Forest Insect Pests in Canada

Editors
J.A. Armstrong and W.G.H. Ives

Published by

Natural Resources Canada
Canadian Forest Service
Science and Sustainable Development Directorate
Ottawa, 1995

Available in Canada through your local bookstore
or by mail from Canada Communication Group – Publishing,
Ottawa, Canada K1A 0S9

Catalogue No. Fo24-235/1995E
ISBN 0-660-15945-7

Production: Paula Irving, Catherine Carmody
Layout and design: Paula Irving, Danielle Monette, Peter Brulé, Francine Langevin
Cover design: Steven Blakeney

Scientific editor: W.G.H. Ives, Natural Resources Canada, Canadian Forest Service,
Northwest Region, Edmonton, Alberta, T6H 3S5
Text editor: E.J. Mullins

Cette publication est aussi disponible en français sous le titre
Insectes forestiers ravageurs au Canada

Canadian Cataloguing in Publication Data

Main entry under title:

Forest insect pests in Canada

Issued also in French under title: Insectes forestiers ravageurs au Canada.
ISBN 0-660-15945-7
Cat. no. Fo24-235/1995E

1. Forest insects — Canada.
2. Trees — Diseases and pests — Canada.
I. Armstrong, J.A., 1929–1991.
II. Ives, W.G.H.
III. Canadian Forest Service. Science and Sustainable Development Directorate.

SB764.C3F67 1995 634.9'6'0971 C95-980076-X

Forestry Canada is now called the Canadian Forest Service and forms part of
the new federal department Natural Resources Canada.

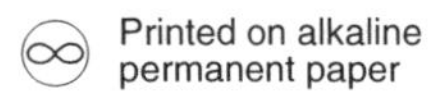

Printed in Canada

Cover: Mature larva of the Douglas-fir tussock moth, *Orgyia pseudotsugata*.

Contents

To Jack Armstrong (1929–1991), his love for entomology, and his perseverance in face of the physical adversity of amyotrophic lateral sclerosis (ALS) for the proposal and development of this book; and to his wife Marion and sons Andrew and John who actively supported him in this significant effort.

Foreword

This book spans 15 years of forest insect pest management in Canada from 1973 to 1988. It was initiated by the late Dr. Jack Armstrong as a sequel to *Aerial Control of Forest Insects in Canada* edited by Dr. M.L.Prebble and published by the Canadian Forestry Service, Environment Canada in 1975. This book has taken a broader view of forest pest management than the previous book as numerous important developments in areas other than aerial control have occurred in recent years. The papers presented in this book cover the range of insect pest management activities in Canada.

Since 1973, we have seen a shift from chemical insecticides to more environmentally benign biological insecticides for operational use. For example, the biological insecticide *Bacillus thuringiensis* (Bt) used in aerial control of spruce budworm has increased in use by forest pest managers from 1% in 1979 to 63% in 1987 of the total area sprayed. As well, the Douglas-fir tussock moth virus is registered for use and has been used in recent operational control programs against this pest. In addition, biological control and silvicultural techniques for managing pests have been implemented more widely.

The impact of insects and the need for pest control have always been crucial considerations in the operational management of our forests, particularly with regard to timber production. Over the past 10 years, insects have annually depleted a volume of wood equal to one third of our annual harvest. When the depletions caused by insects are added to those of diseases and fire, the depletions by all natural causes annually total a volume equivalent to the annual harvest. Similar rates of depletion are anticipated in the future, but may increase if the projections of global warming materialize.

As pressures increase to set aside more forested lands for uses other than timber production and global demand for forest products increases, increased wood production per hectare will be required in the future. In addition, insects affect other forest resources, such as wildlife habitat and aesthetics. Under such circumstances, the management of insect pests will become an increasingly important consideration.

At the same time, insects represent a large and important component of the biodiversity and the food webs in our forests. Therefore, we need to understand the role of insects in forest ecosystems as well as the impacts of operational pest control programs to ensure the sustainable management of forest ecosystems.

The criteria and indicators of sustainable forest management currently being negotiated nationally and internationally include statements on forest condition and biodiversity. Information on insects will be required in both areas.

Within a context of sustainable forest ecosystem management, new technologies for managing insects will be required. Already, there is increased demand for approaches to forest pest management that are environmentally friendly and socially acceptable (for example, nonchemical pest control). Population monitoring techniques for indexing biodiversity as well as for pest management will be required. Other necessary tools include decision support systems, advanced control agents ranging from classical biological controls to biorational insecticides, silvicultural techniques, and insect resistant trees. Biotechnology will be increasingly important in developing biorational insecticides and pest resistant trees.

Another area where entomological expertise will be needed increasingly is that of plant health and quarantine. The possible introduction of exotic insects poses a threat to Canada's forest environment and resources, and has implications for imported goods. Similarly, the possible introduction of insects native to Canada into other areas of the world has implications for Canadian exports. Advice on plant health and quarantine issues, especially regarding nontariff barriers to global trade, will increase in prominence in international negotiations on the trade of forest products. Recognizing the importance of exports of forest products to Canada's economy emphasizes the importance of this issue.

Canadians have demonstrated a strong commitment, at home and internationally, to the sustainable development of our forest resources. Sustainable resource management is essential to the economic, social, and environmental well-being of Canadians, and is an important contribution to a healthy global environment. Forest entomology will assume an increasingly important role in defining and practising sustainable forest ecosystem management in the future.

Yvan Hardy
Assistant Deputy Minister
Canadian Forest Service

Introduction

Canada is a forest nation and the forest sector has long been and continues to be the primary force driving the Canadian economy. Forestry contributes more to our balance in international trade than agriculture, fisheries, fuels, and mining combined and, in 1990, approximately one in every 15 jobs in Canada depended either directly or indirectly on the forest sector. While our forests once appeared to be limitless, it is evident now that they are not. Massive and continuing programs of forest renewal and intensive forest management will be required both to assure sustainable wood supplies and to assure that future generations of Canadians continue to derive all of the many other benefits that flow from a healthy and vigorous forest resource.

No matter how vigorously we attempt to renew our forests, renewal alone will never be able to keep up to the depletion of the resource that occurs in a variety of ways—withdrawal for human uses, and annual losses to fire, disease, destructive forest insects, and competing vegetation. Annual losses to forest insects alone in Canada are estimated to be approximately one third of the total annual harvest and such tremendous losses cannot be allowed to occur if we are to manage our forest resource intelligently. Forest protection must be recognized as an essential and integral component of forest management, and losses, from whatever source, must be managed within acceptable limits that are incorporated in long-term management plans. Failure to do so will result in serious shortfalls in wood supply, expensive disruptions of planned harvesting operations, and less than optimal multiple use of our forests.

This book concerns one aspect of forest protection—the control of forest insects in Canada. It might be looked upon as a sequel to *Aerial Control of Forest Insects in Canada*, edited by Dr. M.L. Prebble and published by Environment Canada in 1975. Since the publication of Prebble's book, significant changes have occurred organizationally within the agencies engaged in forest protection-related research, in provincial policies and procedures relative to forest protection, and in public perception of forest protection practices and their potential to affect human health and environmental quality. All of these have influenced the way in which the control of forest insects is undertaken in Canada and it appears timely now to present an updated overview of this subject and to document what has been achieved since 1975.

The Canadian Forestry Service continued to play a lead role in forest protection-related research both as a part of Environment Canada and more recently as a federal department (Forestry Canada). In 1976, the Canadian Forestry Service decided to physically merge the Insect Pathology Research Institute in Sault Ste. Marie with the Chemical Control Research Institute in Ottawa and to place their programs under one direction. The Forest Pest Management Institute came into being in May, 1977 and staff of the Chemical Control Research Institute moved to the new institute's location in Sault Ste. Marie by the fall of 1978. The program that evolved within this new institute is outlined in detail in the February, 1988 anniversary edition of the Forest Pest Management Institute's newsletter. Essential components of the original programs of both former institutes were maintained, greater emphasis was given to the application of biotechnological techniques to the development of new and improved microbial insecticides, and pesticide formulation and spray atomization and dispersal research was enhanced. The institute's Plantation Pest Management program was strengthened in response to concerns for new pests arising in intensively managed plantations and a new program on biological systems analysis was instituted with beneficial impacts on all research projects. The merger of the microbial control and the chemical control research arms of the Canadian Forestry Service helped remove the competition that existed between these two research arms and paved the way for a more logical approach to the development of integrated pest management techniques.

Throughout the period concerned, the six Canadian Forestry Service (Forestry Canada) regional centres were retained and their programs in insect control-related research and services were maintained much as described by Prebble but with a reduced capacity for insect population dynamics and other entomological research as a result of general resource constraints over the period. Organizationally, provincial forestry departments or special agencies set up by them have remained responsible for the implementation of forest insect control operations. Particularly significant over the period concerned has been the emergence of the New Brunswick Spray Efficacy Research Group in 1981. The group began as a networking of individuals interested in various aspects of aerial spray application technology in New Brunswick (Forest Protection Ltd., New Brunswick Department of Natural Resources and Energy, Forestry Canada—Maritimes Region, University of New Brunswick, and the New Brunswick Research and Productivity Council). It very quickly grew to include researchers from the National Research Council of Canada, Atmospheric Environment Service, and the Forest Pest Management Institute. This research network has expedited significant progress in the aerial application technology of insecticides as described in Part III of this book. Recently, the advantages of such networking in complicated fields such as aerial spray application technology has been realized more broadly and this network has been expanded to include participation of provincial forest authorities from Quebec and Ontario. The network name was changed to Spray Efficacy Research Group to reflect this broader base.

Particularly important since 1975 has been the growing public concern for the environment that has had a profound effect on the forest sector and on our ability to protect and manage the Canadian forest resource adequately. Despite the fact that annually forestry uses only a small fraction of the insecticides that are used in agriculture and around the home, use of these materials, particularly from the air and on Crown forest lands in Canada, has become of major public and political concern for our society. Unfortunately, this concern has been fueled sometimes by less than objective reporting of scientific fact and by news media sensationalism. The result has been confusion on the public's part, in many cases an unsubstantiated public fear of human health and environmental consequences of the use of pesticides in forestry, and political overreaction to the public's perception of pesticide use in forestry. This situation has reached the stage where no chemical insecticides can be applied from the air to Crown forests in Ontario, *Bacillus thuringiensis* (Bt) is used only against spruce budworm in Quebec and Nova Scotia, and the general feeling is that it would be much more politically comfortable to abandon chemicals completely in forestry. The pesticide manufacturing industry, generally, has reacted to this by abandoning its interest in developing forestry use chemical insecticides (for example, Zectran®) and ceasing sustained manufacture of other excellent materials (for example, Matacil®). Unfortunately, Canada is left with only fenitrothion and Bt to use in managing a wide (and growing) range of insect pests destructive to our forest resource. This is a woefully weak arsenal from which to draw weapons for the protection of a resource that is the prime engine of our economy and that contributes in so many other ways to our social well-being.

More positively, this concern for the environment has spurred research and development on microbial control agents and other alternatives to chemical insecticides and has stimulated the application of biotechnological techniques, including genetic engineering, to these ends. It has also stimulated application technology and related research to allow us to formulate pesticides optimally and to deliver them to the target accurately and in the smallest amounts required to achieve the level of protection desired. And the concern is certainly not at odds with attempts to develop integrated pest management techniques.

Impressive advances in research and development in the control of forest insects in Canada have been made since 1975. Generally, these developments have followed the predictions made in the "Outlook" section of the Epilogue of Prebble's book. However, despite the development of nonpersistent chemical insecticides, which have allowed the effective operational use of biological insecticides and resulted in improved methods of aerial application, we are not protecting our forests from destructive forest insects any better now than we were then. Although we have the capability of doing so and with much reduced risk to nontarget components of the forest ecosystem, political reaction to the public perception of pesticides in forestry is preventing our forest managers from using the improved tools and techniques available to them. No amount of additional research is likely to remove this paradox and the situation will remain the same, or worsen, until the public is convinced that forest protection is essential and that it can be practised with little hazard to public health or environmental quality.

This is one of the major problems facing the forest sector today and, until it is resolved, sustainable development of Canada's forest resource will remain an unattainable concept. In his foreword to Prebble's book, R.F. Shaw, then Deputy Minister of Environment Canada, stated "Most of the federal agencies that have been involved in such cooperative effort are now joined in the Department of the Environment, under the broad objectives of undertaking programs to protect and enhance the quality of the environment and to improve the management and sustained economic utilization of the forest, wildlife and aquatic renewable resources of the nation. In pursuing these objectives, I am firmly convinced that forest resource management and environmental management are fully compatible in principle and in practice. They are indeed closer than the proverbial 'two sides of the same coin'. For when we recognize that the forest is more than an assemblage of trees, the objectives of effective forest resource management will neglect neither the trees nor the fauna, nor the associated environmental and social benefits." Shaw recognized the need for integrated renewable resource management and apparently believed it would evolve naturally when most of the federal agencies concerned with such matters were drawn together within Environment Canada. Unfortunately, this has not happened to date and the various renewable natural resource sectors have adopted insular stances instead of joining forces in an integrated approach to resource management. Among other things, this has confused the public even further where renewable resource management (including forest management and protection) is concerned.

It is time now for the renewable resource sectors to come together at the planning table to consider integrated approaches to resource management that will assure sustainable development within all sectors concerned and maintenance of environmental stability. If this can be achieved over the coming years and the public convinced of the value of such an approach, the stage will be set for much improved control of damaging forest insect pests in Canada. The research and development and the operational strategies outlined in this book should provide a stepping stone toward this broader but attainable goal.

Part I

Status of Forest Insect Pests by Region

Chapter 1

Forest Insect Pests in the Newfoundland and Labrador Region

J. Hudak and A.G. Raske

Introduction

Insects are the most numerous component of the forest ecosystem where they serve many functions, ranging from the pollination of some plants to the decomposition of trees. A large number of insect species feed on forest plants, from root hairs to terminal buds, but only a few species become pests by causing severe damage when their populations reach epidemic levels. When insect pests compete with man for a large proportion of the forest resource, they become economically important.

The recent decades in the history of the Newfoundland forests have been characterized by almost continuous, back-to-back outbreaks of the major economic pests: the balsam woolly adelgid, the hemlock looper, and the spruce budworm. These outbreaks have caused extensive tree mortality resulting in imminent shortages of wood supply and in the reduction of aesthetic value of the forests. Forest managers have had to contend with major insect outbreaks for most of the 19 years of this reporting period. The hemlock looper was still causing major damage in 1970 and 1971; the spruce budworm caused serious damage from 1973 to 1984; and the hemlock looper reached serious outbreak levels again from 1984 to 1988. During most of these years the balsam woolly adelgid was also threatening to reach outbreak levels again after subsiding to endemic levels in the early 1960s.

The province of Newfoundland is unique in that it established a royal commission to examine the forest protection and management implications of the spruce budworm outbreak. The appointment of a royal commission to deal with the spruce budworm was evidence of the importance of the forest to the province and of the severity of the impact of a forest pest on forest management.

Several other insect species had become pests by becoming abundant locally in response to more intensive forest management practices, or by threatening to become more important economically. This summary of the status of forest insect pests from 1970 to 1988 has been compiled from publications by the Forest Insect and Disease Survey (FIDS) of the Newfoundland and Labrador Region.

Major Forest Insects, 1973 to 1988

Spruce Budworm

The spruce budworm, *Choristoneura fumiferana*, became the most important economic forest insect pest of the Newfoundland and Labrador Region between 1973 and 1988. Before 1970 only three small infestations had been recorded, and these outbreaks lasted only a few years and caused no appreciable tree mortality.

The recent outbreak started in 1971 in western Newfoundland, and it spread rapidly to almost all of the productive forests by 1979. The size of the infestation, the intensity of defoliation, and the degree of tree mortality were unprecedented in Newfoundland history (Table 1). The infestation primarily extended its boundaries across the island from 1971 to 1974, generally from west to east, but feeding intensity within infested areas of balsam fir and white spruce became extreme from 1976 to 1978. Even extensive areas of black spruce stands and larch trees within infested areas were completely stripped of new foliage for one to several years. Budworm populations collapsed for 3 consecutive years beginning in 1981, with western populations tending to collapse first. Only scattered remnant populations remained at three localities in Newfoundland by 1987 and 1988. One of these remnant populations in the southwestern part of the island occurred in the same locality where the outbreak originated and persisted.

Extensive balsam fir mortality in budworm-infested areas began in 1976, and in that year was confined to stands previously damaged by the hemlock looper or the balsam woolly adelgid. In 1977, balsam fir mortality increased rapidly and continued until 1985. In 1982, black spruce mortality increased suddenly and dramatically, and black spruce continued to die until 1984. Balsam fir and black spruce mortality reached approximately 23 million and 3 million m^3, respectively, comprising 15% of the island's softwood growing stock. The volume of fir and spruce stands with tree mortality totaled 50.8 million m^3, accounting for 30% of the total softwood volume. The total volume of softwoods killed equaled approximately seven times the 1976 estimated annual allowable cut for the province.

Aerial spraying with chemical or biological insecticides did not play a significant role in reducing the damage caused by the outbreak. In the early years, it was felt that the outbreak would terminate naturally without causing significant damage, similar to previous outbreaks. Efforts in subsequent years concentrated on the development of forest access roads to facilitate salvaging of damaged stands. Therefore, no aerial spray program was launched to minimize defoliation until 1977, when a trial spray program was conducted, followed by a small operational program in 1978 after considerable damage had already occurred. Public concern over the safety of aerial spraying with chemicals limited control programs to the treatment of small forest improvement areas and to the use of *Bacillus thuringiensis* (Bt) in 1979 and 1980.

The provincial government appointed a royal commission in 1980 to advise on forest protection and management options to avert the damage caused by the spruce budworm. The commission reported in 1981 and recommended annual control programs with the most effective treatment for as long as necessary to protect the forests from damage caused by budworm feeding. Control operations began in 1981 (the first year of population decline)

Table 1. Area of defoliation by the spruce budworm (*Choristoneura fumiferana*) in productive forests of Newfoundland, area of productive forests with tree mortality, and cumulative volume of trees killed, 1970–1988.

	Area of defoliation class (× 1000 ha)			Cumulative tree mortality (× 1000 m³)		
Year	Light	Moderate and severe	Total[a]	Area[b]	Dead tree volume[c]	Stand volume with dead trees[c]
			Insular Newfoundland			
1970	0	0	0	0	0	0
1971	0.1	0.1	0.2	0	0	0
1972	283	364	647	0	0	0
1973	1 322	91	1 414	0	0	0
1974	1 889	410	2 299	0	0	0
1975	1 448	570	2 018	0	0	0
1976	1 492	1 058	2 550	155	1 100	6 900
1977	1 442	1 288	2 730	175	8 000	13 600
1978	548	794	1 342	300	11 400	22 400
1979	279	973	1 251	427	21 900	38 400
1980	325	926	1 251	427	21 800	41 000
1981	21	379	400	427	21 300	40 000
1982	49	41	90	470	23 200	48 700
1983	36	67	103	484	26 000	50 800
1984	7	15	23	484	27 500	50 800
1985	3	0.5	3.5	484	28 300	50 800
1986	2	2	4	–	–	–
1987	2	4	5	–	–	–
1988	–	0.3	0.3	–	–	–
			Labrador			
1978	4	2	6	0	0	0
1982	–	0.7	0.7	0	0	0
1983	0.1	0.3	0.4	0	0	0

[a] Small discrepancies in totals are due to rounding of numbers.
[b] Area did not increase after 1983.
[c] Mortality after 1983 increased slightly but data not available. Cumulative figures sometimes decreased in time because methods were refined to increase precision.

and continued until 1984 when the outbreak had abated. Operational protection was not provided, therefore, until most of the outbreak had run its course, and the forests had already sustained about 80% of the total tree mortality (Table 1).

Several outbreaks of secondary agents, which normally are not major forest pests, also occurred in stands of fir, spruce, and larch weakened by spruce budworm defoliation. Populations of the foureyed spruce bark beetle, *Polygraphus rufipennis*, increased in weakened trees and contributed substantially to mortality of black spruce. Also, Armillaria root rot, *Armillaria* spp., was abundant in the root systems of both damaged fir and spruce. In the early 1980s, the spruce beetle, *Dendroctonus rufipennis*, caused extensive mortality of white spruce, and the eastern larch beetle, *Dendroctonus simplex*, caused extensive mortality of eastern larch throughout Newfoundland beginning in the late 1970s.

Male budworm moths are being monitored annually by FIDS with pheromone-baited traps at each of 50 permanent sampling sites. The survey began in 1985, and the number of moths trapped annually decreased consistently until 1988. Most moths are being trapped along the west coast of Newfoundland, and many are trapped before local moths have emerged. There is evidence that these moths are transported to Newfoundland from the Maritime provinces by warm air masses. Such transportation of moths to Newfoundland has been documented and it may have contributed to the unprecedented severity of the recent outbreak of the budworm.

Hemlock Looper

The hemlock looper, *Lambdina fiscellaria*, periodically reaches outbreak levels in Newfoundland, and outbreaks that caused tree mortality have been recorded since the early 1900s. Infestations tend to develop rather quickly, cause severe defoliation in any one stand for 2 or 3 years, and then end rather abruptly. An outbreak in a large area, such as Newfoundland, may last 5 to 7 years, with the succession of several infestations.

Large-scale mortality of balsam fir has occurred following 1 to 3 years of severe defoliation, because unlike the spruce budworm this insect feeds indiscriminately and wastefully on both new and old foliage. Therefore the damage to the tree is much greater than the amount of foliage the larvae consumes.

Newfoundland experienced a large outbreak of hemlock looper in the late 1960s, and remnants of this outbreak persisted to 1970 and 1971 (Table 2). Operational spray programs in 1968 and 1969 are generally considered to have prevented considerable tree mortality, although several million cubic metres of timber were killed (mostly in unsprayed areas) by the severe defoliation that occurred in these 2 years.

In the mid- to late 1970s larval numbers increased in several localities, and another outbreak threatened. However, the ongoing spruce budworm outbreak barred looper outbreaks from developing, as budworm larvae consumed fir foliage before the looper emerged in spring. An infestation of the looper developed in the Northern Peninsula in 1979 and 1980, outside of the budworm infestation, and collapsed suddenly. This outbreak caused tree mortality estimated at 200 000 m^3 (Table 2).

Parasites and predators do not play a major role in the population dynamics of the hemlock looper, although a

Table 2. Area of defoliation by the hemlock looper (*Lambdina fiscellaria*) of productive forests of Newfoundland from 1970 to 1988, area of productive forests with tree mortality, and cumulative volume (cubic metres) of trees killed, 1970–1988.

	Area of defoliation class (× 1000 ha)			Cumulative tree mortality (× 1000 m^3)	
Year	Light	Moderate and severe	Total	Area	Volume
1970	33	220	153	73	–
1971	–	–	24	73	–
1979	–	0.9	0.9	0	0
1980	–	–	10	–	195
1981	0	0	0	–	195
1982	0.6	0	0.6	0	0
1983	9	0.6	10	0	0
1984	42	53	95	0	0
1985	79	52	131	0	0
1986	117	215	332	–	3 563
1987	9	152	161	–	4 783
1988	5	13	18	–	–

number of parasites have been introduced into Newfoundland in an attempt to stabilize populations of this pest. Fungal diseases are very important, however, and outbreaks seem to be terminated by epizootics of the fungus *Entomophaga aulicae*, although other species of fungi may be important locally. Fungal diseases build up and spread through looper populations, but not quickly enough to prevent large-scale tree mortality before bringing the outbreak under control.

The recent outbreak started in 1982 in the eastern part of the province. Independent infestations followed in other parts of the island, and many coalesced by 1985 to form the largest and most severe outbreak yet recorded. The outbreak was concentrated along the west coast of the island from the southwestern forests to those of the Northern Peninsula. Commensurate with its size and severity this outbreak has also caused more tree mortality than any previous looper outbreak, and almost 5 million m^3 of timber had been killed by 1987. The infestations in the eastern Avalon Peninsula have been unusual in that they have persisted in the region, but have "moved" annually from area to area.

The province has sprayed portions of the outbreak annually, beginning in 1985.

Balsam Woolly Adelgid

The balsam woolly adelgid, *Adelges piceae*, a serious pest of balsam fir, was accidentally introduced into Newfoundland probably in the early 1920s or 1930s. It was first recorded in the province in 1949, and was causing considerable damage to fir stands in the late 1950s and early 1960s. Populations had collapsed for unknown reasons by the mid- to late-1960s. Since then the adelgid has been slowly spread by wind to other regions of the province and populations have generally increased. This increase is most commonly recognized by damage symptoms that appear after the populations have been high for several years. Surveys now sample for adelgid populations directly before damage appears. Results from such surveys give lead time for forest management decisions. Of special concern is the buildup of adelgid populations and damage in thinned stands of young balsam fir, and research has been initiated to find control measures suitable for such stands.

A damage hazard-rating system has been developed for Newfoundland that is based on site and stand parameters including the elevation of stands and surrounding areas. Areas of greater risk to adelgid damage have been identified and this influences forest management decisions. Defoliating insects caused severe damage in some stands that had already been severely damaged by the balsam woolly adelgid and some of these stands have been cut, burned and planted to black spruce, a non-host species.

Unsuccessful attempts at biological control through the introduction of predators into Newfoundland date back almost 40 years. Similarly, attempts to control the adelgid through chemical sprays have also been unsuccessful. The balsam woolly adelgid therefore maintains its status of an important forest pest because of its potential to cause great economic damage and because of our inability to reduce populations or their damage.

Minor Forest Insects of Conifers 1973 to 1988

Eastern Blackheaded Budworm

High populations of the eastern blackheaded budworm, *Acleris variana*, are common to fir, and sometimes spruce, forests of Newfoundland, and infestations may be local, or cover large areas of more than 100 000 ha. This insect is adapted to northern climates and at times large areas of Labrador are infested with high population levels. Blackheaded budworm outbreaks rarely cause tree mortality, although large numbers of larvae can occur on trees. Larval feeding is decidedly concentrated in the top of crowns, and top mortality is the more common form of severe damage.

The cause of population collapses has not been precisely documented, but FIDS records indicate high levels of parasitism in populations of the blackheaded budworm cause the collapse after 2 and 3 years of infestation.

Infestations in large areas on the island of Newfoundland seem to follow outbreaks of the hemlock looper. Local infestations may occur in any year (Table 3), but large areas were infested from 1973 to 1977 (after the looper outbreak collapsed in 1971 and 1972), and high outbreak levels occurred again in 1987 and 1988 at the collapse of the present looper outbreak (Table 3).

Extensive outbreaks have also occurred periodically in Labrador. The latest outbreak there began in 1973 and peaked in 1976 with over 150 000 ha being infested (Table 3). This outbreak caused little tree mortality and collapsed completely in 1977.

Larch Sawfly

Outbreaks of the larch sawfly, *Pristiphora erichsonii*, are very common in Newfoundland, with populations at high levels locally somewhere on the island or in Labrador in almost every year (Table 4). Outbreaks have been unpredictable, as little is known about the population dynamics of this insect in Newfoundland. Severe larch sawfly defoliation, even for several successive years, does not cause tree mortality, although such damage predisposes larch trees to successful attack by the eastern larch beetle, *Dendroctonus simplex*. Larch is currently not an important commercial forest species in Newfoundland, and the tree is usually only a scattered component within coniferous stands. However, eastern larch or hybrid larch species are sometimes considered for plantations because their high-density fibre content and fast growth on good sites are attractive characteristics.

The masked shrew, *Sorex cinereus*, was introduced into Newfoundland in 1958 as a larch sawfly predator. This introduction still remains a rare example of attempted biological control by introducing a mammal to control an insect. The shrew had dispersed throughout the island by 1970 and sawfly infestations seemed to be less frequent and of shorter

Table 3. Area (hectares) of defoliation by the blackheaded budworm, *Acleris variana*, in various regions of Newfoundland, 1970–1988.

Year	South-west	West-central	Northern Penin.	Central	Eastern	Burin Penin.	Avalon Penin.	Labrador	Total
1970									0
1971	X	X							X
1972	X	X				X			X
1973		X		X	X	X		61 000	61 000+
1974		X	X	X	83 000	19 000	2 000	71 000	175 000+
1975			3 300	X	Y		Y	142 000	222 300+
1976			106 000				56 000	151 000	313 000
1977					1 400		26 200		27 600+
1978									0
1979							X		X
1980		X							X
1981			X						X
1982									0
1983		X							X
1984	X	X							X
1985									0
1986								X	X
1987			3 900				1		3 900
1988			8 100				1		8 100

X=The area of infestation was not determined.
Y=The total area infested in eastern areas and on the Avalon Peninsula is 75 000 ha.

Table 4. Major outbreaks[a] of the larch sawfly, *Pristiphora erichsonii*, in regions of Newfoundland, 1970–1988.

Year	Northern Penin.	West-central	South-west	Central	Avalon Penin.	Labrador
1970	S		L	S		
1971	L		L	L		
1972	L		L			
1973	L		L			
1974	L					
1975			L	S		S
1976			L			S
1977		L	L			S
1978		S	L			S
1979	S		S			S
1980	S		S	L		
1981			S	S	S	
1982				S	L	
1983				L	S	
1984					S	
1985					S	
1986					S	
1987					S	
1988					S	

[a] Light defoliation (L); severe defoliation (S).

duration until the late 1970s (Table 4). Since then several larch sawfly outbreaks have persisted without numerical increase of shrew populations in infested areas.

An attempt to achieve biological control by the introduction of the parasite *Olesicampe benefactor* is currently under way. This parasite has successfully prevented larch sawfly outbreaks in other areas of North America, and is likely to give successful control in Newfoundland as well.

Eastern Larch Beetle

The eastern larch beetle, *Dendroctonus simplex*, is generally considered a secondary insect, successfully attacking only larch trees weakened through damage by other insects, fire, flooding, or drought. However, large beetle populations can overcome tree resistance and kill healthy trees, especially mature ones. In Newfoundland, the first recorded outbreak started in 1976, following severe defoliation of larch by extremely high populations of the spruce budworm. The outbreak gained momentum in the late 1970s, peaked from 1979 to 1981, and populations returned to near endemic levels by 1985. In Labrador, outbreak levels occurred in the mid-1980s following a large outbreak of the larch sawfly.

Estimates of the volume of larch trees killed are difficult to obtain because larch trees are either scattered components of coniferous stands, or clusters of trees surrounded by nonproductive forests on poor sites. Surveys in 1978, 1981, and 1985 in Newfoundland yielded conservative volume estimates of over 100 000 m^3 of larch trees killed. No damage estimates are available for Labrador, but considerably fewer were killed there than on the island.

Foureyed Spruce Bark Beetle

The foureyed spruce bark beetle, *Polygraphus rufipennis*, is an aggressive secondary pest of spruce trees. During the recent outbreak of the spruce budworm, black spruce trees were severely defoliated for 3 or 4 years between 1976 and 1979. Tree mortality of black spruce became extensive in 1981, and the foureyed spruce bark beetle was associated with the sharp increase in spruce mortality in 1982 and 1983. The beetle successfully attacked trees with all degrees of budworm defoliation. However, surviving trees in stands classed as severely damaged by the budworm had a greater probability of being attacked by the bark beetle than did trees with nearly full complements of foliage. The root rot *Armillaria* spp. (*mellea* complex) was also active in the damaged stands. The attack by both of these organisms increased the proportion of trees within stands that died and enhanced the rate and extent of tree mortality. Populations of the bark beetle abated in 1984 and returned to endemic levels by 1985; only single trees have been killed by this beetle since.

Spruce Beetle

In 1979, large numbers of white spruce were being successfully attacked by the spruce beetle, *Dendroctonus rufipennis*, following spruce budworm damage in the late 1970s. Areas of concentrated tree mortality were recorded in 1980 in several areas in western Newfoundland. Populations in other areas of the island increased by 1982, and the outbreak peaked in 1983. Populations of the beetle returned to endemic levels by 1985, because surviving spruce trees had recovered from defoliation damage.

Balsam Fir Sawfly

Populations of the balsam fir sawfly, *Neodiprion abietis* complex, were generally low from 1970 to 1988, but high population levels were recorded in small areas in each year from 1970 to 1976. Before 1970, however, high population levels existed in large areas of balsam fir forests and caused complete defoliation of old growth, but tree mortality usually occurred only in conjunction with feeding damage by other insects. Available information suggests that parasites generally reduce high populations of the balsam fir sawfly. Its larvae are becoming common again, and another outbreak is likely within the next 2 years.

Larch Casebearer

The larch casebearer, *Coleophora laricella*, is an introduced Eurasian insect that was first discovered in Newfoundland in 1941. It now occurs on larch throughout all regions of the island, but outbreaks between 1970 and 1988 have been very small and short-lived. Severe defoliation causes growth loss, and branch mortality may occur during outbreaks, but tree mortality is rare. Two European parasites,

Agathis pumila and *Chrysocharis laricinellae*, were introduced into Ontario, and then collections of these parasites from Ontario were released in Newfoundland between 1944 and 1947, where both species became established. The chalcid, *C. laricinellae*, was the more common parasite until the early 1970s, but since then the braconid, *A. pumila*, has become dominant and is providing successful biological control of the larch casebearer in Newfoundland.

Minor Forest Insects of Hardwoods

Birch Casebearer

The birch casebearer, *Coleophora serratella*, was accidentally introduced into North America from Europe about 1920 and was first discovered in Newfoundland in 1953. It was causing severe defoliation of birch in western Newfoundland in the late 1960s and by 1973 it had spread throughout the island. The birch casebearer feeds on several hardwood species, but species of birch and alder are its principal hosts. Localized or widespread outbreaks occur frequently on the island and high populations cause severe browning of foliage, at times branch mortality, and occasionally the death of trees.

Birch species are not important commercial forest species in Newfoundland, although ecologically they are an important component of the total forest community.

High populations of the birch casebearer may locally be terminated by overwintering mortality or by an unknown egg mortality factor. Attempts at biological control by the introductions of European parasites into Newfoundland seemed unsuccessful until recently. Two species complexes, *Campoplex* spp. and *Dolichogenidea* (= *Apanteles*) spp. (mostly *coleophorae*) were both introduced in the early 1970s into different regions of Newfoundland. Both genera of parasites became established, but parasites of the genus *Campoplex* have not been recovered in recent years. However, *Dolichogenidea coleophorae* has provided good control of the birch casebearer in and near the release area, and this parasite species has been transported from the release site to other areas of the island because it appears to be a promising control agent for this insect.

Satin Moth

The satin moth, *Leucoma salicis*, is an European insect that since its accidental introduction into Newfoundland in 1934 has become common throughout the island. The insect's primary hosts are several species of poplar trees, but willows are also acceptable host plants. Severe but local and short-lived infestations are always present somewhere on the island, but no permanent damage results from its feeding activities. The satin moth is not only a pest in forested areas, but frequently invades urban areas, where it completely strips host trees of foliage, resulting in a loss of their aesthetic value.

In 1936, 1940, and 1942 the parasite *Apanteles melanoscelus* was introduced into Newfoundland and since then has brought high populations of the satin moth under control after the caterpillars have completely stripped the aspen trees for 1 or 2 years.

Mountain-Ash Sawfly

The mountain-ash sawfly, *Pristiphora geniculata*, is always near or at outbreak levels at some locality within the province. Complete defoliation of mountain-ash is common when the insect is abundant, but the host appears to be unharmed even by successive years of defoliation. Defoliation by the mountain-ash sawfly is not important to forestry, but it is of consistent concern in recreational areas and in urban settings.

In 1981, the European parasite, *Olesicampe geniculatae*, was introduced into Newfoundland from Quebec and has become established. The parasite has eliminated the pest near the release area and appears to be spreading into surrounding sawfly populations. It has provided good control of the sawfly in other parts of North America and will likely control it in Newfoundland.

Forest Tent Caterpillar

Male moths of the forest tent caterpillar, *Malacosoma disstria*, periodically appear in Newfoundland in large numbers, but females, eggs, larvae, and feeding damage have never been found. Males were extremely abundant around city lights in 1981 in western Newfoundland, and an annual trapping program began in 1982 with pheromone-baited traps as part of an early detection program. The largest number of males were always trapped in western Newfoundland, but some moths were also trapped in central and eastern areas of the island. No moths have been found in the traps since 1986. The rise and disappearance of this insect in this province coincides with the rise and collapse of the outbreak in 1986 in the Maritime provinces. It seems reasonable to conclude that the male forest tent caterpillar moths trapped in Newfoundland were transported from the Maritime provinces annually, with the aid of warm air-mass movements. Moths had to fly 150 to 400 km over the ocean to land safely. Two things are remarkable; first, the consistency of transport and second, the huge numbers that must have arrived in Newfoundland to result in a trap catch of up to 20 per trap, or several hundred present at each city light. Moths trapped in central and eastern Newfoundland probably first alit in western Newfoundland. The FIDS staff will monitor for the probable permanent arrival of this forest insect to Newfoundland.

Gypsy Moth

The gypsy moth, *Lymantria dispar*, is a forest insect that is not established in Newfoundland. In cooperation with Agriculture Canada, FIDS provides an annual early detection with the use of 250 pheromone-baited traps. The survey began in 1980, and to date three males have been trapped, one at each of the following locations: Corner Brook in 1980, Mummichog Provincial Park in 1985, and St. John's in 1987. These males probably arrived in Newfoundland as hitchhikers on man-made modes of transportation.

Insects of Plantations

Black Army Cutworm

The black army cutworm, *Actebia fennica*, has become a problem in plantations established on recently burned areas. It was first recognized as a forest pest in 1983 when it infested two plantations in central Newfoundland and stripped all the foliage from black spruce seedlings in an area of over 270 ha. Infestations in several other plantations required operational control in 1987 and 1988. Research began in 1988 to find more acceptable and better control methods, because the only registered chemical, a synthetic pyrethroid, caused side effects in aquatic environments.

The cutworm normally feeds on herbaceous plants and hardwood foliage. However, during outbreaks, such as occur on areas burnt within the previous 3 years, the larvae readily defoliate spruce seedlings after other foliage has been consumed.

Yellowheaded Spruce Sawfly

Larval colonies of the yellowheaded spruce sawfly, *Pikonema alaskensis*, are periodically common in young planted stands of black spruce in Newfoundland. Populations do not persist very long in any one stand and all infestations are probably controlled by parasites. Damage to stands usually consists of low levels of defoliation and rarely exceeds sporadic top kill. Infestations were common in several parts of the island in 1971 and 1972, and again in 1983 and 1984. Control measures have not been necessary to date.

European Pine Sawfly

The European pine sawfly, *Neodiprion sertifer*, was accidentally introduced into the St. John's area in 1974 on nursery stock. It quickly became a pest and by 1977 a pine stand near St. John's and many ornamental pines in St. John's were severely defoliated. One year of severe defoliation rarely leads to tree mortality, but consecutive years of defoliation, as would occur without natural controls, would lead to the death of trees.

A pupal parasite, *Pleolophus basizonus*, was introduced for biological control, in infestations around St. John's in each of the 4 years from 1977 to 1980. A larval parasite, *Lophyroplectus oblongopunctatus*, was released annually from 1978 to 1980 and in 1981 a third exotic parasite, *Exenterus abruptorius*, was added to strengthen the parasite complex. Of the three parasites, *P. basizonus* and *L. oblongopunctatus* have been recovered from field collections. In 1983 the emphasis shifted from control by parasites to more efficient control of the pest with a nuclear polyhedrosis virus. Spraying the virus onto early instar larval colonies generally gave complete control in the year of application, and treated trees have not required retreatment to date. However, the virus appears to spread slowly from isolated clusters. Therefore, spraying with the virus has been required annually as new pockets of infestation were discovered.

Warren Root Collar Weevil

The Warren root collar weevil, *Hylobius warreni*, occurs throughout the island at endemic levels. The larvae most commonly feed on the root collars of black spruce but damage to this tree species is usually light. However, population levels increase in pine plantations and pines may be killed by the girdling effect of larval feeding; thus this weevil is hazardous to present and future pine plantations.

In eastern Newfoundland experimental plantings of Scots and lodgepole pine are severely infested and the trees are sometimes killed. Jack pine and Austrian pine are also infested, but tree mortality is less common than on the other two pine species. Soft needle pines are not often attacked by this weevil.

Whitemarked Tussock Moth

The whitemarked tussock moth, *Orgyia leucostigma*, rarely reaches outbreak levels in Newfoundland. A small infestation developed in 1985, however, and over 100 ha of hardwoods and conifers, including seedlings in plantations, were severely defoliated. This tussock moth normally feeds on several hardwoods, but mature larvae can severely damage conifers when other host plants have been denuded. Experimental control was attempted in 1986 and 1987 by spraying the whole infestation with a virus of the Douglas-fir tussock moth, *O. pseudotsugata*, and the virus successfully reduced whitemarked tussock moth populations by 1988.

Other Damaging Insects

Several other insect species have caused minor damage through local abundance. This damage may have been to forests, but more commonly the feeding activity of these insects was conspicuous because it occurred in urban areas, along road sides, in recreational areas, or in plantations. These insects of minor importance are simply listed with the years in which they caused at least moderate damage (Table 5).

Table 5. Insects causing moderate or severe damage to forests or plantations in Newfoundland, 1970–1988.

	Year of reported damage																		
Insect	70	71	72	73	74	75	76	77	78	79	80	81	82	83	84	85	86	87	88
Conifers																			
Balsam twig aphid *Mindarus abietinus*													X	X	X	X	X	X	
European spruce sawfly *Gilpinia hercyniae*	X	X	X									X	X						
Larch budmoth *Zeiraphera improbana*							X	X											
Spruce bud midge *Rhabdophaga swainei*														X	X				
Spruce bud moth *Zeiraphera canadensis*										X				X		X	X		X
Spruce coneworm *Dioryctria reniculelloides*								X	X	X	X								
Strawberry root weevil *Otiorhynchus ovatus*	X																		
White pine needle miner *Ocnerostoma strobivorum*								X											
White pine sawfly *Neodiprion pinetum*																X	X		
	70	71	72	73	74	75	76	77	78	79	80	81	82	83	84	85	86	87	88
Hardwoods																			
Birch leafminer *Fenusa pusilla*	X	X	X	X	X					X	X			X			X	X	
Birch skeletonizer *Bucculatrix canadensisella*		X	X	X															
Dusky birch sawfly *Croesus latitarsus*	X				X				X				X						
European alder leafminer *Fenusa dohrnii*	X	X	X	X	X										X	X	X	X	
European leafroller *Archips rosana*								X			X								
Fall webworm *Hyphantrea cunea*				X	X	X	X	X	X						X			X	X
Large aspen tortrix *Choristoneura conflictana*				X	X	X	X	X	X										X

Table 5. *(Concluded)*

	Year of reported damage																		
Insect	70	71	72	73	74	75	76	77	78	79	80	81	82	83	84	85	86	87	88
Maple bladdergall mite *Vasates quadripedes*					X		X								X				
Mourningcloak butterfly *Nymphalis antiopa*			X	X	X	X										X	X		
Plum webspinning sawfly *Neurotoma inconspicua*						X		X	X	X									
Aspen serpentine leafminer *Phyllocnistis populiella*														X	X		X		X
Poplar-and-willow borer *Cryptorhynchus lapathi*							X												
Rusty tussock moth *Orgyia antiqua*	X				X	X	X	X											
Striped alder sawfly *Hemichroa crocea*	X				X										X	X	X	X	X
Uglynest caterpillar *Archips cerasivorana*							X	X	X	X	X								X
A willow leaf beetle *Chrysomela falsa*							X						X	X	X			X	
A willow sawfly *Nematus limbatu*												X					X		X

Chapter 2
Forest Insect Pests in the Maritimes Region

L.P. MAGASI

Introduction

Not all of the approximately 3 500 species of insects recorded from the forests of the Maritime provinces are forest pests: fewer still are considered pests of major significance. The status of some of these pests is described briefly in this chapter. Special reference is made to the fluctuation in their populations from 1973 to 1987.

Because of changes in forest management practices and the ever-increasing emphasis on plantation forestry, the importance of some forest insects has changed dramatically in recent years. Insects that were economically unimportant members of the ecosystem in the natural forest have suddenly become of major concern in plantations, as they cause significant damage in their new environment—the man-made, single species, even-aged monoculture. Man has always competed with the forest pest for the same resource; now, because of financial investment in plantations and because of increased reliance on the wood expected from the latter, the stakes are much higher. Consequently, attention paid to plantation pests, both old and new, has greatly increased, and to emphasize their importance, additional chapters (Chapters 21–33) are devoted to them.

The information presented here is based on the annual reports of the Forest Insect and Disease Survey (FIDS) in the Maritimes. Liberal use of these reports has been made, including descriptions and tables, without references being noted.

Insects of Natural Forests

Spruce Budworm

The spruce budworm, *Choristoneura fumiferana*, is by far the most economically important forest insect of the balsam fir–spruce forests in the Maritime Region. Because of differences in the history of outbreaks among the three provinces, spruce budworm status is presented separately for each province.

Areas that suffered moderate or severe defoliation during a 20-year period in New Brunswick are tabulated in Table l. The affected areas increased only gradually from the "lows" of the late 1960s until 1972. Then, a dramatic increase in 1973 was followed by further increases in 1974 and 1975. Virtually all susceptible forest in the province was under severe attack during this 3-year period. Infestation decreased in 1976, then gradually increased again until 1983, after which a gradual decrease began—corresponding well with the overall decline of spruce budworm populations in eastern North America—which continues to the present.

The fluctuation in total area affected was accompanied by spatial changes in the most seriously affected areas. In the late 1960s, the outbreaks were most serious in the northern part of New Brunswick. After the 1973 to 1975 period, during which distribution was general, the most serious outbreaks occurred in the southern part of the province. Then, outbreaks expanded northward, especially in western New Brunswick, decreasing in the south and, by the mid-1980s, most of the spruce budworm population existed in northern New Brunswick.

Areas that suffered moderate or severe defoliation during a 20-year period in Nova Scotia are tabulated in Table 2. There was very little spruce budworm damage in the province until 1971, after which populations increased. The size of the area affected fluctuated over the years (reaching a peak in 1976), then declined sharply in 1987, when no visible defoliation could be detected anywhere in Nova Scotia.

Although provincial figures may be useful for comparative purposes, data presented in Table 2 do not adequately represent the spruce budworm situation as it progressed in Nova Scotia. As indicated in Table 3, there were five distinct areas of infestation in the province during the review period.

Table 1. Area (1 000 ha) of moderate and severe defoliation by the spruce budworm, *Choristoneura fumiferana*, in New Brunswick, 1968–1987.

	Area of defoliation by category		
Year	Moderate	Severe	Moderate–Severe
1968	309	80	389
1969	732	247	979
1970	421	153	574
1971	786	633	1 419
1972	968	765	1 733
1973	1 737	1 414	3 151
1974	NA[a]	NA	3 400
1975	NA	NA	3 500
1976	172	226	398
1977	132	342	474
1978	219	450	669
1979	235	1 100	1 335
1980	226	447	673
1981	382	839	1 221
1982	391	811	1 202
1983	355	1 673	2 028
1984	NA	NA	730
1985	NA	NA	1 070
1986	229	698	927
1987	189	241	430

[a]Not available.

Table 2. Area (1 000 ha) of moderate and severe defoliation by the spruce budworm, *Choristoneura fumiferana*, in Nova Scotia, 1968–1987.

Year	Area of defoliation by category		
	Moderate	Severe	Moderate–Severe
1968	NA[a]	NA	22[b]
1969	0	10	10
1970	NA	0	NA
1971	21	161	182
1972	25	115	140
1973	0	89	89
1974	NA	NA	353
1975	187	690	877
1976	530	690	1 220
1977	46	777	823
1978	101	452	553
1979	168	359	527[c]
1980	225	491	716[d]
1981	156	411	567
1982	50	125	175
1983	52	242	294
1984	33	26	59
1985	32	287	319
1986	283	6	289
1987	0	0	0

[a]Not available.
[b]Includes light defoliation.
[c]Does not include 339 000 ha "variable" defoliation.
[d]Does not include 559 000 ha "variable" defoliation.

Table 3. Area (1 000 ha) of moderate and severe defoliation by the spruce budworm, *Choristoneura fumiferana*, in various localities in Nova Scotia, 1971–1987.

Year	Area of moderate and severe defoliation					
	Cumb-erland Co.	Col-chester Co.	Pictou–Antigonish counties	Annapolis Valley	Main-land	Cape Breton
1971	45[a]	0	0	116[a]	161[a]	0
1972	53[a]	0	0	62[a]	115[a]	0
1973	80[a]	0	0	9[a]	89[a]	p
1974	189	0	0	0	189	164
1975	162[a]	42[a]	42[a]	0	204[a]	486[a]
1976	p[b]	p	p	0	0	1220
1977	49	6	6	0	55	769[a]
1978	98	9	1	0	108	648
1979	95	95	24	21	140	726[c]
1980	214[c]	214[c]	65	57	336[c]	936[c]
1981	101	101	13	90	204	363
1982	141	141	30	4[d]	175	0.4[d]
1983	177	177	55	44	276	18
1984	34	34	25	0	59	0
1985	180	180	139[c]	0	319[c]	0
1986	289	289	289	0	289	0
1987	0	0	0	0	0	0

[a]Includes only severe defoliation.
[b]Outbreaks not mapped, but small pockets of defoliation present (p).
[c]Includes "variable" defoliation.
[d]Only moderate defoliation found.

There were four on the mainland; also, all along the northern rim of the province; and one was on Cape Breton Island.

There were two outbreaks in 1971: one in Cumberland County and another in the Annapolis Valley, directly across the Bay of Fundy from heavily infested New Brunswick. The Annapolis Valley outbreak subsided by 1974. The infestation in Cumberland County expanded and, in 1979, merged with part of a smaller outbreak that had existed in parts of Colchester–Pictou counties since 1975. Both the Cumberland–Colchester outbreaks and a smaller one in Pictou–Antigonish counties persisted until 1986, when the outbreak became continuous from the New Brunswick border into western Pictou County. This outbreak, the last remaining area of high infestation in Nova Scotia, collapsed in 1987. There was another small outbreak along the northern rim of the Annapolis Valley from 1979 to 1983, largely the result of a massive moth flight into the area from New Brunswick in 1978.

The largest, most devastating outbreak of spruce budworm ever recorded in the Maritimes occurred on Cape Breton Island. In 1974, defoliation was recorded in a narrow band on the east side of the island. By 1975, severe defoliation was widespread throughout the island and, subsequently, the outbreak expanded and intensified. Spruce budworm populations were extremely high; enough larvae were present to cause severe defoliation many times over. By 1981, more than half (57%) of the balsam fir trees were dead and many more were dying. However, in 1982, spruce budworm populations decreased dramatically, without doubt partly due to lack of food, and the outbreak collapsed in 1983. Tree mortality continued, as trees, stressed beyond the point of no return, succumbed to secondary organisms. By the end of 1986, only 10% of the balsam fir on the Cape Breton Highlands survived and only 6% on the Lowlands.

Areas that suffered moderate or severe defoliation during a 20-year period in Prince Edward Island are tabulated in Table 4. Although spruce budworm populations were present throughout the province and patches of moderate or severe defoliation have been observed since 1970, it was not until 1973 and 1974 that populations really started to increase. The most extensive damage occurred in 1975, when virtually all balsam fir and spruce suffered moderate or severe defoliation. The period of widespread defoliation lasted from the peak year of 1975 until 1978. A drastic reduction occurred in the total area affected in 1979, due to a combination of excellent tree-growing conditions and poor larval survival. Except for 1981, when defoliation covered an area comparable to that

Table 4. Area (1 000 ha) of moderate and severe defoliation by the spruce budworm, *Choristoneura fumiferana*, in Prince Edward Island, 1968–1987.

Year	Area of defoliation by category		
	Moderate	Severe	Moderate–Severe
1968	0	0	0
1969	0	0	0
1970	NA[a]	0	NA[b]
1971	NA	NA	NA[c]
1972	NA	NA	NA[c]
1973	NA	73	73[d]
1974	NA	17	17[d]
1975	NA	NA	243
1976	18	150	168
1977	39	69	108
1978	27	85	112
1979	5	7	12
1980	NA	NA	22[e]
1981	NA	NA	133[e]
1982	3	10	13
1983	14	8	22
1984	7	8	15
1985	13	41	54
1986	64	1	65
1987	0	0	0

[a] Not available.
[b] Moderate defoliation at three locations.
[c] Patches of severe or moderate defoliation.
[d] Includes severe defoliation only.
[e] Includes moderate and severe defoliation.

affected from 1975 to 1978, infestations, although serious, never became general. Until 1984, most of the moderate to severe defoliation occurred in the western part of the province, while in 1985 and 1986, it was recorded in large patches throughout the province. In 1987, all outbreaks collapsed and no visible defoliation was recorded during aerial surveys over Prince Edward Island.

Repeated defoliation of balsam fir caused widespread tree mortality. A survey in 1985 found that 12% of the balsam fir was dead in the western part of the island, 28% in the central part, and 17% in the eastern part. Actual mortality was higher than shown by these figures, which do not account for salvage cuts or old, dead trees no longer standing. Defoliation by the spruce budworm also predisposed trees to successful attack by the spruce beetle.

Spruce Beetle

Spruce beetle, *Dendroctonus rufipennis*, populations started building up in Nova Scotia in 1974 because of widespread blowdown in the eastern and central parts of the province. Contributing to this buildup were the increasing numbers of mature white spruce trees weakened by repeated spruce budworm defoliation.

The first patches of dead spruce trees were observed in central Nova Scotia in 1974, but the outbreak was widespread and by 1978, few areas of the province remained unaffected. A survey in 1979 found that almost half (48%) of the 83 white spruce stands examined were infested. The affected stands were distributed throughout the province but were mainly concentrated on Cape Breton Island, where 73% harbored beetles and, in some stands, more than half of the trees were already dead. By 1980, an estimated 11% of the white spruce in Nova Scotia were affected by the spruce beetle. What made the situation even more serious was that 85% of the affected trees were in the "22 cm+" category, meaning larger trees with the most volume were being lost. Infestations continued to spread and by the end of 1983, no substantial areas on Cape Breton Island remained without severe mortality. In some areas of the Margaree Valley and in the Mabou River valley, an estimated 50% of the white spruce trees were killed in 1983 alone. Overall, 37% of the merchantable white spruce trees that could have produced the highest volume died between the beginning of the outbreak and the end of the year.

There was a dramatic downturn in new tree mortality following the peak year of 1983. On Cape Breton Island, 10% of the remaining white spruce died in 1984 and patches of infestations were still active on the mainland, but the downward trend continued until 1986, when an increase in spruce beetle attacks resulted in tree mortality in some areas in the northeastern mainland. Patchy mortality occurred again in widely separated areas in the province in 1987.

On Prince Edward Island, the first noteworthy mortality of white spruce was reported in 1979 when overmature trees in spruce budworm-damaged stands started to succumb to spruce beetle attack in significant numbers. In 1980, 20% of the white spruce stands were infested in the province and more than half (53%) of these were new infestations. Central Prince Edward Island was worst affected; 61% of the white spruce trees there were successfully attacked. The outbreak continued to intensify and by the end of 1982, an estimated 30% of the merchantable white spruce in the province were dead. Continued attacks in 1983 killed additional trees, but the outbreak peaked in that year. Very few newly killed trees were observed from 1984 to 1986, but there was an increase in 1987 in active locations and numbers of trees affected.

In New Brunswick, the spruce beetle was not known until 1980, when a small stand of white spruce was found to be infested on Grand Manan Island. Additional small infestations found in the southern part of the province brought the number of known infestations to 11 by the end of 1983. In some infestations, spread was limited, but considerable tree mortality occurred; the highest, at 60% mortality, in an area in southeastern New Brunswick. Infestations remained active in 1984 and, in Fundy National Park, where the spruce beetle was first found in 1982, mortality reached 32% by the end of the year. Also in 1984, a significant extension occurred in the known range of the insect when three small infestations were discovered in northwestern New Brunswick. In 1985, a province-wide survey, concentrated in the south, established the presence of spruce beetle at six locations in

Westmorland–Albert counties, five locations in Charlotte County (which included four locations on Grand Manan Island), one location in York County (the only infestation area not known previously), and eight locations in Victoria–Northumberland–Restigouche counties in the area of the three infestations first found in the northwest the previous year. Infestation levels were low, with most represented by a single infested tree. The highest level was in Fundy National Park, where 12% of the trees had been attacked recently.

Some increase in new attacks occurred at a few locations in 1986—52% of the trees were dead and 36% newly attacked in a stand on Grand Manan Island—and infestations remained active in the south during 1987.

Spruce beetle activity appeared to increase in all three provinces at the end of the review period. Even though, for a variety of reasons, the area of mature and, especially, overmature white spruce is rapidly decreasing, there are still many susceptible areas in the Maritimes and another outbreak would cut further into the fast-diminishing supply of lumber-quality wood.

Larch Sawfly

Larch sawfly, *Pristiphora erichsonii*, outbreaks have existed in various parts of the Maritimes since the mid-1950s. Infestations were small originally, but populations gradually built up toward the 1970s. At the beginning of that decade, larch sawfly caused widespread severe defoliation in eastern Prince Edward Island and severe defoliation in many small pockets throughout Nova Scotia; however, there was little activity in New Brunswick.

From 1973 to 1975, the larch sawfly (Table 5) caused very severe defoliation in the eastern half of Prince Edward Island each year, with patches of moderate defoliation scattered throughout. There was some reduction in the area of the severe defoliation in 1975 and the moderately defoliated patches were also absent.

In Nova Scotia during the same period, the areas of severe defoliation increased, patches coalesced, populations grew in areas of moderate defoliation, and by 1975, most of the southern mainland and much of Cape Breton Island suffered uniform, severe defoliation.

Also between 1973 and 1975, small areas of severe and a few areas of moderate defoliation were found in New Brunswick, occurring along the eastern coastal part of the province. The infestation was general north to the Miramichi River. There was also an infestation causing moderate to severe defoliation in the south central part of the province. The size of area affected decreased considerably in 1974 and, except for severe defoliation at St. Andrews in southwestern New Brunswick, only light defoliation was observed in small, widely scattered areas. Severe defoliation in 1975 was somewhat greater than the year before, but was restricted to southern Charlotte County, including Campobello and Deer islands. Generally, the infestations were breaking up.

The outbreak peaked throughout the Maritimes in 1975. In Prince Edward Island, only light defoliation occurred in 1976 with very few small patches of moderate defoliation. Only one area of light defoliation was recorded in 1977 and the larch sawfly was "gone" by 1978. In New Brunswick, where the populations were already on a downward trend before 1975, only three small patches of moderate defoliation could be found in 1977 and none in 1978. In Nova Scotia, infestation levels were greatly reduced in the eastern mainland in 1976, but in the west, defoliation was still extensive and severe. Populations were further reduced in 1977, severe defoliation occurring only in coastal areas in the southwest. Larch sawfly lingered in this area in 1978 and 1979, causing severe or moderate defoliation in isolated patches.

In 1980, larch sawfly populations in the Maritimes were at their lowest level since 1955. This situation remained stable until 1986. During this 7-year period, there were only 12 larch sawfly collections made in the three provinces and in each case only individual trees were affected by a few larvae. This period of very low populations, and the general declines in the late 1970s already noted, may be due to the hymenopterous parasite *Olesicampe benefactor*. Although no follow-up studies were conducted, the parasite was first released in the Maritime provinces in 1967 and has since been recovered several times at various locations.

Populations started increasing in 1987, and roadside trees harbored enough larvae to cause moderate or severe defoliation on a few trees in each of the provinces. Early indications are that a further increase may occur in 1988.

The extensive and prolonged outbreaks of larch sawfly in the 1960s and 1970s, combined with defoliation by the spruce

Table 5. Area (1 000 ha) of moderate and severe defoliation by the larch sawfly, *Pristiphora erichsonii*, in the Maritimes, 1969–1987.

	Area of moderate and severe defoliation		
Year	New Brunswick	Nova Scotia	Prince Edward Island
1969	274	124	57
1970	307	238	243
1971	51	321	274
1972	163	778	337
1973	143	1 546	300
1974	43	1 873	281
1975	53	3 128	240
1976	4	2 101	7
1977	0	442	0
1978	0	221	0
1979	0	147	0
1980	0	0	0
1981	0	0	0
1982	0	0	0
1983	0	0	0
1984	0	0	0
1985	0	0	0
1986	0	0	0
1987	0	0	0

budworm and, in some areas, by whitemarked tussock moth, paved the way for the devastating outbreak of the eastern larch beetle. There were stands, especially in western Nova Scotia and eastern Prince Edward Island, where larch trees exhibited considerable branch and top mortality because of repeated defoliation, but the most serious effect of the larch sawfly was to predispose the trees to beetle attack.

Eastern Larch Beetle

The eastern larch beetle, *Dendroctonus simplex*, normally attacks only weakened, damaged, or recently felled host material. However, when populations are very high, it may infest living and apparently healthy, mature, or overmature trees, and even younger, small diameter trees.

A population buildup was first noticed in Nova Scotia in 1976. This increase in beetle populations followed several years of severe defoliation of larch by the larch sawfly, *P. erichsonii*. In 1977, trees of all ages were succumbing to attack in central Nova Scotia. Beetles were found in 8 of 10 stands examined; tree mortality was 30%, with an additional 36% of the larch trees dying. Twenty-five percent of the larch were affected in eastern Prince Edward Island and one small infestation was found in southern New Brunswick.

The infestations and tree mortality increased rapidly. By 1981, trees were dying in great numbers throughout the Maritimes, except for the northern half of New Brunswick. Both weakened mature trees and healthy young trees were being attacked and killed. Trees infested in the spring most often died by the fall of the same year.

The results of regional surveys in 1978 and 1981, involving the same 97 areas, are shown in Table 6 for comparison and to indicate both the severity and the economic impact of the outbreak. The apparent decrease in mortality in Prince Edward Island is most likely due to the removal of many "old dead" trees.

Although infestations remained active and dying trees were common throughout the region, especially in northwestern New Brunswick and western Nova Scotia (which had the shortest outbreak history), they levelled off in 1982 and a general downward trend in new attacks started in 1984. Cumulative tree mortality has been increasing but at a lower rate than from 1977 to 1981. By 1986, new attacks were observed only in isolated areas of New Brunswick and at one location in western Prince Edward Island.

The outbreak of the eastern larch beetle, which started in 1976, peaked in 1981, and ended by 1986, save for a few spot infestations, left behind a much reduced larch inventory.

Hemlock Looper

The hemlock looper, *Lambdina fiscellaria*, is mainly a defoliator of balsam fir in the Maritimes. It is capable of causing serious damage when populations are high and tree mortality may begin as early as the first year of high infestation.

Hemlock looper populations were low in New Brunswick throughout the review period of 1973 to 1987. No more than a few larvae were present at any given location in any given year and even though in some years they were common, populations always remained at endemic levels.

Occasional outbreaks occurred in Nova Scotia. In 1974, the hemlock looper, along with the spruce budworm and the eastern blackheaded budworm (*Acleris variana*), caused severe defoliation over 8 100 ha on Cape Breton Island. By 1975, the spruce budworm in the area "out-fed" all competition and the hemlock looper disappeared. There were, however, two outbreaks in western Nova Scotia in 1975, covering 3 400 ha and 93 ha, respectively. Both of these collapsed in 1976, but a new area of severe defoliation occurred over 120 ha near Halifax. Tree mortality in the smaller of the 1975 outbreaks was 82%, although most of the trees that died were either understory trees or had been previously weakened by the balsam woolly adelgid. Hemlock looper populations were at endemic levels from 1977 until 1985 when, combined with spruce budworm, they caused 50% defoliation in a 16-ha area in Cumberland County. This infestation persisted in 1986 and was still present in 1987 when, in spite of treatment in both years, moderate or severe defoliation was recorded.

The most serious damage caused by the hemlock looper occurred in Prince Edward Island. In 1975, severe defoliation was recorded in 14 areas of varying sizes (the largest 800 ha) located mostly in the eastern part of the province. The outbreak continued on the 800-ha area in 1976, although populations were somewhat reduced. By the end of the year, 52% of the balsam fir trees (constituting 29% of all trees) were dead. Severe defoliation of balsam fir and hemlock occurred in 1977 and 1978 in the central part of the province. Defoliation occurred in patches of 4–8 ha over 22 000 ha. The outbreak collapsed after 2 years, but tree mortality continued and 80%

Table 6. Spread and magnitude of the eastern larch beetle, *Dendroctonus simplex*, outbreak in the Maritimes, 1978–1981.

Province	Incidence (%)[a]		Mortality (%)[b]		Estimated volume loss[c] (m³)
	1978	1981	1978	1981	
New Brunswick	25	57	3.4	24.0	314 000
Nova Scotia	38	78	22.3	64.0	972 000
Prince Edward Island	50	75	25.0	13.0	11 600

[a] Incidence—percentage of infested locations.
[b] Mortality—percentage of merchantable-size trees.
[c] Volume loss—estimated as of 1981.

Table 7. Condition of conifers affected by the hemlock looper, *Lambdina fiscellaria*, in the central Prince Edward Island outbreak area, based on merchantable volumes (%) at five areas examined in October 1977, 1978, and 1979.

	Balsam fir			Spruce			Hemlock		
Tree condition	1977	1978	1979	1977	1978	1979	1977	1978	1979
Mortality									
Dead less than 2 years	13.2	50.4	65.1	0	0	0	9.8	41.0	91.3
Dead 2 years or more	3.1	3.5	15.1	0	0	0	0.0	0.0	0.0
Total	16.3	53.9	80.2	0	0	0	9.8	41.0	91.3
Defoliation									
Current only	0	0	0.8	21.9	25.1	20.7	0	0	0
Less than 50% complete	0	13.3	12.4	49.9	66.2	70.6	0	0	0
50 to 90% complete	22.6	15.1	6.6	28.2	8.7	8.7	0	17.4	8.7
More than 90% complete	61.1	17.7	0.0	0.0	0.0	0.0	90.2	41.6	0.0
Total	83.7	46.1	19.8	100.0	100.0	100.0	90.2	59.0	8.7

Table 8. Condition of balsam fir trees affected by the balsam woolly adelgid, *Adelges piceae*, in infested stands in the Maritimes in 1980.

						Tree condition in infested stands[b]			
	Stands		No. trees assessed	Dead[a]		Twig attack		Stem attack	
Province	Examined	Infested		Stands	%Trees	Stands	%Trees	Stands	%Trees
New Brunswick	23	6	74	2	14.9	2	9.5	4	29.7
Nova Scotia	30	20	294	4	2.7	12	16.7	16	33.0
Prince Edward Island	7	7	98	0	0	7	31.6	4	9.2
Region	60	33	466	6	4.1	21	18.7	24	27.5

[a]Tree mortality attributed to balsam woolly adelgid attack only. Trees that died from other causes are not included.
[b]The tree conditions are not mutually exclusive; in any given stand, all three conditions assessed may be present.

of the balsam fir and 91% of the hemlock were dead by 1979. The results of fall assessment surveys are presented in Table 7, which show that although spruce budworm also affected the trees in this area, extensive mortality was restricted to tree species susceptible to hemlock looper damage.

After a period of endemic populations, variable defoliation (some severe) occurred in eastern Prince Edward Island over 200 ha in 1985. This outbreak collapsed by 1986. Light defoliation was found that year in another nearby area, but no outbreaks have developed since. However, moth flights have been observed each year and the number of hemlock looper caught in light traps has increased significantly in all parts of the province.

Balsam Woolly Adelgid

The balsam woolly adelgid, *Adelges piceae*, previously known as the balsam woolly aphid, ranked in importance with the spruce budworm during the 1950s and early 1960s as a cause of balsam fir mortality. A succession of winters with below normal temperatures and light snow cover reduced insect populations in 1961–1962 and again in 1970–1971.

Although some infestations, mostly in the coastal areas of Nova Scotia, persisted through the years, "gouty" twigs were the most common manifestation of the insect's presence and stem attacks were reported only in isolated cases. An apparent resurgence of the insect in 1979 prompted a preliminary evaluation. Results showed affected trees in 11% of the areas in New Brunswick, 74% in Nova Scotia, and 25% in Prince Edward Island.

In 1980, a detailed survey found balsam woolly adelgid (or signs of its activity) in 55% of the 60 stands examined in the region: 26% in New Brunswick, 67% in Nova Scotia, and 100% in Prince Edward Island. Infested stands were evenly distributed in Nova Scotia and Prince Edward Island, but five of the six infested stands in New Brunswick were in the eastern part of the province, four of these in coastal areas. A summary of tree conditions in infested stands is presented in Table 8. Tree mortality attributed to chronic attacks by this insect was found in six stands. Mortality ranged from 4% to 100%.

Infestations, and resulting damage, remained generally low and stable, with isolated trees or small groups of trees occasionally exhibiting "gouty" twigs, "umbrella" tops, or "woolly" stems. One of the more serious infestations was noted in Queens County, Nova Scotia, where moderate to severe gouting occurred in a 122-ha Christmas tree plantation in 1981, with the infestation persisting in 1982.

Indications are that the balsam woolly adelgid became more common in New Brunswick in 1987 and that low-level chronic twig attacks generally persisted in Nova Scotia and Prince Edward Island. Experience shows that given the right conditions, populations can increase rapidly and the insect could once again become a major pest.

Whitemarked Tussock Moth

The whitemarked tussock moth, *Orgyia leucostigma*, can kill coniferous trees after a single year of severe defoliation. Although importance remains undiminished in natural forests, the insect may also become a major economic factor in Christmas tree areas, even at relatively light infestations, causing degrade of the trees.

The last major outbreaks in the Maritimes occurred in New Brunswick and Nova Scotia from 1975 to 1978 (Table 9). Populations built up in 1974 in the extreme southeast of New Brunswick and in mixed stands at Portapique, Cumberland County, Nova Scotia. By 1975, the New Brunswick outbreak covered about 25 000 ha and bordered an area of 1 250 ha of defoliation in Nova Scotia; the Portapique infestation of 8 ha in 1974 expanded to cover more than 1 200 ha; an area of almost 1 100 ha was defoliated in Lunenburg County; and smaller outbreaks existed elsewhere in western Nova Scotia. The outbreaks expanded greatly in the same general areas in 1976, but all of them collapsed in 1977 due to high levels of virus infection. A new outbreak of 32 000 ha occurred in 1977 in central Nova Scotia, but that also collapsed after 1 year because of the virus. Populations were at endemic levels throughout the region until 1985 when larvae became more common, and in some areas of southeastern mainland Nova Scotia, egg masses averaged about one per tree in Christmas tree stands. The population buildup continued in 1986 and some defoliation of hardwoods, larch, and balsam fir occurred, but disease was also present in the population and the infestation collapsed without reaching outbreak proportions. Regionally, the whitemarked tussock moth was common throughout in 1987, but populations were low and light defoliation of hardwoods occurred only over 1 000 ha in northeastern Prince Edward Island.

The numbers are present for a quick buildup in populations if the weather conditions favor the insect, and depending on the controlling effects of the virus, outbreaks may occur.

Table 9. Area (hectares) of moderate or severe defoliation by the whitemarked tussock moth, *Orgyia leucostigma*, in New Brunswick and Nova Scotia, 1973–1979.

	Severe or moderate defoliation	
Year	New Brunswick	Nova Scotia
1973	patches	light
1974	patches	8
1975	25 000	3 555
1976	206 400	345 000
1977	0	116 200
1978	0	32 000
1979	0	0

Gypsy Moth

The gypsy moth, *Lymantria dispar*, is the most destructive insect on hardwoods and, to a lesser degree, conifers in the northeastern United States, where it caused severe defoliation over more than 5 million ha in 1981. The northern edge of that infestation was less than 25 km from the New Brunswick border in 1980.

In 1981, gypsy moth egg masses were found in the Maritimes for the first time in about 45 years. In New Brunswick, egg masses were found at four locations, all in the extreme

Table 10. Summary of results of the gypsy moth, *Lymantria dispar*, male adult pheromone trapping program in the Maritime provinces, 1971–1980.

	Maritimes			New Brunswick			Nova Scotia			Prince Edward Island		
	No.	Positive		No.	Positive		No.	Positive		No.	Positive	
Year	traps	No.	%	traps	No.	%	traps	No.	%	traps	No.	%
1971	190	11	5.8	–	2	–	9	–				
1972	400	3	0.7	–	3	–	0	–				
1973	180	7	3.9	110	4[a]	3.6	56	3	5.4	0	0	–
1974	200	2	1.0	60	1	1.7	57	1	1.7	15	0	–
1975	135	1	0.7	58	1	1.7	57	0	0	16	0	–
1976	135	14	10.4	66	10	15.1	56	4	7.1	15	0	–
1977	132	32	24.2	60	17	28.3	43	15[b]	34.9	16	0	–
1978	246	84	34.1	64	17	26.6	167	67	40.1	15	0	–
1979	135	65	48.1	61	43	70.5	59	15	25.4	15	7	46.6
1980	401	240	59.9	199	128	64.3	181	108	59.7	21	4	19.0

[a] First time that two adults were caught in the same trap.
[b] First time that a multiple catch was obtained. Seven of 15 traps were multiple catches, and the highest number of males in a single trap was 10.

Table 11. Average annual catch of male gypsy moths, *Lymantria dispar*, in pheromone traps in southern New Brunswick, 1976–1980.

Area	No. of traps	Average number of male adults per trap				
		1976	1977	1978	1979	1980
Grand Manan Island	10	0.1	0.2	1.0	4.4	34.6
Campobello	5-6	0.2	0.4	0.2	5.6	26.7
Deer Island	4-6	0.5	0.7	0.7	4.5	21.0
St. Andrews	5-10	0.4	0.6	0.6	17.4	33.0
St. Stephen	3-5	0.3	0.3	0.3	8.0	19.0
St. George	5-6	0.4	1.0	0.4	3.0	8.5
Saint John and area	10	0.0	0.2	0.2	1.4	2.5
Fundy National Park	16	0.1	0.1	0.0	1.0	1.1

southern part of Charlotte County. In Nova Scotia, one egg mass was found at Yarmouth in the western end of the province.

The gypsy moth story in the Maritimes goes back to 1969 when FIDS initiated annual gypsy moth pheromone trapping programs in cooperation with the Plant Protection Division of Agriculture Canada. This was in response to concern about the expanding outbreaks in the United States. As those outbreaks increased in size and moved closer to the Maritimes, so did the gypsy moth adult male catches increase. Table 10 summarizes the results of the trapping program from 1971 to 1980, the year before the first egg mass was found.

Traps along the Bay of Fundy (Table 11) showed a steady increase in the number of moths caught and, by 1980, the average number of moths per trap would have justified severe defoliation in the area. However, defoliation did not occur nor were egg masses found. In 1980, to test the idea that male moths originate elsewhere and are brought into the Maritimes by storm fronts, traps at 55 locations were observed daily and moth catches were recorded to indicate the moth invasion pattern. Adults were trapped from July 16 until September 12, but 81% of the catch was taken between July 17 and 28 in Nova Scotia, 84% between July 30 and August 5 in New Brunswick, and 73% between August 1 and 4 in Prince Edward Island. The daily observation trap experiments, which were repeated in 1981 and 1982, involved several hundred traps and gave similar results, which led to altering the traditional trap placement dates in 1983 to minimize interference from large numbers of male moths brought into the region by weather fronts from infested areas of the United States.

Since 1981, gypsy moth egg masses or larvae have been found at 44 locations in New Brunswick and 26 locations in Nova Scotia (Table 12). However, the first noticeable defoliation did not occur until 1987, at Moores Mills in western Charlotte County, New Brunswick, when gypsy moth larvae caused severe defoliation of both hardwoods and conifers in an isolated spot over 5 ha.

At the end of 1987, the status of the gypsy moth in New Brunswick was as follows: only 8 of the 44 "positive" areas were outside Charlotte County, and except for Fredericton, gypsy moth was found only once at any of these locations. In Saint John, an unemerged pupa was found in 1981 and the lack of any activity in that area since then indicates that larvae rather than egg masses were introduced. A similar situation is likely to have occurred in an area near Peel, Carleton County, where 2 years of intensive trapping and searching have failed to confirm the presence of gypsy moth. More than 80% of the known gypsy moth locations are in Charlotte County and all 17 areas where gypsy moth was found in 1987 were in the western half of the county. Consequently, this area is considered at present to sustain a widespread but low-level population, both in the forest and in the urban setting.

Table 12. Summary of new finds and the presence of gypsy moth, *Lymantria dispar*, in New Brunswick and Nova Scotia, 1981–1987. (Life stages other than male moths.)

Year	New Brunswick		Nova Scotia	
	No. new locations	No. locations present	No. new locations	No. locations present
1981	4	4	1	1
1982	7	8	4	5
1983	5	10	7	10
1984	1	8	4	9
1985	8	13	5	15
1986	7	14	4	13
1987	12	17	1	11

In Nova Scotia, the gypsy moth is widespread, having been found at one time or another in all counties in the western half of the province. All 26 locations are urban or semirural and in most of them, the insect population has remained low. The pattern of occurrence suggests that although most of the New Brunswick infestations are likely to have resulted from natural spread from the neighboring United States, in Nova Scotia most of the introductions may have been the result of traffic from infested areas.

In Prince Edward Island, no gypsy moth has yet been found.

Although most gypsy moth infestations are presently low in the Maritimes, the insect is spreading and the potential exists for a serious outbreak.

Forest Tent Caterpillar

The forest tent caterpillar, *Malacosoma disstria*, caused measurable amounts of defoliation in the Maritimes in each year of the review period. However, apart from a few small patches of incidental outbreaks in Nova Scotia between 1974 and 1978, there were only two separate and distinctly different major outbreaks in the region (Table 13).

An outbreak began in western Prince Edward Island in 1973, causing severe defoliation of trembling aspen in several isolated areas. The outbreak expanded and the patches coalesced in 1974. For several years thereafter, the forest tent caterpillar "moved around" in essentially the same general area. After 3 years (1978–1980) of reduced defoliation, populations increased again and the area of defoliation expanded for 4 more years. The outbreak finally collapsed in 1985. Even though disease and parasitism were first noted in the population in the third year of the outbreak and reached 50% by the sixth year, and pupal mortality was as high as 60% in the seventh year, the outbreak lasted for 11 years.

In 1978, a few small patches of trembling aspen were severely defoliated in western New Brunswick, in the Saint John River valley, along the United States border. Thus started the most serious forest tent caterpillar outbreak ever recorded in the Maritimes. In 1979, large areas of severe defoliation occurred along the river, across the center of the province, and in areas along the east coast. By 1980, the entire river valley was covered and this outbreak coalesced with others in central and eastern New Brunswick, forming a crescent-shaped area of defoliation. Besides the "traditional" hosts (trembling aspen, birch, oak, apple, and cherry), sugar maple was defoliated on a large scale for the first time and ash, Manitoba maple, alder, larch, white spruce, colorado blue spruce, ground cover, and garden vegetables were all defoliated. Mass migration of larvae in search of food was followed by mass starvation and early pupation. In 1982, the area affected in the previous year (which was already a "record") almost doubled. The outbreak started to shift southward, disappeared from the northwestern area, but expanded to the south. It also spilled over into western Nova Scotia where the patches of defoliation, the results of large moth flights in 1980, were beginning to combine into sizeable outbreak areas.

During the outbreak, there was a buildup of disease, parasites, and predators, such as the large flesh fly, *Sarcophaga aldrichi*. At the height of the outbreak, mass starvation of larvae occurred at many locations, which resulted in fewer egg masses laid. All of these were factors in weakening the forest tent caterpillar population. In addition, there were 2 consecutive years when the early summer was cool and precipitation was greatly above normal. A weakened forest tent caterpillar population, combined with weather conditions unfavorable to its survival, hastened the almost total collapse of the outbreak in New Brunswick in 1984. Similar weather in the other two provinces was probably a factor in controlling those outbreaks.

The effect of the forest tent caterpillar, along with other stress factors, resulted in a deterioration of the affected hosts, especially trembling aspen and sugar maple. An assessment survey in 1987 found the proportion of healthy trembling aspen to be 53% in Nova Scotia, 26% in Prince Edward Island (a provincial average), and only 3% in New Brunswick.

Table 13. Area (hectares) of forest tent caterpillar, *Malacosoma disstria*, outbreaks in the Maritimes, 1973–1987.

	Area of severe/moderate defoliation		
Year	New Brunswick	Nova Scotia	Prince Edward Island
1973	patches	patches	260
1974	–	–	28 000
1975	–	98	37 700
1976	–	8	37 700
1977	–	8	37 700
1978	patches	8	5 000
1979	37 000	–	5 000
1980	177 000	trace	3 100
1981	775 000	patches	13 800
1982	1 389 000	4 700	18 800
1983	1 119 000	35 000	67 000
1984	94 400	46 000	37 400
1985	–	patches	–
1986	–	200[a]	–
1987	–	200[a]	–

[a] Associated with large aspen tortrix, *Choristoneura conflictana*.

Insects of Plantations

A Seedling Debarking Weevil

A seedling debarking weevil, *Hylobius congener*, has been suspected as the causal agent in the mortality of newly planted coniferous seedlings in central Nova Scotia since the beginning of the 1980s. Its association with the problem was first reported in 1984, when seedling mortality exceeded 85% in some plantations. The insect is present on sites at harvest time or soon afterward. Debarking of the stems of the seedlings occurs from spring to fall, resulting in progressively increasing seedling mortality. The amount of damage is related to forest management practices, such as "hot planting" (that is, the reforesting of newly harvested areas and the proximity of plantations to other recently harvested areas). The concern about this insect by forest managers relates to their increased awareness unexplained plantation failures may have been the result of weevil damage, and the realization that in the absence of practical control methods, the future of large-scale plantation programs may be in jeopardy.

In 1986, a survey of about 130 areas showed that although the most serious damage occurred in eastern mainland Nova Scotia and eastern Prince Edward Island, considerable damage also occurred elsewhere in the region. Table 14 illustrates the percentage of plantations by province in the various damage categories.

In 1987, the weevil continued to damage and kill seedlings in recently established plantations, especially where

Table 14. Frequency of plantations in five categories of seedling mortality in plantations affected by a seedling debarking weevil, *Hylobius congener*, in the Maritimes in 1986.[a]

Mortality range (%)	Plantations (%) New Brunswick	Nova Scotia	Prince Edward Island
None	73	39	41
1-5	20	32	17
6-10	7	14	0
11-20	0	11	17
21-100	0	4	25

[a] Seventeen negative plantations on Cape Breton Island, N.S., have been omitted from the calculations.

planting occurred within two field seasons of the harvesting operation. Damage levels were lower than in 1985 or 1986, apparently because of the extremely dry summer conditions that restricted weevil movement and feeding. However, significant damage still occurred on many sites, resulting in poorly and unevenly stocked plantations. The most severely affected areas remained eastern mainland Nova Scotia and eastern Prince Edward Island. Reports from New Brunswick were scattered and infrequent, but significant mortality levels were encountered in Sunbury County (15%) and Kent County (21%) on black spruce.

Research is in progress to find means of controlling the seedling debarking weevil. Without such control, the insect may continue to jeopardize successful reforestation efforts under certain forest management practices.

Spruce Bud Moths

Spruce bud moth, *Zeiraphera canadensis* (and to a much lesser extent, the purple striped shootworm, *Z. unfortunana*), have been omnipresent forest pests in the Maritimes for as long as FIDS records have been kept, that is, since the late 1930s. Although widespread, insect populations have generally been low except for the occasional flare-up, usually on open-grown white spruce. The last recorded outbreak occurred in New Brunswick in the mid-1960s when spruce in parts of the Southwest Miramichi and the Nashwaak River drainage systems sustained moderate to severe defoliation and also, in the mid-1970s, in Nova Scotia when similar levels of defoliation occurred in areas along the Northumberland Strait and the Fundy Coast.

The spruce bud moth, a not-too-important forest insect in mature forests, became a major pest when, in 1980, it was discovered to be causing defoliation, shoot distortion, and tree deformation in white spruce plantations over large areas of northwestern New Brunswick. In 1982, more than two-thirds of the 180 locations surveyed in the region were infested by spruce bud moth. At more than 40% of these, in both New Brunswick and Prince Edward Island, defoliation and shoot damage were in excess of 10% and were classed as moderate or severe at 10% and 20% of the locations, respectively, in those provinces.

Spruce bud moth was again widespread in 1983 in New Brunswick and in Prince Edward Island, but only sporadic in Nova Scotia. Populations were high and defoliation of white spruce in the spring was considerable over large areas of natural forests in northern New Brunswick. At the same time, in some of these areas, defoliation on balsam fir was minimal. As the season progressed and feeding by the spruce budworm increased on both spruce and balsam fir, the initial difference in defoliation levels on the two hosts became less obvious. Also, spruce bud moth feeding was more difficult to detect on white spruce. By the end of the combined feeding period, the spruce budworm had "out fed" and masked the damage caused by the spruce bud moth in many areas, as witnessed by the abundance of the spruce budworm pupal cases in stands where spruce bud moth was the initial defoliator. The area so affected by spruce bud moth in 1983 has not been determined. During spruce budworm defoliation surveys, 94 000 ha were mapped where detectable foliage discoloration occurred on white spruce without visible defoliation on the balsam fir component. This was likely caused by the spruce bud moth. However, the total area affected was not restricted to this figure.

By 1985, repeated feeding in large areas of white spruce plantations in northwestern New Brunswick resulted not only in shoot distortion but in some instances up to 50 cm of top mortality, involving all of the multiple leaders that developed on the affected trees because of repeated damage. Also in 1985, ground surveys in Nova Scotia indicated that a portion of the 11 600 ha of defoliation mapped during spruce budworm aerial surveys in Antigonish County may have been caused by the spruce bud moth, *Z. canadensis*, rather than the spruce budworm. This observation fits in well with surveys in plantations, indicating that even though on a provincial average spruce bud moth damage remained low (about 10% of shoots affected), there was an increase in populations in certain areas with appropriately higher levels of damage in all three provinces. Although some white spruce plantations remained completely free of the insect, shoot damage levels as high as 96% were recorded in New Brunswick, 55% in Nova Scotia, and 70% in Prince Edward Island.

Of the two species, *Z. canadensis* is the more common, although the proportion of *Z. unfortunana* can reach as high as 21%, as was recorded in a collection from Prince Edward Island in 1987.

The spruce bud moth is a good example of an insect that changed from an insignificant member of the forest ecosystem to a major pest because of changing forest management practices. Even though the insect is not a tree killer, its repeated attacks will distort trees and prevent foresters from realizing the expected maximum value in seriously affected plantations.

European Pine Shoot Moth

The European pine shoot moth, *Rhyacionia buoliana*, was found in the Maritimes for the first time only in 1925, but it spread quickly, probably on infested nursery material, throughout Nova Scotia and Prince Edward Island to become the major forest insect in red pine and Scots pine plantations.

Populations, and the accompanying damage of shoot mortality, stunting, and tree deformation, were high in both Nova Scotia and Prince Edward Island throughout the 1970s, and unaffected plantations, especially in Nova Scotia, were hard to find. Damage was severe, as illustrated by the following examples: a) 73% of the trees had more than 10 shoots infested (or dead) and 26% of the trees had 5–10 shoots infested (and dead) in a 6- to 8-year-old red pine plantation 1 m high; b) 67% of the trees had more than 30% of the shoots killed in another plantation; c) 80% of the shoots were killed on all trees in still another one. In 1980, there was exceptionally good shoot growth and even though insect populations (and damage) were high, apparent damage was less.

Extremely low winter temperatures in 1980–1981 caused very high overwintering larval mortality, almost eliminating the insect from many previously badly damaged plantations. Growth was good in 1981, even in Lunenburg County in southwestern Nova Scotia, where larval survival was highest.

Populations started rebounding in 1982 and damage began to appear. The highest shoot mortality was recorded in Lunenburg County and in parts of central Prince Edward Island. Overall populations of the European pine shoot moth have increased, but they have not reached the pre-1980 levels, although severe damage occurs in many plantations (often similar to that mentioned previously). The insect is considered to be common, but at generally low populations.

Other Damaging Forest Insects

Limitations of space prevent the inclusion of descriptions of the status of other insects that caused severe or moderate damage between 1973 and 1987. These are presented, however, in three tables, one for hardwood insects, another for insects of the natural coniferous forest, and the third for plantation pests. A word of caution is needed on interpreting these tables.

Damage is indicated if it was mentioned at the moderate or severe level in the region, regardless of the size of the problem or the effect of the insect on the forest. For example, a few wild apple trees defoliated by winter moth or thousands of cubic metres of wood killed by the greenstriped mapleworm are indicated in the same way. Please see the appropriate annual report for complete information.

In plantations, a somewhat distorted picture is created because of the restriction of mentioning only moderate or severe damage. Damage may be significant at much lower levels, both in the short term (e.g. 20% of "light" needle loss due to balsam gall midge in a Christmas tree plantation degrades the tree and results in economic loss) and in the long term (e.g. chronic low-level infestations of spruce bud scale result in growth loss). These situations are not indicated in the table.

Mature Forest—Hardwoods

Insect	Prov.	Severe or Moderate Damage														
		73	74	75	76	77	78	79	80	81	82	83	84	85	86	87
Ambermarked birch leafminer	NB									x	x		x			
Profenusa thomsoni	NS															
	PEI															
Oak leaftier	NB															
Psilocorsis quercicella	NS			x	x											
	PEI															
Birch casebearer	NB	x	x	x		x		x	x	x		x	x	x	x	
Coleophora serratella	NS	x		x	x	x	x	x	x	x		x	x	x	x	x
	PEI	x	x	x			x	x	x	x	x	x	x	x	x	x
Birch leafminer	NB	x	x	x	x	x	x	x	x	x	x	x	x	x		x
Fenusa pusilla	NS	x	x	x	x	x	x	x	x	x	x	x	x	x		x
	PEI	x	x	x	x	x	x	x		x	x	x	x	x		x
Birch sawfly	NB															
Arge pectoralis	NS							x								
	PEI															
Birch skeletonizer	NB			x		x								x		
Bucculatrix canadensisella	NS			x	x	x								x	x	x
	PEI			x	x	x								x		
Bruce spanworm	NB		x					x	x		x	x		x		
Operophtera bruceata	NS		x	x												
	PEI															

(Continued)

Mature Forest—Hardwoods *(Continued)*

Insect	Prov.	Severe or Moderate Damage														
		73	74	75	76	77	78	79	80	81	82	83	84	85	86	87
Eastern tent caterpillar	NB								x	x		x	x			
Malacosoma americanum	NS								x	x	x	x	x			
	PEI									x						
Elm leaf beetle	NB						x					x	x			
Xanthogaleruca luteola	NS															
	PEI															
Elm leafminer	NB	x	x	x	x		x	x	x	x	x	x	x	x		x
Fenusa ulmi	NS	x	x	x	x	x	x	x	x	x	x	x	x	x	x	x
	PEI			x	x	x	x	x	x	x	x	x	x	x	x	x
Fall cankerworm	NB		x	x	x	x			x	x	x	x	x	x		
Alsophila pometaria	NS	x	x	x	x			x				x	x	x	x	
	PEI	x	x		x						x	x	x	x	x	
Fall webworm	NB	x	x													
Hyphantria cunea	NS	x	x												x	
	PEI		x					x								
Greenstriped mapleworm	NB					x	x	x								
Dryocampa rubicunda	NS	x	x	x			x	x	x							
	PEI															
Large aspen tortrix	NB	x	x													
Choristoneura conflictana	NS					x									x	x
	PEI		x			x										
Lesser maple spanworm	NB	x	x	x										x		
Itame pustularia	NS															
	PEI															
Maple leafroller	NB	x	x	x	x	x	x	x	x							x
Sparganothis acerivorana	NS		x	x	x	x	x				x					
	PEI	x	x	x					x			x	x	x	x	x
Mountain ash sawfly	NB	x	x						x			x	x	x		x
Pristiphora geniculata	NS	x	x									x	x	x		x
	PEI	x	x									x	x	x	x	x
Oak olethreutid leafroller	NB															
Pseudexentera spoliana	NS					x			x	x	x	x		x	x	x
	PEI													x	x	
Oak leafshredder	NB		x	x	x		x	x	x				x	x		
Croesia semipurpurana	NS	x				x	x									
	PEI	x			x	x			x	x			x			
Orangehumped mapleworm	NB															
Symmerista leucitys	NS											x	x	x		
	PEI															
Poplar leafmining sawfly	NB							x	x	x	x					x
Messa populifoliella	NS															
	PEI															

Mature Forest—Hardwoods *(Concluded)*

Insect	Prov.	73	74	75	76	77	78	79	80	81	82	83	84	85	86	87
		Severe or Moderate Damage														
Spotted aspen leafroller	NB							x								
Pseudosciaphila duplex	NS				x	x		x								
	PEI					x										
Aspen serpentine leafminer	NB				x		x	x	x	x	x	x	x	x	x	x
Phyllocnistis populiella	NS															
	PEI						x									
Redhumped oakworm	NB															
Symmerista canicosta	NS	x	x													
	PEI															
Saddled prominent	NB				x											
Heterocampa guttivitta	NS	x	x									x	x			
	PEI															
Satin moth	NB	x	x	x	x	x	x	x	x	x		x	x	x	x	x
Leucoma salicis	NS	x	x	x	x	x	x	x	x	x		x	x	x		
	PEI	x	x	x	x	x	x	x	x	x		x	x	x	x	x
Spring cankerworm	NB															
Paleacrita vernata	NS	x				x		x		x	x					
	PEI															
Winter moth	NB			x								x				
Operophtera brumata	NS		x	x	x			x	x		x	x	x			
	PEI			x	x			x	x		x	x				

Mature Forest—Coniferous Forests

Insect	Prov.	73	74	75	76	77	78	79	80	81	82	83	84	85	86	87
		Severe or Moderate Damage														
Balsam fir sawfly	NB															
Neodiprion abietis complex	NS			x	x											
	PEI															
Balsam gall midge	NB								x	x	x	x				
Paradiplosis tumifex	NS		x	x	x	x	x				x	x	x			x
	PEI											x				
Balsam shootboring sawfly	NB							x								
Pleroneura brunneicornis	NS															
	PEI															
Balsam twig aphid	NB				x		x				x	x	x			x
Mindarus abietinus	NS				x	x					x	x	x		x	x
	PEI					x						x				
Cedar leafminers	NB															
	NS								x							
	PEI	x	x	x	x	x	x	x	x	x	x	x	x	x	x	x

(Continued)

Mature Forest—Coniferous Forests *(Concluded)*

		Severe or Moderate Damage														
Insect	Prov.	73	74	75	76	77	78	79	80	81	82	83	84	85	86	87
Jack pine budworm	NB				x							x				
Choristoneura pinus	NS															
	PEI															
Larch casebearer	NB							x	x							
Coleophora laricella	NS	x	x	x		x	x		x			x	x	x	x	
	PEI		x	x								x		x		
Pine leaf adelgid	NB															
Pineus pinifoliae	NS													x		x
	PEI															
Spider mites	NB										x					
	NS															
	PEI															
Spruce coneworm	NB															
Dioryctria reniculelloides	NS							x								
	PEI															
White pine sawfly	NB															
Neodiprion pinetum	NS															x
	PEI															

Plantations

Insect	Prov.	73	74	75	76	77	78	79	80	81	82	83	84	85	86	87
Aphids	NB															
Cinara spp.	NS										x					
	PEI															
Eastern spruce gall adelgid	NB								x							
Adelges abietis	NS															x
	PEI															
Jack pine budworm	NB				x					x	x					
Choristoneura pinus	NS															
	PEI															
Northern pitch twig moth	NB											x				
Petrova albicapitana	NS															
	PEI															
Pine tortoise scale	NB						x	x	x	x						
Toumeyella parvicornis	NS						x			x						
	PEI															
Ragged sprucegall aphid	NB				x	x	x									
Pineus similis	NS				x	x										
	PEI					x										
Redheaded jack pine sawfly	NB															
Neodiprion rugifrons	NS												x			
	PEI															

Plantations *(Concluded)*

Insect	Prov.	Severe or Moderate Damage														
		73	74	75	76	77	78	79	80	81	82	83	84	85	86	87
Red pine sawfly	NB															
Neodiprion nanulus nanulus	NS	x	x	x	x		x	x	x	x						
	PEI				x	x										
Root collar weevils	NB															
Hylobius spp.	NS	x	x	x	x	x	x	x	x	x	x	x	x	x	x	x
	PEI	x	x	x	x	x	x	x	x	x	x	x	x	x	x	x
Spruce bud midge	NB									x						
Rhabdophaga swainei	NS								x							
	PEI															
Spruce bud scale	NB		x	x	x	x	x									
Physokermes piceae	NS	x	x	x											x	x
	PEI															
Spider mites	NB									x			x	x	x	
Oligonychus milleri	NS									x					x	
Oligonychus ununguis	PEI															
Swaine jack pine sawfly	NB															
Neodiprion swainei	NS						x	x								
	PEI															
White pine weevil	NB	x	x				x	x	x		x		x		x	x
Pissodes strobi	NS	x	x					x	x		x		x		x	x
	PEI	x						x	x		x		x		x	
Yellowheaded spruce sawfly	NB							x		x		x				x
Pikonema alaskensis	NS	x	x	x	x		x	x	x	x	x					
	PEI				x					x	x	x		x	x	x

Chapter 3
Forest Insect Pests in the Quebec Region

D. Lachance

Introduction

From 1973 to 1987, the spruce budworm, *Choristoneura fumiferana*, continued to be the most serious forest pest in Quebec because of the extent and severity of the infestation and the level of damage. It appears, however, that this insect will approach the endemic level by the early 1990s.

During the period covered by this report, populations of forest tent caterpillar, *Malacosoma disstria*, fluctuated considerably and caused serious damage in several regions of the province. The large aspen tortrix, *Choristoneura conflictana*, was also reported on several occasions over extensive areas of the forest. Finally, there was a substantial increase in spruce bud moth, *Zeiraphera canadensis*, populations in white spruce plantations in the Lower St. Lawrence–Gaspé Peninsula. This phenomenon may be interpreted as a sign of new situations or problems that may arise in the vast, artificially regenerated stands in Quebec and the rest of Canada.

In this chapter, the insects are discussed under three headings: major forest insects, 1973 to 1987; insects attacking plantations; and minor forest insects, 1973 to 1987. Some of the insects that were important in Quebec before 1973 (Prebble 1975) have continued to remain so, while others have caused little recent damage. Other insects that were not discussed in the earlier report have increased considerably in numbers at some point since 1973.

Epidemic levels of insects are discussed, but not the results of annual surveys on their presence or local and normal fluctuations at the endemic level. In addition, control action taken against some insects is occasionally mentioned where appropriate. Most of the data are taken from the regional annual reports prepared by the Forest Insect and Disease Survey (FIDS) of the Laurentian Forestry Centre in cooperation with the Forest Insect and Disease Control branch of the Quebec Department of Energy and Resources, with whom we have been working for approximately 10 years.

Major Forest Insects, 1973 to 1987

Spruce Budworm

The spruce budworm, *Choristoneura fumiferana*, is by far the most damaging insect in Quebec's softwood forests. Three widespread outbreaks have occurred in the province in the present century, one in 1910, another in 1940, and a third in 1967. It is difficult to compare the severity of the three outbreaks, either on a biological basis or on the extent of the damage to the forest, because the condition of the stands affected and the methods available to assess the losses and to control the insect differed considerably from one outbreak to the next. However, the three infestations are reported to have covered 10, 25, and 55 million ha of forest, respectively (Blais 1985).

According to Swaine and Craighead (1924), the first outbreak killed more than 80% of Quebec's balsam fir stands of merchantable age located south of the treeline and extending from the border of Ontario on the west to the Saguenay region on the east. The second infestation resulted in the loss of approximately 85 million cunits (230 million m^3) of balsam fir and spruce. This figure approximates the recent estimate provided by the Quebec Department of Energy and Resources of 235 million m^3 for the third outbreak, which is now ending. This loss is significant in view of the major control efforts made in the 1970s.

The last outbreak began to reach epidemic levels in 1967 and the first surveys of the extent of defoliation were reported in 1968 (Table 1). Like the previous outbreaks, it began in western Quebec. Major pockets of defoliation appeared in the Lower St. Lawrence in 1970 and in the Gaspé Peninsula in 1972. Balsam fir mortality occurred over approximately 250 000 ha in western Quebec in 1973. The infestation reached its peak in 1975 (Figs. 1–5) when areas of moderate to severe defoliation covered more than 28 million ha (Table 1). But it appears that this outbreak cannot extend any farther because virtually all of the forest prone to attack by the budworm has already been attacked. In fact, the infestation began to decline somewhat in extent in 1976, although the areas exhibiting balsam fir and white spruce mortality increased rapidly at that time. In 1976, 1978, 1980, and 1982, tree mortality was reported on 3.8, 6.3, 9.3, and 11.2 million ha, respectively. Beginning in 1986, surveys on the areas exhibiting tree mortality were suspended, partly because of the high level of regeneration that had begun to appear in the dead stands of recent years, and also because of the vast expanses of dead forest that had been felled for salvage.

This outbreak is reported to have been at epidemic levels for 20 years, excluding the few pockets that may have persisted after 1988 in the Gaspé Peninsula and the few small pockets reported in western Quebec in 1965 and 1966.

The control operations against the spruce budworm in Quebec are discussed in Chapter 65.

Forest Tent Caterpillar

Outbreaks of the forest tent caterpillar, *Malacosoma disstria*, a polyphagous insect that feeds openly on hardwoods, generally recur on a 10- to 12-year cycle. Owing to the vast area covered by hardwood forests in Quebec, we will deal with the various outbreaks of the insect on the basis of the area affected. Western Quebec, particularly the census division of Temiscamingue, the Saguenay/Lac-Saint-Jean region, and the south St. Lawrence River will be discussed separately.

Table 1. Areas defoliated by the spruce budworm, *Choristoneura fumiferana*, in Quebec, excluding areas exhibiting tree mortality, from 1968 to 1988, and areas treated annually (data provided by Quebec Department of Energy and Resources).

Year	Defoliation (× 1 000 ha)			Areas of tree mortality (million ha)	Treated areas (× 1 000 ha)
	Light	Moderate/severe	Total		
1968	84	67	151	0	0
1969	577	227	804	0	0
1970	827	1 595	2 242	0	12
1971	1 892	3 385	5 277	0	669
1972	1 740	8 540	10 280	0	760
1973	3 642	6 428	10 070	0.2	3 936
1974	9 356	20 241	29 597	1.1	2 568
1975	4 192	28 108	32 300	3.0	2 887
1976	9 945	19 391	29 336	3.8	3 656
1977	5 489	21 627	27 116	5.7	1 396
1978	3 055	13 203	16 258	6.3	1 246
1979	560	4 751	5 311	8.3	582
1980	1 031	5 191	6 222	9.3	189
1981	632	6 303	6 935	10.2	705
1982	1 062	8 787	9 849	11.2	1 185
1983	946	12 266	13 212	12.0	1 254
1984	3 904	7 139	11 043	12.6	709
1985	2 969	6 291	9 260	12.9	667
1986	752	2 080	2 832	–[a]	51
1987	302	740	1 042	–	198
1988	274	434	708	–	192

[a] Beginning in 1986, areas exhibiting mortality were no longer evaluated.

Table 2. Areas of light and moderate to severe defoliation (in hectares) caused by forest tent caterpillar, *Malacosoma disstria*, in Quebec from 1973 to 1987.

Year	West[a]		South[a]		North[a]		Entire province		
	Light	Mod./severe	Light	Mod./severe	Light	Mod./severe	Light	Mod./severe	Total
1973	–	9 000	–	–	–	–	–	9 000	9 000
1974	–	7 800	–	–	–	–	–	7 800	7 800
1975	–	10 000	–	–	–	3 100	–	13 100	13 100
1976	P[b]	10	–	–	–	4 000	P	4 010	4 010
1977	P	–	–	133	–	4 600	P	4 733	4 733
1978	P	60	–	3 065	1 000	3 600	1 000	6 725	7 725
1979	P	500	500	90 000	–	3 000	500	93 500	94 000
1980	500	500	12 500	750 000	300	30 300	125 800	780 800	906 600
1981	–	400	22 340	718 820	–	17 500	22 340	735 720	758 060
1982	8 000	8 500	49 970	344 471	–	–	42 000	318 500	360 500
1983	–	–	–	–	–	–	–	–	–
1984	–	185	–	–	–	–	–	185	185
1985	8 281	182 344	–	–	–	–	8 281	182 344	190 625
1986	91 563	479 220	–	–	–	–	91 563	479 220	570 783
1987	346 682	561 531	–	–	–	–	346 682	561 531	908 213

[a] West part of the province, including north of Montreal; south of the St. Lawrence River, including the Lower St. Lawrence and the Gaspé Peninsula; north of the St. Lawrence River, including La Mauricie, Saguenay/Lac-Saint-Jean, and the North Shore.
[b] P: defoliation present, extent not evaluated but generally small.

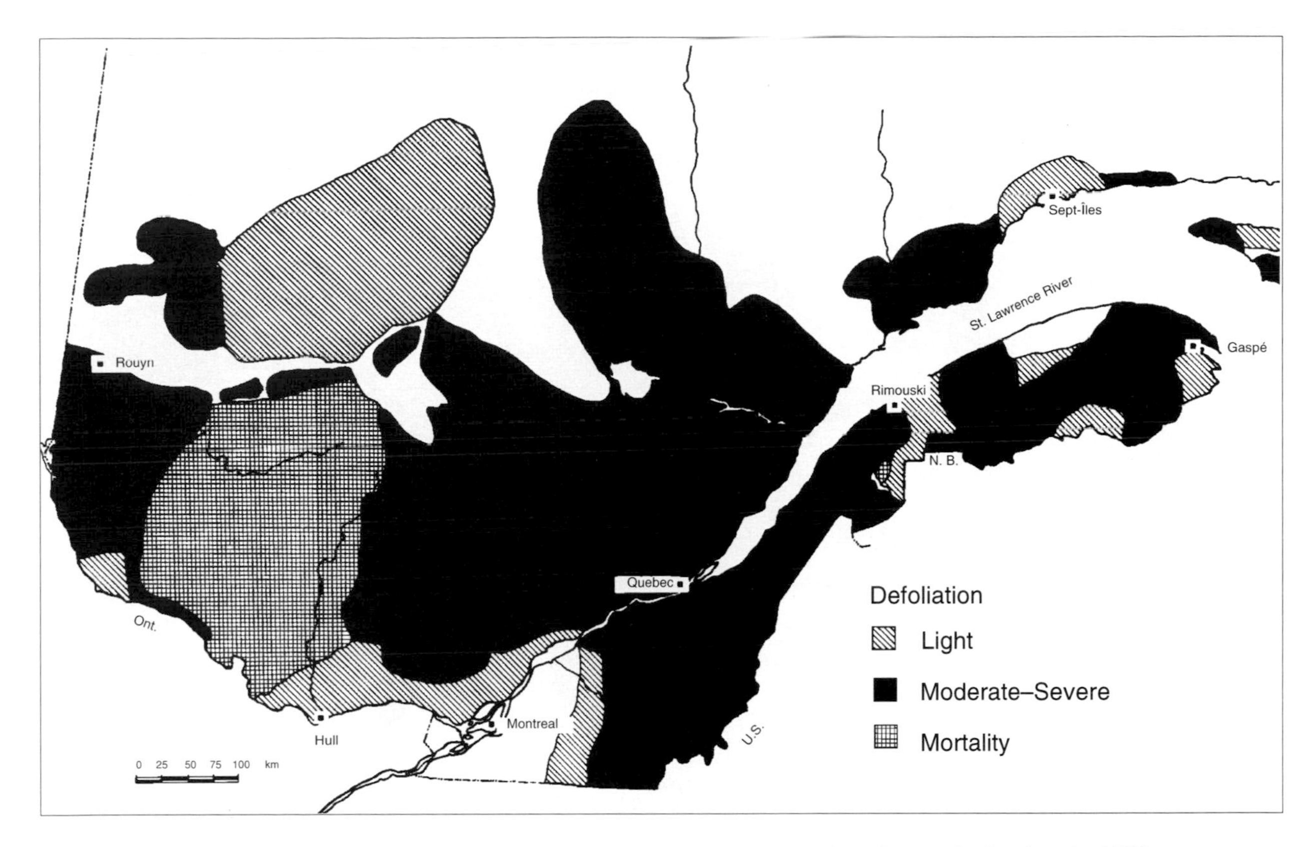

Figure 1. Areas infested by spruce budworm, *Choristoneura fumiferana*, in Quebec in 1975.

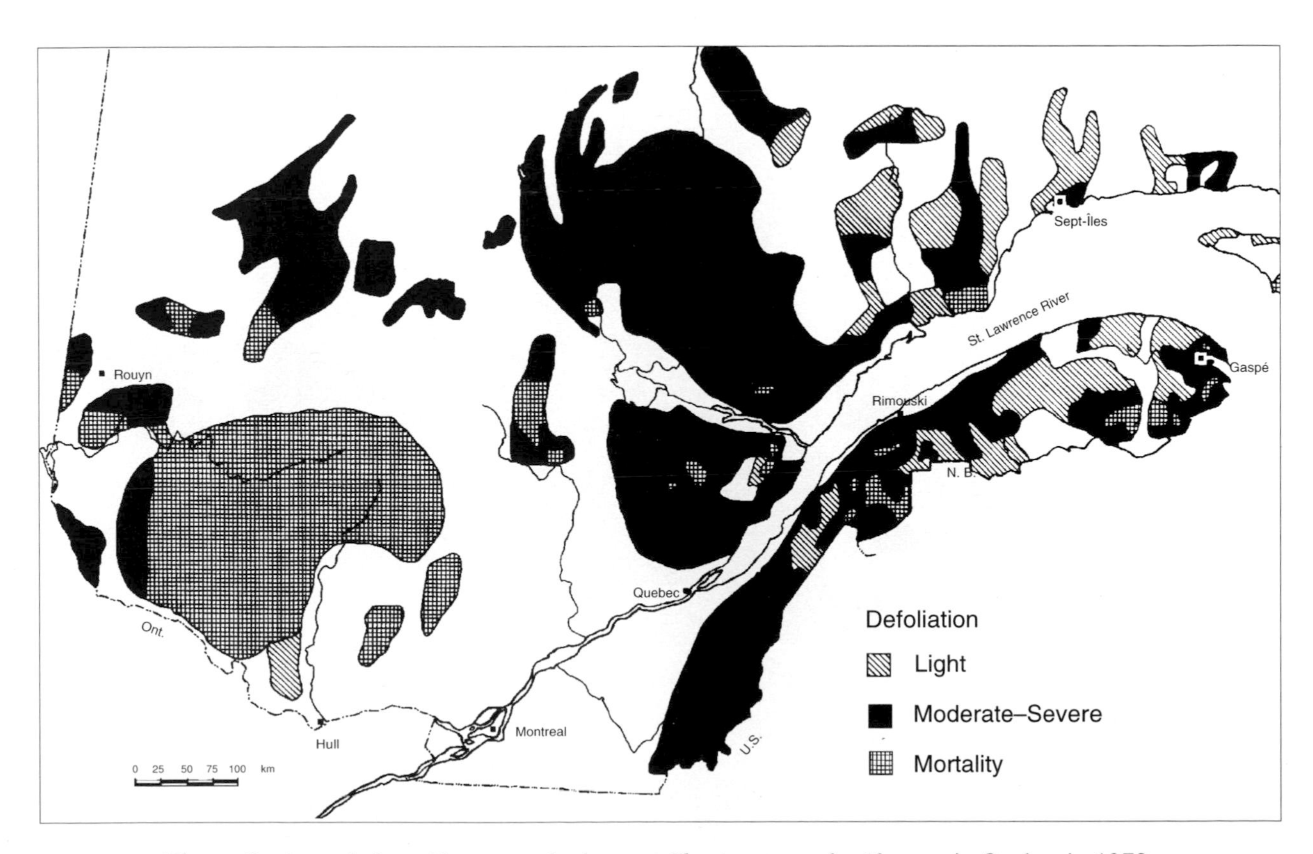

Figure 2. Areas infested by spruce budworm, *Choristoneura fumiferana*, in Quebec in 1978.

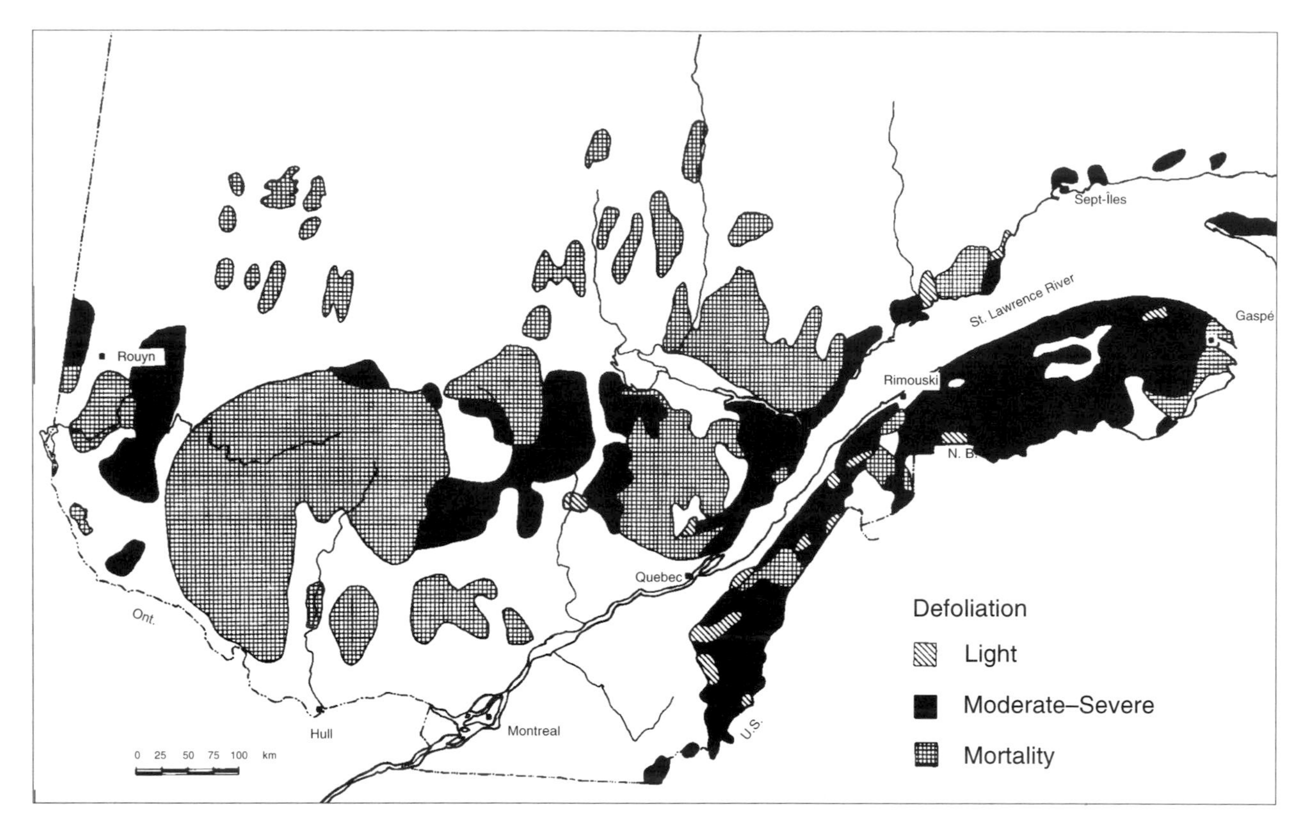

Figure 3. Areas infested by spruce budworm, *Choristoneura fumiferana*, in Quebec in 1981.

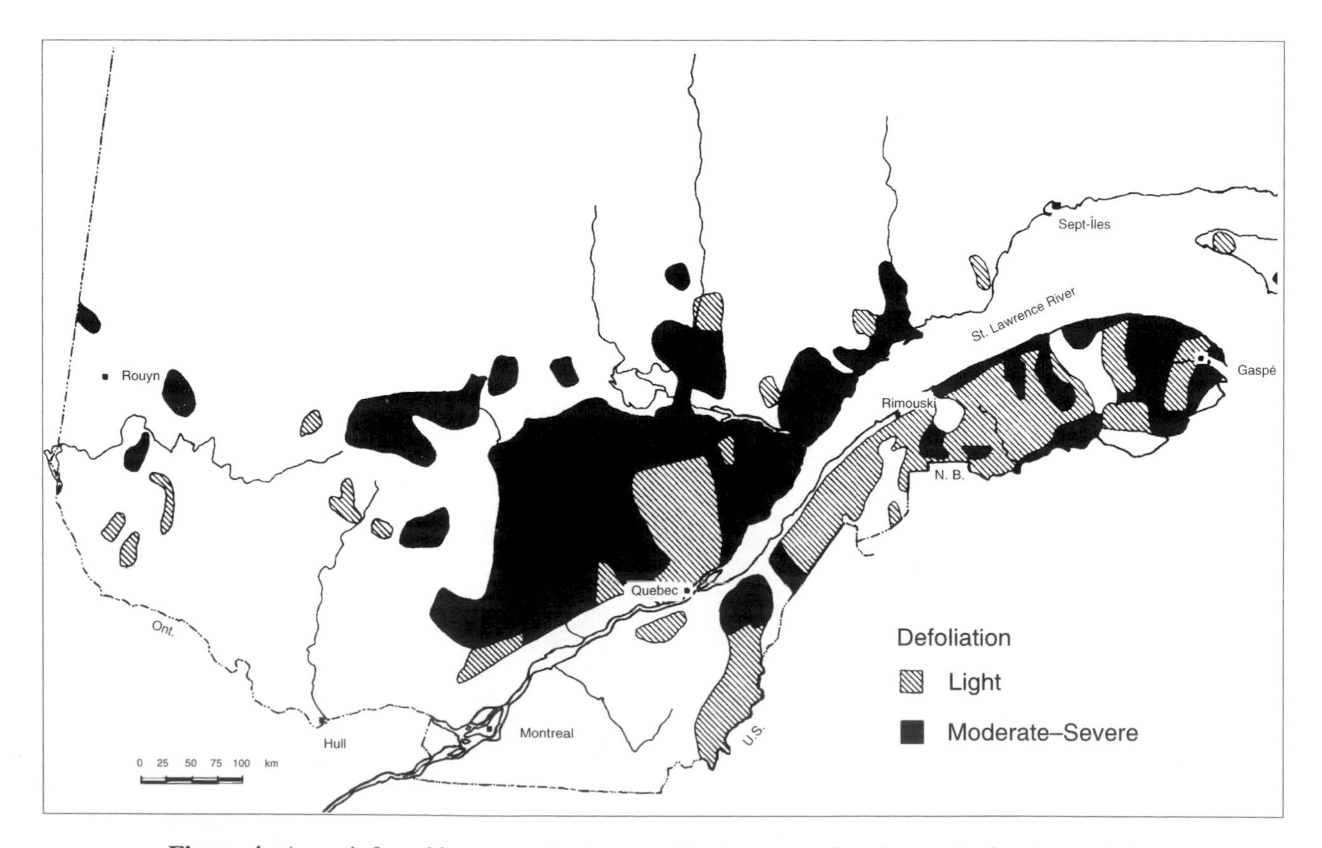

Figure 4. Areas infested by spruce budworm, *Choristoneura fumiferana*, in Quebec in 1984.

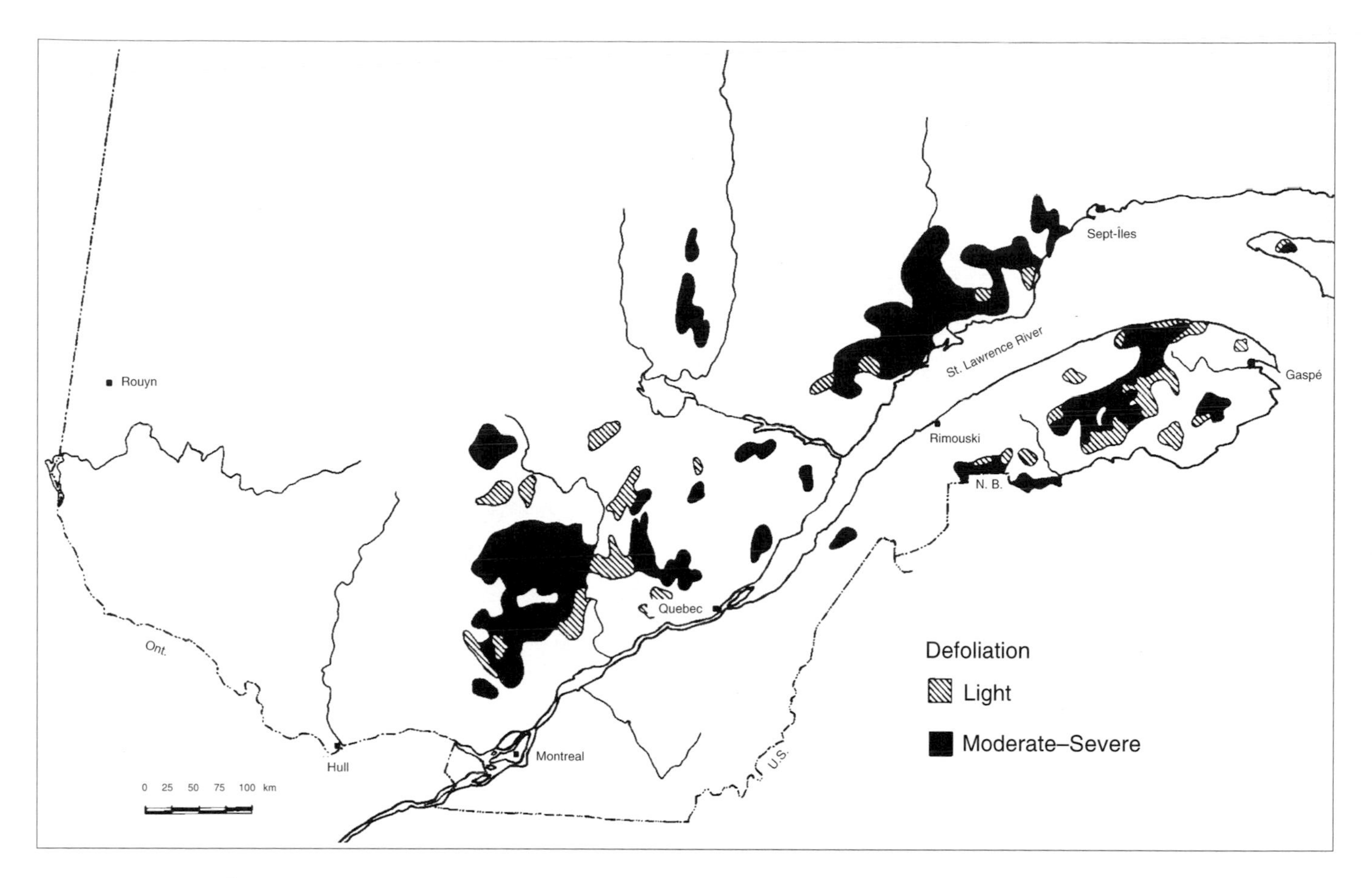

Figure 5. Areas infested by spruce budworm, *Choristoneura fumiferana*, in Quebec in 1986.

The forest tent caterpillar caused some damage primarily to trembling aspen and white birch in western Quebec between 1973 and 1975, particularly near the Ontario border. It almost disappeared in 1976 (Table 2), but reappeared in large numbers in the same area in 1985 and has remained at epidemic levels since. The main area affected is located between the municipalities of Rouyn to the north and Temiscamingue to the south, and stretches from the Ontario border to about 100 km to the east. In 1986, in addition to the trembling aspen, the forest tent caterpillar fed on white birch, speckled alder, balsam poplar, largetooth aspen, and sugar maple. In 1987, areas of severe defoliation declined substantially whereas areas of moderate defoliation increased substantially. However, parasitism of almost 90% was reported in the larvae. Preliminary data for 1988 show that the outbreak declined dramatically in this area.

In the Saguenay/Lac-Saint-Jean area, an infestation that began in 1975 remained at virtually the same level until 1980, when the outbreak of this pest was felt throughout the province. The main sites affected in the region were approximately 36 km^2 of trembling aspen near Notre-Dame-de-la-Doré, and about 10 km^2 of trembling aspen near Sainte-Jeanne-d'Arc. The forest tent caterpillar population fell dramatically in 1981, and almost disappeared in 1982.

The forest tent caterpillar had not been present in southern Quebec in the area south of the St. Lawrence River since 1969 but was reported scattered throughout the area in 1975 and 1976, causing no major damage.

In 1977, pockets of defoliation south and southeast of Montreal were reported over approximately 60% of 133 ha of affected trembling aspen stands. The infestation then spread and intensified, peaking in 1980 and 1981. In 1981, it was estimated that the forest tent caterpillar defoliated 586 000 ha of trembling aspen forests and 172 000 ha of sugar maple, a large percentage of which suffered moderate to severe defoliation (Fig. 6). That same year, the Quebec Department of Energy and Resources successfully conducted an experimental aerial control operation against the forest tent caterpillar, spraying seven maple stands with a formulation of *Bacillus thuringiensis* (Bt) to protect an experiment under way on sugar maple productivity in these areas.

By 1982, the infestation in southern Quebec had declined considerably in severity and extent, decreasing to 397 000 ha of trembling aspen and 14 000 ha of sugar maple, down 33 and 92%, respectively, from the year before. This pest virtually disappeared the following year, owing to unfavorable spring weather conditions and the presence of many parasites and diseases in the population. In 1987, the forest tent caterpillar was still nonexistent in this region.

Bruce Spanworm

The Bruce spanworm, *Operophtera bruceata*, which is native to North America, was collected on 16 hardwood

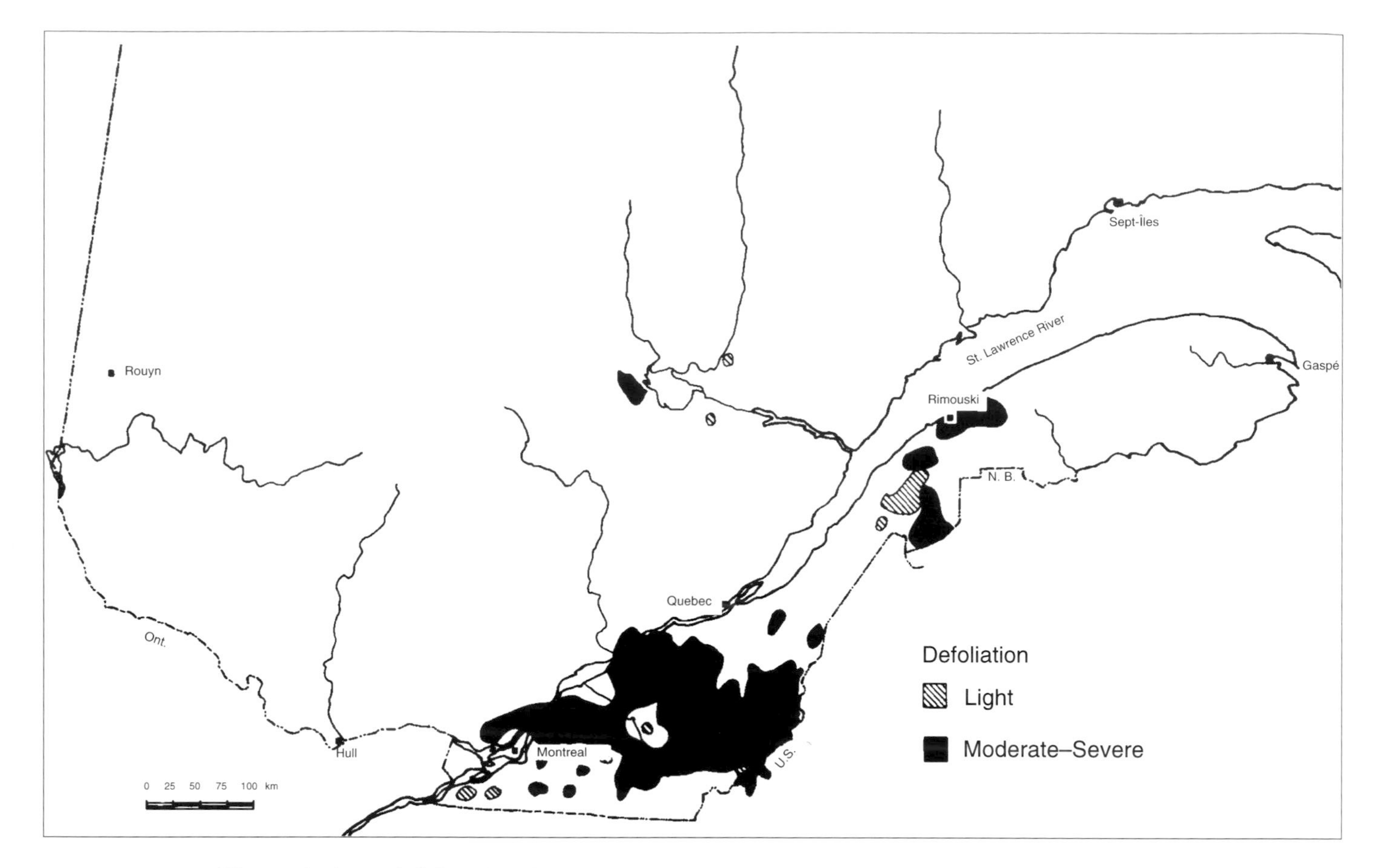

Figure 6. Areas defoliated by forest tent caterpillar, *Malacosoma disstria*, in Quebec in 1981.

species from across Canada. In the east, its preferred hosts are sugar maple and trembling aspen. Major outbreaks in Quebec are relatively recent. The first outbreak was from 1962 to 1965 and a second from 1969 to 1973. No control program was planned at the time of these outbreaks.

A new outbreak occurred throughout Quebec's hardwood forest from 1982 to 1986. In 1982, the Bruce spanworm was found in several sugar maple and trembling aspen stands in the Lower St. Lawrence region. At that time, moderate to severe defoliation occurred in almost 5 000 ha of trembling aspen and 500 ha of sugar maple stands. The following year, the infestation increased considerably, particularly in southern Quebec. It reached its peak in 1984 (Fig. 7) when the area of defoliation totaled almost 72 000 ha in southern and eastern Quebec. This total included areas of moderate and severe defoliation over 21 000 ha of sugar maple and 18 000 ha of trembling aspen stands.

Bruce spanworm populations declined dramatically in these locations beginning in 1985, but increased quickly in western Quebec. In 1986, this insect defoliated 87 000 ha of forest, 83% of which suffered moderate to severe defoliation in the Outaouais area and in the north part of the Montreal area (Fig. 8).

Populations of this insect generally returned to the endemic level the following year.

Large Aspen Tortrix

The last major outbreak of the large aspen tortrix, *Choristoneura conflictana*, in Quebec was from 1968 to 1975. It reached its peak in 1972, at which time an area of almost 10 000 km^2 of forest was defoliated. The insect then attacked trembling aspen throughout the province south of 51° latitude. Moderate to severe defoliation occurred in the following five areas: (1) Lac-Saint-Jean, Saguenay, Montmorency, and Charlevoix; (2) L'Islet, Kamouraska, Rivière-du-Loup, and Rimouski; (3) Compton, Frontenac, Beauce, and Dorchester; (4) Labelle and Gatineau; and (5) Temiscamingue. However, the extent of defoliation was not evaluated.

As is often the case, four other defoliators that feed on the aspen were collected in the same regions. All are associated with the large aspen tortrix, but are not as predominant. They are aspen two leaftier, *Enargia decolor*, birch-aspen leafroller, *Epinotia solandriana*, aspen leafroller, *Pseudexentera oregonana*, and spotted aspen leafroller, *Pseudosciaphila duplex*.

In 1973, the severity of the defoliation declined considerably even though the range of the insect remained virtually unchanged. Its populations continued to fall substantially for the next 2 years, reaching the endemic level in 1976.

An outbreak of large aspen tortrix appears to have begun in 1985. At that time, moderate defoliation occurred in several new stands of trembling aspen, one of which was

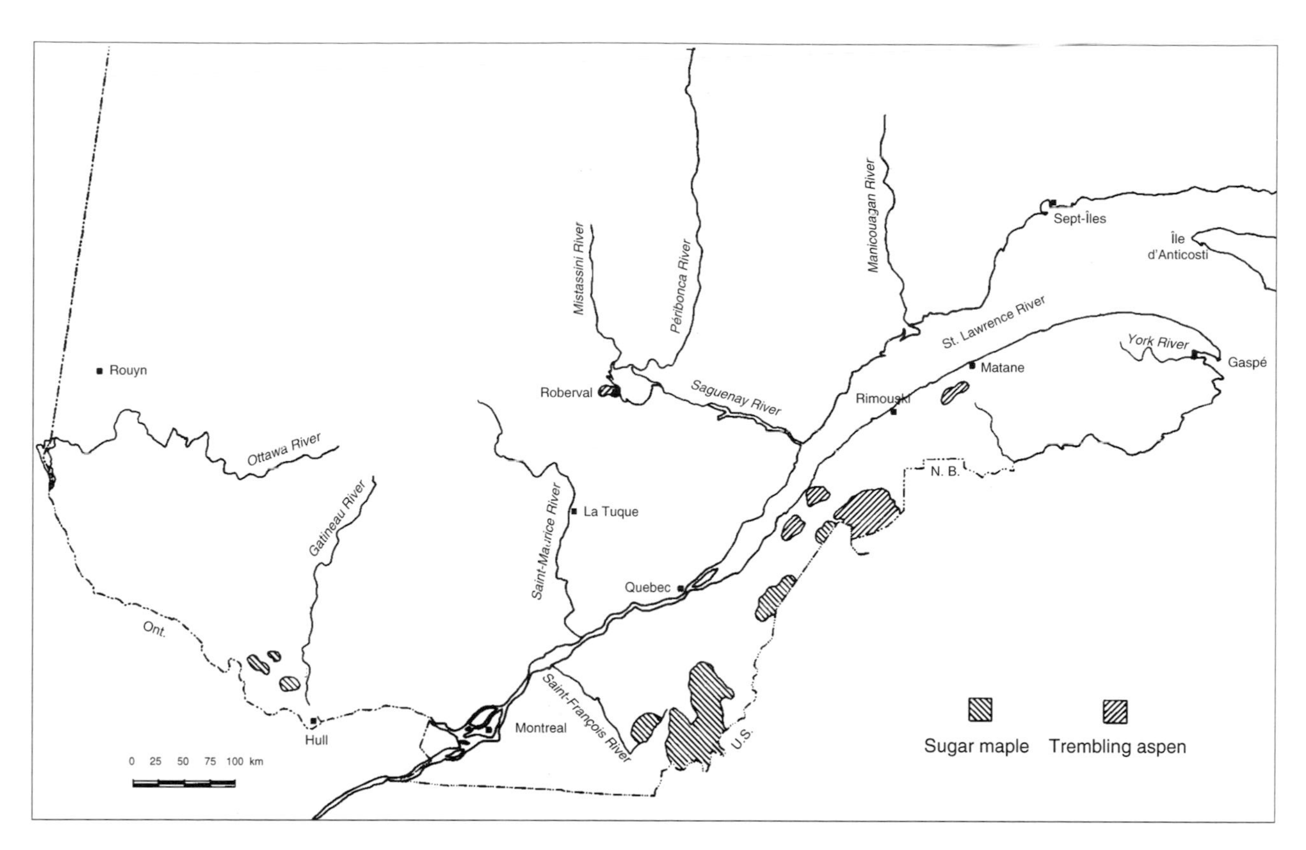

Figure 7. Areas of moderate to severe defoliation of sugar maple and trembling aspen caused by Bruce spanworm, *Operophtera bruceata*, in Quebec in 1984.

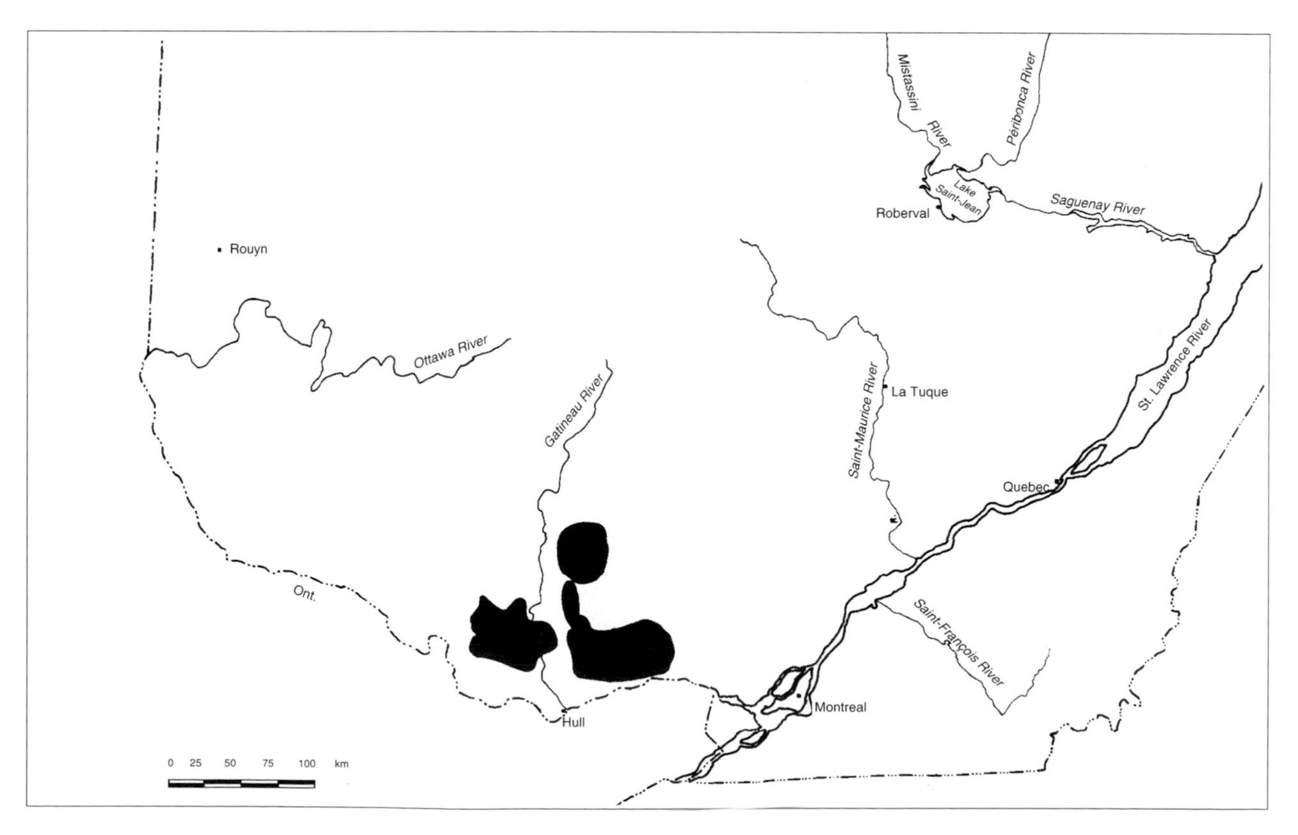

Figure 8. Areas of moderate to severe defoliation caused by Bruce spanworm, *Operophtera bruceata*, in western Quebec in 1986.

100 ha, in the Matane, Kamouraska, and Rivière-du-Loup regions. The widespread presence of the insect was also observed in the Saguenay and Lac-Saint-Jean-Ouest regions. In 1986, approximately 13 000 ha of forest were defoliated to varying degrees on the North Shore of the St. Lawrence River.

In 1987, the insect completely disappeared from the South Shore of the St. Lawrence River but more than 200 000 ha of forest were defoliated on the North Shore and the Saguenay/Lac-Saint-Jean area, nearly 80 000 ha of which suffered moderate to severe defoliation (Fig. 9). This infestation is reported to have declined substantially in 1988.

Gypsy Moth

The gypsy moth, *Lymantria dispar*, was introduced into North America in 1870 and was first detected in Canada in 1924 at three locations in southern Quebec: Henrysburg, Lacolle, and Stanstead. In 1936, it was reported near Saint Stephen, New Brunswick. This pest was eradicated from these localities in 2 years through energetic control operations. In 1956, an egg mass was discovered in the Clarenceville region (Missisquoi), a short distance east of Lacolle, and was destroyed by plant quarantine officials from Agriculture Canada. Subsequently, egg masses were found each year in the same region until 1959, when three major pockets were reported from which the insect later spread (Fig. 10). Measures were taken each year from 1959 to 1965 to completely eliminate the gypsy moth from the several newly discovered pockets, but these efforts were in vain. It is now believed that the gypsy moth is well established in Quebec in the entire area located to the south of the St. Lawrence River, as far east as Quebec City, and including the Chaudière River valley. The insect is also well established in an approximately 20-km band north of the St. Lawrence River that extends from Quebec City to Hull.

The level of infestation varies considerably from year to year on a local basis. However, the insect's general level of activity has decreased dramatically since 1986 and was practically nonexistent in 1988.

The annual populations of this insect have been monitored and forecast through a network of pheromone traps established in 1981 over an area extending from Victoriaville to Lotbinière. This network, which is operated by FIDS of the Laurentian Forestry Centre and maintained in collaboration with Agriculture Canada's Plant Protection Division, was later extended to two adjacent areas. The results showing the moths collected are presented in Table 3.

Aspen Serpentine Leafminer

The aspen serpentine leafminer, *Phyllocnistis populiella*, attacks the lamina of the leaf and can be distinguished from other poplar leafminers by this fact. Although it is present throughout the range of the trembling aspen in Canada, outbreaks of this insect are generally larger in western Canada. The last major outbreak in Quebec began in 1978 when two areas were particularly hard hit: the

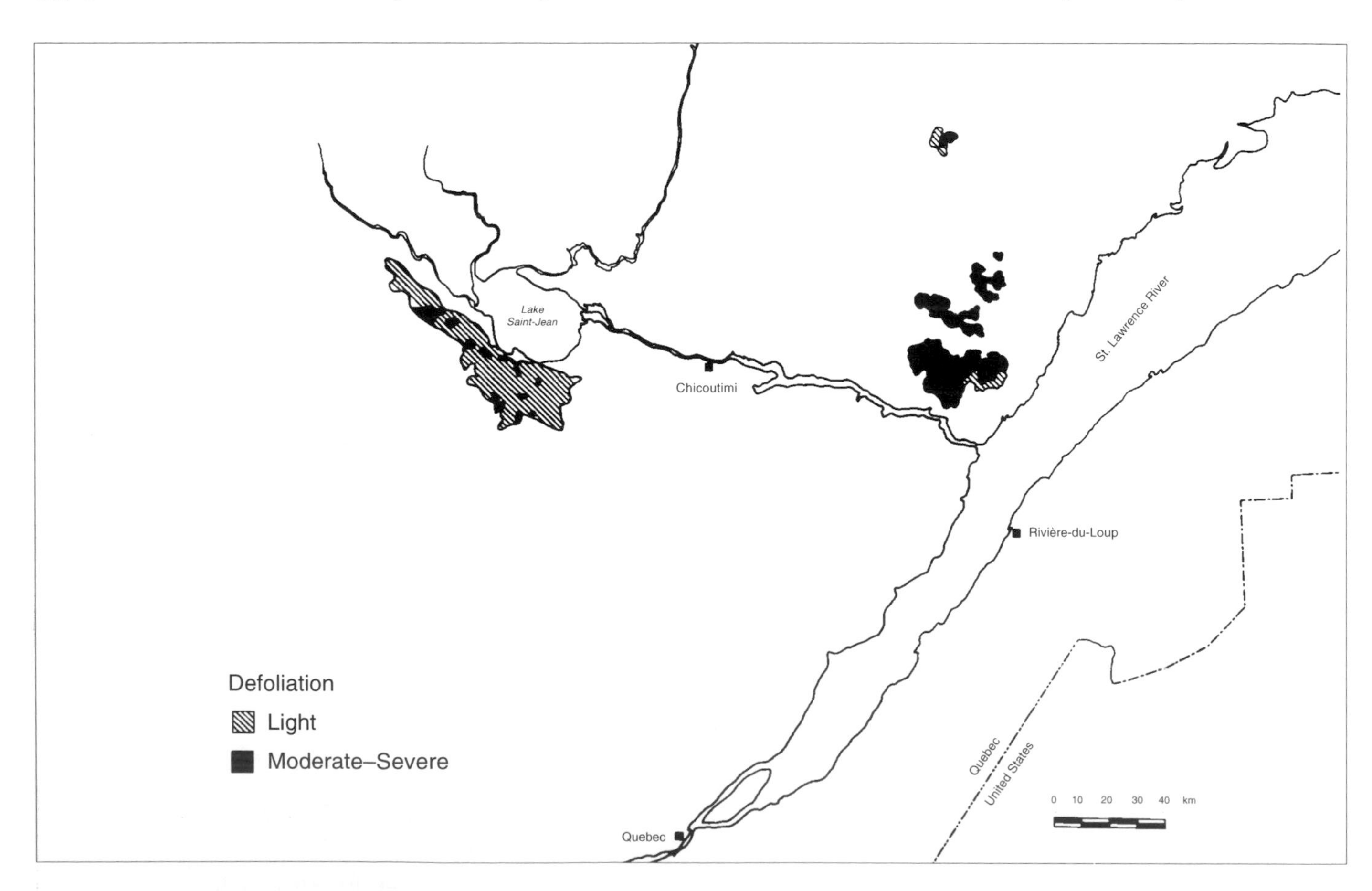

Figure 9. Areas defoliated by large aspen tortrix, *Choristoneura conflictana*, in Quebec in 1987.

Table 3. Male gypsy moths, *Lymantria dispar*, collected in south-central Quebec.

Sector monitored	Year	Number of traps	Total number of moths	Average per trap and maximum per trap ( )	Increase or decrease (%) compared with the previous year
Saint-Antoine-de-Tilley to Inverness	1984	40	681	17.0(29)	–
	1985	44	1 287	19.3(47)	+72.4
	1986	45	376	8.4(22)	-71.3
	1987	45	106	2.3(6)	-72.6
	1988	48	97	2.0(11)	-13.0
Victoriaville to Lotbinière	1981	60	373	6.2(16)	–
	1982	58	871	15.0(38)	+141.9
	1983	57	1 074	18.8(33)	+25.3
	1984	58	1 557	26.8(50)	+42.6
	1985	60	1 989	33.2(68)	+23.9
	1986	60	1 267	21.1(40)	-36.5
	1987	59	986	16.7(37)	-20.9
	1988	60	773	12.9(27)	-22.7
Drummondville	1986	50	1 188	23.8(46)	–
	1987	49	1 164	23.8(40)	0.0
	1988	47	797	17.0(30)	-28.6
Total	1984	98	2 238	22.8(50)	–
	1985	104	3 276	31.5(68)	+38.2
	1986	155	2 831	18.3(46)	-41.9
	1987	153	2 256	14.7(40)	-19.7
	1988	155	1 667	10.8(30)	-26.5

Figure 10. Progression of the outbreak of gypsy moth, *Lymantria dispar*, in Quebec beginning in 1924.

Lower St. Lawrence–Gaspé Peninsula and the North Shore of the St. Lawrence River.

In the Gaspé Peninsula, the infestation became serious in 1981 and reached its peak in 1983, severely damaging more than 1 700 km^2 of poplar forest (Table 4). Virtually all trembling aspen growing in the region were affected (Fig. 11). In 1986, the infestation declined substantially, and a land survey indicated that only 350 km^2 were affected and the damage was generally light. The infestation remained more or less at this level in 1987.

Table 4. Extent of aspen serpentine leafminer, *Phyllocnistis populiella*, infestation in the Lower St. Lawrence–Gaspé Peninsula, from 1981 to 1985, evaluated by aerial surveys.

	Degree of infestation (km^2)			
Year	Light	Moderate	Severe	Total
1981	87	241	148	476
1982	97	465	393	955
1983	337	770	1 709	2 816
1984	586	535	672	1 794
1985	988	884	117	1 990
1986[a]	350	–	–	350

[a] Land survey only.

The aspen serpentine leafminer also caused considerable damage on the North Shore of the St. Lawrence River in the Manicouagan and Toulnustouc river basins from 1982 to 1987. In this region, however, another miner of the trembling aspen, *Lyonetia* sp., was also very active in the same stands and, in some places, caused as much damage as the aspen serpentine leafminer. This infestation appears to have peaked in 1985, affecting more than 1 400 km^2 of forest (Table 5). Land surveys conducted in 1986 and 1987 revealed that this major infestation was still present in the same area as in 1985 but was less severe.

Insects Attacking Plantations

White Pine Weevil

The white pine weevil, *Pissodes strobi*, is a native insect that is present nearly everywhere in Quebec, except on the North Shore of the St. Lawrence River. It continually causes damage, sometimes serious, in local plantations. However, there are no regions in which it is particularly widespread and no major outbreaks have been observed. In 1978, the white pine weevil was reported to have caused light damage,

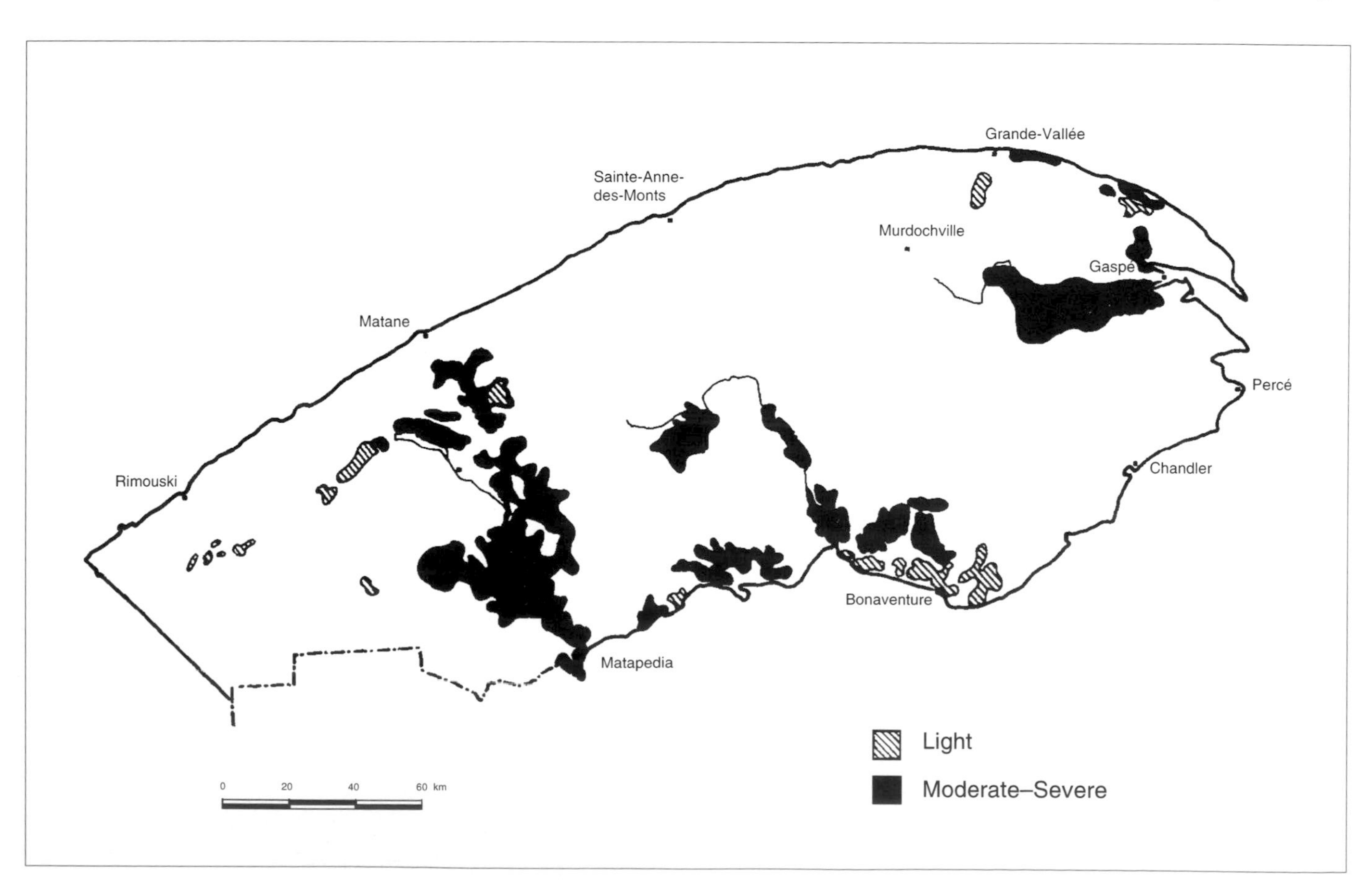

Figure 11. Areas of infestation by aspen serpentine leafminer, *Phyllocnistis populiella*, Lower St. Lawrence–Gaspé Peninsula, Quebec, in 1983.

Table 5. Aerial survey of the areas of trembling aspen affected by the aspen serpentine leafminer, *Phyllocnistis populiella*, and *Lyonetia* sp. on the North Shore from 1982 to 1985.

Year	Degree of infestation (km²)			
	Light	Moderate	Severe	Total
1982	–	–	–	20
1983	8	87	281	376
1984	114	159	242	515
1985[a]	102	711	594	1 407

[a] Aerial surveys were not conducted in 1986 and 1987. However, from land surveys, it is reported that virtually the same areas were affected as in 1985 but less severely.

4% to 10% of the stems affected, in natural black spruce stands in the Abitibi and Temiscamingue regions in the northwest part of the province.

The real potential for damage by the white pine weevil is in pine and spruce plantations, particularly Norway spruce.

Spruce Bud Moth

The spruce bud moth, *Zeiraphera canadensis*, was introduced from Europe in the late 1800s and first reported in Quebec in 1941. It was the subject of much attention in the forestry industry from 1939 to 1958, during the outbreak of spruce budworm, *Choristoneura fumiferana*, when three spruce bud moths were associated with the damage observed. Another insect was the yellow spruce budworm, *Z. fortunana*.

The spruce bud moth has a marked preference for small open-grown spruce trees approximately 2 m high. Therefore, spruce plantations are its favorite site. Although it is present at the endemic level virtually everywhere in Quebec, it began to cause considerable damage in the white spruce plantations of the Lower St. Lawrence–Gaspé Peninsula in the early 1980s.

A qualitative survey conducted in late June 1985 in 136 white spruce plantations revealed the presence of spruce bud moths and damage in 106 plantations and the existence of damage only in 30 plantations. Seven of the 30 plantations suffered particularly severe damage. The insect was present in the entire Lower St. Lawrence–Gaspé Peninsula. The severity of the outbreaks increased somewhat in 1986 but has since remained stable.

According to a survey conducted in 1987, the spruce bud moth was present in 88% of the white spruce plantations in this region and signs of activity were observed on the terminal growth on 32% ± 17.3% of the reforested white spruce trees in this area.

Redheaded Pine Sawfly

The redheaded pine sawfly, *Neodiprion lecontei*, is a native insect that causes noticeable damage mainly in red pine plantations. Although the insect is present in a few plantations each year, it was found more commonly and caused more damage than usual in red pine plantations located in the southern parts of Pontiac and Gatineau regions in southwestern Quebec from 1974 to 1978. In 1978, the Quebec Department of Energy and Resources, in collaboration with the Forest Pest Management Institute of Forestry Canada, carried out a spray operation with a viral suspension in 37 plantations covering an area of more than 700 ha near Fort Coulonge. Excellent control was obtained through caterpillar mortality and foliage protection. In nontreated plantations, damage was more important, but natural control factors eventually brought the insect population down to endemic levels. Since then, in this and other regions, the insect is always present but at endemic levels.

Minor Forest Insects, 1973 to 1987

Jack Pine Budworm

Although there are vast stands of jack pine in Quebec, the jack pine budworm, *Choristoneura pinus*, has caused no major damage to date. It was first reported in the province in 1967, when a small outbreak occurred on about 100 ha on Calumet Island in the Pontiac region. This outbreak disappeared the following year.

Two other small outbreaks have subsequently been reported, one in 1970 near the Baskatong reservoir (Gatineau region) and the other in 1972, a few kilometres northwest of the first outbreak. These two pockets were successfully treated with chemicals and were included in the control programs against the spruce budworm that were then conducted near these sites. In 1970, 1 100 ha were treated and in 1972, 7 500 ha were treated, which included a buffer zone of about 2 500 ha around the area actually affected. Two applications of the insecticide fenitrothion were applied at 0.25 and 0.5 L/ha. The results of the control program were excellent. Since then, the insect has caused no major damage in Quebec.

In 1986 and 1987, because of the probable influence of a major outbreak in Ontario ongoing since 1985, the Forest Insect and Disease Control Service of the Quebec Department of Energy and Resources conducted special research and sampling activities on the insect and collected male moths using pheromone traps. These activities were concentrated along the Ontario–Quebec border, in the Abitibi–Temiscamingue area. In 1986, several larvae were found in 13 natural jack pine stands, but none were collected in 1987. Sampling using 60 pheromone traps located on 20 sites in this region resulted in the collection of only a few moths per site for both years of the survey. This insect has remained at the endemic level in Quebec since 1972.

Swaine Jack Pine Sawfly

Outbreaks of Swaine jack pine sawfly, *Neodiprion swainei*, recur at 8- to 10-year intervals. In Quebec, the last major outbreak ended in 1968. The insect remained at the endemic level until 1980, when local moderate to severe defoliation in areas of jack pine forests began to occur. Tree mortality began to occur in 1982. These relatively isolated infestations were located primarily south of Roberval in the Lac-Saint-Jean-Ouest region and in the northern part of the Champlain region. Moderate to severe defoliation occurred in these locations over areas of approximately 32 000 ha and 3 000 ha, respectively. Aerial chemical spraying was carried out in 1982 and 1983 in the Champlain and Lac-Saint-Jean-Ouest regions, respectively, to prevent high tree mortality. Total control of the insect was obtained both times.

The Swaine jack pine sawfly has continued to attack several stands in these two regions since then, but damage has remained at trace to light levels.

Larch Sawfly

The larch sawfly, *Pristiphora erichsonii*, has been present at the endemic level in Quebec since 1968. A slight increase in its populations was reported from 1974 to 1976 in the regions of Abitibi and Saguenay–Lac-Saint-Jean, where several stands were moderately to seriously affected.

Eastern Larch Beetle

The eastern larch beetle, *Dendroctonus simplex*, occurs primarily in larch stands that have been frequently attacked by larch casebearers, *Coleophora laricella*, or that have been weakened by other causes. Between 1980 and 1983, this insect was observed more often than usual in some stands, and cases of larch mortality were observed in south-central Quebec. A survey was conducted in 1983 in 41 larch stands in that area to briefly assess the effect of this insect. A total of 182 ha were evaluated and 93 ha contained some dead or attacked trees; the beetle was present in all stands except one. After analysis of the data obtained for 7 039 trees, it was estimated that 3 030 m^3 (22.5% of total stand volume) would die in the future. The mean diameter at breast height of all trees in these stands was 19 cm, whereas that of dead or attacked trees was 22 cm. This insect has caused no major damage since.

Balsam Woolly Adelgid

The balsam woolly adelgid, *Adelges piceae*, was first detected in Quebec in 1964 in the Magdalen Islands and in the eastern Gaspé Peninsula; however, it was likely present in these areas for about 15 years (Martineau 1985).

The area of infestation by the balsam woolly adelgid, spread in the 1960s, but no increase has been reported since 1974. A recent survey revealed that about 2 500 km^2 were infested in the Gaspé Peninsula.

Although local populations continue to be active, they cause no new major damage.

Fall Cankerworm

The fall cankerworm, *Alsophila pometaria*, is occasionally present in Quebec on several hardwood species and is often associated with other insects. Its populations have remained at the endemic level in Quebec since 1958. Moderate to severe defoliation occurred only in a few stands in the lower Gatineau region in 1975 and 1976. Elsewhere, only isolated and generally minor cases of defoliation are found.

Oak Leafshredder

Since the relatively major outbreak in the outskirts of Quebec City from 1957 to 1966, the oak leafshredder, *Croesia semipurpurana*, is only occasionally observed in the province. There was a slight increase in its populations from 1973 to 1976 in the southwest part of the province, primarily in the Pontiac, Gatineau, and Papineau regions, where oak trees are more abundant than elsewhere in Quebec. In 1975, defoliation was moderate or severe in 21 of the 26 stands sampled. Since 1977, the insect has been present at the endemic level.

Hemlock Looper

The outbreak of hemlock looper, *Lambdina fiscellaria*, from 1971 to 1973 on Anticosti Island and the North Shore of the St. Lawrence River and the control program carried out there were described in detail in Prebble 1975. Since 1973, no damage has been reported in Quebec, even locally.

Fall Webworm

Outbreaks of fall webworm, *Hyphantria cunea*, which is native to North America, generally recur at approximately 12-year intervals and last an average of 5 years. In Quebec, this insect was considered to be at epidemic levels from 1972 to 1976. The infestation peaked in 1974 and 1975, at which time the webworm was present throughout the range of hardwood species in Quebec.

In 1975, webs were reported to have been observed on 21 different tree species. The main ones were cherry, elm, trembling aspen, birch, and ash. The evaluation of infestation levels in 155 different localities in southern Quebec revealed large populations of the insect in 27% of the localities, moderate populations in 23%, small populations in 28%, and trace levels in 22%. No measures were taken to control this insect.

The fall webworm was again present at the endemic level from 1977 to 1981. From 1982 to 1986, it occurred more frequently in southern Quebec, where it caused local damage. By 1987, it was practically nonexistent.

Larch Casebearer

The larch casebearer, *Coleophora laricella*, was accidentally introduced from Europe into North America in 1886 and was first collected in Canada in Ontario in 1905. Since then, this pest has caused frequent damage to larch trees. In Quebec, the scattered distribution of larch stands and their relatively small sizes may help keep this insect at the endemic level.

However, a major increase in larch casebearer populations occurred in 1979 and 1980. In 1980, the pest was very commonly found from the Gatineau region in the west to the Bellechasse region in the east and from 47° latitude south to the United States border. Outside this area, the larch casebearer was present only at trace levels.

Defoliation in the affected region was generally moderate to severe in stands of approximately 1 or 2 ha each. At Stanstead (Stanstead region), 28 ha were completely defoliated. The most severe outbreak occurred between Montreal and Saint-Michel-des-Saints in the Laurentians, where areas of moderate and severe defoliation were roughly equal. The

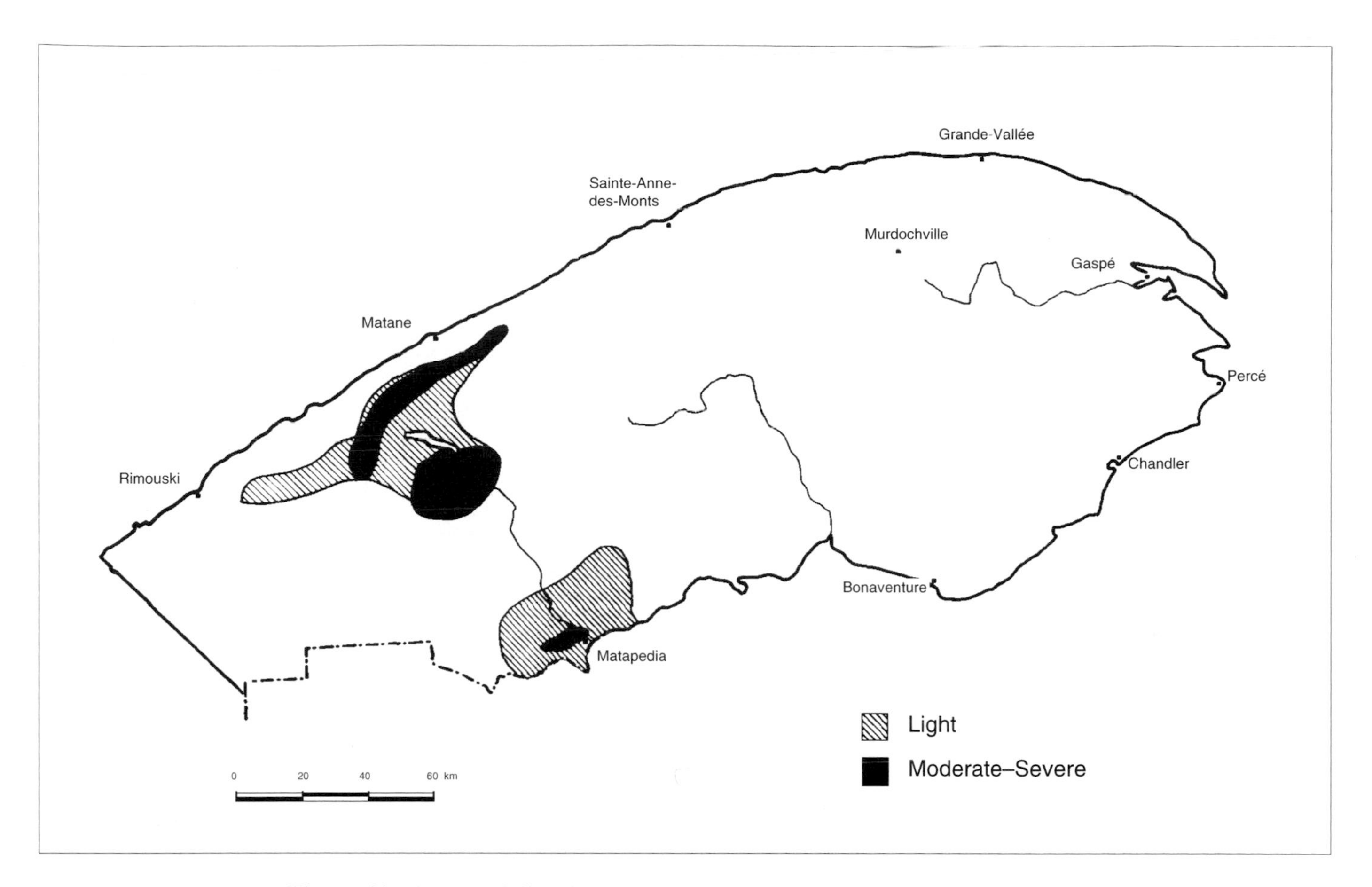

Figure 12. Areas defoliated by the maple leafroller, *Sparganothis acerivorana*, St. Lawrence–Gaspé Peninsula, Quebec, in 1977.

average size of the stands affected were usually about 1 ha. The infestation collapsed completely in 1981.

Maple Leafroller

The maple leafroller, *Sparganothis acerivorana*, feeds primarily on maples but has been reported on 11 hardwood species. Damage by this insect is caused by the caterpillar, which rolls the leaf and then destroys it, which can result in defoliation. Although the maple leafroller is well established in Quebec and is observed annually there, an unusual situation occurred in 1977 when more than 3 000 km^2 of maple forest were affected in the Lower St. Lawrence–Gaspé Peninsula (Fig. 12). The two main species affected were sugar maple and red maple.

About 500 km^2 were severely defoliated in the vicinity of Amqui in the Matapedia region. A second area of about 600 km^2 located closer to the shore of the St. Lawrence River suffered moderate to severe defoliation. Scattered patches of light to moderate defoliation occurred in an area of about 2 000 km^2 adjacent to these two sectors in Rimouski and Matapedia regions. An area of 700 km^2 located between Ascension-de-Matapedia and Saint-Fidèle (Bonaventure) also suffered light to moderate defoliation.

The maple leafroller has caused no similar defoliation since. No control measures were taken against this pest.

References

Blais, J.R. 1985. The Ecology of the Eastern Spruce Budworm: A Review and Discussion. Pages 49-59 in C.J. Sanders, R.W. Stark, E.J. Mullins and J. Murphy, eds. Proc. CANUSA Spruce Budworms Symposium, Bangor, Me.

Blais, J.R.; Benoît, P.; Martineau, R. 1977. Traitements aériens pour combattre la tordeuse des bourgeons de l'épinette au Québec. Pages 125-140 in M.L. Prebble (Ed.), Traitements aériens pour combattre les insectes forestiers au Canada, Environ. Can., Serv. Can. For., Ottawa.

Canadian Forestry Service. Forest Insect and Disease Surveys, Quebec Region 1968 to 1987. Regional Annual Reports. Quebec.

Martineau, R. 1984. Insects harmful to forest trees. Multiscience Publications, Can. For. Serv., Supply Serv. Can. For. Tech. Rep. 32

Prebble, M.L., ed. 1975. Aerial Control of Forest Insects in Canada. Environ. Can., Ottawa.

Swaine, J.M.; Craighead, F.C. 1924. Studies on the spruce budworm, Caocoecia fumiferana (Clem.) Can. Dep. Agric. Tech. Bull. (new series) 37:3-27, Ottawa.

Chapter 4

Forest Insect Pests in the Ontario Region

G.M. Howse

Introduction

Ontario's most important renewable resource is its forests. The forest industry employs, directly or indirectly, approximately 150 000 people, and many communities in Northern Ontario depend on the industry and an adequate supply of wood for survival.

Ontario is a large province. The distance from Cornwall in the east to Kenora in the west is about 1 600 km; from Windsor in the south to Moosonee in the north, more than 1 000 km. The total area of Ontario is 106.8 million ha with productive forest land totaling nearly 40 million ha, most of which (84%) is owned by the Crown. The total volume of wood standing in Ontario is just over 5.1 billion m^3. The most abundant tree species in Ontario is spruce which makes up 40% of the total growing stock volume. Jack pine represents 12.8% of the total volume, 19.8% is poplar, 7.5% is hard maple, and 7.2% is white birch. There are three major forest regions in Ontario. The Deciduous Forest Region in the south is characterized by deciduous tree species such as sugar maple and beech. The Great Lakes–St. Lawrence Forest Region in central Ontario is characterized by a mixture of deciduous and coniferous species. The Boreal Forest Region in the north is the largest of the three and is characterized by black and white spruce, while balsam fir, jack pine, white birch, and trembling aspen are common throughout the region.

The forests of Ontario are continually attacked and damaged by a large variety of destructive insects and other factors. The volume of timber harvested on Crown lands in recent years is about 20 million m^3. The average annual pest-caused losses in Ontario forests was more than 46 million m^3 each year from 1977 to 1981. Insects were responsible for one-third of these losses or about 15.5 million m^3 annually. Knowing the current and expected status of various pest problems is vital to modern forest management and protection. The Forest Insect and Disease Survey (FIDS) has fulfilled the lead role in forest pest surveys in Ontario for more than 50 years. Protection of Crown forests is the responsibility of the Ontario Ministry of Natural Resources; however, protection against pests depends upon reliable detection, identification, evaluation, and prediction of these damaging agents, which in Ontario is the responsibility of the Forest Insect Disease Survey Unit at the Great Lakes Forestry Centre of Forestry Canada. The Unit currently has a staff of 26 people: 7 professional and 19 support staff. Its major function is to provide extensive annual assessments of the status of major forest pests, their effect on the forest, and predictions of expected status. This information is used to assist and improve provincial forest management and protection practices and to determine and evaluate forest pest research needs in Ontario. It is also an essential component in national compilations of the status and effects of forest pests. Other responsibilities of the Unit include assistance to the province on pest control operations, monitoring of acid rain national early warning system plots in Ontario, pest extension services, assistance in plant quarantine activities, and conducting special surveys as necessary. The Unit also conducts research into sampling and damage appraisal methodology and in other areas to improve its own efficiency and effectiveness.

Throughout this chapter, with one exception, geographic locations of insect infestations or damage within Ontario are Ontario Ministry of Natural Resources Administrative regions such as Northwestern Region and North Central Region. Ontario is subdivided into eight such regions and they are depicted in Figure 1. The one exception is for spruce budworm where it was necessary to use broader geographical areas (broader than regions) to describe the situation. This is explained in greater detail in the introductory paragraphs of the spruce budworm section.

The primary source of information in this chapter is the annual national FIDS reports from 1973 to 1987, or in a few instances, regional ranger reports or Survey Bulletins. Ontario Region has been able to maintain a general survey for forest pests, although this general survey has become more focused on specific pests and permanent plot systems in recent years. Virtually all of the information contained in this chapter was collected and published by the Survey Unit, Ontario Region. The author wishes to acknowledge the assistance of several Ontario Region FIDS staff members in compiling this information. They include L.S. MacLeod (retired), W.D. Biggs, H.J. Evans, C.G. Jones, and A.J. Keizer.

Insects of Natural Forests

Spruce Budworm

The spruce budworm, *Choristoneura fumiferana*, is the most destructive insect in Ontario's forests. The present outbreak started in 1967 and since then, sizeable infestations have occurred each year in some part of Ontario to the present (Table 1). The area affected by spruce budworm, such as presented in Table 1, represents the gross area within which stands containing one or more of the major host species (balsam fir, white spruce, or black spruce) show evidence of budworm feeding.

Three distinct regional outbreaks developed in the late 1960s that fell into the broad geographical areas of northwestern, northeastern, and southern Ontario. More specifically in this context: northwestern Ontario is the Northwestern Region plus the western half of the North Central Region or from the Manitoba–Ontario border to Lake Nipigon; Northeastern Ontario is the Northeastern and Northern Regions plus the eastern half of the North Central Region or from Lake Nipigon to the Ontario–Quebec border;

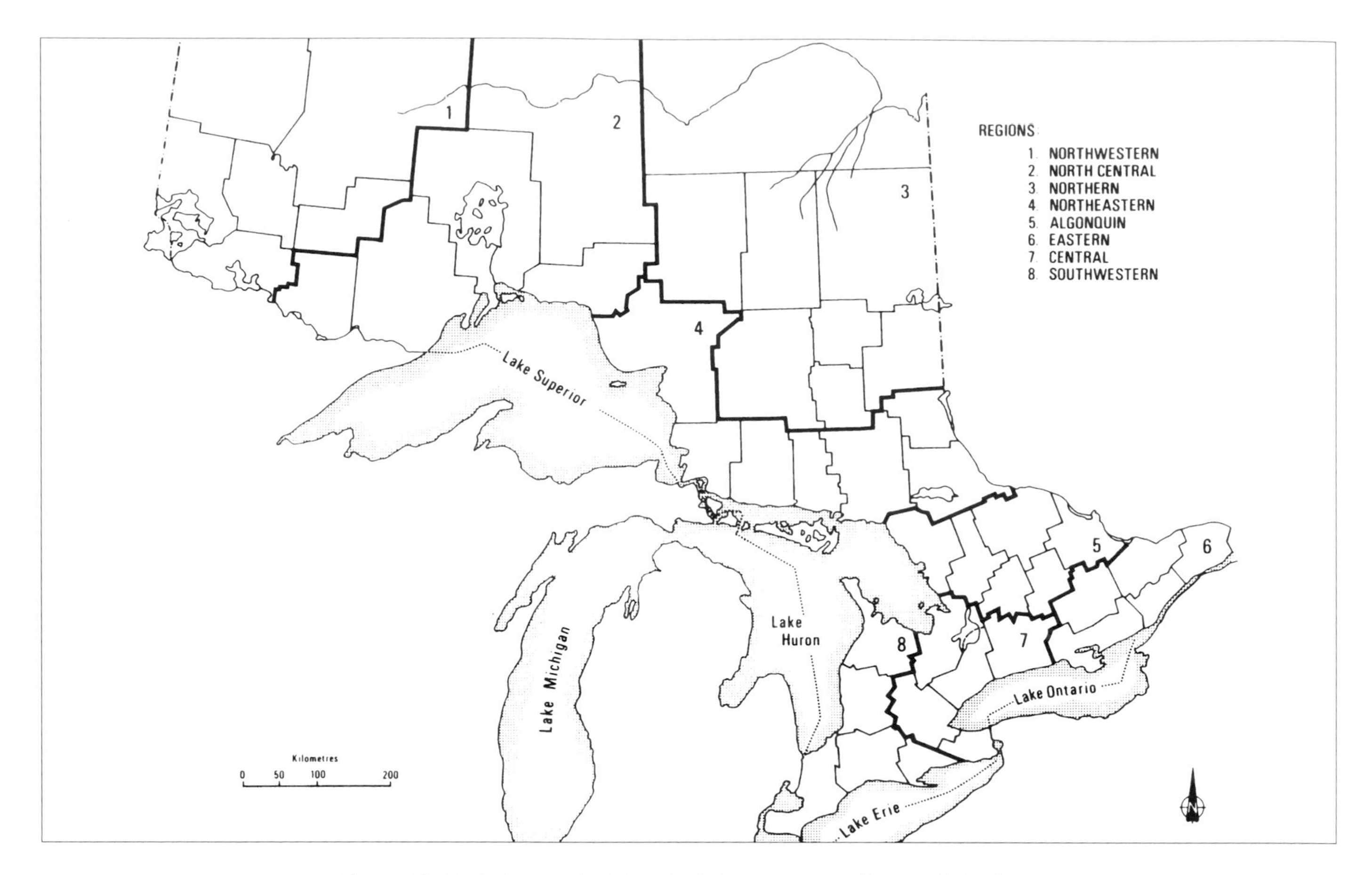

Figure 1. Ontario provincial and administrative regions and districts.

and southern Ontario is the Algonquin, Eastern, Central, and Southwestern regions (Figure 1).

Overall, budworm infestations built up and spread rapidly from the late 1960s to 1980 when the outbreak peaked at a total of 18.8 million ha. During this time, the overall situation, by and large, reflected the outbreak in northeastern Ontario where 90% or more of the total area of budworm infestation was located, particularly from 1976 to 1981. After 1980, infestations in total declined to about 8.7 million ha in 1984, increased to 12.3 million ha in 1985, but declined further in 1986 and 1987 to 7.2 million ha. Budworm-associated host tree mortality was first detected and mapped in 1972. The gross area of tree mortality totaled slightly more than 14 million ha in 1987 (Table 2). However, looking solely at the overall situation can be misleading in some respects. Consideration of subprovincial or regional outbreak patterns will help us better interpret the overall situation.

In northeastern Ontario, scattered infestations totaling 3 000 ha in 1967 increased to more than 17 million ha by 1980. The widespread occurrence of cold temperatures, snow, and freezing rain at the end of May 1972 caused a temporary setback to the budworm in 1973. Another setback in the increase of infestations, cause unknown, occurred in 1977. In 1980, the peak year of infestation, another weather event in the form of freezing temperatures in early June, may have triggered the onset of the decline of the outbreak that occurred in subsequent years. Approximately 1.2 million ha of infestation remained in 1987 (Table 1). Balsam fir mortality was first noted in northeastern Ontario in 1972. The extent of budworm-associated tree mortality totaled 9.6 million ha by 1981, and increased to 11.7 million ha in 1984, and to 11.8 million ha in 1987 (Table 2). The collapse of the outbreak in northeastern Ontario is believed to be attributable to extensive host tree mortality, cold temperatures in 1980, and natural factors, particularly parasites.

In northwestern Ontario, spruce budworm infestations were relatively minor from 1967 to 1976 (Table 1) due to control operations (aerial spraying), conducted by the Ontario Ministry of Natural Resources, that were designed to eliminate infestations or to suppress populations sufficiently to prevent buildup and spread into adjacent areas. These comprehensive control operations were carried out in close collaboration with Forestry Canada which conducted intensive surveys designed to detect and delineate areas of infestation as early as possible. The control program ended in 1976 with about 61 500 ha of infestation in northwestern Ontario. The outbreak then expanded slowly in the late 1970s and early 1980s to 931 000 ha in 1982. Rapid expansion occurred in the next 3 years to 7.2 million ha in 1985. Since then, there has been a decline to slightly less than 6 million ha in 1987 (Table 1). As was the case in northeastern Ontario, severe frosts (cold temperatures) in late May 1987 that killed balsam and spruce foliage as well as spruce budworm larvae, host tree mortality, and natural factors such as parasitism are likely

Table 1. Gross area of moderate to severe defoliation caused by spruce budworm, *Choristoneura fumiferana*, each year for the three regional outbreaks in Ontario, 1967–1988.

	Gross area of moderate to severe defoliation (ha)			
	Southern	Northeastern	Northwestern	Total
1967	60 704	3 035	16 188	79 927
1968	121 408	202 347	0[c]	323 755
1969	310 805	667 746	1 619[c]	980 170
1970	647 511	2 104 411	52 610[c]	2 804 532
1971	1 821 125[b]	3 480 372	52 610[c]	5 354 107
1972	2 347 228	5 422 906[a,b]	28 329[c]	7 798 463
1973	2 428 167	5 058 681	4 047[c]	7 490 895
1974	2 225 820	7 486 847	4 735[c]	9 717 402
1975	2 428 167	11 007 689	18 211[c]	13 454 067
1976	647 511	14 042 898	61 514[c]	14 751 923
1977	407 932	13 468 231	211 979[b]	14 088 142
1978	24 282	14 789 543	342 663	15 156 488
1979	1 001 534	16 939 972	487 873	18 429 379
1980	1 007 000	17 119 000[a]	724 000	18 850 000
1981	601 000	16 958 000	658 000	18 217 000
1982	423 057	6 669 069	931 361	8 023 487
1983	407 494	6 450 910	2 181 458	9 039 862
1984	72 840	4 043 598	4 631 414	8 747 852
1985	1 520	5 130 359	7 200 486	12 332 365
1986	642	1 964 625	6 890 420	8 855 687
1987	350	1 234 272	5 955 141[a]	7 189 763
1988	0	577 219	4 647 515	5 224 734

[a] Unusual weather in late May or early June (snow or frost).
[b] Tree mortality first detected.
[c] Sprayed.

responsible for the decline in northwestern Ontario. However, apparently host tree mortality was not as significant in contributing to the decline as it was in the northeastern part of the province. Tree mortality was first detected in 1977 and slowly increased to a total area affected of 718 000 ha in 1987 (Table 2).

In southern Ontario, the burgeoning budworm outbreak was evident in 1967 in the form of infestations on white spruce in the Ottawa River Valley. Populations increased and spread rapidly in subsequent years, reaching a maximum area of infestation of 2.4 million ha in 1973 and 1975. This was followed by a rapid decline in the next 3 years to less than 25 000 ha in 1978. However, a major reversal in the declining pattern occurred in 1979 and 1980 when slightly more than 1 million ha were infested each year. Since 1981, infestations declined to 1 500 ha in 1985 and 350 ha in 1987 (Table 1). Tree mortality was first noticed in southern Ontario in 1972. The largest increases in tree mortality occurred in 1976 and 1977. The maximum extent of mortality, 1.6 million ha, was reached in 1983 (Table 2).

Overall, gross wood volume losses due to the spruce budworm from 1973 to 1987 total more than 125 million m^3. About 85% of this loss is tree mortality—the other 15% is reduced increment (growth loss) caused by budworm feeding. About 75% of the total loss is balsam fir with the remaining 25% divided between white and black spruce.

Control operations (aerial spraying) have been conducted against spruce budworm in every year from 1968 to 1987. Details of these programs are described in Chapter 66. The area treated varied from year to year but reached a peak in 1986 when 150 000 ha of spruce–fir forest were aerially sprayed. Areas treated in the late 1970s and 1980s included commercial forest and high-value stands such as parks, plantations, wildlife habitat, and seed production areas. The purpose of the spraying was to keep trees alive by minimizing defoliation. Suppression and containment spraying to prevent the spread of budworm into extensive, susceptible forests was carried out in northwestern Ontario from 1968 to 1976.

Jack Pine Budworm

The most important insect problem of jack pine in Ontario (which is the second most important conifer tree species in Ontario) is the jack pine budworm, *Choristoneura*

Table 2. Gross area of budworm-associated tree mortality mapped each year for the three regional outbreaks in Ontario, 1972–1988.

	Gross area of budworm-associated tree mortality (ha)			
	Southern	Northeastern	Northwestern	Total
1967	60 704	3 035	16 188	79 927
1972	?[a]	80 939		80 939
1973	?[a]	202 347		202 347
1974	?[a]	667 746		667 746
1975	121 408	1 214 083		1 335 491
1976	647 511	2 630 514		3 278 025
1977	1 315 257	4 168 353	405	5 484 015
1978	1 347 025	4 734 925	8 095	6 090 045
1979	1 384 055	6 110 886	20 235	7 515 176
1980	1 493 000	6 839 000	24 000	8 356 000
1981	1 550 000	9 572 000	88 000	11 210 000
1982	1 550 000	9 934 000	150 000	11 634 000
1983	1 573 000	10 355 000	191 000	12 119 000
1984	1 573 000	11 702 000	241 000	13 516 000
1985	1 573 000	11 748 592	614 364	13 935 956
1986	1 573 000	11 765 496	647 168	13 985 664
1987	1 573 000	11 776 513	718 325	14 067 838
1988	1 573 000	11 787 356	1 156 119	14 516 475

[a] Mortality present but area not determined.

pinus, a close relative to the spruce budworm. In Ontario, jack pine is the principal host tree attacked by the jack pine budworm but red pine, Scots pine, and eastern white pine are also damaged.

Populations of this insect were generally very low in Ontario from 1973 to 1981. In 1982, several pockets of moderate and severe defoliation totaling 981 ha were mapped in the Georgian Bay area. The pattern of the distribution of the pockets of defoliation raised the possibility that the infestations were the result of moth flights from Michigan, presumably in the summer of 1981.

Populations increased to outbreak proportions in 1983 when a total area of 67 142 ha of infestation were mapped in three widely separated parts of Ontario (Table 3). In 1984, infestations expanded dramatically with moderate to severe defoliation occurring throughout about 1.15 million ha (Table 3).

Major increases occurred in 1985 with 3.66 million ha of defoliation. The largest infestation occurred in the Northeastern and Northern regions where 1.84 million ha of defoliation were mapped followed by 1.76 million ha in the North Central and Northwestern regions (Table 3). In 1986, overall, there was a decline to 1.74 million ha of defoliation. The decline was most pronounced in the northeastern part of Ontario whereas a slight

Table 3. Gross area of current moderate to severe defoliation by the jack pine budworm, *Choristoneura pinus*, in Ontario from 1982 to 1987.

	Area of moderate to severe defoliation (ha)					
Region	1982	1983	1984	1985	1986	1987
Algonquin	981[a]	30 202	25 397	54 034	8 099	
Northeastern		29 970	429 220	887 056	56 627	
Northern			171 595	955 755	93 469	
North Central		6 970	370 568	285 406	31 391	
Northwestern			153 378	1 477 818	1 554 139	504 749
Total	981	67 142	1 150 158	3 660 069	1 743 725	504 749

[a] Included 308 ha in the Southwestern Region and 80 ha in the Central Region.

increase occurred in the Northwestern Region (Table 3). In 1987, populations collapsed everywhere except in the Northwestern Region where slightly more than 500 000 ha of moderate to severe defoliation were mapped (Table 3).

Damage in the form of tree mortality and top kill was common throughout the area infested. Examples of damage to jack pine were as follows: Algonquin Region, 25% tree mortality and 37% top kill; Northeastern Region, 17% mortality and 37% top kill; Northern Region, 2% mortality and 24% top kill; North Central Region, 14% mortality and 10% top kill; Northwestern Region, 24% mortality and 11% top kill. These figures are maximums for the regions and not necessarily representative of average damage conditions. In fact, with a few exceptions, damage is difficult to map in a manner comparable to the spruce budworm, and hence areas of tree mortality cannot be presented for jack pine budworm.

The Ontario Ministry of Natural Resources aerially sprayed about 220 000 ha of jack pine forest in 1985, 482 000 ha in 1986, and 105 000 ha in 1987. Details of these protection programs are described in Chapter 66. The purpose of the spraying programs was to keep trees alive and to minimize damage caused by the feeding larvae.

Forest Tent Caterpillar

The forest tent caterpillar, *Malacosoma disstria*, erupts periodically in Ontario, rising to high population levels and defoliating extensive areas of forest. Outbreaks occur at fairly regular intervals averaging about 10 years and high populations usually persist for 2 to 4 years, but sometimes longer. Outbreaks collapse rapidly, usually in a year or two, as a result of natural factors such as spring frosts, larval starvation, parasites, and diseases.

Recent outbreaks in Ontario include one that lasted from 1949 to 1956 and reached 21 million ha at its peak, one that lasted from 1961 to 1967 and affected nearly 10 million ha at its peak in 1965, and the latest that occurred from 1973 to 1980 and infested almost 18 million ha in 1978.

The principal host of the forest tent caterpillar in Ontario is poplar, particularly trembling aspen, although in southern Ontario the insect also feeds on sugar maple and oak in addition to poplar.

As shown in Table 4, forest tent caterpillar infestations are an annual event in Ontario with the only question being the size of the area affected each year. The 30 600 ha of defoliation in 1983 was the lowest area of infestation from 1973 to 1987. The current outbreak started in 1984 and has expanded steadily since. Based on historical information, this outbreak will probably peak at a total area of about 15–20 million ha in 1990 or 1991.

Table 4. Gross area of moderate to severe defoliation caused by forest tent caterpillar, *Malacosoma disstria*, in Ontario from 1973 to 1987.

Year	Total area defoliated (ha)
1973	377 047
1974	570 000
1975	2 017 340
1976	8 200 000
1977	12 220 000
1978	17 870 000
1979	2 650 000
1980	325 500
1981	228 000
1982	103 700
1983	30 600
1984	124 750
1985	208 942
1986	433 000
1987	1 649 977

A decline of sugar maple trees was reported in the late 1970s in areas defoliated by the forest tent caterpillar in the mid-1970s. The areas affected were around Georgian Bay in the Parry Sound and Bracebridge districts of the Algonquin Region and in the Owen Sound District of the Southwestern Region. By 1980, a total of 28 700 ha of sugar maple forest in Parry Sound and Owen Sound districts were affected and about half of the affected area experienced greater than 25% tree mortality. This represented about 14% of the 202 400 ha of susceptible maple forest in the two districts. The total loss of timber (sugar maple) in the three districts amounted to about 3.26 million m^3 valued at about $32 million. Maples began recovering after the forest tent caterpillar infestation collapsed and since then mortality has been negligible. Some of the dead or dying trees were salvaged but more than half of the mortality was an outright loss.

The Ontario Ministry of Natural Resources conducted aerial spraying operations in 1975, 1977, and 1978 against this pest to reduce larval populations in provincial parks and to minimize foliage loss. Details of these operations are described in Chapter 66.

Bruce Spanworm

The Bruce spanworm, *Operophtera bruceata*, is a hardwood defoliator that was relatively prominent throughout the period covered by this review. In the mid-1970s (1973 to 1978), infestations occurred each year in various parts of Ontario (Table 5). For example, in 1973, light infestations occurred on sugar maple on St. Joseph Island, Sault Ste. Marie District. In 1974, pockets of severe defoliation of sugar maple, beech, and white birch occurred along the southern boundary of Algonquin Provincial Park. In 1975, this outbreak intensified and expanded into the Bracebridge–Lake of Bays area and along the western boundary of Algonquin Park with sugar maple and yellow birch being severely defoliated. In addition, a small, heavy infestation on sugar maple was found north of Sault Ste. Marie. In 1976, the infestations in the Muskoka and Lake of Bays area collapsed whereas infestations along the western boundary of Algonquin Provincial Park continued and intensified. The small infestation on sugar maple north of Sault Ste. Marie persisted. Bruce spanworm infestations were unknown on trembling aspen in Ontario until 1976 when the insect caused severe defoliation of aspen

Table 5. Gross area of moderate to severe defoliation caused by Bruce spanworm, *Operophtera bruceata*, in Ontario from 1973 to 1987.

Year	Region	Area of moderate to severe defoliation (ha)
1973	Northeastern	
1974	Algonquin	
1975	Northeastern and Algonquin	
1976	North Central, Northern, and Algonquin	
1977	North Central	approx. 60 000
1978	North Central	26 700
1979		
1980		
1981		
1982		
1983		
1984	Northeastern	4 286
1985	Northeastern, Algonquin, Central, Southwestern and Eastern	28 764
1986	Northeastern, Algonquin, Central, Southwestern, and Eastern	209 995
1987	North Central, Northeastern, Northern, Algonquin, Central, and Eastern	1 053 717

around Lake Nipigon and light damage around Chapleau. In 1977, the infestation on aspen east and south of Lake Nipigon expanded and intensified. In 1978, infestations south of Lake Nipigon declined to light defoliation but about 26 700 ha of moderate to severe defoliation occurred east of Lake Nipigon. The Nipigon infestations collapsed in 1979, and the insect was not found at damaging levels from 1979 to 1983.

In 1984, two infestations were detected. The larger infestation was on St. Joseph Island, Sault Ste. Marie District, where 4 160 ha of sugar maple were moderately to severely defoliated. The smaller infestation was on Manitoulin Island, Espanola District, where 126 ha of sugar maple were severely defoliated. In 1985, the infestations in the Sault Ste. Marie and Espanola districts expanded to about 28 000 ha of moderate to severe defoliation. New, small, scattered infestations were detected with feeding primarily on understory sugar maple vegetation in sugar maple stands in Pembroke, Bancroft, Minden, and Bracebridge districts, Algonquin Region; on sugar maple and beech (upper canopy and regeneration) in Huronia District, Central Region; on sugar maple in Owen Sound District, Southwestern Region; and in a 20-ha stand of sugar maple in Cornwall District, Eastern Region.

In 1986, infestations continued to expand and new infestations added to a total of 210 000 ha of moderate to severe defoliation. The largest infestations affecting sugar maple were mapped in the Algonquin Park and Pembroke districts. About 29 325 ha were mapped in the Northeastern Region where the Bruce spanworm was feeding on sugar maple, red oak, and trembling aspen in Espanola District on Manitoulin and Cockburn islands and in Sault Ste. Marie District including St. Joseph Island. The 20-ha infestation in the Eastern Region first reported in 1975 reoccurred in 1976. Infestations were also noted in the Huronia District, Central Region, and Owen Sound District, Southwestern Region. In addition to the foregoing, in 1986, varying numbers of Bruce spanworm larvae were found feeding on trembling aspen in conjunction with large aspen tortrix in Chapleau and Gogama districts, Northern Region; Sudbury, Espanola and Sault Ste. Marie districts, Northeastern Region and Thunder Bay District, North Central Region.

In 1987, infestations expanded to a total of 1.05 million ha of moderate to severe defoliation. The largest infestations were recorded in the North Central Region on trembling aspen where about 617 000 ha were mapped in a wide band stretching from Lake Superior across the southern part of Thunder Bay District into Quetico Provincial Park in the southern Atikokan District. In the eastern part of this infestation, forest tent caterpillar and large aspen tortrix fed in conjunction with the Bruce spanworm. In the Northeastern Region, a total of 202 206 ha of moderate to severe defoliation, primarily sugar maple, were mapped in Wawa, Sault Ste. Marie, Blind River, and Espanola districts. Heavy infestations were recorded in Manitoulin and Cockburn islands in Lake Huron and Michipicoten Island in Lake Superior. Infestations on St. Joseph Island collapsed following 3 years of medium to heavy infestations. Infestations also increased in the Algonquin Region where about 233 911 ha of moderate to severe defoliation of sugar maple in the Bracebridge, Minden, Bancroft, Pembroke, and Algonquin Park districts were mapped. Light infestations were also reported from Parry Sound District, Algonquin Region; Lindsay District, Central Region; Carleton Place District, Eastern Region; Chapleau District, Northern Region; and Geraldton, Terrace Bay, and Nipigon districts, North Central Region.

The direct effects of this insect feeding on trees in Ontario are not known, but it is expected that reduced growth, crown dieback, and tree mortality will occur, particularly in sugar maple. Unfortunately, most of the sugar maple stands infested by Bruce spanworm in the Northeastern and Algonquin regions were also infested and severely defoliated by forest tent caterpillar; hence, determining the impact attributable to spanworm will be difficult.

Hemlock Looper

The occurrence of hemlock looper, *Lambdina fiscellaria*, at damaging levels in Ontario has been a rare event in recent years. Localized infestations have occurred in hemlock stands in southern Ontario and other host species in Northern Ontario. Infestations have been short-lived, sometimes lasting only a year or two; nevertheless, trees were killed. Past infestations leading to tree mortality occurred in the Muskoka Lakes area (feeding on hemlock) in 1928–29 and again in the late 1930s; in the Sioux Lookout District in the mid-1940s (feeding on balsam fir and white spruce); on Manitoulin Island feeding on balsam fir from 1949 to 1951; in the Thousand Islands area (St. Lawrence River) feeding on hemlock; at Abitibi Lake feeding on balsam fir, white birch, and poplar in the early 1950s; and in the Lake of Bays, Muskoka Lake, and Parry Sound areas from 1950 to 1955 feeding on hemlock.

No major outbreaks were detected in Ontario from 1956 to 1977. In 1978, moderate to severe defoliation and some mortality of hemlock caused by the looper were observed on islands and lake shores in four adjoining townships in Minden and Bancroft districts in the Algonquin Region. In all, approximately 400 ha were affected; about 60 ha of mature hemlock were moderately to severely defoliated and the remainder was lightly defoliated. At one location, 18% of the hemlock trees had been killed, at another there was 1% mortality. In 1979, populations declined sharply and only one very small pocket of light defoliation was mapped. However, tree mortality increased from 18% to 39% at the first location and from 1% to 13% at the other. Very low numbers of larvae were found in 1980. Since then, this insect has been found in low numbers at several locations throughout the province. Infestations or populations causing damage have not been detected since 1978.

Large Aspen Tortrix

The large aspen tortrix, *Choristoneura conflictana*, is the second most important defoliator of aspen in Ontario. Trembling aspen is the major host but this insect will feed on white birch, balsam poplar, large tooth aspen, willows, alders, and choke cherry during epidemics. Outbreaks of this insect in Ontario have been reported in the vicinity of Toronto, 1912; vicinity of Sault Ste. Marie and Thunder Bay, 1943 to 1946; a large outbreak in north central and northwestern Ontario, 1956 to 1958; and Northern Ontario where up to 10 million ha were affected, 1969 to 1975. Infestations occur for 2 or 3 years in a given location, then collapse abruptly.

Infestations reached a maximum in Ontario in 1973 when 10 million ha of moderate to severe defoliation were mapped (Table 6). A steady decline in populations was recorded annually until 1978 when the insect had virtually disappeared from aspen stands in the province. From 1979 to 1982, small sporadic infestations were reported from widely separated locations in the Northeastern, Algonquin, and Central regions. From 1983 to 1985, populations increased rapidly in the Northeastern Region, mainly in the Espanola District, and along the Ontario–Minnesota border in the Thunder Bay District, North Central Region, where 20 900 ha of moderate to severe defoliation were mapped in 1985. By 1986, the total area of moderate to severe defoliation peaked at 620 899 ha in the North Central, Northern, and Northeastern regions. In 1987, defoliation declined to 121 840 ha with infestations present in the North Central and Northeastern regions. Populations collapsed in the Northern Region.

The impact of large aspen tortrix feeding on aspen trees is not clear. Tree mortality does not seem to be a major consequence of infestation of this insect because significant aspen mortality has not been noted in stands severely defoliated by it. Nevertheless, some branch and tree mortality probably occurs as a result of several years of defoliation, particularly on trees on poor sites or if other stress factors such as drought occurs. The major impact is reduced growth but this has not been measured in Ontario, only estimated. As a result of the 1969 to 1975 outbreak, total aspen growth loss was calculated to be about 12 million m^3.

Table 6. Gross area of moderate to severe defoliation caused by large aspen tortrix *Choristoneura conflictana*, in Ontario from 1973 to 1987.

Year	Region[a]	Area of moderate to severe defoliation (ha)
1973	Northern Ontario	approx. 10 million (estimate)
1974	NW, NC, N, NE, Alg	approx. 1.5 million (estimate)
1975	NE, Alg	scattered pockets
1976	NE	scattered pockets
1977	NE, Alg	light defoliation only
1978	N	pocket of severe defoliation
1979	C	scattered infestations
1980	Alg, NE	2 665
1981	C	510
1982	C, NE	(area data not available)
1983	NE, C	2 690
1984	NE	4 400
1985	NE, NC	27 500
1986	NE, N, NC	620 899
1987	NE, NC	121 840

[a] NW - Northwestern, NC - North Central, N - Northern, NE - Northeastern, Alg - Algonquin, C - Central.

Oak Leafshredder

In recent years, the oak leafshredder, *Croesia semipurpurana*, has been a persistent pest in red oak stands in Ontario. Defoliation by this insect is probably a major predisposing factor in oak decline, dieback, and mortality, and serious damage has occurred in many areas in Ontario with a history of oak leafshredder infestations. Since 1977, the FIDS Unit of the Ontario Region has monitored 12 permanent red oak plots in southern Ontario. By 1987, 9.6% of the original 1 200 trees had died; however, two plots with histories of heavy oak leafshredder infestations had total tree mortalities of 38% and 26%. A general improvement has been observed in the health of the surviving trees for the past several years. This improvement coincided with a general decline in oak leafshredder populations. It has been estimated that about 100 000 m^3 of oak is killed annually as a direct result of insect defoliation or the decline condition.

Numerous widespread areas of moderate to severe defoliation occurred from 1973 to 1975 in the Algonquin, Central, Eastern, and Northeastern regions. Infestations declined from 1976 to 1978, but high populations persisted in the Huronia, Maple, and Niagara districts of the Central Region. A resurgence began in 1979 and peaked in 1980 when approximately 50 000 ha of moderate to severe defoliation were mapped in the Northeastern, Algonquin, Central, and Eastern regions. Another period of decline began in 1981 but high numbers remained in parts of the Central Region until 1983.

Populations remained at relatively low levels until 1987 when increases occurred again in the Huronia District, Central Region, where 560 ha of defoliation were mapped as well as light infestations elsewhere in the region.

The Ontario Ministry of Natural Resources conducted aerial spraying operations in 1973, 1977, 1978, 1979, 1980, and 1983 against this insect to minimize defoliation in high-value oak stands in the Central Region. A total of about 3 700 ha were treated primarily with Sevin-4-oil® or Orthene.® Details of these operations are described in Chapter 66.

Swaine Jack Pine Sawfly

The Swaine jack pine sawfly, *Neodiprion swainei*, is probably the most economically important sawfly threatening jack pine in Ontario (Table 7). Infestations have occurred periodically and have been centered in several apparently highly susceptible stands, which, in many cases, are on rocky lake shores and islands. The insect is present throughout the range of jack pine in Ontario, but the most persistent infestations have occurred in two widely separated areas: (1) the adjoining forest districts of Temagami, Kirkland Lake, Gogama, Chapleau, Sudbury, North Bay, and Parry Sound (northeastern Ontario) and (2) in northwestern Ontario in Fort Frances, Kenora, Dryden, and Atikokan districts. Over the years, infestations have been more widespread, persistent, and damaging in northeastern Ontario than in the northwest. Sporadic infestations have been recorded in the Lady Evelyn Lake and Lake Temagami areas of Temagami District since the early 1940s. Repeated defoliation has killed considerable numbers of trees in this part of Ontario.

Medium to heavy infestations were reported in the early 1970s that continued to persist or increase until 1979 in the Makobe–Lady Evelyn–Temagami lakes area of the Temagami District. Limited salvage cutting was carried out in accessible semimature and mature stands from time to time during the 1970s. In 1980, populations increased and 775 ha of moderate to severe defoliation were mapped. These infestations expanded to 5 700 ha in 1981, declined slightly to 4 650 ha in 1982, declined considerably in 1983 to 518 ha, and collapsed in 1984. The situation in 1981 concerned the Ontario Ministry of Natural Resources because of the threat to extensive jack pine stands in the area. A control program was considered for 1982 but it was decided to take no action because of expected declines in sawfly populations and the option of salvaging damaged timber. Harvesting in damaged areas was conducted from 1980 to 1984. From 1984 to 1987, populations have been

Table 7. Gross area of moderate to severe defoliation caused by Swaine jack pine sawfly, *Neodiprion swainei*, in Ontario from 1973 to 1987.

Year	Location (District)	Area of moderate to severe defoliation (ha)
1973	Temagami	Pockets of heavy infestation near lakes
1974	Temagami	Moderate infestations on islands and shorelines of Lake Temagami
1975	Temagami	Severe damage including tree mortality
	Fort Frances and Kenora	Severe defoliation at Rainy Lake and Lake of the Woods—islands and shorelines
1976	Temagami	Populations remained high
	Fort Frances and Kenora	Population collapsed
1977	Temagami	160 ha of severe defoliation at Banks–Makobe lakes
1978	Temagami	Severe defoliation Banks–Makobe lakes continued, causing considerable tree mortality
1979	Temagami	Banks–Makobe lakes infestations persisted—tree and top mortality, light defoliation in many locations
1980	Temagami	775 ha Banks–Makobe–Big Foot lakes
1981	Temagami	5 700 ha Banks–Makobe–Lady Evelyn lakes
1982	Temagami	4 650 ha Banks–Makobe–Lady Evelyn lakes
1983	Temagami	518 ha—populations declined low populations
1984	Temagami and Kirkland Lake	Low populations
1985	Temagami and Kirkland Lake	Low populations
1986	Temagami	Light defoliation
	Kenora and North Bay	Small numbers of insects collected
1987	Temagami	Varying damage at numerous locations
	Sudbury	Colonies at several locations

low and only trace or light defoliation has been reported for areas where populations had previously been high.

Virtually all Swaine jack pine sawfly activity during the review period occurred in northeastern Ontario except for 1 year, 1975, when severe defoliation occurred at Rainy Lake and Lake of the Woods in the Fort Frances and Kenora districts, Northwestern Region.

Insects of Plantations and High-Value Stands

Seed and Cone Insects

The FIDS Unit, Ontario Region, has conducted annual assessments since 1980 to determine the impact of pests on seed and cone production of the five most important conifer species in Ontario. The first such special study in 1980 looked at black spruce in Northern Ontario. Black spruce flowers and cones were collected in early June and cones in late July. It will be recalled that 1980 was the peak year of the budworm outbreak in Ontario. Flower samples collected in early June in budworm-infested areas were severely infested with spruce budworm, *Choristoneura fumiferana*, and spruce coneworm, *Dioryctria reniculelloides*. In areas free of budworm infestation, the number of insect-infested flowers was low. The second set of samples collected in late July showed that in some heavily budworm-infested areas, high proportions of female flowers aborted and failed to become cones, presumably because of early feeding by lepidopterous larvae. The spruce cone maggot, *Strobilomyia neanthracina*, was the commonest insect although many cones also showed evidence of budworm and coneworm feeding. Overall, 48% of the cones examined showed signs of insect damage (Table 8). In some locations, insects prevented many cones from developing properly and the potential seed production was greatly reduced by insects at other locations where cones had developed successfully. The result in many situations was a virtual failure of black spruce seed and cone production.

In 1981, the impact of pests on white spruce seed and cone production throughout Ontario was studied. More than 70% of the cones examined showed signs of damage (Table 8). Insects such as spruce budworm, spruce bud moth (*Zeiraphera canadensis*), spruce coneworm, spruce seed moth (*Cydia strobilella*), spruce cone maggot, and spruce gall midge (*Mayetiola piceae*), were involved.

Starting in 1982, cones were dissected to determine the number of damaged seeds within damaged cones. From 1982 to 1988, the five conifer species, black spruce, white spruce, red pine, jack pine, and white pine were assessed on either two or three occasions at 3-year intervals. For example, red pine in

Table 8. Damage caused by insects to seed and cones of conifers in Ontario, 1980–1988.

Host	Year	No. of cones examined	Cones damaged (%)	Total seed loss (%)	Principal cause
Black spruce (Northern Ontario)	1980	2 242	48	–	Spruce budworm, spruce coneworm spruce cone maggot
White spruce (Ontario)	1981	1 182	72	–	Spruce budworm, *Choristoneura fumiferana* spruce bud moth, *Zeiraphera canadensis* spruce coneworm, *Dioryctria reniculelloides* spruce seed moth, *Cydia strobilella* spruce cone maggot, *Strobilomyia neanthracina* spruce cone gall midge, *Dasineura canadensis*
Red pine (southern Ontario)	1982 1985 1988	1 344	70	45	Red pine coneworm, *Eucosma monitorana* red pine cone beetle, *Conophthorus resinosae* fir coneworm, *Dioryctria abietivorella* webbing coneworm, *Dioryctria disclusa*
White spruce (Ontario)	1984 1987	3 794	55	29	Spruce cone maggot, spruce cone axis midge, spruce seed moth
Jack pine (Northern Ontario)	1982 1985 1988	2 590	29	11	Unknown lepidopterous larvae, fir coneworm , a midge, *Resseliella* sp.
Black spruce (Northern Ontario)	1983 1986	2 668	33	5	Spruce cone maggot, spruce cone axis midge, spruce cone gall midge
Eastern white pine (southern Ontario)	1983 1986	990	28	3	White pine coneworm, *Eucosma tocullionana* white pine cone beetle, *Conophthorus coniperda* a midge, *Resseliella* sp.

Table 9. Gross area of moderate to severe defoliation by gypsy moth, *Lymantria dispar*, in Ontario from 1981 to 1987.

	Area of moderate to severe defoliation (ha)						
Region	1981	1982	1983	1984	1985	1986	1987
Eastern	1 450	4 800	40 954	80 624	245 987	166 974	11 564
Algonquin	0	0	0	0	330	385	111
Central	0	0	0	0	25	417	888
Southwestern	0	0	0	0	0	0	115
Total	1 450	4 800	40 954	80 624	246 342	167 776	12 678

southern Ontario was studied in 1982, 1985, and 1988. Data for the three samples were combined for presentation in Table 8.

Red pine is subject to the greatest damage. On average, 70% of the cones showed signs of damage. Seed loss in damaged cones averaged about 65%, hence the total seed loss amounted to 45% (Table 8) or almost half of the potential seed production of this species. Total seed loss was 37% in 1982, 42% in 1985 and 50% in 1988, an increasing pattern. Red pine coneworm, *Eucosma monitorana*, and red pine cone beetle, *Conophthorus resinosae*, were the principal causes of loss.

White spruce sustained the second highest losses with an average of 29% for two samples, 1984 and 1987, when losses were 26% and 31%, respectively. Spruce cone maggot, spruce cone axis midge (*Dasineura rachiphaga*), and spruce seed moth were the major insects involved.

Jack pine with seed losses of 11%, black spruce with 5%, and white pine with 3% (Table 8) are not significant in either biological or economic terms except perhaps in some local situations or seed production areas where infestations of cone damaging insects build up and cause heavy losses. During spruce budworm outbreaks, seed and cone production of black spruce and white spruce can be seriously affected, in fact, almost completely eliminated by damage caused by spruce budworm, coneworm, and spruce cone maggot. In contrast, 1985 was the peak year for jack pine budworm in Northern Ontario but there was little evidence that this insect seriously affected jack pine seed and cone production. Perhaps coincidentally, seed loss in 1985 was 22% compared with 9% in 1982 and 5% in 1988.

Plantation Insects

This report will deal separately with each of the five conifer species that were examined during the review period. During the early years of the review period, from 1973 to 1977, special insect surveys of plantations were not carried out in Ontario.

White spruce plantations were examined for a 4-year period. The first formal survey was carried out in 1978 (information from the Survey Bulletin) in the northern portion of Ontario, and subsequent surveys were done across the province in 1981, 1984, and 1987. Spruce budworm *Choristoneura fumiferana*, was the most common insect present during the 4 years; some stands suffered defoliation levels as high as 75%, but most defoliation estimates ranged from 5% to 20%. The spruce coneworm, *Dioryctria reniculelloides*, was present at low levels in 1984 and 1987 with defoliation levels (8%) combined with that of spruce budworm. The yellowheaded spruce sawfly, *Pikonema alaskensis*, was found in 40% of the plantations in Northern Ontario in 1981 and to a lesser degree in both the north and south in 1978 (information from the Survey Bulletin), 1984, and 1987. In addition to the aforementioned insects, the white pine weevil, *Pissodes strobi*, and the spruce bud moth, *Zeiraphera canadensis*, were commonly found, but at very low damage levels.

The province was split geographically for the surveys in 1979 (information from the Survey Bulletin), 1982, 1985, and 1988; red pine was surveyed in the area lying south of a line between the cities of Sault Ste. Marie and North Bay, and jack pine was surveyed north of the line. The importance of any one insect varied with the year in which red pine was examined, but the pine false webworm, *Acantholyda erythrocephala*, and the European pine shoot moth, *Rhyacionia buoliana*, were present for each year, albeit at generally low levels. The European pine sawfly, *Neodiprion sertifer*, and redheaded pine sawfly, *N. lecontei*, were found in some of the years surveyed, but again damage levels were low. In the jack pine survey, the eastern pine shoot borer, *Eucosma gloriola*, and the white pine weevil were two of the most commonly occurring insects, with more than 30% of the stands examined having low damage levels in 1979 (information from the Survey Bulletin) and 1982. Jack pine budworm, *Choristoneura pinus*, was present at low levels in 6–7% of the trees examined in 1985 and 1988.

The surveys of the last two conifer species were carried out in 1980 (information from the Survey Bulletin), 1983, and 1986. The province was split again with eastern white pine surveyed in the southern part and black spruce surveyed in the north. The white pine weevil was the most damaging insect present in the eastern white pine; in 1983, it was found in 60% of the plantations less than 6 m high. Other insects commonly found were the pine bark adelgid (*Pineus strobi*), pine spittlebug (*Aphrophora cribrata*), and the pine false webworm, but generally at very low levels. In the survey of black spruce, spruce budworm was the most important insect pest encountered. In 1980 (information from the Survey Bulletin), low defoliation levels were observed in 47% of the trees examined. Although only present at low levels, the yellowheaded spruce sawfly, spruce coneworm, and white pine weevil were found in each of the 3 years that black spruce was surveyed.

Gypsy Moth

The gypsy moth, *Lymantria dispar*, was first detected in 1969 on Wolfe Island near the city of Kingston. From 1969 to 1980, gypsy moth populations remained relatively low in southern Ontario keeping in mind, however, that the Plant Protection Division, Agriculture Canada, conducted control programs against this insect in Ontario through the 1970s and early 1980s. The extensive aerial spray programs in Ontario conducted by Agriculture Canada were aimed at reducing populations and containing infestations to prevent spread of the insect to the valuable forest stands and recreation areas of eastern and southern Ontario.

In 1981, about 23 pockets of gypsy moth defoliation totaling 1 450 ha were detected in eastern Ontario. Most of the defoliation was found near the town of Kaladar, Tweed District, Eastern Region. Populations increased and spread rapidly during the next 4 years to the point where 246 342 ha of moderate to severe defoliation were mapped in 1985 (Table 9). All but 355 ha of the defoliation was located in the Eastern Region in Tweed, Napanee, Brockville, and Carleton Place districts. In 1986, the infestation declined to less than 170 000 ha and virtually collapsed in 1987 to less than 13 000 ha. Causes of the decline are believed to be an increased incidence of the naturally occurring nuclear polyhedrosis virus and parasites.

To date, there is no evidence of damage, in terms of tree mortality, that would be attributed to gypsy moth.

Pheromone trapping surveys conducted by Forestry Canada indicate that gypsy moth is probably now resident throughout southern Ontario, although some of the catches may be moths from Michigan or from areas south of Lake Erie or Lake Ontario.

Aerial spraying programs were conducted by the Ontario Ministry of Natural Resources in 1982, 1985, 1986, and 1987. More than 100 000 ha were treated in 1986. The purpose of the programs was to minimize defoliation. Details are described in Chapter 66.

Other Noteworthy Insect Pests

In Table 10, brief reports of the occurrence of damage caused by about 41 other insects are listed. These other insects are not classed as major insect pests as is spruce budworm. But most are capable of causing serious injury to or death of living trees, usually on a more sporadic or localized basis than spruce budworm. An insect reported in Table 10, such as birch leafminer, *Fenusa pusilla*, can cause widespread damage annually; however, the host species, except in ornamental situations, is not valued highly.

In a few instances, Table 10 provides information about insects that were mentioned previously such as eastern pine shoot borer, *Eucosma gloriola*, European pine sawfly, *Neodiprion sertifer*, European pine shoot moth, *Rhyacionia buoliana*, pine false webworm, *Acantholyda erythrocephala*, redheaded pine sawfly, Neodiprion lecontei, spruce bud moth, *Zeiraphera canadensis*, white pine weevil, *Pissodes strobi*, and yellowheaded spruce sawfly, *Pikonema alaskensis*. Previous reference to these insects was in a specific context; either seed and cones or plantations and systematic surveys of these high-value situations were initiated only in 1978. Information in Table 10 is more general regarding the host situation and reflects the period of the review, 1973 to 1987.

Table 10. Other noteworthy insects, Ontario, 1973–1987.

Insect	Host	Location	Remarks
Aspen leafroller *Pseudexentera oregonana*	Trembling aspen	Province-wide	Populations began increasing in 1974 and probably peaked in 1976 when there were 175 000 ha of infestation in the Northern Region and a 210 000-ha outbreak west of Thunder Bay. Generally declining numbers with localized areas of moderate to severe defoliation occurred in several areas of the province until 1982 when the population collapsed. In 1987, 9 420 ha were infested in the Hearst District and this recurred in 1988. Also, an outbreak of 1.03 million ha was reported in the Red Lake District in 1988 along with increased numbers at numerous other locations in Northern Ontario .
Balsam fir sawfly *Neodiprion abietis complex*	Balsam fir white spruce black spruce	Province-wide	In 1973, pockets of severe defoliation were present in the Pembroke District; however, populations declined in 1974. A pocket of severe defoliation remained near Arnprior in 1975 as well as moderate defoliation in nine townships in the North Bay District. Pockets of moderate to severe defoliation were present in 1976 and 1977 in the North Bay District. Through the next 6 years varying levels of scattered damage were present in numerous areas across the province. In 1984, defoliation levels

(Continued)

Table 10. Other noteworthy insects, Ontario, 1973–1987. (*Continued*)

Insect	Host	Location	Remarks
Balsam fir sawfly (*Continued*)			ranging from 50% to 75% were found in small groups of trees within an area of 100 000 ha in Kenora District, 6 800 ha in Red Lake District, and 9 800 ha in Ignace District, with heavy defoliation also reported over a 500-ha area in the Algonquin Region. Scattered populations were observed in the Northwestern, Algonquin, and Eastern regions in 1985. The following year moderate to severe defoliation was found scattered over an 11 000-ha area in Northwestern Region and a 1 300-ha area in the Algonquin Region. Populations declined in the remaining 2 years with only scattered occurrences of light defoliation found in both northwestern and southern Ontario.
Beech scale *Cryptococcus fagisuga*	Beech	Central, Eastern, and Southwestern regions	The first occurrence was reported in 1981 in the Central Region. In 1987, low numbers were recorded in the Eastern Region and in 1988 similar population levels were noted in the Southwestern Region. Medium to heavy infestations affected 10% of the trees at one location in 1984 in the Central Region and 18% of the trees in a stand in 1988 in the Eastern Region.
Birch leafminer *Fenusa pusilla*	White birch grey birch	Province-wide	Moderate to severe damage has occurred in every year of the review period. Ornamental damage in urban areas is almost universal. Notable areas of forest tree damage has occurred on white birch in the Latchford–New Liskeard area, Temagami District; in the Thunder Bay area; and on wire birch throughout its range in the Eastern Region.
Birch skeletonizer *Bucculatrix canadensisella*	White birch	Province-wide	In 1973, moderate to severe defoliation was mapped over 23.3 million ha. Infestation collapsed in 1975 and began to accelerate again in 1980. Peaked in 1983 at 8.3 million ha. Collapsed again in 1984. None reported since.
Black army cutworm *Actebia fennica*	Black spruce	Township 239, Hearst District, Northern Region	A major infestation resulted in the loss of 82 000 newly planted trees on a 250-ha burn site in 1984.
	Jack pine	Miramichi Twp, Gogama District and Hill Twp, Chapleau District, Northern Region	High numbers in 1984 and 1985 caused considerable damage to recently planted trees on prescribed burn sites.
Cedar leafminers *Argyresthia aureoargentella* *A. canadensis* *A. thuiella* *Coleotechnites thujaella*	Eastern white cedar	Southern Ontario and Northeastern Region	Widespread infestations occurred in the early 1970s and again from 1979 to 1983. Populations have remained low, except for isolated outbreaks on Manitoulin Island and the Bruce Peninsula, until 1988 when increases were recorded throughout much of the area.
Cherry scallopshell moth *Hydria prunivorata*	Black cherry	Southern Ontario	Populations peaked in 1982 when 100% defoliation of 20 ha occurred in Albemarle Township, Owen Sound District, and on 80 ha in Uxbridge Township, Maple District. Moderate damage was also reported in Lindsay and Bracebridge districts.

(*Continued*)

Table 10. Other noteworthy insects, Ontario, 1973–1987. (*Continued*)

Insect	Host	Location	Remarks
Eastern blackheaded budworm *Acleris variana*	Hemlock	Algonquin Region	Moderate to severe defoliation of hemlock occurred on 14 680 ha in Bracebridge District in 1982. Moderate numbers were reported at several locations in Parry Sound and Minden districts that same year.
		Province-wide	There were light infestations in about one-half of the years from 1973 to 1988. These infestations occurred in numerous areas from Thunder Bay east to Cochrane in Northern Ontario and from the Aylmer District east to the Minden District in southern Ontario.
Eastern pine shoot borer *Eucosma gloriola*	Pine	Province-wide	Numbers of infested leaders reached a high of 67% at one location during the review period and averaged approx. 9% from 1981 to 1988.
Eastern tent caterpillar *Malacosoma americanum*	Cherry and other deciduous species	Southern Ontario and the Northeastern Region	High populations causing widespread, heavy infestations were commonly found from 1973 to 1978 with reductions first noted in 1979 and 1980. No mention of the insect was made for the next 2 years until in 1983 when a general population increase was observed. This trend continued and has resulted in heavy infestations observed in the Eastern, Algonquin, Central, and Northeastern regions throughout the remainder of the review period.
European pine needle midge *Contarinia baeri*	Red pine	Algonquin, Central, Eastern, Northeastern, and Southwestern	Moderate to severe foliar damage to host trees was reported in 6 years in the 16-year review, 1976, 1980, 1982, 1983, 1984, and 1987. Total areas of damage in some years when the information was supplied were as high as 40 ha and the incidence of attack as well as accompanying foliar damage was as high as 90%.
European pine sawfly *Neodiprion sertifer*	Red, Scots, Mugho, and jack pine	Southern Ontario, Northern Region	Populations were generally low with some in the insect's range from 1973 to 1979. High populations were reported in 1980, but declined again the next year and remained low for the next 3 years. Population increases were observed in 1985 and moderate to severe defoliation was observed in the Espanola District in 1986 and 1987. Low numbers were found in 1988.
European pine shoot moth *Rhyacionia buoliana*	Pine	Southern Ontario	Small localized, medium and heavy infestations of plantations were reported in most years, particularly in southwestern Ontario. Notable damage occurred in 1987 at St. Williams Forest Station when an entire compartment was heavily infested with overwintering larvae.
		Northeastern Region	New distribution points have been reported as far west as Sault Ste. Marie District.
Fall webworm *Hyphantria cunea*	Deciduous	Southern Ontario and North Bay District	Widespread medium to heavy infestations were reported from 1973 to 1976 and again from 1983 to 1986 through much of this area with only localized infestations in the intervening years.
		Remainder of Northern Ontario	Pockets of infestations were reported in Thunder Bay District (1975); in the Temagami, Kirkland Lake, and Timmins districts (1984); and in Kenora and Fort Frances districts (1985).

(*Continued*)

Table 10. Other noteworthy insects, Ontario, 1973–1987. *(Continued)*

Insect	Host	Location	Remarks
Grey willow leaf beetle *Tricholochmaea decora*	Willow	Northwestern Region, North Central Region, Northern Region	Periodic infestation caused moderate to severe defoliation to willow over thousands of hectares in these regions from 1985 to 1988.
Greenstriped mapleworm *Dryocampa rubicunda*	Maple	Northeastern Region, Algonquin Region, Northwestern Region, North Central Region	Infestation declined from 11 000 ha in 1973 and 1974 to low numbers by 1978. Remained low with sporadic, small infestations not exceeding 450 ha through 1987. In 1988, small, new infestations were reported in the North Central and Northwestern regions.
Jack pine sawfly *Neodiprion pratti banksianae*	Jack pine	Province-wide	Low populations causing scattered light defoliations were found to have taken place throughout most of the 16 years of the review period. Moderate damage was reported at sites in Owen Sound District in 1973 and no mention was made of the insect in 1974. Severe defoliation was observed in a 19-ha plantation in Pembroke District in 1984 and scattered occurrences of moderate to heavy damage were noted in the Northern Region in 1985.
A jack pine sawfly *Neodiprion pratti paradoxicus*	Jack pine	Algonquin and Eastern Region	Small pockets of moderate and severe defoliation were found in the Eastern Region during the first 3 years of the review period. Population levels were low for the next 2 years and started to increase in southwestern Ontario in 1978. Scattered moderate to severe infestations occurred primarily in the Algonquin Region for the next 4 years. Although only low numbers were found in 1983, occurrences of localized, heavy infestations were common in the Algonquin and Eastern regions for the remainder of the period.
Jack pine tip beetle *Conophthorus banksianae*	Jack pine	Province-wide	First reported in the review period in 1977, then reported every year after. Populations were variable across the province with periodic local outbreaks in 1978, 1980, 1981, 1982, 1983, 1987, and 1988, with most of the damage occurring in the Northern and Northeastern regions. Damage in these years when recorded was typified by an incidence of attack ranging from 25% to 68% with associated terminal damage of up to 9%.
Larch casebearer *Coleophora laricella*	Tamarack European larch	Province-wide	Pockets of moderate to severe defoliation were recorded in the province in each year of the review period. The largest areas of medium to heavy infestation were recorded in Central Region where 500 ha, 260 ha, and 120 ha were affected in 1977, 1978, and 1984, respectively. Light whole-tree mortality was reported in the Algonquin Region in 1981.
Larch sawfly *Pristiphora erichsonii*	Tamarack European larch	Province-wide	From 1974 to 1978 small pockets of moderate to severe defoliation occurred sporadically across Northern Ontario while plantations of exotic larch often sustained severe defoliation in southern Ontario. From 1979 to 1982 defoliation was mostly confined to southern Ontario. From 1982 to 1985 some resurgence was recorded in the North Central Region but numbers have been less since 1986.

(Continued)

Table 10. Other noteworthy insects, Ontario, 1973–1987. *(Continued)*

Insect	Host	Location	Remarks
Linden looper *Erannis tiliaria*	Deciduous	Northeastern Region North Central Region	A massive infestation in Lake Superior Park peaked at 200 000 ha in 1974 and 1975. In 1985, a relatively small infestation (1 050 ha) appeared in the Northeastern Region for 1 year. Low numbers were present until 1988 when 8 835 ha of moderate to severe defoliation occurred in the Terrace Bay District.
Maple leafcutter *Paraclemensia acerifoliella*	Maple	Eastern Region Algonquin Region Central Region	Fluctuating populations of varying proportions from 130 000 ha north of Kingston in 1981 to 1 187 ha in Algonquin Region in 1984 were recorded along with smaller pockets of infestation in these regions.
Oak leafmining sawfly *Profenusa lucifex*	Oak	Lindsay District, Central Region	Moderate to severe foliar damage occurred in an area as large as 8 300 ha in Hamilton and Haldimand townships south of Rice Lake from 1975 to 1980.
Pine engraver *Ips pini*	Red pine	Algonquin Region	In 1983 and 1988 high numbers were found at several points and were associated with drought-stressed trees and log piles.
Pine false webworm *Acantholyda erythrocephala*	Pine	Southern Ontario	A drastic increase was reported in 1976 and this continued until 1981. Medium to heavy infestations were more localized from 1981 to 1986. Populations have been increasing in the past 10 years. Distribution now includes most of the area.
	Northern Ontario		Distribution points have been recorded in North Bay and Hearst districts in the northeast and at Thunder Bay and Lake of the Woods in northwestern Ontario.
Pine tortoise scale *Toumeyella parvicornis*	Jack pine	Northern Region Northeastern Region Algonquin Region	Single and small groups of trees were heavily infested at many locations in these regions during the review period but tree mortality was insignificant.
Red pine sawfly *Neodiprion nanulus nanulus*	Red and jack pine	Province-wide	Noteworthy damage levels did not appear until 1983, when scattered pockets of moderate to severe defoliation were found in the Northwestern, North Central, and Northern regions. Populations remained high the next year in the same regions as well as in the Northeastern Region, but a decline was to follow during the next 2 years. A few exceptions were observed in the Northeastern and Algonquin regions in 1986. Low numbers have been reported on in the remaining 2 years of the review period.
Redheaded jack pine sawfly *Neodiprion rugifrons*	Jack pine	Northern Ontario	Low population levels were reported on during the review period with isolated accounts of high numbers or moderate to severe defoliation taking place in some of the northern districts. Heavy damage was found in Atikokan District in 1974, in the Northern and Northeastern regions in 1980 and 1981; and one pocket of severe defoliation was observed in the Northwestern Region in 1983. Increased insect populations resulted in moderate to severe defoliation at scattered sites in 1988.
Redheaded pine sawfly *Neodiprion lecontei*	Red and jack pine	Southern Ontario and Northeastern Region	During the first 10 years of the review period population fluctuated somewhat, but high numbers of insects causing severe defoliation were present at one place or another in southern

(Continued)

Table 10. Other noteworthy insects, Ontario, 1973–1987. *(Continued)*

Insect	Host	Location	Remarks
Redheaded pine sawfly *(Continued)*			Ontario or Northeastern Region. Low levels of tree mortality were found in the Eastern Region in 1978 and in 1981; 33% mortality levels were reported at one site in the Blind River District. Populations were down in 1983 but have rebounded in southern Ontario throughout the remaining 6 years of the review period.
Saddled prominent *Heterocampa guttivitta*	Sugar maple, beech, and red oak	St. Joseph Island, Sault Ste. Marie District, Northeastern Region	Moderate to severe defoliation occurred in two areas of the island in 1976.
		Huronia and Bracebridge districts Central and Algonquin regions	Light to moderate numbers were reported in several areas in 1981 and 1982.
Saratoga spittlebug *Aphrophora saratogensis*	Red pine	Pembroke District, Algonquin Region	Severe branch and top mortality was reported in 1975 in 13 plantations west and southwest of Pembroke. Aerial spraying of 143 ha was done the following year and was very effective. From 1983 to 1987, relatively low numbers have been reported from this same general area typically causing up to 2% tree mortality and 10% branch mortality.
Satin moth *Leucoma salicis*	Silver poplar, lombardy poplar, willow	Eastern Region	Defoliation was mostly confined to ornamental trees and hedgerows with low numbers affected at each location. Highest numbers were reported in 1982, 1983, and 1986.
Smaller European elm bark beetle *Scolytus multistriatus*	Elm	Southern Ontario	This vector of Dutch elm disease, *Ceratocystis ulmi*, had become well established through much of the area by 1982.
Solitary oak leafminer *Cameraria hamadryadella*	Bur oak, red oak	Central, Eastern, Southwestern, and Northeastern regions	Small areas of medium to heavy infestation were reported in 1983 in the Southwestern and Eastern regions and again in 1986 in the Central and Southwestern regions. In 1988, a 5-ha area of host in the Northwestern Region experienced 80% foliar damage.
Spearmarked black moth *Rheumaptera hastata*	White birch, yellow birch, alder	Northeastern Ontario	An infestation in 1981 caused moderate to severe foliar damage and premature leaf drop on over 772 000 ha mainly in the Wawa and Chapleau districts. Area of damage was confined to Chapleau District and reduced to 410 000 ha in 1982. Populations continued to decline from 1983 to 1985.
Spruce bud moth *Zeiraphera canadensis*	Spruce	All regions except Eastern	Varying degrees of damage to current year's growth of mostly open-grown white spruce trees were recorded throughout the province.

(Continued)

Table 10. Other noteworthy insects, Ontario, 1973–1987. (*Concluded*)

<table>
<tr><th>Insect</th><th>Host</th><th>Location</th><th>Remarks</th></tr>
<tr><td>White pine weevil
Pissodes strobi</td><td>Pine, spruce</td><td>Province-wide</td><td>White pine is the most susceptible species with counts ranging to 99% leader mortality in some plantations. A typical listing of infestation for the province in 1974 is shown below.
<table>
<tr><th>Host</th><th>Leader mortality (%)</th><th>Host</th><th>Leader mortality (%)</th></tr>
<tr><td>wP</td><td>85</td><td>bS</td><td>39</td></tr>
<tr><td>scP</td><td>16</td><td>wS</td><td>18</td></tr>
<tr><td>jP</td><td>9</td><td>nS</td><td>11</td></tr>
</table></td></tr>
<tr><td>Whitespotted sawyer
Monochamus scutellatus</td><td>Jack pine, black spruce, white spruce, eastern white cedar</td><td>Northern Ontario</td><td>No mention was made of this insect from 1973 to 1976 and in the remaining years of the review period damage was caused by adult sawyer beetle feeding, usually adjacent cutovers. Damage to branches with some tree mortality was first seen in 1977 and peaked in 1980 when 200 ha of heavy damage was detected along with 135 ha of tree mortality. Damage to branches and some tree mortality has continued in small localized areas throughout the remainder of the period.</td></tr>
<tr><td>Yellowheaded spruce sawfly
Pikonema alaskensis</td><td>Spruce</td><td>Province-wide</td><td>There have been medium to heavy infestations in each year of the review period. Damage has occurred on roadside, hedgerow, and ornamentals as well as on plantation trees. Particularly heavy defoliation has been recorded in the Englehart–New Liskeard area and has resulted in instances of tree mortality. Tree mortality has also occurred in Cochrane District and in the Northwestern Region. High populations in high-value areas have required control measures on several occasions. One of the largest was in 1975 when 105 ha was aerially sprayed with fenitrothion at Balsam Lake Provincial Park, Lindsay District.</td></tr>
</table>

References

Forest Insect and Disease Conditions in Canada. 1980, 1981, 1982, 1983, 1984, 1985, 1986, 1987. Forestry Canada, Ottawa.

Forest Insect and Disease Survey: Annual Report 1973, 1974, 1975, 1976, 1977, 1978, 1979. Forestry Canada, Ottawa.

Howse, G.M.; Applejohn, M.J. Survey Bulletins, three each year (spring, summer, fall) for Ontario from 1980 to 1987. Forestry Canada, Great Lakes Forestry Centre, Sault Ste. Marie, Ont.

Regional Ranger Reports, eight each year for Ontario from 1973 to 1987. Forestry Canada, Great Lakes Forestry Centre, Sault Ste. Marie, Ont.

Chapter 5

Forest Insect Pests in the Northwest Region

H.F. Cerezke and W.J.A. Volney

Introduction

Relatively few forest insect pests have historically been considered economically important in the Prairie provinces and Northwest Territories (Northwest Region). Information about these pests is obtained primarily from the annual Forest Insect and Disease Survey (FIDS) reports that describe regional outbreaks, insect population trends, and mapped areas of infestation. The criteria for their selection are based on their often widespread occurrence, their known historical pattern of repeated outbreaks, their persistence at high population levels during outbreak periods lasting 2 or more years, and the effect of their primary damage, resulting in reduced volume increment, top kill, and tree mortality. In general, these insect species are fairly host-specific, and although not exclusively so, they are mostly associated with semimature to mature forests. In this chapter, the status and trends in population abundance of the important pest species are described in some detail for the period 1973 to 1990.

Other insect species commonly present but usually of minor economic importance are listed in tabular form, separately for coniferous and deciduous trees. Inclusion of these species helps to pinpoint potentially new pests in localized areas or to recognize those that may develop into outbreaks at relatively infrequent intervals.

The foregoing accounts highlight the pests traditionally associated with natural forests, but new man-made or man-influenced forests that reflect varying levels of forest management practices are emerging across the region. Development of these young forests creates a mosaic of new habitats and opportunities that may favor colonization and expansion of certain endemic insect species normally classed as minor or incidental pests. Some of these may now find new conditions favoring their survival, and thus become threatening pests in localized areas. Species falling into this category and reported since 1973 are also included in the accompanying table.

For more detailed descriptions of the various pests and their outbreak distributions, refer to the annual FIDS reports and other published reference materials listed in the Bibliography.

Insects of Natural Forests

Forest Tent Caterpillar

Historically, the forest tent caterpillar, *Malacosoma disstria*, has been the most spectacular of forest defoliators affecting mainly aspen forests over wide areas across the region. Although its impact on forest productivity is considerable, it is perhaps best known for its direct interaction with people as a nuisance pest in areas of recreation, aesthetics, shelterbelts, and on ornamental tree and shrub plantings. Concern for its overall impact on the aspen forests is increasing, however, because of the current rapidly expanding utilization and management of the aspen and poplar wood and fiber resource.

The forest tent caterpillar periodically erupts into large outbreaks, on average every 10 years or less, that typically result in foliage depletion for 3 to 6 consecutive years over the same forested areas. Across the three Prairie provinces, there were at least four major outbreaks since 1938, when detailed surveys were initiated. Outbreaks appear to develop initially in the Aspen Parkland zone, then spread southward into aspen stands and urban areas within the agricultural zones, and northward into the boreal mixedwood forest. Areas along the lower foothills of Alberta may also be affected as far south as Waterton Lakes National Park. At the peak of outbreaks, most areas of infestation are distributed within the southern two-thirds of the three Prairie provinces, but in Alberta may extend farthest northward almost to the Northwest Territories border.

Throughout the region, a synchrony of outbreak trends is not readily apparent. Since 1973, areas of aspen defoliation have occurred annually to some degree in all three Prairie provinces (Table 1), although infestation pockets tend to shift in location from year to year. During the 18-year period reported (1973 to 1990), population trends in Alberta, Saskatchewan, and probably north-central Manitoba appear to coincide reasonably well, whereas outbreak areas in southern Manitoba appear to be more closely correlated with those in western Ontario.

In Table 1, estimated total outbreak areas are presented for each province. The estimates indicate composite land areas over which moderate to severe levels of defoliation were mapped by ground and aerial surveys and therefore reflect variable areas of actual infested aspen forests. This is because many of the infestations occur throughout the agricultural zone where aspen stands may constitute less than 10% of the land base surveyed. However, as infestations expand, they invariably extend farther into the boreal mixedwood forest and encompass a greater proportion of aspen forests.

The areas of infestation in Table 1 also include defoliation contributed by various other aspen defoliators. Outbreaks of these species often occur concurrently with the forest tent caterpillar and their contributing defoliation cannot usually be separated, except when they occur in the absence of the forest tent caterpillar. The large aspen tortrix, *Choristoneura conflictana*, is the most important of these species and is dealt with in a separate section.

During outbreak periods aspen forests often sustain a severe level of defoliation for 3 or more consecutive years. Mortality of aspen trees rarely occurs as a result of accumulated forest tent caterpillar defoliation, but radial growth losses in the stem and top kill of the upper branch tips may occur after 2 consecutive years of severe defoliation. Further damage may result when defoliation occurs concurrently with

Table 1. Total land areas (hectares) over which infestations of the forest tent caterpillar, *Malacosoma disstria*, caused moderate to severe defoliation annually from 1973 to 1990 in Manitoba, Saskatchewan, and Alberta.

Year	Manitoba	Saskatchewan	Alberta
1973	1 000 000[a]	360 000	100 000[a]
1974	2 150 000	360 000	170 000
1975	3 400 000[a]	300 000[a]	170 000[a]
1976	10 300 000	350 000[a]	300 000[a]
1977	11 900 000	>350 000[a]	3 000 000[a]
1978	71 000	2 200 000[a]	6 400 000[a]
1979	<50	6 800 000	6 400 000
1980	<50	12 800 000	7 500 000
1981	100 000	8 200 000	12 100 000
1982	600 000	2 100 000	13 000 000
1983	600 000	350 000	3 998 000
1984	634 000	3 000	1 335 000
1985	19 500	31 300	2 260 000
1986	17 090	1 771 590	3 160 890
1987	4 403	1 250 000	6 610 700
1988	52 836	4 660 200	13 830 000
1989	306 395	3 953 700	5 899 000
1990	15 178	1 304 610	3 046 360

[a] Areas of infestation are estimates inferred from descriptive information and from aerial survey maps in annual reports of the Forest Insect and Disease Survey.

other stress factors such as spring and summer drought, late spring frost, poor or wet site conditions, stand age, and stem decay. Management of aspen stands for wood and fiber products would therefore have to consider these factors affecting aspen stand dynamics.

An outbreak of the forest tent caterpillar began about 1971 in Manitoba, being first detected in the Interlake region in the vicinity of Lake Winnipeg Narrows. By 1973, numerous patches of moderate to severe defoliation extended over an estimated 1 million ha in southern Manitoba, extending from Whiteshell Provincial Park through the Interlake to Winnipegosis and Dauphin lakes and to Riding Mountain National Park. These patches continued to expand and coalesce during the next few years, reaching a peak in 1976 and 1977 (Table 1). The outbreak area included most of southern Manitoba from about 52o N to within 50 km of the United States border. Expansion extended eastward simultaneously with areas of infestation in western Ontario and also westward into Saskatchewan.

Egg band surveys in 1977 indicated that a general downward trend was likely to occur in 1978. The trend materialized dramatically and continued in 1979. The outbreak collapsed in 1980, except for an infestation in and adjacent to Turtle Mountain Provincial Park which persisted until 1982. The parasitic large flesh fly, *Sarcophaga aldrichi*, was considered to be an important factor during the collapse of the outbreak. In 1982, new infestations appeared in west-central Manitoba, probably as extensions of the existing infestations in Saskatchewan. In 1983, some of these areas broke up, but small and scattered patches extended throughout the Saskatchewan River and Nelson River Forest sections, so that the total outbreak area remained about the same. The outbreak increased to a maximum coverage in 1984, then declined steadily until 1987 (Table 1). Areas of infestation increased again in 1988 and 1989 throughout central and southeastern Manitoba, then declined sharply in 1990.

Before 1973, an outbreak in Saskatchewan and Alberta occurred from about 1958 to 1964 when it collapsed at most locations except in central Alberta near Lake Wabamun. Here, small patches of infestation began to expand slowly after 1966 but remained relatively small for the next 10 years. This area surrounding Wabamun Lake has often been cited as one with long-term chronic infestations that have persisted continuously for more than 20 years.

By 1973, a few scattered infestations were present in both provinces that amounted to less than 500 000 ha. Most of these were concentrated within 200 km north and northwest of Prince Albert, and in aspen forests 50 to 100 km west and southwest of Edmonton. Little change occurred in 1974, but a major spread and intensification was apparent in both provinces by 1975 and 1976, signalling the development of a new outbreak. Expansion of the outbreak continued for the next several years, reaching a peak area of 12.8 million ha in 1980 in Saskatchewan and about 13 million ha in Alberta by 1982. During these peak years of 1980 to 1982, defoliated aspen forests were concentrated throughout the central areas of both provinces, extending from near High Level in northwestern Alberta, south to Red Deer, east to Yorkton, Saskatchewan, and north to beyond Lac La Ronge.

Areas of infestation in central Saskatchewan, west and northwest of Prince Albert, began to break up in 1981 and set a pattern that continued until 1984. In that year only a few small infestations amounting to about 3 000 ha were reported between Hudson Bay and Yorkton. However, in 1985 and 1986, infestations in eastern Alberta, south of the North Saskatchewan River, expanded eastward into Saskatchewan toward Saskatoon. Slight increases also occurred south and southwest of Hudson Bay, suggesting a resurgence of the existing outbreak. In 1987, the infestations in eastern Saskatchewan expanded northward and westward, but declined steadily in 1988, 1989, and 1990 to a few small scattered patches. The large aspen tortrix contributed much of this defoliation in 1989. Infestations in the western part of Saskatchewan declined slightly in 1987 but increased substantially during 1988 and 1989 and declined again in 1990 to about 1 million ha.

A declining trend in outbreak area also occurred in Alberta from 1982 to 1984 but was less dramatic than that occurring in Saskatchewan during the same period. The infestations had broken into numerous patches, extending from the Saskatchewan border near Lloydminster, northwesterly to the British Columbia border, and to about 200 km north of Peace River. In 1985, however, further decreases occurred in northwestern Alberta (Grande Prairie and Peace River Forest districts) but expanded in central and east-central Alberta, reflecting a renewed outbreak trend. This trend of outbreak has continued since 1984 and by 1988 had extended over a composite area of more than 13.8 million ha, including much of the same aspen forests that were infested during the period of 1980 to 1982.

The areas of infestation declined in 1989, and by 1990 most of the defoliated aspen stands were in east-central Alberta and small patches in the Peace River district. Virus infection was commonly observed at many locations in 1988, as well as the large flesh fly, *S. aldrichi*. In 1990, some defoliation injury was contributed by the large aspen tortrix and Bruce spanworm, *Operopthera bruceata*, especially in western Alberta.

Large Aspen Tortrix

The large aspen tortrix, *Choristoneura conflictana*, is a major defoliator of trembling aspen but feeds on other associated broad-leaved species when epidemic. It periodically develops into outbreaks over large areas, though usually not as spectacularly as the forest tent caterpillar. Outbreaks tend to be short-lived, lasting 2 to 3 years, and are often overshadowed by outbreaks of the forest tent caterpillar. Like the forest tent caterpillar, most epidemics of the large aspen tortrix occur throughout the southern two-thirds of the three Prairie provinces, except in its western range where epidemics have extended into the Northwest Territories as far north as the Martin and Ebbutt hills.

Because of its common association with outbreaks of the forest tent caterpillar, the reporting of the large aspen tortrix defoliation damage is less complete, and may in fact be underestimated, especially in northern areas where infestations have been mapped only during aerial surveys.

During outbreaks, aspen forests may be completely defoliated for 1 or 2 years and recover with only some reduction in radial increment and bud damage. However, when the large aspen tortrix occurs in abundance concurrently with the forest tent caterpillar and other aspen defoliators, the damage effects by all species may be additive. The larvae, and especially the large amounts of silk webbing they produce, are considered a nuisance in high-use areas.

Table 2 summarizes the known historical population trends throughout the region from 1973 to 1990. The data reflect a pattern of incomplete coverage and detection throughout the region. However, the highest incidence and frequency of outbreak areas were more apparent in Alberta than in other parts of the region.

Table 2. Years from 1973 to 1990 during which the large aspen tortrix, *Choristoneura conflictana*, was reported at one or more locations in Manitoba, Saskatchewan, Alberta, and the Northwest Territories causing light, light to moderate, or moderate to severe defoliation.

Year	Manitoba	Saskatchewan	Alberta	NWT
1973	c	b	a	-
1974	a	a	c	-
1975	a	0	c	-
1976	a	0	c	-
1977	0	0	c	-
1978	a	c	c	-
1979	0	c	c	-
1980	0	a	c	c
1981	0	0	c	c
1982	0	a	a	c
1983	0	c	c	a
1984	c	c	c	a
1985	c	a	c	0
1986	0	a	a	a
1987	a	c	0	0
1988	c (2 849 ha)	c	b	0
1989	c (18 650 ha)	c	b	0
1990	c (15 540 ha)	b	c	0

a, Present in low abundance or causing light defoliation.
b, Present and causing light to moderate defoliation.
c, Present and causing moderate to severe defoliation.
0, Not found in the surveys.
-, No data available.

An outbreak causing moderate to severe defoliation of aspen forests had occurred in Manitoba from 1971 to 1973, and was extensive from Whiteshell Provincial Park westward through the Interlake to Riding Mountain National Park. Populations were noted to be increasing across the three Prairie provinces in 1972, but declined in 1973, leaving numerous small scattered patches from the Reindeer Lake area to southeastern Manitoba. Populations continued a downward trend in Manitoba and Saskatchewan until 1978, while a new infestation extending for about 50 km developed in northwestern Alberta in 1974 and was extensive in the Peace River and Slave Lake Forest districts in 1977 and 1978.

Low population levels were associated with the forest tent caterpillar in Manitoba in 1978 and then were not reported until 1984 and 1985 when an infestation caused moderate to severe defoliation in Birds Hill Provincial Park. This infestation apparently collapsed in 1986, but low populations were noted in the Interlake area and Duck Mountain Provincial Park in 1987. This infestation expanded during the next 3 years causing moderate to severe defoliation on over 15 000 ha in Duck Mountain Provincial and Riding Mountain National parks in 1990 (Table 2). During 1988 to 1990, low levels of injury were also noted at several widely scattered locations in central and western Manitoba.

In Saskatchewan and Alberta, small localized infestations were interspersed with populations of the forest tent caterpillar and caused severe defoliation in southeastern and west-central Saskatchewan and in central and northwestern Alberta in 1979, 1980 and 1981.

In 1983 and 1984, the large aspen tortrix was the main defoliator of aspen in the Cypress Hills in southern Saskatchewan and Alberta, as well as in the Porcupine Hills and Waterton Lakes National Park in southwestern Alberta. In these areas it accounted for nearly 50 000 ha of moderate to severe defoliation to aspen forests. These outbreaks collapsed in 1985 except for small moderately to severely defoliated patches persisting north of the Porcupine Hills. Scattered low populations were reported in both provinces in 1986, and in 1987, high populations causing moderate to severe defoliation occurred only in central Saskatchewan.

Increased levels of infestation by the large aspen tortrix occurred in 1988 and 1989 in central and eastern Saskatchewan as well as in central and western Alberta. Its distribution generally coincided with that of the forest tent caterpillar in both provinces but contributed to less defoliation injury. In 1990, population levels causing light to moderate defoliation were common throughout central and eastern Saskatchewan, in association with forest tent caterpillar, while in Alberta several areas of moderate to severe defoliation were widespread from the Cypress Hills in the south to the Footner Lake Forest in the northwest.

An outbreak along the Slave River, in the Martin Hills and in Wood Buffalo National Park, Northwest Territories, was first detected in 1980, and by 1981 it had extended over 100 000 ha. It declined into a few small patches in 1982 but persisted at low populations until 1984. Areas of light defoliation were again mapped in the Martin and Ebbutt hills, and north and west of Fort Simpson in 1986, but may have collapsed in 1987. No reports of large aspen tortrix in the Northwest Territories were made during 1988 to 1990.

Spruce Budworm

Outbreaks of the spruce budworm, *Choristoneura fumiferana*, have occurred in the Prairie provinces and the Northwest Territories during the past 18 years and have defoliated considerable areas of forest. Because of the vast area involved it is quite certain that the outbreaks are more extensive than the ones detected by the Forest Insect and Disease Survey of Forestry Canada. The records are incomplete, especially in the 1970s when maps of the outbreaks were seldom provided and areas not reported. The extent and intensity of surveys declined substantially during this period and remote areas were not surveyed. Estimates of the areas defoliated were obtained from the verbal descriptions in the reports and these estimates are probably conservative. The distributions of the spruce budworm in each of the provinces and territories are largely bounded by the extent of host stands within the region. The extent of defoliation is a reflection of the stand and abiotic conditions that permit populations to exceed damaging thresholds.

The gross areas estimated to have been moderately to severely defoliated annually by the spruce budworm in each province and the Northwest Territories for the past 20 years are presented in Table 3. Before and including 1968, there had been extensive outbreaks in the region but all seemed to collapse in 1969. The FIDS reports for the region attribute this sudden collapse to unseasonably late spring frosts that severely damaged the current year's foliage thus interfering with the development of budworm populations in that year. This occurred over a large area, including almost all of the region except southwestern Manitoba. Sudden, dramatic declines in the areas defoliated were to be associated with these unseasonably late spring frosts in two more instances in the region over the period in question.

In Manitoba there were three periods when the areas defoliated by the spruce budworm peaked following the decline of 1969. By 1976, patches of forest over 648 km^2 in the Interlake region were moderately to severely defoliated and populations persisted at high levels in this region for the next 3 years. The peak occurred in 1979. Some of the stands in this area seemed to have sustained moderate to severe defoliation for 11 consecutive years. Apparently unseasonably late spring frosts terminated this outbreak in 1980. The area damaged started to increase in size again in 1982 and peaked in 1984. However, the area most widely affected in the more recent outbreak was east and southeast of Lake Winnipeg. By 1985, tree mortality had reached 25% in some stands with 90% of the balsam fir component killed. The area affected by outbreaks thus shifted southeast from northwestern Manitoba in the 1960s, through the Interlake in the 1970s, and ending in southwestern Manitoba in the 1980s through to 1990.

Following the decline in 1987, areas of infestation increased again in southeastern Manitoba, peaking in 1989. Due to the increasing decline of budworm-damaged forests in this area, the province of Manitoba carried out aerial spray treatments of insecticide for budworm population suppression on over 4 000 ha in 1989 and 1990.

An area in southwestern Manitoba, including the Spruce Woods Provincial Park and the provincial forest, is unique for the persistence of damaging populations. Hardy et al. (1986) reported defoliation in the general vicinity of this area every year from 1938 to 1979. Although the FIDS reports do not indicate defoliation in the area since then, the budworm populations are known to have attained reasonably high densities in 1985 and caused annually some patches of light or moderate defoliation from 1986 to 1990.

Following the collapse of the 1960s outbreak in northeastern Saskatchewan in 1969, no moderate to severe defoliation was reported in that province until 1982. Since then there has been a slow increase in the area in which moderate to severe defoliation has been detected. However, the area involved is south of the previous outbreak close to the Manitoba–Saskatchewan border. These infestations continued at similar intensities to 1990 and prompted increased salvage logging in the area to reduce losses due to accumulated budworm feeding. A new infestation of about 4 500 ha of white spruce–balsam fir forests was discovered west of Prince Albert National Park in 1989 and this has continued in 1990.

Before 1968, there was an outbreak in the Wabasca River drainage that had persisted, based on an interpretation of tree-ring records, from 1941 to 1969 (Stevenson, unpubl. data).[1] Another outbreak in the Chinchaga River drainage also collapsed at the same time. By 1971 no moderate or severe defoliation was reported in Alberta. In 1973, a new outbreak was reported further east in the Athabasca River drainage. This outbreak grew dramatically and by 1974 involved 21 750 million ha of forest in the Athabasca and Clearwater river valleys. The decline of this outbreak was equally as

[1] Stevenson, R.E. 1970. The spruce budworm in northern Alberta: with emphasis on the Wabasca outbreak. Dept. Fish & For., Can. For. Serv., Edmonton, Alta. Int. Rep. A-28.

Table 3. Areas (hectares) of moderate to severe defoliation by the spruce budworm in the Prairie provinces and the Northwest Territories detected between 1968 and 1990.

Year	Manitoba	Saskatchewan	Alberta	NWT	Total
1968	307 600	672 100	262 440	185 100	1 427 240
1969	13 500[a]	0	12 000	19 800	45 300
1970	13 500[a]	0	4 200	26 700[a]	44 400
1971	32 270	0	0	0[a]	33 270
1972	36 100	0	0	1 000[ab]	37 100
1973	38 000	0	9 300	–[c]	47 300
1974	41 000	0	21 750	–[b]	62 750
1975	39 000	0	12 900	–[c]	51 900
1976	111 000	0	13 500	–[c]	124 500
1977	313 000	0	6 400	–[c]	319 400
1978	313 000	0	4 500	–[c]	317 500
1979	520 000	0	0	–[b]	520 000
1980	30 000[a]	0	0	0[b]	30 000
1981	3 000[a]	0	250[d]	0[b]	3 250
1982	31 380	2 000	850[d]	1 000[a]	35 830
1983	40 500	12 700	1 000	11 800	65 600
1984	142 700	15 100	1 000	10 900	169 700
1985	77 500	15 000	1 400	12 500	106 400
1986	34 320	18 500	400	13 000	66 220
1987	15 540	31 600	5 790	2 600	55 530
1988	30 820	31 600	37 290	14 350	114 060
1989	39 870	34 650	52 430	98 600	225 550
1990	18 985	18 780	108 800	107 100	253 665

[a]Estimated as 1 000-ha aggregate defoliation per location with patchy defoliation.
[b]MacKenzie and Liard river drainage not surveyed.
[c]Northwest Territories not surveyed.
[d]Includes 250-ha outbreak near Millet, discovered in 1983 which was at least 3 years old.

dramatic and by 1979 no areas of moderate to severe defoliation were detected. Small localized outbreaks were detected in relic spruce stands within the agricultural zone including Edmonton and south. By 1982, these populations had caused moderate to severe defoliation and persisted until 1987, though populations in Edmonton declined substantially from the late frost in May of that year. A new outbreak in the forests of the Chinchaga River valley, which had been defoliated in the 1960s, was detected in 5 400 ha in 1987, and this area of defoliation has increased annually since then to more than 100 000 ha in 1990. Concurrently, other infestations have developed along the Peace and Athabasca river drainages, indicating a pattern of synchronicity in outbreak development over large areas in the northern half of the province. The progressive expansion of the outbreaks and their effect on the largely spruce forest resource prompted the use of aerially applied insecticide in 1989, the first of its kind in this province, over about 1 000 ha. This strategy was repeated in 1990 when more than 11 000 ha of spruce forests were sprayed.

Information on outbreaks in the Northwest Territories, particularly in the 1970s, is a great deal more sketchy. Outbreaks have been reported from the Mackenzie River valley as far north as Norman Wells (Cerezke 1978). However, during the period concerned the most northerly report of moderate to severe defoliation was reported from the mouth of the Redstone River in 1968. These outbreaks collapsed and no moderate to severe defoliation was detected in 1969. It was not until 1982 that moderate to severe defoliation was again reported in the Mackenzie drainage. This occurred on two islands in the Liard River. In 1986, about 12 500 ha of moderate to severe defoliation was reported in this area but the intensity of this outbreak declined so that the stands were rated as light to moderately defoliated in 1987 and 1988. In 1989, the outbreak along the Liard River drainage expanded to encompass about 98 600 ha of mostly moderately to severely defoliated spruce. Further increases were reported in 1990 as the oubtreak extended northward from the confluence of the Liard River to adjacent areas along the Mackenzie River valley. A second group of outbreaks in the Northwest Territories was reported in Wood Buffalo National Park and in the Slave Lake drainage. These collapsed in 1971. In 1985, a new outbreak of 100 ha was detected in the Little Buffalo and Salt river valleys. The total area of moderate to severe defoliation in the Slave Lake drainage reached 2 600 ha in 1987 and increased slightly in area in 1988. This expanding trend continued in 1989 and 1990 to reach close to 5 000 ha with moderate to severe defoliation.

Recurring budworm outbreaks appear to have influenced the species composition of the susceptible host types in some localized stands. The ability of white spruce to survive prolonged defoliation episodes and the relative vulnerability of balsam fir to this pattern of defoliation may have resulted in almost pure white spruce stands. This process can currently be observed in stands that have colonized old clearings in eastern Manitoba. Most of the balsam fir in these young stands have been killed in the last few years (Moody and Cerezke 1986).

The apparent restriction of spruce budworm infestations to river valleys in Alberta and the Northwest Territories is a reflection of the pattern of distribution of white spruce in this region. Thus white spruce, and consequently spruce budworm outbreaks, are restricted to older stands on mesic upland and moist valley bottom sites along the major river valleys of the region.

Jack Pine Budworm

The jack pine budworm, *Choristoneura pinus*, has been recovered from jack pine in all three Prairie provinces but not from the Northwest Territories. No outbreaks of this insect had been reported in Alberta before 1985 when a small outbreak was first noted in the north-central part of the province. This outbreak has remained static from 1985 to 1989 and has been confined to 70 ha of jack pine forest. It is believed that frequent fires and the consequent juvenility of natural jack pine stands in Alberta have created conditions in which outbreaks of the jack pine budworm are unlikely to persist. The 1985 outbreak suggests that Alberta's jack pine forests are no longer immune to attack by the jack pine budworm.

Table 4 is a summary of the areas moderately to severely defoliated in Manitoba and Saskatchewan from 1968 to 1990. Except for 1980 and 1981, moderate to severe defoliation has been detected somewhere in the two provinces during the period of concern. An outbreak that peaked in 1965 in Saskatchewan and in 1967 in Manitoba declined sharply in the first years of the early 1970s. In both provinces, the area moderately to severely defoliated has reached peaks twice since then: peaks occurred in 1979 and 1985 for Manitoba and in 1977 and 1986 for Saskatchewan.

In a recent analysis of FIDS data on the extent and severity of outbreaks, Volney (1988) found strong correlations between the area damaged and climatic conditions, as indicated by the extent of forest fires in the region, for a period of 4 to 7 years preceding the outbreak. The apparent 10-year cycle of recurring outbreaks during the past 50 years appears to be a direct result of the periodicity in extensive forest fires which indicate prolonged periods without rain in the growing season. Despite these general trends for the region as a whole, there are some notable departures from this pattern. Southeastern Manitoba seems to have a chronic jack pine budworm problem; for 30 of the 50 years in the analysis there was at least moderate defoliation in the area. By contrast, no defoliation was detected in the Nisbet Forest in Saskatchewan from 1940 to 1964 inclusive. Stands in this area were regenerated following fires in the early forest and did not sustain any damage until they were at least 20 years old. Significantly, the total area affected by outbreaks in the two provinces has increased during the period in which records have been kept; more than 2 million ha were defoliated in 1985.

Table 4. Areas (hectares) of moderate to severe defoliation by the jack pine budworm in Manitoba and Saskatchewan detected between 1968 and 1990.

Year	Manitoba	Saskatchewan	Total
1968	177 070	0	177 070
1969	0	0	0
1970	6 480	0	6 480
1971	2 590	0	2 590
1972	11 660	0	11 660
1973	1 040	0	1 040
1974	3 480	0	3 480
1975	35 480	0	35 480
1976	7 000	0	7 000
1977	5 180	22 800	27 980
1978	470 520	60	470 580
1979	508 030	0	508 030
1980	0	0	0
1981	0	0	0
1982	46 000	0	46 000
1983	154 500	0	154 500
1984	761 240	26 800	788 040
1985	2 048 500	130 000	2 178 500
1986	125 356	175 000	300 356
1987	100	2 500	2 600
1988	0	<100	<100
1989	0	0	0
1990	0	0	0

Larch Sawfly

The larch sawfly, *Pristiphora erichsonii*, a major defoliator of tamarack as well as other native and exotic larches, was documented in three major outbreaks in Manitoba and Saskatchewan between 1909 and the early 1970s. Its distribution extends throughout the natural range of tamarack in the Prairie provinces and Northwest Territories and has extended to other southerly locations wherever native and exotic larch species have been planted. During a recent outbreak period that began in southern Manitoba about 1938, which seemingly spread westward and northward through Alberta into the Northwest Territories by about 1960, defoliation damage was extensive and severe. Larch sawfly populations during this outbreak period did not remain high throughout the region but fluctuated widely from year to year and often between stands at the same locations. However, most tamarack stands sustained up to 3 consecutive years of severe defoliation, and almost 20% mortality of tamarack in different stands occurred in Manitoba and Saskatchewan over a 10-year period, while radial increment was reduced to almost nil after 15 years of fluctuating populations. Cumulative damage effects after this outbreak were probably similar in Alberta and the Northwest Territories.

One year of moderate to severe defoliation causes reduced foliage and radial increment production. Continued defoliation results in reduced shoot growth, death of branch tips, further reductions in radial stem growth, and eventually tree death after 3 to 8 years of sustained severe defoliation. Trees stressed by repeated defoliations show a surprising resilience for recovery, but many also succumb to attack by secondary pests, especially the eastern larch beetle, *Dendroctonus simplex*. This bark beetle may have contributed significantly to the widespread mortality of most of the merchantable tamarack stands reportedly to have died during earlier outbreak periods.

Survey records since the early 1970s are less complete for information on larch sawfly infestations across the region. However, a few scattered small areas of moderate to severe defoliation were reported in southeastern Manitoba (Agassiz and Sandilands provincial forests), west-central Manitoba (The Bog area), and in west-central Alberta (near Obed). These infestations remained small with minor fluctuations in population levels from about 1971 to about 1978. From 1975 to 1978, moderate to severe defoliation occurred over about 120 ha in The Bog. Populations remained low for the next several years until 1984 when small areas of moderate to severe defoliation appeared at several locations in west-central Manitoba; only small isolated pockets have persisted up to 1987.

Since the early 1970s, only minor surges in population abundance, causing light defoliation, have been reported in central Saskatchewan in 1987 and again in 1990. In Alberta, the small outbreak near Obed persisted from 1973 to 1982, and then was not reported again until 1987 through to 1990 when light defoliation was noted here and at several other locations in the west-central part of the province.

Survey monitoring in the Northwest Territories was reduced during the 1970s and some infestations may have gone undetected. The first reported infestations causing light defoliation were in 1979 near Fort Smith, and Wood Buffalo National Park. These intensified over large areas covering several hundred square kilometres until 1984 and 1985. The infestations caused light and moderate to severe defoliation at many locations throughout the area bounded by Wrigley, Fort Simpson, Fort Liard, Fort Smith and Yellowknife. Some declines in populations occurred each year from 1983 to 1985. However, there were still several large areas with moderate to severe defoliation persisting each year from 1986 to 1990, especially in the Martin and Ebbutt hills, along sections of the Mackenzie and Liard rivers, and between Hay River and Fort Smith. No tree mortality was observed during this outbreak period in the Northwest Territories.

The trend of larch sawfly abundance since 1973 has, except for infestations in the Northwest Territories, remained largely at endemic or low levels in most tamarack stands across the Prairie provinces. All infestations that reached moderate to severe levels appeared to have remained relatively small and localized. This is in contrast to infestation trends during the previous historical period between 1938 and 1970 and may be accounted for by the possible controlling influence of two introduced parasitic wasps from Europe, *Mesoleius tenthredinis* (a Bavarian strain) and *Olesicampe benefactor*. The first species was released at two locations in Manitoba in 1963 and 1964, whereas the latter species was released in Manitoba in 1961 and subsequently redistributed between 1961 and 1980 to 16 other locations in the three Prairie provinces and Northwest Territories. Post-release monitoring of larch sawfly cocoons, mostly at release sites, indicated that *O. benefactor* may be controlling the larch sawfly populations. Long-term monitoring will help to confirm this.

Mountain Pine Beetle

Numerous outbreaks of the mountain pine beetle, *Dendroctonus ponderosae*, are known to have occurred in British Columbia since 1910, but only two are known to have developed in Alberta. The first occurred between 1939 and 1944 in Banff National Park, following an epidemic in adjacent Kootenay National Park that lasted from 1930 to 1941. The second outbreak began in southwestern Alberta about 1976, presumably from dispersing beetles in adjacent high population sources in Montana and southeastern British Columbia. Small patches of lodgepole pine trees killed by the mountain pine beetle were first identified at 11 locations within Waterton Lakes National Park and northward to the Carbondale River in southwestern Alberta. The infestations subsequently intensified and expanded northward and eastward along the eastern slopes of the Rocky Mountains and attained maximum spread about 130 km north of the Canada–United States border by 1980–1981. During 1979 and 1980, numerous small but scattered infestations were discovered in the Porcupine Hills in southwestern Alberta, and in the Cypress Hills, an area straddling the southern Alberta–Saskatchewan boundary. This area represents the most easterly range in Canada of the beetles main natural host, lodgepole pine. Beetle-attacked trees were also discovered at 38 other locations in urban centers and farm shelterbelts scattered throughout the southern agricultural zone in 1980 and 1981. The main host tree affected at these locations was the planted Scots pine, *Pinus sylvestris*. Farther north, small infestations were discovered in 1982 in the foothills of Alberta, directly east of Banff National Park (Kananaskis area). By 1986, after 10 years at outbreak levels, mountain pine beetle populations had declined to endemic levels at all locations in southern Alberta and southwestern Saskatchewan. In addition to the native lodgepole pine and exotic Scots pine hosts, the mountain pine beetle also killed large numbers of two other native pines in southwestern Alberta, namely whitebark pine, *P. albicaulis*, and limber pine, *P. flexilis*.

During the course of this outbreak in Alberta, total mortality of lodgepole pine has been roughly estimated at 3.48 million trees spread over a composite forested area of at least 2 500 km^2. Data in Table 5 show the yearly tree mortality and volume loss estimates from 1977 to 1987.

The apparent rapid spread of the mountain pine beetle during 1978 to 1980 prompted an intensive control and

Table 5. Summary of the estimated yearly damage caused by the mountain pine beetle, *Dendroctonus ponderosae*, to lodgepole pine during the outbreak in Alberta from 1977 to 1987.

Year	No. of trees killed	No. of infestations	No. of ha infested	Estimated volume killed (m^3)
1977	300[a]	11	–[b]	91
1978	5 500	41	–	1 815
1979	226 000	156	–	74 580
1980	835 000	576	–	275 550
1981	1 250 000	600+	–	412 500
1982	576 000	480+	4 260	190 160
1983	236 000	500+	2 600	79 400
1984	224 700	200+	2 500	75 870
1985	129 600	100+	1 500	42 768
1986	270	<10	<10	89
1987	50	<10	<10	16
Totals	3 483 420			1 152 847

[a]Estimates include tree mortality in Waterton Lakes National Park.
[b]No data available.

salvage program to be established by the provincial forest service in 1980. Simultaneously, detection and control programs involving selective sanitation cutting and burning of infested trees were started in the Cypress Hills as well as in southwestern Alberta, and later in the Kananaskis area. These detection programs have been continued annually to the present date (1990) and have involved the successful deployment of semiochemical baits.

Overall, the control programs have been costly but effective in limiting the spread of the beetle and in reducing infestation centers. Control involved the treatment of about 5 000 lodgepole pine in the Cypress Hills and about 64 500 lodgepole pine in southwestern Alberta. An additional 40 700 infested limber pine were also treated by cutting and burning in the Porcupine Hills and adjacent areas between 1983 and 1986. In 1989, there was a minor resurgence of mountain pine beetle infestations in southwestern Alberta that resulted in mortality of more than 600 lodgepole pine.

Surveys to detect changes in the endemic populations of the mountain pine beetle are being continued in all outbreak areas, including the Rocky Mountain National Parks where a major outbreak still exists in Kootenay National Park. More than 9 000 lodgepole pine were killed in this park by the beetle during 1986 and 1987, and about 4 000 trees were killed annually in 1988 and 1989. Surveys in 1990, however, indicated there were at least 10 000 recently killed lodgepole pine, more than a twofold increase over 1989.

Spruce Beetle

In the Northwest Region, the spruce beetle, *Dendroctonus rufipennis*, attacks Englemann and white spruce throughout their ranges, and rarely attacks black spruce. All known outbreaks of the spruce beetle in this region have occurred in Alberta in mature stands mostly more than 120 years old. All outbreaks have apparently originated from blowdown, cull logs, or from right-of-way logging operations.

An outbreak in southwestern Alberta from 1952 to 1956 caused about 23% mortality of the mature spruce on 490 ha. During the early 1960s in Wood Buffalo National Park, up to 5% of the mature spruce was killed over a 130 km^2 area. Another outbreak in southwestern Alberta from 1966 to 1970 was spread over about 1 820 ha and resulted in more than 59 000 m^3 of beetle-killed timber. This outbreak developed as a result of extensive blowdown in nearby mature spruce stands during the summer of 1964.

The most recent outbreak developed in northwestern Alberta between 1977 and 1983 as a result of extensive blowdown in about 1975 near Rainbow Lake. More than 1 000 ha of mature white spruce in the vicinity of the blowdown were affected where an estimated 64 380 m^3, or about 30% of the total spruce volume, was attacked and killed by 1983. Another area extending over 280 ha sustained patches of tree mortality ranging from 10% to 70%.

Between 1977 and 1984, the accumulated tree mortality ranged between 5% and 70% over a composite land area of at least 100 000 ha in the Footner Lake and Peace River Forest districts. The outbreak had declined by 1984, but numerous scattered patches of dead and dying trees persisted; each patch usually contained one to three affected trees.

The strategy adopted for control of the spruce beetle consisted of first identifying priority stands having 10% or more mortality, followed by salvage logging. During 1983–1984, salvage was undertaken in three main infestation areas in the Footner Lake Forest. Of the total volume of spruce log material harvested, an estimated 19% or 47 500 m^3 had been killed by the spruce beetle. This volume, however, excluded an additional 33 340 m^3 of beetle-killed trees left unharvested because of their advanced deterioration.

In Wood Buffalo National Park, estimated white spruce mortality of 8% to 10% accumulated for 3 or 4 years up to 1984 and extended for several kilometres along the Peace River. Only a few small scattered infestations were noted elsewhere in the province between 1984 and 1990.

Whitespotted Sawyer

The whitespotted sawyer, *Monochamus scutellatus*, occurs throughout the region and is the most economically important of the wood-boring species. Its adults are attracted to dying, freshly cut, or fire-killed coniferous logs. Spruce and pine logs are often more heavily attacked than other coniferous species. The most important damage is contributed by the sawyer beetle larvae that excavate tunnels into the wood, causing the characteristic "worm-hole" defect. Lumber products cut from logs infested by sawyer beetles may be downgraded in value because of the presence of the wormholes, and they also have limited marketability in foreign export markets.

High populations of *M. scutellatus* are mostly associated with fire-killed stands of timber containing a large supply of suitable breeding material. Generally, timber burned during spring or early summer fires is more attractive to the adult whitespotted sawyer than is timber burned in late summer or fall. A major problem arises during the post-fire period when fire-killed timber is salvaged. Early completion of salvage operations and processing the logs into lumber are necessary to minimize log deterioration and losses due to worm-holes.

No reports of major timber losses from sawyer beetle damage occurred from 1973 to 1977. In the spring of 1977, two large fires occurred in central Saskatchewan—one near Montreal Lake, the other in Nipawin Provincial Park. Salvage operations took place during 1978 and 1979. However, high populations of the whitespotted sawyer had built up in the white spruce and jack pine sawlog material and caused extensive downgrading (up to 30% value loss) in the dimension lumber. Losses due to high-density worm-holes were also reported in salvaged rail and fence post materials because the holes increased the risk of moisture penetration and early decay.

The years 1980, 1981, and 1982 were especially high fire-hazard years when large tracts of timber burned in the central and northern parts of the three Prairie provinces and probably also in the Northwest Territories.

A large fire in the Porcupine Hills on the Manitoba–Saskatchewan border burned over several thousand hectares on the Manitoba side, and over 182 000 ha on the Saskatchewan side in 1980. In 1981, another fire burned over 54 000 ha near Kipahigan Lake in northwestern Manitoba. In Alberta, over 1 500 fires destroyed more than 526 000 ha of timber, mostly in the northern part of the province in 1981. In most of the burned areas, large volumes of mature white spruce, black spruce, and jack pine were fire-killed.

Surveys for woodborer attack incidence and value loss estimation were conducted in several major burn areas between 1972 and 1982. Most fire-killed trees became infested with whitespotted sawyer beetles but population levels varied widely depending on factors such as size of burn, tree age, time of the burn, and availability of adult sawyer beetle populations. In a survey in the Porcupine Hills, Manitoba, low to moderate whitespotted sawyer populations on 30% of the trees sampled showed sufficient potential worm-hole damage to cause up to 18% loss in value due to downgrading. A similar survey in the fire near Kipahigan Lake showed high attack densities of *M. scutellatus*, sufficient to cause up to 30% value loss from downgrading. Fire-killed spruce in the Porcupine Hills, Saskatchewan, required extra handling at the mill to sort them into different hazard classes so as to minimize mixing material of different product value. One large mill operating in north-central Alberta estimated losses in fire-killed white spruce lumber products to be between 35% and 40% because of downgrading from worm-holes and checking.

Large salvage operations of fire-killed logs may also dictate extensive millyard stockpiling and storage and result in delays in sawing and processing. This may allow for greater worm-hole damage to accumulate.

In 1983, large volumes of fire-killed timber were salvaged in the Footner Lake Forest, Alberta, but any losses that may have occurred were not reported. No major infestations of the whitespotted sawyer were reported from 1984 to 1990.

Insects of Plantations and Other High-Value Stands

Surveys recently initiated in the Northwest Region in young coniferous plantations and other high-value stands indicate several insect species that contribute significant losses occasionally, often in localized areas. Table 6 summarizes the species considered to be most important. The frequency of collection is only partly indicative of their importance because the intensity of sampling varied considerably for the reporting years from 1973 to 1990.

Other Damaging Forest Insects

Several other common insects that are not usually considered major pests are included in Tables 7 and 8. Outbreaks may develop periodically, however, and may cause damage over relatively small areas. Some may persist for short periods of time at high populations, but at rather infrequent intervals. Still others such as the fall cankerworm may cause significant damage in high-use areas and on urban shade tree plantings but may be classed only as minor forest pests. Other introduced insect species, such as the European spruce sawfly, larch casebearer, and birch leafminers, are also noteworthy because of their present distribution in the region and known potential for causing economic damage. The species are presented separately for those that feed on conifers and those that feed on hardwoods.

Table 6. Important insect pests of young coniferous plantations and other high-value stands in the Northwest Region from 1973 to 1990.

Insects	Province/ territory[a]	Year of reported damage																	
		73	74	75	76	77	78	79	80	81	82	83	84	85	86	87	88	89	90
Eastern pine shootborer	M						X										X		
Eucosma gloriola	S																		
Heinrich	A																		
	NWT																		
Lodgepole terminal weevil	M						X						X	X		X	X	X	
Pissodes terminalis	S						X						X	X	X	X	X	X	X
Hopping	A							X	X				X	X	X	X	X	X	X
	NWT																		
Pine root collar weevil	M												X	X		X		X	
Hylobius radicis	S																		
Buchanan	A																		
	NWT																		
Pitch twig moths	M										X	X	X	X	X	X	X	X	
Petrova albicapitana	S										X	X	X	X	X	X	X		X
(Busck)	A				X	X					X	X	X	X	X	X	X	X	X
P. metallica (Busck)	NWT										X	X	X	X	X	X	X	X	X
Redheaded jack pine sawfly	M													X					
Neodiprion rugifrons	S														X	X			
	A													X		X			
	NWT																		
Spruce bud midge	M																		
Rhabdophaga swainei	S															X	X		X
Felt	A															X	X	X	X
	NWT															X	X		
Warren root collar weevil	M						X			X	X	X						X	
Hylobius warreni	S						X			X	X	X							
Wood	A						X	X		X	X	X				X	X	X	X
	NWT																		
White pine weevil	M	X		X					X		X		X	X	X	X	X	X	X
Pissodes stobi	S	X	X	X	X	X					X	X	X	X	X	X	X	X	
(Peck)	A	X	X	X	X	X			X		X	X	X	X	X	X	X	X	
	NWT																		
Yellowheaded spruce sawfly	M	X		X	X	X	X		X	X	X	X	X	X	X	X	X		
Pikonema alaskensis	S	X	X	X	X	X				X	X	X	X	X	X	X	X	X	X
(Rohwer)	A	X	X	X	X	X				X	X	X	X	X	X	X	X	X	X
	NWT										X	X	X	X	X	X	X	X	X

[a] M = Manitoba; S = Saskatchewan; A = Alberta; NWT = Northwest Territories.

Table 7. Common native or introduced insect pests that periodically cause damage to hardwoods in the Northwest Region from 1973 to 1990.

		Year of reported damage																	
Insects	Province/ territory[a]	73	74	75	76	77	78	79	80	81	82	83	84	85	86	87	88	89	90
Aspen leaf beetles	M	X													X	X			
Chrysomela crotchi Brown	S	X													X	X	X	X	X
C. scripta Fab.	A	X															X	X	X
Gonioctana americana (Sch.)	NWT																	X	X
Aspen leafroller	M	X	X							X			X		X	X	X	X	X
Pseudexentera oregonana	S	X	X				X			X		X			X	X	X	X	
(Wals.)	A	X				X	X	X	X	X			X		X	X	X	X	X
	NWT																		
Birch leafminers	M			X				X	X	X	X	X	X	X	X	X	X		
Fenusa pusilla	S			X	X		X	X	X	X	X	X	X	X	X	X	X	X	
Profenusa thomsoni (Konow)	A			X	X	X	X	X	X	X	X	X	X	X	X	X	X	X	X
	NWT																		
Bruce spanworm	M																X		
Operopthera bruceata (Hulst)	S	X										X	X			X			X
	A					X	X	X	X	X			X	X	X	X	X	X	X
	NWT												X						
Fall cankerworm	M	X	X	X	X	X					X	X	X	X	X	X	X	X	
Alsophila pometaria (Harris)	S	X	X	X	X	X					X	X	X	X	X	X	X	X	X
	A	X										X	X	X	X	X	X	X	X
	NWT																		
Gray willow leaf beetle	M													X					
Tricholochmaea decora decora	S													X		X	X	X	X
(Say)	A															X	X	X	X
	NWT																		X
Mountain ash sawfly	M												X		X				
Pristiphora geniculata	S																		
	A																		
	NWT																		
Poplar borer	M					X		X								X			
Saperda calcarata (Say)	S					X		X							X		X		X
	A	X				X		X							X	X	X		X
	NWT														X	X	X		
Willow leafminer	M		X	X							X	X	X	X					
Micrurapteryx salicifoliella	S		X	X	X						X	X	X	X	X		X		
(Chambers)	A										X	X	X	X	X		X		X
	NWT										X	X	X	X	X		X		X

[a] M = Manitoba; S = Saskatchewan; A = Alberta; NWT = Northwest Territories.

Table 8. Common native or introduced insect pests that periodically cause damage to conifers in the Northwest Region from 1973 to 1990.

Insects	Province/ territory[a]	Year of reported damage																	
		73	74	75	76	77	78	79	80	81	82	83	84	85	86	87	88	89	90
Balsam fir sawfly	M		X	X		X	X								X				
Neodiprion abietis complex	S																		
	A																		
	NWT																		
Douglas-fir beetle	M																		
Dendroctonus pseudotsugae	S																		
Hopkins	A														X	X	X		X
	NWT																		
Eastern blackheaded	M											X	X		X	X	X	X	
budworm	S											X	X	X	X	X	X		
Acleris variana (Fern.)	A											X	X			X	X		
	NWT																		
European spruce sawfly	M								X	X	X	X	X	X	X				
Gilpinia hercyniae	S																		
(Hartig)	A																		
	NWT																		
Fir coneworm	M																		
Dioryctria abietivorella	S														X				
(Grote)	A																		
	NWT																		
Introduced pine sawfly	M											X	X						
Diprion similis	S																		
(Hartig)	A																		
	NWT																		
Jack pine resin midge	M													X	X				
Cecidomyia resinicola	S															X			
(Osten Sacken)	A																		
	NWT																		
Larch casebearer	M								X	X	X	X	X	X	X	X			
Coleophora laricella	S																		
(Hubner)	A																		
	NWT																		
Northern lodgepole needle miner	M																		
Coleotechnites starki	S																X		
(Freeman)	A		X	X				X	X	X	X	X	X		X	X	X	X	X
	NWT																		
Pine engraver	M												X						
Ips pini (Say)	S												X					X	
	A											X	X					X	
	NWT										X	X	X	X					
Spruce bud moth	M																		
Zeiraphera canadensis	S																		
(Mut. and Free.)	A														X				
	NWT																		

(Continued)

Table 8. Common native or introduced insect pests that periodically cause damage to conifers in the Northwest Region from 1973 to 1990. *(Concluded)*

Insects	Province/ territory[a]	Year of reported damage																	
		73	74	75	76	77	78	79	80	81	82	83	84	85	86	87	88	89	90
Spruce cone maggot	M																		
Strobilomyia neanthracina	S																		
(Czerny)	A						X	X											
	NWT																		
Spruce coneworm	M																		
Dioryctria reniculelloides	S																		
(Mut. and Munroe)	A														X				
	NWT																		
Spruce seed moth	M																		
Cydia strobilella (L.)	S																		
	A							X											
	NWT																		
Two-year-cycle budworm	M																		
Choristoneura biennis	S																		
(Freeman)	A													X	X	X			
	NWT																		

[a] M = Manitoba; S = Saskatchewan; A = Alberta; NWT = Northwest Territories.

Bibliography

Beckwith, R.C. 1968. The large aspen tortrix, *Choristoneura conflictana* (Wlkr.), in interior Alaska. U.S. Dept. Agric., For. Serv., Pac. Northwest For. Range Exp. Stn., Portland, Ore. Res. Note PNW-81.

Bright, D.E. 1976. The bark beetles of Canada and Alaska. Part 2. The insects and arachnids of Canada. Res. Branch, Can. Dept. Agric. Publ. 1575. Biosystematics Res. Inst., Ottawa, Ont., 241 p.

Brown, C.E. 1962. The life history and dispersal of the Bruce spanworm, *Operopthera bruceata* (Hulst), (Lepidoptera: Geometridae). Can. Entomol. 94: 1103-1107.

Canadian Forestry Service. 1961 to 1979. Annual reports of the Forest Insect and Disease Survey. Ottawa.

Cerezke, H.F. 1975. White-spotted sawyer beetle in logs. Environ. Can., Can. For. Serv., North. For. Res. Cent., Edmonton, Alta. Inf. Rep. NOR-X-129.

Cerezke, H.F. 1977. Characteristics of damage in tree-length white spruce logs caused by the white-spotted sawyer, *Monochamus scutellatus*. Can. J. For. Res. 7: 232-240.

Cerezke, H.F. 1978. Spruce budworm—how important is it here in the West? In *Forestry Report. Spring 1978*, pp. 4-5. Environ. Can., Can. For. Serv., North. For. Res. Cent., Edmonton, Alta.

Cerezke, H.F.; Emond, F.J. 1989. Forest insect and disease conditions in Alberta, Saskatchewan, Manitoba, and the Northwest Territories in 1987. For. Can., North. For. Cent., Edmonton, Alta. Inf. Rep. NOR-X-300.

Cerezke, H.F.; Emond, F.J.; Gates, H.S. 1991. Forest insect and disease conditions in Alberta, Saskatchewan, Manitoba, and the Northwest Territories in 1990 and predictions for 1991. For. Can., North. For. Cent., Edmonton, Alta. Inf. Rep. NOR-X-318.

Emond, F.J.; Cerezke, H.F. 1989. Forest insect and disease conditions in Alberta, Saskatchewan, Manitoba, and the Northwest Territories in 1988 and predictions for 1989. For. Can., North. For. Cent., Edmonton, Alta. Inf. Rep. NOR-X-303.

Emond, F.J.; Cerezke, H.F. 1990. Forest insect and disease conditions in Alberta, Saskatchewan, Manitoba, and the Northwest Territories in 1989 and predictions for 1990. For. Can., Northwest Region, North. For. Cent., Edmonton, Alta. Inf. Rep. NOR-X-313.

Hardy, D.; Mainville, M.; Schmitt, D.M., 1986. An Atlas of Spruce Budworm Defoliation in Eastern North America, 1938-1980. U.S. Dep. Agric. For. Serv., Cooperative State Res. Serv. Misc. Publ. No. 1449. Washington, D.C.

Hildahl, V.; Reeks, W.A. 1960. Outbreaks of the forest tent caterpillar, *Malacosoma disstria* Hbn., and their effects on trembling aspen in Manitoba and Saskatchewan. Can. Entomol. 92: 199-209.

Hiratsuka, Y.; Cerezke, H.F.; Petty, J. 1980. Forest insect and disease conditions in Alberta, Saskatchewan, Manitoba, and the Northwest Territories in 1979 and predictions for 1980. Environ. Can., Can. For. Serv., North. For. Res. Cent., Edmonton, Alta. Inf. Rep. NOR-X-225.

Hiratsuka, Y.; Cerezke, H.F.; Petty, J.; Still, G.N. 1981. Forest insect and disease conditions in Alberta, Saskatchewan, Manitoba, and the Northwest Territories in 1980 and predictions for 1981. Environ. Can., Can. For. Serv., North. For. Res. Cent., Edmonton, Alta. Inf. Rep. NOR-X-231.

Hiratsuka, Y.; Cerezke, H.F.; Moody, B.H.; Petty, J.; Still, G.N. 1982. Forest insect and disease conditions in Alberta, Saskatchewan, Manitoba, and the Northwest Territories in 1981 and predictions for 1982. Environ. Can., Can. For. Serv., North. For. Res. Cent., Edmonton, Alta. Inf. Rep. NOR-X-239.

Hopping, G.R.; Mathers, W.G. 1945. Observations on outbreaks and control of the mountain pine beetle in the lodgepole pine stands of western Canada. For. Chron. 21: 1-11.

Ives, W.G.H. 1976. The dynamics of the larch sawfly (Hymenoptera Tenthredinidae) populations in southeastern Manitoba. Can. Entomol. 108: 701-730.

Ives, W.G.H. 1981. Environmental factors affecting 21 forest insect defoliators in Manitoba and Saskatchewan, 1945-1969. Environ. Can., Can. For. Serv., North. For. Res. Cent., Edmonton, Alta. Inf. Rep. NOR-X-233.

Ives, W.G.H.; Muldrew, J.A. 1984. *Pristiphora erichsonii* (Hartig), larch sawfly (Hymenoptera: Tenthredinidae). Chap. 64 in J.S. Kelleher and M.A. Hulme (Eds.), Biological control programmes against insects and weeds in Canada 1969-1980. Commonwealth Agric. Bureaux, Slough, Eng.

Moody, B.H.; Cerezke, H.F. 1983. Forest insect and disease conditions in Alberta, Saskatchewan, Manitoba, and the Northwest Territories in 1982 and predictions for 1983. Environ. Can., Can. For. Serv., North. For. Res. Cent., Edmonton, Alta. Inf. Rep. NOR-X-248.

Moody, B.H.; Cerezke, H.F. 1984. Forest insect and disease conditions in Alberta, Saskatchewan, Manitoba, and the Northwest Territories in 1983 and predictions for 1984. Environ. Can., Can. For. Serv., North. For. Res. Cent., Edmonton, Alta. Inf. Rep. NOR-X-261.

Moody, B.H.; Cerezke, H.F. 1985. Forest insect and disease conditions in Alberta, Saskatchewan, Manitoba, and the Northwest Territories in 1984 and predictions for 1985. Can. For. Serv., North. For. Res. Cent., Edmonton, Alta. Inf. Rep. NOR-X-269.

Moody, B.H.; Cerezke, H.F. 1986. Forest insect and disease conditions in Alberta, Saskatchewan, Manitoba, and the Northwest Territories in 1985 and predictions for 1986. Can. For. Serv., North. For. Cent., Edmonton, Alta. Inf. Rep. NOR-X-276.

Muldrew, J.A.; Ives, W.G.H. 1984. Dispersal of *Olesicampe benefactor* and *Mesochorus dimidiatus* in western Canada. Environ. Can., Can. For. Serv., North. For. Res. Cent., Edmonton, Alta. Inf. Rep. NOR-X-258.

Prentice, R.M. 1955. The life history and some aspects of the ecology of the large aspen tortrix, *Choristoneura conflictana* (Wlkr.) (N. Comb.) (Lepidoptera: Tortricidae). Can. Entomol. 87: 461-473.

Safranyik, L. 1971. Spruce beetle. Environ. Can., Can. For. Serv., North. For. Res. Cent., Edmonton, Alta. For. Rep. 1(4): 6.

Safranyik, L.; Raske, A.G. 1970. Sequential sampling plan for *Monochamus* larvae in decked lodgepole pine logs (Coleoptera: Cerambycidae). J. Econ. Ent. 63: 1903-1906.

Safranyik, Y. Shrimpton, D.M.; Whitney, H.S. 1974. Management of lodgepole pine to reduce losses from the mountain pine beetle. Environ. Can., Can. For. Serv., Pac. For. Res. Cent., Victoria, B.C., For. Tech. Rep. 1.

Turnock, W.J. 1972. Geographical and historical variability in population patterns and life systems of the larch sawfly (Hymenoptera: Tenthredinidae). Can. Entomol. 104: 1883-1900.

Volney, W.J.A. 1988. Analysis of historic jack pine budworm outbreaks in the Prairie Provinces of Canada. Can. J. For. Res. 18: 1152-1158.

Wong, H.R. 1974. The identification and origin of the strains of the larch sawfly, *Pristiphora erichsonii* (Hymenoptera: Tenthredinidae), in North America. Can. Entomol. 106: 1121-1131.

Chapter 6

Forest Insect Pests in the Pacific and Yukon Region

G.A. Van Sickle

Introduction

The insect collection maintained at the Pacific Forestry Centre since 1949 is the largest regional collection of forest insects and contains more than 6 000 identified species in 300 families. Fortunately, while they are an important part of the natural ecosystem, most are not significant forest pests. In other cases, being a coastal province, introduced insects can and have become major pest problems in addition to major quarantine concerns. In the original issue of *Aerial Control of Forest Insects in Canada* (Prebble 1975), 16 insect pests involving aerial control were included from this region; however, major pests such as the bark beetles, which are not easily controlled by aerial applications, were not included. Conversely, several of the defoliators that were reported have not been as active in the interim. Previous forest insect pests that have not been recorded at more than endemic levels in British Columbia since 1973 include hemlock needleminer, *Epinotia tsugana*; greenstriped forest looper, *Melanolophia imitata*; saddleback looper, *Ectropis crepuscularia*; and pine butterfly, *Neophasia menapia*. Periodically, pine butterfly adults have been conspicuous in flight, particularly on southern Vancouver Island, but defoliation has not resulted. The phantom hemlock looper, *Nepytia phantasmaria*, only defoliated western hemlock over 100 ha near Coquitlam Lake in 1982 before naturally occurring insect pathogens reduced the populations.

The following brief review includes population and damage fluctuations for those insects that were most active and damaging from 1973 to 1990. In total, more than 25 major and 23 minor insect pests occurring during the 18-year period have been reviewed. Published and file reports of the regional unit of the Forest Insect and Disease Survey (FIDS), which conducts annual overviews of pest conditions, were used extensively to compile the summary tables and for reference. Brief historical descriptions of bark beetle and spruce budworm outbreaks in British Columbia before 1973 are given because this information was not included in *Aerial Control of Forest Insects in Canada* (Prebble 1975). Major control efforts, whether direct aerial applications, salvage harvesting, silvicultural or cultural practices, or population manipulation, are dealt with in other chapters.

Insects of Natural Forests

Mountain Pine Beetle

The mountain pine beetle, *Dendroctonus ponderosae*, continues, as it has for the past decade, to be the most damaging forest insect in British Columbia. Mature lodgepole pine is by far the most commonly killed tree species, followed by western white pine, occasionally ponderosa pine and whitebark pine. Infestations generally persist within individual stands for several years until the pine component is depleted, especially the larger and older trees. The beetles carry blue stain fungi and introduce them into the attacked trees where the fungi develop quickly in the sapwood, interrupting water flow to the crown. The combined action of the beetles and the fungi kill the tree which fades, usually the following spring and summer, to a yellowish red, then brown before the needles fall. Some variation in timing of beetle development and tree response occurs according to temperature and elevation conditions.

Mountain pine beetle outbreaks have been recorded in British Columbia at irregular intervals since at least 1910. Notable early infestations occurred in the Princeton, Okanagan Valley, and Lillooet areas. In the early 1930s, vast areas around Tatla Lake in central British Columbia were infested and included Kootenay, Yoho, and Banff national parks by the 1940s. From 1946 to 1965, infestations were active around Babine Lake in north central British Columbia. The current outbreaks started during the early 1970s. In 1973, a total of only 4 700 ha at about 24 scattered locations were affected, including areas near Nelson, the Okanagan Valley, Hazelton, and the Klinaklini River Valley. By 1976, all the infestations had intensified and expanded tenfold and the infestation started in the Flathead River Valley, adjacent to Glacier National Park in Montana. Infestations continued to increase substantially each year (Fig. 1), culminating in 1984 when red trees (recent faders) were present over more than 482 000 ha in more than 7 700 areas

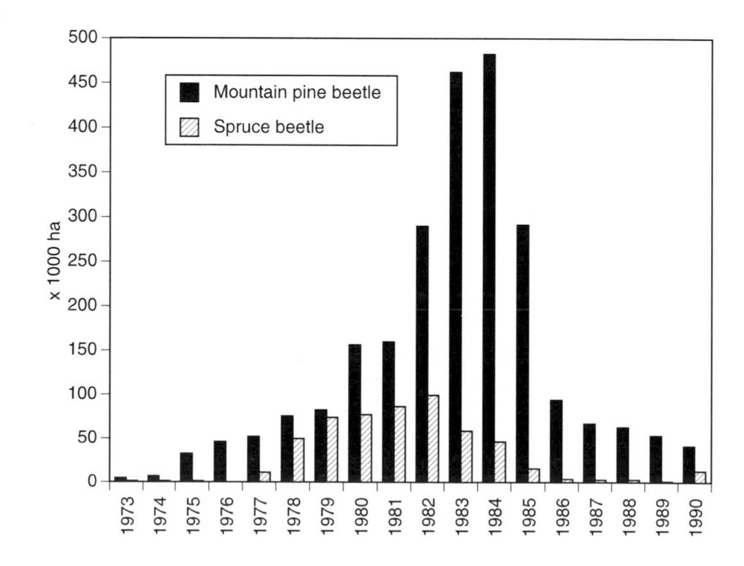

Figure 1. Aerially mapped area (hectares) of trees recently faded (color change) due to mountain pine beetle, *Dendroctonus ponderosae*, and spruce beetle, *D. rufipennis*, in British Columbia, 1973–1990.

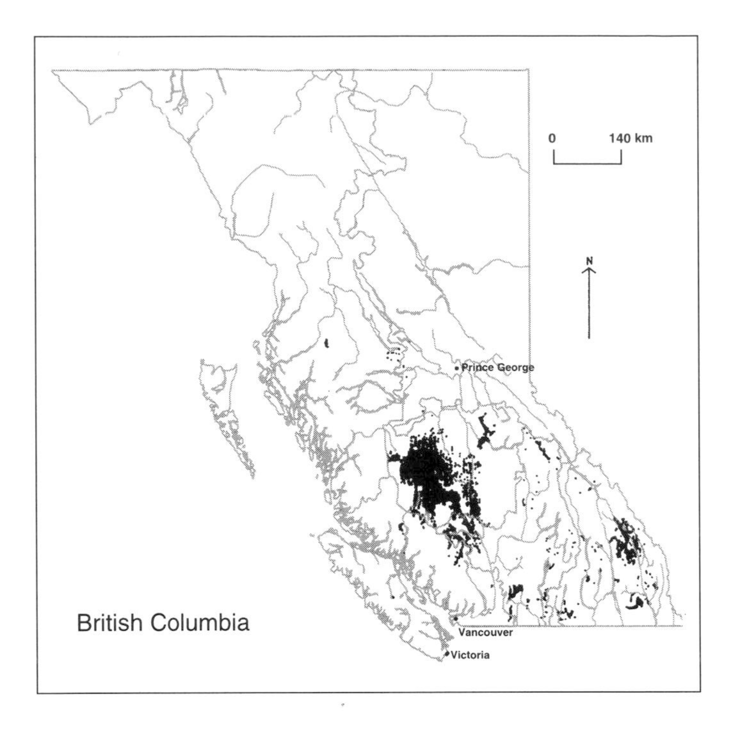

Figure 2. Location of aerially visible damage by the mountain pine beetle, *D. ponderosae*, in British Columbia in 1984 when damage was most concentrated in the Cariboo Forest Region.

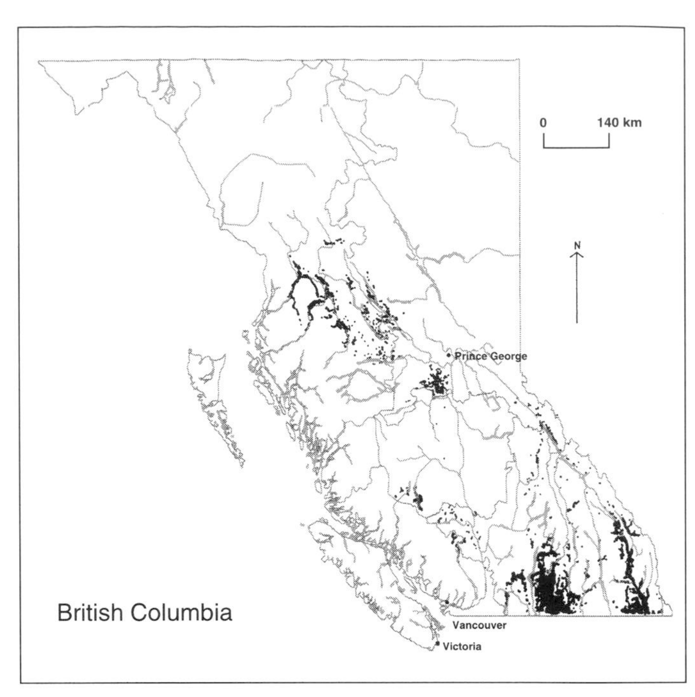

Figure 3. Location of aerially visible damage by the mountain pine beetle, *D. ponderosae*, in British Columbia in 1988.

from the international border in the southeast to north of Prince Rupert (Fig. 2). This was 24 times greater than the area burned by forest fires within the province in 1984. In a single year more than 41 million trees with an estimated volume of 12 million m^3 were killed, a loss equivalent to about 16% of the total annual harvest.

Beetle populations and area infested declined 40% in 1985 and 68% of that again in 1986. The decline was due partly to major control programs and to depletion of susceptible host material in many areas, but primarily to overwintering brood mortality in central British Columbia caused by early, below-normal temperatures in late 1984 and again in 1985. In the Cariboo region, which in 1984 accounted for 80% of the province's infested area, minimum temperatures below –30°C in October and –45°C in December 1984 virtually eliminated beetle populations. Slower development influenced by the cool wet summers of 1983 and 1984 probably also made the populations more susceptible to the early cold snap.

Populations elsewhere were less affected and in 1988 mountain pine beetle remained active over 62 000 ha (Fig. 3) mainly in the Nelson, Okanagan, and Fort St. James areas. A small infestation again developed in the western portion of the Cariboo region in 1988 but declined by 1990.

Since 1973, more than 216 million mature trees have been killed by mountain pine beetle in British Columbia. In many areas the rate of mortality exceeded the local capacity to access and salvage the trees before stain and decay made them unmerchantable. Infestations have forced disruptive and costly changes in forest management planning, killed enough trees to render many stands economically unmerchantable, reduced aesthetic values, increased fire hazard, changed the age and diameter distribution of the pine component, and hastened forest succession. Special bark beetle management programs were funded with emphasis on access construction, photographic and delineation surveys, sanitation harvesting, single tree treatment projects, and the use of aggregating pheromones.

Spruce Beetle

The spruce beetle, *Dendroctonus rufipennis*, is the most destructive insect pest of mature white and Engelmann spruce in western Canada. Periodic outbreaks in British Columbia from the 1940s to the early 1970s killed at least 16 million m^3, and another 22 million m^3 since 1973. The earlier infestations occurred in northwestern British Columbia and southwestern Yukon during the 1940s, at which time up to 90% of the mature spruce were killed. Infestations occurred in southeastern British Columbia from 1952 to 1956, and again from 1967 to 1971, in the Prince Rupert and Prince George regions from 1961 to 1965, and again in the Prince George area in 1967 and 1968.

Spruce beetle adults prefer to attack windfallen, damaged, or stressed trees, and fresh large-diameter slash and high stumps. Healthy mature trees are attacked when populations exceed the capacity of the preferred host material. Trees fade to yellowish green during the winter following successful attacks and, by the second autumn, most of the needles have discolored and may have fallen. This delayed expression of outbreaks is a problem for early detection and control actions.

Spruce beetle populations were generally endemic in the early 1970s but started to increase in 1977 with infestations over 11 100 ha (Fig. 1). An early warm spring and hot summer temperatures accelerated development in many populations in and adjacent to recently logged areas from the usual 2-year cycle to a 1-year cycle, which greatly increased attack potential in 1978. By 1979, infestations covering 74 000 ha were continuing near Babine Lake and Fort St. James and were starting around Prince George. The area of mature white and Engelmann spruce killed peaked at about 100 000 ha in 1982 (Fig. 1) and gradually declined to nearly endemic levels in 1986. The infestations were mostly concentrated around Prince George and the eastern portion of the Prince Rupert region with scattered pockets in and near Bowron Lake Park and in southeastern British Columbia (Fig. 4). Programs of salvage and reduction of beetle populations by the redirection of licensees into infested stands, by the use of nonlethal and lethal-felled trapped trees, by improved utilization standards, and by sanitation logging were widely implemented. Through these efforts and natural depletion of preferred host material, populations and damage declined to about 3 000 ha of scattered infestations in 1986 and 1987, and less than 2 000 ha in 1988. However, windthrow in 1989 led to increased populations and tree mortality over 13 000 ha north of Prince George in 1990.

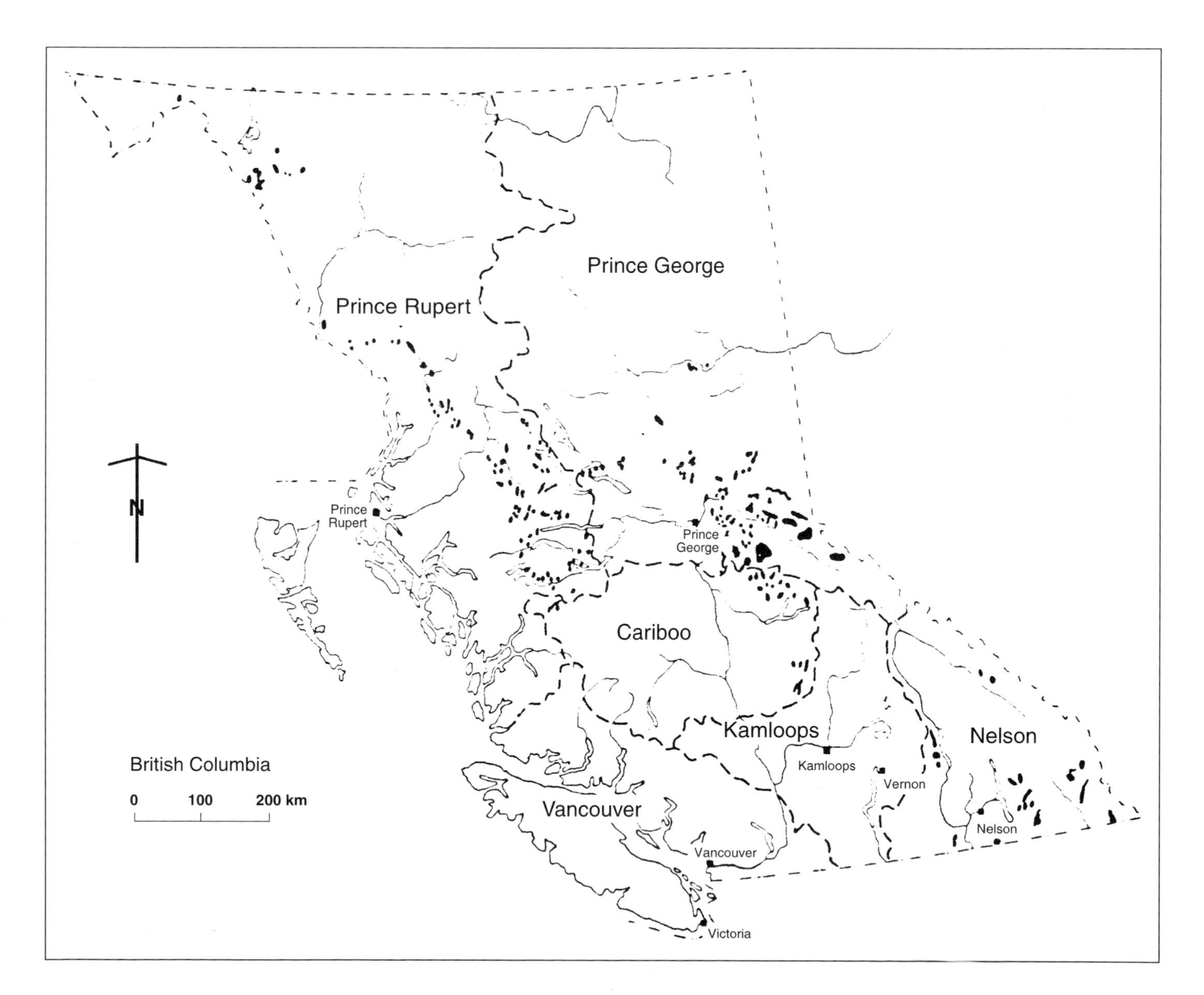

Figure 4. Locations of aerially visible beetle-killed trees at the maximum extent of the spruce beetle, *D. rufipennis*, infestation in 1982.

Douglas-fir Beetle

During the late 1950s and 1960s, more than 500 000 m^3 of mature Douglas-fir in interior British Columbia and 250 000 m^3 in coastal areas were killed by the Douglas-fir beetle, *Dendroctonus pseudotsugae*. The preferred host material includes felled trees, large-diameter slash, high stumps, and decadent and weakened trees. Because of improved utilization and better log inventory and slash disposal, losses since 1973 have been greatly reduced, occurring in small scattered groups or patches primarily in the central interior and often in association with root rots, tussock moth, or other predisposing factors. Warmer and drier weather in the late 1980s predisposed mature trees in scattered patches to beetle attacks that increased throughout interior British Columbia from 1988 to 1990.

Western Balsam Bark Beetle

In association with the lesion-causing fungus, *Ceratocystis dryocoetidis*, the western balsam bark beetle, *Dryocoetes confusus*, kills many large-diameter mature alpine fir throughout British Columbia. Because of the relative inaccessibility of alpine fir and, until recently, its limited commercial value, surveys for damage by this bark beetle have not been consistent or comprehensive. Even so, damage by this bark beetle was recorded before 1900 and frequently observed from 1920 to 1950. From 1957 to 1963, outbreaks west of Salmon Arm and Lillooet extended over an estimated 22 000 ha and a 1963 to 1965 outbreak in the Prince Rupert region killed 30% of the alpine fir over 170 000 ha. However, mortality by this insect–disease complex does not usually fluctuate greatly from year to year; instead, small proportions of susceptible trees are usually killed individually or in small patches each year. Dead trees retain red foliage for several years and the cumulative mortality can be quite conspicuous and significant. Since 1973, tree mortality averaging more than 60 000 m^3/yr has most often been reported, in decreasing order, in the Prince Rupert, northern Kamloops, Nelson, and Prince George regions.

Ambrosia Beetles

Ambrosia beetles, *Trypodendron lineatum* and *Gnathotrichus* spp., continue to be major pests causing degrade of logs, lumber, or veneer and necessitating additional inventory and mill procedures to avoid potential quarantine problems with export products. In British Columbia there are at least five species of ambrosia beetles and most commercial conifer species are susceptible (Shore 1985). Although populations and attacks are greatest in coastal forests, damage occurs periodically throughout the interior of the province. With no chemical pesticide presently registered in Canada for use against ambrosia beetle and an increase in dryland sorting of logs, control procedures have changed from those described by Lejeune and Richmond (1975). Management of log inventories, water misting, and mass trapping of beetles with pheromones or trap logs have been tested and became operational during the 1970s and 1980s (see Chapter 19).

Western Spruce Budworms

Several species of conifer-feeding budworms are now recognized as important defoliators in British Columbia. The western spruce budworm, *Choristoneura occidentalis*, despite the common name, feeds almost exclusively on Douglas-fir in central British Columbia. Spruce budworm, *C. fumiferana*, extends across the northern boreal forest from eastern Canada into northeastern British Columbia where white spruce and alpine fir are the most common hosts. The two-year-cycle budworm, *C. biennis*, occurs on white and Engelmann spruce and alpine fir in north-central British Columbia and at higher elevations in southeastern British Columbia. As the name implies, 2 years are required to complete a life cycle and most of the population is synchronized, with mature larvae feeding most severely and then maturing in even-numbered years. In the odd years, most of the feeding occurs within the buds, although some defoliation may be evident, depending on populations and weather conditions, and some small areas of "off-cycle" populations do exist.

The spruce budworm in British Columbia follows a similar life cycle to that in eastern Canada. Recurrent infestations have been recorded in the Liard River Basin in northeastern British Columbia (Fig. 5) since the late 1950s. Defoliation was greatest during 1964, 1965, and 1969 to 1975, then gradually declined. Defoliation was not reported again until 1984. The affected area reached 94 700 ha in 1986, then declined to 36 000 ha by 1988, but again increased to reach 398 000 ha in 1990 (Table 1, Fig. 6).

The two-year-cycle budworm has been frequently reported in central British Columbia for the past 60 years. Records before 1950 are sketchy due to the vastness and relative remoteness of the region. Infestations south of Prince George have been recorded since 1914, with a 1921 to 1932 infestation peaking in 1926 over about 300 000 ha around Barkerville and major reoccurrences in 1950 and again in 1960. To the west, infestations around Babine Lake occurred from 1950 to 1962 with defoliated areas exceeding 250 000 ha in peak years of 1950, 1956, 1958, and 1960. Infestations occurred in the Kamloops and Nelson regions but were generally less extensive or severe, and of shorter duration. The first epidemic was recorded in 1922 near Fernie; the next, in 1938 north of Kamloops; then periodically from 1940 to 1956 in several areas including Glacier National Park, Vernon, Edgewood, and Nakusp. Large moth flights were observed in 1940 and 1944. Populations were numerous in Kootenay National Park in 1964 to 1969, then in the nearby White River drainage from 1970 to 1974 and reoccurring in 1977 to 1979.

More recently, defoliation occurred in the Bowron Lake–North Thompson River areas during the mid-1970s, then increased in the Prince Rupert and Nelson regions with defoliation mapped over 365 000 ha in 1980 (Figs. 5 and 6) before it gradually declined, although still remaining active, until 1990. Infestations reoccurred around Barkerville and in the Bowron and Willow river drainages in 1974, expanded in 1979, and comprised 95% of the provincial total of 365 000 ha in 1980 before collapsing in 1981. To the northwest, defoliation occurred in 1979 in the Bell-Irving and Kispiox river drainages, expanded over 153 000 ha in 1983, and collapsed in 1984. Concurrently, infestations around Barkerville reoccurred

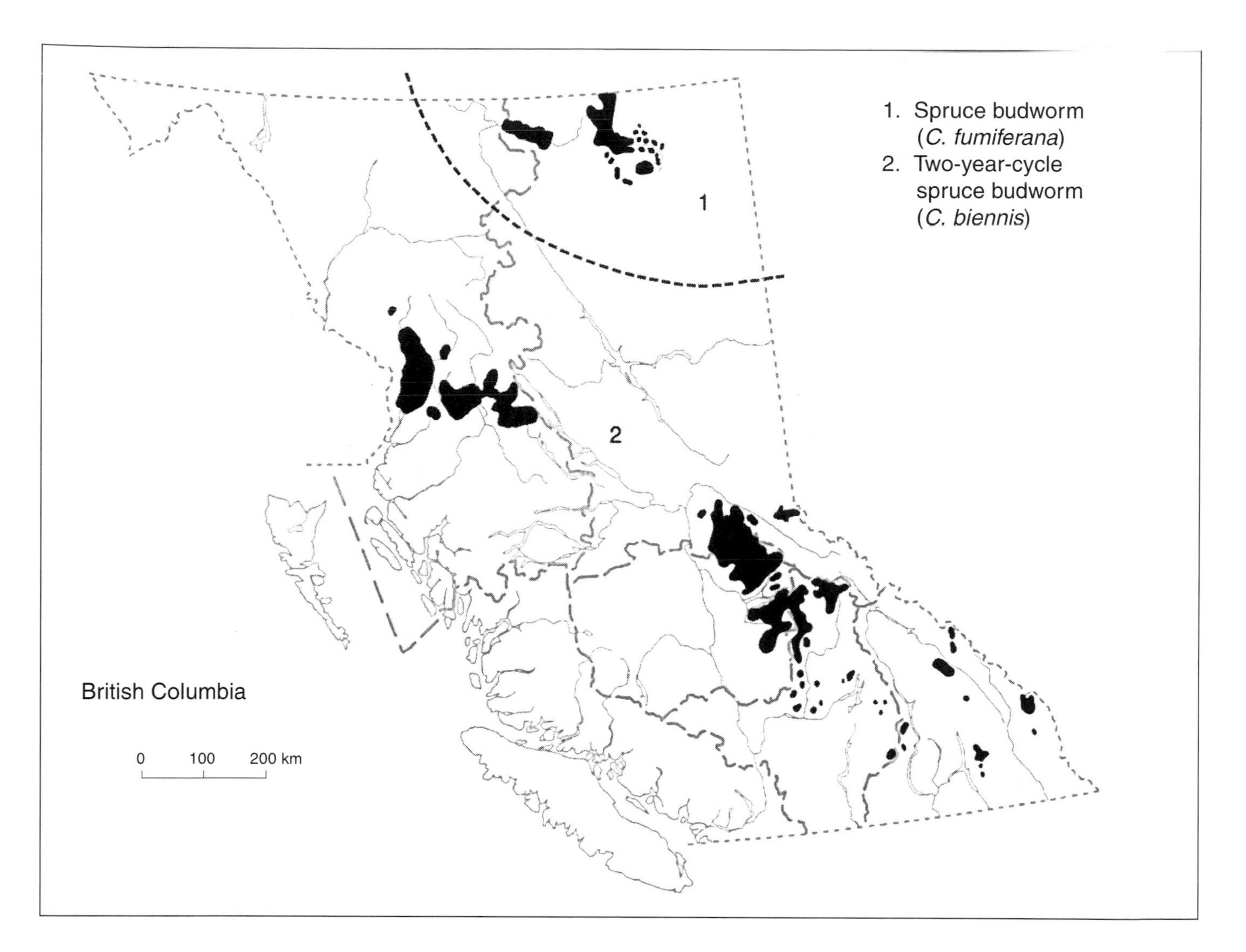

Figure 5. Maximum extent of aerially mapped defoliation by spruce budworm, *Choristoneura fumiferana*, and two-year-cycle budworm, *C. biennis*, in British Columbia between 1973 and 1990.

in 1983, and expanded in 1986 and 1988 along with populations in the North Thompson drainage and scattered locations in the Nelson region, and was recorded for the first time north of Mackenzie in 1989.

Natural control factors have also influenced two-year-cycle budworm populations. Late, cold spring temperatures reduce early-instar survival. Parasitism seldom exceeded 20%. Fungi (*Entomophaga* sp. and *Entomophthora* sp.), entomopox virus, and a microsporidia (*Nosema* sp.) were associated with the 1981 decline in the central interior. In late 1983, the pathogen *Beauveria bassiana* was evident in populations in three-quarters of the trees examined at eight locations throughout the Prince Rupert region and significantly contributed to the collapse of the infestation.

Overall, populations usually fluctuated from year to year, partly because of the 2-year life cycle, and generally declined in each area after 4 to 6 years. Although spruce and alpine fir reproduction was severely damaged and top-kill and growth loss were common, substantial mature tree mortality was not apparent. As more young stands develop following harvest of the higher-elevation spruce–fir types, the potential for control requirements for the two-year-cycle budworm is increasing.

The western spruce budworm is the most widespread and serious defoliator of Douglas-fir in British Columbia. Seven outbreaks have occurred in the province since 1916. The earliest defoliation occurred on Vancouver Island; it occurred again on the island from 1923 to 1930, along with an area east of the Okanagan Valley near Rock Creek. Infestations of 1916 to 1919, 1943 to 1946, and 1948 to 1958 included the Lillooet–Pemberton areas. Defoliation has been mapped each year since 1967, either in the Fraser Canyon or Kamloops areas, with peaks in 1977 and 1987 (Table 1, Fig. 7). The larvae are voracious and wasteful feeders. Tree tops and branch tips are usually the most severely defoliated and bare tops and the loosely webbed browned foliage are conspicuous during late July until washed from the trees by heavy rains. With successive feedings, height and radial growth are reduced, dieback increases, and leader dominance may be lost resulting in forked or crooked

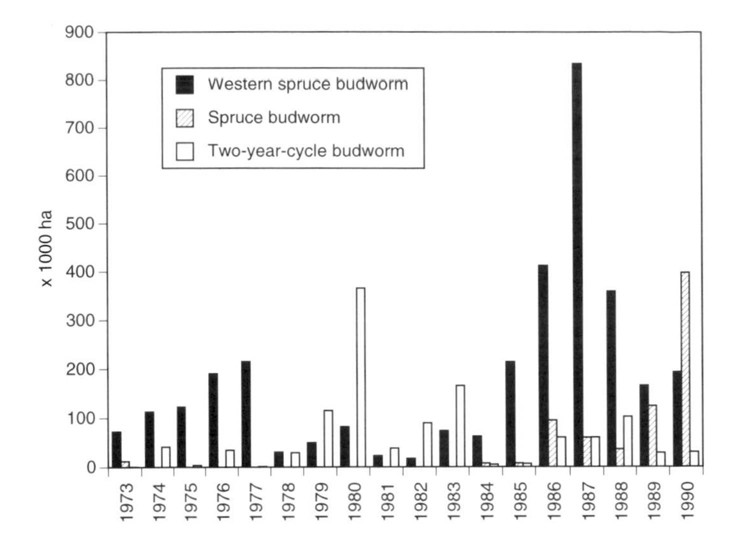

Figure 6. Aerially mapped area (hectares) of defoliation each year in British Columbia by western spruce budworm, *Choristoneura occidentalis*; spruce budworm, *C. fumiferana*; and two-year-cycle budworm, *C. biennis*, 1973–1990.

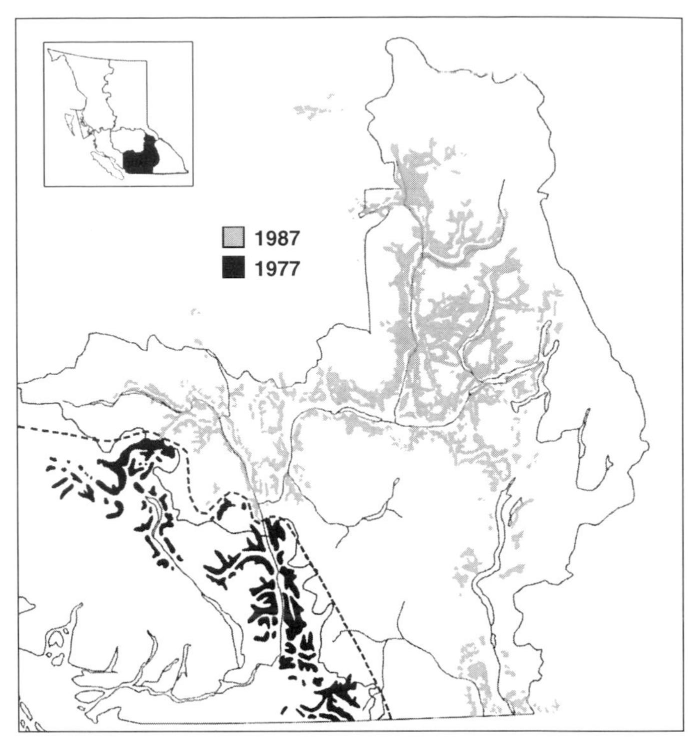

Figure 7. Maximum extent of western spruce budworm, *C. occidentalis*, defoliation during peak infestation years of 1977 and 1987 in British Columbia.

Table 1. Areas (hectares) of aerially visible and mapped defoliation by selected defoliating insects in British Columbia, 1973–1990.

Year	Douglas-fir tussock moth[a]	Western spruce budworm[b]	Two-year-cycle budworm[c]	Spruce budworm[d]	Western black-headed budworm[e]	Western hemlock looper[f]	Western false hemlock looper[g]	Larch sawfly[h]	Forest tent caterpillar[i]
1973	2 050	72 400	500	12 000	121 000	28 250	2 000	50	61 800
1974	3 180	112 700	41 200	less but not mapped	128 000		5 700		2 000
1975	8 100	122 300	3 600		not mapped		700		4 000
1976		190 000	34 500		6 400	10 500		4	7 800
1977		214 300	1 200					2 400	32 000
1978		30 400	29 000					4 500	2 000
1979		49 140	114 420					4 000	
1980		81 500	365 200						
1981	1 060	22 940	38 000				450		
1982	12 820	17 000	89 200			6 500	1 150	12 000	
1983	23 475	73 550	165 000		120	37 250	250	10 400	2 000
1984	160	61 900	4 900	7 300	19 000	13 350		3 000	30 700
1985		213 850	6 550	not mapped	45 600		50		59 025
1986		412 955	60 000	94 700	56 200				93 425
1987		834 255	59 750	59 400	15 205	90		700	24 200
1988		359 700	102 185	35 900	64 400			500	51 870
1989		165 800	17 000	123 750	72 400			100	122 500
1990		193 000	30 000	398 150	630	1 115			206 000

[a] *Orgyia pseudotsugata*; [b] *Choristoneura occidentalis*; [c] *C. biennis*; [d] *C. fumiferana*; [e] *Acleris gloverana*; [f] *Lambdina fiscellaria lugubrosa*; [g] *Nepytia freemani*; [h] *Pristiphora erichsonii*; [i] *Malacosoma disstria*.

stems. Douglas-fir is a very resilient species and although dominant trees are seldom killed, mortality of understory regeneration and suppressed trees can be serious.

In addition to the earlier outbreaks, the western spruce budworm has been continuously active within British Columbia every year since 1967. Localized populations reached epidemic levels by 1970 over 5 000 ha northwest of Pemberton in the coast–interior transition zone. By 1973, this expanded to 72 000 ha in the Lillooet and Birkenhead river valleys, the Fraser Canyon near Boston Bar, and east of Hope. Increases continued each year until in 1977 (Fig. 7) more than 214 000 ha of defoliation were aerially mapped including the Fraser Canyon from Yale to the Nahatlatch River, the Skagit River drainage south of Hope, the Lillooet, Pemberton, Ashcroft areas, with newer areas near Clinton and around Shuswap–Adams lakes. This was the most extensive budworm outbreak recorded to that time and substantial plans were made and then canceled for a control program in the Fraser Canyon (see Chapter 69).

Through a combination of natural control factors including weather, parasitism, asynchrony of bud flush and insect emergence, reduction of suitable foliage, aging populations, and other unknown factors, a major drop in population occurred in 1978 with only 30 000 ha of defoliation mapped. However, with this reduction in the Fraser Canyon–Hope area, there was the beginning of a buildup in the drier interior Douglas-fir forests west of Kamloops near Ashcroft. By 1981, infestations in the Vancouver region had declined but continued around Ashcroft and started near Cache Creek and Clinton. By 1985, infestations again exceeded 200 000 ha extending from Ashcroft–Clinton to north and east of Kamloops, and in 1987, the most extensive outbreak on record covered 834 000 ha with extensions north to Wells Gray Park and south through the Okanagan Valley (Figs. 6 and 7). In 1988, there was a 56% reduction in the area infested to 360 000 ha. The greatest reductions occurred in the Lillooet–North Thompson areas with slight increases in the southern Okanagan and adjacent regions. Further reductions occurred in 1989, but infested areas increased by 15% in 1990, affecting 193 000 ha in three forest regions.

Western Blackheaded Budworm

The biology, earlier outbreaks, and control projects of the western blackheaded budworm, *Acleris gloverana*, mainly a periodic defoliator of western hemlock, true firs, and Sitka and white spruce, were reviewed by Lejeune (1975a). Typical of most blackheaded budworm infestations that rise and decline abruptly, the outbreak that started in 1970 primarily on Vancouver Island peaked during 1972, with 164 000 ha affected, and collapsed during 1973 (Table 1). Heavy rains during early larval development in July were likely the major factor in the decline. Similarly, the infestation in the Prince Rupert–Queen Charlotte Islands area which peaked at 128 000 ha in 1974 occurred in only scattered pockets in 1975 and was absent in 1976. There was a new infestation in 1976 near Blue River which lasted only 1 year, despite numerous eggs in evidence that fall.

Although limited feeding was observed in 1979 and 1983, it was not until 1984 that aerially visible defoliation again occurred over 19 000 ha in the West Kootenay, including Revelstoke and Glacier national parks. This increased in 1985 to include the Queen Charlotte Islands, Kitimat, and the interior wet belt forests near Quesnel Lake. These infestations peaked in 1986 over 56 200 ha (Fig. 8) before declining to 15 000 ha in 1987 and disappearing in these stands in 1988. A new infestation near Holberg on northern Vancouver Island, first evident in 1987, increased to 4 800 ha in 1988 and 7 400 ha in 1989. Only 11 scattered pockets totaling 630 ha remained in 1990. Hemlock sawfly, *Neodiprion tsugae*, was also active along with budworm and followed it on the Queen Charlotte Islands, particularly in young hemlock stands. During this same period, fluctuating sawfly populations lightly defoliated current year shoots on white spruce and alpine fir from east of Prince Rupert to Hazelton and occurred briefly on old growth western hemlock at Harrison Lake near Vancouver and in Wells Gray Provincial Park. Parasitism of larvae has averaged 20 to 40% in older infestations and along with diseases, weather, and starvation are natural factors that contribute to population collapses.

Consecutive years of defoliation have caused some tree mortality with as high as one-third of the mature and two-thirds of the immature trees killed in patches. Top kill and loss of height growth are common and average radial increment in young hemlock averaged 70% less than pre-infestation levels. Significant tree mortality was not recorded in the interior and south coast infestations, which were usually shorter with lighter defoliation.

Douglas-fir Tussock Moth

At least six infestations of the Douglas-fir tussock moth, *Orgyia pseudotsugata*, have been recorded in the dry belt interior of British Columbia since 1916. The infestation in the southern Okanagan Valley that began in 1970 (Lejeune 1975h) peaked in 1972 and was greatly reduced in 1973 by a naturally occurring nuclear polyhedrosis virus. However, new infestations in the north Okanagan Valley, near Salmon Arm and west of Kamloops, developed and expanded. By 1975, populations in the Okanagan Valley had collapsed but infestations west of Kamloops and along the North Thompson River peaked at 8 100 ha (Table 1, Fig. 9) and then collapsed. Only scattered individual shade and forest trees were defoliated in 1989 and 1990.

A 5-year period with only low numbers of moth captures in pheromone traps followed, until 1980 when a few ornamental trees were defoliated east of Kamloops and egg masses were found near Hedley. Defoliation occurred in patches totaling 10 000 ha the following year. By 1983, the area of defoliation doubled that of the previous year to 23 475 ha (Table 1) and virtually collapsed the following year. The main areas affected were east and west of Kamloops and near Hedley, Westwold, Vernon, and Pavilion. About 60% of the larvae sampled from 42 locations contained virus, and parasitism of cocoons was equally high. The numbers of moths captured in pheromone traps in areas selected for frequency of past outbreaks have increased in each of the last 4 years indicating that infestations may be expected in 1991.

Mortality and top kill of trees after successive years of usually severe defoliation are significant. Although extensive salvage logging followed the 1973 to 1975 outbreak, mortality of young trees was evident over more than 1 500 ha in a

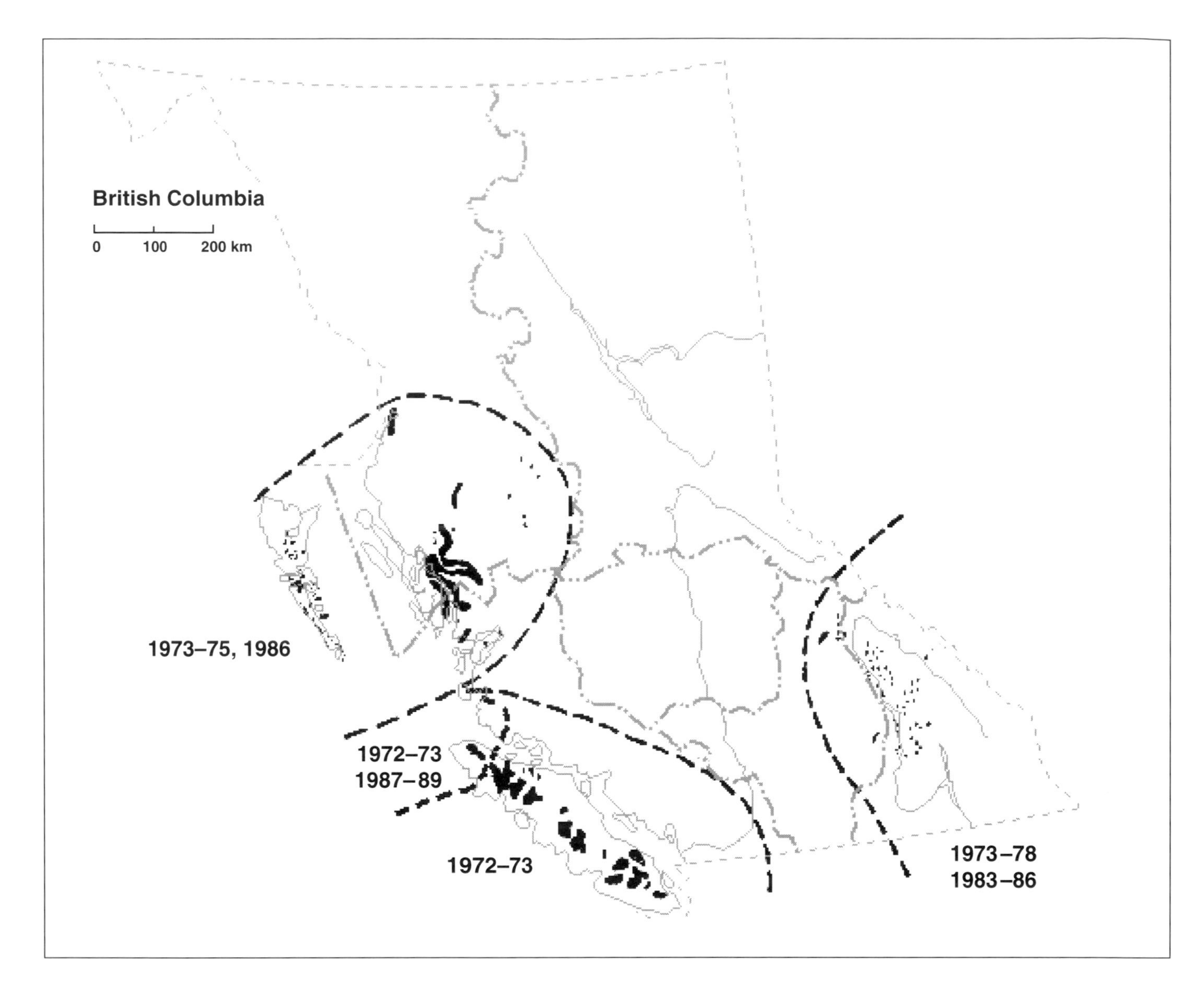

Figure 8. Extent and location of aerially mapped defoliation by western blackheaded budworm, *Acleris gloverana*, infestations in British Columbia, 1972–1989.

1979 aerial survey. On average, 43% of the trees were dead with concentrations occurring on steeper, dry slopes. Similar observations followed the 1981 to 1983 infestation with cumulative mortality of 51% and top kill on 11% of the survivors (Alfaro et al. 1987).

Western Hemlock Looper

Although the western hemlock looper, *Lambdina fiscellaria lugubrosa*, was a frequent and serious defoliator in coastal and interior wet belt forests in earlier years, only two outbreaks have occurred in British Columbia since the last review (Lejeune 1975b). Larvae were common at low numbers causing very light defoliation in the Nelson, Kamloops, and Vancouver regions but the only aerially visible and mapped defoliation occurred in 1973, 1976, from 1982 to 1984, 1987, and in 1990 (Table 1). By 1974, populations along the Columbia River and at Coquitlam Lake had collapsed but were increasing in Wells Gray Provincial Park, where the populations peaked with 10 500 ha infested in 1976 and then collapsed.

Increasing larval numbers were again noted in 1981 and by 1983 defoliation was mapped from Mica to Lower Arrow Lake in the Nelson region, with newer infestations around Shuswap Lake in the Kamloops region and high larval counts near Quesnel Lake. Infestations collapsed in the Nelson region in 1984 and in the Kamloops–Quesnel Lake areas in 1985. A small infestation that severely defoliated western hemlock, western red cedar, and understory shrubs along Jervis Inlet near Vancouver was active only in 1987. In 1990, old-growth western hemlock north of Revelstoke and near Quesnel Lake in eight patches totaling 1 115 ha were lightly defoliated.

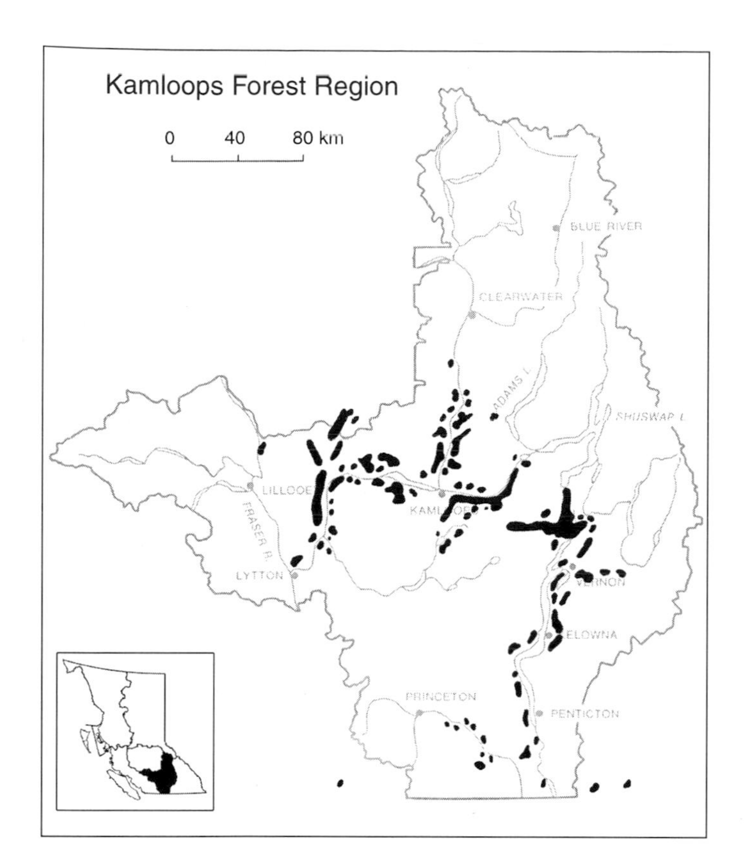

Figure 9. Location of Douglas-fir tussock moth, *Orgyia pseudotsugata*, defoliation during the 1973–1975 and 1981–1984 infestations in British Columbia.

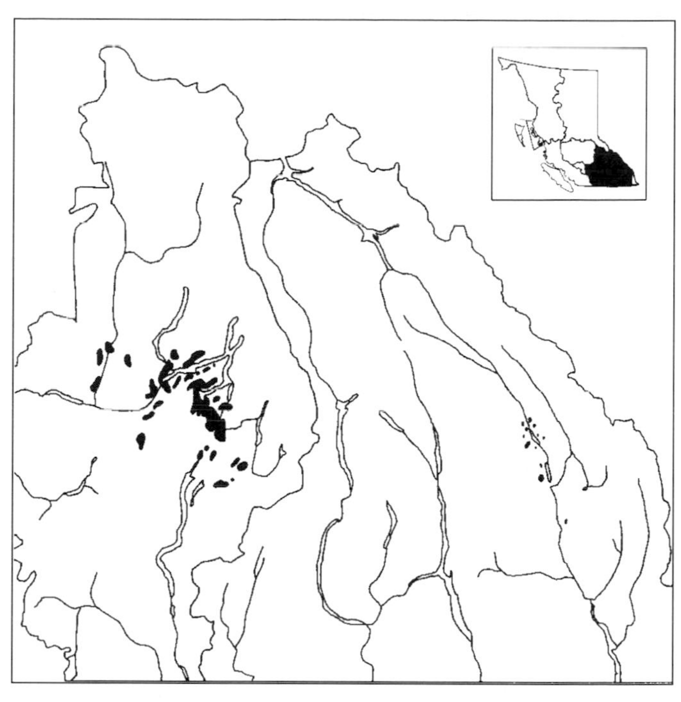

Figure 10. Location of mapped defoliation from western false hemlock looper, *Nepytia freemani*, in British Columbia, 1973–1988.

Egg parasitism as high as 80%, fungal and viral infection, and parasitism of larvae, with cool, wet weather during early larval development appear to be the major natural control factors. Operational controls were not attempted during this period. Mortality of old growth generally averaged less than 5% of the trees, although top kill up to 3 m long was evident on about 25% of the trees.

Western False Hemlock Looper

The infestation of western false hemlock looper, *Nepytia freemani*, near Shuswap Lake which started in 1971 (Lejeune 1975c) expanded and severely defoliated Douglas-fir over about 2 000 ha in 1973 (Table 1). Further defoliation occurred in 1974 but the greatest area of defoliation was near Vernon and along the North Thompson River in association with Douglas-fir tussock moth (Fig. 10). By 1975, the only infestations were 320 ha along the North Thompson and new infestations at Monte Lake and scattered stands in the Windermere Valley. High levels of egg parasitism and a nuclear polyhedrosis virus disease, first isolated in 1975, along with the aerial application of *Bacillus thuringiensis* (Bt) in 1973, caused a decline and only a few larvae were evident in 1976.

Larvae were not again apparent until 1980 with defoliation in 1981 totaling 450 ha, again near Shuswap Lake. This increased to 12 stands totaling 1 150 ha in 1982 before declining in 1983. Small pockets in Christmas tree production areas in the Windermere Valley were defoliated in 1983 and again in 1985. High levels of virus were again factors in the natural declines with damage limited mostly to increment loss, top kill, and some tree mortality, particularly on drier, exposed sites near Sunnybrae.

White Pine Weevil

Previously known in western Canada as the Sitka spruce weevil, *P. sitchensis*, and the Engelmann spruce weevil, *P. engelmannii*, the white pine weevil, *Pissodes strobi*, is a most serious pest in British Columbia not of white pine, but of Sitka spruce. It is common through much of the host range of Sitka spruce except on the Queen Charlotte Islands. Other hosts are white spruce, Engelmann spruce, and Norway spruce.

Attacks that kill the terminal shoots of young trees have been so frequent and widespread that planting of Sitka spruce has been much reduced since the 1970s. A summary of surveys in 310 young spruce stands throughout the province between 1967 and 1982 showed 65% of the stands were damaged in the year of examination. Within stands, trees with currently attacked leaders averaged 9% of white spruce in the Prince George region to about 15% through central and eastern British Columbia, 18% on Sitka spruce in the Vancouver region, and 32% on Sitka spruce in coastal valley bottom sites in the Prince Rupert region. Individual stands can have 85% or more of the leaders currently attacked. The frequency and intensity of attacks are greatest in low-elevation coastal sites on Vancouver Island and the Kitimat Valley, and widespread but usually much less in white or Engelmann spruce throughout the interior. With cumulative attacks over many years, leader dominance, bole form, and considerable volume are lost.

Larch Sawfly

Defoliation of tamarack and western larch by the larch sawfly, *Pristiphora erichsonii*, recorded in the Pacific region since at least the 1940s, has continued periodically in the 1970s and 1980s (Table 1). The damage, however, did not warrant applied control.

Defoliation of tamarack in northern British Columbia was observed in 1973, 1977 to 1978, and 1986 to 1987, and in the Yukon Territory in 1984 and 1987 to 1990. Most occurred in widely scattered bog areas west of Prince George, in the Peace and Liard river drainages, and in northwestern British Columbia near the Yukon border.

Defoliation of western larch in southeastern British Columbia occurred from 1976 to 1979, 1982 to 1984, and 1987 to 1989. Historically, most of the outbreaks have occurred in the East Kootenay near Fernie and north of Kimberley with occasional pockets of defoliation near Grand Forks. The most extensive outbreak developed quickly in 1982 and extended over 12 000 ha in 50 separate areas. A few exotic larch and western larch planted beyond its natural range in Terrace, Prince George, and Vancouver were defoliated from 1988 to 1990.

Sawfly parasites were released in British Columbia for the first and only time in 1978. About 60 adult *Olesicampe benefactor*, reared from sawfly cocoons collected near Winnipeg in 1977, were released at one sawfly-infested western larch site near Sparwood.

Larch Casebearer

The larch casebearer, *Coleophora laricella*, native to Europe and Asia, is an introduced insect pest of western larch in British Columbia. First recorded in the western United States in Idaho in 1957, it likely spread into British Columbia in the early 1960s where it was first collected in 1966. By this time, light populations were observed along the border near Rossland, Creston, along the Salmo and Moyie river valleys, the lower Kootenay Lake, and near Grand Forks. There has been a continued slow spread over most of the southern range of western larch (Fig. 11), with first recordings in Flathead Valley in 1978, and at Cherryville east of Vernon in 1980. Aerial mapping of defoliation was not attempted each year because of the scattered, patchy occurrence of western larch which constitutes only a small proportion of the stand dominated by other tree species.

Defoliation occurred each year during the 1970s, usually at moderate intensity but with some severe patches. During this period some fluctuation of populations occurred (e.g. higher summer temperatures during 1973 caused larval mortality resulting in lighter defoliation in 1974, but by 1975 infested areas were again severely defoliated). Starting in 1969, FIDS imported and released larch casebearer parasites. By 1987, more than 15 000 parasites, mostly *Agathis pumila* or *Chrysocharis laricinellae*, had been released at selected sites. The effectiveness is being evaluated, but early observations are encouraging. Parasites have become established and by 1981 parasitism of collected larvae averaged 59% and defoliation from 1980 to 1982 was very light and patchy. Populations increased in 1983 and in 1984 defoliation occurred over 41 000 ha, although needle cast fungi were also prevalent and careful examination was required to distinguish the damage. Defoliation decreased greatly in 1985 and has been less each year since then with only light defoliation in a few patches in 1988, and either absent or very light at most of the release sites. Populations increased slightly in 1990 and infestations were reported for the first time near Sicamous.

Larch Budmoth

First recorded at epidemic levels in British Columbia in 1965, populations of larch budmoth, *Zeiraphera improbana*, defoliated western larch over 71 500 ha in the western part of the host range in the Nelson and Kamloops regions. The largest single infestation was near Creston over 16 000 ha in the Summit Creek drainage. Most infestations in the Nelson region declined in 1966 to 8 760 ha near Rossland but expanded to 2 100 ha in the adjacent Kamloops region. All declined to endemic levels in 1967 and remained low until 1974.

The first recorded budmoth infestation in the Yukon Territory in 1974 defoliated tamarack north of Watson Lake. Infestations expanded in 1975 throughout most of the host range and continued until 1978 when populations collapsed.

In 1976, new infestations developed in the western part of the host range in southeastern British Columbia and expanded in 1977 to over 2 000 ha in five drainages. In 1978, western larch over 1 900 ha in the west Kootenay were lightly or moderately defoliated but with larval parasitism up to 50%, populations collapsed in 1979. Populations did not occur again until 1983, when high larval populations in the west Kootenay defoliated western larch in 36 patches between 1 200 and 1 500 m elevation, totaling 6 600 ha. Infestations declined in 1984 to 10 areas totaling 1 100 ha, but expanded again in 1985 to 14 800 ha. Larval parasitism in 1986 averaged 31% and was again likely a major factor in the population collapse.

Forest Tent Caterpillar

Forest tent caterpillar, *Malacosoma disstria*, has been reported at varying levels and locations within British Columbia in 14 of the last 18 years. Peak years were 1973, 1977, and 1984 to 1990 (Table 1). Areas in which the most common hosts, trembling aspen and black cottonwood, were most severely defoliated extended from the international border through eastern and central British Columbia, north to Prince George and the Peace River area. Following the 1973 infestation in which more than 60 000 ha were defoliated south and east of Prince George and along the Columbia River, scattered pockets of defoliation reoccurred in the Rocky Mountain Trench. Infestations again peaked near Prince George in 1977, declined substantially in 1978, and populations remained very low for the next 4 years.

Pockets of defoliation occurred in 1983 east of Prince Rupert, in the Salmon River Valley north of Prince George, in the Peace River area, and near Quesnel. In the following years, 70 to 98% of all defoliated areas were concentrated in the Prince George region, either near Prince George or in the Peace River area. The small infestation near Quesnel collapsed in 1984 but pockets of defoliation totaling 1 000 ha occurred near Moricetown, Hazelton, and Kitwanga, and almost 30 000 ha of defoliation were mapped in the Salmon

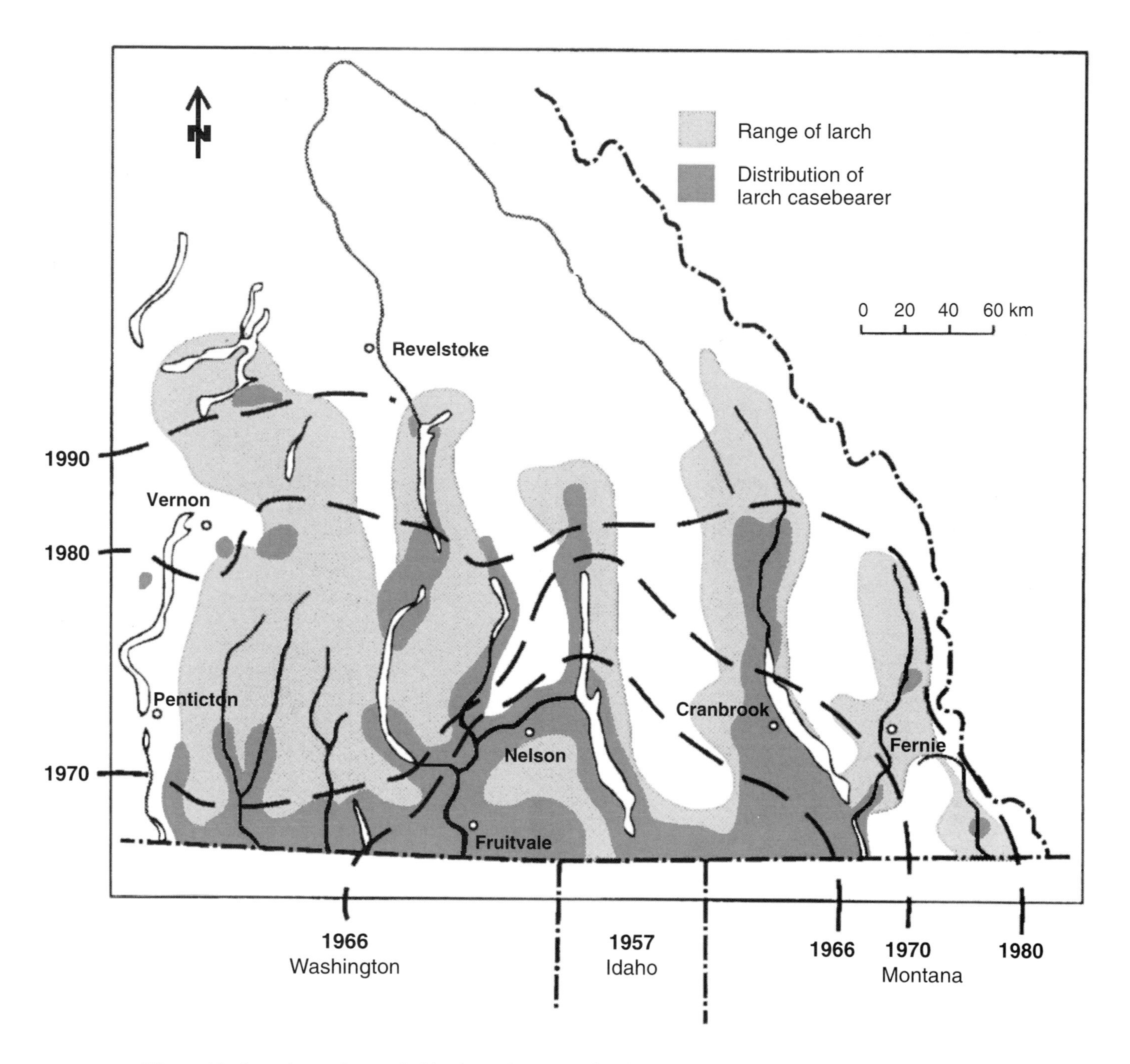

Figure 11. Location and spread of larch casebearer, *Coleophora laricella*, within the natural range of western larch in British Columbia since the insect was introduced in the 1960s.

River Valley and Peace River area. Infestations continued east of Prince Rupert in 1985, increased to over 50 000 ha in the Peace River area, but collapsed along the Salmon River and developed in small pockets near Trail in the West Kootenay. The extent of defoliation almost doubled to 93 400 ha in 1986 (Table 1) largely in the Peace River area with small pockets in the West Kootenay. The small infestation east of Prince Rupert collapsed. After only a 2-year absence, populations reappeared in the Salmon River Valley and near Prince George in 1987, but declined in the Peace River area and increased in the West Kootenay. Cold, wet weather during larval development was likely a factor in the decline. By 1988, the infestation in the Peace River area totaled only 5 000 ha and the major increase continued to be around Prince George and, to a lesser extent, in the Nelson region from Creston to Fernie and north to Donald. Infestations around Prince George, in the Peace River area, and east of 100 Mile House expanded in 1989 and covered more than 200 000 ha in 1990. Infestations in the Nelson region increased slightly in 1989 before declining in 1990.

Northern Tent Caterpillar

The northern tent caterpillar, *Malacosoma californicum pluviale*, a common colonial defoliator of deciduous trees and shrubs, has been recorded in British Columbia at irregular intervals since the 1930s. During the review period, the highest populations occurred in the early 1980s until 1987. In 1984, a small outbreak collapsed in its fourth year in the Bell-Irving and Meziadin areas near Prince Rupert. However, populations continued to increase each year in the Fraser Valley and on the Gulf Islands and southeastern Vancouver Island until 1987 when more than 50% of the young larvae were killed by a nuclear polyhedrosis virus, and defoliation subsided. A first record of this insect from mid-coast British Columbia occurred in 1987 when alder and white birch over 100 ha near Bella Coola were severely defoliated. Colonies were also common and scattered in the Okanagan Valley and the Nelson region until 1989. New infestations, the first in many years, developed in cottonwood stands in the Skeena River Valley and near Kitimat in 1989 and increased in 1990.

Large Aspen Tortrix

The large aspen tortrix, *Choristoneura conflictana*, is another defoliator of trembling aspen, particularly around Prince George and north into the Yukon Territory. Infestations fluctuated around Prince George from 1950 to 1958 and in the Nelson region in 1958 and 1959. More recently, the outbreak that began in 1970 over 9 000 ha near Prince George declined in 1972 and remained endemic until scattered defoliation occurred near Chetwynd in 1975, and from Mackenzie to Fort Ware in 1990. Trembling aspen over more than 38 000 ha near Vanderhoof and Fort Fraser were defoliated in 1980. Aspen along the Alaska Highway were defoliated from 1975 to 1981 and again in 1985 to 1990 with the most widespread and severe defoliation in 1981, when numerous patches from Little Atlin Lake to Dawson City and Snag totaled 50 000 ha. Disease, virus, and parasitism were significant natural control factors in older infestations.

Balsam Woolly Adelgid

The balsam woolly adelgid, *Adelges piceae*, of European origin, is patchily distributed over about 5 000 km^2 of southern Vancouver Island and southwestern British Columbia (Fig. 12). Following its initial discovery near Vancouver in 1958, extensive surveys in the early 1960s established this general distribution. Concern for artificial spread and increased damage led to a voluntary restriction on the importation and movement of *Abies* stock. This was formalized in provincial quarantine regulations in 1966. The initial ban on growing *Abies* nursery stock or ornamentals was amended in 1977 to allow production under annual permit, but only tagged material could be moved and not out of the declared infested zone.

With constraints on movement of susceptible material, the distribution of balsam woolly adelgid has not changed much from that of the initial surveys. It occurs up to about 1 000 m elevation in Lower Fraser River Valley drainages east to Agassiz and Harrison Lake, up the mainland coast to the Sechelt Peninsula, and in Vancouver Island east coast drainages. Several ornamental fir at Oliver and Penticton in interior British Columbia that were found to be heavily attacked in 1967 were immediately sprayed and removed, resulting in no further detections in that vicinity. New areas of infestation, mostly just representing small extensions of the distribution, included Squamish Valley in 1974, Sechelt Peninsula in 1975, Sooke to Jordan River in 1977, the Nanaimo River Valley in 1983, Parksville in 1985, Powell River in 1986, and Hornby Island in 1990. The discovery of severely infested amabilis fir and 30% mortality over 150 ha on West Thurlow Island north of Campbell River in 1987 was the most significant extension. This rather remote, well-established infestation likely started during the mid-1970s.

The insect biology and damage were generally described by Bryant (1975). In British Columbia, populations are most commonly concentrated in the tree crown causing swelling (gout), distortion, and death of twigs and ultimately crown dieback. Heavy stem attacks are less common. It infests all *Abies* species and although alpine fir is the most susceptible to damage, amabilis and grand fir are most frequently infested in coastal British Columbia. Seedlings can be infested and seriously gouted and could, without regulation, easily and rapidly spread the insect throughout the province.

Early infestations in Washington killed an estimated 1.5 billion fbm of timber (Furniss and Carolin 1977). In British Columbia mortality has been lower but is highly variable and patchy. In resurveys of long-term plots in 1987, mortality of mature amabilis fir averaged 15%, ranging from 5% up to 95%. With the discovery in 1983 of surviving populations and

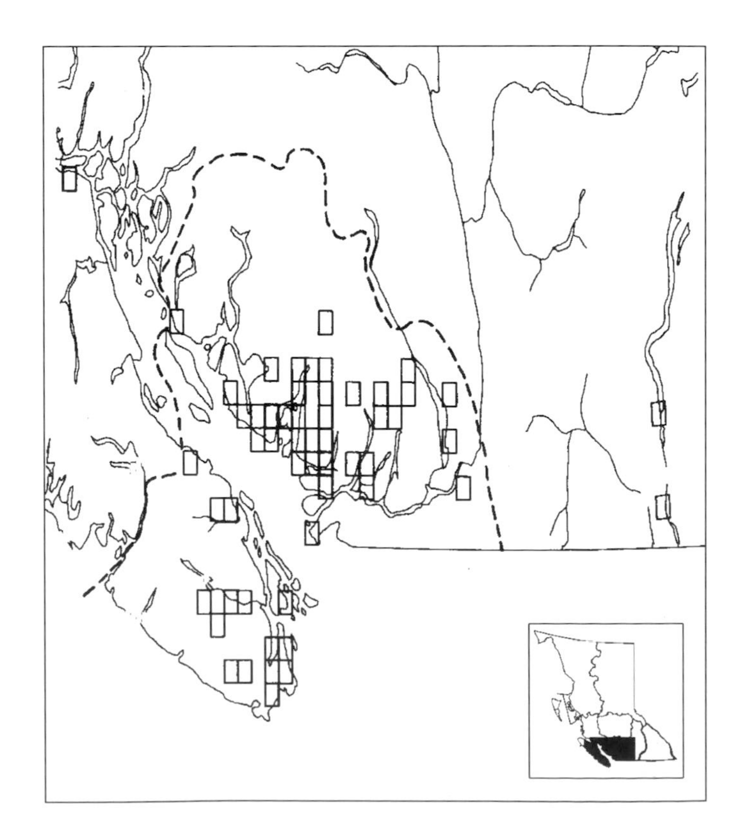

Figure 12. Location of true fir stands infested with the balsam woolly adelgid, *Adelges piceae*, in southwestern British Columbia, and the provincial quarantine boundary, 1988.

damage on alpine fir and grand fir at higher elevations in Idaho, the risk and concern for potential spread into interior British Columbia was re-emphasized. True firs are widely distributed in British Columbia, comprising 20% of the softwood volume and ranking fourth at 13% of the annual harvest. In eastern Canada the adelgid is also a major pest in young Christmas tree plantations. The insect continues to pose a threat to forests and ornamental plantings in British Columbia.

Insects of Plantations and High-Value Stands

Cone and Seed Insects

Research on cone and seed insects has been intensified, and surveys have been emphasized since the early 1970s, when Lejeune (1975j) identified only two significant insects of Douglas-fir cones. Increased harvesting in the interior of the province increased demand for a diversity of tree species and tree seed. By 1986, spruce, true fir, and Douglas-fir each accounted for about 25% of the total seed requirement for the nurseries with 16% pine and 2 to 3% for larch, hemlock, and other species.

The most damaging insects in Douglas-fir cones continued to be the Douglas-fir cone moth, *Barbara colfaxiana*, and the Douglas-fir cone gall midge, *Contarinia oregonensis*, but periodically, severe infestations also include the coneworms *Dioryctria pseudotsugella* and *D. abietivorella*, a seed chalcid, *Megastigmus spermotrophus*, and the Douglas-fir cone scale midge, *Contarinia washingtonensis*. Spruce cones were most frequently and severely infested with spruce seed moth, *Cydia strobilella*, and spruce cone maggot, *Strobilomyia neanthracina* (see also Chapter 33). Despite relatively heavy spruce cone crops and only slightly lighter crops of Douglas-fir in 1979 and 1980, cone and seed infestations were common and frequently severe so that more than 60% of locations sampled were judged unsuitable for cone collection. In 1981, insect infestations were concentrated and more serious in the generally lighter cone crops. Infestation rates showed a small, gradual decline during the next 2 years, but insects were still present in more than 40% of the spruce and Douglas-fir cones and in about one-quarter of the ponderosa pine and alpine fir cones examined in 1983. Cone crops were generally light from 1984 to 1987, and insect damage was scattered and highly variable. For example, in 1987, western spruce budworm, which was widespread at epidemic levels, contributed to a high incidence of damage in two seed orchards. A Douglas-fir coneworm infested most cones examined in the central and southern interior, and Douglas-fir cone moth infested 28% (and up to 85%) of the cones at 34 sites in four regions. Control actions were also implemented in 1987 in most coastal Douglas-fir seed orchards against moderate to high populations of gall midge (90% of cones infested), coneworm (55%), and cone moth (45%). Spruce cone maggot and spruce seed moth infested up to 80% of the white and Engelmann spruce cones at 12 sites in eastern and north central British Columbia. A previously undescribed gall midge, *Chamaediplosis nootkatensis*, infested about 20% of the male and female yellow cedar cones in a seed orchard. Cone crops were up and generally good in 1988, although the seed and cone insects continued to infest about 20% of the cones with considerable variation in the damage among insects, hosts, and locations.

Black Army Cutworm

Although the black army cutworm, *Actebia fennica*, was first collected in British Columbia in 1906 and reported as a pest of agricultural crops, damage to forest plantations was not observed until 1964 when white spruce and all ground cover over 20 ha near Prince George were severely defoliated. Outbreaks, usually 2 years' duration, have occurred most commonly 2 years after harvested forest lands in interior British Columbia have been burned by wild or prescribed fires. The black army cutworm prefers to feed on a wide variety of herbaceous vegetation including fireweed, dandelion, and colt's foot before it attacks conifer seedlings, including white and Engelmann spruce, lodgepole pine, and Douglas-fir. Nocturnal feeding, often in patches, by young larvae occurs shortly after the snow recedes in the spring. Populations can be greatly reduced by animal and bird predation, unfavorable weather or soil conditions, a nuclear polyhedrosis virus, and several parasites. Long-term damage to the seedlings varies greatly according to the distribution of the cutworm, availability of preferred feeding material, natural control factors, and the severity and duration of defoliation. In some stands, partial replanting has been necessary.

Defoliation in young conifer plantations was not reported again until 1974 when severe defoliation of recently planted seedlings occurred in 13 widely separated and previously burned sites, totaling 3 600 ha. Areas affected were east of Prince Rupert, south of Prince George, near Clearwater, and northwest of Golden. These populations collapsed or declined in 1975, but new infested areas occurred near Golden, Revelstoke, and Salmon Arm. These in turn declined in 1976 and, although some larvae were observed feeding mostly on herbaceous vegetation, populations remained low until 1982. Then, more than 32 000 seedlings and other ground cover were defoliated in a dozen recent plantations in the North Thompson area near Kamloops, east of Prince Rupert, and near Fort St. James and Whitehorse. In 1983, populations near Kamloops declined but increased in 35 sites in the Prince George and Prince Rupert regions with about 200 000 lodgepole pine and white spruce seedlings affected. Only six sites were affected in 1984 but feeding was again more common and widespread in these regions in 1985 and 1986 and then declined. In 1987 and 1988, black army cutworm was most active in the Nelson region following the wild fires of 1985. Populations declined in 1989 and remained low in 1990.

On average, less than 20% of those recently burned susceptible stands have become infested, and long-term damage is highly variable and equally unpredictable. Research to improve and calibrate a pheromone-trapping damage index is in progress. Partially because of the difficulty in forecasting outbreaks that are usually of short duration even though defoliation may be severe, damage control actions have focused more on delay of planting than direct chemical treatments. Trials in 1974 tested Diazinon®, Dylox®, Bt, *Cordyceps militaris*, and virus with the first two showing best results (Ross and Ilnytzky 1977). A 1988 British Columbia Forest Service ground spray trial over 16 ha with Ambush® was effective in killing larvae (Moss 1988).

Winter Moth

The winter moth, *Operophtera brumata*, is native to Europe and has been present in the Greater Victoria area since at least 1972; infestations have been expanding since then. Garry oak, maple, willow, and many fruit trees or other deciduous trees and shrubs were frequently and severely defoliated. Defoliation was heaviest in Victoria and on the Saanich Peninsula with populations detected in Duncan in 1983, Sidney and on Saltspring Island by 1984, Nanaimo in 1985, and for the first time on the lower mainland in 1989. Pheromone trap surveys captured male moths in the Tsawwassen and Richmond areas in 1986, although defoliation was not then apparent, and from Sooke to Nanaimo.

After increasing in severity and extent each year for at least 13 years, defoliation declined for the first time in 1985 and by 1987 was reduced to endemic levels. The introduction of more than 18 000 parasitic flies, *Cyzenis albicans*, and 7 000 of a wasp, *Agrypon flaveolatum*, at up to 30 release sites between 1979 and 1982 was apparently successful, as it had been earlier in Nova Scotia. Parasites were recovered by 1981 and by 1984 parasitism averaged 44% with up to 77% at some sites.

Gypsy Moth

The introduced gypsy moth, *Lymantria dispar*, has caused serious, widespread defoliation of many hardwood and softwood shade and forest trees in eastern North America. It is a major public nuisance in recreational areas, and the resulting quarantine regulations restrict movement of nursery and forest products. Although an interception of gypsy moth egg masses was made on nursery stock as early as 1911 in Vancouver, the first trapping on the west coast occurred in California in 1973, followed in 1974 with a capture in Renton, Washington. Increased captures and the 1984 to 1985 Lane County, Oregon, infestation, which prompted a $9.1 million (U.S.) spray program with Bt over 92 000 ha, emphasized the consequence and risk of movement of this insect into British Columbia by travelers from the western states as well as portions of eastern North America. Despite periodic introductions, it has not yet become established in British Columbia largely due to the interagency cooperative detection and treatment programs.

Gypsy moth pheromone trapping surveys were initiated in British Columbia by Agriculture Canada and Forestry Canada in the early 1970s, with emphasis on municipal and recreational areas. With British Columbia Forest Service participation since 1985, more than 8 000 traps are monitored throughout the province each year. In 1978, male moths were trapped and egg masses discovered in the Kitsilano district of Vancouver. Egg masses were destroyed and some properties were treated in 1979 with ground application of carbaryl or insecticidal soap. Concentrated trapping later in 1979 did not recover any moths. Since 1980 moth captures have varied from 1 to almost 200 (Table 2). Except from one location in the Okanagan Valley, the more frequent captures and subsequent discovery of egg masses have been on the lower mainland near Vancouver and four locations on Vancouver Island. With Agriculture Canada's lead, Bt was applied to small areas at Fort Langley in 1984; Courtenay in 1984 and 1985; Chilliwack in 1985, 1986, and 1987; Kelowna, Belmont Park, and Parksville in 1988; Parksville in 1990; and near Victoria airport in 1991. Small numbers of male moths have been captured elsewhere in southern British Columbia without detection of egg masses or resulting defoliation.

Table 2. Gypsy moth, *Lymantria dispar*, in British Columbia 1978–1990.

Location	Total male moths and positive traps										
	1978	1980	1982	1983	1984	1985	1986	1987	1988	1989	1990
Areas sprayed											
Kitsilano	5/1[a]										
Ft. Langley			3/2	26/18[a]							
Courtenay				9/4[a]	25/12						
CFB Chilliwack				1/1	6/3[a]	14/13[a]	8/6[a]				
Kelowna							7/4	198/52[a]	4/4	1/1	
CFB Belmont Park							3/2	6/5[a]	1/1		
Parksville								12/6[a]	2/2	8/8[a]	2/1
Areas not treated											
Tsawwassen/Pt. Roberts		1/1					1/1	4/3	2/2		
Greater Vancouver[b]			6/4	1/1	2/1		1/1		2/2	6/6	11/9
Lower Fraser Valley[b]			1/1		1/1	2/2				3/3	
Central interior B.C.[b]					1/1		3/2		1/1	2/2	1/1
Vancouver Island[b]							1/1	1/1		6/6	107/46[a]
Totals/year	5/1	1/1	10/7	37/24	35/18	16/15	24/17	221/67	12/12	26/26	121/57

[a]Egg masses also found.
[b]Not necessarily the same areas each year.

These locations include lower mainland locations of north, west, and central Vancouver, Sumas, Tsawwassen, Abbotsford; Goldstream Park and Qualicum Beach on Vancouver Island; and Adams River, Clinton, Knutsford, and Sicamous in central British Columbia.

European Pine Shoot Moth

The European pine shoot moth, *Rhyacionia buoliana*, a native to Europe, was first recognized in Canada in 1925 with infested shoots discovered in Ontario, the Maritimes, and at a nursery in Victoria, British Columbia. Until the 1960s, the known distribution in British Columbia was Vancouver and southern Vancouver Island, at which time FIDS emphasized surveys elsewhere in the province. Infested ornamental pines were found in the Kelowna–Penticton area and provincial regulations on movement of pine were imposed from 1968 to 1981. Even so, an infested shipment of Austrian and Mugho pines from Vancouver was discovered in Kelowna in 1972 and landscape plantings of Mugho pine at the Keenleyside Dam near Castlegar were found infested in 1974. From 1976 to 1978, infested shoots were collected throughout the Okanagan Valley from Summerland to Vernon and at Kamloops. Careful clipping and burning of infested shoots, perhaps aided by over-wintering mortality, appear to have controlled the infestation at Castlegar, but even sanitation combined with applications of dimethoate in 1976 has not prevented the establishment of this shoot moth in ornamental plantings in the Okanagan Valley. Further surveys in 1988 indicate that the shoot moth is established in localized urban areas including Victoria to Courtenay, the lower mainland, and the Okanagan Valley. There was no evidence of survival or significant damage in native pines.

Minor Insect Pests

The following information shows the occurrence of selected, usually less serious, forest insect pests in British Columbia from 1973 to 1990.

Insect	73	74	75	76	77	78	79	80	81	82	83	84	85	86	87	88	89	90
Coniferous pests																		
Balsam twig aphid *Mindarus abietinus*						3[a]								2				
Conifer seedling weevil *Steremnius carinatus*										3			3					
Cooley spruce gall adelgid *Adelges cooleyi*	2	2	3	1	2		3						3		2	2	2	2
Gall adelgids *Pineus* spp.		1		2	2	2							3	1	1	1	1	1
Lodgepole pine beetle *Dendroctonus murrayanae*			3										3	3	3	3	3	3
Lodgepole terminal weevil *Pissodes terminalis*	2	2	2			2	3	3					3	2	2	2	2	2
Northern spruce engraver *Ips perturbatus*												3	3	1				1
Phantom hemlock looper *Nepytia phantasmaria*	1									3	1							
Sawflies *Neodiprion* spp.	2	3	3	3	3	3	2				3	3	2	3	3			
Silverspotted tiger moth *Lophocampa argentata*	1				2								1	1				1
Spruce aphid *Elatobium abietinum*	1	3			3	1	2	3	3	1		3	2	1	3	1	1	1
Warren root collar weevil *Hylobius warreni*					2	2		3	2					1	2	3	3	2
Western pine beetle *Dendroctonus brevicomis*				3										2	2		2	
Deciduous pests																		
A birch leafminer *Lyonetia saliciella*		3	3	3	3	3	3			3	3	2	3	2	3	2	3	2

[a] Blank—Not recorded in written reports of that year.
1—Observed or collected but endemic and not damaging to hosts.
2—Common, widespread, but little damage to host(s).
3—High populations, significant damage to host, at least in localized areas.

(Continued)

Minor Insect Pests *(Concluded)*

Insect	73	74	75	76	77	78	79	80	81	82	83	84	85	86	87	88	89	90
Aspen serpentine leafminer *Phyllocnistis populiella*	1			3	3				2					2				
Birch skeletonizer *Bucculatrix canadensisella*					2	3	3		3	3					3		2	2
Fall webworm *Hyphantria cunea*	1	1			2	2		2	2				1	1			1	1
Pacific willow leaf beetle *Tricholochmaea decora carbo*	2		2		2				3				3	1	3			
Satin moth *Leucoma salicis*		3	3	3	3	3					3	3	3	1	2	3	2	1
Striped alder sawfly *Hemichroa crocea*														2				
Uglynest caterpillar *Archips cerasivorana*					2	2												
Western oak looper *Lambdina fiscellaria somniaria*									3	3	2	2	1					
Western winter moth *Erannis tiliaria vancouverensis*														3	3		3	2

[a] Blank—Not recorded in written reports of that year.
1—Observed or collected but endemic and not damaging to hosts.
2—Common, widespread, but little damage to host(s).
3—High populations, significant damage to host, at least in localized areas.

Bibliography

Alfaro, R.I.; Taylor, S.P.; Wegwitz, E.; Brown, R.G. 1987. Douglas-fir tussock moth damage in British Columbia. Forestry Chron. Oct. 1987. pp. 351-355.

British Columbia Plant Protection Advisory Council. 1985. Understanding the gypsy moth threat. Proceedings of an information symposium, Nov. 5, 1985, Vancouver, B.C. 66 p.

Bryant, D.G. 1975. Balsam woolly aphid, *Adelges piceae* (Ratz.). Pages 250-253 *in* M.L. Prebble (Ed.), Aerial control of forest insects in Canada. Environ. Can., Ottawa, Ont.

Canadian Forestry Service. 1973-1979. Annual reports of the Forest Insect and Disease Survey. Ottawa, Ont.

Canadian Forestry Service. 1980-1986. Forest insect and disease conditions in Canada, 1980 to 1986. Ottawa, Ont.

Fiddick, R.L.; Van Sickle, G.A. 1979-1981. Forest insect and disease conditions, British Columbia and Yukon for years 1979 to 1981. Inf. Reps. BC-X-200, -220 and -225. Pac. For. Res. Cent., Victoria, B.C.

Forest Insect and Disease Survey. Maps of major forest insect infestations: Prince Rupert Forest Region 1922-1983. FIDS Rep. 84-3; Vancouver Forest Region 1909-1985. FIDS Rep. 86-8; Kamloops Forest Region 1912-1986. FIDS Rep. 87-8; Cariboo Forest Region 1913-1986. FIDS Rep. 87-9; Nelson Forest Region 1928-1986. FIDS Rep. 87-10; Prince George Forest Region 1944-1986. FIDS Rep. 87-11; Canadian Forestry Service, Pacific Forestry Centre, Victoria, B.C.

Forest Insect and Disease Survey (unnumbered FIDS reports). History of population fluctuations and infestations of important forest insects in the Prince Rupert Forest Region 1914-1981; Vancouver Forest Region 1911-1986; Kamloops Forest Region 1912-1981; Cariboo Forest Region 1913-1982; Nelson Forest Region 1923-1981; Prince George Forest Region 1942-1982; Yukon Territory Forest Region 1952-1983; Canadian Forestry Service, Pacific Forest Research Centre, Victoria, B.C.

Furniss, R.L.; Carolin, V.M. 1977. Western forest insects. USDA For. Serv. Misc. publ. 1339. 654 p.

Harris, J.W.E.; Alfaro, R.I.; Dawson, A.F.; Brown, R.G. 1985. The western spruce budworm in British Columbia 1909-1983. Pac. For. Res. Cent., Victoria, B.C. Inf. Rep. BC-X-257.

Harris, J.W.E.; Dawson, A.F.; Brown, R.G. 1982. The western hemlock looper in British Columbia 1911-1980. Pac. For. Res. Cent., Victoria, B.C. Inf. Rep. BC-X-234.

Harris, J.W.E.; Dawson, A.F.; Brown, R.G. 1985a. The Douglas-fir tussock moth in British Columbia 1916-1984. Pac. For. Cent., Victoria, B.C. Inf. Rep. BC-X-268.

Harris, J.W.E.; Dawson, A.F.; Brown, R.G. 1985b. The eastern false hemlock looper in British Columbia 1942-1984. Pac. For. Cent., Victoria, B.C. Inf. Rep. BC-X-269.

Koot, H.P. 1978. Western blackheaded budworm. Pac. For. Res. Cent., Victoria, B.C. For. Pest Leaflet 4 p.

Lejeune, R.R. 1975a. Western black-headed budworm, *Acleris gloverana* (Wals.). Pages 159-166 *in* M.L. Prebble (Ed.), Aerial control of forest insects in Canada. Environ. Can., Ottawa, Ont.

Lejeune, R.R. 1975b. Western hemlock looper, *Lambdina fiscellaria lugubrosa* (Hulst.). Pages 179-184 *in* M.L. Prebble (Ed.), Aerial control of forest insects in Canada. Environ. Can., Ottawa, Ont.

Lejeune, R.R. 1975c. Western false hemlock looper, *Nepytia freemani* Munroe. Pages 185-187 *in* M.L. Prebble (Ed.), Aerial control of forest insects in Canada. Environ. Can., Ottawa, Ont.

Lejeune, R.R. 1975d. Phantom hemlock looper, *Nepytia phantasmaria* (Stkr.). Pages 188-189 *in* M.L. Prebble (Ed.), Aerial control of forest insects in Canada. Environ. Can., Ottawa, Ont.

Lejeune, R.R. 1975e. Green-striped forest looper, *Melanolophia imitata* (Walker). Pages 190-192 *in* M.L. Prebble (Ed.), Aerial control of forest insects in Canada. Environ. Can., Ottawa, Ont.

Lejeune, R.R. 1975f. Saddle-backed looper, *Ectropis crepuscularia* Schiff. Pages 193-195 *in* M.L. Prebble (Ed.), Aerial control of forest insects in Canada. Environ. Can., Ottawa, Ont.

Lejeune, R.R. 1975g. Pine butterfly, *Neophasia menapia* F. & F. Pages 204-205 *in* M.L. Prebble (Ed.), Aerial control of forest insects in Canada. Environ. Can., Ottawa, Ont.

Lejeune, R.R. 1975h. Douglas-fir tussock moth, *Orgyia pseudotsugata* (McDunnough). Pages 206-207 *in* M.L. Prebble (Ed.), Aerial control of forest insects in Canada. Environ. Can., Ottawa, Ont.

Lejeune, R.R. 1975i. Hemlock needle miner, *Epinotia tsugana* Freeman. Pages 213-214 *in* M.L. Prebble (Ed.), Aerial control of forest insects in Canada. Environ. Can., Ottawa, Ont.

Lejeune, R.R. 1975j. Cone and seed insects. Pages 215-217 *in* M.L. Prebble (Ed.), Aerial control of forest insects in Canada. Environ. Can., Ottawa, Ont.

Lejeune, R.R.; Richmond, H.A. 1975. Striped ambrosia beetle, *Trypodendron lineatum* (Oliv.). Pages 246-249 *in* M.L. Prebble (Ed.), Aerial control of forest insects in Canada. Environ. Can., Ottawa, Ont.

Moss, S. 1988. Black army cutworm and Ambush® - Nelson Forest Region. B.C. Min. For. and Lands, Victoria, B.C. Pest Mgmt. Prog. 7(2): 20.

Prebble, M.L. (Ed.). 1975. Aerial control of forest insects in Canada. Environ. Can., Ottawa, Ont.

Ross, D.A.; Ilnytzky, S. 1977. The black army cutworm in British Columbia. Pac. For. Res. Cent., Victoria, B.C. Inf. Rep. BC-X-154.

Shore, T.L. 1985. Ambrosia beetles. Pac. For. Res. Cent., Victoria, B.C. For. Pest Leaflet 72. 4 p.

Unger, L.S. 1986. Spruce budworms in British Columbia. Pac. For. Cent., Victoria, B.C. For. Pest Leaflet 31. 4 p.

Wood, C.S. 1978. Forest tent caterpillar. Pac. For. Res. Centre, Victoria, B.C. For. Pest Leaflet 17. 3 p.

Wood, C.S.; Van Sickle, G.A. 1982-1990. Forest insect and disease conditions, British Columbia and Yukon for years 1982 to 1988. Pac. For. Cent., Victoria, B.C. Inf. Reps. BC-X-239, -246, -259, -277, -287, -296, -306, -318, and -326.

Chapter 7

Research on the Spruce Budworm, *Choristoneura fumiferana*

C.J. Sanders

The history of research on the spruce budworm, *Choristoneura fumiferana*, reflects the forest industry's level of concern about the damage caused by the budworm. Although evidence of outbreaks has been traced back to the eighteenth century (Packard 1890 [quoted by Swaine and Craighead 1924]; Blais 1968), the first major investigation of the insect's biology was not carried out until the 1910 to 1923 outbreak in Quebec and New Brunswick (Swaine and Craighead 1924). Much basic information on the biology of the insect and the degree and extent of the damage it caused was gathered at that time, but forestry was still in the era of selective logging, and balsam fir, the tree most vulnerable to the budworm, was still considered a weed species. Therefore, there was little concern over the damage the insect caused. Once that outbreak declined, interest in the budworm disappeared, and little further research was carried out until the resurgence of the insect in the 1940s. By this time utilization of the forests had grown dramatically. The anticipated damage was considered unacceptable, and as a result, extensive spray operations began, which have continued to the present. Major research programs were initiated in New Brunswick and Ontario and across the border in the Adirondacks and in Maine.

In New Brunswick, research was centered on the watershed of the Green River under the direction of R. F. Morris. The Green River Project was aimed at explaining the causes of the rise and fall of the outbreak and pioneered the use of life tables in the study of insect population dynamics. The work culminated in a monograph (Morris 1963), which is still a valuable source of information on the biology of the spruce budworm and the factors affecting budworm survival. Elsewhere, the Chemical Control Research Institute was established to improve the effectiveness of chemical insecticides and the Pathology Research Institute to investigate the possibility of using insect pathogens as control agents. Research at the Black Sturgeon Lake and Cedar Lake field stations in northwestern Ontario was noteworthy for the release of numerous species of parasites, mostly western Canadian species, although none of them became established.

The perception of spruce budworm population dynamics at that time was largely influenced by the Green River Project and by the interpretation of the historical records of outbreaks obtained from defoliation maps (Brown 1970) and the analysis of tree-ring growth (Blais 1968). The conclusion was that endemic populations were held in check by predators and parasites and that they were released by several consecutive years of favorable weather. It was hypothesized that favorable weather, in conjunction with extensive areas of mature spruce–fir forest, allowed budworm populations to "escape" the regulating factors and to increase rapidly to epidemic levels. Once populations had escaped in this fashion, it was concluded that outbreaks persisted until extensive tree mortality began, which caused populations to collapse from lack of food.

Following the collapse of budworm outbreaks across eastern North America in the 1950s and 1960s, the research effort also waned, but a nucleus of researchers remained, especially in New Brunswick, where the outbreak persisted. Spruce budworm populations resurged in the late 1960s and 1970s, and the development of new outbreaks was viewed with considerable alarm by the forest industry, which was now using close to the allowable annual cut. Research thinking began to focus more on the integration of spruce budworm control and forest management, a process that has continued and gained momentum up to the present. In an attempt to incorporate our knowledge of budworm biology and its impact on the forest into forest management planning, a computerized model was developed jointly by the Institute of Animal Resource Ecology of the University of British Columbia, under the leadership of C. S. Holling, and Forestry Canada. The structure of the model is discussed in some detail by Régnière and Lysyk (Chapter 8). In summary, it implied that budworm populations in immature forests were held at endemic levels by predation, but that as the trees matured and the food supply increased, the budworm population multiplied faster than the rate of predation, which inevitably led to outbreaks. These persisted until tree mortality was extensive, which caused budworm populations to crash and returned the cycle to the endemic level with populations on immature trees regulated by predation. Qualitatively, this model mimicked reality, and although it is now considered to be based on a false interpretation of the budworm's population dynamics, it is a landmark because it was the first integration of budworm biology and forest dynamics. The model was used as a basis for a review of alternative strategies of forest management in New Brunswick carried out by G. L. Baskerville, which identified the impending shortfall in wood supply due to the budworm in New Brunswick and subsequently in Nova Scotia. Modifications of this model, such as WOSFOP and its derivatives, are now important components in forest management planning in most provinces. Research on the impact of the budworm begun at this time has led to the development of techniques for integrating budworm impact into forest management planning, which are being implemented in New Brunswick (MacLean 1990).

Also in the early 1970s, a major study was begun on the dispersal of female spruce budworm moths. Observations by D. O. Greenbank showed that dispersal was not confined to meteorological events such as the passage of cold fronts, as had previously been thought, but that it occurred every night and involved most of the female moth population. The importance of this, both in population dynamics of the budworm and as a potential method of controlling the budworm by interfering with the dispersal process, led to an integrated research program between 1974 and 1976, which incorporated the use of radar and sophisticated meteorological equipment to monitor budworm movement in the airspace. The results showed that female budworm moths, carrying about half of their egg complement, emigrated every night. On average, they flew about 20 to 50 km downwind, but under favorable conditions they could be carried several hundred kilometres, sufficient to reach Newfoundland from New Brunswick or Cape Breton (Greenbank et al. 1980). These dispersing moths are a tempting target for an aerial control operation, but it is questionable whether the numbers of moths involved have a significant impact on the course of an outbreak. This, and the logistical and environmental problems involved in such a spray operation, have prevented the technique from being developed operationally, as is explained by Kettela (Chapter 10).

The continued increase in the scale of the outbreaks in the 1970s and its potential impact on the forest economies of Canada and the United States led to the two countries signing a memorandum of understanding in 1976 to engage in a joint research program on the spruce budworm and its sibling, the western spruce budworm, *C. occidentalis*, which was causing concern in the west, particularly in the United States. This joint effort came to be known as the CANUSA program and ran from 1977 to 1984. It resulted in a considerable expansion of funding and research in the United States, although in Canada the research effort remained at much the same level as before. CANUSA's impact on the ongoing Canadian research program was limited, therefore, but it generated considerable interaction between the two countries and led to many cooperative research projects, such as acceleration in the development of *Bacillus thuringiensis* (Bt) to an operational level and the use of sex pheromone traps for monitoring budworm populations. It also generated numerous symposia, culminating in a joint research symposium that reviewed progress up to 1984 on both species of spruce budworm (Sanders et al. 1985). In addition, it prompted Forestry Canada (then the Canadian Forestry Service) to review its own budworm research program. A task force was convened and in its subsequent report it emphasized the need for research directed toward the needs of forest managers.

The new awareness of the effect of the spruce budworm on forest management focused attention on the impact of the budworm on tree growth and wood production, and the 1970s saw the beginning of intensive, detailed studies of budworm impact on tree growth in the Maritimes. Increasing concern over the use of chemical insecticides, and unpredictable results with Bt led to more detailed studies on how to improve the efficacy of insecticidal applications. One consequence of this was the formation of the Spray Efficacy Research Group, which has provided valuable information leading to the improved targeting of insecticides and the reduction in drift. Another consequence has been long overdue studies on the behavior of the budworm larvae and how they come into contact with the insecticides, which is described in greater detail by Nigam (Chapter 9).

The end of the CANUSA program coincided with a general decline in spruce budworm populations in both the east and the west. This has resulted in major reductions in the research effort in the United States. However, in Canada, it has been recognized that to improve our ability to cope with the budworm, we must be able to predict when and where outbreaks will occur and how we can interfere with the process. The key to this is an understanding of budworm population dynamics. In the early 1980s, Royama presented his conclusions from a reanalysis of the Green River data. This, coupled with more recent population data from New Brunswick, led him to conclude that the "boom and bust" theory of Morris and Blais, which was the foundation of the Holling model, was erroneous, and that in fact populations oscillate on a 35- to 40-year cycle, regardless of host tree maturity or weather (Royama 1984). Royama was unable to identify the mechanism driving this cycle, but concluded that the key lay in the survival of the larger larvae. He therefore proposed a detailed field study to determine what factors cause larval mortality. As a result, a series of plots have been established in New Brunswick, Quebec, and Ontario, where intensive sampling has been carried out each year, and the larvae are reared to identify the critical factors driving the cycle. The philosophy behind this work, and the results, are discussed in greater depth by Régnière and Lysyk (Chapter 8).

Future Trends

In 1987, Forestry Canada convened a committee under the chairmanship of D. MacLean to review its spruce budworm research program. The review committee recognized that computer-based geographic information systems for handling inventory and site data will offer improved opportunities for incorporating information on spruce budworm into decision-support systems for the forest manager. Such systems can range from short-term plans that involve hazard rating to prioritize stands for harvesting and/or protection through to long-term strategies to reduce budworm numbers and/or to ameliorate damage. The development of such systems will require greater coordination and consultation by Forestry Canada research staff with provincial and industrial foresters, such as is already occurring in New Brunswick, not only to obtain the necessary inventory and growth data, but also to ensure that the output is relevant and practical. The availability of impact data will be crucial to such an exercise, and each region will have to ensure that the information will be available in the required form and detail when necessary. Predictions of when and where damaging populations of budworm will occur will also be crucial. This need can be partially addressed by improved sampling techniques for low-density populations, but the ultimate answer lies in our understanding of population dynamics. Knowing what causes budworm populations to fluctuate will allow us to build realistic process models, so that we can make long-term predictions and also design interventions that will regulate budworm densities.

The need for chemical insecticides will remain, but dependency on the biological insecticide Bt will increase. This will require further sophistication in application techniques to target the insecticide efficiently and increase efficacy. The development of new and improved alternatives must continue. A greater understanding of population dynamics will open up opportunities for enhancing the effects of natural control agents and will improve the chances of successfully introducing new agents. The search for control agents for introduction should be extended to include untapped sources, such as Asia.

Underlying these specific research requirements, there is a need for the development of strategies for managing the spruce budworm as one component of the spruce–fir forest. An important step in this process is the development and evaluation of budworm control tactics in an operational context by establishing trial demonstration areas. Ultimately, the various methods of manipulating and controlling the insect must be integrated with forest management practices into integrated pest management systems.

References

Blais, J.R. 1968. Regional variation in susceptibility of eastern North American forests to budworm attack based on history of outbreaks. For. Chron. 44: 17-23.

Brown, C.E. 1970. A cartographic representation of spruce budworm (*Choristoneura fumiferana* [Clem.]) infestations in eastern Canada, 1909-1966. Can. For. Serv., Ottawa, Rep. 1263.

Greenbank, D.O.; Schaefer, G.W.; Rainey, R.C. 1980. Spruce budworm (Lepidoptera: Tortricidae) moth flight and dispersal: New understanding from canopy observations, radar, and aircraft. Mem. Entomol. Soc. Can. 110, 49 pp.

MacLean, D.A. 1990. Impact of forest pests and fire on stand growth and timber yield: Implications for forest management planning. Can. J. For. Res. 20: 391-404.

Morris, R.F. 1963. The dynamics of epidemic spruce budworm populations. Mem. Entomol. Soc. Can. 31, 332 pp.

Royama, T. 1984. Population dynamics of the spruce budworm *Choristoneura fumiferana*. Ecol. Mono. 54: 429-462.

Sanders, C.J.; Stark, R.W.; Mullins, E.J.; Murphy, J. (Eds.) 1985. Recent advances in spruce budworms research: Proceedings of the CANUSA Spruce Budworms Research Symposium. Bangor, Me. 1984. 527 pp.

Swaine, J.M.; Craighead, F.C. 1924. Studies on the spruce budworm (*Cacoecia fumiferana* Clem.). Part 1. A general account of the outbreaks, injury and associated insects. Can. Dept. Ag., Bull. N.S. 37: 1-27

Chapter 8

Population Dynamics of the Spruce Budworm, *Choristoneura fumiferana*

JACQUES RÉGNIÈRE AND TIMOTHY J. LYSYK

Introduction

The economic impact of the spruce budworm, *Choristoneura fumiferana*, can be minimized by forecasting when, where, and how much damage will occur and planning forestry operations accordingly, as well as by developing control strategies that minimize outbreak intensity and frequency, or terminate and prevent them. These procedures require an accurate understanding of the behavior of spruce budworm populations.

Understanding of the spruce budworm/forest system must not only mimic the system's behavior, it must also provide predictive understanding of the way it will react to management interventions. Royama (1971) discussed the various sources of error encountered in the elaboration of models describing natural systems. One may add that the risk of control failure (noneffect or worsening of a problem) is greater when a wrong model is used in decision making than when random intervention is practised.

The spruce budworm has characteristics similar to those of cyclical outbreak species (Wallner 1987). It is found in a variety of habitats, feeds on several species of conifers, and is not very closely tied to a single plant species. The adult is short-lived, mobile, but is not a typical migratory species as it flies during, rather than before, reproduction. Eggs are laid in masses and the distribution of larvae is highly clumped. This has implications for regulation by natural enemies and sampling. Weather is believed to have a strong influence on its population performance and natural enemies are important in its population dynamics.

Populations of cyclical species are often regulated by two-species interactions, such as predator–prey, parasite– or disease–host systems (Hassell 1978; Anderson and May 1981). In such cases, populations go through more or less regular cycles or oscillations. However, other forms of regulation exist, such as in multiple-equilibrium systems (Berryman 1982). Insects in this category are normally kept in check in a latent state, and their populations are triggered into outbreaks because of a limited capacity of the latent regulator to respond to changes in insect density. The determination of regulation mechanisms is not straightforward, and is often based on hypotheses made from limited data, particularly for periods of low density during which data are difficult to obtain. The population dynamics of the eastern blackheaded budworm, *Acleris variana*, is an example of contrary interpretations of the same data (McNamee 1979; Berryman 1986). This is also the case for spruce budworm.

Two basic questions are addressed by research on spruce budworm population dynamics: the nature of population changes, and mechanisms underlying these changes. The present section is an overview of the field of spruce budworm population dynamics, and a discussion of its integration in the management of budworm-prone forests.

Current State of Knowledge

Nature of Population Changes

There is general agreement that outbreaks of the spruce budworm are cyclical. However, much debate has focused on outbreak frequency and on the behavior of populations between outbreaks. Information available on budworm population changes comes from three different sources of historical records: tree-ring analyses, semiquantitative defoliation or tree mortality records, and more accurate insect survey data. This information must be pieced together and interpreted, although it is difficult to determine underlying causes of general population trends from such records (Régnière 1985).

Radial-growth analysis has been the source of records of outbreaks dating from the 1700s to the early 1900s (e.g. Blais 1981). Although the methods used to obtain and interpret tree-ring information are well developed (Blais 1962), there are basic problems in deriving insect population trends from them. Relationships between density, damage, and radial growth of trees are complex. Changes in population levels below or above certain thresholds cannot be detected by these methods. Also, sampled trees are survivors which would have suffered less damage. As a result, information from tree-ring analyses probably underestimates the extent and severity of past outbreaks. Mild outbreaks may go completely undetected if defoliation is not sufficiently severe. Thus, these records provide little more than presence or absence information about outbreaks.

Defoliation and tree mortality maps, such as those produced by government services (e.g. Hardy et al. 1986), form the bulk of historical information since the mid-1900s. These are also relatively insensitive to population changes occurring at medium or low densities. Changes detected by these methods appear to be very abrupt and give the impression that spruce budworm populations have distinct endemic and epidemic states.

A limited number of truly quantitative population surveys are available for much of New Brunswick since the early 1950s, and detailed information exists for a few specific areas. When presented on an arithmetic scale, these data tend to support a "boom and bust" view of population behavior, with a prolonged endemic period during which populations change little. Part of this impression of abrupt population changes results from the use of standard foliage sampling techniques, which are unreliable when density is very low (Lysyk and Sanders 1987). Nevertheless, when densities are plotted on a logarithmic scale, which places similar emphasis on changes of the same magnitude occurring at low (e.g. a tenfold increase from 1 to 10 larvae/m^2) and high density (e.g. from 10 to 100 larvae/m^2), any apparent endemic period vanishes. Likewise, sudden increases or drops in populations tend to lose importance, and a gradual buildup and decline in population density is emphasized (Fig. 1).

Royama (1984) recognized the approximate nature of historical data and merged available tree-ring analyses, other historical records from New Brunswick and neighboring Quebec, and regional survey data. He concluded that spruce budworm populations have oscillated more or less regularly at a frequency of 30 to 40 years for the past 200 years (Fig. 2). Using more or less the same data, Blais (1983) and others concluded that spruce budworm outbreaks have been

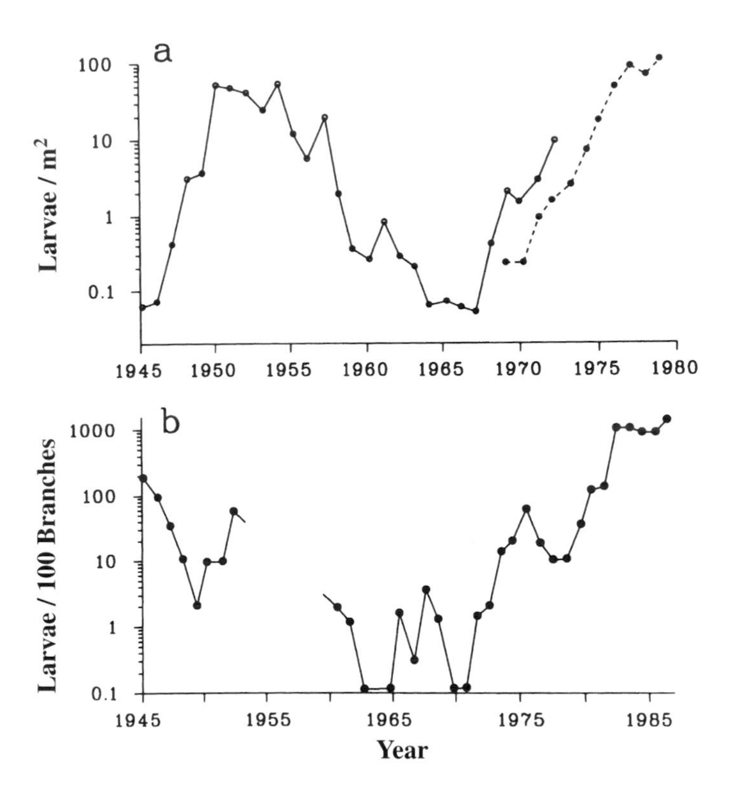

Figure 1. Examples of spruce budworm, *Choristoneura fumiferana*, population density changes in two areas over the past 45 years. (a) Green River, New Brunswick (redrawn from Royama 1984). (b) Black Sturgeon Lake, Ontario (data of Sanders 1988).

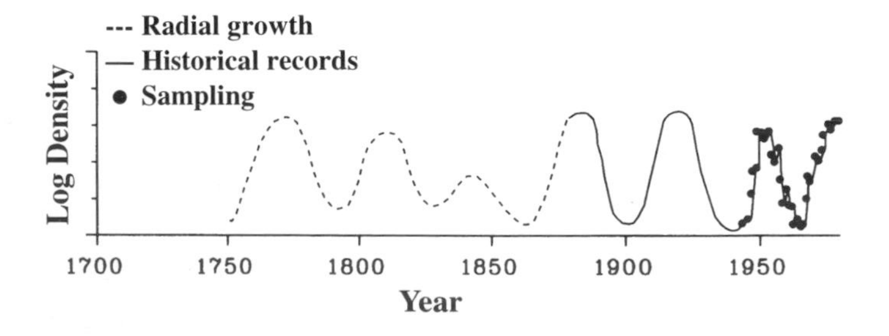

Figure 2. Reconstruction of the spruce budworm, *C. fumiferana*, population trend over the last 250 years in northern New Brunswick (redrawn from Royama 1984).

increasing in frequency over the past century as a result of increasing conifer contents in eastern North American forests. The different opinions of these authors stem from different assumptions concerning the basic processes underlying population changes. Royama's (1984) interpretation allows for mild budworm outbreaks which would leave little or no trace in tree rings (Fig. 1a). However, spruce budworm outbreaks often lead to widespread damage and tree mortality, and Blais (1985) concluded that population decline is a direct result of resource depletion. Thus, Blais' interpretation assumes that all outbreaks are recorded in tree-ring records. Blais also considers outbreaks occurring in different regions of continuous conifer forests as distinct. This tends to multiply the number of outbreaks. Royama sees these outbreaks as generalized phenomena occurring more or less synchronously over large geographical areas. Thus, any increase in frequency of outbreaks is more difficult to substantiate.

Forest composition has been changing as a result of forestry practices, and the overall extent of outbreaks has changed (Blais 1983). Some of the basic processes relating to spruce budworm population dynamics and the forest environment may not be stationary (Fleming 1985), as assumed by Royama (1984).

Interpretation of the broad geographical patterns of spruce budworm outbreaks is also clouded. Many workers have maintained that outbreaks start in epicenters, or foci, from which they spread by moth migration or even larval dispersal into neighboring budworm-free forest (Hardy et al. 1983; Sippell 1984; Blais 1985). There has been disagreement about the nature and location of epicenters, but defoliation and tree-mortality maps tend to support this view. Many workers also believe that an apparent west–east pattern in the spread of spruce budworm outbreaks indicates that spread is at least partly due to downwind migration of moths (Greenbank 1957; Blais 1985). However, as pointed out by Royama (1984) and others, dispersal alone cannot account for the rate of spread. Most authors also believe that moth dispersal can trigger outbreaks only in stands that are ready—"readiness" meaning different things to different people.

Hardy et al. (1983) put forth their version of the epicenter theory based on an analysis of defoliation maps gathered during the most recent budworm outbreak. This theory is particularly interesting because it emphasizes regional differences in the timing, duration and severity of spruce budworm outbreaks, and attempts to interpret these differences ecologically. According to these authors, epicenters, or limited areas where spruce budworm damage was first noticed, have consistently been located in the relatively low-lying and mild Great Lakes–St. Lawrence forest corridor. There, conifer stands are discontinuous and usually of a mixedwood nature, with an important white or red spruce component. Their phytosociological and climatic makeup is in sharp contrast with the stand types presumed by earlier workers to be the foci of spruce budworm outbreaks (Blais 1959). It is significant that damage to trees in these stands is limited and results in little mortality, because the densities reached by the insect are not as extreme and populations decrease sooner there than elsewhere. Epicenter populations would spread, through moth dispersal, to neighboring forests to the north and south, triggering a chain reaction, much as suggested by Greenbank (1957). Spreading outbreaks from adjacent epicenters would coalesce and turn into widespread epidemics. Spread to the south into the Appalachian deciduous forest is of little consequence because of its low conifer content. However, northward spread brings outbreak populations into the boreal forest, which is composed of a high proportion of susceptible balsam fir stands. Southward spread into the high plateau of the Gaspé Peninsula has similar consequences. Populations in these forest types reach very high densities, and consequently damage is severe. Outbreaks also tend to last longer than in epicenter-prone regions, and extensive tree mortality is commonplace. Hardy et al. (1983) also maintain that budworm populations between outbreaks drop to much lower densities in boreal than in meridional forest stands.

Parallel differences in patterns of spruce budworm populations over the province of New Brunswick were observed (Clark and Holling 1979, their fig. 3; Royama 1984, his fig. 2). The red spruce belt, in the central part of the province, is ecologically similar to the coniferous stands within much of the Great Lakes–St. Lawrence forest region (Canada Committee on Ecological Classification 1989). Populations there showed only a slight rise and fall pattern from 1950 to 1980. Those farther to the north and south, in harsher upland areas of the province, underwent more pronounced oscillations. Hardy et al. (1983) interpreted such geographical differences in behavior on the basis of bioclimatic zonation concepts (Cook 1929; Huffaker et al. 1976; Knight and Heikkenen 1980). Presumably, population regulation in the climatically favorable epicenter-prone region is achieved through interactions with natural enemies, while climatic and resource limitations are the factors normally operating on populations in the harsher boreal forests. However, sufficiently detailed information to substantiate or invalidate these hypotheses is mostly lacking.

Other authors suggested that epicenters may be nothing more than the first sites in which budworm populations show recognizable signs of uprise in a given region throughout which they are rising in close synchrony (Stehr 1968; Royama 1984). Greenbank (1957) recognized that increases in spruce budworm populations before the 1949 outbreak in northern New Brunswick were general over much of the area and were not accompanied by significant moth invasions. The existence and nature of epicenters, and their development and spread, have important pest management implications. However, these issues can only be resolved through adequate understanding of the mechanisms underlying population changes.

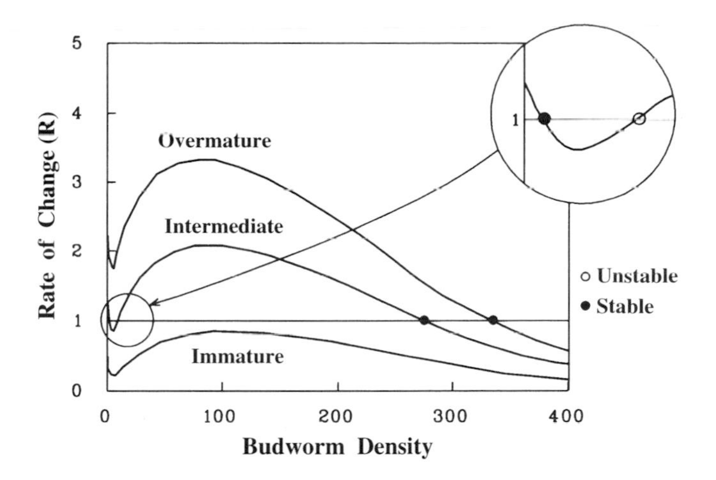

Figure 3. Growth rate curves for spruce budworm, *C. fumiferana*, in different stand types, illustrating the predator pit (insert) of Morris (1963) and multiple equilibria that it would lead to (redrawn from Clark et al. 1979).

Mechanisms of Population Changes

The most extensive set of data available on spruce budworm population dynamics until the early 1980s was gathered in the Green River watershed of New Brunswick from 1947 to 1959. A limited amount of work on low-density budworm populations was continued in later years (Miller and Renault 1976; 1981). There are now two major schools of thought on the topic of budworm population dynamics, both of which have made extensive use of the Green River data.

The philosophy behind the Green River study was that while population events are affected by many factors, they are largely determined by a few key factors (Morris 1959). Variations in survival rates of major age-classes were explained by establishing relationships with natural enemies (parasites and predators), resource availability, and weather. Mott (1963) summarized this analysis in his table 15.1. He found no relationship between egg survival (S_e) and any of the variables examined. Survival of small larvae (S_s), from egg hatch to the establishment of feeding sites in spring, was a decreasing function of average tree diameter, population density, and cumulative defoliation. These findings suggest that there is competition among small larvae, probably during establishment of feeding sites, in severely damaged stands. Survival of larger larvae (S_l) (from peak fourth instar to pupation) was related to stand isolation, weather, phenological synchrony between larvae and shoots, as well as population density.

At densities observed during the Green River study, the relationship between density and S_l was increasing. Watt (1963) hypothesized that predators with a limited numerical response may be responsible for this relationship. He also reasoned that survival should decrease at still higher densities, as a result of severe food shortage, although this was not observed. Extreme weather conditions (i.e. too cold and wet, or too hot and dry) had an adverse effect on S_l, but parasitism had very little impact. Pupal survival (S_p) was directly related to mean maximum air temperature during the period of pupation, and pupal size (an index of fecundity) was a decreasing function of larval density and defoliation history. Mott also showed that variation in generation survival (S_g, from egg to emerged adult) was due mostly to variation in S_l. This led him and his coworkers to conclude that S_l was the major determinant of year-to-year fluctuations in population levels.

Morris (1963) developed a simple key-factor model that related fluctuations of mean maximum air temperature during the period of larval development and current population density (N_t) to the density of the next generation (N_{t+1}). He used this model to simulate population trends from 1925 to 1959. The model produced a gradual increase in population density leading to the outbreak of 1947. Because the density dependence in this model is first-order (meaning that changes in density are related to population density of the current generation only), a decline at the end of an outbreak required either several years of bad weather, or stand mortality. This simulated outbreak was then reduced by high density and poor weather conditions from 1955 to 1959. Royama (1984) suggests that weather did not follow this particular pattern over the entire province of New Brunswick during the Green River period, or during previous outbreaks, although populations generally underwent the same large-scale fluctuations.

Morris himself questioned his regression model's realism. He argued that the effect of natural enemies (especially bird predators) would decrease the rate of population change at intermediate densities as a result of sigmoid functional responses. This, in turn, would make the N_{t+1} to N_t relationship nonlinear and would create a stable, lower equilibrium point around which low-density populations would be held between outbreaks (Fig. 3). Because of a presumed limited capacity of birds to respond numerically to changes in the abundance of any particular prey species, they would constitute an imperfect (saturable) regulator. Thus, increases in prey abundance resulting from high larval survival (e.g. weather) or moth invasion, may not be matched by higher predation and could lead to a release from regulation. Once a population was released, of course, exponential growth would follow, and density would become limited by resource availability. This is a hypothesis that, for the first time, established a biological basis to the boom and bust view of spruce budworm population dynamics.

Morris' ideas were incorporated in a simulation model that accommodated most of the contemporary hypotheses about spruce budworm population dynamics (Jones 1977; Holling et al. 1977). Of particular importance in the structure of this model was the explicit formulation of a close relationship between stand structure, maturity, and larval survival. Bess (1946) and Blais (1952) maintained that this relationship was mostly the result of balsam fir flowering, which enhanced spruce budworm survival by providing an abundance of overwintering sites and a favorable source of succulent food in the spring before the burst of vegetative buds. Blais, in particular, hypothesized that the occurrence of heavy flowering in mature and overmature balsam fir stands was the trigger of spruce budworm outbreaks. However, the Green River data did not support these hypotheses (Greenbank 1963). Rather, it was assumed that a relationship existed between the total surface area of host-plant foliage in a stand and the survival of small larvae during the spring dispersal episode, although no direct evidence of stand-type influence on small larval survival seems to exist (Royama 1984). Under this assumption, increasing survival of small larvae in maturing stands became the driving variable that determined the basic frequency of outbreaks. Indeed, as larval survival increased, the distance between lower equilibrium and release densities decreased, making the eventuality of an outbreak more and more likely (Fig. 3). Weather, which was presumed by Morris to be the key release factor, became a secondary modulator which could only hasten (or delay) the inevitable. In fact, as it stands now, the double-equilibrium theory of spruce budworm population dynamics is essentially deterministic, as it implies that outbreaks are unavoidable consequences of forest maturation (e.g. Clark et al. 1979; Casti 1982).

One of the major arguments against this view of spruce budworm biology is that the interval between outbreaks in a given region has been considerably more variable than the double-equilibrium theory can account for. In addition, outbreaks of the insect have subsided, over large expanses of territory and under many different circumstances, without widespread foliage depletion or tree mortality, particularly at the stand level.

Royama (1984) reinterpreted the Green River Project life tables using a different analytical approach. He emphasized the gradual nature of spruce budworm population buildups and declines, as well as the amount of synchrony in population trends over vast expanses of territory. At the onset, he considered it likely that they were due to common (universal) population regulation processes, resembling such high-order, density-dependent regulation as found in predator–prey, parasite–host systems (Royama 1977). In such systems, population density in a given generation (N_{t+1}) is not only a function of current density (N_t), but also depends on the density of past generations (N_{t-1}, N_{t-2}, ...) because of a delay in the response of major mortality factors to changes in insect density. The simplest function that can produce cyclical density oscillations is a linear and second-order density-dependent process (Royama 1977):

$$N_{t+1} = \alpha \bullet N_t + \beta \bullet N_{t-1} + \xi_t \quad (1)$$

where the term ξ_t represents all density-independent sources of variation in N_{t+1}. Parameters α and β represent density-dependent factors, and determine the basic length and amplitude of population cycles (Fig. 4a). This formulation allows for populations to behave differently when at equal densities N_t but at different points in the outbreak cycle.

The expected value, $E(\xi_t)$, represents the suitability of the biotic and abiotic environment, in terms of mortality and natality, which may vary in time (Fig. 4b). Random fluctuations of ξ_t about $E(\xi_t)$, for example, ε_t, occur in response to deviations of conditions, such as weather, from their average. Royama (1977) addressed only the ε_t term because his main interest was with fluctuations in time, and he assumed $E(\xi_t) = 0$. The influence of ε_t on the behavior of N_{t+1} is twofold; it keeps the population from ever settling, and it makes the exact trajectory of the population trend stochastic (Fig. 4b). Thus, cycles become oscillations, and frequency can be determined precisely only by averaging over a long period. In addition, the severity and duration of outbreaks also become stochastic, and depend on exact sequences of events.

Turchin (1990) detected several cases of forest insect populations that seem to fluctuate according to this type of time-lagged density dependence. But his analysis did not find this in the case of spruce budworm. This analysis was done on a very short time series (50 years) relative to the length of one population oscillation (30–40 years). At any rate, it seems pointless to use a simple second-order density-dependent process as a regression model fitted through Box-Jenkins analysis. The real population processes involved are undoubtedly nonlinear and far more complex. Royama used this type of process merely as an analogy, which he arrived at on the basis of the substantial information contained in the Green River Project life tables.

Royama's view of the nature of density changes in spruce budworm population led to two restrictions in the way density and survival data could be looked at. First, density dependence had to be examined with respect to not only N_t, but also to at least N_{t-1}. Second, because populations may have differed in history, data from different plots could not readily be pooled. Indeed, Royama argued that space was no substitute for time in the analysis of such time-series data. In addition, many mortality factors (such as birds) have the ability to redistribute themselves in response to uneven prey distributions. This type of density dependence does not mean density dependence through time, and caution must be exercised when pooling data from different areas. Both these restrictions were violated by earlier workers (e.g. Watt 1963, fig. 10.5).

In high-order, density-dependent systems, the relative importance of the regulator varies: the lowest mortality rates occur during the rising phase, and the highest during the decline. Royama, therefore, looked for patterns in series of age-interval survival estimates, through time, for separate plots, and attempted to attribute these patterns to various mortality and natality causes in the life tables. Only a few of the Green River Project plots were sampled for a sufficiently long period of time, and most of Royama's arguments are based on data from only four plots.

Royama found that neither egg survival (S_e), nor survival of small larvae (S_s), showed obvious trends in time, although S_s was highly variable. Site-related differences in the levels of S_s existed, but did not support the idea that small larvae fare better in more mature stands. However, S_s was low in severely defoliated stands. Survival of larger larvae (S_l) and, to a lesser extent, pupal survival (S_p), showed a clear decreasing trend during the course of the outbreak. Generation survival (S_g), which is the product of $S_e S_s S_l S_p$, showed a smooth, decreasing trend in all plots (Fig. 5a).

The remaining figure in the life tables is the E/M ratio, which is the density of eggs relative to that of empty pupal cases (apparent realized fecundity). It is determined not only by the sex ratio, fecundity, and oviposition activity of locally emerged females, but also by migratory activity of female moths. The E/M ratio did not show trends over time, but was highly variable from year to year (Fig. 5b). Fluctuations were correlated between plots, and Royama showed a direct effect of weather conditions. Mean levels of the E/M ratio were lower in highly defoliated plots. Royama also demonstrated that the E/M ratio was density dependent, and he concluded that moths tended to emigrate from damaged stands.

Royama concluded that spruce budworm population trends were the result of two major components: a slow, smooth oscillation of large larval survival responsible for the rise-and-fall pattern during an outbreak, and fast, random fluctuations in E/M ratio in response to moth migration. This is at odds with the hypotheses of the double-equilibrium theory, in which slow changes in the survival of small larvae (S_s) caused the rise-and-fall trend, and rapid fluctuations in the survival of large larvae (S_l) determined year-to-year changes in density.

According to this new interpretation, population processes underlying the development and recession of outbreaks are high-order, density-dependent mortality factors operating during the later larval stages. A more detailed examination of causes underlying these patterns is made difficult by limitations in the Green River data. Nevertheless, Royama (1984) concluded that the effects of parasitism, predation, food shortage, and weather were insufficient to explain the observed pattern in S_l, and that singly none could account for the observed population oscillation. He suggested that diseases, which were not well quantified in the Green River study, may be involved. The following is a brief discussion of the possible roles of these classes of mortality in population dynamics of the spruce budworm.

Weather—It has been hypothesized that the influence of weather on the survival of spruce budworm larvae was directly responsible for the initiation (Wellington et al. 1950; Greenbank 1956; Morris 1963) or end of outbreaks (Pilon and Blais 1961; Lucuik 1984). Royama (1984) found a strong correlation between weather and only the E/M ratio, which according to him were mostly due to fluctuations in dispersal activity. Weather is therefore a source of stochastic variation from the basic rise-and-fall pattern in population density.

The stochastic nature of the influence of weather does not exclude the possibility that average climatic conditions, as determined by geographical location and topography, play a role in determining the general properties of population oscillations: frequency, amplitude, and mean level. This effect can be exerted directly on the insect population (ξ_t in equation 1) or through its natural enemies (parameters α and β).

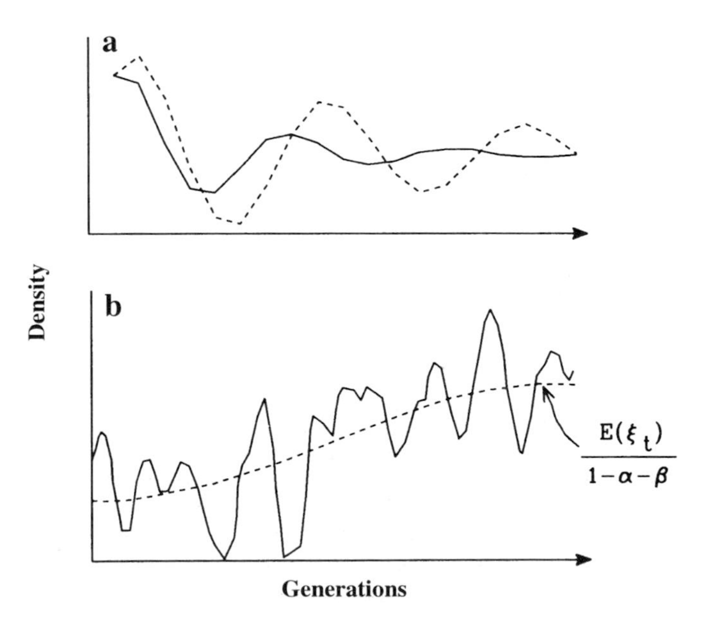

Figure 4. Behavior of equation 1. (a) Response to a single disturbance (ξ_t) at t=0, with different values of α and β. (b) Response to uninterrupted random disturbances (——); equilibrium level corresponding to slowly changing value of $E(\xi_t)$ (.......).

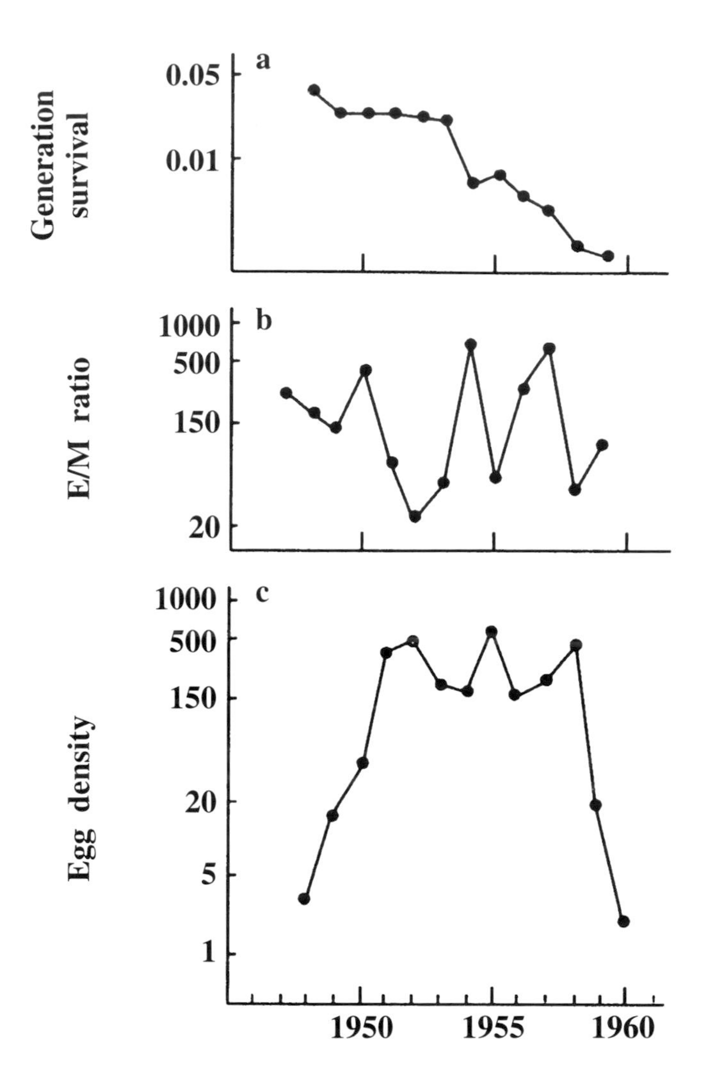

Figure 5. Summary of life table data from Green River plot G4, New Brunswick. (a) Generation survival (from egg to emerging moth). (b) Apparent fecundity (number of eggs laid in stand relative to moths emerged). (c) Resulting population density (per square metre of foliage). Redrawn from Royama (1984).

Food—Resource depletion, in the form of defoliation and tree mortality, is often invoked as the major cause of spruce budworm outbreak collapse (Blais 1985). However, as was observed in Green River and in many other instances, populations of the insect decrease in unison over wide regions despite the patchy nature of stand mortality or severe defoliation. In fact, the survival of large larvae may be higher in defoliated than in undefoliated stands. The Green River data provide good examples of this, where it seems that high larval survival led to severe defoliation. Similar arguments can be made about plant defences as causes of outbreak collapse.

Food availability or quality may impinge on population performance (Bauce and Hardy 1988), but the exact nature and the consequences of such interactions are not clear. Mattson et al. (1991) concluded that natural variation in foliage quality played a minor role in spruce budworm population dynamics, relative to other influences such as natural enemies and weather. However, little is known about tritrophic interactions in this system.

Predators—Royama (1984) argued that birds were too opportunistic in their use of prey to constitute a reliable second-order, density-dependent regulator of this insect. This is supported by experimental data of Miller and Renault (1981). However, recent evidence suggests that birds have a considerable impact on declining spruce budworm populations (Crawford and Jennings 1989; unpubl. data). It is not clear, at this point, how long high larval mortality from these predators is maintained in a low-density population. The importance of other budworm predators is not known.

Parasites—Royama (1984) concluded that a gradual increase in the frequency of parasitism during the Green River outbreak was not large enough to constitute the cause of population oscillations. Nevertheless, several earlier workers have observed considerable increases in the frequency of some parasites, such as *Meteorus trachynotus*, *Winthemia fumiferanae*, *Lypha setifacies*, and *Actia interrupta*, during the final years of outbreaks (Dowden et al. 1948; Jaynes and Drooz 1952; McGugan and Blais 1959; Blais 1960; Miller 1963).

The apparent lack of a sufficient impact of parasites as a major cause of outbreak collapse may be due to inadequate estimates of their frequency. Miller (1955) detailed the methods used in the Green River study. The life cycle of the parasites was considered to consist of 3 distinct periods: (1) attack, (2) development within the host, and (3) exit from the host. Apparent parasitism was calculated from samples collected at several points during a parasite's life cycle. The end of the development period was assumed to occur when the frequency of parasitism decreased. Mortality due to the parasite was calculated on the basis of maximum apparent parasitism and host density at that time. This approach is fine for slow-developing parasites, such as *Apanteles fumiferanae* and *Glypta fumiferanae*, in which attack and exit periods are widely different and percent parasitism reaches a plateau that can be determined easily. However, in most other parasite species, particularly those attacking late larval and pupal stages, development inside the host is brief. This causes two sources of underestimation. First, as pointed out by Royama (1984), a high sampling frequency is required in order not to miss the true peak of parasitism. Even more important is an overlap between the periods of attack and exit from the host. This can lead to gross underestimation of the frequency of parasitism (Van Driesche 1983).

Diseases—Because he could not attribute most of the drop in large larval survival (S_l) during the course of the Green River outbreak to either of the above factors, Royama (1984) hypothesized that a "fifth agent," an unknown complex of viral and protozoan disease, may have been responsible. This may include the microsporidian *Nosema fumiferanae*, which has been underestimated as a mortality factor (Thomson 1958). Viruses have been more sporadic in occurrence, although they seem to be ubiquitous (D. F. Perry, Forestry

Canada, Laurentian Forestry Centre, Sainte-Foy, Quebec, pers. comm.). The Entomophtoralean fungus *Entomophaga aulicae* has caused considerable mortality in localized populations (Perry and Régnière 1986).

Interactions—It is probable that there is not a simple, universal mortality factor regulating populations of the spruce budworm. Rather, this may be accomplished by a group of factors, undoubtedly involving parasites, predators, and diseases which are interacting and competing for budworm as a common resource. Each member of this complex could be limited by specific factors and may therefore play a different role. In fact, it is quite possible that various organisms replace each other in this complex in different habitats or at different points in the life cycle. A simple scenario is as follows. When a spruce budworm population has reached outbreak level, populations of natural enemies exploit this abundant resource as much as possible, given the restrictions imposed by specific limiting factors, such as shelter, alternate hosts (e.g. Maltais et al. 1989) or food sources and environmental conditions. After such a buildup has occurred, any decrease in spruce budworm numbers, in response to fluctuations in environmental conditions or other causes, could lead to a dramatic increase in the impact of these natural enemies. This could easily drive the population into a steep decline, which would stop only after a reduction in the density (or activity) of natural enemy populations as a result of rarified prey.

Biogeographical Considerations

The suitability of environmental conditions to survival and reproduction of a species varies systematically along geographical and topographical axes in response to climatic and edaphic factors. Conditions are conceptually most favorable toward the center of a species' range and degrade toward the periphery. This is expressed by a gradual drop in the intrinsic rate of increase (R), from an optimum in the most favorable part of the range, toward areas where conditions become so unfavorable that populations cannot persist (Fig. 6). This outer edge is ill-defined and may change as a result of fluctuations in the value of key limiting factors. In the case of spruce budworm, conifer stands in the meridional Great Lakes–St. Lawrence forest corridor probably represent the optimum in terms of climate and host plant quality (Hardy et al. 1983). In addition, the increased availability of alternative hosts in these heterogeneous forest stands undoubtedly helps maintain significant populations of important natural enemies during periods when spruce budworm is rare.

In harsher environments, the intrinsic rate of increase of the regulated population is lower; therefore populations recover more slowly from crashes, translating into longer periods of low prey or host abundance. This spells hardship and starvation for regulator populations, which may drop to very low densities or even become extinct locally as a result (Strong 1988). Host or prey populations are then freed to recover in the absence of regulation. This results in slower, but more pronounced population oscillations. Because average environmental conditions are already near the limit for population maintenance, catastrophic deviations are more frequent and severe, and population fluctuations are more erratic (e.g. Pilon and Blais 1961).

The model of Régnière (1984), which simulates regulation of an insect population by a vertically transmitted disease, is used here to illustrate geographical patterns in population changes that can be expected under the above premises. Trends in population density were simulated for an array of sites with different inherent rates of increase along the biogeographical transect illustrated in Figure 6. Populations in the center of the array oscillate at higher frequency and lower amplitude than sites near the edges (Fig. 7a). The lack of cohesion between neighboring populations in this simulation results from the absence of any mechanism by which populations can be synchronized.

There are two mechanisms through which such synchronization of spruce budworm populations is achieved. Weather patterns are normally correlated in space to a degree that diminishes as distances increase. The result of such a correlation may be to synchronize the oscillations of populations that have similar population processes, an idea initially proposed by Moran (1953). Royama (1984) illustrated this idea with simulations using equation 1 as a model. While this is probably true for populations in otherwise identical situations (i.e. for which parameters α and β are the same), it does not hold for populations with even slightly different oscillation frequencies (Fig. 7b).

Migration is another, much more powerful mechanism that links populations geographically and can result in their synchronization (Barbour 1990). A population on the increase sends out migrators. This slows down its rise and accelerates that of neighboring populations. Similarly, a decreasing population receives migrators that can slow the local decline. Under such circumstances, synchronization occurs over the entire array (Fig. 7c). This synchronization is imperfect, however, and the intrinsic oscillatory properties of local populations remain expressed: populations in the most favorable part of the array rise somewhat earlier, reach lower peak densities and drop-off earlier, as do the others. Sites nearer the outer edges of the range show the most severe population fluctuations and prolonged outbreaks.

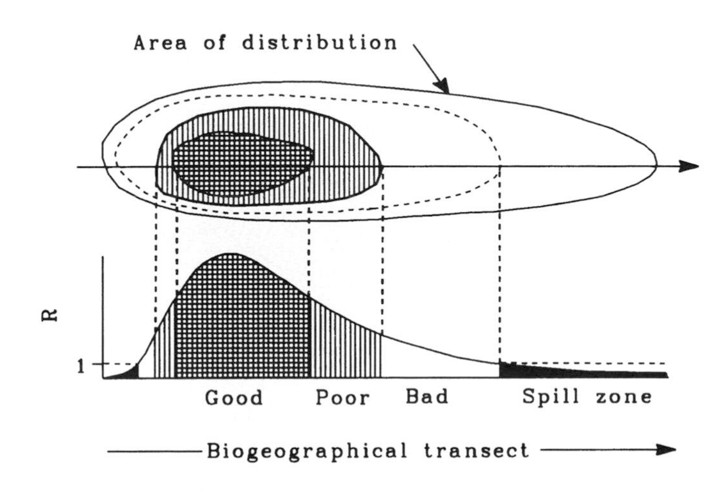

Figure 6. Generalized view of the area of distribution of an organism, and of the intrinsic population growth rates in sites located along a transect across that area.

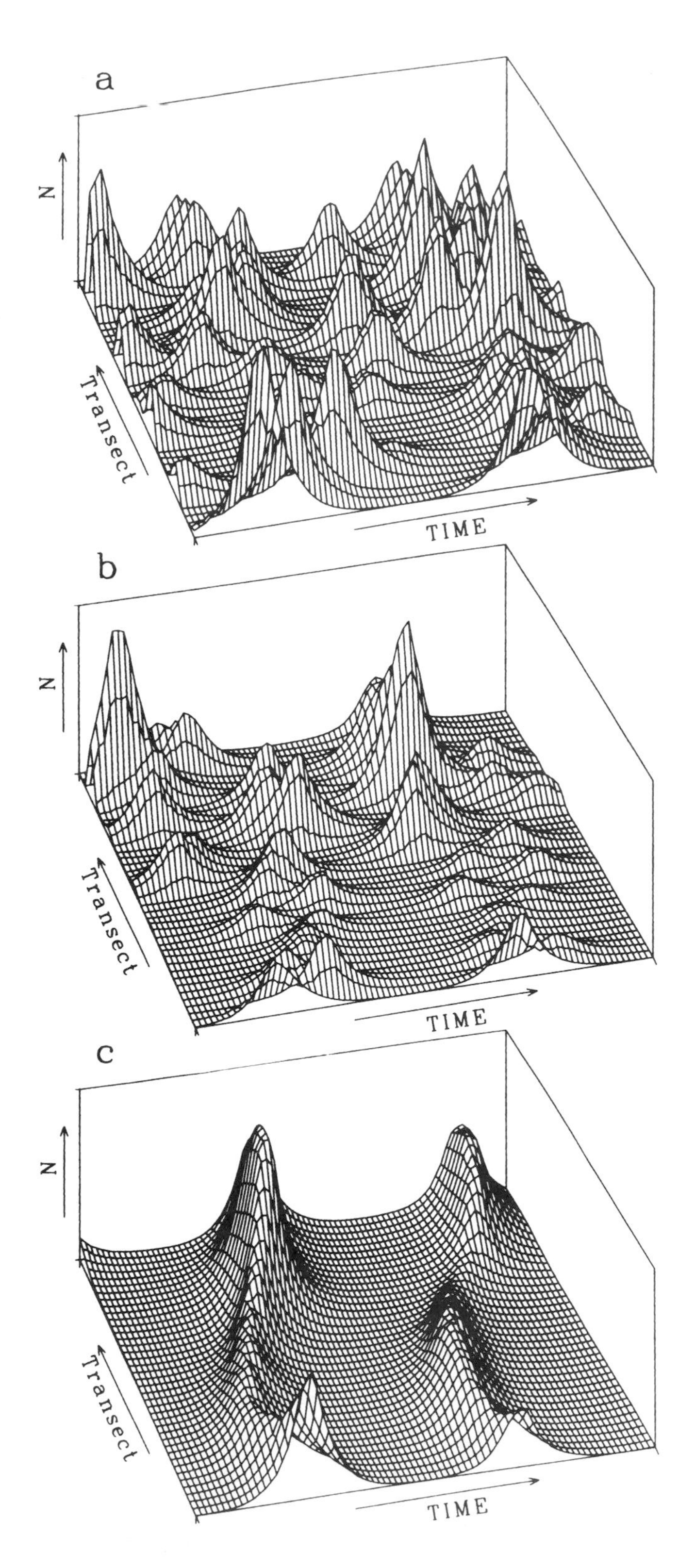

Figure 7. Simulation of population changes in an array of sites located along a biogeographical transect across the area of distribution of an insect. The basic model was described by Régnière (1984). (a) Spatially uncorrelated stochastic disturbances, no moth migration. (b) Spatially correlated disturbances, no moth migration. (c) Spatially correlated disturbances, with moth migration.

These simulations are highly reminiscent of the picture painted by Hardy et al. (1983). However, there is a fundamental distinction. Epicenters in the Hardy et al. hypothesis are physical realities and the cause of outbreaks in outlying areas. In our simulations, epicenters are only the first populations to rise above detection level, much as pointed out by Stehr (1968). In such a case, an epicenter abatement strategy would be pointless, as populations are on the rise, on their own, everywhere.

Population Dynamics and Management of Budworm-Prone Forests

Ideally, management of budworm-prone forests should be based on the integration of a variety of information and decision criteria, ranging from stand inventories to subtle effects of stand structure on population dynamics of the pest. Doing this, however, is a major challenge. Fortunately, these types of activities are readily supported by computer technology, and computer-assisted decision making is becoming an important part of integrated pest management (Welch 1984; Stone et al. 1986; Coulson and Saunders 1987), although no sophisticated systems are yet available for spruce budworm management.

A budworm-forest management decision support system could be centered around a geographical information system (GIS) designed to store, maintain, analyze, and display geographically referenced data (Fig. 8). Forestry GIS data bases currently include spatial information on stands, as well as site-specific statistics such as stand volume, species composition, maturity, topography, and soils. Coupled with growth and yield models, a GIS could be used to update forest inventories and to provide production forecasts and wood flow analyses. With adequate economic and industry-related functions, it may help formulate detailed harvesting, regeneration, and silvicultural plans, taking the geographical location of stands explicitly into consideration.

Spruce budworm management can be done at two levels: foliage protection (crisis management) or population control. The selection of foliage protection methods can be achieved with minimal information about budworm abundance, and models of impact on growth and yield (MacLean and Erdle 1990). Insect abundance and damage information obtained from survey operations, coupled with stand vulnerability indices, provides adequate risk maps. Overlying these with stand-level information, such as standing volume, net worth, or scheduled time of harvest, would form an appropriate basis for forest protection plans (Power 1988). Such a system may facilitate consideration of the costs and benefits of different control options from the points of view of crop protection (Erdle 1989) and environmental impact. Various other aids, such as phenological models (e.g. Régnière 1987), could be used to assist in planning the details of interventions.

Management options that involve forecasting or modifying mid- to long-term spruce budworm population trends require adequate understanding of the mechanisms of population changes. In addition, forecasting implies the availability of early detection methodology to detect rising endemic populations. Data provided by networks of pheromone traps (Allen et al. 1986; Sanders 1988) and adapted sampling

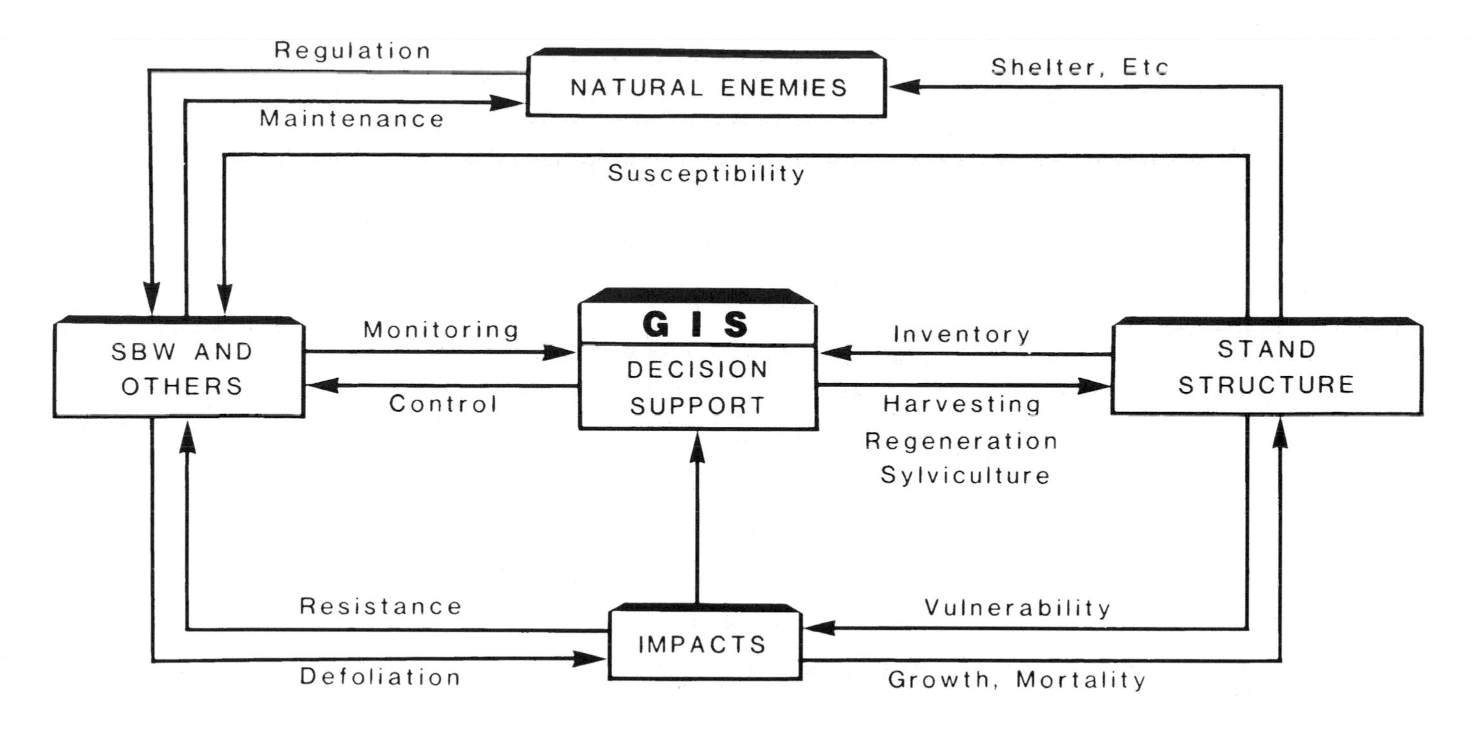

Figure 8. Diagram of a management decision-support system for forests prone to attack by the spruce budworm, *C. fumiferana*.

techniques (Lysyk and Sanders 1987) would be needed. Perhaps the most promising management approaches are aimed at reducing stand suitability to budworm directly through reduced larval survival and oviposition by moths, or indirectly through enhanced action of selected natural enemies. The evaluation of such complex management options requires thorough understanding of the interactions between the stand, the insect, and its enemies. Ongoing research on spruce budworm population dynamics and the biology of the organisms involved is aimed at gathering this knowledge.

References

Allen, D.C.; Abrahamson, L.P.; Eggen, D.A.; Lanier, G.N.; Swier, S.R.; Kelley, R.S.; Auger, M. 1986. Monitoring spruce budworm (Lepidoptera: Tortricidae) populations with pheromone-baited traps. Environ. Entomol. 15: 152-165.

Anderson, R.M.; May, R.M. 1981. The population dynamics of microparasites and their invertebrate hosts. Philo. Trans. Royal Soc. London 291: 451-524.

Barbour, D.A. 1990. Synchronous fluctuations in spatially separated populations of cyclic forest insects. Pages 339-346 *in* A.D. Watt, S.R. Leather, M.D. Hunter and N.A.C. Kidd (Eds.), Population dynamics of forest insects. Intercept. Andover, England.

Bauce, E.; Hardy, Y. 1988. Effects of drainage and severe defoliation on the raw fiber content of balsam fir needles and growth of the spruce budworm (Lepidoptera: Tortricidae). Environ. Entomol. 17: 171-174.

Berryman, A.A. 1982. Biological control, thresholds, and pest outbreaks. Environ. Entomol. 11: 544-549.

Berryman, A.A. 1986. On the dynamics of blackheaded budworm populations. Can. Entomol. 118: 775-779.

Bess, H.A. 1946. Staminate flowers and spruce budworm abundance. Can. Dep. Agric. Bi-Monthly Prog. Rep. 2: 3-4.

Blais, J.R. 1952. The relationship of the spruce budworm (*Choristoneura fumiferana*, Clem.) to the flowering condition of balsam fir (*Abies balsamea* (L. Mill.). Can. J. Zool. 30: 1-29.

Blais, J.R. 1959. Spruce budworm outbreaks and the climax of the boreal forest in eastern North America. Québec Soc. Prot. Plants Rep. 41: 69-75.

Blais, J.R. 1960. Spruce budworm parasite investigations in the lower St. Lawrence and Gaspé regions of Québec. Can. Entomol. 92: 384-396.

Blais, J.R. 1962. Collection and analysis of radial growth data from trees for evidence of past spruce budworm outbreaks. For. Chron. 38: 474-484.

Blais, J.R. 1981. Recurrence of spruce budworm outbreaks for the past two hundred years in western Québec. For. Chron. 57: 273-275.

Blais, J.R. 1983. Trends in the frequency, extent and severity of spruce budworm outbreaks in eastern Canada. Can. J. For. Res. 13: 539-547.

Blais, J.R. 1985. The ecology of the eastern spruce budworm: A review and discussion. Pages 49-59, in C.J. Sanders, R.W. Stark, E.J. Mullins, and J. Murphy (Eds.), Recent advances in spruce budworms research. Proc. CANUSA Spruce Budworms Res. Symp. Bangor, Me. September 16 to 20, 1984.

Canada Committee on Ecological Land Classification (Ecoregions Working Group). 1989. Ecoclimatic regions of Canada (First approximation). Ecological Land Classification Series 23. Sustainable Development Branch. Can. Wildlife Serv., Environ. Can., Ottawa, Canada.

Casti, J. 1982. Catastrophes, control and the inevitability of spruce budworm outbreaks. Ecol. Modelling 14: 293-300.

Clark, W.C.; Holling, C.S. 1979. Process models, equilibrium structures and population dynamics: on the formulation and testing of realistic theory in ecology. Fortsch. Zool. 25: 29-52.

Clark, W.C.; Jones, D.D.; Holling, C.S. 1979. Lessons for ecological policy design. Ecol. Modelling 7: 1-53.

Cook, W.C. 1929. Bioclimatic zonation. Ecology 10: 282-293.

Coulson, R.N.; Saunders, M.C. 1987. Computer-assisted decision-making as applied to entomology. Ann. Rev. Entomol. 32: 415-437.

Crawford, H.S.; Jennings. D.T. 1989. Predation by birds on spruce budworm *Choristoneura fumiferana*: functional, numerical, and total responses. Ecology 70: 152-163.

Dowden, P.B.; Buchanan, W.D.; Carolin, V.M. 1948. Natural control factors affecting the spruce budworm. J. Econ. Enotomol. 41: 457-464.

Erdle, T.A. 1989. Concept and practice of integrated harvest and protection design in the management of eastern spruce-fir forests. Ph.D. Thesis, University of New Brunswick, Fredericton, N.B.

Fleming, R.E. 1985. How should one view the historical record of spruce budworm outbreaks? Pages 136-137 *in* C.J. Sanders, R.W. Stark, E.J. Mullins, and J. Murphy (Eds.), Recent advances in spruce budworms research. Proc. CANUSA Spruce Budworms Res. Symp. Bangor, Me. September 16 to 20, 1984.

Greenbank, D.O. 1956. The role of climate and dispersal in the initiation of outbreaks of the spruce budworm in New Brunswick. I. The role of climate. Can. J. Zool. 34: 453-476.

Greenbank, D.O. 1957. The role of climate and dispersal in the initiation of outbreaks of the spruce budworm in New Brunswick. II. The role of dispersal. Can. J. Zool. 35: 285-403.

Greenbank, D.O. 1963. Staminate flowers and the spruce budworm. Pages 202-218 *in* R.F. Morris (Ed.). The dynamics of epidemic spruce budworm populations. Mem. Entomol. Soc. Can. 31.

Hardy, Y.; Lafond, J.A.; Hamel, L. 1983. The epidemiology of the current spruce budworm outbreak in Quebec. Forest Sci. 29: 715-725.

Hardy, Y.; Mainville, M.; Schmitt, D.M. 1986. A cartographic history of spruce budworm defoliation in eastern North America, 1938-1980. USDA For. Ser. Misc. Publ. 1449. 52 p.

Hassell, M.P. 1978. The dynamics of arthropod predator–prey systems. Princeton Univ. Press. New Jersey.

Holling, C.S.; Jones, D.D.; Clark, W.C. 1977. Ecological policy design: a case study of forest and pest management. Pages 13-90 *in* Proceedings of a conference on pest management. Int. Inst. Appl. Systems Anal. A-2361 Laxenburg, Austria.

Huffaker, C.B.; Simmonds, F.J.; Laing, J.E. 1976. The theoretical and empirical basis of biological control. Pages 41-74 *in* C.B. Huffaker and P.S. Messenger (Eds.), Theory and practice of biological control. Academic Press. N.Y.

Jaynes, L.G.; Drooz, A.T. 1952. The importance of parasites in the spruce budworm infestations in New York and Maine. J. Econ. Entomol. 45: 1057-1061.

Jones, D.D. 1977. The budworm site model. Pages 91-156 *in* Proceedings of a conference on pest management. Int. Inst. Appl. Systems Anal. A-2361 Laxenburg, Austria.

Knight, F.B.; Heikkenen, H.J. 1980. Principles of forest entomology. 5th ed. McGraw-Hill, N.Y. 461 p.

Lucuik, G.S. 1984. Effect of climatic factors on post-diapause emergence and survival of spruce budworm larvae (Lepidoptera: Tortricidae). Can. Entomol. 116: 1077-1083.

Lysyk, T.J.; Sanders, C.J. 1987. A method for sampling endemic populations of the spruce budworm (Lepidoptera: Tortricidae) based on proporation of empty sample units. Can. For. Serv., Sault Ste. Marie, Ontario, Info. Rep O-X-382. 17 p.

McGugan, B.M.; Blais, J.R. 1959. Spruce budworm parasite studies in northwestern Ontario. Can. Entomol. 91: 758-783.

MacLean, D.A.; Erdle, T.A. 1990. Forest inventory updates following insect defoliation: linking stand models with remote sensing of stand condition. Pages 41-51 *in* Division 4, Proceedings of the 19th IUFRO World Congress, Montreal, Quebec (August 1990).

McNamee, P.J. 1979. A process model for eastern blackheaded budworm (Lepidoptera: Tortricidae). Can. Entomol. 111: 55-66.

Maltais, J.; Régnière, J.; Cloutier, C.; Hebert, C.; Perry, D.F. 1989. Seasonal biology of *Meteorus trachynotus* (Hymenoptera: Braconidae) and of its overwintering host *Choristoneura rosaceana* (Lepidoptera: Tortricidae). Can. Entomol. 121: 745-756.

Mattson, J.W.; Haack, R.A.; Lawrence, R.K.; Slocum, S.S. 1991. Considering the nutritional ecology of the spruce budworm in its management. For. Ecol. Manage. 39: 183-210.

Miller, C.A. 1955. A technique for assessing spruce budworm larval mortality caused by parasites. Can. J. Zool. 33: 5-17.

Miller, C.A. 1963. Parasites and the spruce budworm. Pages 228-244 *in* R.F. Morris (Ed.), The dynamics of epidemic spruce budworm populations. Mem. Entomol. Soc. Can. 31.

Miller, C.A.; Renault, T.R. 1976. Incidence of parasitoids attacking endemic spruce budworm (Lepidoptera: Tortricidae) populations in New Brunswick. Can. Entomol. 108: 1045-1052.

Miller, C.A.; Renault, T.R. 1981. The use of experimental populations to assess budworm larval mortality at low densities. Can. For. Serv., Fredericton, N.B. Info. Rep. M-X-115.

Moran, P.A.P. 1953. The statistical analysis of the Canadian lynx cycle. II. Synchronization and meteorology. Australian J. Zool. 32: 302-313.

Morris, R.F. 1959. Single-factor analysis in population dynamics. Ecology 40: 580-588.

Morris, R.F. 1963. The development of predictive equations for the spruce budworm based on key-factor analysis. Pages 116-129 *in* R.F. Morris (Ed.), The dynamics of epidemic spruce budworm populations. Mem. Entomol. Soc. Can. 31.

Mott, D.G. 1963. The forest and the spruce budworm. Pages 189-202 *in* R.F. Morris (Ed.), The dynamics of epidemic spruce budworm populations. Mem. Entomol. Soc. Can. 31.

Perry, D.F.; Régnière, J. 1986. The role of fungal pathogens in spruce budworm population dynamics: frequency and temporal relationships. Pages 167-170 *in* R.A. Samson, J.M. Vlak and D. Peters (Eds.), Fundamental and applied aspects of invertebrate pathology. Wageningen, The Netherlands.

Pilon, J.G.; Blais, J.R. 1961. Weather and outbreaks of the spruce budworm in the province of Quebec from 1939 to 1956. Can. Entomol. 93: 118-123.

Power, J.M. 1988. Decision support systems for the Forest Insect and Disease Survey for pest management. For. Chron. 64: 132-135.

Régnière, J. 1984. Vertical transmission of diseases and population dynamics of insects with discrete generations: a model. J. Theor. Biol. 107: 287-301.

Régnière, J. 1985. Interpreting historical records. Pages 136-137 *in* C.J. Sanders, R.W. Stark, E.J. Mullins and J. Murphy (Eds.), Recent advances in spruce budworms research. Proc. CANUSA Spruce Budworms Res. Symp. Bangor, Me. September 16 to 20, 1984.

Régnière, J. 1987. Temperature-dependent development of eggs and larvae of *Choristoneura fumiferana* (Clem.) (Lepidoptera: Tortricidae) and simulation of its seasonal history. Can. Entomol. 119: 717-728.

Royama, T. 1971. A comparative study of models of predation and parasitism. Res. Popul. Ecol. Suppl. 1. 91 p.

Royama, T. 1977. Population persistence and density dependence. Ecol. Monogr. 47: 1-35.

Royama, T. 1984. Population dynamics of the spruce budworm, *Choristoneura fumiferana*. Ecol. Monogr. 54: 429-462.

Sanders, C.J. 1988. Monitoring spruce budworm population density with sex pheromone traps. Can. Entomol. 120: 175-183.

Sippell, W.L. 1984. Planning now to reduce, postpone or prevent the next spruce budworm outbreak. Pages 57-67 *in* New and improved techniques for monitoring and evaluating spruce budworm populations. USDA For. Serv. Tech. Rep. NE-88.

Stehr, G. 1968. On some concepts in the population biology of the spruce budworm. Proc. Entomol. Soc. Ont. 99: 54-56.

Stone, N.D.; Coulson, R.N.; Frisbie, R.E.; Loh, D.K. 1986. Expert systems in entomology: three approaches to problem solving. Bull. Entomol. Soc. Amer. 32: 161-166.

Strong, D.R. 1988. Parasitoid theory: from aggregation to dispersal. Trends Ecol. Evol. 3: 277-280.

Thomson, H.M. 1958. Some aspects of the epidemiology of a microsporidian parasite of the spruce budworm, *Choristoneura fumiferana* (Clem.). Can. J. Zool. 36: 309-316.

Turchin, P. 1990. Rarity of density dependence or population regulation with lags? Nature 344: 660-663.

Van Driesche, R.G. 1983. Meaning of "percent parasitism" in studies of insect parasitoids. Environ. Entomol. 12: 1611-1622.

Wallner, W.E. 1987. Factors affecting insect population dynamics: differences between outbreak and non-outbreak species. Ann. Rev. Entomol. 32: 317-340.

Watt, K.E.F. 1963. The analysis of the survival of large larvae in the unsprayed area. Pages 52-63 *in* R.F. Morris (Ed.), The dynamics of epidemic spruce budworm populations. Mem. Entomol. Soc. Can. 31.

Welch, S.M. 1984. Developments in computer-based IPM extension delivery systems. Ann. Rev. Entomol. 29: 359-381.

Wellington, W.G.; Fettes, J.J.; Turner, K.B.; Belya, R.M. 1950. Physical and biological indicators of the development of outbreaks of the spruce budworm, *Choristoneura fumiferana* (Clem.) (Lepidoptera: Tortricidae). Can. J. Res. 28: 308-331.

Chapter 9

Response of Spruce Budworm, *Choristoneura fumiferana*, Larvae to Insecticides

P.C. Nigam

Introduction

Research efforts on the response of spruce budworm, *Choristoneura fumiferana*, larvae to insecticides have been increased under the auspices of the New Brunswick Spray Efficacy Research Group (Varty and Godin 1983; Varty and Nigam 1985; Nigam 1987b; Varty and Holmes 1988).

Based on empirical laboratory and field tests, several insecticides were registered for aerial application against spruce budworm larvae (Randall 1969, 1976; Nigam 1971, 1975; Dimond and Morris 1984; Shea and Nigam 1984). In recent years, however, the number of insecticides used in operational protection programs has decreased (Metcalf 1980; Helson 1985; Nigam 1985). Only two insecticides, fenitrothion and *Bacillus thuringiensis* (Bt), were available for use against the spruce budworm in the late 1980s and early 1990s (Nigam 1987b; Carter and Lavigne 1991). To improve the biological efficacy of insecticides, studies on fenitrothion and its formulations were conducted from 1980 to 1987, while studies on Bt formulations have been carried out since 1983.

The effectiveness of aerial application of an insecticide against spruce budworm is expressed as percentage of foliage protected and percent of larval mortality, both of which are largely influenced by the density and size of droplets depositing on the primary target surface (foliage) on or near the larvae (Prebble 1975). The relationship between adequate deposit and efficacy has been expressed as 15–20 droplets per square centimetre on Kromekote® cards or, more recently, mean deposit of 0.5–1 droplets per 1-year-old needle, resulting in approximately 60%–80% larval mortality and 30%–65% foliage protection on average over the years (Randall 1969; Nigam 1980; Wiesner 1984; Varty and Holmes 1988).

The larval response to insecticide contamination of their habitat depends on the characteristics of the droplets, the behavior of larvae of various instars, the host species, and the weather conditions before, during, and after spraying. A basic understanding of how insecticides are transferred to the pest after deposition of various dosages on the host from air or ground application is, however, lacking (Ekblad et al. 1979; Graham-Bryce 1983; Hall 1987).

Spruce budworm larval behavior and growth rate is governed by several intrinsic and extrinsic factors (Wellington 1948a,b, 1949, 1950a,b; Wellington and Cameron 1947; Wellington and Henson 1947; Shepherd 1958; Heron 1965; Eidt 1987). Behavior studies in general show that intrinsic factors like molting, circadian rhythm, and energy requirements (hunger) of larvae have a dominant role in larval activity (Wellington and Cameron 1947; Nigam 1987b). Extrinsic factors like temperature, humidity, and wind speed further modify their activities (Wellington 1950a,b). Little information is currently available on in situ behavior of each instar or on the structure of their microhabitats on different host species that will aid in the planning of operational sprays. This is primarily because the observations on behavior were carried out by naked eye and/or by a hand lens on small and isolated host trees where the larvae were subject to human limitations and interference. To collect qualitative and quantitative data of the larval activities in their natural habitat and without observer interference, a new electronic observation system was developed (Nigam 1987a).

The overall objective was to improve efficacy and efficiency of aerial application of fenitrothion and Bt against spruce budworm larvae by (1) studying the behavior of larvae and droplet characteristics in their microhabitat after experimental or simulated aerial applications under field and laboratory conditions, and (2) elucidating the dose transfer mechanism from initial deposits.

Larval Behavior and Microhabitat on Balsam Fir at Time of Spraying

In New Brunswick, spruce budworm larval populations on balsam fir are mostly in the late third and early fourth instars during the first application of fenitrothion spray. Second applications of fenitrothion are carried out at a minimum interval of 5 days, when most larvae are in the fourth and early fifth instars. Bt is usually sprayed only once, when larvae are in late fourth and early fifth instars (Nigam 1986). Video records of budworm behavior and microhabitat at various stages of larval development were made using a surgical microscope, a time-lapse video cassette recorder, and video and surveillance cameras. The same equipment was used to study droplet and larval interaction. Weather parameters were recorded on a micrologger. Weather data within the canopy were collected using sensors mounted on a meteorological tower located in the middle of the experimental plot and on host tree branches near larval microhabitats (Nigam 1987b).

Budworm larval microhabitats, at the time of first and second spraying of fenitrothion on balsam fir, consist of various types of silk strands and new needles from growing buds (class 2 to 4); Bt is sprayed when buds are in class 4 to 5 (Dorais and Kettela 1982). For maximum efficacy, insecticide droplets must contact larvae and foliage within the microhabitats on which the larvae are feeding. The thickness of the silk strands varies from 2 to 12 μm and they are used for different functions. The budworm larva usually remains within the confines of the silked area at the base of the developing buds during the first spray of fenitrothion. It spins an inner tube of very fine silk among the new needles within the microhabitat as it and the bud grow. This tube is its living chamber and its foraging activity is confined to new needles. The larva eats the needles from inside the developing buds up to the early fifth instar. It is at all times in touch with the silk through its body hairs (setae) or legs. The silk is its lifeline, functioning as a "road system" within its microhabitat. Changes in tension of the silk strands make larvae aware of the outside environment and intruders. Silk strands are also good collectors of dew, rain, and spray droplets (Nigam 1987b).

Third-instar larvae spin their shelters at the base of developing buds and feed by mining into them. Their microhabitat is a tent-like structure made mainly of silk. As buds grow and become less resinous, the larva enters one of three buds on the balsam fir, anchors silk threads to a second bud, and keeps the third bud in reserve. Because buds and larvae are dynamic and are growing actively at this time of the year, larvae spend a great deal of time keeping the shelter together by joining needles with silk within the microhabitat. Fourth-instar larvae usually form a spindle-shaped structure by spinning two buds together and the larvae live inside the spindle, which is kept intact with webbing until the larva molts. The fifth-instar larva usually lives within a single well-developed bud, by tying together alternate needles from midrib to tip of the needle to form a case-like structure. Feeding takes place on the remaining needles inside the case, and the larva usually remains within the case until it is in late fifth instar. If current-year needles are not available, it will feed on old needles but will not consume new needles which are used for structural support of the shelter. Usually, when needles used for foraging within the case are completely consumed, the larva ventures out of the case to collect needles from the reserve bud. If the reserve bud is consumed when the larva reaches late fifth instar, it starts gathering current-year needles from neighboring shoots (Nigam 1986). Before cutting a needle from the shoot, the larva attaches silk strands to it to prevent its loss; then it carries the needle inside the shelter by holding the cut end in its mouthparts. Needles are eaten within their shelter (Nigam 1986, unpubl. video tapes).

Larvae were relatively more active late in the evening (dusk) and very early in the morning (dawn) than in the daytime (Nigam 1987b). General activities such as microhabitat cleaning, excretion, and movements within microhabitat occupy more time than spinning activity, and feeding activity is of short duration compared with the other two activities (Nigam, unpubl. data).

Characteristics of Insecticide Droplets Within Budworm Microhabitat

The microhabitat of spruce budworm larvae on balsam fir at the time of spray operations consists of silk strands or webbing, growing buds or new needles, and usually very few 1-year-old needles. The spray droplets reaching coniferous foliage are 10–60 µm in diameter (Himel 1969; Spillman 1976; Barry et al. 1977; Joyce and Spillman 1978; Barry and Ekblad 1978; Wiesner 1984; Picot et al. 1986; Lambert 1987). Very few papers describe droplet deposition on spruce budworm larvae under field conditions, and such studies were restricted to the western spruce budworm (Himel and Moore 1967; Barry et al. 1977). Droplet deposition on silk strands or webbing was studied by Roberts et al. (1971) under laboratory conditions and by Nigam (1987b) under field conditions. These studies suggest that larval setae can easily be contaminated by droplets of contact and stomach insecticides deposited on silk, as they remain in constant touch with the silk. Particles of a stomach insecticide such as Bt can be transferred from the setae to feeding sites.

Assessments were done to quantify fenitrothion chemically and to characterize the droplet spectra of various formulations of fenitrothion and Bt on webbing and needles during recent field trials (Wiesner 1984; Kettela 1988; Nigam 1987b; Nigam and Holmes 1988, 1989; Varty and Holmes 1988). The droplet behavior and spectra of fenitrothion and Bt sprays and response of the larvae are summarized in the following sections.

Fenitrothion

Fenitrothion was sprayed on balsam fir using a Cessna-188 aircraft and Micronair® AU4000 atomizers in various formulations consisting of 210-g active ingredient per hectare. Emulsion and flowable formulations were applied as 11% active ingredient in water at 1.45 L/ha and ultra-ultra-low volume as 38% active ingredient in Dowanol® at 0.46 L/ha (Nigam 1987b).

Fenitrothion droplets on webbing—The droplets impinging on webbing were chemically analyzed for active ingredient for the ultra-ultra-low volume and emulsion formulations. Approximately 2–10 times more fenitrothion was found in webbing extracted from samples of the former formulation (Nigam 1987b). An attempt was made to characterize droplets impinging the webbing using a surgical microscope, a time-lapse video cassette recorder, and a video camera. The emulsion droplets had evaporated within 10–20 hours after spraying, before most of the samples could be processed. However, emulsion droplets were observed within 3–4 hours of spraying, before evaporation took place. Most of the droplets were less than 15 µm in diameter, and the numbers increased with decreasing drop size. Droplets 6–85 µm in diameter were observed in the ultra-ultra-low volume formulation samples, with a number mean diameter of 37 µm and a volume mean diameter of 42 µm (Nigam 1987b; Nigam and Holmes 1988).

Fenitrothion droplets on needles—The chemical analysis of droplets deposited on balsam fir needles gave a 2.66 ppm mean deposit for ultra-ultra-low volume and a mean deposit of 1.84 ppm for the standard emulsion formulation (Kettela 1988). The droplet spectrum deposited on the needles ranged from 2 to 50 µm; the mode was 10 µm; and the majority of droplets were between 6 and 18 µm in diameter (Wiesner 1984). The peak droplet density averaged nearly one drop per needle (Wiesner 1984), with a range from 0.2 to 1.75 (Varty and Holmes 1988).

Reaction of larvae to fenitrothion droplets within their microhabitat—Larval behavior can be affected by several methods of fenitrothion exposure, namely by direct impaction of droplets on the larvae, inhalation of vapors, contact with deposits on needles, and feeding on needles containing deposits or systemically translocated fenitrothion. Larvae were found in various stages of toxemia. Some larvae were dead inside the webbed shelters, while others came out of the shelters and were hanging head down from their webbing.

The majority of larvae were found on drop trays. Larvae that fell on the drop trays exhibited various stages of toxic symptoms.

Larvae that were found dead inside their shelters or hanging outside the shelters had been exposed to a relatively high dose and died very quickly, compared with those falling on drop trays. Affected larvae showed typical symptoms of nerve poisoning which varied from restlessness, hyperexcitability, convulsions, and tremors to paralysis leading to death. The larvae lose moisture from their cuticle and by regurgitating stomach fluid. Their color changed slowly from light to dark during hyperactivity and convulsions. Eventually, they had tremors in the thoracic region and were dehydrated and shrivelled during and following death (Nigam 1986, unpubl. video tapes). These symptoms are observed generally in larvae treated with fenitrothion in the laboratory or in the field.

Residual toxicity of fenitrothion droplets—Residual toxicity was studied using drop trays in the field and larval mortality (pre- and post-spray) was measured by fallout of larvae from crowns to the drop trays (Nigam 1987b; Varty and Holmes 1988). Results showed a combined effect of interactions among various modes of dose transfer, larval behavior, host development, and weather. Fifth instars were more vulnerable to residues than were earlier instars, as it appears that they had more contact with treated surfaces. The initial high counts of fifth-instar larvae on drop trays decreased with time. Second-instar larvae were less affected initially by fenitrothion residues because the larvae were concealed in 1-year-old needles, but with time and warm weather they emerged and wandered onto the treated needle surfaces. The drop tray counts of second-instar larvae increased with time. Few third- and fourth-instar larvae were found on drop trays because the larvae were protected inside the developing buds; there was no change in the daily rate of fallout up to 4 days after treatment (Nigam 1987b).

Results of bioassay and chemical analysis of foliage carried out for residual toxicity and deterioration of residues after aerial application revealed that residual efficacy of fenitrothion depends largely on initial deposit and the thoroughness of the coverage obtained during spraying. LT_{50} ranged from 3 to 5 days. Mortality due to residual toxicity appears to be highest in the ultra-ultra-low volume formulation and lowest in the flowable formulation (Nigam 1987b). Foliage protection and larval mortality caused by these formulations also appear to be in the same order (E.G. Kettela, pers. comm.).

Bt

Bt experiments were carried out during 1985, 1987, and 1988 to test the efficacy of various formulations on balsam fir. Studies during 1985 and 1987 were exploratory to study behavior of droplets in the budworm microhabitat and to investigate the contamination process occurring in the feeding sites of the larvae. Dyed Thuricide® containing Erio Acid Red dye and Rhoplex sticker was used in 1985 and undiluted Dipel® 6L with 1% Day Glow fluorescent dye was applied in 1987 using four Micronair® AU4000 atomizers on a Cessna-188 aircraft. A surgical microscope, a video camera, and a video cassette recorder were used to record droplets found on needles, webbing, and larvae. Droplets from 10 to 115 µm from both formulations were recorded on needles, webbing, and setae of spruce budworm larvae.

Thuricide® droplets were dark brown, bead- or marble-like solids that maintained their initial size after impaction. They did not break up after impingement on needles or webbing, and maintained a spherical shape on webbing but flattened on impaction on needles. There was no further spreading or evaporation of these drops after deposit due to the Rhoplex sticker. Undiluted Dipel® droplets appeared round and fluid with suspended particulate material. They maintained their shape on webbing, but on needles the carrier oil was absorbed into the tissue and the particulate material was left on the surface. Oil stains from 50 to 700 µm in diameter were observed on the needles. The deposits of particulate material contained varying proportions of dye crystals and a crust-like Bt agglomerate ranging from 15 to 115 µm in diameter. The size of droplets on silk strands ranged from 10 to 90 µm in diameter. Droplets were of three different colors: orange, opaque with milky or silver color, and translucent (like a drop of clear oil).

Most of the droplets on the foliage were found either on old needles or on the new needles at the outer surface of the elongated shoots. However, the larvae were feeding on needles inside the developing shoots or microhabitats at the time of spraying and not on the needles at the outer perimeter, where most of the droplets were found. Larvae are always in contact with the webbing or silk strands of microhabitats, thus droplets are transferred from silk to setae and to the feeding site. It appears from these preliminary studies that setae, particularly those present on the head capsule of budworm larvae, carry Bt spores and crystals from Bt droplets deposited on needles and webs to the feeding site inside their microhabitat. Various droplet sizes were observed on the setae of a few larvae. Bead-like droplets of Dipel® ranging from 5 to 25 µm in diameter were found on the setae.

In 1988, experiments with high potency and undiluted Dipel® 12L were designed to evaluate whether fine atomization would increase coverage or number of droplets in the spruce budworm larval microhabitat and thus increase the efficacy of insecticide. Undiluted Dipel® 12L, with 1% Day Glow dye was applied with a Cessna-188 aircraft equipped with four Micronair® AU4000 atomizers. The droplet spectrum was altered by adjusting the flow rate while atomizing at maximum rotation (approximately 12 000 rpm). Coarse and fine droplet spectra of Dipel® 12L were produced by applying 1.1 L/ha (30 BIU/ha, billion international units per hectare) and 0.55 L/ha (15 BIU/ha), respectively: the effect of atomization on the deposit and efficacy (Nigam and Holmes 1989) is summarized in the following sections.

Dipel® 12L droplets on webbing—The droplets on silk strands were round and had both variable opacity and coloration (orange hues) due to varying amounts of suspended particles of Bt and dye. The range of droplet diameters on the silk strands was from 24 to 130 µm in the 30 BIU plot, and 11 to 100 µm in the 15 BIU plot. Seventy percent of droplets were between 75 and 125 µm in diameter in the 30 BIU plot, and between 15 and 55 µm in the 15 BIU plot. Finer droplets were more common in the 15 BIU plot (0.55 L/ha) than in the 30 BIU plot (1.1 L/ha). The mean droplet density per microhabitat was 0.83 and 1.46 in the 30 and 15 BIU plots, respectively; that is, the finer spectrum (15 BIU) plot had practically twice the density of the coarser spectrum (30 BIU) plot (Nigam and Holmes 1989).

Dipel® 12L droplets on needles—The deposit on needles from the 30 BIU and 15 BIU plots was assessed using ultra-violet light on foliage samples collected from 12 trees from each plot. The droplet spectrum on needles could not be sized due to the nature of droplet composition. The droplets were composed of round white granular agglomerates and particles of orange dye suspended in oil mentioned previously. It was difficult to get a spread factor because suspended particles formed clumps of different shapes and sizes on the needles after the oil was absorbed into leaf tissue. The mean deposits on 1-year-old needles in the 30 BIU and 15 BIU plots were 0.57 and 0.49 drops per needle, respectively. The range of deposit in the 30 BIU plot varied from 0.2 to 1.1 drops per needle. Seven trees out of 12 had more than 0.5 drops per needle. In the 15 BIU plot, most of the deposit was from 0.1 to 0.6 drops per needle, except for two trees that had deposits of more than 1 drop per needle and nine trees with deposits of less than 0.5 drops per needle (Nigam and Holmes 1989).

Reaction of larvae to Bt—Feeding inhibition occurs soon after ingestion of Bt (Heimpel and Angus 1959) but larval death does not occur until hours or, in many cases, days after exposure (Fast and Régnière 1984). Larvae may recover from sublethal doses and start consuming foliage again. They develop septicemia and symptoms of diarrhea, abdominal twitches, and paralysis after ingestion of a lethal dose. They crawl to the base of the microhabitat and eventually drop off the tree and die. It is essential that larvae get a lethal dose at the beginning, otherwise residues may not be effective when they resume feeding because residual toxicity lasts less than 72 hours under field conditions (Nigam and Holmes 1989). Therefore, a high proportion of droplets must contain lethal concentrations of crystals and spores, and this apparently is being achieved by the undiluted application of highly concentrated formulations of Bt.

Residual toxicity of Dipel® 12L—Pre-spray samples for residue bioassay were collected 48 hours before spraying, whereas post-spray sampling was done within 1, 24, and 72 hours after aerial spraying. Twenty to 25 current-year buds were collected from one branch and kept at 0 ± 1° C until bioassayed. Bioassay of foliage from the field for residual efficacy of 15 and 30 BIU/ha sprays was done using laboratory reared fourth-instar larvae. Each plot had 12 samples and 10 larvae were used per sample. The mortality observations were taken after 1, 3, 6, and 9 days of releasing the larvae, and at adult emergence. There was no difference in larval mortality in the 15 and 30 BIU treatments after 9 days of observation. Mortality at adult emergence in 1-hour and 72-hour samples of the 30 BIU plot was significantly higher than that of 15 BIU plot, whereas mortality in the 24-hour samples of both plots was not significantly different. The mean corrected percentage mortality reduced to 9.9 ± 3.9 and 17.4 ± 4.7 from 42.2 ± 8.2 and 36.8 ± 5.8 in 30 and 15 BIU samples, respectively, in 72 hours indicating that residual toxicity lasted for less than 3 days (Nigam and Holmes 1989).

Conclusions

The response of spruce budworm larvae to fenitrothion and Bt varied according to size, number, and concentration of active ingredient of droplets contaminating the larval microhabitat after aerial application. Larvae were more adversely affected by smaller droplets (less than 55 μm in diameter) produced from undiluted, concentrated ultra-ultra-low volume formulations of both insecticides applied at approximately 0.5 L/ha or less, using Micronair® AU4000 atomizers, than the standard application rate of 1.46 to 1.8 L/ha. Microhabitat structures, especially webbing and new needles, favor deposition of smaller droplets. Residual toxicity of fenitrothion and Bt played an important role in efficacy. Both insecticides showed residual toxicity of approximately 3 days.

The mechanism of dose transfer to larvae appears to be from contaminated webbing and needles of the microhabitat. Larvae usually forage on unexposed new needles inside the microhabitat at the time of aerial application of both insecticides. The contamination of foraging sites appears to take place by transfer of insecticide deposits from the outer webbing and needles of microhabitats, through larval setae, while the larvae are carrying out various activities. Setae also help the transfer of contact insecticide droplets from webbing to the cuticle of larvae. The technique for measuring droplet spectra on webbing needs further development. It is difficult to assess the size of droplets depositing on webbing because droplets, especially those of water-based formulations, partially evaporated by the time they are observed.

Fenitrothion-treated larvae exhibited restlessness, convulsions, tremors, and paralysis. Their color changed slowly from light to dark and they dehydrated and shrivelled before death. Larvae die within 24–72 hours after exposure to a lethal dose of fenitrothion. Bt-treated larvae develop septicemia, diarrhea, abdominal twitches, and paralysis, and die 3–6 days after ingesting a lethal dose. Larvae that recover from sublethal doses of Bt resume feeding so it is essential that larvae get a lethal dose at the beginning by spraying high potency formulations.

Spruce budworm larvae feed and spin actively in the evening and early morning, suggesting that evening sprays, if practical, may be more effective than morning applications. In addition, residual life of fenitrothion and Bt may increase due to slow degradation of their deposits at night. Fenitrothion vaporization is reduced due to lower temperatures, and the absence of ultra-violet rays reduces photodegradation of Bt at night.

Efficacy of aerially applied insecticides against spruce budworm larvae can be improved if droplet spectra are produced to maximize impingement on the microhabitat webbing and new foliage of balsam fir, and if ultra-ultra-low volume techniques are practiced using highly potent, fine droplets. Coarser droplets (75–125 μm in diameter) require twice the amount (30 BIU/ha) of Dipel® 12L than finer droplets (15–55 μm) which required only 15 BIU/ha for the same level of residual toxicity and efficacy. The efficiency of fenitrothion and Bt operations can be improved by applying highly concentrated formulations containing a minimum of diluents.

References

Barry, J.W.; Ciesla, W.M.; Tysowski, M.; Ekblad, R.B. 1977. Impaction of insecticide particles on western spruce budworm larvae and Douglas-fir needles. J. Econ. Entomol. 70: 387-388.

Barry, J.W.; Ekblad, R.B. 1978. Deposition of insecticide drops on coniferous foliage. Transactions of the ASAE 21(3): 438-441.

Carter, N.E.; Lavigne, D.R. 1991. Protection spraying against spruce budworm in New Brunswick 1990. Dep. Natural Resources and Energy, Fredericton, N.B., 34 pp.

Dimond, J.B.; Morris, O.N. 1984. Microbial and other biological control. Pages 103-114 *in* D.M. Schmitt, D.G. Grimble and J.L. Searcy (Tech. Coord.), Spruce budworms handbook. USDA For Serv., Agric. Handbook No. 620.

Dorais, L.; Kettela, E.G. 1982. A review of entomological survey and assessment techniques used in regional spruce budworm, *Choristoneura fumiferana* (Clem.), surveys and in the assessment of operational spray programs. Committee for the Standardization of Survey and Assessment Techniques, Eastern Spruce Budworm Council, Que., 43 p.

Eidt, D.C. 1987. Biological interface - insecticides, insects, and post-spray weather. Pages 215-218 *in* G.W. Green (Ed.), Proc. Symp. Aerial Appl. Pest. For. ACAFA, Nat. Res. Coun. Can., Ottawa, Ont. AFA-TN-18 NRC No. 29197.

Ekblad, R.; Armstrong, J.; Barry, J.; Bergen, J.; Millers, I.; Shea, P. 1979. A problem analysis: forest and range aerial pesticide application technology, USDA For Serv., Equip. Devel. Cent. Missoula, Mont. 106 p.

Fast, P.G.; Régnière, J. 1984. Effect of exposure time to *Bacillus thuringiensis* on mortality and recovery of the spruce budworm (Lepidoptera: Tortricidae). Can. Entomol. 116: 123-130.

Graham-Bryce, I.J. 1983. Formulation and application of biologically active chemicals in relation to efficacy and side effects. Pages 463-473 *in* D.L. Whitehead and W.S. Bowers (Eds.), Natural products for innovative pest management. Pergamon Press Ltd.

Hall, F.R. 1987. Parameters governing dose transfer. Pages 259-264 *in* R. Greenhalgh and T.R. Roberts (Eds.), Pesticides science and biotechnology. Blackwell Scientific Publications. Proc. 6th Int. Cong. Pesticide Chem. Ottawa, Ont. 1986.

Heimpel, A.M.; Angus, T.A. 1959. The site of action of crystalliferous bacteria in lepidoptera larvae. J. Inspect Path. 1: 152-170.

Helson, B.V. 1985. Research on chemical insecticides for spruce budworm control at the Forest Pest Management Institute, 1982-present. Page 376 *in* C.J. Sanders, R.W. Stark, E.J. Mullins, and J. Murphy (Eds.), Recent advances in spruce budworms research. Proc. CANUSA Symp., Bangor, Me., Sept. 16-20, 1984. Cat. No. Fo 18-5/1984.

Heron, R.J. 1965. The role of chemotactic stimuli in the feeding of spruce budworm larvae on white spruce. Can. J. Zool. 43(2): 247-269.

Himel, C.M. 1969. New concepts in insecticides for silviculture and old concepts revisited. Proc. Fourth Int. Agric. Aviat. Congr., Kingston, Ont. pp. 275-281.

Himel, C.M.; Moore, A.D. 1967. Spruce budworm mortality as a function of aerial spray droplet size. Science. 156: 1250-1251.

Joyce, R.J.V.; Spillman, J.J. 1978. Discussion of aerial spraying techniques. Pages 13-44 *in* A.V. Holden and D. Bevan (Eds.), Control of pine beauty moth by fenitrothion in Scotland 1978. For. Comm., U.K.

Kettela, E.G. 1988. Review of research and development programs conducted in New Brunswick in 1987 to evaluate pesticides and tactics against the spruce budworm (*C. fumiferana*) and the spruce budmoth (*Z. canadensis*). Pages 10-16 *in* W.A. Sexsmith (compiler). 1987 Rep. Environ., Monitoring of Forest Control Operations (EMOFICO) Dept. of Municipal Affairs and Environ., Fredericton, N.B.

Lambert, M.C. 1987. Quantification of spray deposit in experimental and operational aerial spraying operations. Pages 125-130 in G.W. Green (Ed.), Proc. Symp. Aerial Appl. Pest. For. ACAFA, Nat. Res. Coun. Can., Ottawa, Ont. AFA-TN-18 NRC No. 29197.

Metcalf, R.L. 1980. Changing role of insecticides in crop protection. Ann. Rev. Entomol. 25: 219-256.

Nigam, P.C. 1971. Insecticide evaluation for aerial application against forest insect pests. Proc. Seventh Northeast Aerial Applicators Conf., Ithaca, N.Y. pp. 131-143.

Nigam, P.C. 1975. Chemical insecticides. Pages 8-24 *in* M.L. Prebble (Ed.), Aerial control of forest insects in Canada. Environ. Can., Can. For. Serv., Ottawa, Ont.

Nigam, P.C. 1980. Use of chemical insecticides against eastern spruce budworm in Canada. CANUSA Newsletter ll: 1-3.

Nigam, P.C. 1985. Early laboratory bioassays in Canada. Page 372 *in* C.J. Sanders, R.W. Stark, E.J. Mullins, and J. Murphy (Eds.), Recent advances in spruce budworms research. Proc. CANUSA Symp., Bangor, Me., Sept. 16-20, 1984. Cat. No. Fo 18-5/1984.

Nigam, P.C. 1986. Budworm behaviour and their microhabitat in balsam fir during spray operations (Abstract). Proc. 46th Ann. Meet. Acadian Ent. Soc., Bangor, Me, Apr. 28-30, 1986. 15 p.

Nigam, P.C. 1987a. Video technique for observing spruce budworm larval activity in its natural habitat (Abstract). Proc. 47th Ann. Meet. Acadian Ent. Soc., Charlottetown, P.E.I., Aug. 17-19, 1987, 44 p.

Nigam, P.C. 1987b. Dose transfer and spruce budworm behaviour during operational application of fenitrothion. Pages

281-284 *in* G.W. Green (Ed.), Proc. Symp. Aerial Appl. Pest. For. ACAFA, Nat. Res. Coun. Can., Ottawa, Ont.

Nigam, P.C.; Holmes, S.E. 1988. Progress report of 1987 field experiments on UULV fenitrothion deposition in budworm microhabitat and budworm larval behavior. Internal Rep. to 15th Ann. Forest Pest Control Forum, Ottawa, Ont., Nov. 17-19, 1987. pp. 394-402.

Nigam, P.C.; Holmes, S.E. 1989. Progress report on residual efficacy and dose transfer mechanism of Dipel® 12L sprayed aerially against spruce budworm at Bathurst during 1988. Internal Report to 16th Ann. Forest Pest Control Forum, Ottawa, Ont., Nov. 15-17, 1988. pp. 342-364.

Picot, J.J.C.; Kristmanson, D.D.; Basak-Brown, N. 1986. Canopy deposit and off-target drift in forestry aerial spraying: the effects of operational parameters. Transactions of the ASAE. 29(1): 90-96.

Prebble, M.L. (Ed.). 1975. Aerial control of forest insects in Canada. Environ. Can., Can. For. Serv., Ottawa, Ont. 330 p.

Randall, A.P. 1969. Some aspects of aerial spray experimentation for forest insect control. Proc. Fourth Int. Agric. Aviat. Congr., Kingston, Ont. pp. 308-315.

Randall, A.P. 1976. Insecticides, formulations, and aerial applications technology for spruce budworm control. Pages 77-90 *in* Proc. Symp. on the Spruce Budworm, Nov. 1974, Alexandria, Va. Washington, D.C., USDA For. Ser., Misc. publ. 1327.

Roberts, R.B.; Lyons, R.L.; Page, M.; Miscus, R.P. 1971. Laser halography: its application to the study of the behavior of insecticide particles. J. Econ. Entomol. 64: 2. pp. 533-536.

Shea, P.J.; Nigam, P.C. 1984. Chemical control. Pages 116-132 *in* D.M. Schmitt, D.G. Grimble, and J.L. Searcy (Tech. Coord.). Spruce budworms handbook. USDA For. Serv., Agric. Handbook No. 620.

Shepherd, R.F. 1958. Factors controlling the internal temperatures of spruce budworm larvae, *Choristoneura fumiferana* (Clem.). Can. J. Zool. 36(5): 779-786.

Spillman, J.J. 1976. Optimum droplet sizes for spraying against flying targets. Agric. Aviat. 17: 28-32.

Varty, I.W.; Godin, M.E. 1983. Identification of some of the factors controlling aerial spray efficacy. For. Can.—Maritimes Region Inf. Rep. M-X-142. 29 p.

Varty, I.W.; Holmes, S.E. 1988. Heterogeneity of spray deposit and efficacy within a single swath applied by aircraft over forest infested with spruce budworm. For. Can.—Maritimes Region Inf. Rep. M-X-168. 60 p.

Varty, I.W.; Nigam, P.C. 1985. Spray technology - the biological interface. Page 410 *in* C.J. Sanders, R.W. Stark, E.J. Mullins, and J. Murphy (Eds.), Recent advances in spruce budworms research. Proc. CANUSA Symp., Bangor, Me., Sept. 16-20, 1984. Cat. No. Fo 18-5/1984.

Wellington, W.G. 1948a. The light reactions of the spruce budworm, *Choristoneura fumiferana* Clemens (Lepidoptera: Tortricidae). Can. Entomol. 80(1-12): 56-82.

Wellington, W.G. 1948b. Measurements of the physical environment. Bi-mon. Prog. Rep., For. Ins. Invest., Agric. Can. 4(5): 1-2.

Wellington, W.G. 1949. The effects of temperature and moisture upon the behaviour of the spruce budworm, *Choristoneura fumiferana* Clemens (Lepidoptera: Tortricidae). I. The relative importance of graded temperatures and rates of evaporation in producing aggregations of larvae. Sci. Agric. 29(5): 201-205.

Wellington, W.G. 1950a. Effects of radiation on the temperatures of insectan habitats. Sci. Agric. 30(5): 209-234.

Wellington, W.G. 1950b. Variations in the silk spinning and locomotor activities of larvae of the spruce budworm, *Choristoneura fumiferana* (Clem.), at different rates of evaporation. Trans. Royal Soc. Can. Sect. 5.44 (Series 3): 89-101.

Wellington, W.G.; Cameron, J.M. 1947. Investigations on meteorological and other physical factors. Bi-mon. Prog. Rep., For. Insect Invest., Agric. Can. 3(3): 2.

Wellington, W.G.; Henson, W.R. 1947. Notes on the effects of physical factors on the spruce budworm, *Choristoneura fumiferana* (Clem.). Can. Entomol. 79: 168-170, 195.

Wiesner, C.J. 1984. Droplet deposition and drift in forest spraying. Pages 139-151 *in* W.J.Garner and J. Harvey (Eds.), Chemical and biological controls in forestry. ACS. Symp. Series. No. 238, Chap. 11.

Chapter 10

Attempts to Develop Strategies to Control Spruce Budworm, *Choristoneura fumiferana*, Populations by Spraying Moths

E.G. KETTELA

Introduction

Since the first spray program conducted against spruce budworm, *Choristoneura fumiferana*, larvae in 1952 in New Brunswick, virtually all spray programs have evolved around the philosophy of providing protection to the worst damaged and/or most highly infested areas. Large areas of less infested forest have been left unprotected. This protection strategy by and large provided the results desired by forest managers, but offered little hope of actually controlling the course of infestations or of preventing damage earlier in their development. B. W. Flieger, then President and General Manager of Forest Protection Limited, proposed that treatment of spruce budworm moths might provide a way of controlling populations and would allow forest managers to manage the budworm rather than have the budworm dictate the course of forest management. From 1969 to 1977, a series of trials were conducted in New Brunswick to explore the feasibility of population control by treating the adult spruce budworm. Concurrent with these trials, an intensive research program on spruce budworm moth dispersal was conducted, initially with fairly rudimentary equipment (1971 and 1972) but later, with a significantly higher level of sophistication, using radar (1973 to 1976). This later phase attracted considerable interest, for it offered first, a way to track concentrations of moths; second, a way to determine where they landed, thus providing a target for sprays; and third, the possibility of actually spraying airborne concentrations of moths and killing them before they landed.

Spruce Budworm Moth Dispersal Studies and Their Relationship to Moth Spray Trials

The renewed interest in spruce budworm moth dispersal in 1971 and 1972 and its obvious relationship to spruce budworm population dynamics and modeling, together with the desire to find population control strategies and the availability of radar technology for tracking insect concentrations, all came together in 1973 in a spruce budworm moth flight and dispersal study that wrote a new chapter in spruce budworm biology (Greenbank et al. 1980).

This research program, using radar technology, began in 1973 and was terminated in 1976, when the information and research technology far outstripped the capability of deploying the information realistically and logistically in real-time population control. In retrospect, it is perhaps just as well, for by 1976 the anti-spray movement was gaining prominence and in all likelihood the spray scenarios conceived would have been cancelled. Nonetheless, the study provided considerable insight into the complexities of spruce budworm moth dispersal and its role in the spread of infestations.

The array of electronic equipment, sampling devices, and observations made are summarized in Table 1 (derived from Greenbank et al. 1980). The objectives for each trial by year are also stated.

These studies supported earlier observation made by Henson (1951), Wellington (1954), and Greenbank (1957) on the mass invasion of moths associated with cold fronts and thunderstorms. Further, mass moth movements were associated with most frontal conditions and sea breeze fronts that periodically occurred (Rainey and Haggis 1987). Sea breeze fronts, although they are not frequent, have the potential, if they coincide with moth dispersal, of concentrating the moths and depositing them further inland. Rainey and Haggis (1987) hypothesize that the generation of outbreaks is not "progressive," but that they occur from the emigration, the concentration of moths, and their immigration into nonrelated stands. The implication is that predicting outbreaks is problematic if moth immigration is unknown. They further suggest that the airborne insect-detecting radar is the key element in a search and strike system for controlling airborne moths. However, this hypothesis is not consistent with current theories of spruce budworm population dynamics (see Régnière and Lysyk, Chapter 8) which suggest that moth dispersal is not a key element in the spread of outbreaks.

Moth Spray Trials

The aim of the moth spray trials was to minimize egg deposition. This entailed the treatment of both resident moths and the invading moths that could reinfest protected forest. The spray trials of 1969, 1972, and 1973 (Table 2), which were discussed by Kettela and Miller (1975 a,b), established that it is feasible to kill moths and that phosphamidon is more effective than fenitrothion. The objectives of the trials from 1974 to 1977 were to find the most effective insecticide dosage, time of treatment, number of applications, and type of insecticides. These were all air-to-ground spray trials. There was one "air-to-air" spray trial in which a curtain of spray was used to determine the reaction of moths to the spray cloud and the deposition of insecticide on moths that penetrated the spray cloud.

The moth spray trials were made possible through the services of Forest Protection Limited, who provided all aircraft, insecticide, and considerable ground support. The size of spray operations ranged from 814 000 ha in 1974 to 2 400 ha in 1976. The array of sampling and measuring devices used to determine efficacy included sticky traps baited with pheromone lures to measure male activity; drop trays to collect moths to determine mortality and mating status of female moths; light traps to compare moth abundance in treated and check areas and to detect invasion; observation platforms to determine budworm activity;

Table 1. Summary of methods used to monitor spruce budworm, *Choristoneura fumiferana*, moth dispersal in New Brunswick and the number of nights of operation.[a]

Year	Location	Project study	Observation platform	Night-viewing telescope	Ground-based radar	Airborne radar	Light traps	Meteorological tower	Pilot balloons and rawinsonde ascents	DC-3 windfield study	Aircraft insect collecting net
1971	Acadia	Local flight activity and emigration from heavily infested stands	11					11			
1972	Acadia	Local flight activity and emigration from heavily infested stands	14				14	13			
1972	Upper Blackville	Local flight activity and emigration from moderately infested stand	11								
1972	Greenriver	Local flight activity and emigration from lightly infested stand	14				12				
1973	Chipman	Local flight activity and dispersal	14	5	11		15	12	12	12	4
1974	Chipman	Local flight activity and dispersal	20	8	18		20	14	18		15
	Renous	Moth invasion into area sprayed for spruce budworm moths	21	10	22		22		18	16	16
1975	Chipman	Local flight activity and dispersal	18	13	15	14	20	15	12	18	9
	Juniper	Moth invasion into intensive larval spray area	17		18		21	15	12		7
1976	Acadia	Local flight activity and dispersal	33		33	8+6	41	12	13	12	4

[a] From Greenback et al. 1980.

malaise traps to record moth activity and abundance and to determine mating status and fecundity of emigrating females; daily branch/egg-mass sampling to determine differences between sprays and checks and the rate of egg deposition; high volume spraying of trees (with a hydraulic-truck-mounted sprayer) with insecticide to kill moths daily to determine mating status; and radar (1974, 1976) to measure invasion of moths into selected sites.

Insecticides and Efficacy

Phosphamidon was used in the first trial because it was available (residual left over from larval sprays) and because it was being used against adult insects elsewhere. Subsequent laboratory tests showed that phosphamidon was more toxic to budworm adults than fenitrothion and that its LD_{50} to adult budworms was equivalent to that of fourth instar larvae. There was a 1.64-fold difference between the LD_{50} of male moths and that of the female moths (ml/ml: males, 0.00022; females, 0.00036) (Miller et al. 1973). A summary of spray efficacy for phosphamidon is shown in Table 3. The trials from 1974 to 1976 showed very poor efficacy in preventing egg deposition. These 3 years also coincide with maximum levels of infestation and of moth dispersal in New Brunswick. Treatments with phosphamidon characteristically resulted in an immediate kill of males and old and spent females. Tests with Sevin®-4-oil, aminocarb, and Orthene® in 1976 were disappointing, but the test with aminocarb suggested that gram for gram, it was as toxic or more so than phosphamidon (Miller et al. 1980). Cage experiments indicate that insecticide dose acquisition is probably through moths walking on contaminated foliage. Because gravid female spruce budworm moths are largely sedentary until after they have laid

Table 2. Summary of treatments and objectives of spruce budworm adult spray trials in New Brunswick, 1969–1977.

Year	Treatment	Number		Block size (ha)	Total hectares sprayed	Objectives
		Applications	Blocks			
1969	140 g phosphamidon in 1.46 L/ha. Applied by AgCats (Micronair® 3000)	2X	1	7 000	7 000	Determine whether adult budworm are feasible spray targets
1972	Phosphamidon & fenitrothion 140 g/ha in 1.46 L/ha. Applied by Stearman (B&N)	2X	2	3 000	6 000	To compare efficacy of fenitrothion and phosphamidon
1973	Phosphamidon at 35, 70, and 140 g/ha in 0.88 L/ha. Applied by TBM's (B&N)	1 & 2X	16	8 000	128 000	To compare (1) early and late applications, (2) dosages, and (3) number of applications
1974	70 g/ha of phosphamidon in 0.88 L/ha. Applied by TBM's & DC-6 (B&N)	2X	11 DC-6 33 TBM	29 000 15 000	319 000 495 000 814 000	Operational demonstration spray trial
1975	70 g/ha phosphamidon in 0.88 L/ha. Applied by TBM's (B&N)	2, 3 & 5X	12	700	8 400	To compare replicated trials (a) early and late application (b) 2, 3 & 5 applications
1976	70 g aminocarb 3X, 70 g phosphamidon 3X, 280 g Orthene® 2X, 1 120 g Sevin®-4-Oil 1X & 560 g Sevin®-4-Oil 2X. Applied by AgTrucks (B&N)	3X 3X 2X 1X 2X	10	240	2 400	To compare (a) aminocarb, Orthene®, phosphamidon, and Sevin®-4-oil; (b) to test the persistence of Sevin®-4-oil
1977	70 g phosphamidon in 0.73 L/ha. Applied by AgTrucks	3X	1	3 247	3 247	To test the effectiveness of an early spray to kill males and disrupt mating
	70 g phosphamidon + 1.7 g pyrethrin + 7 g piperonyl butoxide in 0.88 L/ha. Applied by AgTrucks (Micronair® 3000)	1X	1	120	120	To test the effectiveness of phosphamidon + a motor stimulant

50% of their egg complement, they are relatively immune from contact with the insecticide. Therefore, it was hypothesized that the addition of a motor stimulant would activate the moths and expose them to a lethal dose of phosphamidon. Tests with pyrethrins and piperonyl butoxide in the laboratory increased moth activity (Volney and McDougall 1979), but addition of these in the field did not enhance the killing of budworm moths (Miller et al. 1980).

In retrospect, both the spray systems used and the dosages were important factors influencing spray efficacy. From 1969 to 1973, significant control was achieved, but with 140 g/ha rather than the 70 g/ha used from 1974 to 1976. In 1977, the small aircraft used in spray trials were equipped with Micronair® AU 3000 nozzles. Three applications were made, which spanned the emergence period of males and females (Tables 2 and 3). Further, Thomas (1978) showed that susceptibility of female spruce budworm moths, when exposed to sprays of aminocarb and phosphamidon, increased as they deposited eggs and became lighter and more active.

Table 3. Summary of results of spruce budworm moth spraying trials with phosphamidon in New Brunswick, 1969–1977.

Year	Dosage (g/ha)	No. applications	Percent control of egg-mass deposition	Pupa/egg-mass ratio	
				Treated	Check
1969	140	2	72	0.8	2.9
1972	140	2	64	1.0	2.8
1973	140	2	69	1.1	3.4
1974	70	2	23	2.0	2.6
1975	70	2[a]	11	1.6	1.8
	70	3[a]	0	2.4	1.8
	70	5[a]	0	3.1	1.8
1976	70	3[a]	22	2.5	3.2
1977	70	3	44	1.3	2.3

[a] Treatment block size less than 1 000 ha.

The question of the most appropriate timing of spray applications and the number of applications was addressed in trials in 1973 and 1975 to 1977. The conclusion was that a minimum of three applications was necessary and that this should span the early emergence phases of male moths to after 100% female emergence. Even then, an invasion from nonsprayed sites could deposit enough eggs to maintain the status quo, as noted in 1974 and 1975 (Miller et al. 1980). Differential kill of male and female moths was noted throughout these series of trials and was determined to be a key component in understanding the efficacy of sprays on populations and egg deposition. In summary, susceptibility of female moths to aerial sprays is related to chronological age of the moths, with moths less than 2 days old being essentially unaffected by sprays. By the end of the second day, the females have deposited over 50% of their egg complement. Male moths, however, are highly susceptible and over 75% can be killed in one spray. The implication is that an early spray will kill male moths, leaving the females to deposit their eggs, while a late spray will kill females, most of which have deposited their eggs. In either case, the impact on generation survival would tend to be low.

Bibliography

Clark, W.C. 1979. Spatial structure and population dynamics in an insect epidemic ecosystem. Ph.D. thesis, Institute of Animal Resource Ecology, University of British Columbia, Vancouver, 506 p.

Dickison, R.B.B.; Haggis, M.J.; Rainey, R.C. 1983. Spruce budworm moth flight and storms: Case study of a cold front system. J. Clim. Appl. Met. 22, 278, 286.

Dickison, R.B.B.; Haggis, M.J.; Rainey, R.C.; Burns, L.M.D. 1986. Spruce budworm moth flight and storms: Further studies using aircraft and radar. J. Clim. Appl. Met. 25, 1600-1608.

Greenbank, D.O. 1957. The role of climate and dispersal in the initiation of outbreaks of the spruce budworm in New Brunswick, II. The role of dispersal. Can. J. Zool. 35, 385-403.

Greenbank, D.O. 1973. The dispersal process of budworm moths. Can. For. Serv., Marit. For. Res. Cent. Inf. Rep. M-X-39.

Greenbank, D.O.; Schaefer, G.W.; Rainey, R.C. 1980. Spruce budworm moth flight and dispersal: New understanding from canopy observations, radar and aircraft. Mem. Entomol. Soc. Can. 110, 49 p.

Henson, W.R. 1951. Mass flights of the spruce budworm. Can. Entomol. 83, 240.

Kettela, E.G. 1974. Aerial spraying against the spruce budworm in 1974 and a forecast of conditions in the Maritimes region in 1975. *In* Proceedings of the Interdepartmental Committee on Forest Spraying, 1975.

Kettela, E.G.; Miller, C.A. 1975a. Experimental spraying against the spruce budworm in New Brunswick, 1975. *In* Proceedings of the Interdepartmental Committee on Forest Insect Control Operations, 1976.

Kettela, E.G.; Miller, C.A. 1975b. Spray applications against spruce budworm moths. Pages 147-148 *in* M.L. Prebble (Ed.), Aerial Control of Forest Insects in Canada. Dept. of the Environment, Ottawa, Canada.

Miller, C.A.; Stewart, J.F.; Elgee, D.E.; Shaw, D.D.; Morgan, M.G.; Kettela, E.G.; Greenbank, D.O.; Gesner, G.N.; Varty, I.W. 1973. Aerial spraying against spruce budworm adults in New Brunswick. Can For. Serv., Marit. For. Res. Cent. Inf. Rep. M-X-38.

Miller, C.A.; Greenbank, D.O.; Thomas, A.W.; Kettela, E.G.; Volney, W.J.A. 1977. Spruce budworm adult spray tests—1976. Can. For. Serv., Marit. For. Res. Cent. Inf. Rep. M-X-75.

Miller, C.A.; Greenbank, D.O.; Kettela, E.G. 1978. Estimated egg deposition by invading spruce budworm moths (Lepidoptera: Tortricidae). Can. Entomol. 110: 609-615.

Miller, C.A.; Varty, I.W.; Thomas, A.W.; Greenbank, D.O.; Kettela, E.G. 1980. Aerial spraying of spruce budworm moths, New Brunswick, 1972-1977. Can. For. Serv., Marit. For. Res. Cent. Inf. Rep. M-X-110.

Rainey, R.C. 1974. Flying insects as ULV spray targets. Br. Crop Prot. Counc. Monogr. II: 20-28.

Rainey, R.C.; Joyce, R.J.V. 1972. The use of airborne doppler equipment in monitoring windfields for airborne insects. 7th Int. Aerosp. Instrum. Symp. Cranfield 1972: 8.1-8.4.

Rainey, R.C.; Haggis, M.J. 1987. Spruce budworm moth flight: Possible implications of field research findings for population dynamics and management. Proc. Natl. Res. Counc. ACAF Comm. 1987.

Royama, T. 1977. The effect of moth dispersal on the dynamics of a local spruce budworm population. Can. For. Serv. Bi-mon. Res. Notes 33: 43-44.

Royama, T. 1981. Fundamental concepts and methodology for the analysis of animal population dynamics, with particular reference to univoltine species. Ecol. Monogr. 51, 473-493.

Royama, T. 1984. Population dynamics of the spruce budworm. Ecol. Monogr. 54, 429-462.

Sanders, C.J. 1975. Factors affecting adult emergence and mating behaviour of the eastern spruce budworm. Can. Entomol. 107: 967-977.

Sanders, C.J.; Lucuik, G.S. 1975. Effects of photoperiod and size on flight activity and oviposition in eastern spruce budworm (Lepidoptera: Tortricidae). Can. Entomol. 107: 1289-1299.

Schaefer, G.W. 1976. Radar observations of insect flight. Pages 157-197 *in* R.C. Rainey (Ed.), Insect Flight, 7th Symp. Royal Entomol. Soc. London.

Schaefer, G.W. 1979. An airborne radar technique for the investigation and control of migrating insect pests. Phil. Trans. Royal Soc. London. B287, 459-465.

Sippell, W.L. 1984. Planning how to reduce, postpone or prevent the next spruce budworm outbreak. Can. For. Serv., Sault Ste. Marie, Ont., 59-67.

Spillman, J.J. 1980. The design of an aircraft-mounted net for catching airborne insects: Trends in airborne equipment for agriculture and other areas. Proc. Seminar UN/ECE, Warsaw, UN/ECE, Pergamon, 169-180.

Thomas, A.W. 1978. Relationship between oviposition history, current fecundity and the susceptibility of spruce budworm moth (Lepidoptera: Tortricidae) to ULV aerial sprays of insecticides. Can. Entomol. 110: 337-344.

Thomas, A.W.; Miller, C.A.; Greenbank, D.O. 1979. Spruce budworm adult spray tests, New Brunswick, 1977. Can. For. Serv., Marit. For. Res. Centr. Info. Rep. M-X-99, 30 p.

Volney, W.J.A.; McDougall, G.A. 1979. Tests of motor stimulants for eastern spruce budworm moths. Can. Entomol. 111: 237-241.

Wellington, W.G. 1954. Atmospheric circulation processes and insect ecology. Can. Entomol. 86: 312-333.

Chapter 11

Western Spruce Budworm, *Choristoneura occidentalis*

R.F. SHEPHERD, J.C. CUNNINGHAM, AND I.S. OTVOS

Introduction

Evidence in tree ring studies has revealed that six outbreaks of the western spruce budworm, *Choristoneura occidentalis*, have occurred in British Columbia since 1909 (Harris et al. 1985), with defoliation being present somewhere in the province for two-thirds of that time. The last two outbreak periods, 1948–1958 and 1966–present, have been mapped and are well documented. Within outbreak periods, populations increased and waned with no apparent temporal pattern; they can be sustained at high levels or low levels for many years.

Western spruce budworm females lay eggs in imbricated masses on the underside of green needles (Furniss and Carolin 1977). The eggs hatch in August, but the larvae do not feed or grow. They move to protective niches along the branches or tree trunks and spin minute silk shelters in which to hibernate. The larvae emerge the following spring and attempt to mine the swelling buds. If suitable buds are not available, they are forced to mine old needles and survival is reduced. Synchronization with the host is important. Feeding continues within the protective bud until the latter flushes. Survival decreases rapidly thereafter, as the larvae complete their development and pupate among the residual foliage in late June or early July. Moths emerge 10 days later, mate, and the females disperse and oviposit intermittently in different stands.

Douglas-fir, *Pseudotsuga menziesii*, is the primary host and outbreaks have only occurred within its range. Defoliation often occurs in a band on mountain sides apparently related to temperature zones. Generally, the spread of outbreaks has been toward the northeast, following the inflow wind patterns common during the hot July days of adult dispersal (Shepherd 1985). Redistribution of moths within a watershed and dispersal to other areas appear to be main driving components in the dynamics of this pest.

Tree damage characteristically consists of radial and height growth loss, dieback, and deformity (Alfaro 1985). Tree mortality most commonly occurs in the understory below large, heavily infested trees. This mortality is most serious where multiple-age forest management is practised on hot dry sites and the understory provides the next generation of crop trees. If the understory is lost to budworm, the mature trees cannot be removed until a new understory is established, a process requiring many years.

At various times, the use of chemical sprays has been contemplated, but spray operations have never been implemented. It was felt that all budworm populations in a watershed, including those causing only light defoliation, would have to be treated to prevent population resurgence and reinvasion of treated areas. Thereafter, treatments would have to be continued annually to protect the investment of the previous years' control measures. Either way, it was questionable whether there was an economic benefit to be gained from control action. In addition, there is an aversion to spraying chemical insecticides in forests because their use has led to conflicts with different public interest groups. Therefore, research emphasis has been directed toward a solution utilizing biological insecticides, to determine if they can be as effective as chemical insecticides. In addition, long-term population changes following treatments were also monitored to see whether there were significant carry-over effects from one generation to the next.

Experimental Spray Trials Using Viruses

The same viruses that infect the spruce budworm, *C. fumiferana*, also infect the western spruce budworm (Cunningham 1985). The principal viruses studied are both Baculoviridae and most effort has been devoted to the one commonly known as nuclear polyhedrosis virus (NPV); tests have also been conducted with a granulosis virus (GV).

Three field trials have been conducted on western spruce budworm in British Columbia and these are summarized in Table 1. The first was an ad hoc trial in 1976; population reduction due to treatment, calculated using a modified Abbott's formula (Fleming and Retnakaran 1985), was 36%, and the highest incidence of NPV in samples of larvae was 6.8%. This figure fell to 1.3% in 1977. These poor results were attributed to a low dosage of virus applied too late in the season (Shepherd and Cunningham 1984).

The second trial was conducted in 1978, when three 20-ha plots were treated at 750 x 109 polyhedral inclusion bodies per hectare, which was three times the dosage used in 1976. At 15 days post-spray, population reductions due to treatment were 0%, 26%, and 48% in the three plots. The corresponding incidences of larval virus infection were 55%, 87%, and 25%, respectively. Successful adult emergence was lower in treated than in check plots and the male to female ratio was 2:1 in the treated plots compared with 1:1 in the check plots (Hodgkinson et al. 1979). Studies were continued in 1979 and 1980 in two of the treated plots, the insect population having collapsed in the third. Incidence of virus infection decreased over the 2 years following treatment as did population reduction compared to the check plots (Shepherd et al. 1982). This decrease in incidence may have been due to immigration of moths into these small plots. There was no evidence from these trials that NPV can give long-term control of western spruce budworm.

Interest in both nuclear polyhedrosis virus and granulosis virus was intensified when it was found that the western spruce budworm was about 1 700 times (LC50) more susceptible to both viruses than the spruce budworm (Cunningham et al. 1983a). In 1981, NPV and GV were compared at three dosages in a small, ground spray trial on individual, naturally

Table 1. Spray trials with viruses on western spruce budworm, *Choristoneura occidentalis*, in British Columbia.[a]

Year	Virus	Dosage (inclusion bodies/ha)	Tank mix	Peak larval instar at time of application	No. of plots	Total area (ha) treated
1976	NPV	2.5×10^{11}	Water 25% molasses 6% IMC90-001[b] 1% Chevron®[c]	5	1	20.5
1978	NPV	7.5×10^{11}	Water 25% molasses 3% Shade®[d] 0.05% Triton® B-1956[c]	5	3	60
1982	NPV	5.4×10^{11}	25% emulsifiable oil[e] 75% water	4	1	172
1982	GV	1.7×10^{14}	25% emulsifiable oil[e] 75% water	4	1	172

[a] All applications using fixed-wing aircraft with boom and nozzles calibrated to deliver 9.4 L/ha.
[b] UV screening agent supplied by International Minerals Corp.
[c] Stickers.
[d] Same material as IMC90-001 supplied by Sandoz Inc.
[e] Gelled oil vehicle supplied by Abbott Laboratories.

regenerated Douglas-fir trees infested with western spruce budworm larvae. At the two higher dosages, there was little difference in population reduction due to either virus, but at the lowest dosage, GV gave 56% population reduction and the NPV only 6% (Cunningham et al. 1983b). Hence, it appeared that GV had greater potential than NPV as a biocontrol agent for western spruce budworm.

The third trial conducted in 1982 consisted of two large plots (172 ha each) selected for long-term studies of the impact of NPV and GV; it was hoped that the size of these plots would limit the effects of moth immigration. One plot was treated with 9.0 kg of lyophilized, NPV-infected larval powder and the other with 9.0 kg of a GV product (Table 1). In the year of application, results were variable but, on the average, NPV treatment caused a 53% reduction in larval populations compared to only 35% for GV. Results from adult emergence of field-collected, laboratory-reared larvae indicated a greater effect than field population data. The rates of successful adult emergence were 9.4% from the NPV plot, 16.3% from the GV plot, and 52.7% from the untreated check area (Table 2). The same studies and calculations were made during 1983 and 1984, and except for one anomaly (larval population reduction due to GV in 1984), both viruses had the greatest impact in the year of application, less the following year, and still less in the second year following application (Table 2). The epizootic initiated by the spray application decreased over time and did not control the western spruce budworm population (Otvos et al. 1989).

Experimental Spray Trials Using *Bacillus thuringiensis* (Bt)

In conjunction with the NPV trials in 1978, four 40-ha plots were treated with a commercial preparation of Bt called Thuricide® 16B (Sandoz Inc.). Dosage was 20 billion international units (BIU)/ha in a volume of 9.4 L/ha. At 15 days post-spray, population reductions due to treatment were 32%, 66%, 79%, and 91% in the four plots respectively (Hodgkinson et al. 1979). Spruce budworm populations were sampled in these plots during the next 2 years. In 1979, population increases were noted in two of the four treated plots compared with untreated check plots; a small increase was noted in the third and a significant decrease in the fourth plot. In 1980, this trend was reversed with populations decreasing in the check plots and increasing dramatically in the Bt-treated plots (Shepherd et al. 1982). Results were quite variable, and it appears that these small plots also received moth immigration, thus masking treatment and disease carry-over effects.

Laboratory studies have shown that western spruce budworm larvae are more susceptible to both the HD-1 and NRD-12 strains of Bt than are either spruce budworm or jack pine budworm larvae. Western spruce budworm larvae were 3-fold more sensitive than the other two species when fed on host foliage and 8- to 14-fold more sensitive when Bt was incorporated into their diet (van Frankenhuyzen and Fast 1989).

In 1987, the British Columbia Ministry of Forests initiated a large study to determine the efficacy of Bt treatments against the western spruce budworm and of the amount of foliage protection that can be obtained on overstory and understory Douglas-fir trees over a range of sites and forest types. A final report is in preparation.

Logistics of Applying Biological Insecticides

Bt sprays currently appear to be the preferred method for controlling the western spruce budworm in British Columbia, but variability of results and the lack of long-term benefits have prevented the initiation of routine control operations

Table 2. Impact of NPV and GV treatments on western spruce budworm, *Choristoneura occidentalis*, in the year of treatment and the subsequent 2 years assessed by (a) larval population counts from field samples and (b) adult emergence from larval samples reared in the laboratory.

Virus	Year sampled	% Larval population reduction attributed to virus[a]	% Successful adult emergence[b]
NPV	1982	51.8	9.4
	1983	33.7	46.2
	1984	14.4	61.6
GV	1982	34.6	16.3
	1983	14.7	45.3
	1984	25.6	66.3
Checks	1982	–	52.7
	1983	–	65.0
	1984	–	70.6

[a] Calculated using a modified Abbott's formula (Fleming and Retnakaran 1985).
[b] Averages of 2 or 3 samples reported by Otvos et al. 1989.

against this pest. Merchantable Douglas-fir trees are only occasionally killed, so that damage is essentially restricted to growth loss and deformity of future crop trees, and to regeneration loss in multi-aged stands (Alfaro 1985). The cost of one years' reprieve from such damage is difficult to justify in the absence of wholesale stand mortality.

Variability of results can be attributed to many causes, but two stand out: variability in bud flush dates and difficulties in obtaining even and adequate applications in mountainous terrain. Western spruce budworm larvae feed inside the buds and are not accessible until after bud flush. By this time the insects have usually reached the fourth instar and their susceptibility to biological diseases has decreased. In the case of viruses, there is not sufficient time before feeding ceases to obtain good multiplication and transmission of the pathogen. Also, because viruses are slow acting, there is little inhibition of feeding until the larvae are close to pupation, and thus little foliage protection is obtained. Bud phenology varies considerably from tree to tree and from site to site; coordinating aerial applications with bud break over a large area is difficult, if not impossible. The low humidities, unstable weather conditions, and rugged terrain also make applications difficult. Unpredictable winds and changeable lapse conditions result in the loss of finer droplets from the spray deposit, resulting in inadequate and irregular insecticide distributions. All of these factors often lead to compromises in spray operations and to less than satisfactory results.

References

Alfaro, R.I. 1985. Factors affecting radial growth of Douglas-fir caused by western spruce budworm. Pages 251-252 *in* C.J. Sanders, R.W. Stark, E.J. Mullins and J. Murphy (Eds.), Recent advances in spruce budworms research. Can. For. Serv., Ottawa, Ont.

Cunningham, J.C. 1985. Status of viruses as biocontrol agents for spruce budworms. Pages 61-67 in Proc. Symp. Microbial control of spruce budworms and gypsy moths. Windsor Locks, Conn. 1984. U.S.D.A. For. Serv. GTR-NE-100.

Cunningham, J.C.; Kaupp, W.J.; McPhee, J.R. 1983a. A comparison of pathogenicity of two baculoviruses to the spruce budworm and western spruce budworm. Can. For. Serv. Res. Notes 3(2): 9-10.

Cunningham, J.C.; Kaupp, W.J.; McPhee, J.R.; Shepherd, R.F. 1983b. Ground spray trials with two baculoviruses on western spruce budworm. Can. For. Serv. Res. Notes 3(2): 10-11.

Fleming, R.; Retnakaran, A. 1985. Evaluating single treatment data using Abbott's formula with reference to insecticides. J. Econ. Entomol. 78: 1179-1181.

Furniss, R.L.; Carolin, V.M. 1977. Western forest insects. US For. Serv. Misc. Publ. 1339. 654 p.

Harris, J.W.E.; Alfaro, R.I.; Dawson, A.F.; Brown, R.G. 1985. The western spruce budworm in British Columbia, 1909-1983. Can. For. Serv., Pac. For. Cent., Victoria, B.C. Inf. Rep. BC-X-257. 32 pp.

Hodgkinson, R.S.; Finnis, M.; Shepherd, R.F.; Cunningham, J.C. 1979. Aerial applications of nuclear polyhedrosis virus and *Bacillus thuringiensis* against western spruce budworm. B.C. Min. For./Can. For. Serv. Joint Rep. 10. 19 pp.

Otvos, I.S.; Cunningham, J.C.; Kaupp, W.J. 1989. Aerial application of two baculoviruses against the western spruce budworm, Choristoneura occidentalis Freeman (Lepidoptera: Tortricidae), in British Columbia. Can. Entomol. 121: 209-217.

Shepherd, R.F. 1985. A theory on the effects of diverse host-climatic environments in British Columbia on the dynamics of western spruce budworm. Pages 60-70 *in* C.J. Sanders, R.W. Stark, E.J. Mullins, and J. Murphy (Eds.), Recent advances in spruce budworms research. Can. For. Serv., Ottawa, Ont.

Shepherd, R.F.; Cunningham, J.C. 1984. *Choristoneura occidentalis* Freeman, western spruce budworm (Lepidoptera: Tortricidae). Pages 277-279 in J.S. Kelleher and M.A. Hulme (Eds.), Biological control programmes against insects and weeds in Canada 1969-1980. Commonw. Agric. Bureaux, Slough, U.K.

Shepherd, R.F.; Gray, T.G.; Cunningham, J.C. 1982. Effects of nuclear polyhedrosis virus and *Bacillus thuringiensis* on western spruce budworm one and two years after aerial application. Can. Entomol. 114: 281-282.

van Frankenhuyzen, K.; Fast, P.G. 1989. Susceptibility of three coniferophagous *Choristoneura* species (Lepidoptera: Tortricidae) to *Bacillus thuringiensis* var. *kurstaki*. J. Econ. Entomol. 82: 193-196.

Chapter 12

Jack Pine Budworm, *Choristoneura pinus*

B.L. CADOGAN

Introduction

In Canada, the range of jack pine, *Pinus banksiana* (Fig.1), extends from the Mackenzie River valley in western Canada to the Atlantic Ocean; it is one of the major commercial species in the forests of Ontario (Davison 1984). The jack pine budworm, *Choristoneura pinus* (Fig. 2), is the most important insect pest of jack pine in Canada and the lake states of the United States (Clancy et al. 1980; Howse 1984). Although jack pine is its principal host, other species of pine and spruce (*Picea* sp.) are also attacked, especially when they occur as minor components of major jack pine stands (Prebble 1975).

Figure 1. Mature jack pine, *Pinus banksiana*.

Biology

The life history and habits of the jack pine budworm closely resemble those of its close relative, the spruce budworm, *C. fumiferana*, except that the jack pine budworm generally has seven larval instars (Nealis 1987) instead of the six that are standard for spruce budworm. Jack pine budworm moths emerge in July and August and deposit clusters of eggs on the host tree's needles. The eggs hatch in about 6 to 10 days, and the species overwinters, like spruce budworm, as second-instar larvae in hibernacula, and emerges in late May and early June. Jack pine budworm is considered a periodic pest, with outbreaks that usually last 1 to 3 years, occurring primarily in the Prairie provinces (Moody 1986; Frey 1986), in Ontario (Howse 1984, 1986), and infrequently in Quebec (Martineau 1975). In general, the outbreaks in Manitoba coincide with those reported in northwestern Ontario. Although the dynamics of jack pine budworm population epidemics and collapses are not well understood, it appears that population increases are related to the abundance of staminate jack pine flowers.

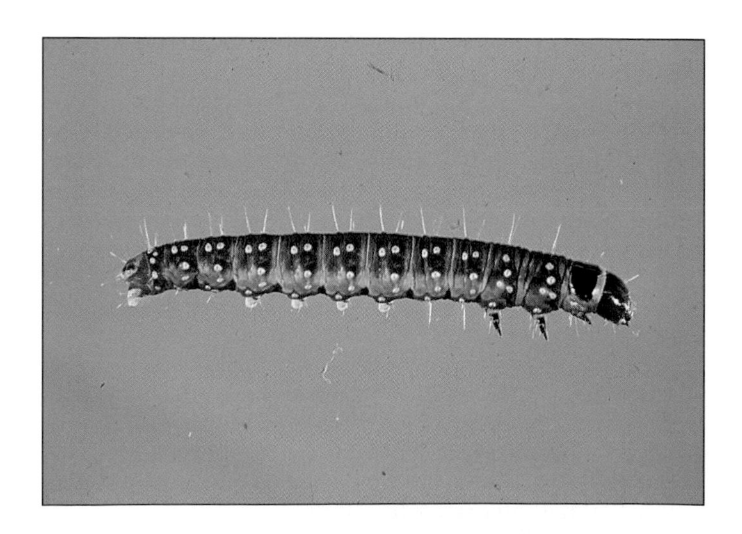

Figure 2. Larval jack pine budworm, *Choristoneura pinus*.

Outbreaks and Damage

Most of the past outbreaks in Canada, although short-lived, generally caused significant economic losses through tree mortality and top kill, but were allowed to run their course without human intervention (Prebble 1975). There were, however, a few epidemics that required suppression and these were satisfactorily treated with DDT, aminocarb, fenitrothion, and Zectran® (DeBoo 1975; Sippell and Howse 1975; Martineau 1975). However, the use of DDT in forests was discontinued in 1973, and Zectran® and aminocarb are no longer available for aerial use against jack pine budworm. There have been few incentives to conduct research in Canada that could lead to registration of insecticides for controlling this pest because most jack pine budworm outbreaks have been sporadic and ephemeral in nature, and the losses incurred during outbreaks have been generally accepted. Most of the research reported since 1973 seems to focus on understanding the dynamics of outbreaks (Howse 1986; Lysyk 1989; Nealis and Lysyk 1988); the biology of the species (Howse 1984; Nealis 1987); or the silvicultural manipulation of the host tree (Volney 1986).

However, by 1983 it was clear that the tree mortality and top kill that are sustained during outbreaks that run their natural course could no longer be routinely accepted. Such losses in wood would cause serious socioeconomic disruptions in those regions of the country that depended almost exclusively on jack pine as a resource. Thus, in 1984–85, a decision was made to intervene in serious budworm outbreaks in northern Ontario and Manitoba. Fenitrothion, the only insecticide then registered for aerial control of this species, was considered to be publicly unacceptable.

Field Research

Facing the above realities, and with the main objective of providing the forest manager with a wider arsenal of insecticides that would facilitate a variety of control strategies, field research was conducted during the budworm outbreak that lasted from 1984 to 1986. This research was primarily intended to determine the efficacy, against the jack pine budworm, of microbial and synthetic chemical insecticides that were already registered for controlling the spruce budworm.

Most jack pine budworm populations in both Ontario and Manitoba were high enough to require insecticidal suppression and adequate for most efficacy trials to be conducted satisfactorily; however, in Manitoba the outbreak was allowed to collapse naturally. In Ontario, several insecticides were assessed for full registration and a significant effort was undertaken to suppress the outbreaks.

In 1985, five formulations of *Bacillus thuringiensis* (Bt), one formulation of a nuclear polyhedrosis virus that is specific to spruce budworm, and two synthetic chemical insecticides, namely Matacil® 180F (aminocarb) and Sumithion® 20F (fenitrothion), were field tested. Both Matacil® 180F and Sumithion® 20F are flowable or suspension concentrate formulations that were intended to be more environmentally acceptable than those aminocarb and fenitrothion formulations that had been used against the spruce budworm. The several Bt formulations ranged in potency from 12 to 16 BIU/L (billion international units per litre). These were new, improved and more potent Bt formulations and the new strategy was to apply them largely undiluted.

There are two basic types of field trials employed in research that use the aerial application of pesticides, and both were conducted against jack pine budworm in Ontario. In 1985, small-scale experimental trials using Bt and the two chemical insecticides were conducted on 50-ha blocks. This type of trial is expected to conform rigidly to scientific principles as they apply to experimental design, implementation, analyses, etc. In 1985 and 1986, semi-operational trials, also called pilot tests, were conducted on blocks of less than 250 ha. Bt was the only material tested. This type of trial is less scientifically rigid than the small-scale experimental trial and usually uses standard operational equipment, methods, and guidelines.

Results from the 1985 small-scale trial (Cadogan et al. 1986) are presented in Table 1. Futura® XLV sprayed once at 30 BIU/ha and Sumithion® 20F sprayed twice at an active ingredient rate of 210 g/ha were very effective in reducing moderately high populations of jack pine budworm, and adequately protected the host tree from defoliation. Matacil® 180F sprayed twice at 70 g/ha was slightly less effective, and Futura® at 20 BIU/ha and nuclear polyhedrosis virus were marginally effective against the pest.

Results from the 1985 semi-operational trials (B.H. McGauley, unpublished 1985 Pest Control Forum Report) indicate that the jack pine budworm populations in the blocks treated with Thuricide® (Table 1) had subsided to levels that were too low to reach definitive conclusions about the product's efficacy. Where the budworm populations were acceptable for research, diluted and undiluted Dipel® sprayed at three volume rates was effective in controlling the pest.

In the 1986 semi-operational trials, Foray® 48B and Dipel® 8AF were both applied undiluted at 20 BIU/ha. Results (J. Meating; G.M. Howse and B.H. McGauley, unpublished 1986 Pest Control Forum Report) indicated that the populations had receded to less than 2.0 larvae per 60-cm branch tip and these levels were such that efficacy assessments were inconclusive.

Conclusions

The results from these 1985 and 1986 trials showed that Bt was capable of controlling jack pine budworm with the same effectiveness as it does the spruce budworm. Large-scale operational spray programs that were conducted with temporarily registered Bt further confirmed the product's efficacy against the jack pine budworm.

True to form, jack pine budworm populations had subsided to endemic levels by 1987. However the research has resulted in several insecticides being registered for aerial application (Table 2). This development will provide the forest manager with additional options with which to confront future epidemics of this species.

Table 1. Insecticides assessed in Ontario during 1985 and 1986 to determine their efficacy in controlling the jack pine budworm, *Choristoneura pinus*.

Location/year	Product name (active ingredient)	Aircraft: Atomizer	Area sprayed (ha)	Appl. rate/ha		Diluent	No. of appl.	Efficacy
				Active ingredient	Litres			
Ontario 1985[a]	Matacil® 180F (aminocarb)	Cessna 188: 4 AU3000	50	70 g	1.5	Water	2	Good
	Sumithion® 20F (fenitrothion)	Cessna 188: 4 AU3000	50	210 g	1.5	Water	2	Very good
	Futura® XLV (Bt)	Cessna 188: 4 AU3000	50	30 BIU[b]	2.1	Undiluted	1	Very good
	Futura® XLV (Bt)	Cessna 188: 4 AU3000	50	20 BIU	1.4	Undiluted	1	Fair
	Nuclear polyhedrosis	Cessna 188: 4 AU3000	50	7.5×10^{11} PIB[c]	9.5	Water	2	Fair
Ontario 1985[d]	Thuricide® 48 LV (Bt)	Cessna: 6 AU5000	240	10 BIU	1.6	Water	1	Inconclusive
	Thuricide® 64 LV (Bt)	Cessna: 6 AU5000	155	20 BIU	1.2	Undiluted	1	Inconclusive
	Thuricide® 64 LV (Bt)	Cessna: 6 AU5000	30	30 BIU	1.8	Undiluted	1	Inconclusive
	Dipel® 132 (Bt)	Cessna: 6 AU5000	75	30 BIU	2.4	Undiluted	1	Good
	Dipel® 176 (Bt)	Cessna: 6 AU5000	155	30 BIU	1.8	Undiluted	1	Very good
	Dipel® 176 (Bt)	Cessna: 6 AU5000	30	30 BIU	3.6	Undiluted	1	Very good

[a] Experiment trials conducted by Forest Pest Management Institute.
[b] Billion international units (BIU).
[c] Polyhedral inclusion bodies (PIB).
[d] Semi-operational trials conducted by Ontario Ministry of Natural Resources and Great Lakes Forestry Centre.

Table 2. Insecticides registered for aerial use against jack pine budworm, *C. pinus*, in Canada and approved for use in woodland[a] and forest[b] management.

Insecticide trade name	Common name	Registrant/ Manufacturer	PCP No.
Dipel® 88	Bt	Abbott Labs Ltd.	16873
Dipel® 132	Bt	Abbott Labs Ltd.	17954
Futura® XLV	Bt	Duphar-BV	20485
Futura® suspension	Bt	Duphar-BV	17778
Bactospeine® F	Bt	Duphar-BV	17782
Thuricide® 32	Bt	Zoecon Corporation, A Sandoz Company	17054
Thuricide® 48LV	Bt	Zoecon Corporation, A Sandoz Company	17980
Novathion® Conc. Liquid	fenitrothion	AS Cheminova	14299
Sumithion® Technical	fenitrothion	Sumitomo Canada Ltd.	11137

[a] Covers treatment of ≤500 ha of the same wooded site.
[b] Covers treatment of >500 ha of a wooded area.

References

Cadogan, B.L.; Zylstra, B.F.; Nystrom, C.; Ebling, P.M.; Pollock, L.B. 1986. Evaluation of a new Futura formulation of Bacillus thuringiensis on populations of jack pine budworm, *Choristoneura pinus pinus* (Lepidoptera: Tortricidae). Proc. Entomol. Soc. Ont. 117: 59-64.

Clancy, K.M.; Gièse, R.L.; Benjamin, D.M. 1980. Predicting jack pine budworm infestations in northwestern Wisconsin. Environ. Entomol. 9: 743-751.

Davison, R.W. 1984. Jack pine management in the boreal forests of Ontario. Pages 169-172 *in* C.R. Smith, and G. Brown (Chairmen), Jack Pine Symposium. Proc. Environ. Can., Can. For. Serv., Sault Ste. Marie, Ont. COJFRC Symp. Proc. 0-P-12. 195 pp.

DeBoo, R.F. 1975. Manitoba Control Projects. 1967. Page 154 *in* M.L. Prebble (Ed.), Aerial control of forest insects in Canada. Environ. Can., Ottawa, Ont.

Frey, G.E. 1986. Jack pine budworm in Saskatchewan. Pages 23-24 *in* Jack pine budworm information exchange. Manitoba Natural Resources, Winnipeg, Man.

Howse, G.M. 1984. Insect pests of jack pine: biology, damage and control. Pages 131–138 *in* Smith C.R. and G. Brown (Chairmen), Jack Pine Symposium. Proc. Environ. Can., Can. For. Serv., Sault Ste. Marie, Ont. COJFRC Symp. Proc. 0-P-12. 195 pp.

Howse, G.M. 1986. Jack pine budworm in Ontario. Pages 47–50 *in* Jack pine budworm information exchange. Manitoba Natural Resources, Winnipeg, Man.

Lysyk, T.J. 1989. A multiple-cohort for simulating jack pine budworm (Lepidoptera: Tortricidae) development under variable temperature conditions. Can. Entomol. 121: 373–387.

Martineau, R. 1975. Jack pine budworm. Quebec Control Projects. Page 157 *in* M.L. Prebble (Ed.), Aerial control of forest insects in Canada. Environ. Can., Ottawa, Ont.

Moody, B.H. 1986. The jack pine budworm history of outbreaks, damage and FIDS sampling and predictions systems in the prairie provinces. Pages 15–22 *in* Jack pine budworm information exchange. Manitoba Natural Resources, Winnipeg, Man. 96 pp.

Nealis, V.G. 1987. The number of instars in jack pine budworm, *Choristoneura pinus pinus* Free. (Lepidoptera: Tortricidae), and the effect of parasitism on head capsule width and development time. Can. Entomol. 119: 773–777.

Nealis, V.G.; Lysyk, T.J. 1988. Sampling overwintering jack pine budworm *Choristoneura pinus pinus* Free. (Lepidoptera: Tortricidae) and two of its parasitoids. (Hymenoptera). Can. Entomol. 120: 1101–1111.

Prebble, M.L. 1975. Jack pine budworm—Introduction. Page 152 *in* M.L. Prebble (Ed.), Aerial control of forest insects in Canada. Environ. Can., Ottawa, Ont.

Sippell, W.L.; Howse, G.M. 1975. Jack pine budworm: Ontario Control Projects 1968-1972. Pages 152–157 *in* M.L. Prebble (Ed.), Aerial control of forest insects in Canada. Environ. Can., Ottawa, Ont.

Volney, W.J.A. 1986. Current jack pine budworm research in Manitoba. Pages 41–43 *in* Jack pine budworm information exchange. Manitoba Natural Resources, Winnipeg, Man. 96 pp.

Chapter 13

Douglas-fir Tussock Moth, *Orgyia pseudotsugata*

I.S. Otvos, J.C. Cunningham, and R.F. Shepherd

Introduction

The Douglas-fir tussock moth, *Orgyia pseudotsugata*, (Lepidoptera: Lymantriidae) is an important defoliator in the interior dry-belt forests of southern British Columbia and the western United States. The primary host tree in British Columbia is Douglas-fir, *Pseudotsuga menziesii*. However, in other areas, heavy feeding can also occur on grand fir, *Abies grandis*, white fir, *A. concolor*, ponderosa pine, *Pinus ponderosa*, and several species of spruce, *Picea* spp.

Outbreaks of the Douglas-fir tussock moth occur periodically at intervals of about 8 to 14 years in British Columbia, Washington, Idaho, Oregon, California, Arizona and New Mexico. Eight outbreaks have been reported since 1916 in susceptible forest stands in British Columbia (Harris et al. 1985; Shepherd and Otvos 1986) (Fig. 1).

Douglas-fir tussock moth population increases are usually first noted in valley bottoms and on open-grown trees near settlements. Infestations usually last for 1 to 4 years in any particular stand and cause growth loss, top kill and tree mortality (Fig. 2) (Wickman 1978; Alfaro et al. 1987). Defoliation occurs in distinct patches in the first year, spreading and coalescing in later years of the outbreak. All age classes and sizes of the host trees may be attacked, and the most severely attacked trees generally die. Large, but less severely defoliated trees are weakened and these may be subsequently attacked and killed by the Douglas-fir bark beetle, *Dendroctonus pseudotsugae*. Outbreaks with noticeable defoliation usually last from 2 to 5 years and are generally terminated by epizootics of a naturally occurring nuclear polyhedrosis virus (NPV) but usually not before severe damage occurs to the infested stands (Dahlsten and Thomas 1969; Mason and Luck 1978; Shepherd and Otvos 1986).

Two morphotypes of NPV virus have been isolated from Douglas-fir tussock moth larvae. In the first type the virus particles are embedded singly (unicapsid) in the protein inclusion-body matrix and in the second type they are embedded in bundles (multicapsid) (Hughes and Addison 1970).

Biology

The Douglas-fir tussock moth has one generation per year. The winged males search out the flightless females and after mating the female lays all her eggs, about 150 to 200 in one mass, on her cocoon in August or September (Wickman

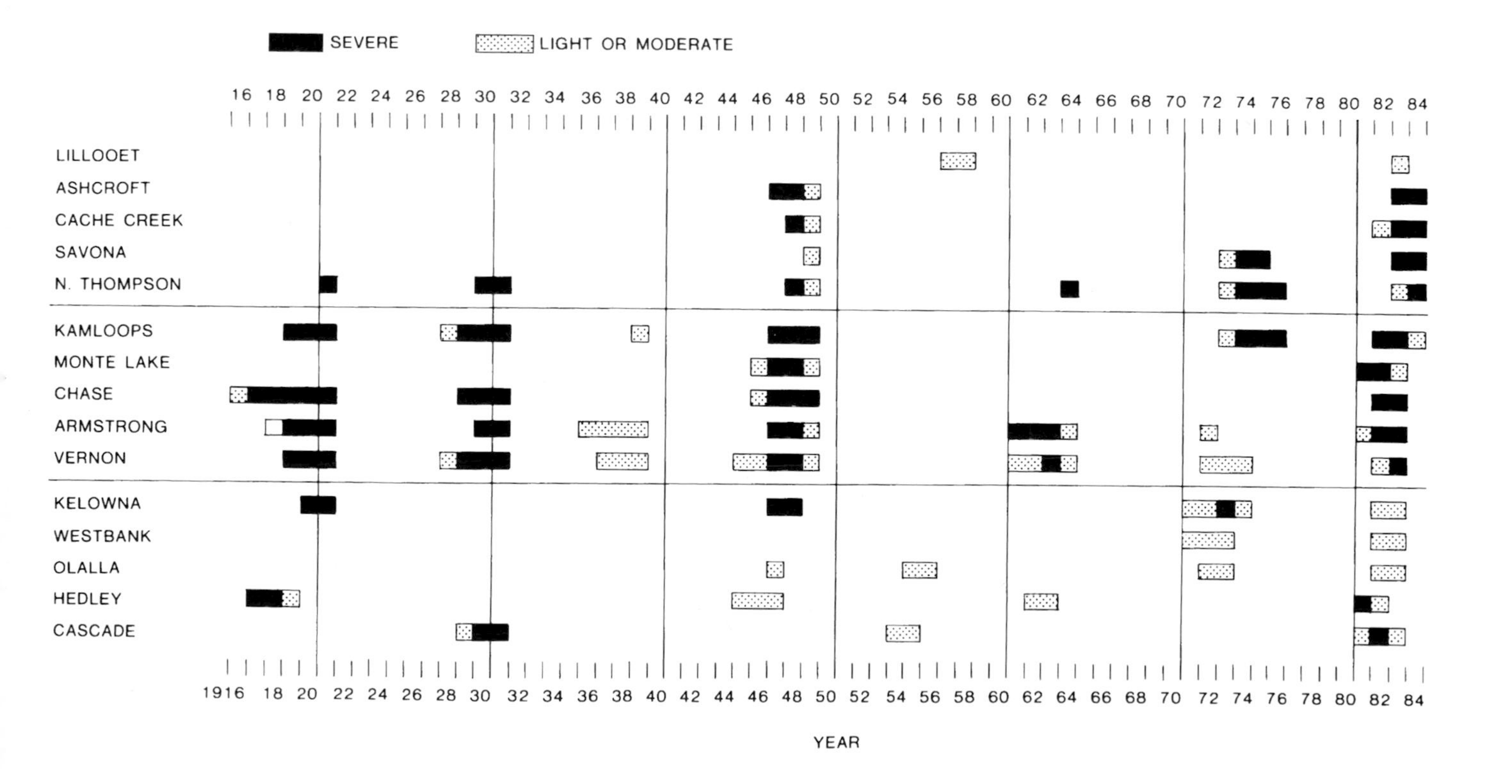

Figure 1. Histogram of Douglas-fir tussock moth, *Orgyia pseudotsugata*, outbreak periods from 1916 to 1984 by geographic locations in British Columbia.

Figure 2. Douglas-fir stands killed by Douglas-fir tussock moth, *O. pseudotsugata*.

Figure 3. Mature Douglas-fir tussock moth, *O. pseudotsugata*, larvae.

and Beckwith 1978; Shepherd et al. 1984a). The egg is the overwintering stage. Eggs hatch just after bud flush from about late May to early June depending on temperature. The young larvae can feed only on new needles of the elongating shoots and can be dispersed by drifting on silken threads carried by wind. Older larvae (Fig. 3) will feed on old foliage if the current year's foliage is depleted. Partly consumed needles dry out and change color, giving a reddish brown appearance to infested trees. Males have five larval instars and females have six. Mature larvae pupate within cocoons in late July or August, usually on the underside of needles and branches; the pupal stage lasts about 2 weeks. At high population densities, cocoons may also be found at many other sites including tree trunks and fence posts.

Early Treatments

During infestations in the 1940s, 1950s, and 1960s, DDT was used operationally from the air to control the tussock moth in Idaho, Oregon, and California (Wickman et al. 1975). Later, Zectran® (carbaryl), Dursban® (organophosphate chloropyplim), and pyrethrin were tested as alternatives to DDT (Wickman et al. 1975; Lejeune 1975). In the early 1970s, a nuclear polyhedrosis virus and *Bacillus thuringiensis* (Bt) were successfully tested in small field plots in Oregon (Stelzer et al. 1975).

These early operational and experimental United States spray treatments were tested in Canada under experimental conditions and the results summarized by Lejeune (1975). Reduced dosages of DDT (0.57 kg/ha, 0.5 lb/acre) and malathion (1.12 kg/ha, 1.0 lb/acre) were tested from the air, each in 9.4 L/ha (1 U.S. gal/acre) of spray formulation, to minimize the undesirable side effects of DDT at the higher dosage (Lejeune 1975). A nuclear polyhedrosis virus was also successfully field-tested by hand spraying on individual trees (Morris 1963).

Operational Field Trials Against the Douglas-fir Tussock Moth in British Columbia, 1975 to 1982

The efficacy of five materials was tested for the control of Douglas-fir tussock moth during 1975 and 1976: acephate, Dimilin®, Bt, and two strains of the naturally occurring nuclear polyhedrosis virus (NPV). The multicapsid strain of NPV was tested in a water and molasses formulation in aerial and ground applications in 1981. In 1982, an oil and water formulation was compared with the molasses and water formulation and reduced dosages of the virus were also tested in oil formulation.

Chemical Insecticides

Acephate (Orthene®) (registered for control of other insect pests) and Dimilin® (not registered for any insects in Canada in 1975) were tested at various dosages of active ingredient in 1975 on 180 ha and acephate was used in operational trials on 8 490 ha in 1976. Acephate was applied at 1.12 kg in a volume of 9.4 L/ha experimentally in 1975 and in operational trials in 1976. Reduced rates of 0.56 and 0.84 kg in 9.4 L/ha were tested in 1976. Dimilin® was applied at 0.28 kg and 0.14 kg/ha. The insecticides were mixed in a 10% aqueous ethylene glycol formulation. The treatment was applied when 90% of the monitored egg masses had hatched and the larvae had moved out to feed on the new shoots. Nine plots, ca. 180 ha in size, were treated in 1975.

Acephate was effective in the operational trials at 1.12 kg/ha and experimentally at both 1.12 and 0.84 kg/ha, but did not provide adequate control at 0.56 kg/ha (Fig. 4). Its effect on tussock moth larvae was rapid and appeared to be complete by 7 days, with little mortality occurring after that time (Shepherd 1980). An experiment on the timing of acephate applications was also conducted on three plots with a total area of 60.7 ha. This application was made when 70% of the egg masses had hatched but the larvae had not necessarily left the egg masses to feed. Effectiveness of this early treatment was reduced because acephate has a short active life and larvae that hatched late because of cool weather escaped exposure.

Dimilin® acted more slowly than acephate but retained its effectiveness for a longer time. Dosages of 0.28 and 0.14 kg/ha gave almost complete mortality, but 0.07 and 0.035 kg/ha did not give acceptable results (Hard et al. 1978). Because of the faster action of acephate, tree defoliation was less (3.4%) than with Dimilin®, (11.9%), although the latter was still acceptable compared with two nontreated areas which suffered 73.7% and 90.8% defoliation. As a result of these trials, Orthene® (acephate) was registered for control of Douglas-fir tussock moth at 1.12 kg active ingredient in 3.8 L water/ha.

The results of the tests with carbaryl (Sevin®) are discussed in Chapter 56.

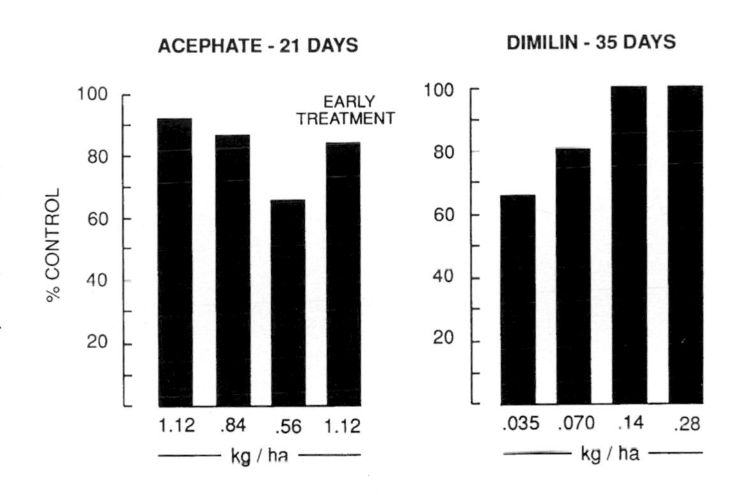

Figure 4. Effectiveness of different concentrations of acephate and Dimilin® and an early treatment using acephate.

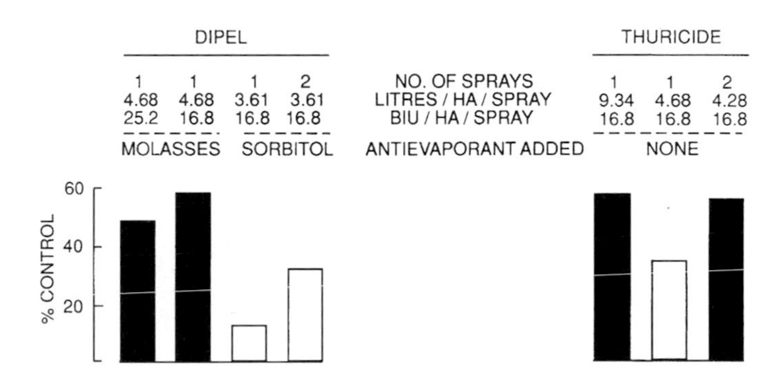

Figure 5. Effectiveness of aerial application of different formulations of Bt against the Douglas-fir tussock moth, *O. pseudotsugata*.

Bt Applications

A total of about 15 260 ha were treated operationally with two Bt products in 1975; Dipel®-SC and Thuricide®-HPC were applied in the Kamloops Forest District in British Columbia using a Cessna AgTruck aircraft equipped with Micronair® AU3000 atomizers. About 8 620 ha were treated with 17.0 billion international units (BIU)/ha of 60% Dipel-SC® mixed with 30% sorbitol and 10% water by volume and applied at 3.6 L/ha. A single application was made over 6 694 ha and double applications were made on 5 245 ha. The remaining 3 320 ha were treated with 17.0 BIU/ha of 80% Thuricide®-HPC mixed with 20% water by volume and applied at 4.7 L/ha. A separate test was made with two applications of Thuricide® for comparison and double and single applications of Thuricide® were also compared. A single application of Thuricide® was made on 3 035 ha and a double application on 283 ha. In 1976, Thuricide® was used operationally along streams and acephate was utilized on the upland sites.

The results of the 1975–76 Bt trials were disappointing; foliage protection and early larval mortality were inconsistent and unacceptably low. Larval mortality in the operational trial spray blocks 21 days after the spray averaged 34% (range: 13% to 57%) (Fig. 5). Larval mortality was determined by counting the number of larvae per unit of branch area on two midcrown branches from 15 to 25 sample trees per plot. Mortality attributed to treatment was calculated from the difference in survival between treated and untreated plots (Shepherd 1980). A double application of both materials and the higher volume of Thuricide® gave 20% better control than a single application of 4.7 L/ha (Shepherd 1980). Satisfactory levels of control in the operational program were not achieved until 35 days after spraying and then only in plots that received double applications. Experiments on increased volumes, concentrations, formulations, and different products indicated that Thuricide® was superior to Dipel®, and molasses was a better additive than sorbitol, but none of the trials resulted in satisfactory foliage protection of the trees.

Virus Applications

The use of virus for control of Douglas-fir tussock moth was considered as early as 1962 in British Columbia (Morris 1963), but it was not until 1975 that a large-scale aerial spray trial was conducted jointly by the British Columbia Forest Service, Forestry Canada, and the United States Forest Service in British Columbia (Stelzer et al, 1977; Shepherd 1980; Cunningham and Shepherd 1984). Three virus stocks were applied aerially to 13 plots with a total area of about 1 062 ha. Two of the stocks were propagated from the multicapsid strain of virus isolated from the Douglas-fir tussock moth in Oregon; one was propagated in the original host and the second in larvae of whitemarked tussock moth, *Orgyia leucostigma*. The third stock was the unicapsid strain of the virus isolated from and propagated in the whitemarked tussock moth from Nova Scotia. All three stocks were tested experimentally on a total of 72 ha at 18.7 L/ha, and in an operational trial on 990 ha at 9.4 L/ha. All three stocks of virus were applied at 2.5×10^{11} polyhedral inclusion bodies (PIB)/ha in an aqueous suspension containing 25% molasses or 50% Sandoz adjuvant V (v/v, concentration is expressed as percent volume over volume). Treatments were applied by Cessna AgTrucks and AgWagons equipped with booms fitted with Tee-jet® 8010 flat fan nozzle tips. Larvae were in the second and third instar at the time of application, and most of the current year's foliage had already been destroyed.

All of the virus treatments resulted in high infection and high larval mortality; no significant differences could be detected between strains or formulations. Larval population reductions in the treated plots ranged from 71% to 93%, 68% to 95%, and 84% to 99% at 21, 28, and 35 days post-spray, respectively. The corresponding larval mortality in the check plots averaged 68%, 76%, and 84%. These population reductions were associated with high levels of naturally occurring virus infection enhanced by the spray application in the treated plots. The high mortality in the check plots was caused by naturally occurring virus; over 50% of the larvae in the check plots were naturally infected 35 days after the date of spraying. This was expected as the outbreak was in its declining phase in that area. Only light defoliation, 11% to 14% of the older foliage, occurred on all but one of the treated plots where spray deposit, measured on Kromekote® cards placed 30 cm above ground level, was poor due to adverse weather during treatment. In contrast, defoliation of the older foliage in the check plots averaged 60%. No egg masses were found in the treated plots, while egg masses were common in the check plots and adjacent untreated areas.

Figure 6. Helicopter spraying against Douglas-fir tussock moth, *O. pseudotsugata*, 1981.

Figure 7. Ground spraying against Douglas-fir tussock moth, *O. pseudotsugata*, 1981.

The results of the 1975 virus trials were promising, but because the treatments were applied in the declining phase of the outbreak the effect of the application on the course of the outbreak could not be evaluated. Following these tests, the multicapsid virus isolated from and produced in Douglas-fir tussock moth larvae was registered by the Environmental Protection Agency in the United States in 1976 under the name TM BioControl-1. The same virus produced in whitemarked tussock moth received temporary registration in Canada in 1983 and full registration in 1987 under the name Virtuss®. TM BioControl-1 was also granted Canadian registration in 1987.

Another Douglas-fir tussock moth population buildup started in 1980 in the Hedley area in southern British Columbia where a previous outbreak had collapsed in 1963. This provided an opportunity to test whether a virus epizootic could be initiated artificially at an early phase of an outbreak before a natural epizootic might occur and before significant tree damage occurred. In 1981, four plots, totalling 20 ha, with moderate to light Douglas-fir tussock moth population densities were aerially sprayed with NPV (Virtuss®) using a helicopter equipped with an 11-m-long boom and nine flat fan nozzles (Tee-jet® 8010); a dosage of 2.2×10^{11} PIB/ha when 60% of the larvae were in the first instar and 40% in the second (Fig. 6) was applied. The aqueous virus suspension, containing 25% molasses, was applied at 11.3 L/ha. Spray deposit, measured on Kromekote® cards, in the four aerially treated plots was poor with 8.4, 8.1, 5.1, and 2.1 droplets/cm^2 (Shepherd et al. 1984b). In addition to the aerial spray trials, trees in two additional plots were sprayed from the ground with a modified orchard-type hydraulic sprayer applying 4.5 L of aqueous suspension/tree (containing 2.4×10^{10} PIB/ha with 25% molasses, v/v) (Fig. 7). An assessment of virus spread was made in one of the ground-treated plots.

An epizootic occurred both in the aerially and ground-treated plots, even at population densities as low as about 40 larvae/m^2 of foliage. Douglas-fir tussock moth populations were reduced and no egg masses were found in any of the treated plots. The occurrence of an epizootic appeared to be inversely related to initial population density; the epizootic occurred sooner, about 2 weeks earlier, in the plots with higher larval densities, and the initial level of virus infection was higher in plots with higher larval densities. By the end of the larval development 100% mortality was recorded. The incidence of viral infection increased slowly over the first 4 weeks and an epizootic developed 5 weeks after treatment in the aerially treated plots. It was surprising to find naturally occurring virus in our check plots because this is not usually prevalent in new outbreaks. However, it could have been introduced by monitoring staff. These experiments showed that the virus can be introduced into a Douglas-fir tussock moth population at an early phase of the outbreak and a viral epizootic can be initiated both by aerial and ground treatment.

In a study of horizontal virus transmission, a line of 15 scattered trees was treated by ground spray and both the treated and the intermediate trees were monitored. Parallel lines of trees at 50 m on either side and one at 100 m on one side were also sampled to investigate spread of the virus. No spray drift was detected on Kromekote® cards placed between the treated line of trees and lines of trees used to detect spread. A line of check trees was located 200 m from the treated trees. Except for the post-spray on the intermediate trees at week 7, the incidence of virus infection in the larvae, collected from the sample trees, followed the expected pattern of spread from an infection source (Shepherd et al. 1984b). Infection appeared to decrease with distance from the treated trees. Thus, treatment of Douglas-fir tussock moth infestations with widely spaced swaths, every 100 to 200 m, may be sufficient to initiate epizootics and protect infested trees.

Virus-killed larvae were not observed in the ground-sprayed plots in the field until 5 to 8 weeks after spraying. Thus, feeding damage continued for most of the larval period and little foliage protection due to treatment could be detected either in ground-treated plots or the aerially treated plots. This indicates that a virus application is most useful when applied early in the outbreak cycle, preferably in the year before significant defoliation takes place, so that infested stands can tolerate defoliation until the treatment takes effects.

In 1981, aqueous tank mixes with 25% molasses (v/v) were used. This mix was widely used in previous tests both in Canada and the United States, but with the low relative humidity in the interior of British Columbia, spray deposits have often been poor. In 1982, the second year of the tussock moth outbreak, Virtuss® was applied in two different tank mixes at another location (Veasy Lake). In an aqueous mix the recommended (label) dosage (2.5×10^{11} PIB/ha) of virus was applied, while in oil mixes (25% Dipel blank carrier vehicle and 75% water) the virus was applied at the recommended dosage as well as at one-third and one-sixteenth of that recommended dosage. Each of these treatments was applied to a 10-ha plot with a fixed-wing aircraft when most larvae were in the first instar. The aircraft was equipped with a boom and Tee-jet® 8005 nozzles calibrated to deliver 9.4 L/ha (Otvos et al. 1987a).

Applications of Virtuss® in the oil mix resulted in population reductions at 6 weeks post-spray which increased with dosage. The full dosage (2.5×10^{11} PIB/ha) resulted in a 95% population reduction; one-third of this gave 91% and one-sixteenth resulted in a 65% reduction. The full dosage of the aqueous mix with molasses reduced the population by 87% (Table 1).

Surveys conducted in the fall of 1982 showed that egg-mass densities in all treated plots were reduced from their spring outbreak values to endemic values or below, while in two of the check plots egg-mass densities remained about the same and doubled in the third. The Virtuss® treatments prevented significant tree mortality in the treated plots. One year after application, less than 1% of sample trees died in the treated plots and 38% died in the check plots. In 1984, two years after the virus application, no tussock moth larvae were

Table 1. Population densities of Douglas-fir tussock moth (*O. pseudotsugata*) larvae and proportion of dead sample trees in plots treated in 1982 with Virtuss® and in matching check plots at Veasy Lake, British Columbia.

Plot no.	Treatment[a] PIB/ha	1982 Pre-spray larval/m^2	% Larval population reduction[b] 1982	% Sample trees killed by Douglas-fir tussock moth 1983	1984
T1	1.6×10^{10}-O	182.8	64.7	0	0
T2	8.3×10^{10}-O	145.8	90.6	2	7
T3	2.5×10^{11}-O	302.0	95.1	0	4
T4	2.5×10^{11}-M	41.8	86.6	0	0
				$\bar{x}$[c] 0.6	2.8
C1	Check	197.5		53	60
C2	Check	136.9		cut for powerline	
C3	Check	360.6		60	62
C4	Check	81.2		0	0
				$\bar{x}$ 37.8	40.7

[a] Treatment: PIB/ha = polyhedral inclusion bodies/hectare; O = oil formulation; M = molasses formulation.
[b] % Reduction was calculated by a modified Abbott's formula (Fleming and Retnakaran 1985).
[c] $\bar{x}$ = Average for the four plots.

found in any of the treated or check plots. Tree recovery, in terms of new foliage production, was good in the treated plots. The number of trees killed increased slightly from the previous year, both in the treated and the check plots (Table 1). Tree recovery continued in 1985 and no additional tree mortality occurred that could be attributed to Douglas-fir tussock moth (Otvos et al. 1987b). It is now considered feasible, as a result of these trials, to reduce the recommended dosage of NPV from 2.5×10^{11} to 8.3×10^{10} PIB/ha and to use either the aqueous or emulsifiable oil formulation for virus applications.

The results of the experiments with NPV over the past few years indicate that the development of Douglas-fir tussock moth outbreaks may be prevented by the application of virus at the beginning of the outbreak. Foliage protection may be poor in the year of application, but will be substantial in the following years. Virus application at reduced dosages makes it economically more acceptable and gives forest managers a viable alternative to the use of chemical insecticides. The cost of virus application may be further reduced by spraying widely spaced swaths, giving only partial coverage of the infested stands. Both the reduced virus dosage and widely spaced swaths will be tested during the next Douglas-fir tussock moth outbreak in British Columbia.

References

Alfaro, R.I.; Taylor, S.P.; Wegwitz, E.; Brown, R.G. 1987. Douglas-fir tussock moth damage in British Columbia. For. Chron. 63: 351-355.

Cunningham, J.C.; Shepherd, R.F. 1984. *Orgyia pseudotsugata* (McDunnough), Douglas-fir tussock moth (Lepidoptera: Lymantriidae). Pages 363-367 *in* J.S. Kelleher and M.A. Hulme (Eds.), Biological Control Programmes against Insects and Weeds in Canada, 1969-1980. Commonw. Agric. Bureaux, Slough, U.K.

Dahlsten, D.L.; Thomas, G.M. 1969. A nucleopolyhedrosis virus in populations of the Douglas-fir tussock moth, *Hemerocampa pseudotsugata*, in California. J. Invertebr. Pathol. 13: 264-271.

Fleming, R.; Retnakaran A. 1985. Evaluating single treatment data using Abbott's formula with reference to insecticides. J. Econ. Entomol. 78: 1179-1181.

Hard, J.S.; Ward, J.D.; Ilnytzky, S.I. 1978. Control of Douglas-fir tussock moth by aerially applied Dimilin (TH6040). USDA For. Serv., Pac. Southw. For. Range Exp. Stn., Res. Paper PSW-130, 6 pp.

Harris, J.W.E.; Dawson, A.F.; Brown, R.G. 1985. The Douglas-fir tussock moth in British Columbia, 1916-1984. Can. For. Serv., Pac. For. Cent., Inf. Rep. BC-X-268, 16 pp.

Hughes, K.M.; Addison, R.B. 1970. Two nuclear polyhedrosis viruses of the Douglas-fir tussock moth. J. Invert. Pathol. 16: 196-204.

Lejeune, R.R. 1975. Douglas-fir tussock moth. Pages 206-207 *in* M.L. Prebble (Ed.), Aerial control of forest insects in Canada. Environ. Can. Cat. No. F023/19/1975.

Mason, R.R.; Luck, R.L. 1978. Population growth and regulation. Pages 41-47 *in* M.H. Brookes, R.W. Stark, and R.W. Campbell (Eds.), The Douglas-fir Tussock Moth: A Synthesis. U.S.D.A. For. Serv. Tech. Bull. 1585.

Morris, O.N. 1963. The natural and artificial control of the Douglas-fir tussock moth. *Orgyia pseudotsugata* (McDunnough), by a nuclear polyhedrosis virus. J. Insect Pathol. 5: 401-414.

Otvos, I.S.; Cunningham J.C.; Friskie, L.M. 1987a. Aerial application of nuclear polyhedrosis virus against Douglas-fir tussock moth, *Orgyia pseudotsugata* (McDunnough) (Lepidoptera: Lymantriidae): I. Impact in the year of application. Can. Entomol. 119: 697-706.

Otvos, I.S.; Cunningham, J.C.; Alfaro, R.I. 1987b. Aerial application of nuclear polyhedrosis virus against Douglas-fir tussock moth, *Orgyia pseudotsugata* (McDunnough) (Lepidoptera: Lymantriidae): II. Impact 1 and 2 years after application. Can. Entomol. 119: 707-715.

Shepherd, R.F. (Ed.). 1980. Operational field trials against the Douglas-fir tussock moth with chemical and biological insecticides. Can. For. Serv., Pac. For. Res. Cent. Inf. Rep. BC-X-201. 19 pp.

Shepherd, R.F.; Otvos, I.S. 1986. Pest management of Douglas-fir tussock moth: procedures for insect monitoring problem evaluation and control actions. Can. For. Serv., Pac. For. Cent., Inf. Rep. BC-X-270, 14 pp.

Shepherd, R.F.; Otvos, I.S.; Chorney, R.J. 1984a. Pest management of Douglas-fir tussock moth (Lepidopetera: Lymantriidae): sampling method to determine egg-mass density. Can. Entomol. 116: 1041-1049.

Shepherd, R.F.; Otvos, I.S.; Chorney, R.J.; Cunningham, J.C. 1984b. Pest management of Douglas-fir tussock moth (Lepidoptera: Lymantriidae): prevention of a Douglas-fir tussock moth outbreak through early treatment with a nuclear polyhedrosis virus by ground and aerial applications. Can. Entomol. 116: 1533-1542.

Stelzer, M.J.; Neisess, J.; Thompson, C.G. 1975. Aerial application of a nuclear polyhedrosis virus and *Bacillus thuringiensis* against the Douglas-fir tussock moth. J. Econ. Entomol. 68: 269-272.

Stelzer, M.J.; Neisess J., Cunningham, J.C.; McPhee, J.R. 1977. Field evaluation of baculovirus stocks against Douglas-fir tussock moth in British Columbia. J. Econ. Entomol. 70: 243-246.

Wickman, B.E. 1978. Tree mortality and top kill related to defoliation by the Douglas-fir tussock moth in the Blue Mountains outbreak. U.S.D.A. For. Serv. Res. Pap. PNW-206. 13 pp.

Wickman, B.E.; Beckwith, R.C. 1978. Life history and habits. Pages 30-37 *in* M.H. Brookes, R.W. Stark and R.W. Campbell (Eds.), The Douglas-fir Tussock Moth: A Synthesis. U.S.D.A. For. Serv. Tech. Bull. 1585.

Wickman, B.E.; Mason, R.R.; Thompson, C.G. 1975. Major outbreaks of the Douglas-fir tussock moth in Oregon and California. USDA For. Serv. Gen. Tech. Rep. PNW-5. 18 p.

Chapter 14

Gypsy Moth, *Lymantria dispar*

L. JOBIN

Introduction

The gypsy moth, *Lymantria dispar*, is a native forest pest of many Eurasian countries. It was introduced into the United States in 1869 and has since become the most important defoliating insect of hardwood forests. Two to three years of consecutive defoliation can result in significant tree mortality, especially in oak forests. The first gypsy moth infestation in Canada was observed in 1924 in southwestern Quebec along the Quebec–United States border. It became established in the same general area during the mid-1960s and is now considered established in Ontario and two of the Maritime provinces, New Brunswick and Nova Scotia.

In Canada, woodland areas of New Brunswick, Quebec, Ontario, and British Columbia are occasionally sprayed with insecticide for eradication or population suppression or containment. Spray programs carried out between 1960 and 1974 are described by Brown (1975). The gypsy moth situation and aerial control programs in Canada from 1975 to 1989 inclusively are summarized in this chapter.

Biology

The gypsy moth has four developmental stages: egg, larva, pupa, and moth (Fig. 1a, b, c, and d, respectively). It has one generation per year and overwinters in the egg stage. Eggs are laid in clusters on trees, stones, logs, and other objects. Each egg cluster contains up to 1 000 eggs which are covered with buff-colored hairs from the abdomen of the female. Tiny larvae emerge from the egg cluster in early May and feed on foliage until July. When full grown, caterpillars are 6.0 to 8.0 cm long, quite hairy, and have a double row of six pairs of red and five pairs of blue dots on their backs. Caterpillars enter the pupal stage in late June and early July. Female pupae are much larger than male and both are dark reddish-brown in color and have scattered yellowish hairs. After about 2 weeks, a moth emerges from the pupa and is active from early July until mid-August. Males are brownish tan with forewings measuring about 4.0 cm. Females are larger than males and are white with black markings on their wings; they are unable to fly and usually lay a single egg mass close to the discarded pupal case.

Preferred host-trees of the gypsy moth in eastern Canada are oak, ironwood, serviceberry, willow, and poplar. Other trees attacked are basswood, cherry, birch, beech, and elm (Lechowicz and Jobin 1983; Mauffette et al. 1983; Mauffette and Lechowicz 1984).

Early Infestation History

In Canada, the gypsy moth was accidentally introduced into Vancouver, British Columbia, in 1912 (Hewitt 1913, 1916), where eight egg masses were collected on cedar shrubs imported from Japan. The second introduction (McLaine 1924, 1925; McLaine and Short 1926; Short 1926; Tessier 1929) occurred in southern Quebec in 1924 and originated from an infestation located in the United States. The infestation, covering 86 ha, was discovered 5 km from the New York State–Quebec border. Successful eradication was attained following spraying with lead arsenate in 1925, 1926, and 1927. Other control methods were also used such as tree banding with burlap, sticky insecticide strips like Tanglefoot®, and burning of shrubby vegetation (McLaine 1927, 1930 and 1939). This introduction of an important forest pest prompted the Canadian Department of Agriculture to pass a Quarantine Act in 1924 that prohibited or restricted the movement of trees likely to harbor the gypsy moth. The Plant Protection Division was responsible for the enforcement of the Quarantine Act, mainly detection and application of control measures against agricultural and forest pests.

Recent Infestation History

A survey program using pheromone-baited traps, initiated in 1954 following an expansion of the gypsy moth-infested areas in neighboring states along the Quebec–United States border, resulted in the discovery in 1956 (Cardinal 1967) of an egg mass in the same general area of Quebec as the first introduction that occurred in 1924. According to Gauthier and Doyle (1957), pockets of defoliation were observed in 1957. Egg mass populations increased from 1956 to 1959 when three isolated infestations were discovered (Canada Department of Agriculture 1959–1981). Ground and aerial spraying operations to eradicate incipient infestations were carried out every year from 1960 to 1974 (Brown 1975), and from 1969, control measures were designed mainly to prevent spread of the insect to other regions in Canada. Since 1972 (Table 1), small, widely scattered patches of light to severe defoliation have been evident in the western area of Quebec. However, a dramatic increase in size of the generally infested area occurred in 1977 (Table 1 and Fig. 3) when light to severe defoliation was observed in an area covering 518 000 ha. The gypsy moth is now established on 42 200 km^2 (Fig. 2) which corresponds to two ecological regions known as Basswood–Maple Forest and Hickory–Maple Forest (Rowe 1972; Jurdant et al. 1977).

The second infestation of the gypsy moth in Canada was detected in southwestern New Brunswick in 1936 (Twinn 1938), and egg masses were found in the same region in 1937 and 1938. Control measures such as application of lead

Figure 1. Developmental stages of the gypsy moth, *Lymantria dispar*: (a) egg masses, (b) larva, (c) pupa, (d) adult.

arsenate and burlap bands to collect and destroy larvae were successful in eradicating the gypsy moth from the province (McLaine 1936-1940, 1939). Egg masses were found again in southern New Brunswick in 1981 and also in four areas of southern Nova Scotia near the State of Maine. In 1982, egg masses were discovered in Nova Scotia, from Yarmouth to Paradise in the Annapolis Valley, and in 1983, in Halifax (Magasi 1984). The insect continued to spread and is now considered to be established in both of these Maritime provinces (Magasi 1988) (Fig. 2). A description of the gypsy moth introduction and dispersal in New Brunswick and Nova Scotia is summarized in annual reports of forest pest conditions in the Maritimes (Magasi 1982–1989).

First detected in two regions of Ontario in 1969, on Wolfe Island near the city of Kingston and in an area adjacent to the Ontario–Quebec border, gypsy moth populations remained relatively low during the next 10 years. Beginning in 1979, pockets of defoliation were located in many hardwood areas in the eastern region of Ontario (Howse 1981). A dramatic population increase (Table 1) occurred in the same region in 1985 (Howse 1987), reaching to the southern part of the Algonquin Region, and oak stands were defoliated in the Pembroke, Bancroft, and Linden Districts. Also, egg masses, larvae, and defoliation were observed in the Niagara District. The three species defoliated there were mainly red oak and mixed hardwood stands containing red oak, birch, and white pine. The gypsy moth is well established in Ontario, and the generally infested area (Fig. 2) covers some 45 500 km^2.

The first gypsy moth infestation in British Columbia was encountered in 1978 when adult moths were captured in pheromone-baited traps, and egg clusters were found in the Kitsilano area of Vancouver (Puritch and Brooks 1981; B.C. Ministry of Forests 1983). Ground applications of insecticides and other control measures were successful in eradicating this outbreak. A small infestation was found again in 1983 in the town of Courtenay, Vancouver Island, and the following year in Chilliwack, some 100 km east of Vancouver. Intensive pheromone-trapping programs to locate and delineate incipient

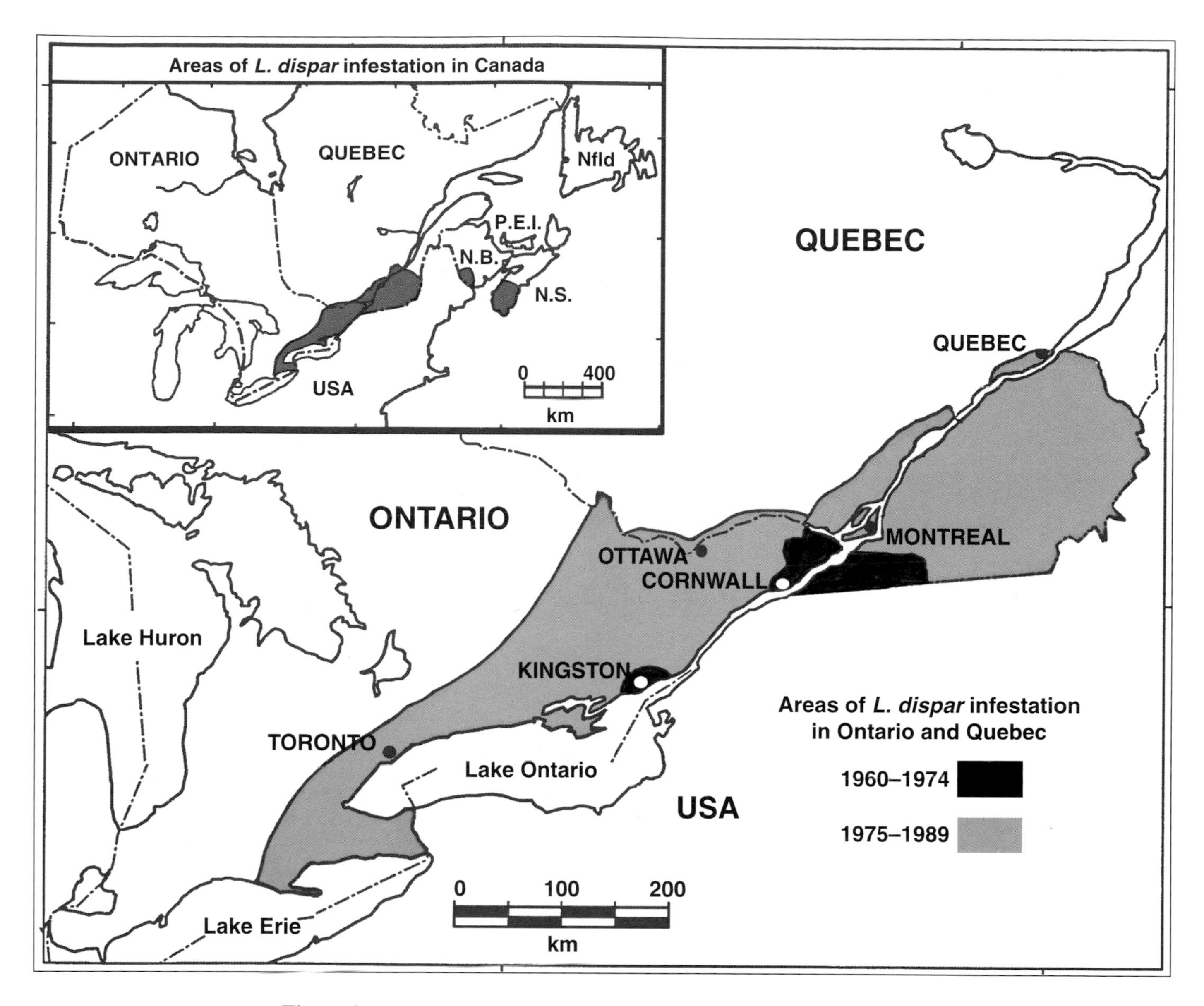

Figure 2. Areas of gypsy moth, *Lymantria dispar*, infestation in Canada.

populations, combined with periodic spraying activities of insecticides, have prevented establishment of the gypsy moth in British Columbia (Wood et al. 1988; Wood and Van Sickle 1987). In Canada, the gypsy moth is considered established in Ontario, Quebec, New Brunswick, and Nova Scotia and has caused light to severe defoliation (Table 1) over a large area of Quebec and Ontario (Fig. 2).

Operational Control

The major expansion of the gypsy moth infestation in 1977 in Quebec and in 1985 in Ontario resulted in 1986 in the largest (103 094 ha) (Table 1) aerial control program against this pest in Canada. A decline in size of the areas with moderate to severe defoliation, attributed to a successful spray program and high natural mortality (Silviculture News 1986),

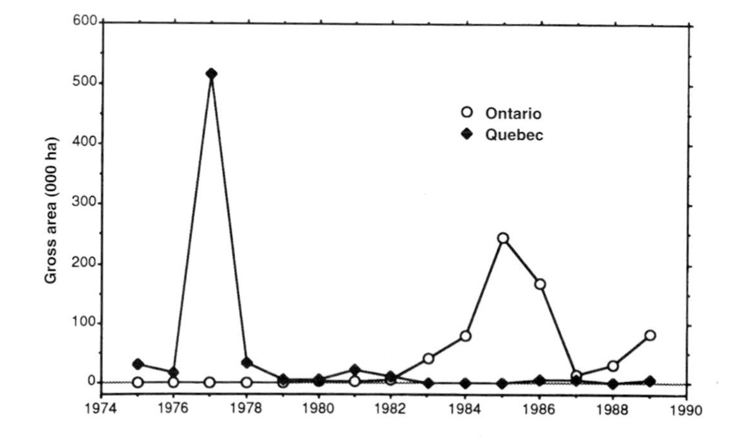

Figure 3. Gross area (hectares) of defoliation by the gypsy moth, *L. dispar*, in Ontario and Quebec from 1975 to 1989.

Table 1. Gross area (ha) of defoliation by the gypsy moth, *L. dispar*, in Ontario and Quebec from 1975 to 1989.

Year	Gross area (ha) of defoliation[a]	
	Ontario[b]	Quebec[c]
1975	L	28 500 L-S
1976	L	15 000 L-S
1977	L	518 000 L-S
1978	L	33 802 L-S
1979	L	4 000 M-S
1980	3 500 L-S	5 200 M-S
1981	1 450 L-S	22 600 L-S
1982	4 800 M-S	9 200 L-S
1983	40 954 M-S	L
1984	80 624 M-S	L
1985	246 342 M-S	200 L-S
1986	167 776 M-S	5 200 L-S
1987	12 678 M-S	4 670 L-S
1988	29 693 M-S	300 L-S
1989	81 640 M-S	5 850 L-S

[a] Small scattered pockets of light defoliation (L); light to severe defoliation (L-S); moderate to severe defoliation (M-S).
[b] Howse and Churcher (1989).
[c] Insectes et maladies des arbres — Québec, 1976-1989.

was observed in Ontario in 1986. Defoliated areas continued to decline in 1987, but a significant increase was observed in 1988 and 1989 (Howse 1989). Distribution of the gypsy moth and aerial control programs in Ontario are described by Howse (1981), Meating et al. (1983), Howse and Applejohn (1985), Howse (1987), and Howse and Churcher (1989).

No aerial spraying has been carried out against the gypsy moth in Quebec since 1981 (Table 2), following a significant population decline observed from 1982 (Table 1).

In the Maritimes, aerial control programs were conducted in 1983 and 1988 by the New Brunswick Ministry of Natural Resources, using the biological insecticide *Bacillus thuringiensis* (Bt). A description of both operations is presented by Magasi (1984, 1989).

Prior to 1979, Agriculture Canada was the only agency responsible for application of ground and aerial control measures against the gypsy moth in Canada. The aerial spray programs in Quebec in 1979, Ontario in 1982, and New Brunswick from 1984 to 1989 were conducted by provincial government departments (Table 2). The operational spray materials used in Ontario and Quebec from the mid-1970s to 1982 were a chemical insecticide Sevin®-4-oil, the insect growth regulator Dimilin® 25W, and different commercial formulations of the biological insecticide Bt. Since 1983, only Bt has been used for the aerial control or eradication of gypsy moth in Canada (Table 1); the dosage most often used is 30 BIU (billion international units)/ha at the rate of one to three applications per treatment.

Aerial spraying operations initiated against the gypsy moth in Quebec in 1960 and Ontario in 1970 (Brown 1975) were directed toward preventing establishment of this insect pest in Canada. As the infested areas became more widespread, population suppression and containment became the objectives of the aerial spray program (Brown 1975). Only a small portion of the generally infested area of the gypsy moth is sprayed in Quebec and Ontario, and the objective of current spray programs is to protect valuable forest and recreational areas from defoliation. Aerial control programs and ground spray programs in British Columbia, New Brunswick, and Nova Scotia are designed to eradicate small incipient infestations and prevent spread of this insect pest to new areas.

Experimental Control

Several registered materials and one biological insecticide were selected for testing of efficacy from the chemical and biological insecticides used in aerial control operations against the gypsy moth from 1975 to 1989 (Table 2).

Aerial field trials were conducted in Quebec in 1977 and 1978 to gather efficacy data for Dimilin®, a compound that kills larvae by interfering with the process of cuticle formation, thus preventing molting from one larval stage to the next. A single treatment of Dimilin® 25W was applied at the rate of 4.68 L/ha using a fixed-wing aircraft. Excellent results based on reduction of larval and egg populations and foliage protection were obtained (Jobin and Caron 1982). The effect of Dimilin® on some of the gypsy moth parasitoids was measured during the 1978 spray trial in Quebec and high parasitoid mortality was reported (Madrid and Stewart 1981).

Results of the 1982 aerial spraying trials carried out in Ontario with Dipel®, Sevin®-4-oil, and Gypchek®, a nuclear polyhydrosis virus developed by the United States Forest Service, are described by Meating et al. 1983. All materials tested gave excellent results in terms of larval and egg mass population reduction and level of leaf damage; they were considered as effective as the insecticide Sevin® or Bt used in an operational control program.

A nuclear polyhydrosis virus named Disparvirus is being developed and tested by Forestry Canada. Aerial field trials were conducted in Ontario in 1988 and 1989 (Table 2) and results indicate that two applications of 5×10^{11} PIB (polyhedral inclusion bodies)/ha^2 is highly effective in reducing gypsy moth populations and defoliation levels (J.C. Cunningham and W.J. Kaupp, pers. comm.).

Table 2. Summary of aerial spraying operations against the gypsy moth, *L. dispar*, in Canada from 1975 to 1989.

Year	Province	Agency[a]	Material (trade name)	Area (ha)	Volume (L/ha)	No. applic.	Dosage BIU/ha[b] PIB/ha[c]
1975	Ontario	Agric. Can.	Sevin-4-oil	1 425	2.40	1	–
	Quebec		Sevin-4-oil	356	2.40	1	–
1977	Ontario	Agric. Can.	Dimilin 25W	1 639	4.68	1	–
	Quebec		Dimilin 25W[d]	178	4.68	1	–
1978	Ontario	Agric. Can.	Dimilin 25W	146	4.68	1	–
	Quebec		Dimilin 25W[d]	498	4.68	1	–
1979	Quebec	MER	Thuricide 32B	1 237	4.68	2	19.8
1980	Ontario	Agric. Can.	Sevin-4-oil	485	2.40	1	–
1981	Ontario	Agric. Can.	Sevin-4-oil	243	2.40	1	–
	Quebec		Dipel 88	29	2.60	1	30
			Sevin-4-oil	445	2.40	1	–
1982	Ontario	Agric. Can.	Dipel 88	7	2.0-3.0	1-2	16.7
		Agric. Can.	Sevin-4-oil[d]	90	2.40	1	–
		Agric. Can.	Dipel 88[d]	133	5.80	2	20
		Agric. Can.	Dipel 88[d]	130	8.75	2	30
		Agric. Can.	Gypchek[e]	63	18.80	2	2.5×10^{11}
1983	New Brunswick	Agric. Can. MNR	Dipel 88	182	–	3	30-40
1984	British Columbia	Agric. Can.	Thuricide HPC	10	35.50	1	30
1985	Ontario	MNR	Dipel 88	170	–	1-2	40
	British Columbia	Agric. Can.	Thuricide 48LV				
			Dipel 132	160	4.80	3	30
1986	Ontario	MNR	Dipel 132	103.094	6.00	2-3	30
	British Columbia	Agric. Can.	Dipel 132	5	12.00	3	30
1987	Ontario	MNR	Dipel132				
			Thuricide 48LV				
			Futura XLV	40.249	2.10-6.10	2-3	30
	British Columbia	Agric. Can.	Dipel 132	25	4.80	3	30
1988	Ontario	MNR	Dipel 132	13.784	2.40	2-3	30
		For. Can.	Disparvirus[e]	64	10.00	2	1.25×10^{12}
	New Brunswick	MNR	Futura XLV	391	2.02	3	30
	British Columbia	Agric. Can.	Dipel 132	112	2.40	3	30
1989	Ontario	MNR	Futura XLV	12.951	2.03	2-3	30
		For. Can.	Disparvirus[e]	90	5.0-10.0	2	5×10^{11}

[a] Agriculture Canada (Agric. Can.), Ministère de l'Énergie et des Ressources du Québec (MER), Ministry of Natural Resources (MNR), and Forestry Canada (For. Can.).
[b] Billion international units per hectare (BIU/ha).
[c] Polyhedral inclusion bodies (PIB/ha).
[d] Operational spray trials.
[e] Experimental spray trials.

References

B.C. Ministry of Forests. 1983. Gypsy moth: a forest threat from the east… and the south. PestTopics no. 9, April 1983. Ministry of Forests, Protection Branch, Victoria, B.C.

Brown, G.S. 1975. Gypsy moth *Porthetria dispar* (L.). Pages 208-212 *in* M.L. Prebble (Ed.), Aerial control of forest insects in Canada. Dep. Environ, Ottawa, Ont. 330 p.

Canada Department Agriculture. 1959. Gypsy moth (*Porthetria dispar* L.)—Québec. Can. Inst. Pest. Rev. 37(7): 238.

Canada Department Agriculture. 1981. Summary of plant quarantine pest and disease situation in Canada, 1980. Ottawa, Ont. June 30, 1981.

Cardinal, J.A. 1967. Lutte contre la spongieuse *Porthetria dispar* L. (Lepidoptera: Lymantriidae) au Québec. Phytoprotection 48(2): 92-100.

Gauthier, G.; Doyle, A. 1957. Insects of the season, Quebec. Can. Inst. Pest Rev. 36(5): 86.

Hewitt, G.C. 1913. Review of entomology relating to Canada in 1912. 43rd Annu. Rep., Entomol. Soc., Ont. (1912) 36: 34-37.

Hewitt, G.C. 1916. The introduction and establishment in Canada of the natural enemies of the brown-tail and gypsy moth. Agric. Gaz. Can. 3(1): 20-21.

Howse, G.M. 1981. Forest insect and disease conditions—Summer 1981. Surv. Bull. pp. 5-6. Can. For. Serv., Great Lakes For. Res. Cent., Sault Ste. Marie, Ont.

Howse, G.M. 1987. Gypsy moth in Ontario. Pages 4-5 *in* Gypsy Moth News no. 15, Nov. 1987. USDA For. Serv., Broomall, Pa.

Howse, G.M.; Applejohn, M.J. 1985. Forest insect and disease conditions in Ontario—Summer 1985. Surv. Bull. pp. 8-11. Can. For. Serv., Great Lakes For. Res. Cent., Sault Ste. Marie, Ont.

Howse, G.M.; Churcher, J.J. 1989. Gypsy moth in Ontario, 1989. 17th Annu. For. Pest Control Forum, Ottawa, Ont. Nov. 14-15, 1989. 14 pp.

Insectes et maladies des arbres—Québec. 1976-1989. Principaux insectes forestiers. Can. For. Serv. and Minist. Énergie et Ressources, Québec. Revue Forêt Conserv. Vol.: 42(4); 43(4); 45(1); 45(4); 46(10); 47(9); 49(10); 51(10); 52(10); 53(10); 54(10) and 56(1).

Jobin, L.; Caron, A. 1982. Results of aerial treatment with Dimilin and *Bacillus thuringiensis* for gypsy moth (*Lymantria dispar* (L.) in Quebec. Can. For. Serv., Ottawa, Ont. Res. Notes 2(3): 18-20.

Jurdant, M.; Bélaire, J.L.; Gérardin, V.; Ducruc, J.P. 1977. Inventaire du capital nature: méthode de classification et de cartographie écologique du territoire. Minist. Pêche et Environ. Canada, Ottawa.

Lechowicz, M.J.; Jobin, L. 1983. Estimating the susceptibility of tree species to attack by the gypsy moth, *Lymantria dispar*. Ecol. Entomol. 8: 171-183.

Madrid, F.J.; Stewart, R.K. 1981. Impact of Diflubenzuron spray on gypsy moth parasitoids in the field. J. Econ. Entomol. 74(1): 1-2.

Magasi, L.P. 1982-1988. Forest pest conditions in the Maritimes. Can. For. Serv., Maritimes For. Res. Cent., Inf. Rep. M-X-135, 141, 149, 154, 159, 161 and 1966. Fredericton, N.B.

Magasi, L.P. 1984. Forest pest conditions in the Maritimes. Can. For. Serv., Maritimes For. Res. Cent., Inf. Rep. M-X-149, Fredericton, N.B. 49 p. (pp. 12-16).

Magasi, L.P. 1985. Forest pest conditions in the Maritimes 1984. Can. For. Serv., Maritimes For. Res. Cent., Inf. Rep. M-X-154, Fredericton, N.B. 49 p. (pp. 13-16).

Magasi, L.P. 1988. Forest pest conditions in the Maritimes 1987. Can. For. Serv., Maritimes For. Res. Cent., Inf. Rep. M-X-166, Fredericton, N.B. 109 p. (pp. 33-38).

Magasi, L.P. 1989. Forest pest conditions in the Maritimes in 1988. For. Can., Maritimes For. Cent. Inf. Rep. M-X-174, Fredericton, N.B. 76 p. (pp. 19-24).

Mauffette, Y.; Lechowicz, M.J.; Jobin, L. 1983. Host preferences of the gypsy moth *Lymantria dispar* (L.) in southern Quebec. Can. J. For. Res. 13:(1): 53-60.

Mauffette, Y.; Lechowicz, M.J. 1984. Differences in the utilization of tree species as larval hosts and preparation sites by the gypsy moth, *Lymantria dispar* (Lepidoptera: Lymantriidae). Can. Entomol. 116: 685-690.

McLaine, L.S. 1924. The outbreak of the gypsy moth in Quebec. 55th Annu. Rep. Entomol. Soc. Ont., 1924: 60-62.

McLaine, L.S. 1925. The outbreak of the gypsy moth in Quebec. 17th Annu. Rep. Que. Soc. Protect. Plants 1925: 32-34.

McLaine, L.S. 1927. The activities of the Division of Foreign Pests Suppression. 57th Annu. Rep. Entomol. Soc. Ont., 1927: 21-22.

McLaine, L.S. 1930. The gypsy moth outbreak in southern Quebec. J. Econ. Entomol. 23(1): 38-41.

McLaine, L.S. 1930-1940. Annual reports of the Division of Foreign Pests Suppression (1936-1937; 1937-1938) and Plan Protection Division (1938-1939; 1939-1940) Agric. Can., Ottawa, Ont.

McLaine, L.S. 1939. Some notes on the gypsy moth eradication campaign in New Brunswick and the Japanese beetle preventive work. 69th Annu. Rep. Entomol. Soc. Ont. 1938: 43-45.

McLaine, L.S.; Short, S.H. 1926. The gypsy moth situation in Quebec. 56th Annu. Rep. Entomol. Soc. Ont., 1925: 67-69.

Meating, J.H.; Lawrence, H.D.; Cunningham, J.C.; Howse, G.M. 1983. The 1982 gypsy moth situation in Ontario: general surveys, spray trials and forecasts for 1983. Can. For. Serv., Sault Ste. Marie, Ont. Inf. Rep. O-X-352. 14 p.

Puritch, G.S.; Brooks, B.C. 1981. Effect of insecticidal soap used in the gypsy moth control program in Kitsilano on insects and vegetation. Can. For. Serv., Pac. For. Res. Cent., Inf. Rep. BC-X-218, Victoria, B.C. 22 p.

Rowe, J.S. 1972. Forest regions of Canada. Environ. Canada, Can. For. Serv., Ottawa, Ont. Publ. no. 1300. 172 p.

Short, S.H. 1926. Invasion de la spongieuse (*Bombya disparate*) à Henryburg, Quebec. 18e Rapp. Ann. Soc. Qué. Prot. Plantes (1925-1926) 136-138.

Silviculture News. 1986. Spraying works in fight against gypsy moth in Ontario. Silviculture Magazine, vol. 1(5), November 1986. p. 9.

Tessier, G. 1929. Quelques ennemis de nos forêts. Nat. Can. 55(11): 241-246.

Twinn, C.R. 1938. A summary of the insect pest situation in Canada in 1937. 68th Annu. Rep. Entomol. Soc. Ont., p. 85.

Wood, C.S.; Van Sickle, G.A. 1987. Forest insect and disease conditions. British Columbia & Yukon 1986. Can. For. Serv., Pac. For. Cent., Inf. Rep. BC-X-287, Victoria, B.C. 35 p. (p. 34).

Wood, C.S.; Van Sickle, G.A.; Humble, L.M. 1988. Forest insect and disease conditions. British Columbia & Yukon 1987. Can. For. Serv., Pac. For. Cent., Inf. Rep. BC-X-296, Victoria, B.C. 40 p. (p. 35).

Wood, C.S.; Van Sickle, G.A. 1989. Forest insect and disease conditions. British Columbia & Yukon 1988. Can. For. Serv., Pac. For. Cent., Inf. Rep. BC-X-306, Victoria, B.C. 33 p. (p. 31).

Chapter 15

Hemlock Looper, *Lambdina fiscellaria*

A.G. Raske, R.J. West, and A. Retnakaran

Introduction

The hemlock looper, *Lambdina fiscellaria*, is a common pest of balsam fir, *Abies balsamea*, and eastern hemlock, *Tsuga canadensis*, forests of eastern Canada, and is distributed as far west as the Rocky Mountains of Alberta. Throughout most of the range of this looper, the principal host is balsam fir, although a large number of conifers and hardwoods are acceptable to late-instar larvae when host foliage has been eaten (Carroll 1956).

Outbreak population levels have occurred periodically in Canada extending from Ontario to Newfoundland, but outbreaks since 1973 have been confined to the Maritime provinces and Newfoundland. These outbreaks are usually associated with bodies of water, such as along the margins of lakes, large rivers, or bogs. Only on Anticosti Island and in Newfoundland are the outbreaks widely distributed and not closely associated with wet habitats.

Infestations tend to develop quickly, cause severe defoliation in any one stand for 2 or 3 years, and then end abruptly (Fig. 1). An outbreak in a large area, such as Newfoundland, may last about 5 to 7 years with the succession of several infestations (Otvos et al. 1971). Large-scale mortality of balsam fir has occurred following 1 to 3 years of severe defoliation in Newfoundland and Anticosti Island. Mortality has also occurred in fir stands along the North Shore of the St. Lawrence River and in hemlock stands in southern Ontario (Otvos et al. 1979; Jobin and Desaulniers 1981).

Biology

The life cycle of the hemlock looper is univoltine. Adults emerge in early to mid-September, mate and oviposit. The moths (Fig. 2) are nocturnal and mate on the boles of trees. Most eggs are laid on the upper boles and in the live crowns of the trees; however, some are laid on the lower boles, on birch bark, in old stumps, or even in the moss layer of the forest floor. Larvae from eggs laid in exposed sites hatch first, and those in cooler microhabitats hatch later. Therefore the larval hatching period may be prolonged (Otvos et al. 1971; Jobin and Desaulniers 1981).

The first two larval instars feed on new needles on the current year's shoot, generally stripping one or both sides of a needle leaving the remainder to shrivel and die (Fig. 3). Third and fourth instar larvae feed wastefully on new and old foliage, usually beginning toward the base of a needle, allowing the remainder to die and drop. The larvae produce silken threads that tend to catch wasted red foliage, giving severely defoliated trees a characteristic dark red-brown appearance. Feeding is generally completed by late July. The larvae then crawl down to pupate on the bole, near the root collar, or in the duff, concealing themselves in or under various places, such as under lichens or debris. Adults are long-lived and are often active till a severe frost kills them in late October or November, but most eggs are laid by mid- to late September (Carroll 1956; Otvos et al. 1971).

Few parasites and predators are closely and consistently associated with the hemlock looper, and none is known to appreciably affect outbreak population levels of this looper. An unidentified braconid parasite was abundant in one sample from a collapsing population in Ontario. A tachinid fly, *Winthemia occidentis*, introduced from British Columbia into Newfoundland, is the most common parasite in Newfoundland, and may infest over 20% of the larvae and pupae of a declining population (Otvos 1973).

Two fungal diseases may influence population levels of the hemlock looper. The fungi *Entomophaga aulicae* and *Erynia radicans* have both been important locally, though the former seems to be the dominant fungus in Newfoundland. This fungus has probably been responsible for the collapse of most populations of the looper since 1950 (Otvos 1973).

Summary of Experimental Control Programs

Only Newfoundland and Nova Scotia have protected forests from the feeding damage of the hemlock looper since 1973. The Nova Scotia Department of Lands and Forests sprayed 136 ha with *Bacillus thuringiensis* (Bt) (Dipel® 132R) in 1986 (T. Smith, Nova Scotia Dep. Lands and Forests, pers. comm.). The Newfoundland Department of Forestry sprayed Bt and fenitrothion annually from 1985 to 1987. In 1985, only fenitrothion was registered for use against the hemlock looper, but Newfoundland received minor use registration for Bt (Dipel® 132R). It was desirable that other materials be made available for operational use against the looper. Therefore, the Newfoundland Forestry Centre and the Forest Pest Management Institute, both of Forestry Canada, in cooperation with the Newfoundland Department of Forestry, tested various materials, dosages, and timings for use against the looper from 1985 to 1987 (Retnakaran et al. 1988; Raske et al. 1986, 1987; West et al. 1987, 1989).

Experimental Program—Bt

Testing of Bt was a major part of the experimental program to obtain field efficacy data required to register new Bt formulations for use against the hemlock looper. Bt was tested in 1985 and 1987 using several commercial formulations, dosages, spray atomizers, timings, and rates of applications (Table 1). In 1985, small planes (Piper Pawnee, Fig. 4) equipped with six Micronair® AU 5000 atomizers

1

2

3

4

5

Figures 1–5

Figure 1. Balsam fir stand severely defoliated by the hemlock looper, *Lambdina fiscellaria*. **Figure 2.** Hemlock looper moth. **Figure 3.** Hemlock looper larva on balsam fir. **Figure 4.** Piper Pawnee airplane used for experimental applications of insecticides against the hemlock looper, and balloon (*upper left*) used to guide the plane. **Figure 5.** Grumman AgCat airplane spraying fenitrothion against the hemlock looper.

were used for experimental sprays, and the same planes equipped with six Micronair® AU 4000 atomizers were used in 1987. The spray deposit was assessed from Kromekote® cards placed near ground level along a cleared line or road through the spray block, and also from mid-crown foliage collected across the spray block. Pre-spray and post-spray larval populations were assessed with beating samples and pupal numbers with the use of burlap traps. Each spray plot was matched with a control plot with a similar pre-spray population level. Population reductions were calculated with Abbott's formula. Larvae and pupae in all plots were sampled to determine levels of parasitism and disease in the population. The percentage of foliage saved was calculated by dividing the absolute difference between the defoliation of the spray plot and its control plot by the defoliation in the control plot and multiplying by 100.

Spray deposits were less for water-based formulations of Bt (sprayed with Micronair® AU 5000s in 1985) than for oil-based formulations (sprayed with Micronair® AU 4000s in 1987) (Table 1). The deposit density on the needles averaged 1 or more droplets per needle for applications of the oil-based formulations, but generally less than 0.5 droplets per needle for applications of water-based formulations.

The Micronair® AU 4000 atomizer provided the best droplet spectra, not only for Bt sprays but also for the other insecticides tested. This atomizer provided the highest percentage of droplets in the desirable range of 30 to 60 µm in diameters and the fewest number of large droplets of over 100 µm in diameter.

The oil-based formulations were more effective against the hemlock looper than water-based formulations. The larval reductions were near 100% and larval mortality occurred within a few days after spraying.

Early applications of the oil-based formulations sharply reduced population levels (Table 2). The single applications reduced larval levels by 74% to 97% and pupal numbers by

Table 1. Details of Bt applications against hemlock looper, *Lambdina fiscellaria*, populations and resulting spray deposit in Newfoundland in 1985 and 1987.

Formulation	Formul. base	Appl. vol. (L/ha)	BIU per ha[a]	Appl. no.	Hemlock looper instar distribution[b]	Post-spray droplet density: K cards[c] dropl./cm²	Post-spray droplet density: Foliage droplets/ needle
1985							
Futura XL	Water	1.4	20	1st	17:75:8:0	4.7	0.38
				2nd	1:18:65:16		0.34
Futura XL	Water	2.1	30	1st	17:75:8:0	5.1	0.08
				2nd	1:18:65:16		0.53
Thuricide 48LV	Water	2.36	30	1st	5:68:27:0	2.7	0.33
				2nd	1:18:65:16		1.34
Thuricide 64B	Water	1.78	30	1st	5:68:27:0	4.2	0.46
				2nd	1:18:65:16		0.70
1987							
Dipel 132	Oil	2.36	30	1st	98:2:0:0	45	1.0
				2nd	34:59:7:0	91	8.1
Dipel 176	Oil	1.78	30	1st	98:2:0:0	44	1.2
				2nd	34:59:7:0	67	2.8
Dipel 176	Oil	2.36	40	1st	34:59:7:0	57	1.6
				–			
Dipel 264	Oil	1.58	40	1st	34:59:7:0	50	1.9
				–			
Dipel 264	Oil	1.18	30	1st	98:2:0:0	43	1.0
				2nd	34:59:7:0	39	1.4

[a] Billion international units per hectare (BIU/ha).
[b] Ratios of instars—first: second: third: fourth.
[c] Kromekote® card.

Table 2. Effect of Bt treatments against hemlock looper, *Lambdina fiscellaria*, populations and foliage saved in Newfoundland in 1985 and 1987.

Treatment	BIU per ha[a]	Population reduction Larval	Pupal	Percent foliage saved[b]
1985				
Futura XL	20	0	0	–
Futura XL	30	10	86	–
Thuricide 48LV	30	34	84	–
Thuricide 64B	30	100	100	–
1987				
Dipel 132	30	98	100	100
Dipel 176	30	99	100	100
Dipel 176	40	85	97	92
Dipel 264	40	97	100	100
Dipel 264	30	96	100	100

[a] Billion international units per hectare (BIU/ha).
[b] Foliage saved = defoliation in control less defoliation in treatment divided by defoliation in control times 100%. Foliage saved in 1985 was near zero for all plots.

nearly 100%. Double applications reduced the larval numbers by more than 95% and pupal numbers by 100%. Samples taken between spraying in the double-application plots indicated reductions of 74% to 82% after the first application; only slightly lower than the 85% and 97% achieved after the second application. Only small differences in population reduction were observed between the three different potencies of Bt formulations tested in 1987.

Several factors including heavy rains and, at times, low population levels made data on foliage protection data difficult to interpret in 1985, but very little foliage protection was observed in the plots. In 1987, however, foliage protection provided by the oil-based Bt sprays was clearly evident and almost all of the foliage was saved.

Oil-based formulations of Bt can give excellent control of the hemlock looper. At application rates of 30 BIU/ha (billion international units), the potency of an oil-based formulation is probably not critical, but the use of higher potency formulations would be more cost effective. Additional experimentation should define the spray window and determine whether one application of Bt will provide good control of the hemlock looper. Further experimentation with water-based formulations is warranted.

Experimental Program—Dimilin®

Dimilin® (=diflubenzuron) is an insect growth regulator and is registered for use against several insect pests in agriculture and forestry. Only field efficacy data were needed to include the hemlock looper on the label of this insecticide. Dimilin® was tested in experimental sprays in 1985, 1986, and 1987.

Methods used for the application of Dimilin® were identical to those used for the Bt, except that in 1986 six Micronair® Au 5000 atomizers were mounted on a Grumman AgCat (Fig. 5) and used for spray delivery. Active ingredient dosages varied from 30 to 70 ha in various spray volumes, but most applications were made with 70g/ha (Table 3). Wettable powder formulations were used except for one application in 1987 when a flowable formulation was tested.

Deposit densities were generally much lower for Dimilin® than those for oil-based formulations of Bt, and slightly lower than those of the water-based formulations of Bt (Tables 1 and 3). Even the flowable formulation, tried in 1987, did not deposit well on cards or on foliage.

Percentage population reduction was acceptable at times (Table 4); however, larvae generally did not die until the late instars and often when they entered the pupal stage. Dimilin® was the only insecticide used against the hemlock looper that caused mortality of emerging moths. Therefore, the total number of emerging moths was about the same for Dimilin® as for fenitrothion in 1986. There was no consistent difference in the efficacy of Dimilin® wettable powder formulations applied at 5.0 L/ha and 2.5 L/ha, but double applications usually caused greater population reductions. One application of the flowable formulation provided better efficacy than one application of the wettable powder, but was not as effective as two applications of the wettable powder.

Foliage protection was usually less than desired, and low for the single applications of wettable powder formulations (Table 4). The foliage of the current shoots was usually severely damaged and the foliage saved was generally that on older shoots.

Water-based formulations of Dimilin® generally did not give adequate foliage deposit. Double applications gave acceptable but delayed population reduction, and therefore foliage protection was less than desired. The flowable formulation was more effective than the wettable powder, but a single application of the flowable formulation still provided inadequate foliage protection. An oil-based formulation may provide a better deposit and be more effective.

Experimental Program—Fenitrothion

Experimental trials with fenitrothion involved two different atomizers, earlier application, reduced dosages of the active ingredient, reduced spray volumes, and the use of water-based formulations.

Formulations of Sumithion® were used for the experimental sprays, and Folethion® served as a benchmark for comparison of experimental treatments of Sumithion® (for Dimilin® and Bt as well). The formulation and dose of Folethion® was identical to that used for the operational spray program in Newfoundland.

The same methods were used for spraying these insecticides as were used for spraying Bt and Dimilin®. The dosage of active ingredient varied from 140 to 210 g/ha in various spray volumes (Table 5). The spray deposits assessed from Kromekote® cards were generally less for the water-based formulations than for the oil-based formulations (Table 5). However, this difference was not evident in the deposits on the

Table 3. Application details of water-based formulations of Dimilin® (=diflubenzuron) against hemlock looper, *Lambdina fiscellaria*, populations and resulting spray deposit in Newfoundland in 1985, 1986, and 1987.

Formulation	Appl. vol (L/ha)	Dosage[a] (g/ha)	Appl. no.	Hemlockz looper instar distribution[b]	Post-spray droplet density[d] K cards[c] droplets/cm²	Foliage (droplets/needle)
1985						
Dimilin 25WP	4.7	35	1st	17:75:8:0	10.1	–
Dimilin 25WP	4.7	70	1st	17:75:8:0	10.6	–
Dimilin 25WP	2.0	30	1st	17:75:8:0	5.6	–
Dimilin 25WP	4.7	70	1st	17:75:8:0	10.2	–
	4.7	70	2nd	1:14:45:40	8.9	–
1986						
Dimilin 25WP	4.7	70	1st	21:79:0:0	–	–
			2nd	10:76:14:0	–	–
Dimilin 25WP	2.5	7	1st	21:79:0:0	–	–
			2nd	10:76:14:0	–	–
1987						
Dimilin Flow.	5.0	70	1st	99:1:0:0	1.5	0.17
Dimilin 25WP	5.0	70	1st	99:1:0:0	2.0	0.33
			2nd	34:59:7:0	7.0	1.12
Dimilin 25WP	2.5	70	1st	99:1:0:0	2.2	0.20
			2nd	34:59:7:0	2.7	0.35
Dimilin 25WP	2.5	70	1st	99:1:0:0	1.1	0.16
Dimilin 25WP	5.0	70	1st	99:1:0:0	2.1	0.42

[a] Dosage expressed as weight of active ingredients.
[b] Ratios of larval instars—first: second: third: fourth.
[c] Kromekote® cards.
[d] Spray deposit not analyzed in 1986 because of equipment failure.

needles because the formulations when sprayed at 0.4 L/ha gave deposits of the active ingredient as high as when they were sprayed at 1.5 L/ha. The spray deposits were not measured in 1986 because of equipment failure.

Early application of the oil-based formulations reduced populations by more than 98% (Table 6) and were the only fenitrothion treatments that gave excellent control. Even a single early application provided over 90% larval reduction as determined from samples taken between the first and second spray. The flowable formulation sprayed at 180 g/ha did not provide adequate larval reduction, but the same amount of active ingredient in an oil-based formulation might have been sufficient. There was no difference in population reduction between the formulations sprayed at 0.4 L/ha and the formulations sprayed operationally at 1.5 L/ha. Water-based formulations were not as effective as the oil-based formulations though the degree of control obtained with the latter may have been improved by the earlier application. Very little foliage was saved in 1985 but the foliage protection provided by late applications in 1986 was adequate. Early applications of the oil-based formulations gave excellent foliage protection in 1987.

In summary, early applications of oil-based formulations of fenitrothion at the very low volume of 0.4 L/ha will give good control of hemlock looper, but even single applications of fenitrothion may offer sufficient control and foliage protection.

Table 4. Effect of Dimilin® treatments against hemlock looper, *Lambdina fiscellaria*, populations and foliage saved in Newfoundland in 1985, 1986, and 1987.

Treatment	Population reduction Larval	Pupal	Percent foliage saved	Percent moth emergence[a]
1985				
Dimilin 25WP	0	0	0	–
Dimilin 25WP	0	52	0	–
Dimilin 25WP	0	92	0	–
Dimilin 25WP	100	100	slight	–
1986				
Dimilin 25WP	162	93	40	18
Dimilin 25WP	126	90	0	7
1987				
Dimilin Flow.	47	62	54	41
Dimilin 25WP	84	99	76	2
Dimilin 25WP	57	96	66	61
Dimilin 25WP	63	88	20	32
Dimilin 25WP	58	44	8	68

[a] Percent moth emergence of larvae that reached the pupal stage.

Table 5. Application details of fenitrothion treatment against hemlock looper, *Lambdina fiscellaria*, populations and resulting spray deposit in Newfoundland in 1985, 1986, and 1987.

Formulation	Formul. base	Appl vol. (L/ha)	Dosage[a] (g/ha)	Appl. no.	Hemlock looper instar distribution[b]	Post-spray droplet density[d] K cards[c] (dropl./cm²)	Foliage[e] (ppm)
1985							
Sumithion Tech.	Water	1.5	210	1st	5:68:27:0	8	12.5
				2nd	1:12:45:42	7	–
Sumithion 20F	Water	1.5	140	1st	4:64:32:0	16	9.2
				2nd	1:12:45:42	20	–
Sumithion 20F	Water	1.5	210	1st	4:54:42:0	4	1.5
				2nd	1:12:45:42	7	–
1986							
Sumithion 20F	Water	0.9	180	1st	20:77:3:0	–	–
				2nd	4:31:65:0	–	–
Folethion Tech.	Oil	1.5	210	1st	20:77:3:0	–	–
				2nd	4:31:65:0	–	–
1987							
Sumithion Tech.	Oil[1]	0.4	210	1st	99:1:0:0	17	4.5
				2nd	34:59:7:0	11	4.2
Sumithion Tech.	Oil[1]	0.4	210	1st	34:59:7:0	7	2.2
				2nd	6:70:24:0	19	5.1
Folethion Tech.	Oil	1.5	210	1st	99:1:0:0	32	5.1
				2nd	34:59:7:0	51	9.0

[a] Dosage expressed as weight of active ingredients.
[b] Ratios of larval instars—first: second: third: fourth.
[c] Kromekote® cards.
[d] Spray deposits not analyzed because of equipment failure.
[e] Deposit density in droplets per needle for 1987: 0.09, 0.52 for treatment 6; 0.23, 0.48 for treatment 7; 0.48, 1.77 for treatment 8.

Table 6. Effect of fenitrothion treatments against hemlock looper, *L. fiscellaria*, populations and foliage saved in Newfoundland in 1985, 1986, and 1987.

Treatment	Population reduction Larval	Pupal	Percent foliage saved
1985			
Sumithion Tech.	63	74	–
Sumithion 20F	9	46	–
Sumithion 20F	0	0	–
1986			
Sumithion 20F	54	92	65
Folethion Tech.	60	99	59
1987			
Sumithion Tech.	99	100	100
Sumithion Tech.	86	79	69
Folethion Tech.	98	100	100

Experimental Program—Matacil®

Matacil® was tested in 1985 at several concentrations, even though preliminary laboratory experiments had indicated that Matacil® would not likely be effective against the hemlock looper.

The same methods and equipment were used for Matacil® as for the other treatments in 1985. Two applications of the flowable (water-based) formulation of Matacil® 180F were sprayed at 90, 135, and 180 g/ha, and each was applied at 1.5 L/ha. The highest rate is twice the dose registered for use against the spruce budworm. The first spray was applied when 5% of the larvae were in the first instar, 60% in the second, and 35% in the third instar.

The spray deposit was similar to that of other water-based formulations with 4 to 8 droplets/cm² on Kromekote® cards. The active ingredient on the foliage averaged 0.55 ppm for the dose of 90 g/ha, 0.80 ppm for 135 g/ha, and 1.22 ppm for 180 g/ha.

Larval and pupal reduction was 0% for the two lower dosages and only near 30% for the highest dose of 180 g/ha. Foliage protection was not evident.

These experiments confirmed that Matacil® is not effective against the hemlock looper and is not recommended.

Conclusions

Early applications of either fenitrothion or Bt will give effective control of the hemlock looper and good foliage protection. The Micronair® AU 4000 nozzle will give the best droplet spectra for all formulations and is recommended for very low-volume aerial treatment in operational forest spray programs. Applications of insecticide during the first larval instar are recommended because later applications often permit considerable defoliation to occur before the spray program can be applied.

References

Carroll, W.J. 1956. History of the hemlock looper, *Lambdina fiscellaria fiscellaria* (Guen.), (Lepidoptera: Geometridae) in Newfoundland, and notes on its biology. Can. Entomol. 88: 587-599.

Jobin, L.J.; Desaulniers, R. 1981. Results of aerial spraying in 1972 and 1973 to control the eastern hemlock looper (*Lambdina fiscellaria fiscellaria* (Guen.)) on Anticosti Island. Environ. Can., Can. For. Serv., Laurentian For. Res. Cent., Québec, Inf. Rep. LAU-X-49E, 30 pp.

Otvos, I.S. 1973. Biological control agents and their role in the population fluctuation in the eastern hemlock looper. Environ. Can., For. Serv., Newfoundland For. Res. Cent., St. John's, Inf. Rep. N-X-102, 34 pp.

Otvos, I.S.; Clark, R.C.; Clarke, L.J. 1971. The hemlock looper in Newfoundland: The outbreak, 1966 to 1971; and aerial spraying, 1968 and 1969. Environ. Can., For. Serv., Newfoundland For. Res. Cent., St. John's, Inf. Rep. N-X-68, 62 pp.

Otvos, I.S.; Clarke, L.J.; Durling, D.S. 1979. A history of recorded eastern hemlock looper outbreaks in Newfoundland. Environ. Can., For. Serv., Newfoundland For. Res. Cent., St. John's, Inf. Rep. N-X-179, 46 pp.

Raske, A.G.; Retnakaran, A. 1987. The effectiveness of diflubenzuron and fenitrothion against the eastern hemlock looper, *Lambdina fiscellaria fiscellaria*, in Newfoundland in 1986. Can. For. Serv., Newfoundland For. Cent., St. John's, Inf. Rep. N-X-263, 30 pp.

Raske, A.G; Retnakaran, A.; West, R.J.; Hudak, J.; Lim, K.P. 1986. The effectiveness of *Bacillus thuringiensis*, Dimilin, Sumithion and Matacil against the hemlock looper, *Lambdina fiscellaria fiscellaria*, in Newfoundland in 1985. Can. For. Serv., Newfoundland For. Cent., St. John's, Inf. Rep. N-X-238, 56 pp.

Retnakaran, A.; Raske, A.G.; West, R.J.; Lim, K.P.; Sundaram, A. 1988. Evaluation of diflubenzuron as a control agent for hemlock looper (Lepidoptera: Geometridae). J. Econ. Entomol. 81: 1698-1705.

West, R.J.; Raske, A.G.; Retnakaran, A.; Lim, K.P. 1987. Efficacy of various *Bacillus thuringiensis* Berliner var. *kurstaki* formulations and dosages in the field against the hemlock looper, *Lambdina fiscellaria fiscellaria* (Guen.) (Lepidoptera: Geometridae), in Newfoundland. Can. Entomol. 119: 449-458.

West, R.J.; Raske, A.G.; Sundaram, A. 1989. Efficacy of oil-based formulations of *Bacillus thuringiensis* Berliner var. *kurstaki* against the hemlock looper, *Lambdina fiscellaria fiscellaria* (Guen.) (Lepidoptera: Geometridae). Can. Entomol. 121: 55-63.

Chapter 16

Oak Leafshredder, *Croesia semipurpurana*

G.G. GRANT, A. RETNAKARAN, AND G.M. HOWSE

Introduction

The oak leafshredder, *Croesia semipurpurana*, is a significant pest of red oak, *Quercus rubra*, and related oak species throughout much of the host range in eastern North America (Beckwith 1963; Cochaux 1968; Sippell 1975; Martineau 1984; Magasi 1987). In Canada, it occurs from Sault Ste. Marie in northern Ontario, through central and southern Ontario and Quebec, and across the Maritimes, except for western Nova Scotia, where the oak olethreutid leafroller, *Pseudexentera cressoniana*, has predominated. The oak leafshredder is capable of causing severe defoliation (more than 95%) over large areas that can lead to tree mortality if it persists for several years or is accompanied by other stress factors such as drought (Nichols 1968; Sterner and Davidson 1981). Defoliation by this insect appears to be a major predisposing factor leading to oak decline and dieback (Staley 1965; Magasi 1987). The insect was first recognized as an important pest in Canada in the mid-1940s when major outbreaks occurred in Ontario and Quebec (Martineau 1984). Serious outbreaks have appeared regularly in these provinces since the early 1950s, and in the Maritimes since the 1970s. These infestations usually last several years although one outbreak near Quebec city persisted for 9 years (Cochaux 1968).

Biology

The life history and biology of the oak leafshredder (known as the oak leaftier in the United States) have been described by Beckwith (1963) for populations in Pennsylvania and Connecticut, by Cochaux (1968) for those in Quebec, and by Sippell (1975) for those in Ontario. Eggs usually hatch in late April and early May in Ontario and Quebec when the leaf buds begin to swell. The yellowish-brown larvae enter the buds, which they mine until the bud flushes, creating a "shot hole" appearance in the open leaves similar to that caused by other oak pests (Martineau 1984). The solitary larvae then feed from folds formed at the leaf margin (Fig. 1) or by tying two leaves together with strands of silk. By mid-June the last (fourth) instar spins to the ground and pupates in the leaf litter. Some pupae can be found in the trees but these are usually parasitized (Cochaux 1968). The pupal stage lasts about 10 days and the small yellow moths, with a wing span of 12 to 15 mm, begin flying around the end of June. The flight season near Sault Ste. Marie and in central Ontario, as indicated by pheromone traps, lasts up to 4 weeks with the peak flight occurring by mid-July. Eggs are about 1 mm long and are laid singly on the apical section of branches, typically on the upper surface around nodes and on the rough bark of older growth (Fig. 2) (Cochaux 1968; Ciesla 1969). Often they become covered with debris and are difficult to find. The insect overwinters in the egg stage.

There is little other biological information available on the oak leafshredder. Insect disease surveys conducted since 1968 by the Forest Pest Management Institute of Forestry Canada and its predecessor, the Insect Pathology Research Institute, have recorded only rare incidences of microsporidian, fungal (*Isaria*), and nuclear polyhedrosis viral infections in larval collections. Records of insect parasites are few, but some 36 species have been reported to attack eggs, larvae, and pupae, although the incidence of attack was generally low (Cochaux 1968; Martineau 1984). There is also some

Figure 1. Feeding damage by the oak leafshredder, *Croesia semipurpurana*.

Figure 2. Eggs of the oak leafshredder, *C. semipurpurana*.

predation by ants and spiders. The impact of these natural mortality factors on population cycles has not been assessed. Thus, the sudden decline of large infestations remains unexplained, but could be related to starvation and loss of habitat due to defoliation (Cochaux 1968) or to abiotic factors.

Population Monitoring

One of the key tools in managing this pest is egg sampling to forecast potential defoliation and population trends so that control operations can be applied in a timely manner. Ciesla (1969) successfully developed an egg-sampling plan that related egg density (from egg counts on branch-tip samples) to subsequent defoliation. This method has been adapted by the Forest Insect and Disease Survey (FIDS) of Forestry Canada in Ontario and requires the collection of two 38-cm branch tips in August from the midcrown of four dominant or codominant trees per stand. Because the eggs are difficult to find, branches must be examined under a microscope in the laboratory. Defoliation estimates of sample plots are taken in June. Although the entire procedure is tedious and time-consuming, it has been successful in identifying stands that may require protection in the following year.

In an attempt to simplify the forecasting process and reduce the effort and costs involved, an experimental pheromone monitoring program is being tested in Ontario to supplement or replace egg sampling. The aim is to compare pheromone trap catches obtained during the summer with egg counts taken later in the year to determine whether there is a statistically reliable relationship between them, one that would indicate that catches in Pherocon® IC sticky traps (Zoecon, Palo Alto, California) provide the same predictive power as egg counts. Ideally, trap catch thresholds will be established that would reflect potential light, moderate, or heavy defoliation in the following year, as has been established for egg counts (Ciesla 1969). The early results are encouraging and show a strong correlation (r^2 = 92%) between the two sampling procedures (Fig. 3). However, further testing is required over a wider range of oak leafshredder populations to validate the relationship before the pheromone trapping program can be fully implemented.

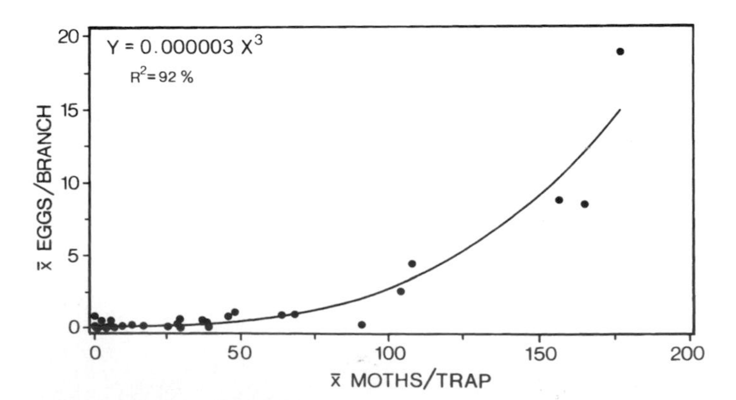

Figure 3. Relationship between average trap catch of oak leafshredder, *C. semipurpurana*, male moths and the average number of eggs per 38-cm branch tip.

Pheromone monitoring as an early warning detection system is also being evaluated in the Maritimes. Because of the efficiency of sampling with pheromone traps, better geographical coverage of the region will be possible and trap catches will also help to differentiate between sites attacked by the oak leafshredder and those by the oak olethreutid leafroller (Grant et al. 1991).

Empirical studies have shown that male moths are attracted to an 85:15 mixture of (E)-11-tetradecenal and (Z)-11-tetradecenal (Weatherston et al. 1978) that is synergized by the addition of acetate analogs of these compounds (Grant et al. 1981). An 85:15:100 mixture of (E)-11-tetradecenal, (Z)-11-tetradecenal and tetradecyl acetate, formulated in PVC (polyvinyl chloride) rods, rubber septa, or polyethylene caps, has been the standard lure used by Forestry Canada for monitoring. Fortuitously, this mixture, particularly at concentrations used for the oak leafshredder, is relatively unattractive to the closely related blueberry leaftier, *Croesia curvalana* (Grant, unpubl. data), although the latter is attracted to a 90: 10 E:Z blend of the aldehydes (Lonergan et al. 1989). *Croesia curvalana* is morphologically very similar to the oak leafshredder and would be difficult to distinguish in traps. Occasionally some spruce budworm, *Choristoneura fumiferana*, moths may be caught when traps are located near conifers because of the similarity of its pheromone (mainly a 95:5 mixture of the E and Z aldehydes) to that of the oak leafshredder. However, the acetate component and the lower E:Z ratio of the aldehydes make the oak leafshredder lure relatively unattractive to the spruce budworm. Recently, chemical identification of the oak leafshredder pheromone has confirmed the presence of the aldehyde components in the female moth (P. Silk, pers. comm.), thus rationalizing the use of the standard lure.

Control

Control of oak leafshredder infestations with pesticides has been warranted in Ontario on several occasions to provide foliage protection and suppress infestations (see Chapters 4 and 66). Orthene® and Sevin® were aerially applied in 1977, 1978, 1979, 1980, and 1983 and were very effective in reducing larval populations and minimizing defoliation. The application of *Bacillus thuringiensis* (Bt) (Dipel® 88) in 1983 also reduced larval populations substantially but afforded less foliage protection than the chemical insecticides. The concealed feeding of the larvae in buds requires application of these pesticides after the buds flush and the insect becomes an open feeder.

An experimental aerial application of an insect growth regulator, diflubenzuron, was conducted in 1982 to assess its efficacy as a potential and more acceptable alternative to traditional insecticides. Unlike most broad-spectrum pesticides, this material has no contact effect and, in general, acts only on larval instars that actively biosynthesize chitin. Upon ingestion, it inhibits chitin formation, which leads to a pronounced weakening of the integument, and the larvae usually die while attempting to moult. Adult insect parasites, predators, and pollinators are generally unaffected since they

Table 1. Effect of aerial application of diflubenzuron to an oak forest infested with the oak leafshredder, *Croesia semipurpurana*, near Barrie, Ontario, in June 1982 (x ± SE).

Plot no.	Treatment	Pre-spray sample (larvae/46-cm branch)	Post-spray sample: Larvae/ 46-cm branch	Post-spray sample: Pupae/ m^2 of litter	Post-spray sample: Adult emergence/ m^2 of litter	% Population reduction[a] due to treatment: Larvae	% Population reduction[a] due to treatment: Pupae	% Population reduction[a] due to treatment: Adult emergence
Check		27.3 ± 4.4	2.7 ± 0.41	73.4 ± 15.82	17.0 ± 4.75			
1	Diflubenzuron, 70 g/ha	10.7 ± 1.5	0.2 ± 0.06	0.6 + 0.13	0	62.2	97.9	100
2	Diflubenzuron, 140 g/ha	13.2 ± 1.6	0.1 ± 0.02	0.6 ± 0.09	0	92.3	98.3	100

[a] % population reduction = $\left[1 - \left(\frac{\text{post-spray density in treatment}}{\text{pre-spray larval density in treatment}} \times \frac{\text{pre-spray larval density in control}}{\text{post-spray density in control}} \right) \right] \times 100$

neither feed on foliage nor form new chitin (Retnakaran and Wright 1987).

Diflubenzuron (25% wettable powder suspended in 7-N-oil) was applied at 70 and 140 g active ingredient (A.I.) in 4.7 L/ha, with a Cessna-AgTruck aircraft equipped with four Micronair® AU-3000 atomizers. The spray was deposited on red oak stands near Barrie, Ontario, when most of the larvae were in the third instar and feeding on opened leaves. Prespray larval population density was estimated from larval counts on 46-cm branch-tip samples taken from the midcrown. Sequential post-spray sampling was conducted to determine (i) the number of larvae per 46-cm branch tip, just before the fourth-instar larvae began to spin down, (ii) the number of pupae per square metre of litter, and (iii) the number of adults emerging from a square metre of litter (Table 1). Defoliation was estimated visually with the aid of binoculars in both the treatment and check blocks after the method of Ciesla (1969). The 140-g dosage resulted in more than 90% reduction in the larval, pupal, and adult populations and provided good foliage protection (Retnakaran and Grant 1985).

An interesting aspect of this study was the use of pheromone monitoring traps to corroborate the post-treatment effects of the diflubenzuron application. This was necessary because of the anticipated difficulty in finding enough late-instar larvae and pupae in the leaf litter to assess following the spray. A comparison of the pheromone trap catches in the check and treated plots demonstrated that survival to the adult stage in the treated plot had been reduced by more than 90% (Fig. 4), confirming the results indicated by the litter sampling data (Table 1). This test suggests that pheromone monitoring could have wider application in forestry as a post-spray assessment tool.

Use of the oak leafshredder sex pheromone for mating disruption may offer another means of suppressing populations, particularly if applied at the start of an outbreak. To date only small plot experiments have been conducted, but

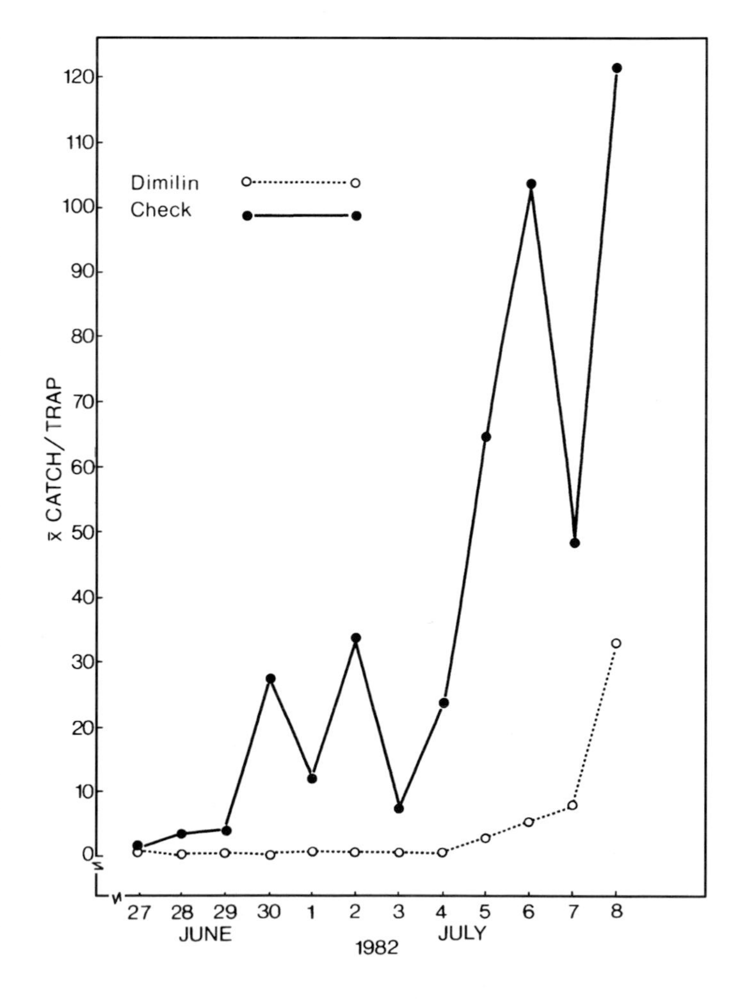

Figure 4. Pheromone trap catches of oak leafshredder, *C. semipurpurana*, moths in a plot treated with diflubenzuron and in the check plot. The total moths caught in the treated plot was reduced by more than 90%.

they have demonstrated that pheromone released from ground emitters can significantly reduce moth catch in pheromone traps, indicating disruption of sexual communication. An experimental aerial application of sex pheromone to study area-wide mating disruption appears to be warranted.

References

Beckwith, R.C. 1963. An oak leaftier, *Croesia semipurpurana* (Lepidoptera: Tortricidae) in Connecticut. Ann. Entomol. Soc. Am. 56: 741-744.

Ciesla, W.H. 1969. Forecasting population trends of an oak leaftier, *Croesia semipurpurana*. J. Econ. Entomol. 62: 1054-1056.

Cochaux, P. 1968. La tordeuse printanière du chêne, *Croesia semipurpurana* (Kft.) (Lepidoptera: Tortricidae) dans la région de Québec. Ann. Entomol. Soc. Que. 13: 98-107.

Grant, G.G.; Frech, D.; MacDonald, L.; Doyle, B. 1981. Effect of additional components on a sex attractant for the oak leafshredder, *Croesia semipurpurana* (Lepidoptera: Tortricidae). Can. Entomol. 113: 449-451.

Grant, G.G.; Pendrel, B.; Slessor, K.N.; Meng, X.Z.; Miller, W.E. 1991. Identification of sex pheromone components for two lepidopteran defoliators, the oak olethreutid leafroller, *Pseudexentera spoliana* (Clemens), and the aspen leafroller, *Pseudexentera oregonana* (Walsingham). Can Entomol. 123: 1209-1218.

Lonergan, G.C.; Ponder, B.M.; Seabrook, W.D.; Kipp, L.R. 1989. Sex pheromone components of the blueberry leaftier moth *Croesia curvalana* Kearfott (Lepidoptera: Tortricidae). J. Chem. Ecol. 15: 2495-2506.

Magasi, L. 1987. Forest Pest Conditions in the Maritimes 1986. Can. For. Serv., Fredericton, N.B., Inf. Rep. M-X-161.

Martineau, R. 1984. Insects Harmful to Forest Trees, Can. For. Serv., Ottawa, Ont., For. Tech. Rep. 32.

Nichols, J.O. 1968. Oak mortality in Pennsylvania—a ten-year study. J. For. 66: 681-694.

Retnakaran, A.; Grant, G.G. 1985. Control of the oak-leaf shredder, *Croesia semipurpurana* (Kearfott) (Lepidoptera: Tortricidae), by aerial application of diflubenzuron. Can. Entomol. 117: 363-369.

Retnakaran, A.; Wright, J.J.E. 1987. Control of insect pests with benzoylphenyl ureas. Pages 205-282 *in* J.E. Wright and A. Retnakaran (Eds.), Chitin and Benzoylphenyl Ureas. Dr. W. Junk Publishers, Dordrecht, the Netherlands.

Sippell, W.L. 1975. Oak leaf shredder *Croesia semipurpurana* (Kft.). Pages 218-219, *in* M.L. Prebble (Ed.), Aerial Control of Forest Insects in Canada. Dep. Environ., Ottawa, Ont.

Staley, J.M. 1965. Decline and mortality of red and scarlet oaks. For. Sci. 11: 2-17.

Sterner, T.E.; Davidson, A.G. 1981. Forest Insect and Disease Conditions in Canada 1980. Can. For. Serv., Ottawa, Ont.

Weatherston, J.; Grant, G.G.; MacDonald, L.M.; Frech, D.; Werner, R.A.; Leznoff, C.C.; Fyles, T.M. 1978. Attraction of various tortricine moths to blends containing *cis*-11-tetradecenal. J. Chem. Ecol. 4: 543-549.

Chapter 17

Bark- and Wood-Boring Insects

L. Safranyik

A large and diverse group of insects, representing many families and species, mine the outer bark, phloem, and wood of forest trees. Living trees of all ages, killed or felled trees, logs, as well as green and seasoned lumber are all affected. Bark- and wood-boring insects are but a small component of the large complex of organisms that influences forest dynamics, and they have important functions in nutrient cycling and in the evolution of forest stands under natural conditions (Wood 1982; Stark 1982).

The primary ecological role of those insect species that kill trees is as thinning agents. They decrease competition among the residual trees for resources essential for growth and development; hasten forest succession; and change stand density, age, and species composition; thus, they aid regeneration of the forest.

Under some circumstances, however, such as during outbreaks of tree-killing bark beetles, high levels of activity by bark- and wood-boring insects cause unacceptably high impacts on economic and social values associated with the forest, such as wood and fiber production, recreation, fish and wildlife, grazing, and hydrology. Any of these can seriously disrupt management plans (Coulson and Stark 1982).

Injury to living trees by bark- and wood-boring insects affects forest productivity in three ways: (a) direct killing, (b) reduction in merchantability, and (c) growth reduction. Damage to dead trees, logs, and lumber affects their value, principally by causing degrade. Such damage may also prevent offshore shipments of such materials.

In general, the most economically important bark- and wood-boring insects are beetles (Coleoptera) in the following families: bark beetles (Scolytidae), roundheaded borers (Cerambycidae), and flatheaded borers (Buprestidae). Only a relatively few species in each of these families attack and kill apparently healthy trees; most attack weakened, dead, or recently felled trees or green lumber. Bark beetles are unique in that they are bark miners or wood borers in both the adult and larval stages whereas this habit is restricted to only the larval stages of most other bark- and wood-boring insects. Most species that are known tree killers mine extensively in the cambium region (e.g. some *Dendroctonus*, *Tetropium*, and *Melanophila*). Others bore deeply in the wood and damage logs and wood products (e.g. *Trypodendron*, *Monochamus*). The main economic impact of bark beetles is direct killing of trees and damage to wood and wood products, whereas the main economic impact of roundheaded and flatheaded borers is only through damage to wood and wood products. Bark beetles are the most injurious among the bark- and wood-boring insects. Consequently, the population biologies and management of the most destructive species have been researched extensively in Canada and elsewhere.

The three basic approaches to reduction of losses from bark- and wood-boring insects are through forest management, biological control, and direct control. These three approaches comprise the following components:

1. Preventive forest management
   - (a) Stand manipulation to reduce tree/stand susceptibility;
   - (b) Sanitary cultural and harvesting practices including:
     - (i) high utilization standards;
     - (ii) prompt removal and processing of windfelled, fire-killed, or otherwise damaged timber;
     - (iii) control of the timing of cultural practices (e.g. thinning and spacing) and harvesting operations; "hot logging," and speedy processing of logs.
2. Biological control
   - (a) Manipulation of beetle populations using behavior-modifying chemicals;
   - (b) Inundative releases of indigenous predators, parasites, and diseases and introduction of exotic natural control agents;
   - (c) Enhancement of the habitat for indigenous natural control agents.
3. Direct control
   - (a) Sanitation salvage logging of infested host materials and burning of bark and slabs;
   - (b) Treating individual trees, logs, and log decks to prevent attack or to kill beetles using chemicals, heat, water, debarking, or other mechanical means; chemical treatment and kiln-drying of lumber;
   - (c) Preparing felled or standing trap trees and trap logs; untreated or treated with insecticide, trap trees and trap logs can be baited with population aggregating pheromone. Following infestation, the insects in untreated trap trees are killed using methods described earlier.
   - (d) Mass trapping using traps baited with population aggregating pheromone.

Normally, under operational conditions, a combination of the three approaches is used. For example, behavior-modifying chemicals of the mountain pine beetle are used extensively in the lodgepole pine, *Pinus contorta* var. *latifolia*, forests of British Columbia for baiting blocks of timber to concentrate the beetles in trees scheduled for harvesting. More detailed discussion of management strategies and tactics, including those that are currently being researched or deserve investigation in the future, are given in conjunction with the following descriptions of the biologies and management of destructive bark- and wood-boring insects and in Chapter 45 on behavior-modifying chemicals.

References

Coulson, R.N.; Stark, R.W. 1982. Integrated management of bark beetles. Pages 315–349 *in* J.B. Mitton and K.B. Sturgeon (Eds.), Bark Beetles in North American conifers. A system for the study of evolutionary biology. University of Texas Press, Austin, Tex. 527 p.

Stark, R.W. 1982. Generalized ecology and life cycle of bark beetles. Pages 21–45 *in* J.B. Mitton and K.B. Sturgeon (Eds.), Bark Beetles in North American conifers. A system for the study of evolutionary biology. University of Texas Press, Austin, Tex. 527 p.

Wood, S.L. 1982. The bark and ambrosia beetles of North and Central America (Coleoptera: Scolytidae), a taxonomic monograph. Great Basin Naturalist Monogr. no. 6. Brigham Young University, Provo, Utah. 1359 p.

Chapter 18

Bark Beetles

L. SAFRANYIK

Introduction

The majority of the estimated 214 species of bark beetles that are known or thought to occur in Canada and Alaska (Bright 1976) attack forest trees, especially conifers. Some species of bark beetles, however, attack fruit and ornamental trees, shrubs, and even nonwoody plants. Most species that attack woody plants breed only in a relatively few host plant species and are often restricted to a certain part of their host plant, such as cones, pith of small branches, the inner bark of the main bole, and xylem tissues. Primary bark beetles are those species that attack and kill trees of apparently normal vigor; those that attack dead, dying, or severely weakened trees are called secondary bark beetles. This classification is of doubtful value, however, since several of the most destructive primary species (e.g. the spruce beetle, *Dendroctonus rufipennis*) actually prefer to attack weakened or felled trees in all phases of their epidemiology, or attack trees of subnormal vigor during endemic periods (e.g. the mountain pine beetle, *D. ponderosae*). Conversely, some secondary species can become killers of apparently healthy trees under certain conditions (e.g. some species of *Ips* following decline of *Dendroctonus* outbreaks).

The most destructive of the true bark beetle species breed in the inner bark and phloem of the main bole. In Canada, all but two species of the most economically important true bark beetles attack and kill coniferous trees (Table 1). The two exceptions are the smaller European elm bark beetle, *Scolytus multistriatus*, an introduced species, and the native elm bark beetle, *Hylurgopinus rufipes*, both of which infest elm trees and are the principal vectors of the Dutch elm disease. Numerous other species could have been included in this list on the basis of occasionally causing locally significant damage (e.g. the western pine beetle, *D. brevicomis*, the foureyed spruce bark beetle, *Polygraphus rufipennis*), or potential for causing significant damage in managed stands (e.g. some species of *Pseudohylesinus*, *Pityogenes*, *Hylastes*). Based on historical records, however, the species listed in Table 1 have caused by far the most significant socioeconomic losses. In the sections following, the life histories, habits, and management of two highly destructive species of true bark beetles and of ambrosia beetles are described.

Table 1. Major bark beetle pests of Canada.

Insect	Canadian distribution	Principal hosts[a]
Dendroctonus ponderosae Mountain pine beetle	British Columbia Alberta	*Pinus contorta* *P. ponderosa* *P. monticola*
D. rufipennis Spruce beetle	Transcontinental	*Picea glauca* *P. engelmannii*
D. pseudotsugae menziesii Douglas-fir beetle	Southern British Columbia Southwestern Alberta	*Pseudotsuga*
D. simplex Eastern larch beetle	Transcontinental	*Larix laricina*
Dryocoetes confusus Western balsam bark beetle	British Columbia Alberta	*Abies lasiocarpa* *Picea engelmannii*
Hylurgopinus rufipes Native elm bark beetle	New Brunswick to Saskatchewan	*Ulmus* sp.
Ips pini Pine engraver	Transcontinental	*Pinus* sp.
Scolytus multistriatus Smaller European elm bark beetle	Western Provinces Southern Ontario Quebec	*Ulmus* sp.
Trypodendron lineatum Striped ambrosia beetle	Transcontinental	All conifers

[a] Hosts in which the insect occurs most commonly and/or on which most of the damage occurs.

Mountain Pine Beetle

The mountain pine beetle, *Dendroctonus ponderosae*, is the most destructive insect enemy of mature pines within the range of its geographic distribution. In recent years, the losses caused by this insect, especially in lodgepole pine, have been devastating (Wood et al. 1985). In stands managed for commercial timber production, the value of losses during epidemics is usually considerably greater than that indicated by the volume loss because proportionately more mortality occurs among the larger-diameter trees.

Canadian Distribution and Host Trees

The insect is native to southwestern Alberta and southern and central British Columbia up to about latitude 56°N (Safranyik et al. 1974). During the late 1970s, mountain pine beetle was recorded for the first time from the Cypress Hills region of southeastern Alberta and adjacent Saskatchewan, causing high mortality in mature lodgepole pine stands. Lodgepole pine, ponderosa pine, and western white pine are the principal hosts, but all native pines and several species of exotic pines are also attacked and killed. Some nonhost trees (e.g. *Picea* spp.) are occasionally attacked and killed, but populations are not usually maintained in such trees.

Signs and Symptoms of Injury

Infested trees are recognized by characteristic symptoms of injury on the bark of the main bole, the inner bark and sapwood, and the foliage (McMullen et al. 1986). Red-brown boring dust, similar in texture to fine sawdust, and pitch tubes on the bark of the main bole are the first signs of attack. Boring dust accumulates in bark crevices at the bases of attacked trees (Fig. 1), but is removed by wind and rain and is therefore only conspicuous for a short time following attack. Sometimes pitch exudes from beetle entry holes and, mixed with boring dust, forms cream- to brown-colored tubes 1 to 3 cm in diameter which are usually most conspicuous on the lower bole but may be found as high as the upper crown (Fig. 2).

Figure 1. Boring dust at the base of a recently attacked lodgepole pine. (Forestry Canada)

The gallery system is the most reliable indicator of mountain pine beetle attacks. On standing trees the female beetles construct vertical egg galleries with slight hooks at the lower ends, in the phloem parallel to the grain. On windfalls the galleries may run toward the top or bottom of the tree, but will always be parallel to the grain. Egg galleries average about 25 cm long and, except for the upper few centimetres, are packed with boring dust. The sides of these egg galleries may contain eggs (Fig. 3) for the first few weeks after attack; later on, egg galleries, larval mines, pupal chambers, and beetles in various developmental stages may be found intermingled under the bark (Fig. 4). Within 2 to 3 weeks after attack, blue stain fungi, carried into the tree by the attacking beetles, begin to form pigments and the infested wood becomes blue-stained (Fig. 5). (The most common blue stain fungi associates of the mountain pine beetle are *Ceratocystis clavigera* and *C. montina*.) The bark of infested trees or logs is often disturbed by woodpeckers searching for beetles or larvae. Small, round emergence holes cut through the bark by emerging beetles can be readily seen and are often numerous on the lower boles of trees from which beetles have emerged.

Figure 2. Pitch tubes formed around entry holes made by female mountain pine beetles, *Dendroctonus ponderosae*, on the bole of a lodgepole pine tree. (Forestry Canada)

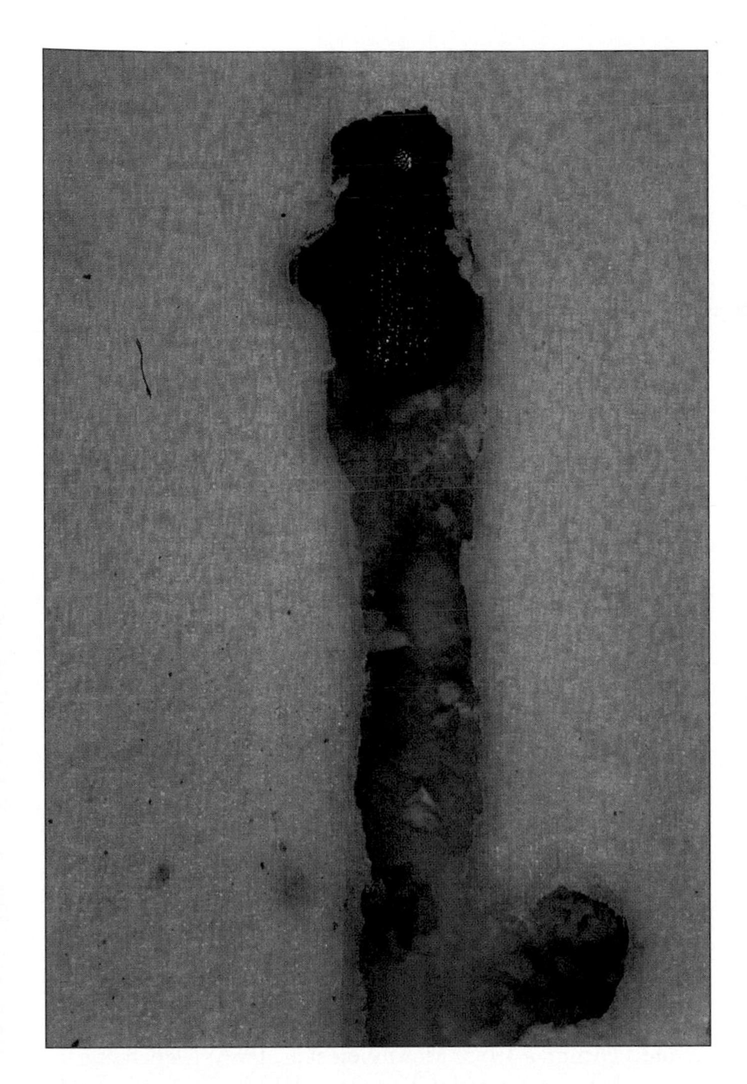

Figure 3. Section of an egg gallery being excavated by female mountain pine beetle, *Dendroctonus ponderosae*. (H.S. Whitney)

Figure 4. Underside of bark showing an egg gallery and larval mines of the mountain pine beetle. (H.S. Whitney)

Figure 5. Blue-stained sapwood. (H.S. Whitney)

Figure 6. Typical fading of lodgepole pine foliage in the spring of the year following attack by mountain pine beetle. (Forestry Canada)

The foliage of infested trees eventually fades and progresses from yellowish green, yellow, and finally to redbrown. Timing of foliage fading varies considerably with tree species and climatic conditions. In lodgepole pine attacked in midsummer, the first observable color change usually occurs in the May–June period following the year of attack (Fig. 6). The red foliage may remain on the tree for several years.

Bionomics

The mountain pine beetle normally has one generation per year. The complete life cycle is spent under the bark of trees, except for a brief period when adults fly to infest new trees during mid- to late summer. The female beetle initiates attack by boring through the outer bark and beginning excavation of an egg gallery in the phloem. As the female begins gallery excavation, she releases population aggregating pheromones[1] which attract more beetles of both sexes to the tree (Borden 1982). Mating takes place in the lower end of the egg gallery. Usually 60 to 80 eggs are laid singly in niches cut in the sides of the egg gallery. Eggs hatch in 1 to 3 weeks and the larvae (white legless grubs with brown heads) feed in the phloem, approximately at right angles to the egg gallery. The larvae pass through four developmental stages (instars) and late-instar larvae become dormant in late October or November. The larvae begin feeding again the next spring and pupate during late spring to early summer. Development is completed in mid-July. The life stages of the mountain pine beetle are shown in Figure 7.

The most common exceptions to the 1-year life cycle occur when many parent females establish two broods in a single dry year, or in very cool years, or at high elevations and latitudes when a proportion of the brood may require more than 1 year to complete development (Amman and Cole 1983).

The insect uses a complex of behavior-modifying chemicals to find and mediate mass attacks on host trees (Borden 1982). This process results in a sequential colonization of nearby host trees surrounding the trees that were attacked first. Trees resist invasion of the inner bark by the beetle blue stain fungi complex with an initial flush of in situ resin ducts into the wound (primary resinosis). This is followed by production, in response to the injury, of new resin in live cells in the vicinity of the wound (secondary resinosis) (Reid et al. 1967; Safranyik et al. 1975). When resinosis is heavy it can repel or kill the invading beetles, isolate the blue stain fungi, and prevent brood establishment. If attacks are to be successful, the beetles and their associated fungi must be present in sufficient numbers to overcome host resistance. Adults carry spores of the blue stain fungi on their bodies and in a special repository on their maxillae (Safranyik et al. 1975). In successfully attacked trees, blue stain fungi quickly colonize and kill tissue and thereby enhance brood development.

During endemic periods, mountain pine beetle populations exist in suppressed, diseased, injured, and decadent trees, and often in trees that have been attacked by secondary bark beetles (Amman 1978; Safranyik et al. 1974). In lodgepole pine, incipient infestations, which characteristically kill small groups of trees, are often found in draws and gullies, along edges of stand openings, or in areas subject to soil compaction or wide fluctuations in the water table. Such stands are usually 80 years old or older with an average diameter (for trees larger than 10 cm) of at least 20 cm. Incipient infestations in ponderosa and western white pine can develop in both mature and immature stands, especially in dense, pure stands on poorer sites and in dense mixed stands.

Just how the transition from endemic to epidemic populations occurs is uncertain. However, all factors that would significantly reduce host resistance and/or increase the size of the beetle population above a level necessary for colonizing at least some of the large-diameter trees could trigger outbreaks (Amman 1978; Amman and Cole 1983; Reid et al. 1967; Safranyik et al. 1974). Outbreaks last from 3 to 20 years and range in size from a few hectares to hundreds of square kilometres, and can recur in a stand at 5- to 20-year intervals until most of the large-diameter pine have been killed.

Management

The two components of forest management that reduce losses from the mountain pine beetle are preventive management and direct control (McMullen et al. 1986).

Preventive management involves long-term planning and forestry practices designed to reduce stand susceptibility, and it requires that stands or areas be rated for their relative potential hazard from outbreaks. Hazard rating can be done using historical records of beetle activity or various hazard rating systems (Amman 1985; McMullen et al. 1986). Depending on management goals and with a clear definition of acceptable risk of loss from mountain pine beetle in the event of an outbreak, one or a combination of the following management options are available (Safranyik et al. 1974): (a) shorter stand rotation, (b) type conversion, (c) formation of mixed stands, (d) formation of age and species mosaics, (e) control of stand structure and stocking.

In British Columbia, all but the last of these management options are used, especially following sanitation logging operations. Spacing and thinning of stands and partial cutting have proven effective in reducing losses in second-growth ponderosa pine (Sartwell and Dolph 1976) as well as in mature lodgepole pine in the United States (Cole and McGregor 1985; Mitchell et al. 1983), but these methods have not been used operationally in Canada.

Direct control does not treat the underlying cause of the problem of stand susceptibility. Experience suggests that the general requirements of successful control operations are as follows (McMullen et al. 1986):

1. Yearly surveillance of infested and susceptible stands;
2. Priorization of stands for treatment, based on socioeconomic considerations;
3. Quick, thorough treatment of incipient infestations;
4. Follow-through with treatment as long as necessary.

[1] Semiochemical is a term used to describe message-bearing chemicals (Law and Regnier 1971). These include pheromones (Karlson and Lüscher 1959), which are chemicals produced by insects to communicate with other members of the same species, and kairomones (Brown et al. 1970), which are chemicals produced by one species (often a host) and used by another species (e.g. a predator) to its advantage.

Figure 7. The life stages of the mountain pine beetle: (a) egg, (b) larva, (c) pupa, (d) adult. (P. Debnam)

Because direct control does not treat the underlying causes of epidemics, repeated treatments are applied until beetle populations are reduced to endemic levels or other management options are implemented. Sanitation logging and individual tree treatments are the two types of operationally applied direct control. Sanitation logging is used to treat the larger or more diffuse infestations. The commonly used individual tree treatments are pile and burn, post-attack injection of trees with the herbicide MSMA (monosodium methane arsenate), and debarking. In high-value or high-impact areas, the lower boles of trees can be treated with an insecticide (2% A.I. carbaryl in water is registered for this use) to protect them from lethal attacks (McMullen et al. 1986). Baiting of trees and stands with population aggregating pheromones to increase both the biological effectiveness and cost-effectiveness of direct control operations has been used increasingly during the past 5 years (Borden et al. 1983) (see Chapter 45).

Current Research and Research Needs

Current research, exclusive of that on behavior-modifying chemicals, is focused on the following areas: (a) evaluation of hazard rating systems for operational use; (b) effects of outbreaks on lodgepole pine stand structure and development; (c) evaluation of the potential of native predators (*Enoclerus* and *Thanasimus* spp.), exotic predators (*Rhizophagus grandis* and *T. formicarius*), and a pathogenic fungus (*Beauveria bassiana*) for controlling mountain pine beetle; (d) short- and long-range dispersal of the mountain pine beetle; (e) forecasting damage trends; (f) beetle population quality and its role in bark beetle epidemiology; (g) determining the effectiveness of partial cutting and spacing in lodgepole pine in reducing losses; and (h) modelling of mountain pine beetle dynamics in lodgepole pine and the spread of infestations. Research topics (a), (e), (g), and (h) are undertaken as part of a joint Canada–U.S. agreement to improve management of mountain pine beetle in lodgepole pine forests. In addition to these topics, more work is urgently needed on mountain pine beetle population dynamics in the endemic state, variation in beetle biology and habits in climatically different regions, role of host tree injury and stress in triggering outbreaks, and incorporation of new knowledge into management guidelines.

Spruce Beetle

The spruce beetle, *Dendroctonus rufipennis*, is one of the most destructive insect pests of mature spruce in Canada. Endemic populations exist in wind-felled trees and logging residue, but injured, diseased, decadent, or otherwise stressed trees may also be attacked and killed. Under favorable conditions, large populations can build up in such host materials and may act as precursors for epidemic infestations. All known outbreaks have occurred following large-scale windthrow, other stand disturbances (such as insect defoliation), severe competition, or large accumulations of human-made slash.

Canadian Distribution and Host Trees

The beetle occurs throughout the spruce forests of Canada. All native species of spruce, *Picea* spp., may be attacked and killed but the major hosts are white spruce, *P. glauca*, Engelmann spruce, *P. engelmannii*, and their hybrids, Safranyik et al. 1981. Under epidemic conditions, some nonhost species (particularly *Pinus* spp.) may also be attacked and killed.

Signs and Symptoms of Injury

The external symptoms of infestation on the bole are similar to those described for mountain pine beetle. On exposed host materials, attacks tend to be concentrated in the shaded areas. No pitch tubes are formed on dead material such as stumps, logs, and broken off or uprooted windfall, and during outbreaks few pitch tubes occur on standing trees. Brown boring dust, similar to that described for mountain pine beetle, often accumulates in large quantities around the bases of infested trees and stumps (Fig. 8) and in crevices in the bark of horizontal host materials.

Egg galleries, averaging about 15 cm long, follow the grain in all host materials. Eggs are deposited in single or double rows into grooves cut by the female beetle in alternate sides of the egg gallery. The larvae tend to feed communally in the early instars and individually in the later stages, a behavior that results in a characteristic larval gallery system (Fig. 9). Oval pupal chambers are formed in the phloem at the ends of larval mines; the pupal chambers often have round exit holes. The sapwood of successfully infested host material is blue-stained within a few weeks after attack.

The foliage of infested trees normally remains green during the first winter. By the following July, the foliage usually fades to greenish yellow (Fig. 10), and it falls shortly thereafter. Sometimes the needles fall without appreciable color change. The bare, red twiglets give a reddish appearance to the crown after most of the needles have dropped.

Figure 8. Boring dust at the base of a spruce stump recently infested by spruce beetle, *Dendroctonus rufipennis*. (Forestry Canada)

Bionomics

The spruce beetle predominantly has a 1-year cycle in the warmer parts of its range but usually requires 2 or rarely 3 years to complete its development in most areas in Canada. In Western Canada, the beetle usually overwinters in the adult stage in areas where the 1-year life cycle predominates and in the larval (1st year) and adult stages (2nd year) elsewhere (Safranyik et al. 1981). Adults emerge to attack new hosts in the late spring to midsummer. The female constructs the egg gallery. Under optimum conditions, an average of about 100 eggs are laid per egg gallery (Schmid and Frye 1977). The larvae are creamy white, legless grubs with brown heads and have two small brown sclerites on the dorsal surfaces of the last two abdominal segments. Fully grown larvae reach about 7 mm long. On a typical 2-year life cycle, the larvae overwinter, resume development in the following spring, pass through a brief pupal stage, and transform into adults by midsummer. The young adults usually remain under the bark throughout the next winter, after which they emerge to attack in the second spring. Some brood adults emerge in late summer and early fall and move to the bases of their brood trees where they re-enter and construct feeding tunnels in which they overwinter, often in groups.

The main factors that affect generation survival are: (a) host resistance, (b) deterioration of the subcortical habitat, (c) climatic events (especially temperature), (d) competition, and (e) natural enemies (especially woodpecker predation). Brood productivity is usually highest in windfall, intermediate in trees, and lowest in logging residue, although high variation in productivity occurs in all host material (Dyer and Taylor 1971; Schmid and Frye 1977).

Host finding, aggregation, and mass attack are mediated through a complex of host- and insect-produced volatiles (Dyer 1973; Borden 1982). Spruce beetles, like most other scolytids, carry into the host with them several microorganisms, including blue stain fungi; the main blue stain fungus associate is *Pesotum*

Figure 9. Typical gallery system for spruce beetle larvae. Note obliteration of individual larval mines in the middle region, due to a commensal feeding habit. (P.M. Hall)

piceae. These fungi kill tissues surrounding the beetle galleries and enhance host colonization (Safranyik et al. 1981).

During outbreaks there is usually a strong direct relationship between the proportion of trees killed and tree diameter (Safranyik 1985). Normally, outbreaks are of short duration (lasting 3–5 years in western Canada), and develop in mature or overmature forests. Outbreaks, however, also develop in some immature stands that may have been stressed by defoliation, competition, or dry, cold soil conditions (Davidson and Prentice 1976; Holsten 1984; Hard and Holsten 1985).

Management

Management approaches must emphasize forestry practices designed to reduce tree and stand susceptibility and prevent accumulations of fresh windfall and logging residue. These practices must be augmented by various techniques of direct control to suppress beetle populations in infested stands.

Figure 10. Typical fading of foliage of infested Engelmann spruce 1 year after attack. (Forestry Canada)

In British Columbia, management of spruce beetle is based on the following principles:

A. Long-term goals
   1. To assign management priorities to stands (based on socioeconomic values and historical beetle pressure in an area);
   2. To plan harvesting of susceptible stands in high-priority areas;
   3. To grow spruce on a short rotation (100 years or less);
   4. To develop mixed stands and stands containing mosaics of age classes and/or tree species.

B. Short-term goals
   1. Windfall and slash management, including:
      (a) prompt salvage of wind-thrown trees
      (b) control of cutting boundary layout to ensure windfirm boundaries
      (c) close utilization standards
      (d) treatment of large-diameter slash, stumps, and residue at landings
      (e) avoidance of extensive soil disturbance near stand edges;
   2. Annual surveys to assess spruce beetle abundance and damage;
   3. Prompt, thorough action on incipient outbreaks (sanitation logging, trap trees, individual tree treatments)

Sanitation logging combined with the felling of trap trees is the most commonly used technique in British Columbia for suppressing spruce beetle populations. Sanitation logging involves the scheduling and placement of logging operations to remove spruce beetle in standing timber and wind-felled trees, as well as the removal of susceptible trees or stands. Trap trees can be of two types: conventional (living, large-diameter spruce felled to attract spruce beetles), and lethal (living, large-diameter trees treated with the silvicide monosodium methane arsenate, MSMA, injected into shallow frills cut near the base 10 to 14 days before felling). Conventional trap trees are felled singly or, more commonly, in small patches. One trap tree is used for each 2 to 10 infested trees, and they are felled during the spring before beetle flight. The main purposes for using trap trees are: (a) to contain emerging beetles in cut blocks before logging, (b) to protect adjacent uninfested timber or leave-blocks, (c) to "mop-up" beetles emerging from stumps and slash following logging, and (d) to suppress beetle populations in local infestations of light to moderate severity. Lethal trap trees are used for the same purposes as conventional trap trees, but mostly in remote areas where subsequent trap tree extraction or treatment are difficult.

Disposal of beetles in individual standing or trap trees is done occasionally by bucking, piling and burning on site, debarking, and by treating the bark with a bark-penetrating insecticide such as 2.4% Dursban®-4E in diesel oil.

During the past 5 years, spruce beetle tree baits, prepared from known semiochemicals, have been used increasingly in British Columbia in all phases of direct control programs from pre-treatment monitoring to containment of active infestations and post-logging "mop-up" (Hodgkinson 1985). Although the strategy of using semiochemicals needs further development, there is already strong evidence to indicate that these chemicals have great potential to significantly increase the efficacy of direct control programs.

Current Research and Research Needs

The focus of current research, excluding that on behavior-modifying chemicals, is on developing a set of management guidelines and a conceptual model of population dynamics. Future work will address beetle survival in various host materials and the effects of windthrow and harvesting practices on beetle epidemiology.

References

Amman, G.D. 1978. Biology, ecology and causes of outbreaks of the mountain pine beetle in lodgepole pine forests. Pages 39-53 *in* A.A. Beeryman, Gene D. Amman, and R.W. Stark (Eds.), Theory and practice of mountain pine beetle management in lodgepole pine forests: Symposium Proceedings: 1978 April 25-27: Pullman, Wash.: Washington State University.

Amman, G.D. 1985. Test of lodgepole pine hazard rating methods for mountain pine beetle infestation in southeastern Idaho. *In* Safranyik, L. (Ed.), Role of the host in the population dynamics of forest insects. Proceedings of IUFRO conference, Banff, Alta., Sept. 4-7, 1983. Can. For. Serv., Pac. For. Cent., Victoria, B.C. 240 p.

Amman, G.D.; Cole, W.E. 1983. Mountain pine beetle dynamics in lodgepole pine forests. Part II: Population dynamics. USDA For. Serv. Int. Res. Stn. Gen. Tech. Rep. INT-145. Ogden Utah. 59 p.

Borden, J.J.H. 1982. Aggregation Pheromones. Pages 74-139 *in* J.B. Mitton and K.B. Sturgeon (Eds.), Bark beetles of North American conifers. A system for the study of evolutionary biology. University of Texas Press, Austin, Tex. 526 p.

Borden, J.H.; Chong, L.J.; Pratt, K.E.G.; Gray, D.R. 1983. The application of behaviour-modifying chemicals to contain infestations of the mountain pine beetle, *Dendroctonus ponderosae* Hopk. For. Chron. 59: 235-239.

Bright, D.E. 1976. The insects and arachnids of Canada. Part 2. The bark beetles of Canada and Alaska. Coleoptera: Scolytidae. Biosyst. Inst. Res. Br., Can. Dept. Agric., Ottawa, Ont., Publ. 1576. 241 p.

Brown, W.L. Jr.; Eisner, T.; Whittaker, R.H. 1970. Allomones and kairomones: transpecific chemical messengers. Biosci. 20: 21-22.

Cole, W.E.; McGregor, M.D. 1985. Reducing and preventing mountain pine beetle outbreaks in lodgepole pine stands by selective cutting *in* Safranyik, L. (Ed.), Role of the host in the population dynamics of forest insects. Proceedings of an IUFRO conference, Banff, Alta., Sept. 4-7, 1983. Can. For. Serv., Pac. For. Cent., Victoria, B.C. 240 p.

Davidson, A.G.; Prentice, R.M. (Eds.). 1967. Important forest insects and diseases of mutual concern to Canada, the United States, and Mexico. Canada. Dept. For. and Rural Devel., Ottawa, Ont. 248 p.

Dyer, E.D.A. 1973. Spruce beetle aggregated by the synthetic pheromone frontalin. Can. J. For. Res. 3: 486-494.

Dyer, E.D.A.; Taylor, D.W. 1971. Spruce beetle brood production in logging slash and wind-thrown trees in British Columbia. Can. For. Serv., Pac. For. Res. Cent., Victoria, B.C. Inf. Rep. BC-X-62 16 p.

Hard, J.S.; Holsten, E.H. 1985. Managing white and Lutz spruce stands in south-central Alaska for increased resistance to spruce beetle, USDA For. Serv. Gen. Tech. Rep. PNW-188, Pac. Northwest For. and Range Exp. Stn., Portland, Ore. 21 p.

Hodgkinson, R.S. 1985. Use of trap trees for spruce beetle management in British Columbia, 1979-1984. B.C. Min. For., Victoria, B.C., Pest Manage. Rep. No. 5. 39 p.

Holsten, E.H. 1984. Factors of susceptibility of spruce beetle attack on white spruce in Alaska. J. Entomol. Soc. B.C. 81: 39-45.

Karlson, P.; Lüscher M. 1959. Pheromones: a new term for a class of biologically active substances. Nature (London) 183: 55-56.

Law, J.H.; Regnier, F.E. 1971. Pheromones. Am. Rev. Biochem. 40: 533-548.

McMullen, L.H.; Safranyik, L.; Linton, D.A. 1986. Suppression of mountain pine beetle infestations in lodgepole pine forests. Can. For. Serv., Pac. For. Cent., Inf. Rep. BC-X-276. 20 p.

Mitchell, R.G.; Waring, R.H.; Pitman, G.B. 1983. Thinning lodgepole pine increases tree vigor and resistance to mountain pine beetle. For. Sci. 29: 204-211.

Reid, R.W.; Whitney, H.S.; Watson, J.A. 1967. Reaction of lodgepole pine to attack by *Dendroctonus ponderosae* Hopkins and blue stain fungi. Can. J. Bot. 45: 1115-1126.

Safranyik, L. 1985. Infestation incidence and mortality in white spruce stands by *Dendroctonus rufipennis* Kirby (Coleoptera: Scolytidae) in central British Columbia, Z. angew. Entomol. 99: 86-93.

Safranyik, L.; Shrimpton, D.M.; Whitney, H.S. 1974. Management of lodgepole pine to reduce losses from the mountain pine beetle. Can. For Serv., Pac. For. Res. Cent., For. Tech. Rep. 1. 24 p.

Safranyik, L.; Shrimpton, D.M.; Whitney, H.S. 1975. An interpretation of the interaction between lodgepole pine, the mountain pine beetle and its associated blue stain fungi in Western Canada. Pages 406-428 *in* David M. Baumgartner (Ed.), Management of lodgepole pine ecosystems symposium: Proc., October 9-13, 1973, Pullman, Wash.: Washington State University, Coop. Ext. Serv.

Safranyik, L.; Shrimpton, D.M.; Whitney, H.S. 1981. The role of host-pest interaction in population dynamics of *Dendroctonus rufipennis* (Kirby) (Coleoptera: Scolytidae). Pages 197-213 *in* A.S. Isaev (Ed.), The role of the host in the population dynamics of forest insects: Proc. Int. MAB/IUFRO Symp., Irkutsk, USSR, August 1981.

Sartwell, C.; Dolph, Jr., R.E. 1976. Silvicultural and direct control of mountain pine beetle in second-growth ponderosa pine. USDA For. Serv. Res. Note PNW-268. 8 p.

Schmid, J.M.; Frye, R.H. 1977. Spruce beetle in the Rockies. USDA For. Serv., Gen. Tech. Rep. RM-49, Rocky Mountain For. and Range Exp. Stn., Fort Collins, Colo. 39 p.

Wood, C.S.; Van Sickle, G.A.; Shore, T.L. 1985. Forest insect and disease conditions. British Columbia and the Yukon, 1984. Can. For. Serv., Pac. For. Cent. Inf. Rep. BC-X-259. 20 p.

Chapter 19

Ambrosia Beetles

T.L. SHORE AND J.A. MCLEAN

Introduction

Ambrosia beetles include all members of the family Platypodidae and many members of the family Scolytidae. The Scolytidae are divided into two main groups, ambrosia beetles and bark beetles. Ambrosia beetles burrow into the sapwood and sometimes the heartwood of their hosts where they and their offspring feed on "ambrosia" fungi. Bark beetles feed on phloem tissue in the inner bark.

In Canada, there are more than 20 species of ambrosia beetle (Table 1). *Platypus wilsoni* is the only ambrosia beetle in the family Platypodidae in this country, the rest are in the family Scolytidae. The majority of ambrosia beetle species cause little economic damage because they attack noncommercial tree species (Table 1), branches, or stumps. In fact, ambrosia beetles can be beneficial because the holes they make and the fungi they introduce into the wood accelerate the decomposition process. However, a few species, primarily the striped ambrosia beetle, *Trypodendron lineatum*, *Gnathotrichus sulcatus*, and *G. retusus*, and to a lesser extent, *G. materiarius*, *Platypus wilsoni*, *Xyleborinus saxeseni*, and *T. rufitarsis*, can cause serious economic damage to logs of commercial forest tree species in Canada. The west coast forest industry is the most sensitive to damage by ambrosia beetles because the larger diameter trees found in this area produce a higher proportion of clear lumber and plywood veneer. Even minor ambrosia beetle damage can reduce the grade of an otherwise clear piece of lumber or plywood face stock. Also, the west coast industry exports more lumber than other parts of Canada to countries such as New Zealand and Australia which do not allow the import of wood damaged by ambrosia beetles.

Table 1. Distribution and hosts of ambrosia beetles in Canada.

Family	Genus	Species	Hosts	Distribution
Platypodidae	*Platypus*	*wilsoni*	Many conifer spp.	British Columbia
Scolytidae[a]	*Trypodendron*	*betulae*	*Betula* spp.	Transcontinental
		lineatum	Many conifer spp.	Transcontinental
		retusum	*Populus*, *Picea* spp.	Transcontinental
		rufitarsis	*Picea*, *Pinus* spp.	Transcontinental
	Corthylus	*columbianus*	Many deciduous spp.	Probably E. Canada[b]
		punctatissimus	Many deciduous spp.	S. Ontario, Quebec
	Gnathotrichus	*materiarius*	Many conifer spp.	E. Canada
		retusus	Conifers, *Alnus*, *Populus*	British Columbia
		sulcatus	Many conifer spp.	British Columbia
	Monarthrum	*fasciatum*	Many deciduous spp.	S. Ontario, Quebec
		mali	Many deciduous spp.	S. Ontario to New Brunswick
		scuttelare	*Quercus* sp.	S. British Columbia
	Xyleborinus	*saxeseni*	Deciduous, *Pinus*, *Tsuga*	S.E. and S.W. Canada
	Xyleborus	*affinis*	Many deciduous spp.	Probably S. Ontario[b]
		celsus	*Carya* sp.	Probably S. Ontario[b]
		dispar	Many deciduous spp.	E. Canada and coastal B.C.
		ferrugineus	Many deciduous spp.	Probably S. Ontario[b]
		obesus	Many deciduous spp.	E. Canada
		pubescens	Many deciduous spp.	S. Ontario
		sayi	Many deciduous spp.	E. Canada
		xylographus	*Quercus* spp.	E. Canada
	Xyloterinus	*politus*	Many deciduous spp.	E. Canada

[a] Sources: Bright (1976), Wood (1982).
[b] Not yet recorded in Canada but present in adjacent states.

Life History

The biology of many of the ambrosia beetle species is not well known. However, the two most damaging species, the striped ambrosia beetle, *T. lineatum* (Fig. 1) and a western species *G. sulcatus*, have been well studied. These two species have quite different life histories. *Trypodendron lineatum* adults overwinter in the forest duff or in bark crevices, while *G. sulcatus* spend the winter in their galleries in the host log. In the spring, when temperatures exceed about 15° C, *T. lineatum* adults fly from the duff in search of host material, which consists of sapwood of coniferous trees that have been dead for at least 3 months, and usually less than 2 years. *Gnathotrichus sulcatus* emerge from logs and fly slightly later than *T. lineatum*. They attack wood felled as recently as 2 weeks before their flight. Both ambrosia beetle species attack most coniferous species in their range. It is believed that suitable wood is identified by its release of "primary attractants" including ethanol and α-pinene, which are perceived by the beetles. In *T. lineatum* the female beetle initiates attack while in *G. sulcatus* it is the male. On finding a suitable host the attacking sex of each species releases a species-specific "secondary attractant" or "aggregation pheromone" that is received by both sexes and results in aggregation and rapid colonization of the host. Following mating the beetles bore into the log and form galleries in the sapwood. They inoculate the gallery walls with ambrosia fungi on which they and their offspring feed as it grows in the wood (Fig. 2).

New ambrosia beetle attacks on logs or green lumber can be identified by piles of white boring dust on the bark or lumber surface. Gallery size and shape can be used to identify the attacking species. *Trypodendron lineatum* has a simpler gallery with a larger diameter (average 1.7 mm) compared with the more complex and narrower (average 1.3 mm) *G. sulcatus* gallery. The galleries of *T. lineatum* penetrate to a depth of about 2.5 to 5.0 cm in Douglas-fir, *Pseudotsuga menziesii*, and slightly deeper in western hemlock, *Tsuga heterophylla*, which has a higher moisture content. *Gnathotrichus sulcatus* galleries penetrate a few centimetres deeper than *T. lineatum* in all host species.

Damage and Loss

Ambrosia beetles cause a serious and continuing loss to the forest industry by degrading lumber and plywood veneer, by increasing sawing time, and by adding to marketing difficulties and costs.

Lumber and Plywood Degrade

The biggest loss is the reduction in grade (degrade) of lumber due to the presence of the dark fungus-stained galleries commonly known as "pinholes" (Fig. 3). These small holes, which the beetles bore in the sapwood, are more of an appearance than a structural defect, but their presence where the most valuable, clear lumber originates makes them particularly damaging. Much of the clear lumber is exported and export grading rules do not allow any pinholes. A piece of otherwise clear lumber having even a few ambrosia beetle holes, in addition to being unexportable, can be degraded from clear to common grade resulting in a substantial drop in value.

Annual degrade losses as a result of ambrosia beetle damage in coastal British Columbia were estimated at around $63 million (McLean 1985). This was a conservative figure based on a lumber recovery factor of 230 fbm/m^3 and degrade value is near $200/Mfbm. In today's market, the recovery factors improved to about 250 fbm/m^3 and degrade value is near $500/Mfbm. At these levels, the total annual degrade losses for coastal British Columbia, if an average of 30% of the logs arriving at the sawmill were infested, would be approximately $173 million.

On plywood veneer, damage by ambrosia beetles often degrades the more valuable face stock to less valuable core stock (Fig. 4). This loss has not yet been quantified but is significant.

Increased Sawing Time

Damage caused by ambrosia beetles to prime lumber and plywood veneer has to be overcome by minimizing the defects through resawing. Boards showing ambrosia beetle attack are often resawn to improve the overall value of the piece by removing pinholes. This resawing of material adds to the amount of time it takes to produce a given amount of lumber as well as to equipment wear. Also, the lumber recovery factor is reduced as a result of the additional cuts.

Marketing Difficulties and Costs

Log marketing can be seriously disrupted by ambrosia beetles and various steps have evolved to manage the problem caused by these insects. These steps include keeping dryland and boom inventories to a minimum during peak beetle flight periods, and processing logs quickly after felling to minimize beetle damage. However, when production levels are geared to factors other than the market, there is a loss in marketing flexibility because it is difficult to respond quickly to changes in the market. If the market is poor, for example, companies may want to maintain their inventory until the market improves, but if this extends over the beetle attack, damage that occurs may offset any gains made by improved market conditions.

A few countries to which Canada exports lumber, most notably Australia and New Zealand, forbid entry of lumber showing ambrosia beetle damage. This is because living beetles may still be in the material and the authorities do not want to import foreign pests. If inspectors find beetle-damaged lumber, the shipment is quarantined and possibly fumigated. This cost is borne by the exporter. For this reason and because a good reputation is required in the competitive international lumber markets, much care is taken to keep beetle-damaged lumber out of all export-based shipments. This care, involving inspection and occasional breakdown and repackaging of loads, is expensive.

Also, a small percentage of raw logs is exported to Asian markets and a relatively high price is paid for these logs. Logs with ambrosia beetle damage have to be sorted out and the export premium is lost. Damaged logs are returned to the domestic log market where the logs are graded at a lower level and, consequently, at a lower value.

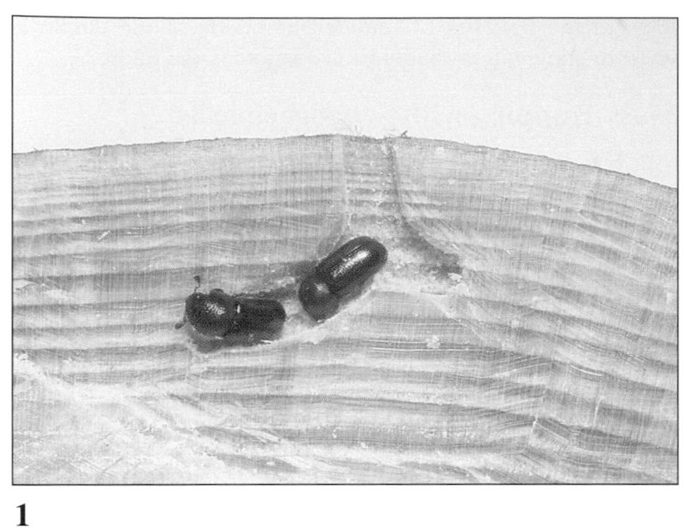

1

2

3

4

5

Figures 1–5

Figure 1. Female and male striped ambrosia beetles, *Trypodendron lineatum*, boring new gallery. (H.A. Moeck) **Figure 2.** *Trypodendron lineatum* larvae feeding on fungi in niches. (H.A. Moeck) **Figure 3.** Lumber showing ambrosia beetle damage. (T.L. Shore) **Figure 4.** A "peeler" log for plywood veneer with ambrosia beetle damage. (H.A. Richmond) **Figure 5.** Sprinkling water on log decks for protection from ambrosia beetles. (W.W. Nijholt)

Figure 6. Mass trapping of ambrosia beetles with a Lindgren funnel trap baited with semiochemicals. (H.A. Moeck)

Control Options

There are several options that have been used or are being used for control of ambrosia beetles: water misting, mass trapping with semiochemicals, careful inventory management, and chemical control. Generally, a pest management system using the first three of these methods is recommended.

Water Misting

Water, applied as a fine mist to logs during the beetle flight period, is an effective method for protecting logs from attack in dryland sorting areas (Richmond and Nijholt 1972; Nijholt 1978) (Fig. 5). The use of this method has been limited, however, because an abundant source of water (preferably fresh) is required and it creates wet working conditions. Difficulties have been caused by debris that frequently clogs intake lines and the equipment is subject to corrosion (especially if salt water is used) and damage by log-moving machinery.

Many dryland sorts are being modernized and include paving and irrigation for dust control. Water sprinkling systems have been installed in many of the newer sorts. In view of the benefits of reduced ambrosia beetle damage, water misting will probably be used more in the future.

Mass Trapping with Semiochemicals

Semiochemical is a term used to describe message-bearing chemicals (Law and Regnier 1971). These include pheromones (Karlson and Lüscher 1959), which are chemicals produced by insects to communicate with other members of the same species, and kairomones (Brown et al. 1970), which are chemicals produced by one species (often hosts) and used by another species (often predators) to their advantage. The semiochemicals used by several ambrosia beetle species to locate suitable hosts and to aggregate other members of their species at these hosts have been identified and are reviewed by Borden and McLean (1979). The striped ambrosia beetle, *Trypodendron lineatum*, *G. sulcatus*, and *G. retusus*, apparently use ethanol and α-pinene, emanating from suitable hosts, as kairomones (Bauer and Vité 1975). The pheromones, racemic sulcatol, S-(+)-sulcatol, and lineatin, have been identified for *G. sulcatus*, *G. retusus*, and *T. lineatum*, respectively. The aggregative response to these pheromones is synergized by the presence of the kairomones (Borden et al. 1980; Shore and McLean 1983a). *Platypus wilsoni* is attracted by the combination of sulcatol, ethanol, and α-pinene (Shore and McLean 1983b).

These semiochemicals, which are now commercially available, can be placed in traps that are then used for survey and mass trapping of ambrosia beetles (Borden and McLean 1979; Lindgren and Borden 1983; Shore and McLean 1985) (Fig. 6). Mass trapping reduces the numbers of ambrosia beetles and thus the damage. Millions of beetles may be trapped in a single dryland sorting area.

Although several types of traps have been used in the past, the Lindgren Funnel Trap (Lindgren 1983) is currently the most common ambrosia beetle trap in Canada. These multiple funnel traps offer the best combination of efficiency, ease of transport, and service (Lindgren et al. 1983). Susceptible logs can also be baited and used as traps, but they have the disadvantages of being a highly variable attraction for the beetle, of saturating when a critical beetle density is reached, and adding to the problem if they are not processed before the attacking beetles have reproduced.

Further research is required to determine whether or not trapping can reduce the population sufficiently to result in economic savings in excess of the cost of the trapping program. In the interim, many forest product companies feel that mass trapping, in conjunction with other preventative methods, contributes to damage reduction and therefore the relatively small cost of such a program is worthwhile.

Careful Inventory Management

The most effective way of reducing ambrosia beetle damage is by minimizing the amount of susceptible wood material available during peak beetle flight periods. Table 2 lists the sources of many ambrosia beetle problems and management opportunities for minimizing damage at these sources.

Chemical Control

Chemical pesticides are rarely used against ambrosia beetles. The beetles preference for attacking the shaded underside of a log makes thorough protective coverage with pesticide sprays or powders extremely difficult. In the past, chemical pesticides such as Lindane and methyl trithion were commonly applied to boomed logs by helicopter to protect the exposed surface against ambrosia beetles (Lejeune and Richmond 1975). However, this practice was discontinued in 1970 because of concerns for the effects of these pesticides on aquatic life, public recreation, and the water supply, and

Table 2. Sources of ambrosia beetle population buildup and management opportunities for minimizing ambrosia beetle damage.

Sources of ambrosia beetle population buildup	Management opportunities for minimizing ambrosia beetle damage
A. Windfalls Provide sources of new host material that maintains endemic population levels.	- Salvage of accessible logs. - Design cutblocks with windfirm boundaries.
B. New Road Right of Ways Logs piled along new roads, together with stumps, provide the first major opportunity for population buildup.	- Cut rights of way only when road-building machinery is available. - Use self-loading trucks to pick up partial loads on new roads as soon as the road base is formed.
C. Felled and Bucked Area Felled timber becomes suitable for attack generally within three months of felling. Suitable logs exposed during the beetles' spring and summer flight will sustain damage. Populations continue to build up in the stumps and slash.	- Keep openings small. Progressive felling encourages the buildup of large populations. - Logs must be moved out promptly after felling with priority given to the most susceptible logs. Avoid cold decking logs in the forest. - Minimize the size of stumps and slash. Burn or break up slash.
D. Dry Land Sorting Area Attack logs transported from the forest will produce brood beetles in late summer. These beetles may fly to the forest margin to overwinter (*T. lineatum*) or attack new logs (*Gnathotrichus* spp.). Resident beetle populations will attack suitable material in the sort. If not processed within two months new brood beetles will begin emerging from these logs adding to the population.	- Sort wood, bundle, and transport to booming grounds or sawmill rapidly. - Avoid stockpiling logs if possible. - Cut back the surrounding forest margin to remove proximity of overwintering sites (*T. lineatum*). - Water misting to further minimize damage. - Mass trapping to reduce resident ambrosia beetle population.
E. Booming Grounds Exposed log surfaces vulnerable to attack; logs already attacked can produce brood that may attack new logs.	- Keep less valuable pulp log booms closer to shore. - Mass trapping to reduce resident ambrosia beetle population.
F. Sawmills and Finished Products Logs stored at mill sites are vulnerable to attack, especially in spring. Logs attacked in the woods or at sorts can produce brood which emerge and overwinter around the mill thus building up a resident. Freshly sawn lumber is vulnerable to attack.	- Maintain minimum inventories. - Move high-value inventories to storage areas remote from overwintering sites (*T. lineatum*). - Water misting to further minimize damage. - Mass trapping to reduce resident ambrosia beetle population. - Kiln drying prevents beetle attack and kills any beetles already in wood.

opposition from woodworkers not wanting to handle insecticide-treated logs. Little effort has been made since then to test the efficacy of, and obtain registration for, other pesticides.

Various chemicals, such as turpentine oils and pine oil, have been tested for repellent effects on ambrosia beetles with mixed results (Nijholt 1973a, 1980; McLean, unpubl. data.). Also, it has been speculated that male striped ambrosia beetles produce an anti-aggregation pheromone to prevent excessive colonization of host material (Nijholt 1970; 1973b; Borden 1973). Further research is required on both repellents and anti-aggregation pheromones.

Current Research

Mass trapping technology for ambrosia beetles is now commercially available to forest companies in British Columbia. Research is being conducted to optimize trap designs, trap numbers and placement, lure designs, and release rates. Work is being done to establish benefit/cost data for mass trapping programs and to develop a reliable method for assessing ambrosia beetle losses. In addition, studies are being conducted to identify the proportion of damage that occurs at each stage of log processing, from the forest to the sawmill. Research is also being conducted on flight behavior responses to semiochemicals in a windtunnel and on dispersal of striped ambrosia beetles, using marked beetles in the forest.

References

Bauer, J.; Vité, J.P. 1975. Host selection by *Trypodendron lineatum*. Naturwissenschaften 62: 539.

Borden, J.H. 1973. Pheromone mask produced by male *Trypodendron lineatum* (Coleoptera: Scolytidae). Can. J. Zool. 52: 533-536.

Borden, J.H.; Lindgren, B.S.; Chong, L. 1980. Ethanol and α-pinene as synergists for the aggregation pheromones of two *Gnathotrichus* species. Can. J. For. Res. 10: 290-292.

Borden, J.H.; McLean, J.A. 1979. Pheromone-based suppression of ambrosia beetles in industrial timber processing areas. Pages 133-154 *in* E.R. Mitchell (Ed.), Management of insect pests with semiochemicals: Concepts and practice. Plenum Press, New York.

Bright, D.J., Jr. 1976. The insects and arachnids of Canada. Part 2: The bark beetles of Canada and Alaska. Coleoptera: Scolytidae. Biosyst. Res. Inst., Can. Dep. Agric., Ottawa, Ont. Publ. 1576. 241 p.

Brown, W.L., Jr.; Eisner, T.; Whittaker, R.H. 1970. Allomones and kairomones: transspecific chemical messengers. Biosci. 20: 21-22.

Karlson, P.; Lüscher, M. 1959. Pheromones: a new term for a class of biologically active substances. Nature (London) 183: 55-56.

Law, J.H.; Regnier, F.E. 1971. Pheromones. Ann. Rev. Biochem. 40: 533-548.

Lejeune, R.R.; Richmond, H.A. 1975. Striped ambrosia beetle *Trypodendron lineatum* (Oliv.). Pages 246-249 *in* M.L. Prebble (Ed.), Aerial control of forest insects in Canada. Can. Dep. Environ., Ottawa, Ont.

Lindgren, B.S. 1983. A multiple funnel trap for scolytid beetles (Coleoptera). Can. Entomol. 115: 229-302.

Lindgren, B.S.; Borden, J.H. 1983. Survey and mass trapping of ambrosia beetles (Coleoptera: Scolytidae) in timber processing areas on Vancouver Island. Can. J. For. Res. 13: 481-493.

Lindgren, B.S.; Borden, J.H.; Chong, L.; Friskie, L.M.; Orr, D.B. 1983. Factors influencing the efficiency of pheromone-baited traps for three species of ambrosia beetles (Coleoptera: Scolytidae). Can. Entomol. 115: 303-313.

McLean, J.A. 1985. Ambrosia beetles: a multimillion dollar degrade problem of sawlogs in coastal British Columbia. For. Chron. 295-298.

Nijholt, W.W. 1970. The effect of mating and the presence of the male ambrosia beetle *Trypodendron lineatum* on "secondary" attraction. Can. Entomol. 102: 894-897.

Nijholt, W.W. 1973a. Ambrosia beetles delayed by turpentine oil. Can. Dep. Fish. and For. Bi-Mon. Res. Notes 29(6): 36.

Nijholt, W.W. 1973b. The effect of male *Trypodendron lineatum* (Coleoptera: Scolytidae) on the response of field populations to secondary attraction. Can. Entomol. 105: 583-590.

Nijholt, W.W. 1978. Evaluation of operational watermisting for log protection from ambrosia beetle damage. Can. For. Serv., Pac. For. Res. Cen. Rep. BC-P-22.

Nijholt, W.W. 1980. Pine oil and oleic acid delay and reduce attacks on logs by ambrosia beetles (Coleoptera: Scolytidae). Can. Entomol. 112: 199-204.

Richmond, H.A.; Nijholt, W.W. 1972. Water misting for log protection from ambrosia beetles in B.C. Can. For. Serv., Pac. For Res. Cent. Inf. Rep. BC-P-4-72.

Shore, T.L.; McLean, J.A. 1983a. A further evaluation of the interactions between the pheromones and two host kairomones of the ambrosia beetles *Trypodendron lineatum* and *Gnathotrichus sulcatus* (Coleoptera: Scolytidae). Can. Entomol. 115: 1-5.

Shore, T.L.; McLean, J.A. 1983b. Attraction of *Platypus wilsoni* Swaine (Coleoptera: Platypodidae) to traps baited with sulcatol, ethanol and α-pinene. Environ. Canada, Can. For. Serv. Res. Notes 3(4): 24-25.

Shore, T.L.; McLean, J.A. 1985. A survey for the ambrosia beetles *Trypodendron lineatum* and *Gnathotrichus retusus* (Coleoptera: Scolytidae) in a sawmill using pheromone-baited traps. Can. Entomol. 117: 49-55.

Wood, S.L. 1982. The bark and ambrosia beetles of North and Central America (Coleoptera: Scolytidae), a taxonomic monograph. Great Basin Naturalist Memoirs Number 6. Brigham Young Univ. 1359 p.

Chapter 20

Wood Borers

L. Safranyik and H.A. Moeck

Introduction

Wood borers are a taxonomically diverse group of insects comprising many species. Some species use the wood for both food and shelter whereas others use it only for shelter. Based on attack preference, wood borers can be divided into three groups: species that attack living trees and recently killed or dying trees and green logs; species that attack only dry wood; and species that attack only older moist wood. Numerous species that attack live and recently dead trees and green logs first mine the phloem region for various time periods before boring into the wood. A few of these species can survive and develop even in dry wood. The most common insects found in wood are beetle larvae (Anderson 1960).

Losses to wood borers as a group were estimated at 1% to 5% of the annual cut (Furniss and Carolin 1977). Much of this damage occurs after trees have been cut or killed and before utilization.

The following groups, listed in order of relative importance to forestry practice within and between groups, contain the economically most important wood-boring insects:

A. Coleoptera:
- (a) ambrosia beetles (Scolytidae, Platypodidae)
- (b) roundheaded and flatheaded wood borers (Cerambycidae, Buprestidae)
- (c) powder-post beetles (Lyctidae, Anobiidae, Bostrichidae, Ptinidae)
- (d) timberworms (Brenthidae, Lymexylidae)
- (e) weevils (Curculiondae)

B. Hymenoptera:
- (a) wood wasps (Siricidae)
- (b) carpenter ants (Formicidae)
- (c) carpenter bees (Anthophoridae: Xylocopinae)

C. Lepidoptera:
- (a) carpenter worms (Cossidae)
- (b) clearwing moths (Sesiidae)

D. Isoptera:
- (a) subterranean termites (Rhinotermitidae)
- (b) dampwood termites (Hodotermitidae)

Coleoptera

Cerambycidae and Buprestidae

The cerambycid and buprestid wood borers are abundant and widely distributed in Canada. Adult cerambycids are also called long-horned or lonicorn beetles because most species have long antennae; adult buprestid beetles are often called metallic wood borers owing to the usually metallic sheen of their exoskeleton. The chief importance of this group of beetles, damaging and beneficial, results from the larval mining in both the sapwood and heartwood of dying, dead, and cut timber (Linsley 1958). Adult emergence holes and larval mines enhance establishment of wood-decaying fungi and contribute to decomposition of wood, a process that causes degrade in wood and wood products. The larval mines or "worm holes" are also graded as defects. Often heavy infestations occur following catastrophic events such as large-scale wind-felling of trees, forest fires (Ross 1960; Gardiner 1957, 1975), or epidemics of tree-killing insects (Belyea 1952; Kinghorn 1954) and may seriously affect the feasibility of salvage operations. Although some species are pests of living trees, few species can kill healthy trees. The larva is the stage encountered most often by foresters and woodworkers.

The adults do not bore into trees; the females lay their eggs under bark scales, in slits cut into the bark, in holes made by other insects, or in cracks or other imperfections in the wood. Based on the type of host material attacked, roundheaded and flatheaded wood borers can be grouped into the following categories: (a) miners of living trees, (b) miners of dying and recently killed trees, and (c) miners of dry wood. The larvae of many species are inner bark wood borers, in that they first mine the inner bark region and later penetrate the wood; a few species mine the inner bark only whereas others mine only the wood.

Winter is usually passed in the larval stage, although some larvae may pupate in the fall and overwinter as adults. The length of the life cycle varies from one to several years, depending on insect species, host condition, and locality; species that mine deeply in the wood in the larval stage usually take longer to complete development than species whose larvae bore only or mainly in the inner bark. Adults are generally sun-loving insects and may appear from early spring to late summer, depending on species and locality, but usually live only for a few weeks. Before laying eggs, adults of some species feed on foliage, tender bark of twigs, fungi, or on pollen, while others do not feed. Adults of several species, especially those in the genus *Monochamus* (Fig. 1), are known or suspected of transmitting the pine wood nematode, *Bursaphelenchus xylophilus* (Steiner and Buhrer 1934) (e.g. Wingfield and Blanchette 1983). Details on the biology, habits, and control of the injurious species of roundheaded and flatheaded borers are found in Anderson (1960), Drooz (1985), and Furniss and Carolin (1977).

Cerambycidae

The tunnels mined by roundheaded borers are broadly oval to circular in cross section; the frass is either distinctly fibrous or coarsely granular, often loose in the tunnel or pushed to the outside through small openings cut through the bark (Fig. 2). The tunnels are long and irregular and increase in size as the larvae grow. Tunnels mined by most

Figure 1. Spotted pine sawyer, *Monochamus mutator* (= *M. maculosus*), adult on lodgepole pine beside emergence hole. (H.A. Moeck)

Figure 2. Frass pile ejected to outside by *Monochamus* larva. (H.A. Moeck)

species range from 1 to 2 cm at the widest point but can range up to 5 cm. The larvae normally are segmented, fleshy, cylindrical, elongate grubs with bodies that taper posteriorly and are several times as long as wide. The thoracic segment is enlarged, with a horny plate on the top of the first segment behind the head. The head is flattened and usually partly embedded in the prothoroax; mouthparts project forward. The larvae are either legless or have short, peglike legs on the thorax. Pupae resemble adults but are usually creamy white, and have wing cases instead of elytra. Adults are medium to large beetles with cylindrical or oblong bodies and antennae that often are longer than the body.

Roundheaded borers that attack living trees are mainly restricted to hardwoods, with a few species attacking softwoods. A few species of inner bark wood miners such as the western larch borer, *Tetropium velutinum* (Kinghorn 1954), can attack and kill weakened trees. The majority of wood borers in this group, however, are known to occasionally cause severe damage to shade, shelterbelt, and forest trees by mining the sapwood and heartwood. In forests managed for wood production, severely infested trees are rendered useless by a high density of grub holes. *Enaphalodes*, *Goes*, and *Saperda* are some of the genera that contain economically important species attacking various species of hardwoods (Berry 1978; Felt 1904 and Joutel; McLeod and Wong 1967; Solomon 1977). *Saperda* attack in aspen is frequently followed by infection by *Hypoxylon* canker (Anderson and Martin 1981; Graham and Harrison 1954).

Numerous species of roundheaded borers attack dying and recently killed trees and green logs. Extensive mining in the sapwood and heartwood and attendant staining and decay may cause serious value loss in lumber produced from infested host materials. *Arhopalus* (Eaton 1959; Eaton and Lyon 1955), *Asemum*, *Ergates*, *Monochamus* (Belyea 1952; Cerezke 1975), *Semanotus* (Gardiner 1970; Wickman 1968a, 1968b), *Xylotrechus* (Furniss and Carolin 1977), *Tetropium* (Kinghorn 1954), attacking various species of softwoods, and Saperda (McLeod and Wong 1967) and *Xylotrechus* (Furniss and Carolin 1977), attacking hardwoods, are some of the genera that contain economically important species. Much of this damage is caused by sawyer beetles, *Monochamus* spp. (Cerezke 1975; Ostaff 1977), especially in fire-killed and wind-felled timber (Gardiner 1975; Prebble and Gardiner 1958; Richmond and Lejeune 1945; Ross 1960). Some species such as the new house borer, *Arhopalus productus*, (Eaton 1959) and the firtree borer, *Semanotus litigiosus* (Gardiner 1970; Wickman 1968), can continue development in lumber cut from infested host materials and, in addition to degrade, also cause damage by emerging through finished surfaces in new buildings.

Several species of roundheaded wood borers infest dry wood. Some species, such as *Callidium* spp. infesting softwoods, only attack recently killed trees and green logs with bark on, but the larvae may continue developing after the wood has dried. The old house borer, *Hylotrupes bajulus* (Moore 1978), an introduced European species, breeds in dry, seasoned sapwood of coniferous trees, especially pines and spruces in eastern North America. Occasionally, serious damage is caused to stored lumber and structural timbers in buildings. This species occurs in Quebec.

Buprestidae

Tunnels excavated by larvae of flatheaded borers are winding, flat, or elliptical, usually several times as wide as deep. The tunnel walls usually have curved grooves cut into them, approximately at a right angle to the long axis of the tunnel. The frass is usually granular and packed into concentric rows and is never ejected to the outside. The larval tunnels, up to 2 cm wide, occur in the inner bark and/or the sapwood or heartwood. When completed, they terminate in elongated pupal chambers that have short, oval exit holes cut to the outside. The larvae of all species that mine the bark and wood have greatly enlarged first and sometimes first to third thoracic segments; the first thoracic segment behind the head bears well-developed sclerotized plates on both the dorsal and ventral surfaces. The head is oval to quadrate and deeply invaginated into the thorax. The body behind the thorax is usually very long in relation to its width and sharply tapered. The larvae have no legs. The pupae are whitish, resemble adults, and have wing cases instead of wings and elytra. The adults are moderate-sized, somewhat flattened, boat-shaped, often brilliantly metallic beetles with slender antennae.

Only a few of the many species of flatheaded borers that mine the inner bark of living trees cause economically important damage. Some of these species mine the inner bark only, while others mine both the inner bark and wood and cause partial or total girdling of the stem; generally, however, they are not primary tree killers. Some species of *Agrilus* attacking various species of hardwood, especially birches, oaks, and poplars (Balch and Prebble 1940; Ball and Simmons 1980; Barter 1957; Wargo 1977), and *Oxypteris* attacking mainly hemlock and spruces in eastern Canada (Balch 1935) and spruce, pines, and firs in the west, can cause serious losses occasionally by extensive killing of trees scorched by fire or weakened by flooding, defoliation, drought, or other causes. Some species of *Oxypteris* such as the black fire beetle, *Oxypteris acuminata*, are known as "firebugs" because they are strongly attracted to sources of infrared radiation such as from still-smouldering trees in forest fires (Evans 1966).

The western cedar borer, *Trachykele blondeli* (Duncan 1979), mines principally in the heartwood of living western red cedars along the Pacific coast and can cause locally serious degrade and cull in trees cut for shingles, poles, and other products. The golden buprestid, *Cypriacis aurulenta* (Duncan 1981), is the most damaging species in this genus and breeds in injured or felled trees and stumps of pines, spruces, true firs, and Douglas-fir in British Columbia. The attacks may occur on trees and logs in the forest, and logs in log storage areas. Larvae continue to develop in lumber milled from infested logs, and under adverse conditions, as in seasoned lumber, several decades may be required for completion of the life cycle (Smith 1962). The damage, consisting of mined boards, timbers, and emergence holes through finished surfaces, is of main concern in buildings; major structural damage rarely occurs. *Chalcophora* spp. mostly breed in the sapwood and heartwood of stumps and logs; sometimes they also attack living trees if there is access to the wood through wounds. Some species, such as the Virginian buprestid, *C. virginiensis*, can severely damage living and recently killed or felled pines, especially following wildfires and wind storms.

Management

Methods for reducing damage caused by roundheaded and flatheaded borers vary with the type of damage, the host species attacked, and the biological characteristics of the insect (Anderson 1960). In general, wood borers that principally mine the inner bark region of living trees are frequently associated with reduced vigor of their hosts. Hence, under forest conditions, damage caused by these insects usually can be reduced significantly by using silvicultural techniques to maintain vigorous growth, stand hygiene, and in some cases, stand closure. The best method for preventing damage from wood borers that attack dying or recently felled trees is the rapid utilization of all such host materials, combined with sanitary harvesting practices. Other measures that may be effective in preventing or reducing the incidence of attacks, depending on the wood-boring species involved, are ponding, peeling, or water sprinkling of logs (Ostaff 1974a; Roff and Dobie 1968); during the attack period, storing of logs in the forest or at log storage areas in shaded locations in high, compact decks with minimum surface area; orienting decks in the north–south direction, with the butt end of logs pointing south, to increase shading within the deck (Cerezke 1975) or covering decks with slash; avoiding bark injuries to trees and logs and reducing the incidence of wane in lumber. Valuable logs and log decks can be protected from grub hole damage by chemical treatment. Lindane is registered in Canada for treatment of wood borer-infested logs (Canadian Pulp and Paper Association 1987; Cerezke 1975; Green 1979; Ostaff 1977). However, special permits may have to be obtained from the responsible provincial organization to use this chemical. Experimental treatments that were shown to be effective were Lindane 0.4 or 1% emulsifiable concentrate (Gardiner 1970; Green 1979; Ostaff 1974a; Ross and Downton 1966; Ross and Geistlinger 1968), and paradichlorobenzene, either as pure crystals or dissolved in tricholoroethylene, with a plastic sheet covering the wood pile (Raske and Robins 1975). Wood borer damage to wood in service can be reduced by ensuring that the lumber is free from bark and grub holes. When lumber is kiln-dried all stages of the insects will be killed (Ostaff and Cech 1978). Sometimes it is feasible to apply an insecticide treatment as discussed for powder-post beetles.

Lyctidae, Anobiidae, Bostrichidae, Ptinidae

The four families of insects that comprise the powder-post beetles collectively include several hundred species in North America. They are distributed widely in Canada, especially in the southeast and British Columbia. Powder-post beetles attack seasoned wood of both softwoods and hardwoods of a large variety of trees, and numerous species are highly destructive to wood in service. Some species, however, can attack only wood in recently killed trees. They are particularly important in mill and lumber yards, warehouses, and in manufactured products in the store and home. Because they usually breed continuously in the same material, if the infestation is not checked the interior of the infested wood may be completely reduced to powder (Smith 1954, 1960). The larvae of all species as well as the adults of some species mine the wood. For details on the biology and habits of the most destructive species, see Anderson (1960), Drooz (1985), Furniss and Carolin (1977), Moore (1979), and Stark and Wood (1962). Damage resembles that of termites but may be distinguished by the following characteristics: (a) damage in dry wood is often in places inaccessible to termites, (b) wood surfaces usually have small, perfectly round emergence holes, (c) the mines are filled with loose or packed powdery dust or tiny pellets.

The Lyctidae are considered to be the most destructive group of powder-post beetles. They attack seasoned and partially seasoned ring-porous hardwoods. The larval tunnels are constructed in the sapwood, usually run with the grain, and are filled with fine flour-like dust; emergence holes are about 1.5 mm in diameter.

Anobiidae breed in the sapwood and heartwood of well-seasoned and usually old hardwoods and softwoods, particularly when such wood is found under constantly humid conditions. Larval tunnels run in all directions with no relation to the grain and are filled with gritty dust; emergence holes are about 3 mm in diameter.

Bostrichidae breed in freshly felled trees with bark on, recently killed hardwoods and softwoods, as well as in processed material. The adult bores through the bark and into the sapwood and then constructs a tunnel across the grain just under the surface of the wood. These tunnels may completely girdle branches and smaller trees. Eggs are deposited along the sides of the tunnel, and the larvae mine mainly in the sapwood and to a lesser extent in the heartwood. Larval mines run parallel with the grain and are filled with fine dust; emergence holes range from 3 to 9 mm in diameter.

Ptinidae are the least important group of the powder-post beetles. They normally attack well-seasoned sapwood and occasionally punky heartwood of both hardwoods and softwoods. Larval galleries run in all directions and are packed with fine boring dust. Typically, infestations occur in boards, timber, and woodwork in old buildings.

Powder-post beetle damage can be minimized using a combination of preventive measures and applied control (Anderson 1960; Ruppel et al. 1975). Susceptible wood should be inspected on acquisition and yearly thereafter for signs of damage. Wood waste and heavily infested wood should be burned, and lightly infested wood treated with an approved insecticide or by kiln drying. Approved insecticidal dips, and seal-coatings of the sapwood against lyctid and bostrichid beetles with linseed oil, or any similar product that closes the pores in the wood, are effective preventive treatments.

Brenthidae, Lymexylidae

The timberworms, except for one species found in cypress, attack only hardwoods (Stark and Wood 1962). They occur in eastern Canada and are known to attack both live as well as dead trees (Drooz 1985). Infestations, however, will not spread to seasoned wood. Members of both families have similar life cycles and damage characteristics. The adults of all species are larger than ambrosia beetles; they are slender, elongated insects with slender legs. The larvae are also slender, with a globular head and a conspicuously armed ninth abdominal segment. Eggs are laid in cracks and the larvae bore deeply into the wood. Some of the burrows may be up to 5 mm in diameter; they may be 1 m or more long and usually cut across the grain. Damage often causes severe losses to the cooperage and ship-building industries.

Curculionidae

Some species of weevils, such as the poplar-and-willow borer, *Cryptorhynchus lapathi*, an introduced insect that attacks mainly willows and poplars (Clark 1976; Cottrell 1959; Harris and Coppel 1967), and members of the genera *Cossonus*, *Hexarthrum*, and *Rhyncolus*, are wood borers in the larval stages. Occasionally these weevils also damage wood products. Overall, the damage is usually of minor importance to forestry.

Hymenoptera

Siricidae

Adult wood wasps (Fig. 3) are medium to large (15–50 mm long) cylindrical insects with thick waists. The head, thorax, and abdomen are of equal width, and the last abdominal segment has a hornlike projection. Female adults have a long stout ovipositor. Adults are usually metallic blue or black or have combinations of black, red, and yellow coloration. They are widely distributed in Canada and attack both hardwoods and softwoods (Morris 1967; Stillwell 1967). The following genera have important wood-boring species: *Sirex*, *Xeris*, *Urocerus*, and *Tremex*. Most species of wood

Figure 3. *Sirex juvencus californicus* ovipositing in ponderosa pine. (H.A. Moeck)

wasps have associated symbiotic fungi such as *Amylostereum areolatum* (Morgan 1968), *A. chailletii* (Stillwell 1966; Talbot 1977), and *Daedalea unicolor* (Stillwell 1967), which also are wood-rotting fungi (Stillwell 1960). The biology of wood wasps in Canada is not well known.

Eggs and spores of the symbiotic fungus are deposited directly into the wood, up to 2.5 cm deep or more by the female. The larvae are white, cylindrical, have three pairs of vestigial thoracic legs, and a pigmented spine at the posterior end. The larvae mine entirely in the wood, packing their circular tunnels with fine borings. The tunnels are usually U-shaped and range up to 20 cm long. It is not known whether the larvae feed entirely on fungi, the wood, or both, but Stillwell (1966) determined that larvae of *Sirex* fail to develop if the fungus is absent. Depending on species, development may take from 1 to 3 years. A few species can attack apparently vigorous trees, but they usually prefer weakened and recently dead trees, wounds in living trees, fire-killed and freshly felled trees. Wood wasps are one of the first groups of insects that invade such host materials (Belyea 1952; Wallis et al. 1974; Wilson 1961).

Extensive damage by wood wasps causes serious degrade to products cut from infested timber. The economically most significant damage is usually caused by infestations in fire-killed and wind-felled timber. Damage by wood wasps can be greatly reduced or prevented by measures such as prompt utilization, immersion into water of infested logs, and kiln-drying of lumber cut from infested trees.

Formicidae

The carpenter ants, large (6–15 mm long), black or red and black ants of the genus *Camponotus*, are cosmopolitan and attack cut timber, decayed and partially decayed trees, wood in service, and even some living trees (Ostaff 1974b; Ruppel, no date b; Smith 1957). They are social insects and do not eat wood but excavate cavities in it for nests in which to rear their young. Carpenter ants infest a wide variety of trees, both hardwoods and softwoods, and cause deterioration and breakage by mining near the base of fire-scarred or butt-rotted standing trees such as western red cedar, Douglas-fir, and balsam fir. Heavily infested trees or logs may become worthless for pulp or lumber. The greatest damage, however, is to house timbers, posts, and poles, especially when the wood is damp or decayed. Their excavations, which are similar to those made by termites but differ in that they run across the grain of the wood and are kept free of frass, are frequently extensive and seriously impair the structural integrity of the infested wood. Buildings are usually invaded from carpenter ant colonies located nearby, especially in wooded areas. Stumps, partially decayed or damaged trees, old logs, and wood waste should be removed from building sites. Sound building construction techniques will reduce infestation by carpenter ants.

Anthophoridae: Xylocopinae

Carpenter bees construct their nests in wood, commonly in pines, cedars, and several hardwoods. Members of the genus *Xylocopa*, which resemble bumblebees, are rated as pests of minor importance. The adults tunnel into the solid wood of beams, posts, and other structural timbers and partition the burrows into cells for rearing their young. The wood is not eaten. The same piece of timber may be attacked for several years; extensive tunnelling may cause structural damage.

Lepidoptera

The wood borers in the Sesiidae and Cossidae cause similar damage. The larvae mine the inner bark and wood of the stem and large branches and can cause serious defects in lumber cut from infested trees. Carpenter worms attack various species of hardwoods while some species of the clearwing moths attack hardwoods and others attack softwoods.

Isoptera

Termites are one of the most destructive groups of insects of dead and dry wood and wood products. However, much of the damage occurs in tropical and subtropical areas where these insects are most abundant. They are social insects, live in colonies in wood or in the ground, and feed on cellulose. In wood, they excavate completely enclosed longitudinal cavities and often mine the wood so extensively that only a thin outer shell remains. These insects are most common in southern British Columbia and the southern part of eastern Canada (Esenther and Gray 1968; Kirby 1967; Ruppel, no date a; Smith 1964; Smith and Johnston 1962; St. George et al. 1960). The dampwood termites usually infest damp and decaying wood, both in the forest as well as in stored wood products and wood in service. They often invade houses, other wood structures such as poles, posts, and wharves, and can cause serious damage. Usually, however, their presence is more of an indicator of wood decay. In general, subterranean termites cause the greatest damage in Canada. They maintain their colonies in the ground and extend their galleries into the wood, principally in the spring wood. Wood that is not in contact with the ground may be reached through earthen tubes constructed by the termites. Prevention is the most effective means of control. Waste lumber in the soil, stumps under buildings, and moist conditions invite termite attack. Contact with the ground of structural timbers should be avoided, and decayed or damaged timbers replaced. Fence posts and trellises should be treated with a preservative before installation.

There is no current research being conducted in Forestry Canada on any of the wood-boring insects described in this chapter.

References

Anderson, G.W.; Martin, M.P. 1981. Factors related to incidence of Hypoxylon cankers in aspen and survival of cankered trees. For. Sci. 27(3): 461-476.

Anderson, R.F. 1960. Forest and Shade Tree Entomology. J. Wiley and Sons, Inc., New York, 428 p.

Balch, R.E. 1935. Notes on the habits of attack of the hemlock borer. Can. Entomol. 67(5): 90-92.

Balch, R.E.; Prebble, J.S. 1940. The bronze birch borer and its relation to the dying of birch in New Brunswick forests. For. Chron. 16(3): 179-201.

Ball, J.; Simmons, G. 1980. The relationship between the bronze birch borer *Agrilus anxius* and birch dieback. J. Arboriculture 6(2): 309-314.

Barter, G.W. 1957. Studies of the bronze birch borer, *Agrilus anxius* Gory, in New Brunswick. Can. Entomol. 89(1): 12-36.

Belyea, R.M. 1952. Death and deterioration of balsam fir weakened by spruce budworm defoliation in Ontario.
Part I. Notes on the seasonal history and habits of insects breeding in severely weakened and dead trees. Can. Entomol. 84(11): 325-335.
Part II. An assessment of the role of associated insect species in the death of severely weakened trees. J. Forestry 50(10): 729-738.

Berry, F.H. 1978. Decay associated with borer wounds in living oaks. USDA For. Serv., Northeast For. Exp. Stn., Broomall, Pa. Res. Note NE-268, 2 p.

Canadian Pulp and Paper Association. 1987. Insecticides registered for forest and woodlands management. Technical Reference Insecticides, 6 p.

Cerezke, H.F. 1975. White-spotted sawyer beetle in logs. Environ. Can., Can. For. Serv., Edmonton, Alta. Inf. Rep. NOR-X-129, 6 p.

Clark, R.C. 1976. The poplar and willow borer. Environ. Can., Can. For. Serv., St. John's, Nfld. For. Note No. 14, 5 p.

Cottrell, C.B. 1959. A brief history of the poplar and willow borer, *Sternochetus lapathi* (L.), in British Columbia. Proceedings of the Entomol. Soc. of B.C. 56: 46-48.

Drooz, A.T. (Ed.). 1985. Insects of Eastern Forests. USDA For. Serv., Washington, D.C. Misc. Publ. 1426.

Duncan, R.W. 1979. Western cedar borer. Environ. Can., Can. For. Serv., Victoria, B.C., Pac. For. Res. Cent., Pest Leaflet FPL 66, 4 p.

Duncan, R.W. 1981. The golden buprestid - a wood boring beetle. Environ. Can., Can. For. Serv., Victoria, B.C. Pac. For. Res. Cent., Pest Leaflet FPL 68, 3 p.

Eaton, C.B. 1959. Observations on the survival of *Arhopalus productus* (LeConte) larvae in Douglas-fir lumber (Coleoptera: Cerambycidae). Pan-Pac. Entomol. 35(2): 114-116.

Eaton, C.B.; Lyon, R.L. 1955. *Arhopalus productus* (LeC.) a borer in new buildings. USDA For. Serv., Calif. For. and Range Exp. Stn., Berkeley, Calif. Tech. Paper No. 11, 11 p.

Esenther, G.R.; Gray, D.E. 1968. Subterranean termite studies in southern Ontario. Can. Entomol. 100(8): 827-834.

Evans, W.G. 1966. Perception of infrared radiation from forest fires by *Melanophila acuminata* De Geer (Buprestidae, Coleoptera). Ecol. 47(6): 1061-1065.

Felt, E.P.; Joutel, L.J. 1904. Monograph of the genus *Saperda*. New York State Museum Bull. 74, Entomol. 20, 86 p.

Furniss, R.L.; Carolin, V.M. 1977. Western Forest Insects. USDA For. Serv., Washington, D.C. Misc. Publ. No. 1339, 654 p.

Gardiner, L.M. 1957. Deterioration of fire-killed pine in Ontario and the causal wood-boring beetles. Canadian Entomol. 89(6): 241-263.

Gardiner, L.M. 1970. New northern Ontario spruce beetle compels May start on log spraying. Can. For. Ind., July, 3 p.

Gardiner, L.M. 1975. Insect attack and value loss in wind-damaged spruce and jack pine stands in Northern Ontario. Can. J. For. Res. 5: 387-398.

Graham, S.A.; Harrison, R.P. 1954. Insect attacks and Hypoxylon infections in aspen. J. For. 52(10): 741-743.

Green, P. 1979. Ont. sawmill perfects method for applying Lindane to logs. Can. For. Ind. 99(9): 28-30, 32.

Harris, J.W.E.; Coppel, H.C. 1967. The poplar-and-willow borer, *Sternochetus* (=*Cryptorhynchus*) *lapathi* (Coleoptera: Curculionidae), in British Columbia. Can. Entomol. 99(4): 411-418.

Kinghorn, J.M. 1954. *Tetropium velutinum* LeConte, a secondary bark-mining cerambycid in western hemlock following the hemlock looper outbreak on Vancouver Island (1948). Can. Dept. Agric. Sci. Serv., For. Biol. Div., For. Zool. Lab. B.C., Interim Tech. Rep. 36 p.

Kirby, C.S. 1967. Termites in Ontario, with particular reference to the Toronto region. Can. Dept. For. and Rural Devel., For. Br., For. Res. Lab., Ont. Region, Inf. Rep. O-X-56, 30 p.

Linsley, E.G. 1958. The role of Cerambycidae in forestry, urban and agricultural environments. Pan-Pac. Entomol. 34(3): 105-124.

McLeod, B.B; Wong, H.R. 1967. Biological notes on *Saperda concolor* LeC. in Manitoba and Saskatchewan (Coleoptera: Cerambycidae). Man. Entomol. 1: 27-33.

Moore, H.B. 1978. The old house borer and update.
Part One. Pest Control 46(3): 14-17, 52-53.
Part Two. Pest Control 46(4): 28, 30, 32
Part Three. Pest Control 46(5): 26-28.

Moore, H.B. 1979. Wood-inhabiting insects in houses: their identification, biology, prevention and control. USDA For. Serv. and Dept. of Housing and Urban Devel. Rep., 133 p., Washington, D.C.

Morgan, F.D. 1968. Bionomics of Siricidae. Ann. Rev. Entomol. 13: 239-256.

Morris, E.V. 1967. Distribution and hosts of some horntails (Siricidae) in British Columbia. J. Entomol. Soc. B.C. 64: 60-63.

Ostaff, D. 1974a. Sawyer beetles. Environ. Can., Can. For. Serv., Eastern For. Products Lab., Ottawa, Ont. Rep. OPX100E, 5 p.

Ostaff, D. 1974b. Carpenter ants. Environ. Can., Can. For. Serv., Eastern For. Products Lab., Ottawa, Ont. Rep. OPX101E, 6 p.

Ostaff, D. 1977. Protect your logs. Don't give wood borers a chance. Env. Can., Can. For. Serv., Eastern For. Products Lab., Ottawa, Ont. Bull. LD8E, 4 p.

Ostaff, D.; Cech, M.Y. 1978. Heat sterilization of spruce-pine-fir lumber containing sawyer beetle larvae (Coleoptera: Cerambycidae, Monochamus sp.). Environ. Can, Can. For. Serv., Eastern For. Products Lab., Ottawa, Ont. Rep. OPX200E, 9 p.

Prebble, M.L.; Gardiner, L.M. 1958. Degrade and value loss in fire-killed pine in Mississagi area of Ontario. For. Chron. 34(2): 139-158.

Raske, A.G.; Robins, J.K. 1975. Wood borer control in spruce logs with p-dichlorobenzene and plastic sheeting (Coleoptera: Cerambycidae). J. Entomol. Soc. Br. Col. 72: 15-17.

Richmond, H.A.; Lejeune, R.R. 1945. The deterioration of fire-killed white spruce by wood-boring insects in northern Saskatchewan. For. Chron. 21(3): 168-192.

Roff, J.W.; Dobie, J. 1968. Water sprinklers check biological deterioration in stored logs. Br. Col. Lumberman 52(5): 1-6.

Ross, D.A. 1960. Damage by long-horned wood borers in fire-killed white spruce, central British Columbia. For. Chron. 36(4): 355-361.

Ross, D.A.; Downton, J.S. 1966. Protecting logs from long-horned wood borers with Lindane emulsion. For. Chron. 42(4): 377-379.

Ross, D.A.; Geistlinger, N.J. 1968. Protecting larch logs from *Tetropium velutinum* LeConte with Lindane emulsion. J. Entomol. Soc. Br. Col. 65: 14-15.

Ruppel, D.H. n.d. a. Termites in British Columbia. Environ. Can., Can. For. Serv., Pac. For. Res. Cent., Victoria, B.C. For. Pest Leaflet No. 57, 6 p.

Ruppel, D.H. n.d. b. Carpenter ants. Environ. Can., Can. For. Serv., Pac. For. Res. Cent., Victoria, B.C. For. Pest Leaflet No. 58, 6 p.

Ruppel, D.H.; Pass, E.C.; Wiens, J.C. 1975. Insects found in and near the home. Environ. Can., Can. For. Serv., Pac. For. Res. Cent., Victoria, B.C. For. Pest Leaflet 29, (rev.), 17 p.

St. George, R.A.; Johnston, H.R.; Kowal, R.J. 1960. Subterranean termites, their prevention and control in buildings. USDA Home and Garden Bull. No. 64, 30 p., Washington, D.C.

Smith, D.N. 1954. Powder-post beetles in structural timber in coastal British Columbia. Can. Dept. Agric. Sci. Serv., For. Biol. Div., Ottawa, Ont. Publ. 903, 8 p.

Smith, D.N. 1957. Carpenter ant infestation and its control. Can. Dept. Agric. Sci. Serv., For. Biol. Div., Ottawa, Ont. Publ. 1013, 5 p.

Smith, D.N. 1960. Strength deterioration of structural timbers in relation to infestation level of Anobiidae. Can. Dept. Agric. Res. Br., For. Biol. Div., For. Biol. Lab., Victoria B.C., Unpubl. Interim Rep. 12 p.

Smith, D.N. 1962. Prolonged larval development in *Buprestis aurulenta* L. (Coleoptera: Buprestidae). A review with new cases. Can. Entomol. 94(6): 586-593.

Smith, D.N. 1964. Controlling termites and preventing losses in British Columbia. Can. Dept. For., For. Entomol. and Pathol. Br., Ottawa, Ont. Leaflet, 12 p.

Smith, V.K. Jr.; Johnston, H.R. 1962. Eastern subterranean termite. USDA For. Serv., Washington, D.C. For. Pest Leaflet 68, 7 p.

Solomon, J.D. 1977. Biology and habits of the oak branch borer (*Goes debilis*). Annals Entomol. Soc. Am. 70(1): 57-59.

Stark, R.W.; Wood, D.L. 1962. Forest Entomology Laboratory Manual. University of California, Berkeley, 106 p.

Steiner, G.; Buhrer, E. 1934. *Aphelenchoides xylophilus*, n.sp., a nematode associated with blue-stain and other fungi in timber. J. Agri. Res. 48: 949-951.

Stillwell, M.A. 1960. Decay associated with woodwasps in balsam fir weakened by insect attack. For. Sci. 6(3): 225-231.

Stillwell, M.A. 1966. Woodwasps (Siricidae) in conifers and the associated fungus, *Stereum chailletii*, in eastern Canada. For. Sci. 12(1): 121-128.

Stillwell, M.A. 1967. The pigeon tremex, *Tremex columba* (Hymenoptera: Siricidae), in New Brunswick. Can. Entomol. 99(7): 685-689.

Talbot, P.H.B. 1977. The *Sirex-Amylostereum-Pinus* association. Ann. Rev. Phytopathol. 15: 41-54.

Wallis, G.W.; Godfrey, J.H.; Richmond, H.A. 1974. Losses in fire-killed timber. Environ. Can., Can. For. Serv., Victoria, B.C. Inf. Rep. BC-X-88, 11 p.

Wargo, P.M. 1977. *Armillariella mellea* and *Agrilus bilineatus* and mortality of defoliated oak trees. For. Sci. 23(4): 485-492.

Wickman, B.E. 1968. Fir tree borer. USDA For. Serv., Washington, D.C., For. Pest Leaflet 168, 6 p.

Wickman, B.E. 1968. The biology of the fir tree borer, *Semanotus litigiosus* (Coleoptera: Cerambycidae), in California. Can. Entomol. 100(2): 208-220.

Wilson, L.F. 1961. Attraction of wood-boring insects to freshly cut pulpsticks. USDA For. Serv., Lake States For. Exp. Stn., Tech. Notes No. 610, 2 p.

Wingfield, M.J.; Blanchette, R.A. 1983. The pine-wood nematode, *Bursaphelenchus xylophilus*, in Minnesota and Wisconsin: insect associates and transmission studies. Can. J. For. Res. 13(6): 1068-1076.

Chapter 21

Control of Insects Affecting Plantations and Managed Stands

D.G. Embree

Introduction

This chapter and others that follow it in this section deal with specific insects and groups of insect pests that threaten the immature forests of Canada. Although these pests are capable of attacking any immature forest, concern is largely focused on immature forests with large amounts of money invested in them, such as plantations or spaced stands. Currently, because of their high "per unit" value and short rotation (7 to 10 years), small plantations (e.g. Christmas tree plantations) are of immediate concern; however, the primary protection effort should be directed at the commercial forest. Consider, for example, that in the 5-year period between 1983 and 1988, more than 2 750 million trees were planted in Canada. This equals 1.6 million ha of plantations costing more than $950 million. The value of such plantations in investment dollars and their role in maintaining a continuing supply of wood for Canada's forest industries require that they be protected.

Plantations and managed young stands are not new in Canada, but the scope of present day operations is unprecedented. Forestry is moving into an era where a significant proportion of the forest resource will consist of plantations and spaced stands. Forest entomologists have predicted increases in the number and variety of insect pests in these so-called "unnatural" tree communities.

Protection Philosophy for the Commercial Forest

Reliance on trees from plantations and spaced stands for an increasingly significant part of our wood supply is a relatively new phenomenon. It is easy to argue that because of the high initial establishment costs, such investments must be protected above all else. Given the pending shortage in wood supply though, a tree is a tree, whether it grows in a plantation or in the wild, and all trees should be equally protected, no matter how hard this is to accomplish. However, the argument that special consideration must be given to plantations hinges on the fact that they have the potential to yield greater value per hectare than equivalent areas of natural forests. Plantations are likely to be readily accessible, fully stocked, of short rotation, and to support superior or genetically improved planting stock. The real advantage to plantation protection, as opposed to that of natural stands, is that it is possible to fashion them (and, in most cases, spaced stands too) in such a way that they can be protected more efficiently than natural stands. The challenge lies in developing attitudes, along with the techniques to back them up, that will make such operations an integral part of forest management planning.

The theoretical guidelines for the design of such plantations are easily arrived at; the opportunities to apply them, however, are extremely limited by the reality of harvesting practices. Aside from the small percentage of plantations on abandoned farm land, most plantations are established on burns and cutovers. Although the size and shape of burns is unpredictable, some control can be exercised over the size and shape of cutovers. However, harvest cuts are controlled to some extent by markets and to a larger extent by the nature of previous harvests and stand type. The best circumstances occur when cutovers are part of an integrated silviculture and harvesting operation, the amount and nature of the regeneration that will follow the harvest is predictable, and the forest nurseries are able to schedule a variety of tree crops to be planted. This level of sophistication in advance planning is a goal of modern forestry not yet achieved. Although the high level of sophistication of present-day forest nurseries means that a variety of planting stock should be available, in many instances planning of plantations follows the harvest, or the burn, and the available planting stock; size, shape, locations and species composition are likely to be preordained by circumstances, not by deliberate planning.

Current Options

Unless a dramatic breakthrough occurs in the biological sciences, the control of insect outbreaks will continue to rely heavily on chemical pesticides for the foreseeable future. Therefore, in deciding options for plantation design (i.e. where not to plant and what to plant where), spray application from the air, because of the efficiency of aerial spraying, has to be a major consideration.

For efficiency and ease of spraying, plantations should be arranged in long rectangles of no more than 400 ha, in consideration of the payload of fixed-wing spray aircraft, and at least 500 m from water courses, given current regulations on the use of pesticides. It may be practical to harvest small unproductive corners adjacent to areas of the main harvest at a loss to establish straight boundaries, making them easier to locate and treat when future protection programs need to be carried out. Greater insurance against complete loss by specific pest species can be achieved by mixing less susceptible tree species with the main crop. But this can only be done between blocks and never within blocks, because of the impracticality of spraying specific trees. The idea of mixing trees within blocks has often been suggested, based on the assumption that susceptible tree species growing in mixed stands are somehow less detectable by or accessible to the pest species. Although this may be the case when dealing with such pests as the balsam woolly adelgid, *Adelges piceae*, it is not a valid assumption for most other pests.

Risk

In the face of a probable insect outbreak, the first question asked is, what risks does a no-spray option pose? In answering this question, the main factor to consider is the age of the plantation. If it is a mature plantation, how close to harvest is it? Is it economical to harvest early, or can it be salvaged following an insect outbreak? Young plantations pose a particular problem. For example, there may be circumstances that would make it cheaper to lose a small number of plantations that can be replanted than to spray a large number of plantations, some of which may not actually need protection. The foregoing statement recognizes that most protection programs cannot be launched at the last minute and that the decision whether or not to spray is based on estimates of populations, usually from the previous generation or at least not from the development stage to be sprayed. It is likely that once a plantation reaches the "free-to-grow" stage, replanting is uneconomical and protection is warranted whenever there is risk. As the plantation reaches maturity the opportunity for salvage becomes a factor.

Although accurate forecasts of likely damage can be made for several forest pests, if the worst case scenario for any age of a plantation means significant loss of money, the decision to spray is more liable to be based on emotion (mainly fear) than on an objective evaluation of risk. Such a reaction is understandable because the prediction of insect outbreaks is, at best, risky and may seem to the forest manager no more reliable than betting on the weather.

Protection Philosophy for Small Plantations

Small plantations, such as most Christmas tree farms, hybrid poplar plantations, seed orchards, and tree-breeding orchards, are akin to agricultural orchards and comparative control options are often sought. After all, the plantations are usually accessible to ground spraying apparatus and should be amenable to protection with agricultural pesticide regimes. However, the registration process still discriminates against forestry-related applications of pesticides. Aerial application of some pesticides is permitted in agricultural orchards, yet these same pesticides cannot be sprayed in an adjacent Christmas tree farm. There is a logical reason for this apparent discrimination. The volume and variety of pesticides used in agriculture dwarfs that of forestry; as a result, efforts to register pesticides for agricultural use began earlier and were more intensive. But over the years, the registration process has gradually become more restrictive and, because of this, forestry has suffered compared with agriculture. A particular problem in forestry is that the intervals between outbreaks of obscure pests can be as long as 20 years, while the outbreak itself may only last as short a time as 2 years. In such circumstances, it is virtually impossible to obtain any sort of permit to spray.

Agricultural orchard protection has relied heavily in recent times on integrated pest management, the objective being more the selective use of pesticides to encourage the survival of beneficial predators, parasitoids, and pollinators, than merely to reduce the use of chemical pesticides. Beneficial insects are vulnerable in orchards because the orchards are veritable islands in a sea of rural landscape usually consisting of a wide variety of agricultural crops. Forest plantations, by contrast, are usually surrounded by other areas of forests, supporting similar tree species, in which beneficial insects abound and are available to reinvade the plantations following spraying.

Control Methods Other Than Pesticides

Because of the relatively small size of plantations and spaced stands, alternatives to pesticides are perhaps more promising for them than can be expected for the large contiguous forest. So far the alternative control options are parasitic nematodes; pheromones; male sterilization; biological pesticides (Bt, viruses, fungi); parasite and predator inundation; and parasite and predator inoculation (classic biological control).

Of these six options, parasitic nematodes, pheromones, and parasite and predator inundation appear to be the most promising for the relatively small-scale operations required by small plantations or spaced stands. The remaining options are equally effective for operations of any size.

Summary

Controlling pest problems in the "human-made" forest will be the challenge of the future. Design of plantations should be such that control techniques can be applied efficiently and safely. Well-integrated harvesting, silviculture, and nursery operations are essential to the achievement of this goal. Many pests of the future can be expected to be new to us, as will be their reaction to the trees and the trees' response to them. Foresters must respond cautiously to each problem, determined to obtain a thorough understanding of the dynamics of the pest and its host. The response should not be to spray first and ask questions later.

Chapter 22

Black Army Cutworm, *Actebia fennica*

ROY F. SHEPHERD AND R.J. WEST

Introduction

Larvae of the black army cutworm, *Actebia fennica*, appear suddenly on sites burned 1 or 2 years previously and begin devouring vegetation, including newly planted seedlings. Severe defoliation results in some seedling mortality and a need for replanting to maintain desired silvicultural stocking standards. Less severe defoliation produces stem deformities and height growth losses. This insect is circumpolar in its distribution; in Canada, outbreaks in the forest have occurred in British Columbia, Ontario, and Newfoundland.

Biology

Eggs are laid in July and August and take about 3 weeks to hatch (Wood and Neilson 1956). There are six larval instars with overwintering occurring in the early instars (Humble et al. 1989). Third and fourth instar larvae can be found actively feeding as soon as the snow melts in spring and the plants begin to grow. At first, the larvae chew small holes in the leaves, but later, they consume all the leafy parts of the plants. On sites where most hardwood and herbaceous vegetation is eaten, conifer seedlings are also consumed. Larvae pupate in the duff in mid- to late June and 3 or 4 weeks later the adults emerge to mate and lay eggs. A highly attractive pheromone has been identified (Struble et al. 1989) and is being used in pheromone traps (Gray et al. 1991).

Outbreaks and Damage

Insect densities change drastically in response to fire. One or 2 years after a prescribed burn or wild fire, thousands of larvae per hectare appear, without apparent warning, and consume nearly all green vegetation. Populations usually remain high for only 1 or occasionally 2 years, then suddenly drop to endemic levels. The distribution of larvae is extremely clumped and the resulting damage to vegetation is patchy. If seedlings have been planted within the infested area, they may be damaged along with the vegetation regrowing in the burned areas. However, certain species of plants are preferred to seedlings; thus, damage is related partly to the types and quantities of alternate vegetation that are available. Timing can also be important; if seedlings have been planted a year or more before defoliation occurs, mortality will be less, compared with seedlings planted concurrent with defoliation.

Mortality cannot be assessed until after bud flush the next growing season as completely defoliated seedlings often regenerate new foliage and survive. More than 60% defoliation was necessary before mortality occurred (Maher and Shepherd 1992). Defoliation blocks root growth and if other factors occur that affect seedling root establishment such as poor planting quality or planting into dry sites, or if planting is followed by drought, there can be synergistic effects that significantly increase mortality.

Height growth also is significantly reduced when defoliation is greater than 60%. If terminal bud destruction accompanies defoliation, average height growth for the year of attack is near zero. On good growing sites that remain moist, growth will recover by the next year. On poor sites, particularly those prone to drought, recovery is much slower and, after 3 years, growth rates are still below normal. In studies of Douglas-fir, *Pseudotsuga menziesii*, and lodgepole pine, *Pinus contorta* var. *latifolia*, the former suffered higher growth losses because of a higher incidence of terminal bud loss at all levels of defoliation.

Control

A pest management strategy has been developed that outlines several steps to follow to avoid severe seedling damage (Shepherd et al. 1992). Forest habitats capable of supporting outbreaks are identified and mapped, and of these, sites that are prone to drought and also have been burned within the last 12 months are identified as being susceptible to an outbreak; severe damage could occur within the next year if a population arises. Moist sites also may have high populations and severe damage, but tree recovery is usually rapid and final damage is negligible. The susceptible sites should be monitored with pheromone traps throughout the moth flight period from July to September to detect areas with moth densities high enough to result in damaging population densities of larvae the next spring. Where risks are high, the planting schedule should be designed to avoid significant damage.

If a pest management strategy is not in place and an outbreak is detected, a useful chemical control method tested in Newfoundland in 1988 (West 1991) may be useful. It is noteworthy that such an application must be applied early enough in the period of larval feeding to prevent damage and thus justify the costs of control. Ambush® 500 EC was applied from the ground at 140 mL in 45 L water/ha (an active ingredient rate of 70 g/ha) with an MB200SK Automatic Mistblower equipped with an AU5000-2 Micronair® air blaster. Larvae were in their third and fourth instars at the time of application. Nine days after treatment, larval density decreased from 28 to less than 1/m^2 in the treated area while it increased from 19 to 33 larvae/m^2 in a nonsprayed area. There was considerable movement of larvae within and around the untreated area; this may account for the increase. The treatment significantly reduced damage to black spruce, *Picea mariana*, seedlings. The nontreated seedlings suffered an average of 49% defoliation and 86% bud damage. In comparison, average defoliation was only 1% in the treated seedlings and only 2% of the buds were damaged.

Ambush® cannot be applied from the air for environmental reasons and alternatives are being sought. The use of a nematode, *Steinernema feltiae*, is under evaluation as a control agent (West 1991). As this species is cold-tolerant, native to Newfoundland soils, and survives passing through a rotary atomizer, it has potential to significantly reduce black army cutworm populations.

References

Gray, T.G.; Shepherd, R.F.; Struble, D.L.; Byres, R.J. 1991. Selection of pheromone trap and attractant load to monitor the black army cutworm, *Actebia fennica* (Tauscher) (Lepidoptera: Noctuidae). J. Chem. Ecol. 17: 309-316.

Humble, L.M.; Shepherd, R.F.; Maher, T.F. 1989. Biology, outbreak characteristics and damage caused by the black army cutworm (Lepidoptera: Noctuidae). Pages 82-88 *in* R.I. Alfaro and S. Glover (Eds.), Insects affecting Reforestation: Biology and Damage. Proc. Symp. IUFRO and Int. Congr. Entomol., July 3-9, 1988, Vancouver, B.C. For. Can., Victoria, B.C. 287 p.

Maher, T.F.; Shepherd, R.F. 1992. Mortality and height growth losses of coniferous seedlings damaged by the black army cutworm. Can. J. For. Res. (In press)

Shepherd, R.F.; Gray, T.G.; Maher, T.F. 1992. Management of black army cutworm. Forestry Canada, Pacific Forestry Centre. Inf. Rep. BC-X-335.

Struble, D.L.; Byers, R.J.; Shepherd, R.F.; Gray, T.G. 1989. Identification of sex pheromone components of the black army cutworm, *Actebia fennica* (Tauscher) (Lepidoptera: Noctuidae) and a sex attractant blend for adult males. Can. Entomol. 121: 557-563.

West, R.J. 1991. Notes on the biology and control of the black army cutworm, *Actebia fennica* (Lepidoptera: Noctuidae), in black spruce plantations. Proc. Entomol. Soc. Ont. 122. (In press)

Wood, G.W.; Neilson, W.T.A. 1956. Notes on the black army cutworm, *Actebia fennica* (Tausch.) (Lepidoptera: Phalaenidae) a pest of low brush blueberry in New Brunswick. Can. Entomol. 88: 93-96.

Chapter 23

Spruce Bud Moth, *Zeiraphera canadensis*

JEAN J. TURGEON, EDWARD G. KETTELA, AND LUC JOBIN

Introduction

Spruce bud moth, *Zeiraphera canadensis*, eggs and larvae have been collected from most North American spruce (*Picea* spp.),[1] although they occur most frequently on white spruce, *P. glauca* (Mutuura and Freeman 1966; Prentice 1965; Turgeon 1986). Populations of spruce bud moth have continually been present in small plantations and natural forests across Canada (MacAndrews 1927; Martineau and Ouellette 1965; Martineau 1985; Mutuura and Freeman 1966; Prentice 1965; Schooley 1983; Magasi 1982) and the northeastern part of the United States (Drooz 1985; Holmes and Osgood 1984). But it was not until about 1980[2] when they infested more than 16 000 ha of white spruce plantation in New Brunswick, which were intensively managed for lumber production, that the spruce bud moth was considered a pest of economic importance. (In Canada, damage resulting from spruce bud moth populations has always been most severe in the Baie des Chaleurs area [MacAndrews 1927; McLeod and Blais 1961; Pilon 1965; Magasi 1982, 1983, 1984; Bonneau et al. 1986].) Feeding by the larvae in these plantations resulted in a significant increase of trees with multiple leaders and inconstant patterns of leader growth (Fig. 1). This damage was expected to reduce the quality and value of the lumber produced (Neilson 1985) and to delay harvesting by several years. This impact appeared critical, especially in provinces such as New Brunswick whose wood supply analysis depends heavily on the performance of new forests (Carrow 1985).

In 1982, studies began on investigating means of controlling the spruce bud moth and preventing its damage to leaders. Because little was known about this insect, its ecology and behavior were reexamined to identify possible control periods, tactics, and strategies. Damage was described and categorized to evaluate its potential impact on growth and form of white spruce; sampling and monitoring tools were developed to detect its presence and assess its population levels; and several experimental and operational control methods were tested under laboratory and field conditions. This chapter provides an overview of this research.

[1] Reports that the spruce bud moth may also infest western hemlock and some firs (Prentice 1965) are doubtful, as no new collections have been reported since the taxonomy of the group has been revised.

[2] Interestingly, the beginning of this outbreak also appears to coincide with the general decline of the spruce budworm, *Choristoneura fumiferana*, in that area (Bonneau et al. 1986; Magasi 1985). Important spruce bud moth damage had been reported in the early 1960s (McLeod and Blais 1961; Pilon 1965; Magasi 1985), immediately following the end of the spruce budworm infestation.

Figure 1. Spruce bud moth, *Zeiraphera canadensis*, damage on white spruce.

Bionomics

After reviewing the taxonomy of North American *Zeiraphera* spp. adults, Mutuura and Freeman (1966) concluded the spruce bud moth, referred to as *Z. ratzeburgiana* by North American authors (MacAndrews 1927; Miller 1950; Mackay 1959; McLeod and Blais 1961; Pilon 1965; Prentice 1965), was a new species, *Z. canadensis*. The first species, *Z. ratzeburgiana*, a pest affecting Norway spruce in Europe (Viktorovskaya 1976), does not occur in Canada (Mutuura and Freeman 1966). The most useful morphological characteristic used to separate lepidopterous larvae that feed on white spruce foliage at the same time as the spruce bud moth is the absence of an anal comb on *Z. canadensis* larvae (Miller 1950; Mackay 1959).

The spruce bud moth is univoltine (a rearing method has been developed and is available from J.J. Turgeon) and overwinters as a diapausing egg (Régnière and Turgeon 1989), between the scales located at the base of the current year growth (Pilon 1965) (Fig. 2a). Egg hatch, which usually occurs in May, is well synchronized with bud burst of white spruce in New Brunswick (Turgeon 1986) and Quebec (L. Jobin, unpubl. data). There are four larval instars (Pilon 1965) which usually complete their development in 20 to 25 days (Turgeon 1985) (Fig. 2b). Fourth-instar larvae drop to the ground and pupate in the litter. Adults emerge about 3 weeks later, and the flight period generally lasts 2 to 3 weeks (Fig. 2c). Detailed information on spruce bud moth biology and behavior can be found in MacAndrews (1927), Pilon (1965), Turgeon (1985), Turgeon et al. (1987), and Hébert (1990).

a

b

c

Figure 2. Spruce bud moth, *Z. canadensis*: (a) eggs, (b) larva, and (c) adult.

The damage caused by each instar has been described by Pilon (1965) and Turgeon (1985). First- and second-instar larvae mine needles of recently burst buds and cause little damage (Fig. 3a). Third- and fourth-instar larvae feed on the proximal end of needles as well as on meristematic tissue (as defined by Graham and Knight 1965), often weakening, and sometimes destroying, the stem (Fig. 3b). (Feeding on the meristematic tissue leaves a characteristic scar that becomes more apparent 2 to 3 weeks after larvae have dropped to the ground to pupate. This scar remains visible for at least 2 years.) Weak stems remain susceptible to breakage by heavy rains, strong winds, and birds until shoot hardening is completed. Although most larvae concentrate their attack in the upper crown and on leaders (Turgeon and Régnière 1987), they do not always destroy the latter, making it difficult to assess the total economic impact of the spruce bud moth on lumber production. Thus, attempts at establishing an economic injury level for lumber production of white spruce should be limited to the damage occurring to the leader and based on loss of form and height growth. A preliminary study revealed that trees with a broken leader, or with evidence of larval feeding on it, are more likely to produce multiple leaders the following year than undamaged trees (J.J. Turgeon, unpubl. data). The position of larval feeding on unbroken leaders was also important in inducing the production of multiple leaders. With this information, damage categories, each with an increasing probability of inducing the production of multiple leaders, were identified and used to assess the impact of the spruce bud moth on growth and form of white spruce. These categories follow: no damage on the leader; larval feeding has occurred on the leader but the scar is at least 5 cm away from the apical buds; larval feeding has occurred within 5 cm of the apical buds; larval feeding has resulted in a broken leader. The relationship between mean larval density of the spruce bud moth in a plantation and the proportion of trees in each damage category is being examined (J.J. Turgeon, unpubl. data). Until results of detailed studies assessing the effects of this insect on white spruce growth, form, and volume are available (D. Quiring, Department of Forest Resources, University of New Brunswick, pers. comm.), these relationships could be used to decide whether control action is warranted. A comparison of the proportion of trees in each of these categories is used to assess the relative efficacy of most control methods being investigated.

It is not only important to know how much and when spruce bud moth damage will occur, but also where. Identifying where is usually the realm of hazard rating (Hedden 1981). Although models that rate stand susceptibility to the spruce moth have not been developed yet, a preliminary analysis of some site, stand, and tree variables associated with differences in spruce bud moth damage intensity has been conducted in Quebec (Boulet 1990; Gagnon et al. 1990). This study has provided valuable information of some factors that may predispose, incite, and contribute to attack of this insect. The next step will be to separate factors that affect tree vigor from those causing stress because both have different characteristics and require different management strategies.

a b

Figure 3. Damage caused by spruce bud moth, *Z. canadensis*: (a) first- and second-instar larvae and (b) third- and fourth-instar larvae.

Sampling and Monitoring

Eggs and Larvae

A 15-cm branch, measured from the scales of a branch's apical growth and taken from the upper crown, has been used as a sampling unit for spruce bud moth eggs and larvae (Turgeon and Régnière 1987). But a new unit may have to be developed because a variable amount of within-tree redistribution of second-instar larvae occurs each year. Helson et al. (1989) used a 45-cm branch that provided less variation in pre- and post-spray samples, but took much longer to process (P. de Groot, Forest Pest Management Institute, pers. comm.). Sequential sampling programs with one (Turgeon and Régnière 1987) and two (Régnière et al. 1988) critical levels of larval densities have also been developed. The Quebec Department of Energy and Resources has used the latter program since 1985 to monitor spruce bud moth changes in geographical distribution, extent of damage, and fluctuations in population levels (Bonneau et al. 1986; Boulet 1987; Bordeleau et al. 1988, 1989). To forecast population levels for the following year, 25 branches per plantation are sampled after completion of oviposition. The scales of each branch are soaked in sodium hydroxide to remove egg clusters which are counted under a microscope (C. Bordeleau, Quebec Department of Energy and Resources, pers. comm.). In the Maritimes, field technicians of Forestry Canada—Maritimes Region use a stand-specific survey to monitor changes in geographical distribution and annual variations in the proportion of shoots and trees infested (Magasi 1982–1989). In each stand, they sample 25 shoots on each of three trees. Population forecasts are not conducted (L. Magasi, Forestry Canada—Maritimes Region, pers. comm.).

Adults

Electroantennogram responses, followed by field screening tests of candidate sex attractants, indicated that (E)-9-tetradecenyl acetate (E9-14:Ac) is attractive to spruce bud moth males (Turgeon and Grant 1988): later this compound was identified as the primary sex pheromone component (Silk et al. 1989). Traps located in the upper crown of white spruce and baited with 10 µg of E9-14:Ac were found to be most effective in trapping males (Turgeon and Grant 1988; Lavallée and Morissette 1988). The influence of climatic conditions on trap captures was also examined (Lavallée et al. 1988). The Forest Pest Management Institute and the Quebec Department of Energy and Resources are investigating the potential of a sex pheromone-based monitoring system. The accuracy of this system in predicting population trends the following year will determine whether it will replace or complement the egg mass survey currently being used in Quebec.

Control

Insecticides

The 16 000 ha of white spruce plantations owned and intensively managed by J.D. Irving Ltd. represented an investment of more than $16 million, and an important component of the province of New Brunswick's wood supply analysis for the next 25 years. To protect this investment, starting in 1980, the company attempted each year to control the spruce bud moth by aerial spraying of both chemical and biological insecticides on third-instar larvae (Table 1), a strategy successful against the spruce budworm, *Choristoneura fumiferana*. These trials, however, were ineffective in reducing spruce bud moth populations and leader damage (E.G. Kettela, unpubl. data; Hartling 1981, 1982), and by 1984, the company had written off their oldest plantations for sawlog production (Neilson 1985). By then, studies on the biology of the spruce bud moth (Turgeon 1985) had revealed third-instar larvae were feeding under needles and scales and thus were not directly exposed to insecticide sprays.

Table 1. Insecticides sprayed aerially[a] to control third-instar larvae of the spruce bud moth, *Zeiraphera canadensis*, in northern New Brunswick, 1981–1986.[b]

Insecticide	Product name	Application	
		Rate	No.
Fenitrothion	Sumithion®	210 g/ha	2
Aminocarb	Matacil®	70 g/ha	2
Trichlorfon	Dylox®	1 120 g/ha	1
Acephate	Orthene®	273 g/ha	1
Bt	Future® FC	30 BIU[c]	1
	SAN 415	20 BIU	1
		25 BIU	1
		30 BIU	1
	Thuricide® 64B	30 BIU	1
	Dipel® 176	30 BIU	1
	Dipel® 132	30 BIU	1
	Futura® FC30	30 BIU	1

[a] Thrush aircraft equipped with a Micronair® atomization system.
[b] E.G. Kettela, unpubl. data; Hartling 1981, 1982.
[c] Billion international units.

There are four periods when spruce bud moths are exposed and therefore vulnerable to insecticide sprays: at egg hatch, when first-instar larvae crawl toward recently burst buds; as second instars, when some larvae reestablish in different shoots; as fourth instars, when larvae drop to the litter to pupate; and as adults (Turgeon 1985). Because redistribution of second-instar larvae varies drastically between years, even in sites with similar densities, and because fourth-instar larvae drop to the ground rapidly without feeding, control efforts have focused mostly on first instars and adults.

Typically, first-instar larvae leave the oviposition site and within 30 minutes crawl on a twig and establish in a suitable shoot (Turgeon 1985). Consequently, to be effective in controlling first-instar larvae, an insecticide requires sufficient toxicity to kill larvae while they crawl toward shoots or shortly afterwards. The insecticide needs to have a relatively long residual activity to ensure control of the larvae during the entire hatching period which usually lasts less than a week (Turgeon 1985).

Using a bioassay where first-instar larvae were allowed to crawl on treated twigs, the toxicity and residual activity of eight chemical insecticides were assessed under laboratory and greenhouse conditions (Table 2). These tests demonstrated that permethrin, a pyrethroid, offered the best potential for control of first-instar larvae of spruce bud moth (Helson et al. 1989).

Among those insecticides tested under field conditions (Table 3), only applications of permethrin at 70 g/ha with

Table 2. Toxicity and residual activity of insecticides screened under laboratory conditions[a] for their potential to control spruce bud moth, *Zeiraphera canadensis*, first-instar larvae crawling on treated white spruce twigs.

Insecticide	Concentration	Relative potency[b]	Relative residual activity
Aminocarb	180 g/L	(0.28) 6	Short
Azinphos-methyl	240 g/L	(0.20) 7	Long
Chlorpyrifos	480 g/L	(1.00) 1	Short
Fenitrothion	1 250 g/L	(0.83) 4	Short
Methomyl	215 g/L	(0.65) 5	Short
Mexacarbate	21.6%	(0.82) 3	Short
Permethrin	50%	(0.94) 2	Long
Thiodicarb	500 g/L	(0.12) 8	Short

[a] Helson et al. 1989. Potter's tower at 5 dosages: 0.005, 0.01, 0.05, 0.1, and 0.5 µg active ingredient/cm².
[b] Potency relative to Chlorpyrifos based on parallel line probit analysis.

fixed-wing aircraft provided an acceptable level of leader protection (less than 10% of the leaders destroyed) (Helson et al. 1989). Although this experiment could not be repeated, an aerial application of 70 g/ha of permethrin with a helicopter and a ground application with a backpack mistblower confirmed these results (B.V. Helson and M. Auger, unpubl. data). Because this insecticide is extremely toxic to aquatic invertebrates and vertebrates, its use for aerial application has to be limited to sites where no aquatic systems are present or where they can be effectively buffered from the effects of the treatment (Kingsbury and Kreutzweiser 1987; Kreutzweiser and Kingsbury 1987). A petition to register permethrin as a ground spray against the spruce bud moth has been filed, but its use would likely be limited to small plantations, leaving owners of larger plantations without any means of preventing damage to current year white spruce leaders. Thus, there is a need to examine alternative tactics.

Aerial application of an insecticide, while adults are flying, was also investigated as a means of providing leader protection for the year following treatment. Optimal efficacy of an adulticide would require the application to be made after most females have emerged but before they begin ovipositing. Between 1985 and 1989, one aerial application of aminocarb (applied dosages of 17.5, 35, 70, and 90 g/ha) and two aerial applications of fenitrothion (70, 105, 140, 210, 280 g/ha) were field-tested (E.G. Kettela, unpubl. data). All sprays were applied with Turbo Thrush or Thrush Commander aircraft equipped with Micronair® AU 3000s or 4000s. The ultra-low volume of fenitrothion (105 g/0.8 L/ha) applied as an adulticide reduced egg lay and subsequent larval populations drastically and provided acceptable levels of protection to trees in the following year (Kettela 1990). Because of this favorable outcome, a petition to register this insecticide at that dosage for spruce bud moth control has been filed.

Timing of insecticide applications for the control of first-instar larvae or adults is critical and is based on accurate prediction of egg hatch and female emergence, respectively. A phenological model was developed and calibrated to simulate the relative abundance of various immature stages under field conditions (Régnière and Turgeon 1989), but for the purpose of planning the timing of control operations, it is likely that the degree-day requirements estimated for 10% egg hatch (146 ± 16 days with a threshold value of 4.4°C) will be favored (Régnière and Turgeon 1989). Helson and Auger (1992) have determined under field conditions that acceptable leader protection is possible by ground spraying trees with permethrin (up to 9 days before the beginning of egg hatch).

The beginning of male emergence can be determined accurately with pheromone traps baited with E9-14:Ac located in the upper crown of white spruce (Turgeon and Grant 1988). Under field conditions, females emerge 1 to 2 days later than males (Turgeon 1985) and begin oviposition 5 to 8 days after emergence (Turgeon et al. 1987). Thus, optimal conditions for adulticidal sprays should occur approximately 6 to 10 days after the first male has been caught in traps. In 1985 and 1986, aerial applications of adulticides in New Brunswick occurred 9 and 8 days, respectively, after the first male had been caught in a pheromone trap (Kettela 1990). On each site, there were three pheromone-baited Multipher or Pherocon IC traps installed 30 m apart. If the application was too early or if adult emergence was longer than usual, a second adulticide spray would have been required (Kettela 1990).

Table 3. Results of aerial spray trials with insecticides for control of spruce bud moth, *Zeiraphera canadensis*, first-instar larvae in New Brunswick.

Insecticide	Product name	Application		Population reduction[a]	% Leader destroyed
		Rate	No.		
1984[b]					
Permethrin	Ambush®	35 g/ha	1	42	28
		70 g/ha	1	81	9
Aminocarb	Matacil®	90 g/ha	2	15	19
		180 g/ha	1	0	33
Control		–	–	–	51
1985[c]					
Mexacarbate	Zectran®	70 g/ha	2	11	27
		150 g/ha	1	20	25
		150 g/ha	2	53	36
		300 g/ha	1	65	22
Control		–	–	–	62

[a]Corrected for natural mortality (Abbott 1925).
[b]Helson et al. 1989.
[c]E.G. Kettela, unpubl. data.

Growth Regulators

The effectiveness of three benzoylphenyl urea insect growth regulators in controlling crawling first-instar spruce bud moth larvae was assessed under field conditions in 1986, in Matapédia, Quebec. These growth regulators were applied with a backpack sprayer and were effective in reducing larval populations and leader growth loss (Table 4). Trials with diflubenzuron and teflubenzuron in 1987, both sprayed by helicopter at rates of 70 and 140 g in 10 L/ha, were not as encouraging because this application was too late and the dosage too low (A. Retnakaran, Forest Pest Management Institute, pers. comm.). The following year, diflubenzuron was similarly applied at rates of 70 and 140 g in 5 L/ha with proper timing, but these trials also failed to reduce larval populations and leader damage. However, when the treated plots were revisited the following year, there was little evidence of spruce bud moths and surprisingly good leader growth. Possible delayed effects of diflubenzuron are currently being examined (A. Retnakaran, pers. comm.).

Attractants

The results of attempts at controlling the spruce bud moth with its sex pheromone are discussed in Chapter 44.

Table 4. Results[a] of ground applications of three insect growth regulators with a backpack mistblower to control first-instar larvae of the spruce bud moth, *Zeiraphera canadensis*, in Matapédia, Quebec.

Treatment	Dosage (g AI[b]/10 L/ha)	Population reduction[c] (%)	Mean leader growth (cm)
Diflubenzuron	200	84	39.8
BASF 153-100-50 wp	400	98	44.4
Control	–	–	27.7
HOE-00522-20w/v flowable	200	73	40.9
Control	–	–	11.4

[a] A. Retnakaran, Forest Pest Management Institute, unpubl. data.
[b] Active ingredient.
[c] Pre-spray densities varied between 1.2 and 2.7 larvae per branch. Population reductions were corrected for natural mortality (Abbott 1925).

Biological Control

Naturally occurring biological control—Natural enemies of the spruce bud moth are listed in Tables 5 and 6. The records used to construct this list were obtained from the reference after each species name, and from collections made near Bathurst and St. Leonard in New Brunswick (J.J. Turgeon, unpubl. data); near St-Léonard de Matapédia, St-Francois d'Assise, and St-Alexis de Matapédia, in Bonaventure County, Quebec (L. Jobin, unpubl. data; Hébert 1990)); and in the Pokiok seed orchard, New Brunswick (D. Quiring, unpubl. data). Older names now in synonymy are given directly below the currently used name.

Over a 3-year period, the proportion of eggs killed by the minute egg parasitoid, *Trichogramma minutum*, in various white spruce plantations near St. Leonard, varied between 3% and 36% (J.J. Turgeon, unpubl. data). During the same period, unknown predators killed up to 30% of spruce bud moth eggs, and this predation appeared constant in each site, irrespective of the density of host eggs. Similar observations were reported by Delucchi et al. (1975), who examined the regulating action of egg predators on populations of the European larch budmoth, *Zeiraphera griseana*.

Most parasitoids of the spruce bud moth attack larvae, but together they cause little mortality. The rates of larval parasitism in southwestern Quebec were 2.8% and 3.6% (L. Daviault, unpubl. data); whereas along the Quebec coast of Baie des Chaleurs in the Gaspé Peninsula, they did not exceed 12.4% (Pilon 1965). In 1983, larval parasitism near Bathurst and St. Leonard was estimated at 2% and 6% (J.J. Turgeon, unpubl. data). The relative contribution of each species to spruce bud moth mortality has not been fully assessed but appears to vary greatly between regions. In the Gaspé Peninsula, *Triclistus podagricus* was considered the most important species (Pilon 1965); whereas in New Brunswick 53% of the parasitized hosts were killed by *Earinus zeirapherae*, 41% by *Enytus montanus*, and the remaining 6% could not be identified (J.J. Turgeon, unpubl. data). Because knowledge of parasitoid guilds of North America *Zeiraphera* spp. appears incomplete when compared with European spp., long-term studies are needed to assess the structure and effects of parasitoid guilds on spruce bud moth population dynamics. Microscopic examination of several hundred larvae of all instars collected in New Brunswick failed to detect the presence of viruses or protozoa (J.J. Turgeon, unpubl. data).

In his study, Pilon (1965) reported spruce bud moth pupal populations were reduced by approximately 80%, and predators were probably responsible for this reduction. In a small experiment conducted in 1984 in New Brunswick over 2 consecutive weeks, two groups of 250 spruce bud moth pupae were left unprotected under white spruce trees for 1 week each. In the first group only 53 pupae were recovered; 14 were recovered from the second group (J.J. Turgeon, unpubl. data). The species of ants and carabids that were observed removing pupae are listed in Table 6. Equally high predation rates were reported in Matapédia (Hébert 1990) and in the Pokiok seed orchard (D. Quiring, pers. comm.). Most of this predation takes place between the time larvae fall from the tree and their disappearance in the litter (D. Quiring, pers. comm.), which typically takes 2 minutes (Hébert 1990). Red mites were also observed on pupae and appeared to be responsible for the death of the pupae. Spruce bud moth adults have been found caught in spider webs (Pilon 1965). A life table is currently being constructed to identify and assess the relative importance of all factors affecting mortality (D. Quiring, pers. comm.).

Applied biological control—Behavioral studies have revealed the microhabitat of spruce bud moth larvae is a high humidity environment, potentially suitable for the survival of host-seeking nematodes. Thus, the possibility of using entomogenous nematodes as biological agents against first- and second-instar larvae was investigated (Turgeon and Finney-Crawley 1991). Larvae feeding in shoots were highly susceptible to *Heterorhabditis bacteriophora* when assayed in petri dishes, but no larval mortality occurred when nematodes were sprayed on trees infested with spruce bud moth, using distilled water as a carrier. The efficacy of nematodes against older instars and pupae is being investigated (D. Eidt, Forestry Canada—Maritimes Region, pers. comm.).

In 1984, 11.5 million females of the minute egg parasitoid *T. minutum* were released in a 1-ha plot infested with a moderate population (seven healthy eggs per branch) of spruce bud moths in New Brunswick. This release and others carried out in 1985 in the same plot resulted in a significant increase in the level of egg parasitism, which reached 65 to 70% (E.G. Kettela, unpubl. data). This tactic may be investigated in more detail (S.M. Smith, University of Toronto, pers. comm.).

Classical biological control—Parasitoid species from *Z. ratzeburgiana*, *Z. diniana*, and other *Zeiraphera* species are currently being collected in Europe and sent to Canada where their compatibility with spruce bud moth is being tested (Mills et al. 1987), and potential candidates for inundative releases are being identified (V. Nealis, Forestry Canada, Ontario Region, pers. comm.).

Table 5. Parasitoids of the spruce bud moth, *Zeiraphera canadensis*, in Canada.

Parasitoid	Stage of host[a] Atk	Kld	Locality[b]	Source of information[c]	Comments
Hymenoptera					
Braconidae					
Apanteles absonus	?	?		Krombein et al. 1979	
Apanteles fumiferanae	?	?		Krombein et al. 1979, Mason 1974	Doubtful host record—overwinters in *C. fumiferana*
Apanteles sp. near *californica*	L	?	1	Pilon 1965	Needle miner parasitoid
Apanteles sp. near *paralechiae*	L	?	4	LJ	New host record—needle miner parasitoid
Apanteles petrovae	?	?		Krombein et al. 1979	*C. fumiferana* parasitoid
Dolichogenidae sp.	L	?	4	LJ	Possibly a *C. fumiferana* parasitoid
Earinus zeirapherae	L	L4	2	JJT, Krombein et al. 1979	
Meteorus pinifolii	L	?		Krombein et al. 1979	
Rogas sp.	L	?	1	Pilon 1965	
Pteromalidae					
Mesopolobus verditer (Norton) (= *Amblymerus verditer* (Norton))	L	P	1	Pilon 1965	
Habrocytus sp.	L	?	4	LJ	
Eulophidae					
Euderus cushmani	L	P	4	LJ	New host record
Tetrastichus anthophilus	L	?	1	Pilon 1965	
Tetrastichus trisulcatus	L	P	4	LJ	New host record
Ichneumonidae					
Enytus montanus	L	L4,P	3,4	JJT, LJ	
Casinaria sp. (= *Horogenes* sp.)	L	?	1	Pilon 1965	Hosts are Lepidoptera that do not conceal themselves when feeding—doubtful host record
Itoplectis conquisitor	L	P	4	LJ	
Scambus brevicornis	L	?	1	Pilon 1965, Bradley 1974, LJ	
Scambus decorus	L	P	4	LJ	Also *C. fumiferana* parasitoid
Scambus hispae	L	P	4	LJ	
Scambus sp.	L	?	1	Pilon 1965	
Triclistus podagricus	L	?	1	Pilon 1965	
Trichogrammatidae					
Trichogramma minutum	E	E	1,2,3	Pilon 1965, JJT, LJ	
Trichogramma sp.	E	E	5	Hebert 1990	

[a] Atk = Attacked, Kld = Killed, ? = Unknown, E = egg, L = larvae, L4 = fourth-instar larvae, P = pupae.
[b] Locality: 1= Bonaventure County (Que.), 1958 & 1959. 2= Bathurst (N.B.), 1984 & 1985. 3= St. Leonard (N.B.), 1984 & 1985. 4= Bonaventure County (Que.), 1987. 5= St-Joseph-de-Matapédia (Que.), 1988.
[c] LJ = L. Jobin, unpubl. data; JJT = J.J. Turgeon, unpubl. data.

Table 6. Predators of the spruce bud moth, *Zeiraphera canadensis*, in Canada.

Predators	Stage of host[a] Atk	Kld	Locality[b]	Source of information[c]	Comments
Hymenoptera					
Formicinae					
Camponotus pennsylvanicus (De Geer)	PP	PP	1	Pilon 1965	Direct observation
Camponotus herculeanus (Linnaeus)	PP,P	PP,P	2	JJT	" "
Camponotus novaeboracensis (Fitch)	PP,P	PP,P	2	JJT	" "
Formica exsectoides Forel	PP,P	PP,P	3	DQ	" "
Formica fusca Linnaeus	PP,P	PP,P	2,3	JJT, DQ	" "
Formica glacialis	PP,P	PP,P	4	Hébert 1990[d]	" "
Formica subnuda Emery	PP,P	PP,P	2	JJT	" "
Myrmicinae					
Myrmica lobicornis fracticornis Emery (= *Myrmica rubra scabrinodis detritinodis* Emery)	PP,P	PP,P	4	Hébert 1990	" "
Myrmica sp.	PP,P	PP,P	2,3	JJT, DQ	" "

[a] Atk = Attacked, Kld = Killed, PP = pre-pupae, P = pupae.
[b] Locality: 1 = Bonaventure County (Que.), 1958 & 1959. 2 = Bathurst (N.B.), 1984 & 1985. 3 = Fredericton (N.B.), 1988. 4 = St-Joseph-de-Matapédia (Que.), 1988.
[c] DQ = D. Quiring, Department of Forest Resources, University of New Brunswick, unpubl. data; JJT = J.J. Turgeon, unpubl. data.
[d] See Hébert (1990) for detailed list of potential predators caught in pitfall traps near St-Joseph-de-Matapédia, 1988.

Crop Plant Resistance

Susceptibility of white spruce leaders to damage by the spruce bud moth was assessed for 72 half-sib families in New Brunswick (Quiring et al. 1991). A significant amount of the variation in damage was attributable to the half-sib family, although environmental factors were more important. Interestingly, trees in less susceptible families displayed better growth than ones in susceptible families, even in the absence of spruce bud moth, suggesting that selection for reduced susceptibility to this insect is compatible with current breeding program objectives. Several mechanisms, such as reduced oviposition and egg survival, asynchrony between egg hatch and bud burst, and perhaps differences in larval survivorship due to plant nutrition and the effect of natural enemies (D. Quiring, unpubl. data), may be involved in reducing susceptibility of white spruce to the spruce bud moth. Potential control tactics include interplanting with susceptible and less susceptible white spruce in a proper ratio (Quiring et al. 1991) or with a cover crop (Auger et al. 1990).

Conclusions

Although research conducted since 1982 has focused almost exclusively on the containment of spruce bud moth populations, with little emphasis on prevention tactics, none of the developed tactics have been registered. Nevertheless, during that period, apparently more efforts have been invested in assessing alternative control strategies and tactics for spruce bud moth than for other pests. Whether this is a reflection of today's more integrated approach to forest pest management rather than the difficulty in finding environmentally and economically acceptable tactics for control of the spruce bud moth is not clear.

A great deal of background information has been generated on the spruce bud moth; however, there is still a need for information leading to the development of models of crop production and pest population dynamics and to the assessment and prediction of pest damage in economic terms. This information is urgently required by forest managers if they are to make informed pest management decisions in line with their management objectives.

References

Abbott, W.S. 1925. A method of computing the effectiveness of an insecticide. J. Econ. Entomol. 18: 265-267.

Auger, M.; Boulet, B.; Gagnon, R.; Hébert, C. 1990. Projet de stratégie de protection contre la tordeuse de l'épinette, *Zeiraphera canadensis* Mut. & Free. Rapport préparé par le Service de la protection contre les insectes et les maladies, Ministère de l'Énergie et des ressources du Québec. 13 p.

Bonneau, G.; Picher, R.; Lachance, D. 1986. Insectes et maladies des arbres, Québec—1985. Forêt Conserv. suppl. 52(10): 32 p.

Bordeleau, C.; Guérin, D.; Innes, L.; Lachance, D.; Picher, R. 1988. Insectes et maladies des arbres, Québec—1987. Forêt Conserv. suppl. 54(10): 32 p.

Bordeleau, C.; Guérin, D.; Innes, L.; Lachance, D.; Picher, R. 1989. Insectes et maladies des arbres, Québec—1988. Forêt Conserv. suppl. 56(1): 32 p.

Boulet, B. 1987. Programme de surveillance dans les plantations. Page 16 *in* Insectes et maladies des arbres, Québec—1986. Forêt Conserv. suppl. 53(10): 32 p.

Boulet, B. 1990. Caractéristiques des arbres et de la station en relation avec les attaques de *Z. canadensis* (Lepidoptera: Tortricidae), sur l'épinette blanche en plantation. Rev. Entomol. Québec. (In press)

Bradley, G.A. 1974. Parasites of forest lepidoptera in Canada. Part 1. Environ. Can., Can For. Serv., Ottawa, Ont. Publ. 1336. 99 p.

Carrow, J.R. 1985. Spruce budmoth—a case history: opportunities. For. Chron. 61: 247-251.

Delucchi, V.; Aeschlimann, J.-P.; Graf, E. 1975. The regulating action of egg predators on the populations of *Zeiraphera diniana* Guénée (Lep. Tortricidae). Bull. Soc. Entomol. Suisse. 48: 37-45.

Drooz, A.T. (Ed.). 1985. Insects of Eastern Forests. USDA, For. Serv., Washington, D.C. Misc. Publ. No. 1426, 608 p.

Gagnon, R.; Pelletier, G.; Boulet, B.; Chabot, M. 1990. Comparaison de deux parties d'une plantation endommagées à différentes intensités par *Z. canadensis* (Lepidoptera: Tortricidae). Rev. Entonoml. Québec. (In press)

Graham, S.A.; Knight, S.B. 1965. Principles of forest entomology. Fourth ed. McGraw-Hill, New York. 417 p.

Hartling, L.K. 1981. A preliminary report on the spruce budmoth (*Zeiraphera* spp.). For. Prot. Ltd. Rep., Fredericton, N.B.

Hartling, L.K. 1982. A summary of the 1982 spray program against the spruce budmoth (*Zeiraphera* spp.). For. Prot. Ltd. Rep., Fredericton, N.B.

Hébert, C. 1990. Biologie et contrôle naturel de *Zeiraphera canadensis* Mut. & Free. (Lepidoptera: Trotricidae) dans la région de Matapédia, Québec. Rapport pour le Ministère de l'Énergie et des Ressources du Québec. 61 p.

Hedden, R.L. 1981. Hazard-rating system development and validation: an overview. Pages 9-12 *in* R.L. Hedden, S.J. Barras and J.E. Coster (Tech. Coords.) Hazard-rating systems in forest insect pest management: symposium proceedings. USDA For. Serv., Gen. Tech. Rep. WO-27. 169 p.

Helson, B.V.; Auger, M. 1992. Timing of permethrin sprays for the control of first-instar larvae of the spruce budmoth, *Zeiraphera canadensis* Mut. & Free. (Lepidoptera: Tortricidae). Can. Entomol. (In press)

Helson, B.V.; de Groot, P.; Turgeon, J.J.; Kettela, E.G. 1989. Toxicity of insecticides to first instar larvae of the spruce budmoth, *Zeiraphera canadensis* (Lepidoptera: Tortricidae): laboratory and field studies. Can. Entomol. 121: 81-91.

Holmes, J.A.; Osgood, E.A. 1984. Chemical control of the spruce budmoth *Zeiraphera canadensis* Mut. & Free. (Lepidoptera: Olethreutidae) on white spruce in Maine. University of Maine at Orono, Maine Agricultural Exp. Stn., Tech. Bull. 112, 21 p.

Kettela, E.G. 1990. Spruce budmoth can be controlled with aerial applications of fenitrothion insecticide to adults. Forestry Canada—Maritimes Region, Tech. Note 225, 7 p.

Kingsbury, P.D.; Kreutzweiser, D.P. 1987. Permethrin treatments in Canadian forests. Part 1: Impact on stream fish. Pestic. Sci. 19: 35-48.

Kreutzweiser, D.P.; Kingsbury, P.D. 1987. Permethrin treatments in Canadian forests. Part 2: Impact on stream invertebrates. Pestic. Sci. 19: 49-60.

Krombien, K.V.; Hurd Jr., P.D.; Smith, D.R.; Burks, B.D. 1979. Catalog of Hymenoptera in America North of Mexico. Smithsonian Institution Press, Washington, D.C.

Lavallée, R.; Morissette, J. 1988. Note sur la comparaison des captures des adultes de *Zeiraphera canadensis* avec trois modèles de piéges Multi-Pher. Phytoprotection 69: 87-89.

Lavallée, R.; Régnière, J.; Morissette, J. 1988. Influence de la température et des précipitations sur le vol des mâles de *Zeiraphera canadensis* en plantations d'épinettes blanches. Phytoprotection 69: 99-103.

MacAndrews, A.H. 1927. Biological notes on *Zeiraphera fortunana* Kft, and *ratzeburgiana* Ratz. (Eucosmidae, Lepid.) Can. Ent. 59: 27-29.

Mackay, M.R. 1959. Larvae of the North American Olethreutidae (Lepidoptera). Can. Entomol. Suppl. 10, 338 p.

Magasi, L. 1982. Forest pest conditions in the Maritimes in 1981. Can. For. Serv., Inf. Rep. M-X-135, 33 p.

Magasi, L. 1983. Forest pest conditions in the Maritimes in 1982. Can. For. Serv., Inf. Rep. M-X-141, 41 p.

Magasi, L. 1984. Forest pest conditions in the Maritimes in 1983. Can. For. Serv., Inf. Rep. M-X-149, 49 p.

Magasi, L. 1985. Forest pest conditions in the Maritimes in 1984. Can. For. Serv., Inf. Rep. M-X-154, 49 p.

Magasi, L. 1986. Forest pest conditions in the Maritimes in 1985. Can. For. Serv.—Maritimes, Inf. Rep. M-X-159, 85 p.

Magasi, L. 1987. Forest pest conditions in the Maritimes in 1986. Can. For. Serv.—Maritimes, Inf. Rep. M-X-161, 67 p.

Magasi, L. 1988. Forest pest conditions in the Maritimes in 1987. Can. For. Serv.—Maritimes, Inf. Rep. M-X-166, 109 p.

Magasi, L. 1989. Forest pest conditions in the Maritimes in 1988. For. Can.—Maritimes Region, Inf. Rep. M-X-174, 76 p.

Martineau, R. 1985. Insectes nuisibles des forêts de l'est du Canada. Rapp. Tech. de Foresterie 32F. Serv. Can. For., Marcel Broquet, 283 p.

Martineau, R.; Ouellette, G.B. 1965. Quebec Region. Ann. rep. Forest Insect and Disease Survey 1965. 38 p.

Mason, W.R.M. 1974. The Apanteles species (Hymenoptera: Braconidae attacking Lepidoptera in the micro-habitat of the spruce budworm (Lepidoptera: Tortricidae). Can. Entomol. 106: 1087-1102.

McLeod, J.M.; Blais, J.R. 1961. Defoliating insects on field spruce in Quebec. Can. Dept. For., Bi-mon. Prog. Rep. 17(1): 2.

Miller, C.A. 1950. A key to some lepidopterous larvae associated with the spruce budworm. Dominion Dept. Agric., Bi-mon. Prog. Rep. 6(1): 1.

Mills, N.J.; Rather, M.; Kruger, K. 1987. Annual Project Report 1987. C.A.B. Intern. Inst. Biol. Control, European Stn. 40 p.

Mutuura, A.; Freeman, T.N. 1966. The North American species of the genus *Zeiraphera* Treitschke (Olethreutidae). J. Res. Lepidop. 5: 153-176.

Neilson, M.M. 1985: Spruce Budmoth—a case history: issues & constraints. For. Chron. 61: 252-255.

Pilon, J.G. 1965. Bionomics of the spruce budmoth *Zeiraphera ratzeburgiana* Ratz., (Lepidoptera: Olethreutidae). Phytoprotection 46: 5-13.

Prentice, R.M. (Comp.) 1965. Forest Lepidoptera of Canada recorded by the Forest Insect Survey. Can. Dept. For. Publ. 1142. pp. 627-629.

Quiring, D.T.; Turgeon, J.J.; Simpson, D.; Smith, A. 1991. Genetically-based differences in susceptibility of white spruce, to spruce budmoth larval feeding. Can. J. For. Res. 21: 42-47.

Régnière J.; Boulet, B.; Turgeon, J.J. 1988. A sequential sampling plan with two critical levels for the spruce budmoth, *Zeiraphera canadensis*, (Lepidoptera: Tortricidae) J. Econ. Entomol. 81: 220-224.

Régnière, J.; Turgeon, J.J. 1989. Temperature-dependent development of the immature stages of *Zeiraphera canadensis* (Lepidoptera: Tortricidae) and simulation of its phenology. Entomol. exp. appl. 50: 185-193.

Schooley, H.O. 1983. Observations of the spruce budmoth and the spruce budmidge on black spruce in Newfoundland. Can. For. Serv., Res. Notes 3(3): 16-17.

Silk, P.J.; Butterworth; E.W.; Kuenen, L.P.S.; Northcott, C.J.; Dunkelblum, E.; Kettela, E.G. 1989. Sex pheromone specificity in two sympatric *Zeiraphera* species: *Zeiraphera canadensis* and *Zeiraphera unfortunana* in New Brunswick. J. Chem. Ecol. 15: 2435-2444.

Turgeon, J.J. 1985. Life cycle and behavior of the spruce budmoth, *Zeiraphera canadensis* (Lepidoptera: Olethreutidae) in New Brunswick. Can. Entomol. 117: 1239-1247.

Turgeon, J.J. 1986. The phenological relationship between larval of the spruce budmoth, *Zeiraphera canadensis* (Lepidoptera: Olethreutidae) and white spruce in northern New Brunswick. Can. Entomol. 118: 345-350.

Turgeon, J.J.; Finney-Crawley, J.R. 1991. Susceptibility of 1st- and 2nd-instar larvae of the spruce budmoth, *Zeiraphera canadensis* (Lepidoptera: Tortricidae), to the entomogenous nematode *Heterorhabditis heliothidis* under controlled conditions. J. Invertebr. Pathol. 57: 126-127.

Turgeon, J.J.; Grant, G.G. 1988. Response of *Zeiraphera canadensis* (Lepidoptera: Tortricidae: Olethreutinae) to candidate sex attractants and factors affecting trap catches. Environ. Entomol. 17: 442-447.

Turgeon, J.J.; Nelson, N.; Kettela, E.G. 1987. Reproductive biology of the spruce budmoth, *Zeiraphera canadensis* (Lepidoptera: Tortricidae: Olethreutinae) in New Brunswick. Can. Entomol. 119: 361-364.

Turgeon, J.J.; Régnière, J. 1987. Development of sampling techniques for the spruce budmoth, *Zeiraphera canadensis*, (Lepidoptera: Tortricidae). Can. Entomol. 119: 239-249.

Viktorovskaya, Ye. A. 1976. Ecological characteristics of *Zeiraphera ratzeburgiana*, Ratz. (Lepidoptera, Tortricidae). Entomol. Rev. 55: 37-39.

Chapter 24

Diprionid Sawflies

D.R. Wallace and J.C. Cunningham

Introduction

Aerial Control of Forest Insects in Canada (Prebble 1975a), which led to the conception of these current reviews, was, as its title implies, much more constrained in its coverage of forest insect pests and research in Canada than the present works. In 1975, only those species against which chemical or biological insecticides had been applied by aerial means were considered. For the diprionid sawflies this included the redheaded pine sawfly, *Neodiprion lecontei*; the European pine sawfly, *N. sertifer*; and the Swaine jack pine sawfly, *N. swainei*. There was no overview of the large body of research on diprionid sawflies, commonly called conifer sawflies, upon which all control investigations and trials or operations depend. We will take a broader perspective in this article to provide a general account of diprionid sawflies in Canada and an entry point to the extensive literature, particularly because no recent comprehensive overview of conifer sawfly research is available. The emphasis will be on work done mainly in Canada, but because of the nature of the sawfly and tree distributions, reference will be made to research on a wider scale. This will be done from a subject matter focus as opposed to a species-by-species treatment, except in the section on Biological Control Using Viruses where there is more detail because of the amount of recent work. Additional information on insect viruses is given by Cunningham and Kaupp in Chapter 35 and on classical biological control by Wallace in Chapter 41. Unfortunately, the extensive non-English literature on the Diprionidae is not as thoroughly covered as warranted and, in this respect particularly, the earliest reference to a specific phenomenon may not be cited, nor is the overall listing of references complete.

The Diprionidae

The Diprionidae is a small group of phytophagous Hymenoptera of the forests of the Northern Hemisphere. Worldwide, some 120 species are recognized, and among those for which the host information is known (most species), all but two are defoliators of coniferous trees. The exceptions are cone-borers. The host association, the flylike appearance of the adults (Figs. 1, 16, 26a, 26b), and the serrated ovipositor (Figs. 2 and 3) used by the females to partially or wholly embed the eggs in the plant tissue (Figs. 4, 5, 17) have given the common name conifer sawflies to the group.

The family is divided into two subfamilies, the Monocteninae comprising 14 species in three genera associated with the tree families Cupressaceae and Taxodiaceae, and the Diprioninae with 105-plus species in eight genera all affiliated with the Pinaceae (Benson 1939, 1945; Smith 1974) (Table 1). Monoctenine sawflies are generally inconspicuous and there are few records of them causing a serious problem, the most notable being the widespread damage to *Cupressus* cones in Sichuan Province, People's Republic of China, in 1982 caused by *Augomonoctenus smithi* with resultant deficiency in seed production (Zeng et al. 1984). On the other hand, the Diprioninae are generally well known because they often cause unacceptable damage in nurseries, plantations (Fig. 13), and natural host tree stands (Fig. 24), as well as to ornamental plantings. The damage may range from unsightliness to tree mortality over large areas.

Because of the importance of the conifer sawflies in forest management and the development of forest entomology in western Europe in the mid-19th century (Schwerdtfeger 1973), about two-thirds of the generally recognized European diprionid species had been described by 1900 and the taxonomy of the major species of the area now has been stable for many years. In North America the group came into prominence during the first half of the 20th century as noted by Swaine (1933) in a paper on insect relationships with the forest. Swaine commented on apparent groups of outbreaks of sawflies in both eastern and western Canada and wondered about the causes. In retrospect, it is impossible to determine whether diprionids did have a period of enhanced abundance in the first half of this century, or whether they became noticed through improved surveys, which in Canada led to the formation of the national Forest Insect Survey (McGugan 1958; Prebble 1956). The development of the Survey was hastened by the discovery of the outbreak of the European spruce sawfly, *Gilpinia hercyniae* (Fig. 27a, b), in eastern Canada in 1930, soon to involve the northeastern United States. This event also stimulated research on diprionids in general, as will be seen in subsequent sections.

Other exotic conifer sawflies became established in eastern North America in this period of rapidly increasing international travel and commerce. The introduced pine sawfly, *Diprion similis* (Fig. 26), was first recorded in Connecticut in 1914 and dispersed rapidly with most damage in this early period being in nurseries and ornamental plantings (Coppel et al. 1974). The nursery pine sawfly, *G. frutetorum* (Figs. 27c, d), was discovered in Massachusetts and Connecticut in 1932 (Smith 1979) and Niagara, Ontario, in 1934 (Gray 1937). The European pine sawfly, *Neodiprion sertifer* (Figs. 1, 2, 4–15), was first found in New Jersey in 1925, but not identified and recognized as an introduced species until 1937 (Schaffner 1939), by which time it was widely distributed. In 1939, it was collected in the Windsor–Sarnia area of southwestern Ontario (Brown 1940). Shaffner (1943) suggests that these European species had been present in North America for many years before being found, perhaps even before the United States federal quarantine regulations were adopted in 1912 (Canada's Destructive Insect and Pest Act was passed in 1910). These introductions showed the need for good taxonomic and biological information on native species to recognize potentially

a

b

c

d

Figure 1. Adult European pine sawfly, *Neodiprion sertifer*: (a) male, dorsal aspect; (b) male, ventral aspect; (c) female, dorsal aspect; (d) female, ventral aspect. (Forestry Canada, Ontario Region)

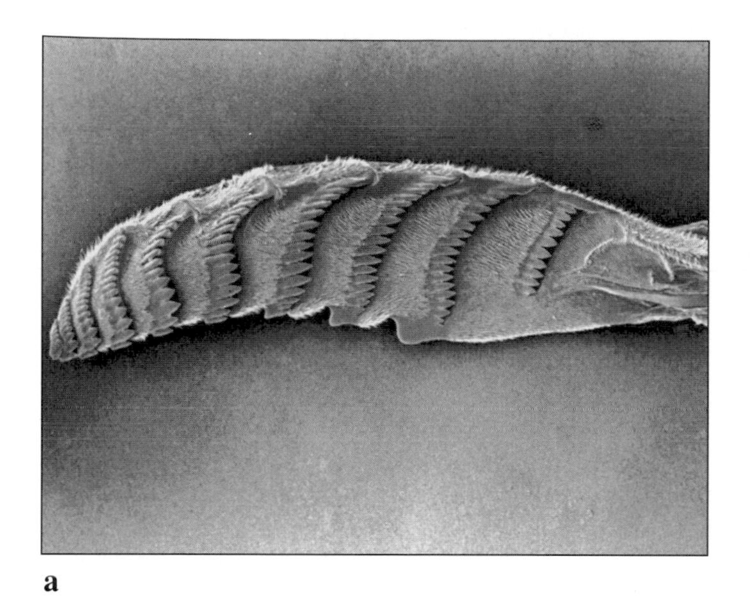

a

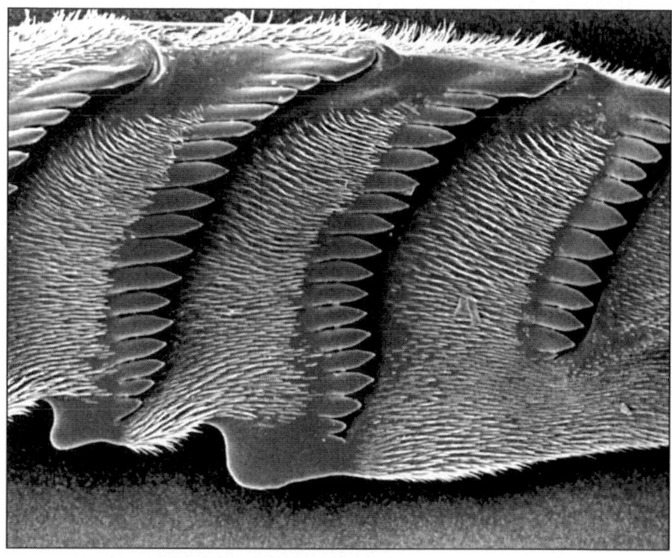

b

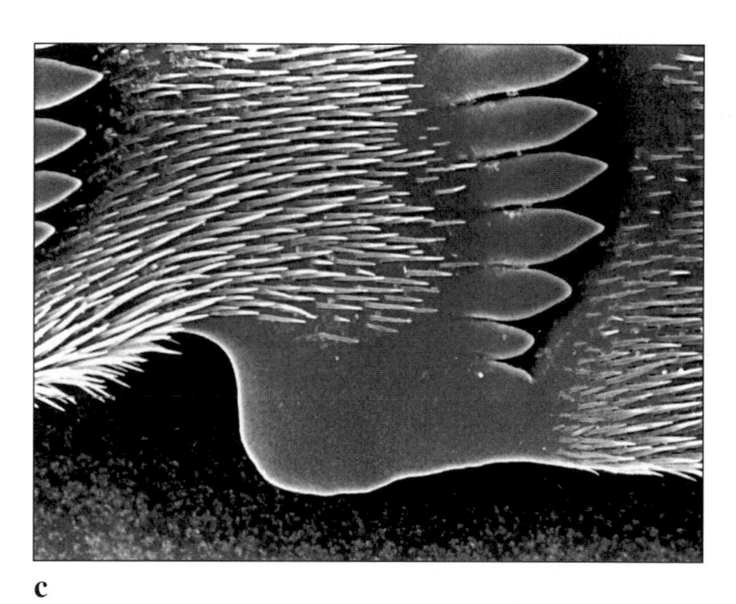

c

Figure 2. Scanning electron micrographs of the European pine sawfly lancet (saw) used by the female to cut slits in pine needles for the insertion of eggs. Characteristics often used for identification: (a) lateral aspect, complete lancet (× 55); (b) basal portion of lancet (× 140); (c) first ventral tooth of lancet (× 335). (C. Davis)

dangerous introductions. One of the key factors in achieving this end is having investigators interested in the group. Fortunately, S.A. Rohwer, Specialist in Forest Hymenoptera with the United States Department of Agriculture, Bureau of Entomology, and his associate W. Middleton provided such a focus by describing many of the new species of diprionids being collected, particularly from the jack pine, *Pinus banksiana*, in the northern United States and adjacent Canada.

In Canada, research on forest insects was developing rapidly under J.M. Swaine's guidance followed by J.J. deGryse (Swaine 1928; deGryse 1947). Field investigations of the pine-feeding diprionids in central Canada were started at Biscotasing, Ontario (Schedl 1931) and at Laniel, Quebec, in 1930. The Biscotasing research was a generalized population study of three species of *Neodiprion* on jack pine (Schedl 1933, 1935, 1937), including two new species. At Laniel, the studies continued for many years and provided the principal biological information and material, along with that coming from wider surveys, that Atwood and Peck used for their systematic study of 12 species of *Neodiprion* sawflies on pines in eastern Canada published in 1943. In western Canada, Hopping and Leech (1936) studied the biology of the hemlock sawfly, *N. tsugae*, in a short-lived outbreak on western hemlock, *Tsuga heterophylla*, on the Queen Charlotte Islands.

Systematics and Evolution

The native North American Diprionidae belong to four genera: *Monoctenus*, *Augomonoctenus*, *Zadiprion*, and *Neodiprion*. Two species of *Gilpinia* and one of *Diprion* have been introduced from Europe as well as one species of *Neodiprion*. The adults may be identified to genus using Smith (1974) and the larvae with Wong and Szlabey (1986). Most species of diprionids in the Western Hemisphere are in the genus *Neodiprion* (Table 1), which has its greatest representation in this region. Of the approximately 50 species of *Neodiprion* worldwide, only one, *N. sertifer*, was known from Eurasia until recently, when six new species were described from the People's Republic of China (Xiao, Huang, and Zhou 1984; Xiao et al. 1984), along with publication of other new records of Diprionidae (Xiao et al. 1983, 1985). Several more species of *Neodiprion* have been described from the southwestern United States, Mexico, Central America, and Cuba since 1984 (Hochmut 1984; Smith and Wagner 1986; Smith 1988), and Smith (1988) indicates that the fauna of the area is "virtually unexplored."

The work of Atwood and Peck (1943) in particular marks the beginning of modern biosystematics of the genus *Neodiprion* and except for possibly only the single, common species *N. lecontei* (Figs. 16–19, 21) which had been well recognized for many years, the beginning of more or less reliable identifications of species involved in outbreaks in eastern Canada, and of experimental material used by investigators. Although the value of a broad range of biological information in species delimitation within the genus *Neodiprion* was clearly demonstrated, most formal species definitions still rely heavily on the anatomy of the female adults and the genitalia (Figs. 2, 3) in particular.

Ross (1955) carried forward the studies of Atwood and Peck and incorporated their findings on eastern Canadian pine-feeding *Neodiprion* species in a complete revision of the genus. Ten new species or subspecies were described and some new biological information was provided. Several species complexes were indicated where the existing information would not resolve species definition. The work of Ross is the most recent complete treatment of the genus. Since 1955 there have been several studies on the application of various techniques to define additional characteristics for species segregation in *Neodiprion* sawflies. Smith (1941) used cytological and reproductive characters for distinguishing the introduced *Gilpinia hercyniae* from *G. polytoma* with which it had been synonymized in Europe. In *G. hercyniae* and *G. polytoma* the search for external distinguishing characters correlated with cytology and parthenogenesis is documented and extended most recently by Goulet (1981). Maxwell (1958) further studied the cytology of sawflies emphasizing the Diprionidae. Although she recognized several unnamed *Neodiprion* entities from western Canada, they were not formally described or named.

Maxwell (1955) in her major comparative study of internal larval anatomy of sawflies provided details of diprionid anatomy, but found that the *Neodiprion* species were "monotonously" similar. An analysis of rectal tooth patterns, which she indicated showed differences between species, was never published.

The use of biochemical data in *Neodiprion* systematics has been investigated in several diverse studies: hemolymph proteins (Whittaker and West 1962); fatty acids (Barlow 1964; Schaefer 1965); hemolymph amino acids (Schaefer and Wallace 1967); chromatographic and serological data (West et al. 1959); and egg pigmentation (Wallace 1964a, 1964b; Wallace and Campbell 1965). Of these different biochemical approaches, egg pigmentation appears to be valuable in the differentiation of closely related species and in studying intraspecific variation.

The sex pheromone chemistry of conifer sawflies has been studied extensively by researchers at the University of Wisconsin (Jewett et al. 1976; Kraemer et al. 1984; Olaifa et al. 1988). This work promises to provide insight into the relationships and species isolation mechanisms among diprionids.

The more recently developed electrophoretic determination of enzyme (isozyme) variation as a proxy for genetic variation in general has been applied to systematic studies of a wide range of plants and animals, including *Neodiprion* species. Kuenzi and Coppel (1986) investigated variation in 13 *Neodiprion* species and *Diprion similis* and found diagnostic profiles for all except three of the entities studied. They concluded that as discrete characters enzyme profiles remain a useful area for more comprehensive research. Woods and Guttman (1987) examined genetic variation in seven species of *Neodiprion* and *D. similis*, particularly with respect to low levels of genetic diversity within the Hymenoptera. Their results also indicate that isozyme research should be expanded. Harvey and Wallace (unpubl. data) also found isozyme differences among about 15 sawfly species examined. The principal problems associated with these studies are that different selections of species and enzyme systems have

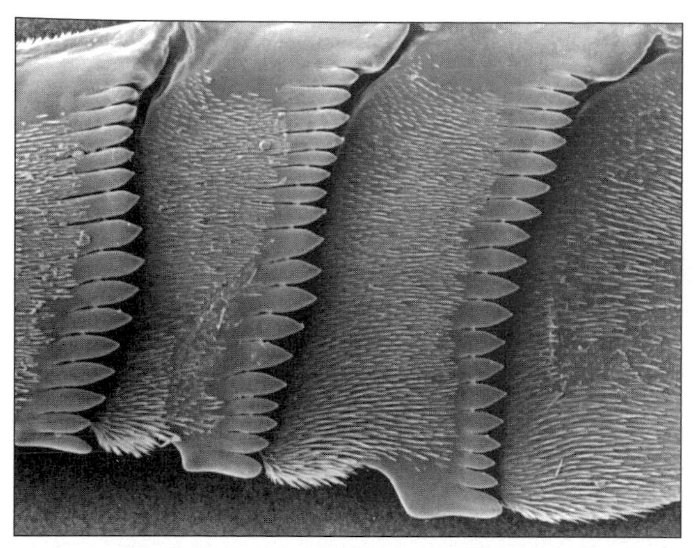
a

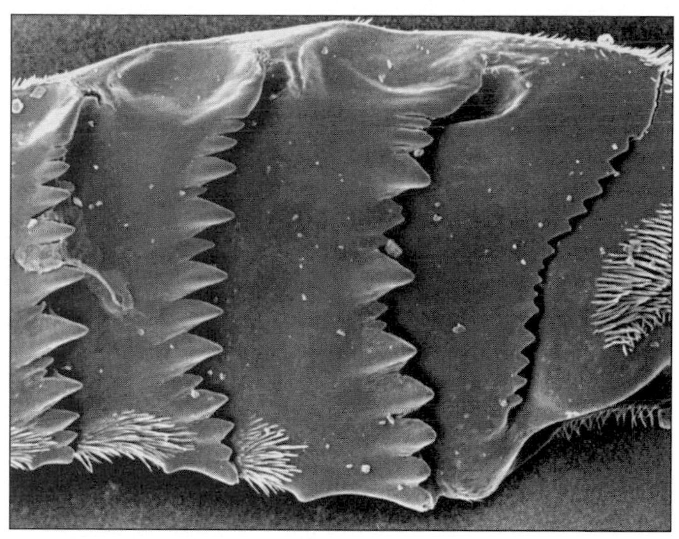
b

c

Figure 3. Scanning electron micrographs of the basal portions of the lancets of three species of diprionid sawflies to show comparison of some characteristics used for identification: (a) redheaded pine sawfly, *Neodiprion lecontei* (× 134); (b) common pine sawfly, *Diprion pini*, from Europe (× 100); (c) a pine sawfly, *Zadiprion* sp., from Mexico (× 134). (C. Davis)

Figure 4. Pine needles showing rows of European pine sawfly (*N. sertifer*) eggs. (Forestry Canada, Ontario Region)

Figure 5. Close-up of European pine sawfly eggs in needle pockets. (Forestry Canada, Ontario Region)

Figure 6. Pine shoot showing typical feeding by newly hatched European pine sawfly larvae. The previous year's needles are partially eaten to produce a characteristic twisted tangle. The new (current year's) foliage is not eaten. (Forestry Canada, Ontario Region)

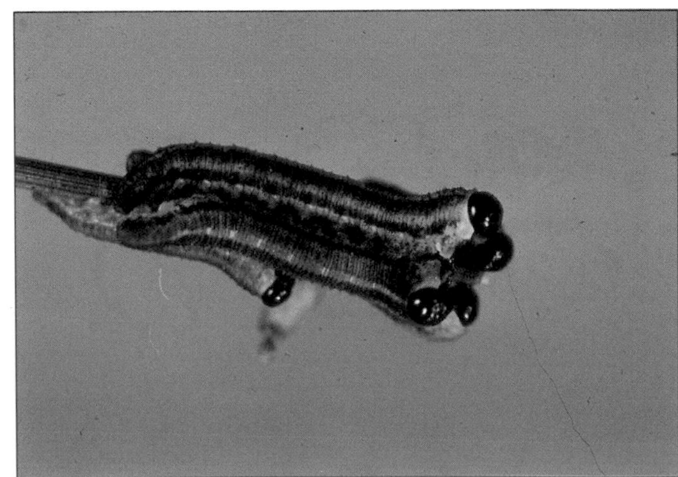

Figure 7. Feeding larvae of the European pine sawfly grouped around the perimeter of a pine needle facing the free end of the needle. (Forestry Canada, Ontario Region)

been used, and representative samples from over wide geographic areas are a necessity.

All of the foregoing studies and some compilations and interpretations of information such as those of Atwood (1961) for Ontario, Knerer and Atwood (1972, 1973), and Knerer (1983) have added to the understanding of diprionid biosystematics. But there have been few investigations of specific taxonomic problems, or of providing descriptions of new entities, particularly in British Columbia and Alberta where *Neodiprion* sawflies are for the most part unknown. Knerer (1984) published an analysis of *N. pratti*, (Fig. 25a, b), in eastern North America, but this study lacks adequate geographical representation and the quantitative methodology necessary to treat the wide distribution and complexity of this entity. As well, it incorporates some apparent misinterpretations of physiological processes.

Much of the fundamental work that is referred to in other sections of this review was done with a broad evolutionary perspective and points to many ways to improve systematic knowledge of the Diprionidae.

Anatomy

There are many descriptions, partial and detailed, of different developmental stages of conifer sawflies and attention will be drawn to only a few sources of information. MacGillivray (1914) presented a general discussion of sawfly anatomy and biology, dealing with various features of many different species. Yuasa (1922) gave an account of sawfly larval anatomy that is a valuable contribution in the field, but difficult to apply because of his use of MacGillivray's largely unaccepted system of nomenclature. Middleton (1921), in his paper on the redheaded pine sawfly, proposed a useful system of nomenclature for the thorax and abdomen of sawfly larvae, one that has had general application. Eliescu (1932) has given us another of the few anatomical works on a diprionid in his paper on the common pine sawfly, *D. pini*, whereas Parker (1934) dealt in more detail with the head of sawfly larvae. Lorenz and Kraus (1957) gave many details in their systematic treatise on European sawfly larvae. A useful extension of species characteristics of European diprionid larvae is provided by Vehrke (1961) in his treatment of three *Gilpinia* species of which *G. hercyniae* is found in Canada. Wallace (1959) has given a general description of the external larval anatomy of *N. swainei* with emphasis on characteristics that are amenable to quantitative analysis of variation. Coppel et al. (1974) have described *D. similis*. The internal larval anatomy of several diprionid species is discussed in Maxwell (1955), which is one of the few major comparative studies of larval features of sawflies.

The description of conifer sawfly adults follows a pattern similar to that shown for the larvae with isolated facets being treated in taxonomic works or detailed studies of a restricted range of structures. Reeks (1937) described the adults of the introduced species *G. hercyniae*, which at that time was causing serious damage to spruce forests in eastern North America and had not yet been distinguished from the very similar European species *G. polytoma*, both of which had been recognized by the German naturalist Hartig in the 1830s and later synonymized. Arora (1953) described the external morphology of *D. pini*, another European species. In Canada, Watson (1961) presented a detailed treatment of adult anatomy and variation in *N. swainei*. Griffiths (1953) carried out a quantitative study on the females of five *Neodiprion* species using 30 sets of measurements and 55 sets of ratios derived from the measurements. No one measurement or ratio would in itself differentiate all five species.

Silk secretion and the structure of cocoons in sawflies, including conifer sawfly species, have received some attention. Rudall and Kenchington (1971) reviewed the status of knowledge on arthropod silks from the standpoint of protein structure. Kenchington (1969) had earlier specifically addressed silk secretion in sawflies, including material from a Canadian population of the *N. abietis* complex. Wong (1951) described the general features of cocoons of some forest defoliating sawflies from Manitoba and Saskatchewan. The fine structure of the cocoons of several sawfly species was studied by Schedl and Klima (1980) using scanning electron microscopy. The European diprionid *G. socia* is among the ones included. In general, it can be concluded that cocoon structure is a fertile area for future comparative work.

There are few studies of conifer sawfly structures at the organ and cellular level. Hinks (1973) has described the neuroendocrine organs of larvae of *N. lecontei*, *N. swainei*, and *G. hercyniae*. Hallberg (1979, 1981) reported on the antennal sensilla of *N. sertifer*, and Hallberg and Lofqvist (1981) described an integral pheromone-producing gland in the abdomen of adults of the same species.

Behavior and Physiology

Another perspective, that of sociobiology, can provide valuable evidence on the evolution of diprionid sawflies and can yield a framework for examining many aspects of their biology. Ghent (1960) discussed the difficulties in associating sawfly behavioral patterns with a stepwise classification of social development that relies fundamentally on the relationship between the female parent and the offspring through one or more generations (Wilson, E.O., 1971, 1975). In the Diprionidae, some species have larvae that are solitary feeders while others form aggregations. Allee (1931) called the type of animal aggregation exhibited by sawfly larvae a sympaedium because the offspring of a female parent form the aggregation without the presence of either parent. This would fit the communal level of social organization among presocial insects, as defined by Wilson (1975), exhibiting cooperation in foraging for food. Wilson (1971) pointed out that behavior and communication among subsocial species may be as complex as that found in eusocial insects. Evidence of this is provided by the conifer sawflies where the research in some areas has gone beyond the descriptive natural history stage to the physiological stage where experimental analyses have been done and there are the beginnings of application in population biology.

Most studies have concentrated on species exhibiting group feeding aggregations, the sympaedia defined earlier. What are the foundations of these aggregations, how are they maintained, and what are their consequences in the population biology of the species? These are questions that the following sections will address.

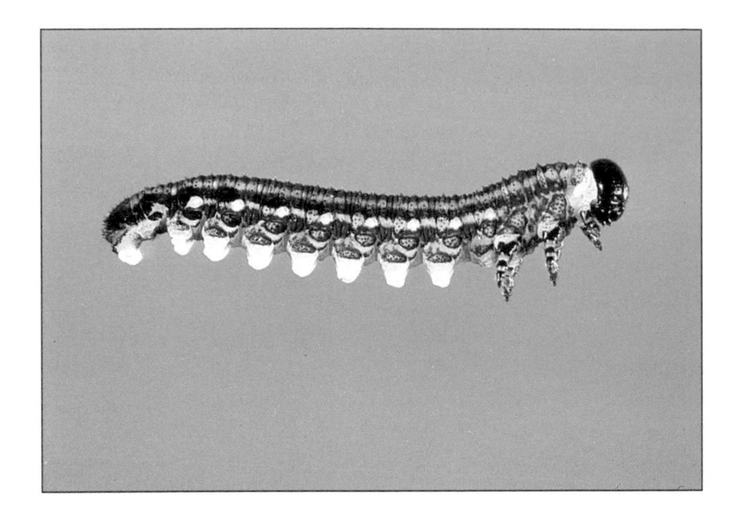

Figure 8. Lateral aspect of a fully developed larva of the European pine sawfly (*N. sertifer*). (Forestry Canada, Ontario Region)

Figure 9. Group of cocoons of the European pine sawfly, normally found in the litter and soil under the trees in which the larvae have fed. (Forestry Canada, Ontario Region)

10a

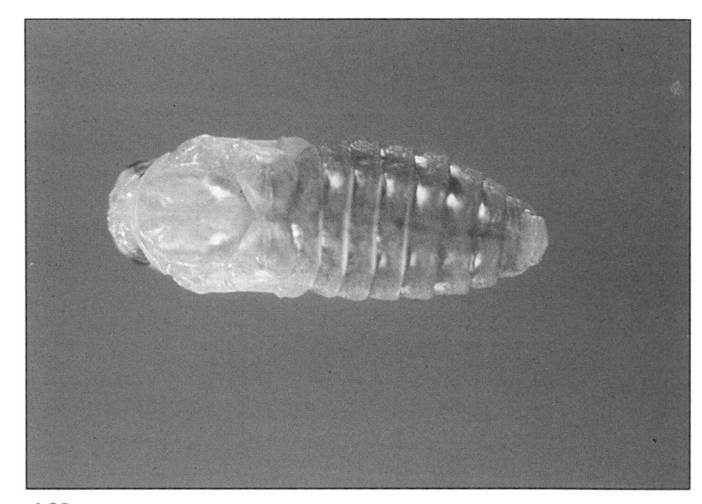

10b

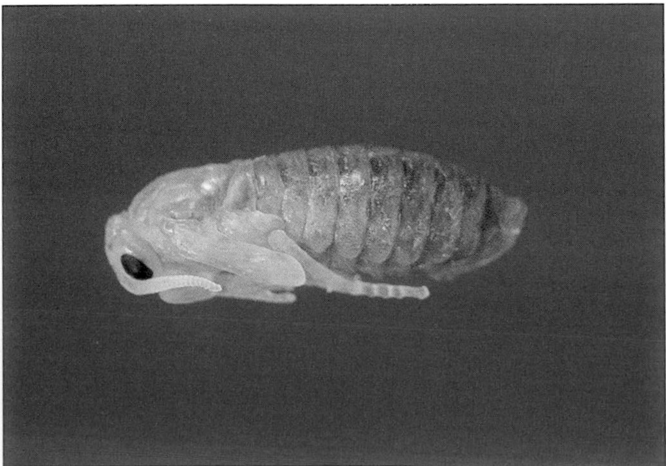

10c

Figure 10. European pine sawfly pupae removed from the cocoon: (a) dorsal aspect, male pupa; (b) dorsal aspect, female pupa; (c) lateral aspect, female pupa. (Forestry Canada, Ontario Region)

Figure 11. Young red pine, *Pinus resinosa*, plantation in the growth stage in which first attacks by the European pine sawfly are usually observed in southern Ontario. (Forestry Canada, Ontario Region)

Figure 12. Feeding by a colony of European pine sawfly larvae on a young red pine. (Forestry Canada, Ontario Region)

Figure 13. Larger Scots pine, *Pinus sylvestris*, that has been defoliated by the European pine sawfly. The feeding takes place early in the summer and the new (current year's) needles have not been eaten. (Forestry Canada, Ontario Region)

Figure 14. European pine sawfly larvae that have been killed by the disease caused by the nuclear polyhedrosis virus. (Forestry Canada, Ontario Region)

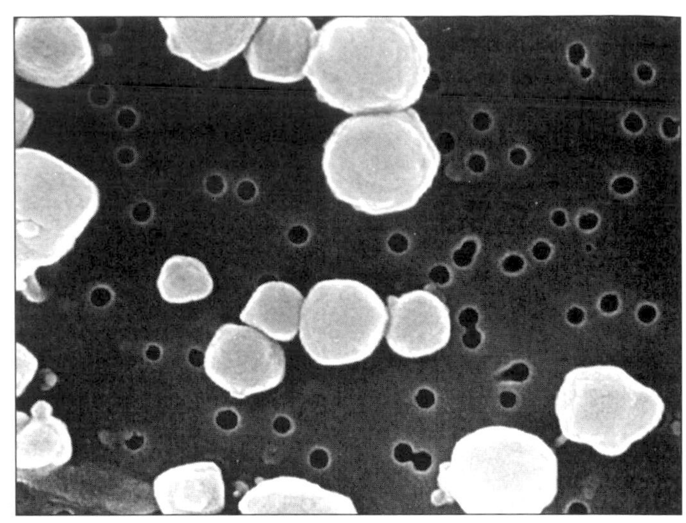

Figure 15. Scanning electron micrograph of the polyhedral inclusion bodies of the European pine sawfly virus. These bodies are formed in larvae killed by the disease; they are released onto the tree and surroundings where they may be ingested by other healthy larvae and in turn infect them with the disease. The polyhedral inclusion bodies shown are resting on a membrane filter with pores about 0.4 μm in diameter (× 7). (C. Davis)

a

b

Figure 16. Female adult redheaded pine sawfly, *Neodiprion lecontei*: (a) dorsal aspect; (b) ventral aspect. (Forestry Canada, Ontario Region)

Figure 17. Female adult redheaded pine sawfly on a pine needle with a row of newly laid eggs. (Forestry Canada, Ontario Region)

Figure 19. Redheaded pine sawfly larva. (Forestry Canada, Ontario Region)

Figure 18. Redheaded pine sawfly larvae feeding. (Forestry Canada, Ontario Region)

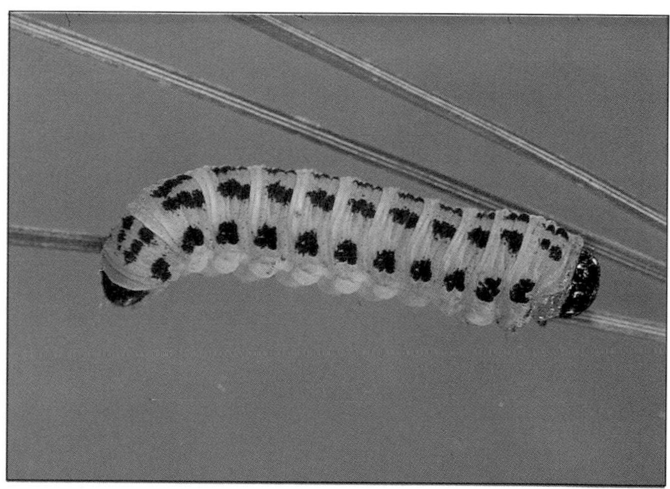

Figure 20. Larva of the white pine sawfly, *Neodiprion pinetum*, a close relative of the redheaded pine sawfly. Note the differences in head color, spot shape and size, and the whitish background color of the larvae of the two species. (Forestry Canada, Ontario Region)

Adult Dispersal, Mate Finding, and Parthenogenesis

It is generally considered that sawflies are poor fliers (e.g. Price et al. 1990, Swaine jack pine sawfly; Benson 1950) and the role that flight plays in species dynamics is an open question. Price et al. (1990) would consider that the poor capability (or lack of stimuli for flight?) of Swaine jack pine sawfly females for flight is a characteristic associated with an eruptive species (one that has low density/low damage and epidemic phases with population levels fluctuating over three-to-five orders of magnitude). Hanski (1987) has suggested that dispersal, that is, behavior leading to solitary larval feeding, is associated with nonoutbreak species of diprionids in Europe. Viktorov and Borodin (1975) found that the number of colonies of European pine sawfly in young pine plantings adjacent to old plantations harboring outbreaks decreased sharply as the distance from the old plantations increased, following a hyperbolic function. The scale of their distances is in metres, maximum 350 to 400 m, but the dispersal was not influenced by the prevailing wind direction. In experimentally infested plantations described by Lyons et al. (1971), this local movement is also recorded, but a much greater scale of distribution is suggested also, once more in keeping with the rapid progression of invasion of southern Ontario reported by Lyons (1964a). C.R. Sullivan (pers. comm.) studied the flight behavior of European pine sawfly adults in the laboratory using flight mill techniques and made field observations of flight initiation. He found that females engage in voluntary and involuntary (from alarm) flight unrelated to copulation when the air temperature is 13° to 15°C, depending on air movement. At low wind speeds the direction of flight is mainly into or across the wind, while at higher wind speeds (more than 15 km/h) it is across or downwind. Male flight is more irregular and associated directly with female location. Gravid females could make continuous flights of up to 9 km at average speeds of 201 cm/s. Males could fly up to 31 km in single flights. Their average flight speed was 217 cm/s. Sullivan concluded that the behavior and activity of the adults made long distance dispersal possible, especially since the ambient weather conditions favoring flight would provide good opportunity for them to be caught in convective air currents and carried downwind. Henson (1965a) proposed that only a small proportion of females fly and that this represents a distinctive behavioral polymorphism. The evidence suggests that adult dispersal has not received wide enough attention.

In many species the majority of adults fly into the crown of the trees under or near which they have spun their cocoons, which are most likely the same trees that served as food during the previous generation. The males usually emerge first and mating takes place shortly after as the females produce a potent sex attractant (Coppel et al. 1960). The general behavior is described in Coppel and Benjamin (1965). Lyons (1976) found that European pine sawfly males may mate several times and that the progeny sex ratios were independent of the number of previous matings by the male parent. Diprionid sawflies are capable of ovipositing without mating and in such cases the offspring are all males. This is termed facultative parthenogenesis, or arrehenotoky. Smith (1941) describes the reproductive cytology and parthenogenesis of conifer sawflies. In the European spruce sawfly, *G. hercyniae*, a very low proportion of males are produced and the species exhibits obligate parthenogenesis (thelytoky). In the European pine sawfly, *N. sertifer*, an arrhenotokous species, in one study by Lyons and Sullivan (1974), the original sex ratio at oviposition was calculated to be 0.573 females. The sex determining mechanism is generally haplodiploidy and thus males are uniparental and females biparental, but in two species, *N. nigroscutum* and *N. pinetum*, biparental (diploid) males were produced when strict inbreeding was carried out for several generations (Smith and Wallace 1971; Smith and Wallace, unpubl. data). The diploid males were not functional.

Oviposition

Adult diprionids are generally short-lived (Pilon 1966), and females oviposit shortly after emergence, especially when mated.

The general distribution of egg clusters within tree crowns and among trees in plantations and natural stands is strongly influenced by the tree growth form and the stand growth stage and condition (Hardy 1971; Hattemer et al. 1969). The distribution of egg clusters has been studied to develop population sampling techniques (Lyons 1964b) and for pest control purposes (Borodin 1972; Tostowaryk and McLeod 1972; Wilson, L.F., 1975), or to understand the ecology of the species better (Stark and Dahlsten 1965). The female adults are photopositive and the egg clusters are generally concentrated at the tops of the trees near the ends of the branches. Needles on the current year's growth (or nodes in the case of trees having multinodal annual growth) may be used for egg deposition depending on the sawfly species and its phenology in relation to the tree growth cycle.

The oviposition habits and characteristics of eggs and egg clusters of many diprionid species such as the redheaded pine sawfly (Fig. 17) have been described in varying degrees of detail for many years (Middleton 1921; Benjamin 1955). The finding by Atwood and Peck (1943) that the characteristics of oviposition can serve as diagnostic traits for at least several pine feeding *Neodiprion* species in eastern Canada, however, has led to a detailed analysis of oviposition and egg clusters in several species (Figs. 4, 5, 17, 22).

Ghent (1955) began the development of methodology for the rigorous analysis of oviposition behavior and the resulting egg clusters. His studies started with the jack pine sawfly, *N. pratti banksianae*, then were extended to the European pine sawfly, *N. sertifer* (Ghent 1959). Needle geometry and the pattern of leg and body movements by the female sawflies were used to explain the observed egg cluster characteristics. Both the above species have row-typed oviposition, also exhibited by the redheaded pine sawfly (Griffiths 1960). In later studies of the Swaine jack pine sawfly, which usually lays single eggs on a needle (Fig. 22) and may display twinned-eggs on adjacent needles of a fascicle, Ghent and Wallace (1958) showed that the growth status of the tree shoot on which the eggs are being laid is also a determining factor in the final configuration of the egg cluster. Ghent and Wallace also suggested that comparative studies of the oviposition behavior based on cluster characteristics can be useful in postulating evolutionary pathways in the genus *Neodiprion*.

Wilkinson (1964, 1971), Wilkinson and Drooz (1979), and Wilkinson and Popp (1989) extended these studies to the blackheaded pine sawfly, *N. excitans* and the slash pine sawfly, *N. merkeli*, in the southeastern United States and Central America where multiple generations of sawflies and multinodal annual growth of the host pines interact with adult behavior and needle geometry to give a rich basis for analyzing sawfly evolution.

The underlying determination of the potential maximum number of eggs available to form clusters lies in the inherent extent to which the females group their eggs on adjacent needles of the host tree. The fecundity of conifer sawflies varies from species to species (Coppel and Benjamin 1965). Fecundity, expressed as the number of eggs laid, is further modified by the food source provided by the host tree, or different host species for polyphagous sawflies. Other factors affecting fecundity and the sawflies involved are as follows: availability of food during the parental larval period for the hemlock sawfly (Hard 1971) and the European pine sawfly (Lister 1980); occurrence of sublethal pathogens for the European pine sawfly (Sullivan and Wallace 1968a, 1968b); and the effects of environmental conditions on the parental generation for the European pine sawfly (Sullivan and Wallace 1967) and the Swaine jack pine sawfly (Lyons 1970; Philogène 1971b). Given these influences on the number of eggs a female has to lay, she emerges from her cocoon ready to mate with her full complement of eggs developed. Warren and Coyne (1958) found that there is little evidence of gradual oocyte development following adult emergence for the loblolly pine sawfly, *N. taedae linearis*, or of resorption of mature oocytes if adult life is protracted. Failure to mate may reduce the number of eggs laid (Lyons 1976), or failure to oviposit at all.

The female sawflies may deposit all the eggs in a tight grouping on adjacent needles of the host tree and so form the basis for large larval feeding groups such as found in the redheaded pine sawfly (Benjamin 1955) (Fig. 18), or at the other extreme, species such as the European spruce sawfly lays eggs singly on needles and the larvae feed singly (Fig. 27a). In *N. nigroscutum*, one of the less common species feeding on jack pine in central Canada and the adjacent United States, the females lay their eggs over several branches or trees (Becker and Benjamin 1967). Females have a fecundity of about 100 eggs but "egg clusters" usually consist of a row of eggs on a single needle. The average number of eggs Becker and Benjamin found in 83 clusters was 8.87. There are other species such as *N. compar* and *N. abbotii* (Tyler and Coppel 1963) that have similar oviposition behavior. Natural mortality during embryonic and larval development results in very small groups of late stage feeding larvae which may have consequences for defence against predators and parasitoids or other factors that may influence population dynamics.

Hatching and Larval Behavior

The establishment of feeding groups and the behavior of feeding larvae have been studied extensively by several authors. Ghent (1955b, 1958, 1960) showed that hatching jack pine sawfly larvae are indifferent to light at first and orient with their heads to the tips of the needles by a "free-end" response, although the strong light reactions demonstrated by older first instar (Green 1954a) may disrupt the behavior. The larvae establish feeding sites near the tips of the needles in groups with their bodies side-by-side parallel to the needle axes (Fig. 7). They feed backward toward the needle bases, at first skeletonizing the needles (Fig. 6) and later consuming the whole needle leaving only a short stub near the basal sheath (Fig. 12). The larvae exhibit strong aggregative tendencies that depend on the presence of foliage. The group is formed through the initiation of feeding by a single larva that is then joined by other larvae up to a maximum determined by the cumulative widths of the larval head capsules in the group in relation to the circumference of the needle. In this way the maximum number of larvae in a group decreases with larval development as head widths increase with each successive molt. Ghent postulated that the larvae initiating feeding might represent a class of larvae particularly adapted to this activity, showing a division of labor in an early stage in the evolution of social organization. The role of such individual differences in insect population dynamics has been discussed by Wellington (1964). The survival advantages of aggregation such as lower early larval mortality, more rapid development, and the concomitant reduction in exposure to natural control agents, larger size, and more effective defensive action have been treated by several authors (Ghent 1960; Henson 1965b; Kalin and Knerer 1977; Lyons 1962; Nakamura 1985; Prop 1959; Trofimov 1974).

Some sawfly larvae exhibit an interesting defensive mechanism against parasitoids and predators by sequestering and regurgitating host-derived material (an allomone, Hendry et al. 1976). Saint-Hilaire (1931) states that this phenomenon had been known for a long time and then describes the oesophagael pouches in the common pine sawfly and the European pine sawfly into which the material is segregated. He further concluded that the regurgitate in these species is pine resin and that the sawfly larvae separate the droplets of resin from the triturated food in the midgut and then transfer them to the storage pouches. The latter have muscle fibres that would promote discharge of the resin. Saint-Hilaire thought that there could not be a defensive function of this process because he had not observed the appropriate larval behavior and to him the resinous drop was not "unpleasant" in odor. His conclusion was that the larvae were dealing with the disposal of a "useless substance in the food". Maxwell (1955) describes the occurrence of the oesophagael pouches in other diprionid species and the even more spectacular analogous structures in pergid sawfly larvae, the repugnatorial glands of Evans (1934) and Tait (1962). Prop (1959) showed that the exudation of resin drops in conjunction with bending, jerking, and stretching movements by conifer sawfly larvae can offer protection against attack by insectan parasitoids such as *Exenterus* spp. and predation by birds. He also concluded that there are marked differences in defensive behavior and success between group feeding and solitary feeding sawfly species. Phillipsen and Coppel (1973) demonstrated the conversion of the foregut (stomodaeum) in the nonfeeding eonymphal introduced pine sawfly larva into a structure to retain resinous material previously sequestered from the food. Eisner et al. (1974) found that eonymphal European pine sawfly larvae can use the resin even within the cocoon to deter an attacker. They also conclusively showed the material

Figure 21. Defoliation by the redheaded pine sawfly. Note that the current year's foliage has been eaten as well as the older foliage. (Forestry Canada, Ontario Region)

Figure 22. Hatched eggs of the Swaine jack pine sawfly, *Neodiprion swainei*, on current year's growth showing the apparent side-by-side location of eggs on the two needles of the fascicle. Mid-stage larvae are feeding on the previous year's needles. (Forestry Canada, Ontario Region)

Figure 23. Swaine jack pine sawfly larvae. (Forestry Canada, Ontario Region)

Figure 24. Defoliated jack pine, *Pinus banksiana*, near Ste. Hedwidge, Quebec (Lac-Saint-Jean region), 1956. (Forestry Canada, Ontario Region)

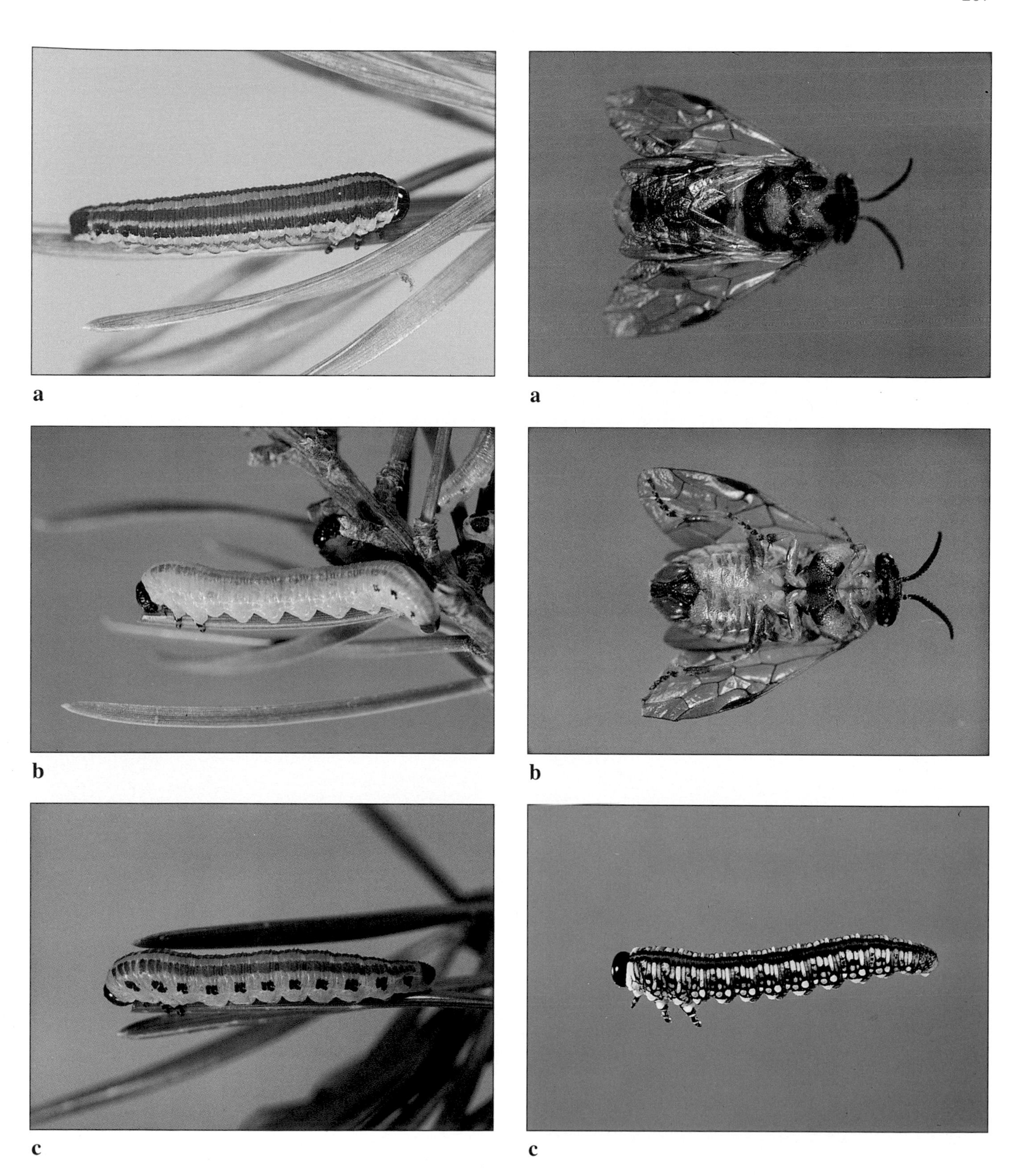

a a

b b

c c

Figure 25. Some *Neodiprion* species larvae on jack pine: (a) *N. pratti*, larval type associated with the name *N. pratti banksianae*; (b) *N. pratti*, larval type associated with the name *N. pratti paradoxicus*; (c) a jack pine sawfly, *N. maurus*. (Forestry Canada, Ontario Region)

Figure 26. The introduced pine sawfly, *Diprion similis*. One of the species of conifer sawfly that has very robust adults (compare with Fig. 1a). (a) Female adult, dorsal aspect; (b) female adult, ventral aspect; (c) larva. (Forestry Canada, Ontario Region)

Figure 27. Two species of *Gilpinia* that have been introduced into North America: (a) larva of the European spruce sawfly, *Gilpinia hercyniae*; (b) facial aspect of the head of the European spruce sawfly larva; (c) larva of the nursery pine sawfly, *G. frutetorum*; (d) facial aspect of the head of the nursery pine sawfly larva. (Forestry Canada, Ontario Region)

is a mixture of needle and branch resins, except in the first instar where the feeding habits of the larvae resulted in only needle resin being present. The separation mechanism in the feeding larvae was found to be very effective in isolating the potentially harmful resin as no resin components could be identified chromatographically in fresh frass or in midgut contents. That no resin was found in the midgut disproves Saint-Hilaire's (1931) belief that the separation takes place in the midgut. This seems logical in considering the risk of damage to the midgut by the resin components. The cuticular lining of the foregut and the oesophagael pouches could retain the plant material for long periods. Ants and spiders were found to be deterred by the pine resin. These defense mechanisms are not completely effective in preventing parasitism and furthermore Green and deFreitas (1955) showed that attack by parasitoids in addition to eliciting the alarm reactions in redheaded and jack pine sawflies caused disruption of normal feeding and would therefore tend to prolong larval development and expose larvae to natural control factors for a longer time.

The dual role of the sequestering of conifer resins by the sawfly larvae, perhaps first to protect the larvae from harmful food components and then to be used in a defensive mechanism, is one that would bear investigation along with other alarm reactions across a range of diprionid species on different host genera. It has already been mentioned that pergid sawflies have an even greater development of the use of plant material, in this case *Eucalyptus* spp. oil, for defence (Carne 1962). McNamee et al. (1981) have proposed that defensive mechanisms against predators are a major factor in population dynamics of defoliators. Benson (1950) in his "Introduction to the Natural History of British Sawflies" provides a more general discussion of the protective mechanisms of sawfly larvae and adults.

Nutrition and Host Tree Interactions

The foregoing sections have emphasized the behavior of the sawflies themselves somewhat apart from the interactions with their host trees, although the behavioral patterns have of course coevolved with their host trees. Here we will consider more directly some aspects of the relationship, that is, tree characteristics as they affect host and food selection, the tree as food, and effects of sawfly feeding on the tree.

The best indicator of host plant range in diprionids is the ability of a tree to support both sawfly oviposition and feeding. Sawfly larvae will accept a broader range of trees as food than can be used by the females for oviposition, although egg laying on an unusual conifer species is observed when foliage of a recognized host and another similar conifer is intermingled or freedom of the ovipositing female is restricted. Some diprionids such as the jack pine sawfly (Schedl 1937) and the white pine sawfly (Rauf and Benjamin 1980) are monophagous, while others such as the redheaded pine sawfly attack a wide range of pine species (Benjamin 1955). Among the pine feeders, most sawfly species do not cross conifer generic boundaries (Table 1). Little is known about the tree characteristics that determine this larger scale host acceptance, but host selection has been studied within some diprionid species.

Wright et al. (1967) investigated the differences among Scots pine varieties in susceptibility to the European pine sawfly using provenances from 108 natural stands from Eurasia. They found that north Eurasian varieties were the least frequently attacked and central European varieties the most. In general, the tallest varieties for their age were attacked most. One variety, *uralensis*, was an exception to this generality and produced slower sawfly larval development. Foliar characteristics for this variety are noted. Hattemer et al. (1969) studied egg deposition by the European pine sawfly on a wide range of species (hard and soft), including Scots pine varieties and hybrid pines. A similar major comparison, also for the European pine sawfly, was done by Henson et al. (1970). In these three works, the major objective was to determine the potential for pest management by selecting for resistance to the sawfly, and the mechanisms of differential attack were not unravelled. In Sweden, Olofsson (1989) found that the introduced lodgepole pine, *Pinus contorta* var. *latifolia*, was more heavily attacked than the native Scots pine.

Once a female diprionid deposits an egg, the association between the insect and the host is very intimate because the eggs are partially or wholly embedded within the host (needle) tissue (Brygider 1952; Breny 1957). The development of the embryo within the egg depends on receiving water from the host tissue. Needle sap several times the dry weight of egg tissue is taken up by European pine sawfly eggs during the spring development. Osmotic regulators in egg-bearing shoots will create sufficient moisture stress to cause mortality in *N. nigroscutum* eggs while the unattacked needles remain healthy (D.R. Wallace, unpubl. data).

Another interaction with the host begins when the first-instar larvae hatch from the eggs and locate a feeding site. Many conifer sawflies feed only on mature needles on the penultimate nodal growth or older and do not voluntarily consume current (new) needles (Fig. 6) (All 1973), especially early in the period of needle growth. This behavior prevails in the Swaine jack pine sawfly, *N. swainei*, where the eggs are deposited in the new foliage (Ghent and Wallace 1958) and also in the redheaded jack pine sawfly, *N. rugifrons*. All (1973) found that the feeding preference is based on the presence of deterrents in the new pine foliage and that the deterrency decreases as the needles mature. Sawfly larvae restricted to feeding on young foliage develop more slowly, weigh less, and may suffer high mortality. Schuh and Benjamin (1984a, 1984b) found that foliar resin acids are the compounds responsible for the feeding deterrency in the *Neodiprion* species they studied. Geri et al. (1988) and Buratti et al. (1988) have investigated the role of needle resin acids in the relationship between Scots pine and the common pine sawfly, but could not show conclusively that feeding deterrency is associated with these compounds.

The quantitative nutritional aspects of diprionid sawfly relationships with the host trees for several North American species were investigated by Fogal (1974). He suggested that factors affecting the utilization of nitrogen in the food are significant in determining the nutritive value of the sawfly food. Other facets of nitrogen metabolism in conifer sawflies are discussed in Fogal and Kwain (1972a, 1972b, 1979), and sterol biosynthesis was investigated by Schaefer et al. (1965). All the nutritional research on diprionid sawflies must employ

Table 1. World checklist of conifer sawflies in the family Diprionidae.[a]

	Host	Distribution	Author and date[b]
Monocteninac (Subfamily)			
Monoctenus (Genus)			
decoratus (Species)	*Cryptomeria*	Japan	Takeuchi 1940
fulvus	*Juniperus*	E. North America	(Norton 1872)
itoi	*Chamaecyparis*	Japan	Okutani ?
juniperi	*Juniperus*	Europe	(Linnaeus 1758)
melliceps		E. North America	(Cresson 1880)
nipponicus	*Juniperus*	Japan	Takeuchi 1940
obscuratus	*Juniperus*	Europe	(Hartig ?)
sadadus	*Juniperus*?	Mexico	Smith 1974
subconstrictus		Sweden	(Thomson ?)
suffusus	*Juniperus*, *Thuja*	E. North America	(Cresson 1880)
Augomonoctenus (Genus)			
libocedri (Species)	*Libocedrus* (cones)	W. Coastal United States	(Rohwer 1918)
pilosus		California	Middlekauff 1967
smithi	*Cupressus funebris* (fruit)	Sichuan, PRC[c]	Xiao et Wu 1983
Rhipidoctenus (Genus)			
cinderellae (Species)		Morocco	Benson 1954
Diprioninae (Subfamily)			
Diprion (Genus)			
butovitchi (Species)	*Pinus*	Sweden	Hedqvist 1967
fukudai	*Pinus*, *Larix*	Japan	Togashi 1964
hutacharernae		Thailand	Smith 1979
liuwanensis	*Pinus massoniana*	Guangxi, Guangdong	Huang et Xiao 1983
	Pinus taiwanensis	Jiangxi, Anhui, PRC	
nanhuaensis	*Pinus yunnanensis*, *armandi*, *massoniana*	Yunnan, Gizhou, PRC	Xiao 1983
	Pinus, *Larix*	Japan	Rohwer 1911
nipponicus	*Pinus*, *Picea*, *Abies*	Eurasia, North Africa	(Linnaeus 1758)
pini		Albania	(Zirngiebl ?)
rufiventris	*Pinus*	Eurasia; E. North America	(Hartig 1834)
similis		(Conn. 1914); Liaoning, PRC	
		Zhejiang, PRC	Zhou et Huang 1983
tianmunicus	*Pinus yunnanensis*	Yunnan, PRC	Xiao et Zhou 1983
wenshanicus			
Prionomeion (Genus)			
gaullei (Species)		Algeria	(Konow ?)
Nesodiprion (Genus)			
biremis (Species)	*Pinus yunnanensis*, *P. khasya*	PRC, Thailand	(Konow 1899)
degenicus	*Pinus densata*	Shaanxi & other, PRC	Xiao et Zhou 1984
huanglongshanicus	*Pinus tabulaeformis*	Shaanxi, PRC	Xiao et Huang 1981
	Pinus armandi		
japonicus	*Pinus*	Japan, Taiwan, Okinawa	(Marlatt 1898)
yananicus	*Pinus tabulaeformis*	Shaanxi, PRC	Huang et Zhou 1981
zhejiangensis	*Pinus*	PRC	Zhou et Xiao 1981

[a] Based on Smith (1974) with updating from literature to September 1990.
[b] Author and date of original description. Parentheses indicate that the species was described originally in a different genus.
[c] People's Republic of China.

(Continued)

Table 1. World checklist of conifer sawflies in the family Diprionidae.[a] *(Continued)*

	Host	Distribution	Author and date[b]
Microdiprion (Genus)			
fuscipennis (Species)	*Abies*	N. Europe	(Forsius 1911)
hakusanus		Japan	Togashi 1970
keteleeriafolius	*Keteleeria evelyiana*	Yunnan, PRC	Xiao et Huang 1984
pallipes	*Pinus*	Eurasia	(Fallen 1808)
Macrodiprion (Genus)			
nemoralis (Species)	*Pinus*	Eurasia	Enslin 1917
Zadiprion (Genus)			
falsus (Species)	*Pinus*	Mexico, Central America	Smith 1988
howdeni		Mexico	Smith 1975
rohweri	*Pinus*	SW United States	(Middleton 1931)
roteus		Mexico	Smith 1975
townsendi	*Pinus*	SW United States	(Cockerell 1898)
Gilpinia (Genus)			
abieticola (Species)	*Picea*	Eurasia	(Dalla Torre 1891)
amamiana		Okinawa	Okutani 1965
baiyinaobaoa	*Picea koraiensis*	Inner Mongolia, PRC	Xiao et Huang 1985
daisetusana		Japan	Takeuchi 1940
disa	*Keteleeria evelyiana*	Yunnan, PRC	Smith 1972
distincta		Japan	Takeuchi 1940
excisa		Germany	Gussakowski ?
fennica	*Picea*	Finland; Heilongjiang, PRC	(Forsius 1911)
frutetorum	*Pinus, Picea*	Eurasia. Introd. E. United States 1932. Niagara, Ont., 1934	(Fabricius 1793)
ghanii	*Picea smithiana*	Pakistan	Smith 1971
hakonensis		Japan	(Matsumura 1912)
hercyniae	*Picea*	Eurasia, Japan, Korea. Introd. Ont. 1922, N.H. 1929. Serious outbreak 1930	(Hartig 1837)
indica	*Cedrus deodara*	Pakistan, N. India	(Cameron 1913)
jinghongensis		Yunnan, PRC	Xiao et Huang 1984
jingxii	*Pinus*	Yunnan, PRC	Xiao et Huang 1984
koreana	*Larix*	Korea, Siberia	(Takagi 1931)
laricis	*Larix*	Europe	(Jurine 1807)
leksawasdii	*Pinus kesiya*	Thailand	Smith 1983
lipuensis	*Pinus massoniana*	Guangxi, PRC	Xiao et Huang 1985
marshalli	*Pinus*	Fuzhou, PRC; Thailand	(Forsius 1931)
nigra	*Picea*?	Japan	Okutani ?
paitooni	*Pinus kesiya*	Thailand	Smith 1983
pallida	*Pinus*	Europe	(Klug 1812)
pindrowi	*Pinus griffithi*	Pakistan	Benson 1961
pinicola	*Pinus koraiensis*	Heilongjiang, PRC	Xiao et Huang 1985
polytoma	*Picea*	Eurasia, Pakistan	(Hartig 1834)

(Continued)

Table 1. World checklist of conifer sawflies in the family Diprionidae.[a] *(Continued)*

	Host	Distribution	Author and date[b]
sachalinensis		Japan	Takeuchi 1940
socia	*Pinus*	Europe	(Klug 1812)
tohi	*Picea*	Japan; Heilongjiang, PRC	Takeuchi 1940
variegata	*Pinus*	Europe	(Hartig 1834)
verticalis	*Pinus*	E. Europe	Gussakovski ?
virens	*Pinus, Picea koraiensis*	Eurasia; Heilongjiang, PRC	(Klug 1812)
yongrenica	*Pinus yunnanensis*	Yunnan, PRC	Xiao et Huang 1984
Neodiprion (Genus)			
abbotii (Species)	*Pinus*	E. North America	(Leach 1817)
abietis (complex)	*Abies, Picea, Tsuga*	Across North America	(Harris 1841)
autumnalis (complex)	*Pinus*	W. United States, Mexico	Smith 1986
bicolor	*Pinus*	Mexico	Smith 1988
burkei	*Pinus*	W. United States	Middleton 1931
chuxiongensis	*Pinus yunnanensis*	Yunnan, PRC	Xiao et Zhou 1984
compar	*Pinus*	E. North America	(Leach 1817)
cubensis	*Pinus*	Cuba	Hochmut 1984
dailingensis	*Pinus koraiensis*	Heilongjiang, PRC	Xiao et Zhou 1985
deleoni	*Abies*	Washington State	Ross 1955
dubiosus	*Pinus*	NE North America	Schedl 1933
edulicola	*Pinus*	SW United States	Ross 1955
equalis	*Pinus*	Mexico	Smith 1988
excitans	*Pinus*	E. United States, Central America	Rohwer 1921
fulviceps	*Pinus*	W. North America	(Cresson 1880)
gillettei	*Pinus*	SW United States, Mexico	(Rowher 1908)
guangxiicus	*Pinus yunnanensis*	Guangxi, PRC	Xiao et Zhou 1985
hetricki	*Pinus*	E. United States	Ross 1955
huizeensis	*Pinus armandi*	Yunnan, PRC	Xiao et Zhou 1984
insularis	*Pinus*	Cuba	(Cresson 1865)
lecontei	*Pinus*	E. North America	(Fitch 1858)
maurus	*Pinus*	North Central North America	Rohwer 1918
merkeli maestrensis	*Pinus*	Cuba	Hochmut 1984
merkeli merkeli	*Pinus*	SE United States, Cuba?	Ross 1961
mundus	*Pinus*	W. United States	Rohwer 1918
nanulus contortae	*Pinus*	W. United States & Canada	Ross 1955
nanulus nanulus	*Pinus*	E. North America to Cordillera	Schedl 1933
nigroscutum	*Pinus*	North Central North America	Middleton 1933
omosus	*Pinus*	Mexico, Central America	Smith 1988
piceae	*Picea wilsonii*	Hebei, PRC	Xiao et Zhou 1985
pinetum	*Pinus strobus*	E. North America	(Norton 1869)
pinusrigidae	*Pinus*	E. United States	(Norton 1868)
pratti banksianae	*Pinus*	NE & Central North America	Rohwer 1925
pratti paradoxicus	*Pinus*	SE Canada, NE United States	Ross 1955
pratti pratti	*Pinus*	E. United States	(Dyar 1899)
rugifrons	*Pinus*	E. Canada, NE United States	Middleton 1933

(Continued)

Table 1. World checklist of conifer sawflies in the family Diprionidae.[a] *(Concluded)*

	Host	Distribution	Author and date[b]
scutellatus	*Pseudotsuga*	W. North America	Rohwer 1918
sertifer	*Pinus*	Eurasia, Japan, Korea Introd. NE North America N.J. 1925, Windsor, Ont. 1939	(Geoffroy 1785)
swainei	*Pinus*	NE North America	Middleton 1931
taedae linearis	*Pinus*	South Central United States	Ross 1955
taedae taedae	*Pinus*	SE United States	Ross 1955
tsugae	*Tsuga, Picea, Abies*	W. United States & Canada	Middleton 1933
ventralis	*Pinus*	Colorado	Ross 1955
virginianus	*Pinus*	E. United States	Rohwer 1918
warreni	*Pinus*	SE United States	Ross 1961
werneri		Arizona	Ross 1955
xiangyunicus	*Pinus yunnanensis*	Yunnan, PRC	Xiao et Zhou 1984
"Unplaced" taxa of *Neodriprion* Rohwer			
abdominalis		North Central United States?	(Say 1824)
edwardsii		California	(Norton 1869)
hypomelas		Colorado	(Rohwer 1908)

[a] Based on Smith (1974) with updating from literature to September 1990.
[b] Author and date of original description. Parentheses indicate that the species was described originally in a different genus.
[c] People's Republic of China.

correlational analyses using natural foliage because there is no artificial diet for these sawflies. Direct testing cannot be used.

More detailed studies on food utilization were reported for the European pine sawfly on Scots pine in Sweden by Larsson and Tenow (1979). Food utilization by the European pine sawfly and the introduced pine sawfly on Scots pine in Canada were studied by Slansky and Fogal (1985). Larsson and Tenow concluded that the European pine sawfly has a high food consumption rate, a low efficiency of dry matter assimilation, and low nitrogen accumulation as adaptations to food low in nitrogen content. They also presented an energy budget for this insect. Slansky (1980) countered that the food consumption rate for diprionid sawflies is similar to that for other sawflies and intermediate for herbivorous insects. Fogal and Kwain (1972c) found that the nutritive value of food influenced the number of larval instars in the introduced pine sawfly. Smirnoff (1973) and Smirnoff and Valero (1975) found that fertilization could increase mortality of Swaine jack pine sawfly, presumably through the influence on foliar nutrients.

In Europe, Schwenke (1964) recognized that nutrition influenced the population dynamics of diprionid sawflies, and this effect on the European pine sawfly was studied in considerable detail by Niemala et al. (1984, 1987). This insect, as well as the common pine sawfly and the European spruce sawfly, have been studied intensively in Scandinavia. A long series of papers on the European spruce sawfly was introduced by Bombosh (1972) and has included work on primary nutritional elements (Lunderstädt et al. 1975; Schopf 1981), secondary components (Schopf 1986), and the interaction effects of insect feeding on the host trees (Lunderstädt 1981; Kulcke 1983).

The most obvious effect of diprionid sawfly attack is the killing of the host trees, which may reach serious levels, as was the case with European spruce sawfly in eastern Canada (Reeks and Barter 1951) and Swaine jack pine sawfly in Quebec (L. Daviault, unpubl. data). Both of these sawflies caused mortality of older trees, but species such as the redheaded pine sawfly may kill young pines (Daviault 1950; Benjamin 1955). There have been several studies of the effects of sawfly defoliation in causing mortality and in reducing growth (Craighead 1940; Eklund 1964; Mott et al. 1957; O'Neil 1962, 1963). Wilson (1966) found that the European pine sawfly seldom kills young pines, but that height and radial growth, especially the latter, are affected. The dynamics of growth in response to European pine sawfly defoliation have been described for Scots pines in Sweden by Ericsson et al. (1980). They found that compensating mechanisms such as improved water status, increased photosynthetic rate, and increased mobilization of starch reserves delayed growth reduction until defoliation was severe. Nutrient dynamics and cycling were also addressed by Rangnekar and Forward (1973), Kimmins (1972), and Fogal and Slansky (1985). Although not dealing with a conifer sawfly, Ierusalimov (1977) found that compensating mechanisms restored the growth loss caused by the previous defoliation.

Just as sawfly defoliation affects the growth of the host trees, other tree characteristics that may alter the suitability of the tree for subsequent attack by the sawfly may be changed. If these latter changes are detrimental to the insect, they are considered defensive mechanisms for the tree (induced defences, Wagner 1988). Thielges (1968) found altered polyphenol metabolism in Scots pine foliage on trees that had been defoliated by the European pine sawfly. Wagner (1986) and Wagner and Evans (1985) working with ponderosa pine, *Pinus ponderosa*, and *N. fulviceps* showed that the damaged foliage was a poorer food for the sawfly, regardless of water stress. Reductions in nitrogen content and tannin compounds were associated with the changes. This aspect of sawfly/host interaction is discussed further in Wagner (1988). Sawfly defoliation also changes the microclimate of the host stand, for example, by allowing the penetration of solar radiation deeper into the canopy or even to the ground surface under the trees. The alterations in habitat temperatures may affect developmental sequences for the sawflies.

Seasonality and Diapause

As a group the conifer sawflies are very widely distributed, from the North polar regions to the tropics, and from sea level to the tree line in mountainous areas. Even individual species such as the European pine sawfly have a broad geographical range, extending latitudinally from close to 70°N latitude in Scandinavia south to about 35°N in both Europe and North America, longitudinally across Eurasia (Kolomiets et al. 1972) and the central region of eastern North America following introduction in the early 1900s (Griffiths 1959), and altitudinally from sea level to elevations of 2 000 m in central Europe (Martinek 1968; Pschorn-Walcher 1965, 1970) and to 3 000 m in the Japanese Alps (Okutani and Ito 1957; Pschorn-Walcher 1965, 1970). Such distributions imply a wide range of seasonal adaptations. These adaptations were first studied as natural history (e.g. Benson 1950) and later by laboratory and field experimentation. Seasonality and the underlying diapause were considered in many early studies simply as means of surviving periods when habitat conditions are not suitable for development. In the case of conifer sawflies, which are so closely tied to northern coniferous forests, this means overwintering. More recently, the important role of diapause as a synchronizing and evolutionary mechanism of populations has gained prominence (Tauber et al. 1986), as well as recognition that population size is often strongly influenced by diapause phenomena. The term "breaking diapause" once was frequently used and many unnatural shock treatments were investigated in attempts to circumvent or terminate the diapause, as though the diapause is something to be avoided. Andrewartha (1952) introduced the concept of the diapause as a developmental process with certain environmental cues for its induction and particular threshold and rate parameters for its termination or fulfillment. This concept has been most important in understanding the seasonal history of sawflies.

In Europe, the European pine sawfly overwinters in the egg stage and is generally considered to have a single generation annually. All other European conifer sawflies overwinter within the cocoon as a prepupal (eonymphal) larva and it has long been recognized that some species such as the common pine sawfly, *D. pini*, can have more than one generation in a summer. In North America, other species of *Neodiprion* were found to overwinter in the egg stage as well. Of the latter, Atwood and Peck (1943) included the red pine sawfly, *N. nanulus*, and the jack pine sawfly, *N. banksianae*, in their work on pine-feeding diprionids. Also, the balsam fir sawfly, *N. abietis* complex, on fir and the hemlock sawfly, *N. tsugae*, on hemlock were described with this characteristic (Bird 1929; Hopping and Leech 1936). Philogène (1971b) has discussed the different types of diapause and the general phenology of many common sawflies including diprionids.

Prebble (1941) conducted pioneering work on sawfly diapause in Canada and clearly showed that the diapause phenomenon goes far beyond being only a way for the European spruce sawfly to deal with the winter. He found that the diapause has a critical role in the epidemiology of the species, the infestations being influenced by the degree of diapause in overwintered cocoons, the duration of diapause, the number of seasonal generations, and survival of cocoons to the time of adult emergence. Prebble and his contemporaries Gobeil (1941) and Brown and Daviault (1942) were able to determine the conditions leading to the fulfillment of diapause and the temperature relationships of postdiapause development in several sawfly species, but were not able to find the factors that influence diapause inception. Prebble states "The only conclusion that can be drawn is that emergent populations are evidently highly sensitive to environmental changes associated with advance of the season."

The experimental investigation of the physiological processes of diapause in insects, which at the time of the authors cited previously was in a rudimentary state, led to the discovery that photoperiodism and the regular cycle of daylength through the year are the key determining factors in diapause inception (Lees 1955; Danilevski 1965), although temperature especially and the quality or quantity of food are strong interacting factors. These last factors themselves are highly linked with daylength progression. In sawflies the understanding of diapause requires first the definition of the different states of development: (a) continuous, uninterrupted development; (b) the "usual" diapause; and (c) intense (prolonged) diapause. Studies like those of Sláma (1958, 1960) on metabolism and respiration can help define the developmental states. With this common language established, the environmental cues, the threshold and rate parameters, the genetic variability among populations in these attributes, and adaptive changes with distributional factors such as latitude and altitude can be more readily studied. Danks (1991) has stressed the importance of constructing system models that take into account all the interactions determining seasonality. Further, Danks (1983) has developed the idea that the individuals in a population exhibiting the extremes of a trait are very important when considering adaptation. Popo (1967) suggested this also in his comparative study of different populations of the European pine sawfly.

The egg-overwintering species such as the European pine sawfly have two diapausing stages. The first is as a young embryo (Brygider 1952) during the late fall and early winter and the second as an eonymphal larva in the cocoon during the summer (Lyons and Griffiths 1962; Wallace and Sullivan 1963). Breny (1957) believed that the arrest in embryogenesis is not a true diapause but a facultative dormancy caused by the unavailability of water in the host needles in the fall. Many other authors have accepted this concept. Experiments by Juutinen (1967) and by Wallace and Sullivan (unpubl. data) in which European pine sawfly oviposition was induced on trees during the period of spring growth and high water availability do not support the host-induction theory. Wallace and Sullivan (1975) were not able to circumvent the embryonic diapause with daylength treatments, but were successful in fulfilling the diapause under total laboratory conditions with exposure to temperatures about 10°C (Wallace and Sullivan 1973). The embryonic diapause must be considered obligate (Zivojinovic 1968). Diapause and cold-hardiness are often associated and this latter attribute of the overwintering eggs has been discussed by Sullivan (1965) and Austarå (1971) in Canada and in Norway, respectively. European pine sawfly eggs may supercool, that is, cool below the normal crystallization point of pure water, to between −26° to −30°C without freezing. Freezing will kill the eggs. Eggs of two other species (the jack pine sawfly, *N. pratti banksianae*, and the red pine sawfly, *N. nanulus nanulus*) from northern regions of Ontario were even more cold hardy than those of the European pine sawfly. Exposure to nonlethal cold temperatures rapidly conditions the eggs to supercool to still lower temperatures, as low as −36°C for the European pine sawfly. Mortality in overwintering egg populations is observed in nature, especially where the eggs occur above the snow cover. In Canada, where the European pine sawfly is principally on young trees, the snow depth in the plantation during the coldest part of the winter is a very important survival factor. Sullivan (1965) discussed the possible role of the natural antifreeze glycerol in the cold hardiness of sawfly eggs.

The feeding and behavior of sawfly larvae and their temperature, humidity, and light relationships as well as their ability to condition thermally, all of which are adaptive to developing in exposed, warm, dry parts of the tree canopy and stand, have been described by Baldwin and House (1954), Green (1954a, 1954b, 1954c), and Green and deFreitas (1955). At the end of the feeding period, the mature larva molts to a nonfeeding prepupal stage which at first is mobile and seeks out a location to spin the cocoons (Fig. 9) (Pilon et al. 1964; Lyons 1964b; McLeod 1966), usually in the organic debris on the ground under the host tree, but in some species like the introduced pine sawfly, the cocoons may be spun on the host trees and on bushy vegetation under the trees as well as in the ground (Coppel et al. 1974). The prepupal larvae in the cocoon become quiescent and may enter diapause as eonymphal larvae.

The length of the summer eonymphal diapause in the European pine sawfly is inversely related to the temperature of the cocooning site (Wallace and Sullivan 1963) and to the date of cocoon spinning (Lyons and Griffiths 1962; Austarå 1966). These two relationships are important in synchronizing adult emergence and egg laying in the fall when the cool conditions favor egg survival by avoiding hot, desiccating conditions and by shortening the period during which eggs may be exposed to natural enemies such as parasitoids. Sullivan and Wallace (1965) showed that the eonymphal diapause is induced by long daylengths, that is, the European pine sawfly is a "short day" insect. Pankevich (1963) had also shown that daylength affects the development time in the cocoon, but left some confusion about the diapause state involved. Wallace and Sullivan (unpubl. data) have shown that the developmental compensation for spinning date is determined by the gradual progression of daylength change over the larval developmental period. The photoperiod–diapause relationship varies among families at a given location (Sullivan and Wallace 1968c), and also between populations (Austarå 1969; Knerer 1983). Populations of the same species living at different latitudes share the same form of photoperiod–diapause relationship, but have the position of the critical photoperiod shifted in order to be adapted to the daylength progression at that latitude. This relationship has been proposed for the European pine sawfly by Wallace and Sullivan (1966).

Under usual circumstances, the European pine sawfly completes its summer diapause and undergoes metamorphosis (Fig. 10) to the adult in the cocoon, emerges, mates, and lays eggs in the fall. The eggs overwinter, larvae hatch in the spring, and feeding is completed in early to midsummer and the cycle repeats. There is only one generation a year, but two variations in the life cycle lead to generation times longer than 1 year. These variations have important adaptive and survival consequences. First, in mountainous regions the European pine sawfly exhibits a 2-year life cycle near the tree line, both in Europe and in Japan (Pschorn-Walcher 1970; Martinek 1968). Insects with a 2-year life cycle overwinter as eggs the first year and prepupal larvae in cocoons the second year. Thus, there are feeding larval populations in alternate years, but adult flight takes place at the normal season as in lowland populations and allows for gene flow among sawfly populations along the ecotone from lowland to tree line. Populations from different alpine locations in a region may not be in phase. This 2-year life cycle adaptation to short warm seasons probably also occurs northward, but the situation is complicated by the occurrence of prolonged diapause.

Prolonged, or intense, diapause in which conymphal larvae remain in the cocoon for periods of years beyond the normal length of time is a phenomenon that has been recognized in sawflies for many years and has been shown to play a significant part in population dynamics. In the European pine sawfly, the prevalence of prolonged diapause is associated with more northern regions (Juutinen 1967) and with cool summers (Lyons et al. 1971; Sullivan and Wallace 1967). Sullivan and Wallace (1967) found that if larvae were exposed to daylengths that would induce development without the normal diapause and were subsequently exposed to low temperature as prepupal larvae within the cocoon, prolonged diapause resulted. In Norway, Austarå (1969) showed that different populations have greatly different inherent capability of entering the prolonged state of arrested development. The conditions that bring about the resumption of development after one to several years are even less well understood than the conditions that bring about the prolonged diapause. Minder (1975) proposed that low temperatures are responsible for the reactivation of development, but the timing of adult emergence was not consistent with events in nature where adults developing after prolonged diapause emerge in synchrony with those from a normal 1-year life cycle. Sullivan and Wallace (1967) found evidence of an inherent annual cycle of reactivation even under laboratory conditions.

Prolonged diapause may act to provide a pool of individuals that can bridge several years of adverse developmental conditions and thus provide survival advantage. This pool of material may explain the sudden, unexpected appearance of a sawfly outbreak if conditions in the soil favor a large emergence of adults. This population survival mechanism is not without cost to individuals. Mortality from natural control factors operating against the sawfly during the cocoon stage increases because the factors have much more time to act. Even though the metabolic rate of individuals in prolonged diapause is very low, many individuals may simply die from exhaustion of body food reserves during the long period of arrested development and the more demanding energy requirements of subsequent metamorphosis. Females emerging after prolonged diapause were found to have a lower fecundity (Sullivan and Wallace 1967). This is an interesting example of a trade-off between survival of the individual and survival of the population (species).

The mechanisms of diapause and development are not as well known for other egg-overwintering diprionids as for the European pine sawfly described in the foregoing. In one population of a member of the *N. abietis* complex on balsam fir, Wallace and Sullivan (1974) found that normal development in the cocoon does not exhibit a photoperiodically determined eonymphal diapause, but that photoperiod does influence the incidence of prolonged diapause and the rate of larval development. Knerer and Atwood (1972) reported that different entities in the *N. abietis* complex have different developmental periods and thus reduced interspecific competition. Wingfield and Warren (1971, 1972) studied the effects of photoperiod and temperature on development in the cocoon in the loblolly pine sawfly, *N. taedae linearis*, in Arkansas and found that this species is a short day insect like the European pine sawfly. This is also true for several other *Neodiprion* species (*N. nanulus nanulus*, *N. pratti banksianae*, and *N. pratti paradoxicus*) that have been investigated by Wallace and Sullivan (unpubl. data), and it is likely that other relationships involved in seasonality are very similar to those of the European pine sawfly.

The species of conifer sawflies that overwinter in the cocoon differ from the foregoing by their propensity to have more than one generation a year depending on their latitudinal distribution range and other factors that determine the length of the warm season. Prebble (1941) delimited zones in eastern North America that support different numbers of generations of the European spruce sawfly. The redheaded pine sawfly, *N. lecontei*, a native, polyphagous North American species, has a single generation a year in southern Canada and adjacent United States, grading up to five generations in Florida (Daviault 1950; Benjamin 1955; Deneve 1968). Middleton (1921) described the occurrence of broods within a generation of this species: when there is more than one generation a year, the seasonal occurrence of feeding larvae is very complicated and has not been thoroughly investigated. Knerer (1983) and Wallace and Sullivan (unpubl. data) found that the eonymphal diapause that occurs within the cocoon in overwintering generations is determined by a long day–short day photoperiod response (an intermediate day insect in the terminology of Danilevski (1965)). Continuous development occurs within a narrow range of daylengths. Knerer found that the photoperiod response is stronger and more precise in northern populations than in southern ones. This is to be expected because the daylength progression at northern latitudes provides a more definitive cue than does the lower amplitude annual cycle that prevails closer to the equator. The occurrence of prolonged diapause in the redheaded pine sawfly is extensive and has been shown to last as long as 5 years (Daviault 1950). The factors inducing and terminating the prolonged diapause are unknown.

The Swaine jack pine sawfly, *N. swainei*, is another species that overwinters in the cocoon. It is monophagous on jack pine and is found principally in Quebec and Ontario and the adjacent American states of the Great Lakes Basin to the southern limit of its host tree (Wallace 1959, 1964a). Details of the life history and phenology have been reported by several authors from both field and laboratory studies beginning with Schedl (1931, 1937). Tripp (1965) discusses seasonal development in Quebec and shows the precarious distributional position that the species occupies, being restricted in the south by its host's range and in the north by the shortness of the warm season (Wallace 1959). The insect undergoes metamorphosis in the cocoon in the spring of the year, the adults emerge and oviposit in midsummer, and larval feeding takes place from late summer into the early fall. The prepupal larvae then spin cocoons and enter diapause which is fulfilled during the winter. Although there is only a single generation a year in nature, diapause induction is under photoperiodic control.

Wallace and Sullivan (1972) showed that the Swaine jack pine sawfly is an intermediate day insect with continuous development taking place when the larvae are exposed to a narrow range of daylengths around 17 to 18 hours. Philogène (1971) found that daylength affecting late-instar larvae has the greatest influence in determining the amount of diapause, and Philogène and Benjamin (1971a) also showed that there is a strong temperature interaction with long daylength. The higher the temperature the greater the proportion of individuals developing without prepupal diapause, given the appropriate photoperiod. Prolonged diapause is also found in the Swaine jack pine sawfly and Philogène and Benjamin (1971a) and McLeod (1972) have presented some evidence on factors that influence its incidence, but there is no clear understanding of its underlying mechanisms.

The photoperiod–diapause relationship has been investigated for several other diprionid sawflies that overwinter in the cocoon and have a diapause in the prepupal stage. All of the species are either long or intermediate day insects with continuous development being induced by long daylengths (*D. similis*—Baker 1970, Wallace and Sullivan, unpubl. data; *G. frutetorum* and *G. hercyniae*—Wallace and Sullivan, unpubl. data; *N. rugifrons*—King and Benjamin 1965, Knerer and Marchant 1973, Knerer 1983, Wallace and Sullivan, unpubl. data; *N. dubiosus* and *N. nigroscutum*—Wallace and Sullivan, unpubl. data). The window of response in the different intermediate day species is of differing widths and in some cases may not provide a clear distinction from a full long day reaction, especially because the experimental evidence may include daylengths that are outside the range of natural daylengths for which the species has undergone selection, for example very long days from 20 hours to continuous light. The diapausing stage within the cocoon is believed to be the eonymphal larva in most species, but arrested development may occur at later stages of morphogenesis as reported by Knerer (1983). In *N. maurus*, a species of restricted distribution in central Canada and adjacent United States, oviposition takes place very early in the spring and the diapause may occur as an adult in the cocoon (Wallace, unpubl. data).

The light level threshold for diapause reactions in diprionid sawflies has not been investigated. Baker (1970) found that light wavelengths at the blue and red ends of the spectrum were less effective in circumventing the diapause in the introduced pine sawfly than wavelengths in the midrange of the visible spectrum. Baker and Atwood (1969) reported that continuous light during rearing of the same species resulted in abnormalities in the thoracic anatomy of adults. Other effects of photoperiod on the immature and adult stages of the Swaine jack pine sawfly have been discussed by Philogène and Benjamin (1971a).

The extensive studies of seasonality and diapause on the common pine sawfly, *D. pini*, in Europe should be indicated. This species is a long day insect (Eichhorn 1976, 1977b; Geri et al. 1984; Geri and Goussard 1988), overwinters in the cocoon, and has a very complicated phenology somewhat analogous to that of the redheaded pine sawfly in North America. There are very marked population differences that Eichhorn (1977a, 1979, 1983) and Sharov and Safonkin (1981) have discussed. Eichhorn differentiates them into lowland and mountain types. Geri et al. (1988) also showed that food influences the occurrence of diapause.

As a final entry in this section, the importance of understanding the mechanisms of diapause and development in determining species limits and in speculating on evolutionary pathways cannot be underestimated. For example, in the European pine sawfly, the observations of Verzhutskii (1966) on the spring flight of adults opens the whole question of this entity being a complex of species in eastern Siberia and other parts of eastern Asia (Suzuki 1934; Okutani and Ito 1957; Lee 1965).

Population Dynamics

In Canada, there have been two major studies with a primary stated purpose of understanding conifer sawfly life systems. These dealt with the Swaine jack pine sawfly, *N. swainei* (McLeod 1970, 1979), and the European pine sawfly, *N. sertifer* (Lyons 1964a; Lyons et al. 1971). Both of these studies were organized with a core population measurement program around which were clustered large numbers of investigations of individual population processes. Research on a third species, the European spruce sawfly, *G. hercyniae*, and long-term population monitoring also allowed an analysis of factors regulating numbers of this sawfly in the Maritime provinces from 1937 to 1963 (Neilson and Morris 1964). In Europe, there have been numerous studies on the European pine sawfly, *N. sertifer* (Hanski 1987) and the common pine sawfly, *D. pini*.

Neilson and Morris (1964) used the key-factor approach to analyze the population data for the European spruce sawfly for 50 consecutive generations over a 25-year period in New Brunswick and found that larval parasitism and a polyhedrosis virus, possibly individually or in combination, are the two factors responsible for the initial collapse of the outbreak and the subsequent regulation at a low level. Cocoon predation by small mammals and wireworms was also identified as an additional density-dependent stabilizing factor. The parasitoids *Drino bohemica* and *Exenterus* spp. were very effective at low sawfly densities, while the polyhedrosis responded to density changes of greater magnitude or longer duration. The cocoon parasitoid, *Dahlbominus fuscipennis*, which was introduced in massive numbers and subsequently found in large numbers during outbreak years (Reeks 1953), did not remain prominent after the outbreak had ended. The minor resurgence of the sawfly in an area where the principal parasitoids were killed by chemical spraying attests to their effectiveness (Neilson and Elgee 1965).

McLeod (1970) discussed the epidemiology of the Swaine jack pine sawfly, in eastern Canada from 1940 to 1967 and described the population studies done in Quebec. He later (1979) analyzed the data using a process simulation model and discussed the results in the context of discontinuous stability of ecological systems. He found that six factors in the model were capable of interacting to mimic many of the qualitative features of population behavior observed in nature. These factors were: (1) competition among groups of animals at different trophic levels for a common limited resource; (2) predation of feeding larvae and adults by insectivorous birds (McLeod 1974) with Type III density response (Holling 1965) which saturates at very low sawfly densities; (3) predation by vertebrates (cinereus shrews) (McLeod 1966) also having a Type III density response, but saturating at prey densities just lower than the level where food limits the sawfly; (4) an upper equilibrium level set by the availability of foliage (food); (5) system time constants caused by lags such as the replacement of foliage following heavy defoliation and especially in the numerical response of synchronous univoltine parasitoids (McLeod 1975); and (6) random inputs from variables such as weather, for example, larval mortality caused by a cold season (Tripp 1965). McLeod also suggested that intervention with a chemical spray early in the increasing phase of the outbreak (the "threat" stage) would likely produce a longer term control of the population below the economic threshold than waiting to act at the "crisis" stage.

Canadian population research on the European pine sawfly was prematurely terminated and there is no detailed published analysis of the overall results. In Europe, there are well-defined, periodic outbreaks that reach moderate densities in well-established stands. However, Lyons et al. (1971) and Lyons (1977) suggest that the pattern was different in North America. The pine plantations in southern Ontario, where the research took place, were more or less isolated and the sawfly outbreaks were unsynchronized independent events. For many years following its introduction, outbreaks of the European pine sawfly affected primarily young plantings as the sawfly dispersed eastward across the province. There was generally an explosive population increase, to densities much higher than in Europe, followed by a decline after 4 to 6 years, with no subsequent outbreaks as the stands matured. This trend repeats the well-recognized pattern of an invading exotic species, whereby the new host resources are exploited before natural enemies become established. Starvation of late-stage larvae, parasitism of prespinning larvae by *Exenterus* spp., and mortality of larvae in the cocoon caused by predators and parasitoids were the principal factors bringing the outbreaks under control. If a newly established plantation was adjacent to, or close enough to, an older plantation that had experienced an outbreak of the European pine sawfly earlier, then the transfer of natural enemies (parasitoids and disease) to the young plantation prevented an outbreak in the new plantation. Viktorov and Borodin (1975) have commented similarly on dispersal patterns in Europe. Adult dispersal was believed by Lyons to be important in determining population trends. In a stand of larger trees the sawfly outbreak was more persistent and irregular than in the young trees and was more damaging. The erratic population behavior in this stand was partly attributable to prolonged (intense) eonymphal diapause which was determined by stand microclimate and its year-to-year variability. The Ontario work ended before it was possible to study the course of invasion and population trends in older natural stands and the effect of a large guild of parasitoids present in indigenous *Neodiprion* species. Nor is there a good understanding of the effect of introduced parasitoids on population regulation.

Hanski (1987) examined the population dynamics of the 11 conifer sawfly species in northern Europe, including the European pine sawfly, *N. sertifer*, and the common pine sawfly, *Diprion pini*, which are the major pest species in the region. Major differences in the occurrence and spatial extent of outbreaks among the 11 species are noted and Hanski concluded that plant defense mechanisms and cocoon predation by generalist predators (shrews and omnivorous small mammals) explained more observations than disease and parasitoid hypotheses. Hanski (1990) further advanced the proposition that European pine sawfly population dynamics are highly influenced by small mammal predation. An interesting association between gregarious feeding behavior and high outbreak frequency was noted. This may be related to differences in the effectiveness of defense mechanisms against parasites and predators in solitary and group-feeding species. Hanski considered that the gregarious species have risk-prone reproductive behavior that increases the variance in density fluctuations and that chance events lead to the start of local population increases and subsequent escape from the regulating factors. No comparable multiple species analysis has been done for North American diprionids where there is a wide range of gregarious behavior, but general experience is that the solitary (in the sense of dispersed egg-laying behavior) species are rare to uncommon while outbreaks occur in the group-feeding species.

The role of host-tree physiological state in the population dynamics of diprionids has been investigated more in Europe than for North American species. Schwenke (1964) concluded that the nutrient (sugar) content of pine needles depends on soil characteristics, especially moisture, and that the sugar-rich needles of pines growing on dry soils promoted the development of outbreaks of the common pine sawfly by promoting faster development, decreased mortality, increased proportion of females, and increased body weight (and therefore fecundity) of the insect. Schwenke also considered that diapause state is the second major determination of population change and that nutrition, through its effect on developmental rate and consequently on daylength exposure, had a reinforcing effect on outbreak development.

The broader questions of population regulation in conifer sawflies have been addressed by McNamee et al. (1981) in the context of the structure and behavior of defoliating insect/forest systems. They recognized that complex process-based simulation models such as those developed for the Swaine jack pine sawfly (McLeod 1979) and the European pine sawfly (D.R. Wallace and K.J. Griffiths, unpubl. data) may provide means of understanding population behavior, but are system-specific and require extensive research for their construction. The authors examined the equilibrium structure of such models to determine a generalized minimum information set required to understand how a defoliator system is regulated. Knowing this, a skeletal model can be used to define the research priorities for a particular system. They suggested that the equilibrium structure of defoliators is determined by five processes: (a) predation by ground-searching predators; (b) predation by crown-searching predators; (c) stochastic conditions such as weather; (d) parasitism or disease infection; and (e) defoliation induced tree mortality. The minimum information required to predict the presence, absence, or form of these processes is the effectiveness of defoliator predator defenses, location of pupation, effects of stochastic conditions such as weather on the defoliator, parasitism synchrony and host specificity, or disease pathogenicity, and probability of tree mortality under a wide range of defoliation.

Natural Control by Parasitoids and Predators and Their Use in Applied Biological Control

The discovery of the European spruce sawfly outbreak in 1930, which was mentioned earlier as stimulating research on the diprionid sawflies in general, also brought about a large biological control program (classical biological control) using parasitoids introduced from Europe. This in turn resulted in a major emphasis in insect pathology in Canada and the specific research on sawfly viruses that is described in the following section. The biological control work was not limited to the European spruce sawfly but included several other diprionid species, especially the European pine sawfly. This work on classical biological control with parasitoids and predators is reported in detail in three volumes published by the Commonwealth Institute of Biological Control (now the CAB International Institute of Biological Control) (McGugan and Coppel 1962; Corbet and Prentice 1971; Kelleher and Hulme 1984) on a case-by-case basis as well as with overall analyses. The reader is referred to these works and to additional comments included in the section on classical biological controls (Chapter 41).

Biological Control Using Viruses

Viruses have been found worldwide in 21 species of Diprionidae (Martignoni and Iwai 1981) of which eight are found in Canada; eight species are listed in Table 2, Chapter 35, and the ninth is from the introduced pine sawfly. Confusion in the taxonomy of the Diprionidae casts doubt on a few host records (Cunningham and Entwistle 1981). These viruses are all nuclear polyhedrosis viruses (NPV) which have many virus particles embedded singly in inclusion bodies (Fig. 15). Unlike NPVs infecting lepidopterous species, viruses infecting sawflies replicate only in midgut epithelial cells, inclusion bodies range in size from 0.5 to 1.5 µm and are generally smaller than inclusion bodies found in Lepidoptera. Sawfly viruses are highly host-specific and it is necessary to develop a different virus for each pest species.

There are no artificial diets for continuous rearing of sawflies and host plant material is required not only for food, but also for oviposition by the females. A cell culture has been developed for redheaded pine sawfly, but it will not sustain viral replication (S.S. Sohi, pers. comm.). Production of sawfly viruses has usually been in the field. Heavily infested plantations are sprayed with a mistblower when larvae are predominantly fourth instar. Dead colonies (Fig. 14) are harvested daily or as often as possible, beginning about a week post-spray. An alternative method has been to transport colonies to a convenient site and infest unwanted trees, such as those found in an abandoned Christmas tree farm. When this method is used, colonies are sprayed directly with a suspension of virus and the length of time to death is more uniform than for virus applied with a mistblower (Cunningham and McPhee 1986).

European Spruce Sawfly

The accidental introduction into Canada of virus affecting the European spruce sawfly, *G. hercyniae*, along with the parasites imported in the late 1930s, has been well documented (Bird and Elgee 1957; Neilson and Morris 1964; Neilson et al. 1971). The last release of this virus in a healthy insect population was in Ontario in 1950 and levels of disease and parasitism were recorded annually until 1959 (Bird and Burk 1961). For the last 30 years, the European spruce sawfly has remained at endemic levels in eastern North America, held in check by virus and parasites. This is an outstanding example of classical biological control. Two instances of local population resurgence were recorded following the application of DDT, but populations returned to endemic levels without human intervention within a few generations following the end of chemical applications (Neilson and Elgee 1965; Neilson et al. 1971; Magasi and Syme 1984). Parasites are effective even at very low host densities and are sensitive to minor population fluctuations, whereas the virus responds to density changes of greater magnitude or to minor increases of prolonged duration. Very effective biological control has been maintained in Canada for about 50 years. Similar successful control was achieved in an outbreak in Wales from 1968 to 1972, although the role of insect parasitoids is not considered as important by British researchers, and birds were a principal agent in virus dispersal (Adams and Entwistle 1981; Entwistle et al. 1983).

Should the situation ever change and European spruce sawfly populations build up to outbreak levels, planned releases of nuclear polyhedrosis virus could be made. European spruce sawfly larvae are solitary, so mass collections for virus production would take longer than collections of gregarious species. However, laboratory culture is simple because females are parthenogenetic and diapause can be avoided by rearing in an 18:6 hour light:dark photoperiod.

European Pine Sawfly

A nuclear polyhedrosis virus was found in dead European pine sawfly, *N. sertifer*, imported from Sweden in 1949 and its development as a biocontrol agent has been described by Wallace et al. 1975. The virus was extensively used in Christmas tree plantations in southern Ontario in the 1950s and 1960s, but no records were kept of areas treated. Since 1973, there have only been two aerial applications of virus (Table 2). The first was in 1975 when 125 ha in Sandbanks Provincial Park were treated with 5.1×10^{10} PIB/ha (polyhedral inclusion bodies per hectare) (Cunningham et al. 1975), and the second was in 1981 when a 19-ha Christmas tree plantation was treated with 3.9×10^{11} PIB/ha. Both these application rates are considerably higher than the currently recommended rate of 5×10^{9} PIB/ha and both operations were completely successful in eradicating European pine sawfly from the treated areas. Each year since 1983, staff of the Newfoundland Forestry Centre have been treating ornamental trees in St. John's, Newfoundland, with virus supplied by the Forest Pest Management Institute.

On a global basis, the European pine sawfly virus is the most widely tested and operationally used insect virus. About 20 000 ha have been treated during the last 30 years with the virus used in 12 countries. Two commercial products are available, Minosarmio-virus from Kemira Oy in Finland and Virox from Oxford Virology in England. Also, a viral insecticide called Virin Diprion is used operationally in the Soviet Union. Toxicological tests of the virus were funded by the U.S. Department of Agriculture Forest Service and a viral insecticide called Neochek-S was registered in 1983 for European pine sawfly control in the United States. A copy of the American registration petition was made available to the Forest Pest Management Institute. A Canadian registration petition, based on the American toxicological studies, was prepared for a product called Sertifervirus and was submitted to Agriculture Canada for evaluation in 1985.

Table 2. Aerial application of European pine sawfly, *Neodiprion sertifer*, nuclear polyhedrosis virus in Ontario, 1973–1991.

Year	Dosage (PIB/ha)[a]	Tank mix	Area treated (ha)	Predominant instar at time of application
1975	5.1×10^{10}	Water 30 g/L IMC 90-001® 0.01% /Chevron® sticker	125	I
1981	3.9×10^{11}	Water 0.04% Rhodamine B dye	19	II, III

[a] Polyhedral inclusion bodies per hectare.

European pine sawfly is currently only a minor pest in Canada, but local outbreaks are reported occasionally. The Canadian product will be available for control of this pest if a registration can be obtained under the Pest Control Products Act. It is recommended that first- and second-instar larvae be treated if a high degree of foliage protection is required.

Redheaded Pine Sawfly

The redheaded pine sawfly, *N. lecontei*, is one of the most damaging insects that attacks young hard pine plantations in Ontario, Quebec, and New Brunswick. In the last 20 years, a great deal of abandoned farm land has been planted with red pine, *Pinus resinosa*, which is one of its principal hosts, although plantations of jack pine, *P. banksiana*, and Scots pine, *P. sylvestris*, can also be attacked. Young trees are at risk until they have grown to the height at which crown closure of the stand occurs.

In 1950, a nuclear polyhedrosis virus affecting the redheaded pine sawfly was found in Ontario by Bird (1961). Ground spray trials were conducted in Ontario (Bird 1971) and a ground spray operation was conducted in Quebec in 1970 (Can. Dep. Fish and Can. For. Serv. 1970; Prebble 1975b). An intensive research project was launched in 1975 at the Insect Pathology Research Institute of Forestry Canada and this viral insecticide was called Lecontvirus. Between 1976 and 1980, 20 plantations with a combined area of 258 ha were aerially sprayed with Lecontvirus (Table 3). Dosages ranged from 1.3×10^9 to 5.5×10^9 PIB/ha, application volumes from 2.4 to 9.4 L/ha, a variety of formulations were tested, and both boom and nozzle and Micronair® spray equipment were used. Larvae were mainly in the second instar at the time of application, but ranged from first to fourth instar. All applications gave good control of redheaded pine sawfly. The lower limits of effective dosage and application volume were not established because all these trials were conducted on private land and the landowners expected effective insect control; hence, testing of extremely low dosages and volumes was not attempted (Cunningham and de Groot 1984; Cunningham et al. 1986).

Lecontvirus was supplied to the Quebec Ministry of Energy and Resources staff who treated a total of 92 plantations with a combined area of 1 051 ha in the Shawville area of Quebec between 1978 and 1980 (Table 4). Most of the virus was applied from the air and the program was an unqualified success (Cunningham and de Groot 1984). Since 1980, Lecontvirus has been supplied to Ontario Ministry of Natural Resources staff; a few applications have also been made by Forestry Canada's Forest Pest Management Institute staff. All these treatments used ground spray equipment and between 1980 and 1990, a total of 478 plantations with a combined area of 3 546 ha were treated (Table 5).

Toxicological tests for Lecontvirus were contracted from the Ontario Veterinary College, Guelph, in the 1970s along with fish and bird studies and a registration petition for this viral insecticide was submitted. It received a temporary registration in 1983 and, after submission of additional data, full registration in 1987. Production costs of Lecontvirus depend on the location of the plantation being used as a virus propagation site, its distance from Sault Ste. Marie and travel costs involved. About 50 virus-infected larvae are required to produce a 1-ha dosage of 5×10^9 PIBs and the estimated cost, based on several years of production, is $2.50/ha. Provided Lecontvirus is applied before

Table 3. Experimental aerial applications of Lecontvirus, a nuclear polyhedrosis virus, by staff of the Forest Pest Management Institute to control infestations of redheaded pine sawfly, *Neodiprion lecontei*, in Ontario.

Year	No. of plantations treated	Total area treated (ha)
1976	3	43
1977	3	52
1978	2	26
1979	4	34
1980	8	103

Table 4. Application of Lecontvirus, a registered nuclear polyhedrosis viral product, by staff of Quebec Ministry of Energy and Resources to control redheaded pine sawfly, *Neodiprion lecontei*, in red pine plantations near Shawville, Quebec.

Year	No. of plantations aerially sprayed	No. of plantations ground sprayed	Total area treated (ha)
1978	36	1	700
1979	42	1	330
1980	–	12	21

Table 5. Ground spray applications of Lecontvirus, a registered nuclear polyhedrosis viral product, by Ontario Ministry of Natural Resources and Forest Pest Management Institute staff to control redheaded pine sawfly, *Neodiprion lecontei*, infestations in Ontario.

Year	No. of districts involved	No. of plantations treated	Total area treated (ha)
1980	5	88	437
1981	5	65	759
1982	8	49	374
1983	7	40	211
1984	4	9	11
1985	7	30	161
1986	7	83	342
1988	7	26	201
1989	4	39	373
1990	6	49	677

most of the larvae reach their fourth instar, successful control is assured and further treatments will not be required in subsequent years (Cunningham et al. 1984).

Swaine Jack Pine Sawfly

A nuclear polyhedrosis virus was found in Swaine jack pine sawfly, *N. swainei*, larvae in northeastern Ontario in 1953 and in Quebec in 1956 (Smirnoff 1962). Following laboratory tests, the virulence of this virus was increased by passage and selection of virus from larvae with the heaviest infections and those that died first. The first aerial spray trial was conducted in 1960. An aqueous and an oil-based formulation were compared at 4.7 and 37.6 L/ha giving dosages of 9.4×10^9 and 7.5×10^{10} PIB/ha, respectively. A fixed-wing aircraft was used to apply the formulations on second-instar larvae. All applications gave excellent results, first mortality was noted 7 days post-spray and cumulative mortality reached 97% (Smirnoff et al. 1962). In 1964, a larger aerial operation involved treating 1 600 ha. Dosages ranged from 1.7×10^{10} PIB/ha to 3.5×10^{10} PIB/ha and the emitted volume was 18.8 L/ha. Unseasonably cold and wet weather after the application reduced insect feeding activity and 25% to 50% of larvae were collected and brought to the laboratory spun cocoons. Surveys were conducted in 1965 and 1966 but transovum transmission of the virus was less than expected; the population eventually collapsed as a result of virus infection and starvation in 1966 (Smirnoff and McLeod 1975).

The initiation of epizootics using spot introductions with low dosages of this virus either aerially or on the ground has been suggested (Finnegan and Smirnoff 1984). The dissemination of infected cocoons has also been suggested. A flagellate, *Herpetomonas swainei*, which co-infects larvae along with the virus, can be used as a marker to distinguish introduced infected adults from naturally occurring virus-infected insects already present (Smirnoff 1972). Although good in theory, these dispersal methods have yet to be proved viable on an operational basis.

No field trials have been conducted since 1964. Swaine jack pine sawfly outbreaks have occurred at irregular intervals and can cause considerable damage and mortality of mature stands of jack pine in Ontario and Quebec. Research on the virus of this species should be reinstated and an evaluation made of the merits of developing it as a biocontrol agent.

Red Pine Sawfly

The red pine sawfly, *N. nanulus nanulus*, is only a minor pest in Canada. Survey records in Sault Ste. Marie document the isolation of a nuclear polyhedrosis virus from this species in 1953. It appears that some laboratory experimentation and small field trials were undertaken with this virus at the Maritimes Forestry Centre of Forestry Canada between 1957 and 1976. Naturally occurring virus was also found in Nova Scotia. The field trials, undertaken by Forest Insect and Disease Survey staff, gave good results, but no publications are available (D.G. Embree, pers. comm.).

In 1986, a small-scale trial was undertaken near Sault Ste. Marie. Individual colonies were collected near Temagami and twigs with colonies were tied on red pine trees. Second- and third-instar larvae were treated with a range of six dosages with 10 mL of each dosage sprayed on three replicated 35-cm shoots with one sawfly colony per shoot. Larvae were examined every second day and first larval mortality and complete mortality of each colony recorded (Table 6). The highest dosage of 10^8 PIB/colony gave complete mortality in 10 days and the lowest dosage of 10^3 PIB/colony gave complete mortality in 22 days. Untreated check colonies pupated in 23.7 ± 0.67 days. These results, combined with the undocumented reports from the Maritimes Forestry Centre, demonstrate that red pine sawfly virus is a potentially useful biocontrol agent that can be developed if the status of this insect changes and it becomes a more serious pest.

Table 6. Effect of nuclear polyhedrosis virus applied on individual colonies of second- and third-instar red pine sawfly, *Neodiprion nanulus nanulus*, in 1986.[a]

PIB/ colony[b]	No. of days until first larval mortality	No. of days until total larval mortality
10^8	8.0	10.0
10^7	10.0 ± 1.20	10.7 ± 0.67
10^6	11.3 ± 0.67	12.0
10^5	12.0	12.7 ± 0.67
10^4	13.3 ± 1.33	16.7 ± 0.67
10^3	15.3 ± 0.67	22.0

[a] Untreated check colonies pupated in 23.7 ± 0.67 days.
[b] Polyhedral inclusion bodies.

Balsam Fir Sawfly

The balsam fir sawfly, *N. abietis* complex, is part of a large, confusing and little studied group of closely related species. It occurs in North America from Mexico to Alaska in the West, across the center of the continent, and down the east coast. Many host species are involved with temporal differences in life histories (Knerer and Atwood 1973) and several distinct species may be involved. In Canada, balsam fir sawfly feeds on the old needles of balsam fir and spruce. Outbreaks are usually of short duration and are probably terminated by naturally occurring nuclear polyhedrosis virus and parasites. The most serious damage is usually reported in conjunction with other pest species such as spruce budworm, *Choristoneura fumiferana*, eastern blackheaded budworm, *Acleris variana*, or balsam woolly adelgid, *Adelges piceae*.

Balsam fir sawfly virus was first recorded by Steinhaus (1949). Samples of virus were collected from the Prairie provinces, Ontario, and Newfoundland during the 1950s and 1960s and were stored at Sault Ste. Marie. In 1973, single-tree experiments were conducted in Ontario. A backpack mistblower was used to treat 2.5-m-tall balsam fir trees at two application times, at late first larval instar and 8 days later at early third larval instar. Concentrations of 10^4, 10^5 and 10^6 PIB/mL applied at 360 mL/tree on first-instar larvae gave 72, 96, and 100% mortality and 10^5, 10^6, and 10^7 PIB/mL on third-instar larvae gave 60, 88, and 100% mortality, respectively (Olofsson 1973).

No further work has been undertaken with this virus since 1973. It appears to be effective for control of balsam fir sawfly if applied on first-instar larvae, but if other pest species are also involved, balsam fir sawfly virus would not be the choice of control agent.

Jack Pine Sawfly

There are problems with the taxonomy of jack pine sawfly, and the *N. pratti* complex may consist of three subspecies or geographical races, although Knerer (1984) contends that these divisions are meaningless. *Neodiprion pratti banksianae* is found in central Canada and the United States and west through Manitoba and Saskatchewan; *N. pratti paradoxicus* in eastern Canada, north central United States, and along the east coast; and *N. pratti pratti* in the Virginias, Carolinas, and Kentucky. Bird (1955) reported a virus in populations of *N. pratti banksianae* near Sault Ste. Marie. In 1953 and 1954, small field trials were conducted and Bird was pessimistic about the value of this virus as a biocontrol agent. However, a subsequent trial using suspensions of 10^3, 10^4, 3×10^4, 3×10^5, and 3×10^6 PIB/mL on first- and second-instar larvae caused 9.7, 55.9, 83.8, 94.6, and 99.1% mortality, respectively (Bird 1961).

Jack pine sawfly is a minor pest in Canada, although it is sometimes found in plantations or natural stands along with other species such as the European pine sawfly and the red pine sawfly. No studies have been undertaken with this virus since the 1950s and reevaluation of its potential as a biocontrol agent would be a worthwhile undertaking.

Brownheaded and Redheaded Jack Pine Sawflies

Brownheaded and redheaded jack pine sawflies, formerly called *N. virginiana* complex, belong to at least two species, *N. dubiosus* and *N. rugifrons*. Laboratory tests and a small-scale field trial were conducted in Ontario by Bird with a nuclear polyhedrosis virus from *N. virginiana* complex larvae in 1966 (Cunningham and Entwistle 1981). Compared with other sawfly viruses, it had low virulence in the laboratory and concentrations of 10^4 to 10^7 PIB/mL on second- and third-instar larvae gave no mortality in the field. It is not known whether the original isolation of the virus was from *N. dubiosus* or *N. rugifrons* and which species was used in the laboratory and field tests. Neither species is a major pest and no further experiments have been conducted with this virus.

Conclusions and Recommendations

Viral insecticides are available from the Forest Pest Management Institute for control of two species of diprionid sawflies in Canada. Lecontvirus for control of redheaded pine sawfly, *N. lecontei*, has been registered since 1983 and the registration package for Sertifervirus, for control of European pine sawfly, *N. sertifer*, was submitted in 1985 and is still under evaluation. Both viruses are highly effective and economical pest control agents. When applied on small, early-instar larvae there is ample time for the virus to spread from diseased to healthy colonies and no further treatment is required. If one larva in a colony becomes infected, the rest of that colony will be infected in a few days. The principal mechanics of virus transmission between colonies are rain and scavenging and predaceous insects, which are often associated with colonies of these sawflies. In Britain, insectivorous birds are considered to be important in the dispersal of the European spruce sawfly virus (Entwistle et al. 1983).

Lecontvirus and Sertifervirus are preparations of lyophilized, virus-infected larvae ground to a fine powder and formulated either as a wettable powder or as a liquid suspension. An emulsifiable oil formulation has been used for distribution to clients, but it has a poor shelf life and it cannot be stored from one season to the next. A liquid formulation is preferable to a wettable powder because it is easier to measure volumes than weights under field conditions. Research is being conducted to develop improved formulations of Lecontvirus and Sertifervirus.

Several other sawfly viruses are possible candidates for development and a cost-benefit analysis is necessary to evaluate whether the expense of toxicological tests and environmental toxicology studies is merited. The current cost of such tests is probably about $250 000. The first candidate on this list is the virus of Swaine jack pine sawfly which is of considerable economic importance. The efficacy of this virus is doubtful when it is compared with Lecontvirus and Sertifervirus and more field trials should be conducted to determine the dosage that gives reproducible and acceptable results. Other possible candidates for development are the viruses of the red pine sawfly, jack pine sawfly, and balsam fir sawfly. Currently, these are only minor pests, but their status can change and they may become serious pests. Likewise, if the pest status of the European spruce sawfly changes from endemic to outbreak, introductions of its virus can be made.

Early literature contains several references to transovum transmission of sawfly viruses (Bird 1955, 1961; Smirnoff 1962). Some doubt was cast on this mode of vertical transmission by Neilson and Elgee (1968) who demonstrated that virus-infected European spruce sawfly adults could contaminate foliage and that contamination of the egg surface is unimportant because sawfly eggs hatch by bursting and larvae do not eat the chorion. These early reports have given some pest managers the impression that sawfly viruses can be applied on late-instar larvae. Although little or no impact is observed in the year of application, it is hoped that a viral epizootic will control the sawfly population the following year. This is not recommended as a pest management strategy and current advice is to wait until the following year and treat small larvae rather than making a late virus application.

Challenges for Research

The family Diprionidae is a small group of organisms as far as insect families are concerned, with many of the included species being well known; yet the group shows a wide range of diversity. These two characteristics make the family particularly suited for additional research into evolutionary mechanisms. In Canada and North America, one of the first needs is a better understanding of the fauna of the western regions of the continent where intensive collecting is required on the wide range of conifer hosts that are found from Alaska to Central America. Other areas of the world where basic exploration is required are across Siberia, in the People's Republic of China, and in Indo-China, although during the last 10 to 15 years there has been progress in the description of new species from the latter two areas. In eastern North America there is still need for study of several species complexes to obtain a better understanding of intraspecific variation.

In many types of research there is a need for comparative studies, for example, to find simple means of identification. The application of pheromone population monitoring techniques currently is hampered by the lack of identification aids for male adults.

Nutrition studies would be facilitated by the development of an artificial food, although propagation of conifer sawflies is unlikely to be divorced entirely from the natural hosts on account of the egg-laying requirements. Even the partial freedom from rearing the sawflies on natural hosts would have practical advantages for virus multiplication, given the general lack of cell cultures from conifer sawflies and the difficulty of virus propagation in cell culture.

Although a great deal is known about the regulation of seasonality in some species, such as the European pine sawfly and the common pine sawfly, more comparative studies of diapause and development are required. The integration of the findings in at least comprehensive conceptual models would assist in the identification of research needs.

Some sawfly nuclear polyhedrosis viruses are among the most virulent viruses known to infect insects. Strains of *Bacillus thuringiensis* that are currently available are ineffective against sawflies, leaving viruses as the only practical microbial agent. Sawfly viruses are highly host-specific and it is necessary to develop a different virus for each pest species. This deters commercial development because the market for each virus is small. Using the latest techniques in genetic engineering, attempts are being made to determine which viral genes influence specificity and host range. If this can be manipulated, it is conceivable that one virus may be developed that can be used to control several species of diprionid sawflies.

References

Adams, P.H.W.; Entwistle, P.F. 1981. An annotated bibliography of *Gilpinia hercyniae* (Hartig), European spruce sawfly. Occasional Paper No. 11, Commonw. For. Inst., Oxford University. 58 p.

All, J.N. 1973. Mechanism of preferential feeding by *Neodiprion swainei* Middleton and *Neodiprion rugifrons* Middleton on *Pinus banksiana* Lambert. Ph.D. Thesis, Univ. of Wisconsin. 257 p.

Allee, W.C. 1931. Animal aggregations. Univ. Chicago Press, Chicago, Ill. 431 p.

Andrewartha, H.G. 1952. Diapause in relation to the ecology of insects. Biol. Rev. 27: 50-107.

Arora, G.L. 1953. The external morphology of *Diprion pini* (L.) (Symphyta-Hymenoptera). East Punjab Univ. Res. Bull. 25: 1-21.

Atwood, C.E. 1961. Present status of the Sawfly Family Diprionidae (Hymenoptera) in Ontario. Proc. Entomol. Soc. Ont. (1960) 91: 205-215.

Atwood, C.E.; Peck, O. 1943. Some native sawflies of the genus *Neodiprion* attacking pines in eastern Canada. Can. J. Res. D 21: 109-144.

Austarå, O. 1966. Laboratory and field investigations of the length of the cocoon stage of *Neodiprion sertifer* (Geoffr.) (Hymenoptera: Diprionidae) in Norway. Medd. f. Det Norske Skogforsφksvesen 21(78), 15 p.

Austarå, O. 1969. Prolonged diapause in *Neodiprion sertifer* (Geoffr.) (Hymenoptera: Diprionidae) from Gaupne in western Norway. Medd. f. Det Norse Skogforsφksvesen 27: 204-213.

Austarå, O. 1971. Cold-hardiness in eggs of *Neodiprion sertifer* (Geoffroy) (Hym., Diprionidae) under natural conditions. Norks. ent. Tidsskr. 18: 45-48.

Baker, W.V. 1970. Effect on photoperiod, wavelength and temperature on larvae of *Diprion similis* Htg. (Hymenoptera: Diprionidae) as expressed by intra-cocoon development and adult emergence. Ph.D. Thesis, Univ. of Toronto.

Baker, W.V.; Atwood, C.E. 1969. The abnormal development of the meso- and meta-thoracic legs of *Diprion similis* (Hymenoptera: Diprionidae) when reared under certain photoperiods. Can. Entomol. 101: 990-994.

Baldwin, W.F.; House, H.L. 1954. Studies on effects of thermal conditioning in two species of sawfly larvae. Can. J. Zool. 32: 9-15.

Barlow, J.S. 1964. Fatty acids in some insect and spider fats. Can. J. Biochem. 42: 1365-1374.

Becker, G.C.; Benjamin, D.M. 1967. The biology of *Neodiprion nigroscutum* (Hymenoptera: Diprionidae) in Wisconsin. Can. Entomol. 99: 146-159.

Benjamin, D.M. 1955. The biology and ecology of the red-headed pine sawfly. U.S. Dept. Agr., For. Serv., Tech. Bull. 1118, 57 p.

Benson, R.B. 1939. On the genera of the Diprionidae (Hymenoptera Symphyta). Bull. Entomol. Res. 30:339-342.

Benson, R.B. 1945. Further note on the classification of the Diprionidae (Hymenoptera Symphyta). Bull. Entomol. Res. 36: 163-164.

Benson, R.B. 1950. An introduction to the natural history of British sawflies. Trans. Soc. for Br. Entomol. 10(2): 45-142.

Bird, F.T. 1955. Virus diseases of sawflies. Can. Entomol. 85: 437-446.

Bird, F.T. 1961. Transmission of some insect viruses with particular reference to ovarial transmission and its importance in the development of epizootics. J. Insect Pathol. 3: 352-380.

Bird, F.T. 1971. *Neodiprion lecontei* (Fitch), red-headed pine sawfly (Hymenoptera: Diprionidae). Pages 148-150 *in* Biological Control Programmes against Insect and Weeds in Canada 1959-1968. Commonw. Inst. Biol. Control. Tech. Comm. No. 4. Commonw. Agric. Bureaux, Slough, England.

Bird, F.T.; Burk, J.M. 1961. Artificially disseminated virus as a factor controlling the European spruce sawfly, *Diprion hercyniae* (Htg.), in the absence of introduced parasites. Can. Entomol. 93: 228-238.

Bird, F.T.; Elgee, D.E. 1957. A virus disease and introduced parasites as factors controlling the European spruce sawfly, *Diprion hercyniae* (Htg.) in central New Brunswick. Can. Entomol. 89: 371-378.

Bird, R.D. 1929. Notes on the fir sawfly, *Neodiprion abietis* Harr. Ann. Rep. Entomol. Soc. Ont. 60: 76-82.

Bombosh, S. 1972. Zur Nahrungsqualität von Fichtennadeln für forstliche Schadinsekten. 1. Zur Problematik der Beeinflussung von Insektenpopulationen durch Veränderung der Nahrungsqualität. Z. ang. Ent. 70: 277-281.

Borodin, A.L. 1972. Method of calculating the numbers of insect in tree crowns using ovipositions of fox-colored sawfly (*Neodiprion sertifer*) as an example. (Original in Russian). Zoologicheskii Zhurnal 51: 738-747. Translation No. 285, Environ. Can., 1973. 27 p.

Breny, R. 1957. Contribution à l'étude de la diapause chez *Neodiprion sertifer* Geoffr. dans la nature. Académie royale de Belgique, Classe des Sciences. Memoires 30(3), 86 p. (No. 1679).

Brown, A.W.A. 1940. A European pine sawfly (*Neodiprion sertifer* Geoff.). Page 22 *in* Annual Report of the Forest Insect Survey 1939. Can. Dept. of Agric., Sci. Serv., Div. Entomol., Ottawa, Ont.

Brown, A.W.A.; Daviault, L. 1942. A comparative study of the influence of temperature on the development of certain sawflies after hibernation in the cocoon. Sci. Agr. 22: 298-306.

Brygider, W. 1952. In what embryonic stage do the eggs of *Neodiprion* enter the winter stage? Can. J. Zool. 30: 99-108.

Buratti, L.; Allais, J.P.; Barbier, M. 1988. The role of resin acids in the relationship between Scots pine and the sawfly *Diprion pini* (Hymenoptera: Diprionidae). I. Resin acids in the needles. Pages 171-187 *in* W.J. Mattson, J. Levieux and C. Bernard-Dagan (Eds.), Springer-Verlag, New York, Berlin, Heidelberg, London, Paris, Tokyo.

Can. Dep. Fish and Can. For. Serv. 1970. Virus trials to control redheaded pine sawfly in Quebec plantations. Inf. Rep. DPC-X-1. Ottawa, Ont. 30 p.

Carne, P.B. 1962. The characteristics and behaviour of the saw-fly *Perga affinis affinis* (Hymenoptera). Australian J. Zool. 10: 1-34.

Coppel, H.C.; Benjamin, D.M. 1965. Bionomics of the Nearctic pine-feeding diprionids. Ann. Rev. Entomol. 10: 69-96.

Coppel, H.C.; Casida, J.E.; Dauterman, W.C. 1960. Evidence for a potent sex attractant in the introduced pine sawfly, *Diprion similis* (Hymenoptera: Diprionidae). Ann. Entomol. Soc. Amer. 53: 510-512.

Coppel, H.C.; Mertins, J.W.; Harris, J.W.E. 1974. The introduced pine sawfly, *Diprion similus* (Hartig) (Hymenoptera: Diprionidae). A review with emphasis on studies in Wisconsin. Univ. of Wisconsin-Madison, College of Agriculture and Life Sciences, Research Division, Res. Bull. R2393, 74 p. + 18 plates.

Corbet, P.S.; Prentice, R.M. (Compilers). 1971. Biological control programmes against insects and weeds in Canada 1959-1968. Tech. Comm. No. 4, Commonwealth Institute of Biological Control, Trinidad. Commonw. Agric. Bureaux, Slough, England. 286 p.

Craighead, F.C. 1940. Some effects of artificial defoliation on pine and larch. J. Forestry 38: 885-888.

Cunningham, J.C.; de Groot, P. 1984. *Neodiprion lecontei* (Fitch), redheaded pine sawfly (Hymenoptera: Diprionidae). Pages 323-329 *in* J.S. Kelleher and M.A. Hulme. Biological control programmes against insects and weeds in Canada 1969-1980. 410 p. Commonwealth Agricultural Bureaux, Farnham Royal, Slough, England.

Cunningham, J.C.; de Groot, P.; Kaupp, W.J. 1986. A review of aerial spray trials with Lecontvirus for control of redheaded pine sawfly, *Neodiprion lecontei* (Hymenoptera: Diprionidae), in Ontario. Proc. Entomol. Soc. Ont. 117: 65-72.

Cunningham, J.C.; de Groot, P.; McPhee, J.R. 1984. Lecontvirus: a viral insecticide for control of redheaded pine sawfly, *Neodiprion lecontei*. Can. For. Serv., For. Pest Man. Inst., Sault Ste. Marie, Ont. Tech. Note No. 2. 5 p.

Cunningham, J.C.; Entwistle, P.F. 1981. Control of sawflies by baculovirus. Pages 379-407 *in* H.D. Burgess (Ed.), Microbial control of pests and plant diseases 1970-1980. Academic Press, London, England.

Cunningham, J.C.; Kaupp, W.J.; McPhee, J.R.; Sippell, W.L.; Barnes, C.A. 1975. Aerial application of a nuclear polyhedrosis virus to control European pine sawfly. Can. For. Serv. Ottawa, Ont. Bi-mon. Res. Notes 31: 39-40.

Cunningham, J.C.; McPhee, J.R. 1986. Production of sawfly viruses in plantations. Can. For. Serv., For. Pest Manage. Inst., Sault Ste. Marie, Ont. Tech. Note No. 4, 4 p.

Danilevski, A.S. 1965. Photoperiodism and seasonal development of insects. Oliver and Boyd, Edinburgh and London, 283 p.

Danks, H.V. 1983. Extreme individuals in natural populations. Bull. Entomol. Soc. Amer. 29(1): 41-46.

Danks, H.V. 1991. Life cycle pathways and the analysis of complex life cycles in insects. Can. Entomol. 123: 23-40.

Daviault, L. 1950. La tenthrède de Leconte (*Neodiprion lecontei* (Fitch)) dans la province de Québec. 32e Rapport de la Société de Québec pour la Protection des Plantes, 1-20.

de Gryse, J.J. 1947. Noxious forest insects and their control. The Canada Year Book 1947, Can. Dept. of Trade and Comm., Dominion Bureau of Statistics, Reprinted, 10 p.

Deneve, R.T. 1968. Biological and morphological studies of five species of pine defoliating sawflies in Florida. M.Sc. Thesis, Univ. of Florida. 55 p.

Eichhorn, O. 1976. Dauerzucht von *Diprion pini* L. (Hym.: Diprionidae) im Laboratorium unter Berücksichtigung der Fotoperiode. Anz. Schädlingskde., Pflanzenschutz, Umweltschutz 49: 38-41.

Eichhorn, O. 1977a. Autökologische Untersuchungen an Populationen der gemeinen Kiefern-Buschhornblattwespe *Diprion pini* (L.) (Hym.: Diprionidae). I. Herkunftsbedingte Unterschiedde im Schlüpfverlauf und Diapause. Z. Angew. Entomol. 82: 395-414.

Eichhorn, O. 1977b. Autökologische Untersuchungen an Populationen der gemeinen Kiefern-Buschhornblattwespe *Diprion pini* (L.) (Hym.: Diprionidae). III. Laborzuchten. Z. Angew. Entomol. 84: 264-282.

Eichhorn, O. 1979. Autökologische Untersuchungen an Populationen der gemeinen Kiefern-Buschhornblattwespe *Diprion pini* (L.) (Hym.: Diprionidae). IV. Generations- und Schlüpfwellenfolge. Z. Angew. Entomol. 88: 378-398.

Eichhorn, O. 1983. Dormanzverhalten der Gemeinen Kiefern-Buschhornblattwespe (*Diprion pini* L.) (Hym., Diprionidae) un ihrer Parasiten. Z. Angew. Entomol. 95: 482-498.

Eisner, T.; Johessee, J.S.; Carrel, J.; Hendry, L.B.; Meinwald, J. 1974. Defensive use by an insect of a plant resin. Science 184: 996-999.

Eklund, B. 1964. On the effect of the damage caused by the European pine sawfly as measured by the diameter growth of pine at breast height. (Original in Swedish). Norrlands Skogs. Forb. Tidskr. 205-208. Translation No. 65, Can. Dept. For., Ottawa, Ont. 1966, 14 p.

Eliescu, G. 1932. Beiträge zur Kenntnis der Morphologie, Anatomie und Biologie von *Lophyrus pini* L. Z. Angew. Entomol. 19: 22-26, 188-205.

Entwistle, P.F.; Adams, P.H.W.; Evans, H.F.; Rivers, C.F. 1983. Epizootiology of a nuclear polyhedrosis virus (Baculoviridae) in European spruce sawfly (*Gilpinia hercyniae*): spread of disease from small epicentres in comparison with spread of vaculovirus diseases in other hosts. J. Appl. Ecol. 20: 473-489.

Ericsson, A.; Helqvist, J.; Hillerdal-Hagstromer, K.; Larsson, S.; Mattson-Djos, E.; Tenow, O. 1980. Consumption and pine growth - Hypotheses on effects on growth processes by needle-eating insects. *In* T. Persson (Ed.), Structure and function of northern coniferous forests. An ecosystem study. Ecol. Bull. (Stockholm) 32: 537-545.

Evans, J.W. 1934. Notes on the behavior of larval communities of *P. dorsalis* (Leach). Trans. R. Entomol. Soc. Lond. 82: 455-460.

Finnegan, R.J.; Smirnoff, W.A.; 1984. Neodiprion swainei (Middleton), Swaine Jack Pine Sawfly (Hymenoptera: Diproinidae). Chapter 59, pages 341-348 *in* J.S. Kelleher and M.A. Hulme (Eds.), Biological control programmes against insects and weeds in Canada 1969-1980. Commonwealth Agricultural Bureaux, Farnham Royal, Slough, England. 410 p.

Fogal, W.H. 1974. Nutritive value of pine foliage for some diprionid sawflies. Proc. Entomol. Soc. Ont. 105: 101-118.

Fogal, W.H.; Kwain, M.-J. 1972a. Uric acid excretion by midgut epithelium in *Gilpinia hercyniae* (Htg.). Can. J. Zool. 50: 143-145.

Fogal, W.H.; Kwain, M.-J. 1972b. The effects of temperature on the occurrence of melanotic inclusions and related physiology in a sawfly, *Gilpinia hercyniae*. J. Insect Physiol. 18: 1545-1564.

Fogal, W.H.; Kwain, M.-J. 1972c. Host plant nutritive value and variable number of instars in a sawfly, *Diprion similis*. Israel J. Entomol. 7: 63-72.

Fogal, W.H.; Kwain, M.-J. 1979. Total protein and amino acid content of the gut of *Neodiprion sertifer* (Hymenoptera: Diprionidae) during metamorphosis. Proc. Entomol. Soc. Ont. 110: 104-106.

Fogal, W.H.; Slansky, F. Jr. 1985. Contribution of feeding by European pine sawfly larvae to litter production and element flux in Scots pine plantations. Can. J. For. Res. 15: 484-487.

Geri, C.; Buratti, L.; Allais, J.P. 1988. The role of resin acids in the relationship between Scots pine and the sawfly, *Diprion pini* (Hymenoptera: Diprionidae). II. Correlations with the biology of *Diprion pini*. Pages 189-201 *in* W.J. Mattson, J. Levieux, and C. Bernard-Dagan (Eds.), Springer-Verlag, New York, Berlin, Heidelberg, London, Paris, Tokyo.

Geri, C.; Goussard, F. 1988. Incidence de la photophase et de la température sur la diapause de *Diprion pini* L. (Hym., Diprionidae). J. Appl. Entomol. 106: 150-172.

Geri, C.; Goussard, F.; Allais, J.P.; Buratti, L. 1988. Incidence de l'alimentation sur le développement et la diapause de *Diprion pini* L. (Hym., Diprionidae). J. Appl. Entomol. 106: 451-464.

Geri, C.; Goussard, F.; Liger, A. 1984. Influence de la photopériode sur les arrêts de développement de *Diprion pini* L. (Hymenoptera, Diprionidae) C. R. Acad. Agric. Fr. 70: 245-251.

Ghent, A.W. 1955a. Oviposition behavior of the jack-pine sawfly, *Neodiprion americanus banksianae* Roh., as indicated by an analysis of egg clusters. Can. Entomol. 87: 229-238.

Ghent, A.W. 1955b. Light reactions of newly-hatched larvae of the jack-pine sawfly, *Neodiprion americanus banksianae* Roh. Can. Dept. Agric., Sci. Serv., For. Biol. Div., Ottawa, Ont. Bi-mon. Prog. Rep. 11(2): 2.

Ghent, A.W. 1958. Studies of the feeding orientation of the jack pine sawfly, *Neodiprion pratti banksianae* Roh. Can. J. Zool. 36: 175-183.

Ghent, A.W. 1959. Row-type oviposition in *Neodiprion* sawflies as exemplified by the European pine sawfly, *N. sertifer* (Geoff.). Can. J. Zool. 37: 267-281.

Ghent, A.W. 1960. A study of the group-feeding behavior of larvae of the jack pine sawfly, *Neodiprion pratti banksianae* Roh. Behavior 16: 110-148.

Ghent, A.W.; Wallace, D.R. 1958. Oviposition behavior of the Swaine jack-pine sawfly. For. Sci. 4: 264-272.

Gobeil, A.R. 1941. La diapause chez les tenthrèdes. Can. J. Res. (D) 19: 363-416.

Goulet, H. 1981. New external distinguishing characters for the sawflies *Gilpinia hercyniae* and *G. polytoma* (Hymenoptera: Diprionidae). Can. Entomol. 113: 769-771.

Gray, D.E. 1937. Notes on the occurrence of *Diprion frutetorum* (Fabr.) in southern Ontario. 68th Rep. Entomol. Soc. Ont., p. 50.

Green, G.W. 1954a. Some laboratory investigations of the light reactions of the larvae of *Neodiprion americanus banksianae* Roh. and N. lecontei (Fitch) (Hymenoptera: Diprionidae). Can. Entomol. 86: 207-222.

Green, G.W. 1954b. Humidity reactions and water balance of larvae of *Neodiprion americanus banksianae* Roh. and *L. lecontei* (Fitch) (Hymenoptera: Diprionidae). Can. Entomol. 86: 261-274.

Green, G.W. 1954c. The functions of the eyes and antennae in the orientation of adults of *Neodiprion lecontei* (Fitch). Can. Entomol. 86: 371-376.

Green, G.W.; deFreitas, A.S. 1955. Frass-drop studies of larvae of *Neodiprion americanus banksianae* Roh. and *N. lecontei* (Fitch) (Hymenoptera: Diprionidae). Can. Entomol. 87: 427-440.

Griffiths, K.J. 1953. Variations in morphological characters of some sawflies of the Family Diprionidae. M.A. Thesis, Univ. of Toronto. 47 p.

Griffiths, K.J. 1959. Observations on the European pine sawfly, *Neodiprion sertifer* (Geoff.), and its parasites in southern Ontario. Can. Entomol. 91: 501-512.

Griffiths, K.J. 1960. Oviposition of the red-headed pine sawfly, *Neodiprion lecontei* (Fitch). Can. Entomol. 92: 430-435.

Hallberg, E. 1979. The fine structure of the antennal sensilla of the pine saw fly *Neodiprion sertifer* (Insecta: Hymenoptera). Protoplasma 101: 111-126.

Hallberg, E. 1981. Johnston's organ in *Neodiprion sertifer* (Insecta: Hymenoptera). J. Morph. 167: 305-312.

Hallberg, E.; Lofqvist, J. 1981. Morphology and ultrastructure of an integral pheromone gland in the abdomen of the pine sawfly *Neodiprion sertifer* (Insecta: Hymenoptera): a potential source of sex pheromones. Can. J. Zool. 59: 47-53.

Hanski, I. 1987. Pine sawfly population dynamics: patterns, processes, problems. Oikos 50: 327-355.

Hanski, I. 1990. Small mammal predation and the population dynamics of *Neodiprion sertifer*. Pages 253-264 *in* A.D. Watt, S.R. Leather, M.D. Hunter, and N.A.C. Kidd. Population dynamics of forest insects. Intercept, Hampshire.

Hard, J.S. 1971. Effects of semistarvation during larval stages on survival and fecundity of the hemlock sawfly in the laboratory. USDA For. Serv., Pac. Northwest For. and Range Exp. Stn., Portland, Ore., Res. Note PNW-157, 8 p.

Hardy, Y.J. 1971. Relationship between the structure and location of Scots pine needles and the oviposition of *Neodiprion sertifer* (Geoff.) (Hymenoptera: Diprionidae). Ph.D. Thesis, State Univ. College of Forestry at Syracuse Univ. Syracuse, N.Y. 106 p.

Hattemer, H.H.; Henson, W.R.; Mergen, F. 1969. Some factors in the distribution of European pine sawfly egg clusters in an experimental plantation of hard pines. Theor. and Appl. Genetics 39: 280-289.

Hendry, L.B.; Kostelc, J.G.; Hindenlang, D.M.; Wichmann, J.K.; Fix, C.J.; Korzeniowski, S.H. 1976. Chemical messengers in insects and plants. Pages 351-384 *in* J.W. Wallace and R.L. Mansell. Biochemical interaction between plants and insects, Vol. 10, Recent advances in phytochemistry. Plenum Press, New York and London.

Henson, W.R. 1965a. Individual variation and the dispersive capacity of *Neodiprion sertifer* (Hymenoptera). Proc. XII. Int. Congr. Entomol. London, England. 1964.

Henson, W.R. 1965b. Individual rearing of the larvae of *Neodiprion sertifer* (Geoffroy) (Hymenoptera: Diprionidae) Can. Entomol. 97: 773-779.

Henson, W.R.; O'Neil, L.C.; Mergen, F. 1970. Natural variation in susceptibility of *Pinus* to *Neodiprion* sawflies as a basis for the development of a breeding scheme for resistant trees. Yale Univ., School of Forestry Bull. No. 78, 71 p.

Hinks, C.F. 1973. The neuroendocrine organs of larvae of *Neodiprion lecontei*, *N. swainei*, and *Diprion hercyniae* (Hymenoptera: Diprionidae). Can. Entomol. 105: 725-731.

Hochmut, R. 1984. El genero *Neodiprion* Rohwer, 1918 (Hymenoptera: Diprionidae) en Cuba. Poeyana No. 263, 16 p.

Holling, C.S. 1965. The functional response of predators to prey density and its role in mimicry and population regulations. Mem. Entomol. Soc. Can. 45, 60 p.

Hopping, G.R.; Leech, H.B. 1936. Sawfly biologies. I. *Neodiprion tsugae* Middleton. Can. Entomol. 68: 71-79.

Ierusalimov, Ye. N. 1977. Restoration of needle biomass and increment of a pine stand damaged by the pine moth. (Original in Russian). Lesovedeniye (6): 79-85. Translation No. 1650, Canada Dept. of Fish. and Environ., Ottawa, Ont. 1978, 12 p.

Jewett, D.M.; Matsumura, F.; Coppel, H.C. 1976. Sex pheromone specificity in the pine sawflies: interchange of acid moieties in an ester. Science 192: 51-53.

Juutinen, P. 1967. Bionomics and occurrence of the fox-colored sawfly (*Neodiprion sertifer* Geoffr.) in Finland in the years 1959 to 1965. (Original in German). Communicationes Instituti Forestalis Fenniae 63(5), 129 p. Translation No. 1310, Environ. Can., Ottawa, Ont. 1977. 184 p.

Kalin, M.; Knerer, G. 1977. Group and mass effects in diprionid sawflies. Nature 267: 427-429.

Kelleher, J.S.; Hulme, M.A. (Eds.). 1984. Biological control programmes against insects and weeds in Canada 1969-1980. Commonw. Agric. Bureaux, Slough, England. 410 p.

Kenchington, W. 1969. Silk secretion in sawflies. J. Morph. 127: 355-362.

Kimmins, J.P. 1972. Relative contributions of leaching, litter-fall and defoliation by *Neodiprion sertifer* (Hymenoptera) to the removal of cesium-134 from red pine. Oikos 23: 226-234.

King, L.L.; Benjamin, D.M. 1965. The effect of photoperiod and temperature on the development of multivoltine populations of *Neodiprion rugifrons* Middleton. Proc. North Cent. Br. Entomol. Soc. Amer. 20: 139-140.

Knerer, G. 1983. Diapause strategies in diprionid sawflies. Naturwissenschaften 70: 203-205.

Knerer, G. 1984. Morphological and physiological clines in *Neodiprion pratti* (Dyar) (Symphyta, Diprionidae) in eastern North America. Z. Angew. Entomol. 97: 9-21.

Knerer, G.; Atwood, C.E. 1972. Evolutionary trends in the subsocial sawflies belonging to the *Neodiprion abietis* complex (Hymenoptera: Tenthredinoidea). Am. Zool. 12: 407-418.

Knerer, G.; Atwood, C.E. 1973. Diprionidae sawflies: polymorphism and speciation. Science 179: 1090-1099.

Knerer, G.; Marchant, R. 1973. Diapause induction in the sawfly *Neodiprion rugifrons* Middleton (Hymenoptera: Diprionidae). Can. J. Zool. 51: 105-108.

Kolomiets, N.G.; Stadnitskii, G.V.; Vorontsov, A.I. 1972. The European pine sawfly. Distribution, biology, damage, natural enemies, and control. (Original in Russian). Publishing House "Nauka", Siberian Branch, Novosibirsk, 148 p. Translation No. 390, Environ. Can., Ottawa, Ont. 1974, 197 p.

Kraemer, M.E.; Coppel, H.C.; Matsumura, F.; Kikukawa, T.; Benoit, P. 1984. Field and electroantennogram responses to sex pheromone isomers by monophagous jack pine sawflies (Hymenoptera: Diprionidae). J. Chem. Ecol. 10: 983-995.

Kuenzi, F.M.; Coppel, H.C. 1986. Isozymes of the sawflies *Neodiprion* and *Diprion similis*: Diagnostic characters and genetic distance. Biochem. Syst. Ecol. 14: 423-429.

Kulcke, J. 1983. Zur Nahrungsqualität von Fichtennadeln für forstliche Schadinsekten. 21. Vergleich der Larventwicklung und Auswertung von Nadelonhaltsstoffen aus identischen Pflanzenmaterial von Fichte (*Picea abies* Karst.) durch gleichzeitig angestezte Larven von *Gilpinia hercyniae* (Hym., Diprion.) und *Lymantria monacha* bzw. *Lymantria dispar* (Lep., Lymantr.). Z. Angew. Entomol. 95: 390-405.

Larsson, S.; Tenow, O. 1979. Utilization of dry matter and bioelements in larvae of *Neodiprion sertifer* Geoffr. (Hym., Diprionidae) feeding on Scots pine (*Pinus sylvestris* L.). Oecologia (Berl.) 43: 157-172.

Lee, S.O. 1965. Studies on the ecology of pine sawfly, *Neodiprion sertifer* Geoffroy. (Hymenoptera: Diprionidae). (Original in Korean). Research Reports of the Office of Rural Development (Suwon) 8(2): 127-132. Translation No. 302, Can. Dept. For., Ottawa, Ont. 1969. 11 p.

Lees, A.D. 1955. The physiology of diapause in arthropods. Cambridge Monog. in Exp. Biol., no. 4, 151 p. The University Press, Cambridge.

Lister, S.A. 1980. The effect of food supply on development, mortality and fecundity of *Neodiprion sertifer*. B.Sc. Thesis in Forestry, Lakehead Univ., School of Forestry, Thunder Bay, Ont. 60 p.

Lorenz, H.; Kraus, M. 1957. Die Larvalsystematik der Blattwespen. Akademie-Verlag, Berlin. 339 p., 435 Textfiguren.

Lunderstädt, J. 1981. The role of food as a density-determining factor for phytophagous insects with reference to the relationship between Norway spruce (*Picea abies* Karst) and *Gilpinia hercyniae* Htg. (Hymenoptera, Diprionidae). For. Ecol. and Mgmt. 3 (1980/1981): 335-353.

Lunderstädt, J.; Schwarz, U.; Dinish, K.N. 1975. Über Zusammenhange zwischen dem Zuckerstoffwechsel bei Larven von *Gilpinia hercyniae* (Hym., Diprionidae) und dem Kohlenhydratgehalt ihrer natürlichen Nahrung, den Nadeln von Fichte (*Picea abies* Karst). Z. Angew. Entomol. 77: 258-262.

Lyons, L.A. 1962. The effect of aggregation on egg and larval survival in *Neodiprion swainei* Midd. (Hymenoptera: Diprionidae). Can. Entomol. 94: 49-58.

Lyons, L.A. 1964a. The European pine sawfly, *Neodiprion sertifer* (Geoff.) (Hymenoptera: Diprionidae). A review with emphasis on studies in Ontario. Proc. Entomol. Soc. Ont. 94: 5-37.

Lyons, L.A. 1964b. The spatial distribution of two pine sawflies and methods of sampling for the study of population dynamics. Can. Entomol. 96: 1373-1407.

Lyons, L.A. 1970. Some population features of reproductive capacity in *Neodiprion swainei* (Hymenoptera: Diprionidae). Can. Entomol. 102: 68-84.

Lyons, L.A. 1976. Mating ability in *Neodiprion sertifer* (Hymenoptera: Diprionidae). Can. Entomol. 108: 321-326.

Lyons, L.A. 1977. On the population dynamics of *Neodiprion* sawflies. *In* H.M. Kulman and H.C. Chiang (Eds.). Insect Ecology - Papers presented in the A.C. Hodson Lectures. Univ. Minn. Agric. Exp. Stn. Tech. Bull. 310: 48-55.

Lyons, L.A.; Griffiths, K.J. 1962. Observations on the development of *Neodiprion sertifer* (Geoff.) within the cocoon (Hymenoptera: Diprionidae). Can. Entomol. 94: 994-1001.

Lyons, L.A.; Sullivan, C.R. 1974. Early differential mortality in *Neodiprion sertifer* (Hymenoptera: Diprionidae). Can. Entomol. 106: 1-10.

Lyons, L.A.; Sullivan, C.R.; Wallace, D.R.; Griffiths, K.J. 1971. Problems in the management of forest pest populations. Proc. Tall Timbers Conf. on Ecological Animal Control by Habitat Management No. 3: 129-140.

MacGillivray, A.D. 1914. The immature stages of the Tenthredinoidea. 44th Ann. Rep. Entomol. Soc. Ont. (1913): 54-75.

Magasi, L.P.; Syme, P.D. 1984. *Gilpinia hercyniae* (Hartig), European spruce sawfly (Hymenoptera: Diprionidae). Chapter 52, Pages 295-297 *in* J.S. Kelleher and M.A. Hulme. Biological control programmes against insects and weeds in Canada 1969-1980. Commonwealth Agricultural Bureaux, Farnham Royal, Slough, England. 410 p.

Martignoni, M.E.; Iwai, P.J. 1981. A catalogue of viral diseases of insects, mites and ticks. Pages 897-911 *in* H.D. Burgess (Ed.), Microbial control of pests and plant diseases 1970-1980. Academic Press, London, England.

Martinek, V. 1968. A contribution to the bionomy of the European pine sawfly *Neodiprion sertifer* (Geoff.) in the Ore Mountains and in the Giant Mountains. (Original in Czechoslovakian). Opera Corcontica 5: 175-199. Translation No. 103, Can. Dept. Fish. and For., Ottawa, Ont. 1970. 36 p.

Maxwell, D.E. 1955. The comparative internal larval anatomy of sawflies (Hymenoptera: Symphyta). Can. Entomol. v. 87, Suppl. 1, 132 p.

Maxwell, D.E. 1958. Sawfly cytology with emphasis upon the Diprionidae (Hymenoptera: Symphyta). Proc. 10th Int. Congr. of Entomol., 1956. 2: 961-978.

McGugan, B.M. 1958. The Canadian Forest Insect Survey. Proc. Tenth Int. Congr. Entomol. 1956. 4: 219-232.

McGugan, B.M.; Coppel, H.C. 1962. Part 2. Biological control of forest pests, 1910-1958. Pages 35-216 *in* J.H. McLeod, B.M. McGugan and H.C. Coppel. A review of the biological control attempts against insects and weeds in Canada. Tech. Comm. No. 2, Commonw. Inst. of Biol. Control, Trinidad. Commonw. Agric. Bureaux, Farnham Royal, England.

McLeod, J.M. 1966. The spatial distribution of cocoons of *Neodiprion swainei* in a jack pine stand. A cartographic analysis of cocoon distribution with reference to predation by small mammals. Can. Entomol. 98: 430-447.

McLeod, J.M. 1970. The epidemiology of the Swaine jack-pine sawfly, *Neodiprion swainei* Midd. For. Chron. 46: 1-8.

McLeod, J.M. 1972. Mass rearing techniques for larvae of the Swaine jack pine sawfly, *Neodiprion swainei* Middleton and effects on normal and prolonged diapause. Environ. Can. For. Serv. Laurentian For. Res. Cent., Quebec. Inf. Rep. Q-X-26, 21 p.

McLeod, J.M. 1974. Bird population studies in the Swaine jack pine sawfly life system. Cent. de Rech. For. des Laurentides, Quebec. Can. Dept. Environ. Inf. Rep. LAU-X-10, 86 p.

McLeod, J.M. 1975. Parasitoid evaluation: the monitoring problem. Melsh. Entomol. Series 17: 1-12.

McLeod, J.M. 1979. Discontinuous stability in a sawfly life system and its relevance to pest management strategies. Pages 68-81 *in* W.E. Waters (Ed.), Current topics in forest entomology. Selected papers from the XVth Int. Congr. of Entomol., Washington, D.C., Aug. 1976. USDA For. Serv., Gen. Tech. Rep. WO-8.

McNamee, P.J.; McLeod, J.M.; Holling, C.S. 1981. The structure and behaviour of defoliating insect/forest systems. Res. Popul. Ecol. 23: 280-298.

Middleton, W. 1921. Leconte's sawfly, an enemy of young pines. J. Agric. Res. 20(10): 741-760, 3 plates.

Minder, N.F. 1975. The effect of temperature on the rate of reactivation and the development of the larvae of the fox-colored sawfly (*Neodiprion sertifer*) with prolonged diapause. (Original in Russian.) Zoologicheskii Zhurnal 54(6): 944-951, Translation No. 1107, Environ. Can., Ottawa, Ont. 1976. 17 p.

Mott, D.G.; Nairn, L.D.; Cook, J.A. 1957. Radial growth in forest trees and effects of insect defoliation. For. Sci. 3: 286-304.

Nakamura, H. 1985. Ecological studies on the aggregation of the European pine sawfly, *Neodiprion sertifer* (Geoffroy) (Hymenoptera: Diprionidae). (Original in Japanese.) Bull. Joto Gakuen Women's Junior College, No. 15, p. 41-154. Translation No. 2636249, Agr. Can., Ottawa, Ont. 1987, 278 p.

Neilson, M.M.; Elgee, D.E. 1965. An unusual increase of spruce sawfly numbers in New Brunswick. Bi-mon. Prog. Rep. 21(2): 1, Can. Dept. For., Ottawa, Ont.

Neilson, M.M.; Elgee, D.E. 1968. The method and role of vertical transmission of a nucleopolyhedrosis virus in the European spruce sawfly, *Diprion hercyniae*. J. Invert. Pathol. 12: 132-139.

Neilson, M.M.; Martineau, R.; Rose, A.H. 1971. *Diprion hercyniae* (Hartig), The European spruce sawfly (Hymenoptera: Diprionidae). Pages 136-143 *in* Biological Control Programmes against Insects and Weeds in Canada 1959-1968. Commonw. Inst. of Biol. Control. Tech. Comm. No. 4. Commonw. Agric. Bureaux, Slough, England.

Neilson, M.M.; Morris, R.F. 1964. The regulation of European spruce sawfly numbers in the Maritime Provinces of Canada from 1937 to 1963. Can. Entomol. 96: 773-784.

Niemela, P.; Rousi, M.; Saarenmaa, H. 1987. Topographical delimitation of *Neodiprion sertifer* (Hym., Diprionidae) outbreaks on Scots pine in relation to needle quality. J. Appl. Entomol. 103: 84-91.

Niemela, P.; Tuomi, J.; Mannila, R.; Ojala, P. 1984. The effect of previous damage on the quality of Scots pine foliage as food for Diprionid sawflies. Z. Angew. Entomol. 98: 33-43.

Okutani, T.; Ito, T. 1957. On a pine sawfly in high mountains of Japan. The New Entomologist 6: 1-3.

Olaifa, J.I.; Matsumura, F.; Kikukawa, T.; Coppel, H.C. 1988. Pheromone-dependent species recognition mechanisms between *Neodiprion pinetum* and *Diprion similis* on white pine. J. Chem. Ecol. 14: 1131-1144.

Olofsson, E. 1973. Evaluation of a nuclear polyhedrosis virus as an agent for the control of balsam fir sawfly, *Neodiprion abietis* (Harr.), Can. For. Serv. Ottawa, Ont. Inf. Rep. IP-X-2. 30 p.

Olofsson, E. 1989. Oviposition behavior and host selection in *Neodiprion sertifer* (Geoffr.) (Hym., Diprionidae). J. Appl. Entomol. 107: 357-364.

O'Neil, L.C. 1962. Some effects of artificial defoliation on the growth of jack pine (*Pinus banksiana* Lamb.). Can. J. Bot. 40: 273-280.

O'Neil, L.C. 1963. The suppression of growth rings in jack pine in relation to defoliation by the Swaine jack-pine sawfly. Can. J. Bot. 41: 227-235.

Pankevich, T.P. 1963. Effect of light on the duration of development of eonymphal stage of the European pine sawfly (*Neodiprion sertifer*). Proc. Acad. Sci. Byelorussian S.S.R., Ser. Biol. Sci. No. 1, 110-114.

Parker, H.L. 1934. Notes on the anatomy of tenthredinid larvae, with special reference to the head. Boll. Lab. Zool. Portici. 28: 159-191.

Phillipsen, W.J.; Coppel, H.C. 1973. Changes in the mature gut of the introduced pine sawfly, *Diprion similis* (Hartig) (Hymenoptera: Diprionidae) during transformation. Ann. Entomol. Soc. Am. 66: 1360-1362.

Philogène, B.J.R. 1971a. Détermination du stade larvaire sensible à la lumière chez *Neodiprion swainei* Midd. (Hyménoptère: Diprionide) Can. J. Zool. 49: 449-450.

Philogène, B.J.R. 1971b. Revue dcs travaux sur les formes de diapause chez les Tenthrédionoïdes (Hyménotpères: symphites) les plus communs. Ann. Entomol. Soc. Québec 16: 112-119.

Philogène, B.J.R.; Benjamin, D.M. 1971a. Diapause in the Swaine jack pine sawfly, *Neodiprion swainei*, as influenced by temperature and photoperiod. J. Insect Physiol. 17: 1711-1716.

Philogène, B.J.R.; Benjamin, D.M. 1971b. Temperature and photoperiod effects on the immature stages and adults of *Neodiprion swainei* (Hymenoptera: Diprionidae). Can. Entomol. 103: 1705-1715.

Pilon, J.G. 1966. Influence of temperature and relative humidity on the life-span of adults of the Swaine jack-pine sawfly, *Neodiprion swainei* Midd. (Hymenoptera: Diprionidae). Can. Entomol. 98: 789-794.

Pilon, J.G.; Tripp, H.A.; McLeod, J.M.; Ilnitzky, S.L. 1964. Influence of temperature on prespinning eonymphs of the Swaine jack-pine sawfly, *Neodiprion swainei* Midd. (Hymenoptera: Diprionidae). Can. Entomol. 96: 1450-1457.

Popo, A. 1967. Comparative ecological investigations of different populations of the European pine sawfly *Neodiprion sertifer* Geoffr. (Hymenoptera, Diprionidae). (Original in German.) Thesis for Obtaining the Degree of Doctor at the Forestry Faculty of the Georg-August University, Gottingen in Hann. Munden. 145 p. Translation No. 264, Can. Dept. For. and Rural Dev., Ottawa, Ont. 1968. 171 p.

Prebblc, M.L. 1941. The diapause and related phenomena in *Gilpinia polytoma* (Hartig). Can. J. Res. (D) 19: 295-322, 323-346, 350-362, 417-436, 437-454.

Prebble, M.L. 1956. Entomology in Canada up to 1956. Forest entomology. Can. Entomol. 88: 350-363.

Prebble, M.L. (Ed.) 1975a. Aerial Control of Forest Insects in Canada. Dept. Environ. Ottawa, Ont., 330 p.

Prebble, M.L. 1975b. Red-headed pine sawfly *Neodiprion lecontei* (Fitch). Pages 220-223 *in* M.L. Prebble (Ed.), Aerial Control of Forest Insects in Canada. Dept. Environ. Ottawa, Ont.

Price, P.W.; Cobb, N.; Craig, T.P.; Fernandes, G.W.; Itami, J.K.; Mopper, S.; Preszler, R.W. 1990. Insect herbivore population dynamics on trees and shrubs: New approaches relevant to latent and eruptive species and life table development. Pages 1-38 *in* E.A. Bernays (Ed.), Insect-Plant Interactions. CRC Press, Boca Raton, Ann Arbor, Boston.

Prop, N. 1959. Protection against birds and parasites in some species of tenthredinid larvae. Arch. Neerlandaises de Zoologie 13(3): 1-68.

Pschorn-Walcher, H. 1965. The ecology of *Neodiprion sertifer* (Geoff.) (Hym., Diprionidae) and a review of its parasite complex in Europe. Tech. Bull. Commonw. Inst. Biol. Cont. 5: 33-97.

Pschorn-Walcher, H. 1970. Studies on the biology and ecology of the alpine form of *Neodiprion sertifer* (Geoff.) (Hym.: Diprionidae) in the Swiss Alps. Z. Angew. Entomol. 66: 64-83.

Rangnekar, P.V.; Forward, D.F. 1973. Foliar nutrition and wood growth in red pine: effects of darkening and defoliation on the distribution of ^{14}C photosynthate in young trees. Can. J. Bot. 51: 103-108.

Rauf, A.; Benjamin, D.M. 1980. The biology of the white pine sawfly, *Neodiprion pinetum* (Hymenoptera: Diprionidae) in Wisconsin. Great Lakes Entomol. 13: 219-224.

Reeks, W.A. 1937. The morphology of the adult of *Diprion polytomum* (Hartig). Can. Entomol. 69: 257-264.

Reeks, W.A. 1953. The establishment of introduced parasites of the European spruce sawfly (*Diprion hercyniae* (Htg.) (Hymenoptera: Diprionidae) in the Maritime Provinces. Can. J. Agric. Sci. 33: 405-429.

Reeks, W.A.; Barter, G.W. 1951. Growth reduction and mortality of spruce caused by the European spruce sawfly, *Gilpinia hercyniae* (Htg.) (Hymenoptera: Diprionidae). For. Chron. 27: 1-16.

Ross, H.H. 1955. The taxonomy and evolution of the sawfly genus *Neodiprion*. For. Sci. 1: 196-209.

Rudall, K.M.; Kenchington, W. 1971. Arthropod silks: The problem of fibrous proteins in animal tissues. Ann. Rev. Entomol. 16: 73-96.

Saint-Hilaire, K. 1931. Über Vorderdarmanhange bei *Lophyrus*-Larven und ihre Bedeutung. Zeit. fur dic Morphologie und Okologie der Tiere 21: 608-616.

Schaefer, C.H. 1965. Fatty acids of the Virginia pine sawfly, *Neodiprion pratti* Dyar. Can. Entomol. 97: 941-945.

Schaefer, C.H.; Kaplanis, J.N.; Robbins, W.E. 1965. The relationship of the sterols of the Virginia pine sawfly, *Neodiprion pratti* Dyar, to those of two hosts plants, *Pinus virginiana* Mill and *Pinus rigida* Mill. J. Insect Physiol. 11: 1013-1021.

Schaefer, C.H.; Wallace, D.R. 1967. Evaluation of larval hemolymph amino acids as a taxonomic character for *Neodiprion* sawflies (Hymenoptera: Diprionidae). Can. Entomol. 99: 574-578.

Shaffner, J.V., Jr. 1939. *Neodiprion sertifer* (Geoff.), a pine sawfly accidentally introduced into New Jersey from Europe. J. Econ. Entomol. 32: 887-888.

Shaffner, J.V., Jr. 1943. Sawflies injurious to conifers in the Northeastern States. J. For. 41: 580-588.

Schedl, K. 1931. Notes on jack pine sawflies in Northern Ontario. 61st Ann. Rep. Entomol. Soc. Ont. 1930, 75-79.

Schedl, K. 1933. Statistische Untersuchungen über die Kopfkapsel breiten bei Blattwespen. Z. Angew. Entomol. 20: 449-460.

Schedl, K. 1935. Zwei neue Blattwespen aus Kanada (Hym., Tenthr.). Mitt. Detsch. Entomol. Ges. 6: 39-44.

Schedl, K. 1937. Quantitative Freilandstudien an Blattwespen der *Pinus banksiana* mit besonderer Berucksichtigung der Methodik. Z. Angew. Entomol. 24: 25-70, 181-215.

Schedl, W.; Klima, J. 1980. Rasterelektronenoptische Untersuchungen zur Feinstruktur von Blattwespen-Kokons (Hymenoptera: Symphyta). Z. Angew. Entomol. 89: 34-42.

Schopf, R. 1981. Untersuchungen zum Aminosaürenstoffwechsel der Ficten-nadeln fressenden Blattwespe *Gilpinia hercyniae* Ht. (Hym., Diprionidae). Z. Angew. Entomol. 92: 84-92.

Schopf, R. 1986. The effect of secondary needle compounds on the development of phytophagous insects. For. Ecol. Manage. 15 (1986): 55-64.

Schuh, B.A.; Benjamin, D.M. 1984a. The chemical feeding ecology of *N. dubiosus* Schedl, *N. rugifrons* Midd., and *N. lecontei* (Fitch) on jack pine (*Pinus banksiana* Lamb.). J.Chem. Ecol. 10: 1071-1079.

Schuh, B.A.; Benjamin, D.M. 1984b. Evaluation of commercial resin acids as feeding deterrents against *Neodiprion dubiosus*, *N. lecontei*, and *N. rugifrons* (Hymenoptera: Diprionidae). J. Econ. Entomol. 77: 802-805.

Schwenke, W. 1964. Grundzüge der Populationsdynamik und Bekämpfung der gemeinen Kiefernbuschhorn-Blattwespe, *Diprion pini* L. Z. Angew. Entomol. 54: 101-107.

Schwerdtfeger, F. 1973. Forest entomology. Pages 361-386 *in* R.F. Smith, T.E. Mittler and C.N. Smith (Eds.), History of entomology. Ann. Rev., Palo Alto, Calif.

Sharov, A.A.; Safonkin, A.F. 1981. Seasonal developmental dynamics of eggs and larvae of *Diprion pini* L. (Hymenoptera, Diprionidae). Entomol. Rev. 59(1): 64-69.

Sláma, K. 1958. The total metabolism of insects. 8. The respiratory metabolism of the prepupal stages and pupae in sawflies. (Original in Czechoslovakian). Acta Societatis Entomologicae Bohemoslovenicae 22: 334-346. Translation No. 72525, Agric. Can., Ottawa, Ont. 16 p.

Sláma, K. 1960. Metabolism during diapause and development in sawfly metamorphosis. Pages 195-201 *in* The Ontogeny of Insects. Acta symposii de evolutione insectorum, Praha 1959.

Slansky, F., Jr. 1980. High consumption rate by *Neodiprion sertifer*? A comment on a paper by Larsson and Tenow. Oecologia(Berl.) 46: 133-134.

Slansky, F., Jr.; Fogal, W.H. 1985. Utilization of dry matter and elements by larvae of the sawflies *Neodiprion sertifer* (Geoff.) and *Diprion similis* (Hartig) feeding on Scots pine needles. Can. J. Zool. 63: 506-510.

Smirnoff, W.A. 1962. Trans-ovum transmission of virus of *Neodiprion swainei* Middleton (Hymenoptera: Tenthredinidae), J. Insect Pathol. 4: 192-200.

Smirnoff, W.A. 1972. Promoting virus epizootics in populations of the Swaine jack pine sawfly by infected adults. Bioscience 22: 662-663.

Smirnoff, W.A. 1973. Increased mortality of the Swaine jack pine sawfly, and foliar nitrogen concentrations after urea fertilization. Can. J. For. Res. 3: 112-121.

Smirnoff, W.A.; McLeod, J.M. 1975. Swaine jack pine sawfly, *Neodiprion swainei* Middleton. Pages 235-240 *in* M.L. Prebble (Ed.), Aerial Control of Forest Insects in Canada. Dept. Environ. Ottawa, Ont.

Smirnoff, W.A.; Fettes, J.J.; Haliburton, W. 1962. A virus disease of Swaine's jack pine sawfly, *Neodiprion swainei* Midd. sprayed from an aircraft. Can. Entomol. 94: 477-486.

Smirnoff, W.A.; Valero, J. 1975. Effets à moyen terme de la fertilisation par urée ou par potassium sur *Pinus banksiana* L. et le comportement de ses insectes dévastateurs: tel gue *Neodiprion swainei* (Hymenoptera, Tenthredinidae) et *Toumeyella numismaticum* (Homoptera, Coccidae). Can. J. For. Res. 5: 236-244.

Smith, D.R. 1974. Conifer sawflies, Diprionidae: Key to North American genera, checklist of world species, and new species from Mexico (Hymenoptera). Proc. Entomol. Soc. Wash. 76: 409-418.

Smith, D.R. 1979. Symphyta. Pages 3-137 *in* K.V. Krombien, P.D. Hurd Jr., D.R. Smith, and B.D. Burks (Compilers). Catalog of Hymenoptera in America north of Mexico. Smithsonian Institution Press, Washington, D.C.

Smith, D.R. 1988. A synopsis of the sawflies (Hymenoptera: Symphyta) of America south of the United States: introduction, Xyelidae, Pamphiliidae, Cimbicidae, Diprionidae, Xiphriidae, Siricidae, Orussidae, Cephidae. Syst. Entomol. 13: 205-261.

Smith, D.R.; Wagner, M.R. 1986. Recognition of two species in the pine feeding "*Neodiprion fulviceps* complex" (Hymenoptera: Diprionidae) of western United States. Proc. Entomol. Soc. Wash. 88: 215-226.

Smith, S.G. 1941. A new form of spruce sawfly identified by means of its cytology and parthenogenesis. Sci. Agric. 21: 245-305.

Smith, S.G.; Wallace, D.R. 1971. Allelic sex determination in a lower hymenopteran, *Neodiprion nigroscutum* Midd. Can. J. Genet. Cytol. 13: 617-621.

Stark, R.W.; Dahlsten, D.L. 1965. Notes on the distribution of eggs of a species in the *Neodiprion fulviceps* complex (Hymenoptera: Diprionidae). Can. Entomol. 97: 550-552.

Steinhaus, E.A. 1949. Principles of Insect Pathology. McGraw-Hill, New York. 757 p.

Sullivan, C.R. 1965. Laboratory and field investigations on the ability of eggs of the European pine sawfly, *Neodiprion sertifer* (Geoffroy) to withstand low winter temperatures. Can. Entomol. 97: 978-993.

Sullivan, C.R.; Wallace, D.R. 1965. Photoperiodism in the development of the European pine sawfly, *Neodiprion sertifer* (Geoff.). Can. J. Zool. 43: 233-245.

Sullivan, C.R.; Wallace, D.R. 1967. Interaction of temperature and photoperiod in the induction of prolonged diapause in *Neodiprion sertifer*. Can. Entomol. 99: 834-850.

Sullivan, C.R.; Wallace, D.R., 1968a. Inclusions in adults of the European pine sawfly, *Neodiprion sertifer* (Geoff.). Can. J. Zool. 46: 959-963.

Sullivan, C.R.; Wallace, D.R. 1968b. Reductions in fecundity associated with inclusions in *Neodiprion sertifer* (Geoff.) Proc. Entomol. Soc. Ont. 99: 57-60.

Sullivan, C.R.; Wallace, D.R. 1968c. Variations in the photoperiodic response of *Neodiprion sertifer*. Can. J. Zool. 46: 1082-1083.

Suzuki, A. 1934. On *Neodiprion sertifer* Geoffroy. (Original in Japanese). Oyo-Dobutsugaku-Zasshi (Tokyo) 6(1): 254-273. Translation No. 4, Can. Dept. For., Ottawa, Ont. 1964. 29 p.

Swaine, J.M. 1928. Forest entomology and its development in Canada. Can. Dept. of Agric., Ottawa, Ont., Pamphlet 97, 20 pp.

Swaine, J.M. 1933. The relation of insect activities to forest development as exemplified in the forests of Eastern North America. Sci. Agric. 14: 8-31.

Tait, N.N. 1962. The anatomy of the sawfly *Perga affinis affinis* Kirby (Hymenoptera: Symphyta). Austral. J. Zool. 10: 652-683.

Tauber, M.J.; Tauber, C.A.; Masaki, S. 1986. Seasonal adaptations of insects. Oxford University Press, New York, Oxford, 411 p.

Thielges, B.A. 1968. Altered polyphenol metabolism in the foliage of *Pinus sylvestris* associated with European pine sawfly attack. Can. J. Bot. 46: 724-725.

Tostowaryk, W.; McLeod, J.M. 1972. Sequential sampling for egg clusters of the Swaine jack pine sawfly, *Neodiprion swainei* (Hymenoptera: Diprionidae). Can. Entomol. 104: 1343-1347.

Tripp, H.A. 1965. The development of *Neodiprion swainei* Middleton (Hymenoptera: Diprionidae) in the Province of Quebec. Can. Entomol. 97: 92-107.

Trofimov, S.B. 1974. A study of the effect of the group in larvae of the European pine sawfly (*Neodiprion sertifer*) (Hymenoptera: Diprionidae). (Original in Russian). Zoologicheskii Zhurnal 53(3): 368-375. Translation No. 963. Environ. Can., Ottawa, Ont. 1975, 17 p.

Tyler, G.M.; Coppel, H.C. 1963. Preliminary notes on the biology of *Neodiprion abbotii* (Leach) in Wisconsin. Proc. N. Cent. Br., Entomol. Soc. Am. 28: 63.

Vehrke, H. 1961. Zur Unterscheidung der Larven vor *Gilpinia abieticola* (D.T.), *G. polytoma* (Htg.) und *G. hercyniae* (Htg.) (Hymenoptera: Diprionidae) nach dem Zeichnungsmuster der Kopfkapseln. Z. Angew. Entomol. 48: 176-185.

Verzhutskii, B.N. 1966. Sawflies of the Baikal Region. (Original in Russian). Akademiya Nauk, Moscow. 163 p. Translation No. 81, Can. Dept. Fish. and For., Ottawa, Ont. 390 p.

Viktorov, G.; Borodin, A.L. 1975. Migration patterns of fox-colored sawfly (*Neodiprion sertifer*) from old to young pine plantations. (Original in Russian). Zoologicheskii Zhurnal 54(7): 1092-1095. Translation No. 1271, Environ. Can., Ottawa, Ont. 1977, 9 p.

Wagner, M.R. 1986. Influence of moisture stress and induced resistance on ponderosa pine, *Pinus ponderosa*, on the pine sawfly, *Neodiprion fulviceps*. For. Ecol. Manage. 15: 43-53.

Wagner, M.R. 1988. Induced defenses in ponderosa pine against defoliating insects. Pages 141-155 *in* W.J. Mattson, J. Levieux and C. Bernard-Dagan (Eds.), Springer-Verlag, New York, Berlin, Heidelberg, London, Paris, Tokyo.

Wagner, M.R.; Evans, P.D. 1985. Defoliation increases nutritional quality and allelochemics of pine seedlings. Oecologia 67: 235-237.

Wallace, D.R. 1959. Occurrence of the Swaine jack pine sawfly and external anatomy of the mature, feeding larvae. Ph.D. Thesis. Dept. Entomol. and Plant Path., McGill Univ., Montreal, Que. 173 p.

Wallace, D.R. 1964a. Egg pigmentation, a new criterion for use in diprionid sawfly taxonomy. Ph.D. Thesis, Dept. Entomol. and Plant Path., McGill Univ., Montreal, Que. 142 p.

Wallace, D. R. 1964b. Spectrophotometry in the taxonomy of conifer sawflies. Can. Entomol. 96: 162.

Wallace, D.R.; Cameron, J.M.; Sullivan, C.R. 1975. European pine sawfly *Neodiprion sertifer* (Geoff.). Pages 224-229 *in* M.L. Prebble (Ed.), Aerial Control of Forest Insects in Canada. Dept. Environ. Ottawa, Ont.

Wallace, D.R.; Campbell, I.M. 1965. Method for characterizing sawfly egg pigmentation. Nature 207: 1363-1364.

Wallace, D.R.; Sullivan, C.R. 1963. Laboratory and field investigations of the effect of temperature on the development of *Neodiprion sertifer* (Geoff.) in the cocoon. Can. Entomol. 95: 1051-1066.

Wallace, D.R.; Sullivan, C.R. 1966. Geographic variation in the photoperiodic reaction of *Neodiprion sertifer* (Geoff.). Can. J. Zool. 44: 147.

Wallace, D.R.; Sullivan, C.R. 1972. Some effects of photoperiod on the development of *Neodiprion swainei* Midd. Can. J. Zool. 50: 1055-1061.

Wallace, D.R.; Sullivan, C.R. 1973. First successful total laboratory development of eggs of the European pine sawfly [*Neodiprion sertifer* (Geoff.)]. Can. Dept. Environ., Canadian For. Serv., Ottawa, Ont. Bi-mon. Res. Notes 29(1): 3.

Wallace, D.R.; Sullivan, C.R. 1974. Photoperiodism in the early balsam strain of the *Neodiprion abietis* complex (Hymenoptera: Diprionidae). Can. J. Zool. 52: 507-513.

Wallace, D.R.; Sullivan, C.R. 1975. Effects of daylength over one complete and a partial succeeding generation on development in *Neodiprion sertifer* (Hymenoptera: Diprionidae). Entomol. Exp. Appl. 18: 399-411.

Warren, L.O.; Coyne, J.F. 1958. The pine sawfly, *Neodiprion taedae linearis* Ross, in Arkansas. Agric. Exp. Stn., Univ. of Arkansas, Fayetteville. Bull. 602, 23 p.

Watson, W.Y. 1961. Studies on the Diprionidae (Hymenoptera). I. Bibliography of the Genus *Neodiprion* in eastern North America. II. Adult anatomy and variation in *Neodiprion swainei* Midd. Interim Rep. 1960-3, Can. Dept. For., For. Biol. Div., For. Insect Lab., Sault Ste. Marie, Ont. 33 p. + 27 plates.

Wellington, W.G. 1964. Qualitative changes in natural populations during changes in abundance. Can. Entomol. 96: 436-451.

West, A.S.; Horwood, R.H.; Fourns, T.R.; Hudson, A. 1959. Systematics of *Neodiprion* sawflies. I. Preliminary report on serological and chromatographic studies. Ann. Rep. Entomol. Soc. Ont. (1958) 89: 58-68.

Whittaker, J.R.; West, A.S. 1962. A starch gel electrophoretic study of insect hemolymph proteins. Can. J. Zool. 40: 655-671.

Wilkinson, R.C. 1964. Development in *Neodiprion excitans* Rohwer as related to oviposition and pine needle growth. Can. Entomol. 96: 1142-1147.

Wilkinson, R.C. 1971. Slash-pine sawfly *Neodiprion merkeli*. 1. Oviposition pattern and description of egg, female larva, pupa, and cocoon. Ann. Entomol. Soc. Am. 64: 241-247.

Wilkinson, R.C.; Drooz, A.T. 1979. Oviposition, fecundity, and parasites of *Neodiprion excitans* from Belize, C.A. Environ. Entomol. 8: 501-505.

Wilkinson, R.C.; Popp, M.P. 1989. Oviposition behavior of *Neodiprion merkeli* (Hymenoptera: Diprionidae) in two-needle and three-needle fascicles of slash pine. Environ. Entomol. 18: 678-682.

Wilson, E.O. 1971. The insect societies. The Belknap Press of Harvard University Press, Cambridge, Mass. 548 p.

Wilson, E.O. 1975. Sociobiology. The new synthesis. The Belknap Press of Harvard University Press, Cambridge, Mass. and London, England. 697 p.

Wilson, L.F. 1966. Effects of different population levels of the European pine sawfly on young Scots pine trees. J. Econ. Entomol. 59: 1043-1049.

Wilson, L.F. 1975. Spatial distribution of egg clusters of the European pine sawfly *Neodiprion sertifer* (Geoff.), in young pine plantations in Michigan. Great Lakes Entomol. 8: 123-134.

Wingfield, M.; Warren, L.O. 1971. The effect of temperature on the cocoon stage of the sawfly *Neodiprion taedae linearis*. J. Kansas Entomol. Soc. 44: 491-500.

Wingfield, M.; Warren, L.O. 1972. The effect of photoperiod on the development of the sawfly *Neodiprion taedae linearis*. J. Kansas Entomol. Soc. 45: 1-6.

Wong, H.R. 1951. Cocoons of some sawflies that defoliate forest trees in Manitoba and Saskatchewan. Ann. Rep. Entomol. Soc. Ont. 82: 61-67.

Wong, H.R.; Szlabey, D.L. 1986. Larvae of the North American genera of Diprionidae (Hymenoptera: Symphyta). Can. Entomol. 118: 577-587.

Woods, P.E.; Guttman, S.I. 1987. Genetic variation in *Neodiprion* (Hymenoptera: Symphyta: Diprionidae) sawflies and a comment on low levels of genetic diversity within the Hymenoptera. Ann. Entomol. Soc. Am. 80: 590-599.

Wright, J.W.; Wilson, L.F.; Randall, W.K. 1967. Differences among Scots pine varieties in susceptibility to European pine sawfly. For. Sci. 13: 175-181.

Xiao, G.R.; Huang, X.Y.; Zhou, S.Z. 1983. Sawflies of the genus *Diprion* from China (Hymenoptera, Symphyta, Diprionidae). Scientia Silvae Sinicae 19(3): 277-282.

Xiao, G.R.; Huang, X.Y.; Zhou, S.Z. 1984. Sawflies of the family Diprionidae from China (Hymenoptera, Symphyta). Scientia Silvae Sinicae 20(4): 366-371.

Xiao, G.R.; Huang, X.Y.; Zhou, S.Z. 1985. Sawflies of the family Diprionidae from China (Hymenoptera, Symphyta) (continued). Scientia Silvae Sinicae 21(1): 30-41.

Xiao, G.R.; Zhou, S.Z.; Huang, X.Y. 1984. Seven new species of the family Diprionidae from Yunnan, China (Hymenoptera, Symphyta). Entomol. Taxon. Bull. 6: 141-150.

Yuasa, H. 1922. A classification of the larvae of the Tenthredinoidea. Illinois Biol. Monogr. 7(4), 172 p.

Zeng, C.H.; Jin, M.; Liu, M.Q.; Su, X.G. 1984. On the bionomics and control of *Augomonoctenus smithi* Xiao et Wu (Hymenoptera, Diprionidae). Scientia Silvae Sinicae 20: 332-335.

Zivojinovic, D. 1968. Contribution to the study of the diapause of the European pine sawfly (*Neodiprion sertifer*). (Original in Serbian). Zastita bilja 89-92: 349-356. Translation No. 137, Environ. Can., Ottawa, Ont. 1972. 15 p.

Chapter 25

Terminal Weevils

A. RETNAKARAN AND JOHN W.E. HARRIS

Introduction

Terminal weevils (Coleoptera: Curculionidae) feed and oviposit in the leader or terminal shoot of pine and spruce up to pole size across Canada, adversely affecting the growth and shape of trees. The larvae feed in the terminal shoot and kill it, thus eliminating a year or more of growth each time the tree is attacked, which may be as often as every second year and sometimes may even be in two successive years. Also, as a result, the growth of the tree is no longer straight. The major damaging species belong to the genus *Pissodes*.

In eastern Canada, the principal terminal weevil is the white pine weevil, *Pissodes strobi*, which attacks several pine species. It is also an important economic pest in the western provinces, where it attacks spruce. We will primarily address this species as it has the greatest economic importance and has had the most work done on it.

Another species, the lodgepole terminal weevil, *P. terminalis*, attacks pines in the west and as far east as Manitoba. In addition, the weevils *Magdalis gentilis* and *Cylindrocopturus* sp. attack the terminals of pine (and also apparently boles) much in the same way as the lodgepole terminal weevil (Kovacs 1988).

Ecology

White pine weevil occurs across most of Canada (Little 1971). In the east, it has a preference for eastern white pine, *Pinus strobus*; Scots pine, *P. sylvestris*; jack pine, *P. banksiana*; Norway spruce, *Picea abies*; and white spruce, *P. glauca*, but these are by no means the complete list of hosts (Wallace and Sullivan 1985). In the west, it attacks Sitka spruce, *P. sitchensis*; Engelmann spruce, *P. engelmannii*; and white spruce, *P. glauca*.

Smith and Sugden (1969) separated the white pine weevil into three species. Their conclusions were supported by VanderSar et al. (1977) who agreed, based on feeding experiments, that there may be three different ecotypes of the white pine weevil: the engelmannii and sitchensis types in the west, and the strobi type in the east, each with it own host preference.

Adult white pine weevils typically overwinter in the duff around the base of host trees (Stevenson 1967; Belyea and Sullivan 1956), although on the west coast of British Columbia some of them may spend the winter within the leader (Silver 1968). They emerge in early spring to feed on the bark of the terminal leader and lateral branches (Fig. 1). They are poor fliers and usually do not migrate far (McMullen and Condrashoff 1973). Soon after emergence they mate and oviposit as many as 100 eggs in the feeding punctures made in the previous year's leader growth. The eggs hatch and the larvae feed on the phloem, girdling the leader. There may be as many as 50 or more larvae in a leader and they typically mine downwards.

As the larvae are developing through their later instars, the current year's developing terminal shoot becomes deformed and dies. Pupation occurs within the leader. In early fall, pre-diapause adults may emerge or they may remain within the leader. Those which emerge, feed and enter the duff in late fall and overwinter under the host tree. Natural mortality may be very high during the overwintering stage and ranges from 50% to 80% (Dixon et al. 1979; Dixon and Houseweart 1982; VanderSar 1977). The weevil is univoltine with one life cycle during a year. In the east, it seldom survives more than a year (Wallace and Sullivan 1985), whereas McMullen and Condrashoff (1973) found that adult weevils in the west lived up to 5 years.

When the terminal shoot is killed, a lateral from the first whorl below the damage begins to assume dominance. The resulting deformity from repeated weevil attack can reduce the value of logs (Stiell 1979; Alfaro 1989) cut from attack trees. Spruce weevil-caused losses in volume for sawlog sizes can range from 22% to 63% and the associated reductions in lumber value average 25% owing to lower grades (Brace 1972). More recently, large-scale white pine weevil infestations in white spruce and jack pine plantations have seriously affected the quality of pulp logs in northern Ontario (A. Retnakaran, unpubl. data).

The lodgepole terminal weevil typically infests the current year's terminal growth of lodgepole pine, *Pinus contorta*, var. *latifolia*, and jack pine. There may be several larvae in a shoot and they normally mine upwards. Overwintering is generally in the larval stage. There is uncertainty in current taxonomic work; the insect may also attack pine boles.

Control Strategies

Adult weevils feed openly during the summer, and can be controlled as they are exposed on the leader. Fall control of pre-diapause spruce weevils can be considered successful only if it substantially exceeds the normal natural overwintering mortality (Dixon et al. 1979; Dixon and Houseweart 1982; VanderSar 1977), which can reach up to 80%. Nevertheless, the escape of only a few weevils out of a potential of 50 to 100 per leader can maintain a static level of continuing attack.

In general, spruce are considered susceptible to attack when they are 5 or more years old and 2 to 15 m or more high, and when the terminal leader is 4 mm or more in diameter. However, younger and smaller trees can be attacked, especially when the available trees are limited in number.

A simulation model of white pine weevil on spruce in western Canada was developed by McMullen et al. (1987). This model provides the background necessary for understanding the interaction of the spruce weevil and its host and for examining the relative merits of various control strategies.

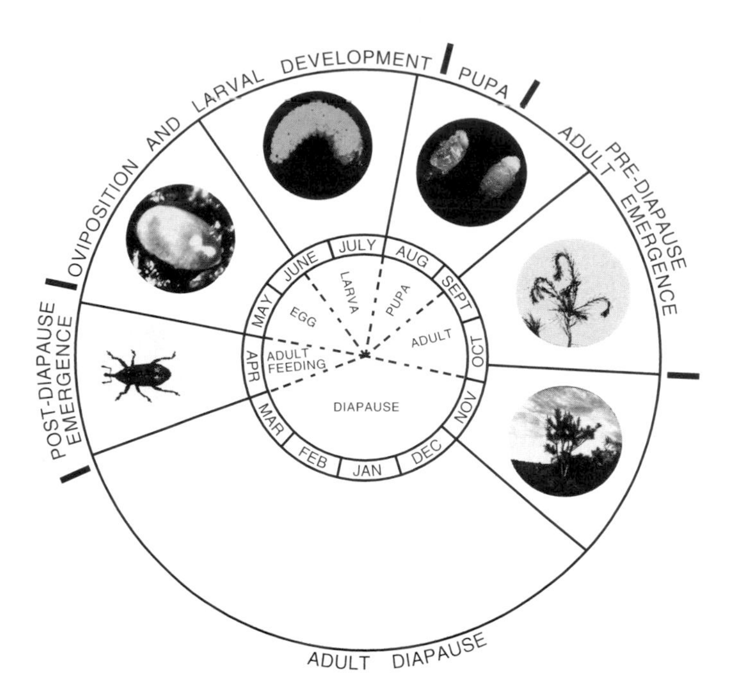

Figure 1. Life history of the white pine weevil, *Pissodes strobi*.

Several different approaches to terminal weevil control on spruce have been attempted and they can be broadly classified into two groups. The same principles should apply to the other terminal weevil species should it be necessary to apply controls. The prophylactic methods include those that prevent weevil attack, whereas the therapeutic methods include those where the weevil population is vigorously managed by control agents (Table 1) (de Groot 1985).

Prophylactic Control Methods

This approach is basically to prevent weevils from ovipositing in the terminal leader, thereby preventing the internal larval feeding that causes the leader to die.

Resistant trees—It is theoretically possible to select host trees resistant to weevil attack. There has been some limited success in detection of resistant types but such varieties are not available for large-scale planting yet (Zsuffa 1985).

Shading the host trees—When trees are grown under the shade of other trees, as an understory, the chances of weevil attack are reduced. This method is less effective with hardwoods, because the weevils are active before the hardwoods leaf out in the spring. Such shading can also result in growth loss (Berry and Stiell 1976; Droska 1982; Sullivan 1961). Stevenson and Petty (1968) reported that the incidence of lodgepole terminal weevil attack is greatest in open-grown stands of regeneration lodgepole pine and jack pine.

Moist overwintering sites—Increasing the density of a stand results in the site being more moist, which acts as a deterrent to overwintering weevils. But such a dense stand on a moist site often favors white pine blister rust, a serious fungal pathogen of the white pine.

Leader dimensions—Leaders that are longer and larger in diameter are clearly preferred by the lodgepole terminal weevil (Maher 1982).

Clipping and burning infested leaders—This is actually a mechanical sanitation method that is labor-intensive and nonpolluting. When done with care so that virtually all the attacked leaders are removed before the weevils can escape, this method can be helpful in reducing weevil damage. Leaders that are clipped must be treated or burned because populations can complete their life cycle and emerge as adults after clipping (Maher 1982). It is best if surrounding areas that harbor the weevil and act as reservoirs are also treated.

Infested spruce leaders are topped by deformed "shepherd's crooks" (the new, developing current year's growth which dies due to weevil mining in last year's leader below it) and should be clipped and burnt in July or early August after parasitism but before the weevils emerge. Pine leaders infested by lodgepole terminal weevils must be clipped by early spring. Kovacs (1988) suggested clipping in February because most of the attacked leaders have changed color by then and snow in the plantations makes it easy to get around and reach the tops of higher trees.

Unfortunately, few studies have compared clipped and unclipped stands over several years; most studies consist of counts made only in clipped stands. Also, it is difficult to find two or more comparable stands.

The major disadvantage of the clipping method is that it destroys the weevils' natural enemy complex while eliminating much of the weevil population. An improvement on this method, protecting the natural enemies, is discussed under Biological Control. Another problem with clipping is that it is often not done thoroughly enough: some leaders invariably are missed because the attack is overlooked, or it is not visible at clipping time, or it is too high to reach with the tools available. In addition, weevils may fly in from unclipped areas and some weevils overwinter more than one year, which is significant at least at the beginning; those already at large when clipping is done will not be removed by clipping. Thus, there will always be some weevils in a stand to attack enough leaders to become new foci of infestation if they are not dealt with promptly each year.

Pheromones—Many insects produce an aggregation pheromone, which is used as a cue for mating. It is possible to use pheromone-baited traps to trap-out such insects or to confuse them by spraying an entire area with pheromone. Thus, they will be unable to locate the source, which will prevent them from aggregating and mating. It has been difficult to demonstrate the existence of a pheromone system in these weevils. It is conceivable that host chemicals play a more important role in weevil aggregation, and consequently mating, than pheromones (J.H. Borden, pers. comm.).

Host-chemicals—Several different chemicals, especially terpenes, from the host plant attract or deter the weevil to or from the host. Alfaro et al. (1979, 1980) determined that chemicals that induce feeding constitute a complex mixture of volatile and nonvolatile compounds in the phloem and on the surface

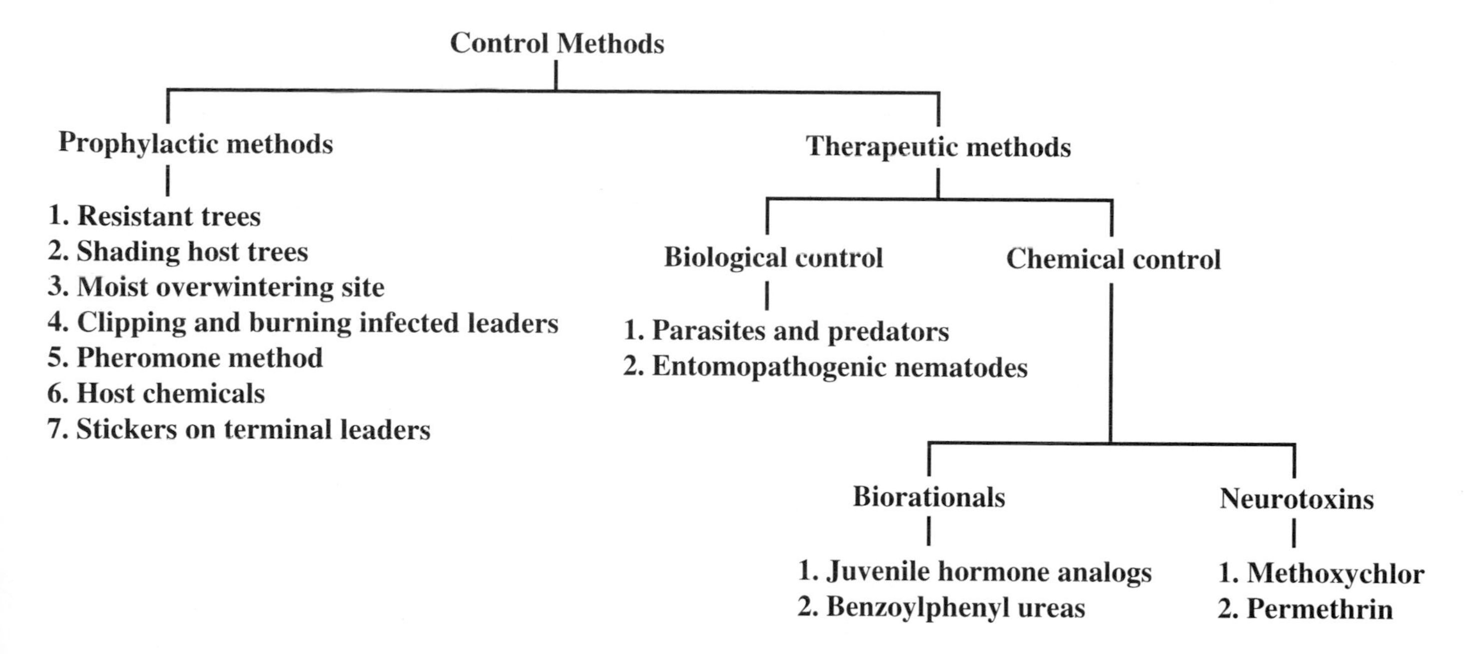

Figure 2. Methods of weevil control attempted.

of the bark and needles. Possibly a method could be developed to interfere with the host-selection process. Alfaro et al. (1981) determined that leaf oils from western red cedar, *Thuja plicata*, and pine oils (Alfaro et al. 1984) contained feeding deterrents for the whitepine weevil.

Alfaro and Borden (1985) investigated the constituents of Sitka spruce leaders, finding chemicals that were feeding stimulants, deterrents, and repellants. Many terpenes were attractive at low concentrations but deterred feeding at high concentrations. This information could be used in the development of weevil-resistant varieties or pest management strategies based on chemical feeding deterrents. Preliminary efforts at spraying these compounds in the field showed very little effect, but more work is warranted. Alfaro (1988) tested whether host preference in feeding experiments could be changed by rearing the weevils in the nonhost lodgepole pine. Because progeny emerged from lodgepole pine still preferred their host Sitka spruce, Alfaro (1988) concluded that host preference in this insect was probably genetically determined. It was felt that resistance might be introduced by crossing resistant with otherwise more desirable species of nonresistant spruce.

Stickers on terminal leaders—Certain stickers can be applied as a spray to the terminal leader early in spring prior to weevil emergence to substantially reduce weevil attacks. The results have been inconsistent but warrant further research (A. Retnakaran, unpubl. data).

Therapeutic Control Methods

These methods actively control weevil populations by destroying certain stages of the weevil and can be classified as biological control and chemical control.

Biological Control

This approach includes the use of living organisms for weevil control. A complex of predators and parasitoids occurs naturally with the terminal weevils and in some cases may significantly reduce their numbers.

In eastern Canada, few natural enemies are associated with the white pine weevil. Attempts to infect the different stages of the weevil with certain strains of entomopathogenic nematodes have been unsuccessful. Perhaps there are other strains that may be infective. For instance, the clover root weevil, *Sitona hispidulus*, was successfully infected with two different strains of nematodes (Jaworska and Wiech 1988).

Biological control in western Canada has involved attempts to enhance the naturally occurring native insect enemies of the white pine weevil that occur in varying numbers wherever the weevils are found. The white pine weevil, *P. strobi*, lodgepole terminal weevil, *P. terminalis*, and a terminal weevil, *M. gentilis*, are native to North America and do not occur elsewhere; hence, one cannot search for their natural enemies abroad. However, there has been some work on importing enemies of similar bark and wood-boring insects from outside the continent. Attempts are being made to import *Coeloides pissodis* from *P. strobi* from eastern Canada to British Columbia.

Native predators and parasites of the white pine weevil in the west (coastal British Columbia) were summarized by Alfaro et al. (1985). The principal species are a predator, *Lonchaea corticis* (Diptera: Lonchaeidae), and a number of hymenopteran parasites: *Bracon pini*, *Allodorus crassigaster* (Braconidae); *Dolichomitus terebrans* (Ichneumonidae); *Rhopalicus pulchripennis* (Pteromalidae); and *Eurytoma pissodis*, *E. cleri*, *E. picea* (Eurytomidae). Stevenson (1967) and VanderSar (1978) had identified the above same species as occurring in interior British Columbia. The principal

Table 1. Results of 1987–88 ground spray trials of Dimilin® against the white pine weevil, *Pissodes strobi*.

Treatment	Location & area	Formulation & dosage	Time of application	Tree species	Damage
1	Madoc, Ont. 5.0 ha	Control	–	White pine 2 m	23%
2	Madoc, Ont. 1.9 ha	25% WP in water 263 g AI in 158 L/ha	Fall 1987	White pine 2 m	7%
3	Madoc, Ont. 6.0 ha	ODC-45 in 7N oil 500 g AI in 500 L/ha	Spring 1988	White pine 2 m	3%
4	Kirkwood, Ont. 1.0 ha	Control	–	Scots pine 3 m	47%
5	Kirkwood, Ont. 1.0 ha	25% WP in water 375 g AI in 100 L/ha	Fall 1987	Scots pine 3 m	12%
6	Kirkwood, Ont. 1.0 ha	25% WP in water 375 g AI in 150 L/ha	Spring 1988	Scots pine 3 m	2%
7	Kirkwood, Ont. 0.25 ha	SC-48 in water 500 g AI in 40 L/ha	Spring 1988	Scots pine 3 m	36%
8	Kirkwood, Ont. 0.25 ha	ODC-45 in '585' oil 500 g AI in 40 L/ha	Spring 1988	Scots pine 3 m	14%
9	Beauceville, Que. 0.8 ha	Control	–	Norway spruce 4.5 m	40%
10	Beauceville, Que. 0.4 ha	25% WP in water 250 g AI in 200 L/ha	Fall 1987	Norway spruce 4.5 m	22%
11	Beauceville, Que. 0.4 ha	25% WP in water 250 g AI in 200 L/ha	Spring 1988	Norway spruce 4.5 m	36%
12	St. Narcisse, Que. 0.4 ha	Control	–	Norway spruce 3 m	21%
13	St. Narcisse, Que. 0.4 ha	ODC-45 in 7N oil 800 g AI in 100 L/ha	Spring 1988	Norway spruce 3 m	0%

natural enemy appears to be *L. corticis* (Alfaro and Borden 1980; Hulme and Harris 1989).

All of the above, except *Lonchaea* and *Allodorus* but with the addition of a *Brachistes* sp., also attack the lodgepole terminal weevil, *P. terminalis* (Stevenson and Petty 1968). Kovacs (1988) studied a complex of *P. terminalis*, *M. gentilis*, and *Cylindrocopturus* sp. He found *R. pulchripennis* to be the dominant parasite, with two species of *Eurytoma* ranking second in abundance. He also found the ichneumonid *D. terebrans* and a group of other Hymenoptera: *Allodorus* sp., *Mesopolobus* sp. (Pteromalidae), *Hyssopus* sp. (Eulophidae), *Telenomus* sp. (Scelionidae), and *Coeloides rufovariegatus* (Braconidae). Kovacs (1988) suggested the growing of *Lupinus* sp. in association with pine plantations to provide attractive feeding sites for some of the parasitoids. Maher (1982) also listed several other parasitoid species of the lodgepole terminal weevil reported in the literature from the United States.

While much of the work to date has involved identifying, counting, and studying the biology of the principal native enemies, some of this research, particularly that on the white pine weevil, has included attempts to cage-rear these native enemies with mass, inundative releases in mind. To date they have proven difficult to rear because of the necessity to also rear weevils as hosts; both *Pissodes* and their parasitoids and predators have proven to be fairly host-specific and cannot be raised in any great numbers on simple-to-rear alternate hosts or artificial media.

A modification of spruce leader clipping is currently being studied in British Columbia. This system combines mechanical and biological control. It involves clipping as before, but with the added refinement of releasing the natural enemies back into the stand, rather than killing them along with the weevils. Hulme et al. (1986, 1987) studied ways of doing this by cold treatments (the weevil is more sensitive to low temperatures than the natural enemies) and by enclosing the clippings in a screened cage with a mesh size that confines the weevils but releases the natural enemies. This enhancement of the weevil's natural enemies and terminal leader clipping technique is being tested.

Kovacs (1988) suggested that the screen technique would not work with the complex of pine terminal weevils because *Cylindrocopturus* was too small and would escape through any screen designed to release *Pissodes* parasitoids and predators. The cold-hardiness of *P. terminalis*, *M. gentilis* and *Cylindrocopturus* has not been tested.

The one attempt at introducing a *Pissodes* parasitoid into Canadian forests was made in 1950 when a *Coeloides* sp. from Europe was released in southern Quebec (McGugan and Coppel 1962). It was later determined that this parasite closely resembled a native species, *C. pissodis*.

Cooperation for recent western introductions has been arranged with the Commonwealth Institute of Biological Control in Europe. For the past several years, small numbers of three species, *Eubazus* (*Allodorus*) *semirugosus*, *Coeloides sordidator*, and *Scambus sudeticus*, from Scots pine cones that have been attacked by *Pissodes validirostris* have been imported and cage-released at the Pacific Forestry Centre in British Columbia. All have attacked and been reared through to adults on the white pine weevil. There were no free releases in this round of introductions.

Chemical Control

Classical chemical control involves the application of insecticides by aircraft or ground means to kill weevil adults overwintering in the ground or feeding and ovipositing on branches, or to kill eggs, larvae, and pupae which are under the bark or within the wood of the leader shoot.

The chemicals that have been used to control the weevil fall under the two following categories: biorationals and related compounds—the so-called insect growth regulators that act on a unique insect or arthropodan system; and broad-spectrum insecticides—neurotoxins that act as cholinesterase inhibitors.

Insect Growth Regulators

Juvenile hormone analogs—The pre-diapause weevils that are ready to enter the duff to overwinter in early fall lack juvenile hormone. They are sexually inactive and the sex organs are not fully developed. When these insects are treated with a juvenile hormone analog, the sex organs, especially the ovaries in the female, become fully developed; the weevils become sexually active, mate, and lay fertile eggs that do not survive the winter. The reserve fats in the fat bodies of the adults, stored for overwintering, are mobilized and the hormone-treated pre-diapause weevils are thus unprepared for overwintering (Retnakaran 1974). Similar treatment of adults in the spring resulted in retardation of progeny growth rates (McMullen and Sahota 1974).

Chitin synthesis inhibitors—Benzoylphenyl ureas selectively inhibit chitin synthesis in arthropods and are environmentally acceptable (Retnakaran and Wright 1987). Curculionids show a characteristic ovicidal response that was initially discovered on the boll weevil, *Anthonomus grandis grandis* (Wright and Villavaso 1983). The eggs get contaminated with diflubenzuron (Dimilin®) either directly from the female or indirectly from the male during sperm transfer. The embryo inside the egg fails to molt, normally resulting in mortality. Such an ovicidal effect could be effective against the white pine weevil, which is univoltine and lays only one clutch of eggs. Field trials conducted with several different formulations of diflubenzuron have shown that the weevils are well controlled with little or no damage to the trees if (i) the insecticide is applied early in the spring before the weevils emerge from the duff, and (ii) the coverage is good, with a substantial number of small droplets on the terminal shoot (Table 1).

In the west, Sahota and McMullen (1979) investigated some insect growth regulators, concluding that both diflubenzuron and precocene, a juvenile hormone antagonist, significantly reduced progeny production in the white pine weevil but the effect was not sufficient to warrant field trials.

Conventional Insecticides (Neurotoxins)

The other type of chemical control is the use of conventional or broad-spectrum insecticides that are, for the most part, neurotoxic and act on the cholinesterase enzyme system.

Early work in western Canada on the white pine weevil was done by Silver (1968), who achieved satisfactory control with DDT and phosphamidon with a hand-sprayer but concluded that repeated applications would be necessary to protect a plantation throughout the critical years. More recent work on this same pest was done by Ilnytzky (1973), who tested several insecticides; he concluded that fenitrothion and methyl trithion were the most promising compounds, effective for control at 2% minimum concentrations. Further testing of ground and aerial applications under field conditions was recommended. A list of the literature on chemical control of the weevil was prepared by Sutherland and DeBoo (1973).

The chemical most commonly used in recent years was methoxychlor, which has been applied at rates as high as 2.5 kg/ha (de Groot 1985; Deboo and Campbell 1971, 1972a, 1972b, 1974a, 1974b; Sundaram 1973, 1974, 1975, 1977; Sundaram and LeCompte 1975; Sundaram et al. 1972). Proper timing of applications and complete coverage were essential for success. Aerial applications did not appear to be as persistent as ground applications. Synthetic pyrethroids such as permethrin are effective but the lack of persistence is a problem.

No significant work on silvicultural or chemical control has been done on the lodgepole terminal weevil as its impact on plantations has been much less than that of the white pine weevil (Stevenson and Petty 1968).

Summary

The cryptic life style of the damaging larval stages of spruce and pine terminal weevils, particularly *Pissodes strobi*, together with the general ineffectiveness of systemic insecticides on conifers, makes them extremely difficult insect pests to control. While considering various options for control, it is important to take a holistic approach. For instance, growing the host tree under a canopy might lessen weevil damage but favors disease incidence and limits tree growth. At present, the mechanical procedure of clipping infested leaders, possibly with natural enemy augmentation, and early treatment with methoxychlor, when warranted, appear to be practical options. Control with diflubenzuron promises to be an exciting possibility in the near future. Use of host volatiles and entomopathogenic nematodes, *inter alia*, are certainly worth examining.

References

Alfaro, René I. 1988. Laboratory feeding and colonization of non-host lodgepole pine by two populations of *Pissodes strobi* (Peck) (Coleoptera: Curculionidae). Can. Entomol. 120: 167-173.

Alfaro, René I. 1989. Stem defects in Sitka spruce induced by Sitka spruce weevil, *Pissodes strobi* (Peck.). Pages 177-185 *in* René I. Alfaro and S.G. Glover (Eds.), Insects affecting reforestation: biology and damage. Proc. IUFRO working group on insects affecting reforestation (S2. 07-03), XVIII Int. Congr. of Entomol., Vancouver, Canada, July 3-9, 1988. For. Can. Victoria, B.C. 256 p.

Alfaro, R.I.; Borden, J.H. 1980. Predation by *Lonchaea corticis* (Diptera: Lonchaeidae) on the white pine weevil, *Pissodes strobi* (Coleoptera: Curculionidae). Can. Entomol. 112: 1259-1270.

Alfaro, R.I.; Borden, J.H. 1985. Factors determining the feeding of the white pine weevil (Coleoptera: Curculionidae) on its coastal British Columbia hosts, Sitka spruce. Proc. Entomol. Soc. Ont. Suppl. 116: 63-66.

Alfaro, R.I.; Borden, J.H.; Harris, L.J.; Nijholt, W.E.; McMullen, L.H. 1984. Pine oil, a feeding deterrent for the white pine weevil, *Pissodes strobi* (Coleoptera: Curculionidae). Can. Entomol. 116: 41-44.

Alfaro, R.I.; Hulme, M.A.; Harris, J.W.E. 1985. Insects associated with the Sitka spruce weevil, *Pissodes strobi* (Col.: Curculionidae) on Sitka spruce, *Picea sitchensis* in British Columbia, Canada. Entomophaga 30(4): 415-418.

Alfaro, R.I.; Pierce, Jr., H.D.; Borden, J.H.; Oehlschlager, A.C. 1979. A quantitative feeding bioassay for *Pissodes strobi* Peck (Coleoptera: Curculionidae). J. Chem. Ecol. 5(5): 663-671.

Alfaro, R.I.; Pierce, Jr., H.D.; Borden, J.H.; Oehlchlager, A.C. 1980. Role of volatile and nonvolatile components of Sitka spruce bark as feeding stimulants for *Pissodes strobi* Peck (Coleoptera: Curculionidae). Can. J. Zool. 58(4): 626-632.

Alfaro, R.I.; Pierce, Jr., H.D.; Borden, J.H.; Oehlchlager, A.C. 1981. Insect feeding and oviposition deterrents from western red cedar foliage. J. Chem. Ecol. 7(1): 39-48.

Belyea, R.M.; Sullivan, C.R., 1956. The white pine weevil: a review of current knowledge. For. Chron. 32(1) 58: 67.

Berry, A.B.; Stiell, W.M. 1976. Control of white pine weevil damage through manipulation of stand climate: preliminary results. Petawawa For. Exp. Stn., Chalk River, Ont. Inf. Rep. PS-X-61. 8 p.

Brace, L.G. 1972. Weevil control could raise value of white pine by 25%. Can. For. Ind. 92: 42-45.

Deboo, R.R.; Campbell, L.M. 1971. Plantation research: IV. Field evaluation of insecticides for control of white-pine weevil (*Pissodes strobi*) in Ontario, 1971. Chem. Cont. Res. Inst., Ottawa, Ont. Inf. Rep. CC-X-11. 22 p.

DeBoo, R.F.; Campbell, L.M. 1972a. Plantation research: VI. Hydraulic sprayer applications of insecticides for control of white pine weevil (*Pissodes strobi*) in Ontario. Dept. Environ., Can. For. Serv., Ottawa, Ont. Inf. Rep. CC-X-24. 15 p.

DeBoo, R.F.; Campbell, L.M. 1972b. Plantation research: VII. Experimental aerial applications of methoxychlor for control of white-pine weevil (*Pissodes strobi*) in Ontario, 1972. Chem. Cont. Res. Inst., Ottawa, Ont. Inf. Rep. CC-X-25. 31 p.

DeBoo, R.F.; Campbell, L.M. 1974a. Plantation research: X. Experimental aerial applications of methoxychlor and Gardona for control of white pine weevil (*Pissodes strobi*) in Ontario, 1973. Chem. Cont. Res. Inst., Ottawa, Ont. Inf. Rep. CC-X-68. 23 p.

DeBoo, R.F.; Campbell, L.M. 1974b. Plantation research: XI. Experimental aerial applications of methoxychlor and carbaryl for control of white pine weevil (*Pissodes strobi*) in Ontario, 1974. Chem. Cont. Res. Inst., Ottawa, Ont. Inf. Rep. CC-X-87. 30 p.

de Groot, P. 1985. Chemical control of insect pests of white pine. White Pine Symposium Suppl. Proc. Entomol. Soc. Ont. 116: 67-71.

Dixon, W.N.; Houseweart, M.W. 1982. Life tables of the whitepine weevil, *Pissodes strobi*, in central Maine. Environ. Entomol. 11: 555-564.

Dixon, W.N.; Houseweart, M.W.; Sheffer, S.M. 1979. Fall temporal activity and overwintering sites of the white pine weevil, *Pissodes strobi*, in central Maine. Ann. Entomol. Soc. Am. 72: 840-844.

Droska, J.S. 1982. Responses of the white pine weevil to selected environmental factors in the fall. M.Sc. Thesis, University of Maine, Orono. 60 p.

Hulme, Michael A.; Harris, John W.E. 1989. How *Lonchaea corticis* Taylor may impact broods of *Pissodes strobi* (Peck) in *Picea sitchensis* (Bong.) Carr. Pages 161-166 *in* René I. Alfaro and S.G. Glover, (Eds.), Insects affecting refraction: biology and damage. Proc. IUFRO working group on insects affecting reforestation (S2.07-03), XVIII Int. Congr. of Entomol., Vancouver, Canada, July 3-9, 1988. For. Can., Victoria, B.C. 256 p.

Hulme, Michael A.; Dawson, Allan F.; Harris, John W.E. 1986. Exploiting cold-hardiness to separate *Pissodes strobi* (Peck) (Coleoptera: Curculionidae) from associated insects in leaders of *Picea sitchensis* (Bong.) Carr. Can. Entomol. 118: 1115-1122.

Hulme, Michael A.; Harris, John W.E.; Dawson, Allan F. 1987. Exploiting adult girth to separate *Pissodes strobi* (Peck) (Coleoptera: Curculionidae) from associated insects in leaders of *Picea sitchensis* (Bong.) Carr. Can. Entomol. 119: 751-753.

Ilnytzky, S. 1973. Evaluation of residual toxicity of six insecticides for control of Sitka spruce weevil. Can. For. Serv. Bi-mon. Res. Notes 29(2): 9-10.

Jaworska, M.; Wiech, K. 1988. Susceptibility of the clover root weevil, *Sitona hispidulus* F. (Col., Curculionidae) to *Steinernema feltiae S. bibionis* and *Heterorhabditis bacteriophora*. J. Appl. Entomol. 106: 373-376.

Kovacs, Ervin. 1988. Terminal weevils of lodgepole pine and their parasitoid complex in British Columbia. Master's Thesis, University of British Columbia. 107 p.

Little, E.L., Jr. 1971. Atlas of United States trees. Vol 1. Conifers and important hardwoods. USDA For. Serv. Misc. Publ. 1146.

McGugan, B.M.; Coppel, H.C. 1962. Part II. Biological control of forest insects—1910-1958. Pages 40, 53 *in* J.H. McLeod, B.M. McGugan, and H.C. Coppel, A review of the biological control attempts against insects and weeds in Canada. Tech. Comm. 2, Commonw. Inst. Biol. Cont., Trinidad. Commonw. Agric. Bureaux, Eng. 216 p.

McMullen, L.H.; Condrashoff, S.F. 1973. Notes on dispersal, longevity and overwintering of adult *Pissodes strobi* (Peck) (Coleoptera: Curculionidae) on Vancouver Island. J. Entomol. Soc. B.C. 70: 22-26.

McMullen, L.H. Sahota, T.S. 1974. Effect of a juvenile hormone analogue on developmental rate and growth rate of progeny in *Pissodes strobi* (Coleoptera: Curculionidae). Can. Entomol. 106: 1015-1018.

McMullen, L.H.; Thomson, A.J.; Quenet, R.V. 1987. Sitka spruce weevil (*Pissodes strobi*) population dynamics and control: a simulation model based on field relationships. Can. For. Serv. Inf. Rep. BC-X-288. 20 p.

Maher, Thomas Francis. 1982. The biology and impact of the lodgepole terminal weevil in the Cariboo Forest Region. Master's Thesis, University of British Columbia. 84 p.

Retnakaran, A. 1974. Induction of sexual maturity in the white pine weevil, *Pissodes strobi* (Coleoptera: Curculionidae), by some analogues of juvenile hormone. Can. Entomol. 106: 831-834.

Retnakaran, A.; Wright, J.E. 1987. Control of insect pests with benzoylphenyl ureas. *In* Chitin and benzoylphenyl ureas. Dr. W. Junk Publishers, Dordrecht, the Netherlands.

Sahota, T.S.; McMullen, L.H. 1979. Reduction in progeny production in the spruce weevil, *Pissodes strobi* Peck, by two insect growth regulators. Can. For. Serv. Bi-mon. Res. Notes 35(6): 32-33.

Smith, S.G.; Sugden, B.A. 1969. Host trees and breeding sites of native North American *Pissodes* bark weevils, with a note on synonymy. Ann. Entomol. Soc. Am. 62(1): 146-148.

Silver, G.T. 1968. Studies on the Sitka spruce weevil, *Pissodes sitchensis*, in British Columbia. Can. Entomol. 100(1): 93-110.

Stiell, W.M. 1979. Releasing unweeviled white pine to ensure first-log quality of final crop. For. Chron. 55: 142-143.

Stevenson, Robert E. 1967. Notes on the biology of the Engelmann spruce weevil, *Pissodes engelmanni* (Curculionidae: Coleoptera) and its parasites and predators. Can. Entomol. 99(2): 201-213.

Stevenson, R.E.; Petty, J.J. 1968. Lodgepole terminal weevil (*Pissodes terminalis* Hopkins) in the Alberta/Northwest Territories Region. Can. For. Rural Dev. Bi-mon. Res. Notes 24(1): 6.

Sullivan, C.R. 1961. The effect of weather and physical attributes of white pine leaders on the behavior and survival of the white pine weevil, *Pissodes strobi* Peck, in mixed stands. Can. Entomol. 93: 721-741.

Sundaram, K.M.S. 1973. Persistence studies of insecticides: I. Aerial application of methoxychlor for control of white pine weevil in Ontario, 1973. Chem. Cont. Res. Inst., Ottawa, Ont. Inf. Rep. CC-X-57. 34 p.

Sundaram, K.M.S. 1974. Persistence studies of insecticides: II. Degradation of Gardona on white pine leaders (*Pinus strobus* L.) after aerial application for control of white pine weevil (*Pissodes strobi* Peck) in Ontario, 1973. Chem. Cont. Res. Inst., Ottawa, Ont. Inf. Rep. CC-X-62. 31 p.

Sundaram, K.M.S. 1975. Persistence studies of insecticides: VI. Degradation of methoxychlor on white pine leaders (*Pinus strobus* L.) after aerial application for control of white pine weevil (*Pissodes strobi* Peck) in Ontario, spring 1974. Chem. Cont. Res. Inst., Ottawa, Ont. Inf. Rep. CC-X-99. 18 p.

Sundaram, K.M.S. 1977. A study on the comparative deposit levels and persistence of two methoxychlor formulations used in white pine weevil control. Chem. Cont. Res. Inst., Ottawa, Ont. Inf. Rep. CC-X-142. 14 p.

Sundaram, K.M.S.; LeCompte, P.E. 1975. Persistence studies of insecticides: V. Degradation of carbaryl on white pine leaders (*Pinus strobus* L.) after aerial application for control of white pine weevil (*Pissodes strobi* Peck) in Ontario, 1974. Chem. Cont. Res. Inst., Ottawa, Ont. Inf. Rep. CC-X-98. 16 p.

Sundaram, K.M.S.; Smith, G.G.; O'Brien, W.; Bonnet, D. 1972. A preliminary report on the persistence of methoxychlor for the control of white pine weevil in plantations. Chem. Cont. Res. Inst., Ottawa, Ont. Inf. Rep. CC-X-31. 27 p.

Sutherland, B.G.; DeBoo, R.F. 1973. White pine weevil bibliography. Chem. Cont. Res. Inst., Ottawa, Ont. Inf. Rep. CC-X-40. 26 p.

VanderSar, T.J.D. 1977. Overwintering survival of *Pissodes strobi* (Peck) (Coleoptera: Curculionidae). J. Entomol. Soc. B.C. 74: 37.

VanderSar, T.J.D. 1978. Emergence of predator and parasites of the white pine weevil, *Pissodes strobi* (Coleoptera: Curculionidae) from Engelmann spruce. J. Entomol. Soc. B.C. 75: 14-18.

VanderSar, T.J.D.; Borden, J.H.; McLean, J.A. 1977. Host preference of *Pissodes strobi* Peck (Coleoptera: Curculionidae) reared from three native hosts. J. Chem. Ecol. 3(4): 377-389.

Wallace, D.R.; Sullivan, C.R. 1985. The white pine weevil, *Pissodes strobi* (Coleoptera: Curculionidae): A review emphasizing behavior and development in relation to physical factors. Proc. Entomol. Soc. Ont., suppl. to Vol. 116: 39-62. White Pine Symp., Petawawa Natl. For. Inst., Sept. 14, 1984.

Wright, J.E.; Villavaso, E.J. 1983. Boll weevil sterility. Pages 153-177 *in* E.P. Lloyd, R.L. Ridgeway, and W.C. Cross (Eds.), Cotton insect management with special reference to the boll weevil. Plenum Press, New York.

Zsuffa, L. 1985. The genetic improvement of eastern white pine in Ontario. White Pine Symp. Suppl. Proc. Entomol. Soc. Ont. 116: 91-94.

Chapter 26

Yellowheaded Spruce Sawfly, *Pikonema alaskensis*

PETER DE GROOT

Introduction

The yellowheaded spruce sawfly, *Pikonema alaskensis*, is a native insect that is found throughout most of the range of spruce in Canada. Detailed information on the biology and on damage caused by this sawfly can be found in Rose and Lindquist (1977) and Martineau (1984, and references therein). The sawfly attacks the following spruces: white, *Picea glauca*; black, *P. mariana*; red, *P. rubens*; Sitka, *P. sitchensis*; blue, *P. pungens*; Norway, *P. abies*; and Engelmann, *P. engelmannii*. The yellowheaded spruce sawfly is primarily a defoliator (Fig. 1) of young open-grown trees in plantations, shelterbelts, ornamental plantings, hedges, nurseries, and areas of natural regeneration, and rarely of mature forests. From 3 to 4 years of moderate to severe defoliation usually kills the tree, and most trees die after complete defoliation. Trees that survive several years of attack will show a reduction in annual growth, may have dead branches, and may become predisposed to attack from other insects or diseases.

Figure 1. Defoliation of white spruce caused by the yellowheaded spruce sawfly, *Pikonema alaskensis*. (Forestry Canada, Northwest Region)

Biology

The yellowheaded spruce sawfly has one generation per year. The adults (Fig. 2) emerge from the soil during late May to mid-June, depending on local weather conditions. The adult flight period coincides with the opening of the spruce buds. Eggs are laid in a shallow slit in the current year's needles, usually one per needle, at the base. Occasionally, eggs are deposited in the bark between needles. After about a week, the eggs hatch and the young larvae begin to feed on the edges of the needles. As the larvae mature they consume entire needles (Fig. 3). When the new growth is exhausted, they continue to feed on older needles. Larvae feed for 30 to 40 days in loose groups; when feeding is completed, usually in July, the larvae drop to the ground and construct a cocoon in the soil where they overwinter as prepupae. The sawflies pupate about 2 weeks prior to emergence as adults.

The primary sex pheromone of the sawfly has been identified as (Z)-10-Nonadecenal (Bartelt and Jones 1984; Bartelt et al. 1982a, b), and the secondary pheromone is (Z)-5-Tetradecen-1-ol (Bartelt et al. 1984). Morse and Kulman (1985) used the primary sex pheromone of the sawfly and the pheromone of *Syndipnus rubiginosus*, an ichneumonid parasite, to assess the relationship between sawfly defoliation and trap catches of the two insects. Based on their findings, they suggested that surveys using the sex pheromones of both insects may be useful in estimating the extent of natural control of the pest. The sex pheromones of the sawfly have not been used in pest management programs in Canada.

Natural Control

There are numerous hymenopterous and dipterous parasites of the sawfly. Houseweart et al. (1984) compiled a list of 41 species of Hymenoptera and 9 species of Diptera that are known to be parasites or hyperparasites of yellowheaded spruce sawfly. Raizenne (1957) reported 13 species of parasites from Ontario. Twelve species of sawfly parasites attacked 46.7% of the larvae in Nova Scotia (Thompson and Kulman 1980). In Minnesota, Houseweart and Kulman (1976) observed 17 parasitic species; parasitism of the late-instar larvae ranged from 1.5% to 19.6%. Overwintering predation by insects (primarily by elaterid and carabid larvae) and small mammals (predominately voles and shrews) killed about 66% of the sawfly cocoons (Houseweart and Kulman 1976). There are at least 16 insect parasites of the sawfly in Newfoundland, and these appear to be a major factor in regulating outbreaks and preventing appreciable tree mortality (Clark et al. 1974). Natural enemies alone may have little impact, however, in quickly suppressing a population when conditions are favorable for a rapid increase in the sawfly population. A manipulated biological control program aimed at increasing egg and early larval parasitism appears to be the most feasible way to use natural enemies effectively (Houseweart and Kulman 1976). Biological control using insect parasites has not been attempted in Canada.

Entomopathogenic fungi (*Aspergillus flavus*, *Aspergillus* sp. and *Beauveria bassiana*) and one microsporidian parasite (*Thelohania pristophorae*) have been recorded as pathogenic

Figure 2. Adult yellowheaded spruce sawfly, *P. alaskensis*. (Forest Insect and Disease Survey, Ontario Region)

Figure 3. Yellowheaded spruce sawfly, *P. alaskensis*, larva feeding on white spruce. (Forestry Canada, Northwest Region)

organisms of the yellowheaded spruce sawfly in Quebec (Smirnoff and Juneau 1973). Insect disease surveys conducted by the Forest Pest Management Institute have found sawflies infected with a *Beauveria* sp. and a microsporidian (*Pleistophora* sp.). At the institute many samples of the sawfly have been examined over the years in an attempt to find a nuclear polyhedrosis virus in the population, but to date none has been found. Nuclear polyhedrosis viruses are generally species-specific, and a screening program of sawfly virus isolates available at the institute (Cunningham and Kaupp, Chapter 35) has not been undertaken; only the virus attacking the European spruce sawfly, *Gilpinia hercyniae*, has been tested on yellowheaded spruce sawfly. Third- and fourth-instar larvae on white spruce foliage in lantern globes were sprayed with a suspension of virus containing 10^7 PIB/mL (polyhedral inclusion bodies per millilitre); all larvae pupated normally and no mortality was recorded (J.R. McPhee and F.T. Bird, unpubl. data).

Despite the large number of mortality factors, artificial control of this insect may be necessary, particularly for high-value trees.

Artificial Control

When a few small trees are infested, the sawfly can be controlled easily by manually removing and destroying the larvae. If several trees are infested, or if the larvae are too numerous to remove, an insecticide application may be necessary. In Canada, experimental and operational control trials for yellowheaded spruce sawfly have used chemical insecticides exclusively. A list of the chemical insecticides tested in Canada since 1973 is given in Table 1.

Drouin and Kusch (1973–1976, 1978–1980) conducted ground spray trials to evaluate various chemical insecticides, formulations, and application methods for the control of sawfly larvae. Their aim was to gather the necessary technical data to support new Canadian registrations of insecticides and to develop suitable and low-hazard insecticide application techniques. They evaluated systemic and contact insecticides with ground spray equipment. Systemic insecticides were applied to the soil in early spring, before the sawflies began to feed. For each insecticide, the dosage was based on the tree diameter and was applied into holes dug underneath the tree crown. The contact insecticides were applied either with a hydraulic sprayer, back pack mistblower, knapsack sprayer, or an ultra-low volume (Turbair) sprayer when the sawflies were feeding. The assessment of insecticide efficacy was usually done within 2 or 3 days of application, and was based on the presence of remaining larvae. In general, most of the insecticides that Drouin and Kusch tested were effective in controlling the sawfly. As a result of the efforts by Drouin and Kusch, several insecticides were registered for control of the sawfly (Drouin and Kusch 1979). Several small-scale operational control programs have been conducted in Canada, often using malathion applied with ground spray equipment. These programs have usually been successful.

Both systemic and contact insecticides can be used to control the sawfly, but contact insecticides are preferred because they are easier to apply. Malathion is often the contact

Table 1. Experimental field trials with chemical insecticides for control of yellowheaded spruce sawfly, *Pikonema alaskensis*, in Canada, 1972 to 1979.

Insecticide		Type	Application method[a]	Application rate	Percent control	Reference[b]
acephate	SP	systemic	soil drench	120 g/0.16 kg	98	Drouin and Kusch 1976
aldicarb	G	systemic	soil drench	306 g/3.06 kg	95	Drouin and Kusch 1975
aldicarb	G	systemic	soil drench	7.8 mL/2.5 cm	80	Drouin and Kusch 1974
carbaryl	EC	contact	hydraulic sprayer	114 g/227 L	100	Drouin and Kusch 1975
carbaryl	WP	contact	mist blower	3.10 g/4.5 L	100	Drouin and Kusch 1976
carbaryl	EC	contact	mist blower	23.7 mL/3.8 L	95	Drouin and Kusch 1974
carbaryl	EC	contact	mist blower	96.2 mL/3.8 L	100	Drouin and Kusch 1973
carbofuran	G	systemic	soil drench	100 g/1.0 kg	98	Drouin and Kusch 1976
carbofuran	G	systemic	soil drench	227 g/2.27 kg	90	Drouin and Kusch 1975
carbofuran	G	systemic	soil drench	7.8 mL/2.5 cm	80	Drouin and Kusch 1974
carbofuran	EC	systemic	mist blower	29.6 mL/3.8 L	95	Drouin and Kusch 1974
chordimeform	EC	systemic	soil drench	110 g/0.22 L	64	Drouin and Kusch 1976
chlorpyriphos	WP	contact	mist blower	22.2 mL/3.8 L	95	Drouin and Kusch 1974
cypermethrin	EC	contact	mist blower	1.0 mL/L	100	Drouin and Kusch 1980
diazinon	EC	contact	hydraulic sprayer	114 g/227 L	95	Drouin and Kusch 1975
diazinon	EC	contact	mist blower	3.10 g/4.5 L	100	Drouin and Kusch 1976
diazinon	EC	contact	soil drench	125 g/0.25 L	0	Drouin and Kusch 1976
DM[c]	ULV	contact	ULV	3.4 g/56.8 mL	51	Drouin and Kusch 1976
DM	ULV	contact	ULV	1.78 mL/29.6 mL	70	Drouin and Kusch 1974
DDM[d]	ULV	contact	ULV	56.8 mL	98	Drouin and Kusch 1976
dimethoate	EC	systemic	hydraulic sprayer	114 g/227 L	100	Drouin and Kusch 1975
dimethoate	EC	systemic	mist blower	3.10 g/4.5 L	100	Drouin and Kusch 1976
dimethoate	EC	systemic	soil drench	110 g/0.22 L	98	Drouin and Kusch 1976
dimethoate	EC	systemic	mist blower	11.8 mL/3.8 L	100	Drouin and Kusch 1973
fenvalerate	EC	contact	hydraulic sprayer	0.013 mL/L	100	Drouin and Kusch 1979
fenvalerate	EC	contact	mist blower	1.0 mL/L	100	Drouin and Kusch 1980
fenitrothion	EC	contact	aerial spray	-	100	FIDS 1976
CGA 12223	G	contact	soil drench	90 g/0.90 kg	47	Drouin and Kusch 1976
isothioate	G	contact	soil drench	71 g/1.78 kg	0	Drouin and Kusch 1976
K 840	EC	systemic	mist blower	14.8 mL/3.8 L	20	Drouin and Kusch 1974
malathion	EC	contact	mist blower	88.8 mL/3.8 L	100	Drouin and Kusch 1973
malathion	ULV	contact	ULV sprayer	53.3 mL/0.946 L	95	Drouin and Kusch 1974
malathion	EC	contact	hydraulic sprayer	6.3 to 31.5 mL/L	100	Drouin and Kusch 1979
malathion	EC	contact	mist blower	3.10 g/4.5 L	100	Drouin and Kusch 1976
malathion	ULV	contact	ULV sprayer	1.0 g/56 mL	95	Drouin and Kusch 1976
malathion		contact	aerial sprayer	-	100	Applejohn and Weir 1978
methidathion	EC	contact	mist blower	1.6 mL/L	100	Drouin and Kusch 1980
methidathion	ULV	contact	ULV sprayer	28.4 mL/L	100	Drouin and Kusch 1980
methidathion	EC	contact	soil drench	112 g/0.28 L	0	Drouin and Kusch 1976
methomyl	L	contact	mist blower	3.10 g/4.5 L	100	Drouin and Kusch 1976
methomyl	WP	contact	mist blower	26.6 mL/3.8 L	100	Drouin and Kusch 1974
mexacarbate	EC	contact	mist blower	23.7 mL/3.8 L	100	Drouin and Kusch 1973
M 3726	EC	contact	mist blower	3.2 mL/L	100	Drouin and Kusch 1978
oxamyl	EC	contact	mist blower	1.6 mL/L	100	Drouin and Kusch 1980
permethrin	EC	contact	knapsack sprayer	0.2 mL/L	95	Drouin and Kusch 1980
permethrin	EC	contact	mist blower	3.2 mL/L	100	Drouin and Kusch 1978
permethrin	EC	contact	mist blower	1.6 mL/L	100	Drouin and Kusch 1980
phorate	G	systemic	soil drench	7.8 mL/L	25	Drouin and Kusch 1974
phosmet	EC	systemic	hydraulic sprayer	114 g/227 L	100	Drouin and Kusch 1975
phosmet	WP	systemic	mist blower	3.10 g/4.5 L	100	Drouin and Kusch 1976
phoxim	G	systemic	soil drench	79 g/0.79 kg	0	Drouin and Kusch 1976
PP 484	EC	systemic	mist blower	7.4 mL/3.8 L	90	Drouin and Kusch 1974
PP 505	G	systemic	soil drench	102 g/1.02 kg	90	Drouin and Kusch 1975
PP 505	G	systemic	soil drench	79.4 g/0.79 kg	67	Drouin and Kusch 1976
propoxur	EC	contact	soil drench	120 g/0.80 L	98	Drouin and Kusch 1976
resmethrin	ULV	contact	ULV spray	0.10 g/14 mL	100	Drouin and Kusch 1975
resmethrin	ULV	contact	ULV spray	0.40 g/0.057 L	91	Drouin and Kusch 1976
WL 43467	EC	contact	mist blower	3.2 mL/L	100	Drouin and Kusch 1978
WL 43479	EC	contact	mist blower	3.2 mL/L	100	Drouin and Kusch 1978
WL 43775	EC	contact	mist blower	3.2 mL/L	100	Drouin and Kusch 1978

[a] Active ingredient/unit of measure.
[b] FIDS = Forest Insect and Disease Survey Report.
[c] DM = Dichlorvos + Methoxychlor.
[d] DDM = Dimethoate + Dicofol + Methoxychlor.

insecticide of choice because it is effective, easily available, and relatively safe to use. Most applications of insecticides are made with ground spray equipment, but aerial applications are used when large areas have to be treated quickly. For example, in 1975, the Ontario Ministry of Natural Resources conducted a control operation with fenitrothion on a 105-ha white spruce plantation in Balsam Lake Provincial Park near Minden, Ontario. Excellent control was achieved (Forest Insect and Disease Survey Annual Report 1975, published 1976). In comparison with many other forest insect pests, control of the yellowheaded spruce sawfly is not difficult. Insecticide applications are easy to target because the sawfly is an open-feeding insect usually found on small and accessible trees.

References

Applejohn, M.J.; Weir, H.J. 1978. Forest insect and disease surveys in the central region of Ontario, 1977. Can. For. Serv. Great Lakes Forest Research Centre, Sault Ste. Marie, Ontario. Unpublished report.

Bartelt, R.J.; Jones, R.L. 1984. (Z)-10-Nonadecanal: a pheromonally active air oxidation product of the (Z,Z)-9,10 dienes in the yellowheaded spruce sawfly. J. Chem. Ecol. 9: 1333-1341.

Bartelt, R.J.; Jones, R.L.; Krick, T.P. 1984. (Z)-5-Tetradecen-1-ol: a secondary pheromone of the yellowheaded spruce sawfly, and its relationship to (Z)-10-nonadecenal. J. Chem. Ecol. 9: 1343-1352.

Bartelt, R.J.; Jones, R.L.; Kulman H.M. 1982a. Evidence for a multicomponent sex pheromone in the yellowheaded spruce sawfly. J. Chem. Ecol. 8: 83-94.

Bartelt, R.J.; Jones, R.L.; Kulman H.M. 1982b. Hydrocarbon components of the yellowheaded spruce sawfly sex pheromone: a series of (Z,Z)-9,19 dienes. J. Chem. Ecol. 8: 95-114.

Clark, R.C.; Clarke, L.J.; Pardy, K.E. 1974. The sawflies and horntails of Newfoundland. Can. For. Serv., Newfoundland and Labrador Region, St. John's. Inf. Rep. N-X-128.

Drouin, J.A.; Kusch, D.S. 1973. Summary of insecticide field trials on shade and shelterbelt trees in Alberta, 1972. Can. For. Serv., Northwest Region, Edmonton, Alta. File Report NOR-Y-66.

Drouin, J.A.; Kusch, D.S. 1974. Insecticide field trails on shade and shelterbelt trees in Alberta and Saskatchewan, 1973. Can. For. Serv., Northwest Region, Edmonton, Alta. Inf. Rep. NOR-X-81.

Drouin, J.A.; Kusch, D.S. 1975. Pesticide field trials on shade and shelterbelt trees in Alberta and Saskatchewan, 1974. Can. For. Serv., Northwest Region, Edmonton, Alta. Inf. Rep. NOR-X-131.

Drouin, J.A.; Kusch, D.S. 1976. Pesticide field trials on shade and shelterbelt trees in Alberta, 1975. Can. For. Serv., Northwest Region, Edmonton, Alta. Inf. Rep. NOR-X-150.

Drouin, J.A.; Kusch, D.S. 1978. Pesticide field trials on shade and shelterbelt trees in Alberta, 1977. Can. For. Serv., Northwest Region, Edmonton, Alta. Inf. Rep. NOR-X-205.

Drouin, J.A.; Kusch, D.S. 1979. Pesticide field trials on shade and shelterbelt trees in Alberta, 1978. Can. For. Serv., Northwest Region, Edmonton, Alta. Inf. Rep. NOR-X-213.

Drouin, J.A.; Kusch, D.S. 1980. Pesticide field trials on shade and shelterbelt trees in Alberta, 1979. Can. For. Serv., Northwest Region, Edmonton, Alta. Inf. Rep. NOR-X-227.

Forest Insect and Disease Survey. 1976. Pages 58 and 59, *in* Annual report of the Forest and Disease Survey, 1975, Ottawa.

Houseweart, M.W.; Kulman, H.M. 1976. Life tables of the yellowheaded spruce sawfly, *Pikonema alaskensis* (Rohwer) (Hymenoptera: Tenthredinidae) in Minnesota. Environ. Entomol. 5: 859-867.

Houseweart, M.W.; Kulman, H.M.; Thompson, L.C.; Hansen, R.W. 1984. Parasitoids of two spruce sawflies in the genus *Pikonema* (Hymenoptera: Tenthredinidae). College of Forest Resources Research Bulletin 4, Maine Agricultural Experiment Station, Misc. Rep. 294.

Martineau, R. 1984. Insects harmful to forest trees. Forestry Technical Report 32, Can. For. Serv., Ottawa. 261 p.

Morse, B.W.; Kulman, H.M. 1985. Monitoring damage by yellowheaded spruce sawflies with sawfly and parasitoid pheromones. Environ. Entomol. 14: 131-133.

Raizenne, H. 1957. Forest sawflies of southern Ontario and their parasites. Can. Dep. Agric., Biol. Div. Publ. 1009. Ottawa. 45 p.

Rose, A.H.; Lindquist, O.H. 1977. Insects of eastern spruces, fir, and hemlock. Forestry Technical Report 23, Can. For. Serv., Ottawa. 159 p.

Smirnoff, W.A.; Juneau, A. 1973. Quinze années de recherches sur les micro-organismes des insectes forestiers de la province de Québec (1957-1972). Ann. Soc. Entomol. Qué. 18: 147-181.

Thompson, L.C.; Kulman, H.M. 1980. Parasites of the yellowheaded spruce sawfly, *Pikonema alaskensis* (Hymenoptera: Tenthredinidae), in Maine and Nova Scotia. Can. Entomol. 112: 25-29.

Chapter 27

Pine False Webworm, *Acantholyda erythrocephala*

D. Barry Lyons

Introduction

The pine false webworm, *Acantholyda erythrocephala*, is endemic to the Palearctic Region, and occurs in Great Britain, in central and northern Europe as far as Lapland, and as far east as the Caucasus and western Siberia, Korea, and Japan (Middlekauff 1958; Charles and Chevin 1977). This species was first reported in North America in 1925 from Pennsylvania (Wells 1926) and has since been reported in New Jersey (Soraci 1938), New York (Middlekauff 1938), Connecticut (Plumb 1945), and the Lake States (Wilson 1977). The pine false webworm was first reported in Ontario in 1961 in Scarborough Township (Eidt and McPhee 1963). It was not until the late 1970s that the extent of its distribution and of the defoliation that it caused were realized. Damage that had been previously ascribed to other pamphiliids was correctly attributed to this insect when adults were collected and identified (Howse et al. 1982). It has been reported from throughout most of southern Ontario and in the Lake of the Woods region in the northwest (Syme 1981). Isolated populations have subsequently been detected at Thunder Bay, and in Arnott and Bastedo Townships in northern Ontario (Forestry Canada, Ontario Region, Forest Insect and Disease Survey, unpubl. data). The species was detected for the first time in western Canada during 1989, while defoliating Scots pine at Edmonton, Alberta (Edmond and Cerezke 1990).

Host Plant Relationship and Impact

Host plants of the pine false webworm are species of *Pinus*. In North America, red pine, *Pinus resinosa*, and eastern white pine, *P. strobus*, are the most favored hosts, while Austrian pine, *P. nigra*, is the least favored host (Middlekauff 1958). Scots pine, *Pinus sylvestris*, is preferred to Austrian pine in mixed stands in Europe (Jahn 1967; Schmutzenhofer 1975).

This insect has become more common in recent years in Ontario (Syme 1981). Although not widespread, it has been locally abundant in pine plantations (Fig. 1). Up to 96% defoliation has been reported in red pine plantations (Moody 1990). Even in low numbers, the sawfly is especially destructive in Christmas tree plantations (Fig. 2), where the webs and defoliation reduce the market value of the trees (Syme 1981). In New York, the species has become a severe and persistent problem and host trees have been killed (Hofacker et al. 1989).

Biology

In spring adult males emerge before females from earthen cells in the soil (Schwerdtfeger 1944). Females mate soon after emergence. Slits are cut in the flattened surface of 1-year-old pine needles (Griswold 1939) by the female's ovipositor and eggs are inserted by a crease of the chorion into these slits (Fig. 3) in contiguous rows (Middlekauff 1958). The eggs obtain moisture from the vascular system of the plant (Schwerdtfeger 1941).

Newly hatched larvae descend the needles to the twigs, where they spin silk around the base of the needles and feed gregariously on the sides above the fascicle. While maturing, the larvae feed on the basal ends of the needles that they pull toward themselves (Griswold 1939). Silk, frass, exuviae, and uneaten needles form webs (Fig. 4) that enclose the larvae. From 1 to 35 (average 6–10) larvae inhabit silken tubes within these webs (Schewerdtfeger 1944), which are multifunctional. In addition to thermoregulation, the silk-spinning habit of the larvae serves to facilitate mobility (Kolomietz 1967) and attachment to the food plant. Larvae are seldom seen on branches outside of the webs, unless they have completely defoliated the branch or until they are ready to drop. The larvae move along the substrate (i.e. a needle, a branch, or other surface) on their backs by moving their heads from side to side and attaching a silk ladder under which they travel. This behavior has been observed on numerous occasions in all but the ultimate instar. Larvae are unable to grasp their food, but bite off the needles at the base and feed as the needles are drawn into the web. Disturbed larvae retract into the web, away from the food, and need a method of securing the needles. Examination of several webs containing intact needles revealed that the larvae had guyed the needles with strands of silk before feeding on them (Lyons 1988).

Male larvae pass through five instars and females through six during their arboreal development (Schwerdtfeger 1941). In one study (Lyons 1988), the size-class intervals for instars (Fig. 5) occurred at about the same head-capsule widths for the 3 years for which frequency distributions were compared. However, the head-capsule sizes for North American populations were larger than those of a European population (Schwerdtfeger 1941).

Ultimate-instar larvae fall to the ground and burrow into the soil on completion of feeding in early summer (Middlekauff 1958). The larvae, at this stage referred to as eonymphs (Fig. 6), mold earthen cells and enter summer diapause. Most eonymphs emerge from diapause in autumn and transform to pronymphs, which have characteristic pupal eyes visible through the larval skin (Schwerdtfeger 1941). Some eonymphs may remain in diapause for several years (Schwerdtfeger 1944). This insect is essentially univoltine (Griswold 1939); pronymphs and, less commonly, eonymphs in prolonged diapause are the overwintering stages. Eonymphs were found below the humus layer at depths of 5 to 11 cm in Europe (Schwerdtfeger 1941; Jahn 1967). In Ontario, they were found from just below the soil surface to a depth of 9 cm (Lyons 1988). The mean depth

Figure 1. Red pines defoliated and webbed by the pine false webworm, *Acantholyda erythrocephala*. (D.B. Lyons)

Figure 3. Female pine false webworm ovipositing on red pine needles. (D.B. Lyons)

Figure 2. Scots pine Christmas tree plantation severely defoliated by the pine false webworm. (W.D. Biggs)

Figure 4. Web of the pine false webworm. (D.B. Lyons)

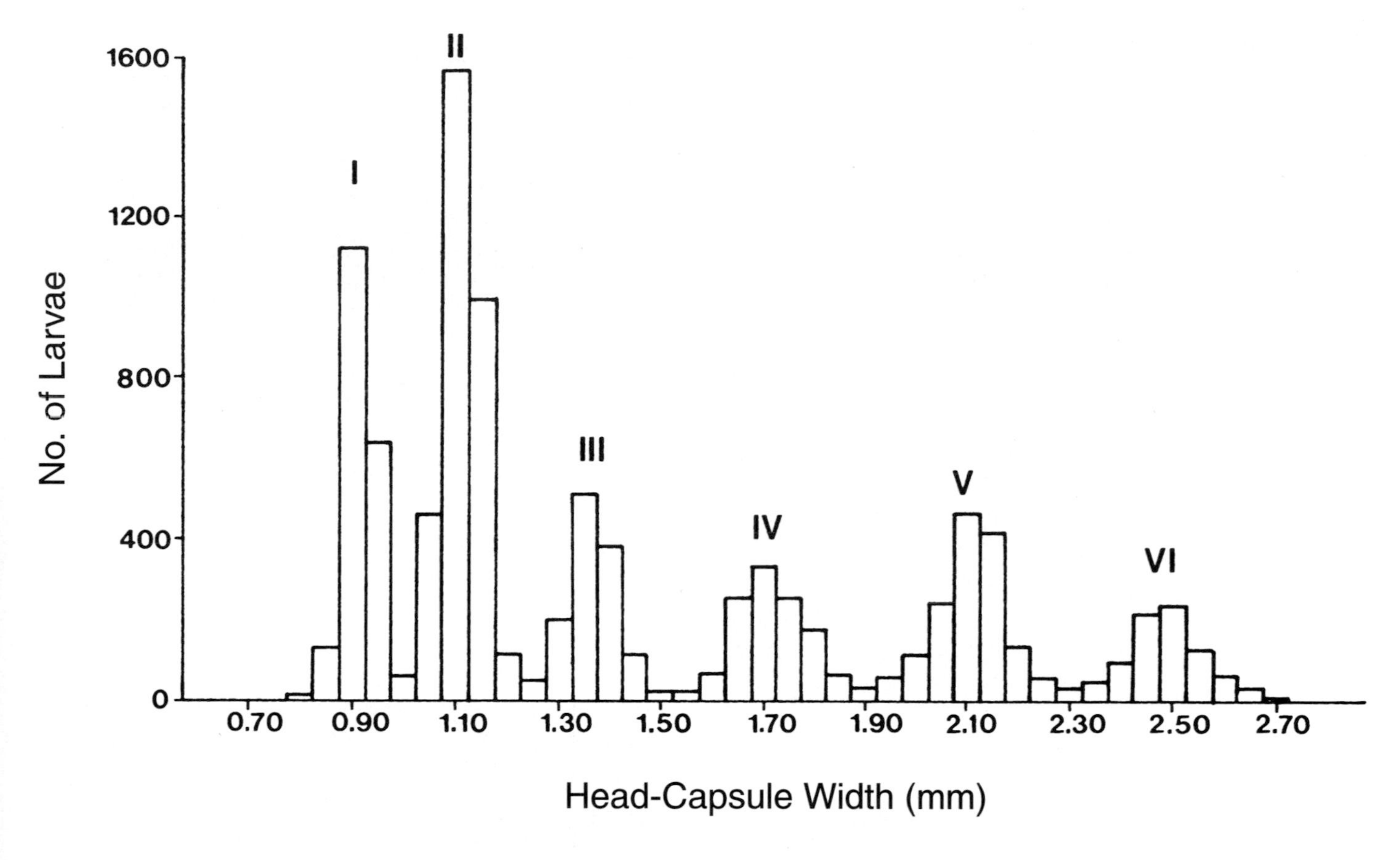

Figure 5. Head-capsule distribution of larvae of the pine false webworm, *A. erythrocephala*, from Lakehurst, Ontario, in 1986. Roman numerals indicate instar.

was less than the 5-cm minimum suggested in earlier studies. Pupation (Fig. 7) occurs in the spring when soil temperatures increase.

Females deposit small numbers of large, high-quality eggs. Rapid deposition of eggs, even at low temperatures, means that if females die prematurely, the majority of eggs has already been deposited. As a result, survival of the females has little effect on survival of their generation. Although most eggs are mature and ready to be oviposited when the female emerges from the soil, a few immature oocytes are always observed during dissections of newly emerged females. Maturation of these eggs occurs after the emergence of the females.

The degree of defoliation of the host trees resulted in variable potential fecundities. Mean (standard error) potential fecundities from heavily, moderately, and lightly defoliated zones in the same plantation were 18.7 (0.77), 25.9 (0.68), and 35.0 (0.55) eggs per female, respectively (Lyons 1988). A few females contained no mature eggs on emergence. Defoliation affects fecundity by reducing egg numbers and thus acts as a feedback system in regulating population levels.

The flight period of the pine false webworm is reported to extend from 21 days (Rumphorst and Goossen 1960) to 1 month (Jahn 1967). Adults are diurnal and peak flight activity occurs between 1 100 and 1 500 hours on sunny, calm days; activity is minimal on cool, rainy, or windy days (Rumphorst and Goossen 1960). At constant temperatures between 14.9° and 26.6°C, females begin to oviposit during daylight hours within the first 24 hours after emergence, indicating a negligible preoviposition period (Lyons 1988).

The variable sex ratio observed in field populations suggests differential mortality between the sexes or differential mechanisms of sex determination. In a European population, females generally predominated in less heavily infested habitats, and where the insects were more numerous, the sex ratio was even or favored males (Jahn 1967). For three defoliation zones examined in Ontario (Lyons 1988), males predominated where the insects were most abundant, and females were more numerous where the sawflies were least abundant. Where densities were intermediate, the sex ratio was approximately 1:1. These differences may have resulted from the failure of some females to mate before ovipositing at high densities.

Natural Control

Special insect parasitoids attack the pine false webworm in Europe (Schwerdtfeger 1941, 1944; Rumphorst and Goossen 1960; Kolomietz 1967; Schmutzenhofer 1975). The biology and descriptions of some of these species can be found in Eichhorn (1988). In North America, few species of

Figure 6. Eonymph of the pine false webworm, *A. erythrocephala.* (E.R. Rayner)

Figure 7. Pupae of the pine false webworm. (E.R. Rayner)

Figure 8. Adult *Sinophorus megalodontis*, a parasitoid of the pine false webworm. (E.R. Rayner)

Figure 9. Adult female of *Olesicampe* sp., a parasitoid of the pine false webworm. (E.R. Rayner)

parasitoids have been reared from this host. The ichneumonid *Ctenopelma erythrocephalae* was described from females collected while ovipositing in eggs of the sawfly at Oakland, New Jersey (Barron 1981).

Two additional species of ichneumonids were reared from this host in Ontario (Lyons 1988). *Sinophorus megalodontis* (Fig. 8) belongs to a species group that attacks web-spinning sawflies (Sanborne 1984). Another member of the group, *S. crassifemur*, attacks larvae of this and related species in the Palearctic Region (Schwerdtfeger 1944). The other ichneumonid reared from the pine false webworm is an undescribed species (H. Townes, American Entomological Institute, Gainesville, Fla., pers. comm.) of the poorly known genus *Olesicampe* (Fig. 9). One species of this very large cosmopolitan genus is responsible for significant mortality of pamphiliid populations (Billany et al. 1985), and some, such as *O. benefactor* and *O. geniculatae*, have been introduced successfully as biological control agents against sawflies (Muldrew 1967; Quednau and Lim 1983).

Sinophorus megalodontis and *Olesicampe* sp. are solitary endoparasitoids of the webworm and like their hosts are univoltine. *Olesicampe* sp. overwinters in the host larvae as late-instar larvae. Parasitoid larvae prevent the hosts from transforming into pronymphs. Development is completed in the spring, when the larvae emerge from the host cadavers. The parasitoid larva spins a cocoon in the earthen cell, pupates, and emerges as an adult. *Sinophorus megalodontis* emerges from the host in the fall and overwinters in its own cocoon within the host's earthen cell.

Females of both species oviposit into all instars of the host larvae. The eggs hatch soon after deposition, as evidenced by the presence of parasitoid larvae in first-instar host larvae. Parasitoid larvae remain as first instars until the last instar of the host. Adults mate and begin ovipositing soon after emergence, which is indicated by the coincidence of the onset of emergence and the occurrence of parasitized host larvae. Parasitoid attacks are distributed equally between the sexes of the host. Thus, the parasitoids are not responsible for the variable sex ratios observed in webworm populations. The increasingly greater incidence of parasitism in later-dropping host larvae suggests that the development of the larvae may be retarded by parasitism (Lyons 1988).

Females of *Olesicampe* sp. are apparently unable to discriminate between previously parasitized hosts (i.e. superparasitism) and hosts attacked by *S. megalodontis* (i.e. multiparasitism). Encapsulation of the parasitoid larvae by hemolymph cells of the host is common.

The transcontinental distribution of *S. megalodontis* (Sanborne 1984) suggests that this species is endemic to North America. Unidentified species of *Olesicampe* and *Sinophorus* have been reported to be parasitoids of species of *Cephalcia* in Canada (Eidt 1969). It is also likely that the species of *Olesicampe* discussed here originated in the New World. Even though members of these genera attack the pine false webworm in the Old World, these Nearctic species are poorly adapted to their recently introduced host. These parasitoids, as a result of a lack of synchronization with the host, poor searching ability, and encapsulation, do not have much impact on the host populations.

In a moderately infested plantation, both parasitoid species were most abundant in the upper half of the canopy, excluding the leader and first whorl of branches, and were less abundant in the lower canopy. The distribution of parasitoids is influenced by the distribution of host larvae. In a heavily defoliated plantation, host larvae were equally distributed in the high and low strata and the parasitoids were equally distributed among the hosts. Parasitized hosts were distributed uniformly around the trees.

In 1990, parasitoids were reared from eggs of the pine false webworm collected near Minden, Ontario. The wasps were identified as belonging to the *Trichogramma minutum* complex (J.D. Pinto, University of California, pers. comm.).

A virus was reported in a European population of pine false webworm, but its occurrence remains unconfirmed (Jahn 1967). Wilson (1984) successfully infected larvae with the microsporidian *Pleistophora schubergi* but because of rearing problems was unable to assess the impact on the host.

Insect predators of pine false webworm reported in the European literature include a coccinellid, *Anatis ocellata*; a carabid, *Calosoma sycophanta* (Jahn 1967); and hemipterans, snakeflies, ants (Schwerdtfeger 1944), dragonflies, and assassinflies (Fedoryak and Usol'tsev 1978).

In 1981, the services of the Commonwealth Institute of Biological Control were enlisted to review the European literature on the species and undertake surveys of the parasitoid complex of European populations of the sawfly and related species with the intention of exporting potential biological control agents to Canada. These investigations were continued until 1987, when low population densities of pamphiliids in Europe made the work impractical.

Chemical Control

There is currently no insecticide registered in Canada for use against this insect (Syme 1981). Recently, laboratory bioassays of potential insecticidal control agents, targeted against early instars of the pine false webworm, have been conducted (Lyons and Helson, unpubl. data). Field trials of the most promising of these chemicals and an assessment of their efficacy associated with insecticide spray drift were also conducted. All chemicals tested in the laboratory are common insecticides used in forestry and registered against several pest species, namely, Ambush® 500EC, Baygon® LC, Cygon® 4E, Diazinon® 50EC, Dursban® 4E, Malathion® 500EC, Methoxychlor EC, Orthene®, Sevin® XLR, and Sumithion® 20F. None of the insecticides inhibited egg hatch. Of the chemicals tested, Ambush® (permethrin) and Sevin® (carbaryl) showed the most promise at reducing larval numbers at low concentrations and were therefore selected as the best candidates for field trials. Ambush® and Sevin® are both registered for control of sawflies on pine.

Field trials were conducted in a heavily infested red pine plantation in Parry Sound District, Ontario. Insecticides were applied to individual treatment rows of approximately 1.6-m-tall trees using a Solo backpack mistblower. Treatment rows were oriented north-south and sprays were applied by walking along the west side of the trees in a southerly direction. Sevin® was applied at application rates of 125, 250, and

500 g/ha of active ingredient, while Ambush® was tested at 35 to 70 g/ha. In addition, double applications of Sevin® were made at 125 g/ha and Ambush® at 35 g/ha. Sprays were timed to coincide with the termination of adult emergence. Assessment of efficacy was done using mean mortality per treatment defined as one minus the number of larvae on a branch tip divided by the initial number of eggs per branch tip.

The best results (more than 70% mortality) were obtained with application rates of 250 g/ha or greater of Sevin® or the double applications of either insecticide. Ambush® at 70 g/ha and Sevin® at 125 g/ha were significantly better than the controls, but Ambush® at 35 g/ha showed no effect. For application rates of Sevin® at 250 and 500 g/ha, mean mortality was determined for branches in each cardinal direction, with branches in upper and lower strata pooled. For both application rates the results were remarkably similar, with a higher mortality of larvae on north- and west-facing branches compared with east- and south-facing branches. The implication from these data is that better control was achieved on the side of the trees on which the insecticides were applied, even using a mistblower on relatively small trees. Control of the webworm was hindered by spray distribution problems and probably by the protracted phenology of the insects.

To determine the effect of insecticide drift on mortality in adjacent rows, a plot was set up consisting of 11 rows by about 10 trees. Sevin® was applied to row 0 at 500 g/ha. Mortality was greatest on treatment rows and declined exponentially with distance. Tree rows adjacent to treatment rows received significant doses, which suggests that not all rows must be treated to achieve control.

References

Barron, J.R. 1981. The Nearctic species of *Ctenopelma* (Hymenoptera, Ichneumonidae, Ctenopelmatinae). Nat. Can. 108: 17-56.

Billany, D.J.; Winter, T.G.; Gauld, I.D. 1985. *Olesicampe monticola* (Hedwig) (Hymenoptera: Ichneumonidae) redescribed together with notes on its biology as a parasite of *Cephalcia lariciphila* (Wachtl.) (Hymenoptera: Pamphiliidae). Bull. Entomol. Res. 75: 267-274.

Charles, P.J.; Chevin, H. 1977. Note sur le genre *Acantholyda* (Hyménoptères - Symphytes-Pamphiliidae), et plus particulièrement sur *Acantholyda hieroglyphica* (Christ.). Rev. For. Fr. 29: 22-26.

Edmond, F.J.; Cerezke, H.F. 1990. Forest insect and disease conditions in Alberta, Saskatchewan, Manitoba, and the Northwest Territories in 1989 and predictions for 1990. For. Can., Northwest Region. Inf. Rep. NOR-X-313, 23 p.

Eichhorn, O. 1988. Studies on the spruce webspinning sawflies of the genus *Cephalcia* spp. Panz. (Hym., Pamphiliidae). II. Larval- and nymph parasites. J. Appl. Entomol. 105: 105-140.

Eidt, D.C. 1969. The life history of a web-spinning sawfly of spruce, *Cephalcia fascipennis* (Cresson) (Hymenoptera: Pamphiliidae). Can. Entomol. 97: 148-153.

Eidt, D.C.; McPhee, J.R. 1963. *Acantholyda erythrocephala* (L.) new in Canada. Can. Dep. For., Ottawa, Ont. Bi-mon. Prog. Rep. 19(4): 2.

Fedoryak, V.E.; Usol'tsev, V.A. 1978. Modelling of the population dynamics of *Acantholyda stellata* Christ and *A. erythrocephala* L. during the period of the development of eggs and larvae in tree crowns. Lesovedenie 4: 104-107. [transl. from Russian]

Griswold, C.L. 1939. Introduced pine sawfly. J. Econ. Entomol. 32: 474-478.

Hofacker, T.H.; Loomis, R.C.; Worrall, A.J. 1989. Forest insect and disease conditions in the United States 1988. USDA For. Serv., For. Pest Manage., Washington, D.C.

Howse, G.M.; Syme, P.D.; Gross, H.L. 1982. Ontario region. Pages 39-54 *in* Annual reports of the Forest Insect and Disease Survey 1978 and 1979. Dep. Environ., Can. For. Serv., Ottawa, Ont.

Jahn, E. 1967. On a population outbreak of the pine false webworm, *Acantholyda erythrocephala* Chr. in the Steinfeld, lower Austria, in the years 1964-1967. Anz. Schaedlingskd. 39: 145-152. [transl. from German]

Kolomietz, N.G. 1967. Sawfly weaver (distribution, biology, damage, natural enemies, control). Nauka Novosibirsk, USSR. [transl. from Russian]

Lyons, D.B. 1988. Phenology and biometeorology of pine false webworm (Hymenoptera: Pamphiliidae) and its parasitoids in Ontario. Ph.D. Thesis, Univ. of B.C., Vancouver. 176 p.

Middlekauff, W.W. 1938. Occurrence of a European sawfly *A. erythrocephala* (L.) in N.Y. State. J. New York Entomol. Soc. 46: 438.

Middlekauff, W.W. 1958. The North American sawflies of the genera *Acantholydae, Cephalcia*, and *Neurotoma* (Hymenoptera, Pamphiliidae). Univ. Calif. Publ. Entomol. 14: 51-174.

Moody, B.H. (Comp.). 1990. Forest insect and disease conditions in Canada 1988. For. Can., Ottawa, Ont.

Muldrew, J.A. 1967. Biology and initial dispersal of *Olesicampe* (*Holocremnus*) sp. nr. *nematorum* (Hymenoptera: Ichneumonidae), a parasite of the larch sawfly recently established in Manitoba. Can. Entomol. 99: 312-321.

Plumb, G.H. 1945. Miscellaneous insect notes. Connecticut State Entomol., 44th Rep., Conn. Agric. Exp. Stn. Bull. (New Haven) 488: 420-421.

Quedneau, F.W.; Lim, K.P. 1983. *Olesicampe geniculatae*, a new palaearctic ichneumonid parasite of *Pristiphora geniculata* (Hymenoptera: Tenthredinidae). Can. Entomol. 115: 109-113.

Rumphorst, H.; Goossen, H. 1960. Contribution to the biology and control of the pine false webworm (*Acantholydae erythrocephala*). Anz. Schädlingskd. 33: 149-154. [transl. from German]

Sanborne, M. 1984. A revision of the world species of *Sinophorus* (Ichneumonidae). Mem. Amer. Entomol. Inst. (Ann Arbor, Mich.) 38, 403 p.

Schmutzenhofer, H. 1975. Sawflies injurious to forests in Austria. 1. *Acantholyda erythrocephala* L. Cbl. Ges. Forstwesen 92: 1-8. [transl. from German]

Schwerdtfeger, F. 1941. A contribution to the knowledge of the pine-plantation web-spinning sawfly (*Acantholyda erythrocephala*). Z. Angew. Entomol. 28: 125-156. [transl. from German]

Schwerdtfeger, F. 1944. Further observations on the biology of the pine false webworm, *Acantholyda erythrocephala* L. Z. Angew. Entomol. 30: 364-371. [transl. from German]

Soraci, F.A. 1938. Distribution of the sawfly *Acantholyda erythrocephala* in New Jersey. J. New York Entomol. Soc. 46: 444.

Syme, P.D. 1981. Occurrence of the introduced sawfly *Acantholyda erythrocephala* (L.) in Ontario. Dep. Environ., Can. For. Serv., Ottawa, Ont. Res. Notes 1: 4-5.

Wells, A.B. 1926. Notes on tree and shrub insects in southwestern Pennsylvania. Entomol. News 37: 254-258.

Wilson, G.G. 1984. Infection of the pine false webworm by *Pleistophora schubergi* (Microsporida). Dep. Environ., Can. For. Serv., Ottawa, Ont. Res. Notes 4: 7-8.

Wilson, L.F. 1977. A guide to insect injury of conifers in the Lake States. USDA Agric. Handbook No. 501.

Chapter 28

Root Collar Weevils, *Hylobius* and *Steremnius* spp.

H.F. Cerezke and B.A. Pendrel

Introduction

Several forest insects have become prominent pests in Canada since the early 1950s, largely in response to human-influenced habitat conditions created during the process of forest removal and the early establishment of the new forest. The cut stumps and logging slash left behind after harvesting provide abundant attractive breeding materials for several closely related weevil species in the genera *Hylobius* and *Steremnius*. Upon emerging from the breeding sites, the adult weevils are attracted to the newly planted or naturally grown seedlings on which they feed. Their subsequent debarking damage on the stems results in reduced growth, disfigured stems, and tree mortality. The important species in this group include *Hylobius congener* (Martin 1962, 1964), the pales weevil, *H. pales* (Finnegan 1959), and a close relative, *Steremnius carinatus* (Condrashoff 1968; Lejeune 1962) (Table 1).

Three other *Hylobius* species are also important pests in young coniferous stands, although they also attack older trees.

Table 1. Summary of tree species damaged, breeding sites, and type of injury caused by *Hylobius* and *Steremnius* weevil species in Canada.

Species	Tree species damaged	Breeding sites	Type of injury
Hylobius congener	All conifers	Inner bark of fresh-cut stumps and roots; occasionally logs or large diameter slash	Girdling of stem of 1–4-year-old planted seedlings
Hylobius pales Pales weevil	Eastern white pine Jack pine Red pine Scots pine	Inner bark of fresh cut stumps and roots	Girdling of stem of mostly 1–2-year-old seedlings
Hylobius pinicola Couper collar weevil	Balsam fir Black spruce Tamarack White spruce	Inner bark and cambium of roots and root collar of live trees	Girdling of roots and root collar area of young to mature trees
Hylobius radicis Pine root collar weevil	Jack pine Lodgepole pine Ponderosa pine Red pine Scots pine	Inner bark and cambium of root collar and main lateral roots of live trees	Girdling of root collar and main lateral roots of trees, mostly up to 25 years old
Hylobius warreni Warren root collar weevil	Jack pine Lodgepole pine Scots pine Western white pine White spruce	Inner bark and cambium of root collar and main lateral roots of live trees	Girdling of root collar and main lateral roots of trees, mostly up to 25 years old
Steremnius carinatus	Amabilis fir Douglas-fir Sitka spruce Western hemlock	Inner bark of cut stumps, roots, and felled trees	Girdling of lower stem of 1–3-year-old seedlings

These species include the Couper collar weevil, *H. pinicola* (Smerlis 1961; Warren 1960b; Warren and Parrott 1965; Whitney 1961; Wood 1957), the pine root collar weevil, *H. radicis* (Finnegan 1962a; Prentice 1955; Wilson and Millers 1983), and the Warren root collar weevil, *H. warreni* (Cerezke 1970c, 1973a, 1974; Finnegan 1962b; Warren 1956a, 1956b; Warren and Parrott 1965; Wood and Van Sickle 1986) (Table 1). The larvae of these species attack the root collar and main roots of live conifers and thus are most damaging during the larval feeding stages. The larvae bore into the bark, scoring the cambium and outer sapwood, and cause various degrees of root and stem girdling. Completely girdled trees are killed, while less severely damaged trees may suffer reduced growth, or are made less windfirm because of weakened roots and stems. In addition, because the wounds may accumulate over the life of the tree, they can provide points of entry for root and stem decay fungi (Smerlis 1957, 1961; Warren and Whitney 1951; Whitney 1952, 1961, 1962; Wilson and Millers 1983).

General Description and Taxonomy

Adults of *Hylobius* and *Steremnius* are small (e.g. *H. congener*, 5.8–9.0 mm long; *H. pales*, 5.8–11.3 mm long; *S. carinatus*, 6.2–9.0 mm long) to large robust weevils (e.g. *H. radicis*, 9.4–13.0 mm long; *H. pinicola*, 11.3–14.3 mm long; *H. warreni*, 11.7–15.1 mm long) (Brown 1966; Condrashoff 1968; Finnegan 1961; Wood 1957). Two of the species are known to be flightless (*H. warreni* and *S. carinatus*), and hence most of their dispersal, attacks, and breeding occur in the vicinity of origin (Condrashoff 1968; Warren 1960b). Their larvae are white, legless, and falciform when full grown.

Morphological descriptions of the species have been provided by Condrashoff (1966), Warren (1960b), Watson (1955), and Wood (1957), and taxonomic keys for the species have been prepared by Brown (1966), Condrashoff (1966), Finnegan (1961), Goyer and Hertel (1970), Thomas (1964), Warner (1966), and Wood (1957). Wilson et al. (1966), described an external sex character useful for distinguishing male and female adults of all North American *Hylobius* species.

Early confusion existed in the identification of *H. pinicola* and *H. warreni* because the two species resemble each other closely and were often found in the same habitats (Finnegan 1962b; Warren 1956a; Whitney 1961). The insects referred to in early studies in Manitoba (Warren 1956a, 1956d) consisted of these two species combined under one taxonomic entity, namely as *Hypomolyx piceus* (Warren 1960b). This was later clarified by Wood (1957), who invalidated the name *Hypomolyx*, erected *Hylobius warreni* as a new species, and also provided a taxonomic key to separate the two North American species, *H. pinicola* and *H. warreni*, from the Eurasian species, *Hylobius piceus*. According to Wood, *H. pinicola* most closely resembles *H. piceus*, and suggested that the two may in fact be subspecies. Martineau (1984) includes a description of the "large spruce weevil," *Hylobius piceus*, in Quebec, but the species is more likely that of *H. warreni* (Warren 1958; Wood 1957).

Distribution and Hosts

Of the five *Hylobius* species, *H. congener*, *H. pinicola*, and *H. warreni* occur throughout most of the forested areas of Canada, from Newfoundland to coastal British Columbia, and at least into southern parts of the Northwest Territories (Cerezke 1969; Grant 1966; Martin 1964; Nord et al. 1984; Warner 1966; Warren 1956b; Warren and Parrott 1965; Wood 1957). *Hylobius pinicola* appears to extend farthest north to central Yukon where several adult collections have been made (Grant 1966).

The distributions of *H. pales* and *H. radicis* are limited to southern Canada, and extend from southeastern Manitoba to Nova Scotia (Finnegan 1956, 1959, 1962a; Nord et al. 1984; Prentice 1955; Wallace 1954; Warner 1966; Warren 1956c). Wilson and Millers (1983) noted that *H. radicis* occurred eastward into Newfoundland, based upon early identifications (Lewis 1954; Reeks et al. 1949). However, Warren and Parrott (1965) found no evidence to support this statement and suggested that in early identifications, records of *H. radicis* were likely misidentified *H. warreni*.

Steremnius carinatus occurs only in western Canada, extending along the mainland coast of British Columbia, on Vancouver Island and the Queen Charlotte Islands, and in the interior southeastern wet belt of the province (Brown 1966).

The tree species on which economic damage has been sustained are listed in Table 1. However, the host species on which broods are produced for *H. congener*, *H. pales*, and *S. carinatus* may differ somewhat from those that are damaged. In Ontario, *H. pales* adults were reared from stumps of red, jack, eastern white, and Scots pines (Finnegan 1959). Stumps, slash, and cut logs of Scots, red, and eastern white pines were commonly used as breeding materials for *H. congener* (Martin 1964; Welty and Houseweart 1985). Additional breeding hosts of *H. congener* reported in the Maritime provinces include white, red, and red–black hybrid spruces (Butterworth and Pendrel, unpubl. data). Host selection studies indicated that *S. carinatus* favored Douglas-fir and Sitka spruce for brood establishment, but this species also laid eggs on western hemlock and amabilis fir (Condrashoff 1968).

Some geographical variations in the host preferences of *H. pinicola* and *H. warreni* have been noted. *Hylobius pinicola* has been most commonly associated with white spruce and tamarack (Wood 1957). However, Smerlis (1961) observed extensive root wounding on balsam fir by this species in Quebec, but found no damage on the white spruce within the same stands. Ecological studies of root collar weevils in Manitoba (Warren 1956a, 1956d, 1960b) and of root diseases in Saskatchewan (Whitney 1961) indicated that white spruce was a major host of *H. pinicola* and *H. warreni* on moist to wet sites. Jack pine is also cited as a major host of *H. warreni* in central and eastern Canada (Howse 1984; Warren 1958). In the foothills of Alberta and in British Columbia, lodgepole pine is the preferred host for *H. warreni* on all sites (Cerezke, 1969, 1970a; Reid 1952; Stark 1959), whereas western white pine is attacked in some localized areas (Ross 1955).

Life Histories and Behavior

The six weevil species can be conveniently divided into two groups for discussion on their life histories and general behavior. The first group includes *H. congener*, *H. pales*, and *S. carinatus*.

The life histories of these three species vary slightly in length of development and in seasonal occurrence, but all three appear to require one or more years to complete development from egg to adult. Life history data are known for *H. congener* only from southern Ontario (Martin 1962, 1964), and Pendrel (unpubl. data) suggested there were many informational gaps in reference to recent infestations in the Maritime provinces. In southern Ontario, *H congener* adults emerged from their overwintering sites after mid-May, fed for about 2 weeks, then dispersed by flight for 1–2 weeks to areas with fresh-cut logs and stumps. Here they commenced mating and oviposition by late May or early June, often favoring fresh cut logs over stumps as breeding sites. The eggs hatch after 10 days and the larvae excavate tunnels mostly in the inner phloem, completing their development by mid-August to late September. They then burrow into the wood to a depth of about 5 mm before turning at right angles and excavating along the grain for several centimetres, filling the tunnel with fine wood chips behind them. The larvae remain in the tunnels throughout the winter. Pupation and transformation to adults are not completed until the following July and August. The young adults emerge to feed intermittently on the inner bark of logs and slash before entering the forest duff to overwinter. After emergence in the spring they feed for some time before taking flight. The longevity of the adults is unknown. A 2-year generation period seems likely in southern Ontario (Martin 1964), in Maine (Welty and Houseweart 1985), and in Nova Scotia as well (Pendrel, unpubl. data).

Some *H. pales* also seem to require 2 years to complete each generation in southern Ontario, but most require only 1 year, which makes its life history more complex (Finnegan 1959). Adults may be found throughout the year and are known to overwinter at least twice. Overwintered adults emerge from the soil during late April and early May, then feed for about 2 months before selecting oviposition sites on the roots and root collar areas of cut stumps in July. They then cut niches in the bark and deposit up to 30 eggs per female during their first year as overwintered adults and an additional 18 eggs per female during their second year. Eggs hatch within 10 days and larval development (most often on the roots) is completed by early September. About 70% of these larvae construct chip cocoons in which they pupate at this time and emerge as new adults by early October. The remaining 30% overwinter as larvae, and emerge as new adults from mid-June to August in the following year (Finnegan 1959). These adults may lay a few eggs in the fall, before overwintering and merging with the other portion of the adult population.

Adults of *S. carinatus* may overwinter up to three times and establish new broods each year (Condrashoff 1968). Adults that have overwintered mate and lay eggs from spring to late summer. The eggs are deposited in niches under the bark scales on roots and stumps, and in slash of dead coniferous material. Brood development from egg to adult varies with temperature conditions, and may require about 1.5 years in open sites and about 2 years in shaded sites. Pupation occurs in chip cocoons excavated in the bark or outer sapwood and new adults may appear from April to July. The adults are omnivorous feeders and may select plant species other than conifers to feed on.

All three species share several common behavioral patterns, in addition to selecting dead host materials for breeding sites. All pupate in chip cocoons (*H. congener* varies slightly), tend to be long-lived in the adult stage, and are largely nocturnal and terrestrial during their feeding and oviposition periods. Feeding, dispersion, and diurnal and seasonal behavior patterns have been described for adult *H. pales* (Corneil and Wilson 1984a, 1984b, 1984c). Adult trapping studies of *H. congener* suggested that they were initially attracted to stand edges of new cut areas (Martin 1964; Welty and Houseweart 1985). This behavioral pattern was described also for *Hylobius abietis*, a similar seedling debarking weevil in Europe (Eidmann 1979).

The second group of weevil species includes *H. radicis*, *H. pinicola* and *H. warreni*. The time required to complete one generation from egg to adult is about 2 years for *H. radicis* (Finnegan 1962a; Raffa and Hunt 1989; Wilson and Millers 1983), and is probably similar for the other two species (Cerezke 1970a; Reid 1952; Stark 1959). Warren (1956a) also suggested a 2-year cycle for *H. pinicola* and *H. warreni*, although he noted that the larval stage could extend 3 to 4 years for *H. warreni*, based upon artificial rearing studies (Warren 1960a). All three species probably lay eggs from May to September, with a peak in July. Most larval feeding is confined to the roots and root collar areas, between the mineral soil and the upper surface of the forest duff. The larvae overwinter at least once in the case of *H. radicis*, and probably twice in the case of the other two species. Larval development is completed in late spring when a hardened pupal cell is formed from soil and bark particles mixed with resin. Here they develop into a short pre-pupal stage in June, then pupate and form adults. *Hylobius radicis* adults emerge during July and August, while *H. pinicola* and *H. warreni* may emerge somewhat later in the fall and in the following spring. Because the adults live and reproduce for 2 or more years, there may be three or more overlapping generations, and all stages can be found during the spring to fall period. *Hylobius radicis* adults usually live for 2 years and produce eggs in both years. Insectary-reared adults laid an average total of 32 eggs per female (Finnegan 1962a), whereas laboratory-reared adults laid an average of 48 eggs (Schaffner and McIntyre 1944).

Hylobius warreni females appeared to lay few if any eggs during their first year, but continued to lay eggs during each of the following 3 years of adult life. Individual females laid up to 36 eggs in one season but averaged only 12 per female in one season under field-reared conditions (Cerezke 1969). Recent studies of *H. radicis* indicate that this species also requires an overwintering period to initiate oviposition, and that the stimulus to oviposit may be partly mediated by kairomones such as host-produced volatiles and ethanol (Hunt and Raffa 1989).

The general behavior of the three species appears to be similar. All are largely nocturnal and tend to remain inactive in the duff at or near tree bases during daylight hours, dispersing between trees and up stems to feed on branches and twigs at night (Cerezke 1969; Wilson and Millers 1983). Flight has been observed only rarely for *H. radicis* (Wilson and Millers 1983). Mark-recapture studies of *H radicis* and of *H. warreni* indicate they move several metres between trees during a single night, but the movement is all pedestrian, and dispersal during the summer is limited because their movements are random (Cerezke 1969; Wilson 1968; Wilson and Millers 1983). Some differences in the daily and seasonal activity patterns were observed between the sexes of *H. warreni* that relate to feeding, oviposition, resting, and dispersion (Cerezke 1969). Males and females of newly formed adults of *H. radicis* and *H. warreni* tend to occur in about equal numbers (Cerezke 1969; Raffa and Hall 1988; Raffa and Hunt 1989; Wilson and Millers 1983).

Methods for Estimating Damage and Population Levels

The assessment of seedling damage caused by *H. congener*, *H. pales* and *S. carinatus* relies on various general-type surveys designed to record incidence of mortality and debarking damage on surviving seedlings. Debarking damage may be grouped into four or five classes that express the degree of partial girdling (Condrashoff 1968; Lynch 1984; Welty and Houseweart 1985; Wood and Van Sickle 1986). A variety of techniques has been used to estimate adult abundance, including pit-fall traps, insecticidal-treated baits, light traps, flight-intercept traps, tree shaking, sifting duff, split-pine bolt traps, and various pine host and synthetic host attractants (Condrashoff 1968; Corneil and Wilson 1984b; Lynch 1984; Magasi 1988; Norlander 1987, 1988; Welty and Houseweart 1985; Witcosky et al. 1987). Raffa and Hunt (1988, 1989) used pit-fall traps, baited with ethanol and turpentine, to successfully monitor *H. pales* adult populations; male and female adults were equally attracted. Similar field trials with semiochemical attractants in Nova Scotia have shown that these materials are potentially useful for monitoring populations of *H. congener*, and for risk-rating sites before planting (Magasi 1988), as was done in Europe for *H. abietis* (Norlander 1987). Semiochemical studies of *H. abietis* in Europe and *H. pales* in North America suggest that both host kairomone activity and weevil-produced pheromone activity are involved (Kalo 1986; Norlander 1987; Selander 1978; Phillips et al. 1988; Tilles et al. 1988).

Surveys to detect damage caused by *H. radicis*, involving aerial and ground checking, are detailed in Wilson and Millers (1983) and apply equally for *H. warreni*. These authors also describe survey methods to estimate immature weevil populations and amount of tree injury, based upon the percentage of trees with larvae, pupae, and callow adults in pupal cells. A relationship was established between percentage trees infested and average number of weevils per tree.

A sequential-type survey method was adapted for estimating *H. warreni* abundance in lodgepole pine stands of different age classes and was field-tested over a large area in western Alberta (Cerezke 1970b, 1970c). This method is also used for determining estimates of the average number of weevils per tree, from which absolute numbers of the immature population per unit area can be estimated. A more detailed population census of immature forms of *H. warreni* (exclusive of eggs) was described by Cerezke (1973a).

A survey method that relies on searching the root collar base was developed to estimate adult populations of *H. radicis*, but this method has time and tree size limitations (Wilson and Millers 1983). *Hylobius warreni* adults were successfully collected in a crawl-in trap fixed to the lower tree stem. The trap appears useful for estimating adult numbers by mark-recapture techniques (Cerezke 1969). A similar crawl-in trap was deployed to capture *H. radicis* (Raffa and Hall 1988).

Damage appraisal methods and tree and stand damage indices have been developed for *H. radicis* and for *H. pinicola–H. warreni* complex (Warren 1956a; Wilson and Millers 1983). Warren's method provides a six-point stand damage rating based on percentage girdling classes tallied on individual white spruce trees. Warren later (unpubl. report) applied the tree and stand damage index method to both white spruce and jack pine that were attacked by *H. warreni*. The technique described by Wilson and Millers utilizes measurements of girdled outer and inner bark on the root collar of trees and provides a composite stand damage index.

Host Injury

The host injury caused by adult feeding and larval tunneling will each be discussed separately, and the seedling damage caused by adult feeding will be further subdivided by insect species.

Seedling Damage Caused by Adult Feeding

Hylobius congener—Feeding by *H. congener* was first observed in the late 1950s in southern Ontario on logs, stumps, and slash, in association with the decline of old-field red pine plantations (Martin 1962, 1964), and more recently (since 1981), seedling damage has been observed in the Maritime provinces (Magasi 1986, 1987, 1988, 1989). Virtually all conifer species are attacked (Table 1). *Hylobius congener* adults are attracted to recently harvested sites by the odors of host tree volatiles released from fresh-cut conifer logs and stumps (Martin 1964; Welty and Houseweart 1985). Problems caused by the debarking of seedlings by the adult weevils have developed because most plantations are now established 1 to 3 years after harvesting. Adults that are attracted to and fly into the harvest site, as well as adults emerging from the cut stumps and roots, utilize the planted conifer seedlings as a food source along with natural regeneration (Fig. 1). Most seedling losses occur within 1 to 2 years after planting, but some losses can continue for up to 4 years; some sites may even require replanting or postponement of planting (Magasi 1988, 1989). Surveys conducted at 130 locations in 1986 and at 171 locations in 1988 in the Maritime provinces showed that *H. congener* was distributed widely in all three provinces, but was causing the most severe damage in eastern Nova Scotia including Cape Breton Island and in central and eastern Prince Edward Island (Magasi 1987, 1988, 1989;

Figure 1. Adult *Hylobius congener* feeding on the stem of a red pine seedling. (Bruce Pendrel, Forestry Canada)

Pendrel, unpubl. data). Mortality exceeded 85% in some plantations and was the cause of some plantation failures in the past.

The adult weevils feed mostly on the stem near ground level. The feeding scars are at first irregularly shaped holes in the bark, but these coalesce as feeding intensifies. Debarking damage occurs from May to October, but there is a tendency to peak in June (Magasi 1987, 1988; Pendrel, unpubl. data; Welty and Houseweart 1985). The stems may be partly or completely girdled. Seedlings with completely girdled stems are killed immediately, but some of the partly girdled seedlings may survive. Those that do may be susceptible to diseases but this has not been investigated.

Hylobius pales—Damage by *H. pales* has thus far been reported only in southern Ontario and Quebec, mostly in association with Christmas tree plantations, where selective cutting on an annual basis provides a continuous supply of breeding material (Finnegan 1959; Martineau 1984). Like *H. congener*, the adults of *H. pales* are attracted to the cut stumps for breeding sites and feed mostly on conifer seedlings. They debark the stems of young seedlings as well as the branches of larger trees, but the latter is of little consequence once the trees are over 3 m tall. Although some feeding damage may occur throughout the summer, two distinct feeding peaks have been observed in Christmas tree plantations: one in mid-September when new adults begin emerging from the stumps, and another in the spring by adults that have overwintered in the duff (Finnegan 1959). Pierson (1921) had noted earlier that when stands were cut in a single operation, the freshly cut stumps attracted immigrant adults into the area in the spring time; seedling damage could also occur at this time.

Damage to seedlings may occur at or above ground level, and also below ground level when roots are loosely packed. On saplings, twig mortality of up to 40% may occur on 5- to 10-year-old Christmas trees, and sometimes buds are killed. The problem subsides after 1 to 2 years when the stumps start deteriorating (Nord et al. 1984). Finnegan (1959) noted that damage to pines by *H. pales* in Ontario had never been as extensive or severe as in some parts of eastern United States, partly because there had been no large-scale cutting operations of pine forests to allow for population buildup. In Michigan, the estimated injury and profit losses in Christmas tree stands caused by the pales weevil depended largely upon the marketing procedure but ranged up to 19% (Corneil and Wilson 1980, 1986).

Steremnius carinatus—This insect has been reported on several occasions since 1961 as a pest in areas planted with coniferous seedlings following large-scale continuous logging and road clearance (Condrashoff 1968; Lejeune 1962). Severe damage generally occurs in plantations established 2 to 3 years after timber harvesting and on seedlings up to 2 years old. The seedlings are usually considered safe from damage after one full growing season. Most feeding occurs in the spring and the fall or during wet periods. Debarking damage occurs on the main stem from 1 cm below to 2 to 3 cm above ground level. Over 40% of planted Douglas-fir seedlings have been killed or damaged on some logged and burned sites along the west coast of Vancouver Island (Condrashoff 1968; Lejeune 1962). More recently, in 1983, an average 6% of seedlings were killed in 17 of 30 plantation sites, including 12% of the amabilis fir in six areas and 10% of western hemlock in 18 areas. A resurvey of four of the plantations later in the same year revealed that mortality had increased substantially on the same species and included 5% mortality of Sitka spruce (Wood et al. 1984). Similar surveys in 1985 showed that up to 9% seedling mortality occurred in 12 plantations of six major conifer species: amabilis fir, Douglas-fir, western hemlock, western red cedar, Sitka spruce, and grand fir, and that additional seedlings were partly girdled at the root collar (Wood and Van Sickle 1986). Studies in disturbed Douglas-fir stands in Oregon indicated that *S. carinatus* is one of a complex of root bark beetle vectors of *Leptographium wageneri*, the causal agent of black-stain root disease (Witcosky 1989; Witcosky et al. 1986, 1987).

Root Collar Damage Caused by Larval Girdling

The most severe and extensive damage caused by root collar weevils in Canada has generally been attributed to *H. radicis*, *H. pinicola,* and *H. warreni* (Warren and Parrott 1965). Important damage by *H. radicis* has not been reported in natural stands but has occurred since the early 1950s in southern Ontario and southeastern Manitoba in Scots and lodgepole pine plantations, growing mainly on well-drained sandy soils (Finnegan 1956; Prentice 1955; Wallace 1954; Warren 1956c; Wilson and Millers 1983). Pure stands of exotic pines such as Austrian, lodgepole, mugho, and Scots seemed particularly susceptible to attacks by the weevil and suffered higher mortality than did the native red and jack pines. Up to 25% mortality occurred in Scots and lodgepole pine plantations in Manitoba and 90% mortality in Scots and red pine plantations in southern Ontario (Finnegan 1962a; Prentice 1955).

The amount of tree mortality depends partly on the size of the tree when first attacked and upon the density of the larval population. Scots pine attacked when 4 to 5 years old died 3 to 4 years later, whereas trees attacked when 10 to 15 years old died after 6 to 8 years (Finnegan 1962a). Trees that survive various degrees of partial girdling produce shorter shoots, and leader length is reduced in proportion to the percentage of root collar girdled (Wilson and Millers 1983). The feeding in the root collar and root crown may extend over several years, resulting in partially or completely severed cambium around the stem. Trees that survive several years of feeding damage form callous tissues that overgrow the wounds, and the root collar becomes fluted. The stem growth is often constricted at the wound site, leaving the tree susceptible to windthrow (Finnegan 1962a; Wilson and Millers 1983). The incidence of wood-staining and decay organisms associated with the wounds was insignificant, but the weevil may be a potential vector of stain-causing fungi such as *Leptographium* (Raffa and Hall 1988). Trees weakened by stem girdling may become susceptible to secondary pests such as the eastern pine weevil, *Pissodes nemorensis*, and engraver beetles such as the pine engraver, *Ips pini* (Wilson and Millers 1983).

In even-aged, well-stocked plantations, damage and mortality appear to occur randomly and decline after crown closure. Trees growing along the margins of stands or at the edges of openings supported higher larval populations and were more severely damaged than interior trees (Millers 1965). Hunt and Raffa (1989) reported that the numbers of *H. radicis* captured in pitfall traps within a Scots pine plantation were significantly related to the percentage incidence of trees classed with declining foliar symptoms.

Tree injury caused by *H. pinicola* has been reported in association with that of *H. warreni*, on the same or adjacent white spruce trees within the same stands (Warren 1956a, 1960b; Whitney 1961). It has not yet been reported causing significant damage in upland plantations. However, little is known about its potential, and it may become a pest of young black spruce and tamarack on wet sites where it is relatively abundant (Warren and Parrott 1965).

Injuries attributed to *H. pinicola* on balsam fir in Quebec occurred mostly on the root systems of dominant, codominant, and intermediate tree classes (Smerlis 1957, 1961). The feeding wounds formed open galleries that exposed the xylem to wood-destroying fungi, completely girdling some roots. A relationship between the larval wounds and root- and butt-rotting fungi was established. The larval wounds were the most important infection courts on dominant and codominant trees; 56% of the total number of root and butt rots present in 36- to 55-year-old stands were associated with the larval wounds (Smerlis 1961).

In natural and planted stands of lodgepole pine, initial attack by *H. warreni* has occurred when trees are 5 to 6 years old, and continues through to stand maturity (Cerezke 1969). In even-aged stands of all ages this weevil selectively attacks the larger dominant and codominant trees. As in the case of *H. radicis*, various degrees of partial and complete girdling occur on the root collar and roots. In natural stands of lodgepole pine, mortality usually occurs in stands less than 30 years old and seldom exceeds 10% (Cerezke 1969, 1970c, 1974). Feeding damage on young trees is concentrated around the root collar, but most of the feeding on older, larger trees occurs on the lateral roots, thus reducing the risk of stem girdling. Individual trees can be attacked repeatedly throughout their development, so that by age 40 to 60 years, 80% to 100% of all trees in stands have accumulated old weevil wounds that often cause partial girdling. These wounds cause periodic growth reductions on a large proportion of the trees during stand development, but attempts to quantify the effects of partial girdling have been only partly successful and apply mostly to trees younger than 25 years (Cerezke 1970c, 1972, 1974). Leader growth is reduced on trees with partially girdled stems. Partial root collar girdling restricts nutrient distribution into some lateral roots, reducing their radial increment and thereby affecting the stability of the root system (Cerezke 1974; Långström and Hellqvist 1989). Up to 17% reduction in radial increment of the stem and up to 11.5% reduction of leader length occurred on trees after sustaining 50% girdling of the root collar (Cerezke 1970c). Warren (unpubl. data, 1960) showed that leader growth of white spruce saplings with a 40% to 50% damage index was reduced up to 35%.

The size of *H. warreni* populations in natural stands was positively correlated with tree diameter and depth of duff material around tree bases, and open-grown trees were more susceptible than trees within stands (Cerezke 1969).

Various surveys of *H. warreni* damage have been reported in plantations of native and exotic pine species across Canada; in some cases the exotic species have suffered more severely. Examples include 63% mortality of plantation-grown Scots pine in Quebec (Finnegan 1962b); up to 29.2% mortality of pine species in plantations in Newfoundland (Warren and Parrott 1965); up to 20% of lodgepole pine in British Columbia (Wood and Van Sickle 1986); and over 25% of planted lodgepole pine in Alberta (Cerezke, unpubl. data).

Little evidence was found in lodgepole pine stands in Alberta attacked by *H. warreni* that the larval wounds contributed significantly as courts of entry for disease organisms, except in one mature stand (Baranyay and Stevenson 1964; Cerezke 1969; Nordin 1956; Stark 1959). However, Armillaria root rot has been associated with the wounds of this weevil in British Columbia (Wood et al. 1984). In white spruce stands in Manitoba and Saskatchewan, *Hylobius* (mostly *H. warreni*) wounds provided an important avenue of infection for root-rotting and staining fungi (Warren and Whitney 1951; Whitney 1952, 1961) and were a significant factor in the development of stand opening disease (Whitney 1962).

Management Strategies and Control

Site factors that contribute to high-risk of debarking damage by *H. congener* have been investigated in Maine and in the Maritime provinces; several factors have been identified but require further site evaluation (Pendrel, unpubl. data; Welty and Houseweart 1985).

Various methods developed in Europe for control of *H. abietis* are being tested and adapted for control of *H. congener* in the Maritime provinces (Pendrel, unpubl. data). These include delayed planting for up to 3 years after clearcutting, soil scarification using a disk trencher, a Bracke and an anchor-chain scarifier (reduced damage by 29%, 25%, and 21%, respectively), placement of physical barriers around seedlings such as plastic collars and nylon nets (reduced damage by 31%), and spraying seedlings with poly-butenes such as "Tree Tanglefoot" (reduced damage by 34%). Dipping seedlings in water-based insecticides may also have application, as used in Europe (Eidmann 1979; Heritage et al. 1989), but this requires further field testing.

Delayed planting for 1 to 2 years after cutting and treating seedlings with insecticide before planting have been the two most effective methods of reducing *H. pales* damage (Lynch 1984; Nord et al. 1984). However, delaying planting may have costly side-effects in controlling competing vegetation and in site preparation. *Hylobius pales* adults are not attracted to sites that had a previous nonconiferous cover, but high populations have occurred in large clearcuts that were adjacent to areas clearcut 1 year earlier (Nord et al. 1984). The spacing and timing of adjacent clearcuts are important factors to consider in reducing *H. pales* and *H. congener* immigration and spread into new areas.

In Christmas tree plantations, additional control measures for *H. pales* included removal of stumps in the spring, treatment of stumps and live branches and foliage with insecticide, and retaining a whorl of live branches on stumps to keep them alive and unattractive for oviposition sites (Corneil and Wilson 1980, 1984c; Nord et al. 1984). Planting container-grown seedlings in summer avoided the heavy spring feeding period and appeared to be an alternative to insecticidal treatment. Attempts to trap out adult weevil populations have been unsuccessful (Nord et al. 1984).

Biological agents to control *H. pales* in particular have received some attention but with limited success, and further development work is warranted. These have included parasitic nematodes, entomogenous fungi, and a gregarine protozoan. Several species of mites are associated with *H. pales* adults and larvae but their relationship is not understood (Nord et al. 1984). The terrestrial and nocturnal habits of adults of all *Hylobius* species and of *S. carinatus* provide an opportunistic link in the life cycle to explore methods of potential microbial control.

Condrashoff (1968) suggested several remedial measures to reduce damage caused by *S. carinatus*: these included planting immediately after logging and slash-burning (seedlings are considered immune after one growing season); planting older larger seedlings on high-risk sites and in the spring rather than in the fall; and monitoring to assess adult population levels prior to planting. In Oregon, Witcosky (1989) developed a crop production–pest management system for management of black-stain root disease in plantations of Douglas-fir. The scheme is designed to weaken the links between intensive forest practices and the epidemiology of black-stain root disease and its insect vectors.

Areas established for pine plantations can be risk-rated for potential damage by *H. radicis* by using four parameters: stand location relative to zones with cool summers and cold winters; proximity to weevil brood sources; tree species susceptibility; and tree species mixtures (Raffa and Hunt 1989; Wilson and Millers 1983). There is little known about the site factors influencing *H. pinicola* other than its host preferences and moist site requirements. White spruce site conditions of high-risk to *H. warreni* in Manitoba have been partly described by Warren (1956a, 1956d, 1960b). In lodgepole pine forests in Alberta, highest levels of *H. warreni* populations occurred on rich, well-drained moist growing sites of high productivity classes (Cerezke 1969, 1970a). No *H. warreni* infestations were found in lodgepole pine stands at elevations above 1 600 m (Cerezke 1970b).

Detailed guidelines were developed for the control and management of *H. radicis* in Michigan and have application elsewhere (Wilson and Millers 1983). These include both preventative and control strategies compatible with current pine management practices. They incorporate site evaluation for risk and pine species selection, site preparation and planting, plantation maintenance, and plantings for windbreaks and as ornamentals.

Silvicultural treatments consisting of pruning branches from the lower 60 cm of the stem and removing the litter and top soil from around the tree base reduced *H. radicis* populations satisfactorily in red pine plantations (Wilson and Millers 1983). Warren (1956a) suggested similar soil removal treatment to help control the *H. pinicola–H. warreni* complex in Manitoba. This treatment was also applied recently for control of *H. warreni* in lodgepole pine plantations in Alberta, but its long-term effects have not been evaluated. Removal of the litter from around the tree base destroys the adult weevil habitat, while pruning the lower branches decreases moisture and increases temperatures; both factors are important to weevil survival.

The effects of clearcutting mature lodgepole pine stands on *H. warreni* population development and survival were investigated in Alberta over a 5-year period (Cerezke 1973a). The study showed that larvae could complete their development in the cut stumps for 1 and 2 years after cutting, but that an estimated 88.4% of the larval population apparently died. More adults were produced in the clearcut areas in the second year than in the adjacent uncut areas. While clearcutting alone provided a high level of control, it was hypothesized that additional mortality of all stages would be enhanced when clearcutting was followed by scarification or prescribed burning treatments (Cerezke 1970a, 1973a).

In mature stands infested with *H. warreni* the risk of damage and tree mortality is often high in the new stands subsequently established after harvesting (Emond and Cerezke 1989; Wood and Van Sickle 1989). This may be because some weevils can survive in an area for as long as 6 to 7 years after the mature trees are harvested; larvae in the stumps may require 2 years to complete their development, and the adults live 4 to 5 years (Cerezke 1969, 1970a, 1973a). Young trees established in the clearcuts, either by planting or by natural regeneration, immediately after harvesting can be old and large enough (i.e. over 5 years) to be attractive as oviposition sites for surviving adults (Cerezke 1970a). Hence, trees can be attacked from the resident population already distributed throughout the clearcut, rather than by adults that must first immigrate into the area from adjacent areas. The period between harvesting and establishment of the new stand therefore provides the most opportune time to restrict reestablishment of the weevil and its numerical increase in young stands. Some strategies to reduce *H. warreni* damage and its population increase in young stands following clearcutting have been described (Cerezke 1970a, 1973a).

Thinning young naturally grown lodgepole pine stands infested with *H. warreni* had the initial effect of concentrating weevil populations on the residual trees and of increasing the proportion of residual trees with damage (Cerezke, unpubl.

data). This treatment needs to be evaluated over several years and in relation to weevil abundance and stand characteristics.

A provenance test plantation of Scots pine grown in Michigan from seed collected in 108 natural stands in Europe and Asia was assessed for genetic differences among varieties showing resistance to *H. radicis* damage (Wright and Wilson 1972). Varietal susceptibility was related to the percentage of trees killed and was generally highest in varieties from central Europe, Scandinavia, and Siberia. The most resistant varieties had short needles, were dark green, were moderately fast growing, and were native to southern Europe. Several characteristics were examined to identify the possible mode of resistance but none of the traits measured consistently correlated with *H. radicis* damage.

Hylobius warreni damage has been reported in provenance test plantations of jack, lodgepole, and Scots pines (including six varieties of the latter species) in Ontario, but no varietal preference was established (Foster and Hook 1972; Foster et al. 1968).

Several early attempts were made to control *H. radicis* with insecticides having long residual effect, many of which are no longer registered in Canada (Finnegan and Stewart 1962; Wilson and Millers 1983). Control of larvae was often variable because of their protection in the soil, bark, and resinous galleries, whereas adults were more readily killed. Attempts to control *H. warreni* larvae with systemic insecticides applied as soil drenches also gave highly variable results (Drouin and Kusch 1975, 1978).

Biological control of *H. radicis, H. pinicola*, and *H. warreni* offer some potential, but few experimental trials have been undertaken (Schmiege 1963; Wilson and Millers 1983). While excess moisture in the pupal cell appeared to be one of the most important mortality factors affecting mature larvae, pre-pupae, pupae and teneral adults (Cerezke 1973b; Finnegan 1962a), the controlling influence of the known biological agents is largely unknown.

Parasitic insect species include a braconid that attacks *H. radicis* (Finnegan 1962a; Wilson and Millers 1983), and an ichneumonid was identified that attacks *H. pinicola* and accounts for up to 5% mortality of *H. warreni* (Cerezke 1973b).

Other parasites include mites, nematodes and the entomogenous fungus, *Beauveria bassiana*. Some entomophilic nematodes that inhabit the soil show potential for control applications to root weevils (Eidt, pers. comm.). Certain spider and carabid species are known to prey on *H. radicis* adults and shrews and small rodents may also contribute to adult losses (Cerezke 1973b; Wilson and Millers 1983).

References

Baranyay, J.A.; Stevenson, G.R. 1964. Mortality caused by Armillaria root rot, peridermium rusts, and other destructive agents in lodgepole pine regeneration. For. Chron. 40: 350-361.

Brown, W.J. 1966. The species of *Steremnius* Schoenherr (Coleoptera: Curculionidae). Can. Entomol. 98: 586-587.

Cerezke, H.F. 1969. The distribution and abundance of the root weevil, *Hylobius warreni* Wood, in relation to lodgepole pine stand conditions in Alberta. Ph.D. Thesis, Univ. B.C. i-xvii + 221 p.

Cerezke, H.F. 1970a. Biology and control of Warren's collar weevil, *Hylobius warreni* Wood, in Alberta. Dept. Fish. and For., Can. For. Serv., Calgary, Alta. Int. Rep. A-27.

Cerezke, H.F. 1970b. Survey report of the weevil, *Hylobius warreni* Wood, in the foothills of Alberta. Dept. Fish. and For., Can. For. Serv., Edmonton, Alta. Int. Rep. A-38.

Cerezke, H.F. 1970c. A method for estimating abundance of the weevil, *Hylobius warreni* Wood, and its damage in lodgepole pine stands. For. Chron. 46: 392-396.

Cerezke, H.F. 1972. Effects of weevil feeding on resin duct density and radial increment in lodgepole pine. Can. J. For. Res. 2: 11-15.

Cerezke, H.F. 1973a. Survival of the weevil, *Hylobius warreni* Wood, in lodgepole pine stumps. Can. J. For. Res. 3: 367-372.

Cerezke, H.F. 1973b. Some parasites and predators of *Hylobius warreni* in Alberta. Environ. Can., Ottawa, Ont. Bi-mon. Res. Notes 29: 24-25.

Cerezke, H.F. 1974. Effects of partial girdling on growth in lodgepole pine with application to damage by the weevil *Hylobius warreni* Wood. Can. J. For. Res. 4: 312-320.

Condrashoff, S.F. 1966. A description of the immature stages of *Steremnius carinatus* (Boheman) (Coleoptera: Curculionidae). Can. Entomol. 98: 663-667.

Condrashoff, S.F. 1968. Biology of *Steremnius carinatus* (Coleoptera: Curculionidae), a reforestation pest in coastal British Columbia. Can. Entomol. 100: 386-394.

Corneil, J.A.; Wilson, L.F. 1980. Pales weevil - rational for its injury and control. Mich. Christmas Tree J., Fall 1980: 16-17.

Corneil, J.A.; Wilson, L.F. 1984a. Some light and temperature effects on the behavior of the pales weevil, *Hylobius pales* (Coleoptera: Curculionidae). Great Lakes Entomol. 17: 225-228.

Corneil, J.A.; Wilson, L.F. 1984b. Dispersal and seasonal activity of the pales weevil, *Hylobius pales* (Coleoptera: Curculionidae), in Michigan Christmas tree plantations. Can. Entomol. 116: 711-717.

Corneil, J.A.; Wilson, L.F. 1984c. Live branches on pine stumps deter pales weevil breeding in Michigan (Coleoptera: Curculionidae). Great Lakes Entomol. 17: 229-231.

Corneil, J.A.; Wilson, L.F. 1986. Impact of feeding by pales weevil (Coleoptera: Curculionidae) on Christmas tree stands in southeastern Michigan. J. Econ. Entomol. 79: 192-196.

Drouin, J.A.; Kusch, D.S. 1975. Pesticide field trials on shade and shelterbelt trees in Alberta, 1974. Environ. Can., Can. For. Serv., Edmonton, Alta. Inf. Rep. NOR-X-131.

Drouin, J.A.; Kusch, D.S. 1978. Pesticide field trials on shade and shelterbelt trees in Alberta, 1977. Environ. Can., Can. For. Serv., Edmonton, Alta. Inf. Rep. NOR-X-205.

Eidmann, H.H. 1979. Integrated management of pine weevil (*Hylobius abietis*) populations in Sweden. *In* W.E. Waters (Ed.). Current Topics in Forest Entomology. Selected Papers from the XVth International Congress of Entomology, Washington, D.C., Aug. 1976.

Emond, F.J.; Cerezke, H.F. 1989. Forest insect and disease conditions in Alberta, Saskatchewan, Manitoba, and the Northwest Territories in 1988 and some predictions for 1989. For. Can., Can. For. Serv., North. For. Cent., Edmonton, Alta. Inf. Rep. NOR-X-303.

Finnegan, R.J. 1956. Weevils attacking pines in southern Ontario. Agric. Can., Sci. Serv., For. Biol. Div., Ottawa, Ont. Bi-mon. Prog. Rep. 12: 3.

Finnegan, R.J. 1959. The pales weevil, *Hylobius pales* (Hbst.), in southern Ontario. Can. Entomol. 91: 664-670.

Finnegan, R.J. 1961. A field key to the North American species of *Hylobius* (Curculionidae). Can. Entomol. 93: 501-502.

Finnegan, R.J. 1962a. The pine root collar weevil, *Hylobius radicis* Buch., in southern Ontario. Can. Entomol. 94: 11-17.

Finnegan, R.J. 1962b. Serious damage to Scots pine by the weevil, *Hylobius warreni* Wood. Can. Dept. For., For. Ent. and Path. Br., Ottawa. Ont. Bi-mon. Prog. Rep. 18: 2.

Finnegan, R.J.; Stewart, K.E. 1962. Control of the pine root collar weevil, *Hylobius radicis*. J. Econ. Entomol. 55: 482-486.

Foster, H.R.; Foreman, F.; MacLeod, L.S.; Ingram, W. 1968. Annual district reports of the Forest Insect and Disease Survey in Ontario, 1967, Northern Forest Region. Can. Dept. Fish. and For., For. Br., Great Lakes For. Res. Cent., Sault Ste. Marie, Ont. 15 p.

Foster, H.R.; Hook, J. 1972. Forest insect and disease surveys in the Northern survey region, 1971. Environ. Can., Can. For. Serv., Great Lakes For. Res. Cent., Sault Ste. Marie, Ont. Inf. Rep. 0-X-161.

Goyer, R.A.; Hertel, G.D. 1970. A field key to the pine-infesting *Hylobius* weevils of Wisconsin and their allies. Madison, Wis., Univ. of Wisconsin. Univ. Wisc. For. Res. Note 152.

Grant, J. 1966. The hosts and distribution of the root weevils *Hylobius pinicola* (Couper) and *H. warreni* Wood in British Columbia. J. Entomol. Soc. B.C. 63: 3-5.

Heritage, S.; Collins, S.; Evans, H.F. 1989. A survey of damage by *Hylobius abietis* and *Hylastes* spp. in Britain. Pages 36-42 *in* R.I. Alfaro and S.G. Glover (Eds.) Insects affecting reforestation: biology and damage. Proc. IUFRO working group on insects affecting reforestation (S2.07-03), Vancouver, B.C., July 3-9, 1988.

Howse, G.M. 1984. Insect pests of jack pine: biology, damage and control, Pages 131-138 *in* Proc. of Jack Pine Symp., Environ. Can., Can. For. Serv., Great Lakes For. Res. Cent., Sault Ste. Marie, Ont. COJFRC Symp. Proc. O-P-12.

Hunt, D.W.A.; Raffa, K.F. 1989. Attraction of *Hylobius radicis* and *Pachylobius picivorus* (Coleoptera: Curculionidae) to ethanol and turpentine in pitfall traps. Environ. Entomol. 18: 351-355.

Kalo, P. 1986. Structure elucidation of behaviourally active compounds in the large pine weevil, *Hylobius abietis* L. by microanalytical methods. Ann. Acad. Sci. Fenn. Serv. A2 208: 1-48.

Långström, B.; Hellqvist, C. 1989. Effects of defoliation, decapitation, and partial girdling on root and shoot growth of pine and spruce seedlings, Pages 89-100 *in* R.I. Alfaro and S.G. Glover (Eds.). Insects affecting reforestation: biology and damage. Proc. IUFRO working group on insects affecting reforestation (S2.07-03), Vancouver, B.C., July 3-9, 1988.

Lejeune, R.R. 1962. A new reforestation problem by a weevil *Steremnius carinatus* Boh. Can. Dept. For., For. Ent. and Path. Br., Ottawa, Ont. Bi-mon. Prog. Rep. 18: 3.

Lewis, H.S. 1954. Forest plantations in Newfoundland. Prog. Rep. NF-5, Can. Dept. Northern Affairs and Nat. Resources, St. John's, Nfld.

Lynch, A.M. 1984. The pales weevil, *Hylobius pales* (Herbst): a synthesis of the literature. J. Georgia Entomol. Soc. 19: 1-34.

Magasi, L.P. 1986. Forest pest conditions in the Maritimes 1985. Can. For. Serv., Maritimes, Fredericton, N.B. Inf. Rep. M-X-159.

Magasi, L.P. 1987. Forest pest conditions in the Maritimes 1986. Can. For. Serv., Maritimes, Fredericton, N.B. Inf. Rep. M-X-161.

Magasi, L.P. 1988. Forest pest conditions in the Maritimes 1987. Can. For. Serv., Maritimes, Fredericton, N.B. Inf. Rep. M-X-166.

Magasi, L.P. 1989. Forest pest conditions in the Maritimes in 1988. For. Can., Maritimes, Fredericton, N.B. Inf. Rep. M-X-174.

Martin, J.L. 1962. *Hylobius congener* Dalla Torre on *Pinus* spp. in Ontario. Can. Dept. For., For. Ent. and Path. Br., Ottawa, Ont. Bi-mon. Prog. Rep. 18: 1-2.

Martin, J.L. 1964. The insect ecology of red pine plantations in central Ontario. II. Life history and control of Curculionidae. Can. Entomol. 96: 1408-1417.

Martineau, R. 1984. Insects harmful to forest trees. Agric. Can., Can. For. Serv., Min. Supply and Services, Ottawa, Ont. 261 p.

Millers, I. 1965. The biology and ecology of the pine root collar weevil, *Hylobius radicis* Buchanan, in Wisconsin. Dissertation. Univ. Wisc., Madison. 151 p.

Nord, J.C.; Ragenovich, I.; Doggett, C.A. 1984. Pales weevil. U.S. Dept. Agric., For. Serv., Southern Region. For. Insect and Dis. Leaflet 104.

Nordin, V.J. 1956. Insects associated with fire-scarred lodgepole pine in Alberta. Agric. Can., Sci. Serv., For. Biol. Div., Ottawa, Ont. Bi-mon. Prog. Rep. 12: 3.

Norlander, G. 1987. A method for trapping *Hylobius abietis* (L.) with a standardized bait and its potential for forecasting seedling damage. Scand. J. For. Res. 2: 199-213.

Norlander, G. 1988. The use of artificial baits to forecast seedling damage caused by *Hylobius abietis* (Coleoptera:

Curculionidae). Abstract *in* Proc. XVIII Int. Congr. of Entomol., Vancouver, B.C. 1988.

Phillips, T.W.; Wilkening, A.J.; Atkinson, T.H.; Nation, J.L.; Wilkinson, R.C.; Foltz, J.L. 1988. Synergism of turpentine and ethanol as attractants for certain pine-infesting beetles (Coleoptera). Environ. Entomol. 17: 456-462.

Pierson, H.B. 1921. The life history and control of the pales weevil (*Hylobius pales*). Harvard For. Bull. 3. Cambridge, Mass., Harvard Univ. 33 p.

Prentice, R.M. 1955. Pine root-collar weevil in Manitoba. Agric. Can., Sci. Serv., For. Biol. Div., Ottawa, Ont. Bi-mon. Prog. Rep. 11: 1-2.

Raffa, K.F.; Hall, D.J. 1988. Seasonal occurrence of pine root collar weevil, *Hylobius radicis* (Coleoptera: Curculionidae), in red pine stands undergoing decline. Great Lakes Entomol. 21: 69-74.

Raffa, K.F.; Hunt, W.A. 1988. Use of baited pitfall traps for monitoring pales weevil, *Hylobius pales* (Coleoptera: Curculionidae). Great Lakes Entomol. 21: 123-125.

Raffa, K.F.; Hunt, D.W.A. 1989. Microsite and interspecific interactions affecting emergence of root-infesting pine weevils (Coleoptera: Curculionidae) in Wisconsin. Ann. Entomol. Soc. Am. 82: 438-445.

Reeks, W.A.; Forbes, R.S.; Cuming, F.C. 1949. Forest Insect Survey Annual Report, Maritime Provinces. Agric. Can., Div. of Entomol., Ottawa, Ont.

Reid, R.W. 1952. *Hypomolyx piceus* (DeG.) on lodgepole pine. Agr. Can., For. Biol. Div., Ottawa. Ont. Bi-mon. Prog. Rep. 8: 2-3.

Ross, D.A. 1955. A weevil of white pine roots. Agric. Can., Sci. Serv., For. Biol. Div., Ottawa, Ont. Bi-mon. Prog. Rep. 11: 4.

Schaffner, J.V., Jr.; McIntyre, H.I. 1944. The pine root collar weevil. J. For. 42: 269-275.

Schmiege, D.C. 1963. The feasibility of using a Neoplectanid nematode for control of some forest insect pests. J. Econ. Entomol. 56: 427-431.

Selander, J. 1978. Evidence of pheromone mediated behavior in the large pine weevil, *Hylobius abietis*. Ann. Entomol. Fenn. 44: 105-112.

Smerlis, E. 1957. *Hylobius* injuries as infection courts of root and butt rots in immature balsam fir stands. Agric. Can., For. Biol. Div., Ottawa, Ont. Bi-mon. Prog. Rep. 13: 1.

Smerlis, E. 1961. Pathological condition of immature fir stands of *Hylocomium-Oxalis* type in the Laurentide Park, Quebec. For. Chron. 37: 109-115.

Stark, R.W. 1959. *Hylobius warreni* Wood in Alberta. Agric. Can., Res. Br., For. Biol. Div., Ottawa, Ont. Bi-mon. Prog. Rep. 15: 2.

Thomas, J.B. 1964. A key to the larvae and pupae of three weevils (Coleoptera: Curculionidae). Can. Entomol. 96: 1417-1420.

Tilles, D.A.; Eidmann, H.H.; Solbreck, B. 1988. Mating stimulant of the pine weevil, *Hylobius abietis*. J. Chem. Ecol. 14: 1495-1503.

Wallace, D.R. 1954. The pine root-collar weevil in Ontario. Agric. Can., Sci. Serv., For. Biol. Div., Ottawa, Ont. Bi-mon. Prog. Rep. 10: 2.

Warner, R.E. 1966. A review of *Hylobius* of North America, with a new species injurious to slash pine (Coleoptera: Curculionidae). Coleop. Bull. 20: 65-81.

Warren, G.L. 1956a. The effect of some site factors on the abundance of *Hypomolyx piceus* (Coleoptera: Curculionidae). Ecol. 37: 132-139.

Warren, G.L. 1956b. Root injury to conifers in Canada by species of *Hylobius* and *Hypomolyx* (Coleoptera: Curculionidae). For. Chron. 32: 7-10.

Warren, G.L. 1956c. Observations on the origin of infestations of the pine root collar weevil in plantations. Can. Agric., Sci. Serv., For. Biol. Div., Ottawa, Ont. Bi-mon. Prog. Rep. 12(3): 2-3.

Warren, G.L. 1956d. pH and incidence of attack of *Hypomolyx piceus* (DeG.). Can. Agric., Sci. Serv., For. Biol. Div., Ottawa, Ont. Bi-mon. Prog. Rep. 12(4): 2-3.

Warren, G.L. 1958. A method of rearing bark- and cambium-feeding beetles with particular reference to *Hylobius warreni* Wood (Coleoptera: Curculionidae). Can. Entomol. 90: 425-428.

Warren, G.L. 1960a. Progress in the artificial rearing of *Hylobius warreni* Wood. Can. Agric., Res. Br., For. Biol. Div., Ottawa, Ont. Bi-mon. Prog. Rep. 16(6): 2-3.

Warren, G.L. 1960b. External anatomy of the adult of *Hylobius warreni* Wood (Coleoptera: Curculionidae) and comparison with *H. pinicola* (Couper). Can. Entomol. 92: 321-341.

Warren, G.L.; Parrott, W.C. 1965. Status of root weevils, *Hylobius* spp. in coniferous plantations in Newfoundland. For. Can., Cornerbrook, Nfld. Inf. Rep. N-X-1.

Warren, G.L.; Whitney, R.D. 1951. Spruce root borer (*Hypomolyx* sp.), root wounds, and root diseases of white spruce. Can. Agric., Sci. Serv., For. Biol. Div., Ottawa, Ont. Bi-mon. Prog. Rep. 7(4): 2-3.

Watson, W.Y. 1955. A description of the full grown larvae of *Hylobius radicis* Buch. (Coleoptera: Curculionidae). Can. Entomol. 87: 37-41.

Welty, C.; Houseweart, M.W. 1985. Site influences on *Hylobius congener* (Coleoptera: Curculionidae), a seedling debarking weevil of conifer plantations in Maine. Environ. Entomol. 14: 826-833.

Whitney, R.D. 1952. Relationship between entry of root-rotting fungi and root wounding by *Hypomolyx* and other factors in white spruce. Can. Agric., Sci. Serv., For. Biol. Div., Ottawa, Ont. Bi-mon. Prog. Rep. 8(1): 2.

Whitney, R.D. 1961. Root wounds and associated root rots of white spruce. For. Chron. 37: 401-411.

Whitney, R.D. 1962. Studies in forest pathology. XXIV *Polyporus tomentosus* Fr. as a major factor in stand-opening disease of white spruce. Can. J. Bot. 40: 1631-1658.

Wilson, L.F. 1968. Habits and movements of the adult pine root collar weevil in young red pine plantations. Ann. Entomol. Soc. Amer. 61: 1365-1369.

Wilson, L.F. and Millers, I. 1983 Pine root collar weevil– its ecology and management. U.S. Dept. Agric., For. Serv., North Cent. Exp. Stn., East Lansing, Mich. Tech. Bull. No. 1675.

Wilson, L.F.; Waddell, C.D.; Millers, I. 1966. A way to distinguish sex of adult *Hylobius* weevils in the field. Can. Entomol. 98: 1118-1119.

Witcosky, J.J. 1989. Root beetles, stand disturbance, and management of black-stain root disease in plantations of Douglas-fir. Pages 58-70 *in* R.I. Alfaro and S.G. Glover (Eds.). Insects affecting reforestation: biology and damage. Proc. IUFRO working group on insects affecting reforestation (S2.07-03). Vancouver, B.C., July 3-9, 1988.

Witcosky, J.J.; Schowalter, T.D.; Hansen, E.M. 1986. *Hylastes nigrinus* (Coleoptera: Scolytidae), *Pissodes fasciatus*, and *Steremnius carinatus* (Coleoptera: Curculionidae) as vectors of black-stain root disease of Douglas-fir. Environ. Entomol. 15: 1090-1095.

Witcosky, J.J.; Schowalter, T.D.; Hansen, E.M. 1987. Host-derived attractants for the beetles *Hylastes nigrinus* (Coleoptera: Scolytidae) and *Steremnius carinatus* (Coleoptera: Curculionidae). Environ. Entomol. 16: 310-313.

Wood, S.L. 1957. The North American allies of *Hylobius piceus* (De Geer) (Coleoptera: Curculionidae). Can. Entomol. 89: 37-43.

Wood, C.S.; Van Sickle, G.A.; Shore, T.L. 1984. Forest insect and disease conditions, British Columbia & Yukon 1983. Environ. Can., Can. For. Serv., Pac. For. Res. Cent., Victoria, B.C. Inf. Rep. BC-X-246.

Wood, C.S.; Van Sickle, G.A. 1986. Forest insects and disease conditions, British Columbia & Yukon 1985. Environ. Can., Can. For. Serv., Pac. For. Cent., Victoria, B.C. Inf. Rep. BC-X-277.

Wood, C.S.; Van Sickle, G.A. 1989. Forest insect and disease conditions, British Columbia & Yukon 1988. For. Can., Pac. For. Cent., Victoria, B.C. Inf. Rep. BC-X-306.

Wright, J.W.; Wilson, L.F. 1972. Genetic differences in Scotch pine resistance to pine root collar weevil. Michigan State Univ., Agric. Exp. Stn., East Lansing. Res. Rep. 159: 6 p.

Chapter 29

Cecidomyiidae Gall Midges

L.M. Humble and R.J. West

Introduction

Flies of the family Cecidomyiidae, commonly referred to as gall midges, rank sixth in the order Diptera in number of included species. The majority of cecidomyiids induce galls in plant shoots, buds, leaves, and cones, but several species are zoophagous or fungivorous (Mani 1964).

Adults are tiny, fragile flies, less than 5 mm long with reduced wing venation. Larvae pass through three instars and are often brightly colored. The larvae of many species have a distinct "breastplate" or sternal spatula that distinguishes them from other dipterans.

Although many cecidomyiids feed on trees (Barnes 1951), few are considered important forest pests. None of the recorded species in Canada currently cause extensive damage to immature forests; however, with the increase in area of young stands, their status as pests may increase. Those species known to attack regeneration are discussed below; cecidomyiids associated with cones are reviewed in Chapter 33.

Gall Midges of Pines

The jack pine resin midge, *Cecidomyia resinicola*, and the jack pine midge, *C. piniinopis*, are widely distributed in North America. In Canada, both species range from British Columbia east to Quebec. The jack pine resin midge attacks jack pine, *Pinus banksiana*, pitch pine, *P. rigida*, and lodgepole pine, *P. contorta* var. *latifolia*, while the jack pine midge attacks jack pine, lodgepole pine, and ponderosa pine, *Pinus ponderosa* (Gagné 1978; Ives and Wong 1988; Reeks 1960).

Jack pine resin midge adults emerge from late May to early June and live 3 to 4 days. Eggs are laid singly on the bark or needles of lightly resinous current shoots. They hatch in about 6 days and newly hatched larvae crawl to pitch masses and embed themselves. Larvae usually occur singly in small pitch masses but may be gregarious in larger masses. Although the larvae apparently feed only on the resin, larval feeding causes damage to the cambium, stimulating additional pitch flow. The immature larvae overwinter in pitch masses on the twigs. In early spring, larvae complete feeding and pupate during May in delicate cocoons within the pitch (Gagné 1978; Ives and Wong 1988; Reeks 1960).

The life histories of both midges appear to be similar. Jack pine midge adults emerge about 4 to 5 days later than those of the jack pine resin midge, and the females oviposit during the period of rapid shoot elongation. The larvae usually feed singly in pitch droplets or sometimes in pockets at the base of needles. They overwinter within a cavity on the shoot near the base of a bud. In mid-May, mature larvae leave the pitch-filled cavities and pupate in silken cocoons spun on the bark or needles (Gagné 1978; Ives and Wong 1988; Reeks 1960).

Although both species of pine pitch midges have been present during past outbreaks, only the jack pine resin midge has been abundant (Reeks 1960). Since 1960, at least three outbreaks have occurred on jack pine in plantations in central Quebec and south-central Ontario and in natural stands in western Quebec (Martineau 1984). In the Prairie provinces, damage has occurred on trees weakened by jack pine budworm, *Choristoneura pinus*, attack. Although heavy infestations can occur on older trees growing on poor sites, the most important damage occurs on regeneration. Larval feeding causes mortality of new shoots, with up to 75% of the shoots being killed during severe infestations. The jack pine resin midge may cause leader mortality and loss of height growth in young jack pine, but infestations are rarely severe enough to affect radial increment or cause mortality (Ives and Wong 1988; Martineau 1984; Reeks 1960).

No chemical control measures are available for pine pitch midges. Martineau (1984) notes that these species may be eliminated from isolated trees or small plantations by clipping and burning all infested shoots.

The introduced European pine needle midge, *Contarinia baeri*, was first identified in Canada after premature needle loss was observed in Christmas tree plantations of Scots pine, *Pinus sylvestris*, in Quebec in 1968 (Martineau and Ouellette 1969; Canadian Forestry Service 1973). Defoliation by this midge occurred annually up to 1972 and from 1975 through 1977 in Scots and red pine, *Pinus resinosa*, plantations in Quebec and was reported for the first time from Christmas tree plantations in Ontario in 1976 (Canadian Forestry Service 1973, 1976, 1978, 1981a). Moderate to severe leader damage occurred in Ontario in Scots and red pine plantations during 1980, 1983, and 1984 (Canadian Forestry Service 1981b, 1984, 1985). This insect has also been recovered from red pine in New Brunswick where it has been present since at least 1953 (Reeks 1954; DeBoo et al. 1973).

The European pine needle midge is univoltine in Canada. Larvae pupate in the spring and adults emerge from late June to early July. Females lay their eggs in the needle sheaths among the upper portion of the crown. After hatching, larvae bore through the inner side of a needle, at the needle base, and feed on needle tissue until mid- to late August. Mature larvae spin cocoons within the damaged needle cluster and overwinter within the litter (DeBoo et al. 1973; Martineau 1984).

Larval feeding within the sheath causes attacked needles to wither and droop away from healthy ones. By late July infested foliage dies and by early August it drops from the tree. Needle loss is generally confined to the lateral and terminal shoots in the upper third of the crown in sheared Christmas tree plantations, but in ornamental plantings damage may extend throughout the crown. DeBoo et al. (1973) noted only slight mortality of shoots, even after severe defoliation by this midge. Top-kill or tree mortality have not been observed.

Chemical control trials against this midge in Scots pine plantations were conducted by DeBoo et al. in 1973. Ground applications of granular aldicarb, and soil drenches of lindane and disulfoton were tested as controls for adults emerging from cocoons before oviposition. Foliar applications of fenthion, malathion, dimethoate, trichlorfon, disulfoton, demeton, and propoxur with knapsack mistblowers and of fenthion, malathion, and dimethoate with knapsack compressed air sprayers were tested as controls for larval populations. Granular applications of aldicarb (2.24 kg/ha) or foliar applications of dimethoate (10 g/L at 258 L/ha for a mistblower; 1 g/L at 853 L/ha or 2 g/L at 831 L/ha for a compressed air sprayer), fenthion (9 g/L at 168 L/ha for a mistblower; 2 g/L at 899 L/ha or 4 g/L at 809 L/ha for a compressed air sprayer) and malathion (25 g/L at 180 L/ha or 191 L/ha for a mistblower; 3 g/L at 1 146 L/ha or 6 g/L at 1 168 L/ha for a compressed air sprayer) provided good protection during outbreaks.

Periodic outbreaks of the red pine needle midge, *Thecodiplosis piniresinosae*, have lasted 1 to 2 years in plantations in Ontario and Quebec, causing premature needle drop of red pine (Canadian Forestry Service 1978, 1981a, 1981b, 1982). Kearby and Benjamin (1964) studied the biology of and damage caused by the red pine needle midge. Overwintering larvae pupate in spring and adults emerge from late May through early June as the shoots and needles of red pine are elongating. Eggs are laid singly or in clusters on either side of the bases of the developing needle fascicles during the first 2 weeks of June. Larvae hatch in less than a week and mine into the base of a needle fascicle feeding either between the needles within the sheath or on the needle surface beneath the sheath. Infested needle fascicles become enlarged and there is a reduction in needle length. Most fascicles contain two larvae, although up to 12 may be present. Feeding is completed by late September. Mature larvae cut an exit hole in the needle sheath and drop to the ground in early October. They apparently overwinter in the soil without forming a cocoon (Kearby and Benjamin 1964).

Attacked needles first droop, then turn brown, and drop in the fall of the year of attack. In heavy infestations, terminal and lateral shoot mortality is often observed when more than 75% of the needles are lost. Stem deformation and reduced leader growth can result in plantations with persistent outbreaks. Outbreaks persist in open, slow-growing plantations about 10 to 17 years old. Vigorously growing plantations with rapidly elongating shoots and needles are not severely infested.

No chemical control methods have been developed for this insect.

Gall Midges of Douglas-fir

Three species of gall midges, *Contarinia pseudotsugae*, *C. constricta*, and *C. cuniculator*, cause leaf galls on the current year's foliage of Douglas-fir, *Pseudotsuga menziesii*, in British Columbia. Outbreaks of gall midges have been noted since 1935 and can have an economic impact on the Christmas tree industry. *Contarinia pseudotsugae* is found through most of the range of Douglas-fir in British Columbia. *Contarinia constricta* is restricted to the south-central areas (interior) of the province, while *C. cuniculator* occurs primarily in the interior and is present only rarely on the coast. The greater abundance and periodic outbreaks of *C. pseudotsugae* in the interior dry-belt areas of the province make it the most damaging of the three species (Condrashoff 1961, 1962a, 1962b). The biology of this gall midge and the economic damage caused by it are outlined below. Those of the outer two species are similar.

Contarinia pseudotsugae is univoltine and overwinters as a mature larva in the soil or litter. Pupation occurs in late April or early May. Adults emerge, mate, and females lay eggs on newly opened buds or under the scales of swollen buds by mid-May. Larvae hatch in about 3 days and bore directly into the lower epidermis of new needles. Larvae feed within the needles throughout the summer and early fall. Mature larvae exit the needles through the lower epidermis from early October through mid-December and drop to the ground to overwinter.

This insect causes heavy needle losses in severe infestations. Continuation of an infestation for 2 or 3 years can result in severe defoliation, especially on younger trees. Trees heavily attacked for several years can also sustain twig dieback. Tree mortality has not been attributed to these midges; however, foliar loss makes trees unacceptable for the Christmas tree market (Condrashoff 1962b).

Ross and Arrand (1963) and Ross et al. (1964) conducted trials to determine the efficacy of various formulations and concentrations of endosulfan, DDT, and lindane against Douglas-fir midges (*Contarinia* spp.) attacking commercially grown Christmas trees in British Columbia. Hand-sprayed applications, to the point of run-off, of endosulfan (EC, 1 g/L to 3.75 g/L) and DDT (EC, 3.12 g/L) at budflush were found to give satisfactory control. Wettable powder formulations of lindane (1.2 g/L), endosulfan (2.0 g/L), and DDT (2.0 g/L) were ineffective.

Gall Midges of Spruce

The spruce gall midge, *Mayetiola piceae*, causes clusters of hemispherical galls on the new shoots of white spruce, *Picea glauca*, and red spruce, *P. rubens* (Smith 1952). Its range extends from Yukon and Alberta to the Atlantic provinces. High populations were reported causing browning of natural regeneration in the Yukon during 1968 and 1969 (Canada Department of Fisheries and Forestry 1969, 1970). Damage is generally restricted to ornamental and shelterbelt trees in the Prairies. Surveys in the Maritimes in 1982 found measurable populations at only 4 of the 72 locations examined (Canadian Forestry Service 1983).

The spruce gall midge is univoltine, overwintering as mature larvae within the galls. Larvae pupate in the spring and adults emerge in late May or June. Eggs are laid at the base of needles on new shoots and newly hatched larvae bore into the shoots and form cells. Gall tissues develop, resulting in shoots that are about twice the normal diameter. Heavily infested shoots curl and often die; however, the midge is generally not abundant enough to severely affect tree health.

Less severely attacked shoots may survive and continue to grow in a curled fashion (Smith 1952; Ives and Wong 1988).

No control attempts have been made against this minor pest in Canada.

The spruce bud midge, *Rhabdophaga swainei*, occurs throughout the boreal forest, attacking native white spruce, red spruce, and black spruce, *P. mariana*, as well as introduced Norway spruce, *P. abies*, and blue spruce, *P. pungens* (Cerezke 1972; Clark 1952).

There is one generation a year. Newly hatched larvae enter bud primordia of flushing shoots in late spring and gall tissues develop (West 1989). The larvae feed within a chamber in the galled buds throughout the summer and mature by late fall. Larvae overwinter within the buds and pupate from late April to early May. Adults emerge from the buds about 3 weeks later and females lay eggs singly between the needles or basal scales of elongating shoots.

Most infested buds are found in the upper crown on the terminals of main branches and stems on open-grown spruce trees 2 to 4 m high. Height loss and stem distortion (multiple leaders), associated with terminal bud loss, are the main results of spruce bud midge attack, but even these effects are minimal (Cerezke 1972; West 1989).

No control attempts have been made against this pest in Canada. However, Hard et al. (1988) conducted field trials of fenvalerate (0.25 g/L, 0.125 g/L, and 0.625 g/L) and acephate (10 g/L, 5 g/L, and 2.5 g/L) against the spruce bud midge on black and white spruce seedlings in south-central Alaska. Formulations were applied by backpack hydraulic sprayer until run-off, after shoot elongation had begun. The only treatment that provided significant protection was the highest concentration of fenvalerate tested. The authors recommended that further insecticide trials be conducted using both proven systemics, such as dimethoate and metasystox, and additional rainfast compounds, such as carbaryl and permethrin. They recommended against testing higher concentrations of fenvalerate.

Gall Midges of True Firs

Two species of gall midges, the introduced false balsam gall midge, *Dasineura balsamicola*, and the balsam gall midge, *Paradiplosis tumifex*, are associated with galling of the needles of balsam fir, *Abies balsamea*, throughout its range in Canada. Populations of balsam gall midges have been noted almost annually in eastern Canada since 1938. In the Maritimes, outbreaks occurred in 1938 and 1939, 1945 and 1946, from 1956 to 1960, and from 1965 to 1968. Since then populations have remained relatively low, with localized outbreaks noted from 1975 to 1978 and from 1980 to 1986. Localized outbreaks, lasting only a few years, have occurred in Quebec since 1939, and in Ontario, severe outbreaks were reported between 1959 and 1963 (Canadian Forestry Service 1981b, 1983–1987; Martineau 1984; Smith and Forbes 1962).

Smith and Forbes (1962) first identified the presence of two species of midges on galled needles in the Maritimes. However, until Osgood and Gagné (1978) demonstrated that the balsam gall midge caused gall formation, damage was attributed to the introduced false balsam gall midge, an inquiline in galls of *P. tumifex*. Aspects of the biology of one or both species of gall midges have been studied by Osgood and Gagné (1978), Giese and Benjamin (1959), Smith and Forbes (1962), West and Shorthouse (1982), and Shorthouse and West (1986). Little difference is evident in the life histories of the two species. Both are univoltine. Adults are present during the latter half of May. Eggs are laid in the newly opening buds during late May and early June and hatch in 2 to 3 days. The first instar of both species is completed before the end of June. Larvae begin molting to the third instar by the end of August but the balsam gall midge, *P. tumifex*, does not advance beyond the second instar when the false balsam gall midge, *D. balsamicola*, is present. The third instar of the balsam gall midge matures in late September, about 2 weeks before the introduced species. At maturity, larvae of both species leave the galls to overwinter in the duff or mineral soil. Both gall midges pupate in early May; however, only the introduced species pupates in a cocoon.

The balsam gall midge causes the formation of galls at the base of the needles of the current year's foliage on mature and immature balsam fir. The heaviest infestations generally occur on open slow-growing immature trees less than 8 m high. In dense stands, attack is restricted to stand margins (Giese and Benjamin 1959; Smith and Forbes 1962). Galled needles gradually turn yellow and drop from the twigs during late October through November. Since the balsam gall midge causes premature foliage loss, it is of economic concern because severely attacked trees are unacceptable for the Christmas tree market. However, damage is seldom permanent. If infested trees are not cut for 3 to 4 years after the collapse of an outbreak, they will have recovered enough to be marketable (Smith et al. 1981).

Control of balsam gall midge is required only when it is necessary to preserve the appearance of trees used for ornamental purposes or for the Christmas market. Smith et al. (1981) provided chemical control recommendations and application rates for mist blowers (56 L/ha) and hydraulic sprayers (1 125 L/ha). They recommended the application of diazinon (12.5 g/L for mistblowers; 0.62 g/L for hydraulic sprayers) in early June for control of the midge. Dimethoate (19.9 g/L for mistblowers; 1.0 g/L for hydraulic sprayers) and malathion (24.9 g/L for mistblowers; 1.25 g/L for hydraulic sprayers) were also noted as providing fair protection against midge damage.

Gall Midges of Yellow Cypress

In 1987, a new species of gall midge was discovered causing galls on the foliage of yellow cypress, *Chamaecyparis nootkatensis*, in a seed orchard in British Columbia (Wood et al. 1987). Subsequent surveys have shown the insect to be widely distributed in coastal areas. This pest, initially reported as *Contarinia* n. sp., was recently described by Gagné and Duncan (1990) as *Chamaediplosis nootkatensis*.

Some research has been completed on the biology of this midge (R. Duncan, Forestry Canada, pers. comm.). At low elevations, two overlapping generations are completed annually. Peak adult flights occur in April and in June. In cooler natural stands, where only one generation occurs annually, flight occurs in July. Eggs are laid singly near the base of scale leaves. First-instar larvae bore into the meristematic tissue near a growing tip or into the base of a scale leaf and cause a swelling to develop. Galling of the terminal growth is very evident by the time the second instar is reached. The gall increases in size through the third instar, and averages 3 mm in diameter at maturity. Pupation occurs within the gall. Before adult eclosion the pupa pushes itself through the gall wall. Galling usually occurs at the tip of vegetative shoots or in the developing megasporangia and microsporangia. Galled structures are killed back to the nearest crotch. Thus, attack by this pest has the potential to affect cone production in commercial seed orchards. Damage in natural stands is very low; however, in seed orchards it can exceed 15% of the available shoots and immature cones.

Research trials are currently under way to identify suitable chemical control options for this pest in seed orchards (D. Summers, B.C. Forest Service, pers. comm.).

References

Barnes, H.F. 1951. Gall midges of economic importance. Vol. V. Gall midges of trees. Crosby Lockwood and Son, Ltd. London. 270 p.

Canada Department of Fisheries and Forestry. 1968. Annual report of the Forest Insect and Disease Survey 1969. Ottawa, Ont.

Canada Department of Fisheries and Forestry. 1969. Annual report of the Forest Insect and Disease Survey 1970. Ottawa, Ont.

Canada Department of Fisheries and Rural Development. 1968. Annual report of the Forest Insect and Disease Survey 1967. Ottawa, Ont.

Canadian Forestry Service. 1973. Annual report of the Forest Insect and Disease Survey, 1972. Environ. Can. Ottawa, Ont.

Canadian Forestry Service. 1976. Annual report of the Forest Insect and Disease Survey, 1975. Environ. Can. Ottawa, Ont.

Canadian Forestry Service, 1978. Annual report of the Forest Insect and Disease Survey, 1976. Environ. Can. Ottawa, Ont.

Canadian Forestry Service. 1981a. Annual report of the Forest Insect and Disease Survey, 1977. Environ. Can. Ottawa, Ont.

Canadian Forestry Service. 1981b. Forest insect and disease conditions in Canada 1980. Environ. Can. Ottawa, Ont.

Canadian Forestry Service. 1982. Annual report of the Forest Insect and Disease Survey, 1978 and 1979. Environ. Can. Ottawa, Ont.

Canadian Forestry Service. 1983. Forest insect and disease conditions in Canada 1982. Environ. Can. Ottawa, Ont.

Canadian Forestry Service. 1984. Forest insect and disease conditions in Canada 1983. Environ. Can. Ottawa, Ont.

Canadian Forestry Service. 1985. Forest insect and disease conditions in Canada 1984. Environ. Can. Ottawa, Ont.

Canadian Forestry Service. 1986. Forest insect and disease conditions in Canada 1985. Agric. Can. Ottawa, Ont.

Canadian Forestry Service. 1987. Forest insect and disease conditions in Canada 1986. Agric. Can. Ottawa.

Cerezke, H.F. 1972. Observations on the distribution of the spruce bud midge (*Rhabdophaga swainei* Felt) in black and white spruce crowns and its effect on height growth. Can. J. For. Res. 2: 69-72.

Clark, J. 1952. The spruce bud midge, *Rhabdophaga swainei* Felt (Cecidomyiidae: Diptera). Can. Entomol. 84: 87-89.

Condrashoff, S.F. 1961. Three new species of *Contarinia* Rond. (Diptera: Cecidomyiidae) in Douglas-fir needles. Can. Entomol. 93: 123-130.

Condrashoff, S.F. 1962a. Bionomics of three closely related species of *Contarinia* Rond. (Diptera: Cecidomyiidae) from Douglas-fir needles. Can. Entomol. 94: 376-394.

Condrashoff, S.F. 1962b. Douglas-fir needle midges—pests of Christmas trees in British Columbia. Can. Dept. For., Ottawa 5 p.

DeBoo, R.F.; Campbell, L.M.; Laplante, J.P.; Daviault, L.P. 1973. Plantation research: VIII. The pine needle midge, *Contarinia baeri* (Diptera: Cecidomyiidae), a new insect pest of Scots pine. Environ. Can., Chem. Cont. Res. Inst., Ottawa, Ont. Inf. Rep. CC-X-41.

Gagné, R.J. 1978. A systematic analysis of the pine pitch midges, *Cecidomyia* spp. (Diptera: Cecidomyiidae). U.S. Dept. Agric., Sci. Educ. Admin., Washington, D.C. Tech. Bull. 1575.

Gagné, R.J.; Duncan, R.W. 1990. A new species of Cecidomyiidae (Diptera) damaging shoot tips of yellow cypress, *Chamaecyparis nootkatensis*, and a new genus for two gall midges on Cupressaceae. Proc. Entomol. Soc. Wash. 92(1): 146-152.

Giese, R.L.; Benjamin, D.M. 1959. The biology and ecology of the balsam gall midge in Wisconsin. For. Sci. 5(2): 193-208.

Hard, J.; Shea, P.; Holsten, E. 1988. Field trials of fenvalerate and acephate to control spruce bud midge, *Dasneura swainei* (Diptera: Cecidomyiidae). J. Entomol. Soc. Brit. Columbia. 85: 40-44.

Ives, W.G.H.; Wong, H.R. 1988. Tree and shrub insects of the Prairie Provinces. Can. For. Serv., Northern For. Cent., Edmonton, Alta. Inf. Rep. NOR-X-292.

Kearby, W.H.; Benjamin, D.M. 1964. The biology and ecology of the red-pine needle midge and its role in fall browning of red pine foliage. Can. Entomol. 96: 1313-1322.

Mani, M.S. 1964. Ecology of plant galls. Junk, The Hague, the Netherlands. 434 p.

Martineau, R. 1984. Insects harmful to forest trees. Can. For. Serv., Environ. Can., Ottawa, Ont. For. Tech. Rep. 32. Multiscience Publications.

Martineau, R.; Ouellette, G.B. 1969. Une cécidomyie sur le pin sylvestre. Pages 15-16 *in* Rapport Annuel—1969. Relevé des Insectes et des Maladies des Arbres, Région de Québec. Can. Dept. Fish. and For., Quebec. Inf. Rep. Q-F-X-6.

Osgood, E.A.; Gagné, R.J. 1978. Biology and taxonomy of two gall midges (Diptera: Cecidomyiidae) found in galls on balsam fir needles with description of a new species of *Paradiplosis*. Ann. Entomol. Soc. Am. 71: 85-91.

Reeks, W.A. 1954. Damage to red pine by a midge. Can. Dept. Agric., Ottawa, Ont. Bi-mon. Prog. Rep. 10(3): 1-2.

Reeks, W.A. 1960. Observations on the life history, distribution, and abundance of two species of *Cecidomyia* (Diptera: Cecidomyiidae) on jack pine in Manitoba and Saskatchewan. Can. Entomol. 82: 154-160.

Ross, D.A.; Arrand, J. 1963. Preliminary insecticide tests against the Douglas-fir needle midges. *Contarinia* spp., Larkin, B.C., 1962. Proc. Entomol. Soc. Brit. Columbia. 60: 32-33.

Ross, D.A.; Arrand, J.; Geistlinger, N.J. 1964. Further insecticide tests against the Douglas-fir needle midges, *Contarinia* spp. Proc. Entomol. Soc. Brit. Columbia. 61: 20-22.

Shorthouse, J.D.; West, R.J. 1986. Role of the inquiline, *Dasineura balsamicola* (Diptera: Cecidomyiidae), in the balsam fir needle gall. Proc. Entomol. Soc. Ont. 117: 1-7.

Smith, C.C. 1952. The life-history and galls of a spruce gall midge, *Phytophaga picea* Felt (Diptera: Cecidomyiidae). Can. Entomol. 84: 272-275.

Smith, C.C.; Forbes, R.S. 1962. Gall midges on Balsam fir needles in the Maritime Provinces. Can. Dept. For., Ottawa, Ont. Bi-mon. Prog. Rep. 18(3): 1.

Smith, C.C.; Newell, W.R.; Renault, T.R. 1981. Common insects and diseases of balsam fir Christmas trees. Environ. Can., Can. For. Serv., Ottawa, Ont. 60 p.

West, R.J. 1989. The biology, damage and within-tree distribution of the spruce bud midge, *Rhabdophaga swainei* Felt (Diptera: Cecidomyiidae), on black spruce in Newfoundland. Pages 213-223 *in* R.I. Alfaro; S. Glover (Eds.), Insects affecting reforestation: biology and damage. Proc. IUFRO Symp., Vancouver, B.C. July 3-9, 1988.

West, R.J.; Shorthouse, J.D. 1982. Morphology of the balsam fir needle gall induced by the midge *Paradiplosis tumifex* (Diptera: Cecidomyiidae). Can. J. Bot. 60(2): 131-140.

Wood, C.S.; Van Sickle, G.A.; Humble, L.M. 1987. Forest insect and disease conditions British Columbia & Yukon 1987. Can. For. Serv., Pac. For. Cent., Victoria, B.C. Inf. Rep. BC-X-296.

Chapter 30

European Pine Shoot Moth, *Rhyacionia buoliana*, and Other Olethreutid Shoot Borers and Tip Moths

P.D. Syme, G.G. Grant, and T.G. Gray

Introduction

The European pine shoot moth, *Rhyacionia buoliana*, has been known as a destructive pest of hard pine plantations in North America since it was identified in 1914 in the New York area (Busk 1914). It was discovered in Ontario and Nova Scotia in 1925 and was first recorded in western Canada at Victoria, British Columbia, in the same year (McLaine 1926). The first recorded outbreak in the west did not occur until 1938 (Mathers 1938) on native lodgepole pine, *Pinus contorta* var. *latifolia*, planted as ornamentals in South Vancouver. Its history and biology in Canada before 1969 have been the subject of extensive investigations and are well summarized elsewhere (McGugan and Coppel 1962; Pointing and Green 1962; Miller 1967; Pointing and Miller 1967; Syme 1971, 1984; Rose and Lindquist 1984; Martineau 1984). The status of the shoot moth has been under continuous surveillance and has been recorded annually in the reports of the Forest Insect and Disease Survey.

In most of eastern Canada, infestations have remained at a low ebb since the early 1980s. Changes in the uses of pine have probably played a role in the changing attitude of foresters, but the insect is not now considered as great a threat as in the past. However, in Nova Scotia and Prince Edward Island, there seems to be a population increase and there is concern for the future of young red pine, *P. resinosa*, and Scots pine, *P. sylvestris*, in these provinces (L. Magasi, pers. comm.).

In British Columbia and the Yukon, the shoot moth is the most important shoot-boring insect of pines. From 1968 to 1981 a quarantine to protect native pines was implemented in British Columbia under the Plant Protection Act to prohibit the transportation of infested pines from the Vancouver Forest Region. Currently, there are no provincial regulations restricting the movement of ornamental or nursery stock from coastal nurseries to the interior of the province. However, a federal regulation requires that all imported stock entering British Columbia bear a phytosanitary certificate stating it to have been free of European pine shoot moth for at least 2 years before the importation date. These regulations are being changed and likely they will be abandoned.

Since 1973, negligible research has been done in Canada on the European pine shoot moth or its parasites, except for the pheromone work described below.

Biology

In eastern Canada red pine is most seriously damaged by the shoot moth. Scots pine, Austrian pine (*P. nigra*), ponderosa pine (*P. ponderosa*), and ornamental mugho pine (*P. mugo mughus*) are moderately susceptible, whereas pitch pine (*P. rigida*), Virginia pine (*P. virginiana*), jack pine (*P. banksiana*), and eastern white pine (*P. strobus*) are relatively resistant. In the Pacific and Yukon regions, the European pine shoot moth is considered a potential threat to native ponderosa and lodgepole pines (Evans 1973; Pointing and Miller 1967), and it is the most damaging insect of ornamental pines on southern Vancouver Island and in the Lower Fraser Valley (Harris and Ross 1975).

The shoot moth has one generation per year and adults appear in the field over a 4-week period that usually begins in early June. The small moths have a light orange-yellow head and thorax with gray abdomen and light reddish-orange forewings with silvery white markings and grayish hindwings. Moths generally emerge at sunrise but are not active until the early evening when virgin females "call," emitting a sex pheromone to attract males. After mating, about 60 flat, oval, 1-mm long, cream-colored eggs are laid singly, or in small groups, on the twigs or needle sheaths of the current year's growth. Development is temperature dependent and usually requires about 2 weeks in the field. First- and second-instar larvae, which are yellowish with black heads and thoracic shields, spin small webs between the needle bases and the twig (de Gryse 1932; Pointing and Green 1962). They mine needle bases on new growth from within this protection. The needles discolor and die, but this apparently has little effect on the vigor of the host. After mining several needle-pairs, the larvae move up to the buds and spin a larger web, lined with resin, between the buds and nearby needles. They bore into the buds and kill them. The half-grown larvae cease feeding by mid- to late August to overwinter within the buds.

Feeding on the terminal bud seriously injures the main stem and reduces height growth. Repeated destruction of the entire terminal cluster during the summer months initially causes a profusion of shoot growth (witches'-broom) and eventually a spike top when the leader dies (Fig. 1). Various degrees of crookedness result from lateral branches or buds assuming dominance over terminal shoots. The insect occasionally attacks cones (Chapter 33).

In the spring, the spinning of new webs and molting precede the resumption of feeding, which coincides with the onset of tree growth (Pointing and Miller 1967). Throughout the spring, larval movements to the upper whorls are marked by new webs. Here, the larger buds, and frequently the tree leader are attacked (Pointing 1963). The most noticeable injury caused by this insect occurs when only one side of the main shoot is damaged and continued growth forms a posthorn, or crook. The yellow-brown pupae are formed in silk-lined chambers in mined shoots or large buds from late May to mid-June and moths emerge in 2 or 3 weeks.

Figure 1. "Spike top" on young red pine tree caused by repeated destruction of terminal buds by the European pine shoot moth, *Rhyacionia buoliana*.

Other Shoot Borers and Tip Moths

Native shoot borers, shoot moths, and tip moths in the genera *Rhyacionia* and *Eucosma* attack pines (Table 1) and like the European pine shoot moth can cause loss of form and incremental growth because they characteristically damage new apical shoots, including the terminal leader. Except for the pine shoot borer, *R. busckana*, occurring in British Columbia as mentioned below, the life cycles of native *Rhyacionia* species differ from that of European pine shoot moth in several respects (Powell and Miller 1978; Grant et al. 1985). Larvae of native species complete their development in the same year the eggs are laid and they pupate in the soil rather than in the larval gallery on the tree. Adults emerge the following spring in April and May, much earlier than the European pine shoot moth, and often fly when snow is still on the ground. There is considerable host and geographical overlap between native species and the European pine shoot moth, but the native species range considerably farther north, at least in the east.

The pine tip moth, *Rhyacionia adana*, known in Canada only in Ontario on red, jack, and Scots pines, attacks trees about 1 m or less high in plantations, nurseries, and natural stands (Martin 1960). Larvae mine the shoots and kill the buds before pupating on the root collar just below the soil surface. Often the pine tip moth and the European pine shoot moth concurrently attack plantations of less than 5 to 6 years old. They can be separated by the following differences: their respective larval size, because of the timing of their larval development; the timing of their entry into shoots; and their pupation sites. Dead shoots with needle growth of about 2 cm indicate damage caused by the pine tip moth, whereas those without needle growth are caused by European pine shoot moth. The pine tip moth is the only *Rhyacionia* species other than the European pine shoot moth known to cause serious commercial damage to pine plantations (Rose and Lindquist 1984; USDA Forest Service 1985; Wilson 1977).

The yellow jack pine shoot borer, *Rhyacionia sonia*, has been found at scattered locations throughout Ontario, upper Michigan, and southeastern Manitoba (Miller 1967; Powell and Miller 1978; Rose and Lindquist 1984). The yellowish larvae mine the twigs of new jack pine shoots in late June and July, producing diagnostic silk and excreta among the needles in the dead shoot tip (USDA Forest Service 1985; Powell and Miller 1978; Rose and Lindquist 1984). Serious injury has not been recorded.

The pine shoot borer, *R. busckana*, is a transcontinental species that may encompass several unknown species (Powell and Miller 1978). In the east, the red jack pine shoot borer, *R. granti*, was recently recognized as a long-confused sibling species, segregated initially from the pine shoot borer by a difference in its response to sex attractants (Table 1). Subsequent studies discerned morphological differences (Miller 1985; Grant et al. 1985). Pine shoot borer larvae are a minor pest of red and Scots pine, whereas the red jack pine shoot borer is found only on jack pine (Miller 1985). In British Columbia, the pine shoot borer is found on ponderosa pine but it is bionomically distinct from its eastern counterpart in that it overwinters as a larva, pupates on the tree in the larval feeding chamber, and flies from July to September (Evans 1982). Thus, it could be another distinct species. Its sex attractant is unknown.

A ponderosa pine tip moth, *Rhyacionia zozana*, is a pest of young ponderosa pines in British Columbia. It also attacks lodgepole pine, Jeffrey pine (*P. jeffreyi*), sugar pine (*P. lambertiana*), Digger pine (*P. sabiniana*), and a variety of other pines (Furniss and Carolin 1977). The larvae mine young growing shoots of open-grown trees less than 2 m tall and thereby decrease growth. The insect does not seriously harm native pines.

The western pine shoot borer, *Eucosma sonomana*, whose eggs are laid singly on swelling buds (Evans 1982), is a pest of ponderosa pines in British Columbia (Furniss and Carolin 1977; Grant 1958). From 36 to 75% of the trees in a 20-ha site near Cascade, British Columbia, had 1 to 17 damaged tips per tree over a 4-year period in the early 1960s. Because the insect prefers terminal leaders, damage can lead to significant reduction in annual height growth (Stozek 1973).

The eastern pine shoot borer, *Eucosma gloriola*, is a widely distributed eastern species recognized as taxonomically and geographically distinct from the western pine shoot borer, *E. sonomana* (DeBoo et al. 1971). It occurs throughout the natural range of eastern white pine (Fowells 1965) except in the extreme southern part of this range and in the Atlantic provinces (DeBoo et al. 1971). It has not been a problem in natural mixed stands, but it is a potentially serious pest of young pine plantations, attacking jack pine in the northern part of its range and white or red pine in the southern part (DeBoo et al. 1971; Rose and Lindquist 1984). Infested trees become bushy with repeated attack and when the terminal leader is damaged the trees are forked and stunted (USDA Forest Service 1985). Often the leaders break in high winds because the larvae girdle the inside of the shoot (Wong and Campbell 1967).

The life history and biology of the eastern pine shoot borer have been discussed by DeBoo et al. (1971). Moths emerge from overwintering cocoons in May, overlapping the earlier flight of native *Rhyacionia* species (Grant et al. 1985). Where the eggs are laid is not definitely known. Hatched larvae bore directly into the pith, usually a few centimetres above the base of the new shoots, and feed toward the base of the shoots. Near the end of the feeding period in July, the larvae reverse direction and widen the original tunnel within the pith zone. The fully grown larvae, almost 13 mm long with pale bodies and yellow-brown heads, finally cut a hole to the outside and drop to the ground to pupate within cocoons in the soil.

Control of the European Pine Shoot Moth

Red pine in particular is able to outgrow extensive European pine shoot moth damage (Syme 1976) and the level of control required may not be as critical as once supposed (Syme 1971; Pointing and Green 1962). In Scots pine Christmas tree plantations, shearing and pruning usually reduce populations to acceptable levels. Because cold winter temperatures limit the northward spread of the European pine shoot moth in Canada (Beique 1960; Green 1962), pruning lower branches of susceptible pine trees to eliminate or at least decrease the proportion of the crown that will be under the snow will help to reduce the surviving population during a severe winter.

Insecticides

Control of the European pine shoot moth and other *Rhyacionia* and *Eucosma* species is often difficult because the larvae are internal feeders and the damage they cause is usually not evident until after they have finished feeding or are close to pupation. Therefore, timing of chemical control is difficult and the availability of effective insecticides is limited.

Table 1. Summary of hosts and sex pheromones (P) or sex attractants (A) for shoot moths, tip moths, and shoot borers attacking *Pinus* species in Canada.

Species	Host	Attractive compounds		Optimum ratio	Reference[a]
Rhyacionia buoliana	Most hard pines	(E)-9-dodecenyl acetate + (E)-9-dodecenol	(P)	97:3	1
R. adana	*P. resinosa, P. banksiana, P. sylvestris*	(E)-9-dodecenyl acetate + (E)-9-dodecenol	(A)	98:2 - 80:20	2
R. busckana (eastern)[b]	*P. resinosa, P. sylvestris*	(E)-9-dodecenyl acetate + (E)-9-dodecenol	(A)	98:2 - 80:20	2
R. busckana (western)[b]	*P. ponderosae*	unknown	(A)	-	3
R. granti	*P. banksiana*	(E,E)-8,10-dodecadienyl acetate	(A)	-	2
R. sonia	*P. banksiana*	(E)-9-dodecenyl acetate[c]	(A)	-	2
R. zozana	*P. ponderosa*	(E)-9-dodecenyl acetate + (E)-9-dodecenol	(P)	70:30 - 95:5	4
Eucosma gloriola	*P. resinosa, P. banksiana*	(Z)-9-dodecenyl acetate + (E)-9-dodecenyl acetate	(A)	85:15	2
E. sonomana	*P. ponderosae*	(Z)-9-dodecenyl acetate + (E)-9-dodecenyl acetate	(A)	2:1	5

[a]References: 1, Gray et al. 1984; 2, Grant et al. 1985; 3, Evans 1982; 4, Niwa et al. 1987; 5, Sower et al. 1979.
[b]Because of differences in their bionomics, these insects may not be the same species (see text).
[c]Only one moth caught with this compound.

Chemical methods of control of the European pine shoot moth have been investigated (Winter and Scott 1977; Pree and Saunders 1972; Carolin et al. 1962; Miller and Haynes 1961; Butcher and Haynes 1959, 1960). Although fumigation can be detrimental to young seedlings, methyl bromide fumigation of dormant seedlings in the spring (Flink and Brigham 1959), of dormant bare root and container-grown seedlings in both the spring and fall (Ilnytzky and Sweeten 1976), or phosdrin dip or spray in the fall (Butcher and Haynes 1958) give satisfactory control in nurseries. Although the larva spends most of its life within the tree, it wanders from needle to needle, or from bud to bud, but remains close to the bases of the needles. A contact insecticide, therefore, would have to drench the trees to reach the needle bases and only a small proportion of larvae would be affected. Similarly, stomach poisons would also have to be applied at a high rate to penetrate to the buds, but even then, the larva does not consume the outer parts of the bud (Pointing 1962). The high rate of application of these insecticides would create a danger to wildlife and to natural predators and parasites that tend to suppress the shoot moth populations.

Systemic poisons have been tried with varying degrees of success in Europe and North America. Carbofuran applied as granules (Pree and Saunders 1972) or as a soil drench (Boyd et al. 1968) appears to offer the most potential with the least risk of phytotoxicity. Its effectiveness decreases as the tree size increases, however (Pree and Saunders 1972). Generally, the cost of systemics is prohibitive for large-scale use and few are registered because of their toxicity.

The cost of insecticides and the complications attending their use preclude their use in most commercial situations. Pheromone traps may be useful in improving the timing of insecticide applications (Gargiullo et al. 1983), but this has not been investigated for European pine shoot moth in Canada. Protection of high-value ornamental trees with costly insecticides may be acceptable in some circumstances, but hand pruning and killing infested shoots in the fall or spring is usually an effective alternative, except in large plantations where it is not practical. Within an integrated pest management program, chemical control is best applied in the spring to treat the susceptible, wandering stage of the larvae and cause the least damage to *Orgilus obscurator*, the most common and effective introduced parasite of the European pine shoot moth (Syme 1971).

Parasites and Predators

In Ontario, studies of the parasites of the European pine shoot moth can be grouped conveniently into three phases. Before 1962, they focused primarily on the biology, life histories, and morphology of several native and introduced parasites (Juillet 1959, 1960a,b,c, 1961; Juillet and Laviolette 1966; Arthur 1961, 1962, 1963; Arthur et al. 1964; Leius 1961a,b, 1963, 1967). During the mid- and late 1960s, emphasis was placed on the impact of the parasites on host survival. Starting in 1970 the emphasis once again changed, and studies centered on the effects of wild carrot, *Daucus carota*, and other flowering plants on the survival and effectiveness of one of the most efficient parasites, *O. obscurator* (Syme 1984).

Early studies indicated that parasites were essentially ineffective (Pointing and Miller 1967), but later studies showed that under favorable conditions of both weather and adult food (nectar) sources, parasites could have devastating effects on shoot moth populations (Syme 1971 1984). Indeed, in one case (Syme 1971), where 133 females were introduced in 1960, in the presence of wild carrot, parasitism by *O. obscurator* alone rose to 92% by 1964, with the complete eradication of the host population by 1965.

The method whereby the probability of control can be enhanced by manipulating the environment into which a parasite is introduced, thus favoring its survival and effectiveness, has been promoted. This approach represents a new working principle that requires emphasis in future control attempts and it has been put into effect in Nova Scotia (Syme 1984).

Pheromones—Survey Detection, Mating Disruption

In 1974, mugho, Austrian, and Scots pines were heavily infested with shoot moth near Castlegar in the interior of British Columbia. These nonnative pines had been brought from a coastal nursery and planted in 1968. To support the 1968 quarantine mentioned earlier, detection and monitoring of European pine shoot moth populations had been implemented, and an extensive pheromone trapping program was initiated by the Forest Insect and Disease Survey of Forestry Canada in 1974. Monitoring of moth populations was facilitated by the isolation and identification of the sex pheromone by Smith et al. (1974). This trapping program continued until the quarantine was lifted in 1981. During this period, moderate to heavy infestations of European pine shoot moth were monitored in the interior of the province at Kelowna, Vernon, Kamloops, Castlegar, and Trail, as well as in the Vancouver area.

During the course of the monitoring program, unexpected null trap catches were encountered. Field observations revealed that male moths were attracted to traps but failed to enter. The synthetic sex pheromone obtained from chemical supply companies was suspected and a study was initiated to determine why the attractant failed to lure moths into the traps. A reinvestigation of the sex pheromone (Gray et al. 1984) revealed the presence of three components, (E)-9-dodecenol (E9-12:OH), dodecyl acetate, and dodecanol, in addition to the previously identified (E)-9-dodecenyl acetate (E9-12:Ac). The absence of E9-12:OH from the synthetic attractant used for trapping explained why occasionally no moths were captured in traps. An effective lure, containing E9-12:Ac and E9-12:OH in a ratio of 97:3 embedded separately in polyvinylchloride (5% by weight) and cut into 2.5 × 3 mm diameter rods, was formulated for the detection and monitoring of the moths.

Since the removal of the quarantine in 1981 the provincial and federal authorities have cooperatively utilized pheromone lures and Pherocon 1C sticky traps (Ecoyoos Consulting Services, Osoyoos, British Columbia, V0H 1V0) to detect infestations of the European pine shoot moth. Endemic populations are controlled by clipping and burning infested shoots. Because of these efforts, areas previously containing moderate to high infestations in the interior of British Columbia and on

Figure 2. Scots pine tree with butcher string and inhibitor lures attached to prevent males from mating with female moths.

Vancouver Island currently possess low populations of shoot moth. Detection traps deployed in southwestern Alberta from 1981 to 1985 failed to catch any adults, indicating an absence of European pine shoot moth in that area.

Sex attractants useful for monitoring other species of shoot moths and shoot borers occurring in Canada are known (Table 1). The native *Rhyacionia* species are sympatric and have overlapping flight periods. At least three species are attracted to the same compounds as the European pine shoot moth, but since they fly earlier in the spring they would not show up in survey traps for the European pine shoot moth (Grant et al. 1985). *Eucosma* species are attracted to similar chemicals but the attractant blends are composed predominantly of the (Z)-isomer of the compounds (Table 1).

Selective control of insect populations by mating disruption is most effective against low populations over limited areas (Beroza 1976). Because shoot moths and borers cause noticeable damage at relatively low populations within plantations, they are ideal targets for control by this means. Considerable success in suppressing populations of the western pine shoot borer and reducing damage has been demonstrated in the western United States (Daterman 1982; Sartwell et al. 1983). Studies with the European pine shoot moth, *Rhyacionia buoliana*, and other *Rhyacionia* species in that area (Daterman et al. 1975; Berisford 1982) and in England (Baker and Longhurst 1981) indicate that mating disruption with sex pheromone is also potentially useful for controlling shoot moth in Canada, but there have been no attempts to do so. However, an inhibitor of the sex pheromone, (Z)-9-dodecenyl acetate (Z9-12:Ac) (Smith et al. 1974), was used in 1979 in a small study conducted at Richmond, British Columbia, to try to prevent males from locating females and mating: Z9-12:Ac (2.5% by weight) was imbedded in polyvinylchloride, cut into 5 × 3 mm diameter rods, then placed inside Beem embedding capsules (J.B. EM Services Inc., Dorval, Quebec, H9R 4S8), and attached to butcher string at 1-m intervals. The string was then wound around infested Scots pine so there was a capsule every metre over the trees' foliar surface (Fig. 2). Although mating was disrupted, causing a 67% reduction in the following year's population, this was not sufficient to reduce it to a manageable level (Gray, unpubl. data).

References

Arthur, A.P. 1961. The cleptoparasitic habits and immature stages of *Eurytoma pini* Bugbee (Hymenoptera: Chalcidae), a parasite of the European pine shoot moth, *Rhyacionia buoliana* (Schiff.) (Lepidoptera: Olethreutidae). Can. Entomol. 93: 655-660.

Arthur, A.P. 1962. Influence of host tree on abundance of *Itoplectis conquisitor* (Say) (Hymenoptera: Ichneumonidae), a polyphagous parasite of the European pine shoot moth, *Rhyacionia buoliana* (Schiff.) (Lepidoptera: Olethreutidae). Can. Entomol. 94: 337-347.

Arthur, A.P. 1963. Life histories and immature stages of four ichneumonid parasites of the European pine shoot moth *Rhyacionia buoliana* (Schiff.) in Ontario. Can. Entomol. 95: 1078-1091.

Arthur, A.P.; Stainer J.E.R.; Turnbull, A.L. 1964. The interaction between *Orgilus obscurator* (Nees) (Hymenoptera: Braconidae) and *Temelucha interruptor* (Grav.) (Hymenoptera: Ichneumonidae), parasites of the pine shoot moth, *Rhyacionia buoliana* (Schiff.) (Lepidoptera: Olethreutidae). Can. Entomol. 96: 1030-1034.

Baker, R.; Longhurst, C. 1981. Chemical control of insect behaviour. Phil. Trans. R. Soc. Lond. B 295: 73-82.

Beique, R. 1960. The importance of the European pine shoot moth *Rhyacionia buoliana* (Schiff.) in Quebec City and vicinity. Can. Entomol. 92: 858-862.

Berisford, C.W. 1982. Pheromones of *Rhyacionia* spp.: Identification, function and utility. J. Ga. Entomol. 17: 23-30.

Beroza, M. Ed. 1976. Pest management with insect pheromones. Am. Chem. Soc. symposium series. Washington, D.C. 191 p.

Boyd, J.P.; Barton, R.L; Walton, R.R. 1968. Control of Nantucket pine tip moth or pinews in ornamental or small plantations by systemic insecticides. Oklahoma Agric. Exp. Stn. Bull. B-661 39 p.

Busk, A. 1914. A destructive pine moth introduced from Europe. J. Econ. Entomol. 7: 340-341.

Butcher, J.W.; Haynes, D.L. 1958. Fall control of European pine shoot moth on pine seedlings. Quart. Bull. Mich. Agric. Exp. Stn. 41: 264-268.

Butcher, J.W.; Haynes, D.L. 1959. Experiments with new insecticides for control of European pine shoot moth. Quart. Bull. Mich. Agric. Exp. Stn. 41: 734-744.

Butcher, J.W.; Haynes, D.L. 1960. Influence of timing and insect biology on the effectiveness of insecticides applied for control of European pine shoot moth, *Rhyacionia buoliana*. J. Econ. Entomol. 53: 349-354.

Carolin, V.M.; Klein, W.H.; Thompson, R.M. 1962. Eradicating European pine shoot moth on ornamental pines with methyl bromide. USDA For. Serv., Pac.

Northwest For. and Range Exp. Stn., Portland, Ore. Res. Paper. 47, 16 p.

Daterman, G.E. 1982. Control of western pine shoot borer damage by mating disruption - a reality. *In* A.F. Kydonieus and M. Beroza (Eds.), Insect Supression with Controlled Release Pheromone Systems,Vol II., CRC Press, Boca Raton, Fla.

Daterman, G.E.; Doyle Daves, G. Jr.; Smith, R.G. 1975. Comparison of sex pheromone versus an inhibitor for mating disruption of pheromone communication in *Rhyacionia buoliana*. Environ. Entomol. 4: 944-946.

DeBoo, R.F.; Sippell, W.L.; Wong, H.R. 1971. The eastern pine-shoot borer, *Eucosma gloriola* (Lepidoptera: Tortricidae), in North America. Can. Entomol. 103: 1473-1486.

de Gryse, J.J. 1932. Note on the early stages of the European pine shoot moth. Can. Entomol. 64: 169-173.

Evans, D. 1973. Establishment and survival of European pine shoot moth on container-grown 1-0 lodgepole pine. Can. For. Serv., Pac. For. Res. Cent., Victoria, B.C. Inf. Rep. BC-X-79. 7 p.

Evans, D. 1982. Pine shoot insects common in British Columbia. Can. For. Serv., Pac. For. Res. Cent., Victoria, B.C. Inf. Rep. BC-X-233. 56 p.

Flink, P.R.; Brigham, P.M. 1959. Preliminary report on red pine nursery stock fumigation. J. For. 57: 662-663.

Fowells, H.A. 1965. Silvics of forest trees of the United States. USDA For. Serv. Washington, D.C. Agric. Handb. 271. 762 p.

Furniss, R.L.; Carolin, V.M. 1977. Western Forest Insects. USDA For. Serv., Washington, D.C. Misc. Publ. 1339.

Gargiullo, P.M.; Berisford, C.W.; Canalos, C.G.; Richmond, J.A. 1983. How to time dimethoate sprays against the Nantucket pine tip moth. Georgia For. Res. Paper 44, 10 p.

Grant, G.G.; MacDonald, L.; Frech, D.; Hall, K.; Slessor, K.N. 1985. Sex attractants for some eastern species of *Rhyacionia*, including a new species, and *Eucosma gloriola* (Lepidoptera: Tortricidae). Can. Entomol. 117: 1489-1496.

Grant, J. 1958. Observations on a pine shoot moth, *Eucosma sonomana* Kft. (Lepidoptera: Olethreutidae) Proc. Entomol. Soc. B.C. 55: 25-27.

Gray, T.G.; Slessor, K.N.; Shepherd, R.F.; Grant, G.G.; Manville, J.F. 1984. European pine shoot moth, *Rhyacionia buoliana* (Lepidoptera: Tortricidae): Identification of additional pheromone components resulting in an improved lure. Can. Entomol. 116: 1525-1532.

Green, G.W. 1962. Low winter temperatures and the European pine shoot moth, *Rhyacionia buoliana* (Schiff.) in Ontario. Can. Entomol. 94: 314-346.

Harris, J.W.E.; Ross, D.A. 1975. European pine shoot moth. Can. For. Serv. Pacif. For. Res. Cent., Victoria, B.C. For. Pest Leafl. 18. Rev. Apr. 1975. 4 p.

Ilnytzky, S.; Sweeten, J.R. 1976. Fumigation of bareroot and container-grown lodgepole pine seedlings for European pine shoot moth control. Tree Planters' Notes, (USFS) 27 (4): 21-22, 33.

Juillet, J.A. 1959. Morphology of immature stages, life history, and behaviour of three hymenopterous parasites of the European pine shoot moth, *Rhyacionia buoliana* (Schiff.) (Lepidoptera: Olethreutidae). Can. Entomol. 91: 709-719.

Juillet, J.A. 1960a. Immature stages, life histories and behaviour of two hymenopterous parasites of the European pine shoot moth, *Rhyacionia buoliana* (Schiff.) (Lepidoptera: Olethreutidae). Can. Entomol. 92: 342-346.

Juillet, J.A. 1960b. Resistance to low temperatures of the overwintering stages of two introduced parasites of the European pine shoot moth, *Rhyacionia buoliana* (Schiff.) (Lepidoptera: Olethreutidae). Can. Entomol. 92: 701-704.

Juillet, J.A. 1960c. Some factors influencing the flight activity of hymenopterous parasites. Can J. Zool. 38: 1057-1061.

Juillet, J.A. 1961. Observations on arthropod predators of the European pine shoot moth, *Rhyacionia buoliana* (Schiff.) (Lepidoptera: Olethreutidae), in Ontario. Can. Entomol. 93: 195-198.

Juillet, J.A.; Laviolette, R. 1966. Notes sur la présence de *Rhyacionia buoliana* (Schiff.) à Sherbrooke et dans la région. Phytoprotection 47: 114-115.

Leius, K. 1961a. Influence of food on fecundity and longevity of adults of *Itoplectis conquisitor* (Say) (Hymenoptera: Ichneumonidae). Can. Entomol. 93: 771-780.

Leius, K. 1961b. Influence of various foods on fecundity and longevity of adults of *Scambus buolianae* (Htg.) (Hymenoptera: Ichneumonidae). Can. Entomol. 93: 1079-1084.

Leius, K. 1963. Effects of pollens on fecundity and longevity of adult *Scambus buolianae* (Htg.) (Hymenoptera: Ichneumonidae). Can. Entomol. 95: 202-207.

Leius, K. 1967. Food sources and preferences of adults of a parasite, *Scambus buolianae* (Hym.: Ichn.), and their consequences. Can. Entomol. 99: 865-871.

Martin, J.L. 1960. Life history of the pine tip moth, *Rhyacionia adana* Heinrich, in Ontario (Lepidoptera: Olethreutidae). Can. Entomol. 92: 724-728.

Martineau, R. 1984. Insects harmful to forest trees. For. Tech. Rep. 32, Can. For. Serv., Ottawa. 261 p.

Mathers, W.G. 1938. Can. Dept. Agric., Ann. Rept. of the Vancouver Forest Insect Lab., pp. 58-63.

McGugan, B.M.; Coppell, H.C. 1962. Biological control of forest insects, 1910-1958. Pages 35-127 *in* A review of the biological control attempts against insects and weeds in Canada. Commonw. Inst. Biol. Cont. Tech. Comm. 2. Slough, Eng.

McLaine, L.S. 1926. A preliminary announcement of the outbreak of the European pine shoot moth. 56th Ann. Rept. Entomol. Soc. Ont. (1925) pp. 71-72.

Miller, W.E. 1967. The European pine shoot moth - ecology and control in the Lake States. For. Sci. Monogr. 14, 72 p.

Miller, W.E. 1985. Nearctic Rhyacionia pine tip moths: A revised identity and a new species (Lepidoptera: Tortricidae). Great Lakes Entomol. 18: 119-122.

Miller, W.E.; Haynes, D.L. 1961. Experiments with concentrated DDT sprays for European pine shoot moth suppression in forest plantations. J. Econ. Entomol. 54: 1014-1018.

Niwa, C.G.; Sower, L.L.; Daterman, G.E. 1987. Chemistry and field evaluation of the sex pheromone of ponderosa pine tip moth, *Rhyacionia zozana* (Lepidoptera: Tortricidae). Environ. Entomol. 16: 1287-1290.

Pointing, P.J. 1962. The effectiveness of a microbial insecticide against larvae of the European pine shoot moth, *Rhyacionia buoliana* (Schiffermuller). J. Insect Path. 4: 484-497.

Pointing, P.J. 1963. The biology and behaviour of the European pine shoot moth, *Rhyacionia buoliana* (Schiff.), in southern Ontario. II. Egg, larva and pupa. Can. Entomol. 95: 844-863.

Pointing, P.J.; Green, G.W. 1962. A review of the history and biology of the European pine shoot moth, *Rhyacionia buoliana* (Schiff.) (Lepidoptera: Olethreutidae) in Ontario. Proc. Entomol. Soc. Ontario 92: 58-69.

Pointing, P.J.; Miller, W.E. 1967. European pine shoot moth. Pages 162-166 *in* Important forest insects and diseases of mutual concern to Canada, the United States and Mexico. Can. Dept. For. Rur. Dev., Ottawa, Publ. 1180.

Powell, J.A.; Miller, W.E. 1978. Nearctic pine tip moths of the genus Rhyacionia: Biosystematic review (Lepidoptera: Tortricidae, Olethreutinae). USDA For. Serv., Washington, D.C.Agric. Handb. 514. 51 p.

Pree, D.J.; Saunders, J.L. 1972. Chemical control of the European pine shoot moth. J. Econ. Entomol. 65: 1081-1085.

Rose, A.H.; Lindquist, O.H. 1984. Insects of eastern pines (Rev.) Dept. Environ. Publ. 1313. Ottawa. 127 p.

Sartwell, C.; Daterman, G.E.; Overhulser, D.L.; Sower, L.L. 1983. Mating disruption of western pine shoot moth borer (Lepidoptera: Tortricidae) with widely spaced releasers of synthetic pheromone. J. Econ. Entomol. 76: 1148-1151.

Smith, R.G.; Daterman, G.E.; Daves, G.D. Jr.; McMurtrey, K.D.; Roelofs, W.L. 1974. Sex pheromone of the European pine shoot moth: Chemical identification and field tests. J. Insect Physiol. 20: 661-668.

Sower, L.L.; Daterman; G.E.; Sartwell, C.; Cory, H.T. 1979. Attractants for the western pine shoot borer, *Eucosma sonomana*, and *Rhyacionia zozana* determined by field screening. Environ. Entomol. 8: 265-267.

Stozek, K.J. 1973. Damage to ponderosa pine plantations by the western pine shoot borer. J. For. 71: 701-705.

Syme, P.D. 1971. *Rhyacionia buoliana* (Schiff.), European pine shoot moth (Lepidoptera: Olethreutidae). Pages 194-205 *in* Biological control programmes against insects and weeds in Canada, 1959-1968. Commonw. Inst. Biol. Cont. Tech. Comm. 4. Slough, Eng.

Syme, P.D. 1976. Red pine and the European pine shoot moth in Ontario. Dept. Environ., Can. For. Serv., Great Lakes For. Res. Cent., Sault Ste. Marie, Ont. Int. Rep. O-X-244. 17 p.

Syme, P.D. 1984. *Rhyacionia buoliana* (Schiff.), European pine shoot moth, (Lepidoptera: Tortricidae). Pages 387-394 *in* J.S. Kelleher, and M.A. Hulme (Eds.), Biological control programmes against insects and weeds in Canada 1969-1980. Commonw. Agric. Bureaux, Slough, Eng.

USDA Forest Service. 1985. Insects of eastern forests. Washington, D.C. Misc. Publ. 1426. 608 p.

Wilson, L.F. 1977. A guide to insect injury of conifers in the lake states. USDA For. Serv., Washington, D.C. Agric. Handbook No. 501. 218 p.

Winter, T.G. and T.M. Scott. 1977. Chemical control of the pine shoot moth, *Rhyacionia buoliana* (Dennis and Schiffermuller) (Lepidoptera: Tortricidae) in seed orchards in Britain. Forestry 50: 161-164.

Wong, H.R. and A.E. Campbell, 1967. The larval feeding habits of the eastern pine shoot borer, *Eucosma gloriola* Heinrich (Lepidoptera: Tortricidae) in jack pine regeneration in Manitoba. Manitoba Entomol. 1: 42-46.

Chapter 31

Pitch-Blister Moths, *Petrova* spp.

H.R. Wong, J.A. Drouin, and C.L. Rentz

Introduction

Pitch-blister moths are a group of insects whose larvae construct hollow resinous nodules with silk and frass over the feeding site on pine shoots. Heinrich (1923) placed this group of insects along with others described in several other genera in his genus *Petrova*, and indicated there were 12 species and 1 subspecies in North America. Since then several new species have been described (McDunnough 1938; Miller 1959, 1978) and changes have been made in names and status of other species (Brewer and Stevens 1973; Miller 1978; Hodges et al. 1983). Miller (1978) noted that 8 of the 14 species of *Petrova* in North America are pitch-blister moths, and 5 species are present in Canada (Table 1). Although *P. pallipennis*, described by McDunnough (1938), is present in Canada and resembles the pitch twig moth, *P. comstockiana*, in outward appearance, it is not a pitch-blister moth like the latter (Miller 1978; Butcher and Hodgson 1949).

Biological information relating to the pitch-blister moths in Canada is presented in Table 1. The northern pitch twig moth, *P. albicapitana*, and the Virginia pitch-nodule moth, *P. wenzeli*, according to Miller (1978) may be color morphs of one biennial population. The former species predominates in northern North America and the latter in east-central North America. Miller (1978) indicated that the specimens of a metallic pitch nodule moth, *P. metallica*, reported by McLeod and Tostowaryk (1971) in Quebec were a new species, *P. mafica* on jack pine (*Pinus banksiana*). The life cycle of *P. comstockiana*, noted by Miller and Neiswander (1956) as annual in Ohio, is probably biennial in Canada. This is suggested by the life cycle of *P. metallica* which is annual in Minnesota and South Dakota (Miller 1978), but biennial in Canada (Stark 1957; Wong et al. 1985). The life cycle of *P. mafica* is probably biennial in Canada, because the known life cycles of the species of *Petrova* in this country are biennial and adult specimens of *P. mafica* were noted as emerging with *P. albicapitana* from nodules on jack pine in Quebec (McLeod and Tostowaryk 1971).

The two most important species in Canada are the northern pitch twig moth, *P. albicapitana* and a metallic pitch nodule moth, *P. metallica*. Several papers have reported on the life history of *P. albicapitana* (Turnock 1953; McLeod and Tostowaryk 1971; Tracy and Osgood 1981), but only sketchy accounts on the biology of *P. metallica* are available (Stark 1957; Hopping 1961; Wong et al. 1985). The life history of *P. metallica* and a comparison of the differences between the two major species of *Petrova* in Canada are discussed in this report.

Damage and Abundance

The northern pitch twig moth is distributed from Nova Scotia to British Columbia and attacks the twigs, branches, and leaders of jack pine (*Pinus banksiana*), lodgepole pine (*P. contorta* var. *latifolia*), and Scots pine (*P. sylvestris*). This species was most abundant on jack pine from 0.3 to 1.5 m high in Manitoba plantations (Turnock 1953), and in several mature stands of jack pine 50 to 60 years old in Quebec (McLeod and Tostowaryk 1971). *Petrova metallica* is distributed from the Cypress Hills in Saskatchewan to British Columbia and attacks usually the twigs and branches and occasionally the

Table 1. Some aspects of the biology of *Petrova* spp. in Canada.

Species	Distribution	*Pinus* host	Location of nodule	Life cycle
P. albicapitana	Nova Scotia to British Columbia	*P. banksiana* *P. contorta* *P. sylvestris*	internodal and nodal	biennial (Turnock 1953)
P. comstockiana	Ontario	*P. rigida* *P. sylvestris* *P. resinosa*	internodal	Annual in Ohio (Miller and Neiswander 1956), unknown in Canada
P. mafica	Ontario, Quebec	*P. banksiana*	internodal	unknown
P. metallica	Cypress Hills, Sask., to British Columbia	*P. contorta*	internodal	biennial (Stark 1957, Wong et al.1985)
P. wenzeli	Ontario	*P. banksiana*	internodal and nodal	biennial (Miller and Altmann 1958)

leader of small seedlings of lodgepole pine. Pitch-blister nodules of this species were observed on trees from 0.3 to 25 m high and the most severe attacks occur on trees 3.0 m high. Field studies in 1982 to 1984 with a synthetic sex attractant[1] (Wong et al. 1985) revealed that *P. metallica* was much more abundant than previously reported (Prentice 1965). The numbers of *P. metallica* captured were greater than *P. albicapitana* (Table 2). The abundance of each species captured appears to be related to elevation. Both species are common at 900 to 1 300 m. Above 1 300 m the number of captured males of *P. albicapitana* generally decreases whereas the number of *P. metallica* increases. No *P. metallica* were captured in areas below 800 m elevation. The capture of more males of *P. metallica* than expected at Whitecourt (900 m) and Swan Hills (800 m) in Alberta may be the result of the males having been carried into these areas by prevailing winds, which regularly occur from nearby higher elevations to the northwest. Field observations failed to detect *P. metallica* pitch blister nodules or damage in the vicinity of the traps.

The larvae of both species feed on the new and old growth of the stem and branches of pines. They combine the pitch exuding from canals punctured by them during feeding with

[1] 7Z, 9E-dodecadienyl-acetate, supplied by the Prairie Regional Laboratory, National Research Council of Canada, Saskatoon, Saskatchewan.

Table 2. Capture of *Petrova albicapitana* and *P. metallica* males by traps baited with sex attractant at various elevations and areas in Alberta.

Location	Elevation (m)	Stand	No. of baits	Total *P. albicapitana*	Total *P. metallica*
1982					
McLeod[a]	1 300	Lodgepole	21	276	495
McLeod[a]	1 300	Lodgepole	21	134	316
Edson	900	Lodgepole	8	10	2
Devon	850	Lodgepole	6	30	0
1983					
Berland[a]	1 500	Lodgepole	16	5	90
Athabasca[a]	1 300	Lodgepole	16	49	204
Marlboro[a]	1 200–1 300	Lodgepole	16	19	49
McLeod[a]	1 300	Lodgepole	16	142	181
Grande Prairie	700	Lodgepole	8	4	0
Slave Lake	800	Lodgepole	8	6	0
Whitecourt	900	Lodgepole	8	1	17
Nojack	800	Jack and lodgepole	8	5	0
1984					
Berland[a]	1 500	Lodgepole	13	8	80
Athabasca[a]	1 300	Lodgepole	13	40	172
Marlboro[a]	1 200	Lodgepole	13	22	79
McLeod[a]	1 300	Lodgepole	13	100	106
Greg River[a]	1 700	Lodgepole	13	8	87
Banff	1 400	Lodgepole	13	0	394
Simonette	1 200	Lodgepole	13	0	64
Cypress Hills	1 400	Lodgepole	13	0	763
Whitecourt	1 000	Lodgepole	13	0	15
Swan Hills Tower	1 200	Lodgepole	13	0	77
Swan Hills	800	Lodgepole	13	13	19
Florida Hills	1 000	Lodgepole	13	0	31
Waterton Park	1 500	Lodgepole	13	0	150
Clear Hills	1 300	Lodgepole	13	0	0
Porcupine Hills	1 300	Lodgepole	13	0	52
Bruderheim	700	Jack	13	42	0
Hondo	700	Jack	13	28	0
Caslan	700	Jack	13	15	0
Nojack	800	Jack and lodgepole	13	8	0

[a] Names of areas designated by Weldwood of Canada Ltd. in its woodlands operations near Hilton, Alberta.

silk and frass to form round nodules (Figs. 5, 6). Damage by *P. albicapitana* becomes noticeable after the larva has migrated to the node where it builds a second-year nodule. The larva feeding beneath the bark frequently girdles and kills the stem or shoot (Fig. 9). The foliage acquires a reddish-brown color producing a "flagging" appearance (Fig. 14). Damage to growing terminal shoots by second-year larvae could result in a forked trunk or dead leaders, but more frequently in crooked and weakened trunks (Turnock 1953) and height reduction in the upper crown (McLeod and Tostowaryk 1971). Larvae of *P. metallica* do not girdle the cortical tissue, but tunnel in the xylem and pith (Fig. 10) and do not normally kill the shoot. *Petrova metallica* causes "flagging" only when it attacks slender shoots or leaders of small seedlings causing breakage by wind and snow. The second-year larvae do not migrate from nodules on branches or stems to the leader so that dead, crooked, or weakened trunks do not usually occur.

Bella (1985) observed that the thinning of lodgepole pine stands in west-central Alberta to improve the growth and yield increased the incidence of attack by *P. albicapitana*, *P. metallica*, and the lodgepole terminal weevil, *Pissodes terminalis*, to 25% from 16% in unthinned stands. The most common type of damage by pitch-blister moths is reduced height growth and breakage (Figs. 6, 14) resulting in lower merchantable volume production. Because only a small proportion of the total volume of foliage is affected, and the damage impact is expected to lessen with increasing age, pitch-blister moths will not likely cause tree mortality (McLeod and Tostowaryk 1971; Bella 1985).

Life History of *Petrova metallica*

This insect was first described by Busck (1914) as *Evetria metallica* and later placed in the genus *Petrova* by Heinrich (1923). The earliest indication that this species is a common insect attacking lodgepole pine was reported by Stark (1957) in Alberta. Hopping (1961) indicated that *P. metallica*, *P. albicapitana*, and *P. luculentana* caused serious injury to twigs and terminals of lodgepole pine in the same province. The latter species, described by Heinrich (1920) as *Evetria luculentana*, was placed by Miller (1978) as a junior synonym of *P. metallica*.

The adults of *P. metallica* appear at least 2 weeks earlier than the northern pitch-twig moth, and at higher elevations (1 500 m), much earlier than the latter species (Figs. 1 and 2).

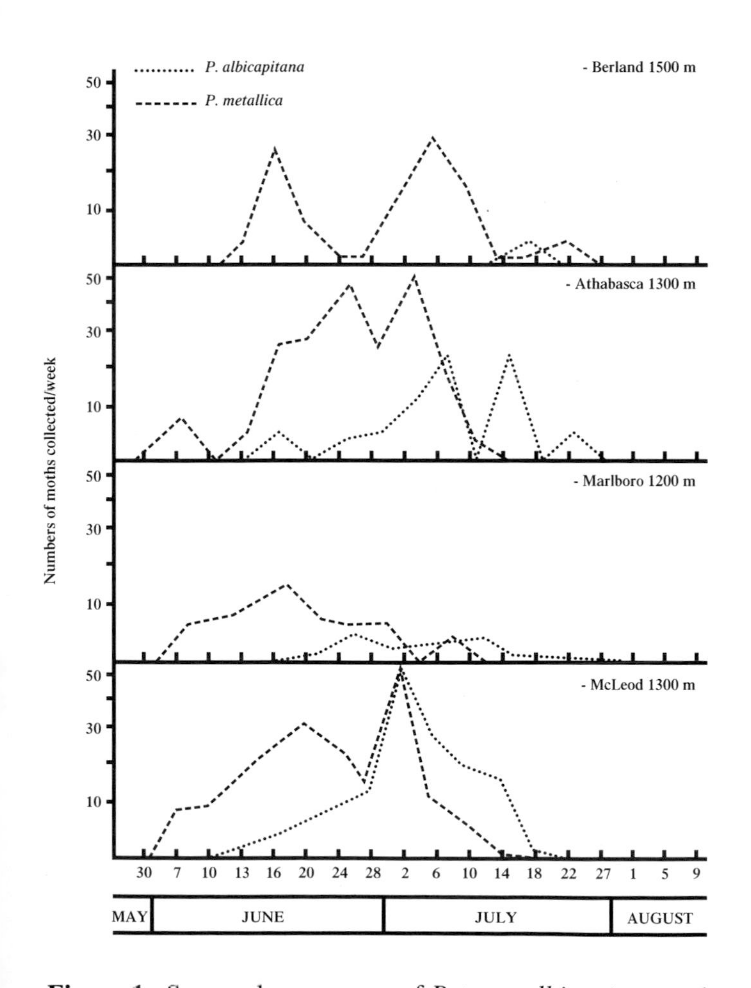

Figure 1. Seasonal occurrence of *Petrova albicapitana* and *P. metallica* as indicated by males captured per week in pheromone traps near Hinton, Alberta, 1983.

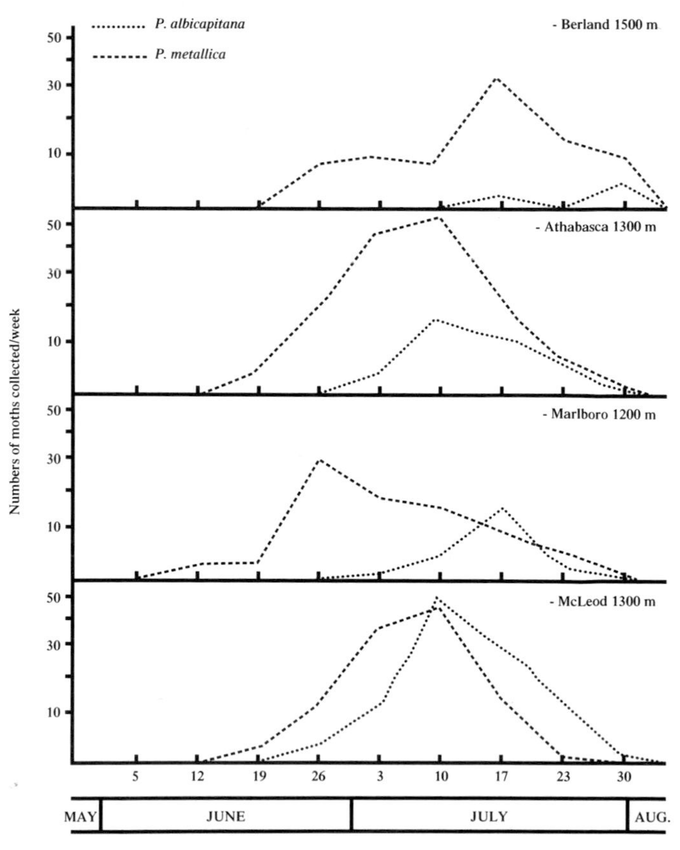

Figure 2. Seasonal occurrence of *Petrova albicapitana* and *P. metallica* as indicated by males captured per week in pheromone traps near Hinton, Alberta, 1984.

In 1983, they were captured from late May to mid-July at elevations of 900 to 1 300 m, but at 1 500 m emergence extended from early June to late July. Cold and wet weather during the spring and early summer will result in the presence of more than one peak emergence period (Fig. 1). When the season was about 2 weeks later, as in 1984 (Fig. 2), the emergence period for *P. metallica* was extended and ended about the same time as that of *P. albicapitana*.

Eggs were not observed in the field, and attempts to have moths oviposit in caged lodgepole pine were not very successful. Only one egg was deposited by a captive female in late July near the apical bud on the base of a needle sheath located on a lateral shoot about 7 mm in diameter. The yellowish egg was about 0.7 mm and flat and oval. The first-instar larva penetrates the needle sheath and feeds on the needle base before entering the bark and wood. The presence of two or more dead needles on the new growth and a small silk tent with pitch and frass are the first sign of damage by *P. metallica* on lodgepole pine (Fig. 11). This type of damage was observed on shoots 4 to 15 mm in diameter, which were mainly laterals and occasionally leaders of small seedlings. There appears to be a size restriction on shoots suitable for oviposition. When the laterals of small seedlings were unsuitable in size, the female was forced to oviposit on the thicker leader. In the wood, the larva tunnels in the xylem and pitch, enlarging the silk-lined reddish-brown nodule (Fig. 10). It forms a silk-lined chamber in the tunnel to overwinter. The larva becomes active in early May, tunnelling in the pith, enlarging the nodule and building compartments (Fig. 12). The feeding damage causes the area of the shoot behind the gall to swell (Fig. 8); the diameter of the swellings vary from 5 to 12 mm. The larva passes the second winter in the enlarged nodule. In the third season, it continues feeding in the xylem and pitch for about 2 weeks and the length of the tunnel may vary from 10 to 27 mm. The larva prepares for pupation by building a silk-lined pupal chamber with a thin outer wall made of pitch and silk near the apical end of the nodule. The pupa is oriented with its head toward the outer wall. Before the adult emerges, the pupa forces itself through the exit and protrudes about half way (Fig. 13) or completely out of the nodule suspended by silken threads.

Comparison Between *Petrova albicapitana* and *P. metallica*

The damage and biology of the northern pitch twig moth, *P. albicapitana*, and a metallic pitch nodule moth, *P. metallica*, are similar in many respects. Females oviposit on the needle sheaths of the new growth. The larvae, which closely resemble one another (MacKay 1959), enter the wood and construct resinous nodules of silk and frass. Both species have a biennial life history and their pupae protrude from the nodules before adults emerge. It has been generally assumed that nodules on the nodes of the main stem (Fig. 5) were made by larvae of *P. albicapitana* migrating from smaller nodules on the internodes of branches, and those second-year and third-year nodules on the internodes of branches were constructed by nonmigrating larvae of *P. metallica* (Fig. 6). Wong et al. (1985) noted that nodules of *P. albicapitana* could occur internodally on the main stem, and those of *P. metallica* could also occur on the main stem just below the junction of a branch of lodgepole pine saplings.

It has been suggested that the migrating larvae of pitch-blister moths have survival value (Miller and Altmann 1958). Newly hatched larvae can only penetrate the bark of currently developing shoots and feed on the cortical tissue. Because the larvae grow faster than the small shoots, they migrate to crotches of larger shoots to meet their needs for more food and better shelters. Those larvae that do not migrate, feed on the xylem and pith and cause that area of the shoot to hypertrophy, thus providing the additional food and bigger shelters for the growing larvae.

There are several differences in the appearance and habits that can be used to separate the two species (Table 3). The most

Table 3. Comparison between *Petrova metallica* and *P. albicapitana*.

Character	*P. metallica*	*P. albicapitana*
Color of forewing	Grayish brown sprinkled with rust, banded by metallic and white, sometimes forming an indistinct checkerboard pattern (Miller 1978)	Rusty, sprinkled with black, banded by shining white or gray (Miller 1978)
Color of hindwing	Grayish brown (Miller 1978)	Gray (Miller 1978)
Part of shoot attacked	Xylem and pith (Miller 1978, Wong et al. 1985)	Cortical tissue (Turnock 1953, Miller 1978, Wong et al. 1985)
Type of damage	Tunneling the shoot (Wong et al. 1985)	Girdling the shoot (Turnock 1953, Wong et al. 1985)
Larva migrating from internode to node	No (Wong et al. 1985)	Yes (Turnock 1953, Wong et al. 1985)
Texture of nodule	Verrucose with pale areas of dried resin (Wong et al. 1985)	Generally smooth and unicolorous (Wong et al. 1985)
Swelling on shoot caused by feeding damage	Yes (Wong et al. 1985)	No (Wong et al. 1985)

Table 4. Parasites of pitch-blister moths.

Species	Hosts (*Petrova*)[a]
Hymenoptera	
Ichneumonidae	
Exeristes comstockii	AL,[d j k l] CO,[d g] HO,[f] ME[l]
Exochus turgidus	AL[d]
Campoplex conocola	AL[d]
Campoplex sp.	ME[l]
Lissonata conocola	AL,[d] ME[l]
Lissonata folii	ME[l]
Scambus dioryctriae	AL,[d] ME[l]
Scambus sp.	AL[k]
Sinophorus sp. nr. *sulcatellus*	AL[k]
Aphidiidae	
Aphidius sp.	AL[k]
Braconidae	
Agathis pini	CO,[e g] HO,[e f] WE[h]
Agathis sp.	AL[k]
Apanteles petrovae	AL[e j k]
Apanteles sp.	AL[k]
Bracon sp.	AR[c]
Chelonus (*Microchelonus*) *recurvariae*	AL[e]
Macrocentrus cuniculus	AL,[e k l] AR,[c e] ME[e l]
Orgilus lateralis	AL[e]
Eulophidae	
Elachertus sp.	AL[l]
Hyssopus benefactor	AL,[b k] CO,[b] HO,[f] ME,[b i] WE[b]
Hyssopus evetriae	AL,[b k] WE[b h]
Hyssopus thymus	AL,[j] CO,[b g] WE[b h]
Diptera	
Tachinidae	
Lixophaga sp.	AL,[k] ME[l]
Phrynofrontina sp.	AL,[j] ME[i]

[a] AL = *albicapitana*, AR = *arizonensis*, CO = *comstockiana*, ME = *metallica*, HO = *houseri*, WE = *wenzeli*.
[b] From Burks 1971; [c] Brewer and Stevens 1973; [d] Carlson 1971; [e] Marsh 1971; [f] Miller 1963; [g] Miller and Neiswander 1956; [h] Miller and Altmann 1958; [i] Stark 1957; [j] Tracy and Osgood 1981; [k] Turnock 1953; [l] from present study.

evident are the rusty color of the forewing and gray hindwing in *P. albicapitana* (Fig. 3) and grayish-brown color of the forewing and hindwing in *P. metallica* (Fig. 4). The nodules of *P. albicapitana* are generally smooth and unicolorous (Fig. 5), and those of *P. metallica* are verrucose with pale areas of dried resin (Fig. 13). Larvae of *P. albicapitana* feed in the bark, cambium, and outer xylem and partly girdle the shoot (Fig. 9). They usually migrate to a crotch on the main stem in the second season and do not cause the shoot to swell (Fig. 7). Larvae of *P. metallica* feed in the xylem and pith, tunnelling downward beyond the nodule (Fig. 10) and causing a slight swelling of the shoot (Fig. 8). They do not migrate and complete their development in the same area of the shoot.

Table 5. Percent parasitism of *Petrova metallica* in Alberta.

Year	Location	Date	No. nodules examined	No. nodules with parasites	Percent parasitism
1984	Hinton	March 15	100	28	28
1985	Hinton	March 15	100	21	21
	Hinton	July 15	100	7	7
	Berland	August 15	100	10	10
1986	Hinton	July 15	100	7	7
	Hinton	August 15	150	15	10

Natural Control

The species of parasites reared from the nodules of pitch-blister moths are recorded in Table 4. Mortality caused by these parasites was generally low. In populations of the Virginia pitch-nodule moth, *P. wenzeli*, Miller and Altmann (1958) noted in Ohio and Maryland that parasitism of 14% in the spring and 7% in the winter was nearly all caused by *Hyssopus thymus* or *H. evetriae*. The parasites recovered from the pitch twig moth, *P. comstockiana*, by Miller and Neiswander (1956) in Ohio were *Agathis pini*, *H. thymus*, *Exeristes comstockii*, and *Perilampus fulvicornis* (hyperparasite of *A. pini*). They indicated that *A. pini* kept populations of *P. comstockiana* at endemic levels in the forests, but not in ornamental plantings. *Hyssopus thymus* accounted for 7% of the mortality of the young larvae. *Agathis pini* was observed by Miller (1963) to have caused about two-thirds of the 48% to 54% parasitism of the shortleaf pitch-blister moth, *P. houseri*, in Washington Co., Ohio. The other parasites were *E. comstockii*, *H. benefactor*, and a Braconid sp. The life cycles of *A. pini*, *P. fulvicornis*, and *H. thymus* are detailed by Miller (1955). Brewer and Stevens (1973) reported that only 7 out of 281 nodules of the pinyon pitch-nodule moth, *P. arizonensis*, they observed in Colorado were infested by *Bracon* sp. and *Macrocentrus cuniculus*. Parasitism of the northern pitch twig moth, *P. albicapitana*, varied from 2% to 16.7% in Manitoba and Saskatchewan according to Turnock (1953), and the parasites most commonly recovered were *E. comstockii* and *M. cuniculus*. Low parasitism of *P. albicapitana* by *Phrynofrontina* sp., *Apanteles petrovae*, *E. comstockii*, and *H. thymus* was recorded also by Tracy and Osgood (1981) in Maine. Parasitism of the metallic pitch-nodule moth, *P. metallica*, in Alberta varied from 7% to 28% from 1984 to 1986 (Table 5). The most common species were *M. cuniculus*.

Predation by birds on the larvae and pupae of pitch-blister moths is not a common occurrence. The only report of birds preying on an unusually large proportion of pupae and larvae was by Miller and Altmann (1958), who suspected that the Carolina chickadee, *Parus carolinensis*, caused 56% predation of *P. wenzeli*. Brewer and Stevens (1973) observed that birds had apparently preyed on a few nodules of *P. arizonensis* but did not indicate the birds involved. Signs of predation on the larvae and pupae of *P. metallica* were apparent in lodgepole pine stands in Alberta. The gray jay, *Perisoreus canadensis* (Salt 1986), was the only bird actually

3

4

5

Figures 3–8.

Wings, pitch nodules, and damage by *Petrova* spp.

Figure 3. Wings of *Petrova albicaptana*. **Figure 4.** Wings of *P. metallica*. **Figure 5.** Nodule constructed by *P. albicapitana*. **Figure 6.** Nodule constructed by *P. metallica* showing broken apical end of stem. **Figure 7.** Opposite view showing unswollen stem infested by *P. albicapitana*. **Figure 8.** Opposite view showing swollen stem infested by *P. metallica* and broken apical end.

6

7

8

9

11

10

Figures 9–14.
Pitch nodules and damage by *Petrova* spp.

Figure 9. Opened nodule of *Petrova albicapitana* showing damage from larval feeding. **Figure 10.** Opened nodule of *P. metallica* showing damage from larval feeding. **Figure 11.** Dead needles and young nodule indicate where *P. metallica* entered stem. **Figure 12.** Opened nodule showing compartments constructed by *P. metallica* larva. **Figure 13.** Empty pupa of *P. metallica* protruding from nodule. **Figure 14.** Dead broken shoot caused by feeding damage by *P. albicapitana*.

12

13

14

observed extracting larvae from pitch-blister nodules, but other species are probably involved.

The Mackenzie red squirrel, *Tamiasciurus hudsonicus preblei*, and the British Columbia red squirrel, *T. hudsonicus columbiensis* (Soper 1964), were observed attacking nodules of *P. albicapitana* and *P. metallica* in Alberta.

Acknowledgments

We are indebted to Dr. E.W. Underhill and his colleagues of the Plant Biotech. Institute, National Research Council of Canada, Saskatoon, Saskatchewan, for preparing and supplying the synthetic sex attractants for *Petrova albicapitana* and *P. metallica*. Our appreciation is extended to present and past members of Biosystematics Research Canada, Ottawa, for kindly identifying the parasites. Weldwood of Canada Ltd. is acknowledged for permission to study insects in its lodgepole pine stands.

References

Bella, I.E. 1985. Pest damage incidence in natural and thinned lodgepole pine in Alberta. For. Chron. 61: 233-235.

Brewer, J.W.; Stevens, R.E. 1973. The pinyon pitch nodule moth in Colorado. Ann. entomol. Soc. Am. 66: 789-792.

Busck, A. 1914. Descriptions of new Microlepidoptera of forest trees. Proc. Entomol. Soc. Wash. 16: 143-150.

Burks, B.D. 1971. Eulophidae. Pages 967-1022 *in* Krombein, K.V., P.D. Hurd, Jr., D.R. Smith and B.D. Burks (Eds.), Catalog of Hymenoptera in America North of Mexico, Vol. 1. Smithsonian Institution Press, Washington, D.C.

Butcher, J.W.; Hodgson, A.C. 1949. Biological and ecological studies on some lepidopterous bud and shoot insects of jack pine (Lepidoptera Olethreutidae). Can. Entomol. 81: 161-173.

Carlson, R.W. 1971. Ichneumonidae. Pages 315-740 *in* Krombein, K.V., P.D. Hurd, Jr., D.R. Smith and B.D. Burks (Eds.), Catalog of Hymenoptera in America North of Mexico, Vol. 1. Smithsonian Institution Press, Washington, D.C.

Heinrich, C. 1920. On some forest Lepidoptera with descriptions of new species, larvae and pupae. Bull. U.S. Natl. Mus. Washington, D.C. 57: 56-57.

Heinrich, C. 1923. Revision of the North American moths of the subfamily Eucosminae of the family Olethreutidae. Bull U.S. Natl. Mus. Washington, D.C. 123: 1-297.

Hodges, D.W.; Dominick, T.; Davis, D.R.; Ferguson, D.C.; Franclemont, J.G.; Munroe, E.G.; Powell, J.A. 1983. Check list of the Lepidoptera north of Mexico including Greenland. E.W. Classey Ltd. and The Wedge Entomol. Res. Found. London, England.

Hopping, G.R. 1961. Insects injurious to lodgepole pine in the Canadian Rocky Mountain Region. Pages 77-87 *in* Smithers, L.A. (Ed.), Lodgepole pine in Alberta. Can. For. Ser. Dep. Fish. For. Edmonton, Alta. 380 p.

MacKay, M.R. 1959. Larvae of the North American Olethreutidae (Lepidoptera). Can. Entomol. 91. Supp. 10.

Marsh, P.M. 1971. Braconidae. Pages 144-295 *in* Krombein, K.V., P.D. Hurd, Jr., D.R. Smith and B.D. Burks (Eds.), Catalog of Hymenoptera in America north of Mexico, Vol. 1. Smithsonian Institution Press, Washington, D.C.

McDunnough, J. 1938. Some apparently new Eucosmidae (Lepid.). Can. Entomol. 70: 90-100.

McLeod, J.M.; Tostowaryk, W. 1971. Outbreaks of pitch nodule makers (*Petrova* spp.) in Quebec jack pine forests. Can. For. Serv., Laurentian For. Res. Centr. Quebec City. Inf. Rep. Q-X-24, 5 p.

Miller, W.E. 1955. Notes on the life cycles of three parasites of the pitch twig moth. Ohio J. Sci. 55: 317-319.

Miller, W.E. 1959. *Petrova houseri*, a new pitch-nodule moth from eastern North America. Ohio. J. Sci. 59: 230-232.

Miller, W.E. 1963. The shortleaf pitch-blister moth, *Petrova houseri* Miller. Ohio J. Sci. 63: 297-301.

Miller, W.E. 1978. *Petrova* pitch-blister moths of North America and Europe: two new species and synopsis (Olethreutidae). Ann. Ent. Soc. Am. 71: 329-340.

Miller, W.E.; Altmann, S.A. 1958. Ecological observations on the Virginia pitch-nodule moth, *Petrova wenzeli* (Kearfott), including a note on its nomenclature. Ohio J. Sci. 58: 273-281.

Miller, W.E.; Neiswander, R.B. 1956. The pitch twig moth and its occurrence in Ohio. Ohio Agric. Exp. Stn. Res. Bull. 779, 24 p.

Prentice, R.M. (Comp.). 1965. Forest Lepidoptera of Canada. Vol. 4. Microlepidoptera. Can. Dep. For. Ottawa. Publ. 1142. 840 p.

Salt, W.R. 1986. The birds of Alberta. Queen's Printer, Edmonton, 498 p.

Soper, J.D. 1964. The mammals of Alberta. Queen's Printer, Edmonton, 402 p.

Stark, R.W. 1957. Pitch nodule maker in Banff National Park. Can. Dept. Agric. Ottawa. Bi-mon. Prog. Rep. 13(4): 2-3.

Tracy, R.A.; Osgood, E.A. 1981. Two new parasite records and notes on *Petrova albicapitana* (Busck) (Lepidoptera: Olethreutidae) on jack pine, *Pinus banksiana* Lamb. in Maine. Ent. News. 92: 101-105.

Turnock, W.T. 1953. Some aspects of the life history and ecology of the pitch nodule maker, *Petrova albicapitana* (Busck) (Lepidoptera: Olethreutidae). Can. Entomol. 85: 233-243.

Wong, H.R.; Drouin, J.A.; Rentz, C.L. 1985. *Petrova albicapitana* (Busck) and *P. metallica* (Busck) (Lepidoptera: Tortricidae) in *Pinus contorta* Dougl. stands in Alberta. Can. Entomol. 117: 1463-1470.

Chapter 32

Woolly Adelgids

John W.E. Harris and Wade W. Bowers

Introduction

Members of the family Adelgidae are minute phytophagous insects that feed exclusively on conifers. The group is morphologically and biologically allied to two other families in the superfamily Aphidoidea, the Aphididae and Phylloxeridae. Some species exhibit various waxy, thread-like or powdery exudations; the Adelgidae are usually covered with a waxy "wool," giving a white, matted surface to bark and twigs whenever many individuals are present. Adelgids differ from true aphids in having shorter antennal segments, a reduced wing venation, no cornicles, and all females are oviparous. Phylloxerids are distinguished from adelgids by the form of their wings and antennae, and all phylloxerids feed on dicotyledonous plants. The Adelgidae contain two genera, *Adelges* and *Pineus*, having five and four distinct pairs of lateral abdominal spiracles, respectively. Both genera are represented in Canada by species with highly variable and complex behavior and seasonal life history patterns.

Biology

For those species where intensive biological studies have been carried out, several generalizations can be made concerning adelgid biology. As for all Homoptera, adelgids undergo simple metamorphosis. The life cycle is typically pentamorphic with three heterogeneous generations, sexuales, fundatrix, and alata migrans, on the primary host spruce (*Picea* spp.), and two heterogeneous generations, the sistens and sexupara, on the secondary host (various conifers). Although alternation of hosts is common, some parthenogenetic forms are confined to a single host. Sistens (dormant forms) on the secondary host tree may also produce nondiapause forms. However, these progrediens (nondormant forms) usually account for a small segment of the population.

Adelgids feed by inserting their sucking mouthparts, which include four piercing stylets, into the phloem tissue of bark and needles. Trees harboring feeding adelgids exhibit a variety of symptoms including chlorosis, necrosis, and pathological resin flow. Depending on adelgid abundance, swelling, galls, and defoliation may occur. These structural abnormalities are often accompanied by desiccation and shoot inhibition.

Status as Pest Insects

Except for balsam woolly adelgid, *Adelges piceae*, and Cooley spruce gall adelgid, *A. cooleyi*, little research has been directed to the Adelgidae on forest trees in Canada. The Forest Insect and Disease Survey (FIDS) of Forestry Canada has for over half a century monitored adelgid distribution patterns and their damage on forest and shade trees. However, there is still a paucity of information on the bionomics and population dynamics of many species of the genera *Adelges* and *Pineus*. Most species have little or no economic impact on their hosts, so the need for operational pest management strategies is minimal or nonexistent. However, as forestry practices intensify, adelgid problems are expected to increase.

Minor Pests

Species of *Adelges* known to contribute to tree losses in Canada include the balsam woolly adelgid, the Cooley spruce gall adelgid, and the eastern spruce gall adelgid, *A. abietis*. The former will be discussed later in detail. The latter two species cause gall formations on spruce and occasionally inflict significant losses in nurseries, plantations, and on ornamental trees (Martineau 1984; USDA 1985). In western Canada, the Cooley spruce gall adelgid uses Douglas-fir (*Pseudotsuga menziesii*) as an alternate host and is known to discolor foliage of seedlings and ornamentals, including Christmas trees. Research to control the eastern spruce gall adelgid and the Cooley spruce gall adelgid has been largely limited to insecticide and sanitation trials. Two other adelgids, the spruce gall adelgid, *A. lariciatus*, and an introduced European species, the pale spruce gall adelgid, *A. laricis*, attack spruce and larch across Canada; two western adelgids, *A. oregonensis* and *A. tsugae*, attack larch and hemlock. These species seldom require control and, consequently, studies have been limited to detection and survey.

Adelgids of the genus *Pineus* include the ragged spruce gall adelgid, *P. similis*, the red spruce adelgid, *P. floccus*, the pine leaf adelgid, *P. pinifoliae*, the pine bark adelgid, *P. strobi*, as well as *P. abietinus*, *P. coloradensis*, and *P. sylvestris*. Little damage has been attributed to members of the genus *Pineus*; however, surveys in the Maritimes from 1942 to the mid-1960s disclosed some damage by *P. pinifoliae*. Magasi (1988) reported a significant increase in populations of this adelgid on eastern white pine (*Pinus strobus*) in western Nova Scotia in 1987. Nevertheless, it is generally accepted that as for most species of *Adelges*, species of *Pineus* are controlled by natural mortality factors. Accordingly, the group has commanded little research attention.

Major Pests

In contrast to the majority of the above species, considerable effort has been expended in Canada to research and develop reliable and effective control strategies for balsam woolly adelgid. This pest was introduced into Canada from Europe early in the century, occurs on both coasts, and has caused extensive tree damage. An annotated bibliography prepared by Schooley and Oldford (1981) lists about

353 publications dealing with the pest. Detailed accounts of the biology and life history of *Adelges piceae* are provided by Annand (1928), Balch (1952), Mitchell (1960), and Schooley and Bryant (1978). Discussed elsewhere are infestation trends (Aldrich and Drooz 1967; Amman and Fedde 1971), sampling methodology (Amman 1969; Talmon de l'Armée 1972; Bryant 1972, 1976), and the role of climate on *A. piceae* populations (Atkins and Hall 1969; Greenbank 1970). Some of the most intensive efforts in balsam woolly adelgid research were aimed at elucidating adelgid–host relationships (Balch 1955; Foulger 1968; Amman 1970; Schooley 1976; Puritch 1977; Puritch and Johnson 1971; Fedde 1973; Mullick 1975; Page 1975; Hudak 1976; Sagio 1976). Efforts at control began in eastern Canada shortly after the insect was first reported in Nova Scotia in 1929. The voluminous literature on control precludes a comprehensive review. Rather, the following synopsis will focus on research highlights and recent developments.

Biological Control

Classical biological control programs aimed at population regulation have added much new knowledge but appear to have had minimal impact on the problem. Schooley et al. (1984) provides the most recent review of predators introduced into Canada to control the adelgid. Eight species of introduced predators have become established, including *Adalia ronina*, *Aphidecta obliterata*, *Pullus impexus*, *Laricobius erichsonii*, *Aphidoletes thompsoni*, *Cremifania nigrocellulata*, and two *Leucopis* spp. However, work conducted over the past 55 years has not resulted in a predator or predator complex capable of keeping adelgid populations at acceptable levels. In fact, no attempt at predator release has been made in eastern Canada since 1968 and in western Canada since 1969 (Harris and Dawson 1979). Furthermore, there are no known parasites of balsam woolly adelgid, and although several fungi have been reported to attack it (Harris et al. 1966; Smirnoff 1970), none have appeared promising as control agents. Other entomopathogenic microorganisms, including bacteria, viruses, protozoa, and nematodes may offer potential, but have not been found as naturally occurring entomopathogens or been developed as microbial insecticides.

One reason for the apparent failure of biotic mortality agents used against balsam woolly adelgid is that historically biological control has been a tactical procedure, practised largely on an empirical basis. As a result, important theoretical considerations that relate to agent selection and release are sometimes ignored or not fully understood. For example, given the widespread and variable crown and trunk distribution of the adelgid, only natural enemies with efficient searching abilities are likely to succeed. Other factors that may have limited the efficacy of adelgid natural enemies include the lack of temporal synchronization, limited fecundity that may leave many adelgids unattacked, or a response threshold of prey density that is too high. Indeed, the suitability of biological control for balsam woolly adelgid may be in question because there is evidence in Europe, where the adelgid infests silver fir, *Abies alba*, that tree resistance, not the predator complex, appears to be the factor responsible for preventing outbreaks.

Chemical Insecticides

A great deal of interest has been shown in the development and application of chemical insecticides to control balsam woolly adelgid (Nigam 1972). When applied at high concentrations and high volumes, contact and systemic insecticides have proven effective on ornamental and nursery trees. Unfortunately, it has not been possible to control populations of the pest in natural stands.

Initial attempts in the Maritimes to control the insect with nicotine sulfate and miscible oils met with little success (Nigam 1967). In 1963, the Chemical Control Research Institute screened new organic compounds including organophosphate and carbamate insecticides (Hopewell and Bryant 1966). Ground spray applications with diazinon, Baygon,® and Dursban® resulted in 72%, 75%, and 85% adelgid mortality, respectively (Hopewell and Bryant 1969), but failed to provide acceptable total population reduction. An experimental airspray then was conducted in 1968 at Deer Lake, Newfoundland. Test plots 16.2 ha (40 acres) in size were treated with each of diazinon, Furadan® (carbofuran), Dursban® (chlorpyrifos), and Baygon® (propoxur). Although the distribution of spray throughout the plots was rated as acceptable, there was no evidence of significant adelgid mortality. During the 1968–69 spray program in Newfoundland against hemlock looper, *Lambdina fiscellaria*, two applications of fenitrothion appeared to reduce adelgid crawlers. However, populations were not reduced enough to prevent twig deformation (Bryant 1975).

Tests of chemicals in western Canada were more limited than in the east and most were done in cooperation with eastern efforts. Early experiments (Randall et al. 1967) indicated propoxur was a potentially useful contact insecticide. In further tests carbofuran and propoxur were investigated for systemic activity. Carbofuran was applied with and without nitrogenous fertilizer as granules to the soil for translocation through the roots, and both chemicals were injected directly into tree stems but results were not encouraging (Nigam 1976).

Carrow et al. (1977) reported a 30% to 50% increase in adelgid mortality in tree crowns when carbofuran was combined with urea fertilizer; however, the treatment did not prevent adelgid recolonization. The potential of host fertilization as a single tactic to indirectly control adelgid populations has also been investigated by applying nitrogen fertilizers to soil and to foliage. Adelgid populations decreased following foliar application with ammonium nitrate but apparently increased after foliar application of urea (Carrow and Graham 1968; Carrow and Betts 1973). Despite evidence that some forms of fertilizer may depress adelgid populations by affecting the nutritional value of trees, the exact mechanism by which individuals are affected has not been established. Coupled with economic considerations there is little likelihood that fertilizers will be rigorously pursued as a control alternative.

Control of adelgid on seedlings became of particular importance in British Columbia after the establishment of quarantine areas and legislation requiring treatment to eradicate adelgids on stock being moved. Various dips and sprays were tested (Puritch et al. 1980) including insecticidal soaps, preferred because of their low toxicity to workers handling the seedlings (Puritch and Talmon de l'Armée 1974;

Puritch 1975). Recent changes in quarantine areas have renewed interest in direct control efforts. In British Columbia, the Ministry of Forests has obtained control with permethrin and insecticidal soap combined and sprayed to run-off. Schooley et al. (1984) noted that of 51 insecticides tested in laboratory and greenhouse experiments, none was effective under field conditions.

The failure to control outbreaks over large areas is not fully understood; however, the cryptic feeding habit of the insect and the lower concentrations and volumes necessary for wide-scale insecticide applications appear to be limiting factors. It seems, therefore, that difficult and innovative research will be necessary before chemical insecticides can be developed for effective control of the balsam woolly adelgid.

Other Control Methods

Several other methods of control have been investigated. The feasibility of using entomopathogenic nematodes was investigated by Finney and Bennett (1984). Experiments to determine the susceptibility of the balsam woolly adelgid to *Neoaplectana bibionis* resulted in low infection rates; additional research on steinernematid nematodes as biological control agents has been proposed.

In 1987, a cooperative project was initiated between the Newfoundland Forestry Centre and the Forest Pest Management Institute, both of Forestry Canada, to assess the potential use of insect growth regulators (IGRs) against the balsam woolly adelgid. These compounds were tested in Newfoundland in 1987 and 1988 during field trials at Bottom Brook and Crabbes River, respectively. Test materials were diluted with water and applied as aqueous sprays to the point of run-off from a backpack mist blower. Results indicated that some potential exists for growth regulators; work is continuing on dose–response relationships, timing of application, and biological effects.

Silvicultural methods have been used to reduce or contain the spread of infestations. For example, in Newfoundland, cutting of stands infested with high levels of adelgid has met with some success in reducing aphid populations at Terra Nova National Park and near Corner Brook (Schooley and Bryant 1978). Furthermore, several thousand hectares of balsam fir severely affected by adelgid have been converted at considerable cost to black spruce (*Picea mariana*).

Discussion

Although present populations of the balsam woolly adelgid in Canada are low, the insect remains a potential threat, particularly in areas damaged in the past, including Newfoundland, New Brunswick, Nova Scotia, the Gaspé region of Quebec, and southwestern British Columbia. In Newfoundland, balsam woolly adelgid continues to cause extensive growth loss and mortality in balsam fir stands. The most recent damage is of particular concern to forest managers as it affects fir stands that were thinned at high cost. The occurrence of the pest in areas of high-value stands has given impetus to new research designed to evaluate the role of the adelgid in thinned stands. Studies are being carried out to determine whether thinning juvenile balsam fir promotes population buildup. Also, in 1987, a 5-year study to assess growth impact was initiated along with research to investigate the seasonal history, bionomics, and chemical ecology of the adelgid in thinned and unthinned stands.

In British Columbia a quarantine to restrict the movement of seedlings and logs remains in effect. Nursery stock cannot be moved from within an infestation, and even within infested areas, stock must be treated with insecticides. A recently discovered northern extension of the infestation now includes nurseries that were previously outside the infestation boundaries. Consequently, there has been renewed testing of pesticide formulations for treating nursery stock (G. Shrimpton and R. Sturrock, pers. comm.). Of these, an insecticidal soap formulation sprayed to run-off before moving stock into the field appears to be the most promising treatment.

References

Aldrich, R.C.; Drooz, A.T. 1967. Estimated Fraser fir mortality and balsam woolly aphid infestation trend using aerial color photography. For. Sci. 13: 300-313.

Amman, G.D. 1969. A method of sampling the balsam woolly aphid on Fraser fir in North Carolina. Can. Entomol. 101: 883-889.

Amman, G.D. 1970. Distribution of redwood caused by the balsam woolly aphid in Fraser fir of North Carolina. U.S. Dept. Agric., For. Serv., Washington, D.C. Res. Note SE-135. 4 p.

Amman, G.D.; Fedde, G.F. 1971. Infestation trends of the balsam woolly aphid in an *Abies alba* plantation in North Carolina. U.S. Dept. Agric., For. Serv., Washington, D.C. Res. Note SE-148. 6 p.

Annand, P.N. 1928. A contribution towards a monograph of the Adelginae (Phylloxeridae) of North America. Stanford Univ. Publ. Biol. Sci. 6: 7-146.

Atkins, M.D.; Hall, A.A. 1969. Effect of light and temperature on the activity of balsam woolly aphid crawlers. Can. Entomol. 101: 481-488.

Balch, R.E. 1952. Studies of the balsam woolly aphid *Adelges piceae* (Ratz.) and its effect on balsam fir, *Abies balsamea* (L.) Mill. Can. Dept. Agric., Ottawa, Ont. Pub. No. 867, 76 p.

Balch, R.E. 1955. Effects of balsam woolly aphid on the quality of pulp. Can. Dept. Agric., Ottawa, Ont. Bi-mon. Prog. Rep. 11(4): 1.

Bryant, D.G. 1972. The measurement of population density of the balsam woolly aphid, *Adelges piceae* (Ratz.) (Homoptera: Phylloxeridae), a highly aggregate insect. Ph.D. Dissertation. Yale University. 169 p.

Bryant, D.G. 1975. Balsam woolly aphid. Pages 250-254 *in* M.L. Prebble (Ed.), Aerial Control of Forest Insects in Canada. Can. Dept. of the Environ., Ottawa, Ont.

Bryant, D.G. 1976. Sampling populations of *Adelges piceae* (Homoptera: Phylloxeridae) on balsam fir, *Abies balsamea*. Can. Entomol. 108: 1113-1124.

Carrow, J.R.; Betts, R.E. 1973. Effects of different foliar-applied nitrogen fertilizers on balsam woolly aphid. Can. J. For. Res. 3: 122-139.

Carrow, J.R.; Graham, K. 1968. Nitrogen fertilization of the host tree and population growth of the balsam woolly aphid, *Adelges piceae*. Can. Entomol. 100: 478-485.

Carrow, J.R.; Puritch, G.S.; Nigam, P.C. 1977. Field test of Furadan and Baygon against balsam woolly aphid in British Columbia. Can. Forest. Serv., Ottawa, Ont. Bi-mon. Res. Notes 33(2): 10.

Fedde, G.F. 1973. Impact of the balsam woolly aphid (Homoptera: Phylloxeridae) on cones and seed produced by infested Fraser fir. Can. Entomol. 105: 673-680.

Finney, J.R.; Bennett, G.F. 1984. The use of steinernematid nematodes for control of the hemlock looper, balsam woolly aphid, larch and spruce bark beetles in insular Newfoundland. Prog. Rep., Can. For. Serv., Nfld. For. Cent., St. John's, Nlfd. 10 p.

Foulger, A.N. 1968. Effect of aphid infestation on properties of grand fir. For. Prod. J. 18(1): 43-47.

Greenbank, D.O. 1970. Climate and ecology of the balsam woolly aphid. Can. Entomol. 102: 546-578.

Harris, J.W.E.; Allen, S.J.; Collis, D.C.; Harvey, E.G. 1966. Status of the balsam woolly aphid, *Adelges piceae* (Ratz.) in British Columbia. Can. Dept. For., Victoria, B.C. Inf. Rep. BC-X-5. 12 p.

Harris, J.W.E.; Dawson, A.F. 1979. Predator release program for balsam woolly aphid, *Adelges piceae* (Homoptera: Adelgidae), in British Columbia, 1960-1969. J. Entomol. Soc. B.C. 76: 21-26.

Hopewell, W.W.; Bryant, D.G. 1966. Tests of various insecticides for chemical control of the balsam woolly aphid in Newfoundland. Can. Dept. For., Ottawa, Ont. Bi-mon. Prog. Rep. 22(2): 1.

Hopewell, W.W.; Bryant D.G. 1969. Chemical control of *Adelges piceae* (Homoptera: Adelgidae) in Newfoundland, 1967. Can. Entomol. 101: 1112-1114.

Hudak, J. 1976. Microbial deterioration of balsam fir damaged by the balsam woolly aphid in western Newfoundland. Ph.D. Dissertation. Coll. Environ. Sci. Stat. Univ., New York.

Magasi, L.P. 1988. Forest pest conditions in the Maritimes. Can. For. Serv., Fredericton, N.B. Inf. Rep. M-X-166. 109 p.

Martineau, R. 1984. Insects harmful to forest trees. Multisci. Publ. Ltd. 261 p.

Mitchell, R.G. 1960. The biology of the balsam woolly aphid, *Chermes piceae* Ratz. in Oregon and Washington and the identification and evaluation of its native predators. Ph.D. Dissertation. Oregon State College, Corvallis, Oregon.

Mullick, D.B. 1975. A new tissue essential to necrophylactic periderm formation in the bark of four conifers. Can. J. Bot. 53: 2443-2457.

Nigam, P.C. 1967. Chemical control trials against the balsam wooly aphid in New Brunswick. Chem. Cont. Res. Inst., Ottawa, Ont. Internal Rep. CC-3, Mar., 19 p.

Nigam, P.C. 1972. Summary of toxicity of insecticides and chemical control studies against balsam woolly aphid. Can. For. Serv., Ottawa, Ont. Inf. Rep. CV-X-26. 7 p.

Nigam, P.C. 1976. Summary of chemical control studies against balsam woolly aphid in British Columbia, 1973-74. Can. For. Serv., Ottawa, Ont. Inf. Rep. CC-X-123. 6 p.

Page, G. 1975. The impact of balsam woolly aphid damage on balsam fir stands in Newfoundland. Can. J. For. Res. 5: 195-209.

Puritch, G.S. 1975. The toxic effects of fatty acids and their salts on the balsam woolly aphid, *Adelges piceae* (Ratz.). Can. J. For. Res. 5: 515-522.

Puritch, G.S. 1977. Distribution and phenolic composition of sapwood and heartwood in *Abies grandis* and effects of the balsam woolly aphid. Can. J. For. Res. 7: 54-62.

Puritch, G.S.; Johnson, R.P.C. 1971. Effects of infestation by balsam woolly aphid, *Adelges piceae* (Ratz.) on the ultrastructure of bordered-pit membranes of grand fir, *Abies grandis* (Doug.) Lindl. J. Exp. Bot. 22: 953-958.

Puritch, G.S.; Nigam, P.C.; Carrow, J.R. 1980. Chemical control of balsam wooly aphid (Homoptera: Adelgidae) on seedlings of *Abies amabilis*. J. Entomol. Soc. B.C. 77: 15-18.

Puritch, George S.; Talmon de l'Armée, M. 1974. Biocidal effect of fatty acids and soaps on the balsam woolly aphid. Can. For. Serv., Ottawa, Ont. Bi-mon. Res. Notes 30(6): 35-36.

Randall, A.P.; Hopewell, W.W.; Nigam, P.C. 1967. Chemical control studies on the balsam woolly aphid (*Adelges piceae* (Ratz.)) Can. Dept. For. and Rural Dev., Ottawa, Ont. Bi-mon. Res. Notes 23(3): 18-19.

Sagio, R. 1976. Anatomical changes in the secondary phloem of grand fir (*Abies grandis*) induced by the balsam woolly aphid. Can. J. Bot. 54: 1903-1910.

Schooley, H.O. 1976. Effect of balsam woolly aphid on cone and seed production by balsam fir. For. Chron. 52: 237-239.

Schooley, H.O.; Bryant, D.G. 1978. The balsam woolly adelgid in Newfoundland. Can. For. Serv., St. John's, Nfld. Inf. Rep. N-X-160. 72 p.

Schooley, H.O.; Harris, J.W.E.; Pendrel, B. 1984. *Adelges piceae* (Ratz.), Balsam Woolly Adelgid (Homoptera: Adelgidae). Pages 229-234 *in* J.S. Kelleher and M.A. Hulme (Eds.), Biological control programs against insects and weeds in Canada 1969-1980. Commonw. Agric. Bureaux, Slough, England.

Schooley, H.O.; Oldford, L. 1981. An annotated bibliography of the balsam woolly aphid (*Adelges piceae* (Ratzeburg)). Can. For. Serv., Nfld. For. Res. Cent., St. John's, Nfld. Inf. Rep. N-X-196. 97 p.

Smirnoff, W.A. 1970. Fungus diseases affecting *Adelges piceae* in the fir forest of the Gaspé Peninsula, Quebec. Can. Entomol. 102: 799-805.

Talmon de l'Armée, M. 1972. A tree platform for crown sampling. Can. J. For. Res. 2: 166-167.

USDA. 1985. Insects of Eastern Forests. U.S. Dept. Agric., Washington, D.C. Misc. Publ. No. 1426, Dec. 1985. 608 p.

Chapter 33

Cone and Seed Insects

G.E. Miller, J.J. Turgeon, and P. de Groot

Introduction

Cone and seed insect pests adversely affect seed production in many conifers, but only those attacking Douglas-fir and spruces have been studied in detail. However, the establishment of increased numbers of seed orchards and the commitment to reforest with superior seed have prompted a realization of the potential impact and significance of cone and seed insects and of the need for management systems to minimize losses.

Most of the important cone and seed insects in Canada have already been identified from information collected primarily by the Forest Insect and Disease Survey (FIDS) of Forestry Canada. The greatest number of records exists for British Columbia where Douglas-fir has been surveyed for 30 years, several other conifer species for 15 to 20 years.

Recognition of insect pests and the identification of potential insect problems by orchard managers are critical first steps in pest management. These steps have been aided by several extension publications (e.g. Ruth 1980; Ruth et al. 1982; Churcher et al. 1985). Following insect identification, the basic components of a pest management system include (i) knowledge of the biology, behavior, and population dynamics of both the host and pest species; (ii) ability to predict crop abundance as well as the impact of the pest population; and (iii) development of direct and supportive control tactics (Huffaker and Smith 1980).

The development of insect pest management systems is particularly relevant to seed orchards. Crop values are high so a wide range of management activities are feasible; orchards cover small, well-defined areas; and trained staff are usually on site allowing for frequent monitoring and quick action if control measures are warranted. The area committed to seed orchards in Canada has increased dramatically during the last decade from 364 ha in 1981 to an expected 2 036 ha in 1988 (Morgenstern 1985). A pest management system is now operational for Douglas-fir in coastal British Columbia (Miller 1984a), and systems for other conifers are in various stages of development. Ultimately, development of these systems will assist orchard managers in selecting optimal control measures for dealing with insect problems and should minimize the need for applications of pesticides.

In this chapter, we identify the major insect pests of conifer seeds in Canada as well as features of their life cycles and population dynamics that could influence pest management strategies. We also discuss damage appraisal and insect monitoring techniques, and review cone and seed insect control programs (both operational and research) that have been conducted in Canada.

Insects Infesting Cones and Seeds of Canadian Conifers and Their Biologies

A broad range of insect genera and species from several families is known to infest the reproductive structures (pollen- and seed-cones) of commercially important conifers in Canada (Table 1). Only a few of these insects are known to significantly reduce seed production; the remainder have either a low or unknown economic importance. Insects may affect the production of seeds directly by feeding on conelets, cones, and seeds or indirectly by damaging foliage, twigs, and branches that could bear reproductive organs or by reducing tree vigor (Miller 1980). Among those that have a direct impact, some feed exclusively on reproductive structures and cannot develop with them; we will refer to these insects as obligatory. Other insects, which are not dependent on these structures for their survival, but will feed on them opportunistically, will be referred to as facultative. Defoliators such as the spruce budworm, *Choristoneura fumiferana*, and western spruce budworm, *C. occidentalis*, are good examples of facultative pests. In this section, our emphasis has been placed primarily on obligatory insects and on those aspects of their biologies with direct relevance to pest management.

Some information on the known distributions, general descriptions, and life histories and habits of most of these insects is already available (Hedlin et al. 1980), and the literature on flies and cone beetles attacking North American conifers has recently been compiled and annotated (de Groot 1986a, 1986b). Nevertheless, the taxonomy and biology of many of these insects are not always well understood. Indeed, several species are known only to the genus level. This is particularly true for seed, gall, and scale midges, as well as cone moths. A clearer understanding of the taxonomy and host specificity of cone-infesting anthomyiid flies has been gained only recently (Michelsen 1988).

Some features of the insect life cycles that are important in the identification of pest management strategies are listed in Table 2. Some generalities can be made, although their biologies, behaviors, and impacts differ. Virtually all insects are univoltine and attack at discrete stages of cone development. In most conifers, in which seed-cones mature over one summer, initial attacks occur when seed-cones (conelets or "flowers") are open to receive pollen, but a few species oviposit later when cones are about half their mature length. However, in pines, in which cones mature over two summers, the initial attack occurs mostly in the spring of the second year of cone development when the cones begin to elongate; only a few species are known to damage the first-year conelets (de Groot 1986c). Sclerotic cones can also be infested by anobiid beetles 1 year or more after they have matured. In all cases, feeding is completed over one summer.

Table 1. Insects infesting cones of Canadian conifers and their importance.

Species	Common name[a]	Firs						Larches			Spruces					Pines									Cedar	Hemlocks		
		amabilis	balsam	white	grand	alpine	California red	alpine	tamarack	western	Engelmann	white	black	red	Sitka	jack	lodgepole	western white	ponderosa	red	pitch	eastern white	Scotch	Douglas-fir	western red	eastern	western	mountain
Coleoptera																												
Conophthorus coniperda	White pine cone beetle	.	.	.	.	.	.	.	.	.	.	.	.	.	.	.	.	.	.	.	.	S[b]	.	.	.	.	.	.
Conophthorus resinosae	Red pine cone beetle	.	.	.	.	.	.	.	.	.	.	.	.	.	.	?	.	.	.	.	S	.	.	.	.	.	.	.
Conophthorus ponderosae	Ponderosa pine cone beetle	.	.	.	.	.	.	.	.	.	.	.	.	.	.	.	L	L	S	.	.	.	.	.	.	.	.	.
Ernobius bicolor		.	.	.	.	.	.	.	.	.	.	.	L	.	.	.	.	.	.	.	.	.	.	.	.	.	.	.
Ernobius mollis		.	.	.	.	.	.	.	.	.	.	.	L	.	.	.	.	.	.	.	.	.	.	.	.	.	.	.
Ernobius nigrans		.	.	.	.	.	.	.	.	.	.	.	L	.	.	.	.	.	.	.	.	.	.	.	.	.	.	.
Ernobius pallitarsis		.	.	.	.	.	.	.	.	.	.	.	.	.	.	.	.	.	L	.	.	.	.	.	.	.	.	.
Ernobius punctulatus	Douglas-fir cone beetle	.	.	.	.	.	.	.	.	.	.	.	.	.	.	.	.	.	.	.	.	.	.	L	.	.	.	.
Tenebroides sp.	Cadelle	.	.	.	.	.	.	.	.	.	.	.	U	.	.	.	.	.	.	.	.	.	.	.	.	.	.	.
Diptera																												
Asynapta hopkinsi	Cone resin midge	.	.	L	L	?	L	.	.	.	?	?	.	.	.	.	L	.	L	?	.	.	.	.	.	.	.	?
Camptomyia pseudotsugae		.	.	.	.	.	.	.	.	.	.	.	.	.	.	.	.	.	.	.	.	.	.	L	.	.	.	.
Contarinia oregonensis	Douglas-fir cone gall midge	.	.	.	.	.	.	.	.	.	.	.	.	.	.	.	.	.	.	.	.	.	.	S	.	.	.	.
Contarinia washingtonensis	Douglas-fir cone scale midge	.	.	.	.	.	.	.	.	.	.	.	.	.	.	.	.	.	.	.	.	.	.	O	.	.	.	.
Dasineura abiesemia	A fir seed midge	S	.	U	S	S	U	.	.	.	.	.	.	.	.	.	.	.	.	.	.	.	.	.	.	.	.	.
Dasineura canadensis	Spruce cone gall midge	.	.	.	.	.	.	.	.	.	.	U	?	.	.	.	.	.	.	.	.	.	.	.	.	.	.	.
Dasineura rachiphaga	Spruce cone axis midge	.	.	.	.	.	.	.	.	L	L	L	.	L	.	.	.	.	.	.	.	.	.	.	.	.	.	.
Dasineura sp.		O	L	.	S	S	.	.	.	.	.	.	.	.	.	.	.	.	.	.	.	.	.	.	.	.	.	.
Earomyia abietum	Fir seed maggot	S	.	L	S	S	L	.	.	.	.	.	.	.	.	.	.	.	.	.	.	.	.	.	.	.	.	.
Earomyia barbara		.	.	.	.	.	.	.	.	.	.	.	.	.	.	.	.	.	.	.	.	.	.	L	.	.	.	.
Earomyia aquilonia		.	.	.	.	U	.	.	U	.	.	.	.	.	.	.	.	.	.	.	.	.	.	U	.	.	.	.
Earomyia longistylata		U	.	U	U	.	U	.	.	.	.	.	.	.	.	.	.	.	.	.	.	.	.	.	.	.	.	.
Earomyia sp.		.	L	.	O	.	.	.	.	.	.	.	.	.	.	.	.	.	.	.	.	.	.	.	.	.	.	.
Mayetiola carpophaga	Spruce seed midge	.	.	.	.	.	.	.	.	.	?	O	L	.	?	.	.	.	.	.	.	.	.	.	.	.	.	.
Mayetiola thujae	Western red cedar cone midge	.	.	.	.	.	.	.	.	.	.	.	.	.	.	.	.	.	.	.	.	.	.	S	.	.	.	.
Resseliella sp.		.	.	.	.	.	.	.	L	.	.	.	L	.	.	L	.	.	L	.	.	.	.	.	.	.	.	.
Strobilomyia abietis	A fir cone maggot	.	?	U	.	U	.	.	.	.	.	.	.	.	.	.	.	.	.	.	.	.	.	.	.	.	.	.
Strobilomyia carbonaria	A balsam fir cone maggot	.	U	.	.	.	.	.	.	.	.	.	.	.	.	.	.	.	.	.	.	.	.	.	.	.	.	.
Strobilomyia laricis	A larch cone maggot	.	.	.	.	.	.	.	U	O	.	.	.	.	.	.	.	.	.	.	.	.	.	.	.	.	.	.
Strobilomyia macalpine		.	.	.	.	.	.	U	U	.	.	.	.	.	.	.	.	.	.	.	.	.	.	.	.	.	.	.
Strobilomyia neanthracina	Spruce cone maggot	.	.	.	.	.	.	.	.	.	S	S	S	?	S	.	.	.	.	.	.	.	.	.	.	.	.	?
Strobilomyia viaria	A larch cone maggot	.	.	.	.	.	.	.	O	.	.	.	.	.	.	.	.	.	.	.	.	.	.	.	.	.	.	.

(Continued)

Table 1. Insects infesting cones of Canadian conifers and their importance. *(Continued)*

Species	Common name[a]	Firs						Larches			Spruces					Pines									Cedar	Hemlocks		
		amabilis	balsam	white	grand	alpine	California red	alpine	tamarack	western	Engelmann	white	black	red	Sitka	jack	lodgepole	western white	ponderosa	red	pitch	eastern white	Scotch	Douglas-fir	western red	eastern	western	mountain
Hemiptera																												
Leptoglossus occidentalis	Western conifer seed bug	.	.	.	U	.	.	.	.	.	.	.	.	.	.	.	U	U	U	.	.	.	.	O	.	.	.	.
Tetyra bipunctata	Shieldbacked pine seed bug	.	.	.	.	.	.	.	.	.	.	.	.	.	.	L	.	.	.	U	U	U	.	.	.	.	.	.
Homoptera																												
Adelges cooleyi	Cooley spruce gall adelgid	.	.	.	.	.	.	.	.	.	.	L	.	.	.	.	.	.	.	.	.	.	.	L	.	.	.	.
Adelges lariciatus	Spruce gall adelgid	.	.	.	.	.	.	.	U	.	.	.	.	.	.	.	.	.	.	.	.	.	.	.	.	.	.	.
Adelges laricis	Pale spruce gall adelgid	.	.	.	.	.	.	.	U	.	.	.	.	.	.	.	.	.	.	.	.	.	.	.	.	.	.	.
Adelges piceae[c]	Balsam woolly adelgid	U	U	.	U	U	.	.	.	.	.	.	.	.	.	.	.	.	.	.	.	.	.	.	.	.	.	.
Mindarus abietinus	Balsam twig aphid	.	.	.	.	.	.	.	.	.	.	.	U	.	.	.	.	.	.	.	.	.	.	.	.	.	.	.
Hymenoptera																												
Megastigmus albifrons	A ponderosa pine seed chalcid	.	.	.	.	.	.	.	.	.	.	.	.	.	.	.	.	.	U	.	.	.	.	.	.	.	.	.
Megastigmus atedius	Spruce seed chalcid	.	.	.	.	.	.	.	.	.	U	U	U	.	.	.	.	.	.	.	.	.	U	.	.	.	.	.
Megastigmus laricis	Larch seed chalcid	.	.	.	.	.	.	.	U	.	.	.	.	.	.	.	.	.	.	.	.	.	.	.	.	.	.	.
Megastigmus lasiocarpae		L	.	.	.	L	.	.	.	.	.	.	.	.	.	.	.	.	.	.	.	.	.	.	.	.	.	.
Megastigmus milleri		.	.	.	U	.	.	.	.	.	.	.	.	.	.	.	.	.	.	.	.	.	.	.	.	.	.	.
Megastigmus pinus	Fir seed chalcid	O	.	O	L	O	.	.	.	.	.	.	.	.	.	.	.	.	.	.	.	.	.	.	.	.	.	.
Megastigmus rafni		.	.	U	L	.	U	.	.	.	.	.	.	.	.	.	.	.	.	.	.	.	.	.	.	.	.	.
Megastigmus specularis	Balsam fir seed chalcid	.	U	.	.	.	.	.	.	.	.	.	.	.	.	.	.	.	.	.	.	.	U	.	.	.	.	.
Megastigmus spermotrophus	Douglas-fir seed chalcid	.	.	.	.	.	.	.	.	.	.	.	.	.	.	.	.	.	.	.	.	.	.	O	.	.	.	.
Megastigmus sp.		.	.	.	.	.	.	.	U	.	.	.	.	.	.	.	.	.	.	.	.	.	.	.	.	.	.	.
Megastigmus tsugae		.	.	.	.	.	.	.	.	.	.	.	.	.	.	.	.	.	.	.	.	.	.	.	.	.	L	L
Xyela alberta		.	.	.	.	.	.	.	.	.	.	.	.	.	.	.	U	.	.	.	.	.	.	.	.	.	.	.
Xyela alpigena		.	.	.	.	.	.	.	.	.	.	.	.	.	.	.	.	.	.	.	.	.	U	.	.	.	.	.
Xyela bakeri		.	.	.	.	.	.	.	.	.	.	.	.	.	.	.	.	.	U	.	.	.	.	.	.	.	.	.
Xyela cheloma		.	.	.	.	.	.	.	.	.	.	.	.	.	.	.	.	.	U	.	.	.	.	.	.	.	.	.
Xyela minor	Pine flower sawfly	.	.	.	.	.	.	.	.	.	.	.	.	.	.	.	.	.	U	.	.	.	.	.	.	.	.	.
Xyela obscura		.	.	.	.	.	.	.	.	.	.	.	.	.	.	U	.	.	U	.	.	.	.	.	.	.	.	.

[a] As given by P. Benoit (1985) in *Insect Names in Canada*.

[b] The impact of the pest is either of low (L), significant (S), occasionally significant (otherwise low) (O), and unknown or undetermined (U) importance. "?" indicates a possible host.

[c] Its impact on Fraser fir is significant.

(Continued)

Table 1. Insects infesting cones of Canadian conifers and their importance. *(Continued)*

		Firs						Larches			Spruces					Pines									Cedar	Hemlocks		
Species	Common name[a]	amabilis	balsam	white	grand	alpine	California red	alpine	tamarack	western	Engelmann	white	black	red	Sitka	jack	lodgepole	western white	ponderosa	red	pitch	eastern white	Scotch	Douglas-fir	western red	eastern	western	mountain
Lepidoptera																												
Acleris variana	Eastern blackheaded budworm	.	.	.	.	.	.	.	.	.	.	.	L	.	.	.	.	.	.	.	.	.	.	.	.	.	.	.
Archips packardianas	Spring spruce needle moth	.	.	.	.	.	.	.	.	.	.	.	L	.	.	.	.	.	.	.	.	.	.	.	.	.	.	.
Archips alberta		.	.	.	.	.	.	.	.	.	.	.	L	.	.	.	.	.	.	.	.	.	.	.	.	.	.	.
Barbara colfaxiana	Douglas-fir cone moth	.	.	.	.	.	.	.	.	.	.	.	.	.	.	.	.	.	.	.	.	.	.	S	.	.	.	.
Barbara mappana		.	L	.	.	.	.	.	.	.	.	L	.	.	.	.	.	.	.	.	.	.	.	.	.	.	.	.
Barbara sp.	A fir cone moth	O	.	.	O	.	O	.	.	.	.	.	.	.	.	.	.	.	.	.	.	.	.	.	.	.	.	.
Choristoneura occidentalis	Western spruce budworm	.	.	O	O	O	.	.	.	O	O	O	.	.	.	.	.	.	.	.	.	.	.	O	.	.	.	.
Choristoneura fumiferana	Spruce budworm	.	O	.	.	.	.	.	O	.	.	O	O	O	.	.	.	.	.	.	.	.	.	.	.	U	.	.
Choristoneura pinus pinus	Jack pine budworm	.	.	.	.	.	.	.	.	.	.	.	.	.	.	O	L	.	.	L	.	.	.	.	.	.	.	.
Choristoneura rosaceana	Obliquebanded leafroller	.	.	.	.	.	.	.	.	.	.	.	L	.	.	.	.	.	.	.	.	.	.	.	.	.	.	.
Coleotechnites atrupictela		.	.	.	.	.	.	.	.	.	.	.	L	.	.	.	.	.	.	.	.	.	.	.	.	.	.	.
Coleotechnites blastovora		.	.	.	.	.	.	.	.	.	.	.	L	.	.	.	.	.	.	.	.	.	.	.	.	.	.	.
Coleotechnites laricis	Orange larch tubemaker	.	.	.	.	.	.	.	L	.	.	.	.	.	.	.	.	.	.	.	.	.	.	.	.	.	.	.
Coleotechnites piceaella	Orange spruce needleminer	.	.	.	.	.	.	.	.	.	.	.	L	.	.	.	.	.	.	.	.	.	.	.	.	.	.	.
Cydia miscitata		.	.	.	.	.	.	.	.	.	.	.	.	.	.	.	.	.	L	.	.	.	.	.	.	.	.	.
Cydia piperana	Ponderosa pine seedworm	.	.	.	.	.	.	.	.	.	.	.	.	.	.	.	.	.	S	.	.	.	.	.	.	.	.	.
Cydia strobilella	Spruce seedworm	.	.	.	.	.	.	.	.	.	S	S	O	L	O	.	.	.	.	.	.	.	.	.	.	.	.	.
Cydia toreuta	Eastern pine seedworm	.	.	.	.	.	.	.	.	.	.	.	.	.	.	L	L	.	.	L	.	.	.	.	.	.	.	.
Dioryctria abietivorella	Fir coneworm	L	O	O	L	O	O	.	.	?	.	L	O	?	.	L	L	O	L	L	.	L	L	O	.	.	L	.
Dioryctria auranticella	Ponderosa pine coneworm	.	.	.	.	.	.	.	.	.	.	.	.	.	.	.	.	.	O	.	.	.	.	.	.	.	.	.
Dioryctria cambiicola		.	.	.	.	.	.	.	.	.	.	.	.	.	.	U	.	U	.	.	.	.	.	.	.	.	.	.
Dioryctria disclusa	Webbing coneworm	.	.	.	.	.	.	.	.	.	.	?	.	.	.	O	.	.	.	O	O	.	O	.	.	.	.	.
Dioryctria pentictonella		.	.	.	.	.	.	.	.	.	.	.	.	.	.	U	.	U	.	.	.	.	.	.	.	.	.	.
Dioryctria pseudotsugella		.	.	.	.	.	.	.	.	.	.	.	.	.	.	.	.	.	.	.	.	.	.	L	.	.	.	.
Dioryctria reniculelloides	Spruce coneworm	L	L	.	L	L	.	.	?	.	L	O	O	L	L	?	L	.	.	.	.	.	.	L	.	.	.	L
Dioryctria rossi		.	.	.	.	.	.	.	.	.	.	.	.	.	.	.	.	.	U	.	.	.	.	.	.	.	.	.
Eucosma monitorana	Red pine cone borer	.	.	.	.	.	.	.	.	.	.	.	.	.	.	.	.	.	.	O	.	.	.	.	.	.	.	.
Eucosma ponderosa	A ponderosa pine cone borer	.	.	.	.	.	.	.	.	.	.	.	.	.	.	.	.	.	L	.	.	.	.	.	.	.	.	.
Eucosma recissoriana	Lodgepole pine cone borer	.	.	.	L	L	.	.	.	.	.	.	.	.	.	.	L	L	.	.	.	.	.	.	.	.	.	.
Eucosma tocullionana	White pine cone borer	.	L	.	.	.	.	.	.	.	.	.	.	.	.	.	.	.	.	.	.	L	.	.	.	.	.	.
Eupithecia albicapitata		.	U	.	.	.	.	.	.	.	U	U	U	.	.	.	?	.	.	.	.	.	.	U	.	.	.	U
Eupithecia columbrata		U	.	.	.	.	.	.	.	.	.	U	.	.	U	.	?	.	.	.	.	.	.	.	.	.	.	U
Eupithecia mutata	Spruce cone looper	.	U	.	.	.	.	.	.	.	.	U	.	.	.	.	.	.	.	.	.	.	.	.	.	.	.	.

(Continued)

Table 1. Insects infesting cones of Canadian conifers and their importance. *(Concluded)*

		Firs						Larches			Spruces					Pines									Cedar	Hemlocks		
Species	Common name[a]	amabilis	balsam	white	grand	alpine	California red	alpine	tamarack	western	Engelmann	white	black	red	Sitka	jack	lodgepole	western white	ponderosa	red	pitch	eastern white	Scotch	Douglas-fir	western red	eastern	western	mountain
Eupithecia spermaphaga	Fir cone looper	L	.	L	.	.	L	.	.	.	.	.	.	.	.	.	.	L	L	.	.	.	.	L	.	.	.	L
Exoteleia nepheos	Pine candle moth	.	.	.	.	.	.	.	.	.	.	.	.	.	.	.	.	.	.	L	.	.	L	.	.	.	.	.
Henricus fuscodorsana	Cone cochylid	.	.	.	.	.	.	.	.	L	.	L	.	.	L	.	.	.	.	.	.	.	.	L	.	.	.	.
Herculia thymetusalis		.	.	.	.	.	.	.	.	.	.	.	L	.	.	.	.	.	.	.	.	.	.	.	.	.	.	.
Holocera augusti		.	.	.	.	.	.	.	.	.	.	.	.	.	.	.	.	.	.	.	.	.	.	.	L	.	.	.
Holcocera immaculella		.	.	.	.	.	.	.	.	.	.	.	L	.	.	L	L	.	.	L	.	.	.	L	.	.	.	.

[a] As given by P. Benoit (1985) in *Insect Names in Canada*.

[b] The impact of the pest is either of low (L), significant (S), occasionally significant (otherwise low) (O), and unknown or undetermined (U) importance. "?" indicates a possible host.

[c] Its impact on Fraser fir is significant.

The feeding stages of obligatory species are usually spent entirely within the protected environment that cones afford, except for the mobile seed bugs that feed on the seeds from the surface of the cones. Facultative cone feeders can damage cones at most stages of cone development and may feed outside the cones for part of their development.

Damage by obligatory insects is often difficult to detect. Infested cones cannot always be separated from healthy cones by the outward appearances as symptoms of attack are often not visible before cone harvest. The presence of eggs on the cone surface between scales at the time of oviposition is the only sign of attack. Adult cone beetles in the genus *Conophthorus* kill the young cones before ovipositing in them, and attack by the spruce cone maggot, *Strobilomyia neanthracina*, reduces cone size (Prevost et al. 1988). Exit holes of this insect from the cones become visible in mid- to late July. The symptoms of attacks by other insects are visible only if infestations are heavy enough to cause necrosis of cones or parts of cones (Hedlin et al. 1980).

The individual impact of most species of defoliators and needle miners on cone production is usually small, except during pest outbreaks. The damage caused by these facultative pests is usually easy to detect visually. It appears in the form of surface feeding or twisted cones. Because of the general nature of the damage inflicted by facultative pests, all but coneworm damage is extremely difficult to associate with a given species. Attacks by coneworms in the genus *Dioryctria* are typified by large amounts of coarse frass inside or on the outside of damaged cones. The seeds are not always destroyed but the cones are usually deformed sufficiently to prevent extraction of seeds (Prevost et al. 1988).

The number of seeds destroyed per insect varies among species (Table 2). Larvae of seed chalcids and seed midges each consume one seed during their development. Larvae of gall midges may occur singly or in groups within a gall and cause the loss of one or both seeds on the affected scale. Infestations by insects that mine the cones do not necessarily result in the loss of all the viable seeds. Single spruce cone maggots, *Strobilomyia neanthracine*, may destroy 65% (Tripp and Hedlin 1956) to 75% (J.J. Turgeon, unpubl. data) of the seeds in white and black spruce cones, and a single larva of the Douglas-fir cone moth, *Barbara colfaxiana*, can destroy about 60% of a cone's seed (Hedlin 1960). The filled seeds remaining in these cones germinate successfully (Hedlin and Ruth 1977; Prevost et al. 1988). Female cone beetles, *Conophthorus* spp., destroy all seeds because the stem of the cone is girdled during the attack (Lyons 1957).

Associations between cone and seed insects and their parasitoids and predators are difficult to determine because of the cryptic nature of these insects. Consequently, there is a paucity of information on the natural enemies that attack pests of seeds and cones, and their potential value as control agents in pest management is difficult to evaluate. Apparent rates of parasitism have been reported to be high for the Douglas-fir cone moth (Hedlin 1960), but low for midges (Miller 1984b). In a few instances where natural enemies have specifically been studied (e.g. Hedlin 1960), their long-term impact on dynamics of the host populations is not well known.

Fluctuations in cone and seed crop size have a tremendous influence on the population dynamics of obligatory cone and seed insects. In natural stands, the intervals between large cone crops are variable, depending on the conifer species and on environmental factors. The number of cones produced annually during these intervals is drastically reduced, thus limiting the population densities of obligatory cone and seed feeders (Lyons 1957; Mattson 1971; Miller et al. 1984). Where cone production is more consistent, such as in seed orchards, losses to insects are usually greater and more consistent (Mattson 1971; Hedlin et al. 1980).

Another factor of particular importance in the population dynamics of cone and seed insects is the ability of obligatory insects to remain dormant for 1 year or more (Table 2). For example, some individuals of the Douglas-fir cone moth, which diapause as pharate adults, have emerged 3 years after entering dormancy and others appeared to be still alive after 5 years (G.E. Miller, unpubl. data). This mechanism, referred to as "prolonged diapause," is used by obligatory insects to compensate for the irregular production of cones and allows them to survive through periods of reduced cone production. Temperature has been implicated in the induction and termination of prolonged diapause (Hedlin et al. 1982; Miller and Ruth 1986), although the biological mechanisms involved are poorly understood. Other endogenous and exogenous factors are likely involved as well (Roques 1989). Prolonged diapause promotes survival of local populations in two ways: (i) by maintaining a reserve of insects over time, even though heavy mortality is known to occur in diapausing insects (Miller and Ruth 1986; Roques 1988), and (ii) by preventing the loss of the population in years of light cone production when the carrying capacity of the cones would be exceeded if all the insects emerged. Even with a large proportion of a local population remaining in prolonged diapause, small cone crops have been killed by heavy insect infestations (G.E. Miller, unpubl. data).

Insects overwinter either in cones (both on trees and on the ground) or in the duff layer of the soil (Table 2). Except for the spruce cone maggot, the red pine cone beetle (*Conophthorus resinosae*), and possibly some coneworms, insects leaving cones to overwinter do not vacate the cones until about the time of seed fall. This feature of the life cycle of cone and seed insects has a significant effect on the development of pest management strategies. Indeed, by harvesting cones just before seed fall, most insects are removed with the cones and, subsequently, orchard infestations result exclusively from the migration of pests from nearby stands, not from the buildup of on-site populations. Removal of the local population could be particularly important as a control measure in seed orchards where crops are harvested annually, especially if the orchard phenology is out of synchrony with wild stands or if nearby stands are freed of the tree species present in the orchard. The removal of all Douglas-fir trees within a radius of 3 to 5 km combined with the annual removal of all cones from a seed orchard has been suggested as a method to limit the impact of the Douglas-fir seed chalcid, *Megastigmus spermotrophus*, in seed orchards (Roques 1986a).

The females of many insect species release pheromones to attract males, and these chemicals provide a potential method for monitoring pest populations. In the last decade, a sex pheromone or attractant has been identified for the following obligatory pests: Douglas-fir cone moth (Hedlin et al. 1983); a cone moth, *B. mappana* (Reed and Chisholm

Table 2. Key features of the bionomics of cone and seed insects important for insect control (summarized from Hedlin et al. 1980 and references therein).

Insect pest	Host phenology when attacked	Overwintering site (stage[a])	Leave cones in (stage[a])	Prolonged diapause	Structure fed upon	Seeds/cone destroyed per insect
Coleoptera: Anobiidae						
Ernobius sp.	mature cone	cone (L)	summer (A)	Unknown	scales and seeds	undetermined
Coleoptera: Scolytidae						
Conophthorus coniperda	2nd year	cone (A)	spring (A)	No	cones	100% (2-4 cones/female)
Conophthorus resinosae	2nd year	bud (A)	summer (A)	No	cones	100% (15-20 cones/female)
Diptera: Anthomyiidae						
Strobilomyia sp.	pollination	duff (P)	summer (L)	Yes	scales and seeds	50-65% of wS[c] seeds
Diptera: Cecidomyiidae						
Anysapta sp. (pines)	2nd year	duff (P)	summer (L)	Unknown	scales	undetermined
Contarinia oregonensis	pollination	duff (L)	autumn (L)	Yes	scales (galls)	1 or 2
Contarinia washintonensis	post pollination	duff (L)	autumn (L)	Yes	scales	usually none
Camptomyia pseudotsugae	post pollination	duff (L)	autumn (L)	Yes	scales	usually none
Dasineura abiesemia	pollination	seed in duff (L)	autumn (L)	Yes	seeds	1
Dasineura canadensis	pollination	cone (L)	spring (A)	Yes	scales (galls)	no known impact
Desineura rachiphaga	pollination	cone (L)	spring (A)	Yes	cone axis	no known impact
Mayetiola carpophaga	pollination	seed in duff (L)	autumn (L)	Yes	seeds	1
Mayetiola thujae	pollination	cone (P)	spring (A)	Yes	scales	100% of seeds
Resseliella sp.	post pollination	cone (P)?	autumn (L)	Yes	scales	undetermined
Diptera: Chloropidae						
Earomyia sp.	pollination	duff (P)	summer (L)	Yes	seeds	undetermined
Hemiptera: Coreidae						
Leptoglossus occidentalis	post pollination	bark (A)	n.a.	No	seeds, conelets, pollen	undetermined
Hemiptera: Pentatomidae						
Tetyra bipunctata	post pollination	bark (A)	n.a.	No	seeds	undetermined
Hymenoptera: Torymidae						
Megastigmus sp.	post pollination	seed in duff (L)	autumn (L)	Yes	seeds	1 seed
Hymenoptera: Xyelidae						
Xyela sp.	pollen devp.	duff (PP)	n.a.	Yes	pollen	n.a.
Lepidoptera: Cochylidae						
Henricus fuscodorsanus	post pollination	duff (P)	autumn (L)	No	scales and seeds	undetermined
Lepidoptera: Gelecheiidae						
Coleotechnites sp.	post pollination	?	?	No	scales and seeds	undetermined
Exoteleia nepheos	pollen devp.	needle (L)	n.a.	No	pollen	n.a.
Lepidoptera: Geometridae						
Eupithecia sp.	post pollination	cone (P)	spring (A)	No	scales and seeds	10%
Lepidoptera: Pyralidae						
Dioryctria abietivorella	post pollination	cone (L)	spring (A)	No	scales and seeds	undetermined
Dioryctria reniculelloides	bud burst[b]	tree (L)	summer (A)	No	scales and seeds	> 1 cone
Dioryctria disclusa	post pollination	tree (L)	summer (A)	No	scales and seeds	1 or 2 cones
Lepidoptera: Tortricidae						
Barbara colfaxiana	pollination	cone (P)	spring (A)	Yes	scales and seeds	60% of seeds
Barbara sp.	pollination	cone (P)	spring (A)	Yes	scales and seeds	undetermined
Eucosma sp.	2nd year	duff (P)	summer (L)	?	scales and seeds	1/2 to 2 cones
Cydia sp. (pines)	2nd year	cone (L)	spring (A)	Yes	seeds	undetermined
Cydia sp. (spruces)	pollination	cone (L)	spring (A)	Yes	seeds	seeds 34% of wS[c] seeds
Choristoneura sp.	bud burst[b]	tree (L)	summer (L)	No	scales and seeds, pollen	> 1 cone

[a] L = larva; A = adult; P = pupa; PP = pre-pupa.
[b] Feeding continues throughout the remainder of cone development.
[c] wS = white spruce.
n.a. Not applicable.

1985); Ponderosa cone borer, *Eucosma ponderosa*, and a cone borer, *E. recissoriana* (Stevens et al. 1985); Ponderosa pine seedworm, *Cydia piperana* (Sartwell et al. 1985; Stevens et al. 1985); spruce seed moth, *C. strobilella* (Roelofs and Brown 1982; Weatherston et al. 1977; Grant et al. 1989); eastern pine seedworm, *C. toreuta* (Chisholm et al. 1985; Katovich et al. 1989); and a webbing coneworm, *Dioryctria disclusa* (Meyer et al. 1982). Sex pheromones are also available for important facultative pests such as the spruce coneworm, *Dioryctria reniculelloides* (Grant et al. 1987), and the spruce budworm (Sanders and Weatherston 1976). The Douglas-fir cone gall midge, *Contarinia oregonensis*, is known to utilize a sex pheromone (Miller and Borden 1981), as are other cone midges (G.E. Miller, unpubl. data), but the chemical identity of the pheromones has not been determined for this important group of cone and seed pests. Other ongoing Canadian research into pheromones includes studies on cone beetles, *Conophthorus* spp. (de Groot et al. 1991), and coneworms, *Dioryctria* spp.

Monitoring Seed Losses and Cone and Seed Insects

Recent studies have emphasized the identification of the various mortality factors affecting the reproductive structures of Canadian conifers and their respective impacts on crops (Amirault 1984; Amirault and Brown 1986; de Groot 1986c; Miller et al. 1984; Prevost 1986; Prevost et al. 1988; West 1986). This information combined with knowledge of seed losses, which have occurred historically in forest stands, indicates the impact that can be expected in seed orchards and seed production areas.

Overall losses, expressed as damage ratings, caused by insects in Canada are listed in Table 3. More specific information on losses to individual insect species are available in references listed in Table 3, in most reports of insecticide trials (Table 4), and other publications (e.g. Radcliffe 1952; Hedlin 1964, 1966, 1967; Miller et al. 1984). From this information, insects clearly do not cause the same levels of seed loss to all conifers. For example, surveys conducted in British Columbia indicate that on average, losses to insects in Douglas-fir, true firs, and white and Engelmann spruces equalled or exceeded the numbers of filled seeds produced (Miller and Ruth 1989). Hemlocks and lodgepole pine did not lose significant amounts of seed; ponderosa pine consistently lost a moderate amount of seed; and western red cedar, Sitka spruce, western larch, and western white pine suffered light losses, although heavy losses occurred occasionally.

In conifers known to suffer frequent heavy seed losses to insects, the losses can vary dramatically from year to year. In British Columbia, losses in Douglas-fir seed orchards have ranged from less than 5% to more than 95% (G.E. Miller, unpubl. data). Historical data also indicate that damage in a given year is influenced by the size of the cone crop in the previous year (Miller et al. 1984; Gagnon 1986). Generally, crops (even heavy crops) produced the year following a heavy cone crop are heavily attacked, whereas heavy crops preceded by light cone crops will normally be lightly damaged. This extreme variability in damage emphasizes the need for practical pest monitoring systems.

In seed orchards, distribution of seed losses among clones or families can also be important. Most of the seed currently produced in Douglas-fir orchards in British Columbia is borne on trees originating from a relatively small number of clones or families (M. Crown, British Columbia Forest Service, Coastal Seed Orchards, Duncan, B.C., pers. comm.). Hedlin and Ruth (1978) found statistically significant variations in damage by cone and seed insects among clones, although the differences were not considered of practical significance. Seed loss to insects could aggravate such imbalanced seed production if clones, which produce seed less frequently, are particularly susceptible to insects. Furthermore, if heavily producing clones were particularly susceptible to cone and seed insects, pest control operations would likely have to be implemented more frequently (Schowalter and Haverty 1989).

Seed losses by cone and seed insects are difficult to estimate precisely because other factors, such as pollination and seed abortion, also affect seed set. Seed chalcids and seed midges consume only filled seeds so estimating their damage is relatively straightforward. However, most obligate insects are general cone feeders, consuming seed, scale, and bract tissue as they mine cones. A mine through a seed site indicates that a potential seed, but not necessarily a filled seed, was lost. The bagging of cones to exclude the presence of other insects is one approach that has been used in the United States for determining the true impact of specific insects on seed production, but this approach has not yet been used in Canadian studies. Insecticides can also be used to estimate the seed potential of cones (Prevost et al. 1988); however, seldom is an insecticide 100% effective and some systemic insecticides may be phytotoxic to seeds.

Losses can also be difficult to estimate for facultative cone and seed insects that destroy reproductive buds, because their damage is not as readily ascertained as that of insects in cones, and few studies have examined the seed-destroying capacity of facultative pests. Young larvae of the spruce budworm are known to favor reproductive buds as feeding sites (Powell 1973; Schooley 1978, 1980; Amirault and Brown 1986), and thus have caused greater seed loss than has been recognized in cone surveys. Similarly, insects that feed on twigs may also be important because their feeding causes the death of twigs bearing reproductive buds. Shoot moths in the genus *Eucosma* are recognized as major pests of pine seed orchards in the southern United States for this reason (DeBarr and Barber 1975; Goyer and Nachod 1976). Heavy infestations of balsam woolly adelgid, *Adelges piceae*, are also known to reduce cone production (Schooley 1975).

No verified methods have specifically been developed for estimating seed losses to insects at a stand or seed orchard level with specified precision and accuracy. Kozak (1964) developed a sequential sampling procedure for estimating seed losses on individual Douglas-fir trees but it was not considered usable at the stand level because of the large sample size required. However, a method for estimating seed production in Douglas-fir seed orchards has been developed (Bartram and Miller 1988) and it could be used to estimate losses to insects. A similar procedure has been developed for southern pines in the United States (Bramlett and Godbee 1982).

Table 3. Damage caused by obligate cone and seed insect complexes on conifers in Canada.

Conifer	References	Damage ratings[a]
Douglas-fir	FIDS (1978–1987); Allen and Ruth (1979); Ruth et al. (1980); Miller and Ruth (1989)	nil to heavy
Fir		
balsam	Fye and Wylie (1968); Lachance et al. (1985)	light to heavy
grand	Miller and Ruth (1989)	nil to heavy
Pacific Silver	Miller and Ruth (1989)	nil to heavy
subalpine	Ruth et al. (1980); Miller and Ruth (1989)	nil to heavy
Western red cedar	Ruth et al. (1980); Miller and Ruth (1989)	nil to moderate
Hemlock		
mountain	Miller and Ruth (1989)	nil to light
western	Ruth et al. (1980); Miller and Ruth (1989)	nil to light
Larch		
tamarack	Lachance et al. (1985)	moderate
western	Ruth et al. (1980); Lachance et al. (1985); Miller and Ruth (1989)	moderate
Pine		
jack	Howse et al. (1983); FIDS (1986); de Groot (1986c); Anonymous (1987)	Light to heavy
lodgepole	Ruth et al. (1980); Miller and Ruth (1989)	nil to light
ponderosa	Ruth et al. (1980); Miller and Ruth (1989)	nil to heavy
red	Lyons (1957); FIDS (1983, 1986); Howse et al. (1983); FIDS (1986); Anonymous (1987)	nil to heavy
Scots	Anonymous (1987)	nil to light
white, eastern	Howse et al. (1981); FIDS (1984, 1987)	nil to moderate
white, western	Miller and Ruth (1989)	nil to moderate
Spruce		
black	Fye and Wylie (1968); Haig and McPhee (1969); FIDS (1981, 1984, 1987); Howse et al. (1981); Syme (1981); Howse et al. (1982); Lachance et al. (1985); Anonymous (1987); Miller and Ruth (1989)	nil to heavy
Engelmann	FIDS (1978–87); Allen and Ruth (1979); Ruth et al. (1980); Miller and Ruth (1989)	nil to heavy
Norway	Lachance et al. (1985); Anonymous 1987	light to moderate
red	Lachance et al. (1985)	light
Sitka	Miller and Ruth (1989)	nil to moderate
white	FIDS (1978-1987); Fye and Wylie (1968); Allen and Ruth (1979); Ruth et al (1980); Howse et al. (1982); Magasi (1983); Lachance et al. (1985); Cerezke and Holmes (1986); Anonymous (1987); Miller and Ruth (1989)	nil to heavy

[a] nil = no attack; light = <20% seeds or <30% cones damaged; moderate = 20% to 50% seed or 31% to 69% cones damaged; heavy = >50% seed or >70% cones damaged.

Table 4. Insecticide field trials for control of obligatory cone and seed insects in Canada.

Conifer	Insecticide			Pest	Reference
	Type	Active ingredient	Application method		
Douglas-fir	systemic	acephate	foliar spray	cone gall midge, chalcid	Summers and Miller (1986)
		acephate	injection	cone gall midge, chalcid	Summers and Miller (1986)
		bidrin	foliar spray	midges, cone moth, seed chalcid	Hedlin (1966)
		dimethoate	foliar spray	midges, cone moth, seed chalcid	Hedlin (1966)
		dimethoate	foliar spray	cone gall midge, chalcid	Summers and Miller (1986)
		oxydemeton-methyl	foliar spray	midges, cone moth, seed chalcid	Hedlin (1966)
		oxydemeton-methyl	foliar spray	cone gall midge, chalcid	Summers and Miller (1986)
		oxydemeton-methyl	injection	cone gall midge, chalcid	Summers and Miller (1986)
	contact	carbaryl	foliar spray	cone gall midge	Miller (1983a)
		fatty acids	foliar spray	cone gall midge	Miller (1982c)
		fenitrothion	foliar spray	midges, cone moth, seed chalcid	Hedlin (1966)
Jack pine	systemic	carbofuran	granules to soil	webbing coneworm	Brown and Amirault (1985)
		dimethoate	foliar spray	webbing coneworm	Brown and Amirault (1985)
	contact	fenitrothion	foliar spray	webbing coneworm	Brown and Amirault (1985)
		permethrin	foliar spray	webbing coneworm	Brown and Amirault (1985)
Black spruce	systemic	acephate	implants	maggot, coneworm, seed moth	West (pers. comm.)
		carbofuran	liquid to soil	maggot, coneworm, seed moth	Fogal et al. (1988)
		carbofuran	granules to soil	maggot, coneworm, seed moth	Fogal et al. (1988)
		dimethoate	foliar spray	midge	Haig and McPhee (1969)
		oxydemeton-methyl	foliar spray	midge	Haig and McPhee (1969)
White spruce	systemic	acephate	foliar spray	midge, maggot, coneworm, seed moth	Fogal and Lopushanski (1985)
		carbofuran	granules to soil	midges, cone maggot, seed moth	Cerezke and Holmes (1986)
		dicrotophos	foliar spray	midges, maggot, seed moth, chalcid	Hedlin (1973)
		dicrotophos	injection	midge, maggot, coneworm, seed moth	Fogal and Lopushanski (1984)
		dimethoate	foliar spray	midges, maggot, seed moth, chalcid	Hedlin (1973)
		dimethoate	foliar spray	cone maggot	Miller and Hutcheson (1981)
		dimethoate	foliar spray	midge, maggot, coneworm, seed moth	Fogal and Lopushanski (1985)
		formothion	foliar spray	midges, maggot, seed moth, chalcid	Hedlin (1973)
		methomyl	foliar spray	midge, maggot, coneworm, seed moth	Fogal and Lopushanksi (1985)
		oxydemeton-methyl	foliar spray	midges, maggot, seed moth, chalcid	Hedlin (1973)
		oxydemeton-methyl	injection	midge, maggot, coneworm, seed moth	Fogal and Lopushanski (1984)
	biological	*Beauveria bassiana*	dust to soil	cone maggot	Fogal (1986)
		Beauveria bassiana	dust to cones	cone maggot, seed moth	Fogal et al. (1986)
Tamarack	systemic	carbofuran	granules to soil	midge, maggot, chalcid	Amirault and Brown (1986)
		dimethoate	foliar spray	midge, maggot, chalcid	Amirault and Brown (1986)

Cones infested by internal feeders cannot be separated from healthy cones by outward appearances so the cones must be dismantled to estimate insect damage. The proportions of seeds, filled and damaged, along the face of a cone cut in half were similar to those for whole cones in Douglas-fir (Johnson and Heikkenen 1958; DeMar 1964) and ponderosa pine (Schmid et al. 1985). Cone slicing has been used operationally and in research as a method for indexing insect damage and is currently being evaluated for other Canadian conifers. Slicing cones to index insect damage is attractive because it takes one tenth to one twentieth the amount of time as cone dissections.

Development of methods to predict seed losses is a major thrust of current cone and seed insect research in Canada. For predictions of seed losses to be of practical significance, they must occur before any windows of opportunity for operational controls. Miller (1986a, 1986b) found that for the Douglas-fir cone gall midge, the key pest in Douglas-fir seed orchards in coastal British Columbia, counts of egg-infested scales in conelets in the spring were highly correlated with damage at cone harvest. Conelets are sampled and anticipated damage estimated just before cone development reaches the optimum stage for treatment with systemic insecticides. As a result, a predictive system based on this correlation was developed and is now in operational use.

Similar systems are currently being developed for use in spruce, larch, and western red cedar seed orchards. Systems already exist for predicting defoliation by the spruce budworms based on egg counts (Carolin and Coulter 1972; Dorais and Kettela 1982), and modification of these to predict seed loss in seed orchards should be possible. However, egg counting is not reasonable for all cone and seed pests. Seed chalcids oviposit their tiny eggs directly into seeds, making counting impractical. Predicting damage by insects that attack cones when they are about half their mature length, or later via egg counts, is of limited value because there is no effective control method for the insects at this stage of crop development.

Adult trapping is useful for timing insecticide applications and offers a potential tool for predicting damage if adult catches are related to the damage at cone harvest. Light traps are currently used in pine seed orchards in the southeastern United States to capture night-flying moths (Yates and Ebel 1975), but are not used as an operational method of monitoring Douglas-fir seed pests in British Columbia because of their lack of specificity. The most specific monitoring tools currently available are sex pheromones and related chemical attractants. These compounds typically attract single species or a few closely related species and are currently being used to monitor pest populations in deciduous fruit orchards (Madsen et al. 1975). The relationships between seed losses and trap catches of spruce seed moth and Douglas-fir cone moth have recently been quantified.

Traps mimicking stimuli originating from cones also offer potential as monitoring tools. In Europe, color traps have been shown to attract anthomyiid flies (Roques 1986b) and seed wasps (Roques 1986c). In Canada, research is being conducted to evaluate the potential of color traps as monitoring tools for the Douglas-fir seed chalcid, the spruce cone maggot, and the Douglas-fir cone gall midge. Volatile chemicals produced by cones could also be attractive to insects and are being tested for activity in the Douglas-fir cone moth (Benn and Ruth, unpubl. data). Such host-mimicking traps, known to catch both sexes of cone maggots in Europe (Roques 1986b) and the Douglas-fir cone moth (Benn and Ruth, unpubl. data), may prove to predict damage better than sex pheromone-baited traps that catch only males.

The development of insect monitoring systems requires the following knowledge: spatial and statistical distribution of insects and cones within seed production areas or seed orchards and trees; the relationships between insect numbers and damage; mortality rates likely to occur during the period between time of sampling and cone harvest. But the expected level of insect damage is not the only factor that must be considered when deciding whether or not insect control is justified (Miller 1983b). Situation-specific variables such as seed value (the most important variable), crop size, and the cost and efficacy of control operations must also be included. With this information, realistic threshold and injury levels can be established and control decisions facilitated.

Insect Management

Insect pest management strategies for seed orchards in Canada have been discussed by Miller (1980, 1982a, 1983b, 1984a). The strategies depend on several factors: the insects causing concern, their life histories, habits, and population dynamics; seed orchard location; tree species; and the tactics available for use. Currently, insecticides are the only practical tools available for operational insect control in the year of attack. Cultural practices are useful in long-term management of seed orchards to prevent or reduce insect infestations but are of no use in controlling insects after the attacks have occurred.

Crop Management Practices, Insect Infestations, and Control

Seed losses to insects can vary dramatically among sites. Historical seed losses to insects at potential seed orchard sites should be one factor considered in the site selection process, with a view to minimizing future pest problems.

Crop management practices, such as the use of prescribed fire and the destruction of unwanted cone crops, can be specifically aimed at pest control in seed orchards, while others, such as delaying reproductive bud burst and tree topping, are primarily aimed at the other objectives, but have benefits in pest management. On the other hand, certain practices, such as tree spacing and stimulation of crops in otherwise poor crop production years, can promote pest problems.

The use of prescribed fire for the control of the red pine cone beetle, *Conophthorus resinosae*, was evaluated in New Brunswick (Mellish 1987). An experimental burn in the early spring failed to reduce populations of the beetles, probably because many of the beetles had emerged before the burn. Timing of a spring fire may be difficult in some years, and it is suggested that a late fall burn be investigated as an alternative time to control the cone beetle. Prescribed fire shows promise for use in white pine seed orchards in the United States (Wade et al. 1989).

The removal and destruction of all mature (including infested) cones either from the tree or ground is a sound cultural practice that can prevent a buildup of pest populations on site. The early collection of cones and the quick extraction of seed should help prevent cone losses to anobiid beetles that attack mature cones (Schooley 1983). Destruction of unwanted cone crops can also be an effective tool in pest management in orchards. Unwanted crops include those too small to manage as a productive crop as well as cones on root stocks in clonal orchards.

Reproductive bud burst is delayed by misting trees with cold water in the spring in some British Columbia Douglas-fir seed orchards, primarily to prevent pollen contamination (Fashler and Devitt 1980). Miller (1983c) evaluated the effectiveness of cold-water misting to prevent or reduce Douglas-fir cone gall midge infestations. He found that the greatest reduction in damage (up to 99.2% in one seed orchard) occurred when pollination was delayed by 10 days in combination with the earliest flowering trees which were the most heavily attacked. No reductions occurred when the delay was 5 days or less. It has not been possible to predict the efficacy of misting because the delay in bud burst is determined by weather, which varies from year to year, and because the synchrony between the presence of the adult midge and the susceptible host stage is not consistent.

Tree topping is sometimes carried out in Douglas-fir orchards to facilitate supplemental pollination, cone collection, and insecticide applications. It also allows use of a wider range of application equipment options.

Orchard managers originally thought that stimulation of cone crops in years when cone production in forest stands was poor would avoid insect problems, but just the opposite has occurred. If crops are poor in nearby forest stands, insects migrate and can cause heavy damage in orchards because infestations become intensified on whatever cones are available. Application of insecticides has been necessary to control heavy pest damage in Douglas-fir seed orchards that had been stimulated by root pruning to obtain manageable crops in generally poor crop years.

Spacing trees is a practice necessary to promote good cone production, but it is also known to increase infestations by cone and seed insects (Kraft 1968; Schenk and Goyer 1967). Species considered unimportant or rare in forests may become significant pests in orchards, because orchards may provide more suitable environments than forest stands for some insect species. For example, the western conifer seed bug is considered rare in natural stands of Douglas-fir (Hedlin 1974) but has periodically caused serious damage (up to 36% seed loss) in Douglas-fir orchards (G.E. Miller, unpubl. data).

Chemical Control

Two major types of chemical insecticides have been tested for efficacy in the protection of cones and seed—systemic and contact. Systemic insecticides are absorbed into tree tissues and translocated, so they can be used to kill insects already inside young cones, whereas contact insecticides, to be effective, must contact the insect directly and are applied when the insect is, or is about to become, exposed. Both types of insecticides have roles in management of cone and seed insects, but only three insecticides are registered for use against obligatory cone and seed insects in Canada. Oxydemeton-methyl (a systemic) and azinphosmethyl (contact) are registered for use on Douglas-fir and dimethoate (systemic) is registered for use on both Douglas-fir and white spruce.

Numerous insecticides have been tested against cone and seed insects in Canada, particularly the facultative cone pests. Insecticide trials against obligatory cone insects are listed in Table 4. The materials are not likely to be registered solely for cone and seed insects, because seed orchards offer a very limited market for insecticide use. Future trials should therefore center on insecticides registered for other uses because with validation of efficacy, such insecticides can also be registered for use in seed orchards or seed production areas through the minor-use clauses of the Pest Control Products Act.

So far, because of the young ages of seed orchards, insecticides have been used operationally only in Douglas-fir orchards in British Columbia for control of obligatory pests. Both dimethoate and oxydemeton-methyl have been applied as sprays until runoff at 0.5 to 1.0% active ingredient and 0.5% active ingredient, respectively.

Application methods and equipment—The most practical application method and equipment will depend on the conditions in the orchard (e.g. tree size, spacing, and terrain), the target pest, and the insecticide to be used. Contact insecticides can be applied as foliar sprays whereas systemic insecticides can be sprayed on foliage, injected or implanted in trunks, brushed onto bark as topical applications, or incorporated into the soil. Painting systemics onto bark was ineffective on Douglas-fir (Hedlin 1966), but sprays, injections, and soil incorporations are effective. With sprays of systemic insecticides, maximum cone and seed insect control is obtained by spraying cones and surrounding foliage until runoff occurs.

Sprays can be applied with backpack sprayers, hydraulic sprayers, airblast sprayers, and mistblowers from the ground, or with aerial application equipment. Each ground sprayer has its own characteristics in terms of droplet size produced and speed of application (see de Groot, Chapter 54). Fogal and Lopushanski (1988) describe modifications of commercial equipment for the incorporation of granular and liquid formulations into the soil in coniferous stands.

Lejeune (1975) reported on an earlier aerial application for control of cone and seed insects on Douglas-fir. In the only recent aerial application trial, dimethoate was applied to white spruce in British Columbia via three techniques: broadcast application, single tree application with a short horizontal boom, and single tree application with an "A"-frame boom (Miller and Hutcheson 1981). The dimethoate was applied at 2% active ingredient at 83.5 L/ha in the broadcast application, 0.8% active ingredient at 12.3 L per tree with the horizontal boom, and 0.8% active ingredient at 2.3 L per tree with the "A"-frame boom. Populations of spruce cone maggot were reduced by 68% in the broadcast spray, 100% with the horizontal boom, and 87% with the "A"-frame boom. The number of seeds per cone was increased by 22% and 43% for the broadcast and horizontal booms, respectively. Aerial application is now a widespread practice in the southeastern United States because many trees have become too tall for efficient treatment with ground spray equipment, and aerial application can be done very quickly in the short time available for effective control.

Very few direct comparisons of application technologies have been made in Canada and more are needed to develop recommendations for their use. Results to date have indicated that hydraulic sprays are better than airblast sprays on Douglas-fir (Summers and Miller 1986); that double applications (hydraulic sprayer) of dimethoate 1 week apart are no more effective than single applications against the Douglas-fir cone gall midge (Summers and Miller 1986); that foliar sprays of oxydemeton-methyl (0.25 to 1.0% active ingredient, soaking spray) are more effective than injections (Mauget® injectors, 50% active ingredient) at one unit per 15 or 30 cm dbh on Douglas-fir (Summers and Miller 1986); and that soil incorporation of liquid carbofuran may be better than that of a granular formulation on black spruce (Fogal et al. 1988).

Efficacy trials—Systemic insecticides that are effective as foliar sprays against at least one targeted pest include the following: dimethoate on Douglas-fir (Hedlin 1966; Summers and Miller 1986; Summers and Ruth 1987), black spruce (Haig and McPhee 1969; Prevost et al. 1988), white spruce (Hedlin 1973; Fogal and Lopushanski 1985), and tamarack (Amirault and Brown 1986); oxydemeton-methyl on Douglas-fir (Hedlin 1966; Summers and Miller 1986) and white spruce (Hedlin 1973; Fogal and Lopushanski 1985); bidrin on Douglas-fir (Hedlin 1973); and methomyl (Fogal and Lopushanski 1985) and formothion (Hedlin 1973) on white spruce. Acephate gave inconsistent results on Douglas-fir; it was very effective in some years but was ineffective in others (Summers and Miller 1986). The reasons for the variability are unknown. Dicrotophos was not effective as a foliar spray on white spruce (Hedlin 1973). The application rates tested in trials to date have been in the 0.25% to 2.0% active ingredient range.

Trunk injections of oxydemeton-methyl were effective on Douglas-fir (Summers and Miller 1986) and white spruce (Fogal and Lopushanski 1984) as were injections of dicrotophos on white spruce (Fogal and Lopushanski 1984). Implants of acephate were effective on black spruce (R. West, Forestry Canada, Newfoundland Forestry Centre, St. John's, Nfld., pers. comm.) and white spruce (W. Fogal, Forestry Canada, Petawawa National Forestry Institute, Chalk River, Ont., pers. comm.). Generally, soil incorporations of carbofuran at rates below 10 g/cm dbh have resulted in a low level of effectiveness on white spruce (Cerezke and Holmes 1986), jack pine (Brown and Amirault 1985), and black spruce (Brown and Amirault 1985; Fogal et al. 1988). Active ingredient rates above 10 g/cm dbh have shown more effectiveness but also considerably more phytotoxicity (Brown and Amirault 1985; Fogal et al. 1981).

Effective contact insecticide trials include fenitrothion against the spruce budworm on black spruce, fenitrothion and permethrin against jack pine budworm, *Choristoneura pinus* (Brown and Amirault 1985), and permethrin and malathion against the western conifer seed bug, *Leptoglossus occidentalis*, on Douglas-fir (Summers and Ruth 1987). Ineffective contact insecticides include fatty acid derivatives ("soaps") (Miller 1982c), carbaryl (Miller 1983a), and diatomaceous earth (Summers and Ruth 1987) on Douglas-fir; fenitrothion and permethrin on larch; and permethrin on black spruce (Brown and Amirault 1985).

All of the effective insecticides, except for bidrin because of its high mammalian toxicity, have potential use in the control programs if registrations are obtained. In direct comparisons, dimethoate was superior to oxydemeton-methyl, formothion, and fenitrothion on white spruce (Hedlin 1973), and achieved similar results to oxydemeton-methyl on Douglas-fir (Summers and Miller 1986).

Effectiveness of insecticides has varied among insect species. For example, dimethoate was effective against a cone maggot but not against the spruce budworm, or a cone resin midge, or a seed chalcid on larch (Amirault and Brown 1986). Dimethoate was effective against lepidopterans on black spruce but was ineffective against the spruce cone maggot, *Strobilomyia neanthracina*, and the spruce cone axis midge, *Dasineura rachiphaga* (Prevost et al. 1988). In Douglas-fir seed orchards, dimethoate applied primarily for control of the Douglas-fir cone gall midge has proven erratic in the control of Douglas-fir seed chalcid (Summers and Miller 1986). Such variability may well result from the timing of the sprays rather than from differences in susceptibility among insects. Seed chalcids and resin midges attack cones about a month after the pollination period. If systemic insecticides are applied to cones when they are turning down, there may not be enough residual insecticide left in the cone tissues to cause consistent mortality rates in insects that oviposit later in the season.

All trials have been primarily targeted at insects that attack conelets open to receive pollen or, in the case of pines, young conelets in the spring of the second year of their development. Preventative sprays are likely to be the most effective for insects attacking half-grown cones, and for insects attacking pine cones in their second year of development (based on experiences in the southeastern United States), but no trials have been specifically targeted at these insects in Canada.

The effectiveness of the insecticides indicated previously is based on their significant reductions of insect populations. Unfortunately, a reduction in insect density does not always result in significant increases in seed production because of the adverse effect of some other factor, such as poor pollination. For example, no increase in seed production was observed with oxydemeton-methyl on Douglas-fir (Summers and Miller 1986) or carbofuran on white spruce (Fogal et al. 1981). The lack of increased yield may have been the result of phytotoxicity of the insecticide to ovules or developing seeds. This is difficult to determine, however, as seed production is frequently so variable among trees or among cones within a tree that treatment effects are masked.

Timing—Timing of insecticide applications is critical to their efficacy. Contact insecticides are used primarily to prevent damage and must be applied when the target pest is active. For many cone and seed insects (particularly the internal cone feeders) the exposure time is short, which presents problems for the proper timing of the application. However, the use of pheromone traps or degree-day models may circumvent the problem. In spite of this difficulty, contact insecticides will be essential in the management of some insects, particularly budworms and coneworms.

Systemic insecticides were originally thought to be applicable at any time during cone development. We now know that timing applications of systemic insecticides is more critical and affects insecticidal efficacy for insects that oviposit in open conelets. In Douglas-fir, the efficacy of dimethoate sprays applied to conelets that had been pendant for 1 week was dramatically lower than that of sprays applied to conelets beginning to turn down (Summers and Miller 1986). Although timing was not specifically studied, Miller and Hutcheson (1981) observed that more damage by the spruce cone maggot occurred on trees treated after the cones were between the horizontal and pendant positions than on trees treated when the cones were just starting to turn down. Timing of injections and implants has not been studied in Canada, but studies in the United States indicate that time can affect efficacy.

Phytotoxicity—Systemic insecticides can be phytotoxic to foliage, cones, and seeds. Granular carbofuran is phytotoxic to white spruce at rates of 9 g/cm dbh or higher (Fogal et al. 1981, 1988; Cerezke and Holmes 1986). Miller (1982b) reported that Dimethoate 4E applied operationally as a foliar spray caused moderate to heavy foliage burn on about 25% of the treated Douglas-fir trees, the effects apparently being associated with certain clones. The cone abortion rate was slightly higher on trees sprayed with dimethoate (Cygon 2E) than on untreated trees (15% vs. 5%, respectively) but the differences were not statistically significant (Miller 1983a). Although some systemic insecticides may control insects without increasing seed yields, there has not yet been a clear demonstration that these insecticides are, in fact, phytotoxic to seeds. Selective phytotoxicity can be operationally useful: potassium oleate (an insecticidal soap) causes conelet abortion when applied to conelets open to receive pollen (Miller 1983c) and has been used to abort unwanted cone crops.

Because a tree (or clone) is susceptible to damage by one systemic insecticide does not mean that it will be susceptible to another. For example, trees damaged by dimethoate were not adversely affected when treated with Meta-Systox-R. (D. Summers, British Columbia Forest Service, Silviculture Branch, Victoria, B.C., pers. comm.).

Biological Control

There are several biological control agents that can be used for suppression of cone and seed insects, including parasites, predators, viruses, bacteria, and fungi, but only the last two agents have been tested. *Bacillus thuringiensis* (Bt), a bacterial insecticide, has been widely used to protect foliage in operational programs for control of the budworms and coneworms and may be useful for other cone and seed insects.

The fungus *Beauvaria bassiana* is pathogenic to a wide variety of insects and has been tested in Canada for the control of some cone and seed insects. Timonin et al. (1980) found that under laboratory conditions, larvae and puparia of the spruce cone maggot, the spruce cone axis midge, and the spruce cone gall midge, also larvae of the spruce seed moth, were very susceptible to preconditioned conidia of *B. bassiana*. The conidia were applied directly to the insect by a brush or by dipping the insect in a conidial suspension. In tests against the spruce cone maggot, 42% mortality due to treatment occurred under low moisture conditions, but none was observed in a high moisture environment (Fogal 1986). Under seminatural field conditions, 21% insect mortality was achieved when larvae were added to soil treated with conidia at a rate of 2.6×10^6 conidia per gram of humus plus 1.5×10^6 conidia per square centimetre layered on the soil surface.

In a field study, Fogal et al. (1986) dusted individual conelets of white spruce with approximately 9.5×10^6 conidiospores each on May 18 and May 21 after the cone scales had closed. The treatments had no significant effect on cone abortion or on damage by the spruce cone maggot and the spruce seed moth. A significant increase (55%) in the number of sound seeds was recorded in the conelets treated May 21, but not in those treated May 18. Although reasons for the observed increase were not known, it was concluded that successful control depends on the application of viable conidiospores at or near the peak of oviposition of adult insects.

Several parasites and predators have been reared from cone and seed insects, but they offer little protection to developing crops in seed orchards because most of these natural enemies kill pests after they have completed their damage. A notable exception are egg parasitoids, which because they kill eggs, can reduce infestations. Hulme and Miller (1988) released *Trichogramma* sp. against Douglas-fir cone moth and obtained low levels of parasitism. Factors limiting the effectiveness of the wasp included the poor quality of the parasitoid stock, a poor release method, and ant predation.

Seed Treatments

Seed chalcids and seed midges are not always removed during seed cleaning and sorting, and control of these insects may be required when infested seed is to be exported. Heating Douglas-fir seeds at 45° C for 33 hours killed all seed chalcids without significantly affecting seed germinability (Ruth and Hedlin 1974). A similar treatment was effective for controlling seed chalcids and seed midges in true fir seeds (Miller and others, in preparation).

Future Directions

The five phases used to classify the status of crop protection in cotton agricultural ecosystems (Smith 1969) are also applicable to cone and seed crops (Cameron 1984):

Phase	Characteristics
Subsistence	• Seeds come from natural stands or seed production areas. • Little crop management, and insecticides are rarely used.
Exploitation	• Establishment of seed orchards. • Intensive crop management, and pesticides are used as preventive sprays on fixed schedule.

Crisis	• Insects that never caused problems before become primary pests. • Use of more frequent pesticide sprays.
Disaster	• Excessive use of pesticides. • Production costs exceed value of crop. • Pesticide residues reach intolerable levels.
Pest Management	• Control programs use ecological factors. • Control is optimized rather than maximized.

Currently, managers of southern pine seed orchards are heading for the crisis phase, as secondary insects begin to appear due to the frequent use of pesticides. Through additional research, they are attempting to reach the pest management phase while avoiding the disaster phase (Cameron 1984). It is becoming increasingly difficult to register new pesticides in Canada, and the semiurban location of many seed orchards in Canada inhibits the use of pesticides. Most Canadian seed orchards are still at the early stages of the exploitation phase, and the current research emphasis on the development of improved, ecologically oriented pest management programs that optimize the cost to benefit ratio of crop protection (Huffaker and Smith 1980) should continue to avoid both the disaster and crisis phases and to reduce our dependence on pesticides.

Specific information needs for the development of sound pest management systems in Canada include the following:

- The identity of insects affecting seed production in most conifers as well as their seed-destroying capacity. Good databases exist only for Douglas-fir, white spruce, and Engelmann spruce in British Columbia.
- Development of methodologies for estimating crop sizes and for monitoring pest populations; in particular, development of damage prediction systems wherever possible. Currently only one damage prediction system exists, that for Douglas-fir cone gall midge.
- Very few new registrations of pesticides are likely for cone and seed insect control in the foreseeable future. Use of the minor-use registration clauses of the Pest Control Products Act represents the only viable route of registration because of the small size of the seed orchard market for insecticides. In Canada, insecticides are registered for use on Douglas-fir, spruce, and western red cedar. None are registered for use on pines, true firs, and larches. Conifers in this latter group can suffer heavy seed losses, at least occasionally, to insects. If seed orchards of these species become infested, orchard managers have no tools to protect their investment.
- The uptake and phytotoxicity of systemic insecticides by conifers is not well understood. Given the variability reported in efficacy trials, more information in this area is needed if more consistent results are to be achieved.
- A better understanding of insect population dynamics is essential, particularly in relation to the effects of cone crop size and periodicity on population levels. This knowledge will offer seed orchard managers the possibility of manipulating the abundance of cones and the frequency of production to limit the impact of insects and optimize crop yield. Further knowledge of the natural enemies of cone and seed insects may also provide future alternatives to insecticides.

References

Allen, S.J.; Ruth, D.S. 1979. Cone and seed insects, Cariboo Forest Region. Can. For. Serv., Pac. For. Res. Cent., Victoria, B.C. Pest Rep., 3 p.

Amirault, P.A. 1984. An investigation of the cone and seed insects of eastern larch, *Larix laricina* (Du Roi) K. Koch, and attempts to control damage using chemical insecticides. M.Sc.F. Thesis, Univ. of New Brunswick, Fredericton.

Amirault, P.A.; Brown, N.R. 1986. Cone and seed insects of tamarack, *Larix laricina* (Du Roi) K. Koch, and attempts to control damage using chemical insecticides. Can. Entomol. 118: 589-596.

Anon. 1987. Insectes et maladies des arbres, Québec, 1986. Gouvernement du Canada, Service canadien des forêts et Gouvernement du Québec, ministère de l'Énergie et des Ressources, Service de la protection contre les insectes et les maladies, Ste-Foy (Québec). 33 p.

Bartram, C.; Miller, G. 1988. Estimation of seed orchard efficiencies by means of multistage variable probability sampling. Can. J. For. Res. 18: 1397-1404.

Bramlett, D.L.; Godbee, J.F., Jr. 1982. Inventory-monitoring systems for southern pine seed orchards. Ga. For. Comm. Res. Pap. No. 28.

Brown, N.R.; Amirault, P.A. 1985. Studies on the biology and control of cone and seed insects of selected conifers in the Maritime Provinces. Report submitted under contract 08SC.KH209-3-0137 to the Canadian Forestry Service—Maritimes.

Cameron, R.S. 1984. History, status, and future needs for entomology research in southern forests. Pages 40-46 Proc. *in* 10th Anniversary East Texas For. Entomol. Seminar. Kurth Lake, Tex.

Carolin, V.M.; Coulter, W.K. 1972. Sampling populations of western spruce budworm and predicting defoliation on Douglas-fir in eastern Oregon. USDA. For. Serv., Corvallis, Ore. Res. Pap. PNW-149.

Cerezke, H.F.; Holmes, R.E. 1986. Control studies with carbofuran on seed and cone insects of white spruce. Can. For. Serv., Edmonton, Alta. Inf. Rep. NOR-X-280, 10 p.

Chisholm, M.D.; Reed, D.W.; Underhill, E.W.; Palaniswamy, P.; Wong, J.W. 1985. Attraction of tortricid moths of subfamily Olethreutinas to field traps baited with dodecadienes. J. Chem. Ecol. 11: 217-230.

Churcher, J.J.; Furderer, R.E.; McGauley, B.H. 1985. Insects affecting seed production of spruces in Ontario. Ont. Min. of Nat. Resources, Toronto. Pest Cont. Rep. 21.

DeBarr, G.L.; Barber, L.R. 1975. Mortality factors reducing the 1967-69 slash pine seed crop in Baker County, Florida—a life table approach. USDA. For. Serv., Washington, D.C. Res. Paper SE-131, 16 p.

de Groot, P. 1986a. Diptera associated with cones and seeds of North American conifers: an annotated bibliography. Can. For. Serv., For. Pest. Manage. Inst., Sault Ste. Marie, Ont. Inf. Rep. FPM-X-76.

de Groot, P. 1986b. Cone & twig beetles (Coleoptera: Scolytidae) of the genus Conophthorus: An annotated bibliography. Can. For. Serv., For. Pest Manage. Inst., Sault Ste. Marie, Ont. Inf. Rep. FPM-X-76.

de Groot, P. 1986c. Mortality factors of jack pine, *Pinus banksiana*, Lamb., strobili. Pages 39-52 *in* A. Roques (Comp. and Ed.), Proc. Cone and Seed Insects Working Party 2nd Conf. (IUFRO S2.07-01), Briancon, Fr.

de Groot, P.; DeBarr, G.L.; Birgersson, G.O.; Pierce, H.D.; Borden, J.H.; Berisford, Y.C.; Berisford, C.W. 1991. Evidence for a female-produced pheromone in the white pine cone beetle, *Conophthorus coniperda* (Shwarz) and in the red pine cone beetle, *C. resinosae* Hopkins (Coleoptera: Scolytidae). Can. Entomol. 123: 1057-1064.

DeMar, C.J. 1964. Predicting insect caused damage to Douglas-fir seed from samples of young cones. USDA. For. Serv., Washington, D.C. Res. Note PSW 40.

Dorais, L.; Kettela, E.G. 1982. A review of entomological survey and assessment techniques used in regional spruce budworm, *Choristoneura fumiferana*, (Clem.) surveys and in the assessment of operational spray programs. A report of the Committee for the Standardization of Survey and Assessment Techniques, Eastern Spruce Budworm Council. Ministère de L'Énergie et des Ressources, Que. 43 p.

Fashler, A.M.K.; Devitt, W.J.B. 1980. A practical solution to Douglas-fir seed orchard pollen contamination. For. Chron. 56: 237-241.

FIDS (Forest Insect and Disease Survey). 1979-1987. Annual reports. Can. For. Serv., Ottawa, Ont.

Fogal, W.H. 1986. Applying *Beauveria bassiana*, (Bals.) Vuill. to soil for control of the spruce cone maggot *Lasiomma anthracina*, (Czerny). Pages 257-266 *in* A. Roques (Comp. and Ed.), Proc. Cone and Seed Insects Working Party 2nd Conf. (IUFRO S2.07-01), Briancon, Fr.

Fogal, W.H.; Lopushanski, S.M. 1984. Stem injection of systemic insecticides for control of white spruce seed and cone insects. Pages 157-167 *in* H.O. Yates III (Comp. and Ed.), Proc. Cone and Seed Insects Working Party 1st Conf. (IUFRO S2.07), Athens, Ga.

Fogal, W.H.; Lopushanski, S.M. 1985. A test of foliar-applied insecticides to prevent damage to white spruce cones by insects. For. Chron. 61: 499-502.

Fogal, W.H.; Lopushanski, S.M. 1988. Prototype equipment for soil incorporating granular and liquid formulations of insecticides in conifer seed stands. Proc. Entomol. Soc. Ont. 119: 79-81.

Fogal, W.H.; Lopushanski, S.M.; MacLeod, D.A.; Winston, D.A. 1988. Soil incorporation of carbofuran for protecting black spruce seed trees from insects. Proc. Entomol. Soc. Ont. 119: 69-78.

Fogal, W.H.; Thurston, G.S.; Chant, G.D. 1986. Reducing seed losses to insects by treating white spruce conelets with conidiospores of *Beauveria bassiana*. Proc. Entomol. Soc. Ont. 117: 95-98.

Fogal, W.H.; Winston, D.A.; Lopushanski, S.M.; MacLeod, D.A.; Willcocks, A.J. 1981. Soil application of carbofuran to control spruce budworm, *Choristoneura fumiferana*, (Lepidoptera: Tortricidae), in a managed white spruce seed production area. Can. Entomol. 113: 949-951.

Fye, R.E.; Wylie, W.D. 1968. Notes on insects attacking spruce and fir cones at Black Sturgeon Lake, Ontario, 1963-4. Can. Dept. For. and Rural Develop., Ottawa, Ont. Bi-mon. Prog. Rep. 24(6): 47-48.

Gagnon, D. 1986. Les insectes ravageurs des cones et des graines au Québec: La situation actuelle. Memoire de fin d'études. Faculté de Foresterie et de Géodesie, Université Laval, Que.

Goyer, R.A.; Nachod, L.H. 1976. Loblolly pine conelet, cone, seed losses to insects and other factors in a Louisiana seed orchard. For. Sci. 22: 386-391.

Grant, G.G.; Fogal, W.H.; West, R.J.; Slessor, K.N.; Miller, G.E. 1989. A sex attractant for the spruce seed moth, *Cydia strobilella* (L.) and the effect of lure dosage and trap height on capture of male moths. Can. Entomol. 121: 691-697.

Grant, G.G.; Prevost, Y.H.; Slessor, K.N.; King, G.G.S.; West, R.J. 1987. Identification of the sex pheromone of the spruce coneworm *Dioryctria reniculelloides* (Lepidoptera: Pyralidae). Environ. Entomol. 16: 905-909.

Haig, R.A.; McPhee, H.G. 1969. Black spruce cone insect control trials, Longlac, Ontario, 1967-68. Can. Dep. Fish. and For., Great Lakes For. Res. Cent., Sault Ste. Marie, Ont. Inf. Rep. 0-X-110.

Hedlin, A.F. 1960. On the life history of the Douglas-fir cone moth, *Barbara colfaxiana* (Kft.) (Lepidoptera: Olethreutidae) and one of its parasites, *Glypta evetriae* Cush. (Hymenoptera: Ichneumonidae) Can. Entomol. 92: 826-834.

Hedlin, A.F. 1964. Results of a six-year plot study on Douglas-fir cone insect population fluctuations. For. Sci. 10: 124-128.

Hedlin, A.F. 1966. Prevention of insect-caused seed loss in Douglas-fir with systemic insecticides. For. Chron. 42: 76-82.

Hedlin, A.F. 1967. Cone insects of grand fir, *Abies grandis* (Douglas) Lindley, in British Columbia. J. Entomol. Soc. B.C. 64: 40-44.

Hedlin, A.F. 1973. Spruce cone insects in British Columbia and their control. Can. Entomol. 105: 113-122.

Hedlin, A.F. 1974. Cone and seed insects of British Columbia. Can. For. Serv. Pac. For. Res. Cent., Victoria, B.C. Inf. Rep. BC-X-90.

Hedlin, A.F.; Miller, G.E.; Ruth, D.S. 1982. Induction of prolonged diapause in *Barbara colfaxiana*, (Lepidoptera: Olethreutidae): correlations with cone crops and weather. Can. Entomol. 114: 465-471.

Hedlin, A.F.; Ruth D.S. 1977. Comparison of germinability of seed from insect-infested and uninfested cones. Can. For. Serv., Ottawa, Ont. Bi-mon. Res. Notes 33(5): 34.

Hedlin, A.F.; Ruth, D.S. 1978. Examination of Douglas-fir clones for differences in susceptibility to damage by cone and seed insects. J. Entomol. Soc. B.C. 75: 33-34.

Hedlin, A.F.; Weatherston, J.; Ruth, D.S.; Miller, G.E. 1983. Chemical lure for male Douglas-fir cone moth, *Barbara colifaxiana*, (Lepidoptera: Olethreutidae). Environ. Entomol. 12: 1751-1753.

Hedlin, A.F.; Yates, H.O., III; Cibrian Tovar, D.; Ebel, B.H.; Koerber, T.W.; Merkel, E.P. 1980. Cone and seed insects of North American conifers. Can. For. Serv./USDA For. Serv./Secretaria de Agricultura y Recursos Hidraulicos, Mex. Victoria, B.C. 122 p.

Howse, G.M.; Gross, H.L.; Syme, P.D.; Myren, D.T.; Meating, J.H.; Applejohn, M.J. 1982. Forest insect and disease conditions in Ontario, 1981. Can. For. Serv., Great Lakes For. Res. Cent., Sault Ste. Marie, Ont. Inf. Rep. 0-X-339.

Howse, G.M.; Syme, P.D.; Gross, H.L.; Myren, D.T.; Applejohn, M.J. 1981. Forest insect and disease conditions in Ontario, 1980. Can. For. Serv., Great Lakes For. Res. Cent., Sault Ste. Marie, Ont. Inf. Rep. 0-X-327, 50 p.

Howse, G.M.; Syme, P.D.; Gross, H.L.; Myren, D.T.; Meating, J.H.; Applejohn, M.J.; Smith, K.L. 1983. Forest insect and disease conditions in Ontario, 1982. Can. For. Serv., Great Lakes For. Res. Cent., Sault Ste. Marie, Ont. Inf. Rep. 0-X-350, 39 p.

Huffaker, C.B.; Smith, R.F. 1980. Rationale, organization and development of a national integrated pest management project. Pages 1-35 *in* C.B. Huffacker (Ed.), New Technology of Pest Control. John Wiley & Sons, N.Y.

Hulme, M.A.; Miller, G.E. 1988. Potential for control of *Barbara colfaxiana* (Kearfott) (Lepidoptera: Olethreutidae) using *Trichogramma*, sp. Int. Symp. of Trichogramma, Colloq. IRNA 43: 483-488.

Johnson, N.E.; Heikkenen, H.J. 1958. Damage to the seed of Douglas-fir by the Douglas-fir cone gall midge. For. Sci. 4: 274-282.

Katovich, S.A.; Swedenborg, P.D.; Giblin, M.; Underhill, E. 1989. Evidence for (*E*, *Z*) -8, 10-Dodecadienyl acetate as the major component of the sex pheromone of the eastern pine seedworm, *Cydia toreuta* (Lepidoptera: Tortricidae). J. Chem. Ecol. 15: 581-590.

Kozak, A. 1964. Sequential sampling for improving cone collection and studying damage by cone and seed insects in Douglas-fir. For. Chron. 40: 210-218.

Kraft, K.J. 1968. Ecology of the cone moth *Laspeyresia toreuta*, in *Pinus banksiana* stands. Ann. Entomol. Soc. Am. 61: 1452-1462.

Lachance, D.; Benoit, P.; Laflamme, G.; Bonneau, G.; Picher, R. 1985. Insectes et maladies des arbres, Québec, 1984. Gouvernement du Canada, Service canadien des forets et Gouvernement du Québec, Ministère de l'Énergie et des Ressources, Service d'entomologie et de pathologie, Ste-Foy (Québec). 33 p.

Lejeune, R.R. 1975. Cone and seed insects. Pages 215-217 *in* M.L. Prebble (Ed.), Aerial Control of Forest Insects in Canada. Environ. Can., Ottawa, Ont. 330 p.

Lyons, L.A. 1957. Insects affecting seed production in red pine. Part IV. Recognition and extent of damage to cones. Can. Entomol. 89: 264-271.

Magasi, L.P. 1983. Forest pest conditions in the Maritimes 1982. Can. For. Serv., Maritimes For. Res. Cent., Fredericton, N.B. Inf. Rep. M-X-141.

Madsen, H.F.; Peters, H.F.; Vakenti, J.M. 1975. Pest management: experience in six British Columbia apple orchards. Can. Entomol. 107: 873-877.

Mattson, W.J. 1971. Relationship between cone crop size and cone damage by insects in red pine seed production areas. Can. Entomol. 103: 617-621.

Mellish, S.B. 1987. The use of prescribed fire for control of the red pine cone beetle (*Conophthorus resinosae* Hopkins) (Coleoptera: Scolytidae) in a New Brunswick red pine (*Pinus resinosa* Aiton) seed production area. M.Sc. Thesis, Univ. of New Brunswick, Fredericton. 120 p.

Meyer, W.L.; DeBarr, G.L.; Berisford, C.W.; Barber, L.R.; Roelofs, W.L. 1982. Identification of the sex pheromone of the webbing coneworm moth. *Dioryctria disclusa* (Lepidoptera: Pyralidae). Environ. Entomol. 11: 986-988.

Michelson, V. 1988. A world revision of *Strobilomyia* gen. *in* the anthomyiid seed pests of conifers (Diptera: Anthomyiidae). Syst. Entomol. 13: 271-314.

Miller, G.E. 1980. Pest management in Douglas-fir seed orchards in British Columbia: a problem analysis. Simon Fraser Univ. Pest Manage. Paper 22. 138 p.

Miller, G.E. 1982a. Strategies for control of insect pests in seed orchards. Part 2. Pages 49-59 *in* Proc. 18th Annual Meeting Can. Tree Improv. Assoc., Duncan, B.C., Aug. 17-20, 1981.

Miller, G.E. 1982b. Phytotoxicity of Dimethoate 4ER to Douglas-fir seed orchards. Can. For. Serv., Victoria, B.C. Res. Notes 2: 7-8.

Miller, G.E. 1982c. Phytotoxicity to Douglas-fir megastrobili and efficiency against Douglas-fir cone gall midge of five fatty acid derivatives. Can. J. For. Res. 12: 1021-1024.

Miller, G.E. 1983a. Biology, sampling and control of the Douglas-fir cone gall midge, *Contarinia oregonensis* Foote (Diptera: Cecidomyiidae), in Douglas-fir seed orchards in British Columbia. Ph.D. Thesis, Simon Fraser Univ., Burnaby, B.C. 192 p.

Miller, G.E. 1983b. When is controlling cone and seed insects in Douglas-fir seed orchards justified? For. Chron. 59: 304-307.

Miller, G.E. 1983c. Evaluation of the effectiveness of cold-water misting of trees in seed orchards for control of Douglas-fir cone midge (Diptera: Cecidomyiidae). J. Econ. Entomol. 76: 916-919.

Miller, G.E. 1984a. Pest management in Douglas-fir seed orchards in British Columbia. Pages 179-185 *in* H.O. Yates III (Ed.), Proc. Cone and Seed Insect Working Party 1st Conference (IUFRO S2.07-01), Athens, Ga.

Miller, G.E. 1984b. Biological factors affecting *Contarinia oregonensis* infestations in Douglas-fir seed orchards on Vancouver Island, British Columbia. Environ. Entomol. 13: 873-877.

Miller, G.E. 1986a. Distribution of *Contarinia oregonensis* Foote (Diptera: Cecidomyiidae) eggs in Douglas-fir seed orchards and a method of estimating egg density. Can. Entomol. 118: 1291-1295.

Miller, G.E. 1986b. Damage prediction for *Contarinia oregonensis* Foote (Diptera: Cecidomyiidae) in Douglas-fir seed orchards. Can. Entomol. 118: 1297-1306.

Miller, G.E.; Borden, J.H. 1981. Evidence for a sex pheromone in the Douglas-fir cone gall midge. Can. For. Serv., Victoria, B.C. Res. Notes 1(2): 9-10.

Miller, G.E.; Hedlin, A.F.; Ruth, D.S. 1984. Damage by two Douglas-fir cone and seed insects: correlation with cone crop size. J. Entomol. Soc. B.C. 81: 46-50.

Miller, G.E.; Hutcheson, D.W. 1981. Aerial spraying for control of the spiral spruce-cone borer. *Hylemyia anthracina* (Diptera: Anthomyiidae). J. Entomol. Soc. B.C. 78: 3-6.

Miller, G.E.; Ruth, D.S. 1986. Effect of temperature during May to August on termination of prolonged diapause in the Douglas-fir cone moth (Lepidoptera: Tortricidae). Can. Entomol. 118: 1073-1074.

Miller, G.E.; Ruth, D.S. 1989. Relative importance of cone and seed insect species on commercially important conifers in British Columbia. *In* G.E. Miller (Ed.), Proc. Cone and Seed Insect Working Party 3rd Conference (IUFRO S2.07-01), 1988 June 26-20, Victoria, B.C.

Morgenstern, E.K. 1985. Status and future of seed orchards in Canada. Part 1:248 *in* Proc. 20th Annual Meeting, Can. Tree Improv. Assoc., Québec City, Que., Aug. 19-22, 1985.

Powell, G.R. 1973. The spruce budworm and megasporangiate strobili of balsam fir. Can. J. For. Res. 3: 424-429.

Prevost, Y. 1986. The relationship between the development of cones of black spruce, *Picea mariana* (Mill.) B.S.P., and their insect fauna. Ph.D. Thesis, Univ. of Guelph, Guelph, Ont.

Prevost, Y.; Laing, J.E.; Haavisto, V.F. 1988. Seasonal damage by insects and squirrels to female reproductive structures of black spruce, *Picea mariana* (Mill.) B.S.P. Can. Entomol. 120: 1113-1121.

Radcliff, D.N. 1952. An appraisal of seed damage by the Douglas-fir cone moth, in British Columbia. For. Chron. 28: 19-24.

Reed, D.W.; Chisholm, M.D. 1985. Attraction of moth species of Tortricidae, Gelechiidae, Geometridae, Drepanidae, Pyralidae, and Gracillariidae families to field traps baited with conjugated dienes. J. Chem. Ecol. 11: 1645-1657.

Reid, R.W. 1956. Coniferous seed and cone insects found in Alberta and the Rocky Mountain National Parks. Can. Dept. Agric., For. Biol. Div., Ottawa, Ont. Bi-mon. Prog. Rep. 12(4): 3.

Roelofs, W.L.; Brown, R.L. 1982. Pheromones and evolutionary relationships of tortricidae. Ann. Rev. Ecol. Sys. 13: 395-422.

Roques, A. 1986a. Dynamique d'infestation des nouveaux vergers à graines de Douglas du sud de la France par le chalcidien ravageur *Megastigmus spermotrophus* Wachtl. (Hymenoptera: Torymidae). Pages 685-694 *in* Proc. 18th IUFRO World Congress, Div. 2, Vol. II. Ljubljama, Yugoslavia.

Roques, A. 1986b. Response des adultes de *Lasiomma melania*, ravageur des cones de *Larix decidua*, à des pièges colores de differents types. Entomol. exp. & appl. 40: 177-187.

Roques, A. 1986c. Interaction between visual and olfactory signals in cone recognition by insect pests. Pages 153-160 *in* Labeyrie, Fabres and Lachaise (Eds.), Proc. 6th Int. Symp. Insect-Plant Relationships, Pau, Fr.

Roques, A. 1988. The larch cone fly in the French Alps. Pages 1-28 *in* A.A. Berryman (Ed.), Dynamics of forest insect populations. Plenum Publishing Corp.

Ruth, D.S. 1980. A guide to insect pests in Douglas-fir seed orchards. Can. For. Serv., Pac. For. Cent., Victoria, B.C. Inf. Rep. BC-X-204.

Ruth, D.S.; Hedlin, A.F. 1974. Temperature treatment of Douglas-fir seeds for control of the seed chalcid, *Megastigmus spermotrophus* Wachtl. Can. J. For. Res. 4: 441-445.

Ruth, D.S.; Miller, G.E.; Sutherland, J.R. 1982. A guide to common insect pests & diseases in spruce seed orchards in B.C.. Can. For. Serv., Pac. For. Cent., Victoria, B.C. Inf. Rep. BC-X-231.

Ruth, D.S.; Senecal, M.A.; Carlson, J.A. 1980. Cone and seed pests 1980. Can. For. Serv., Pac. For. Res. Cent., Victoria, B.C. Pest Rep. 16 p.

Sanders, C.J.; Weatherston, J. 1976. Sex pheromone of the eastern spruce budworm (Lepidoptera: Tortricidae): Optimum blend of trans- and cis-11-tetradecenal. Can. Entomol. 108: 1285-1290.

Sartwell, C.; Daterman, G.E.; Sower, L.L. 1985. A synthetic attractant for male moths of a biotype in the *Cydia piperana* complex (Lepidoptera: Tortricidae). Can. Entomol. 117: 1151-1152.

Schenk, J.A.; Goyer, R.A. 1967. Cone and seed insects of western white pine in northern Idaho: distribution and seed losses in relation to stand density. J. For. 65: 186-187.

Schmid, J.M.; Mata, S.A.; Mitchell, J.C. 1985. Estimating sound seeds in ponderosa pine cones from half-face cones. USDA For. Serv., Washington, D.C. Res. Note RM-459. 3 p.

Schooley, H.O. 1975. Cone production of balsam fir damaged by balsam woolly aphid. For. Chron. 51: 1-3.

Schooley, H.O. 1978. Effects of spruce budworm on cone production by balsam fir. For. Chron. 54: 298-301.

Schooley, H.O. 1980. Damage to Black Spruce cone crops by the spruce budworm. Can. For. Serv., Nfld. For. Cent., St. John's, Nfld. Inf. Rep. N-X-187, 15 p.

Schooley, H.O. 1983. A deathwatch cone beetle (Anobiidae: *Ernobius bicolor*) reduces the natural storage of black spruce seed in Newfoundland. For. Chron. 59: 139-142.

Schowalter, T.D.; Haverty, M.I. 1989. Influence of host genotype on Douglas-fir seed losses to *Contarinia oregonensis* (Diptera: Cecidomyiidae) and *Megastigmus spermotrophus* (Hymenoptera: Torymidae) in Western Oregon. Environ. Entomol. 18: 94-97.

Smith, R.F. 1969. The new and the old in pest control. Proc. Acad. Nazion. Lincei, Rome (1968) 366: 21-30.

Stevens, R.E.; Sartwell, C.; Koerber, T.W.; Powell, J.A.; Daterman, G.E.; Sower, L.L. 1985. Forest tortricids trapped using *Eucosma* and *Rhyacionia* synthetic sex attactants. J. Lepidop. Soc. 39: 26-32.

Summers, D.; Miller, G.E. 1986. Experience with systemic insecticides for control of cone and seed insects in Douglas-fir seed orchards in coastal British Columbia, Canada. Pages 267-283 *in* A. Roques (Ed.), Proc. Cone and Seed Insect Working Party 2nd Conf. (IUFRO S2.07-01), Briancon, Fr.

Summers, D.; Ruth, D.S. 1987. Effect of diatomaceous earth, malathion, dimethoate and permethrin on *Leptoglossus occidentalis* (Hemiptera: Coreidae): a pest of conifer seed. J. Entomol. Soc. B.C. 84: 33-38.

Syme, P.D. 1981. Black spruce cone and seed insects—a special report. Can. For. Serv., Great Lakes For. Res. Cent., Sault Ste. Marie, Ont. For. Res. Newsletter, pp. 1-2.

Timonin, M.T.; Fogal, W.H.; Lopushanski, S.M. 1980. Possibility of using white and green muscardine fungi for control of cone and insect pests. Can. Entomol. 112: 849-854.

Tripp, H.A.; Hedlin, A.F. 1956. An ecological study and damage appraisal of white spruce cone insects. For. Chron. 32: 400-410.

Wade, D.D.; DeBarr, G.L.; Bartier, L.R.; Manchester, E. 1989. Prescribed file—A cost effective control for the white pine cone beetle. Pages 117-121 *in* Proc. 10th Conference on Fire and Forest Meteorology. April 17-21, 1989. Ottawa, Canada.

Weatherston, J.; Hedlin, A.F.; Ruth, D.S.; MacDonald, L.M.; Leznoff, C.C.; Fyles, T.M. 1977. Chemical and field studies on the sex pheromones of the cone and seed moths *Barbara colfaxiana* and *Laspeyresia youngana*. Experientia 33: 723-724.

West, R.J. 1986. Seasonal incidence of cone pests of black spruce in Newfoundland. Can. For. Serv., Nfld. For. Cent., St. John's, Nfld. Inf. Rep. N-X-244.

Yates, H.O., III; Ebel, B.H. 1975. A light-trapping guide to seasonal occurrence of pine seed- and cone-damaging moths of the Georgia Piedmont. USDA For. Serv., Washington, D.C. Res. Note SE-210.

Part II
Pest Management Technology

Introduction

In this section of the book, prominent Forestry Canada, university, and provincial researchers and managers working in various aspects of forest pest management describe advances that have been made in the development of microbial and chemical insecticides, parasites and predators, insect pheromones, and silvicultural techniques for use in the management of forest insect pests. These authors describe the development of a wide range of control agents, the tools that the forest manager must have available to combat destructive forest insect pests. Other authors in this section will discuss techniques for the detection and assessment of pest populations and the development of computer models and expert systems that assist the forest manager in making sound pest management decisions. Later in this book, others will describe research and development in spray application technology aimed at delivering these tools to the target effectively and efficiently. Still others will describe research and development on the environmental impact of these control agents. All of this information, plus detailed knowledge of the biology, behavior, and population dynamics of the insect pest is essential for the development of pest management technologies that are effective, efficient, and environmentally conscious and that can be used in an integrated manner as essential components of good forest management.

The contents of this section of the book indicates a marked decline since 1975 in the development and registration of new neurotoxic chemical insecticides and in the availability of previously registered and effective ones (for example, Matacil® and Zectran®). The decline in this class of control agents has been due partly to very high costs associated with the development and registration of such materials and to reluctance by the pesticide manufacturing industry to develop and promote materials that when used for forestry purposes are, rightly or wrongly, the subjects of very negative public concern. Recently, there has been renewed interest in the development of new insect growth regulators, particularly the benzoylphenol ureas, juvenile hormone analogs, and others, with some very promising materials showing up that might help to fill this gap. There has been renewed interest in insect feeding deterrents also, but none of these are sufficiently developed for use against major forest insect pests, at least for the immediate future.

The period concerned has seen increased research and development activity in microbial insecticides. Bt is now used widely in the control of major lepidopteran forest insect pests. Research on its mode of action, the development of more concentrated formulations, improvements in application technology, and increased competition between Bt manufacturers have improved its operational effectiveness and narrowed the cost differential between it and chemical insecticides in the control of major pests. Over the same period, research on insect baculoviruses in Canada has begun to pay off in practical terms and has shown that several nuclear polyhedrosis viruses are extremely effective control agents. At the time of writing, Forestry Canada's Forest Pest Management Institute holds Canadian registrations for Virtuss, a very effective virus against the Douglas-fir tussock moth, *Orgyia pseudotsugata*, and other closely related tussock moths, and for Lecontvirus, an extremely potent virus against the redheaded pine sawfly, *Neodiprion lecontei*. The institute has submitted registration petitions for viruses for the European pine sawfly, *N. sertifer*, and the gypsy moth, *Lymantria dispar*, and is developing improved application technology for the latter. Research on fungal pathogens has concentrated on the Entomophthorales; a great deal is now known about their mode of action and host–pathogen interactions. A major stumbling block to their development as useful control agents, however, is the lack of efficient techniques for mass producing them in forms suitable for application to pest populations in the field. For Protozoa, research over the period has concentrated on the microsporidia. A great deal is now known about these organisms but there appears to be little opportunity to use them as direct control agents and, consequently, research on them from that particular perspective has waned over recent years. Research interest in nematodes is increasing but much remains to be done before their usefulness in Canadian forestry can be determined.

Increased research and development interest is once more being aimed at parasites and predators as control agents for forest insect pests in Canada. Significant success has been achieved by the import and release of parasites against introduced insects but there has been not nearly as much success with releases against native pests. Research interest is now turning to the potential of inundative releases, to an understanding of the role of parasites and predators in the regulation of pest populations, and to integrated pest management techniques that will allow the use of other control agents with minimal impact on parasites and predators.

Research and development on insect pheromones has received continuing attention in Canada and has been simplified through the identification and production of synthetic pheromones. Significant success has been attained in the use of pheromones in detecting the occurrence of specific insects and in monitoring their population fluctuations (for example, gypsy moth and Douglas-fir tussock moth), while others are approaching the operational use stage. As control agents, pheromones are now being used successfully against bark- and wood-boring beetles and good potential exists for their development and use against other pests, particularly those with rather localized populations.

Because of these advances, one might ask, Do we have a suitable range of pest control products available in Canada today to combat important forest insect pests?

The direct answer is that we are woefully short of registered and readily available pest control products to adequately handle either existing pest problems or new ones that will surely develop with intensive forest management and as climate change imposes additional stress on our forest resource.

Few, if any, classical neurotoxic insecticides are being developed by the pesticide industry for forestry use as stated earlier, and the environmental acceptability of currently registered and avialable materials in this class is being challenged.

Bt is finally coming of age and is being widely used in forestry in Canada. Continuing research and development will, undoubtedly, increase its effectiveness significantly. However, we must not rely too much on Bt and exclude other options or use Bt when it is not the most appropriate choice.

There are some very promising new insect growth regulators in the developmental stream that may be more acceptable to the general public than neurotoxic chemicals, but, to date, obtaining registration for them has been difficult.

Extensive advances have been made with insect viruses and two of these are already registered for use in Canada as noted earlier. However, these materials are so species-specific and, to date at least, so labor intensive to produce that commercial producers have shown little interest in them. The same may be said for entomogenous fungi which generally are at a significantly less advanced stage of development than the viruses.

Significant advances have been and are being made with parasites, predators, nematodes, and pheromones. Apparently, they will be used mostly with other agents and/or techniques in integrated pest management programs when these are developed in the future.

Biotechnological approaches will undoubtedly provide new and improved microbial control agents, some of which can probably be tailor-made for specific situations. However, this will take time and, as yet, registration protocols for these next generation pest control products have not been promulgated.

Impressive advances have been made in the development of forest pest management tools and techniques over the past few decades and the potential exists for even more spectacular advances in the future. All of these advances may be looked upon as steps toward the development and institution of integrated forest pest management practices within the broader concepts of integrated forest management and sustainable renewable resource development. Desirable as such goals may be and as impressive as the advances toward them to date might seem, they will not be achieved overnight and, in the interim, less than completely acceptable forest pest management techniques will have to be implemented if we wish to protect our forests adequately. Along the way, forest managers will require ready access to all registered pest control products and, together with their colleagues from other renewable natural resource sectors, will have to determine and implement the protection strategy that optimizes the many and varied benefits that can and should flow from a healthy forest resource.

Chapter 34

Development and Current Status of *Bacillus thuringiensis* for Control of Defoliating Forest Insects[1]

K. VAN FRANKENHUYZEN

Introduction

The past decade of large-scale spraying against defoliating forest Lepidoptera in Canada is characterized by a gradual increase in the use of microbial insecticides based on the bacterium *Bacillus thuringiensis* Berliner (commonly abbreviated as Bt). Aerial application of Bt is now considered a viable and effective strategy for providing foliage protection. It took more than 25 years of concerted collaboration between researchers, forest managers, and industry in both Canada and the United States to reduce the major obstacles of inconsistent efficacy and high cost. Many field tests were conducted between initial availability of commercial products in the early 1960s and operational acceptance in the mid-1980s. Most of those tests have been reviewed extensively (Morris et al. 1975; Smirnoff and Morris 1982; Morris 1982b; Cunningham 1985). This review describes events that made a critical contribution to the operational use of Bt in forestry, with emphasis on achievements in the last 8 years. An overview of its current status (1988) is presented and prospects for further optimization are discussed.

Background

Bacillus thuringiensis is a naturally occurring bacterium that is pathogenic to larvae of many insect species. It was first recognized as a disease agent in silkworm in Japan and flour moths in Germany (Berliner 1915) and has since been isolated from various species around the world (Krieg and Langenbruch 1981), including many forest insects (Morris 1982b). It now appears to be a worldwide and ubiquitous component of soil microbiota (Delucca et al. 1981; Martin and Travers 1989).

Insecticidal activity of the bacterium is mainly associated with a parasporal protein body, often referred to as the crystal, which is formed during sporulation. After ingestion by a susceptible insect, the crystal is converted by digestive juices to a toxic protein that destroys the cells lining the gut (Percy and Fast 1983). The larva stops feeding and dies within a few days if a lethal dose is ingested, but recovers and resumes feeding if the dose is sublethal (Fast and Régnière 1984; Retnakaran et al. 1983; van Frankenhuyzen and Nystrom 1987). Germination of ingested spores and subsequent septicemia by multiplication of vegetative cells contribute to larval death in many species (Heimpel and Angus 1959; Smirnoff 1974).

[1]Originally published in *The Forestry Chronicle*, October 1990, pages 498–507.

Bacillus thuringiensis is a collective name for a complex of more than 30 currently known subspecies, each with unique insecticidal properties (Dulmage 1981). Toxicity is exhibited specifically against Lepidoptera (e.g. subsp. *kurstaki*), Diptera (e.g. subsp. *israelensis*), or Coleoptera (e.g. subsp. *tenebrionis*). There is a high level of specificity within each subspecies. For example, only 200 species of predominantly foliage-feeding Lepidoptera are known to be susceptible to the *kurstaki* subsp., which is used in commercial formulations available for defoliator control. Because most other insect orders are unaffected (Krieg and Langenbruch 1981), Bt is a narrow-spectrum insecticide that is safe to natural enemies and other beneficial and nontarget organisms (Morris 1982b).

Commercial production is achieved by fermentation (Dulmage and Rhodes 1971). When cultured in an appropriate nutrient broth, vegetative cells sporulate and eventually lyse, releasing spores and crystals into the medium. The spore–crystal complex forms the active ingredient of commercial formulations. Bt was first available commercially in the late 1930s in France as a product called Sporeine. Commercial production in North America did not begin until the early 1950s (Hall 1963) and was primarily directed against agricultural pests.

Early Use

The first experimental aerial applications of Bt in forestry were conducted in 1960 against spruce budworm, *Choristoneura fumiferana*, in New Brunswick (Mott et al. 1961) and western blackheaded budworm, *Acleris gloverana*, in British Columbia (Kinghorn et al. 1961). Further tests in 1962 and 1969 yielded results that were encouraging but far from adequate (Morris et al. 1975). Two events in the 1960s were of particular significance to the development of Bt. First was the discovery of HD-1 (Dulmage 1970), a *kurstaki* isolate that was more toxic to lepidopterans than the one that was initially used in commercial products. HD-1 was quickly adopted in North America for commercial production. Second was the establishment of an international system for standardizing the potency of formulations. Early formulations were standardized on the basis of spore counts, which did not relate directly to insecticidal activity. Suggestions by Bonnefoi et al. (1958) and Burgerjon (1959) to express potency in biological units based on insect bioassays resulted in the adoption of a European Bt standard in 1966 (Burges 1967), which was later replaced by HD-1 as the North American standard (Dulmage et al. 1971). Insecticidal activity is now routinely expressed as the number of international units per unit weight or volume, as determined in bioassays against the cabbage looper, *Trichoplusia ni*, in parallel with the HD-1 standard.

Availability of formulations based on HD-1 and with standardized potency started a period of intensive field testing between 1971 and 1973 in Ontario and Quebec. Formulations containing about 4 BIU/L (billion international units per litre) were tested at 10 to 20 BIU/ha in 4.7 to 18.9 L. Although results were highly inconsistent, the good showing in some trials encouraged the expectation that Bt could indeed provide adequate foliage protection and provided the basis for continued and intensified testing in the remainder of the decade (Morris et al. 1975).

Potential and Constraints

Variable results of trials in the early 1970s (Smirnoff and Morris 1982; Morris 1982b; Dorais 1985) were attributed to inconsistent spray deposition, and consequently efforts were primarily directed at increasing the delivery of droplets to the target by improving both formulation and application technology. Formulations were altered to improve spray deposition by using spray additives (Morris et al. 1980) and to improve effectiveness by adding small amounts of the enzyme chitinase (Smirnoff et al. 1973; Smirnoff 1974; Morris 1976) or low concentrations of chemical insecticides (Morris and Armstrong 1974; Morris 1977). Various aircraft and delivery systems were tested, ranging from helicopters and single-engine biplanes equipped with rotary atomizers to four-engine aircraft fitted with boom and nozzles (Smirnoff and Morris 1982). Although effectiveness of Bt showed marked improvement during those trials, results remained inconsistent with cost of treatment up to four times higher than chemical insecticide treatment (Blais 1976).

Improvements in cost effectiveness were achieved in the late 1970s when commercial formulations became more concentrated, that is, when potencies increased from 4.2 BIU/L in early products (e.g. Thuricide® 16B) to 8.4 BIU/L (e.g. Thuricide® 32LV, Dipel® 88, Novabac® 3) (Table 1). More concentrated products reduced shipping costs (bulk transport), permitted application of lower volumes, and improved spray plane productivity (Dorais 1985). By the end

Table 1. Bt products currently registered or pending registration for aerial application in forestry.

Year	Trade Name	Type of formulation[a]	Potency (BIU/L)[b]	Registrant (supplier)
1973	Thuricide 16B	FC	4.2	Zoecon
1978	Novabac-3	FC	8.6	Biochem
1980	Dipel 88	S	8.4	Abbott
1981	Thuricide 32B	FC	8.4	Sandoz (Zoecon)
	Thuricide 32LV	FC	8.4	Sandoz (Zoecon)
1984	Thuricide 32F	FC	8.3	Sandoz (Zoecon)
	Bactospeine	FC	9.7	Duphar
	Futura	FC	14.4	Duphar (Chemagro)
	Thuricide 48LV	FC	12.7	Sandoz (Zoecon)
	Dipel 132	S	12.7	Abbott
1985	Envirobac-ES	FC	8.4	Pfizer
1988	Dipel 176 (8L)	S	16.9	Abbott
	Futura XLV	FC	14.4	Duphar (Chemagro)
	Dipel 48AF[c]	FC	12.7	Abbott
	Dipel 64AF[c]	FC	16.9	Abbott
	Foray 48B[c]	FC	12.7	Novo
	Dipel 254[c]	S	25.4	Abbott
	Futura XLV-HP[c]	FC	33.0	Duphar (Chemagro)
	Ecodart[c]	FC	16.9	CIL
	Condor[c]	FC, S	NA[d]	Ecogen

[a] Flowable concentrate (FC); emulsifiable suspension (S).
[b] Billion international units per litre.
[c] Pending registration.
[d] Expressed as concentration of toxin protein.

Source: Agriculture Canada, Pesticides Directorate, Ottawa.

of the decade, performance of Bt for spruce budworm control had improved to the point that it was recommended for use in environmentally sensitive areas (Canada–U.S. guidelines, Morris 1980), and was considered an operational alternative to chemical insecticides in Quebec (Dorais 1985). As a result, Bt was used operationally on about 100 000 ha in 1979 and 1980 (Table 2). However, variable efficacy and a three- to fourfold higher cost of treatment (Table 3) remained major obstacles to wider acceptance. Other constraints that put severe restrictions on spray program managers and undermined their confidence in Bt included reported reduced efficacy against populations exceeding 25 larvae per 45-cm branch and a window of application that was restricted to about 10 days after onset of bud flush (Dorais 1985).

Reducing the Constraints

Use of Bt in forestry accelerated rapidly in the early 1980s. By 1985, Bt had become a fully operational insecticide that was widely used for control of spruce budworm (Table 2) as well as other defoliators (Table 4). This surge in progress can be attributed to a confluence of several critical developments. The use of higher dosages and lower volumes, operational use before cost parity was achieved, better formulations, new knowledge on effective droplet sizes and dosage requirements, and corresponding improvements in application technology, all contributed to reducing the obstacles of inconsistent efficacy and high costs, while a concurrent shift in political climate favoring the use of biologicals pushed Bt into operational use before uncertainties were completely resolved.

Application Rate

Inconsistent efficacy arose partially from the use of marginal dosages and from the difficulty of obtaining adequate spray coverage. The dosage of 20 BIU in 4.7 L/ha, recommended in the 1978 Canada–U.S. guidelines (Morris 1980), was essentially a compromise between available product potency, the minimum emitted volume considered necessary to obtain adequate coverage, and the desire to reduce treatment costs. Most of the experimentation throughout the 1970s was conducted within the recommended rate of 20 BIU/ha, even though the minimum effective dosage rate had never been determined. Inadequacy of this application rate was first suspected when Fast (1976) compared the expected dose on the target foliage with estimated dose requirements for spruce budworm larvae, and was confirmed between 1979 and 1981 when the guidelines were tested in extensive field trials under the auspices of the Canada–United States Spruce Budworm Program (CANUSA) (Morris 1980, 1981; Cunningham 1985). It was not until 1981 and 1982 that field tests were designed specifically to determine optimum dosage and volume application rates (Morris 1984; Lewis et al. 1984). The general conclusion was that 20 BIU/ha was effective only under optimum conditions and against low populations, whereas 30 BIU/ha produced much more consistent results, even against populations exceeding 30 larvae per 45-cm branch (Morris 1982a; Carrow 1983; Dorais 1985).

Availability of concentrated formulations permitted application of the higher dosage in a lower volume. The initial call for concentrated products was based on the desire to reduce shipping done by Smirnoff, who developed a concentrated formulation (Futura) that could be applied at 20 BIU/ha in a final volume of 2.5 L. Trials at that rate were conducted as early as 1978 (Smirnoff 1980). Consistently good results between 1980 and 1983 (Smirnoff and Valero 1983; Smirnoff 1985) resulted in the registration of Futura® in 1984 (Table 1). Commercial formulations containing 12.7 BIU/L became available for experimental use in the early 1980s. Application of such formulations at 30 BIU/ha in 2.4 L provided excellent results (Dimond 1982) and was as effective as application in 4.7 or 9.4 L/ha (Morris 1984). These findings were reflected in

Table 2. Operational use of Bt for control of spruce budworm, *Choristoneura fumiferana*, 1979–1988: number of hectares (ha) sprayed with Bt and percentage of total area treated (%).

	Ontario		Quebec		New Brunswick		Newfoundland		Nova Scotia		Total	
Year	ha	%	ha	%	ha	%	ha	%	ha	%	ha	%
1979	3 454	17	17 030	3	205	1	5 870	100	556	100	27 115	1
1980	4 374	44	22 180	12	10 500	1	11 761	100	25 670	100	74 485	4
1981	6 576	67	15 001	2	0	0	1 900	1	31 194	100	54 671	2
1982	3 068	90	31 877	2	4 000	1	4 725	10	19 153	100	62 823	2
1983	2 763	87	45 627	4	10 300	1	0	0	20 726	100	79 416	3
1984	2 688	82	296 568	42	37 300	4	3 110	12	20 537	100	360 660	20
1985	29 370	100	512 155	73	81 000	12	3 450	100	49 719	100	675 363	45
1986	150 663	100	32 789	50	111 500	22	0	0	56 155	100	351 107	45
1987	76 689	100	197 992	100	91 300	19	0	0	31 080	100	397 191	51
1988	14 023	100	208 064	100	210 500	49	0	0	0	0	432 587	63
Total	**293 668**		**1 379 283**		**556 605**		**30 816**		**254 790**		**2 515 418**	

Sources: Carrow (1983); Annual Forest Pest Control Forum Reports, Canadian Forestry Service, Ottawa.

Table 3. Relative cost[a] of treatment with Bt and chemical insecticides in New Brunswick and Quebec, 1980–1988.

Year	Quebec			New Brunswick[b]		
	Bt	Chem.	Ratio	Bt	Chem.	Ratio
1980	31.31	8.59	3.6			
1981	27.94	6.24	4.5			
1982	18.16	5.79	3.1			
1983	23.54	7.67	3			
1984	18.63	10.16	1.8	16.38	11.52	1.4
1985	22.52	12.99	1.7	18.71	11.82	1.6
1986	17.37			16.6	11.5	1.4
1987	23.24			21.54	16.42	1.3
1988	23.02			21.3	16.98	1.2

[a] Cost of operation excluding assessment and surveys, administrative overhead, and research costs in $/ha.
[b] Small spray planes only, not including TBM.

Source: N. Carter, NB Dept. Natural Resources; L. Dorais, Quebec Department of Energy and Resources.

updated guidelines for operational use (Morris et al. 1984; Dimond and Morris 1984).

Operational Experience

Operational use before cost and efficacy were competitive with chemical insecticides catalyzed significant cost reductions and provided the experience needed to improve efficacy. From 1979 to 1983, Bt was applied as a small but consistent proportion of the spruce budworm control program (Table 2) at 20 to 30 BIU in 2.4 to 7.0 L/ha (Carrow 1983). Increased use and competitive bidding by suppliers forced the price down. The average cost per BIU dropped from $0.68 in 1980 to $0.32 in 1983 and the costs per hectare fell from $13.18 to $6.96 in spite of a general increase in dosage from 20 to 30 BIU/ha (Carrow 1983). By 1983, the differential in application costs between Bt and chemical insecticides had narrowed to a two- to threefold difference, depending on the jurisdiction. Practical experience gained from operational use between 1979 and 1983 also reduced the problem of reliability: users agreed that foliage protection at 30 BIU/ha was generally as good as that obtained with chemical insecticides (Carrow 1983). Early operational experience thus precipitated essential cost reductions while increasing confidence in the product, which opened the door for much wider use in subsequent years.

Undiluted Application

The trend of the late 1970s to increase product potency continued in the 1980s. Initial problems with handling of the highly viscous formulations were eliminated with the introduction of low-viscosity formulations. Potency of registered products increased from 8.4 BIU/L in 1980 to 12.7 BIU/L in 1984 and 16.9 BIU/L in 1988 (Table 1). The major benefit of the high-potency formulations was that they could be applied undiluted because they were designed for optimum atomization and reduced droplet evaporation. Application of undiluted formulations had yielded promising results in trials in 1981 and 1982 (Dimond 1982; Morris 1984) and was increasingly seen as the ultimate way to reduce treatment costs (Carrow 1983; Dimond and Morris 1984). The undiluted product was first used operationally in 1982 in New Brunswick on 3 200 ha (Kettela 1983, 1985), and became the standard for spruce budworm control after 1984, when formulations containing 12.7 BIU/L were approved for operational use. From then on, Bt was routinely applied at 20 or 30 BIU/ha in 1.6 or 2.4 L. Successful application of such low volumes hinged on concurrent improvements in application technology that were based on a better understanding of effective droplet sizes and dose requirements (see next section).

Undiluted high-potency products undoubtedly made the most important contribution to reducing the constraints of high cost and unreliable efficacy. The use of low-spray volumes reduced application costs by increasing spray plane productivity. Elimination of on-site mixing further increased plane productivity by allowing more loads per spray period. For example, productivity of four-engine spray planes in the Quebec spray program increased from 42 740 ha per year for application of 20 BIU in 4.7 L/ha (diluted) to 125 320 ha for undiluted application of the same dose in 1.6 L/ha (Dorais

Table 4. Operational use of Bt against defoliating forest insects, 1985–1988.

Target species	Number of hectares sprayed				Total
	1985	1986	1987	1988	
Spruce budworm	675 694	351 107	397 061	432 587	1 856 449
Jack pine budworm	248 676	482 032	105 463		836 171
Hemlock looper	2 365	5 420	4 183	23 788	35 756
Gypsy moth	170	103 094	40 249	13 784	157 297
Total	926 905	941 653	546 956	470 159	2 885 673

Source: Annual Forest Pest Control Forum Reports, Forestry Canada, Ottawa.

Table 5. Comparative efficacy of Bt and chemical insecticides for operational control of spruce budworm in Quebec, 1984–1986.

Year	Product	Dosage/ha	Volume (L/ha)	Total area (ha)	No. larvae/ 45-cm tip	% Mortality	% Reduction in defoliation
1984	Matacil	2 × 52 g a.i.[a]	1.4	141 521	7.4	77.0	43
	Thuricide 48LV	1 × 30 BIU[b]	2.4	182 346	7.4	73.0	35
1985	Fenitrothion	2 × 210 g a.i.	1.4	162 676	8.0	81.2	40
	Thuricide 48LV	1 × 30 BIU	2.4	456 639	8.8	83.0	46
1986	Fenitrothion	2 × 210 g a.i.	1.4	33 038	13.8	93.5	33
	Thuricide 48LV	1 × 30 BIU	2.4	18 160	14.4	86.1	31

[a] Grams active ingredient.
[b] Billion international units.

Source: L. Dorais, M. Auger, Quebec Department of Energy and Resources.

1985). Furthermore, handling time and costs were reduced and problems associated with instability of mixed formulations were avoided, thus facilitating overall program management. The net effect of those factors was a further narrowing of the cost differential between Bt and chemicals (Table 3). In the state of Maine, total treatment cost showed a threefold reduction between 1980 and 1985 (Irland and Rumpf 1987). The use of high-potency products also increased efficacy and reliability. In the 1984 to 1986 spray programs in Quebec, when both chemical insecticides and Bt were used on a large scale, single applications of Bt at 30 BIU/ha in 2.4 L generally performed as well as double applications of chemicals (Table 5). However, this was not the case in New Brunswick in 1988 (Carter 1988).

Effective Droplet Size and Dose Requirements

Implementation of undiluted formulations depended critically on improvements in the application of chemical insecticides, in particular the use of small droplets. In early Bt applications, spray systems were calibrated to deliver droplets in the 100- to 300-μm size range to optimize spray deposition as measured on Kromekote® cards (Morris 1980; Morris et al. 1984). Data generated by the New Brunswick Spray Efficacy Research Group in the early 1980s (Picot et al. 1985, 1986) and work on the pine beauty moth, *Panolis flammea*, in Scotland (Holden and Bevan 1978) provided overwhelming evidence that droplets less than 100 μg impinge more effectively on coniferous foliage than large droplets. The use of such small droplets requires efficient atomization of the spray formulation. For aerial application this is best achieved with rotary atomizers. Availability of nonvolatile, low-viscosity formulations permitted direct translation of this concept to the application of Bt. Rotary atomizers are capable of generating a high proportion of droplets below 100 μm from small volumes of undiluted high-potency formulations (Yates and Cowden 1986) and are the key to success of undiluted applications.

Because Bt has to be ingested to be effective, success of a spray treatment is determined by the interaction between insect feeding activity, the number of spray droplets per unit feeding area, and the dose in those droplets. The importance of droplet density has long been recognized. Spray volumes were chosen and aircraft were calibrated to obtain a deposit of at least 25 droplets per square centimetre on Kromekote® cards placed on the ground (Smirnoff and Morris 1982; Morris 1980) because lower deposits were associated with poor efficacy (Grimble and Morris 1983). The dose delivered to the target foliage was not assessed as a critical determinant of efficacy until 1981 when Fast and Sundaram (unpubl. data) explored the interaction between size, density, and toxicant concentration of spray droplets in determining spruce budworm mortality. Using a droplet size of 30 to 50 μm, they demonstrated that increasing the product concentration from 2.1 to 12.7 BIU/L increased mortality from 20% to more than 90%. The number of droplets required to obtain 50% mortality decreased from 10 droplets per needle for a 2.1 BIU/L concentration to less than 1 droplet for 8.4 BIU/L. In a subsequent field trial in 1983, a deposit of 1 droplet per needle of a 6.3 BIU/L spray mixture resulted in a 50% reduction in larval survival (Fast et al. 1985a). Work in Quebec in 1984 established that 0.5 droplets per needle of a 12.7 BIU/L formulation was required to obtain 50% foliage protection or more (Lambert 1987).

The pronounced effect of concentration on the density of spray droplets was further demonstrated in field trials in 1984 and 1985. Fast et al. (1985b, 1986) related estimates of larval survival and defoliation to density of spray droplets (mostly in the 25- to 75-μm size range) on balsam fir foliage for formulations containing 8.4 to 16.9 BIU/L after undiluted application at 30 BIU/ha. Higher potency products provided better efficacy, particularly at lower deposits. Similar relationships were established in subsequent years for various high-potency products (e.g. Wiesner and Kettela 1987). In general, formulations with 12.7 BIU/L or more provided at least 50% reduction in defoliation at a deposit of 0.5 droplets per needle, while higher droplet densities were required for less concentrated formulations. A high concentration of Bt in the spray droplets thus played a key role in the increase in efficacy associated with undiluted application.

Volume versus Concentration

The work by Fast et al. (1985b, 1986b) was significant because it demonstrated the dominant role of product potency in determining efficacy. Their findings suggest that the more concentrated the formulation is, the fewer droplets are required to deliver an efficacious dose to the target. Thus, optimum dosage requirements should be considered in terms of BIU/L rather than BIU/ha. If the potency is high enough, acceptable efficacy should be obtainable at lower application rates as long as the applied volume can be atomized into a sufficient number of droplets to deliver an efficacious dose to the target. That thinking stimulated the development of products with even higher potencies, currently up to 33.0 BIU/L (Table 1), which can be applied in volumes as low as 0.7 or 0.9 L/ha, and provided the basis for the development of the so-called enhanced atomization application technique.

Enhanced Atomization Application

Droplets 15 to 55 µm in diameter appear optimal for spruce budworm-insecticide contact on coniferous foliage (Picot et al. 1985, 1986). They also impinge effectively on the silk strands of budworm feeding shelters, which may be an important mechanism of dose transfer to early instars (Nigam 1987). Droplet volume in that size range generated by rotary atomizers can be enhanced by reducing the flow rate. Van Vliet and Picot (1987) demonstrated in wind tunnel studies that the Micronair® AU4000 atomizer produced more than 80% of the emitted volume in the 15- to 55-µm size range at flow rates below 2 L/min. Such flow rates allow maximum rotational speed and enhance atomization of the spray liquid compared with the standard flow rates of 5 to 7 L/min, thus producing more droplets in the effective size range. Enhanced atomization trials in New Brunswick in 1987 demonstrated that undiluted application of a 12.7 BIU/L formulation at 15 BIU/ha using a flow rate of 2 L/min per atomizer was as effective against spruce budworm as the standard application of 30 BIU/ha using 6.7 L/min. Availability of products containing 33 BIU/L in 1988 enabled application of 15 BIU/ha in 0.5 L. The deposition of droplets and efficacy were comparable with conventional applications of 30 BIU/ha using 5.0 L/min (E.G. Kettela, unpubl. data). Work is continuing to further test this promising technique.

Operational Use

The proportion of the area treated with Bt increased from 1% to 4% between 1979 and 1983 to 20% in 1984 and 45% to 50% in subsequent years (Table 2). This is primarily due to increasing use in Quebec, and in Ontario, following the provincial government's decisions in 1985 to stop aerial use of chemical insecticides in public forests. Chemical insecticides remained the main tool in New Brunswick until 1988 when Bt was used in almost half of the program. Although increasing public pressure against the use of chemical insecticides in forests played a key role in the switch to Bt, the switch was possible because improvements in efficacy and cost effectiveness achieved in the early 1980s gave politicians a viable, albeit more expensive, alternative.

Operational use after 1984 involved predominantly undiluted applications of formulations with 12.7 or 14.4 BIU/L at 20 or 30 BIU in 1.6 to 2.4 L/ha. The main products used were Dipel® 132, Thuricide® 48LV, and Futura® (Fig. 1). First operational use of a 16.9 BIU/L product (Dipel® 176) at 30 BIU in 1.8 L/ha occurred in 1988 in Newfoundland against hemlock looper, *Lambdina fiscellaria*. Formulations containing 25 to 33 BIU/L have been available for experimental use since 1986 and were successfully tested in volumes of 0.9 to 1.2 L/ha against the spruce budworm (Wiesner and Kettela 1987; E.G. Kettela, unpubl. data), and the hemlock looper (West et al. 1989). Super-high potency formulations thus allow application volumes that are comparable with or below those for chemical insecticides and are expected to become the mainstay of forest protection programs as soon as registrations are completed (Table 1).

Because of increasing restrictions on chemical insecticides in recent years, the use of Bt was extended to control several other defoliating species (Table 4). A recent outbreak of jack pine budworm, *Choristoneura pinus*, in Ontario and Manitoba resulted in extensive control operations between 1985 and 1987. Because laboratory bioassays indicated a similar level of susceptibility for jack pine budworm as

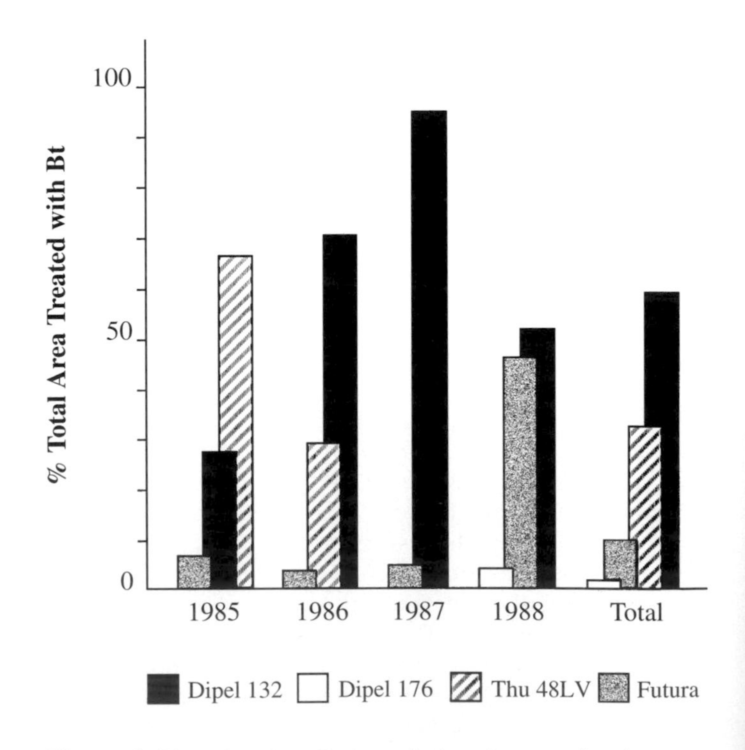

Figure 1. Use of various Bt formulations in operational spraying against defoliators in all jurisdictions, 1985–1988.

the spruce budworm (van Frankenhuyzen and Fast 1989), application of undiluted high-potency formulations was recommended and successfully implemented. Formulations with 12.7 BIU/L applied at 20 BIU in 1.6 L/ha generally provided satisfactory foliage protection.

Operational control of the gypsy moth, *Lymantria dispar*, in Ontario started in 1985 and mushroomed to 100 000 ha in 1986. High-potency formulations were diluted and applied at 30 BIU in 5 to 6 L/ha because a high volume application rate was considered necessary to obtain adequate coverage of the foliage throughout the hardwood canopy. However, studies in 1987 and 1988 demonstrated that application of undiluted Bt provided better canopy penetration and equivalent larval mortality at lower spray deposits than application of diluted Bt in higher volumes (van Frankenhuyzen et al. 1989). The 1988 program was consequently conducted with the undiluted product applied at 30 BIU/ha in 2.4 L and was the most effective gypsy moth control program conducted to date (G.M. Howse, unpubl. data).

Undiluted application of high-potency products was also successful in experimental programs against the hemlock looper in Newfoundland (West et al. 1987, 1989) and operational use increased from 2% to 6% between 1985 and 1987 to 34% in 1988. Foliage protection was generally equivalent to that of fenitrothion (Table 6). Many other defoliators are susceptible to Bt including occasional outbreak species such as forest tent caterpillar, *Malacosoma disstria*, eastern tent caterpillar, *M. americanum*, western spruce budworm, *Choristoneura occidentalis*, blackheaded budworms, *Acleris* sp., and whitemarked tussock moth, *Orgyia leucostigma*. Field tests are needed to determine the feasibility of using undiluted Bt for control of these species.

Current Constraints and Future Prospects

After almost 30 years of research and development, Bt has become a major tool in forest protection programs. Operational use of high-potency formulations applied undiluted in low volumes has reduced but not eliminated the initial constraints of high treatment costs and inconsistent efficacy. Operational costs are still higher than for chemical insecticides (Table 3), but the main drawback is that efficacy depends more on proper timing of application and favorable post-spray weather conditions, which results in a narrower window of application and less predictable efficacy. The 1988 experience in New Brunswick is a good example of this: reduction in defoliation was 61% for fenitrothion compared with 24% for Bt (Table 7). Current research to reduce these constraints is focused on the development of higher potency products with greater foliar persistence, and product improvement by the selection of more effective strains and by biotechnological manipulation.

Improved Formulations

The development of formulations with higher potency and greater residual toxicity is considered the key to short-term improvement of Bt efficacy. The benefit of such formulations is severalfold. Variable tree and insect phenology within large spray blocks make proper timing of a single spray application difficult, particularly when using Bt which has no contact toxicity and limited residual toxicity. Reliability of Bt might be increased by the adoption of a split application scheme, as is used for chemical insecticides, but high treatment costs have previously prohibited this. However, the registration of formulations in the near future with much higher potencies

Table 6. Comparative efficacy of Bt and fenitrothion for operational control of the hemlock looper in Newfoundland, 1986–1988.

Year	Product	Dosage/ha	Volume (L/ha)	Area (ha)	% area with defoliation			
					nil	light	mod.	severe
1986[a]	Fenitro[b]	2 × 210 g a.i.[c]	1.5	79 028	84	3	3	10
	Dipel 132	1 × 30 BIU[d]	2.4	5 420	76	0	0	24
1987[e]	Fenitro	2 × 210 g a.i.	1.5	164 362	84	5	3	8
	Dipel 132	1 × 30 BIU	2.4	4 183	78	6	1	15
1988	Fenitro	2 × 210 g a.i.	1.5	45 138	91	6	<1	2
	Dipel 176	1 × 30 BIU	1.8	6 473	91	7	<1	7
	Futura XLV	1 × 30 BIU	2.0	3 704	94	3	<1	3
	Dipel 176	2 × 30 BIU	1.8	1 372	100	-	-	-
	Dipel 176+	1 × 30 BIU	1.8					
	+Fenitro	1 × 210 g a.i.	1.5	10 410	97	2	<1	<1

[a] About 1.5% of area received single application of fenitrothion.
[b] Fenitrothion.
[c] Grams active ingredient.
[d] Billion international units.
[e] About 9.6% of area received double application of fenitrothion.

Source: H. Crummey, Newfoundland Department of Forest Resources and Lands.

Table 7. Comparative efficacy of Bt and fenitrothion for operational control of the eastern spruce budworm in New Brunswick, 1988.

Product	Dosage/ha	Volume (L/ha)	Total area (ha)	No. larvae/ 45-cm tip	% Mortality	% Reduction in defoliation
Fenitrothion[a]	2 × 210 g a.i.[a]	1.4	231 000	21.6	72	61
Futura XLV	1 × 30 BIU[b]	2.0	210 500	26.0	50	24

[a] Grams active ingredient.
[b] Billion international units.

Source: Carter 1988.

should permit operational use of reduced dosages at 0.5 L/ha (see previous section, "Enhanced Atomization Application"). This could reduce treatment costs to where split applications become competitive. Higher product potency is also expected to decrease the dependence on favorable weather for treatment success after application by ensuring rapid ingestion of a lethal dose (van Frankenhuyzen 1990). Another prerequisite to improving reliability of Bt efficacy is better persistence of spray deposits on the foliage. Residual toxicity of present formulations is limited to a few days, particularly under rainy conditions (van Frankenhuyzen and Nystrom 1989). Increased persistence of spray deposits will minimize adverse effects of post-spray precipitation, thereby allowing more latitude in timing of spray application.

Strain Selection and Modification

An alternative approach to reducing cost and increasing effectiveness of Bt is the commercialization of strains with specific toxicity to forest insects. This can be achieved by identifying natural isolates with the desired activity range or by enhancing natural strains using conventional genetic techniques or genetic engineering techniques.

Of the naturally occurring strains, the HD-1 isolate of current formulations is one of literally thousands of possible candidates (Martin and Travers 1989), but efforts to identify more potent strains have yielded little benefit for forest insect control to date (Morris and Moore 1983; Dubois 1985; Dubois et al. 1989). New strains are continually being isolated (Travers et al. 1987), however, and the search for more effective ones is actively pursued. One example of possible success in forestry is commercialization of the strain A20 (Biodart, Table 1), although superior performance in the field has yet to be demonstrated.

Genetic enhancement of toxicity is an approach that was used by Ecogen to produce Condor (Table 1). Toxin genes of two strains were combined by using a native conjugal plasmid transfer system to improve toxicity to spruce budworm and gypsy moth. Condor is currently undergoing field tests for registration. Condor is also an example of a new approach for standardization of product potency, based on quantification of toxin protein content. The proliferation of target-specific products is making the use of international units obsolete and cumbersome. Because activity against cabbage looper is often not meaningful, each product needs to be labelled with different international units for each target pest. Examples are the recent introduction of units for the Colorado potato beetle, *Leptinotarsa decemlineata* (Zehnder and Gelernter 1989) and Spodoptera for standardization of products developed specifically against these targets.

The great diversity of natural strains, each with a different activity spectrum, offers opportunities for the application of genetic engineering techniques. Because genes that code for crystal proteins are located on plasmids, they are highly amenable to modification by using recombinant DNA techniques. Modifications include increased toxin yield, altered range of insecticidal activity, or enhanced environmental persistence (Lüthy 1986; Wilcox et al. 1986; Whiteley and Schnepf 1986). Biotechnological improvement of Bt is enjoying worldwide commercial interest, but most efforts are focused on the larger agricultural market. In Canada, a research network was established in 1984 with the goal of tailoring an improved Bt product for forest insect control. The network, called Biocide, involves researchers from Forestry Canada, National Research Council, and several universities. Toxin genes from various strains were cloned and sequenced (Hefford et al. 1987) and researchers are now elucidating molecular determinants of toxin activity and specificity. Site-directed mutagenesis and chemical modification will then be used to manipulate these determinants to achieve maximum benefit for forestry applications. In addition, a mechanism of sunlight inactivation of the crystal protein has been established (Pozsgay et al. 1987) and ways of improving sunlight stability are being explored. Achievements to date will accelerate the development of novel, target-specific microbial insecticides based on recombinant Bt.

The future of Bt for forest insect control looks bright. Now that operational use is well established, manufacturers are increasingly confident that they face a promising market. Increasing competition (from three main producers in 1985 to seven potential contenders in 1988) is expected to accelerate product improvement and result in further cost reductions in the short term. Genetic engineering offers exciting opportunities in the long term, provided risks associated with the release of genetically engineered microorganisms (Pimentel et al. 1989) can be minimized and regulatory requirements can be worked out. After 30 years of research and development, we are just beginning to understand and exploit the real potential of Bt for control of forest insects.

References

Berliner, E. 1915. Uber die Schlaffsucht der Mehlmottenraupe (*Ephestia Kuhniella* Zell) und ihren Erreger, *Bacillus thuringiensis*. Z. Angew. Entomol. 2: 29-56.

Blais, J.R. 1976. Can *Bacillus thuringiensis* replace chemical insecticides in the control of spruce budworm? The Forestry Chronicle 52: 57-60.

Bonnefoi, A.; Burgerjon, A.; Grison, P. 1958. Titrage biologique des préparations de spores de *Bacillus thuringiensis*. C.R. Acad. Sci. 247: 1418-1420.

Burges, H.D. 1967. The standardization of products based on *Bacillus thuringiensis*. Pages 306-338 *in* Proc. Int. Colloq. Insect Pathol. and Microb. Control, Wageningen, the Netherlands, 1966. North Holland Publ. Co., Amsterdam.

Burgerjon, A. 1959. Titrage et définition d'une unité biologique pour les préparations de *Bacillus thuringiensis* Berliner. Entomophaga 4: 201-206.

Carrow, J.R. 1983. Bt and the spruce budworm -1983. New Brunswick Department of Natural Resources, Fredericton, N.B.

Carter, N. 1988. Protection spraying against spruce budworm in New Brunswick, 1988. Department of Natural Resources and Energy, Fredericton, N.B.

Cunningham, J.C. 1985. Biorationals for control of spruce budworms. Pages 320-349 *in* Sanders et al. (Eds.), Recent advances in spruce budworms research. Proc. CANUSA Spruce Budworms Research Symposium, Bangor, Me., Sept. 1984. Canadian Forestry Service, Ottawa, Ont.

Delucca, A.J.; Simonson, J.G.; Larson, A.D., 1981. *Bacillus thuringiensis* distribution in soils of the United States. Can. J. Microbiol. 27: 865-870.

Dimond, J.B. 1982. Effects of aerial sprays of undiluted *Bacillus thuringiensis* formulations on spruce budworm. Maine Life Sci. Agric. Exp. Station, Univ. Maine at Orono. Misc. Rep. No. 274.

Dimond, J.B.; Morris, O.N.; 1984. Microbial and other biological control. Pages 104-114 *in* Spruce budworms handbook: Managing the srpuce budworm in eastern Northern America. USDA Forest Service, Agriculture Handbook No. 620, Washington, D.C.

Dorais, L. 1985. Four-engine aircraft experience in the application of *Bacillus thuringiensis* against the spruce budworm in Quebec. Pages 13-15 *in* Grimble, D.G., and F.B. Lewis (Eds.), Symp. Proc. Microbial Control of Spruce Budworms and Gypsy Moths. Windsor Locks, Conn. April 1984. USDA For. Serv., Broomall, Pa. GTR-NE-100.

Dubois, N.R. 1985. Selection of new more potent strains of *Bacillus thuringiensis* for use against gypsy moth and spruce budworm. Pages 99-102 *in* Grimble, D.G. and F.B. Lewis (Eds.), Symp. Proc. Microbial Control of Spruce Budworms and Gypsy Moths. Windsor Locks, Conn. April 1984. USDA For. Serv. Broomall, Pa. GTR-NE-100.

Dubois, N.R.; Huntley, P.J.; Newman, D. 1989. Potency of *Bacillus thuringiensis* against gypsy moth and spruce budworm larvae: 1980-86. USDA Forest Service, Northeastern Forest Experiment Station, Broomall, Pa. GTR-NE-131.

Dulmage, H.T. 1970. Insecticidal activity of HD-1, a new isolate of *Bacillus thuringiensis* var. *alesti*. J. Invert. Pathol. 15: 232-239.

Dulmage, H.T. 1981. Insecticidal activity of isolates of *Bacillus thuringiensis* and their potential for pest control. Pages 193-222 *in* Burges, H.D. (Ed.), Microbial Control of Pests and Plant Diseases 1970-1980. Academic Press, London.

Dulmage, H.T.; Boening, O.P.; Rehnborg, C.S.; Hansen, G.D. 1971. A proposed standardized bioassay for formulations of *Bacillus thuringiensis* based on the International Unit. J. Invert. Pathol. 18: 240-246.

Dulmage, H.T.; Rhodes, R.A. 1971. Production of pathogens in artificial media. Pages 507-540 *in* Burges, H.D. and N.W. Hussey (Eds.), Microbial control of insects and mites. Academic Press, London.

Fast, P.G. 1976. Some calculations relevant to field applications of *Bacillus thuringiensis*. Can. For. Serv., Ottawa. Bi-monthly Res. Notes 32: 21.

Fast, P.G.; Régnière, J. 1984. Effect of exposure time to *Bacillus thuringiensis* on mortality and recovery of the spruce budworm (Lepidoptera: Tortricidae). Can. Entomol. 116: 123-130.

Fast, P.G.; Kettela, E.G.; Wiesner, C.J. 1985a. Measurement of foliar deposits of Bt and their relation to efficacy. Pages 148-149 *in* D.G. Grimble and F.L. Lewis (Eds.), Symp. Proc. Microbial Control of Spruce Budworms and Gypsy Moths. Windsor Locks, Conn. April 1984. USDA For. Serv., Broomall, Pa. GTR-NE-100.

Fast, P.G.; Kettela, E.G.; Wiesner, C.J. 1985b. Assessment of the influence of concentration and foliar deposition on the efficacy of *Bacillus thuringiensis*. New Brunswick Research and Productivity Council Report No. C/85/047, Fredericton, N.B.

Fast, P.G.; Kettela, E.G.; Wiesner, C.J. 1986. Assessment of efficacy of *Bacillus thuringiensis* against the spruce budworm. New Brunswick Research and Productivity Council Report No. C/86/005, Fredericton, N.B.

Grimble, D.G.; Morris, O.N. 1983. Regional evaluation of Bt for spruce budworm control. Agric. Info. Bull. 458, USDA For. Ser., Washington, D.C.

Hall, I.M. 1963. Microbial control. Pages 477-517 *in* Steinhaus, E.A. (Ed.), Insect pathology, an advanced treatise. Academic Press, New York.

Hefford, M.; Brousseau, R.; Prefontaine, G.; Hanna, Z.; Condie, J.; Lau, P.C.K. 1987. Sequence of a Lepidopteran toxin gene of *Bacillus thuringiensis* subsp. *kurstake* NRD-12. J. Biotech. 6: 307-322.

Heimpel, A.M.; Angus, T.A. 1959. The site of action of crystalliferous bacteria in Lepidoptera larvae. J. Insect Pathol. 1: 152-170.

Holden, A.V.; Bevan, D. 1978. Control of pine beauty moth by fenitrothion in Scotland, 1978. Forestry Commission Report, Farnham, Surrey, England.

Irland, L.C.; Rumpf, T.A. 1987. Cost trends for *Bacillus thuringiensis* in the Maine spruce budworm control program. Bull. Ent. Soc. Am. 33: 86-90.

Kettela, E.G. 1983. Operational use of Bt in New Brunswick. Part 2. Efficacy. Pages 46-53 *in* Carrow, J.R. (Ed.), Bt and the spruce budworm - 1983. New Brunswick Dept. Nat. Resource, Fredericton, N.B.

Kettela, E.G. 1985. Review of foliage protection spray operations against the spruce budworm with *Bacillus thuringiensis kurstaki* from 1980 to 1983 in Nova Scotia and New Brunswick, Canada. Pages 19-22 *in* Grimble, D.C. and F.B. Lewis (Eds.), Symp. Proc. Microbial Control of Spruce Budworms and Gypsy Moths. Windsor Locks, Conn. April 1984. USDA For. Serv., Broomall, Pa. GTR-NE-100.

Kinghorn, J.M.; Fisher, R.A.; Angus, T.A.; Heimpel, A.M. 1961. Aerial spray trials against the black headed budworm in British Columbia. Department of Forestry, Ottawa, Ont. Bimonthly Progress Report 17: 3-4.

Krieg, A.; Langenbruch, G.A. 1981. Susceptibility of arthropod species to *Bacillus thuringiensis*. Pages 837-896 *in* Burges, H.D. (Ed.), Microbial Control of Pests and Plant Diseases 1970-1980. Academic Press, London.

Lambert, M. 1987. Quantification of spray deposits in experimental and operational aerial spraying operations. Pages 125-129 *in* Green, G.W. (Ed.), Proc. Symp. Aerial Application of Pesticides in Forestry. Assoc. Committee on Agriculture and Forestry Aviation, NRC, Ottawa, Ont. AFA-TN-18, NRC No. 29197.

Lewis, F.B.; Walton, G.S.; Dimond, J.B.; Morris, O.N.; Parker, B.; Reardon, R.C. 1984. Aerial application of Bt against spruce budworm: 1982 Bt cooperative field tests-combined summary. J. Econ. Entomol. 77: 999-1003.

Lüthy, P. 1986. Genetics and aspects of genetic manipulation of *Bacillus thuringiensis*. Pages 97-110 *in* Krieg, A. and A.M. Huger (Eds.), Symposium in memorium Dr. Enrst Berliner anlasslich des 75. Jahrestage der Erstbeschreibung von *Bacillus thuringiensis*. Mitt. Aus der Biologischen Bundesanstalt fur Land- und Forstwirtschaft. Heft 223, Berlin-Dahlem.

Martin, P.A.W.; Travers, R.S. 1989. Worldwide abundance and distribution of *Bacillus thuringiensis* isolates. Appl. Env. Microbiol. 55: 2437-2442.

Morris, O.N. 1976. A two-year study of the efficacy of Bt-chitinase combinations in spruce budworm control. Can. Entomol. 108: 225-233.

Morris, O.N. 1977. Long term study of the effectiveness of aerial application of *Bacillus thuringiensis*-acephate combinations against the spruce budworm, *Choristoneura fumiferana* (Lepidoptera: Tortricidae). Can. Entomol. 109: 1239-1248.

Morris, O.N. 1980. Report of the 1979 CANUSA cooperative *Bacillus thuringiensis* spray trials. Canadian Forestry Service, Sault Ste. Marie, Ont. Inf. Rep. FPM-X-40.

Morris, O.N. 1981. Report of the 1980 cooperative *Bacillus thuringiensis* spray trials. Canadian Forestry Service, Sault Ste. Marie, Ont. Inf. Rep. FPM-X-48.

Morris, O.N. 1982a. Report of the 1981 cooperative *Bacillus thuringiensis* spray trials. Canadian Forestry Service, Sault Ste. Marie, Ont. Inf. Rep. FPM-X-58.

Morris, O.N. 1982b. Bacteria as pesticides: forest applications. Pages 239-287 *in* Kurstak, E. (Ed.). Microbial and viral pesticides. Marcel Dekker, New York.

Morris, O.N. 1984. Field response of the spruce budworm, *Choristoneura fumiferana* (Lepidoptera: Tortricidae) to dosage and volume application rates of commercial *Bacillus thuringiensis*. Can. Entomol. 116: 983-990.

Morris, O.N.; Angus, T.A.; Smirnoff, W.A. 1975. Field trials of *Bacillus thuringiensis* against the spruce budworm, 1960-1973. Pages 129-133 *in* Prebble, M.L. (Ed.), Aerial control of forest insects in Canada, Dept. of Environment, Ottawa, Ont.

Morris, O.N.; Armstrong, J.A. 1974. Aerial application of *Bacillus thuringiensis*-Orthene combinations against the spruce budworm. Canadian Forestry Service, Ottawa, Ont. Inf. Rep. CC-X-71.

Morris, O.N.; Dimond, J.B.; Lewis, F.B. 1984. Guidelines for operational use of *Bacillus thuringiensis* against the spruce budworm. USDA Forest Service, Agriculture Handbook No. 621, Washington, D.C.

Morris, O.N.; Hildebrand, M.J.; Armstrong, J.A. 1980. Preliminary field studies on the use of additives to improve deposition rate and efficacy of commercial formulations of *Bacillus thuringiensis* applied against the spruce budworm. Canadian Forestry Service, Sault Ste. Marie, Ont. Inf. Rep. FPM-X-32.

Morris, O.N.; Moore, A. 1983. Relative potencies of 50 isolates of *Bacillus thuringiensis* for larvae of the spruce budworm, *Choristoneura fumiferana* (Lepidoptera: Tortricidae). Can. Entomol. 115: 815-822.

Mott, D.G.; Angus, T.A.; Heimpel, A.M.; Fisher, R.A. 1961. Aerial application of Thuricide against the spruce budworm in New Brunswick. Department of Forestry, Ottawa, Ont. Bi-monthly Progress Report 17: 2.

Nigam, P.C. 1987. Dose transfer and spruce budworm behaviour during operational application of fenitrothion. Pages 281-284 *in* Green, G.W. (Ed.), Proc. Symp. Aerial Application of Pesticides in Forestry Assoc. Committee on Agriculture and Forestry Aviation, NRC, Ottawa, Ont. AFA-TN-18, NRC No. 29197.

Percy, J.; Fast, P.G. 1983. *Bacillus thuringiensis* crystal toxin: ultrastructural studies of its effect on silkworm midgut cells. J. Invert. Pathol. 41: 86-98.

Picot, J.J.C.; Bontemps, X; Kristmanson, D.D. 1985. Measuring spray atomizer droplet spectrum down to 0.5 μm size. Trans. ASAE 28: 1367-1370.

Picot, J.J.C.; Kristmanson, D.D.; Basak-Brown, N. 1986. Canopy deposit and off-target drift in forestry aerial spraying: the effects of operational parameters. Trans. ASAE 29: 90-96.

Pimentel, D.; Hunter, M.S.; LaGro, J.A.; Efroymson, R.A., 1989. Benefits and risks of genetic engineering in agriculture. BioScience 39: 606-614.

Pozsgay, M.; Fast, P.; Kaplan, H.; Carey, P.R. 1987. The effect of sunlight on the protein crystals from *Bacillus thuringiensis* varieties kurstaki HD-1 and NRD-12: a Raman spectroscopic study. J. Invert. Path. 50: 246-253.

Retnakaran, A.; Lauzon, H.; Fast, P.G. 1983. *Bacillus thuringiensis* induced anorexia in the spruce budworm, *Choristoneura fumiferana*. Ent. exp. et appl. 34: 233-239.

Smirnoff, W.A. 1974. Three years of aerial field experiments with *Bacillus thuringiensis* plus chitinase formulation against the spruce budworm. J. Invert. Pathol. 24: 344-348.

Smirnoff, W.A. 1980. Deposit assessment of *Bacillus thuringiensis* formulations applied from an aircraft. Can. J. Microbiol. 26: 1364-1366.

Smirnoff, W.A. 1985. Field tests with a highly concentrated *Bacillus thuringiensis* formula against spruce budworm. Pages 55-59 *in* Grimble, D.G. and F.B. Lewis (Ed.), Symp. Proc. Microbial Control of Spruce Budworms and Gypsy Moths. Windsor Locks, Conn. April 1984. USDA For. Serv., Broomall, Pa. GTR-NE-100.

Smirnoff, W.A.; Fettes, J.J.; Desaulniers, R. 1973. Aerial spraying of a *Bacillus thuringiensis*-chitinase formulation for control of the spruce budworm. Can. Entomol. 105: 1535-1544.

Smirnoff, W.A.; Morris, O.N. 1982. Field development of *Bacillus thuringiensis* in eastern Canada. Pages 238-247 *in* Kelleher, J.S. and M.A. Hulme (Eds.), Biological control programmes against insect and weeds in Canada 1969-1980. Commonwealth Agricultural Bureaux, Slough, England.

Smirnoff, W.A.; Valero, J.R. 1983. Characteristics of a highly concentrated *Bacillus thuringiensis* formulation against spruce budworm, *Choristoneura fumiferana* (Lepidoptera: Tortricidae). Can. Entomol. 115: 443-444.

Travers, R.S.; Martin, P.A.; Reichelderfer, C.F. 1987. Selective process for efficient isolation of soil Bacillus spp. Appl. Env. Microbiol. 53: 1263-1266.

van Frankenhuyzen, K. 1990. Effect of temperature and exposure time on toxicity of *Bacillus thuringiensis* spray deposits to spruce budworm, *Choristoneura fumiferana* (Lepidoptera: Tortricidae). Can. Entomol. 122: 69-75.

van Frankenhuyzen, K.; Fast, P.G. 1989. Susceptibility of three coniferophagous Choristoneura species (Lepidoptera: Tortricidae) to *Bacillus thuringiensis* var. *kurstaki*. J. Econ. Entomol. 82: 193-196.

van Frankenhuyzen, K.; Howard, C.; Churcher, J.; Howse, G.; Lawrence, D. 1989. Deposition and efficacy of Dipel 8AF applied diluted and undiluted against the gypsy moth in southeastern Ontario. Forestry Canada, Forest Pest Management Institute, Sault Ste. Marie, Ont. Info. Rep. FPM-X-84.

van Frankenhuyzen, K.; Nystrom, C.W. 1987. Effect of temperature on mortality and recovery of spruce budworm (Lepidoptera: Tortricidae) exposed to *Bacillus thuringiensis*. Can. Entomol. 119: 941-954.

van Frankenhuyzen, K.; Nystrom, C.W. 1989. Residual toxicity of a high-potency formulation of *Bacillus thuringiensis* to spruce budworm (Lepidoptera: Tortricidae). J. Econ. Entomol. 82: 868-872.

van Vliet, M.W.; Picot, J.J.C. 1987. Drop spectrum characterization for the Micronair AU4000 aerial spray atomizer. Atom. Spray Technol. 3: 123-134.

West, R.J.; Raske, A.G.; Retnakaran, A.; Lim, K.P. 1987. Efficacy of various *Bacillus thuringiensis* var. *kurstaki* formulations and dosages in the field against the hemlock looper, *Lambdina fiscellaria fiscellaria* (Guen.) (Lepidoptera: Geometridae) in Newfoundland. Can. Entomol. 119: 449-458.

West, R.J.; Raske, A.G.; Sundaram, A. 1989. Efficacy of oil-based formulations of *Bacillus thuringiensis* var. *kurstaki* against hemlock looper, *Lambdina fiscellaria fiscellaria* (Guen.) (Lepidoptera: Geometridae). Can. Entomol. 121: 55-63.

Whitely, H.R.; Schnepf, H.E. 1986. The moleculecular of parasporal crystal body formation in *Bacillus thuringiensis*. Ann. Rev. Microbiol. 40: 549-576.

Wiesner, C.J.; Kettela, E.G. 1987. 1986 Dipel deposit-efficacy studies and bioassay validation. New Brunswick Research and Productivity Council Report C/86/110, Fredericton, N.B.

Wilcox, D.R.; Shivakumar, A.G.; Melin, B.E. 1986. Genetic engineering of bioinsecticides. Pages 395-413 *in* Inouye, M. and R. Sarma (Eds.), Protein Engineering: Applications in Science, Medicine, and Industry. Academic Press, Inc.

Yates, W.E.; Cowden, R.E. 1986. Drop size spectra for Beecomist, Micronair, and 8006 flat fan nozzle with Dipel 8L and Thuricide 48LV. USDA Forest Service, Forest Pest Management, Davis, Ca. FPM 86-4.

Zehnder, G.W.; Gelernter, W.D. 1989. Activity of the M-One formulation of a new strain of *Bacillus thuringiensis* against the Colorado potato beetle (Coleoptera: Chrysomelidae): Relationship between susceptibility and insect life stage. J. Econ. Entomol. 82: 756-761.

Chapter 35

Insect Viruses

J.C. Cunningham and W.J. Kaupp

Introduction

Several groups of viruses isolated from insects are recorded by the International Committee on Taxonomy of Viruses (Matthews 1982); these have been summarized by Evans and Entwistle (1987) who describe 15 distinct groups or sub-groups. For forest pest control in Canada, research has been limited to viruses that have inclusion bodies. These are protein bodies in which the virus particles are occluded; when ingested by susceptible insect larvae, the inclusion bodies dissolve in the alkaline gut juice and release infectious virus particles which penetrate gut cells and initiate the cycle of infection. Of particular interest are Baculoviridae sub-group A, better known as nuclear polyhedrosis viruses (NPV) and sub-group B, known as granulosis viruses (GV) (Figs. 1a and 1b). Research has also been conducted using cytoplasmic polyhedrosis viruses (CPV) which are classified as Reoviridae and entomopoxviruses (EPV) classified as Poxviridae. Inclusion bodies range in size from about 0.5 µm to 10 µm; they can be detected and enumerated using a light microscope.

Interest in viruses for biological control of forest insect pests in Canada was triggered in the late 1930s by the discovery that a polyhedrosis virus, introduced accidentally from Europe along with imported parasites, was largely responsible for the collapse of a major outbreak of European spruce sawfly, *Gilpinia hercyniae*, (Balch and Bird 1944; Cameron 1975a). At that time, European spruce sawfly was the most important forest insect pest in Canada. Following the collapse of the European spruce sawfly outbreak, spruce budworm, *Choristoneura fumiferana*, ranked first as the economically most important forest insect pest. Attention has now been focused on it for about four decades. It was hoped that a virus could be used to terminate outbreaks in the same manner as the one that terminated the outbreak of the European spruce sawfly. Although several viruses have been found to infect spruce budworm, their artificial dissemination has, to date, failed to initiate the spectacular epizootics observed in European spruce sawfly populations.

All insect viruses must be ingested to cause infection. NPVs and GVs are highly host specific; in some instances one virus will infect several species in the same genus, but this is not always the case and it is usually necessary to develop one virus for each major insect pest. This host specificity is highly advantageous from an environmental standpoint, but it can deter commercialization of viral insecticides. All the research and development of viral insecticides in Canada have taken place in government laboratories.

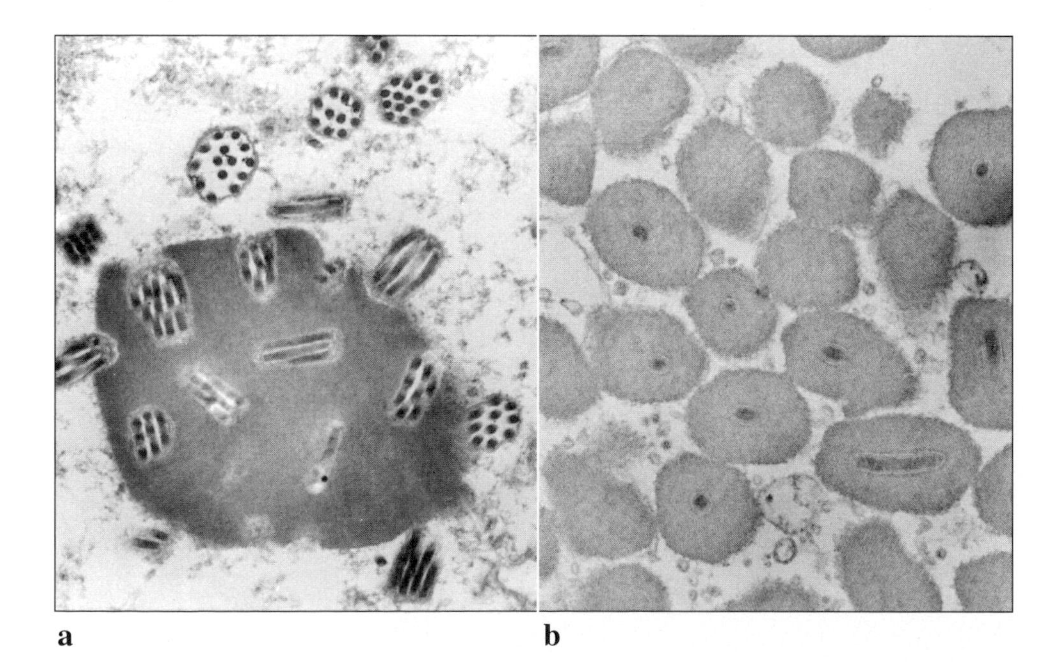

Figure 1. (a) Ultrathin section of the formation of an inclusion body of a nuclear polyhedrosis virus that has bundles of virus particles occluded in the protein. × 29 000; (b) ultrathin section of granulosis virus inclusion bodies that contain one virus particle per occlusion. × 45 000.

Research designed to elucidate the interrelationships between the insect population, the virus, and the environment is of paramount importance for effective use of viruses as biocontrol agents (Entwistle 1986). Epizootiological studies have indicated that use strategies and application methods, which increase foliar persistence of spray deposits of viral insecticides and increase the quantity of virus inoculum produced by virus-killed larvae in epizootics, have the highest probability of increasing the effectiveness of a viral pathogen. Persistence of the virus in the host population, the rate of horizontal transmission (from insect to insect), and the probability of vertical transmission (from generation to generation) are influenced in part by these two factors, especially where forest insects are concerned.

Ideally, a virus should be sprayed soon after all eggs have hatched and larvae are actively feeding because small larvae are more susceptible than larger larvae. NPVs and GVs are slow acting under field conditions and take 7 to 20 days to kill host insect larvae. When these small larvae die, their bodies release more virus inoculum onto the host plant. If this inoculum remains viable for sufficient time to be ingested by other larvae, horizontal transmission occurs and a secondary cycle will begin. Various biotic factors, such as birds and predacious, parasitic and scavenging insects, and abiotic factors, such as wind and rain, can disperse the virus inoculum. These factors either increase horizontal transmission or impede it by removing virus from the foliage into the soil. Although virus in the soil is virtually eliminated from the host insect's environment, it can remain viable for many years and may be an important reservoir for virus inoculum between insect outbreaks (Thompson et al. 1981).

Vertical transmission of virus can occur by several methods. Virus-killed larvae and pupae can remain over winter on trees and contain active virus. Virus also overwinters in such protected places as needle bases of conifers and in cracks and crevices in bark (Kaupp 1983). Larvae of the next generation are infected when they ingest foliage contaminated with virus leached from these reservoirs. Sub-lethally infected adults can also contaminate egg surfaces and foliage during oviposition and eggs may be contaminated by rain splash from environmental reservoirs of virus.

This chapter deals with various stages in the development of viral insecticides. The current status of viruses for forest insect pest control in Canada is listed in Tables 1 and 2. Most attention is focused on developments between 1973 and the present. Insect species are discussed individually; some are

Table 1. Viruses investigated in Canada for control of lepidopterous forest insect pests.

Species	Viruses tested[a]	Ground/ aerial spray[b]	Registration status
Spruce budworm *Choristoneura fumiferna*	NPV, GV CPV, EPV	G & A	
Western spruce budworm *C. occidentalis*	NPV, GV	G & A	
Jack pine budworm *C. pinus*	NPV	A	
Douglas-fir tussock moth *Orgyia pseudotsugata*	NPV	G & A	Virtuss® & TM BioControl-1 registered 1983
White-marked tussock moth *O. leucostigma*	NPV	G & A	
Gypsy moth *Lymantria dispar*	NPV	A	Disparvirus submitted 1990
Forest tent caterpillar *Malacosoma disstria*	NPV, CPV	G & A	
Bruce spanworm *Operophtera bruceata*	NPV	G	
Winter moth *O. brumata*	NPV	G	
Hemlock looper *Lambdina fiscellaria*	NPV	G	
Black army cutworm *Actebia fennica*	NPV	G	

[a]Nuclear polyhedrosis viruses (NPV); granulosis viruses (GV); cytoplasmic polyhedrosis viruses (CPV); and entomopoxviruses (EPV).
[b]Ground (G) and/or aerial (A) spray applications.

Table 2. Nuclear polyhedrosis viruses investigated in Canada for control of hymenopterous forest insect pests.

Species	Ground/ aerial spray[a]	Registration status
European spruce sawfly *Diprion hercyniae*	G	
European pine sawfly *Neodiprion sertifer*	G & A	Sertifervirus submitted 1985
Redheaded pine sawfly *N. lecontei*	G & A	Lecontvirus registered 1983
Balsam fir sawfly *N. abietis* complex	G	
Swaine jack pine sawfly *N. swainei*	G & A	
Red pine sawfly *N. nanulus*	G	
Jack pine sawfly *N. pratti banksianae*	G	
Redheaded jack pine sawfly *N. rugifrons*	G	

[a]Ground (G) and/or aerial (A) spray applications.

dealt with in greater detail in other sections of this book. When a promising viral insecticide is identified it must be produced in sufficient quantity to conduct a field trial. If a viral insecticide is shown to be effective and economical, the next step is its registration under the Pest Control Products Act (Canada). The final step should be technology transfer and commercialization, but this has yet to be achieved in Canada.

Virus Production

Viruses replicate in living cells; they can only be mass produced in susceptible cell cultures or in insect larvae. The former method is technically feasible for some insect/virus systems (Fig. 2) but, at present, the cost is too high for production of viruses to be used in field trials. The price of a medium to support cell growth is the main cost factor and over the past two decades extensive research in industrial, government, and university laboratories has made considerable progress toward alleviating this problem. The second method, using insect larvae, is the only practical means currently available for propagating insect viruses in the quantities required for field trials.

Few companies wish to get involved with rearing insects, and commercialization of insect viruses has made little progress compared to *Bacillus thuringiensis* (Bt), which can be grown in liquid culture in fermenters. At the Forest Pest Management Institute, henceforth referred to as the Institute, several viruses have been produced in sufficient quantities to treat areas ranging from one hundred to several thousand hectares. These include spruce budworm NPV, GV and EPV; Douglas-fir tussock moth, *Orgyia pseudotsugata*, NPV and gypsy moth, *Lymantria dispar*, NPV. Production in insect larvae reared on artificial diet is labor-intensive, although it is possible that the handling of some insect systems can be mechanized to a certain extent. All operations involve processing large numbers of larvae, which are reared to the correct size for infection, transferred to diet treated with virus, harvested at the appropriate time when either dead or heavily infected and then frozen. Further processing at the Institute involves lyophilization of the virus-infected cadavers and grinding this material to a fine powder suitable for formulating in a liquid to be applied with aerial or ground spray equipment.

Entirely different techniques are used to produce NPVs of colonial species of diprionid sawflies (Cunningham and McPhee 1986). There are no artificial diets for sawflies; they have to be reared on fresh foliage in the laboratory. For large-scale virus production, laboratory rearing is out of the question. Hence, these viruses are produced by spraying heavily infested plantations with virus when larvae are mainly in their fourth instar and later clipping foliage with dead and diseased colonies. Dead larvae are picked off the foliage and processed in the same manner as laboratory-produced virus material.

All viral insecticides are bioassayed to determine their potency and checked for the presence of undesirable bacterial contaminants (Podgwaite and Bruen 1978). Cost of viral insecticides basically depends on dosage, which in turn depends on the number of larvae which have to be reared, infected, harvested, and processed to treat a given area. The most economical are sawfly viruses, such as the NPVs of European pine sawfly, *Neodiprion sertifer*, and redheaded pine sawfly, *N. lecontei*, produced in plantations; about 50 virus-infected larvae are required to treat 1 ha at a cost of about $2.50. Laboratory-produced viral insecticides require considerably more handling than field-produced sawfly viruses and are consequently more expensive.

Virus Registration

In the 1950s and 1960s, little thought was given to registration of viruses under the Pest Control Products Act because no material was sold, and viral insecticides were generally regarded as safe. However, in the 1970s, guidelines for safety testing and registration were issued by several agencies. The most detailed were those produced by the American Environmental Protection Agency for registration of biorational pesticides and their guidelines have generally been followed in Canada. New Canadian guidelines have been drafted for microbial pesticides and are now under review. The only registrations for viral insecticides in Canada are those held by the Institute. Temporary registrations were obtained for Virtuss (Douglas-fir tussock moth NPV) and Lecontvirus (redheaded pine sawfly NPV) in 1983 and those were upgraded to full registrations in 1987. At that time, a third viral insecticide, TM BioControl-1 (a USDA Forest Service Douglas-fir tussock moth NPV product), also got full registration in Canada.

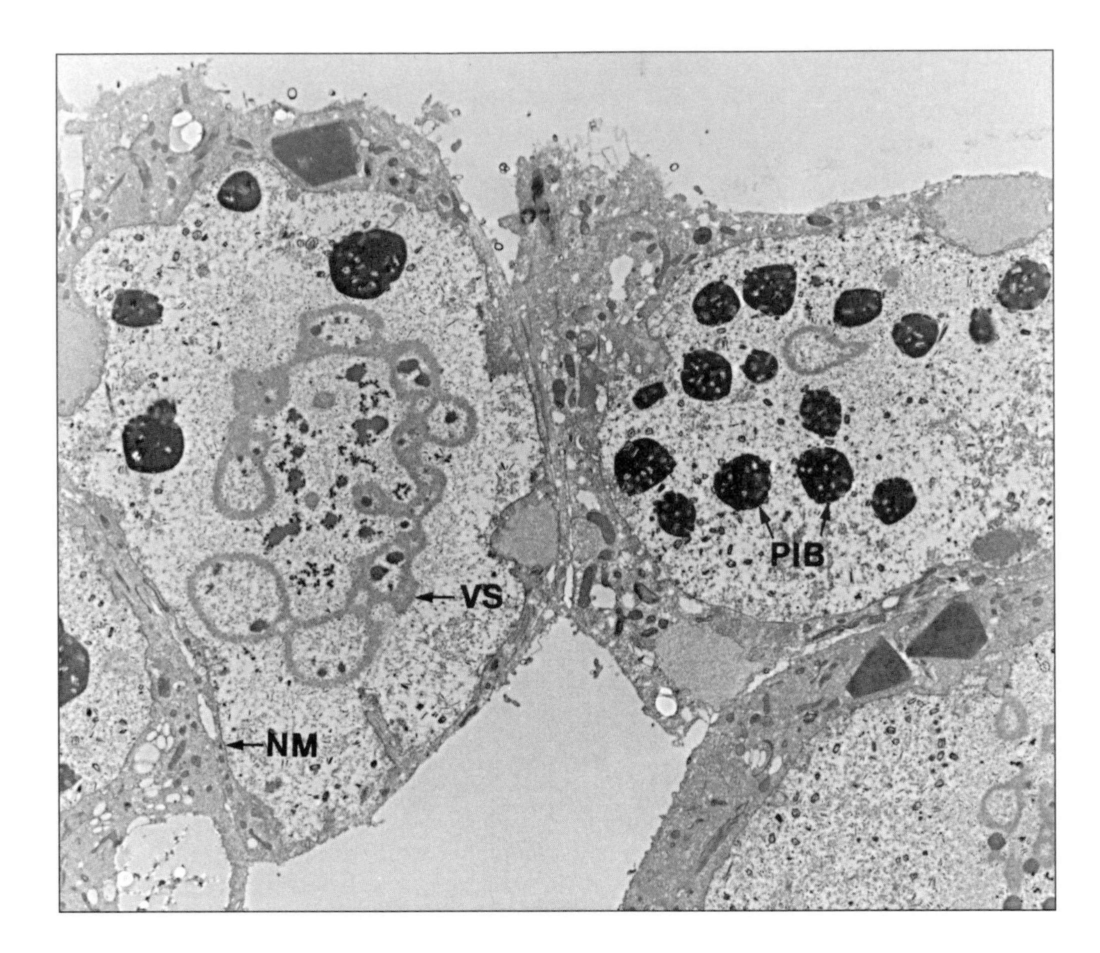

Figure 2. Spruce budworm nuclear polyhedrosis virus growing in cell culture. VS = virogenic stoma where virus particles are formed. PIB = polyhedral inclusion bodies. NM = nuclear membrane. × 4 860.

Safety testing of viral insecticides in Canada began in 1974 when registration of spruce budworm NPV was contemplated. Mammalian toxicology studies and fish and wildlife testing for this NPV were undertaken, but its efficacy in field trials did not merit further development and registration. Following spruce budworm NPV toxicology tests, a series of contracts was arranged for the safety testing of redheaded pine sawfly NPV. While this work was being undertaken in Canada, the USDA Forest Service funded toxicology tests of Douglas-fir tussock moth NPV, gypsy moth NPV, and European pine sawfly NPV. These three viral insecticides, named TM BioControl-1, Gypchek, and Neochek-S obtained registrations in the United States from the Environmental Protection Agency in 1976, 1978 and 1983, respectively. American authorities made their registration petitions available to Forestry Canada; their toxicology data were used in the Virtuss® registration petition, and TM BioControl-1 itself was registered in Canada to facilitate importation of the American product by the British Columbia Forest Service. The Neochek-S data package was reworked into a Canadian petition for Sertifervirus, which was submitted in 1985 and is being evaluated. The Gypchek data package was reworked for a Canadian gypsy moth NPV product called Disparvirus and submitted in 1990 to Agriculture Canada for evaluation.

Research on Insect Viruses, 1974 to 1988

Spruce Budworm

Because of the economic importance of spruce budworm, considerably more research has been conducted on viruses infecting this species than any other Canadian forest insect pest. From 1971 to 1973, 28 plots with a total area of 1 130 ha were treated with NPV, CPV, and EPV (Cameron 1975b). It was found that contrary to expectations based on laboratory studies, NPV had considerably greater potential for spruce budworm control than EPV, so research was concentrated on NPV (Cunningham and Howse 1984; Cunningham 1985). Subsequently, except for 2 plots with a combined area of 6 ha treated with GV, all trials were conducted using NPV and between 1974 and 1983, 37 plots with a total area of 1 510 ha were treated. Dosages of NPV have ranged from 2.5×10^{10} PIB/ha (polyhedral inclusion bodies per hectare) to a double treatment with 3.4×10^{12} PIB/ha followed by 2.3×10^{12} PIB/ha, although the dosage most commonly applied has been 7.5×10^{11} PIB/ha, which, in some tests, gave satisfactory levels of population reduction but negligible foliage protection (Cunningham and Howse 1984). The LD_{50} dosages of NPV on third-, fourth-, fifth-, and sixth-instar budworm larvae were established as 455, 1 141, 13 016, and 36 544 PIBs, respectively (Kaupp and Ebling 1990). Most sprays were applied on fourth- or fifth-instar larvae at budflush, but four attempts were made to treat second-instar larvae as they emerged from hibernacula and mined into needles and buds. Emitted volumes have ranged from 4.5 to 28.2 L/ha, with most applications at 9.4 L/ha. Both boom and nozzle and Micronair® rotary spray atomizers have been used and a variety of tank mixes and stickers tested. Most tank mixes have been aqueous and contained 25% v/v molasses, but an emulsifiable oil was also tested. A few trials have involved application of chemical insecticides, either fenitrothion or acephate, followed by a virus.

The aim of most of these spruce budworm virus trials was to initiate an epizootic that would eventually regulate the insect population; naturally occurring virus epizootics have never been observed to terminate a spruce budworm outbreak. Following the first aerial application of NPV in 1971 on two isolated mature white spruce, *Picea glauca*, stands, insect populations and incidence of virus infection were monitored until 1977 when the insect population declined to a very low level. Over this period, the NPV had a considerable impact on the spruce budworm population, and because reinvasion of these isolated stands did not occur, reduced defoliation due to lower insect population densities prevented tree mortality (Cunningham et al. 1975; Cunningham and Howse 1984). Subsequent efforts to repeat this experiment were unsuccessful. Although NPV has been detected at low levels in the year following the year of application, the impact of vertically transmitted virus was always much less than the actual spray application.

The most recent field trial with spruce budworm NPV was conducted in Quebec in 1983 (Kaupp et al. 1990). Field trials with spruce budworm viruses will continue and a search is being made for a new strain of virus that will initiate and maintain an epizootic. Genetic engineering of spruce budworm NPV is a high research priority at the Institute and attempts are being made to identify the genes that control such functions as host specificity and virulence. Research is also being conducted to determine why horizontal and vertical transmission of spruce budworm NPV occur less with spruce budworm than with some other NPVs that infect forest insect pests. The question arises whether this is a function of virulence of the virus or the behavior and ecology of the host insect.

Western Spruce Budworm

Western spruce budworm, *C. occidentalis*, is infected by the same viruses as the spruce budworm, but it is 2 000-fold more susceptible to both NPV and GV (Cunningham et al. 1983). Aerial spray trials were conducted with NPV in 1976, 1978, and 1982 with a total area of 252.5 ha treated and with GV in 1982 with 172 ha treated. These trials are reported in greater detail in Chapter 11. In spite of the greater susceptibility of western spruce budworm, attempts to initiate epizootics were unsuccessful (Otvos et al. 1989).

Jack Pine Budworm

Jack pine budworm, *C. pinus*, is a cyclical pest. An outbreak between 1984 and 1986 presented the opportunity to test spruce budworm NPV on this species. Both insects are equally susceptible to the virus in the laboratory. In 1985, a 50-ha plot in Gogama District was aerially sprayed with 7.5×10^{11} PIB/ha in an emitted volume of 9.5 L/ha when larvae were at the peak of their fourth instar. The virus caused 40.6% mortality in larvae collected 6 days post-spray and reared individually on artificial diet. Larval population reduction due to treatment (Abbott's formula) was 61%, pupal reduction 66%, and reduction in the adult emergence was 74%, but no foliage protection was achieved. In 1986, the incidence of vertical transmission of the NPV was assessed by individually rearing larvae collected from the treated plot and an untreated area. Dead larvae were examined microscopically to determine the cause of death; 8.1% died from NPV in the treated plot and 1.5% in the untreated. Successful adult emergence was almost identical from the two plots with 48.4 and 51.1%, respectively (J.C. Cunningham and B.L. Cadogan, unpubl. data).

Douglas-fir Tussock Moth

Two morphologically distinct NPVs affect Douglas-fir tussock moth; one has virus particles occluded singly in polyhedral inclusion bodies (SNPV) and the other has them multiple-embedded in bundles (MNPV) (Hughes and Addison 1970). Outbreaks of this cyclical pest in British Columbia and the western United States are invariably terminated by epizootics of naturally occurring NPV, but not before considerable damage and tree mortality have occurred (Shepherd and Otvos 1986). In both Canada and the United States, viruses were considered prime candidates for development as biocontrol agents and the first field trial was conducted in Canada when individual trees were treated using a pressurized hand sprayer (Morris 1963).

In the United States, the homologous host insect, Douglas-fir tussock moth, has been used for virus production and, in Canada, whitemarked tussock moth, *O. leucostigma*, is preferred. The latter is easier to rear, has no obligatory diapause, and goes through its life cycle faster than Douglas-fir tussock moth. Aerial spray trials were conducted in British Columbia in 1974 (Cunningham and Shepherd 1984), in 1975 in collaboration with USDA Forest Service personnel (Stelzer et al. 1977), in 1976 (Ilnytzky et al. 1977), in 1981 (Shepherd et al. 1984), and in 1982 (Otvos et al. 1987a, 1987b). A summary of these trials, in which a total of 24 plots with an area of 1 136 ha was treated, is given in Table 3, with more complete details in Chapter 13. Generally, virus applications were slow to control Douglas-fir tussock moth populations because only a small proportion of the population became infected by ingesting the spray deposit. However, these larvae died and released more virus inoculum, which caused secondary infection by horizontal transmission. By the onset of pupation, a major population reduction is achieved, although saving of foliage may be negligible. In the year after a virus application, there are very few surviving egg masses and a single virus treatment virtually eradicates the insect population.

The USDA Forest Service arranged safety testing of the MNPV and obtained registration for their viral insecticide called TM BioControl-1 in 1976. The American safety data were used to prepare a petition for the Canadian product called

Table 3. Aerial spray trials with NPV on first- and second-instar Douglas-fir tussock moth, *Orgyia pseudotsugata*, larvae in British Columbia: application was with fixed-wing aircraft fitted with boom and nozzle equipment.

Year	Virus (host)[a]	Dosage (PIB/ha)[b]	Application volume (L/ha)	Tank mix	Plots	Area treated (ha)	% population reduction due to treatment[c]
1974	MNPV (*O.l.*)	2.5×10^{11} (double)	18.8	3% IMC 90-001[d] 0.1% Chevron sticker	1	10	97
1975	MNPV (*O.p.*)	2.5×10^{11}	18.8	25% molasses 6% Sandoz Shade	3	24	92
	MNPV (*O.l.*)	2.5×10^{11}	18.8	50% Sandoz adjuvant V	3	24	78
	SNPV (*O.l.*)	2.5×10^{11}	18.8	50% Sandoz adjuvant V	3	24	81
	MNPV (*O.l.*)	2.5×10^{11}	9.4	50% Sandoz adjuvant V	1	80	91
	MNPV (*O.p.*)	2.5×10^{11}	9.4	50% Sandoz adjuvant V	1	80	89
	MNPV (*O.p.*)	2.5×10^{11}	9.4	25% molasses 6% Sandoz Shade	1	510	95
	SNPV (*O.l.*)	2.5×10^{11}	9.4	25% molasses	1	320	66
1976	SNPV (*O.p.*)	1.25×10^{11}	18.8	25% molasses	1	2	82
	SNPV (*O.l.*)	1.25×10^{11}	18.8	25% molasses	1	2	90
1981[e]	MNPV (*O.l.*)	2.2×10^{11}	11.3	25% molasses 0.2% Chevron sticker	4	20	76
1982	MNPV (*O.l.*)	2.5×10^{11}	9.4	Emulsifiable oil[f]	1	10	95
	MNPV (*O.l.*)	8.3×10^{10}	9.4	Emulsifiable oil	1	10	91
	MNPV (*O.l.*)	1.6×10^{10}	9.4	Emulsifiable oil	1	10	65
	MNPV (*O.l.*)	2.5×10^{11}	9.4	25% molasses	1	10	87

[a] Virus produced in *O.l.* = *Orgyia leucostigma*, *O.p.* = *O. pseudotsugata*.
[b] Polyhedral inclusion bodies per hectare.
[c] Calculated using a modified Abbott's formula on samples collected from 28 to 42 days post-spray.
[d] Sunscreening agents.
[e] Helicopter used for application.
[f] Abbott Laboratories' blank carrier vehicle for Dipel® 88.

Virtuss® (containing the same MNPV), which received a temporary registration in 1983 and full registration in 1987. The dosage on the label of both products is 2.5×10^{11} PIB/ha or its equivalent in activity units. The British Columbia Ministry of Forests purchased sufficient TM BioControl-1 from the USDA Forest Service to treat 8 000 ha. A Canadian registration was obtained for the American product to facilitate its importation into Canada. There is also sufficient Virtuss® in storage to treat about 1 000 ha when the next outbreak occurs in British Columbia. A pest management strategy is in place and it is hoped that the outbreaks will be detected early by monitoring with pheromone traps and will then be suppressed rapidly with applications of NPV (Shepherd et al. 1984; Shepherd and Otvos 1986). Use strategies for NPV are well established, but some refining of methodologies could well reduce costs and enhance efficacy. Plans are being made to apply NPV to 200 ha in 1991 following survey results from pheromone traps and from egg mass counts.

Whitemarked Tussock Moth

The whitemarked tussock moth can be a serious forest pest in Nova Scotia where it feeds on both balsam fir, *Abies balsamea*, and hardwoods; it can be particularly damaging to balsam fir Christmas tree plantations. It is a pest in urban areas throughout eastern Canada where it feeds on a variety of hardwood shade trees. Outbreaks, like those of the Douglas-fir tussock moth, are ultimately controlled by an NPV but in the case of whitemarked tussock moth, the virus is an SNPV.

In a small ad hoc trial in 1975, this SNPV was aerially sprayed on 18 ha of infested mature balsam fir trees in

Nova Scotia. The aqueous tank mix contained 50% Sandoz Adjuvant V, the dosage was 2.2×10^{11} PIB/ha, the emitted volume was 22.6 L/ha, and larvae were mainly in the first instar. Population reduction due to treatment was 85% and the treated area was clearly visible from the air because it suffered only light defoliation compared with surrounding untreated areas (Embree et al. 1984). In 1976, two applications of virus suspensions were dispersed in exploding canisters fired from a homemade mortar. A volume of 1.9 L was dispersed over 1 000 to 2 000 m^2, but virus was already in the insect population so it was not possible to assess the impact of these applications (Embree et al. 1984).

The MNPV of Douglas-fir tussock moth, registered as Virtuss® in Canada and TM BioControl-1 in the United States, also infects whitemarked tussock moth and rusty tussock moth, *O. antiqua* (Hughes 1976), and it would be advantageous to add these species to the Virtuss® label. A ground spray trial was undertaken in 1986 in Newfoundland. The virus was applied on small white birch, *Betula papyrifera*, and balsam fir trees in two 0.24-ha plots at 2.5×10^{11} PIB/ha in 20 L of water per hectare when larvae were mostly in the first instar. Beating samples from balsam fir showed 59% and 78% population reduction due to treatment in the two plots at 5 weeks post-spray. At this time 62% and 69% of larvae were infected with NPV in the treated plots and 18% and 21% in the check plots. The following year it was found that the insect population had collapsed in the whole area (West et al. 1987). In 1987, a Virtuss® aerial spray trial was undertaken in Newfoundland. A 25-ha plot containing predominantly white birch and balsam fir was treated using a fixed-wing aircraft fitted with Micronair® atomizers when larvae were in the second and third instars. The aqueous tank mix contained 60 g/L Orzan LS (an ultraviolet protectant), dosage was 2.5×10^{11} PIB/ha, and emitted volume was 9.4 L/ha. The insect population collapsed in both the treated and check plot (West et al. 1989). In both the ground and aerial spray trial in Newfoundland, it is presumed that the virus spread from the treated to untreated check areas, although the role of naturally occurring NPV cannot be discounted. However, there is no doubt that Virtuss® is effective for controlling whitemarked tussock moth.

Gypsy Moth

Gypsy moth NPV has been extensively studied in Europe and the United States, but only recently in Canada. This virus is prevalent in older infestations of gypsy moth and is a key factor, along with parasites, in causing populations to collapse. In the United States, the first ground spray trial was conducted in 1963 (Rollinson et al. 1965) and the first aerial spray trial in 1974 (Yendol et al. 1977). An American gypsy moth NPV product called Gypchek was registered by the Environmental Protection Agency in 1978. The label calls for a double application of dosages ranging from 2.5×10^{11} to 1.25×10^{12} PIB/ha, 7 to 10 days apart. A registration petition for a Canadian product called Disparvirus was submitted in 1990. Much of the data was obtained from the USDA Forest Service submission for Gypchek.

Gypsy moth has been found in Quebec along the American border for many years. Cardinal and Smirnoff (1973) attempted to initiate epizootics of gypsy moth NPV by handspraying egg masses and treating 8-m oak trees with a mechanized sprayer; larvae emerging from treated egg masses died in 3 weeks. In Ontario, gypsy moth was first detected on Wolfe Island, near Kingston, in 1969, but populations remained low until 1981 when about 1 000 ha of defoliation were reported in the Kaladar area. In 1982, experimental spray trials were conducted with several agents including Gypchek supplied by the USDA Forest Service. Details of virus spray trials conducted in Ontario are given in Table 4. In two plots treated with a double application of 2.5×10^{11} PIB/ha (total 5×10^{11} PIB/ha), there was a 90% and 95% reduction in fall egg mass density compared with spring egg mass density. There was a 55% reduction in one check plot and a 324% increase in a second (Meating et al. 1983). In 1985, Disparvirus production began at the Institute using the same strain of NPV as Gypcheck. In 1986, two dosages of Disparvirus were tested near Sharbot Lake, a double application of 2.7×10^{11} PIB/ha (total 5.4×10^{11} PIB/ha), and almost tenfold that dosage with a double application of 2.2×10^{12} PIB/ha (total 4.4×10^{12} PIB/ha). There was a 99% decline from spring to fall egg mass density in the plot receiving the higher dosage and a 52% decline in the plot receiving the lower dosage. However, major declines of 80% and 95% were also recorded in two check areas (Kaupp et al. 1988).

Intensive field testing of Disparvirus began in 1988 in order to develop it as an operational control agent for gypsy moth in Ontario. Considerable progress has been made during the last 3 years, but some refinements are still required before it is suitable for operational use. In 1988, three plots in Lindsay District were treated with a double application of 1.25×10^{12} PIB/ha for a total of 2.5×10^{12} PIB/ha. The emitted volume was 10 L/ha and the tank mix contained 25% molasses and 6% Orzan LS. The two applications, 3 days apart, were on predominantly first-instar larvae. A further three plots were selected as untreated checks. Pre-spray gypsy moth egg mass densities ranged from 1 430 to 8 520/ha in the six plots. Defoliation of red oak, *Quercus rubra*, was estimated at 14% in two of the treated plots compared with 82% and 90% in the corresponding check plots. The third treated plot suffered 46% defoliation, which was attributed to oak leafshredder, *Croesia semipurpurana*, Bruce spanworm, *Operophtera bruceata*, and forest tent caterpillar, *Malacosoma disstria*, as well as gypsy moth; defoliation in the corresponding check plot was 31%. From spring to fall, egg mass densities in the treated plots were reduced by 76%, 93%, and 98%. In corresponding check plots an increase of 56% and reductions of 50% and 70% were recorded.

Disparvirus should not cost much more than Bt if it is going to be used operationally. It requires about 1 000 gypsy moth larvae to produce the 1-ha dosage applied in 1988. The virus was applied in an emitted volume of 10.0 L/ha, whereas Bt is routinely applied in volumes of less than 2.0 L/ha. The molasses and Orzan tank mix, although effective, is difficult to mix and unsuitable for use by commercial applicators.

In 1989, some of these problems were addressed when both a reduced dosage and emitted volume of Disparvirus were tested in Lindsay District. A double application of 5×10^{11} PIB/ha (total 10^{12} PIB/ha) was tested in emitted volumes of 10.0 L/ha and 5.0 L/ha. The tank mix contained

Table 4. Aerial spray trials with NPV on gypsy moth, *Lymantria dispar*, in Ontario.

Year	Total dosage PIB/ha[a] (double application)	Tank mix	Spray equipment	Volume L/ha	Predominant instar treated (first/second application)	Days between applications	No. of plots	Area treated (ha)
1982	5×10^{11}	Water 25% emulsifiable oil	Fixed-wing B&N[b]	18.8	2/3	6	2	63
1986	4.4×10^{12}	Water 25% molasses 6% Orzan LS	Helicopter Micronair®	9.4	1&2/2	5	1	10
1986	5.4×10^{11}	As above	Helicopter Micronair®	9.4	1&2/2	5	1	88
1988	2.5×10^{12}	As above	Fixed-wing Micronair®	10.0	1/1	3	3	64
1989	10^{12}	Water 25% molasses 10% Orzan LS 2% Rhoplex B60A	As above	10.0 & 5.0	1/1	3	6	90
1990	10^{12}	As above	As above	5.0 & 2.5	1/1	5	6	60
1990	10^{12}	Water 25% emulsifiable oil	As above	5.0	1/1	8	3	30

[a]Polyhedral inclusion bodies per hectare.
[b]Boom and nozzle.

25% molasses, 10% Orzan LS, and 2% Rhoplex B60A sticker. The two applications, 3 days apart, were on predominantly first-instar larvae. A further three plots were selected as untreated checks. Pre-spray gypsy moth egg mass densities ranged from 2 180/ha to 11 900/ha in six treated plots and from 2 000 to 6 380/ha in three check plots. Defoliation was assessed on both red oak and white oak, *Q. alba*. There was significantly less defoliation in all the treated plots compared with the check plots where defoliation ranged from 75% to 82%. One plot in the 10.0 L/ha treatment suffered 46% defoliation, but the other five plots ranged from 19% to 38%. In the 10.0 L/ha treatments, reduction in egg mass densities between spring and fall were 61%, 62%, and 83%. In the 5.0 L/ha treatments, reduction in egg mass densities were 71%, 81%, and 98%. In the three check plots, egg mass densities fell 9% in one plot and increased by 12% and 82% in the other two. Based on these findings, it was recommended that the dosage be reduced to a double application of 5×10^{11} PIB/ha and that the emitted volume be reduced to 5.0 L/ha (Cunningham, Kaupp, and Brown, unpubl. data).

Following analysis of the 1989 field trial, the emitted volume was further reduced in 1990 and tested at 2.5 L/ha compared with 5.0 L/ha. It was also decided to test an emulsifiable oil tank mix which is easier to mix than molasses and Orzan. Sufficient Disparvirus was produced at the Institute to spray 60 ha, and Gypchek, obtained from the USDA Forest Service, was used to test the oil formulation. The oil was the blank carrier vehicle used for Dipel® 176, which was supplied by Abbott Laboratories, and a 25% oil, 75% water tank mix was used. The double application of 5×10^{11} PIB/ha and the aqueous tank mix used in 1990 were the same as those applied in 1989. The three treatments were applied on three replicated plots in Simcoe District and a further four plots were selected as checks. The aqueous tank mix was applied at 5.0 L/ha and 2.5 L/ha and the emulsifiable oil tank mix at 5.0 L/ha. Pre-spray egg mass densities ranged from 2 280/ha to 8 900/ha in the nine treated plots and from 2 390 to 6 690/ha in the four check plots. For the third year in succession, larvae were mainly in the first instar at the time of application. The treatments with the aqueous tank mix were 5 days apart and those with the emulsifiable oil tank mix were 8 days apart.

In the 2.5 L/ha aqueous tank mix treatments, the plots suffered 30%, 34%, and 46% defoliation, respectively. Reductions in egg mass densities from spring to fall were 38%, 59%, and 64%. In the plots treated at 5.0 L/ha with the aqueous tank mix, reductions in egg mass densities were 79%, 88%, and 89% and defoliation ranged from 29% to 38%. The Gypchek in the emulsifiable oil tank mix at 5.0 L/ha gave excellent results with reductions in egg mass densities of 82%, 88%, and 89%. Defoliation of the four check plots ranged from 46% to 93% and increases in egg mass densities of 43%, 65%, 125%, and 245% were recorded (Cunningham, Kaupp, and Brown, unpubl. data).

A double application of 5×10^{11} PIB/ha (total 10^{12} PIB/ha) is effective for controlling moderate density populations of gypsy moth (2 000 to 6 000 egg masses/ha). An emitted volume of 5.0 L/ha appears to be adequate and 2.5 L/ha appears to be too low when using the aqueous tank mix. Encouraging results were obtained with the emulsifiable oil tank mix at 5.0 L/ha, in spite of the lack of a sun screening agent. This tank mix is easy to handle compared with the aqueous tank mix. A commercial source capable of supplying the quantities of Disparvirus needed for widespread use is essential if this product is to be used operationally.

Forest Tent Caterpillar

The forest tent caterpillar, *Malacosoma disstria*, has a long history of periodic outbreaks in Canada in all provinces except Newfoundland. It attacks a wide variety of trees, but trembling aspen, *Populus tremuloides*, is the preferred host. As early as 1915, "wilt disease" was described in forest tent caterpillar (Chapman and Glaser 1915), although some population studies have indicated that the role of NPV is unimportant in population dynamics (Hodson 1941; Smirnoff 1968; Witter et al. 1972). Field trials with NPV have been comprehensively reviewed by Ives (1984a). In 1963, ground spray trials with a backpack mistblower were conducted in Ontario (Stairs 1964) and another small-scale spray was conducted in New Brunswick (Forbes et al. 1972). A series of trials was conducted in Alberta from 1976 to 1980. Egg masses were sprayed in 1976, and a mistblower was used in 1977 (Ives and Muldrew 1978). In 1978, 1979, and 1980, a helicopter was used to treat a total of 17 small plots with dosages ranging from 4.4×10^9 to 1.8×10^{13} PIB/ha (Ives et al. 1982). Virus transmission from 1 year to the next was evaluated in all these tests. On the basis of these trials in Alberta, it was concluded that NPV is not particularly effective in controlling forest tent caterpillar and only high dosages, applied before egg hatch or on small larvae, are effective.

Bruce Spanworm

A polyhedrosis virus has been recorded in populations of Bruce spanworm in British Columbia, the Maritime provinces, and Quebec. It was credited with terminating outbreaks in eastern Canada in 1963 and 1964 (Ives and Cunningham 1980). In a small ground spray trial, conducted in 1979 in Alberta, trembling aspen were sprayed to run-off with a backpack mistblower. Encouraging results were obtained with 10^7 PIB/mL on first-instar larvae and 10^8 PIB/mL on second- and third-instar larvae, and foliage protection was obvious (Ives and Cunningham 1980; Ives 1984b). These ground spray trials indicated that Bruce spanworm NPV is a potentially useful biocontrol agent. However, it is inefficient to produce this NPV in its homologous host, which only yields about 10^8 PIB/larva. If an extensive research program is contemplated, a search should be made for an insect host that gives a better yield of virus.

Winter Moth

Winter moth, *Operophtera brumata*, was accidentally introduced into Nova Scotia from Europe, probably in the mid-1930s. It is not a major forest pest and although it favors shade trees such as oaks, *Quercus* spp., and lindens, *Tilia* spp., it is principally a problem in apple orchards. Interestingly, it is causing damage to Sitka spruce, *Picea sitchensis*, in Britain and this is the first record of winter moth on a coniferous host. An NPV was found in this species in Nova Scotia (Neilson 1965). Because no viable Canadian virus was available, a sample of NPV from Britain was tested on winter moth in a small ground spray trial in 1979 in an abandoned apple orchard in Victoria, British Columbia. Three concentrations of virus, 10^6, 10^7, and 10^8 PIB/mL, were applied at 1 L per tree just after larval hatch and another series of trees was treated 8 days later when buds had flushed. Best results were obtained with 10^8 PIB/mL applied early which gave 46% population reduction due to the treatment and significant foliage protection. Surveys in 1980 revealed negligible generation to generation transmission of the NPV (Cunningham et al. 1981). It required 357 larvae to produce the 1 L of spray to treat one tree, so without having an impact over more than 1 year, this NPV is not a good candidate for a biocontrol agent. However, should winter moth become a pest of coniferous trees in Canada, the epizootiology of this NPV may be entirely different from that on deciduous trees.

Hemlock Looper

The hemlock looper, *Lambdina fiscellaria*, is a particularly destructive pest of balsam fir in Newfoundland. An NPV has been isolated from the hemlock looper and NPVs isolated from western hemlock looper, *L. fiscellaria lugubrosa*, and western oak looper, *L. fiscellaria somniaria*, also infect hemlock looper (Cunningham 1970). Western oak looper NPV was propagated in hemlock looper larvae and both purified and crude preparations of NPV were compared in a small ground spray trial in Newfoundland in 1970. Concentrations ranging from 2×10^5 to 10^7 PIB/mL were applied with a backpack mistblower on first- to third-instar larvae at 1 500 to 6 000 L/ha. First mortality was observed at 20 days post-spray; larval mortality averaged 20% in the treated plots and all trees were heavily defoliated. Pupal mortality due to treatment was 66% (Cunningham 1982). No further virus research has been undertaken on this species since 1970. It is difficult to mass produce this virus and the NPV did not appear particularly effective in the one, very limited field trial.

Black Army Cutworm

The black army cutworm, *Actebia fennica*, can be a serious pest of young conifer seedlings planted in burned-over areas. It occurs across Canada from Newfoundland to British Columbia and overwinters as a first- or second-instar larva. Feeding on sprouting vegetation begins as soon as the snow melts. An NPV isolated from bertha armyworm, *Mamestra configurata*, was found to infect first-instar black army cutworm (Burke 1974). Virus was propagated in bertha armyworm and shipped to British Columbia for a small field trial near Golden in 1974. Three 405-m^2 plots were sprayed with a dosage of 1.6×10^{12} PIB/ha in 150 L/ha using a portable mistblower. The spray was applied in the evening, which is peak feeding time for larvae, and allowed 8-hours exposure with no virus inactivation from sunlight. Larvae sampled at 14 days post-spray showed no sign of virus infection. Hence, this virus is either ineffective for black army cutworm control or most of the treated larvae moved away from the sprayed plots (Ross and Ilnytzky 1977).

Three more viruses from noctuid larvae have been found to infect black army cutworm; these are from alfalfa looper, *Autographa californica*, the armyworm, *Pseudaletia unipuncta*, and the cabbageworm, *M. brassicae*. The alfalfa looper NPV has a wide host range and there has been some interest in producing it commercially. The cabbageworm NPV also has a wide host range and it is produced commercially by Calliope in France under the name Mamestrin. There is interest in developing a viral insecticide for control of this species, but the cyclical and sporadic nature of outbreaks makes it difficult to organize a research program.

Diprionid Sawflies

The most effective and economical viral insecticides developed to date in Canada are those used to control a few species of sawflies. A more detailed description of these viruses is given in Chapter 24. The NPV of European spruce sawfly is a landmark in Canadian pest management and a textbook example of classic biological control. The last study on this virus in Canada was conducted in 1950 when transmission and spread were monitored in Ontario (Bird and Burke 1961). The European spruce sawfly population in eastern Canada is currently at an endemic level, held in check by NPV and parasites.

The introduction to Canada in 1949 of the NPV of the European pine sawfly and subsequent field trials are well documented (Bird 1953; Wallace et al. 1975; Cunningham and Entwistle 1981). As a result of widespread application of this NPV and release of parasites in the 1950s and 1960s, European pine sawfly is currently only a minor pest. In 1975, 125 ha in Sandbanks Provincial Park, Ontario, were aerially sprayed with NPV to control European pine sawfly and, in 1981, a 19-ha Christmas tree plantation in Ontario was aerially sprayed. The virus has been used annually since 1983 to treat ornamental trees in St. John's, Newfoundland. In spite of widespread use of this NPV, it was not until 1985 that a registration petition was submitted for the Institute product called Sertifervirus.

The NPV of redheaded pine sawfly was found in Ontario in 1950 and reported by Bird (1961). Several ground spray trials were conducted between 1950 and 1970, and between 1975 and 1980 a series of aerial spray trials was conducted when 20 plantations with a combined area of 258 ha were treated (Cunningham et al. 1986). Between 1978 and 1980, the Quebec Ministry of Energy and Resources treated 78 plantations from the air and 14 from the ground with a combined area of 1 051 ha. Between 1980 and 1990, the Ontario Ministry of Natural Resources treated 478 plantations with a combined area of 3 546 ha using ground spray equipment. The Institute product, called Lecontvirus, received a temporary registration in 1983 and full registration in 1987.

An NPV was found in Swaine jack pine sawfly, *N. swainei*, in northeastern Ontario in 1953 and in Quebec in 1956 (Smirnoff 1962). The first aerial spray trial with this virus was conducted in 1960 (Smirnoff et al. 1962) and in 1964, 1 600 ha were treated (Smirnoff and McLeod 1975). Unfortunately, the 1964 operation was a failure due to cold weather following the application and no further field trials have been conducted with this virus. The dissemination of virus-infected cocoons has been suggested as a means of initiating epizootics (Smirnoff 1972), but this method has yet to be proved viable on an operational scale. This NPV may still be worthwhile developing as a viral insecticide because of the economic importance of Swaine jack pine sawfly. Several other sawfly viruses have been tested in ground spray trials, but it is doubtful if the economic importance of these pests merits development of viral insecticides. NPVs are available for jack pine sawfly, *N. pratti banksianae*, and the redheaded jack pine sawfly, *N. rugifrons*, but no tests have been conducted in the last 20 years. Ground spray tests in 1973 indicated that the NPV of balsam firm sawfly, *N. abietis* complex, has considerable potential as a biological control agent (Olofsson 1973). An NPV has been recorded from red pine sawfly, *N. nanulus*, and known for some time. It was used successfully to control this insect in plantations in the Maritimes in the 1970s although these operations were not documented (D.G. Embree, pers. comm.). In 1986, a small field trial was conducted in Ontario on individual colonies using a range of concentrations of this NPV and preliminary results indicated that it could be a promising candidate for development.

Conclusions

Insect viruses, particularly NPVs, are a viable alternative to Bt and synthetic chemical insecticides for control of some forest insect pests in Canada. Viruses that initiate epizootics have an obvious advantage over other control agents that merely act as stomach or contact poisons. Research into the mechanisms of persistence and spread of insect viruses is essential when it comes to improving strategies for their use as pest management tools. Viruses are highly effective for control of European pine sawfly and redheaded pine sawfly and are strongly recommended over synthetic chemical insecticides for these pests in plantations. Available strains of Bt are ineffective against hymenopterous pests. Chemical treatments for sawflies usually need to be repeated annually, whereas a correctly timed virus application will give protection for several years, long enough, in fact, for trees to have grown sufficiently that they are no longer at risk. Demand for Sertifervirus and Lecontvirus is not sufficient to merit commercialization of these products and the Institute distributes them to clients free of charge. As shown in Table 2, there are six more sawfly NPVs that could be developed as viral insecticides in Canada. Currently, only one of these, Swaine jack pine sawfly, can be considered to have significant economic importance. Of course, the status of potentially damaging insects can change dramatically and some minor pest species may become major pests in the future.

Commercial interest in viral insecticides has lagged behind interest in Bt and this is a problem if they are to be extensively used on an operational basis. Their high degree of specificity, which is so attractive from an environmental standpoint, deters commercialization. Also, not many companies want to produce viruses in insect larvae, although, if production in cell culture becomes economically feasible, there will be much greater commercial interest. Gypsy moth NPV is a good candidate for commercialization because it is not a cyclical pest and because there are markets for this product in both Europe and North America. Douglas-fir tussock moth NPV has proved to be a highly effective pest management tool and its use is recommended in preference to Bt or synthetic chemical insecticides which may, in fact, prolong rather than terminate outbreaks by delaying progress of naturally occurring NPV epizootics. Douglas-fir tussock moth is a cyclical pest that is another deterrent to commercial production of its NPV, but considerable quantities of viral insecticide are currently held by government agencies both in Canada and the United States. Future outbreaks will provide an opportunity to validate the pest management model based on pheromone trapping, egg mass counting, and application of NPV early in the outbreak cycle (Shepherd and Otvos 1986).

Recombinant DNA technology has revolutionized the industrial, agricultural, and medical use of microorganisms including viruses. Viral insecticides are obvious candidates for improvement by genetic engineering, and speculation on the possibilities is limitless. Alfalfa looper NPV grown in a cabbage looper, *Trichoplusia ni*, cell line is the best studied system and several viral genes have been identified physically and functionally. One such gene codes for the inclusion body protein and it is possible to remove it and replace it with a gene coding for a foreign protein. This baculovirus has been widely used for expressing foreign genes and now more than 140 proteins with medical and veterinary applications have been produced in this baculovirus/insect cell system including interferon and insulin. Commercial interest in baculoviruses as expression vectors is currently much greater than in baculoviruses as insecticides; however, research is being conducted into the genes affecting host range and virulence. There is also a possibility of inserting foreign genes with insecticidal properties into baculoviruses (Kirschbaum 1985). The genome of spruce budworm NPV has been characterized and a physical map constructed (Arif and Doerfler 1983, 1984). Physical maps have also been constructed for the genomes of Douglas-fir tussock moth NPV (Leisy et al. 1984) and gypsy moth NPV (McClintock and Dougherty 1988; Smith et al. 1988). A field release of alfalfa looper NPV with a genetic marker incorporated in the genome was made in the United Kingdom in 1986 (Bishop 1986). Progress is relatively slow, but it is expected that genetically improved viral insecticides will be available within the next decade. They will, however, still have to be produced in susceptible insect larvae unless the problem of mass production in cell culture is resolved.

The most exciting and positive aspect of using viruses for insect pest control is the ability of some viruses to initiate epizootics in some ecological situations with some insect species. Forests are an ideal ecosystem for virus epizootics because of the long-standing, undisturbed nature of the crop and the large areas involved. Also, biological persistence of NPVs and GVs appears to increase with increasing complexity of plant architecture, so it is much greater on trees than annual field crops (Entwistle et al. 1988). Even if some viruses are costly to produce, provided that long-term insect population regulation can be achieved, the initial cost of the application can be amortized over several years. Under these circumstances, the cost will appear reasonable and competitive with other pest control options; the concept is certainly more appealing than annual applications of Bt or synthetic chemical pesticides.

References

Arif, B.M.; Doerfler, W. 1983. Characterization of the DNA from *Choristoneura fumiferana* nuclear polyhedrosis virus. Zbl. Bakt. Hyg. Abt. Orig. A. 254: 147-148.

Arif, B.M.; Doerfler, W. 1984. Identification and localization of reiterated sequences in the *Choristoneura fumiferana* MNPV genome. EMBO J. 3: 525-529.

Balch, R.E.; Bird, F.T. 1944. A disease of the European spruce sawfly, *Gilpinia hercyniae* (Htg.), and its place in natural control. Sci. Agric. 25: 65-80.

Bird, F.T. 1953. The use of a virus disease in the biological control of the European pine sawfly, *Neodiprion sertifer* (Geoffr.). Can. Entomol. 85: 437-445.

Bird, F.T. 1961. Transmission of some insect viruses with particular reference to ovarial transmission and its importance in the development of epizootics. J. Insect Pathol. 3: 532-580.

Bird, F.T.; Burke, J.M. 1961. Artificially disseminated virus as a factor controlling the European spruce sawfly, *Diprion hercyniae* (Htg.), in the absence of introduced parasites. Can. Entomol. 89: 371-378.

Bishop, D.H.L. 1986. UK release of genetically marked virus. Nature 323: 496.

Burke, J.M. 1974. Infection tests with a virus of the Bertha armyworm. Environ. Can., For. Serv., Ottawa, Ont. Bi-mon. Res. Notes 30(5): 29.

Cameron, J.M. 1975a. Biological insecticides. Pages 25-33 *in* M.L. Prebble (Ed.), Aerial Control of Forest Insects in Canada. Dept. Environ., Ottawa, Ont.

Cameron, J.M. 1975b. Field trials of spruce budworm viruses, 1971-1973. Pages 134-137 *in* M.L. Prebble (Ed.), Aerial Control of Forest Insects in Canada. Dept. Environ., Ottawa, Ont.

Cardinal, J.A.; Smirnoff, W.A. 1973. Introduction expérimentale de la polyédrie nucléaire de *Porthetria dispar* L. (Lépidoptera: Lymantriidae) en forêt. Phytoprotection 54: 48-51.

Chapman, J.W.; Glaser, R.W. 1915. A preliminary list of insects which have wilt, with a comparative study of their polyhedra. J. Econ. Entomol. 8: 140-149.

Cunningham, J.C. 1970. Pathogenicity tests of nuclear polyhedrosis viruses infecting the eastern hemlock looper, *Lambdina fiscellaria fiscellaria*. Can. Entomol. 102: 1534-1539.

Cunningham, J.C. 1982. Field trials with baculovirses: control of forest insect pests. Pages 335-386 *in* E. Kurstaki (Ed.), Microbial and Viral Pesticides. Dekker, New York.

Cunningham, J.C. 1985. Status of viruses as biocontrol agents for spruce budworms. Pages 61-67 *in* Proc. Symp. Microbial Control of Spruce Budworms and Gypsy Moths. Windsor Locks, Conn. 1984. USDA For. Serv. Rep. GTR-NE-100.

Cunningham, J.C.; de Groot, P.; Kaupp, W.J. 1986. A review of aerial spray trials with Lecontvirus for control of redheaded pine sawfly, *Neodiprion lecontei* (Hymenoptera: Diprionidae), in Ontario. Proc. Ent. Soc. Ont. 117: 65-72.

Cunningham, J.C.; Entwistle, P.F. 1981. Control of sawflies by baculovirus. Pages 379-407 *in* H.D. Burges (Ed.), Microbial Control of Pests and Plant Diseases 1970-1980. Academic Press, London, England.

Cunningham, J.C.; Howse, G.M. 1984. *Choristoneura fumiferana* (Clemens), spruce budworm (Lepidoptera: Tortricidae). Pages 277-279 *in* J.S. Kelleher and M.A. Hulme (Eds.), Biological Control Programmes against Insects and Weeds in Canada 1969-1980. Commonw. Agric. Bureaux, Slough, England.

Cunningham, J.C.; Howse, G.M.; Kaupp, W.J.; McPhee, J.R.; Harnden, A.A.; White, M.B.E. 1975. The persistence of nuclear polyhedrosis virus and entomopoxvirus in populations of spruce budworm. Can. For. Serv., Insect Path. Res. Inst., Sault Ste. Marie, Ont. Inf. Rep. IP-X-10. 32 p.

Cunningham, J.C.; Kaupp, W.J.; McPhee, J.R. 1983. A comparison of pathogenicity of two baculoviruses to the spruce budworm and western spruce budworm. Can. For. Serv., Ottawa, Ont. Res. Notes 3(2): 9-10.

Cunningham, J.C.; McPhee, J.R. 1986. Production of sawfly viruses in plantations. Can. For. Serv., For. Pest Manage. Inst., Sault Ste. Marie, Ont. Tech. Note No.4. 4 p.

Cunningham, J.C.; Shepherd, R.F. 1984. *Orgyia pseudotsugata* (McDunnough), Douglas-fir tussock moth (Lepidoptera: Lymantriidae). Pages 363-367 *in* J.S. Kelleher and M.A. Hulme (Eds.), Biological Control Programmes against Insects and Weeds in Canada 1969-1980. Commonw. Agric. Bureaux, Slough, England.

Cunningham, J.C.; Tonks, N.V.; Kaupp, W.J. 1981. Viruses to control winter moth *Operophtera brumata* (Lepidoptera: Geometridae). J. Entomol. Soc. B.C. 78: 17-24.

Embree, E.G.; Elgie, D.E.; Estabrooks, G.F. 1984. *Orgyia leucostigma* (J.E. Smith), whitemarked tussock moth (Lepidoptera: Lymantriidae). Pages 359-361 *in* J.S. Kelleher and M.A. Hulme (Eds.), Biological Control Programmes against Insects and Weeds in Canada 1969-1980. Commonw. Agric. Bureaux, Slough, England.

Entwistle, P.F. 1986. Epizootiology and strategies of microbial control. Pages 257-278 *in* J.M. Franz and M. Lindaur (Eds.), Biological Plant and Health Protection. Fortsch. Zool. Vol. 32. Fishcher Verlag, Stuttgart.

Entwistle, P.F.; Cory, J.S.; Doyle, C. 1988. An overview of insect baculovirus ecology as a background to field release of a genetically manipulated nuclear polyhedrosis virus. Pages 72-80 *in* W. Klingmuller (Ed.), Risk Assessment for Deliberate Releases. Springer-Verlag, Berlin.

Evans, H.F.; Entwistle, P.F. 1987. Viral diseases. Pages 257-322 *in* J.R. Fuxa and Y. Tanada (Eds.), Epizootiology of Insect Diseases. Wiley, New York.

Forbes, R.S.; Underwood, G.R.; van Sickle, G.A. 1972. Maritimes Regions. Pages 19-33 *in* Annual Report of the Forest Insect and Disease Survey, 1971. Can. For. Serv., Ottawa, Ont.

Hodson, A.M. 1941. An ecological study of the forest tent caterpillar, *Malacosoma disstria* Hbn., in northern Minnesota. Univ. of Minn. Agric. Exp. Stn. Tech. Bull. 148.

Hughes, K.M. 1976. Notes on the nuclear polyhedrosis viruses of tussock moths of the genus *Orgyia* (Lepidoptera). Can. Entomol. 108: 479-484.

Hughes, K.M.; Addison, A.B. 1970. Two nuclear polyhedrosis viruses of the Douglas-fir tussock moth. J. Invertebr. Pathol. 16: 196-204.

Ilnytzky, S.; McPhee, J.R.; Cunningham, J.C. 1977. Comparison of field-propagated nuclear polyhedrosis virus from Douglas-fir tussock moth with laboratory produced virus. Can. Dept. Fish. and For., Ottawa, Ont. Bi-mon. Res. Notes 33: 5-6.

Ives, W.G.H. 1984a. *Malacosoma disstria* Hubner, forest tent caterpillar (Lepidoptera: Lasiocampidae). Pages 311-319 *in* J.S. Kelleher and M.A. Hulme (Eds.), Biological Control Programmes against Insects and Weeds in Canada 1969-1980. Commonw. Agric. Bureaux, Slough, England.

Ives, W.G.H. 1984b. *Operophtera bruceata* (Hulst), Bruce spanworm (Lepidoptera: Geometridae). Pages 349-351 *in* J.S. Kelleher and M.A. Hulme (Eds.), Biological Control Programmes against Insects and Weeds in Canada 1969-1980. Commonw. Agric. Bureaux, Slough, England.

Ives, W.G.H.; Cunningham, J.C. 1980. Application of nuclear polyhedrosis virus to control Bruce spanworm (Lepidoptera: Geometridae). Can. Entomol. 112: 741-744.

Ives, W.G.H.; Muldrew, J.A. 1978. Preliminary evaluations of the effectiveness of nucleopolyhedrosis virus sprays to control the forest tent caterpillar in Alberta. Can. For. Serv., Northern For. Res. Cent., Edmonton, Alta. Inf. Rep. NOR-X-204. 10 p.

Ives, W.G.H; Muldrew, J.A.; Smith, R.M. 1982. Experimental aerial application of forest tent caterpillar baculovirus. Can. For. Serv., Northern For. Res. Cent., Edmonton, Alta. Inf. Rep. NOR-X-240. 9 p.

Kaupp, W.J. 1983. Persistence of *Neodiprion sertifer* (Hymenoptera: Diprionidae) nuclear polyhedrosis virus on *Pinus contorta* foliage. Can. Entomol. 115: 869-873.

Kaupp, W.J.; Cunningham, J.C.; Cadogan, B.L. 1990. Aerial application of high dosages of nuclear polyhedrosis virus to early instar spruce budworm, *Choristoneura fumiferana* (Clem.). For. Can., For. Pest Manage. Inst., Sault Ste. Marie, Ont. Inf. Rep. FPMI-X-85. 9 p.

Kaupp, W.J.; Cunningham, J.C.; Meating, J.H.; Howse, G.M.; Denys, A. 1988. Aerial spray trials with Disparvirus in Ontario in 1986. Can. For. Serv., For. Pest Manage. Inst., Sault Ste. Marie, Ont. Inf. Rep. FPM-X-82. 7 p.

Kaupp, W.J.; Ebling, P.M. 1990. Response of third-, fourth-, fifth- and sixth-instar spruce budworm, *Choristoneura fuminferana* (Clem.), larvae to nuclear polyhedrosis virus. Can. Entomol. 122: 1037-1038.

Kirschbaum, J.B. 1985. Potential implications of genetic engineering and other biotechnologies to insect control. Annu. Rev. Entomol. 30: 51-70.

Leisy, D.J.; Rohrmann, G.F.; Beaudreau, G.S. 1984. Conservation of genome organization in two multicapsid nuclear polyhedrosis viruses. J. Virol. 52: 699-702.

Matthews, R.E.F. 1982. Classification and nomenclature of viruses. 4th Rep. Int. Comm. Taxonomy Viruses. Intervirology 17: 200 p.

McClintock, J.T.; Dougherty, E.M. 1988. Restriction mapping of *Lymantria dispar* nuclear polyhedrosis virus DNA: localisation of the polyhedrin gene and identification of four homologous regions. J. Gen. Virol. 69: 2303-2312.

Meating, J.H.; Lawrence, H.D.; Cunningham, J.C.; Howse, G.M. 1983. The 1982 gypsy moth situation in Ontario: general surveys, spray trials and forecasts for 1983. Can. For. Serv., Great Lakes For. Res. Cen., Sault Ste. Marie, Ont. Inf. Rep. O-X-352. 14 p.

Morris, O.N. 1963. The natural and artificial control of the Douglas-fir tussock moth, *Orgyia pseudotsugata* McDunnough, by a nuclear polyhedrosis virus. J. Insect Pathol. 5: 401-414.

Neilson, M.M. 1965. A new nuclear polyhedrosis virus of the winter moth, *Operophtera brumata* (L.). Can. Dept. For., Ottawa, Ont. Bi-mon. Prog. Rep. 21: 1.

Olofsson, E. 1973. Evaluation of a nuclear polyhedrosis virus as an agent for the control of balsam fir sawfly, *Neodiprion abietis* (Harr.). Can. For. Serv., Insect Pathol. Res. Inst., Sault Ste. Marie, Ont. Inf. Rep. IP-X-2. 30 p.

Otvos, I.S.; Cunningham, J.C.; Friskie, L.M. 1987a. Aerial application of nuclear polyhedrosis virus against Douglas-fir tussock moth, *Orgyia pseudotsugata* (McDunnough) (Lepidoptera: Lymantriidae): I. Impact in the year of application. Can. Entomol. 119: 697-706.

Otvos, I.S.; Cunningham, J.C.; Alfaro, R.I. 1987b. Aerial application of nuclear polyhedrosis virus against Douglas-fir tussock moth, *Orgyia pseudotsugata* (McDunnough) (Lepidoptera: Lymantriidae): II. Impact 1 and 2 years after application. Can. Entomol. 119: 707-715.

Otvos, I.S.; Cunningham, J.C.; Kaupp, W.J. 1989. Aerial application of two baculoviruses against the western spruce budworm, *Choristoneura occidentalis* Freeman (Lepidoptera: Tortricidae), in British Columbia. Can. Entomol. 121: 209-217.

Podgwaite, J.D.; Bruen, R.B. 1978. Procedures for the microbiological examination of production batch preparations of the nuclear polyhedrosis virus (Baculovirus) of the gypsy moth, *Lymantria dispar*. USDA For. Serv., Broomall, Pa. Gen. Tech. Rep. NE-38. 8 p.

Rollinson, W.D.; Lewis, F.B.; Waters, W.E. 1965. The successful use of a nuclear polyhedrosis virus against the gypsy moth. J. Invertebr. Pathol. 7: 515-517.

Ross, D.A.; Ilnytzky, S. 1977. The black army cutworm *Actebia fennica* (Tauscher) in British Columbia. Dept. of Fish. and Environ., For. Serv., Pac. For. Res. Cent., Victoria, B.C. Inf. Rep. BC-X-154. 23 p.

Shepherd, R.F.; Otvos, I.S. 1986. Pest management of Douglas-fir tussock moth: procedures for insect monitoring, problem evaluation and control actions. Can. For. Serv., Pac. For. Res. Cent., Victoria, B.C. Inf. Rep. BC-X-270. 14 p.

Shepherd, R.F.; Otvos, I.S.; Chorney, R.J.; Cunningham, J.C. 1984. Pest management of Douglas-fir tussock moth (Lepidoptera: Lymantriidae): prevention of an outbreak through early treatment with a nuclear polyhedrosis virus by ground and aerial applications. Can. Entomol. 116: 1533-1542.

Smirnoff, W.A. 1962. Trans-ovum transmission of virus of *Neodiprion swainei* Middleton (Hymenoptera: Tenthredinidae). J. Insect Pathol. 4: 192-200.

Smirnoff, W.A. 1968. Microorganisms isolated from *Malacosoma americanum* and *Malacosoma disstria* in the Province of Quebec. Dept. of For. and Rural Dev., For. Branch, Ottawa, Ont. Bi-mon. Res. Notes 24: 4.

Smirnoff, W.A. 1972. Promoting virus epizootics in populations of the Swaine jack pine sawfly by infected adults. Bioscience 22: 662-663.

Smirnoff, W.A.; Fettes, J.J.; Haliburton, W. 1962. A virus disease of Swaine's jack-pine sawfly, *Neodiprion swainei* Midd., sprayed from an aircraft. Can. Entomol. 94: 477-486.

Smirnoff, W.A.; McLeod, J.M. 1975. Swaine jack-pine sawfly *Neodiprion swainei* Middleton. Pages 235-240 *in* M.L. Prebble (Ed.), Aerial Control of Forest Insects in Canada. Dept. of Environ., Ottawa, Ont.

Smith, I.R.L.; van Beek, N.A.M.; Podgwaite, J.D.; Wood, H.A. 1988. Physical map and polyhedrin gene sequence of *Lymantria dispar* nuclear polyhedrosis virus. Gene 71: 97-105.

Stairs, G.R. 1964. Dissemination of nuclear polyhedrosis virus against the forest tent caterpillar, *Malacosoma disstria* (Hubner) (Lepidoptera: Lasiocampidae). Can. Entomol. 96: 1017-1020.

Stelzer, M.; Neisess, J.; Cunningham, J.C.; McPhee, J.R. 1977. Field evaluation of baculovirus stocks against Douglas-fir tussock moth in British Columbia. J. Econ. Entomol. 70: 243-246.

Thompson, C.G.; Scott, D.W.; Wickman, B.E. 1981. Long-term persistence of the nuclear polyhedrosis virus of the Douglas-fir tussock moth, *Orgyia pseudotsugata* (Lepidoptera: Lymantriidae), in forest soil. Environ. Entomol. 10: 254-255.

Wallace, D.R.; Cameron, J.M.; Sullivan, C.R. 1975. European pine sawfly, *Neodiprion sertifer* (Geoff.). Pages 224-230 *in* M.L. Prebble (Ed.), Aerial Control of Forest Insects in Canada. Dept. of Environ., Ottawa, Ont.

West, R.J.; Cunningham, J.C.; Kaupp, W.J. 1987. Ground spray applications of Virtuss®, a nuclear polyhedrosis virus, against white-marked tussock moth larvae at Bottom

Brook, Newfoundland in 1986. Can. For. Serv., Nfld. For. Res. Cent., St. John's, Nfld. Inf. Rep. N-X-257. 10 p.

West, R.J.; Kaupp, W.J.; Cunningham, J.C. 1989. Aerial application of Virtuss®, a nuclear polyhedrosis virus, against whitemarked tussock moth in Newfoundland in 1987. For. Can., Nfld. and Lab. Region, St. John's, Nfld. Inf. Rep. N-X-270. 10 p.

Witter, J.A.; Kulman, H.M.; Hodson, A.C. 1972. Life tables for the forest tent caterpillar. Ann. Entomol. Soc. Am. 65: 25-31.

Yendol, W.G.; Hedlund, R.C.; Lewis, F.B. 1977. Field investigations of a baculovirus of the gypsy moth. J. Econ. Entomol. 70: 598-602.

Chapter 36

Fungal Pathogens

D.F. PERRY, D. TYRRELL, AND D. STRONGMAN

Introduction

Because so little was known about fungal pathogens, many early studies concentrated on different aspects of basic research, including the identification of species pathogenic to forest insects; the biological description of important species and their laboratory manipulation; and the elucidation of the influence of abiotic factors on fungal development, and on fungal morphology, epizootiology, and biochemistry.

In Canada and throughout the world, research on fungal insect pathogens has greatly advanced during the past 15 years. A fungal-based product had not been commercialized for any purpose at the time of Prebble's (1975) review of Canadian forestry activity. In the late 1970s and early 1980s, several products were developed for agricultural use, but none is currently registered for forestry use. Although small-scale trials against several insects have been successful in implanting mycosis, a lack of information on the epizootiological aspects of insect disease has led to variable results.

Biological control research on fungal pathogens includes the applied aspects of mass production on industrial media, bioassays and laboratory screening of pathogens, and small-scale trials. Initially, a fungus is considered to be a promising candidate if it has been grown on artificial medium in submerged culture and has a high degree of pathogenicity. Secondly, spore production under industrial-type fermentation conditions is investigated. Finally, it is necessary to master the germination processes that lead to infection in the field.

During the early 1980s, the Canadian research effort further concentrated on the fungal insect pathogens on forest pests within the framework of population dynamics studies, most notably, of the spruce budworm, *Choristoneura fumiferana*, in the belief that increased knowledge of the mechanisms by which mortality is determined would lead to strategic advances in application methodology. Although studies of fungal pathogens of other insects have continued and have shown the spectacular mortality that fungi can cause, they will not be reviewed here due to lack of space. Readers interested in background research and in studies on other insects should refer to the Bibliography.

Development and Selection of Fungi for Biocontrol

Fungi and Spruce Budworm

The spruce budworm is host to eight fungal pathogens, four of which occur frequently (Table 1). Mortality due to Entomophthorales species, primarily *Erynia radicans* and *Entomophaga aulicae*, can exceed 90% (Perry and Régnière 1986), although levels of up to 25% are more common (Fig. 1). The two most common species of Fungi Imperfecti, *Paecilomyces farinosus* and *Hirsutella longicolla*, have been observed to kill up to 40%, respectively (Fig. 2). Strategic use of either group continues to appear promising, and current efforts to develop biological controls have concentrated on *Entomophaga aulicae* and *P. farinosus*. Life cycle characteristics and recent applied research accomplishments justify this choice of species.

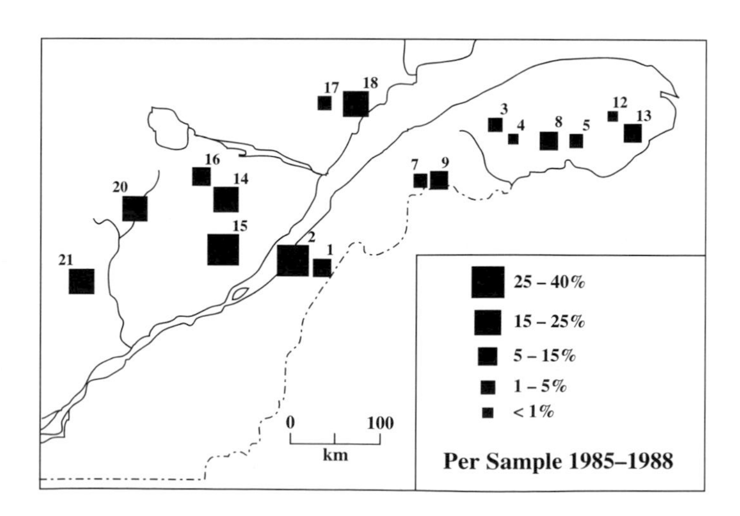

Figure 1. Maximum mortality due to *Paecilomyces farinosus* in Quebec.

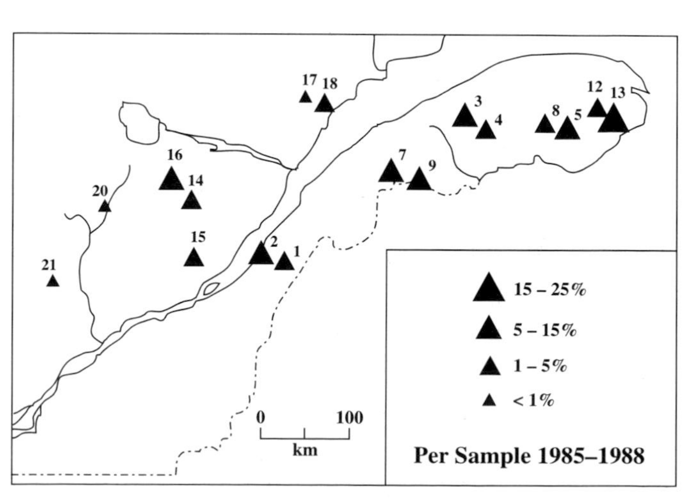

Figure 2. Maximum mortality due to *Entomophaga aulicae* in Quebec.

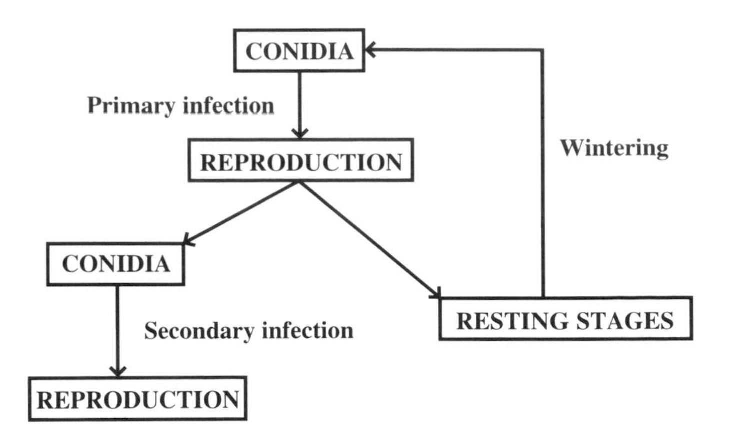

Figure 3. The fungal life cycle.

Table 1. Fungal pathogens of the spruce budworm, *Choristoneura fumiferana*.

Frequently encountered species	Less common species
Entomophthorales	Entomophthorales
Erynia radicans	*Conidiobolus* sp.
Entomophaga aulicae	
Fungi Imperfecti	Fungi Imperfecti
Paecilomyces farinosus	*Verticillium lecanii*
Hirsutella longicolla	*Metarhizium anisopliae*
	Beauveria bassiana

Table 2. Species of fungi screened against the spruce budworm, *Choristoneura fumiferana*.

Entomophthorales	Fungi Imperfecti
Erynia radicans	*Paecilomyces farinosus*
Entomophaga aulicae	*Paecilomyces fomusoroseus*
Conidiobolus sp.	*Paecilomyces canadensis*
	Beauveria bassiana
	Metarhizium anisopliae
	Hirsutella gigantae
	Hirsutella thompsonii
	Nomuraea rileyi
	Verticillium lecanii

Table 3. Current treatment strategies.

Entomophaga aulicae
Conidial application in highly fluctuating populations at second-instar emergence, peak fourth instar, or at egg hatch. Resting spore application before second-instar emergence (spring or late summer).
Paecilomyces farinosus
Conidial application in high resilient populations, at second-instar emergence, peak fourth instar, or at egg hatch.

The Fungal Life Cycle

The fungi are similar to other insect pathogens in many ways in that specialized stages adapted to particular purposes are formed at different times (Fig. 3). Infection stages lead to reproductive phases within the host and the disease is subsequently transmitted horizontally. Both infectious propagules (conidia) and survival stages (resting spores or sclerotia) are found outside the host and are thus responsible for transmission and persistence. Environmental conditions and genetic background both influence sporulation and spore form. Synchronous development of pathogen and host depends on suitable day-length or light intensity and temperature, because both of these factors influence the spore germination process that leads to infection (Payandeh et al. 1978; Perry and Fleming 1989a, b; Tyrrell 1988; van Roermund et al. 1984; Wallace et al. 1976). Selecting isolates to test for biological control purposes is difficult because of the genetic variability associated with these pathogens and the dramatic differences between individual isolates in their responses to external conditions. Forestry Canada currently manages a culture collection of more than 5 000 fungi pathogenic to forest insects and in particular to the spruce budworm. Field studies have contributed to the establishment of this culture bank and allowed us to more effectively introduce fungal pathogens of budworm into the forest ecosystem. Laboratory studies have continued to screen as many isolates as possible in relation to their potential control use (Table 2). The criteria for selection include temperature and light adaptations, inoculum performance, and persistence.

Treatment Goals in Relation to Fungal Biology

The long-term nature of epizootiological studies has been a limiting factor in proceeding to application trials, although useful in the derivation of new application strategies (Table 3). The majority of infectious inocula survive in the soil through the formation of resistant stages, or, as in *P. farinosus*, colonization of senescent balsam fir needles (Harney 1988). *Paecilomyces farinosus* attacks throughout the larval period (Fig. 4), whereas infection by *Entomophaga aulicae* is delayed until the fifth budworm instar, due to resting spore germination and larval movement.

Fungal infection patterns—Two different patterns in fungal infection of spruce budworm occur (Fig. 5). Rapid, spectacular mortality characterizes *E. aulicae*, whereas *P. farinosus* maintains moderate levels of infection for longer periods of time. Although both species have an LT_{50} of 3 to 4 days, they differ in their ability to kill, sporulate, and transmit the disease: *E. aulicae* sporulates in less than 24 hours, but *P. farinosus* requires several days; *E. aulicae* actively ejects conidia, and consequently disseminates more rapidly than *P. farinosus*, which does not; *E. aulicae* produces fewer conidia than *P. farinosus*, but they are more virulent; and finally, *E. aulicae* is easier to screen for highly pathogenic strains while *P. farinosus* is easier to screen for high conidia production. Isolates of both species also differ in their host specificity: some isolates can attack only spruce budworm, while others can attack several alternate hosts.

Development of epizootics—Epizootics of Entomophthorales species, as exemplified by *Entomophaga aulicae*, require a rapid and abundant production of conidia on budworm cadavers. Conodial producing isolates dominate until the end of an epizootic, whereupon resting spore producers increase in frequency. Many more resting spore producers are also found when the percentage mycosis is low. Abiotic conditions subtly influence the performance of an isolate, but an introduced strain must have the capacity to compete with background infections from naturally occurring pathogens. Experimental applications to characterize isolate performance in relation to indigenous populations are impractical, however, as part of a major screening process. In response to this dilemma, population genetic studies have been initiated in an attempt to quantify variability and gene flow among isolates (D.F. Perry, unpubl. data; D. Tyrrell, unpubl. data). Although a single genome can dominate a given population, other genetic patterns are present and genetic composition of populations varies with time. The nature of clonal varieties or dominant genomes may be implied from these observations, but conclusive evidence necessitates the study of hundreds of isolates.

A broad hypothesis for use in planning genetic studies of populations can be proposed, however, based on the previously mentioned observations. At any given time, a ratio of individuals exists in a population of fungi, some strains more adapted to infection and others more adapted to survival. Their relative proportions increase or decrease with time (the age of an infestation). Characteristics of the budworm population favorable to epizootics, including larval density and defoliation stimulating increased movement, will at some point coincide with a fungal population dominated by highly pathogenic individuals that rapidly produce conidia on the cadavers. Because of the lower survival rate of this isolate type, and the possible collapse of the budworm population, the relative frequency of survival forms increases the following year and lower infection rates are observed.

Imperfect fungi are somewhat similar in their epidemiology. *Paecilomyces farinosus* is rare in budworm populations until the infestation is at high density. Due to a lower pathogenicity, a large accumulation of infectious conidia is required for mortality to occur. This fungal proliferation occurs on senescent or dead balsam fir needles and is strongly associated with stand age. Differences in pathogenicity between isolates from needles and those obtained from dead budworms are not consistent nor representative of enzymatic profiles, because there is high within-group variation in pathogenicity and in the amount of conidia produced (Harney 1988; D.F. Perry, unpubl. data.).

Screening of pathogenic fungi—Screening of naturally occurring budworm pathogens and other fungi that are effective against a variety of insect pests has continued throughout the 1980s. The budworm is susceptible to a broad group of fungi (Jones 1985), even though many do not naturally infest the budworm in the Canadian forest. The degree of pathogenicity and the ability of a fungus to perform under northern forest conditions vary greatly within a single species. Isolates that unite the desired characteristics, both from a technical and an ecological point of view, are required. In *E. aulicae*, pathogenicity and epizootic potential are high, whereas persistence in a population is low. These characteristics point to its use as a prescribed treatment in highly fluctuating populations of budworm. In contrast, in *P. farinosus*, pathogenicity and epizootic potential are comparatively low, favoring an application in stable resilient populations. Both strategies imply rapid reductions in larval populations or the reduction of populations.

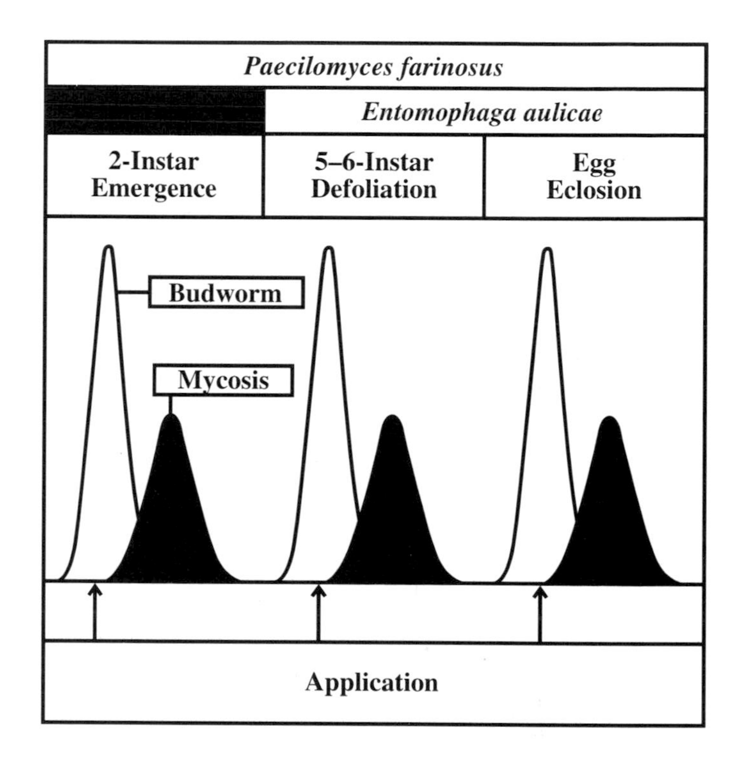

Figure 4. Natural mycosis and treatment strategies.

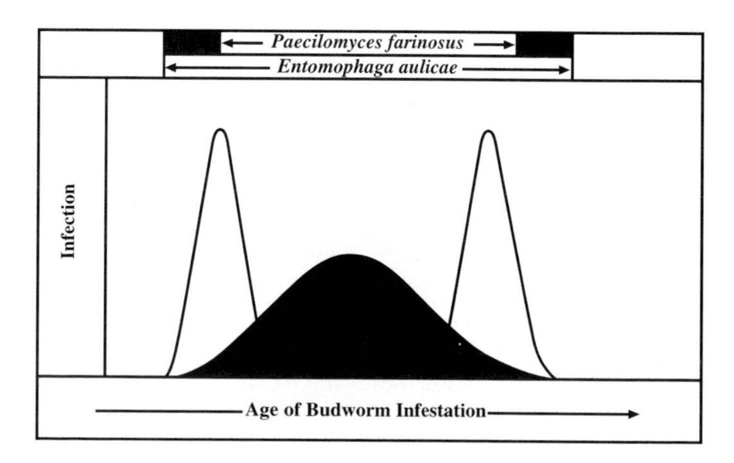

Figure 5. Long-term patterns in mycosis.

Experimental applications at early, mid-, and late season periods take advantage of larval movement and maximize disease introduction. It may be advantageous to combine isolates of like or different species.

Planning field trials—The success of field trials depends on the germination of resting spores and conidia, the persistence of infectivity, and the influence of the introduction on budworm dynamics under operational conditions (high damage risk infestations). Planning of such trials is now possible because sufficient observational and quantitative biological data are available, along with a synthesis of the life cycle processes of fungal pathogen species (Fleming 1985; Fleming and Perry 1986; Perry and Whitfield 1984). Corroborative field results emphasize the relationship between spore germination and the timing of infection (Perry and Fleming 1989a, b; D.F. Perry, unpubl. data). Additional information on the persistence and survival of spore forms in a treatment model has aided experimental field trials. Laboratory selection of isolates that are adapted to particular temperature and humidity regimes and are highly pathogenic has been accomplished. Fungi that are most persistent on balsam fir foliage have been identified for delivery at a time when the budworm is most accessible. Although extreme levels of solar radiation and relative humidity may hinder the process of infection, meteorological data obtained since the early 1980s suggest that these factors will generally not be limiting. Protective formulations can be developed to enhance treatment efficacy, and thus partially overcome the adverse effects of unfavorable weather.

Selecting isolates for field trials—*Entomophaga aulicae* isolates can be divided into those that produce resting spores and those that do not. Resting spores are highly resistant to external conditions and are an appropriate form to apply. Research has been oriented toward finding the culture conditions that would stimulate resting spore formation. Parallel studies on the pathogenicity of the two types of isolates indicate that resting spore producers are consistently less pathogenic than those isolates that yield only conidia (Papierok and Wilding 1979; D. Tyrrell, unpubl. data.). Resting spores germinate, producing infectious conidia that persist over a 1 to 2 month period, but conidia germinate in 24 hours and are faster acting. The selection of an isolate or spore form must consider a variety of ecological adaptations that will favor a given treatment strategy, such as germination rates of resting spores and conidia. The proposed treatments must therefore take these patterns into account. In selecting spore forms and isolates for biological control, many advantages can be gleaned from these phenomena. Conidia can be best used in situations where immediate larval mortality is desirable, within a timing margin associated with persistence on foliage. Resting spores can be applied well in advance of a target mortality date to take advantage of the natural course of events.

Operational control requirements—Operational control of the spruce budworm attempts to reduce larval populations before annual defoliation, requiring treatments to take effect before peak fifth-instar occurrence. Large tracts of forested land are aerially sprayed, requiring mass-produced, stable, and concentrated formulations. Droplet size and distribution must be optimized to achieve maximum delivery to the target. Fungal formulations based on those currently used with Bt may suffice, but the actual droplet spectrum for optimal use will depend on spore size and dosage values. *Entomophaga aulicae* spores are 40 µm in diameter, while those of *P. farinosus* are only 1 to 2 µm in diameter. Shear forces, created upon exit of small droplets from nozzles, and evaporation rates of these small droplets are important parameters that must be studied to determine whether the application of fungi is compatible with existing technologies. Fungi are readily produced in batch fermentation processes, but formation of the desired spore type may necessitate some modification of traditional methods.[1]

Resting spores are produced in a single fermentation, optimized for growth and spore yield. By mixing the active growing phase before sporulation with an inert material, the fermentation period can be shortened from 7 to 5 days. Similarly, conidial production as a two-stage process consists of growth optimization in submerged culture and the induction of sporulation on an inert material. Stabilization of live spores, using current methodologies, allows for a shelf life of 12 months, but this can be extended with cold storage. Progress in the formulation of fungi will require increased research and is subject to the vagaries of strain selection. Standardization of production for specific high performance isolates may be revised once a particular strain proves effective.

Insecticidal Metabolites

Recently, research efforts investigating fungal-produced toxins for use against insect pests have increased (Burges 1981; Clay et al. 1985; Carroll 1986; Carroll and Carroll 1978; Clark 1988; Dunphy and Nolan 1982; Grove and Pople 1980; Miller et al. 1985; Murakoshi et al. 1978; Strongman et al. 1988). Phylloplane fungi (epiphytic and endophytic species) are isolated and studied for their ability to produce toxins active against forest pests. Application of the fungus or a toxic compound and use as a vector for genes associated with the production of Bt toxin are being explored. The natural role of these metabolites and their mode of action are currently under investigation, thus adding a new dimension to the use of fungi in biological control. Toxins from the insect pathogens are also being studied.

Discussion

The past 25 years of research on fungal pathogens of insects has added to our basic knowledge of how fungi behave and has led to the resolution of many technical problems that impeded fungal use in biological control programs. It is perhaps time to add fungi to the arsenal with which operational programs attack insect pests. The registration of the first fungal-based product for forestry use in Canada will undoubtedly occur at the beginning of the next decade.

[1] Latgé, J.P.; Perry, D.F. 1980. Perfectionnements apportés aux procédés de préparation de spores durables d'entomophthorales pathogènes d'insectes, préparation de spores ainsi obtenus et compositions phytosanitaires contenant les dites préparations. Patent 80-24-769 (France). McCabe, D.E.; Soper, R.S. 1985. Preparation of an entomopathogenic fungal insect control agent. U.S. Patent 4,530,834.

Bibliography

Burges, H.D. 1981. Microbial control of pests and plant diseases. Academic Press, London, New York. 949 p.

Carroll, G.C. 1986. The biology of endophytism in plants with particular reference to woody perennials. Pages 205-222 *in* N.N. Fokkema and J. Van Den Neuvel (Eds.), Microbiology of the Phyllosphere. London, Eng. 392 p.

Carroll, G.C.; Carroll, F.E. 1978. Studies on the incidence of coniferous needle endophytes in the Pacific Northwest. Can. J. Bot. 56: 3034-3043.

Clark, C.L. 1988. Toxicity of endophytic fungi from conifer needles to the Eastern spruce budworm *Choristoneura fumiferana*. M. Sc. Thesis, Univ. of New Brunswick. 66 p.

Clay, K.; Hardy, T.N.; Hammond, A.M. 1985. Fungal endophytes of grasses and their effects on an insect herbivore. Oecologia (Berl.) 66: 1-6.

Dunphy, G.B.; Nolan, R.A. 1977. Regeneration of protoplasts of *Entomophthora egressa*, a fungal pathogen of the eastern hemlock looper. Can. J. Bot. 55: 107-113.

Dunphy, G.B.; Nolan, R.A. 1979. Effects of physical factors on protoplasts of *Entomophthora egressa*. Mycologia 71: 1589-602.

Dunphy, G.B.; Nolan, R.A. 1980a. A study of the surface proteins of *Entomophthora egressa* protoplasts and of larval spruce budworm hemocytes. J. Insect Pathol. 38: 352-361.

Dunphy, G.B.; Nolan, R.A. 1980b. Response of eastern hemlock looper hymocytes to selected stages of *Entomophthora egressa* and other foreign particles. J. Insect Pathol. 36: 71-84.

Dunphy, G.B.; Nolan, R.A. 1982. Cellular immune responses of spruce budworm larvae to *Entomophthora egressa* protoplasts and other test particles. J. Invertebr. Pathol. 39: 81-92.

Dunphy, G.B.; Nolan, R.A.; Otvos, I.S. 1977a. Ninhydrin-positive substance analysis of larval hemolymph of the eastern hemlock looper, *Lambdina fiscellaria fiscelleria* (Lepidoptera: Geometridae) and growth of *Entomophora egressa* protoplasts. Can. Entomol. 109: 341-346.

Dunphy, G.B.; Keough, K.M.W.; Nolan, R.A. 1977b. Fatty acid composition and lipid content of late larval and pupal stages of the eastern hemlock looper, *Lambdina fiscelleria fiscelleria* (Lepidoptera: Geometridae). Can. Entomol. 109: 347-350.

Fleming, R.A. 1985. Infectious diseases as part of the group of mortality factors affecting spruce budworm population dynamics. Pages 110-111 *in* C.J. Sanders, R.W. Stark, E.J. Mullins and J. Murphy (Eds.), Recent advances in spruce budworm research. Proc. CANUSA Spruce Budworms Res. Symp. Bangor, Me., Sept. 16-20, 1984. Can. For. Serv., Ottawa, Ont.

Fleming, R.A.; Perry, D.F. 1985. Simulation of phenological development to determine strategies for implanting entomophthoralean fungi in populations of eastern spruce budworm. Abstract *in* Proc. Annu. Meeting. Invertebr. Pathol, Sault Ste. Marie, Ont.

Fleming, R.A.; Perry, D.F. 1986. Simulation as an aid to developing strategies for forest pest management. Pages 694-698 *in* G.C. Vansteenkisste, E.J.H. Kerckhotts, L. Dekker, and J.C. Zuideerat (Eds.), Proc. 2nd. European Simulation Congr., Antwerp, Belgium, Sept. 9-12, 1986. Soc. Computer Simulation, San Diego, Calif.

Grove, J.R.; Pople, M. 1980. The insecticidal activity of Beauverin and the enniatin complex. Mycopathologia 70: 103-105.

Harney, S. 1988. Ecological and physiological aspects of *Paecilomyces farinosus* (Holm ex S.F. Gray) Brown and Smith, a potential biological control agent of the spruce budworm, M.Sc. Thesis, Concordia Univ. 83 p.

Jones, B.M. 1985. Effects of temperature on development and infectivity of anamorphic fungi with potential for control of eastern spruce budworm. M.Sc. Thesis, Univ. of Waterloo. 154 p.

Latgé, J.P.; Perry, D.F.; Prevost, M.C.; Samson, R.A. 1989. Ultrastructural studies of primary spores of *Conidiobolus, Erynia* and related Entomophthorales. Can. J. Bot. 67: 2576-2589.

Latgé, J.P.; Soper, R.S.; Madore, C.D. 1986. Media suitable for the industrial production of *Entomophthora virulenta* zygospores. Biotechnol. Bioeng. 19: 1269-1284.

Lim, K.P.; Perry, D.F. 1983. Field text of the entomopathogenic fungi, *Zoophthora radicans* and *Entomophthora egressa*, against the spruce budworm in western Newfoundland, 1982. Can. For. Serv., Nfld. For. Cent. 3281 (1982-1983a). St. John's, Nfld.

Lim, K.P.; Raske, A.G.; Tyrrell, D.; Perry, D.F. 1981. The 1981 field trials of the entomopathogenic fungi, *Zoophthora radicans* and *Entomophthora egressa*, to control spruce budworm in Newfoundland. Can. For. Serv., Nfld. For. Cent. 3281 (1981-1982a). St. John's, Nfld.

MacLeod, D.M.; Cameron, J.W.M.; Soper, R.S. 1966. The influence of environmental conditions on epizootics caused by entomogenous fungi. Rev. Roum. Biol. 11: 125-134.

MacLeod, D.M.; Muller-Kogler, E. 1970. Insect pathogens: species originally described from their resting spores as *Tarichium* species (Entomophthorales: Entomophthoraceae). Mycologia 65: 823-893.

MacLeod, D.M.; Muller-Kogler, E.; Wilding, N. 1986. Entomophthora species with *E. muscae*-like conidia. Mycologia 67: 1-29.

MacLeod, D.M.; Tyrrell, D.; Soper, R.S. 1979. *Entomophthora canadensis* n.sp., a fungus pathogenic on the woolly pine needle aphid, *Schizolachnus piniradiatae*. Can. J. Bot. 57: 2663-2672.

May, B.D.; Roberts, D.W.; Soper, R.S. 1979. Intra-specific genetic variation in laboratory strains of *Entomophthora* as determined by enzyme electrophoresis. Exp. Mycologia 3: 289-297.

McCabe, D.E.; Humber, R.A.; Soper, R.S. 1984. Observation and interpretation of nuclear reductions during maturation and germination of entomophthoalean resting spores. Mycologia 76: 1104-1107.

Miller, J.D.; Strongman, D.; Whitney, N.J. 1985. Observations on fungi associated with spruce budworm infested balsam fir needles. Can. J. For. Res. 15: 896-901.

Murakoshi, S.; Ichino, M.; Suzuki, A.; Kanaoka, M.; Isogai, A.; Tamura, S. 1978. Presence of toxic substances in fungus bodies of the entomopathogenic fungi *Beauveria bassiana* and *Verticillium lecanii*. Appl. Entomol. Zool. 13: 97-102.

Murrin, F.; Nolan, R.A. 1987. Ultrastructure of the infection of spruce budworm larvae by the fungus *Entomophaga aulicae*. Can. J. Bot. 65: 1694-1706.

Nolan, R.A.; Dunphy, G.B.; MacLeod, D.M. 1976. In vitro germination of *Entomophthora egressa* resting spores. Can. J. Bot. 54: 1131-1134.

Papierok, B.; Wilding, N. 1979. Étude de comportement de plusieurs souches de *Conidiobolus obscurus* (Zygomycetes: Entomophthoraceae) vis-à-vis des pucerons *Acyrthosiphon pisum* et *Sitobion avenae* (Homoptera: Aphididae). Entomophaga 26: 241-249.

Payandeh, B.; MacLeod, D.M.; Wallace, D.R. 1978. Germination of *Entomophthora aphidis* resting spores under constant temperatures. Can. J. Bot. 56: 2328-2333.

Payandeh, B.; MacLeod, D.M.; Wallace, D.R. 1980. An empirical regression function suitable for modelling spore germination subject to temperature threshold. Can. J. Bot. 58: 936-941.

Perry, D.F. 1985. Epizootic development of entomophthoralean fungi. Page 107 *in* C.J. Sanders, R.W. Stark, E.J. Mullins and J. Murphy (Eds.), Recent advances in spruce budworms research. Proc. CANUSA Spruce Budworms Res. Symp. Bangor, Me., Sept. 16-20, 1984. Can. For. Serv., Ottawa, Ont.

Perry, D.F. 1988. Germination of *Erynia bullata* resting spores. J. Insect Pathol. 51: 161-162.

Perry, D.F.; Fleming, R.A. 1989a. *Erynia crustosa* zygospore germination in relation to mycosis of *Choristoneura fumiferana*. Mycologia 81: 154-158.

Perry, D.F.; Fleming, R.A. 1989b. The timing of *Erynia radicans* resting spore germination in relation to mycosis of *Choristoneura fumiferana*. Can. J. Bot. 67: 1657-1663.

Perry, D.F.; Régnière, J. 1986. The role of fungal pathogens in spruce budworm population dynamics: frequency and temporal relationships. Pages 167-170 *in* J.M. Vlak and D. Peters (Eds.), Fundamental and applied aspects of invertebrate pathology. Wageningen, the Netherlands.

Perry, D.F.; Tyrrell, D.; Delyzer, A.J. 1982. The mode of germination of *Zoophthora radicans* zygospores. Mycologia 74: 549-554.

Perry, D.F.; Whitfield, G.H. 1984. The interrelationship between microbial entomopathogens and insect hosts: a systems' study approach with particular reference to the Entomophthorales and the Eastern spruce budworm. Pages 307-331 *in* J.M. Anderson, A.D.M. Rayner and D.W.H. Walton (Eds.), Invertebrate-microbial Interactions. Cambridge Univ. Press, New York.

Prebble, M.L. 1975. Aerial control of forest insects in Canada. Environ. Can., Ottawa, Ont.

Otvos, I.S.; MacLeod, D.M.; Tyrrell, D. 1983. Two species of *Entomophthora* pathogenic to the eastern hemlock looper (Lepidoptera: Geometridae) in Newfoundland. Can. Entomol. 105: 1435-1441.

Smirnoff, W.A. 1970. Fungus diseases affecting *Adelges piceae* in the fir forest of the Gaspé peninsula, Quebec. Can. Entomol. 102: 799-805.

Soper, R.S. 1963. *Massospora levispora*, a new species of fungus pathogenic to the cicada, *Okanagana rimosa*. Can. J. Bot. 41: 875-878.

Soper, R.S. 1974. The genus *Massospora*, entomopathogenic for cicadas, Part I, Taxonomy of the genus. Mycotaxon 1: 13-40.

Soper, R.S.; DeLyzer, A.J.; Smith, L.F.R. 1976a. The genus *Massospora*, entomopathogenic for cicada, Part II, Biology of *Massospora levispora* and its host *Okanagana rimosa*. With notes on *Massospora cicadina* on the periodical cicadas. Ann. Entomol. Soc. Am. 69: 89-95.

Soper, R.S.; Smith, L.F.R.; DeLyzer, A.J. 1976b. Epizooiology of *Massospora levispora* in an isolated population of *Okanagana rimosa*. Ann. Entomol. Soc. Am. 69: 275-283.

Soper, R.S.; MacLeod, D.M. 1981. Descriptive epizooiology of an aphid mycosis. U.S. Dept. Agric., Washington, D.C. Publ. 1632. 17 p.

Soper, R.S.; May, B.; Martinelli, B. 1983. *Entomophaga grylli* enzyme polymorphism as a technique for pathotype identification. Environ. Entomol. 12: 720-723.

Strongman, D.B.; Strunz, G.M.; Giguère, P.; Yu, C.M.; Calhoun, L. 1988. Enniatins from *Fusarium avenaceum* isolated from balsam fir foliage and their toxicity to spruce budworm larvae, *Choristoneura fumiferana* (Clem.) (Lepidoptera: Tortricidae) J. Chem. Ecol. 14: 753-764.

Tyrrell, D. 1967. The fatty acid composition of 17 *Entomophthora* isolates. Can. J. Microbiol. 13: 755-760.

Tyrrell, D. 1969. Biochemical systematics of fungi. Bot. Rev. 35: 305-316.

Tyrrell, D. 1977. Occurrence of protoplasts in the natural life cycle of *Entomophthora egressa*. Exp. Mycologia 1: 259-263.

Tyrrell, D. 1988. Survival of *Entomophaga aulicae* in dried insect larvae. J. Invertebr. Pathol. 52: 187-188.

Tyrrell, D.; MacLeod, D.M. 1972. Spontaneous formation of protoplasts by a species of *Entomophthora*. J. Insect Pathol. 19: 354-360.

Tyrrell, D.; Peberdy, J.F. 1979. Spontaneous and enzyme-induced protoplasts from *Entomophthora* species. Adv. Proto. Res. 5: 229-234.

Vandenberg, J.D.; Soper, R.S. 1975. Isolation and identification of *Entomophthora* spp. Fres. (Phycomycetes: Entomophthorales) from the spruce budworm, *Choristoneura fumiferana* Clem. (Lepidoptera: Tortricidae.) J. N.Y. Entomol. Soc. 83: 254-255.

Vandenberg, J.D.; Soper, R.S. 1978. Prevalence of Entomophthorales mycosis in populations of spruce budworm, *Choristoneura fumiferana*. Environ. Entomol. 7: 847-853.

Vandenberg, J.D.; Soper, R.S. 1979. A bioassay technique for *Entomophthora sphaorosperma* on the spruce budworm, *Choristoneura fumiferana*. J. Insect Pathol. 33: 148-154.

van Roermund, H.; Perry, D.F.; Tyrrell, D. 1984. Influence of temperature, light, nutrients and pH in determination of the mode of conidial germination in *Zoophthora radicans*. Trans. Br. Mycol. Soc. 82: 31-38.

Wallace, D.R.; MacLeod, D.M.; Sullivan, C.R.; Tyrrell, D.; DeLyzer, A.J. 1976. Induction of resting spore germination in *Entomophthora aphidis* by long day light conditions. Can. J. Bot. 54: 1410-1418.

Whitney, H.S. 1982. Relationship between bark beetles and symbiotic organisms. Pages 183-211 *in* J.B. Mitton and K.B. Sturgeon (Eds.), Bark beetles in North American conifers, a system for the study of evolutionary biology. Univ. Texas Press, Austin, Texas.

Chapter 37

Protozoa and Nematodes

G.G. Wilson, J.R. Finney-Crawley, and D.C. Eidt

Introduction

Protozoa and nematodes have probably received the least amount of attention of all the parasitic organisms as control agents against Canadian forest pest insects; however, as research advances, the potential of these organisms will surely be enhanced.

Protozoans are a group of naturally occurring insect pathogens that often have adverse effects on the survival and reproduction of forest insect pests. The microsporidia, in particular, play a significant role in the natural suppression of some insect populations (Henry 1981; Wilson 1982a). Protozoa generally produce chronic rather than acute diseases, and their effects as control agents are therefore not as dramatic as some of the other pathogenic microbes. Interest in microsporidia as potential biological control agents has been steadily increasing, probably because current pest management concepts emphasize conservation and augmentation of existing natural control agents, but much basic research is still needed.

The only insect parasitic nematodes that have been seriously considered as biological control agents of forest insect pests in Canada belong to the genera *Steinernema* and *Heterorhabditis*. Many species and strains of these nematodes have been isolated, some of which differ from each other in characteristics such as temperature tolerance, resistance to desiccation, and host-seeking ability. In many parts of the world these nematodes are available and continue to be evaluated for use in diverse environments for the control of a range of insect pests of economic importance (Gaugler 1981; Gaugler and Kaya 1990; Kaya 1985; Nickle 1984; Webster 1980). Although available in Canada, their potential for forest insect pest control under this specific environmental regime has yet to be exploited.

The available information on protozoa and nematodes infecting major Canadian forest insect pests is reviewed in this chapter and their value as control agents is also discussed. The discussion for protozoa is limited to microsporidia, as this is the most studied group of protozoa in forest insects.

General Considerations

Other than brief mention in a few reports (Brooks 1980; Henry 1981; and Wilson 1982a), no general review has been presented on protozoan pathogens of Canadian forest insect pests. Reports by Smirnoff and Juneau (1973), Wilson (1981a), and general insect disease surveys conducted by the Forest Pest Management Institute of Forestry Canada indicate that numerous Canadian forest insects are subject to infection by various species of microsporidia. The spruce budworm, *Choristoneura fumiferana*, is the only major forest pest that has received much attention, as far as the microsporidia are concerned (Wilson 1981b; Wilson et al. 1984). Limited studies have been conducted of microsporidia infecting the forest tent caterpillar and the larch sawfly, and species infecting other insects have been identified.

Steinernematid and heterorhabditid nematodes possess many attributes that support their use as control agents in place of or integrated with chemical insecticides. The infective stage of the nematodes possesses a distinct advantage over chemical control agents in that some are attracted to and will actively seek out an insect host (Gaugler et al. 1980; Pye and Burman 1981; Schmidt and All 1978, 1979), whereas chemical control agents have to be brought into contact with or ingested by the target insect. Therefore, nematodes can be considered for control of feeding and nonfeeding stages of an insect and can be particularly useful for the control of insects in cryptic habitats (Dutky 1959; Finney and Walker 1977; Lindegren et al. 1978, 1981; Miller and Bedding 1982; Moore 1970).

The infective nematode may gain access to a host via the mouth, anus, spiracles, or directly through the cuticle. Once inside the host it must reach the haemocoel before further development can occur. Here the nematode releases a plug of symbiotically associated bacteria that multiply in the insect, causing a fatal septicemia (usually within 24 to 48 hours). Subsequent to host mortality, the nematode feeds on the bacteria and multiplies until the food source is exhausted. At this time the infective stage nematodes are produced and these escape from the cadaver into the environment. The number of larvae produced depends on the susceptibility and size of the insect host but may be in the hundreds of thousands.

The high reproductive potential of the nematode has been used for its mass production. A wide range of insect hosts has been used in small-scale in vivo methodologies (Dutky et al. 1964; Lindegren et al. 1979); however, these tend to be labor intensive and costly. Recent advances in vitro culture have led to the production of nematodes on solid and liquid media on a more commercially viable scale (Popiel 1987). The nematodes can be stored and transported in water, on matrices such as a damp sponge, or in a partially dehydrated or "anhydrobiotic" condition. Application in the field can be carried out using any of the conventional methods currently used in the application of chemicals (Dutky 1959), and indeed the nematodes could be applied with some of the commercially available chemical insecticides or microbial agents (Hara and Kaya 1982, 1983a,b; Kamionek 1979; Lam and Webster 1972). The nematodes have undergone a series of safety tests and have proved to be of no threat to any mammal (Gaugler and Boush 1979; Poinar 1972). They did not affect nontarget soil or aquatic arthropods in a variety of turfgrass, crop, and stream situations (Georgis et al. 1991).

The susceptibility of a range of forest insect pests to steinernematid and heterorhabditid nematodes have been tested in the laboratory, in most cases with positive results

(Table 1); however, these results have rarely been duplicated in the field (Finney and Walker 1979; Kaya and Reardon 1982; Kaya et al. 1981; Pye and Pye 1985; Reardon et al. 1986). The reason lies in the environmental sensitivity of these nematodes, particularly to moisture, temperature, and ultraviolet light. The presence of moisture is essential to survival of the infective stage nematode in the field. Application of the infective stage nematodes to exposed surfaces results in their rapid drying, which is lethal. Survival may be enhanced somewhat at high humidities, but in the forest environment foliar applications without anti-desiccant protection are usually ineffective.

Compared with foliar application, field tests in which the nematodes were applied to the soil have had far more favorable results. Although rapid drying on an exposed surface is lethal to the infective stage of nematodes, they can survive even prolonged drying in soil due to the slow and gradual nature of the drying process. Nematodes survive at soil moisture values well below the permanent wilting point of plants and readily revive when higher moisture levels return (Simons and Poinar 1973). The only restrictive effect of low moisture levels in the soil is to limit nematode movement during these periods.

Temperature affects both infectivity and ability of infective stage nematodes to cause host mortality. The first steinernematids and heterorhabditids that were isolated and subsequently used in laboratory and field experiments had reduced infectivity below 17°C (Finney and Bennett 1984a). Optimum activity occurred at 25°C with no development above 33°C or below 10°C (Kaya 1977). More recently, steinernematids that are infective and lethal to hosts at temperatures as low as 4°C have been isolated (Finney 1984).

Solar radiation adversely affects steinernamatid and heterorhabditid nematodes. Natural sunlight reduces the pathogenicity of infective stage larvae, while irradiation with ultraviolet light for very short periods of time impairs nematode pathogenicity, development, and reproductive capacities (Gaugler and Boush 1978).

Because particular problems are associated with the use of nematodes for the control of specific pests, the key to successful use of these nematodes lies in the choice of an appropriate nematode for use in a certain environment against a particular stage(s) of development of the target insect.

Among the susceptible insects listed in Table 1, we will discuss the potential use and specific field-related problems that could have an impact on effective control of the spruce budworm, *Choristoneura fumiferana*, the spruce bud moth, *Zeiraphera canadensis*, and the seedling debarking weevil, *Hylobius congener*.

Spruce Budworm

Microsporidia

Nosema fumiferanae is probably the most widespread and common pathogen of the spruce budworm. It was first reported by Graham (1948) and described in detail by Thomson (1955), who reported that although the main site of infection is the mid-gut cells, most other tissues eventually become infected. Observations based on laboratory tests and examinations of field-collected budworm show that *N. fumiferanae* retards larval and pupal development, lowers pupal weight, and reduces adult fecundity (Thomson 1958; Wilson 1977, 1983). Wilson (1985) reported an LD_{50} of 2.8×10^6 *N. fumiferanae* spores for fourth-instar spruce budworm larvae with an LT_{50} of about 17 days. The greatest number of spores in an infected budworm was 2.0×10^7/mg of body weight and this occurred after ingestion of 5×10^6 spores. Bauer and Nordin (1988), using a different method of dosing (synthetic diet instead of balsam fir needles), reported an even lower LD_{50} of 2.0×10^5 *N. fumiferanae* spores, for fourth-instar spruce budworm with an LT_{50} range of 6 to 19 days.

Two studies of wild populations show that the percentage of infection increases with the age of infestation. In a study of the Uxbridge forest of Ontario, Thomson (1960a) reported that infections in overwintering larvae from 1955 to 1959 were 36.4, 45.7, 56.1, 69.1, and 81.3%, respectively, and stated, "Great reductions in egg numbers in 1959 and the subsequent low larval population were probably due to direct larval mortality and a reduction in fertility of adults, both caused by microsporidia." The levels of *N. fumiferanae* infections in increasing and decreasing spruce budworm populations were studied during a 4-year period (Wilson 1987). Infections in the increasing budworm population increased from 18% to more than 30% during the period, while infections in the declining population increased from 52% to 62%. Levels of infection are therefore significantly higher in the older (declining) infestations.

Nosema fumiferanae is readily transmitted perorally and transovarially into its host, the spruce budworm. Infected females but not infected males readily transmit the pathogen to their offspring (Wilson 1982b). A study of the relationship between infected females and their offspring showed that an average of 90% of their offspring were infected (G.G. Wilson, unpubl. data). Increasing the spore counts in infected adults caused an increase in the spore counts of the larvae, but at a declining rate.

The spruce budworm is also readily infected by the microsporidium *Pleistophora schubergi* (Wilson 1984a), which occurs in field populations of this insect (Wilson 1975). The LD_{50} for fourth-instar larvae is 6.3×10^4 spores, which is considerably lower than that of *N. fumiferanae*. This microsporidium also reduces budworm weight, adult longevity, and fecundity. Transovarial transmission of the pathogen could not be demonstrated (Wilson 1984b).

Preliminary field spray trials have been carried out to test the effectiveness of *N. fumiferanae* and *P. schubergi* against the spruce budworm (Wilson and Kaupp 1976). A packsack-type mist blower was used to apply suspensions of microsporidian spores on individual white spruce, *Picea glauca*, and balsam fir, *Abies balsamea*, trees infested with spruce budworm. The aqueous suspension contained 25% (v/v) molasses, 30 g/L of IMC 90-001 (Sandoz Wonder, Inc., Homestead, Florida) sunlight protectant, and the desired spore concentration to provide 5×10^{10} spores per tree. Levels of infection for *N. fumiferanae* in budworm 19 days after treatment were 44% on white spruce and 66% on balsam fir; 25 days after treatment the values were 60% on white spruce and 73% on balsam fir. The levels of *N. fumiferanae* infection

Table 1. Susceptibility of some forest insect pests to steinernematid and heterorhabditid nematodes.

Insect species	Larvae	Pre/pupa	Adult	References
Lepidoptera				
Actebia fennica	+			Finney-Crawley, unpubl. data
Choristoneura fumiferana	+	+-		Schmiege 1963
	+	+-		Finney et al. 1982
	+			Finney and Bennett 1984a
Choristoneura occidentalis	+			Kaya et al. 1981
Choristoneura pinus	+	+-		Schmiege 1963
Coleophora serratella	+-			Finney and Bennett 1984b
Lambdina fiscellaria		+		Finney and Bennett 1984b
Leucoma salicis	+-			Finney and Bennett 1984b
Lymantria dispar	+			Reardon et al. 1986
	+			Shapiro et al. 1985
Ryacionia buoliana	+			Schmiege 1963
Symmerista albifrons	+-			Schmiege 1963
Zeiraphera canadensis	+	+	+	Finney-Crawley, unpubl. data
Coleoptera				
Dendroctonus frontalis	+			Moore 1970
Dendroctonus ponderosae	+	+	+-	MacVean and Brewer 1981
Hylobius abietis	+-			Pye and Burman 1977, 1978
Hylobius congener	+		+	Finney-Crawley, unpubl. data
Hylobius pales	+			Thomas 1970
Hylobius radicis	+			Jackson and Moore 1969
			+-	Schmiege 1963
Monochamus scutellatus	+-	+-		Schmiege 1963
Pissodes strobi			-	Schmiege 1963
Xanthogaleruca luteola	+	+		Kaya et al. 1981
Scolytus multistriatus	+	+	+	MacVean and Brewer 1981
Scolytus scolytus	+		+	Finney and Walker 1976
	+			Finney and Walker 1977
Hymenoptera				
Cephalcia lariciphila		+		Georgis 1981
Diprion similis	+			Schmiege 1963
Neodriprion lecontei	+			Schmiege 1963
Neodriprion sertifer	+	+		Finney and Bennett 1983
Pikonema alaskensis	+			Schmiege 1963
Pristiphora erichsonii	+	+-		Schmiege 1963
	+			Finney and Bennett 1983
	+			Webster and Bronskill 1968
Pristiphora geniculata	+			Finney and Bennett 1983

\+ = very susceptible.
+- = moderately susceptible.
\- = resistant.
Blank = not tested.

in the check plots ranged from 25% to 38%. Application of *P. schubergi* spores resulted in infection rates of 65% for budworm on white spruce and 96% for those on balsam fir 19 days after treatment. There was no infection in larvae from the check area. These results were based on data recorded when larvae were treated predominantly in the fourth and fifth instars. *Nosema fumiferanae* infection in the treated area remained higher than in the check area for 3 years after treatment. Although initial infection by *P. schubergi* was higher than that of *N. fumiferanae*, there was no carry-over of *P. schubergi* to the next year. Application of *N. fumiferanae* to spruce trees when budworms were in the needle-mining stage did not increase the levels of infection. This probably occurred because of the precise timing that would be necessary for the spray application to contact larvae moving from their hibernacula to the foliage, and larvae in needles would not be exposed to the microsporidian spores.

Nematodes

The first three larval instars of the spruce budworm occupy cryptic habitats to which delivery of chemical insecticides is a problem. Control measures are, therefore, usually directed at the fourth-, fifth-, and sixth-instar larvae which feed on the foliage before pupation. The microhabitats of the needle-mining and bud-mining first-, second-, and third-instar larvae are environments with high humidity, which can support nematode survival once access is gained to those sites. The nematode of choice, therefore, is one that is able to actively seek out the target insects at the temperature at which these larval instars are developing.

Laboratory study (Finney and Bennett 1984a) showed that due to its seeking habit, *Heterorhabditis bacteriophora* could penetrate the webbing of the hibernacula and kill first and second instars in them. The third-instar larva was also susceptible in petri dish experiments. However, the infectivity of *H. bacteriophora* decreased with temperature and was completely reduced under the temperature regime of 4° to 12°C at which the first-, second-, and third-instar larvae develop in Newfoundland. To attempt control of the early instars in the field, a low temperature tolerant nematode with a good host-seeking capacity such as *Steinernema feltiae* or *Steinernema* sp. Newfoundland strain No. 1 would have to be used.

Finney et al. (1982) found that the fourth-, fifth-, and sixth-instar larvae of the spruce budworm plus the pupal stage were all susceptible to *H. bacteriophora*. In this case, *H. bacteriophora* can be applied as a control measure because it would be infective over the temperature range at which these larval stages develop. Furthermore, the effective success of an application could be extended by the demonstrated mass production of the infective stage in infected hosts. Depending on the timing of the original application, the infective stages could infect later instars, pupae, or possibly early instars of the next budworm generation once released from the cadavers into the environment.

The use of more than one nematode increases the time frame over which nematode applications can be made in the field. However, unless humidity is high at the time of application, it seems likely that an anti-desiccant would have to be added to the spray formulation to ensure survival of the nematodes until infection of the target host occurred. Kaya and Reardon (1982) evaluated foliar applications of *Steinernema carpocapsae* in three different tank mixes for control of the western spruce budworm, *Choristoneura occidentalis*, and determined that Norbak, a water-absorbent polymer, and Methocoel J75MS, a methlycellulose polymer, extended nematode survival longest but could not obtain significant nematode infections of larvae or pupae in either of these formulations. Research directed at extending nematode persistence in the field during the post-treatment period is essential to effective foliar application of nematodes for control of this insect.

Forest Tent Caterpillar

Microsporidia

The forest tent caterpillar, *Malacosoma disstria*, is commonly infected in nature by the microsporidium *Nosema disstriae*. This pathogen attacks primarily the silk glands and the midgut epithelium but other tissues are also infected (Thomson 1959). Spores of *N. disstriae* fed to larvae of the tent caterpillar adversely affected pupal weights, adult fecundity, and longevity (Wilson 1979). Wilson (1984c) demonstrated that the forest tent caterpillar is also susceptible to *Pleistophora schubergi* and *Vairimorpha necatrix*. Wilson (1984c) tested third- and fifth-instar larvae of the tent caterpillar against all three microsporidia. At the time of treatment, larvae received doses of 5×10^3, 5×10^4, 5×10^5, 5×10^6, and 5×10^7 spores. Trembling aspen, *Populus tremuloides*, leaf disks (5 mm in diameter) were treated with 5 µL of the required spore suspension. More than 80% mortality occurred for all three microsporidia when third-instar larvae ingested 5×10^4 spores. The relative susceptibility of the tent caterpillar to the three pathogens was indicated by the LD_{50} values for spores to kill fifth-instar larvae; these were *V. necatrix*, 1.4×10^3; *P. schubergi*, 2.3×10^4; and *N. disstriae*, 2.3×10^5.

The only field trial using microsporidia against the forest tent caterpillar was conducted by Wilson and Kaupp (1977). These tests consisted of a combination of two microsporidian pathogens, *N. disstriae* and *P. schubergi*. Treated trees (trembling aspen) were sprayed with 150 mL of an aqueous formulation consisting of 50% (v/v) molasses and 30 g/L of IMC 90-001 sunlight protectant applied with a packsack-type mist blower. Each tree received a total of 1.8×10^{11} spores. Spraying with 1:1 mixture of *N. disstriae* and *P. schubergi* spores resulted in a significant increase in levels of *P. schubergi* but not *N. disstriae*. The average level of infection for *P. schubergi* was 76% in treated larvae compared with 0% in larvae from the check plots. A second area was treated with a 96:4 mixture of *P. schubergi* and *N. disstriae* spores. The average level of infection in this case was 85% for *P. schubergi* and 25% for *N. disstriae* with 0% and 23%, respectively, in the check area. As was the case for spruce budworm treated with *P. schubergi*, there was no carry-over of the pathogen in forest tent caterpillar (Wilson 1980).

Spruce Bud Moth

Nematodes

The spruce bud moth, *Zeiraphera canadensis*, is particularly difficult to control by conventional means because it is rarely exposed (Turgeon 1985). Exposure occurs only after hatching when the first instars migrate to the expanding bud, when the fourth-stage larvae leave the foliage and drop to the ground, and when the adult stage is reached. The second- and third-instar larvae feed on needle tips under the bud cap. Laboratory investigations showed that *Heterorhabditis bacteriophora* could penetrate the buds and kill the larvae inside (Turgeon and Finney-Crawley 1991) and that the fourth instars, pupae, and adults were susceptible to the nematode (D.C. Eidt and J.R. Finney-Crawley, unpubl. data).

In this situation, there are two options for use of nematodes in a control program against this pest: a foliar or a ground application. Foliar application against all pre-pupal instars is subject to the nematode temperature tolerance requirements and the formulation of the nematode applications to enhance survival on the foliage, a methodology that requires further research. The alternative, a nematode spray directly into the leaf litter at the base of the trees later in the season would be more conducive to nematode survival. Such an experiment was carried out in 1988 (Eidt and Dunphy 1991).

Three species of nematodes in suspensions were applied to the soil beneath white spruce trees in split-tree treatments in which one half was the control. The treatments were applied in late evening to avoid rapid evaporation and desiccation of the nematodes. The results indicated that *H. bacteriophora* was not an effective control agent (D.C. Eidt and J.R. Finney-Crawley, unpubl. data). However, only small doses were applied (0.1×10^4 nemas/m^2). *Steinernema* sp. Newfoundland strain No. 1 (*feltiae* in Eidt and Dunphy 1991) reduced emergence by 71% even though the rate applied was low (0.13×10^6 nemas/m^2). *Steinernema carpocapsae* was effective at all rates: 53% reduction (4.0×10^6 nemas/m^2), 79% reduction (20×10^6 nemas/m^2), and 67% reduction in emergence (40×10^6nemas/m^2). There was evidence, however, of nematode migration from the treated to the untreated side of the trees.

In 1989, further soil treatments, in which whole tree plots were treated, gave clearer results. However, the highest population reductions, measured as moths emerging, were 68% using nearly 28 million *S. carpocapsae* per square metre, and 76% using over 55 million per square metre (Eidt and Dunphy 1991).

On the assumption that proprietary Bt formulations contain appropriate, biologically benign anti-desiccants and wetting agents, Eidt and Dunphy (1991) devised a spray mixture of *S. carpocapsae* and Futura XLV. They successfully reduced populations by 82% in 1989 when they sprayed trees to runoff using fewer than 10 000 infective juveniles and 125 mL Futura per litre of aqueous mix. The results were repeated in 1990 and 1991 using Futura XLV and Foray 48B for adjuvants, and at the same time the amounts of Bt formulation and total spray volume were substantially reduced.

Larch Sawfly

Microsporidia

A microsporidian species was first described from the larch sawfly, *Pristiphora erichsonii*, by Smirnoff (1966) and assigned the name *Thelophania pristiphorae*. Another microsporidium, thought to be *Pleistophora* sp., was identified from larch sawfly in Sault Ste. Marie in 1973 collections from Uxbridge Township, Ontario. Percy et al. (1982) described this pathogen in some detail and indicated that infection levels up to 25% occurred in natural populations of the larch sawfly; also, the insect is susceptible to the microsporidium *P. schubergi* (Wilson 1981c). When first- and second-instar larvae were fed in the laboratory on foliage that had been sprayed with aqueous suspensions that contained 4×10^6 and 4×10^8 *P. schubergi* spores per millilitre, the resulting infection levels were 45% and 65%, respectively. An aqueous suspension of *P. schubergi* containing 5×10^8 spores/mL and 30 g/L of IMC-90-001 sunlight protectant was also sprayed on branches of larch trees naturally infested with colonies of the larch sawfly. This treatment resulted in infection levels of 40% compared with 10% in larvae from the check area.

Seedling Debarking Weevil

Nematodes

Hylobius congener, a seedling debarking weevil, is an important forest pest. The species colonizes clearcut areas and the adults feed extensively on the bark of planted seedlings. As they tend to be concealed feeders, control by conventional means is difficult, but successful field trials with nematodes for the control of a similar pest have been carried out in Sweden. Burman et al. (1979) showed that the larval populations of the large pine weevil, *H. abietis*, decreased sharply after *Steinernema carpocapsae* Leningrad strain was applied to tree stumps. This method, however, was labor intensive. In 1985, Pye and Pye (1985) dipped young seedlings in soil into a suspension of the infective stage of *S. carpocapsae* Umeå strain before planting. Mortality among the nematode-treated plants due to gnawing by adults of *H. abietis* was reduced to 21% compared with 64% mortality in controls. The larval populations were reduced by 89%. Seedlings were also sprayed with nematodes in suspensions with and without anti-desiccant. These treatments were less effective in protecting seedlings.

Laboratory tests showed that the larval and adult stages of *H. congener* were susceptible to *Steinernema* sp. Newfoundland strain No. 1 (J.R. Finney-Crawley, unpubl. data). On the basis of these tests, a field trial was carried out in Nova Scotia in 1988 (D.C. Eidt and J.R. Finney-Crawley, unpubl. data) whereby the roots of red pine, *Pinus resinosa*, seedlings in multipots were dipped in a suspension of the infective stage nematodes before planting. Although some cross-infection and reinfestation occurred due to the mobility of the weevils, comparison of treated and untreated seedlings suggested that the nematode treatment increased seedling survival.

In 1989 and 1990, experiments were carried out in field cages with introduced weevil populations (D.C. Eidt, S. Zervos, and A.E. Pye, unpubl. data). It was clearly demonstrated that either contamination of controls by infected, hyperactive weevils, invasion of controls by weevils from adjacent untreated areas, or both were responsible for the lack of clearly defined results from the 1988 field experiment.

In 1991, entire clearcuts were treated by applying an aqueous suspension with a watering can to seedlings in multipots the evening before they were to be planted by conventional means (D.C. Eidt, S. Zervos, and A.E. Pye, unpubl. data). *Steinernema carpocapsae* was applied at 400 000 Umeå strain and 500 000 All strain per tree. Damage assessment in the fall showed that seedling losses were reduced from 37% and 47% in two controls to 5% and 7% in the treatments. Losses up to 10% are tolerated before replanting or fill-in planting is deemed necessary. The experimental design did not permit statistical analysis of the results, so it will have to be demonstrated that the results can be repeated before a recommendation can be made.

Other Insects

Microsporidia

Table 2 lists some of the Canadian forest insects from which protozoa have been isolated or have proved to be susceptible in laboratory tests. Samples submitted to the disease survey in Sault Ste. Marie have shown, however, that other forest insects have also been infected with microsporidia. For example, during 1986 and 1987, the following insects were found to be infected with as yet undescribed species of microsporidia: spruce coneworm, *Dioryctria reniculelloides*; fir coneworm, *D. abietivorella*; Bruce spanworm, *Operophtera bruceata*; mountain-ash sawfly, *Pristiphora geniculata*; eastern blackheaded budworm, *Acleris variana*; and linden looper, *Erannis tiliaria*.

Nematodes

The black vine weevil, *Otiorhynchus sulcatus*, is frequently a pest in forest nurseries, particularly of trees held longer than 1 year, such as those used for grafting. For this species, one of the principal pests of outdoor potted plants, ornamental conifers, and small fruit crops, entomopathogenic nematodes are widely recommended in North America and Europe. S. Zervos (pers. comm.) demonstrated that *S. carpocapsae* at 440 nemas/cm^2 (100 000/4L pot), applied to the soil surface, can reduce numbers of grubs and pupae more effectively than lindane at 0.15 g/pot. Many treated trees produced new roots and regained vigor. Questions of timing of applications, treatment of trees in multipots, and persistence of the nematodes remain to be worked out.

Other forest insect pests against which entomopathogenic nematodes are being considered or tested in Canada are the balsam gall midge, *Paradiplosis tumifex*, in balsam fir Christmas tree plantations (D.C. Eidt and C.A.A. Weaver, unpubl. data); elm leaf beetle, *Xanthogaleruca luteola*, on shade trees (D.C. Eidt and S. Zervos, unpubl. data); winter moth, *Operophtera brumata*, on shade trees (D.C. Eidt, S. Walde, and S. Zervos, unpubl. data); black army cutworm, *Actebia fennica*, in newly planted clearcuts (R. West, pers. comm.); spruce cone maggots, *Strobilomyia* spp., in seed orchards (J. Sweeney, pers. comm.); strawberry root weevil,

Table 2. Some Canadian forest insects from which protozoa have been isolated or that have been infected in the laboratory.

Host	Pathogen[a]	Reference
Acantholyda erythrocephala	*Pleistophera schubergi*	Wilson 1984d
Archips cerasivorana	*Nosema cerasivoranae* *Pleistophora* sp. *Thelophania* sp.	Thomson 1960b Wilson and Burke 1978 Smirnoff 1965
Choristoneura conflictana	*Nosema thomsoni*	Wilson and Burke 1971
Choristoneura occidentalis	*Nosema* sp.	Sweeney and McLean 1987
Choristoneura pinus	*Nosema* sp.	Thomson 1959 Wilson 1986
Neodriprion swainei	*Herpetomonas swainei*	Smirnoff and Lipa 1970
Pissodes strobi	*Nosema* sp.	Wilson 1984e
Pristiphora erichsonii	*Thelophania pristiphorae*	Smirnoff 1966

[a] All are microsporidia except for *Herpetomonas swainei*, which is a flagellate.

Otiorhynchus ovatus, and other root weevils in nurseries and Christmas tree plantations (M. Hubbes, pers. comm.); and a web-spinning sawfly, *Acantholyda erythrocephala* in pine plantations (B. Lyons, pers. comm.).

Discussion and Conclusion

Microsporidia

Many Canadian forest insect pests are naturally infected with one or more species of microsporidia, some of which have yet to be identified. Most of the research has been conducted on the spruce budworm, and we have really only begun to examine the ecology of microsporidia in forest insects in general. Studies are needed to determine the role of microsporidia in integrated control; timing and methods for introducing microsporidia into field populations; and the importance of self-perpetuation of microsporidia in regulating pest populations. Vertically transmitted diseases such as *Nosema fumiferanae* may be crucial elements in the population dynamics of many insect pests, even if their virulence is low (Régnière 1984). If ecologically acceptable control procedures are to be developed for forest pests, the microsporidia must be considered. Although some species may have potential in direct control, most are chronic debilitating agents. The latter may predispose the host insect to attack by other factors, including chemical insecticides and other pathogens, thus increasing their effectiveness.

Nematodes

Steinernematid and heterorhabditid nematodes are not suitable for use as control agents against every forest insect pest. To implement them successfully, the entire host–nematode–environment complex must be studied. The susceptibility of all stages of the insect should be checked against a range of nematodes because not all stages of development are equally susceptible to all nematodes. Further choice of nematode will depend on the inherent characteristics of different species and strains. Research to date has shown that they may differ in their host-seeking ability (Bedding and Miller 1981), temperature tolerance (Finney 1984; Finney and Bennett 1984a), and resistance to desiccation (Cleary and Finney-Crawley, unpubl. data); meanwhile, all available nematodes have not yet been fully characterized and this is an area where further research is needed. Finally, it is necessary to determine whether the habitat of the target host is conducive to nematode survival in the post-treatment phase. Recent work with Bt–nematode combinations indicates that for some species of defoliators on some host plant species, formulated nematodes will have a role to play in forest insect control. However, effective treatment of insects on exposed foliage requires further research in formulation of these nematodes with anti-desiccants, evaporetardants, and photoprotectants that are not phytotoxic or lethal or detrimental to the nematodes. Soil, the natural habitat of the infective stage larvae, and cryptic habitats offer environments, however, that are especially suited to nematode survival. Because nematodes may become established in these habitats, repeated application of them may not be required. If the use of steinernematid and heterorhabditid nematodes for control of forest insect pests in Canada is to be exploited to its fullest potential, immediate research should focus on reduction of those insect pests that inhabit these environments during the life cycles and thereby provide the forest industry with yet another tool for insect pest control.

References

Bauer, L.S.; Nordin, G.L. 1988. Pathogenicity of *Nosema fumiferana* (Thomson) (Microsporidia) in spruce budworm. *Choristoneura fumiferana* (Clemens), and implications of diapause conditions. Can. Entomol. 120: 221-229.

Bedding, R.A.; Miller, L.A. 1981. Disinfecting blackcurrent cuttings of *Synanthedon tipuliformis* using the insect parasitic nematode, *Neoaplectana bibionis*. Environ. Entomol. 10: 449-453.

Brooks, W.M. 1980. Production and efficacy of protozoa. Biotechnol. and Bioeng. 22: 1415-1440.

Burman, M.; Pye, A.E.; Nojd, N.O. 1979. Preliminary field trial of the nematode *Neoaplectana carpocapsae* against larvae of the large pine weevil, *Hylobius abietis* (Coleoptera: Curculionidae) Ann. Entom. Fennici. 45: 88.

Dutky, S.R. 1959. Insect microbiology. Adv. Appl. Microbiol. 1: 175-200.

Dutky, S.R.; Thompson, J.V.; Cantwell, G.E. 1964. A technique for the mass propagation of the DD-136 nematode. J. Insect Pathol. 6: 417-422.

Eidt, D.C.; Dunphy, G.B. 1991. Control of spruce budmoth, *Zeiraphera canadensis* Mut. and Free., in white spruce plantations with entomopathogenic nematodes, *Steinernema* spp. Can. Entomol. 123: 379-385.

Finney, J.R. 1984. Alternative sources of steinernematid nematodes for use as biocontrol agents against insect pests in Newfoundland. CANUSA Newsletter. 34: 5, Ottawa, Ont.

Finney, J.R.; Bennett, G.F. 1983. The susceptibility of some sawflies (Hymenoptera: Tenthrediadae) to *Heterorhabditis heliothidis* (Nematoda: Rhabditida) under laboratory conditions. Can. J. Zool. 61: 1177-1180.

Finney, J.R.; Bennett, G.F. 1984a. Susceptibility of early instars of the spruce budworm (Lepidoptera: Tortricidae) to *Heterorhabitis heliothidis* (Nematoda: Rhabditida). Can. Entomol. 116: 285-286.

Finney, J.R.; Bennett, G.F. 1984b. *Heterorhabditis heliothidis*: a potential biocontrol agent of agricultural and forest pests in Newfoundland. J. Agric. Entomol. 1 (3): 287-295.

Finney, J.R.; Lim, K.P.; Bennett, G.F. 1982. The susceptibility of the spruce budworm *Choristoneura fumiferana* (Lepidoptera: Tortricidae) to *Heterorhabditis heliothidis* (Nematode: Heterorhabditidae) in the laboratory. Can. J. Zool. 60: 958-961.

Finney, J.R.; Mordue, W. 1976. The susceptibility of the elm bark beetle *Scolytus scolytus* to the DD-136 strain of *Neoaplectana* sp. Ann. Appl. Biol. 83: 311-312.

Finney, J.R.; Walker, C. 1977. The DD-136 strain of *Neoaplectana carpocapsae* as a potential biological control agent for the European elm bark beetle, *Scolytus scolytus* (Fab). J. Invertebr. Pathol. 29: 7-9.

Finney, J.R.; Walker, C. 1979. Assessment of a field trial using the DD-136 strain of *Neoaplectana* sp. for the control of *Scolytus scolytus*. J. Invertebr. Pathol. 33: 239-241.

Gaugler, R. 1981. Biological control potential of Neoaplectanid nematodes. J. Nematol. 13(3): 241-249.

Gaugler, R.; Boush, G.M. 1978. Effects of ultraviolet radiation and sunlight on the entomogenous nematode, *Neoaplectana carpocapsae*. J. Invertebr. Pathol. 32: 291-296.

Gaugler, R.; Boush, G.M. 1979. Non-susceptibility of rats to the entomogeneous nematode, *Neoaplectana carpocapsae*. Environ. Entomol. 8: 658-660.

Gaugler, R.; Kaya, H.K. 1990. Entomopathogenic nematodes in biological control. CRC Press, Boca Raton, Florida. 365 pp.

Gaugler, R.; LeBeck, L.; Nakagaki, B.; Boush, G.M. 1980. Orientation of the entomogeneous nematode *Neoaplectana carpocapsae* to carbon dioxide. Environ. Entomol. 9: 649-652.

Georgis, R. 1981. Studies on *Neoaplectana carpocapsae* in the web-spinning larch sawfly *Cephalcia lariciphila*. Ph.D. Thesis, University of Reading, Reading, England.

Georgis, R.; Kaya, H.K.; Gaugler, R. 1991. Effect of steinernematid and heterorhabditid nematodes (Rhabditida: Steinernematidae and Heterorhabditidae) on nontarget arthropods. Environ. Entomol. 20: 815-822.

Graham, K. 1948. Insect pathology. Can. Dept. Agric., Ottawa, Ont. Bi-mon. Prog. Rep. 4: 2.

Hara, A.H.; Kaya, H.K. 1982. Effects of selected insecticides on the *in vitro* development of the entomogeneous nematode *Neoaplectana carpocapsae*. J. Nematol. 14: 486-491.

Hara, A.H.; Kaya, H.K. 1983a. Toxicity of selected organophosphate and carbamate pesticides to infective juveniles of the entomogeneous nematode *Neoaplectana carpocapsae* (Rhabditida: Steinernamatidae). Environ. Entomol. 12: 496-510.

Hara, A.H.; Kaya, H.K. 1983b. Development of the entomogeneous nematode *Neoaplectana carpocapsae* (Rhabditida: Steinernamatidae), in insecticide-killed beet armyworm (Lepidoptera: Noctuidae). J. Econ. Entomol. 76: 423-426.

Henry, J.E. 1981. Natural and applied control of insects by protozoa. Ann. Rev. Ent. 26: 49-73.

Jackson, G.J.; Moore, G.E. 1969. Infectivity of nematodes *Neoaplectana* species, for the larvae of the weevil *Hylobius pales*, after rearing in species isolation. J. Invertebr. Pathol. 14: 194-198.

Kamionek, M. 1979. Influence of pesticides on the mortality and effectiveness of *Neoaplectana carpocapsae* Weiser. Pages 87-88 *in* J. Weiser (Ed.), Progress in Invertebrate Pathology: proceedings of the International Colloquium of Invertebrate Pathology and XIth annual meeting of the Society for Invertebrate Pathology, Sept. 1978, Agricultural College Campus, Prague, CSSR.

Kaya, H.K. 1977. Development of DD-136 strain of *Neoaplectana carpocapsae* at constant temperatures. J. Nematol. 9: 346-349.

Kaya, H.K. 1985. Entomogeneous nematodes for insect control in IPM systems. Pages 283-302 *in* M.A. Hoy and D.C. Herzog (Eds.), Biological Control in Agriculture IPM Systems, Academic Press, Orlando.

Kaya, H.K.; Hara, A.H.; Reardon, R.C. 1981. Laboratory and field evaluation of *Neoaplectana carpocapsae* (Rhabditida: Steinernematidae) against the elm leaf beetle (Coleoptera: Chrysomelidae) and the western spruce budworm (Lepidoptera: Tortricidae). Can. Entomol. 113: 787-793.

Kaya, H.K.; Reardon, R.C. 1982. Evaluation of *Neoaplectana carpocapsae* for biological control of the western spruce budworm, *Choristoneura occidentalis*: ineffectiveness and persistence of tank mixes. J. Nematol. 14(4): 595-597.

Lam, A.B.Q.; Webster, J.M. 1972. Effect of the DD-136 nematode and of a B-Exotoxin preparation of *Bacillus thuringiensis* var. *thuringiensis* on leatherjackets, *Tipula paludosa* larvae. J. Invertebr. Pathol. 20: 141-149.

Lindegren, J.E.; Curtis, C.E.; Poinar, G.O., Jr. 1978. Parasitic nematode seeks out navel orangeworm in almond orchards. Calif. Agric. 32: 10-11.

Lindegren, J.E.; Hoffman, D.D.; Collier, S.S.; Fries, R.D. 1979. Propagation and storage of *Neoaplectana carpocapsae* using *Anyelios transitella* (Walker) adults. USDA/SEA, Washington, D.C. Adv. Agric. Technol. Western Series No. 3.

Lindegren, J.E.; Yamashita, T.T.; Barnett, W.W. 1981. Parasitic nematodes may control carpenterworm in fig trees. Calif. Agric. 35: 25-36.

MacVean, C.M.; Brewer, J.W. 1981. Suitability of *Scolytus multistriatus* and *Dendroctonus ponderosae* as hosts for the entomogeneous nematode *Neoaplectana carpocapsae*. J. Econ. Entomol. 74: 601-607.

Miller, L.A.; Bedding, R.A. 1982. Field testing of the insect parasitic nematode, *Neoaplectana bibionis* (Nematoda: Steinernematidae) against currant borer moth *Synanthedon tipuliformis* (Lepidotera: Sesiidae) in blackcurrants. Entomophaga 27 (1): 109-114.

Moore, G.E. 1970. *Dendroctonus frontalis* infection by the DD-136 strain of *Neoaplectana carpocapsae* and its bacterium complex. J. Nematol. 2: 341-344.

Nickle, W. 1984. Plant and Insect Nematodes. Marcel Dekker Inc., New York and Basel. 925 p.

Percy, J.; Wilson, G.; Burke, J. 1982. Development and ultrastructure of a microsporidian parasite in midgut cells of the larch sawfly, *Pristiphora erichsonii* (Hymenoptera: Tenthredinidae). J. Invertebr. Pathol. 39: 49-59.

Poinar, G.O., Jr. 1972. Nematodes as facultative parasites of insects. Annu. Rev. Entomol. 17: 103-122.

Popiel, I. 1987. Nematodes parasitic in insects. Pages 379-393 *in* A.E.R. Taylor and J.P. Baker (Eds.), *In Vitro* Methods for Parasite Cultivation. 3rd ed. Academic Press, Toronto.

Pye, A.E.; Burman, M. 1977. Pathogenicity of the nematode *Neoaplectana carpocapsae* (Rhabditida: Steinernematidae) and *Hylobius abietis* (Coleoptera: Curculionidae). Annales Entomnologici Fennic. 43: 115-119.

Pye, A.E.; Burman, M. 1978. *Neoaplectana carpocapsae*: infection and reproduction in large pine weevil larvae, *Hylobius abietis*. Exp. Parasitol. 46: 1-11.

Pye, A.E.; Burman, M. 1981. *Neoaplectana carpocapsae*: nematode accumulations on chemical and bacterial ingredients. Exp. Parasitol. 51: 13-20.

Pye, A.E.; Pye, N.L. 1985. Different applications of the insect parasitic nematode *Neoaplectana carpocapsae* to control the large pine weevil, *Hylobius abietis*. Nematologica. 31(1): 109-116.

Reardon, R.C.; Kaya, H.K.; Fusco, R.A.; Lewis, F.B. 1986. Evaluation of *Steinernema feltiae* and *S. bibionis* (Rhadbditidae: Steinernematidae) for suppression of *Lymantria dispar* (Lepidoptera: Lymantriidae) in Pennsylvania, U.S.A. Agriculture, Ecosystems and Environment. 15: 1-10.

Régnière, J. 1984. Vertical transmission of diseases and population dynamics of insects with discrete generations: a model. J. Theor. Biol. 107: 287-301.

Schmidt, J.; All, J.N. 1978. Chemical attraction of *Neoaplectana carpocapsae* (Nematoda: Steinernematidae) to insect larvae. Environ. Entomol. 7: 605-607.

Schmidt, J.; All, J.N. 1979. Attraction of *Neoaplectana carpocapsae* (Nematoda: Steinernematidae) to common excretory products of insects. Environ. Entomol. 8: 55-61.

Schmiege, D.C. 1963. The feasibility of using neoaplectanid nematode for control of some forest insect pests. J. Econ. Ent. 56(4): 427-431.

Shapiro, O.; Poinar, G.O., Jr.; Lindegren, E. 1985. Suitability of *Lymantria dispar* (Lepidoptera: Lymantriidae) as a host for the entomogeneous nematode, *Steinernema feltiae* (Rhabditida: Steinernematidae). J. Ecol. Entomol. 78: 342-345.

Simons, W.R.; Poinar, G.O., Jr. 1973. The ability of *Neoaplectana carpocapsae* (Steinernematidae: Nematodea) to survive extended periods of desiccation. J. Invertebr. Pathol. 22: 228-230.

Smirnoff, W.A. 1965. The occurrence of *Nosema* and *Plistophora* microsporidians on *Archips cerasivoranus* (Fitch) in Québec. Ann. Soc. Ent. du Québec 10: 121-124.

Smirnoff, W.A. 1966. *Thelophania pristiphorae* sp. n., microsporidian parasite of the larch sawfly, *Pristiphora erichsonii* (Hymenoptera: Tenthredinidae). J. Invertebr. Pathol. 8: 360-364.

Smirnoff, W.A.; Juneau, A. 1973. Quinze années de recherches sur les micro-organisms des insectes forestiers de la Province de Québec (1957-1972). Ann. Soc. Ent. 18: 147-181.

Smirnoff, W.A.; Lipa, J.J. 1970. *Herpetomonas swainei* sp. n., a new flagellate parasite of *Neodiprion swainei* (Hymenoptera: Tenthredinidae). J. Invertebr. Pathol. 16: 187-195.

Sweeney, J.D.; McLean, J.A. 1987. Effect of sublethal infection levels of *Nosema* sp. on the pheromone-mediated behavior of the western spruce budworm, *Choristoneura occidentalis* Freeman (Lepidoptera: Tortricidae). Can. Entomol. 119: 587-594.

Thomas, 1970. Neoaplectanid nematodes as parasites of the pales weevil larva, *Hylobius pales*. Entomol. News. 81: 91.

Thomson, H.M. 1955. *Perezia fumiferanae* n. sp., a new species of microsporidia from the spruce budworm *Choristoneura fumiferana* (Clem.). J. Parasitol. 41: 1-8.

Thomson, H.M. 1958. Some aspects of the epidemiology of a microsporidian parasite of the spruce budworm, *Choristoneura fumiferana* (Clem). Can. J. Zool. 36: 309-316.

Thomson, H.M. 1959. A microsporidian parasite of the forest tent caterpillar, *Malacosoma disstria* Hbn. Can. J. Zool. 37: 217-221.

Thomson, H.M. 1960a. The possible control of a budworm infestation by a microsporidian disease. Can. Dep. Agric., Ottawa, Ont. Bi-Mon. Prog. Rep. 16: 1.

Thomson, H.M. 1960b. *Nosema cerasivoranae* n. sp., a microsporidian parasite of the ugly-nest caterpillar, *Archips cerasivorana* (Fitch). Can. J. Zool. 38: 643-644.

Turgeon, J.J. 1985. Life cycle and behavior of the spruce budmoth, *Zeiraphera canadensis* (Lepidoptera: Olethreutidae) in New Brunswick. Can. Entomol. 117: 1239-1247.

Turgeon, J.J.; Finney-Crawley, J.R. 1991. Susceptibility of first and second instar larvae of the spruce budmoth, *Zeiraphera canadensis* (Lepidoptera: Tortricidae) to the entomogenous nematode *Heterorhabditis heliothidis* under controlled conditions. J. Invertebr. Pathol. 57: 126-127.

Webster, J.M. 1980. Biocontrol: the potential of entomophilic nematodes in insect management. J. Nematol. 12: 270-278.

Webster, J.M.; Bronskill, J. 1968. Use of Gelgard M and an evaporation retardant to facilitate control of larch sawfly by a nematode-bacterium complex. J. Econ. Entomol. 61(5): 1370-1373.

Wilson, G.G. 1975. Occurrence of *Thelohania* sp. and *Pleistophora* sp. (Microsporida: Nosematidae) in *Choristoneura fumiferana* (Lepidoptera: Tortricidae). Can. J. Zool. 53: 1799-1802.

Wilson, G.G. 1977. The effects of feeding microsporidian (*Nosema fumiferanae*) spores to naturally infected spruce budworm (*Choristoneura fumiferana*). Can. J. Zool. 55: 249-250.

Wilson, G.G. 1979. Effects of *Nosema disstriae* (Microsporida) on the forest tent caterpillar, *Malacosoma disstria* (Lepidoptera: Lasiocampidae). Proc. Ent. Soc. Ont. 110: 97-99.

Wilson, G.G. 1980. Persistence of microsporidia in populations of the spruce budworm and forest tent caterpillar. Can. For. Serv., For. Pest Manage. Inst., Sault Ste. Marie, Ont. Inf. Rep. FPM-X-39.

Wilson, G.G. 1981a. The potential of *Pleistophora schubergi* in microbial control of forest insects. Can. For. Serv., For. Pest Manage. Inst., Sault Ste. Marie, Ont. Inf. Rep. FPM-X-49.

Wilson, G.G. 1981b. "Microsporidia" *in* J. Hudak and A.G. Raske (Eds.), Review of the spruce budworm outbreak in Newfoundland—its control and forest management implications. Can. For. Serv., Nfld. For. Res. Cent., St. Johns, Nfld. Inf. Rep. N-X-205.

Wilson, G.G. 1981c. Susceptibility of the larch sawfly to *Pleistophora schubergi* (Microsporidia). Can. For. Serv., Ottawa, Ont. Res. Notes 1: 1.

Wilson, G.G. 1982a. Protozoa for insect control. Pages 587-600 *in* E. Kurstak (Ed.), Microbial and Viral Pesticides. Marcel Dekker, Inc., New York.

Wilson, G.G. 1982b. Transmission of *Nosema fumiferanae* (Microsporida) to its host *Choristoneura fumiferana* (Clem.). Z. Parasitenkd 68: 47-51.

Wilson, G.G. 1983. A dosing technique and the effects of sublethal doses of *Nosema fumiferanae* (Microsporidia) on its host the spruce budworm, *Choristoneura fumiferana*. Parasitol. 87: 371-376.

Wilson, G.G. 1984a. Dose-mortality response of *Choristoneura fumiferana* (Lepidoptera: Tortricidae) to a microsporidium, *Pleistophora schubergi*. Proc. Ent. Soc. Ont. 115: 93-94.

Wilson, G.G. 1984b. The transmission and effects of *Nosema fumiferanae* and *Pleistophora schubergi* on *Choristoneura fumiferana* (Lepidoptera: Tortricidae). Proc. Ent. Soc. Ont. 115: 71-75.

Wilson, G.G. 1984c. Pathogenicity of *Nosema disstriae, Pleistophora schubergi* and *Vairimorpha necatrix* (Microsporida) to larvae of the forest tent caterpillar, *Malacosoma disstria*. Z. Parasitenkd. 70: 763-767.

Wilson, G.G. 1984d. Infection of the pine false webworm by *Pleistophora schubergi* (Microsporida). Can. For. Serv., Ottawa, Ont. Res. Note 4: 7-8.

Wilson, G.G. 1984e. Observations of a microsporidian parasite in the white pine weevil; *Pissodes strobi* (Peck) (Coleoptera: Curculionidae). Can. For. Serv., Ottawa, Ont. Res. Note 4: 33-35.

Wilson, G.G. 1985. Dosage-mortality response of *Choristoneura fumiferana* (Clem.) to the microsporidium, *Nosema fumiferanae*. Can. For. Serv., For. Pest Manage. Inst., Sault Ste. Marie, Ont. Inf. Rep. FPM-X-68. 7 p.

Wilson, G.G. 1986. A comparison of the effects of *Nosema fumiferanae* and a *Nosema* sp. (Microsporida) isolated from jack pine budworm on *Choristoneura fumiferana* (Clem.) and *Choristoneura pinus pinus* Free. Can. For. Serv., For. Pest Manage. Inst., Sault Ste. Marie, Ont. Inf. Rep. FPM-X-77. 10 p.

Wilson, G.G. 1987. Observations on the level of infection and intensity of *Nosema fumiferanae* (Microsporida) in two different field populations of the spruce budworm, *Choristoneura fumiferana*. Can. For. Serv., For. Pest Manage. Inst., Sault Ste. Marie, Ont. Inf. Rep. FPM-X-79. 15 p.

Wilson, G.G.; Burke, J.M. 1971. *Nosema thomsoni* n. sp., a microsporidian from *Choristoneura conflictana* (Lepidoptera: Tortricidae). Can. J. Zool. 49: 786-788.

Wilson, G.G.; and Burke, J.M. 1978. Microsporidian parasites of *Archips cerasivoranus* (Fitch) in the district of Algoma, Ontario. Proc. Ent. Soc. Ont. 109: 84-85.

Wilson, G.G.; Kaupp, W.J. 1976. Application of *Nosema fumiferanae* and *Pleistophora schubergi* (Microsporida) against the spruce budworm in Ontario, 1976. Can. For. Serv., Ins. Path. Res. Inst., Sault Ste. Marie, Ont. Inf. Rep. IP-X-15. 14 p.

Wilson, G.G.; Kaupp, W.J. 1977. Application of *Nosema disstriae* and *Pleistophora schubergi* (Microsporida) against the forest tent caterpillar in Ontario, 1977. Can. For. Serv., For. Pest Manage. Inst., Sault Ste. Marie, Ont. Inf. Rep. FPM-X-5. 9 p.

Wilson, G.G.; Tyrrell, D.; Ennis, T.J. 1984. Applications of microsporidia and fungi, and of genetic manipulations. Pages 260-266 *in* J.S. Kelleher and M.A. Hulme (Eds.), Biological Control Programmes against Insects and Weeds in Canada 1969-1980. Commonw. Agric. Bureaux, London, England.

Chapter 38

Neurotoxic Insecticides

B.V. Helson and P.C. Nigam

Introduction

Chemical insecticides have been the major tools used to control forest insects in Canada for several decades. The insecticides that are currently registered for this purpose belong to four major classes: organophosphates, carbamates, organochlorines, and pyrethroids (Table 1). All are neurotoxins, which exert their lethal effects via the insect's nervous system. The general attributes of such neurotoxic insecticides include high toxicity to insects, selectivity for mammals, quick action, ease of application to large areas at low volumes, and economical cost.

Over the past two decades, chemicals with this neurotoxic mode of action have been a major source of potential insecticides for forest insect control. Scientists at the Forest Pest Management Institute of Forestry Canada (formerly the Chemical Control Research Institute) have been actively testing such chemicals to identify new, highly effective, and selective insecticides in an effort to improve the efficiency and safety of forest control operations and to increase the number of control options available for forest insect pests.

In this chapter, we will review the progress made in the laboratory at the Forest Pest Management Institute since 1973 in determining the toxicological properties of new, neurotoxic insecticides and formulations on forest insect pests and in developing insecticides for the control of newly recognized pests. The ground-based field tests that have been conducted with insecticides during this period will also be included. Experimental aerial applications are reported in other sections of this book.

Product Development Approach

New insecticides are selected for testing at the Forest Pest Management Institute based on several criteria. The insecticide must be potentially highly toxic to at least Lepidoptera larvae because the major forest pests in Canada are in this group. Only those insecticides that do not appear to pose an unacceptable risk to human health or the environment are selected for development. The chemical company must be committed to the development of the product for forestry use, and it is a distinct advantage if the insecticide already has been or potentially will be registered for agricultural or other major uses. The agricultural market is a large and stable one, while the forestry market is relatively small and periodical in nature. Recently, some experimental chemicals have been developed that have novel modes of action. Selected ones are screened to assess their spectrum of insecticidal activity against an array of representative forest pests belonging to different orders and families of insects. This provides valuable information for deciding on toxicological grounds if the compound has potential for any forestry use.

A sequential testing procedure that advances from the laboratory to the field is used to evaluate the potential of new chemical insecticides for the control of spruce budworm, *Choristoneura fumiferana*, and other forest insect pests. Initially, the contact toxicity of candidate insecticides to sprayed spruce budworm larvae is determined relative to standard insecticides such as fenithrothion and aminocarb, which have been used extensively for spruce budworm control in Canada. Each insecticide is also tested for its toxicity to larvae on sprayed host foliage. These tests provide essential information on the toxicity of the compound by different exposure routes that could occur during spray applications in the field, namely, direct impingement of the spray on larvae, indirect contact with deposits on foliage, and ingestion of deposits on foliage. Because the latter two routes are important, the persistence of the insecticide on foliage could also influence its potential efficacy against the spruce budworm. Consequently, the residual toxicity of promising insecticides is measured under natural weathering conditions. Experience has shown that insecticides with high contact toxicity but poor residual toxicity are not very successful in field trials.

By comparing the toxicity of a new insecticide with that of a standard insecticide with similar toxicological properties, potential field dosages can be estimated based on the known effective dosage rates for the standard. Before promoting the insecticide to field trials, information on its mammalian, wildlife, and environmental toxicology is carefully examined in relation to the estimated field dosages to assess its relative safety for large-scale forestry operations. This toxicological information is obtained from the pesticide company or other external sources. Cost of the chemical, potential availability, and its use pattern in agriculture are also considered. If the chemical still has potential for use in forestry after these investigations and if the pesticide company is still interested in further development, the insecticide is recommended for field trials. Between 1973 and 1978, simulated aerial sprays and/or mistblower trials were the next step in the sequence of events in product development, followed by experimental aerial applications on small plots as exemplified by the development of Orthene® (acephate) (Nigam 1973, 1975c; Nigam and Hopewell 1973; Hopewell and Nigam 1974; Hopewell 1975, 1977b; Armstrong and Nigam 1975). More recently, new products have been taken directly to aerial trials on small plots.

Experimental Techniques

The use of a modified Potter's tower for determining the contact toxicity of insecticides to forest insect pests has been described by Nigam (1975c). Briefly, insects are sprayed in this tower after calibrating it to deliver volumes of dyed

Table 1. Insecticides registered for forest insect control in Canada.

Insecticide	Class[a]	Trade name(s)	Manufacturer(s) Supplier(s)	Application method[b]	Target pests
Acephate	OP	Orthene	Chevron Chemical (Canada) Ltd.	G,A	Spruce budworm larvae (all instars), western spruce budworm, oak leafshredder, Douglas-fir tussock moth, gypsy moth larvae, tent caterpillars, aphids, bagworm, fall webworm, Nantucket pine tip moth, sawflies, tussock moths, mites
Aminocarb	C	Matacil	Chemagro Ltd.	A	Spruce budworm
Azinphosmethyl	OP	Guthion APM Azinphosmethyl	Chemagro Ltd., Chipman Inc., Solcoor (Canada) Ltd.	G	Aphids, scale insects, cone midges, cone moths, cone worms, seed worms, European pine shoot moth, Nantucket pine tip moth
Carbaryl	C	Sevin	May & Baker Inc., Sanex Chemicals Ltd.	G,A	Gypsy moth, Douglas-fir tussock moth, mountain pine beetle, spruce budworm
Chlorpyrifos	OP	Dursban	Dow Chemical Canada Inc.	G	Elm bark beetles, tent caterpillars, sawflies
Diazinon	OP	Diazinon	Chipman Inc.	G	Leaf miners, bagworms, scale insects, aphids, caterpillars, thrips, European pine shoot moth
Dimethoate	OP	Cygon	Niagara Chemical, Pfizer Company Ltd., Cyanamid Canada Ltd.	G,A	Spruce budworm, pine shoot moth, seed & cone insects, aphids, bagworm, Nantucket pine tip moth, pine needle scale, Zimmerman pine moth, black-headed pine sawfly, birch leafminer, spider mites, scale insects, spruce needle miner, Sitka spruce weevil, forest tent caterpillar
Fenitrothion	OP	Folithion Novathion Sumithion	Chemagro Ltd., A/S Cheminova, Sumitomo Canada Ltd., BASF Canada Ltd.	A,G	Spruce budworm, hemlock looper, western hemlock looper, sawflies, canker worm, jack pine budworm, Swaine jack pine sawfly, spruce bud moth
Lindane	OC	Lindane	Ditchling Corp. Ltd., Niagara Chemical, Sanex Chemicals Ltd.	G	Pine bark beetles, sawyer beetles, ambrosia beetles, cereal leaf beetles, pinehole borer
Methomyl	C	Lannate, Nudrin	Dupont Canada Inc., Pecten Chemicals Inc.	G	Spruce budworm
Methoxychlor	OC	Methoxychlor	Green Cross, Chipman Inc.	G	Elm bark beetles
Oxydemeton-methyl	OP	Metasystox-R	Chemagro	G,A	Balsam twig aphid, cone and seed insects
Permethrin	P	Ambush	Chipman Inc.	G	Douglas-fir tussock moth, white marked tussock moth, spruce budworm, open-feeding sawflies, eastern tent caterpillar, forest tent caterpillar, fall webworm, eunonymus webworm, open-feeding aphids, spruce coneworm, black army cutworm
Trichlorfon	OP	Dylox, Danex	Chemagro Ltd., Solcoor (Canada) Ltd.	G,A	Spruce budworm, gypsy moth, forest tent caterpillar

[a]Class of insecticide: OP = organophosphate; C = carbamate; OC = organochlorine; and P = pyrethroid.
[b]Application method: G = ground; and A = aerial.

insecticide mixes equivalent to deposits between 1 and 10 L/ha. The mortality of insects is determined 72 hours after treatment or longer, corrected for any mortality in a control group (Abbott 1925), and subjected to probit analysis (Finney 1962; Bickle 1968). From this analysis, an LC_{50} value is obtained as a standard measure of the toxicity of an insecticide. Before 1982, oil-based mixtures of technical grade insecticides or formulated products were used. Since then, aqueous-based mixtures of emulsifiable concentrate or flowable formulations have been developed because of the recent, increased emphasis on evaluating new formulations of existing insecticides and the preference by spray operators for water as a diluent. To determine the toxicity of insecticide deposits on host foliage, larch foliage was sprayed in the modified Potter's tower and untreated insects were exposed to this foliage, typically for 72 hours. Larch was used because it is available for much longer than balsam fir or white spruce buds. Buds from these latter two species are now being used because methods have been developed to flush potted trees throughout most of the year in the greenhouse after a cold treatment.

Nigam (1975c) has also described the general procedures for determining the residual toxicity of insecticides. Basically, potted trees are sprayed with measured dosages of an insecticide and placed outdoors under natural weather conditions, which are continuously monitored on site. At selected intervals after treatment, usually 4 hours, 1, 3, 5, and 10 days, foliage is clipped from these trees and insects are exposed to this foliage in the laboratory for 72 hours.

The spray chamber now being used for these tests is different from the one illustrated in Nigam (1975c). The current chamber is a rectangular box (2.45 m high × 1.14 m wide × 0.68 m deep). The potted tree is placed on an adjustable shelf in the chamber 2 m below the spray nozzle. The tree is not rotated on a turntable. An electrically driven, spinning disk nozzle from a Flak, handheld, ultra-low-volume sprayer is used instead of Potter's nozzles that were used in the former chamber. Spray distribution, droplet sizes, and numbers can be varied as desired by changing the rotation speed of the spinning disk with a rheostat. As with the previous chamber, the insecticide is delivered to the nozzle from a Manostat digipet driven by a gear motor that is operated for the desired period by a timer.

To determine the residual toxicity of insecticides to spruce budworm larvae, the past standard procedure was to conduct three separate tests over 3 years on both balsam fir and white spruce trees at one selected dosage of active ingredient, typically 224 g/ha. Recently, some insecticides have been tested at several dosages to compare their residual toxicities not only at the same dosage but also at dosages providing similar initial toxicities.

For simulated aerial sprays, a device developed and described by Hopewell (1974) was used to treat individual trees 2 to 3 m tall in the field. Photographs of the apparatus are provided in Hopewell and Nigam (1974) and Hopewell (1977a). A portable, woodframed plastic shelter enclosing a 4.5-m^2 area is erected around a tree. A measured amount of insecticide solution is placed in the syringe of the spraying unit and delivered to a spinning disk by battery-operated motors at a rate of about 1 mL/minute. An operator moves the unit systematically over the tree by means of a right-angled pole with a 2.1-m horizontal arm and a 2.4-m vertical shaft. The insecticide concentrations and volumes that were normally used in aerial sprays for forestry at the time could be applied with this device and it produced realistic spray drop densities and size spectra (Hopewell 1974). Spray deposition was apparently very efficient with this technique because measured deposits in glass dishes adjacent to the trees were typically similar to the applied dosages.

An FMC Rotomist® 100HT sprayer was used for mistblower trials (DeBoo and Campbell 1974b; DeBoo 1980).

Spruce Budworm

From 1973 to 1988, close to 70 new insecticides and formulations were tested on spruce budworm, *Choristoneura fumiferana*, larvae in the laboratory by the authors (Helson 1985a, 1985b; Nigam 1981, 1985). Most of these insecticides were neurotoxins belonging to three major classes: the organophosphates, carbamates, and pyrethroids. The major advance in insecticide chemistry during this period was the development of more photostable and, hence, persistent pyrethroids. Since 1975, representatives of this group including cypermethrin, fenvalerate, deltamethrin, and particularly permethrin have been studied in depth to assess their potential for spruce budworm control. Many new organophosphates were tested between 1973 and 1980, while relatively few new carbamates were introduced during the past 15 years. Since 1980, most emphasis has been on evaluating new formulations of fenitrothion, aminocarb, carbaryl, and mexacarbate. Recently, some novel products have been tested including hydramethylnon, an amidinohydrazone compound; ethofenprox, a pyrethroid-like insecticide with relatively low fish toxicity; and ∞-terthienyl, a naturally occurring secondary metabolite that is found in members of the plant family, Asteraceae.

Larvae—New Insecticides

Many of the insecticides that were tested against spruce budworm larvae over the last 15 years are listed in Table 2 along with their contact toxicity to laboratory-reared fifth-instar larvae.

The pyrethroid insecticides exhibited exceptional toxicity to spruce budworm larvae. The most toxic ones are one to two orders of magnitude more active than the organophosphates and carbamate insecticides tested. Several new organophosphates were up to five times more toxic than fenitrothion, while the few new carbamate insecticides tested were no more toxic than aminocarb.

The toxicological properties of the more promising insecticides were investigated further in the laboratory (Table 3). In addition to their generally high contact toxicity, these insecticides were very toxic to budworm larvae exposed to deposits on foliage. Most of these insecticides possessed relatively long residual effectiveness, indicating that they had good potential for controlling spruce budworm larvae by the major routes of exposure discussed above. These insecticides were selected or recommended for field tests.

Table 2. Selected insecticides tested against the spruce budworm, *Choristoneura fumiferana*, 1973–1988.

Insecticides	Larvae 72 h LC_{50} ($\mu g/cm^2$)[a]	Slope	Toxicity ranking[b] Pupae	Toxicity ranking[b] Adults
Organophosphates				
A5675A (Ciba Geigy)[c]	0.178	4.80		
A41286 (Abbott)	0.125	5.31		
acephate	0.349	3.77	4	≤4[g]
azamethiphos	0.015	5.94		
azinphos-methyl	–	–		3
Bay 78182 (Chemagro)	0.089	5.26		
carbophenothion	–	–	4	
CGA 19795 (Ciba Geigy)	0.066	6.99		
chlorphoxim	0.033	3.14		
chlorpyrifos-methyl	0.061	5.41		
diazinon	0.581	5.01		
dimethoate	0.228[d]	3.65	4	
fenitrothion	0.096[d]	6.14	4	3
iodofenphos	–[d]	–		3
leptophos	0.266	5.95		
malathion	0.222	5.92		
methidathion	0.089	3.53	3	
MV770 (Stauffer)	0.117	5.19		
phosmet	0.064	5.85	3	
phosphamidon	0.088[d]	7.29	3	3
phoxim	0.056	4.65	4	
primidophos	0.308	5.80		
profenofos	0.051	9.39		
prothiofos	0.198	5.40		
RH0994 (Rohm and Haas)	0.103	5.13		
RHC 367 (Rohm and Haas)	0.058	3.60		
SAN 279 (Sandoz-Wander)	0.037	6.15		
sulprofos	0.019	6.44		3
trichlorfon	0.172	5.40	4	
trifenofos	0.049	6.92		
Carbamates				
aminocarb	0.023[d]	4.28	4	3
(Matacil® 180F)	0.057[e]	4.25		
Dimetilan	–	–	5	
Carbamates *(Continued)*				
methomyl	0.017	2.07	3	
mexacarbate	0.027[d]	3.61	3	
(Zectran® UCZF 19)	0.067[e]	4.91		
thiodicarb	0.134[e]	3.09		
U56295 (Upjohn)	0.130	2.31		
UC72987 (Union Carbide)	0.038	4.78		
Pyrethroids				
ABG 6010 (Abbott)	0.131	5.99		
ABG 6070 (Abbott)	0.142	3.21		
bioethanomethrin	0.013	4.09		
bioethanomethrin (racemic mix)	0.018	2.63		
cismethrin	0.007	4.03		
cypermethrin	0.003	3.92		≤2[g]
deltamethrin	0.0005	2.97		
ethofenprox	0.042[e]	7.87		
fenpropanate	0.011	3.92		
fenvalerate	0.011	3.98		
FMC 45497 (FMC)	0.004	8.94		
FMC 45812 (FMC)	0.011	6.53		
flucythrinate	0.005	3.81		
NRDC 168-S (Procida)	0.0003	2.55		
permethrin	0.015	4.72		3
permethrin (Ambush)	0.035[e]	5.73		
resmethrin	0.026	4.27		
Organochlorines				
A 47170 (Abbott)	3.535	2.16		
A 47171 (Abbott)	4.174	3.22		
Miscellaneous				
AC 268-962 (American Cyanamid)	0.106[f]	2.57		
hydramethylon	0.061[f]	3.25		

[a] Contact toxicity to laboratory-reared fifth-instar larvae.
[b] The following numbers represent the range in LC_{50} ($\mu g/cm^2$) values corresponding to each toxicity ranking (in parentheses): <0.001(1); 0.002–0.01(2); 0.02–0.1(3); 0.2–1(4); and >2(5).
[c] Manufacturer of numbered experimental insecticides.
[d] Toxicity to field-collected larvae reported in Nigam (1975c).
[e] H_2O used as carrier with formulated product.
[f] Toxicity on treated foliage. See Table 3 for comparative values. Contact toxicity of these compounds low.
[g] Equal to or one rank less than the indicated one (≤).

Several were evaluated in simulated aerial spray tests (Nigam and Hopewell 1973; Hopewell and Nigam 1974; Hopewell 1975; Hopewell 1977b) or mistblower trials (DeBoo and Campbell 1974b; DeBoo 1980). The results of these field tests are summarized in Tables 4 and 5, where all dosages are expressed as weights of active ingredient per unit of area.

In a preliminary simulated aerial spray test, phoxim appeared to be at least as effective as fenitrothion, although spray deposits of fenitrothion were much lower than phoxim. In a subsequent test, phoxim did not provide as good control as fenitrothion at similar dosages. The residual toxicity of phoxim was low in laboratory tests under natural weathering conditions (Table 3). In these tests, the addition of an ultraviolet light inhibitor actually reduced phoxim's potency (Hopewell and Nigam 1974). This insecticide was effective in mistblower trials at 280 g/ha.

In other tests (Table 4), leptophos at 280 g/ha (330–460 g/ha measured dye deposit) was similar in effectiveness to fenitrothion at 280 g/ha (205–295 g/ha measured dye deposit). A split application of leptophos at 140 g/ha did not provide as good budworm mortality or foliage protection as the single applications.

Acephate was tested extensively over 4 years (Table 4). At applied dosages of 140 g/ha or more in volumes as low as 1.5 L/ha, both technical acephate and Orthene® 75% Soluble Powder in water provided larval reductions of 75% or more and foliage protection greater than 60% when timed against fourth and fifth instars. In early applications against second- and third-instar larvae, acephate was generally not very effective, although some foliage protection was achieved. This insecticide was also very effective in controlling spruce budworm larvae in a mistblower application at 550 g/ha (Table 5).

Permethrin was selected for intensive investigation of its potential for spruce budworm control, as a representative of the new group of more photostable pyrethroids. In initial tests in 1975 (Table 4), permethrin was effective in reducing budworm populations and defoliation at an applied dosage as low as 18 g/ha. In further tests the following year, applications of 63 to 64 g/ha usually provided good control against fourth- and fifth-instar larvae. According to Hopewell (1977b), concentration of active ingredient or droplet density did not seem to influence effectiveness; the controlling factor appeared to be the dosage of permethrin. Applications of permethrin early in the spring against second-instar larvae, emerging from

Table 3. Insecticides selected or recommended for field evaluation of their efficacy to spruce budworm, *Choristoneura fumiferana*, larvae, 1973–1988.

	Toxicity to fifth-instar larvae									
	Direct contact		Treated larch foliage		Residual toxicity					
	72 h LC_{50}		72 h LC_{50}		Dosage	72 h percent mortality after weathering for				
Insecticide	(μg/cm^2)	Slope	(μg/cm^2)	Slope	(g/ha)[a]	0	1	3	5	10 days
Leptophos	0.266	5.95	0.099	3.77	224	100	94	47	34	25
Phoxim	0.056	4.65	–	–	224	77	13	12	0	0
Acephate	0.349	3.77	0.042	3.40	560	99	55	16	32	22
Permethrin	0.015	4.72	0.033	3.84	224	96	96	79	69	20
Chlorpyrifos-methyl	0.061	5.41	0.028	5.06	224	99	91	85	78	60
Azamethiphos	0.015	5.94	0.094	2.35	224	93	74	18	4	3
Sulprofos	0.019	6.44	0.066	3.80	224	98	87	57	43	37
Cypermethrin	0.003	3.92	0.012	1.93	28	99	100	97	87	65
Thiodicarb[b]	0.134	3.09	0.028	2.51	200	97	92	88	60	43
Mexacarbate[b]	0.067	4.91	0.028	4.34	200	98	78	70	47	34
Standards										
Fenitrothion	0.096	6.14	0.070	3.90	224	96	65	22	5	4
Fenitrothion[b]	0.388	4.43	0.113	5.11	224	99	47	31	16	6
Aminocarb	0.030	5.04	0.021	3.32	224	100	89	92	75	58
Aminocarb[b]	0.057	4.25	0.025	3.33	200	98	92	69	50	36

[a] Grams active ingredient per hectare.
[b] Mixed in water; the other insecticides were mixed in an oil-based carrier.

Table 4. Insecticides tested against spruce budworm, *Choristoneura fumiferana*, larvae by simulated aerial spray on white spruce trees, 1973–1976.

Insecticide	Year	Instar	Applied rate g/ha[a]	Applied rate L/ha	% SBL[b] reduction	% Defoliation Treated	% Defoliation Check	% Defoliation reduction	Reference[c]
Phoxim	1973	5,6	360	3.6	77				1
"	1974	4	280	2.9	50-56	6-9	24		2
Leptophos	1975	4,5	2 × 140	2 × 1.5	70	32	41	30	3
"	1975	4,5	280	2.9	81	11–18	41	60–76	3
Acephate	1973	5,6	360	3.6	76				1
technical	1974	4	280	2.9	86	5	24		2
"	1974	4	420	4.4	96	3	24		2
"	1974	4	630	6.6	95	5	24		2
"	1975	2	280	2.9	60–72	20–50	60	17–67	3
"	1975	4,5	140	1.5	75–77	23–24	60	61–64	3
"	1975	4,5	280	2.9	93–94	8–16	60	73–87	3
Acephate	1975	4,5	140	1.5	87	19	60	68	3
75% S.	1975	4,5	280	2.9	92	18	60	71	3
"	1976	4,5	70	1.4	69			45	4
"	1976	2	70	1.4	11			30	4
"	1976	2	140	1.4	40			52	4
"	1976	2	290	2.9	48			38	4
Permethrin	1975	4,5	7	1.5	49	31	41	33	3
"	1975	4,5	18	1.5	81	16	41	66	3
"	1975	4,5	35	1.5	68	20	41	56	3
"	1975	5,6	70	1.5–2.9	47–70	26–39	41	26–44	3
"	1975	5,6	140	2.9	70	28	41	40	3
Permethrin	1976	4	19	1.1	0–35			0–35	4
"	1976	4,5	36–39	1.1–4.5	0–56			0–64	4
"	1976	4,5	63–64	1.1–4.5	6–86			0–77	4
"	1976	2	27	3.4	62			66	4
"	1976	2	38	1.1–2.3	62–77			58–70	4
"	1976	2	48	3.4	52			81	4
"	1976	2	63	1.1–2.3	53–76			55–76	4
Chlorpyrifos-	1976	5	110	1.1	71			80	4
methyl	1976	5	230	2.3	52			74	4
"	1976	5	450	4.5	79			85	4
Fenitrothion	1973	5,6	360	3.6	53–60				1
"	1974	4	280	2.9	57–67	8–10	24		2
"	1975	5,6	280	2.9	71–83	12–13	45	71–73	3
"	1976	5	290	2.9	52			26	4
"	1976	5	200	1.1	57			67	4

[a] Grams active ingredient per hectare.
[b] Spruce budworm larvae.
[c] References: 1. Nigam and Hopewell 1973; 2. Hopewell and Nigam 1974; 3. Hopewell 1975; 4. Hopewell 1977b.

hibernaculae, were also effective at applied dosages of 27 to 64 g/ha. In these, and in the previous mid-season tests, the measured dye deposits in petri dishes were normally about twice that of the applied dosages. In mistblower trials, permethrin was one of the most effective insecticides evaluated by DeBoo (1980) and DeBoo and Campbell (1972, 1974a, 1974b). Treatments of 18 to 35 g/ha were very effective in reducing population densities of third- to fifth-instar larvae and provided high levels of foliage protection.

In preliminary tests (Table 4), chlorpyrifos-methyl was at least as effective as fenitrothion for budworm control at similar deposits. Treatments of carbaryl and methomyl by mistblower provided very good control at the dosages tested (Table 5).

Except for leptophos, these insecticides were also tested in small-scale aerial trials during the review period. The toxicity of leptophos to spruce budworm was not particularly high and the company was not interested in further development. Two other insecticides listed in Table 3, azamethiphos and mexacarbate (Zectran®), proceeded directly to such aerial trials. Zectran® had been previously tested, registered, and used for spruce budworm control in Canada (Nigam 1972, 1975c). Its Canadian registration lapsed after 1973 but interest in this insecticide was renewed when Union Carbide Agricultural Products obtained the rights to Zectran® from Dow Chemical in 1981. The results of the aerial trials with these insecticides are reported elsewhere in this book. Three other promising insecticides, sulprofos, thiodicarb, and cypermethrin, have not been field tested because the respective companies decided not to pursue development of these products for forestry use.

Of the insecticides reviewed previously, acephate has received a registration for aerial and ground use against spruce budworm larvae (Table 1). Carbaryl is also registered for such use. Methomyl and permethrin are registered for ground application only. The impact of permethrin on aquatic arthropods was concluded to be too great to recommend the large-scale aerial use of this insecticide against spruce budworm (Kingsbury and Kreutzweiser 1987; Kreutzweiser and Kingsbury 1987). The other insecticides have not been registered for various reasons such as unsatisfactory efficacy at tested dosages, or poor market potential.

Larvae—New Formulations

Since 1980, scientists in the Chemical Control Agents Program at the Forest Pest Management Institute were involved primarily with the development and evaluation of new formulations and tank mixes of insecticides currently or previously registered for spruce budworm control in Canada (Table 6) (Helson 1985b; Nigam 1985).

Matacil® 180F is a flowable formulation containing 180 g of aminocarb per litre. This formulation consists of fine particles of technical aminocarb suspended with special additives in an agricultural-type spray oil. It can be mixed directly with an oil carrier such as ID585 or with water plus an emulsifier such as Atlox® 3409F or Triton® X114. Matacil® 180F was developed to replace Matacil® 1.8D, which contained nonylphenol as the primary solvent. Concern had been expressed about the toxicity of nonylphenol to certain aquatic organisms, which were more sensitive to this solvent than to aminocarb itself. The contact, foliar, and residual toxicities of Matacil® 180F to spruce budworm larvae is comparable to other aminocarb formulations (Table 6). Matacil® 180F proceeded

Table 5. Insecticides tested against spruce budworm, *Choristoneura fumiferana*, larvae by mistblower on white spruce trees, 1973–1978.

Insecticide	Year	Instar	Applied rate g/ha[a]	Applied rate L/ha	% SBL[b] reduction	% Defoliation Treated	% Defoliation Check	Reference[c]
Phoxim	1974	4	300	152	79	8	29–61	1
Acephate	1974	4	550	140	98	6	29–61	1
Permethrin	75–78	3–5	7	130–200	59	35	63	2
"	"	"	9	"	43	12	63	2
"	"	"	12	"	82	7	63	2
"	"	"	18	"	65	24	63	2
"	"	"	24	"	96	10	63	2
"	"	"	35[d]	"	91	11	71	2
"	"	"	70[d]	"	92	10	71	2
Carbaryl	1974	4	665	150	89	13	29–61	1
Methomyl	1974	4	260	160	85	11	29–61	1
Fenitrothion	1973	4,5	420	190	51	>50	>50	3

[a] Grams active ingredient per hectare.
[b] Spruce budworm larvae.
[c] References: 1. DeBoo and Campbell 1974b; 2. DeBoo 1980; 3. DeBoo and Campbell 1974a.
[d] White spruce and balsam fir combined.

to aerial field trials and is now registered for spruce budworm control in Canada.

Sumithion® 20F is a novel, emulsion-type, flowable formulation containing liquid technical fenitrothion suspended in water in the presence of dispersing and thickening agents (Tsuji and Fuyama 1982). This formulation can be mixed directly with water without the addition of an emulsifier or can be sprayed undiluted. Sumithion® 20F is as toxic to spruce budworm larvae as other aqueous-based fenitrothion formulations and its residual activity on foliage is also comparable (Table 6). Aerial trials were conducted with this formulation to determine its efficacy in the field and it is now registered for spruce budworm control in Canada.

A new flowable formulation of carbaryl, Sevin® FR, was introduced by Union Carbide and underwent preliminary testing in the insecticide evaluation program at the Forest Pest Management Institute. Sevin® FR is a suspension of microfine carbaryl in water, which can be mixed directly with water as a carrier. Its contact toxicity to spruce budworm larvae was similar to Sevin®-4-oil, the conventional oil-based formulation and to Sevin® XLR, a formulation used in agriculture. Its residual toxicity also appeared to be at least as high as the other formulations.

Zectran® UCZ15 is an aqueous-based flowable formulation of mexacarbate developed by Union Carbide. Its longevity under natural weathering conditions and its toxicity to spruce budworm larvae were comparable to the new formulation of mexacarbate, Zectran® UCZF19. Development of Sevin® FR and the flowable formulation of Zectran® for forestry use was discontinued by the company before these formulations were tested in the field.

Pupae and Adults

In contact toxicity tests, spruce budworm pupae were susceptible to several insecticides including methidathion, phosmet, phosphamidon, and methomyl (Table 2). Pupae generally appeared to be less susceptible than fifth-instar larvae to most insecticides, particularly aminocarb and mexacarbate.

Control measures aimed at spruce budworm adults have been considered as a potential strategy for altering the course of future infestations. Phosphamidon was evaluated against moths in aerial applications in New Brunswick (Kettela and Miller 1975). These field trials raised several questions about the effectiveness of different insecticides, the relative susceptibility of males versus females, and virgin versus mated females, and effects of insecticide treatments on oviposition. Generally, those insecticides tested in the laboratory were quite toxic to spruce budworm adults, particularly permethrin and cypermethrin (Table 2). Males were approximately twice as susceptible as females to most insecticides. The toxicities of fenitrothion and phosphamidon were similar to virgin and mated females.

Table 6. New insecticide formulations evaluated against spruce budworm, *Choristoneura fumiferana*, larvae in the laboratory, 1973–1988.

		Toxicity to fifth-instar larvae									
		Direct contact		Treated larch foliage		Residual toxicity					
		72 h LC_{50}		72 h LC_{50}		Dosage	72 h percent mortality after weathering for				
Formulations	Carrier	(μg/cm^2)	Slope	(μg/cm^2)	Slope	(g/ha)[a]	0	1	3	5	10 days
Matacil® 180F	ID585	0.027	4.02	0.025	3.33	224	100	87	88	67	58
Matacil® 1.8D	ID585	0.025	3.84	–	–	224	100	95	72	67	40
Matacil® tech	ID585	0.030	5.04	0.021	3.32	224	100	89	92	75	58
Sumithion® 20F	H_2O	0.294	3.61	0.087	3.67	224	99	43	25	0	4
Sumithion® tech	H_2O[b]	0.388	4.43	0.113	5.11	224	99	47	31	16	6
Sevin® FR	Dowanol TPM	0.566	2.44	–	–	560	–	–	89	75	31
Sevin®-4-Oil	ID585	0.695	2.63	–	–	560	93	91	61	40	5
Sevin® XLR	H_2O	0.768	3.07	–	–	560	100	64	61	42	3
Zectran® UCZF 15	H_2O	0.056	4.67	0.023	3.17	100	98	78	25	25	4
Zectran® UCZF 19	H_2O[b]	0.067	4.91	0.028	4.34	100	97	52	50	27	12

[a]Grams active ingredient per hectare.
[b]Water plus emulsifier.

Other Forest Insects

From 1973 to 1988, various new and registered insecticides have been tested on about 20 species of pests from different regions of Canada by the authors (Nigam 1973, 1975a, 1975b; Helson et al. 1989) (Table 7). A major recent thrust in this research has been the assessment of the potential of pyrethroids, particularly permethrin, for the control of certain forest pests on small areas such as plantations where no water is present, both by ground and aerial applications.

Spruce Bud Moth Larvae

The spruce bud moth, *Zeiraphera canadensis*, has recently become an important pest of white spruce plantations in New Brunswick and Quebec. Because no control measures were available for this pest, laboratory investigations were initiated in 1982 to identify potential insecticides for its control. Biological studies indicated that the only larval stage exposed to contact insecticides may be newly hatched larvae that crawl along branches from the eggs to the spruce buds (Turgeon 1985). Consequently, bioassay techniques were developed to evaluate the effects of insecticides on first-instar larvae by direct exposure to sprays and by exposure to insecticide deposits on white spruce twigs, the primary route of exposure. The residual toxicities of selected insecticides were also determined by spraying potted white spruce seedlings and placing larvae on these seedlings at periodic intervals up to 5 days after treatment. The insecticides tested on spruce bud moth larvae are listed in Table 7. Although several insecticides are toxic to spruce bud moth larvae, permethrin has the best potential for controlling newly hatched larvae because of its consistently high toxicity and longevity (Helson et al. 1989). Aerial trials by fixed-wing aircraft and helicopter, and mistblower tests in New Brunswick and Quebec white spruce plantations have confirmed that permethrin at 70 g/ha of the active ingredient is effective in controlling *Z. canadensis* and reducing damage to tree leaders (Helson et al. 1989; Helson and Auger 1991).

Black Army Cutworm Larvae

The black army cutworm, *Actebia fennica*, has been a periodic pest of newly planted conifer seedlings in burned areas in British Columbia, Ontario, and Newfoundland. The contact toxicity of several insecticides to fourth-instar larvae has been determined in the laboratory (Table 7). Pyrethroid insecticides, including permethrin, are toxic to this species. Next to the pyrethroids, the most toxic insecticides are chlorpyrifos, methomyl, and endosulfan. Most other insecticides possess relatively low toxicity and do not have much potential for the control of this pest. Permethrin is registered for the control of black army cutworm by ground application (Table 1).

Bruce Spanworm Larvae

The contact toxicities of seven insecticides to larvae of the Bruce spanworm, *Operophtera bruceata*, a periodic defoliator of trembling aspen and sugar maple, have been determined in the laboratory (Table 7). Permethrin is the most toxic, followed by fenitrothion and aminocarb.

Rusty and Douglas-fir Tussock Moth Larvae

Of the six insecticides tested on rusty tussock moth, *Orgyia antiqua*, permethrin and fenvalerate were the most toxic to fourth-instar larvae (Table 7). Aminocarb, mexacarbate, acephate, and fenitrothion have been tested on third- and fourth-instar Douglas-fir tussock moth, *O. pseudosugata*, larvae. The LC_{50} values ranged from 0.1 to 0.8 μg/cm^2, with the carbamates being most toxic and fenitrothion least toxic to this species. Acephate, carbaryl, and permethrin are registered for Douglas-fir tussock moth larvae (Table 1).

Gypsy Moth Larvae

The gypsy moth, *Lymantria dispar*, has been present in parts of Canada, particularly Ontario and Quebec, for many years; the infestation in Ontario began to grow dramatically in 1981. Bioassay methods have been developed for evaluating the toxicity of promising new insecticides in the laboratory both by direct contact with the larvae and by exposure to insecticide deposits on discs of red oak leaves (Table 7). Permethrin and ethofenprox are very toxic to second-instar larvae. Carbaryl, the standard insecticide, and two other carbamates, aminocarb and mexacarbate, are also toxic to gypsy moth larvae. Acephate, carbaryl and trichlorfon are registered for this pest in Canada (Table 1).

Hemlock Looper Larvae

In 1984, a new outbreak of the hemlock looper, *Lambdina fiscellaria*, a serious pest of balsam fir, occurred in Newfoundland. Because technical fenitrothion was the only insecticide registered for the aerial control of hemlock looper (Table 1), research was initiated to determine the potential of the new formulations of aminocarb, mexacarbate, and fenitrothion, Matacil® 180F, Zectran® UCZF19, and Sumithion® 20F, respectively, for controlling this pest. Zectran® and Sumithion® 20F are promising products. Zectran® had up to triple the toxicity of a standard emulsion of technical fenitrothion to third-instar larvae by both direct contact exposure and exposure to deposits on balsam firm foliage. Sumithion® 20F was up to twice as toxic as the fenitrothion emulsion. The toxicity of Matacil® 180F was similar to that of technical fenitrothion. Permethrin was also found to be very toxic to hemlock looper larvae.

Jack Pine Budworm Larvae

In 1984, a new outbreak of jack pine budworm, *Choristoneura pinus*, an important pest of jack pine, occurred in Ontario. The only registered insecticide for this insect was technical fenitrothion. Therefore, Matacil® 180F, Zectran® UCZF19, and Sumithion® 20F were tested against this budworm. They are all good candidates for control of this species because their toxicities to jack pine budworm larvae, by direct contact and on treated foliage, were equal to or greater than a standard emulsion of technical fenitrothion.

Sawfly Larvae

The larval stages of the redheaded pine sawfly, *Neodiprion lecontei*, Swaine jack pine sawfly, *N. swainei*, jack

Table 7. Insecticides tested on selected forest insect pests in the laboratory.

Insecticides	Contact toxicity rank[a] for designated insect species[b]										
	SBM	BAC	BS	RTM	GM	RPS	SJS	JPS	LS	MAS	WPW
Organophosphates											
acephate	≥4[c]	4	5	4	4	4	4	4	4	4	5
azamethiphos						2	2				
azinphos-methyl	3	4									4
carbophenothion								4			≥4[c]
chlorphoxim						3	3		3		4
chlorpyrifos	3	4				3[e]		3	3[e]	3	4[e]
chlorpyrifos-methyl						3	3	3	3		4
diazinon		4								3	
dimethoate	5[d]	5	4			≤3[c,e]	≤3[c,e]		3[e]	3	4
etrimfos						3	2	3			4
fenitrothion	3	5	4		≤5[c]	3	3	3	3		4
leptophos						4		3			4
malathion		5	4	4		≤3[c]	3	≥2[c]	3	3	
methidathion						2	2	2	3		4
methyl Trithion											4
phosmet						3					4
phosphamidon			4			3	3[e]	3	3[e]		4[e]
phoxim							3	3			4
profenofos						3	3				
prothiofos							≥3[c]		≤4[c]		
sulprofos	3						3	3	3		≥4[c]
tetrachlorvinphos						3	3[e]		3[e]		4[e]
trichlorfon	≥4[c]	5		4	4	3[e]		4[e]	3[e]		
trifenofos						3	3	3	3		
Carbamates											
aminocarb	4	5	4		3	2[e]	3[e]	3	3[e]		5
carbaryl	5[d]	5		3	3			3		≤3[c]	
Dimetilan						3					5
methomyl	3	4				2[e]	2[e]	2	3[e]		4[e]
mexacarbate	3	4			3	≤3[c,e]	3[e]	3[e]	3[e]		4[e]
propoxur						3[e]	2[e]		3[e]		
thiodicarb	4	5			4						
Pyrethroids											
allethrin								4	4		
bioethanomethrin						3	2		2		
cismethrin						2	2	2	3		
cypermethrin	2[d]	3				2	2	2		2	3
deltamethrin	2[d]	2				2	2	1		3	2
ethofenprox		3			2						4
fenpropanate						3	3	3			
fenvalerate		3		2		3	3	3	3		4
flucythrinate								3			
permethrin	3	3	2	2	2	2	3	3	3	3	3
resmethrin						3	3	3	2		4
Organochlorines											
endosulfan		4									
methoxychlor	5[d]	5							5[e]	4	5
Miscellaneous											
Baygon MEB						5		5			
fenarimol						≥4[c]	5				
thiocyclam								5			4

[a] The following numbers represent the range in LC_{50} (μg/cm^2) values corresponding to each contact toxicity ranking (in parentheses): <0.001(1); 0.002-0.01(2); 0.02-0.1(3); 0.2-1(4); and >2(5).

[b] Species and stage tested: SBM = spruce bud moth first-instar larvae; BAC = black army cutworm fourth-instar larvae; BS = Bruce spanworm fourth-instar larvae; RTM = rusty tussock moth fourth-instar larvae; GM = gypsy moth second-instar larvae; RPS = redheaded pine sawfly fourth- and fifth-instar larvae; SJS = Swaine jack pine sawfly fourth-instar larvae; JPS = jack pine sawfly fourth-instar larvae; LS = larch sawfly fourth- and fifth-instar larvae; MAS = mountain-ash sawfly fourth- and fifth-instar larvae; and WPW = white pine weevil adults.

[c] Equal to, or one rank more or less than the indicated one, respectively (≥, ≤).

[d] Rank based on toxicity to fourth-instar spruce bud moth larvae.

[e] Based on toxicity values in Nigam 1975c.

pine sawfly, *N. pratti banksianae*, larch sawfly, *Pristiphora erichsonii*, and mountain-ash sawfly, *P. geniculata*, are generally very susceptible to many insecticides (Table 7). Pyrethroids such as deltamethrin and cypermethrin were the most toxic insecticides. The toxicity of azamethiphos was comparable to these pyrethroids. In fact, several insecticides in three different classes were very toxic, with LC_{50} values ranging from 0.01 to 0.03 μg/cm^2 for most species. These include the organophosphates, methidathion, etrimfos, chlorpyrifos-methyl, tetrachlorvinphos, phoxim, malathion, fenitrothion, and diazinon; the pyrethroids, permethrin, fenvalerate, and resmethrin; and the carbamates, methomyl, aminocarb, mexacarbate, and propoxur. The contact toxicities of acephate and methoxychlor to sawfly larvae were relatively low. The susceptibilities of the different species to insecticides were normally similar.

In simulated aerial spray tests against the redheaded pine sawfly, fenitrothion at 145 g/ha, permethrin at 22 g/ha, phosphamidon at 36 g/ha, aminocarb at 145 g/ha, and propoxur at 145 g/ha all gave good to excellent foliage protection but acephate at 145 g/ha was not effective (Hopewell 1977a). In Alberta, field trials using mistblowers and hydraulic sprayers have been conducted on shade and shelterbelt trees to determine the effectiveness of fenvalerate, malathion, permethrin, cypermethrin, diazinon, methomyl, carbaryl, phosmet, dimethoate, acephate, carbofuran, propoxur, resmethrin, chlordimeform, isothioate, methidathion, phoxim, oxamyl, aldicarb, and other experimental insecticides, or mixtures of insecticides against yellowheaded spruce sawfly and larch sawfly larvae (Drouin and Kusch 1974, 1975, 1976, 1977, 1978, 1979). Acephate, chlorpyrifos, dimethoate, fenitrothion, and permethrin are registered for use against these pests (Table 1).

White Pine Weevil

The white pine weevil, *Pissodes strobi*, is an economically important pest of several conifer species including white pine, jack pine, Sitka spruce, and Engelmann spruce. Because adults are the only stage exposed to contact insecticides, control measures have been directed against them, either during the spring when females are feeding and laying eggs on the leaders, or in the fall after the new generation of adult weevils has emerged and is feeding on the trees. Currently, the only insecticide used for white pine weevil control in Canada is methoxychlor, applied at high dosages with ground spray equipment. Aerial applications with methoxychlor have not been successful. Dimethoate is registered for Sitka spruce weevil, now considered to be synonymous with the white pine weevil (Table 1). Consequently, new insecticides have been tested in an attempt to find effective products to control this pest (Table 7). Most insecticides are not very toxic to adults but some of the pyrethroids show promise. The potential of permethrin is being evaluated extensively by examining its toxicity by direct contact, on white pine branches and on potted white pine trees.

A Seedling Debarking Weevil

Adults of a seedling debarking weevil, *Hylobius congener*, have recently been recognized as a pest of conifer seedlings in Nova Scotia. A potential control strategy is one that is used in Europe to control a similar pest, *H. abietis*: seedlings are treated with an insecticide before or after planting to prevent damage for up to two seasons. Preliminary tests were conducted to determine the relative toxicity of permethrin, chlorpyrifos, fenitrothion, lindane, carbaryl, and methoxychlor to adults on branches of white pine or white spruce dipped in solutions of these insecticides. The first four were most toxic and are being evaluated further for their effectiveness and longevity on seedlings dipped in or sprayed with selected concentrations, planted in pots, and placed outdoors for up to 2 seasons.

Nontarget Aquatic Arthropods

Permethrin and other pyrethroids are some of the most toxic and effective insecticides for controlling many forest insect pests including spruce bud moth, black army cutworm, hemlock looper, gypsy moth, and white pine weevil. Because permethrin also has low toxicity to mammals and birds, it could be very useful for managing such insect pests. However, it is highly toxic to fish, and more importantly, to aquatic insects and crustaceans. This has severely curtailed its potential use in forest pest management to date. No direct fish mortality has been observed in forest streams after experimental aerial applications of permethrin at dosages of active ingredient between 8.8 and 70 g/ha. However, these applications severely disturbed invertebrate communities in streams with resultant secondary effects on fish including altered diets, temporary growth rate reductions, and decreases in fish densities, presumably due to emigration (Kingsbury and Kreutzweiser 1987; Kreutzweiser and Kingsbury 1987). Consequently, these authors have recommended that aerial applications of permethrin at these dosages should only be considered appropriate in situations where no productive fishery habitat is present or where they can be effectively buffered. These conditions are met in many plantations that have been planted recently and permethrin would be a very useful tool for the control of several pests attacking this valuable resource.

Since 1983, scientists at the Forest Pest Management Institute have been investigating the size of buffer zones necessary to minimize significant mortality of fish food organisms due to permethrin sprays. A technique has been developed using bioassays with two indicator species, *Gammarus pseudolimnaeus* and *Aedes aegypti*, larvae in artificial aquatic habitats to assess the impact of permethrin drift downwind of spray applications (Helson et al. 1986). Children's plastic wading pools, 500-ml Mason jars or 9-L aluminum pans containing stream water are placed in replicate at selected distances downwind of the spray site. The amphipods or mosquito larvae are either placed directly in these containers before or immediately after the application, or in water samples collected from the pools and pans after the spray. These bioassays are very sensitive with LC_{50} values of 0.25 to 0.37 ppb for *G. pseudolimnaeus* and 0.69 to 1.85 ppb for *A. aegypti* larvae obtained in concurrent standards. The mortality–concentration relationships from the standards can be used to estimate the quantity of permethrin deposited with distance downwind of the application and to predict the mortality at different distances based on drift models developed from measured deposits on artificial collection surfaces. Predicted mortalities can then be compared with actual mortalities at downwind sites to test the adequacy of these models. These techniques have been used successfully in studies to develop and test models for predicting the size of buffer zones required for mistblower and aerial applications of permethrin (Payne et al. 1986, 1988). Besides being relatively simple to perform and inexpensive, such bioassays can be repeated on the same site in the same year and eliminate the need to contaminate natural bodies of water and avoid the tedious sampling required for such systems. Research is continuing with these bioassays at the Forest Pest Management Institute to determine the size of buffers required for small-scale aerial sprays of permethrin in an effort to generate the information needed to register this insecticide for specific uses in plantations.

References

Abbott, W.S. 1925. A method of computing the effectiveness of an insecticide. J. Econ. Entomol. 18: 265-267.

Armstrong, J.A.; Nigam, P.C. 1975. The effectiveness of the aerial application of Orthene® against spruce budworm at Petawawa Forest Experiment Station during 1974. Environ. Can., Can. For. Serv., Chem. Cont. Res. Inst., Ottawa, Ont. Inf. Rep. CC-X-82. 28 p.

Bickle, A. 1968. S103 Probit Single Line and Parallel Line Analysis. A computer program written by Statistical Research Section, Engineering and Statistical Research Institute, Research Branch, Agric. Can, Ottawa, Ont.

DeBoo, R.F. 1980. Experimental applications of permethrin by mistblower for control of spruce budworm in Quebec, 1975-78. Can. For. Serv., Ottawa, Ont. Bi-mon. Res. Notes 36(5):23-24.

DeBoo, R.F., Campbell, L.M. 1972. Plantation research: V. Mistblower applications of dilute insecticide solutions for control of *Choristoneura fumiferana* on white spruce in Quebec, 1972. Environ. Can., Can. For. Serv., Chem. Cont. Res. Inst., Ottawa, Ont. Inf. Rep. CC-X-21. 17 p.

DeBoo, R.F.; Campbell, L.M. 1974a. Evaluation of commercial preparations of *Bacillus thuringiensis* with and without chitinase against spruce budworm. C. Assessment of effectiveness by mistblower and aerial application, Spruce Woods, Manitoba. Environ. Can., Can. For. Serv., Chem. Cont. Res. Inst., Ottawa, Ont. Inf. Rep. CC-X-59. 31 p.

DeBoo, R.F.; Campbell, L.M. 1974b. Plantation research: XII. Experimental applications of insecticides by mistblower for control of *Choristoneura fumiferana* on white spruce in Quebec, 1974. Environ. Can., Can. For. Serv., Chem. Cont. Res. Inst., Ottawa, Ont. Inf. Rep. CC-X-88. 24 p.

Drouin, J.A.; Kusch, D.S. 1974. Insecticide field trials on shade and shelterbelt trees in Alberta and Saskatchewan, 1973. Environ. Can., Can. For. Serv., North For. Res. Cent., Edmonton, Alta. Inf. Rep. NOR-X-81. 40 p.

Drouin, J.A.; Kusch, D.S. 1975. Pesticide field trials on shade and shelterbelt trees in Alberta and Saskatchewan, 1974. Environ. Can., Can. For. Serv., North For. Res. Cent., Edmonton, Alta. Inf. Rep. NOR-X-131. 30 p.

Drouin, J.A.; Kusch, D.S. 1976. Pesticide field trials on shade and shelterbelt trees in Alberta, 1975. Environ. Can., Can. For. Serv., North For. Res. Cent., Edmonton, Alta. Inf. Rep. NOR-X-150. 29 p.

Drouin, J.A.; Kusch, D.S. 1977. Pesticide field trials on shade and shelterbelt trees in Alberta, 1976. Environ. Can., Can. For. Serv., North For. Res. Cent., Edmonton, Alta. Inf. Rep. NOR-X-184. 20 p.

Drouin, J.A.; Kusch, D.S. 1978. Pesticide field trials on shade and shelterbelt trees in Alberta, 1977. Environ. Can., Can. For. Serv., North For. Res. Cent., Edmonton, Alta. Inf. Rep. NOR-X-205. 16 p.

Drouin, J.A.; Kusch, D.S. 1979. Pesticide field trials on shade and shelterbelt trees in Alberta, 1978. Environ. Can., Can. For. Serv., North For. Res. Cent., Edmonton, Alta. Inf. Rep. NOR-X-213. 16 p.

Finney, D.J. 1962. Probit Analysis. 2nd. ed. Cambridge University Press. 318 p.

Helson, B.V. 1985a. Chemical insecticides for spruce budworm. Pages 131-135 *in* D. Schmidt (Ed.), Spruce-Fir Management and Spruce Budworm. Society of American Foresters Region VI Technical Conference. Apr. 24-26, 1984. Burlington, Ver. Gen. Tech. Rep. NE-99. USDA For. Serv., NE For. Expt. Stn., Broomall, Pa. 1985. 217 p.

Helson, B.V. 1985b. Research on chemical insecticides for spruce budworm control at the Forest Pest Management Institute, 1982-present. Page 376 *in* C.J. Sanders, R.W. Stark, E.J. Mullins, and J. Murphy (Eds.), Recent Advances in Spruce Budworms Research. Proc. CANUSA. Spruce Budworms Res. Symp., Bangor, Me., Sept. 16-20, 1984. 527 p.

Helson, B.V.; Auger, M. 1991. Efficacy and timing of permethrin mistblower sprays for the control of first-instar larvae of the spruce bud moth, *Zeiraphera canadensis* Mut. and Free. (Lepidoptera: Tortricidae). Can. Entomol. 123: 1319-1326.

Helson, B.V.; de Groot, P.; Turgeon, J.J.; Kettela, E.G. 1989. Toxicity of insecticides to first instar larvae of the spruce budmoth, *Zeiraphera canadensis* Mut. and Free. (Lepidoptera: Tortricidae): laboratory and field studies. Can. Entomol. 121: 81-91.

Helson, B.V.; Kingsbury, P.D.; de Groot, P. 1986. The use of bioassays to assess aquatic arthropod mortality from permethrin drift deposits. Aquatic Toxicol. 9: 253-262.

Hopewell, W.W. 1974. Evaluation of commercial preparations of *Bacillus thuringiensis* with and without chitinase against spruce budworm. B. Simulated aerial sprays in a young white spruce plantation, Shawville, Quebec. Environ. Can., Can. For. Serv., Chem. Cont. Res. Inst., Ottawa, Ont. Inf. Rep. CC-X-59. 11 p.

Hopewell, W.W. 1975. Field evaluation of Orthene®, Phosvel®, FMC 33297 and TH 6040 against *Choristoneura fumiferana*, applied as simulated aerial spray. Environ. Can., Can. For. Serv., Chem. Cont. Res. Inst., Ottawa, Ont. Inf. Rep. CC-X-115, 24 p. + Fig.

Hopewell, W.W. 1977a. Field evaluation of eight insecticides for control of *Neodiprion lecontei* on red pine, *Pinus resinosa*, by simulated aerial spray. Environ. Can., Can.

For. Serv., For. Pest Manage. Inst., Sault Ste. Marie, Ont. Inf. Rep. FPM-X-7. 20 p.

Hopewell, W.W. 1977b. Field evaluation of the pyrethroid NRCD-143, compared with fenitrothion, acephate and chlorpyrifos-methyl as simulated aerial spray deposit for control of the spruce budworm. *Choristoneura fumiferana* Clem. Environ. Can., Can. For. Serv., Chem. Cont. Res. Inst., Ottawa, Ont. Inf. Rep. CC-X-132. 31 p.

Hopewell, W.W.; Nigam, P.C. 1974. Field evaluation of Orthene®, phoxim and fenitrothion against spruce budworm (*Choristoneura fumiferana*) applied as a simulated aerial spray. Environ. Can., Can. For. Serv., Chem. Cont. Res. Inst., Ottawa, Ont. Inf. Rep. CC-X-83. 14 p.

Kettela, E.G.; Miller, C.A. 1975. Spray applications against spruce budworm moths. Pages 147-148 *in* M.L. Prebble (Ed.), Aerial Control of Forest Insects in Canada. Can. For. Serv., Dept. Environ. Ottawa, Ont. 330 p.

Kingsbury, P.D.; Kreutzweiser, D.P. 1987. Permethrin treatments in Canadian forests. Part 1: Impact on stream fish. Pestic. Sci. 19: 35-48.

Kreutzweiser, D.P.; Kingsbury, P.D. 1987. Permethrin treatments in Canadian forests. Part 2: Impact on stream invertebrates. Pestic. Sci. 19: 49-60.

Nigam, P.C. 1972. Contact and residual toxicity studies of Zectran against 18 species of forest insect pests. Can. Dept. Environ., Can. For. Serv., Chem. Cont. Res. Inst., Ottawa, Ont. Inf. Rep. CC-X-29. 6 p.

Nigam, P.C. 1973. Summary of contact, stomach and residual toxicity of insecticides against forest insect pests during 1973. Environ. Can., Can. For. Serv. Chem. Cont. Res. Inst., Ottawa, Ont. Inf. Rep. CC-X-46. 9 p.

Nigam, P.C. 1975a. Summary of contact, stomach and residual toxicity of insecticides against forest insect pests during 1974. Environ. Can., Can. For. Serv., Chem. Cont. Res. Inst., Ottawa, Ont. Inf. Rep. CC-X-100. 10 p.

Nigam, P.C. 1975b. Summary of laboratory evaluation of insecticides against various species of forest insect pests during 1975. Environ. Can., Can. For. Serv., Chem. Cont. Res. Inst., Ottawa, Ont. Inf. Rep. CC-X-124. 9 p.

Nigam, P.C. 1975c. Chemical insecticides. Pages 8-24. *in* M.L. Prebble (Ed.), Aerial Control of Forest Insects in Canada. Can. For. Serv., Dept. Environ., Ottawa, Ont. 330 p.

Nigam, P.C. 1981. New chemical insecticides. Pages 127-128 *in* J. Hudak and A.G. Raske, (Eds.), Review of the Spruce Budworm Outbreak in Newfoundland—Its Control and Forest Management Implications. Can. For. Serv., Nfld. For. Res. Cent., St. John's, Nfld. Inf. Rep. N-X-205. 280 p.

Nigam, P.C. 1985. Early laboratory bioassays in Canada. Page 372 *in* C.J. Sanders, R.W. Stark, E.J. Mullins and J. Murphy (Eds.), Recent Advances in Spruce Budworm Research. Proc. CANUSA Spruce Budworm Res. Symp., Bangor, Me., Sept. 16-20, 1984. 527 p.

Nigam, P.C.; Hopewell, W.W. 1973. Preliminary field evaluation of phoxim and Orthene® against spruce budworm on individual trees as simulated aircraft spray. Environ. Can., Can. For. Serv., Chem. Cont. Res. Inst., Ottawa, Ont. Inf. Rep. CC-X-60. 15 p.

Payne, N.J.; Helson, B.V.; Sundaram, K.M.S.; Fleming, R.A. 1988. Estimating buffer zone widths for pesticide applications. Pestic. Sci. 24: 147-161.

Payne, N.; Helson, B.; Sundaram, K.; Kingsbury, P.; Fleming, R.; de Groot, P. 1986. Estimating the buffer required around water during permethrin applications. Can. For. Serv., For. Pest. Manage. Inst., Sault Ste. Marie, Ont. Inf. Rep. FPM-X-70. 26 p.

Turgeon, J.J. 1985. Life cycle and behavior of the spruce budmoth, *Zeiraphera canadensis* (Lepidoptera: Olethreutidae), in New Brunswick. Can. Entomol. 117: 1239-1247.

Tsuji, K.; Fuyama, H. 1982. Development of emulsion-type flowable formulation. Pages 361-366 *in* R. Greenhalgh and N. Drescher (Eds.), Pesticide Chemistry: Human Welfare and the Environment. Vol. 4. Pesticide Residues and Formulation Chemistry. Pergamon Press, Toronto. 429 p.

Chapter 39

Insect Growth Regulators

ARTHUR RETNAKARAN

Introduction

Insect growth regulators are biochemicals that adversely interfere with the normal growth and development of insects. Their spectrum of activity is restricted, in most instances, to either the class insecta or phylum arthropoda. In general, they are environmentally acceptable, having negligible effects on nontarget species. Because they act on unique biochemical systems in the insect, they are also called biorational control agents. They can be synthetic compounds, natural products, or analogs of naturally occurring chemicals.

Insect growth regulators include a list of compounds that are mostly chemically unrelated, but have a common effect on some aspect of growth and development in insects. By the very nature of the definition, new groups and new compounds are constantly being added. A broad classification that includes the different types of insect growth regulators currently being developed is given in Table 1.

Hormone Analogs

The sequential polymorphism observed during metamorphosis is orchestrated by the brain through its neurosecretions, much like the tropic functions of the pituitary gland in vertebrates. The two principal hormones involved in the growth and development are the juvenile hormone and 20-hydroxy ecdysone or molting hormone. The secretion of juvenile hormone by the corpora allata is controlled by allatotropin and allatohibin secreted by the neurosecretory cells in the brain. The secretion of the molting hormone is controlled by the prothoracicotropic hormone secreted again by the neurosecretory cells in the brain (Retnakaran et al. 1985; Steel and Davey 1985). Several hormone analogs have been tested as agents and the chief ones will be highlighted.

Juvenile Hormone Analogs

Carroll Williams, who did much of the pioneering research on juvenile hormones, suggested in 1967 that they could be used as insect-specific control agents and heralded them as "third generation insecticides" following the inorganic pesticides and the chlorinated hydrocarbons (Williams 1967). Among its functions, the juvenile hormone preserves the larval state and in its absence, the larva metamorphoses into the adult. When the last larval instar is treated with this hormone, many of the larval characters are retained in the pupa. This larval–pupal intermediate seldom reaches the adult stage. The larval stadium is often prolonged.

Several hundred juvenile hormone analogs, some of them many times more active than the native material, have been synthesized but only a few have had commercial success. Basically these compounds are useful as control agents only when the adult stage is the one that causes the damage, such as mosquitoes, fleas, cockroaches, and flies. If the larval stage is the one responsible for the damage, then treatment with the material prolongs the larval stage increasing the damage, even though population control can be achieved. In the context of forestry, promising results have been obtained against the balsam woolly adelgid, *Adelges piceae* (Bowers and Retnakaran 1988).

Anti-Juvenile Hormone Compounds

Some compounds such as the naturally occurring precocenes (Bowers et al. 1976) or synthetic compounds like fluoromevalonate (Staal et al. 1981) act as antagonists to juvenile hormones. When young larvae are treated with these materials, they tend to pupate prematurely, abbreviating the larval stage. Unfortunately, at present, none of the available compounds have proved to be very effective on forest pest species.

Table 1. Classification of insect growth regulators.

I Hormone analogs	II Enzyme inhibitors	III Natural products
(a) Juvenile hormone analogs e.g. methoprene	(a) Chitin synthesis inhibitors e.g. diflubenzuron or Dimilin®	(a) General inhibitors of growth and development e.g. azadirachtin
(b) Anti-juvenile hormone compounds e.g. precocene		
(c) Ecdysone analogs e.g. benzoyl hydrazines		
(d) Neuropeptides e.g. prothoracicotropic hormone		

Ecdysone Analogs

The hormone 20-hydroxy ecdysone or edcysterone is responsible for the induction of the molting process in insects. Therefore, it is theoretically possible to adversely interfere with molting using analogs of this hormone. Several azasterols showed activity (Walker and Svoboda 1973) but because of their steroid nature they tended to be difficult to synthesize and proved to be labile. Recently, certain benzoyl hydrazines have been found to have potent molting hormone activity (Wing 1988; Wing et al. 1988). Preliminary studies have indicated that these nonsteroidal agonists show potent activity against several forest insect pests. Within 24 hours after ingestion, the larvae stop feeding and enter into an abnormal molting phase where the head capsule slips and the old cuticle separates with an ill-formed new cuticle underneath.

Neuropeptide Analogs

Neurohormones such as the prothoracicotropic hormone, allatotropin, and allatohibin are a few of the known tropic hormones that regulate other endocrine activity responsible for growth and development. A slight alteration in the titer and time of release would have serious deleterious effects. Since many of these hormones are peptides and are produced by single genes, it is theoretically possible to insert these genes into microbial agents and produce genetically engineered control agents (Keely and Hayes 1987). By altering the DNA sequence, neuropeptide analogs can possibly be produced, and some of these may have agonistic or antagonistic activity or may irreversibly bind to the receptor site. Such activity that adversely affects the pest can be developed into control agents. Research in this area has been initiated by Forestry Canada as well as in other laboratories.

Enzyme Inhibitors

Unique biochemical pathways that have critical roles in the growth and metamorphic processes can be upset if some of the rate-limiting enzymes are inhibited. One such discovery is the benzoylphenyl ureas that inhibit chitin synthesis in insects (Retnakaran and Wright 1987). A compound belonging to this class that has been well studied is diflubenzuron or Dimilin®. Upon ingestion, it inhibits chitin formation in the cuticle and as a result the insect dies during molting. Dimilin® has been shown to be very effective against many forest insects. In the United States, its use has steadily increased and in 1988 more than 65% of the spray operations conducted against the gypsy moth, *Lymantria dispar*, was with this material (Retnakaran 1989). Numerous benzoylphenyl urea analogs are currently being tested for activity.

Natural Products

Several natural products have been shown to have insect growth regulator activity. Some of them, for example, juvabione (Bowers et al. 1966) and juvocimenes (Bowers and Nishida 1980), show juvenile hormone activity; precocenes show anti-juvenile hormone activity (Bowers et al. 1976); ponasterone and other phyto-ecdysteroids show molting hormone activity (Horn and Bergamasco 1985). There are others that are insect growth regulators but do not fit into the hormone category. For example, caffeine and aminophylline inhibit adult development (McDaniel and Berry 1974); sclerin, a metabolite produced by the fungus *Sclerotinia libertiana*, affects ecdysis (Shinoda et al. 1977); azadirachitin affects molting (Schmutterer 1987). Many of these naturally occurring insect growth regulators have the potential to be developed as control agents.

References

Bowers, W.; Retnakaran, A. 1988. The effectiveness of Insect Growth Regulators against the balsam woolly aphid in Newfoundland in 1987. Report of the 15th Annual Forest Pest Control Forum, Nov. 17-19, 1987. Forestry Canada, Ottawa, Ont.

Bowers, W.S.; Fales, H.M.; Thompson, M.J.; Uebel, E.C. 1966. Juvenile hormone: Identification of an active compound from balsam fir. Science 154: 1020-1021.

Bowers, W.S.; Nishida, R. 1980. Juvocimenes: Potent juvenile hormone mimics from sweet basil. Science 209: 1030-1032.

Bowers, W.S.; Ohta, T.; Cleere, J.S.; Marsella, P.A. 1976. Discovery of anti-juvenile hormones in plants. Science 193: 542-547.

Horn, D.H.S.; Bergamasco, R. 1985. Chemistry of Ecdysteroids. Pages 185-248 *in* G.A. Kerkut and L.I. Gilbert (Eds.), Comprehensive Insect Physiology, Biochemistry and Pharmacology. Vol. 7. Ch. 6. Pergamon Press, Oxford.

Keely, L.L.; Hayes, T.K. 1987. Speculations in biotechnology applications for insect neuroendocine research. Insect Biochem. 17: 639-651.

McDaniel, C.N.; Berry, S.J. 1974. Effects of caffeine and aminophylline on adult development of the cecropia moth. J. Insect Physiol. 20: 245-252.

Retnakaran, A. 1989. Gypsy moth control with Dimilin in Ontario. For. Can., For. Pest. Manage. Inst., Sault Ste. Marie, Ont. Inf. Rep.

Retnakaran, A.; Granett, J.; Ennis, T. 1985. Insect Growth Regulators. Pages 529-601 *in* G.A. Kerkut and L.I. Gilbert (Eds.), Comprehensive Insect Physiology, Biochemistry and Pharmacology. Vol. 12. Ch. 15. Pergamon Press, Oxford.

Retnakaran, A.; Wright, J.E. 1987. Control of insect pests with Benzoylphenyl ureas. Pages 205-282 *in* J.E. Wright and A. Retnakaran (Eds.), Chitin and Benzoylphenyl Ureas. Dr. W. Junk Publishers, Dordrecht, Netherlands.

Schmutterer, H. 1987. Insect growth-disrupting and fecundity-reducing ingredients from the neem and chinaberry trees. Pages 119-170 *in* E.D. Morgan and N.B. Mandaava (Eds.), Handbook of Natural Pesticides. Vol. 3. C.R.C. Press, Boca Raton, Fla.

Shinoda, S.; Kamada, A.; Asans, S.; Taniguchi, M.; Satomura, Y. 1977. Effect of sclerin on imaginal ecdysis of the silkworm, *Bombyx mori* L. Agric. Biol. Chem. 41: 2483-2484.

Staal, G.B.; Henrick, C.A.; Bergot, B.J.; Cerf, D.C.; Edwards, J.P.; Kramer, S.J. 1981. Relationships and interactions between JH and anti-JH analogues in Lepidoptera. Pages 323-340 *in* F. Schnal, A. Zabza, J.J. Menn, and B. Cymborowski (Eds.), Regulation of Insect Development and Behavior. Part. I. Wroclaw Technical University Press, Wroclaw, Poland.

Steel, C.G.H.; Davey, K.G. 1985. Integration in the Insect Endocrine System. Pages 1-35 in G.A. Kerkut and L.I. Gilbert (Eds.), Comprehensive Insect Physiology Biochemistry and Pharmacology. Vol. 8. Ch. 1. Pergamon Press, Oxford.

Walker, W.F.; Svoboda, J.A. 1973. Suppression of Mexican bean beetle development by foliar applications of azasterols and reversal by cholesterol. Ent. Exp. Appl. 16: 422-426.

Williams, C.M. 1967. Third generation pesticides. Sci. Amer. 217: 13-17.

Wing, K.D. 1988. RH-5849, a Nonsteroidal Ecdysone Agonist: Effects on a Drosophila cell line. Science 241: 467-469.

Wing, K.D.; Slawecki, R.A.; Carlson, G.R. 1988. RH-5849, a Nonsteroidal Ecdysone Agonist: Effects on Larval Lepidoptera. Science 241: 470-472.

Chapter 40

Natural Antifeedants

GEORGE M. STRUNZ

Secondary Metabolites as Plant Defenses

It is generally accepted that secondary plant metabolites emerged as defenses for plants in their co-evolution with phytophagous insects (Rhoades 1979). In time, some insects were able to gain a competitive advantage by evolving mechanisms that enabled them to overcome the adverse effects of the defensive chemicals of a particular plant or group of plants. The plants, with their defenses thus breached, could become hosts to the adapted insects. There were insects that took further advantage of these phytochemicals, developing special receptors for them, and using their presence as "token stimuli" for recognition of their host plants (Rees 1969; Schoonhoven 1967). Notwithstanding such adaption by many insects, the production of toxic, repellent, or antifeedant metabolites by a plant generally provides it with effective protection against predation by the majority of insect herbivores (cf. Jermy 1966: Bernays 1983). The manifest function of naturally occurring antifeedants in defending plants against phytophagous insects has provided impetus for exploring their potential as nontoxic, environmentally acceptable agents for crop protection.

Antifeedants

The terms "antifeedant," "feeding deterrent," and occasionally "rejectant" are used synonymously in the literature to describe substances that when contacted prevent or interrupt feeding activity (Schoonhoven 1982). The expression "repellent" is reserved for substances that induce movement away from the source of stimulus (Schoonhoven 1982).

Toxins and phytochemicals whose assimilation requires high expenditure of energy may produce some effects similar to those of antifeedants. Careful observation and suitable bioassay design will usually indicate whether such effects as reduced food intake, retarded development, or diminished insect weight truly result from antifeedant activity.

Electrophysiological studies of insect chemosensory systems have shed some light on the mechanisms involved in antifeedant perception (Schoonhoven 1982).

Most research on insect antifeedants has focused on compounds from higher plants, but fungi and bacteria also produce metabolites with antifeedant properties (e.g. Rowan et al. 1986). Several synthetic substances, including certain organometallic fungicides (Ascher 1979), have similarly been found to possess significant antifeedant activity against insects.

Probably the most potent natural antifeedant described to date is azadirachtin (Fig. 1) found in leaves and berries of the neem tree, *Azadirachta indica*, which grows commonly in India and parts of Africa (Butterworth and Morgan 1971; Warthen 1979; Krause et al. 1987, and references cited therein; Rembold 1989). Azadirachtin shows spectacular antifeedant activity against the desert locust, *Schistocerca gregaria*. Feeding by this insect is inhibited by a solution containing 40 μg/L azadirachtin absorbed on filter paper, which corresponds to about 1 ng/cm^2 of the antifeedant on the substrate (Butterworth and Morgan 1971). Azadirachtin is not a universal antifeedant and displays a range of activity against other insects. The compound has also been reported to possess powerful growth regulating properties (Kubo and Klocke 1982; Rembold 1989). A synthetic hydroxy dihydrofuran acetal, embodying features of a portion of the complex azadirachtin structure, has been found to possess potent antifeedant activity (Ley et al. 1987).

Recent literature is replete with reports on natural products displaying antifeedant properties against a variety of economically important insect pests. Although a few structure–activity relationships have begun to emerge (e.g. Ley et al. 1987; Simmonds et al. 1990; Blaney et al. 1990), the molecular basis for antifeedant activity remains obscure (cf. D'Ischia et al. 1982). The wide variety of structural types that are endowed with antifeedant properties may be illustrated by reference to a random sampling of known feeding deterrents. Warburganal (Fig. 2) from bark of *Warburgia* spp. is active against the African armyworm, *Spodoptera exempta*, at 0.1 ppm (Kubo et al. 1976a). The furocoumarin isopimpinelline (Fig. 3) exhibits antifeedant activity against the tobacco cutworm, *Spodoptera litura*, at 5 ppm (Yajima et al. 1977), whereas 50 ppm of the less potent neolignan antifeedant piperenone (Fig. 4) is required to inhibit feeding by this insect (Matsui and Munakata 1976). Ajugarin 1 (Fig. 5), one of three closely related clerodane antifeedants from foliage of *Ajuga remota*, is active against *S. exempta* at 100 ppm (Kubo et al. 1976b), and xylomolin (Fig. 6) from unripe fruit of *Xylocarpus moluccensis* inhibits feeding by this insect at the same concentration (Kubo et al. 1976c). Citrus limonoid by-products such as obacunone (Fig. 7) have been found to reduce feeding and lower the growth rate in the fall armyworm, *Spodoptera frugiperda*, and the cotton earworm, *Heliothis zea* (vide infra) (Klocke and Kubo 1982). It is no surprise that many alkaloids, in addition to their other bioactivity, have revealed potent antifeedant properties against insects (e.g. Bentley et al. 1982).

Secondary Metabolites from Host Trees and Spruce Budworm Feeding

Although detailed information on the feeding behavior of spruce budworm larvae, *Choristoneura fumiferana*, has been available since the 1950s (Blais 1952; McGugan 1954; Blais 1957; Greenbank 1963), the first approach to a systematic examination of chemical aspects of the spruce budworm–host tree interaction can be attributed to Heron (1965). Heron studied the feeding responses of fifth-instar spruce budworm larvae

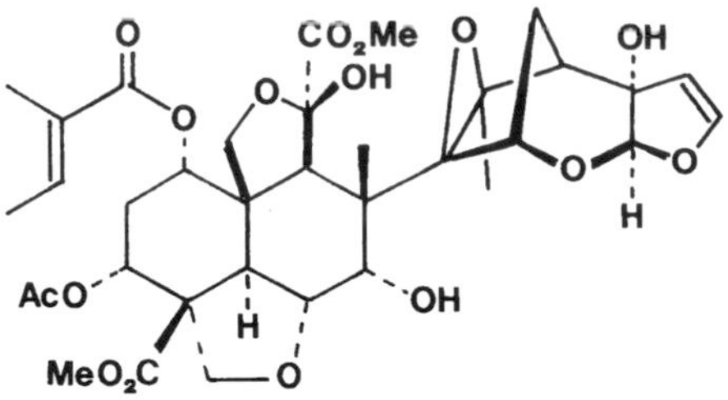

Fig. 1

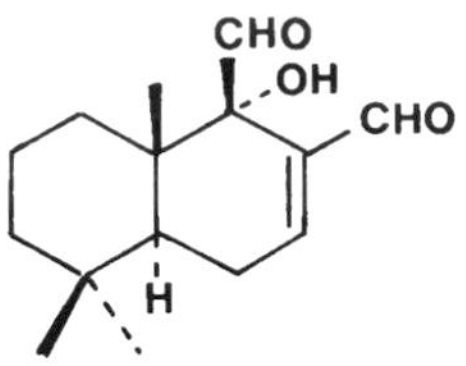

Fig. 2

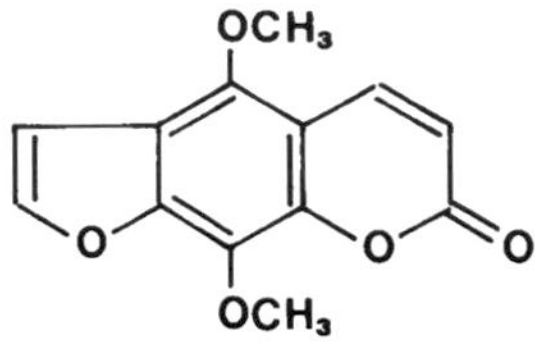

Fig. 3

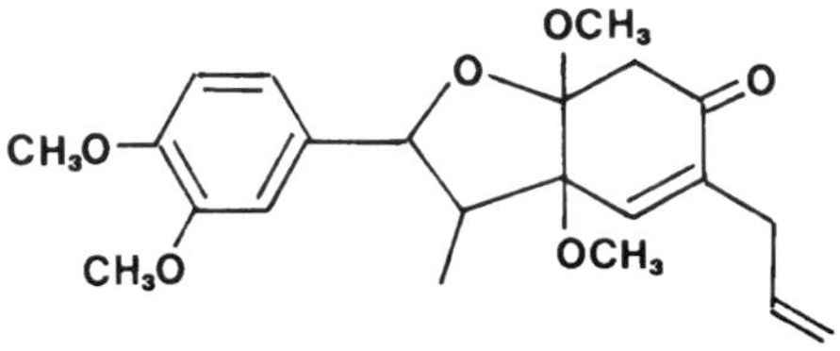

Fig. 4

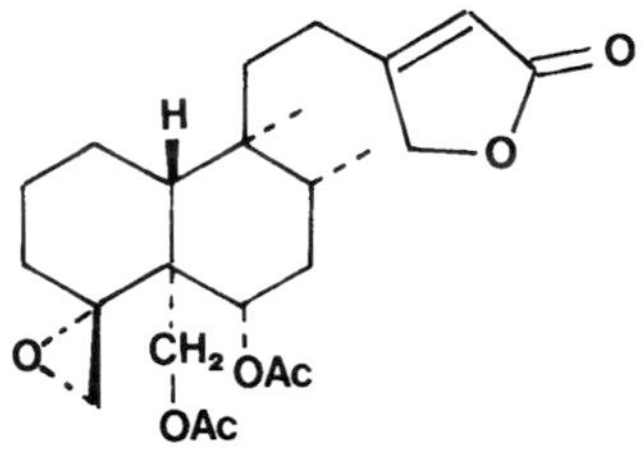

Fig. 5

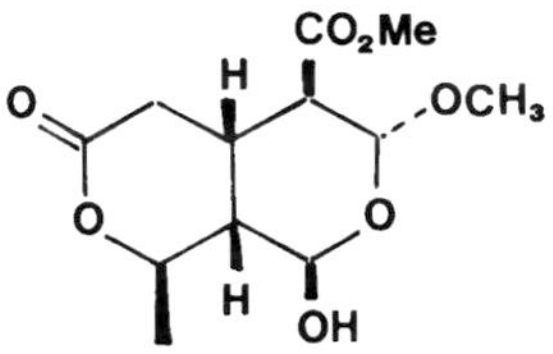

Fig. 6

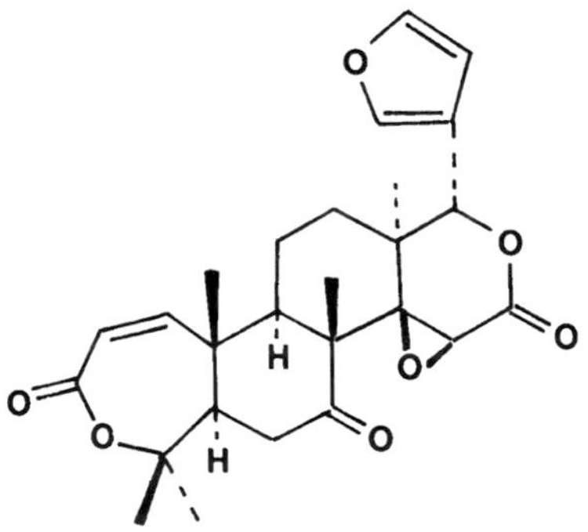

Fig. 7

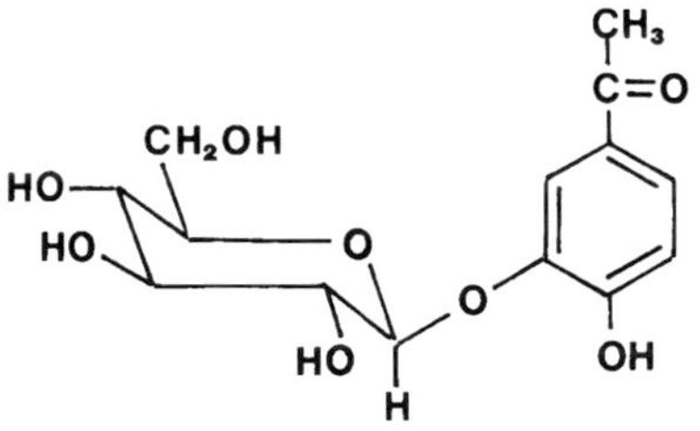

Fig. 8

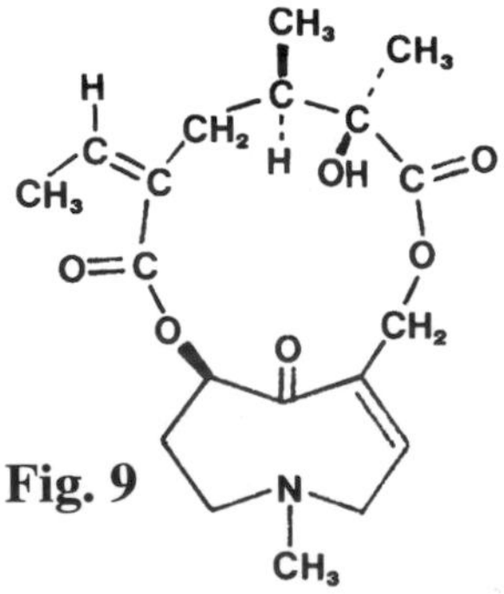

Fig. 9

Figures 1–18. Chemical structures of some natural insect antifeedants.

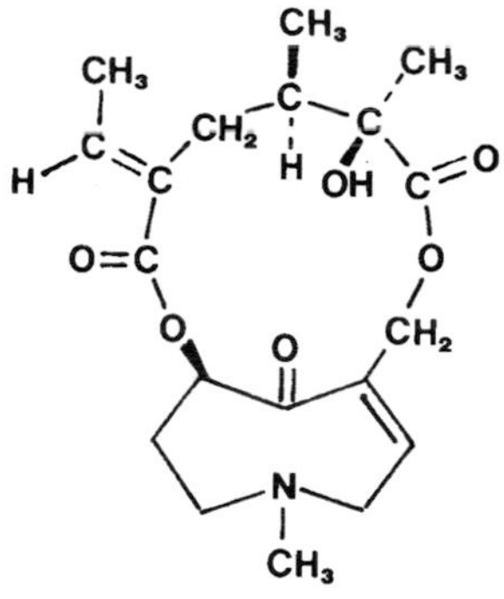

Fig. 10

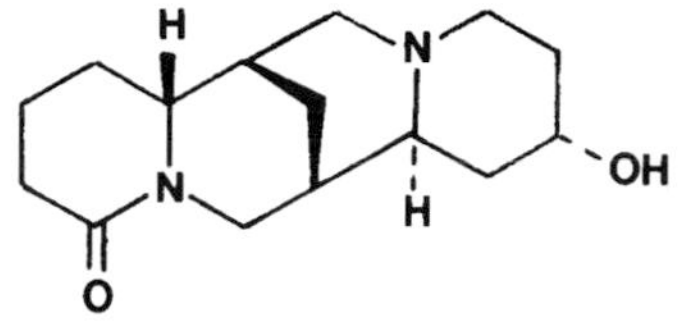

Fig. 11

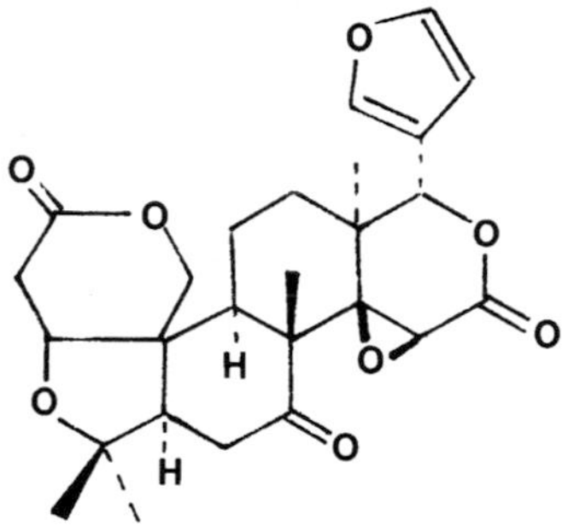

Fig. 12

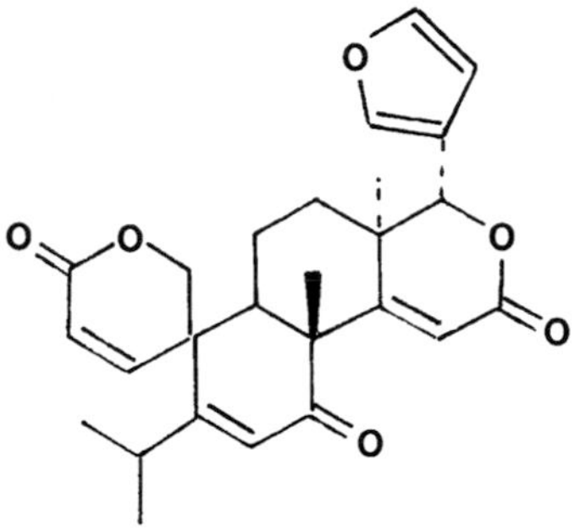

Fig. 13

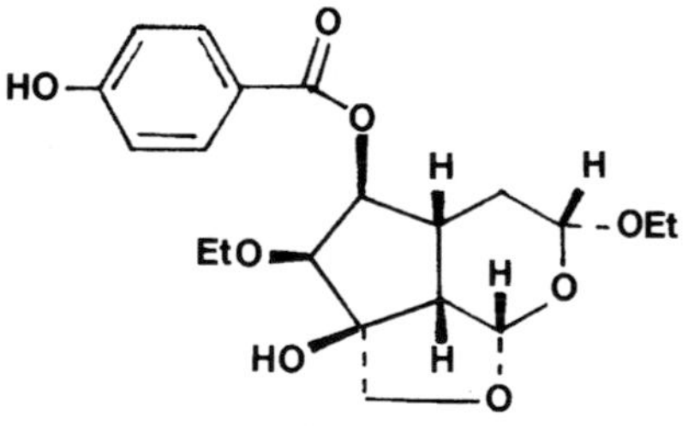

Fig. 14

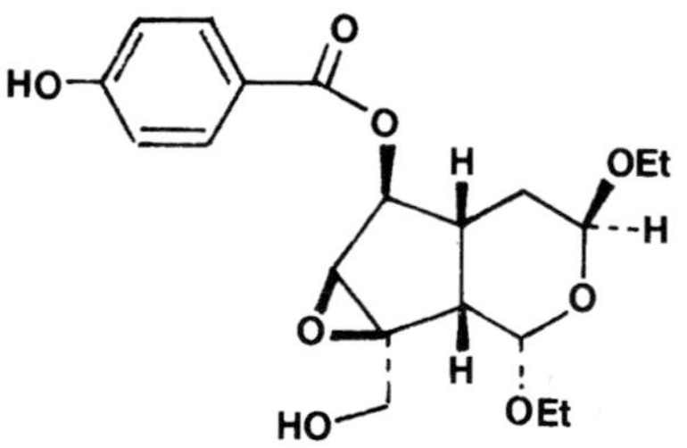

Fig. 15

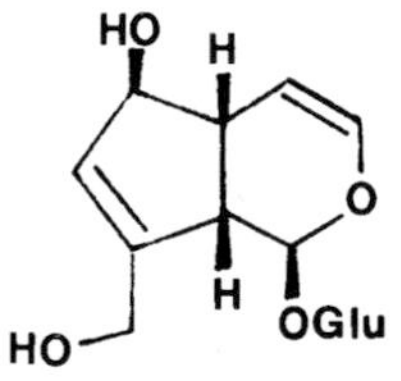

Fig. 16

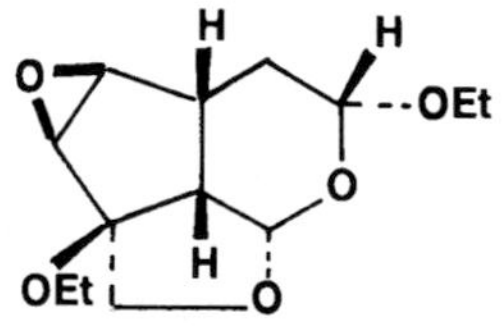

Fig. 17

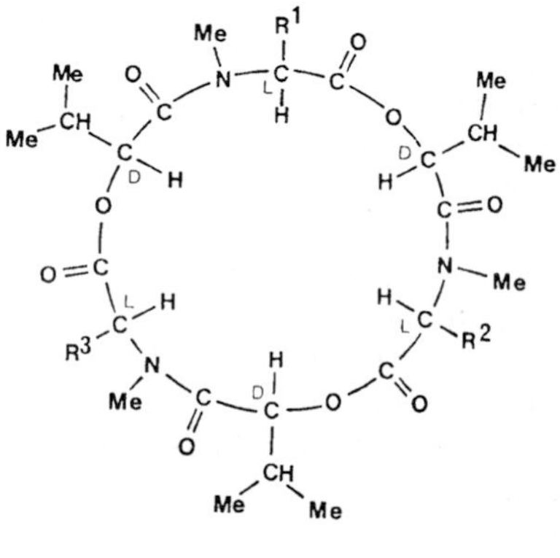

Fig. 18

toward extracts of staminate flowers, new vegetative shoots, and mature needles of white spruce, *Picea glauca*. The effects of several pure compounds, generally polar conifer constituents, were also examined. Heron found that sucrose and the amino acid L-proline, both constituents of the highly acceptable staminate flowers of *P. glauca*, showed pronounced phagostimulant effects in the bioassays. The phenolic glucoside pungenin (Fig. 8) is present at high concentrations in mature foliage of various *Picae* species (Neish 1957, 1958) but is virtually absent in the new shoots. Heron (1965) reported that fifth-instar spruce budworm larvae, when offered a choice between disks of Japanese elder pith impregnated with the phagostimulant sucrose and identical disks also treated with a 1% solution of pungenin, showed a marked preference for the former. This finding appeared to provide a plausible, if only partial, explanation why new vegetative shoots, devoid of pungenin, are much more acceptable to the larvae than mature foliage in which the glucoside is abundant (Heron 1965).

The suggestion that pungenin might be an important feeding deterrent for spruce budworm seems to have been accepted by entomologists during the ensuing two decades (e.g. Mattson et al. 1983). Little additional research on this compound was reported until 1986, when Strunz et al. described a chemical synthesis of pungenin from the commercially available precursor acetovanillone. Synthetic pungenin, as well as natural material isolated from white spruce according to Neish (1957), was used to reinvestigate the effects of the glucoside on spruce budworm feeding (Strunz et al. 1986). The results can be summarized as follows: spruce budworm larvae developed successfully from second to sixth instar on synthetic diet (McMorran 1965) containing pungenin at a concentration of 5%, a level higher than that normally present in mature foliage of white spruce at the time the insects are feeding (cf. Neish 1958). No evidence was detected of increased mortality or retarded development among these insects. The performance of a separate group of sixth-instar larvae fed a diet containing 4.8% pungenin was assessed on the basis of weights of diet consumed, frass produced, and adult moths. No significant differences were observed between the performances of male test and control budworm, but the corresponding weights for the (larger) female test insects were about 30% lower than those of the control females. Although pungenin exhibits deterrency in a choice situation, it can't be considered a potent antifeedant. Feeding by spruce budworm on foliage containing abnormally high concentrations of pungenin, however, could extract a toll from these insects.

Terpenes are prominent foliar constituents of all the host trees of spruce budworm and undoubtedly have a protective function for these trees against many species of herbivores. Their presence is a significant factor in the spruce budworm–host interaction (e.g. Städler 1974), but their role in feeding is subtle and remains poorly understood. Although it appears unlikely that terpenoid host constituents have important antifeedant properties against spruce budworm, the disposal by the insect of at least some of these compounds is accomplished at significant metabolic cost. Thus, Mattson et al. (1983) observed that several terpenes in both balsam firm and white spruce, including alpha- and beta-pinenes, camphene, beta-phellandrene, bornyl acetate, and terpinolene, were negatively correlated with weight gain of spruce budworm. Cates and his co-workers have made related observations in studying the interaction between the insect's western relative, the western spruce budworm, *Choristoneura occidentalis*, and Douglas-fir, *Pseudotsuga menziesii* (e.g. Redak and Cates 1984). The possibility that terpene composition could provide a potential resistance mechanism against the depredations of the spruce budworm remains a moot question (Wagner and Tinus 1985).

The Quest for Antifeedants for the Spruce Budworm

It is logical to turn to nonhost plants in searching for natural antifeedants for protection of host-trees against defoliation by the spruce budworm. Mass screening of plant extracts for antifeedant effects requires a bioassay that is operationally simple, fast, reproducible, and modest in its requirements for test material. An assay system developed by Bentley et al. (1979) appeared to meet these criteria, albeit at the expense of sensitivity. In the test, sixth-instar larvae are induced to feed on a filter paper substrate impregnated with the synergistic phagostimulants sucrose and L-proline (cf. Heron 1965). Frass resulting from ingestion of this material is easily recognizable and counts of frass pellets after 48 hours indicate the quantity consumed. The effects on feeding of adding plant extracts and test chemicals can thus be observed readily. Using this bioassay, extracts of more than 110 nonhost plants, as well as about 60 naturally occurring chemicals, were tested for their effects on spruce budworm feeding (Bentley et al. 1982). Among the extracts, surprisingly, only six showed deterrency greater than 75% relative to controls. All plants in this most active group contain alkaloids and, in each case, the greatest activity was observed to be localized in the basic fractions. Among the pure alkaloids tested, including representatives of the solanum, pyrrolizidine, quinolizidine, berberine, and strychnos groups, fewer than 25% displayed deterrent activity in the most potent category.

Although the nature of the bioassay used in this investigation did not address the responses of starving insects or the effects on the insects' performance of consuming deterrents with modest activity, the study revealed some leads that merit further investigation. (Pungenin on paper diet disks at concentrations up to 5% was not observed to cause significant reduction of feeding by sixth-instar *C. fumiferana* larvae [Strunz et al. 1986].)

More detailed accounts of spruce budworm feeding responses to alkaloids in the pyrrolizidine (Bentley et al. 1984a), lupine (Bentley et al. 1984b), and solanum groups (Bentley et al. 1984c), besides indicating wide variability in deterrence within a structural class, highlight some interesting structure–activity relationships. For example, the results indicate the importance of an α, β-unsaturated lactone or ester moiety with appropriate (Z) double bond geometry, in conferring (spruce budworm) deterrent properties on the pyrrolizidine alkaloids. Thus, while senkerkine (Fig. 9) is highly deterrent, neosenkerkine (Fig. 10), which differs only in the (E) configuration of the Δ^{15} double bond, is inactive, as is otosenine, the epoxide of senkerkine (Bentley et al. 1984a).

Of a dozen lupine alkaloids assayed for deterrent effects on spruce budworm feeding, only esters of 13-hydroxylupanine with conjugated carboxylic acids displayed significant activity. The alcohol, 13-hydroxylupanine (Fig. 11) was not deterrent.

Five steroidal *Solanum* alkaloids, tomatine, α-solanine, their aglycones, tomatidine and solanidine, as well as α-chaconine, all showed significant feeding inhibition at 10^{-3} M.

Some shortcomings of the paper disk bioassay system have been alluded to previously, and the use of a more nutritious diet generally gives more meaningful quantitative results. In testing some citrus limonoids as potential antifeedants for spruce budworm, Alford and Bentley (1986) homogenized these compounds into artificial diet and monitored the amount of diet consumed by sixth-instar larvae in choice and no-choice situations. In addition, they studied the effects of the chemicals on survival and development time from the fourth instar to the pupal stage. Although limonin (Fig. 12) and some other citrus limonoids are known to reduce feeding and retard the growth rate of the fall armyworm, *S. frugiperda*, and the corn earworm, *H. zea* (Klocke and Kubo 1982), none of the limonoids, limonin, deoxylimonin, or citrolin (Fig. 13) suppressed feeding by the spruce budworm at the high concentration of 1 000 ppm. Citrolin at 500 ppm did extend the larval development time by about 40% relative to that of the controls, but neither limonin nor its deoxy derivative showed significant effects on budworm development (Alford and Bentley 1986).

A preliminary study of the response of sixth-instar spruce budworm larvae to the redoubtable antifeedant-growth regulator azadirachtin (Fig. 1) was conducted by A.W. Thomas and the author, using bioassay methology developed by Thomas (Forestry Canada—Maritimes). At a concentration of 47 ppm in an artificial diet, azadirachtin (provided by Professor J.T. Arnason, University of Ottawa) caused a mean feeding depression of about 75%, calculated using the same equation as Alford and Bentley (1986). The effects of azadirachtin as a growth regulator are, in fact, evident at much lower concentrations, and this is its principal mode of action against many insects (Saxena 1989; Rembold 1989), including the spruce budworm. The potential of a commercial neem seed extract preparation for operational control of spruce budworm is currently being investigated at the Forestry Canada—Maritimes laboratory in collaboration with scientists at McGill University (Thomas et al. 1992).

Chang and Nakanishi (1983) reported on the screening of extracts from foliage of 40 nonhost trees for antifeedant activity against the spruce budworm. When incorporated into diet at 50 to 100 ppm, antifeedant properties were attributed to an iridoid, designated specionin, from leaves of *Catalpa speciosa*. Specionin was assigned the structure shown in Figure 14 (Chang and Nakanishi 1983), but this was subsequently amended to that shown in Figure 15 on the basis of synthetic and spectroscopic investigation (Van der Eycken et al. 1986).

Chan et al. (1987) considered that a totally synthetic approach to specionin was unlikely to produce quantities of the antifeedant sufficient for large-scale bioassays, much less for field trials or operational use. The approach adopted by these researchers was to prepare analogs of specionin for testing from abundantly available iridoid natural products. Aucubin (Fig. 16), an iridoid glucoside constituting 1% of fresh plant weight of *Aucuba japonica*, was initially selected as a suitable precursor and was chemically modified to give several products more or less structurally related to specionin (Chan et al. 1987). These compounds, at 0.2% wet weight in McMorran diet, were tested for their effects on spruce budworm. In one of the assays, larvae were reared from the second instar on treated diet and any retardation of development relative to controls was recorded after 15 days. In the other bioassay, feeding inhibition was determined by differences in dry weights of frass, as well as moths, between test and control insects. Of seven compounds tested, including catalposide, also from *C. speciosa*, only that shown in Figure 17 was significantly active on both bioassays (Chan et al. 1987). The activity of this analog, though modest, illustrates the validity of the approach and provides encouragement for further research along these lines.

As stated earlier, fungi and bacteria also produce substances that deter insect feeding. Retnakaran et al. (1983) reported that sublethal doses of *Bacillus thuringiensis* (Bt) suppress feeding by the spruce budworm. The mechanism of this temporary feeding inhibition is uncertain, but the authors suggest that cytotoxic effects of Bt on the midgut epithelium may be responsible. This example illustrates that it may sometimes be difficult to distinguish between antifeedant and toxic effects. Thomas and Wiesner have used the suppression of spruce budworm feeding as the basis for a rapid and effective method for quantitative determination of foliar residues of Bt (A.W. Thomas, pers. comm.).

Thomas (1987), in studying the effect of age of current-year shoots of white spruce on the performance of spruce budworm larvae, observed that foliage collected in mid-June, at the time larvae are feeding in the field, was highly acceptable and nutritionally adequate for the insects. He made the surprising observation that larvae offered foliage sampled from the same trees less than 3 weeks later ate little and suffered greater than 90% mortality, but this effect diminished in foliage collected subsequently. Initial attempts at extraction of the principle(s) responsible for the mortality of insects fed with the early July foliage of white spruce were unsuccessful. Analysis of the decline in nitrogen content of the needles in relation to the mortality data indicated that this was unlikely to be the lethal factor. The hypothesis that high concentrations of pungenin or tannins might be responsible was rejected when the characteristics of pungenin were examined (Strunz et al. 1986), and the author and Thomas subsequently determined that the tannin content of the acceptable needles was higher than that of the lethal foliage. Information about the substance that caused the death of the insects remains scant, other than it is heat labile (Thomas 1987). Although a surge in concentration of certain foliar chemicals could have been responsible, the lethal factor was more likely to have been of microbial origin.

A growing literature attests to increasing interest in the role of plant–fungal interactions in providing some protection to the plant against the depredations of phytophagous insects and other herbivores (references in Strongman et al. 1988). Miller et al. (1985) observed that infestation of balsam fir by spruce budworm had a marked effect on the foliage mycoflora, and that some fungi from infested trees were toxic to spruce budworm larvae when ingested.

One of the more toxic epiphytes isolated from spruce budworm-infested balsam fir foliage was identified as *Fusarium avenaceum*. The major toxic principle produced by cultures of this fungus was identified as enniatin complex, rich in enniatins A and A_1 (Strongman et al. 1988). The enniatins are a group of cyclohexadepsipeptide ionophores (Fig. 18) whose pronounced bioactivity, including antibiotic effects, results from their ability to disrupt ion transport across membranes (Shemyakin et al. 1965). Besides its toxicity to spruce budworm larvae, preliminary evidence based on a choice bioassay suggests that the enniatin A/A_1 complex deters feeding by these insects (Strongman et al. 1988 and pers. comm.). Other insecticidal metabolites of fungi from spruce budworm host trees, currently under investigation, may possess antifeedant properties in addition to their toxicity.

Conclusion

The problems associated with continued widespread use of synthetic broad spectrum pesticides have lent urgency to the search for alternative, environmentally acceptable approaches to crop protection. Among other possible strategies, the concept of suppressing feeding by phytophagous insect pests on crop plants through the agency of nontoxic biodegradable substances, has much to commend it. Although substances that inhibit feeding by insect herbivores are abundant in the plant kingdom, the development of antifeedant compounds for operational use in crop protection is no trivial objective. Such a compound, besides being effective at acceptable concentrations, needs to be endowed with several additional attributes. It must be abundantly available from renewable natural sources or have a chemical structure sufficiently simple to permit economical large-scale production by synthesis. Whatever its origin, its cost must be realistic in the context of the proposed use. It should be substantially less toxic to nontarget organisms than conventional insecticides. It should ideally be taken up systemically by the crop plant to ensure that fresh leaf surfaces exposed to the insect herbivore remain protected. It should be sufficiently stable to avoid the necessity of repeated seasonal applications but should not pose problems of long-term environmental persistence. It should be relatively selective but have a sufficient spectrum of activity to be commercially viable.

Of the natural antifeedants considered for operational use in protecting agricultural crops, neem preparations containing azadirachtin and other feeding inhibitors come closest to meeting these criteria and have received the greatest attention (Pradhan et al. 1962; Butterworth and Morgan 1971; Warthen 1979; Saxena 1986, 1989; Rembold 1989). Establishment of a series of International Neem Conferences attests to the widespread interest in these substances. Although the application of antifeedants in agriculture and forestry has thus far been very limited, it has been demonstrated under field conditions using neem preparations (e.g. Ladd et al. 1978; Saxena 1986; Zehnder and Warthen 1988; Saxena 1989), as well as synthetic organotin antifeedants (Jermy 1961; Murbach 1967; Kamel et al. 1970; All and Benjamin 1976; Ascher 1979), that inhibition of insect feeding is, indeed, a viable approach to crop protection. It appears certain that from existing secondary plant metabolites, conservatively estimated at 400 000 (Swain 1977), research will produce several antifeedants possessing appropriate qualities for crop protection.

The prospect of substituting natural antifeedants for conventional insecticides in attempts to control spruce budworm over large areas of forest certainly appears remote now. However, antifeedants could well have a role in integrated pest management programs, and their potential for use in forest nurseries and small plantations is as great as it is for any agricultural crop.

Success reported in the insertion of a gene coding for Bt endotoxin into tobacco plants using recombinant DNA technology (Meusen and Zabeau 1986) suggests that the genetic engineering of trees to produce effective antifeedants in their own foliage may be a productive area of research within the next few years.

References

Alford, A.R.; Bentley, M.D. 1986. Citrus limonoids as potential antifeedants for the spruce budworm (Lepidoptera: Tortricidae). J. Econ. Entomol. 79: 35-38.

All, J.N.; Benjamin, D.M. 1976. Potential of antifeedants to control larval feeding of selected *Neodiprion* sawflies (Hymenoptera: Diprionidae). Can. Entomol. 108: 1137-1144.

Ascher, K.R.S. 1979. Fifteen years (1963-1978) of organotin antifeedants—a chronological bibliography. Phytoparasitica 7: 117-137.

Bentley, M.D.; Leonard, D.E.; Bushway, R.J. 1984c. *Solanum* alkaloids as larval feeding deterrents for spruce budworm, *Choristoneura fumiferana* (Lepidoptera: Tortricidae). Ann. Entomol. Soc. Am. 77: 401-403.

Bentley, M.D.; Leonard, D.E.; Leach S.; Reynolds, E.; Stoddard, W.; Tomkinson, B.; Tomkinson, D.; Strunz, G.; Yatagai, M. 1982. Effects of some naturally occurring chemicals and extracts of non-host plants on feeding by spruce budworm larvae (*Choristoneura fumiferana*). Tech. Bull. 107. Life Sci. and Agric. Exp. Stn., Univ. of Maine at Orono. 23 p.

Bentley, M.D.; Leonard, D.E.; Mott, D.G. 1979. Spruce budworm feeding deterrents. Paper presented at ESA National Meeting, Denver, Colo., Nov. 29, 1979.

Bentley, M.D.; Leonard, D.E.; Reynolds, E.K.; Leach, S.; Beck, A.B.; Murakoshi, I. 1984b. Lupine alkaloids as larval feeding deterrents for spruce budworm, *Choristoneura fumiferana* (Lepidoptera: Tortricidae). Ann. Entomol. Soc. Am. 77: 398-400.

Bentley, M.D.; Leonard, D.E.; Stoddard, W.F.; Zalkow, L.H. 1984a. Pyrrolizidine alkaloids as larval feeding deterrents for spruce budworm, *Choristoneura fumiferana* (Lepidoptera: Tortricidae). Ann. Entomol. Soc. Am. 77: 393-397.

Bernays, E.A. 1983. Antifeedants in crop pest management. Pages 254-273 *in* D.L. Whitehead and W.S. Bowers (Eds.), Natural Products for Innovative Pest Management. Pergamon Press, New York.

Blais, J.R. 1952. The relationship of the spruce budworm (*Choristoneura fumiferana* (Clem.)) to the flowering condition of balsam fir (*Abies balsamea* (L) Mill.). Can. J. Zool. 30: 1-29.

Blais, J.R. 1957. Some relationships of the spruce budworm to black spruce. For. Chron. 33: 364-372.

Blaney, W.M.; Simmonds, M.S.J.; Ley, S.V.; Anderson, J.C.; Toogood, P.L. 1990. Antifeedant effects of azadirachtin and structurally related compounds on lepidopterous larvae. Entomol. Exp. Appl. 55: 149-160.

Butterworth, J.H.; Morgan, E.D. 1971. Investigation of the locust-feeding inhibition of the seeds of the neem tree, *Azadirachta indica*. J. Insect Physiol. 17: 969-977.

Chan, T.H.; Zhang, Y.J.; Sauriol, F.; Thomas, A.W.; Strunz, G.M. 1987. Studies in iridoid chemistry and spruce budworm (*Choristoneura fumiferana*) antifeedants. Can. J. Chem. 65: 1853-1858.

Chang, C.C.; Nakanishi, K. 1983. Specionin, an iridoid insect antifeedant from *Catalpa speciosa*. J. Chem. Soc. Chem. Commun. 605-606.

D'Ischia, M.; Prota, G.; Sodano, G. 1982. Reaction of polygodial with primary amines: an alternative explanation to the antifeedant activity. Tetrahedron Lett. 23: 3295-3298.

Greenbank, D.O. 1963. Staminate flowers and the spruce budworm. Pages 202-218 *in* R.F. Morris (Ed.), The Dynamics of Epidemic Spruce Budworm Populations. Mem. Entomol. Soc. Can. No. 31, Ottawa, Ont.

Heron, R.J. 1965. The role of chemotactic stimuli in the feeding behavior of spruce budworm larvae on white spruce. Can. J. Zool. 43: 247-269.

Jermy, T. 1961. The rejective effect of some inorganic salts on the Colorado beetle (*Leptinotarsa decemlineata* Say) adults and larvae. Növényyéd Kut. Inter. Evk. 8: 121-130.

Jermy, T. 1966. Feeding inhibitors and food preference in chewing phytophagous insects. Entomol. Exp. Appl. 9: 1-12.

Kamel, A.A.M.; Mitri, S.H.; Abo El Ghar, M.; Zake, M.M. 1970. Field trials for the use of two promising antifeedants against the cotton leaf worm. Bull. Ent. Soc. Egypt. Econ. Ser. 4: 35-41.

Klocke, J.A.; Kubo, I. 1982. Citrus limonoid by-products as insect control agents. Entomol. Exp. Appl. 32: 299-301.

Krause, W.; Bokel, J.; Bruhn, A.; Cramer, R.; Klaiber, I.; Klenk, A.; Nagl, E.; Pöhnl, H.; Sadlo, H.; Vogler, B. 1987. Structure determination by nmr of azadirachtin and related compounds from *Azadirachta indica* A. Juss. (Meliaceae). Tetrahedron 43: 2817-2830.

Kubo, I.; Lee, Y.-W.; Pettei, M.; Pilkiewicz, F.; Nakanishi, K. 1976a. Potent army worm antifeedants from the East African *Warburgia* plants. J. Chem. Soc. Chem. Commun. 1013-1014.

Kubo, I.; Lee, Y.-W.; Balogh-Nair, Y.; Nakanishi, K.; Chapya, A. 1976b. Structure of ajugarins. J. Chem. Soc. Chem. Commun. 949-950.

Kubo, I.; Miura, I.; Nakanishi, K. 1976c. The structure of xylomollin, a secoiridoid hemiacetal acetal. J. Am. Chem. Soc. 98: 6704-6705.

Kubo, I.; Klocke, J.A. 1982. Azadirachtin, insect ecdysis inhibitor. Agric. Biol. Chem. 46: 1951-1954.

Ladd, T.L.; Jacobson, M.; Buriff, C.R. 1978. Japanese beetles: extracts from neem tree seeds as feeding deterrents. J. Econ. Entomol. 71: 810-813.

Ley, S.V.; Santafianos, D.; Blaney, W.M.; Simmonds, M.S.J. 1987. Synthesis of a hydroxy dihydrofuran acetal related to aradirachtin: a potent insect antifeedant. Tetrahedron lett. 28: 221-224.

Matsui, K.; Munakata, K. 1976. Insect antifeeding substances in *Parabenzoin preacor* and *Piper futokadzura*. Agric. Biol. Chem. 40: 1045-1046.

Mattson, W.J.; Slocum, S.S.; Koller, C.N. 1983. Spruce budworm (*Choristoneura fumiferana*) performance in relation to foliar chemistry of its host plants. Pages 55-65 *in* R.L. Talerico and M. Montgomery (coordinators), Proceedings: Forest Defoliator-Host Interactions: A Comparison Between Gypsy Moth and Spruce Budworms. U.S. Dept. Agric., For. Ser., Northeast Stn., Broomall, Pa. Gen. Tech. Rep. NE-85.

McGugan, B.M. 1954. Needle-mining habits and larval instars of the spruce budworm. Can. Entomol. 86: 439-454.

McMorran, A. 1965. A synthetic diet for the spruce budworm, *Choristoneura fumiferana* (Clem.) (Lepidoptera: Torticidae). Can. Entomol. 97: 58-62.

Meusen, R.L.; Zabeau, M. 1986. Engineering of Bt endotoxin genes into plants: potential for producing insect resistant seed. Abst. Sixth Int. Congr. Pest. Chem., IUPAC, Ottawa, Ont., Aug. 10-15, 2F-03.

Miller, J.D.; Strongman, D.; Whitney, N.J. 1985. Observations on fungi associated with spruce budworm infested balsam fir needles. Can. J. For. Res. 15: 896-901.

Murbach, R. 1967. Effet en plein champ, de fongicides à base de fentin-acétate, de manébe et d'oxychlorure de cuivre sur la densité de population du doryphore de la pomme de terre (*Leptinotarsa decemlineata* Say). Schweiz. Landwirt. Forsch. 6: 345-357.

Neish, A.C. 1957. Pungenin: a glucoside found in leaves of *Picea pungens* (Colorado spruce). Can. J. Biochem. Physiol. 35: 161-167.

Neish, A.C. 1958. Seasonal changes in metabolism of spruce leaves. Can. J. Bot. 36: 649-662.

Pradhan, S.; Jotwani, M.G.; Rai, B.K. 1962. The neem seed deterrent to locusts. Indian Farming 12: 7-11.

Redak, R.A.; Cates, R.G. 1984. Douglas-fir (*Pseudotsuga menziesii*)—spruce budworm (*Choristoneura occidentalis*) interactions: The effect of nutrition, chemical defenses, tissue phenology, and tree physical parameters on budworm success. Oecologia 62: 61-67.

Rees, C.J.C. 1969. Chemoreceptor specificity associated with choice of feeding site by the beetle *Chrysolina brunsvicensis* on its foodplant, *Hypericum hirsutum*. Entomol. Exp. Appl. 12: 565-583.

Rembold, H. 1989. Azadirachtins: their structure and mode of action. Chapter 11 *in* J.T. Arnason, B.J.R. Philogène, and

P. Morand (Eds.), Insecticides of Plant Origin. ACS Symp. Series 387, pp. 150-163. Am. Chem. Soc., Washington, D.C.

Retnakaran, A.; Lauzon, H.; Fast, P. 1983. *Bacillus thuringiensis* induced anorexia in the spruce budworm, *Choristoneura fumiferana*. Entomol. Exp. Appl. 34: 233-239.

Rhoades, D.F. 1979. Evolution of plant chemical defense against herbivores. Pages 3-54 *in* G.A. Rosenthal and D.F. Janzen (Eds.), Herbivores; their interaction with secondary plant metabolites. Academic Press, New York.

Rowan, D.R.; Hunt, M.B.; Gaynor, D.L. 1986. Peramine, a novel insect feeding deterrent from ryegrass infected with the endophyte *Acremonium loliae*. J. Chem. Soc. Chem. Commun. 935-936.

Saxena, R.C. 1986. Neem seed oil—a potential antifeedant against insect pests of rice. Abst. Sixth Int. Congr. Pest. Chem. IUPAC, Ottawa, Ont., Aug. 10-15. 2D/E-20.

Saxena, R.C. 1989. Insecticides from neem. Chapter 9 *in* J.T. Arnason, B.J.R. Philogène and P. Morand (Eds.), Insecticides of Plant Origin. ACS Symp. Series 387, pp. 110-135. Am. Chem. Soc., Washington, D.C.

Schoonhoven, L.M. 1967. Chemoreception of mustard oil glucosides in larvae of *Pieris brassicae*. Kon. Ned. Akad. Wetensch. Proc. Ser. C. 70: 556-568.

Schoonhoven, L.M. 1982. Biological aspects of antifeedants. Entomol. Exp. Appl. 31: 57-69.

Shemyakin, M.M.; Vinogradova, E.I.; Feigina, M.Y.; Aldanova, N.A.; Loginova, N.F.; Ryabova, I.D.; Pavlenka, I.A. 1965. The structure—antimicrobial relation for valinomycin depsipeptides. Experientia 21: 548.

Simmonds, M.S.J.; Blaney, W.M.; Ley, S.V.; Anderson, J.C.; Toogood, P.L. 1990. Azadirachtin: structural requirements for reducing growth and increasing mortality in lepidopterous insects. Entomol. Exp. Appl. 55: 169-181.

Städler, E. 1974. Host plant stimuli affecting oviposition behavior of the eastern spruce budworm. Ent. exp. appl. 17: 176-188.

Strongman, D.B.; Strunz, G.M.; Giguère, P.; Yu, C.-M.; Calhoun, L. 1988. Enniatins from *Fusarium avenaceum* isolated from balsam fir foliage and their toxicity to spruce budworm larvae, *Choristoneura fumiferana* (Clem.) (Lepidoptera: Tortricidae). J. Chem. Ecol. 14: 753-763.

Strunz, G.M.; Giguère, P.; Thomas, A.W. 1986. Synthesis of pungenin, a foliar constituent of some spruce species, and investigation of its efficacy as a feeding deterrent for spruce budworm (*Choristoneura fumiferana* (Clem.)). J. Chem. Ecol. 12: 251-259.

Swain, T. 1977. Secondary plant compounds as protective agents. Ann. Rev. Plant Physiol. 28: 479-501.

Thomas, A.W. 1987. The effect of age of current-year shoots of *Picea glauca* on survival, development time, and feeding efficiency of 6th-instar larvae of *Choristoneura fumiferana*. Entomol. Exp. Appl. 43: 251-260.

Thomas, A.W.; Strunz, G.M.; Chiasson, M.; Chan, T.H. 1992. Potential of Margosan-O, an azadirachtin-containing formulation from neem seed extract, as a control agent for spruce budworm, *Choristoneura fumiferana*. Entomol. Exp. Appl. (In press)

Van der Eycken, E.; De Bruyn, A.; Van der Eycken, J.; Callant, P.; Vandervalle, M. 1986. Iridoids: the structure elucidation of specionin based on chemical evidence and 1H NMR analysis. Tetrahedron 42: 5385-5396.

Wagner, M.R.; Tinus, R.W. 1985. Mechanism of tree resistance to spruce budworms—is there a terpene connection? Pages 121-123 *in* C.J. Sanders, R.W. Stark, E.J. Mullins and J. Murphy (Eds.), Recent Advances in Spruce Budworm Research. Proc. CANUSA Spruce Budworms Res. Symp., Bangor, Me.

Warthen, J.D. 1979. *Azadirachta indica*: A source of insect feeding inhibitors and growth regulators. USDA Agric. Rev. and Manage. ARM-NE-4, 22 p.

Yajima, T.; Kato, N.; Munakata, K. 1977. Isolation of insect antifeeding principles in *Orixa japonica* Thunb. Agric. Biol. Chem. 41: 1263-1268.

Zehnder, G.; Warthen, J.D. 1988. Feeding inhibition and mortality effects of neem-seed extract on the Colorado potato beetle (Coleoptera: Chrysomelidae). J. Econ. Entomol. 81: 1040-1044.

Chapter 41

Classical Biological Control

D.R. Wallace

Introduction

Classical biological control may be defined as the regulation of a pest population by importing natural enemies that are not indigenous to the territory where the target species is a pest. Often the pest is itself an exotic which in its new environment has become more damaging than in its homeland. Nordland (1984) uses the term "inoculative" for parasitoid releases made to implement classical biological control. Inoculative implies that only small numbers of the natural enemies may be released, and that to achieve the control of the pest, the introduced agent must multiply and be self-sustaining. This principle of being self-sustaining is the most important characteristic of classical biological control that makes its application in forestry particularly valuable, but it also makes the assessment of biological control programs difficult.

The first use of the name "classical biological control" is unclear. Caltagirone (1981) suggested that the term "classical" was adopted because the introduction strategy described above was used in the remarkable early successes in pest control with natural enemies, which included the case of the cottony cushion scale, *Icerya purchasi*, in California by importation of the vedalia beetle, *Rodolia cardinalis*, from Australia in 1888 (Doutt 1964). The early successes, coupled with the fact that many of the pests plaguing the development of agriculture and forestry in North America were of foreign origin, especially European, resulted in a very heavy reliance on the classical, or inoculative, method. In other parts of the world, biological control had been practised for many centuries, usually by increasing the effectiveness of native parasites and predators through crop cultural methods, by local relocations, or by artificially multiplying native organisms to augment those already present. The emphasis on classical introductions continued in North America until the whole practice of biological control lost a great deal of appeal following the development of synthetic chemical insecticides in the early 1940s. The recent recognition of the environmental risks associated with chemical pesticides and the serious problem of pest resistance to chemicals has brought about a resurgence in interest in biological control. The main thrust of the renewal is still classical introductions, but the ecological basis for conservation and protection of the environment is turning attention more toward other aspects of biological control (i.e. enhancement of native parasitoids and predators through crop management practices) (Chapter 43) and artificially increasing the numbers of native organisms (Chapter 42).

Suitability of Classical Introductions to Forestry

A useful way of examining the suitability of classical biological control for application in forest management is to compare some of the basic characteristics of forest and agricultural ecosystems and relevant cultural practices (Table 1). The differences between the two systems (Nealis and Wallace 1991) emphasize the need for low-cost pest control methods for forestry because the value of the crop per unit area is low, the crop areas affected by pests are often very large, and pest outbreaks may last for several years and reoccur more than once in a crop rotation. Because the pest control treatments may be many years removed from harvest, the market value of the crop is not necessarily assured. There is little or no evidence that conventional pesticide applications in forestry control populations, but may simply provide a temporary respite for the trees which may or may not outlast the natural course of the pest outbreak. There is argument in some cases, such as the spruce budworm, that extensive pesticide application may lengthen the pest outbreak and reinforce the need for repeated treatments.

Table 1. Characteristics of forest and agricultural crop systems and management practices that influence pest management strategies.[a]

Forestry	Agriculture
Long rotation	Short rotation
Complex ecosystem, resilient to disturbance	Simpler ecosystem, easily disturbed
Less disturbance	Frequently and severely disturbed
Extensive areas	Smaller crop areas
Crop unit area value low Low annual accrual	Crop unit area value high
More damage tolerance	Low damage threshold
Long/large pest outbreaks	Smaller units for treatment
Access may be difficult and pests cryptic	More amenable to treatment
Market for pesticides not stable	Generally a stable pesticide market
Environment pristine in the public image	Public sees agricultural areas as highly altered

[a] Based in part on Nealis and Wallace (1991)

It is clearly not practical on economic and logistic grounds to treat more than a small fraction of an area affected by forest pests with conventional pesticides, and neither is it environmentally acceptable or desirable to do so even if broad spectrum microbial pesticides are used. Forest areas that have received expensive silvicultural treatments or that bear high-value stands are treated, however, and large residual areas of the pest outbreak are left "uncontrolled." General forest management trends are that more and more forest production will be concentrated on smaller areas of better growing sites close to manufacturing and consuming areas, and that cultural practices on these sites will be intensive and expensive. Under these circumstances, pest outbreaks in the large, less intensively managed surrounding forests may spill over and cause serious damage in the prime production areas, or it may not be possible to suppress pest populations in these managed areas in the midst of a surrounding outbreak. A management system that provides inexpensive, safe pest control over large areas is therefore required.

Forest trees support many insects that depend on the trees for their food supply. Syme and Nystrom (1988) list about 1 600 species as tree feeding associates found in the forests in Ontario by the Forest Insect and Disease Survey (FIDS) over a period of more than 50 years. This probably represents only a fraction of the true number because the FIDS methodology tends to emphasize foliavores and miss cryptic feeders. Out of all these species over the 50-year period, only about 40 species are considered major pests and another 90 are recognized to cause sporadic or localized damage. It is evident, therefore, that most forest herbivores never affect the trees to an extent that we can detect an economically important loss. The reasons that so few of the many forest-feeding insects become pests are complex, but it depends on the reproductive characteristics of the insects and the interaction with their hosts and the influence of natural enemies which include parasitoids and predators. In many cases, the parasitoids and predators prevent the herbivores from becoming pests and this "invisible" biological control can be demonstrated by intensive study and experimentation.

Frequently, when a herbivore that is not a pest in the region where it has evolved is accidentally introduced into a new region (e.g. from Europe to North America), the transfer involves small numbers of the herbivore and the natural enemies are left behind. Native natural enemies in the new country may not transfer to the introduced species. Without the controlling influence of natural enemies, the herbivore may become a pest if the climate and food plants in the new territory are suitable. Classical biological control programs have sought to reassociate major elements of the natural control complex in the originating region with the pest in its new home. Where parasitoids and predators are involved, usually only small numbers of the organisms are released and must become established, multiply, and disperse on their own. Not all releases result in establishment of the exotic parasites and predators, but in many instances the introduction programs have led to good control with a high benefit-to-cost ratio, although the very nature of classical biological control makes benefit–cost evaluation extremely difficult. Wagge (1991) considers that the success rate is about 25%. The level of success varies from case to case but, in the best circumstances, the status of the introduced species is reduced from a pest to an unnoticed component of the fauna. An example of this type of result in Canada is the long-term control of the European spruce sawfly which erupted in a serious outbreak in eastern North America in the 1930s. In other species the severity and frequency of outbreaks/damage may be reduced. DeBach (1964) has expressed the foregoing very succinctly: "...the end result of an outstanding example of biological control is not spectacular and is likely to go unnoticed and unappreciated because the formerly abundant organism has been reduced to a rare species which is attacked by rare natural enemies. It is easy to overlook the results and to forget a problem when it has disappeared."

Comparing the properties of insect pest control most appropriate for forestry with the characteristics of classical biological control clearly indicates the appropriateness of the method for forest pest management.

Canadian Historical Overview and Evaluation of Control Programs

There is an overwhelming amount of literature on biological control of pests, much of it relating to the classical introduction method. General treatments that outline the development of the science include Sweetman (1936, 1958), DeBach (1964), Huffaker and Messenger (1976), and Mackauer et al. (1989). These texts are international in scope, although somewhat North American in focus. McClay (1991) provides a current general synthesis for Canada, including forest pests (Nealis and Wallace 1991). The history of classical biological control in Canada was given by Baird (1956) in a general history of entomology compiled by Glen (1956).

The earliest introduction of a parasitoid recorded is in 1882 when William Saunders, the first director of the Dominion Experimental Farms, obtained *Trichogramma minutum* from New York State for release against the imported currantworm, *Nematus ribesii*, in Ontario gardens. Today, it would be considered that *T. minutum* would have been present already in the area of release. The first classical biological control project for a forest pest in Canada was organized by C.G. Hewitt in 1909 when he was appointed Dominion entomologist. He came to this posting from England where he had been investigating the natural enemies of the larch sawfly, *Pristiphora erichsonii*. This species had destroyed most of the merchantable tamarack, *Larix laricina*, in eastern Canada by the time of his arrival in the country, but he began a program of parasitoid introduction that was continued for several decades. This case history is discussed in a later section to illustrate features of classical introductions. Another early introduction program was organized to combat the browntail moth, *Euproctis chrysorrhoea*, which had invaded New Brunswick and Nova Scotia from the New England States from 1909 to 1911. This project marked the beginning of intensive Canadian involvement in international biological control, at first with the United States Bureau of Entomology, particularly the staff of the Gypsy Moth and Brown-Tail Moth Laboratory at Melrose Highlands, Massachusetts.

In England, the Imperial Bureau of Entomology founded the Farnham House Laboratory in 1927 to respond to the needs for classical biological control throughout the British Empire, with a very strong influence and cooperation from Canada (Thompson 1936). The cooperation between Canadian biological control specialists and the Farnham House Laboratory has continued through many successions of change and reorganization to the current program with the Commonwealth Agricultural Bureaux International Institute of Biological Control. Dr. W.R. Thompson, a Canadian specialist in applied biological control, in the mathematical relationships between parasitoids and their hosts, and in the systematics and biology of the Diptera, became director of the Farnham House Laboratory in 1928 and retired as director of the Commonwealth Institute of Biological Control in 1958. The activities of the Farnham House Laboratory were transferred to Canada at the Dominion Parasite Laboratory in Belleville, Ontario, in 1940 for the duration of World War II. From 1948 to 1962 the institute headquarters were in Ottawa (Anstey 1986; Downes 1972).

The first biological control laboratory in Canada was established by federal authorities at the University of New Brunswick in Fredericton in 1912 as a center for the browntail moth studies. Work continued here on the forementioned and other species until 1922 when parasitoid studies were relocated to Ottawa. A new parasitoid laboratory was set up in Chatham, Ontario, in 1925 to investigate biological control of agricultural pests, but the redistribution of larch sawfly parasitoids was also handled. The Dominion Parasite Laboratory was established in 1929 at Belleville and new facilities were opened in 1936 (Anon. 1936), including a quarantine and rearing building for mass rearing of imported parasitoids. The successful effort to control the European spruce sawfly beginning about 1933 was a major factor in the expansion of biological control. The Belleville laboratory became The Research Institute for Biological Control in 1959. It was closed in 1972 and the staff was dispersed to other centers in a reorganization of Agriculture Canada. The laboratory at Belleville was responsible for importations and research on parasitoids for forest pest management, while research on the biology and population dynamics of the pests, including the monitoring of parasitoid releases, was now done at the forest entomology laboratories (Prebble 1956; Wallace 1990). In 1954, more responsibility for biological control of forest pests was transferred to forestry laboratories with mainly aspects of quarantine regulation and importation remaining with organizations in Agriculture Canada.

There are numerous reviews of Canadian biological control work from a wide variety of perspectives. The most basic are three volumes published by the Commonwealth Agricultural Bureaux that cover case histories of biological control programs until 1980 (McLeod et al. 1962; Corbet 1971; Kelleher and Hulme 1984). Table 2 shows the forest pest species that are included in these compilations. Relatively few additions to the species being treated have occurred since 1958, and several of these are more involved with insect pathology and microbial control than with parasitoids and predators. There are also general overviews and assessments of the biological control research and application included in the publications cited above (Biological Control of Forest Insects, 1910–1958, McGugan and Coppel 1962; Current Approach to Biological Control of Forest Insects, Reeks and Cameron 1971; Status and Potential of Biological Control in Canada, Munroe 1971; Biological Control of Forest Insect Pests in Canada 1969–1980; Retrospect and Prospect, Hulme and Green 1984).

Turnbull and Chant (1961) gave what they called a "stocktaking" of biological control in Canada that includes forestry applications, although the underlying sentiment is more agricultural. These authors take a very critical and somewhat negative view of biological control in Canada and perhaps in retrospect were too alarmist. However, their concerns about safeguards against introductions of biological control agents that ultimately prove harmful will be returned to in a later section. They did recognize that 12 of the 31 projects they reviewed were successes.

Beirne (1975) also reviewed biological control attempts by classical introductions in Canada and concluded that the strong restrictions advocated by Turnbull and Chant were not warranted by the record of unexpected harmful effects of parasitoid introductions and that failures in establishment and control were more related to inadequate procedures than to inherent characteristics of the method. He believed that biological control had much unrealized potential for use in Canada. Beirne (1973) had earlier discussed the evolution of biological control in Canada from a very personal viewpoint and covered the interaction of forestry and agricultural influences.

Hulme (1988) evaluated the recent Canadian record in applied biological control of forest pests and concluded that one third of 21 attempts have been successful, the pests being almost permanently controlled. Partial control has been recorded in another one third, and in a few other cases, it was too soon to make an evaluation. Just under one half of the examples involved the use of parasitoids or predators through inoculative introductions. Table 3 lists forest pests that have been controlled by parasitoids or for which parasitoids have had a proven or probable role in achieving control. Note that all are introduced pests while many species listed in Table 2 are native pests.

An economic evaluation of classical introductions in Canada is more difficult. Many of the reasons for this difficulty are inherent in the strategy. The disappearance of successfully controlled pests results in a lack of support for long-term studies and quantitative measurements to determine the mechanisms of population regulation and the degree of regulation that has been provided by the applied biological control program. This shortcoming has been discussed in many of the reviews of Canadian programs. The result is a serious loss of information that could assist in developing a fundamental understanding of biological control leading to improvements in future programs. An example of this is the question of the mechanism of successful control that has been achieved with the winter moth (Embree 1966, Roland 1988). An additional effect of the premature termination of intensive studies and impact evaluation of both the pests on their hosts and the parasitoids on the pests is that the benefit–cost data needed to justify continued application of classical introductions is unavailable or lacks authority. Hulme (1988) examined the return on investment for this type of pest

Table 2. Case histories of Canadian biological control programs in forestry reviewed in Commonwealth Institute of Biological Control Publications indicated.[a]

Species[b]	1910–58	1959–68	1969–80
Homoptera			
Balsam woolly adelgid, *Adelges piceae*	X	X	X
Pine bark adelegid, *Pineus strobi*	X		
Western fruit lecanium, *Lecanium tiliae*		X	
Juniper scale, *Carulaspis juniperi*	X		
Coleoptera			
White pine weevil, *Pissodes strobi*	X		
Spruce beetle, *Dendroctonus rufipennis*	X		
Lepidoptera			
Satin moth, *Leucoma salicis*	X	X	X
Browntail moth, *Euproctis chrysorrhoea*	X		
Forest tent caterpillar, *Malacosoma disstria*	X	X	X
Winter moth, *Operophtera brumata*	X	X	X
Hemlock looper, *Lambdina fiscellaria*	X		
Western oak looper, *Lambdina fiscellaria somniaria*			X
European pine shoot moth, *Rhyacionia buoliana*	X	X	X
Spruce seed moth, *Cydia strobilella*	X		
Spruce budworm, *Choristoneura fumiferana*	X	X	X
A lodgepole needleminer, *Coleotechnites starki*	X		
Pine needleminer, *Exoteleia pinifoliella*	X		
Larch casebearer, *Coleophora laricella*	X	X	X
Western spruce budworm, *Choristoneura occidentalis*			X
Birch casebearer, *Coleophora serratella*			X
Gypsy moth, *Lymantria dispar*			X
Bruce spanworm, *Operophtera bruceata*			X
Whitemarked tussock moth, *Orgyia leucostigma*			X
Douglas-fir tussock moth, *Orygia pseudotsugata*			X
Hymenoptera			
Introduced pine sawfly, *Diprion similis*	X		
European pine sawfly, *Neodiprion sertifer*	X	X	X
Nursery pine sawfly, *Gilpinia frutetorum*	X		
European spruce sawfly, *Gilpinia hercyniae*	X	X	X
Striped alder sawfly, *Hemichroa crocea*	X		
Larch sawfly, *Pristiphora erichsonii*	X	X	X
Mountain-ash sawfly, *Pristiphora geniculata*	X		X
Redheaded pine sawfly, *Neodiprion lecontei*		X	X
Swaine jack pine sawfly, *Neodiprion swainei*		X	X
Birch leafminer, *Fenusa pusilla*			X
Balsam fir sawfly, *Neodiprion abietis complex*			X

[a]1910–1958, McGugan and Coppel (1962); 1959–1968, Corbet (1971); 1969–1980, Kelleher and Hulme (1984).
[b]Includes work on biological control using microbials.

Table 3. Tree pests in Canada that have been controlled by inoculative releases of parasitoids or ones for which parasitoids have had a proven or probable role in regulation (modified from Hulme 1988).

Satin moth, *Leucoma salicis*	
Browntail moth, *Euproctis chrysorrhoea*	(possible role only)
Winter moth, *Operophtera brumata*	
European pine shoot moth, *Rhyacionia buoliana*	
Larch casebearer, *Coleophora laricella*	
European pine sawfly, *Neodiprion sertifer*	(possible role + baculovirus)
European spruce sawfly, *Gilpinia hercyniae*	+ baculovirus
Larch sawfly, *Pristiphora erichsonii*	
Mountain-ash sawfly, *Pristiphora geniculata*	

management and concluded that although lack of data prevents a rigorous national analysis, the evidence available shows a large return on investment. In Australia, Marsden et al. (1980) showed a 10:1 benefit overall for 12 control attempts, of which 8 were successful. Tisdell (1990) has addressed the general issue of economic impact and also concluded that high rates of return on investment have been achieved with introductions of exotic biological control agents.

Methodology of Classical Introductions

Pschorn-Walcher (1977) reviewed the methodology of biological control with particular reference to forest insects and has illustrated the typical sequence of steps with examples from historical cases, many in the context of Canadian forestry programs. These steps have been summarized by Ehler (1990b) under three phases: the pre-introductory phase, the introductory phase, and the post-introductory phase. He also tabulates many of the questions that need to be addressed in a well-planned classical introduction project and provides references relating to them. Other general discussions of methodology are presented by Ehler (1990a) and Wagge (1991).

The pre-introductory phase begins with an evaluation of the pest as a proposed target for releases of exotic natural enemies. This evaluation should be based on thorough population studies of the pest in the region of concern, including an economic analysis of the potential return on investment from a control program. The economic analysis may be very difficult in forestry applications because of the long crop rotations. The amenability of the pest and its habitat to classical biological control is assessed and a preliminary idea of the control strategy formed. As this process advances, the search for potential source regions comes into play based on ecological similarity of target and source habitats, potential hosts and prey of parasitoids and predators in the foreign areas, and the introduction strategy that seems appropriate. These considerations will be influenced strongly by whether the pest is itself an exotic in the target area or whether it is an indigenous organism. Foreign exploration and population research begins at this time to study the dynamics of the source system at least to a rudimentary extent, if this information is not already known, and to determine the feasibility and method of obtaining the natural enemies for study and possible release. In the late stages of the pre-introductory phase, selected natural enemies are studied for compatibility with the target pest and its habitat, competitive relationships in the target system, and safety concerns relating to introductions.

Research may also include developing methods of propagating the natural control agents if the abundance of the parasitoid or predator in the source region is low. The investigation of potential agents may take place in the region of origin or in quarantine facilities in the receiving region. Personnel who will be involved in making releases and monitoring the results should be directly involved in the exploration and subsequent research.

The transition between the pre-introductory and introductory phases of a classical biological control project is not sharp, but the introductory phase certainly begins when the decision has been taken to release one or more exotic organisms. Release areas and methods are determined and the plans for determining establishment and dispersal made. Foreign exploration is increased to provide sufficient material for the proposed releases, or artificial multiplication of the parasitoids or predators is in place to provide the requirements. The releases are made and the establishment success or failure determined. This may take several years and also includes estimating the initial spread of the organism.

The post-introductory phase includes defining the longer term dispersal of the introduced exotic and its impact on the target pest. This is a continuation of the population dynamics studies that were part of the pre-introductory phase. Not only should the effectiveness of the introduced agent be determined but also the mechanism of regulation (or failure to control) should be elucidated. Host damage assessment must be included so that an economic evaluation of the project can be made. The post-introductory phase of a classical biological control program may cover many years and will obviously decrease in intensity over time depending on the circumstances. Populations of introduced organisms often exhibit marked changes in behavior over time as they interact with the native fauna and flora. These changes are important in developing a framework for applied biological control. In addition, if the pest still exhibits cycles of abundance after the introductions have been made, in which the damage to the host may exceed the economic threshold, it is necessary to understand the changing interactions of the exotic control agent at the different population levels.

The process of classical biological control is long term and requires an expenditure of effort for many years, depending on the degree to which there is a commitment to sound preparatory and follow-up research. Omitting work in a casual approach to rapid introductions of exotics increases the risk of causing harmful effects and increases the probability of failure of the program.

A Case History in Canadian Forestry Application of Classical Introductions

One of the most interesting cases in Canadian forest pest biological control using parasitoids is that of the larch sawfly which was alluded to earlier (McGugan and Coppel 1962; Turnock and Muldrew 1971; Ives and Muldrew 1984). It illustrates ultimate success but it also provides insights into pitfalls and complications.

One hundred years ago tamarack was a useful and commercially important species in the northeastern and north central regions of North America. There was a large volume of merchantable timber, much of it in mature stands. Between about 1880 and 1920, huge numbers of tamaracks died after severe defoliation by the larch sawfly over several years. The tamarack has never regained its position as a commercially valuable species, although today it and other species of larch, *Larix* spp., may be considered for short rotation planting and therefore loss of increment brought about by the larch sawfly would be of concern.

The sudden irruption of severe outbreaks of the sawfly in the late 19th century and its apparent rapid spread westward led scientists to believe that the pest had been introduced recently from Europe (Hewitt 1912; Coppel and Leius 1955) where it had been known as a minor defoliator of larch that appeared sporadically and only rarely killed trees (Pschorn-Walcher and Eichhorn 1963; Pschorn-Walcher and Zinnert 1971; Turnock 1972). C.G. Hewitt, the first Dominion entomologist, who had studied the natural control of the larch sawfly in an outbreak in the Lake District in England before his appointment in Canada, believed that the larch sawfly was introduced from Europe. He had the parasitoid *Mesoleius tenthredinis*, an effective regulatory agent in England, imported and released in Ontario, Quebec, and Manitoba between 1910 and 1913. The parasite became established and dispersed rapidly as well as being relocated in several regions in later years. During the two or three decades following establishment in Canada, *M. tenthredinis* became a major parasitoid of the sawfly in North America and apparently checked sawfly outbreaks to a large extent. Turnock (1972) examined the geographical and historical variability in population dynamics of the larch sawfly, showing that in North America the species had a small natural control complex in contrast to the diversity of biotic mortality factors that are found in the regions of Eurasia where population levels are low and have small fluctuations in amplitude. This difference may account for the severe early outbreaks observed in North America. The introduction of a single effective parasitoid into the system changed the population behavior.

In the late 1930s the larch sawfly again reached outbreak population levels in central Canada and the United States. It was found that larvae were not being killed by *M. tenthredinis* in the high proportion previously observed. A renewed and intensified research program on the larch sawfly was initiated in the 1940s and it was found that the reduction in effectiveness of *M. tenthredinis* resulted from encapsulation of the parasitoid eggs by blood cells of the host, a type of immune response (Muldrew 1953). The loss of effectiveness of this key introduced parasitoid allowed the sawfly to escape from regulation.

The reason for the change in behavior of the sawfly/*M. tenthredinis* system was eventually discovered and demonstrates two important issues in classical biological control programs. In studies of variation in the larch sawfly across its Holarctic range, Wong (1974) found that the species is composed of several strains and the early North American outbreaks represented strains that he postulated had reached the continent from Eurasia by natural dispersal across the Bering Land Bridge, possibly in the Miocene. These strains did not bring with them the parasitoid *M. tenthredinis* or its progenitors, nor did they have or develop the capability of encapsulating this parasitoid. In the process of making the releases of *M. tenthredinis*, larch sawfly cocoons were set out in the forest for the parasites to emerge without precautions to prevent the introduction of foreign larch sawfly females (the species has obligatory parthenogenesis, Smith 1955) or hyperparasites and other unwanted organisms (Criddle 1928; Graham 1953). This process allowed a strain of the larch sawfly capable of encapsulating *M. tenthredinis* to be introduced from England and in about 30 years it built up to dominance in the Canadian Prairies and produced the new series of outbreaks. Without the application of fundamental research on systematics and evolution, the details of these events would not have come to light. The second issue that is graphically illustrated is the need for quarantine and release protocols to prevent the introduction of organisms other than the desired species. Current practices would not permit the method of release used from 1910 to 1913.

Further research in Europe and Canada identified a strain of *M. tenthredinis*, the Bavarian strain, that does not induce the encapsulation reaction of the resistant larch sawfly strain. The Bavarian strain was introduced in 1963–64, and increases in the effectiveness of *M. tenthredinis* have been observed (Turnock and Muldrew 1971). Attempts to relocate this strain artificially from its original introduction and dispersal area in Manitoba have been complicated by the inability to distinguish the different *M. tenthredinis* strains morphologically.

Another parasitoid, *Olesicampe benefactor*, also introduced from Europe into Manitoba (Muldrew 1967) between 1961 and 1963, has dispersed rapidly and has been extensively relocated by entomologists into various regions of North America. The evidence to date is that *O. benefactor* has reestablished regulation of larch sawfly populations in many areas. However, there are further complications that could affect the long-term effectiveness of the parasitoid. The larch sawfly can encapsulate *O. benefactor* and the parasitoid is attacked by a native North American hyperparasitoid, *Mesochorus globulator*, that was rarely found before the introduction and multiplication of *O. benefactor*. The ultimate stabilization state of the system with the addition of the newest European introduction is still to come. A sidelight to the effectiveness of *O. benefactor* is the recent successful control of another European sawfly species in Canada (Forbes and Daviault 1964), the mountain-ash sawfly, *Pristiphora geniculata*, by the related species *O. geniculatae* in a program described by Quednau (1990). The similarities of the two sawfly/parasitoid systems have not been compared.

Other parasitoid releases and relocations against the larch sawfly have been unsuccessful, as has been the use of insect pathogens. One other classical introduction against the sawfly bears comment. The masked shrew, *Sorex cinereus*, was introduced into Newfoundland in 1958 (Buckner 1966; Warren 1970). Newfoundland has a restricted small mammal fauna, shrews were absent, and the masked shrew had been found to be an important predator of sawfly cocoons. The introduction of the shrew did not lead to the elimination of sawfly outbreaks on the island and the effects on the fauna are not known. Buckner noted changes in the characteristics of the shrew in Newfoundland when compared with populations on the mainland. Warren noted some social consequences of the introduction and colonization, although he concluded that the introduction was beneficial to the regulation of several forest pests and had by 1970 shown no harmful effects. The absence of intensive baseline studies and follow-up to such a major alteration of the fauna of a large island like Newfoundland leaves the questions of desirability of this type of introduction in the balance.

Continued research on the larch sawfly and its population regulation is required to determine the outcome of the series of introductions described, although lapses in the records will prevent a full analysis of the processes involved, an analysis that would add to the understanding of classical introductions.

Changing Targets for Classical Introductions

Until recent years, the possibility of introducing exotic biological control agents against native pest species was given little consideration, or when releases were made the attempts were not treated in the same determined fashion as projects against introduced pests. Pimentel (1963) brought together information on the use of parasitoids and predators to control native pests. He proposed the concept of new associations between parasitoids and predators as being more effective in the context of introductions of exotic natural enemies than associations that have coevolved, based on the development of homeostatic mechanisms in the latter instance. This principle was not supported by other authors such as Huffaker et al. (1971, 1976) and Wagge (1991), and there is still an active debate about the validity of the proposal. Hokkanen and Pimentel (1984) reasserted the argument, and then in 1989 they redefined the idea of a new association to mean "an association between an exploiter and its victim... when the exploitation or attack strategy of this exploiter differs from the strategy of the native exploiters that have played a major role in the evolutionary history of the victim." This definition is harder to apply than if the exploiter is simply being drawn from a different habitat or from the natural enemy guild of another herbivore than the target pest. The conclusion at this time appears to be that old associations offer more opportunities for inoculative introductions but that new associations are sufficiently promising to be applied.

Of particular relevance to classical biological control of native Canadian pests are a general review of the topic by Carl (1982) and a specific examination by Mills (1983) of the possibilities for biological control of the spruce budworm using natural enemies from Europe where *Choristoneura fumiferana* is not found. The latter paper suggests ways of avoiding the shortcomings of earlier similar attempts and provides guidance in the selection of suitable parasitoids for introduction. The general review by Carl concludes that in principle there is no reason to avoid the application of introductions of exotic biocontrol agents to the management of native pests. The major drawback is the need for more extensive research and experimentation to assure compatibility between the exotic and its target and the suitability for release of the organism in the new environment, both in terms of successful establishment and the possibility of harmful effects to nontarget systems.

Changing Attitudes Toward Biological Control

The early settlers were free to bring any living plants and animals to the New World that would survive transportation. It was not until the early 1900s that quarantine regulations were established in Canada (1910) and the United States (1912). This also meant that early introductions of biological control agents were not strictly regulated. The regulatory aspects of importation and release of exotic parasitoids and predators in Canada are discussed by Byrne (1991) and at present are relatively simple. To obtain a permit for importation the end use (e.g. quarantine or other research, contained release [cage], free release) must be stated and a reasonable protocol for testing proposed. In other countries such as Australia and New Zealand where the indigenous fauna and flora have suffered irreparable harm from both accidental and deliberate introductions of foreign organisms, the regulations affecting classical biological control are much more stringent (Palmer 1989). There is currently an international effort (Wagge 1991) to assure a higher level of environmental safety in the use of exotic natural control organisms by promoting more uniform regulations. The risks involved in classical biological control are not restricted by national boundaries if the adjoining countries share identical or similar ecological zones, because once established the exotic multiplies and disperses. For Canada this implies close cooperation with the United States. Protocols for the United States are described in Coulson and Soper (1989).

There is little evidence to support the concerns of Turnbull and Chant (1961) that general safety precautions of releases in Canada were very lax; however, the increasing recognition by the public and researchers alike of the need for environmental protection, and the public's especially critical approach to pest management in forestry, both dictate that biological control researchers and practitioners must actively work with environmental interests and regulatory agencies to formulate sound rules for introductions that address risk adequately and do not block the application of a valuable pest management tool. Hoy et al. (1991) have suggested that biological control practitioners for commercial releases of arthropod natural enemies be licensed just as pesticide applicators are. Because most classical introductions are made by researchers, this could imply ultimate licensing for them as well, a high level of regulation.

Current Canadian Forestry Projects in Classical Biological Control

Current projects in Canada using introductions of exotic natural control agents for forest pests are being carried out principally or entirely in Forestry Canada laboratories and depend on research and exploration in Europe done by the Commonwealth Agricultural Bureaux International Institute of Biological Control. The European work is centered at the field station at Delémont, Switzerland.

The main target pests of the current programs are gypsy moth, *Lymantria dispar*; white pine weevil, *Pissodes strobi*; spruce bud moth, *Zeiraphera canadensis*; and spruce budworm, *Choristoneura fumiferana*.

Several other pests are also being examined in the pre-introductory phase to determine the desirability of initiating a more intensive program or waiting for adequate populations for obtaining natural control agents to be located.

Introductions of the tachinid parasitoid *Ceranthia samarensis* are being made against the gypsy moth using both material collected in Europe and propagated at the Quebec Regional Laboratory of Forestry Canada. In Europe, the recent exploration and collecting has concentrated on the natural enemy complex of scarce populations in contrast to most of the earlier work which involved outbreak populations. The methodology depends on exposure and retrieval of cohorts of host larvae in areas where the pest is known to occur but is rare at the present time (Mills 1990; Mills and Nealis 1991). Exploitation of this approach may offer greater availability of exotic parasitoids for other pests as well as the gypsy moth (Fig. 1).

Figure 1. Mating flies of the gypsy moth parasitoid *Ceranthia samarensis* which has been imported from Europe and propagated for release in Canada. (W. Quednau, Forestry Canada)

Three of the four current main projects are targeted against native pests and the majority of the other pests being considered for classical introductions are also indigenous North American species. Although native species were looked at in the past (Table 2), the concentration of effort on them represents a new view.

Future of Classical Introductions

The record of successes in classical biological control of forest pests in Canada and the worldwide trend toward use of biological methods of pest management, in the light of the risks and problems associated with the application of conventional synthetic chemical insecticides, should indicate that classical introductions will become a more important component of pest management in the future. This is especially true because the public views biological control as desirable and, providing that care is taken by practitioners to follow proper safety practices, it should be possible to avoid stifling regulation of classical biological control.

Van Lenteren (1980) questioned the likelihood that theoretical population ecology would soon develop methods to predict the outcome of introducing natural enemies. He believed that the practice of biological control is more of an art than a science. Other authors have taken the extreme view that no amount of planning and basic research will replace empirical trials by introducing the natural enemies in the new environment. Yet ecologists, for example, Murdock (1989), contend that some general principles from population theory can provide guidance in the selection of targets and natural enemies for biological control, and he calls for experimental analyses of both successful and unsuccessful biological control attempts to provide a better link between practice and theory. There is ample evidence to argue in either direction. Some confounding issues, however, suggest that it is important to seek a balance between fundamental research and practical application: for instance, whether to release a single natural enemy or a complex of species; intraspecific variability that results in establishment and successful control with strains of a natural enemy from one source area but not another; and the record that natural enemies showing little promise of being effective from studies in the source area perform well when introduced. The successes of classical introductions in forestry in Canada have come from such a balance. It could be said "Art Based on Science."

References

Anon. 1936. Conference on biological methods of controlling insect pests, Belleville, Ontario, June 25 and 26, 1936. Dept. Agric., Entomol. Br., Ottawa, Ont. 80 p.

Anstey, T.H. 1986. One hundred harvests. Research Branch Agriculture Canada 1886-1986. Res. Br. Agric. Can., Ottawa, Ont. Historical Series No. 27.

Baird, A.B. 1956. Biological control. Pages 363-367 *in* R. Glen (Compiler), Entomology in Canada up to 1956. A review of developments and accomplishments. Can. Entomol. 88.

Beirne, B.P. 1973. Influence on the development and evolution of biological control in Canada. Bull. Entomol. Soc. Can. 5(3): 85-89.

Beirne, B.P. 1975. Biological control attempts by introductions against pest insects in the field in Canada. Can. Entomol. 107: 225-236.

Buckner, C.H. 1966. The role of vertebrate predators in the biological control of forest insects. Ann. Rev. Entomol. 11: 449-470.

Byrne, J. 1991. Registration and regulation in biological control. Pages 113-118 *in* A.S. McClay (Ed.), Proceedings of the Workshop on Biological Control of Pests in Canada, October 11-12, 1990, Calgary, Alberta. Alta. Environ. Cent., Vegreville, Alta. AECV91-P1.

Caltagirone, L.E. 1981. Landmark examples in classical biological control. Ann. Rev. Entomol. 26: 213-232.

Carl, K.P. 1982. Biological control of native pests by introduced natural enemies. Commonw. Inst. of Biol. Cont. Biocontrol News and Inf. 3(3): 191-200.

Coppel, H.C.; Leius, K. 1955. History of the larch sawfly, with notes on origin and biology. Pages 103-111 *in* Approaches to the study of forest insects with special reference to the larch sawfly, *Pristiphora erichsonii* (Htg.). Can. Entomol. 87.

Corbet, P.S. (Ed.). 1971. Biological control programmes against insects and weeds in Canada 1959-1968. Commonw. Agric. Bureaux, Slough, Eng. Commonw. Inst. Biol. Cont. Tech. Commun. No. 4. 266 p.

Coulson, J.R.; Soper R.S. 1989. Protocols for the introduction of biological control agents in the U.S. Pages 1-35 *in* R.P. Kahn (Ed.), Plant Protection and Quarantine. Vol. III, Special Topics. CRC Press, Inc. Boca Raton, Fla.

Criddle, N. 1928. The introduction and establishment of the larch sawfly parasite, *Mesoleius tenthredinis* Morley, into southern Manitoba. (Hymen.). Can. Entomol. 60: 51-53.

DeBach, P. (Ed.). 1964. Biological control of insect pests and weeds. Reinhold Publishing Co., New York. 844 p.

DeBach, P. 1964. The scope of biological control. Pages 3-20 *in* P. DeBach (Ed.), Biological control of insect pests and weeds. Reinhold Publishing Co., New York.

Doutt, R.L. 1964. Historical development of biological control. Pages 21-42 *in* P. DeBach (Ed.), Biological control of insect pests and weeds. Reinhold Publishing Co., New York.

Downes, J.A. 1972. William Robin Thompson 1887-1972 (with Bibliography prepared by D. McClymont). Bull. Entomol. Soc. Can. 4(3): 38-47.

Ehler, L.E. 1989. 6. Introduction strategies in biological control of insects. Pages 111-134 *in* M. Mackauer; L.E. Ehler; J. Roland (Eds.), Critical issues in biological control. Intercept, Andover, Hants.

Ehler, L.E. 1990. 1. Some contemporary issues in biological control of insects and their relevance to the use of entomopathogenic nematodes. Pages 1-19 *in* R. Gaugler; H.K. Kaya (Eds.), Entomopathogenic nematodes in biological control. CRC Press, Boca Raton, Ann Arbor, Boston.

Embree, D.G. 1966. The role of introduced parasites in the control of the winter moth in Nova Scotia. Can. Entomol. 98: 1159-1168.

Forbes, R.S.; Daviault, L. 1964. The biology of the mountain-ash sawfly, *Pristiphora geniculata* (Htg.) (Hymenoptera: Tenthredinidae), in eastern Canada. Can. Entomol. 96: 1117-1133.

Glen, R. (Compiler). 1956. Entomology in Canada up to 1956. A review of developments and accomplishments. Can. Entomol. 88: 285-371.

Graham, A.R. 1953. Biology and establishment in Canada of *Mesoleius tenthredinis* Morley (Hymenoptera: Ichneumonidae), a parasite of the larch sawfly, *Pristiphora erichsonii* (Hartig) (Hymenoptera: Tenthredinidae). 35th report of the Quebec Soc. Protection Plants. pp. 61-75.

Hewitt, C.G. 1912. The large larch sawfly (*Nematus erichsonii*) with an account of its parasites, other natural enemies and means of control. Can. Dept. Agric., Entomol. Bull. 5. 42 p.

Hokkanen, H.; Pimentel, D. 1984. New approach for selecting biological control agents. Can. Entomol. 116: 1109-1121.

Hokkanen, H.M.T.; Pimentel, D. 1989. New associations in biological control: theory and practice. Can. Entomol. 121: 829-840.

Hoy, M.A.; Nowierski, R.M.; Johnson, M.W.; Flexner, J.L. 1991. Issues and ethics in commercial releases of arthropod natural enemies. Am. Entomol. 37(2): 74-75.

Huffaker, C.B.; Messenger, P.S. (Eds.). 1976. Theory and practice of biological control. Academic Press, New York, San Francisco, London. 788 p.

Huffaker, C.B.; Messenger, P.S.; DeBach, P. 1971. The natural enemy component in natural control and the theory of biological control. Pages 16-67 *in* C.B. Huffaker (Ed.), Biological control. Plenum, New York.

Huffaker, C.B.; Simmonds, F.J.; Laing, J.E. 1976. The theoretical and empirical basis of biological control. Pages 41-78 *in* C.B. Huffaker; P.S. Messenger (Eds.), Theory and practice of biological control. Academic Press, New York, San Francisco, London.

Hulme, M.A. 1988. The recent Canadian record in applied biological control of forest insect pests. For. Chron. 64: 27-31.

Ives, W.G.H.; Muldrew, J.A. 1984. *Pristiphora erichsonii* (Hartig), Larch sawfly (Hymenoptera: Tenthredinidae). Pages 369-380 *in* J.S. Kelleher; M.A. Hulme (Eds.), Biological programmes against insects and weeds in Canada 1969-1980. Commonw. Agric. Bureaux, Slough, Eng.

Kelleher, J.S.; Hulme, M.A. (Eds.). 1984. Biological programmes against insects and weeds in Canada 1969-1980. Commonw. Agric. Bureaux, Slough, Eng. 410 p.

Mackauer, M.; Ehler, L.E.; Roland, J. (Eds.). 1989. Critical issues in biological control. Intercept, Andover, Hants. 330 p.

Marsden, J.S.; Martin, G.E.; Parham, D.J.; Ridsill Smith, T.J.; Johnston, B.G. 1980. Returns on Australian agricultural research: joint IAC-CSIRO benefit cost study of the CSIRO Division of Entomology. Commonw. Scientific and Industrial Res. Organ., Canberra.

McClay, A.S. (Ed.). 1991. Proceedings of the Workshop on Biological Control of Pests in Canada, October 11-12, 1990, Calgary, Alberta. Alta. Environ. Cent., Vegreville, Alta. AECV91-P1. 136 p.

McGugan, B.M.; H.C. Coppel. 1962. Part II — Biological control of forest insects, 1910-1958. Pages 35-216 *in* J.H. McLeod; B.M. McGugan; H.C. Coppel. A review of the biological control attempts against insects and weeds in Canada. Commonw. Agric. Bureaux, Bucks, Eng. Tech. Commun. No. 2.

McLeod, J.H.; McGugan, B.M.; Coppel, H.C. 1962. A review of the biological control attempts against insects and weeds in Canada. Commonw. Agric. Bureaux, Bucks, Eng. Tech. Commun. No. 2. 216 p.

Mills, N.J. 1983. Possibilities for the biological control of *Choristoneura fumiferana* (Clemens) using natural enemies from Europe. Commonw. Inst. Biol. Cont., Biocontrol News and Inf. 4(2): 103-125.

Mills, N.J. 1990. Are parasitoids of significance in endemic populations of forest defoliators? Some experimental observations from gypsy moth, *Lymantria dispar* (Lepidoptera: Lymantriidae). Pages 265-274 *in* Population dynamics of forest insects. Intercept Ltd., Andover, Hampshire.

Mills, N.J.; Nealis, V.G. 1991. European field collections and Canadian releases of *Ceranthia samarensis* (Diptera: Tachinidae); a parasitoid of the gypsy moth. Entomophaga. In press.

Muldrew, J.A. 1953. The natural immunity of the larch sawfly (*Pristiphora erichsonii* (Htg.)) to the introduced parasite *Mesoleius tenthredinis* Morley, in Manitoba and Saskatchewan. Can. J. Zool. 31: 313-332.

Muldrew, J.A. 1967. Biology and initial dispersal of *Olesicampe* (*Holocremnus*) Sp. Nr. *nematorum* (Hymenoptera: Ichneumonidae), a parasite of the larch sawfly recently established in Manitoba. Can. Entomol. 99: 312-321.

Murdoch, W.W. 1989. 1. The relevance of pest-enemy models to biological control. Pages 1-24 *in* M. Mackauer, L.E. Ehler, J. Roland (Eds.), Critical issues in biological control. Intercept, Andover, Hants.

Nealis, V.G.; Wallace, D.R. Biological control of forest pests by insect parasitoids. Pages 15-22 *in* A.S. McClay (Ed.), Proceedings of the Workshop on Biological Control of Pests in Canada, October 11-12, 1990, Calgary, Alberta. Alta. Environ. Cent., Vegreville, Alta. AECV91-P1.

Norlund, D.A. 1984. Biological control with entomophagous insects. J. Georgia Entomol. Soc. 2nd Suppl. 19: 14-27.

Palmer, W.A. 1989. Recent legislation in Australia affecting biocontrol. International Organization for Biological Control, Nearctic Regional Section, Newsletter No. 30: 5-9.

Pimentel, D. 1963. Introducing parasites and predators to control native pests. Can. Entomol. 95: 785-792.

Prebble, M.L. 1956. Forest entomology. Pages 350-359 in R. Glen (Compiler). Entomology in Canada up to 1956. A review of developments and accomplishments. Can. Entomol. 88.

Pschorn-Walcher, H. 1977. Biological control of forest insects. Ann. Rev. Entomol. 22: 1-22.

Pschorn-Walcher, H; Eichorn, O. 1963. Investigations on the ecology and natural control of the larch sawfly (*Pristiphora erichsonii* (Htg.)) (Hym., Tenthredinidae) in central Europe. I. Abundance, life history and ecology of *P. erichsonii* and other sawflies on larch. Commonw. Inst. Biol. Cont. Tech. Bull. 3: 51-82.

Pschorn-Walcher, H.; Zinnert, K.D. 1971. Investigations on the ecology and natural control of the larch sawfly (*Pristiphora erichsonii* (Htg.)) (Hym., Tenthredinidae) in central Europe. II. Natural enemies: their biology and ecology and their role as mortality factors in *P. erichsonii*. Commonw. Inst. Biol. Cont. Tech. Bull. 14: 1-50.

Quednau, F.W. 1990. Introduction, permanent establishment, and dispersal in eastern Canada of *Olesccampe geniculatae*. Quednau and Lim (Hymenoptera: Ichneumonidae), an important biological control agent of the mountain ash sawfly, *Pristiphora geniculata* (Hartig) (Hymenoptera: Tenthredinidae). Can. Entomol. 122: 921-934.

Roland, J. 1988. Decline of winter moth populations in North America: direct versus indirect effect of introduced parasites. J. Animal Ecol. 57: 523-531.

Smith, S.G. 1955. Cytogenetics of obligatory parthenogenesis. Pages 131-135 *in* Approaches to the study of forest insects with special reference to the larch sawfly, *Pristiphora erichsonii* (Htg.). Can. Entomol. 87.

Syme, P.D.; Nystrom, K.L. 1988. Insects and mites associated with Ontario trees. Classification, common names, main hosts and importance. Environ. Can., Can. For. Serv., Great Lakes For. Res. Cent. Inf. Rep. O-X-392, 131 p.

Sweetman, H.L. 1936. The biological control of insects. Comstock Publ. Co., Ithaca, N.Y. 461 p.

Sweetman, H.L. 1958. The principles of biological control. Interrelation of hosts and pests and utilization in regulation of animal and plant populations. Wm. C. Brown Co., Dubuque, Iowa. 560 p.

Thompson, W.R. 1936. The work of the Farnham House Laboratory. Pages 5-13 *in* Anonymous. Conference on biological methods of controlling insect pests, Belleville, Ont., June 25 and 26, 1936. Dept. Agric., Entomological Br., Ottawa, Ont.

Tisdell, C.A. 1990. 15. Economic impact of biological control of weeds and insects. Pages 301-316 *in* M. Mackauer; L.E.

Ehler; J. Roland (Eds.), Critical issues in biological control. Intercept, Andover, Hants.

Turnbull, A.L.; Chant, D.A. 1961. The practice and theory of biological control of insects in Canada. Can. J. Zool. 39: 697-753.

Turnock, W.J. 1972. Geographical and historical variability in population patterns and life systems of the larch sawfly (Hymenoptera: Tenthredinidae). Can. Entomol. 104: 1883-1900.

Turnock, W.J.; Muldrew, J.A. 1971. Part III. Forest insects. 45. *Pristiphora erichsonii* (Hartig), larch sawfly (Hymenoptera: Tenthredinidae). Pages 175-194 *in* P.S. Corbet (Ed.), Biological control programmes against insects and weeds in Canada 1959-1968. Commonw. Agric. Bureaux, Slough, Eng. Commonw. Inst. Biol. Cont. Tech. Commun. No. 4.

van Lenteren, J.C. 1980. Evaluation of control capabilities of natural enemies: does art have to become science? Netherlands J. Zool. 30: 369-381.

Wagge, J.K. 1991. Biological control: the old and the new. Pages 105-112 *in* A.S. McClay (Ed.), Proceedings of the Workshop on Biological Control of Pests in Canada, October 11-12, 1990, Calgary, Alberta. Alta. Environ. Cent., Vegreville, Alta. AECV91-P1.

Wallace, D.R. 1990. Forest entomology or entomology in the forest? Canadian research and development. For. Chron. 66: 120-125.

Warren, G.L. 1970. Introduction of the masked shrew to improve control of forest insects in Newfoundland. Proc. Tall Timbers Conf. on Ecological Animal Control by Habitat Management, February 26-28, 1970. pp. 185-202.

Wong, H.R. 1974. The identification and origin of the strains of the larch sawfly, *Pristiphora erichsonii* (Hymenoptera: Tenthredinidae), in North America. Can. Entomol. 106: 1121-1131.

Chapter 42

Inundative Releases

D.R. WALLACE AND S.M. SMITH

Introduction

Biological control, the planned use of beneficial organisms for pest management, arose naturally from the observations and actions of farmers and naturalists around the world as an integral part of the cultural history of the respective region. Many different approaches to biological control have been used depending on the local conditions and requirements. DeBach (1964) consolidated much of the earlier material, especially relating to the development and practice of biological control in America, and recognized several general approaches, including augmentative releases, both inoculative and inundative.

Flanders (1930) coined the name "inundative" for those methods that rely solely or largely on the control action of the individuals that have been released. The effect is immediate and "non-sustaining," and in the case of parasitoids and predators, does not depend on the released organisms proliferating in the environment. This terminology has the connotation of releasing large numbers of natural enemies, but as Stinner (1977) suggested, in practice the distinction between inundative and inoculative releases is often uncertain. For example, a single release of a relatively small number of individuals into a low-level population of a pest gradation may effect immediate control, while in another instance an inoculative release may involve very large numbers of the natural enemy.

Ridgway, King, and Carrillo (1977) in Ridgway and Vinson (1977) suggested the more general term "augmentation" to describe the artificial increase of natural enemies. They point out that many activities, including "inoculative releases, seasonal colonization, supplemental releases, strategic releases, programmed releases, or inundative releases" can be included in this definition, and that the use of supplemental foods and behavior-altering chemicals may also be considered under augmentation. When only naturally occurring parasitoids are affected, the latter two categories seem more aptly called conservation techniques because they involve manipulation of the environment to favor the natural control agents rather than direct supplementation of the parasitoids or predators. A comprehensive system of clearly defined categories may not be possible; therefore, for the purpose of this paper, we will use Nordlund's (1984) classification in which he defines three types of releases:

1. Inoculative release—liberation of an imported exotic parasitoid or predator, often in small numbers, in a country where it is not native (often to reassociate the natural enemies of an introduced pest).

2. Augmentative release—liberation of a parasitoid or a predator in a country where it is present (indigenous or introduced before) to reinforce its potential; these releases may be periodical and eventually involve large numbers.

3. Inundative release—liberation of a parasitoid or predator, indigenous or exotic, generally in large numbers, to protect limited areas; it is an actual biological treatment.

We will deal with the last category in the review. An historical perspective of inundative releases worldwide will be presented with emphasis placed on studies conducted in Canada or of particular relevance to Canada. The next section will address the primary agent used in recent work in Canada, the egg parasitoid *Trichogramma minutum*, outlining the studies conducted in Quebec, Maine, and Ontario involving the spruce budworm, *Choristoneura fumiferana*. The final section deals with the prospects for using inundative releases for managing Canadian forest insect pests.

Historical Perspective

International Developments

Although the concept of inundative release is very old (Flanders [1930] cites Enock for noting in 1895 the potential for the mass propagation of *Trichogramma*), interest in its application has been periodic. In the early part of the 1900s into the 1930s, there was much activity in Europe and the United States in studying both the fundamental aspects of *Trichogramma* biology and the practical use of these parasitoids in inundative releases. Field results were unpredictable, and in North America and Western Europe the development of cheap and very effective synthetic chemical insecticides, beginning in the late 1930s, and their widespread utilization after 1945 resulted in a general loss of interest in biological control. The recognition that chemical pesticides have serious local and global environmental risks came into prominence in the 1960s and has led to more emphasis on biological control. In addition, the rise of pest resistance to chemical insecticides coupled with the uncertainty and extremely high costs of developing new materials, further shifted the interest to biological methods. Finally, more open scientific exchange with the Soviet Union and the People's Republic of China (henceforth referred to as China) brought to the forefront their advances in the application of biological control (Coulson 1981; McFadden et al. 1981; Cock 1985), in particular the concept of inundation.

Worldwide, over 28 species of insects and mites are used in inundative releases to control greenhouse, agricultural, and forestry pests (King et al. 1985). In the United States alone, over 8 species of predators and 20 species of parasitoids are reared and released. Although 9 of the 20 parasitoid species are used to control orchard pests, to date none has been used operationally against forest insects. There are currently over 40 commercial producers of biological agents for use in

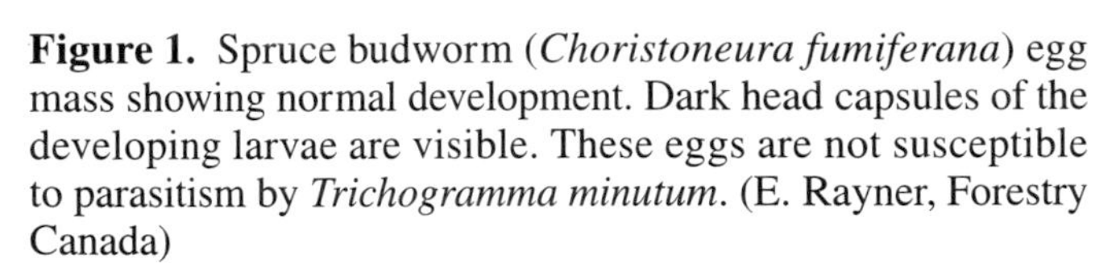

Figure 1. Spruce budworm (*Choristoneura fumiferana*) egg mass showing normal development. Dark head capsules of the developing larvae are visible. These eggs are not susceptible to parasitism by *Trichogramma minutum*. (E. Rayner, Forestry Canada)

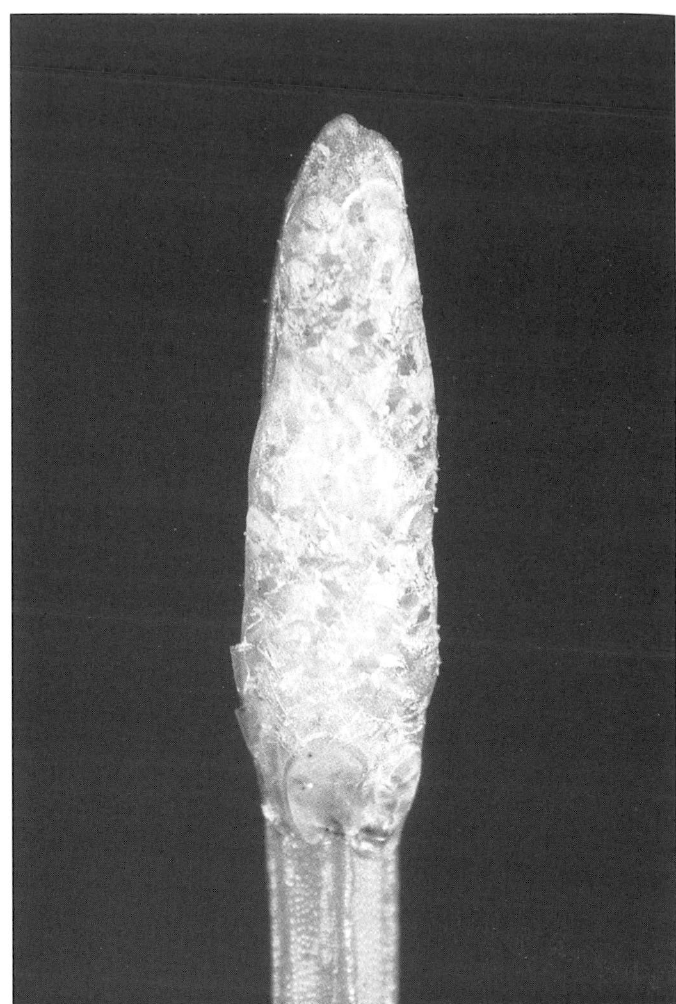

Figure 2. Budworm egg mass from which larvae have hatched. A few yellowish, infertile eggs remain. (E. Rayner, Forestry Canada)

inundative releases throughout the world. Over 43 species of natural enemies are propagated for use against more than 25 target pests in 9 arthropod orders (Anon. 1989).

The literature suggests that there are three major groups of parasitoids that have been used in large-scale inundative releases: tachinid flies; and two groups of egg parasitoids, scelionid wasps and trichogrammatid wasps. Grenier (1988) reported that one species of tachinid fly had been released in Florida to control *Diatraea saccharalis* on sugarcane (*Saccharum officinarum*), between 1973 and 1976. Effectiveness was limited when over 1500 flies per hectare were released on 80 ha in 1975 and 1400 flies per hectare on 48 ha in 1976. The lack of success was considered to have resulted from dispersal of the flies and local climatic/geographical conditions.

Scelionid wasps have also been used worldwide in inundative releases against the egg stage of insect pests. Orr (1988) reported releases made in several countries, including Japan, Iran, India, and the Soviet Union, and in Africa. The majority of these releases were against agricultural pests with only one record of using *Telenomus terebrans* against the forest defoliator *Dendrolimus sibiricus* in the Soviet Union between 1950 and 1960.

Several reviews are available that discuss the use of trichogrammatid wasps in inundative releases against the egg stage of pest insects (Stinner 1977; Beglyarov and Smetnik 1977; Huffaker 1977; Ridgway et al. 1977). Many aspects of research on *Trichogramma* and its use are included in the symposium volume "*Trichogramma and other egg parasites*" edited by Voegele et al. (1988). Species of *Trichogramma* are reared commercially in over 12 countries and used against pests on over 11 different crops and for research purposes in at least 21 countries as shown in Ridgway and Morrison (1985). They state that the principal target pests are on crops such as corn (seven countries), cotton (five countries), sugar-

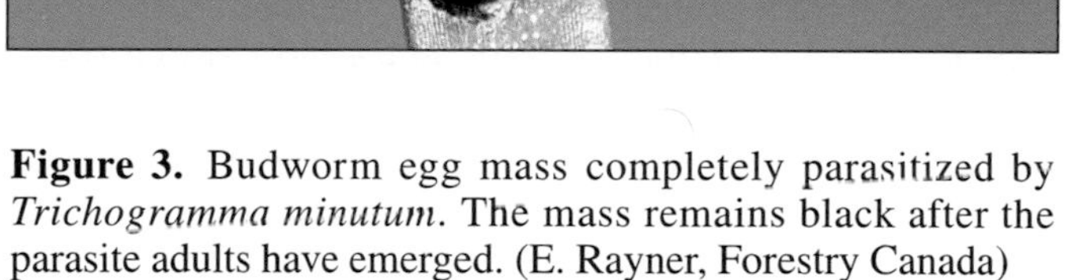

Figure 3. Budworm egg mass completely parasitized by *Trichogramma minutum*. The mass remains black after the parasite adults have emerged. (E. Rayner, Forestry Canada)

Figure 4. *Trichogramma minutum* female on a laboratory-reared budworm egg mass. (University of Guelph)

cane (four countries), sugarbeet (two countries), cabbage (two countries), and rice. Aside from Europe and the USA, *Trichogramma* have been used extensively in the Soviet Union and China where the socioeconomic factors of labor availability and cost, as well as the high cost of chemical pesticides, have favored biological control methods. Gusev and Lebedev (1988) reported that in 1985 the Soviet Union had extensive facilities devoted to the production of these wasps for use on more than 15 million ha of crops. Although *Trichogramma* have been used primarily against agricultural pests in the Soviet Union and China, some species have been released inundatively against forest pests such as *Dendrolimus* spp. attacking pines (Voronin and Grinberg 1981; Cock 1985). Sun and Yu (1988) reported that since 1970, *T. dendrolimi* had been used for the control of the pine-feeding moths, *Dendrolimus* spp. and *Clostera anachoreta*, on 1 million ha in China. They also list several other forest pests against which *Trichogramma* have been used. The usefulness of the method, especially in lower density populations, is indicated and they suggest problems requiring research. Other lepidopterous pests, including the European silver-fir budworm, *Choristoneura murinana*, the cone and seed feeder, *Dioryctria ebeli*, the European pine shoot moth, *Rhyacionia buoliana*, and the teak skeletonizer, *Pyrausta machaeralis*, have been treated with *Trichogramma* (Belmont and Habeck 1983; Tsankov et al. 1980; Patil and Thontadarya 1984).

In other forestry applications, Blumenthal et al. (1981) reviewed the results of a small number of tests using inundative releases of parasitoids against the gypsy moth, *Lymantria dispar*, in the United States and one in Spain. Several species of parasitoids were involved: *Anastatus disparis*, *Apanteles melanoscelus*, *A. liparidis*, *A. porthetriae*, and *Brachymeria intermedia*. They found that the releases were generally unsuccessful in reducing post-release egg mass density. More recent developments in the use of augmentative parasitoid releases in integrated programs against the gypsy moth will be covered in a later section.

The Canadian Record

Detailed compilations of biological control work in Canada (McLeod et al. 1962; Corbet and Prentice 1971; Kelleher and Hulme 1984) show that up to 1980, inundative releases of parasitoids and predators were largely untried here in the control of both agricultural and forest insect pests, excluding the application of parasitoids and predators to manage greenhouse pests. Kelleher and Hulme (1984) calls the latter an integration method. Van Lenteren and Woets (1988) apply the term "seasonal inoculative" to the principal release strategy used in greenhouses and also state that the inundative method is used for minor pests. Biological and integrated greenhouse pest control is being applied successfully in Canada (Elliott et al. 1987). There are currently two rearing facilities, one in Ontario and one in British Columbia, that produce natural enemies of greenhouse pests commercially (Anon. 1989).

In other aspects of agriculture, McLeod et al. (1962) concluded that "the mass production of biotic agents is of little value except under special circumstances." The special circumstances that he gave did include the saturation of a pest population with biotic agents if there is no satisfactory chemical control to prevent economic loss. He had stated already that the release of 30 million individuals of several species of *Trichogramma* against the oriental fruit moth, *Grapholitha molesta*, in the Niagara Peninsula of Ontario from 1928 to 1933 had had no evident effect on the fruit moth population. Laing (1987) pointed out that this evaluation rests on an erroneous assumption about the purpose and mechanism of inundative releases. Recently, there has been a renewed interest in *Trichogramma* in Ontario apple orchards (Yu et al. 1984; Hagley 1986) and its interaction with pesticides (Hagley and Laing 1989). *Trichogramma* material produced at The University of Guelph has been used in release trials against the European corn borer, *Ostrinia nubilalis*, in Ontario (Laing and Eden 1990).

In forestry, there is little, if any, early work that relates to the inundative approach (Hulme 1982). Mention should be made of the massive and more recently considered controversial introduction of the parasitoid *Dahlbominus fuscipennis* against the European spruce sawfly, *Gilpinia hercyniae*. Beginning in 1934 and continuing until 1948 (but principally in the late 1930s and early 1940s), 882 million of this species were used against the European spruce sawfly and another nearly 5 million against other sawfly species. Most of this material came from a large-scale propagation program at the Dominion Parasite Laboratory in Belleville, Ontario, but many millions were obtained from European collections. McGugan and Coppel (1962) use the term "saturation releases" in connection with *D. fuscipennis*, and Baird (1937) suggested that "to secure quick results in heavily infested areas, liberations in units of 10,000 should be made at intervals of a quarter mile or less." Although this sounds like an inundative approach, only inoculation was intended and there are no analyses to show local, direct suppression of the sawfly in association with these massive releases. Reeks (1953) discussed the establishment, distribution, and effectiveness of *D. fuscipennis* in the Maritime provinces. Although it became established and achieved a high level of parasitism in outbreak sawfly populations, it is not one of the biotic agents credited with long-term control of the European spruce sawfly (Bird and Elgee 1957; Neilson and Morris 1964).

A detailed analysis of all the records on parasitoid propagation and introduction over the very active period into the 1950s undoubtedly would identify situations and agents that would suggest the suitability of inundative techniques. Such a study is beyond the scope of this review. At the time, Canadian biological control work was focused exclusively on inoculative methods.

A Canadian Forestry Mission to China (Anon. 1975) reported on the emphasis on biological control in forestry and on the heavy reliance on the use of inundative releases of *Trichogramma* spp. Two other Canadian forestry delegations, one to the Soviet Union in 1977 (Angus et al. 1979) and the other to China in 1984 (Ennis et al. 1984) dealt with biological control specifically. These contacts raised Canadian interest in inundative parasitoid releases in general and in *Trichogramma* in particular. In addition to the work on the spruce budworm that will be covered in the next section of this paper, there have been recent studies on the possibility of using *Trichogramma* against the Douglas-fir cone moth *Barbara colfaxiana* in British Columbia (Hulme and Miller 1988) and the spruce bud moth, *Zeiraphera canadensis*, in New Brunswick (Laing and Eden 1990).

Inundative Releases with *Trichogramma* and the Spruce Budworm

There have been several analyses of natural parasitism of spruce budworm eggs by *Trichogramma*. The most prominent feature of these studies is the great variability of the level of parasitism by *T. minutum* (Daviault 1950; Miller 1953; Thomas 1966; Jaynes and Drooz 1952; Wilkes et al. 1948; McGugan and Blais 1959; Anderson 1976) in both western and eastern North America. Thomas (1966) did not find any relationship between parasitism level and budworm host density, but Kemp and Simmons (1978) found that higher parasitism occurred in forest stands with increased diversity of non-budworm host tree species, possibly related to more alternate hosts being present. More recently, Meating et al. (1985) found a geographical trend in parasitism rate in Ontario that may result from this host diversity effect. Miller (1953, 1963) and Thomas (1966) concluded that *T. minutum* is not generally an important natural control agent and that its influence on the spruce budworm population tends to be small. The main reason identified is that the parasite needs alternate hosts to maintain itself and to multiply. The usefulness of *T. minutum* in biological control seems limited to inundation.

Quebec

In 1970, Quednau (pers. comm.) initiated a study at the Laurentian Forestry Centre of Forestry Canada to apply inundative releases of *Trichogramma minutum* to the control of the spruce budworm. Earlier, Quednau had published several papers on the biosystematics and ecology of *Trichogramma* species (Flanders and Quednau 1960; Quednau 1960) and

5

7

8

6

Figures 5–8

Figure 5. Rearing *T. minutum* on *Sitotroga cerealella* at the Biological Control Laboratory, University of Guelph. Each vertical crib holds 10 kg of wheat as food for the grain moth. Grain moth adults from these cribs lay eggs that are then parasitized by *T. minutum*. (University of Guelph) **Figure 6.** "Window-box" parasitization unit used to obtain uniform, high level of parasitization of the grain moth eggs by *T. minutum* by controlling the movement of the parasites with light. (University of Guelph) **Figure 7.** One million parasitized *Sitotroga* eggs being prepared for application. (University of Guelph) **Figure 8.** Aerial view of one of the test plots in Rogers Township, northwest of Hearst, Ontario. Note the gray, defoliated balsam fir trees. Some of the deposit sampling stations (white), a pheromone trap for budworm male adults, and a red guidance balloon are shown. (G. Howse, Forestry Canada)

was therefore well prepared to carry out this investigation. He worked on the budworm–*Trichogramma* host–parasitoid relationship and began work on the propagation of a local strain of *T. minutum* from the spruce budworm, using *Anagasta kuehniella* (=*Ephestia kuehniella*) as the factitious host. He started large-cage experiments to determine release rates of *Trichogramma* that would be required at various budworm densities. The work at the Laurentian Centre ran from 1970 through 1973. Quednau planned to do field trials using 1 million *Trichogramma* per acre in 1974. The work was stopped because Quednau experienced health problems associated with the handling and propagation of the parasitoids, and his plan to use *Sitotroga cerealella* for mass rearing was not possible because the Angoumois grain moth was not considered established in Canada and importation was prohibited (Quednau, pers. comm.). The only public record of this work was a paper presented at the Northeastern Forest Insect Work Conference, University of Maine, Orono, Maine, April 4, 1973.

Maine

From 1978 to 1981, a research program on the use of *Trichogramma minutum* against the spruce budworm was carried out in Maine by the University of Maine and the USDA Forest Service, Northeastern Forest Experiment Station, Orono, Maine, as part of the Canada–United States (CANUSA) Spruce Budworms Program. Houseweart (1985) summarizes the findings of this work and provides references to the relevant publications. Laboratory studies on mass rearing and parasite quality compared the characteristics of *T. minutum* reared on spruce budworm with adults from *Sitotroga cerealella*, a suitable factitious host for main propagation. Parasitoids from budworm were superior and passage through budworm for at least one generation before release was suggested, although this would not likely be practical. It was also shown that progeny production by *T. minutum* is concentrated in the first 2 days of oviposition and that the budworm eggs are more susceptible to parasitism for a few days just after deposition. These factors favor multiple, closely timed releases. Field studies in 1979 and 1980 showed that budworm egg deposition in Maine occurred over a period of almost 1 month, and the deposition curve approximated a normal distribution around a central peak. Using laboratory developmental data for the parasitoid and field temperatures for Maine, it was calculated that budworm eggs parasitized in the early part of the deposition period would produce adults in time to attack budworm eggs deposited late.

Houseweart and his associates found no evidence of *T. minutum* overwintering in budworm eggs, but in a study of alternate hosts for the parasitoid did find it overwintering in the forest tent caterpillar, *Malacosoma disstria*. Houseweart et al. (1984) do not provide convincing evidence, however, that the tent caterpillar is the principal overwintering host. They did discover a new species of *Trichogramma*, sp. nr. *nubilale* (USDA Insect Identification and Beneficial Insect Introduction Institute Culture No. 630-22-X-1), on the budworm, the forest tent caterpillar, and several other lepidopterous hosts. This species is characterized by a long ovipositor that could be advantageous in attacking multiple-layered egg masses and protected or hidden eggs.

Several point-source and broadcast releases were made against epidemic spruce budworm populations in Maine. Local *T. minutum* proved superior to a California commercial strain and an aerial, broadcast release in 1981 gave significantly increased parasitism over control plots, but did not suppress the very high level population.

Ontario

The encouraging results from the Maine trials against the spruce budworm and the emphasis elsewhere internationally on the use of inundative releases for agriculture, and to a lesser extent for forestry, led to the initiation of a major project to investigate more fully the use of *T. minutum* against the spruce budworm in 1981–1982. The program was developed by R.J. Carrow of the Ontario Ministry of Natural Resources in cooperation with the Biological Control Laboratory, University of Guelph, the Faculty of Forestry, University of Toronto, and the Great Lakes Forestry Centre of Forestry Canada. There were three primary objectives: (1) to develop the technology for mass production of *Trichogramma* in Canada; (2) to develop the technology for handling and releasing the parasitoid in the forest environment; and (3) to determine whether inundative releases of *Trichogramma* could significantly increase budworm egg parasitism and decrease subsequent larval populations.

The mass production objective became the responsibility of the University of Guelph group, handling and releasing technology was addressed by the Pest Control Section of the Ontario Ministry of Natural Resources, and the ground operations were a joint undertaking of the University of Toronto and Forestry Canada with the assistance of the Ontario Ministry. The trials took place in the Hearst area of northern Ontario each year from 1982 to 1986, inclusive. Concurrently with the earlier part of the field work, additional research on aspects of parasitoid strainal characteristics and identification were under way at the University of Toronto (Smith and Hubbes 1986a, 1986b, 1986c).

The full account of the "Ontario Project" can be found in Smith et al. (1990) and Figures 1 to 16 present a brief pictorial account of the work. The primary goals were all met. Methodology for production, handling, application, and field assessment was developed and tested. Tests over the 5-year period permitted the construction of a simple dose response curve (Fig. 17). The relationship indicates that parasitism levels of 60% can be achieved with a double release of 4 million females per hectare, and that over 80% parasitism will result from increasing the application rate to 10 million females each release. Higher rates produced little parasitism increase. Further, it was shown that larval populations the following season were reduced proportionally to egg parasitism (Table 1).

The *Trichogramma* production at Guelph from 1983 to 1985 totalled 212.5 million females and supported several smaller studies in addition to the spruce budworm research already described.

Prospects for Inundative Releases in Forestry

Conceptually, inundative releases are attractive, because it is recognized that some natural enemies can become abun-

Table 1. Parasitism of natural spruce budworm eggs and reductions in larval populations on plots receiving various rates of *Trichogramma minutum* near Hearst, Ontario, from 1982 to 1985, inclusive.[a]

Year	Treatment plot	Plot	Parasite source	Parasitoid release: Date	Parasitoid release: Total (10^6/ha)[d]	No. SBW[b] eggs per 45-cm branch[c]	Viable egg parasitized (%)	No. SBW[b] larvae per 45-cm branch	Population reduction (%)[e]	Control (%)[f]	Expected defoliation (%)[g]
1982	Control	C1	-	-	-	240.5a[i]	0.3	-	-	-	-
	Release	1	BCL[h] & Biogenesis	14 July	0.6 (1)	228.6a	1.0	-	-	-	-
1983[j]	Control	C1	-	-	-	247.7a	0.1	22.4a[i]	91.0	-	80
	Release	1	Rincon Vitova	14 July	8.8 (1)	168.2b	15.9	7.6b	95.5	50	45
1984	Control	C2	-	-	-	78.0a	1.1	34.9a	55.3	-	94
	Release	2	BCL	10 & 16 July	22.7 (2)	75.4a	79.5	6.1b	91.9	82	41
		3	BCL+ Bt	10 & 16 July	22.7 (2)	80.5a	83.4	2.4c	97.0	93	20
		4	Rincon Vitova	10 & 16 July	30.9 (2)	12.0b	14.1	-	-	-	-
1985	Control	C3 & C4	-	-	-	100.7a	0.2	11.9a	88.2	-	60
	Release	5 & 6	BCL	9 & 19 July	4.2 (2)	94.9a	15.6	12.0a	87.4	0	60
		7 & 8	BCL	9 & 19 July	8.4 (2)	119.7a	22.2	8.2b	93.2	42	47
		9	BCL	9 & 19 July	16.9 (2)	87.6a	30.4	5.7b	93.5	45	40

[a] Source: Smith et al. 1990.
[b] Spruce budworm.
[c] Based on unpublished data from the site, includes conversion for a mean of 14.6 eggs in each SBW egg mass (Smith 1985).
[d] Total number of female *T. minutum* released; number of releases is indicated in parentheses.
[e] Represents the change in SBW density from eggs per 45-cm branch to larvae per 45-cm branch within the same generation.
[f] Percent control = $1 - \frac{\text{post-release density in treatment}}{\text{pre-release density in treatment}} \times \frac{\text{pre-release density in control}}{\text{post-release density in control}} \times 100$ from Fleming and Retnakaran (1985).
[g] Predicted level of defoliation based on 130 ± 23 buds per branch (field data) and the relationship between the number of fourth-instar spruce budworm larvae per bud per 45-cm branch described by Dorais and Kettela (1982).
[h] Biocontrol Laboratory, University of Guelph.
[i] Means followed by the same letter within each year and column are not significantly different at the $P \leq 0.05$ level (Duncan's New Multiple 1955 Range Test).
[j] Data provided by the Forest Insect and Disease Survey, Forestry Canada, Sault Ste. Marie, Ont.

dant enough to prevent a pest from causing unacceptable damage, and that these same insects can be reared easily with simple materials. Over the years, the majority of the natural enemies chosen for such work have been native species and therefore have posed minimal risk when released back into their natural environments. In practice this relatively straightforward approach to biological control is far from simple, and the development of models of increasing complexity (Knipling and McGuire 1968; Barclay et al. 1985; Smith and You 1990) can assist greatly in the planning of proposed inundative releases and the prediction of the effect on the pest. Ridgway and Vinson (1977) show how augmentation (inundation) is particularly suited to the severe disturbance and ecosystem simplicity associated with modern agriculture, conditions that work against classical or inoculative biological control. Forest renewal and management may also develop in an analogous manner with smaller areas of intensively treated tree monocultures, and thus present situations more amenable to inundation than the extensive "natural forests."

The potential use of inundative releases in forestry, either with *Trichogramma* or other predators or parasitoids, will be constrained by several factors. Primarily, the mass production of sufficient numbers of natural enemy at a reasonable cost will determine whether this approach will ever be developed beyond the research stage. Production of the large numbers that may be needed to treat extensive forest pest outbreaks may be prohibitive. The mass production of any biological

9

11

10

Figures 9–11

Figure 9. Metering hopper device mounted inside the helicopter cockpit. A tube from this apparatus carries the parasitized eggs to the "slinger" unit mounted under the helicopter. (D. Wallace, Forestry Canada) **Figure 10.** Slinger unit mounted under the helicopter. This is a modified Brohm-style centrifugal dispenser initially developed for the distribution of jack pine seed. (D. Wallace, Forestry Canada) **Figure 11.** Helicopter making an application in the early evening when low wind speeds were optimum for preventing unwanted dispersal of the parasitized material. A guidance balloon (*lower right*) and a deposit monitoring station are visible. (D. Wallace, Forestry Canada)

12

14

13

Figures 12–14

Figure 12. Stand and platform used to raise deposit monitoring cards into openings in the forest canopy. (D. Wallace, Forestry Canada) **Figure 13.** Sticky monitoring card, 25 × 25 cm. These cards were positioned just before the application and removed immediately following. The exposed cards were covered with clear plastic which allowed storage and counting. (D. Wallace, Forestry Canada) **Figure 14.** Funnel collecting trap used to obtain material from the application for quality testing such as parasite emergence, parasite development times, sex ratio, and ability to attack budworm eggs. (D. Wallace, Forestry Canada)

Figure 15. Rope system used to position sentinel budworm egg clusters at different levels in the branch tip area of budworm host trees. The diagonal loops ran between a "pig-tail" stuck in the ground and a ring in the horizontal rope running from tree to tree. These loops allowed the rapid changing of the sentinel clusters. (D. Wallace, Forestry Canada)

Figure 16. Small twig of balsam fir bearing a sentinel budworm egg cluster. The twig is attached to one of the diagonal loops. The clusters were changed at 3-day intervals to ensure that the budworm eggs were fresh and susceptible to *Trichogramma* attack. (D. Wallace, Forestry Canada)

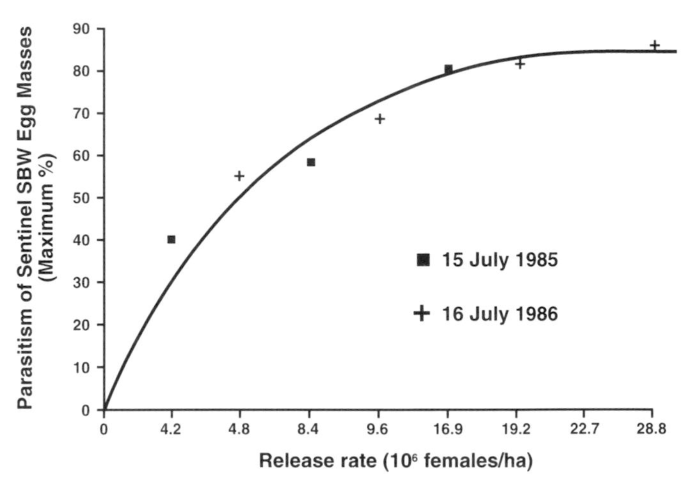

Figure 17. Effect of release rate of *Trichogramma minutum* on the maximum parasitism level achieved on sentinel egg masses attached to balsam fir and white spruce trees near Hearst, Ontario, in 1985 and 1986 (Smith et al. 1990).

agent for field use in Canada will be restricted by the biological capability for propagating or storing the large numbers for use during the few summer months when susceptible pest stages are present. It will also depend on how much the production can be automated on a large scale, thereby reducing the cost per unit of product. Although a limited host range is desirable from the perspective of impact on non-target organisms, the commercialization of species-specific agents is much less attractive to the investor and more difficult for the applicator.

The development of inundative releases, both in agriculture and forestry, will also depend on research to assure proper and effective application strategies and techniques. The technology for releasing large numbers of predators or parasitoids must be improved to meet the needs of many different pest situations. As well, a registration process for the application of these biotic agents needs to be established so that acceptance of biological control by the public can be maintained. These regulations must be pragmatic and recognize the complex nature of biological control.

During the 1980s, the Ontario *Trichogramma* project in Canada achieved its main goals. The results of this study, however, suggested the need to develop a full-scale production facility that could produce high quality parasitoids in large numbers at a low cost on a regular and continuous basis. The project also emphasized the need to develop strategies and application techniques that would optimize the use of *Trichogramma* to minimize the dosage required. Finally, the need for more research on parasitoid strain selection, quality control, and the identification of other pest targets was recognized. All of these needs are being addressed by a 5-year program announced by the Ontario Premier's Council Technology Fund in November, 1989. Under this project CIBA-GEIGY Canada will develop a *Trichogramma* production unit in Ontario, and research on commercial production, parasitoid improvement, release methods, and application strategies will be conducted by the University of Toronto, the University of Guelph, and the Ontario Ministry of Natural Resources.

Although the work outlined above deals only with *Trichogramma* to be used against the spruce budworm, it represents a start into the large-scale commercial production of parasitoids for forestry in Canada without which many research avenues are closed. Diversification into the use of other insect biological control agents and a broader range of target pests is at least a possibility. This will allow a balanced development of biological control using an appropriate selection of conservation, inoculative, and inundative techniques within generalized forest management systems. Examples of programs of this type for the gypsy moth in urban- and suburban-forested areas in Maryland and Virginia are discussed by Reardon et al. (1987) and Ticehurst and Finley (1988). The development of these integrated pest management approaches will be aided by the capability of including inundative parasitoid releases.

References

Anderson, J.F. 1976. Egg parasitoids of forest defoliating Lepidoptera. Pages 223-249 *in* J.F. Anderson and H.K. Kaya (Eds.), Perspectives in Forest Entomology. Academic Press. New York, San Francisco, London.

Angus, T.A.; Carrow, J.R.; Mcdonald, D.R. 1979. Report of the visit of the Canadian Delegation on Forest Pest Control to the USSR: 1977. Environ. Can., Can. For. Serv., Ottawa, Ont. 13p. + Appendix I, II.

Anonymous. 1975. Report of the Canadian Forestry Mission to the Peoples Republic of China, October, 1974. Environ. Can., Can. For. Serv., Ottawa, Ont. 46 p.

Anonymous. 1989. Directory of producers of natural enemies of common pests. IPM Practitioner 6(4): 15-17.

Baird, A.B. 1937. Biological control of the spruce sawfly, *Diprion polytomum* Hartig. Pulp & Paper Mag. Canada 38: 311-315.

Barclay, H.J.; Otvos, I.S.; Thomson, A.J. 1985. Models of periodic inundation of parasitoids for pest control. Can. Entomol. 117: 705-716.

Beglyarov, G.A.; Smetnik, A.I. 1977. Seasonal colonization of entomophages in the USSR. Pages 283-328 *in* R.L. Ridgway and S.B. Vinson (Eds.), Biological Control by Augmentation of Natural Eenemies. Plenum Press, New York. 480 p.

Belmont, R.A.; Habeck, D.H. 1983. Parasitoids of *Dioryctria* spp. (*Pyralidae: Lepidoptera*) coneworms in slash pine seed production areas of north Florida, U.S.A. Florida Entomol. 66: 399-407.

Bird, F.T.; Elgee, D.E. 1957. A virus disease and introduced parasites as factors controlling the European spruce sawfly, *Diprion hercyniae* (Htg.), in central New Brunswick. Can. Entomol. 89: 371-378.

Blumenthal, E.M.; Fusco, R.A.; Reardon, R.C. 1981. Parasite augmentation. *In* C.D. Doane and M.J. McManus (Eds.), The gypsy moth: Research toward integrated pest management. Forest Service, Science and Education Agency, Animal and Plant Health Inspection Service. USDA, Washington, D.C. Tech. Bull. 1584: 402-408.

Cock, M.J.W. 1985. The use of parasitoids for augmentative biological control of pests in the People's Republic of China. Commonw. Inst. Biol. Cont., Biocontrol News and Information 6(3): 213-223.

Corbet, P.S.; Prentice, R.M. (Compilers). 1971. Biological control programmes against insects and weeds in Canada 1959-1968. Tech. Comm. No. 4, Commonw. Inst. Biol. Cont., Trinidad. Commonw. Agric. Bureaux, Slough, England. 286 p.

Coulson, J.R. (Ed.). 1981. Proceedings of the Joint American-Soviet Conference on Use of Beneficial Organisms in the Control of Crop Pests, Washington, D.C., U.S.A., August 13-14, 1979. Entomol. Soc. Am., College Park, Md. 62 p.

Daviault, L. 1950. Les parasites de la tordeuse des bourgeons de l'épinette (*Choristoneura fumiferana* Clem.) dans la province de Québec. 31st Report of the Quebec Society for the Protection of Plants. 1948-1949. pp. 41-45.

DeBach, P. (Ed.). 1964. Biological Control of Insect Pests and Weeds. Reinhold Publishing Corp. New York. 844 p.

Dorais, L.; Kettela, E.G. 1982. A review of entomological survey and assessment techniques used in regional spruce budworm, *Choristoneura fumiferana* (Clem.), surveys and in the assessment of operational spray programs. Eastern Spruce Budworm Council Rep. 43 p.

Elliott, D.; Gilkeson, L.A.; Gillespie, D. 1987. The development of greenhouse biological control in western Canadian vegetable greenhouses and plantscapes. Bull. SROP 10(2): 52-56.

Ennis, T.J.; Hulme, M.A.; Otvos, I.S.; Sanders, C.J. 1984. Report on Scientific Mission on Biological Control in Forestry in Peoples Republic of China, 1984. Agric. Can., Can. For. Serv. 18 p. + Appendix I-XIII.

Flanders, S.E. 1930. Mass production of egg parasites of the Genus *Trichogramma*. Hilgardia 4: 465-501.

Flanders, S.E.; Quednau, W. 1960. Taxonomy of the Genus *Trichogramma* (*Hymenoptera, Chalcidoidea, Trichogrammatidae*). Entomophaga 5: 285-294.

Fleming, R.; Retnakaran, A. 1985. Evaluating single treatment data using Abbott's formula with reference to insecticides. J. Econ. Entomol. 78: 1179-1181.

Grenier, S. 1988. Applied biological control with tachinid flies (*Diptera, Tachinidae*): A review. Anz. Schadlingskde, Pflanzenschutz, Umweltschutz 61: 49-56.

Gusev, G.V.; Lebedev, G.I. 1988. Present state of *Trichogramma* application and research. Pages 477-481 in J. Voegele, J. Waage, and J. Van Lenteren (Eds.), *trichogramma* and other egg parasites. Les trichogrammes et autres parasitoïdes oophages. 2nd International Symposium, Guangzhou (Chine), 10-15 Novembre, 1986. Les colloques de l'INRA, No. 43, 644 p.

Hagley, E.A.C. 1986. Occurrence of *Trichogramma* spp. (*Hymenoptera: Trichogrammatidae*) in apple orchards in southern Ontario. Proc. Entomol. Soc. Ont. 17: 79-82.

Hagley, E.A.C.; Laing, J.E. 1989. Effect of pesticides on parasitism of artificially distributed eggs of the codling moth, *Cydia pomonella* (Lepidoptera: Tortricidae) by *Trichogramma* spp. (Hymenoptera: Trichogrammatidae). Proceedings Entomol. Soc. Ont. 120: 25-33.

Houseweart, M.W. 1985. *Trichogramma* vs. the Spruce Budworm. Pages 395-396 *in* C.J. Sanders, R.W. Stark, E.J. Mullins, and J. Murphy (Eds.), Recent advances in spruce budworm research. Proceedings of the CANUSA Spruce Budworms Research Symposium, Bangor, Maine. September 16-20, 1984.

Houseweart, M.W.; Jennings, D.T.; Pease, S.H.; Lawrence, R.K. 1984. Alternate insect hosts and characteristics of forest stands supporting native populations of *Trichogramma minutum* Riley. Cooperative Forestry Research Unit, College of Forest Resources, Maine Agricultural Experiment Station, University of Maine at Orono. CFRU Research Bull 5, 32 p.

Huffaker, C.B. 1977. Augmentation of natural enemies in the People's Republic of China. Pages 329-340 *in* R.L. Ridgway and S.B. Vinson (Eds.), Biological Control by Augmentation of Natural Enemies. Plenum Press, New York. 480 p.

Hulme, M.A. 1982. Biological control in the Canadian Forestry Service. Environ. Can., Can. For. Serv., Ottawa, Ont. Rep. DPC-X-11. 45 p.

Hulme, M.A.; Miller, G.E. 1988. Potential for control of *Barbara colfaxiana* (Kearfott): (Lepidoptera: Olethreutidae) using *Trichogramma* sp. *In* J. Voegele, J. Waage, and J. Van Lenteren (Eds.), *Trichogramma* and other egg parasites. Les trichogrammes et autres parasitoides oophages. 2nd International Symposium, Guangzhou (Chine), 10-15 Novembre, 1986. Les colloques de l'INRA, No. 43: 483-488.

Jaynes, H.A.; Drooz, A.T. 1952. The importance of parasites in the spruce budworm infestations in New York and Maine. J. Econ. Entomol. 45: 1057-1061.

Kelleher, J.S.; Hulme, M.A. (Eds.). 1984. Biological control programmes against insects and weeds in Canada 1969-1980. Commonw. Agric. Bureaux, Slough, England. 410 p.

Kemp, W.P.; Simmons, G.A. 1978. The influence of stand factors on parasitism of spruce budworm eggs by *Trichogramma minutum*. Environ. Entomol. 7: 685-688.

King, E.G.; Bull, D.L.; Bouse, L.F.; Phillips, J.R. 1985. Introduction: Biological control of *Heliothis* spp. in cotton by augmentative releases of *Trichogramma*. Southwest Entomol. Suppl. 8: 1-10.

Knipling, E.F.; McGuire Jr., J.U. 1968. Population models to appraise the limitations and potentialities of *Trichogramma* in managing host insect populations. USDA Tech. Bull. No. 1387 Washington, D.C. 44 p.

Laing, J.E. 1987. Augmentation and inundation Pages 60-63 *in* Biological Control in Canada. *In* Report based on the Biological Control Workshop, Winnipeg, 9-10 October 1986. Res. Br., Agric. Can.

Laing, J.E.; Eden, G.M. 1990. Mass production of *Trichogramma minutum* Riley on factitious host eggs. *In* S.M. Smith, J.R. Carrow, and J.E Laing (Eds.), Inundative release of the egg parasitoid, *Trichogramma minutum* (Hymenoptera: Trichogrammatidae), against forest insect pests such as the spruce budworm, *Choristoneura fumiferana* (Lepidoptera: Tortricidae): The Ontario Project 1982-1986. Mem. Entomol. Soc. Can. No. 153, 87 p.

McFadden, M.W.; Dahlsten, D.L.; Berisford, C.W.; Knight, F.B.; Metterhouse, W.W. 1981. Integrated pest management in China's forests. J. For. 79: 723-726 + photo page.

McGugan, B.M.; Blais, J.R. 1959. Spruce budworm parasite studies in northwestern Ontario. Can. Entomol. 91: 758-783.

McGugan, B.M.; Coppel. H.C. 1962. Part 2. Biological control of forest pests, 1910-1958. *In* J.H. McLeod, B.M. McGugan, and H.C. Coppel (Eds.), A review of the biological control attempts against insects and weeds in Canada. Technical Communication No. 2, Commonw. Inst. of Biological Control, Trinidad. Commonw. Agric. Bureaux, Farnham Royal, Slough, England. 35-216.

McLeod, J.H.; McGugan, B.M.; Coppel, H.C. 1962. A review of the biological control attempts against insects and weeds in Canada. Part 1—Biological control of pests of crops, fruit trees, ornamentals, and weeds in Canada up to 1959, by J.H. McLeod. Part 2—Biological control of forest pests, 1910-1958, by B.M. McGugan and H.C. Coppel. Tech. Comm. No. 2, Commonw. Inst. of Biological Control, Trinidad. Commonw. Agric. Bureaux, Farnham Royal, Slough, England. 216 p.

Meating, J.H.; Howse, G.M.; Syme, P.D.; Smith, K.L. 1985. Spruce budworm parasitoid surveys in Ontario. Pages 105-106 *in* C.J. Sanders, R.W. Stark, E.J. Mullins, and J. Murphy (Eds.), Recent advances in spruce budworm research. Proceedings of the CANUSA Spruce Budworms Research Symposium, Bangor, Maine. September 16-20, 1984.

Miller, C.A. 1953. Parasitism of spruce budworm eggs by *Trichogramma minutum*. Can. Dept. Agric., Sci. Serv., Div. For. Biol., Ottawa, Ont. Bi-mon. Prog. Rep. 9(4): 1.

Miller, C.A. 1963. Parasites of the spruce budworm. *In* R.F. Morris (Ed.), The dynamics of epidemic spruce budworm populations. Mem. Entomol. Soc. Can. 31: 228-244.

Neilson, M.M.; Morris, R.F. 1964. The regulation of European spruce sawfly numbers in the Maritime Provinces of Canada from 1937 to 1963. Can. Entomol. 96: 773-784.

Nordlund, D.A. 1984. Biological control with entomophagous insects. J. Georgia Entomol. Soc. 2nd Suppl. 19: 14-27.

Orr, D.B. 1988. Scelionid wasps as biological control agents: A review. Florida Entomol. 71: 506-528.

Patil, B.V.; Thontadarya, T.S. 1984. Efficiency of egg parasite, *Trichogramma* spp. in parasitizing the eggs of the teak skeletonizer, *Pyrausta machaeralis* Walker (Lepidoptera: Pyralidae). Indian For. 110: 413-418.

Quednau, W. 1960. Über die Identität der Trichogramma-Arten und einiger ihrer Ökotypen (Hymenoptera, Chalcidoidea, Trichogrammatidae). Mitt. Biol. Bundesanst. Berl. Dahlem H. 100: 11-50.

Reardon, R.; McManus, M.; Kolodny-hirsch, D.; Tichenor, R.; Raupp, M.; Schwalbe, C.; Webb, R.; Meckley, P. 1987. Development and implementation of a gypsy moth integrated pest management program. J. Arboric. 13: 209-216.

Reeks, W.A. 1953. The establishment of introduced parasites of the European spruce sawfly (*Diprion hercyniae* Htg.) (Hymenoptera: Diprionidae) in the Maritime Provinces. Can. J. Agric. Sci. 33: 405-429.

Ridgway, R.L.; King, E.G.; Carillo, J.L. 1977. Augmentation of natural enemies for control of plant pests in the Western Hemisphere. Pages 379-416 *in* R.L. Ridgway and S.B. Vinson (Eds.), Biological Control by Augmentation of Natural Enemies. Insect and Mite Control with Parasites and Predators. Plenum Press. New York and London.

Ridgway, R.L.; Morrison, R.K. 1985. Worldwide perspective on practical utilization of *Trichogramma* with special reference to control of Heliothis on cotton. Southwestern Entomol. Suppl. 8: 190-197.

Ridgway, R.L.; Vinson, S.B. (Eds.). 1977. Biological Control by Augmentation of Natural enemies. Insect and Mite Control with Parasites and Predators. Plenum Press. New York and London. 480 p.

Smith, S.M. 1985. Feasibility of using the egg parasitoid, *Trichogramma minutum* Riley, for biological control of the spruce budworm. Ph.D. dissertation, University of Toronto. 221 p.

Smith, S.M.; Carrow, J.R; Laing, J.E. (Eds.). 1990. Inundative release of the egg parasitoid, *Trichogramma minutum* (Hymenoptera: Trichogrammatidae), against forest insect pests such as the spruce budworm, *Choristoneura fumiferana* (Lepidoptera: Tortricidae): the Ontario Project 1982-1986. Mem. Entomol. Soc. Can. No. 153, 87 p.

Smith, S.M.; Hubbes, M. 1986a. Isoenzyme patterns and biology of *Trichogramma minutum* Riley as influenced by rearing temperature and host. Entomol. Exp. Appl. 42: 249-258.

Smith, S.M.; Hubbes, M. 1986b. Strains of *Trichogramma minutum* Riley. I. Biological and biochemical characterization. Z. Angew. Entomol. 101: 223-239.

Smith, S.M.; Hubbes, M. 1986c. Strains of *Trichogramma minutum* Riley. II. Utilization for release against the spruce budworm. Z. Angew. Entomol. 102: 81-93.

Smith, S.M.; Wallace, D.R.; Howse, G.; Meating, J. 1990. 3.5 Suppression of spruce budworm populations by *Trichogramma minutum* Riley, 1982-1986. *In* S.M. Smith, J.R. Carrow, and J.E. Laing (Eds.), Inundative release of the egg parasitoid, *Trichogramma minutum* (Hymenoptera: Trichogrammatidae), against forest insect pests such as the spruce budworm, *Choristoneura fumiferana* (Lepidoptera: Tortricidae): The Ontario Project 1982-1986. Mem. Entomol. Soc. Can. No. 153, 87 p.

Smith, S.M.; You, M. 1990. A life system simulation model for improving inundative releases of the egg parasite, *Trichogramma minutum* (Hymenoptera: Trichogrammatidae) against the spruce budworm (Lepidoptera: Tortricidae). Ecol. Model. 51: 123-142.

Stinner, R.E. 1977. Efficacy of inundative releases. Ann. Rev. Entomol. 22: 515-531.

Sun, X.; Yu, E. 1988. Use of *Trichogramma dendrolimi* in forest pest control in China. Pages 591-596 *in* J. Voegele, J. Waage, and J.Van Lenteren (Eds.), Trichogramma and other egg parasites. Lestrichogrammes et autres parasitoides oophages. 2nd Int. Symp., Guangzhou (Chine), 10-15 Novembre, 1986. Les colloques de l'INRA, No. 43, 644 p.

Thomas, H.A. 1966. Parasitism by *Trichogramma minutum* (Hymenoptera: Trichogrammatidae) in the spruce budworm outbreak in Maine. Ann. Entomol. Soc. Am. 59: 723-725.

Ticehurst, M.; Finley, S. 1988. An urban forest integrated pest management program for gypsy moth: an example. J. Arboric. 14: 172-175.

Tsankov, G.; Chernev, T.; Laktarieva, L.; Bochev, N.; Ficheva, E. 1980. The eastern Rhodope strain of *Trichogramma* isolated from *Rhyacionia buoliana*, and the possibility of using it in biological control. Gorskostop. Nauka 17: 51-56.

Van Lenteren, J.C.; Woets, J. 1988. Biological and integrated pest control in greenhouses. Ann. Rev. Entomol. 33: 239-269.

Voegele, J.; Waage, J.; Van Lenteren, J. (Eds.). 1988. *Trichogramma* and other egg parasites. Les trichogrammes et autres parasitoides oophages. 2nd Int. Symp., Guangzhou (Chine), 10-15 Novembre, 1986. Les colloques de l'INRA, No. 43,644 p.

Voronin, K.E.; Grinberg, A.M. 1981. The current status and prospects of *Trichogramma* utilization in the USSR. Pages 49-51 *in* Coulson, J.R. (Ed.), Proc. Joint Amer.-Soviet Conf. on Use of Beneficial Organisms in the Control of Crop Pests. Entomol. Soc. Am., College Park, Md. 62 p.

Wilkes, A.; Coppel, H.C.; Mathers, W.G. 1948. Notes on the insect parasites of the spruce budworm *Choristoneura fumiferana* (Clem.) in British Columbia. Can. Entomol. 80: 138-155.

Yu, D.S.K.; Hagley, E.A.C.; Laing, J.E. 1984. Biology of *Trichogramma minutum* (Riley) collected from apples in southern Ontario. Environ. Entomol. 13: 1324-1329.

Chapter 43

Role of Native Parasitoids and Predators in Control Programs

V.G. Nealis and M.A. Hulme

Introduction

The role of parasitoids (parasitic insects) and predators in controlling introduced forest insect pests has been discussed in the preceding chapters. Many of these introduced insects are pests in their new environment primarily because the enemies that normally attack them in their native range are not present in the new habitat. Successful introduction of key parasitoids or predators can produce dramatic, long-lasting control. Demonstrating the role of native parasitoids and predators in the natural regulation of insect populations, however, is much more difficult because methods for incorporating their effect on pest populations into pest management programs are not well-developed. Nevertheless, many (e.g. Knight and Heikkenen 1980) consider the encouragement of indigenous natural enemies to be an important aspect of biological control.

The utilization of native parasitoids and predators of Canadian forest pests is an almost unexplored tactic of pest control, despite the fact that the tactic has extensive potential. This chapter is devoted to the development of a better understanding of native natural enemies. This is an indispensable starting point for the development of a sound integrated approach to forest pest management.

Natural Enemies and Forest Pest Management

There are several characteristics of forest pests that make better utilization of indigenous natural enemies a sensible option for pest managers:

1) Native parasitoids and predators of forest pests are abundant. Many of our most damaging forest pests are native insects that already have a diverse and well-developed natural enemy fauna. Although these natural enemies may not maintain pest populations below economic damage levels in all stands at all times, they do regularly inflict substantial mortality on pest populations. Efficient pest management should complement, not hinder, this naturally occurring mortality.

2) Forest pest problems in Canada are typically extensive. The periodic irruption of populations of conifer-feeding budworms over millions of hectares of Canadian forests is an all-too-familiar example. The application of insecticides throughout the affected area thus becomes prohibitively expensive. We are forced to rely on a natural end to the epidemics, hopefully without extensive tree mortality. Moreover, because we are only able to treat a proportion of the infested area, it is important that our spray programs be deployed where they are most badly needed. A knowledge of the expected effects of natural enemies would assist in making these strategic decisions.

3) Many forest pests are physically inaccessible. Bark beetles, tip weevils, and shoot and bud moths are examples of pests that have cryptic habits. This characteristic makes them difficult to reach with conventional controls. The potential for manipulating indigenous natural enemies is good, however, since these already have access to the target pest.

4) Several forest pests damage trees without always killing them. Most pest control operations in Canadian forests are aimed at defoliating insects where pest populations reach very high densities for short periods of time. Because the generation time of trees is long, relative to the duration of these infestations, attacked trees may recover from even severe defoliation. The significance of this resilience of trees for pest management is that pest populations need not always be lowered as drastically as is often required in agriculture. Natural enemies may be able to provide sufficient protection for most years of an infestation.

Impact of Natural Enemies on Populations of Forest Pests

Pest managers have several reasons to be interested in when, where, and how much mortality is inflicted on pest populations by natural enemies. There are, for example, repeated references in Prebble (1975) on the confounding effect of mortality caused by natural enemies on the interpretation of results from chemical spraying. Richmond (1986) concluded that in every chemical spray program carried out against forest defoliators in British Columbia, infestations declined similarly in both the treated and untreated areas. Without reliable estimates of natural mortality, unambiguous assessment of chemical spray efficacy will continue to be difficult.

Prediction of certain patterns of natural mortality, such as consistent increases in parasitism as pest outbreaks persist in an area (Blais 1965; Miller 1966; Witter and Kulman 1979) may help explain the inability of some forecasting schemes to reliably predict damage. Pest density thresholds that prompt control measures should vary with the expected amount of natural mortality. Unless expected mortality due to natural enemies can be included in these forecasts, unnecessary control action or, conversely, the substantial risks of not intervening, will continue to pose a dilemma. Should certain population patterns become reasonably predictable (as in the case of population cycles of short duration such as those of the forest tent caterpillar, *Malacosoma disstria*), then pest control action can be restricted to those few areas where any loss would be intolerable. Elsewhere, the natural course of events can be permitted to occur.

Several experimental methods have been devised for assessing the impact of natural enemies on pest populations (DeBach et al. 1976; Luck et al. 1988). Unfortunately, these methods have been sparsely utilized in forest ecosystems. The approaches are broadly grouped as methods of addition, of exclusion, and of interference. The first method refers to inoculative or inundative releases of natural enemies. This aspect is covered in Chapter 42 of this book. The second method excludes natural enemies from the pest environment, usually with physical barriers such as cages or sticky traps. Comparison with areas where indigenous natural enemies are unrestricted has clearly shown the impact of parasitoids (Price and Tripp 1972), ants (Campbell and Torgerson 1982), and birds (Torgerson and Campbell 1982) upon pest populations. The third method is related to the second; it generally comprises a partial removal of natural enemies from the pest ecosystem, most effectively by the use of broad-spectrum chemical insecticides. In forestry, two inadvertent examples, involving the European spruce sawfly (Neilson et al. 1971) and hemlock scale insects (McClure 1978), have shown that dramatic pest resurgence can follow the elimination of the natural enemy fauna by spray programs.

Apart from direct experimental manipulation, census data for forest insect pests prepared as life tables document the prevalence of natural enemies in populations of forest pests. Significant mortality caused by the combined action of several natural enemies, mostly parasitoids, has been associated with major declines in many pest populations: spruce budworm, *Choristoneura fumiferana* (Blais 1960; Miller 1963); jack pine budworm, *C. pinus* (Dixon 1961; Allen et al. 1969; Nealis 1991); eastern blackheaded budworm, *Acleris variana* (Miller 1966); elm spanworm, *Ennomos subsignaria* (Kaya and Anderson 1974); and forest tent caterpillar (Witter and Kulman 1979). Even with insects such as the Douglas-fir tussock moth, *Orgyia pseudotsugata*, where there is strong evidence that population declines are the direct result of virus epizootics, predators and parasitoids may play a supporting role (Torgersen and Campbell 1978).

In addition to the impact of natural enemies on outbreak populations of forest insect pests, there is general recognition that natural enemies are instrumental in the maintenance of insect populations below damaging levels. Birds may be particularly important in this role (Holmes et al. 1979; Otvos 1979; Crawford et al. 1983). If this is the case, then studies of forest pest populations during low to moderate density periods will be necessary to understand the conditions under which natural control of populations is achieved and to predict when and where natural mortality will not be sufficient to dampen an incipient outbreak.

Enhancement of Populations of Natural Enemies

In the enhancement of natural enemy populations, there is active manipulation of the environment to favor certain beneficial species. One method of enhancement is augmentation of natural enemy populations by release of laboratory-reared predators or parasitoids, a tactic already discussed by Wallace and Smith (in Chapter 42). There are many other ways that the existing beneficial fauna can be enhanced. The number of possibilities is limited only by our knowledge of the biological requisites of natural enemies and by our imagination in exploiting this knowledge.

Food sources have been added to attract and maintain adult parasitoid populations in an area (Syme 1976). Providing nesting sites for bird predators is a commonly used method of encouraging bird populations in Europe and Asia but has not been extensively used in North America (Takekawa et al. 1982). Langelier and Garton (1986) suggest several specific guidelines for increasing bird populations in forest habitats including silvicultural techniques that encourage vertical and horizontal diversity in the forest profile, and species composition and harvesting methods that create open strips and some slash. Many of these practices would also encourage generalist arthropod predators such as solitary wasps (Jennings and Houseweart 1984). The challenge is to harmonize these practices with other forestry activities.

Certain specific chemicals that modify behavior could also be employed. Parasitoids and predators sense a variety of specific chemicals termed kairomones in the environment to locate their hosts and prey. These chemicals could be identified and used to manipulate the behavior of natural enemies in ways that facilitate pest control. The potential of this approach is promising (Lewis et al. 1975) and work is now in progress that combines our knowledge of the behavior of insects to kairomones with measurement of rates of parasitism or predation in the field.

There are also several less direct approaches to enhancing populations of natural enemies. One is to subsidize pest populations when they fall to particularly low levels. At first glance locally increasing pest populations may seem counterproductive, but such a strategy could help to maintain a critical number of natural enemies that could quickly respond to future increases in pest density. Equivalently, pest managers could apply insecticides at a rate that would reduce local pest populations to moderate but not low levels.

Many indigenous natural enemies can attack insects related to the pest insect. The manipulation of alternative host or prey populations to maintain a diverse natural enemy complex may also be considered. This aspect could be addressed through silvicultural means: establishing mixed stands or encouraging a herbaceous layer that supports alternative hosts or prey of natural enemies.

Conservation of Natural Enemies

The conservation of natural enemies is another active approach to their use in managing forest pest populations. Here, we accept that insecticides will continue to have a role in forest pest management. The question becomes how to utilize these insecticides in such a way that they complement, rather than reduce, the effects of natural enemies.

The use of chemical insecticides in forestry can disrupt natural enemy complexes in many ways. Most obviously, they may kill the beneficial fauna. Some chemical insecticides are particularly undesirable because they are toxic to parasitoids (Croft and Brown 1975; Franz et al. 1980) and birds (Peakall and Bart 1983). This detrimental effect of insecticides on the beneficial fauna may be relatively short-lived, or may persist throughout the season (Freitag and Poulter 1970) or for several years after the spray (Price 1972). Besides the direct toxic effects of insecticides on beneficial insects and birds, there may be more subtle secondary effects such as mating or nesting disruption (Peakall and Bart 1983) or the elimination of innocuous, nontarget forest pests that support natural enemy populations between outbreaks of the pest insect (Croft and Brown 1975; Martinat et al. 1988).

The result of these negative effects of chemical insecticides on natural enemies on the control program can range from a reduced effectiveness of the spray due to compensatory relief of predation pressure on the pest population, to the resurgence of pest populations to even higher levels than before the protection program was initiated (Waage et al. 1985). There are many examples of this latter situation from the agricultural pest literature. In forestry, resurgence of the pest species due to the elimination of natural enemies following insecticide application has been documented for European spruce sawfly (Neilson et al. 1971) and for scale insects on hemlock (McClure 1978).

There are several options to ameliorate these negative consequences of spraying. The inclusion of natural enemy abundance in pest population forecasts may, in some insstances, identify areas where sprays will not be necessary due to the presence of relatively high populations of beneficial species. Spruce and jack pine budworms, for example, are attacked by at least two species of parasitoids that overwinter within the budworm. Forecast methods that rely on estimates of the number of overwintering budworms can be easily extended to estimate the rate of parasitism. Where parasitism is high, pest managers may want to reconsider their spray target zones to conserve these natural enemies (Benjamin 1965; Nealis and Lysyk 1988). A related tactic refers to a previous point regarding the general increase in parasitism as pest populations persist in an area. If the history of an infestation is known, it may be possible to conclude that a population collapse is about to occur and that spraying may actually extend the infestation by retarding increases in natural enemy populations.

When a spray program is considered necessary, choosing insecticides that are selectively more toxic to the pest insect than to beneficial species is clearly a desirable strategy. Certain microbial insecticides such as Bt have the advantage of being most toxic to herbivorous insects and relatively harmless to predators and parasitoids (Hassan et al. 1987). In the case of Bt, the action of parasitoids may actually be amplified by the use of Bt due to the differential survival of parasitized individuals (Nealis and van Frankenhuyzen 1990).

Consideration of the timing of insecticide application in the forest is another way natural enemies can be conserved during a spray program. Strategic questions regarding the effect of different timings of insecticide application and their combination with natural control have been addressed theoretically (Barclay 1982; Hassell 1984), but there is a need for experimental investigation of this tactic for each forest pest. Waage et al. (1985) have illustrated the various scenarios. One approach to the problem is the development of phenology models for both the pest and its natural enemies that permit a spray schedule that minimizes the impact on natural enemies by applying the insecticide at a time when the natural enemies are least vulnerable (Lysyk and Nealis 1988).

A knowledge of the biology of natural enemies can also be used to advantage in the case of cultural control practices. Hulme et al. (1986, 1987) exploited the relative cold-hardiness and size of natural enemies of a tip weevil to treat spruce leaders in such a way that natural enemies were conserved. The treated leaders could then be returned, with their indigenous natural enemies, to the infested areas.

The conservation of natural enemies is truly an integrated approach to forest pest management because it incorporates the action of beneficial organisms into existing control options.

Conclusion

Considering all the factors discussed in this section as part of a pest management strategy is clearly a complex task. Such an approach to pest management, however, is essential as we shift our emphasis from short-term control objectives and the protection of current-year foliage, to the development of pest management programs designed to maintain a sustained resource yield. We believe that an appreciation of the importance of natural enemies can contribute significantly to this more comprehensive view of forest pest management.

There is already an extensive body of knowledge on the biology and ecology of native parasitoids and predators which could be incorporated into pest management strategies. In addition to minor modifications in chemical spray programs, such as refraining from spraying areas where natural enemies are abundant or the use of more specific insecticides and modified spray schedules, certain basic tenets of pest control may need to be examined if we are to make most effective use of this knowledge. For example, control tactics previously regarded as ineffective may merit reconsideration when used to support the action of natural enemies. Or, these less effective methods may even become preferred as they reduce pest populations to levels that are economically tolerable but still abundant enough to support a natural enemy fauna that can respond quickly to increases in pest densities.

The application of research in ecology to the management of forest pest populations requires two important components. From research we need more specific information on the population biology and impact of both the pest and its natural enemies. From pest managers we need the willingness and imagination to employ the resultant information in the development of more comprehensive pest management strategies.

References

Allen, D.C.; Knight, F.B.; Foltz, J.L.; Mattson, W.J. 1969. Influence of parasites on two populations of *Choristoneura pinus* (Lepidoptera: Tortricidae) in Michigan. Ann. Entomol. Soc. Am. 62: 1469-1475.

Barclay, H.J. 1982. Models for pest control using predator release, habitat management and pesticide release in combination. J. Appl. Biol. 19: 337-348.

Benjamin, D.M. 1965. Evaluation of outbreak populations of the jack pine budworm, *Choristoneura pinus* Freeman (Lepidoptera). Proc. XIIth Int. Cong. Entomol. 696-697.

Blais, J.R. 1960. Spruce budworm parasite investigations in the lower St. Lawrence and Gaspé regions of Quebec. Can. Entomol. 92: 384-396.

Blais, J.R. 1965. Parasite studies in two residual spruce budworm (*Choristoneura fumiferana* (Clem.) outbreaks in Quebec. Can. Entomol. 97: 129-136.

Campbell, R.W.; Torgersen, T.R. 1982. Some effects of predaceous ants on western spruce budworm pupae in north central Washington. Environ. Entomol. 11: 111-114.

Crawford, H.S.; Titterington, R.W.; Jennings, D.T. 1983. Bird predation and spruce budworm populations. J. Forest. 81: 433-435.

Croft, B.A.; Brown, A.W.A. 1975. Responses of arthropod natural enemies to insecticides. Ann. Rev. Entomol. 20: 285-336.

DeBach, P.; Huffaker, C.B.; MacPhee, A.W. 1976. Evaluation of the impact of natural enemies. Pages 255-285 *in* Huffaker C.B. and P.S. Messenger (Eds.), Theory and Practice of Biological Control. Academic Press, New York.

Dixon, J.C. 1961. The biology and ecology of jack pine budworm in Wisconsin with special reference to insect parasites. Ph.D. thesis, University of Wisconsin. ix + 164 pp.

Franz, J.M.; Bogenshutz, H.; Hassan, S.A.; Huang, P.; Naton, E.; Suter, H.; Viggiani, G. 1980. Results of a joint pesticide test programme by the working group: pesticides and beneficial arthropods. Entomophaga. 25: 231-236.

Freitag, R.; Poulter, F. 1970. The effects of the insecticides Sumithion and phosphamidon on populations of five species of carabid beetles and two species of lycosid spiders in northwestern Ontario. Can. Entomol. 102: 1307-1311.

Hassan, S.A.; Albert, R.; Bigler, F.; Blaisinger, P.; Bogenschutz, H.; Boiler, E.; Brun, J.; Chiverton, P.; Edwards, P.; Englert, W.D.; Huang, P.; Inglesfield, C.; Naton, E.; Domen, P.A.; Overmeer, W.P.J.; Rioeckmann, W.; Samsoe-Petersen, L.; Staubli, A.; Tuset, J.J.; Viggiani, G.; Vanwetswinkel, G. 1987. Results of the third joint pesticide testing programme by the IOBC/WPRS-working group "Pesticides and beneficial organisms." J. Appl. Entomol. 103: 92-107.

Hassell, M.P. 1984. Insecticides in host-parasitoid interactions. Theor. Pop. Biol. 26: 378-386.

Holmes, R.T.; Shultz, J.C.; Northnagle, P. 1979. Bird predation on forest insects; an exclosure experiment. Science 206: 462-463.

Hulme, M.A.; Dawson, A.F.; Harris, J.W.E. 1986. Exploiting cold-hardiness to separate *Pissodes strobi* (Peck) (Coleoptera: Curculionidae) from associated insects in leaders of *Picea sitchensis* (Bong.) Carr. Can. Entomol. 118: 1115-1122.

Hulme, M.A.; Harris, J.W.E.; Dawson, A.F. 1987. Exploiting adult girth to separate *Pissodes strobi* (Peck) (Coleoptera: Curculionidae) from associated insects in leaders of *Picea sitchensis* (Bong.) Carr. Can. Entomol. 119: 751-753.

Jennings, D.T.; Houseweart, M.W. 1984. Predation by eumenid wasps (Hymenoptera: Eumenidae) on spruce budworm (Lepidoptera: Tortricidae) and other lepidopterous larvae in spruce-fir forests in Maine. Ann. Entomol. Soc. Am. 77: 39-45.

Kaya, H.K.; Anderson, J.F. 1974. Collapse of the elm spanworm outbreak: role of *Doencrytus* sp. Environ. Entomol. 3: 659-663.

Knight, F.B.; Heikkenen, H.J. 1980. Principles of Forest Entomology. 5th ed. McGraw-Hill, New York. xiii + 461 pp.

Langelier, L.A.; Garton, E.O. 1986. Management guidelines for increasing populations of birds that feed on western spruce budworm. USDA Agric. Hdbk. No. 653, 19 pp.

Lewis, W.J.; Jones, R.L.; Nordlund, D.A.; Sparks, A.M. 1975. Kairomones and their use for management of entomophagous insects: I. Evaluation for increasing rates of parasitization by *Trichogramma* spp. in the field. J. Chem. Ecol. 3: 343-347.

Luck R.F.; Shepard, B.M.; Kenmore, P.E. 1988. Experimental methods for evaluating arthropod natural enemies. Ann. Rev. Entomol. 33: 367-391.

Lysyk, T.J.; Nealis, V.G. 1988. Temperature requirements for development of the jack pine budworm (Lepidoptera: Tortricidae) and two of its parasitoids (Hymenoptera). J. Econ. Entomol. 81: 1045-1051.

Martinat, P.J.; Coffman, C.C.; Dodge, K.; Cooper, R.J.; Whitmore, R.C. 1988. Effect of diflubenzuron in a central Appalachian forest. J. Econ. Entomol. 81: 261-267.

McClure, M.S. 1978. Resurgence of the scale, *Florinia externa* (Homoptera: Diaspididae) on hemlock following insecticide application. Environ. Entomol. 6: 480-484.

Miller, C.A. 1963. Parasites and the spruce budworm. Pages 228-241 *in* R.F. Morris (Ed.), The dynamics of epidemic spruce budworm populations. Mem. Entomol. Soc. Canada.

Miller, C.A. 1966. The black-headed budworm in eastern Canada. Can. Entomol. 98: 592-613.

Nealis, V.G. 1991. Parasitism in sustained and collapsing populations of the jack pine budworm, *Choristoneura pinus pinus* Free. (Lepidoptera: Tortricidae), in Ontario, 1985-1987. Can. Entomol. 123: 1065-1075.

Nealis, V.G.; van Frankenhuyzen, K. 1990. Interactions between *Bacillus thuringiensis* Berliner and *Apantales fumiferanae* Vier. (Hymenoptera: Braconidae), a parasitoid of the spruce budworm, *Choristoneura fumiferana* (Clem.) (Lepidoptera: Torticidae). Can. Entomol. 122: 585-594.

Nealis, V.G.; Lysyk, T.J. 1988. Sampling overwintering jack pine budworm, *Choristoneura pinus pinus* Free. (Lepidoptera: Tortricidae), and two of its parasitoids (Hymenoptera). Can. Entomol. 120: 1101-1111.

Neilson, M.M.; Martineau, R.; Rose, A.H. 1971. *Diprion hercyniae* (Hartig) european spruce sawfly in biological control programmes against insects and weeds in Canada 1959-1968. Tech. Comm. 4 Comm. Inst. Biol. Control, Comm. Agric. Bureau, Slough, England. pp. 136-143.

Otvos, I.S. 1979. The effects of insectivorous bird activities in forest ecosystems: an evaluation. Pages 341-374 *in* Dickson, J.G., R.N. Conner, R.R. Fleet, J.C. Kroll and J.A. Jackson (Eds.), The Role of Insectivorous Birds in Forest Ecosystems. Academic Press, New York.

Peakall, D.B.; Bart, J.R. 1983. Impacts of aerial applications of insecticides on forest birds. CRC Crit. Rev. Environ. Control 13: 117-165.

Prebble, M.S. (Ed.). 1975. Aerial Control of Forest Insects in Canada, Environ. Can., Ottawa. 330 p.

Price, P.W. 1972. Immediate and long-term effects of insecticide application on parasitoids in jack pine stands in Quebec. Can. Entomol. 104: 263-270.

Price, P.W.; Tripp, H.A. 1972. Activity patterns of parasitoids on the Swaine jack pine sawfly, *Neodiprion swainei* (Hymenoptera: Diprionidae), and parasitoid impact on the host. Can. Entomol. 104: 1003-1016.

Richmond, H.A. 1986. Forest entomology: from pack horse to helicopter. Pest Management Rep. No. 8. B.C. Ministry of Forestry and Lands, Victoria. 44 pp.

Syme, P.D. 1976. Red pine and the European pine shoot moth in Ontario. Can. For. Serv., Ontario Region, Sault Ste. Marie. Inf. Rep. O-X-244. 17 pp.

Takekawa, J.Y.; Garton, E.O.; Langelier, L.A. 1982. Biological control of forest insect outbreaks: The use of avian predators. North Am. Wildl. Conf. 47: 393-409.

Torgersen, T.R.; Campbell, R.W. 1978. Natural mortality. Pages 47-53 *in* Brookes, M.H., R.W. Stark, R.W. Campbell (Eds.), The Douglas-fir Tussock Moth: A Synthesis. USDA For. Serv. Tech. Bull. 1585, Washington D.C.

Torgersen, T.R.; Campbell, R.W. 1982. Some effects of avian predators on the western spruce budworm in north central Washington. Environ. Entomol. 11: 429-431.

Waage, J.K.; Hassell, M.P.; Godfray, H.C.J. 1985. The dynamics of pest-parasitoid-insecticide interactions. J. Appl. Ecol. 22: 825-838.

Witter, J.A.; Kulman, H.M. 1979. The parasite complex of the forest tent caterpillars in northern Minnesota. Environ. Entomol. 8: 723-731.

Chapter 44

Pheromones

C.J. Sanders, G.G. Grant, R.F. Shepherd,
P.J. Silk, E.W. Butterworth, and E.G. Kettela

Introduction

Pheromones are defined as chemicals produced by individuals of one species to communicate with others of the same species. The most widely known are sex pheromones; however, there are also aggregation and alarm pheromones. Among the defoliating Lepidoptera and sawflies, sex pheromones are produced by sexually mature females when they are ready to mate. When a receptive male detects the pheromone, it flies upwind, maintaining contact with the pheromone plume, enabling it to contact the "calling" female. Among bark beetles and weevils the signals are more complex. Both aggregation and the location and identification of the opposite sex are involved, and pheromones may be produced by both sexes. Frequently host-plant odors are also implicated. In addition, the meaning of the message may vary with the ratios of the constituent compounds and with the physiological state of the recipient insect.

Parasitic Hymenoptera and many Diptera also use sex pheromones to locate sexual partners, and female parasitoids depend upon odors to locate their prey. Such chemicals, which benefit the receiver to the detriment of the emitter, are referred to as "kairomones." Research into kairomones has not received the same attention that pheromones have, and none have been tested in Canadian forestry. However, it is recognized that kairomones have potential for manipulating parasitoids to enhance their effectiveness, and research in this area is expanding.

Because of their species specificity, high potency, and low environmental impact, pheromones have received considerable attention from pest managers, both for detecting and monitoring populations and for regulating populations by mass trapping or by disrupting normal mating behavior. Registration is not required for pheromones or related compounds when used as lures in traps, but registration is required for insecticides placed in the traps to kill the insects, and Vaportape II plastic strips impregnated with dichlorvos (HealthChem Corp., New York) have recently received registration for this purpose.

The use of pheromones in pest management of Lepidoptera and defoliating Hymenoptera (sawflies) has focused entirely on sex pheromones emitted by the adult female insects. In some instances, use has been made of the natural pheromone extracted from the tip of the abdomens of the female, where the pheromone gland is located, but this is cumbersome, and extensive use of pheromones has become possible only after the identification and manufacture of synthetic pheromones. For the first identification, that of the silkworm, 500 000 abdomen tips were required to provide sufficient material for the analysis (Butenandt et al. 1959). Improved technology has now enabled identifications to be made from a sample of air that has passed over a single pheromone gland. Initially, many pheromones were identified as single compounds, but more stringent bioassay techniques have shown that the original identification was often incomplete and that additional, minor components may be necessary for a complete level of response. Synthetic sex pheromones of Lepidoptera and sawflies are under development in forestry with two major objectives in mind: to lure insects to traps or to disrupt normal mating behavior.

Trapping

In theory, traps can be used to capture sufficient males to decrease mating success of the females and consequently reduce subsequent egg production. Attempts have been made in this way to eradicate localized, discrete populations of gypsy moths in both the United States and Canada, but, in general, the use of traps for population regulation is impractical in forestry where outbreaks are extensive and often inaccessible. The greatest potential value of sex pheromone traps is for detection and monitoring. Because of the potency of the pheromones, significant numbers of insects can be caught in traps even when population densities are so low that they cannot be detected by conventional sampling techniques without a disproportionate sampling effort. Traps placed beyond the known limits of an insect's range can be used to detect population spread, as with the gypsy moth throughout North America and the larch casebearer in British Columbia. Pheromone traps have also proved useful for obtaining specimens at low population densities for taxonomic studies and range definition. Alternatively, in species that cycle from low to high density, traps can be used to monitor changes in density to provide early warning of an impending outbreak. If the relationship between catch and population density is sufficiently reliable, then threshold catches can be determined to indicate the need for more intensive surveys. The earliest trap designs utilized sticky surfaces to catch the insects. These still have a role to play in detection, but for quantitative work and for mass trapping, saturation of the sticky surface with insects becomes a problem; this has led to the development of higher capacity bucket traps in which the insects are killed by insecticide.

Pheromone trapping systems are now in use or are under development in Canada to detect, survey, and monitor a variety of Lepidoptera and sawflies that attack coniferous and deciduous forests, plantations, seed orchards, nurseries, and forest regeneration sites. Attractants and known pheromones for these are listed in Table 1 (courtesy G.G. Grant). In Table 1, detection refers to surveys that are designed to locate insect infestations, determine their extent, and track their geographical spread. The term "monitoring" indicates that the trap catch is used to estimate insect abundance, follow

Table 1. Known sex pheromones and attractants for common Lepidopteran and Hymenopteran pests of forests in Canada.

Family species common name	Attractive compounds[a]	Ratio	Chemical status[b]	Forestry Canada usage	References chemical identification[c]
	LEPIDOPTERA				
ARCTIIDAE					
Hyphantria cunea Fall webworm	(Z,Z)-9,12-18:Ald (Z,Z,Z)-9,12,15-18:Ald (Z,Z)-3,6-cis-9,10-epoxy-21:Hy	1:6:27	P		Hill et al. 1982
COLEOPHORIDAE					
Coleophora laricella Larch casebearer	Z5-10:OH		A	Detection (developmental)	Priesner et al. 1982
COSSIDAE					
Prionoxystus robiniae Carpenter worm	(Z,Z)-3,5-14:Ac		A		Doolittle et al. 1976
GEOMETRIDAE					
Operophtera bruceata Bruce spanworm	(Z,Z,Z)-1,3,6,9-19:Hy		A P		Roelofs et al. 1982 Underhill et al. 1987
Operophtera brumata Winter moth	(Z,Z,Z)-1,3,6,9-19:Hy		P	Detection	Roelofs et al. 1982
Operophtera occidentalis	(Z,Z,Z)-1,3,6,9:Hy		A		Roelofs et al. 1982
Alsophila pometaria Fall cankerworm	(Z,Z,Z)-3,6,9-19:Hy (Z,Z,Z,E)-3,6,9,11-19:Hy (Z,Z,Z,Z)-3,6,9,11-19:Hy	12:27:7	P		Wong et al. 1984
LASIOCAMPIDAE					
Malacosoma disstria Forest tent caterpillar	(Z,E)-5,7-12:Ald (Z,E)-5,7-12:OH	1:10	P	Detection (semi-operational)	Chisholm et al. 1980
Malacosoma americanum Eastern tent caterpillar	(Z,E)-5,7-12:OH (E,Z)-5,7-12:Ald		P		Roelofs et al. 1986
Malacosoma californicum pluviale Northern tent caterpillar	(E,Z)-5,7-12:Ald		P		Underhill et al. 1980
LYMANTRIIDAE					
Lymantria dispar Gypsy moth	2-Me-cis-7,8-epoxy-18:Hy (+)-enantiomer		P A	Detection/Control (operational)	Bierl et al. 1970 Cardé et al. 1977
Orgyia pseudotsugata Douglas-fir tussock moth	Z6-11-21:Kt		P	Monitoring (operational)	Smith et al. 1975
Orgyia antiqua Rusty tussock moth	Z6-11-21:Kt		A		Daterman et al. 1976
Dasychira plagiata Northern conifer tussock moth	Z6-11-21:Kt		A		Grant 1977
Dasychira pinicola Pine tussock moth	Z6-11-21:Kt		P		Slessor & Grant, unpubl. data

(Continued)

Table 1. Known sex pheromones and attractants for common Lepidopteran and Hymenopteran pests of forests in Canada. *(Continued)*

Family species common name	Attractive compounds[a]	Ratio	Chemical status[b]	Forestry Canada usage	References chemical identification[c]
NOCTUIDAE					
Actebia fennica	Z7-12:Ac	1:20	A	Detection	Steck et al. 1979
Black army cutworm	Z11-14:Ac		P	(developmental)	Struble et al. 1989
Peridroma saucia	Z11-16:Ac	2:1	A	Detection	Struble et al. 1976
Variegated cutworm	Z9-14:Ac			(developmental)	
PYRALIDAE					
Dioryctria reniculelloides	Z9-14:Ac	3:0.15	P	Detection	Grant et al. 1987
	Z7-12:Ac			(developmental)	
Spruce coneworm					
Dioryctria disclusa	Z9-14:Ac		P		Meyer et al. 1982
Webbing coneworm					
TORTRICIDAE (OLETHREUTIDAE)					
Barbara colfaxiana	Z9-12:OH	75:25	A	Detection	Hedlin et al. 1983
Douglas-fir cone moth	Z9-12:Ac			(developmental)	
Barbara mappana	E9-12:Ac		A		Reed and Chisholm 1985b
Cydia piperana biotype	E9-12:Ac	1:1	A		Stevens et al. 1985
	Z9-12:Ac				
or	E9-12:Ac	7:3	A		Sartwell et al. 1985
	Z9-12:Ac				
Cydia populana	(E,E)-8,10-12:Ald		A		Chisholm et al. 1985
or	(E,E)-8,10-12:Ac		A		Stevens et al. 1985
Cydia strobilella	E8-12:Ac		A	Detection	Roelofs & Brown 1982
Spruce seed moth				(developmental)	Grant et al. 1988
Cydia toreuta	(E,Z)-8,10-12:Ac		A		Chisholm et al. 1985
Eastern pine seedworm					
Epinotia nanana	E8-12:Ac	1:1	A		Booij and Voerman 1984
European spruce needle miner	E10-12:Ac				
Epinotia criddleana	E9-12:Ac		A		Reed and Chisholm 1985a
Eucosma bobana	Z9-12:Ac		A		Stevens et al. 1985
A Pinyon cone borer					
Eucosma gloriola	Z9-12:Ac	9:1	A		Grant et al. 1985
Eastern pine shoot borer	E9-12:Ac				

[a] Compounds listed include only those that contribute to attraction or have proven behavioral effects. Abbreviations follow patterns typically found in the literature (e.g. Arn et al. 1986). Position and geometry of double bonds are indicated in the usual fashion and chain length and functional groups are usually separated by a colon: (e.g. (Z,E)-5,7-12:Ald = (Z,E)-5,7-dodecadienal). Functional groups are abbreviated as follows: Ac = acetate, OH = alcohol, Ald = aldehyde, epoxy = epoxide, Pr = proprionate, Kt = ketone, Hy = hydrocarbon chain, Me = methyl group. The chirality of asymmetrical isomers of sawfly pheromones is indicated by the usual R/S system.

[b] P = sex pheromone (i.e. compounds chemically identified); A = sex attractant, compounds found by empirical field testing.

[c] Appropriate reference on same line as letter (A or P) indicating chemical status.

(Continued)

Table 1. Known sex pheromones and attractants for common Lepidopteran and Hymenopteran pests of forests in Canada. *(Continued)*

Family species common name	Attractive compounds[a]	Ratio	Chemical status[b]	Forestry Canada usage	References chemical identification[c]
Eucosma ponderosae A western pine cone borer	Z9-12:Ac		A		Stevens et al. 1985
Eucosma recissoriana Lodgepole pine cone borer	Z9-12:Ac		A		Stevens et al. 1985
Eucosma sonomona Western pine shoot borer	Z9-12:Ac E9-12:Ac	2:1	A		Sower et al. 1979
Petrova albicapitana Northern pitch twig moth	(Z,E)-7,9-12:Ac		A P		Wong et al. 1985 Dix and Underhill 1988
Petrova metallica A metallic pitch nodule moth	(Z,E)-7,9-12:Ac		A P		Wong et al. 1985 Dix and Underhill 1988
or	E9-12:Ac		A		Stevens et al. 1985
or	E9-12:Ac Z9-12:Ac	7:3	A		Sartwell et al. 1985
Proteoteras crescentana	Z8-12:Ac Z8-12:OH	4:1	P		Underhill (in Roelofs & Brown 1982)
Proteoteras willingana Boxelder twig borer	Z8-12:Ac Z8-12:OH	1:9	P		Underhill (in Roelofs & Brown 1982)
Rhyacionia adana	E9-12:Ac E9-12:OH	9:1	A		Grant et al. 1985
Rhyacionia buoliana European pine shoot moth	E9-12:Ac E9-12:OH		P	Detection (operational)	Gray et al. 1984 Smith et al. 1974
Rhyacionia busckana	E9-12:Ac E9-12:OH	9:1	A		Grant et al. 1985
Rhyacionia granti	(E,E)-8,10-12:Ac		A		Grant et al. 1985
Rhyacionia zozana A ponderosa pine tip moth	E9-12:Ac E9-12:OH	70:30–95:5	P		Niwa et al. 1987
Zeiraphera canadensis Spruce bud moth	Z9-14:Ac		A P	Monitoring (developmental)	Turgeon & Grant 1988 Silk et al. 1989
Zeiraphera unfortunana Purplestriped shootworm	E9-12:Ac		P	Detection (developmental)	Silk et al. 1988
TORTRICIDAE (TORTRICIDAE)					
Acleris emargana	E11-14:Ac		A		Weatherston et al. 1974
Archips semiferana Oak leafroller	E11-14:Ac Z11-14:Ac	2:1	P		Miller et al. 1976
Sparganothis acerivorana Maple leaf roller	Z11-14:Ald E11-14:Ald Z11-14:Ac	65:35:10	P		Grant and Slessor 1983
Choristoneura biennis Two-year cycle budworm	E11-14:Ac		A		Sanders et al. 1974

(Continued)

Table 1. Known sex pheromones and attractants for common Lepidopteran and Hymenopteran pests of forests in Canada. *(Continued)*

Family species common name	Attractive compounds[a]	Ratio	Chemical status[b]	Forestry Canada usage	References chemical identification[c]
Choristoneura carnana	E11-14:Ald Z11-14:Ald	92:8	A		Liebhold and Volney 1985
Choristoneura conflictana Large aspen tortrix	Z11-14:Ald		A P	Monitoring (developmental)	Weatherston et al. 1976 Grant and Slessor, unpubl. data
Choristoneura fumiferana Spruce budworm	E11-14:Ald Z11-14:Ald	95:5	P	Monitoring (semi-operational)	Weatherston et al. 1971 Silk et al. 1980
Choristoneura occidentalis Western spruce budworm	E11-14:Ald Z11-14:Ald	90:10	P	Monitoring (developmental)	Silk et al. 1982 Cory et al. 1982
Choristoneura orae Annual spruce budworm	E11-14:Ac Z11-14:Ac E11-14:OH	82:9:9	P	Detection (developmental)	Gray et al. 1984
Choristoneura pinus Jack pine budworm	E11-14:Ac Z11-14:Ac E11-14:OH Z11-14:OH	22.8:3.7:2.6:0.4	P	Detection (developmental)	Silk et al. 1985
Choristoneura retiniana	E11-14:Ac Z11-14:Ac E11-14:OH Z11-14:OH	83:7:0.9:0.1	P		Daterman et al. 1984
Croesia semipurpurana Oak leafshredder	E11-14:Ald Z11-14:Ald 14:Ac	85:15 85:15 85:15:100	A P A	 Monitoring (developmental)	Weatherston et al. 1978 P. Silk, pers. comm. Grant et al. 1981
Raphia frater Yellowmarked caterpillar	Z7-12:OH		A		Weatherston et al. 1974
HYMENOPTERA					
DIPRIONIDAE					
Diprion similis Introduced pine sawfly	3,7-diMe-2-15:Pr (2S,3S,7S)+(2S,3R,7R) isomers	5:1	P		Olaifa et al. 1988
Gilpinia frutetorum Nursery pine sawfly	3,7-diMe-2-15:Ac (2S,3R,7R) isomer		A		Kikukawa et al. 1982

[a] Compounds listed include only those that contribute to attraction or have proven behavioral effects. Abbreviations follow patterns typically found in the literature (e.g. Arn et al. 1986). Position and geometry of double bonds are indicated in the usual fashion and chain length and functional groups are usually separated by a colon: (e.g. (Z,E)-5,7-12:Ald = (Z,E)-5,7-dodecadienal). Functional groups are abbreviated as follows: Ac = acetate, OH = alcohol, Ald = aldehyde, epoxy = epoxide, Pr = proprionate, Kt = ketone, Hy = hydrocarbon chain, Me = methyl group. The chirality of asymmetrical isomers of sawfly pheromones is indicated by the usual R/S system.

[b] P = sex pheromone (i.e. compounds chemically identified); A = sex attractant, compounds found by empirical field testing.

[c] Appropriate reference on same line as letter (A or P) indicating chemical status.

(Continued)

Table 1. Known sex pheromones and attractants for common Lepidopteran and Hymenopteran pests of forests in Canada. *(Concluded)*

Family species common name	Attractive compounds[a]	Ratio	Chemical status[b]	Forestry Canada usage	References chemical identification[c]
Neodiprion lecontei Redheaded pine sawfly	3,7-diMe-2-15:Ac (2S,3S,7S) isomer		P		Jewett et al. 1976 Matsumura et al. 1979
Neodiprion nanulus nanulus Red pine sawfly	3,7-diMe-2-15:Ac (2S,3S,7S) isomer		P		Olaifa et al. 1987
Neodiprion pinetum White pine sawfly	3,7-diMe-2-15:Ac (2S,3S,7S)+(2S,3R,7R) isomers	1:2	P		Olaifa et al. 1988
Neodiprion pratti banksianae Jack pine sawfly	3,7-diMe-2-15:Ac (2S,3S,7S)+(2S,3R,7R) isomers	5:1	A		Olaifa et al. 1984
Neodiprion sertifer European pine sawfly	3,7-diMe-2-15:Ac (2S,3S,7S)+(2S,3R,7R) or (2S,3S,7S)+(2S,3R,7R/S)	5:0.003 5:0.003	P		Jewett et al. 1976
Neodiprion swainei Swaine jack pine sawfly	3,7-diMe-2-15:Pr		A		Jewett et al. 1978
TENTHREDINIDAE					
Pikonema alaskensis Yellowheaded spruce sawfly	Z10-19:Ald	-	P		Bartelt and Jones 1983
	Z5-14:OH				Bartelt et al. 1983

[a] Compounds listed include only those that contribute to attraction or have proven behavioral effects. Abbreviations follow patterns typically found in the literature (e.g. Arn et al. 1986). Position and geometry of double bonds are indicated in the usual fashion and chain length and functional groups are usually separated by a colon: (e.g. (Z,E)-5,7-12:Ald = (Z,E)-5,7-dodecadienal). Functional groups are abbreviated as follows: Ac = acetate, OH = alcohol, Ald = aldehyde, epoxy = epoxide, Pr = proprionate, Kt = ketone, Hy = hydrocarbon chain, Me = methyl group. The chirality of asymmetrical isomers of sawfly pheromones is indicated by the usual R/S system.

[b] P = sex pheromone (i.e. compounds chemically identified); A = sex attractant, compounds found by empirical field testing.

[c] Appropriate reference on same line as letter (A or P) indicating chemical status.

population trends, or predict outbreaks and future damage. Only two of the trapping systems can be considered fully operational at this time: the survey system used to detect the spread of the gypsy moth and the monitoring system used to locate imminent infestations of the Douglas-fir tussock moth, and these will be discussed in more detail. Several other trapping systems are in the developmental stage but require further evaluation on the relationship between trap catch and population density. Of these, the monitoring system for spruce budworm will be discussed in more detail. In addition, brief mention will be made of pheromone trapping systems that are under development for other species. In only one instance is a pheromone being used operationally for the control of a lepidopterous pest; a trapping-out program has been used for the gypsy moth to assist in eradicating isolated populations in the Maritimes.

Gypsy Moth

The use of sex pheromone traps to detect the presence and spread of the gypsy moth, *Lymantria dispar*, and to delimit infestations began in 1956. Since 1983, traps have also been used to eradicate isolated gypsy moth populations in the Maritimes. The gypsy moth is still classed as "quarantinable" and therefore falls under the jurisdiction of the Plant Health Directorate of Agriculture Canada, henceforth referred to as Agriculture Canada. Strategies for controlling gypsy moth have differed in different regions and have involved Agriculture Canada, Forestry Canada, provincial agencies and municipalities, often in cooperative programs. The same agencies are responsible for deploying traps for detection and monitoring gypsy moth populations. Summaries of both the control programs and detection and monitoring are available in the

Table 2. Numbers of pheromone traps deployed in Canada in 1988 to monitor populations of the gypsy moth.[a]

Province	Detection surveys	Delimitation surveys	Eradication programs	Leading edge surveys	Export certification	Total traps placed	Total traps recovered
Newfoundland	240					240	237
Prince Edward Island	320					320	314
Nova Scotia	361	639	2 436		202	3 638	3 297
New Brunswick	307		2 101		22	2 430	2 290
Quebec	155			185		340	325
Ontario	803			429		1 232	1 187
Manitoba	132					132	127
Saskatchewan	58					58	55
Alberta	163					163	154
British Columbia	4 979	126	1 852			7 557	7 219
Totals	7 518	765	6 389	614	224	16 110	15 205

[a] Information provided by C. Slight, Plant Protection Division, Agriculture Canada, Ottawa, Ont.

annual "Summary of Plant Quarantine Pest and Disease Situations in Canada" by Agriculture Canada and also in the annual regional reports on forest pest conditions put out by the Forest Insect and Disease Survey (FIDS) Branch in each Forestry Canada regional establishment.

The traps and lures are obtained from the United States Department of Agriculture and are distributed by Agriculture Canada. For the first 10 years of the program, cardboard cups coated inside with Tanglefoot® and stapled horizontally on a tree were used as traps, baited with extracts of female abdomen-tips. From 1964 on, delta traps have been used, baited with synthetic attractant, first Gyplure®, then racemic Disparlure®, and, since 1982, (+) Disparlure®. The number of traps deployed has increased steadily over the years, from a total of 1 856 in 1977, to 6 882 in 1980 when the program was first extended to cover all 10 provinces, to 11 400 in 83 and 16 110 in 1988. To provide an overview of the scale and complexity of the program, a breakdown of trap use and deployment across Canada in 1988 is shown in Table 2 (courtesy Agriculture Canada). Since 1983, a significant number of traps has been used in New Brunswick and Nova Scotia to suppress outbreaks by trapping-out male moths before they mate. From 1983 through 1986, an independent mass-trapping program was carried out in the vicinity of Québec City using Multi-pher®, nonsaturating bucket traps. For detection surveys the density of traps recommended by Agriculture Canada is 1 per 40 ha or about 1 per 1.6 km of rural road. This is increased up to about 14 per ha for delimiting the extent of outbreaks. For mass trapping a density of 15 to 25 per ha is recommended.

Douglas-fir Tussock Moth

Pheromone traps are used to monitor population changes of the Douglas-fir tussock moth, *Orgyia pseudotsugata*, as an essential component of an integrated pest management program. Traps are used to provide 1 to 2 years advance warning of when and where outbreaks will occur. This allows time for production and application of the naturally occurring, highly infectious polyhedrosis virus before serious defoliation is anticipated.

Defoliation by the Douglas-fir tussock moth occurs in specific, identifiable ecozones representing the driest and hottest part of the range of Douglas-fir. Population fluctuations are usually synchronized over distances of at least 50 km. As density at the most favored site increases rapidly over a number of generations, there is a slower, but still positive increase in density in surrounding areas. Consequently, even if a trap site is not situated at the center of an impending outbreak, an upward trend will be reflected in the moth trap catches (Shepherd et al. 1985). Permanent monitoring stations are established at about 30-km intervals. Currently, the FIDS of Forestry Canada maintains 19 of these stations in British Columbia. At each station six delta-shaped, sticky, pheromone traps are deployed each year for the duration of the moth flight. Each trap is baited with a polyvinyl chloride pellet containing 0.1% (weight/weight) of the synthetic pheromone, *Z*-6-heneicosen-11-one (Shepherd et al. 1985).

Three years of upward trend in moth catch, or an average catch exceeding 25 moths at any station, is taken as an indication that defoliation may occur somewhere in the watershed the following year. Two years before defoliation (i.e., after

2 years of increasing catch), an auxiliary detection survey is carried out with single traps placed at 1-km intervals along all access roads that surround the monitoring station to define more precisely where populations are erupting. Stands near traps with high catches are ground-surveyed for egg masses. Sequential egg-mass sampling provides predictions of defoliation and determines the need for virus treatment (Shepherd et al. 1984). Detection surveys are continued each year to locate newly erupting populations until a general collapse occurs.

The pheromone monitoring system is conservative; 3 years of increasing catch are not always accompanied by high egg-mass densities. However, if a threatening population does arise, the trapping system will detect it.

Spruce Budworm

Research into the development of pheromone traps for monitoring spruce budworm, *Choristoneura fumiferana*, populations began following the identification of the major pheromone components (Sanders and Weatherston 1976) and received additional impetus during the years of the Canada–United States (CANUSA) Spruce Budworms Program, when funding was provided for a cooperative program in both the United States and Canada. This culminated in recommendations for a monitoring program using Multi-pher® traps (Jobin 1985) containing a plastic strip impregnated with dichlorvos, and polyvinyl chloride pellets containing 0.03% (weight/weight) of the pheromone—a 95:5 blend of (E:Z)-11-tetradecenals. Deployment of traps at head height in a triangular layout was recommended, with 40 m between traps (Allen et al. 1986). Subsequent experience indicated that the release rate from the recommended lures was too low, resulting in many zero catches at low population densities. This, and questions as to the reliability of the polyvinyl chloride lures led to a recommendation for a switch to Biolures (Consep Membranes Inc., Bend, Oregon).

Trap catches have, on a broad scale, reflected population trends, but inconsistencies have occurred. The relationship between trap catch and population density has varied from region to region and from year to year. This makes the trap catches unsuitable for accurately predicting population densities in the following year. Possibly these discrepancies are due to differences in climate and weather, and with more research these factors could be allowed for. Nevertheless, experience thus far suggests that the traps have great potential for monitoring long-term trends in low-density populations to determine when and where populations are on the increase (Sanders 1988). With data from a few more years, it should be possible to establish thresholds to indicate when more intensive assessment is required.

Other Cases

In addition to the above examples, pheromone monitoring and detection systems are being developed for other forest pests to supplement or replace traditional sampling methods that generally are more time consuming or costly, particularly when the target population is at a low density and sampling locations are widely separated. For example, the prediction of population trends of the oak leafshredder, *Croesia semipurpurana*, and forecasts of potential defoliation are currently dependent on egg sampling, which is a tedious procedure involving the use of pole clippers to obtain midcrown branch samples that are subsequently examined under a microscope to count the eggs. By contrast, pheromone traps are easy to deploy and counting insects is less work. At low and moderate population densities, catches in sticky traps baited with an 85:15:100 mixture of (E:Z)-11-tetradecenals and tetradecyl acetate (Grant et al. 1981) correlate strongly with egg-mass counts. However, comparisons involving higher population densities are needed to demonstrate that pheromone trapping can reliably replace egg surveys (see Chapter 16). Pheromone monitoring of oak leafshredder populations has been successfully used as a post-treatment assessment tool in the evaluation of Dimilin® for control of this insect (Retnakaran and Grant 1985).

Pheromone trapping of the forest tent caterpillar, *Malacosoma disstria*, is used to monitor low-density populations in the Maritimes and is considered an accurate means of predicting future population trends. Consequently, it has replaced egg-mass sampling, saving considerable work and expense in conducting annual surveys (Magasi 1988). Pheromone trapping is used in Newfoundland to warn of invasion by forest tent caterpillar moths from the mainland.

Pheromone trapping of the spruce bud moth, *Zeiraphera canadensis*, is under study as a potential replacement or supplement for egg-mass sampling to predict population trends (see Chapter 23) in a cooperative study between the Forest Pest Management Institute and the Ministère de l'Énergie et des Ressources du Québec. Pheromone trapping is effective for determining the initiation of adult flight (Turgeon and Grant 1988), which could be of value for timing adulticide sprays, a potential option for control of this insect (Turgeon 1985).

In British Columbia, pheromone traps are of value in detecting new infestations of the European pine shoot moth, *Rhyacionia buoliana*, and were formerly vital to the implementation of quarantine regulations prohibiting the inland shipment of infested ornamental pines from coastal nurseries (see Chapter 30). Trapping is also important in the Maritimes to monitor populations of the European pine shoot moth in susceptible young pine plantations. A pheromone detection and monitoring system for the black army cutworm, *Actebia fennica*, is currently being tested in British Columbia (R.F. Shepherd, pers. comm.). This insect invades recently burned regeneration sites and, after depleting its preferred host species, attacks planted coniferous seedlings. Its appearance in an area is sudden, and because the larval and adult behavior are cryptic, populations are easily overlooked, leading to significant damage with little advance notice. The aim of pheromone trapping is to provide early warning of moth invasion and to monitor population buildups so that timely control actions can be initiated if needed.

A pheromone-based monitoring system has been tested for the large aspen tortrix, *Choristoneura conflictana*, a pest of *Populus* species, especially trembling aspen, *P. tremuloides*. A recent 3-year study compared moth catches with larval and egg-mass densities in 18 aspen stands located in northern Ontario (Grant 1989, unpubl. data). Results indicate that three or four Multi-pher® traps in each plot are adequate, each baited with a rubber septum containing 10 μg of (Z)-11-tetradecenal, the major sex pheromone component of the female moth (Weatherston et al. 1976). Catches are well correlated with larval density in both the current and subsequent year, indicating the potential value for monitoring population fluctuations. Relating trap catches to defoliation estimates was not useful because damage in the sampling sites was often caused by other defoliators, such as the forest tent caterpillar, or the aspen twoleaf tier, *Enargia decolor*, which confounded any comparisons involving levels of defoliation. Pheromone trapping of the large aspen tortrix can replace other sampling methods for this insect and is best suited for low to moderate populations. High populations are usually self-evident and there is probably little merit in trapping in this situation. In addition, larval parasitism is often high in these populations, which decreases the size of the subsequent moth population and causes lower than expected trap catches; this could be misleading if interpreted alone.

Pheromone components of the major sawfly pests in Canada have been identified (Table 1), but they are not being used currently for monitoring or detection purposes. The reasons for this are the lack of specificity of the available pheromone components for target species of interest, combined with the impossibility of morphologically separating male sawflies of different species caught in the same traps. Thus, where sawfly species overlap, or potentially do so, the survey trap catches cannot be used reliably to assess the target species. This problem should be overcome when the specific enantiomers utilized by the various sawfly species are chemically resolved.

Mating Disruption

The value of pheromones for monitoring forest insects is now well accepted, but their use for population regulation is proving more problematical.

The strategy of mating disruption assumes that high concentrations of synthetic pheromone will obscure the natural pheromone plumes emitted by the female moths or cause adaptation of the male insects, although the exact mechanisms are not understood. Adequate levels of the synthetic pheromone are required in the atmosphere throughout the insect's flight period. This necessitates special formulations that protect the chemicals from degradation and release them at an adequate rate for periods of several weeks. Some, such as hollow fibers and plastic flakes, require specialized equipment for their application, but have given promising results. Others, such as microcapsules, can be sprayed with conventional boom and nozzle equipment, which gives them a practical advantage. Pheromones have proved commercially successful for a number of agricultural pests, but their application in forestry has been very limited.

Formulations have been registered in the United States for aerial use against the gypsy moth, and Hercon® flakes

Table 3. Field trials of the aerial application of sex pheromones to disrupt spruce budworm mating behavior, 1975–1988.

Year	Location	Formulation (+ sticker)	Dosage[a] (g)	Area treated	References
1975	Ontario	NCR microcapsules	7	12 ha	Sanders 1976
1977	Ontario	Conrel fibres	1.5	10 ha	Sanders 1979
		Conrel flakes	9	250 ha	
		Conrel flakes	14	10 ha	
1978	New Brunswick/ Nova Scotia	Conrel fibres	0.1–25	8 plots × 10 ha	Miller 1979 Sanders and Silk 1982
1980	Maine	Hercon flakes (+ Pherotec)	25	145 & 30 ha	Dimond et al. 1984
		Hercon flakes	250	9 ha	
1981	Ontario	Hercon flakes (+ Pherotec)	70	30 ha	Sanders and Silk 1982
1983	New Brunswick	Hercon flakes (+ Pherotec)	250	cages in 2-, 3-ha plots	Seabrook and Kipp 1986
1984	New Brunswick	Hercon flakes (+ AR 1990)	250	cages in 2-, 5-ha plots	Seabrook and Kipp 1986
1985	New Brunswick	Hercon flakes (+ AR 1990)	250	5 ha	Seabrook and Kipp, report ESA 1987 meeting
1986	New Brunswick	Hercon flakes (+ AR 1990)	250	5 ha	Seabrook and Kipp, report ESA 1987 meeting
1987	New Brunswick	ICI[b] microcapsules	100	16 ha	Seabrook and Kipp, pers. comm.
1988	New Brunswick	ICI microcapsules	20	25 ha	Seabrook and Kipp, pers. comm.
			90	25 ha	

[a] Active ingredients per hectare.
[b] ICI Agrochemicals.

have been used operationally against outlying infestations with apparent success (Webb et al. 1988). Both hollow fiber and flake formulations of the pheromone of the western pine shoot borer, *Eucosma sonomana*, are registered in the United States and have proven to be effective in reducing damage (Sower and Overhulser 1986). Hand-applied Hercon® flakes are also effective in reducing populations of a ponderosa pine tip moth, *Rhyacionia zozana* (Niwa et al. 1988). Trials with hollow fiber and flake formulations of the Douglas-fir tussock moth pheromone have been carried out in the United States (Sower et al. 1979) and in Canada in British Columbia (Sower et al. 1983). Results were encouraging, with 70% or higher reduction in numbers of fertile egg-masses, and registration for a hollow fiber formulation is now being sought in the United States (L.L. Sower, pers. comm.).

No pheromones have yet been registered for use against forestry pests in Canada. Extensive aerial trials have been carried out against the spruce budworm and, more recently, trials have been initiated against the spruce bud moth. Small-scale field trials using hand-placed formulations have also been carried out against the white-marked tussock moth, *Orgyia leucostigma* (Grant 1978). Because of its extensive range and highly mobile adults, the spruce budworm is not a good candidate for pheromone disruption. Cryptic or less mobile insects such as the spruce bud moth, are better candidates.

Spruce Budworm

Trials involving the aerial application of pheromone to disrupt the mating behavior of the spruce budworm are summarized in Table 3. The first trial was carried out in 1975, shortly after the identification of the major pheromone components (Sanders and Weatherston 1976). The objective was limited to determining if the presence of the pheromone would reduce the ability of males to locate virgin female moths in traps. Reduction in trap catch averaged over 97% (Sanders 1976), indicating a profound biological effect. Trials in 1977 through 1981 were designed to determine if reduction in trap catch resulted in fewer matings and a reduction in oviposition. All trials resulted in major reductions in trap catch, but assessment of mating and oviposition was confounded by lack of information on the performance of the formulations and by the possibility of mated, egg-bearing females invading the plots from untreated areas.

The problem of dispersal has been circumvented in New Brunswick by the use of large cages into which known numbers of budworm are released (Seabrook and Kipp 1986). Trials using these cages were conducted from 1983 through 1986. The area in which the cages were erected was sprayed each year with a 95:5 blend of (E:Z)-11-tetradecenal formulated in Hercon® laminated plastic flakes. Reductions in mating of up to 80% were recorded (Seabrook and Kipp 1986). It was also concluded that disruption is less effective at high population densities. At the same time, techniques have been developed and refined for measuring the concentrations of pheromone in the atmosphere; it is now possible to measure the amount of naturally produced pheromone present in the air space within a forest stand. In 1987 and 1988, trials were resumed on uncaged, natural populations in New Brunswick. An ICI Agrochemicals microcapsule formulation was used, but results were again confounded by high population densities and the possibility of invasion by female moths from outside the treated area (W.D. Seabrook and L.R. Kipp, pers. comm.).

Spruce Bud Moth

The cryptic larval feeding behavior of the spruce bud moth, *Zeiraphera canadensis*, and other members of the genus *Zeiraphera*, make them difficult to control with conventional larvicides, and population reduction using the mating-disruption technique with sex pheromones is currently being intensively studied. The sex pheromone of the spruce bud moth has been identified as E-9-tetradecenyl acetate (Silk et al. 1989) and has been found to be effective in trapping males of this species (Turgeon and Grant 1988). In 1987, a series of trap-disruption experiments in New Brunswick in high populations of the spruce bud moth showed that both Hercon® plastic flakes and ICI Agrochemicals microcapsules gave 100% reductions in trap catch over the entire adult flight period when applied at rates of 1, 10, and 100 g pheromone/ha. In 1988, two plots were aerially treated with the microcapsules at 10 and 50 g pheromone/ha using "raindrop" nozzles. In comparison with the check plot, a higher percentage of females were unmated during the first 2 to 3 days post-spray in the 10 g/ha plot, egg deposition appeared to start later, and there was a 21% reduction in actual egg counts. Interpretation of results at the 50 g/ha dosage was confounded by a large decrease in egg densities in the associated check plot, which was attributable to adjacent adulticide spray tests.

Further testing is being planned, but either further refinement of the pheromone blend or improvements in formulation are required before more field trials are carried out.

References

Allen, D.C.; Abrahamson, L.P.; Eggen, D.A.; Lanier, G.N.; Swier, S.R.; Kelley, R.S.; Auger, M. 1986. Monitoring spruce budworm (Lepidoptera: Tortricidae) populations with pheromone-baited traps. Environ. Entomol. 15: 152-165.

Arn, H.; Toth, M.; Priesner, E. 1986. List of sex pheromones of Lepidoptera and related attractants. Intern. Organ. for Biol. Cont. 123 p.

Baker, R.; Longhurst, C.; Selwood, D.; Billany, D. 1983. Ortho-aminoacetophenone: a component of the sex pheromone system of the web-spinning larch sawfly, *Cephalcia lariciphila* Wachtl. Experientia 39: 993-994.

Bartelt, R.J.; Jones, R.L. 1983. (Z)-10-nonadecenal: a pheromonally active air oxidation product of (Z,Z)-9,19 Dienes in yellowheaded spruce sawfly. J. Chem. Ecol. 9: 1333-1341.

Bartelt, R.J.; Jones, R.L.; Krick, T.P. 1983. (5)-tetradecen-1-ol: a secondary pheromone of the yellowheaded spruce sawfly, and its relationship to (Z)-10-nonadecenal. J. Chem. Ecol. 9: 1343-1352.

Bartelt, R.J.; Jones, R.L.; Kulman, H.M. 1982. Evidence for a multicomponent sex pheromone in the yellowheaded spruce sawfly. J. Chem. Ecol. 8: 83-84.

Bartelt, R.J.; Jones, R.L.; Kulman, H.M. 1983. Hydrocarbon components of the yellowheaded spruce sawfly sex pheromone: a series of (Z,Z)-9-19 Dienes. J. Chem. Ecol. 9: 95-114.

Bierl, B.A.; Beroza, M.; Collier, C.W. 1970. Potent sex attractant of the gypsy moth: its isolation, identification, and synthesis. Sci. 170: 87-89.

Bobb, M.L. 1972. Influence of sex pheromones on mating behavior and populations of Virginia pine sawfly. Environ. Entomol. 1: 78-80.

Booij, C.J.H.; Voerman, S. 1984. New sex attractants for 35 tortricid and 4 other lepidopterous species, found by systematic field screening in the Netherlands. J. Chem. Ecol. 10: 135-144.

Booij, C.J.H.; Voerman, S. 1984. (Z)-11-hexadecenyl compounds as attractants for male microlepidoptera of the subfamilies Argyerthunae, Glyphipteryginae, and Crambinae. Entomol. Exp. Appl. 36: 47-53.

Butenandt, A.; Beckman, R.; Stamm, D.; Hecker, E. 1959. Uber den Sexual-Lockstoff des Seidenspinners, *Bombyx mori*. Reindarstellung and Konstitution. Z. Naturforsch. 146: 283-284.

Cardé, R.T.; Doane, C.C.; Baker, T.C.; Iwaki, S.; Marumo, S. 1977. Attractancy of optically active pheromone for male gypsy moths. Environ. Entomol. 6: 768-772.

Casida, J.E.; Coppel, H.C.; Watanabe, T. 1963. Purification and potency of the sex attractant from the introduced pine sawfly, *Diprion similis*. J. Econ. Entomol. 56: 18-24.

Chisholm, M.D.; Underhill, E.W.; Steck, W.; Slessor, K.N.; Grant, G.G. 1980. (Z)-5,(E)-7-dodecadienal and (Z)-5,(E)-7-dodecadien-1-ol, sex pheromone components of the forest tent caterpillar, *Malacosoma disstria*. Environ. Entomol. 9: 278-282.

Chisholm, M.D.; Palaniswamy, P.; Underhill, E.W. 1982. Orientation disruption of male forest tent caterpillar, *Malacosoma disstria* (Hubner) (Lepidoptera: Lasiocampidae), by air permeation with sex pheromone components. Environ. Entomol. 11: 1248-1250.

Chisholm, M.D.; Reed, D.W.; Underhill, E.W.; Palaniswamy, P.; Wong, J.W. 1985. Attraction of tortricid moths of subfamily Olethreutinae to field traps baited with dodecadienes. J. Chem. Ecol. 11: 217-230.

Cory, H.T.; Daterman, G.E.; Daves, Jr., G.D.; Sower, L.L.; Shepherd, R.F.; Sanders, C.J. 1982. Chemistry and field evaluation of the sex pheromone of western spruce budworm, *Choristoneura occidentalis*, Freeman. J. Chem. Ecol. 8: 339-350.

Daterman, G.E.; Peterson, L.J.; Robbins, R.G.; Sower, L.L.; Daves, Jr., G.D.; Smith, R.G. 1976. Laboratory and field bioassay of the Douglas-fir tussock moth pheromone, (Z)-6-heneicosene-11-one. Environ. Entomol. 5: 1187-1190.

Daterman, G.E.; Cory, H.T.; Sower, L.L.; Daves, Jr., G.D. 1984. Sex pheromone of a conifer-feeding budworm, *Choristoneura retiniana*, Walsingham. J. Chem. Ecol. 10: 153-160.

Dimond, J.B.; Mott, D.G.; Kemp, W.P.; Krall, J.H. 1984. A field test of mating-suppression using the spruce budworm sex pheromone. Maine Agric. Exp. Stn., Univ. of Maine, Tech. Bull. 113, 14 p.

Dix, M.E.; Underhill, E.W. 1988. Sex pheromone of *Retinia metallica* (Busck) (Lepidoptera: Tortricidae): Identification and field studies. Can. Entomol. 120: 721-726.

Doolittle, R.E.; Roelofs, W.L.; Solomon, J.D.; Cardé, R.T.; Beroza, M. 1976. (Z,E)-3,5-tetradecadien-1-ol acetate sex attractant for the carpenterworm moth, *Prionoxystus robiniae* (Peck) (Lepidoptera: Cossidae). J. Chem. Ecol. 2: 399-410.

Eller, F.J.; Bartelt, R.J.; Jones, R.L.; Kulman, H.M. 1984. Ethyl(Z)-9-Hexadecenoate: a sex pheromone of *Syndipnus rubiginosus*, a sawfly parasitoid. J. Chem. Ecol. 10: 291-300.

Grant, G.G. 1977. Interspecific pheromone responses of tussock moths and some isolating mechanisms of eastern species. Environ. Entomol. 6: 739-742.

Grant, G.G. 1978. Field trials on disruption of pheromone communication of tussock moths. J. Econ. Entomol. 71: 453-457.

Grant, G.G. 1991. Development and use of pheromones for monitoring lepidopterous forest defoliators in North America. For. Ecol. Manage. 39: 153-162.

Grant, G.G.; Frech, D.; MacDonald, L.; Doyle, B. 1981. Effect of additional components on a sex attractant for the oak leafshredder, *Croesia semipurpurana* (Lepidoptera: Tortricidae). Can. Entomol. 113: 449-451.

Grant, G.G.; MacDonald, L.; Frech, D.; Hall, K.; Slessor, K.N. 1985. Sex attractants for some eastern species of *Rhyacionia*, including a new species, and *Eucosma gloriola* (Lepidoptera: Tortricidae). Can. Entomol. 117: 1489-1496.

Grant, G.G.; Prévost, Y.H.; Slessor, K.N.; King, G.G.S.; West, R.J. 1987. Identification of the sex pheromone of the spruce coneworm, *Dioryctria reniculelloides* (Lepidoptera: Pyralidae). Environ. Entomol. 16: 905-909.

Grant, G.G.; Slessor, K.N. 1983. Pheromone of the maple leaf roller *Cenopis acerivorana* (Lepidoptera: Tortricidae). Can. Entomol. 115: 189-194.

Gray, T.G.; Slessor, K.N.; Grant, G.G.; Shepherd, R.F.; Holsten, E.H.; Tracey, A.S. 1984. Identification and field testing of pheromone components of *Choristoneura orae* (Lepidoptera: Tortricidae). Can. Entomol. 116: 51-56.

Gray, T.G.; Slessor, K.N.; Shepherd, R.F.; Grant, G.G.; Manville, J.F. 1984. European pine shoot moth, *Rhyacionia buoliana* (Lepidoptera: Tortricidae): identification of additional pheromone components resulting in an improved lure. Can. Entomol. 116: 1525-1532.

Hedlin, A.F.; Weatherston, J.; Ruth, D.S.; Miller, G.E. 1983. Chemical lure for male Douglas-fir cone moth, *Barbara colfaxiana* (Lepidoptera: Olethreutidae). Environ. Entomol. 12: 1751-1753.

Hill, A.S.; Kovalev, B.G.; Nikolaeva, L.N.; Roelofs, W.L. 1982. Sex pheromone of the fall webworm moth, *Hyphantria cunea*. J. Chem. Ecol. 8: 383-396.

Jewett, D.M.; Matsumura, F.; Coppel, H.C. 1976. Sex pheromone specificity in the pine sawflies: interchange of acid moieties in an ester. Sci. 192: 51-53.

Jewett, D.; Matsumura, F.; Coppel, H.C. 1978. Preparation and use of sex attractants for four species of pine sawflies. J. Chem. Ecol. 4: 277-287.

Jobin, L. 1985. Development of a large capacity pheromone trap for monitoring forest insect pest populations. Pages 243-244 *in* C.D. Sanders, R.W. Stark, E.J. Mullins, and J. Murphy (Eds.), Recent Advances in Spruce Budworm Research. Proc. CANUSA Spruce Budworms Symp. Bangor, Me., Sept. 16-20, 1984.

Jonsson, S.; Bergstrom, G., Lanne, B.S., Stensdotter, U. 1988. Defensive odor emission from larvae of two sawfly species, *Pristiphora erichsonii* and *P. wesmaeli*. J. Chem. Ecol. 14: 713-721.

Kikukawa, T.; Matsumura, F.; Kraemer, M.; Coppel, H.C.; Akira, Tai. 1982. Field attractiveness of chirally defined synthetic attractants to males of *Diprion similis* and *Gilpinia frutetorum*. J. Chem. Ecol. 8: 301-314.

Kochansky, J.; Tette, J.; Taschenberg, E.F.; Cardé, R.T.; Kaissling, K.E.; Roelofs, W.L. 1975. Sex pheromone of the moth *Antheraea polyphemus*. J. Insect Physiol. 21: 1977-1983.

Kraemer, M.E.; Coppel, H.C.; Kikukawa, T.; Matsumura, F.; Thomas, H.A.; Thompson, L.C.; Mori, K. 1983. Field and electroantennogram responses to sex pheromone optical isomers by four fall-flying sawfly species (Hymenoptera: Diprionidae, *Neodiprion*). Environ. Entomol. 12: 1592-1596.

Kraemer, M.E.; Coppel, H.C.; Matsumura, F.; Kikukawa, T.; Mori, K. 1979. Field responses of the white pine sawfly, *Neodiprion pinetum*, to optical isomers of sawfly sex pheromones. Environ. Entomol. 8: 519-520.

Kraemer, M.E.; Coppel, H.C.; Matsumura, F.; Wilkinson, R.C.; Kikukawa, T. 1981. Field and electroantennogram responses of the red-headed pine sawfly, *Neodiprion lecontei*) (Fitch), to optical isomers of sawfly sex pheromones. J. Chem. Ecol. 7: 1063-1072.

Kuenzi, F.M.; Coppel, H.C. 1986. Isozymes of the sawflies *Neodiprion* and *Diprion similis*: diagnostic characters and genetic distance. Biochem. System. and Ecol. 14: 423-429.

Liebhold, A.M.; Volney, W.J.A. 1985. Effects of attractant composition and release rate on attraction of male *Choristoneura retiniana*, *C. occidentalis*, and *C. carnana* (Lepidoptera: Tortricidae). Can. Entomol. 117: 447-457.

Longhurst, C.; Baker, R.; Mori, K. 1980. Response of the sawfly *Diprion similis* to chiral sex pheromones. Experientia 36: 946-947.

Magasi, L.P. 1988. Forest pest conditions in the Maritimes 1987. Can. For. Serv., Fredericton, N.B.. Inf. Rep. M-X-166.

Matsumura, F.; Akaira, T.; Coppel, H.C.; Imaida, M. 1979. Chiral specificity of the sex pheromone of the red-headed pine sawfly, *Neodiprion lecontei*. J. Chem. Ecol. 5: 237-249.

Meyer, W.L.; Debarr, G.L.; Berisford, C.W.; Barber, L.R.; Roelofs, W.L. 1982. Identification of the sex pheromone of the webbing coneworm moth, *Dioryctria disclusa* (Lepidoptera: Pyralidae). Environ. Entomol. 11: 986-988.

Miller, C.A. 1979. Report of spruce budworm pheromone trials, Maritimes 1978. Int. Rep., Maritimes For. Res. Cent., Can. For. Serv., Fredericton, N.B. 149 p.

Miller, J.R.; Baker, T.C.; Cardé, R.T.; Roelofs, W.L. 1976. Reinvestigation of oak leaf roller sex pheromone components and the hypothesis that they vary with diet. Sci. 192: 140-143.

Morse, B.W.; Kulman, H.M. 1985. Monitoring damage by yellowheaded spruce sawflies with sawfly and parasitoid pheromones. Environ. Entomol. 14: 131-133.

Niwa, C.G.; Daterman, G.E.; Sartwell, C.; Sower, L.L. 1988. Control of *Rhyacionia zozana* (Lepidoptera: Tortricidae) by mating disruption with synthetic sex pheromone. Environ. Entomol. 17: 593-595.

Niwa, C.G.; Sower, L.L.; Daterman, G.E. 1987. Chemistry and field evaluation of the sex pheromone of the pine tip moth, *Rhyacionia zozana* (Lepidoptera: Tortricidae). Environ. Entomol. 16: 1287-1290.

Olaifa, J.I.; Kikukawa, T.; Matsumura, F.; Coppel, H.C. 1984. Response of male jack pine sawfly, *Neodiprion pratti banksianae* (Hymenoptera: Diprionidae), to mixtures of optical isomers of the sex pheromone, 3,7-Dimethylpentadecan-2-ol. Environ. Entomol. 13: 1274-1277.

Olaifa, J.I.; Matsumura, F.; Coppel, H.C. 1987. Field responses and gas-liquid chromatograph separation of optically active synthetic and natural pheromones in two sympatric Diprionid sawflies, *Neodiprion nanulus nanulus* and *Neodiprion sertifer* (Hymenoptera: Diprionidae). J. Chem. Ecol. 13: 1395-1408.

Olaifa, J.I.; Matsumura, F.; Kikukawa, T.; Coppel, H.C. 1988. Pheromone-dependent species recognition mechanisms between *Neodiprion pinetum* and *Diprion similis* on white pine. J. Chem. Ecol. 14: 1131-1144.

Priesner, E.; Altenkirch, W.; Battensweiler, W.; Bogenschütz, H. 1982. Evaluation of (Z)-5-decan-1-ol as an attractant for male larch casebearer moths, *Coleophora laricella*. Z. Naturforsch. 37c: 953-966.

Ramaswamy, S.B.; Cardé, R.T. 1982. Non-saturating traps and long-life attractant lures for monitoring spruce budworm males. J. Econ. Entomol. 75: 126-129.

Reed, D.W.; Chisholm, M.D. 1985a. Field trapping of three *Epinotia* species with (Z,Z)-7,9-dodecadienyl acetate (Lepidoptera: Tortricidae). J. Chem. Ecol. 11: 1389-1398.

Reed, D.W.; Chisholm, M.D. 1985b. Attraction of moth species of Tortricidae, Gelechiidae, Geometridae, Drepanidae, Pyralidae, and Gracillariidae families to field traps baited with conjugated dienes. J. Chem. Ecol. 11: 1645-1657.

Retnakaran, A.; Grant. G.G. 1985. Control of the oak leafshredder, *Croesia semipurpurana* (Kearfott), by aerial application of diflurbenzuron. Can. Entomol. 117: 363-369.

Roelofs, W.L.; Brown, R.L. 1982. Pheromones and evolutionary relationships of Tortricidae. Ann. Rev. Eco. Syst. 13: 359-422.

Roelofs, W.L.; Hill, A.S.; Linn, C.E.; Meinwald, J.; Jain, S.C.; Herbert, H.J.; Smith, R.F. 1982. Sex pheromone of the winter moth, a geometrid with unusually low temperature precopulatory responses. Science. 217: 657-659.

Sanders, C.J. 1976. Disruption of sex attraction in the eastern spruce budworm. Environ. Entomol. 5: 868-872.

Sanders, C.J. 1978. Evaluation of sex attractant traps for monitoring spruce budworm populations (Lepidoptera: Tortricidae). Can. Entomol. 110: 43-50.

Sanders, C.J. 1979. Spruce budworm mating disruption trials using synthetic attractant in Conrel fibres (Ontario, 1977). Can. For. Serv., Sault Ste. Marie, Ont. Inf. Rep. O-X-285, 32 p.

Sanders, C.J. 1986. Evaluation of high-capacity, non-saturating traps for monitoring population densities of spruce budworm (Lepidoptera: Tortricidae). Can. Entomol. 118: 611-619.

Sanders, C.J. 1988. Monitoring spruce budworm population density with sex pheromone traps. Can. Entomol. 120: 175-183.

Sanders, C.J.; Daterman, G.E.; Shepherd, R.F.; Cerecke, H. 1974. Sex attractants for the two species of western budworm, *Choristoneura biennis* and *C. viridis* (Lepidoptera: Tortricidae). Can. Entomol. 106: 157-159.

Sanders, C.J.; Meighen, E.A. 1987. Controlled-release sex pheromone lures for monitoring spruce budworm populations. Can. Entomol. 119: 305-313.

Sanders, C.J.; Seabrook, W.D. 1982. Disruption of mating in the spruce budworm, *Choristoneura fumiferana* (Clemens). Page 312 *in* A.F. Kydonieus and M. Beroza (Eds.), Insect Suppression with Controlled Release Pheromone Systems. Vol.II. C.R.C. Press Inc., Boca Raton, Fla.

Sanders, C.J.; Silk, P.J. 1982. Disruption of spruce budworm mating by means of Hercon plastic laminated flakes, Ontario 1981. Can. For. Serv. Inf. Rep. 0-X-335, 22 p.

Sanders, C.J.; Weatherston, J. 1976. Sex pheromone of the eastern spruce budworm (Lepidoptera: Tortricidae): Optimum blend of *trans*- and *cis*-11-tetradecenal. Can. Entomol. 108: 1285-1290.

Sartwell, C.; Daterman, G.E.; Sower, L.L. 1985. A synthetic attractant for male moths of a biotype in the *Cydia piperana* complex (Lepidoptera: Tortricidae). Can. Entomol. 117: 1151-1152.

Schonherr, J.; Weir, T.; Langguth, C.; Reulecke, T. 1979. Factors influencing the female sex attractiveness in pine sawflies Diprion pini and Neodiprion sertifer (Hym. Diprionidae). Mitt. Schweiz. Entomol. Ges. 52: 217-222.

Seabrook. W.D.; Kipp, L.R. 1986. The use of a two component blend of the spruce budworm sex pheromone for mating suppression. Proc. 13th Int. Symp. on Controlled Release of Bioactive Materials. Controlled Release Soc. Inc.

Shepherd, R.F.; Gray, T.G.; Chorney, R.J.; Daterman, G.E. 1985. Pest management of Douglas-fir tussock moth, *Orgyia pseudotsugata* (Lepidoptera: Lymantriidae): monitoring endemic populations with pheromone traps to detect incipient outbreaks. Can. Entomol. 117: 839-838.

Shepherd, R.F.; Otvos, I.S. 1986. Pest management of Douglas-fir tussock moth: procedures for insect monitoring, problem evaluation and control actions. For. Can., Pac. Region, Victoria, B.C. Inf. Rep. BC-X-270, 14 p.

Shepherd, R.F.; Otvos, I.S.; Chorney, R.J. 1984. Pest management of Douglas-fir tussock moth (Lepidoptera: Lymantriidae): a sequential sampling method to determine egg mass density. Can. Entomol. 116: 1041-1049.

Silk, P.J.; Butterworth, E.W.; Kuenen, L.P.S.; Northcott, C.J.; Dunkelblum, E.; Kettela, E.G. 1989. Identification of a sex pheromone component of the spruce budmoth, *Zeiraphera canadensis*. J. Chem. Ecol. 15: 2435-2444.

Silk, P.J.; Butterworth, E.W.; Kuenen, L.P.S.; Northcott, C.J.; Kettela, E.G. 1988. Sex pheromone of purplestriped shootworm, *Zeiraphera unfortunana* Powell. J. Chem. Ecol. 14: 1417-1425.

Silk, P.J.; Kuenen, L.P.S.; Tan, S.H.; Roelofs, W.L.; Sanders, C.J.; Alford, A.R. 1985. Identification of the sex pheromone components of jack pine budworm, *Choristoneura pinus pinus* Freeman. J. Chem. Ecol. 11: 159-167.

Silk, P.J., Tan, S.H.; Wiesner, C.J.; Ross, R.J.; Lonergan, G.C. 1980. Sex pheromone chemistry of the eastern spruce budworm, *Choristoneura fumiferana*. Environ. Entomol. 9: 640-644.

Silk, P.J.; Wiesner, C.J.; Tan, S.H.; Ross, R.J.; Grant, G.G. 1982. Sex pheromone chemistry of the western spruce budworm *Choristoneura occidentalis* Free. J. Chem. Ecol. 8: 351-362.

Smith, R.G.; Daterman, G.E.; Daves, Jr., G.D. 1975. Douglas-fir tussock moth: Sex pheromone identification and synthesis. Sci. 188: 63-64.

Smith, R.G.; Daterman, G.E.; Daves, Jr., G.D.; McMurtrey, K.D.; Roelofs, W.L. 1974. Sex pheromone of the European pine shoot moth: chemical identification and field tests. J. Insect Physiol. 20: 661-668.

Sower, L.L.; Daterman, G.E.; Funkhouser, W.; Sartwell, C. 1983. Pheromone disruption controls Douglas-fir tussock moth (Lepidoptera: Lymantriidae) reproduction at high densities. Can. Entomol. 115: 965-969.

Sower, L.L.; Daterman, G.E.; Orchard, R.D.; Sartwell, C. 1979. Reduction of Douglas-fir tussock moth reproduction with synthetic sex pheromone. J. Econ. Entomol. 72: 739-742.

Sower, L.L.; Daterman, G.E.; Sartwell, C.; Cory, H.T. 1979. Attractants for the western pine shoot borer, *Eucosma sonomana*, and *Rhyacionia zozana* determined by field screening. Environ. Entomol. 8: 265-267.

Sower, L.L.; Overhulser, D.L. 1986. Recovery of *Eucosma sonomana* (Lepidoptera: Tortricidae) populations after mating disruption treatments. J. Econ. Entomol. 79: 1645-1647.

Steck, W.F.; Chisholm, M.D.; Bailey, B.K.; Underhill, E.W. 1979. Moth sex attractants found by systematic field testing of 3-component acetate-aldehyde candidate lures. Can. Entomol. 111: 1263-1269.

Stevens, R.E.; Sartwell, C.; Koerber, T.W.; Powell, J.A.; Daterman, G.E.; Sower, L.L. 1985. Forest tortricids trapped using *Eucosma* and *Rhyacionia* synthetic sex attractants. J. Lepid. Soc. 39: 26-32.

Struble, D.L.; Byers, J.R.; Shepherd, R.F.; Gray, T.G. 1989. Identification of sex pheromone components of the black army cutworm, *Actebia fennica* (Tauscher) (Lepidoptera: Noctuidae), and a sex attractant blend for adult males. Can. Entomol. 121: 557-563.

Struble, D.L.; Swailes, G.E.; Steck, W.F.; Underhill, E.W.; Chisholm, M.D. 1976. A sex attractant for the adult males of variegated cutworm, *Peridroma saucia*. Environ. Entomol. 5: 988-990.

Turgeon, J.J. 1985. Life cycle and behavior of the spruce budmoth, *Zeiraphera canadensis* (Lepidoptera: Olethreutidae), in New Brunswick. Can. Entomol. 117: 1239-1247.

Turgeon, J.J.; Grant, G.G. 1988. Response of *Zeiraphera canadensis* (Lepidoptera: Tortricidae: Olethreutinae) to candidate sex attractants and factors affecting trap catches. Environ. Entomol. 17: 442-447.

Underhill, E.W.; Chisholm, M.D.; Steck, W. 1980. (E)-5,(Z)-7-dodecadienal, a sex pheromone component of the western tent caterpillar, *Malacosoma californicum* (Lepidoptera: Lasiocampidae). Can. Entomol. 112: 629-631.

Underhill, E.W.; Millar, J.G.; Ring, R.A.; Wong, J.W.; Barton, D.; Giblin, M. 1987. Use of a sex attractant and an inhibitor for monitoring winter moth and Bruce spanworm populations. J. Chem. Ecol. 13: 1319-1330.

Weatherston, J.; Davidson, L.M.; Siminini, D. 1974. Attractants for several male forest Lepidoptera. Can. Entomol. 106: 781-782.

Weatherston, J.; Grant, G.G.; MacDonald, L.M.; Frech, D.; Werner, R.A.; Leznoff, C.C.; Fyles, T.M. 1978. Attraction of various Tortricine moths to blends containing *cis*-11-tetradecenal. J. Chem. Ecol. 4: 543-549.

Weatherston, J.; Percy, J.E.; MacDonald, L.M. 1976. Field testing of *cis*-11-tetradecenal as attractant or synergist in Tortricinae. Experientia. 32: 178-179.

Weatherston, J.; Roelats, W.; Comeau, A.; Sanders, C.J. 1971. Studies of physiologically active arthropod secretions. X. Sex pheromone of the eastern spruce budworm, *Choristoneura fumiferana* (Lepidoptera: Tortricidae). Can. Entomol. 103: 1741-1747.

Webb, R.E.; Tatman K.M.; Leonhardt, B.A.; Plimmer, J.R.; Boyd, J.K.; Bystrak, P.G.; Schwalbe, C.P.; Douglas, L.W. 1988. Effect of aerial application of racemic disparlure on male trap catch female mating success of gypsy moths (Lepidoptera: Lymantriidae). J. Econ. Entomol. 81: 268-273.

Wilkinson, R.C.; Chappelka III, A.H.; Kraemer, M.E.; Coppel, H.C.; Matsumura, F. 1982. Field responses of redheaded pine sawfly males to a synthetic pheromone and virgin females in Florida. J. Chem. Ecol. 8: 471-475.

Wilkinson, R.C.; Chappelka III, A.H.; Kraemer, M.E.; Coppel, H.C.; Matsumura, F. 1987. Effect of height on responses of redheaded pine sawfly (Hymenoptera: Diprionidae) males to synthetic pheromone and virgin females. Environ. Entomol. 16: 1152-1156.

Wong, H.R.; Drouin, J.A.; Rentz, C.L. 1985. *Petrova albicapitana* (Busck) and *P. metallica* (Busck) (Lepidoptera: Tortricidae) in *Pinus contorta* Dougl. stands of Alberta. Can. Entomol. 117: 1463-1470.

Wong, J.W.; Palaniswamy, P.; Underhill, E.W.; Steck, W.F.; Chisholm, M.D. 1984a. Novel sex pheromone components from the fall cankerworm moth, *Alsophila pometaria*. J. Chem. Ecol. 10: 463-473.

Wong, J.W.; Palaniswamy, P.; Underhill, E.W.; Steck, W.F.; Chisholm, M.D. 1984b. Sex pheromone components of fall cankerworm moth, *Alsophila pometaria* synthesis and field trapping. J. Chem. Ecol. 10: 1579-1596.

Chapter 45

Development and Use of Semiochemicals Against Bark and Timber Beetles

JOHN H. BORDEN

Introduction

Since Anderson's (1948) discovery that mass attack by pine engravers, *Ips pini*, was associated with odors produced by male beetles, interest in the semiochemical-mediated behavior of bark and timber beetles has grown phenomenally.[1] Most of these insects are in the family Scolytidae. In this family, more than any other, insect-produced semiochemicals may serve several roles and may in turn be supplemented by semiochemicals produced by other insect species or even their host trees. For example, one semiochemical may serve at low concentration as an aggregations pheromone and act synergistically with another pheromone, as well as a kairomone from the host. At high concentration the same semiochemical may serve as a spacing or antiaggregation pheromone. At all concentrations it may serve as an allomone protecting an occupied host from attack by a competing species. Finally, the same semiochemical may be used as a kairomone by a predator or parasite seeking to dine or oviposit on its bark beetle host.

This complexity has presented a considerable challenge to scientists seeking to understand chemical communication in the Scolytidae. At the same time, these complex systems may have many points of vulnerability (Borden 1989), and these have been profitably exploited in the development of semiochemical-based pest management programs.

Isolation and Identification of Semiochemicals

A standard protocol (Silverstein et al. 1967) has characteristically been used to isolate and identify semiochemicals used by bark and timber beetles. It relies on the prior discovery of semiochemical-mediated phenomena (e.g. aggregation) in the target species. Each of the steps in the protocol is monitored by bioassays for the phenomenon in question.

[1]Nordlund's (1981) terminology for chemical-releasing stimuli is followed herein with particular reference to pheromones, allomones, and kairomones, as follows: "*Semiochemical.* A chemical involved in the interaction between organisms. *Pheromone.* A substance that is secreted by an organism to the outside and causes a specific reaction in a receiving organism of the same species. *Allomone.* A substance produced or acquired by an organism that, when it contacts an individual of another species in the natural context, evokes in the receiver a behavioral or physiological reaction that is adaptively favorable to the emitter but not to the receiver. *Kairomone.* A substance produced or acquired by an organism that, when it contacts an individual of another species in the natural context, evokes in the receiver a behavioral or physiological reaction that is adaptively favorable to the receiver but not to the emitter."

The first step is to produce large amounts of starting material containing the semiochemical. In the first successful isolations, huge quantities of bark or timber beetle frass were collected (Wood et al. 1966; Silverstein et al. 1966, 1968; MacConnell et al. 1977). As chemical and biological methodology became increasingly sophisticated, microextractions of single beetles (Slessor et al. 1985), or capturing the volatiles produced by individual beetles (Gries et al. 1988), became possible. In one European program, the beetles were treated with a juvenile hormone analog to enhance the production of pheromones (Francke et al. 1977).

Extracted or captured chemicals are separated chromatographically (Brand et al. 1979), using such techniques as silica gel and preparative gas chromatography. Isolated compounds of verified activity are purified and derivatized, if necessary, into easily identified materials. Many spectrometric analyses (Brand et al. 1979) are used, including coupled gas chromatography–mass spectrometry, and nuclear magnetic resonance spectrometry. After it was discovered that many bark beetle pheromones were optically active, new techniques were developed to resolve the enantiomers of a compound (Schurig et al. 1983; Slessor et al. 1985). The identification of isolated compounds is confirmed by rational synthesis and spectrometric comparison of isolated and synthetic materials. Finally, the newly identified compounds are synthesized or purchased in amounts sufficient for field testing, and in anticipation of operational use, commercial syntheses are devised. Some of the research on the isolation, identification, and synthesis of semiochemicals used by bark and timber beetles has been done in Canada, and the semiochemicals used by several Canadian species are known (Table 1).

Vulnerability of Semiochemical-Mediated Systems

A generalized mass-attack sequence by scolytid beetles (Fig. 1) typically starts with host-seeking beetles in a dispersal flight and ends with the termination of attack on a fully utilized resource. Two principal strategies have been developed to exploit the beetles' dependence on semiochemicals in this natural sequence (Borden 1989).

The first strategy exploits the attraction that occurs in nature when a host tree or log is undergoing mass attack (Fig. 1). In a responsive state primed by flight exercise (Borden et al. 1986), these beetles are vulnerable to artificial attractive sources that mimic those that are produced in nature by either or both sexes, as well as their hosts. Application of this strategy may capitalize on the beetles' visual response to host form (e.g. by baiting large-diameter host trees with semiochemicals, or by

Table 1. Semiochemicals isolated and/or identified in Canadian bark and timber beetles: their source and biological activity.

Family, species, and common name (if any)	Semiochemical[a]	Source and biological activity
Curculionidae		
Pissodes approximatus Northern pine weevil	grandisol grandisal	Identified from hindgut extracts and captured volatiles from males. In field tests, grandisal with odor of red pine attractive to both sexes, grandisol inactive. Mixture of both compounds attractive; when odor of red pine present, response increases to level induced by four or five males on red pine (Booth et al. 1983).
Pissodes strobi White pine weevil	grandisol grandisal	Isolated from hindgut extracts and captured volatiles from males (Booth et al. 1983). Not attractive in field tests; response of *P. approximatus* inhibited by unknown allomone (Phillips and Lanier 1986).
Scolytidae		
Dendroctonus brevicomis Western pine beetle	frontalin *exo*-brevicomin *trans*-verbenol verbenone [ipsdienol]	*Exo*-brevicomin (Silverstein et al. 1968) and myrcene (Silverstein 1970) isolated and identified from female frass, frontalin from male hindguts (Kinzer et al. 1969). All three attractive in field (Bedard et al. 1969), myrcene and *exo*-brevicomin mainly to males and frontalin to females (Vité and Pitman 1969). Enantiomeric composition primarily (-)-frontalin and (+)-*exo*-brevicomin (Stewart et al. 1977). 1*S*,5*R*-(-)-frontalin and 1*R*,5*S*,7*R*-(+)-*exo*-brevicomin are the active isomers in field tests (Wood et al. 1976). Verbenone and *trans*-verbenol identified from hindguts of males and females, respectively (Renwick 1967). Verbenone proposed as antiaggregation pheromone (Renwick and Vité 1970). Verbenone and *trans*-verbenol inhibit response (Wood 1972; Bedard et al. 1980a,b). (+)-Ipsdienol detected in guts of males exposed to myrcene vapors (Byers 1982), or in fresh hosts (Byers et al. 1984); inhibited response alone (Byers 1982) or in combination with verbenene (Paine and Hanlon 1991) to aggragation pheromones with myrcene in the field.
Dendroctonus ponderosae Mountain pine beetle	*trans*-verbenol *exo*-brevicomin *endo*-brevicomin frontalin verbenone ipsdienol α-pinene myrcene terpinoline	*Trans*-verbenol isolated and identified from females (Pitman et al. 1968), attractive to both sexes in field with α-pinene (Pitman 1971). (-)-*trans*-Verbenol active (Borden et al. 1987a). Brevicomins identified from hindgut or abdominal extracts, or beetle volatiles (Pitman et al. 1969; Rudinsky et al. 1974a). In British Columbia lodgepole pine populations, *exo*-brevicomin attractive with *trans*-verbenol and a monoterpene in traps (Conn et al. 1983; Borden et al. 1987a) and on trap trees (Borden et al. 1983a). In British Columbia and Oregon, *exo*-brevicomin at high concentrations inhibits response (Rudinsky et al. 1974a; Ryker and Rudinsky 1982; Borden et al. 1987a); at low concentrations unattractive in Oregon (Libbey

[a]Chemical names for pheromones (but not monoterpene kairomones) listed alphabetically by common or trivial name (if any) are as follows:

exo-brevicomin: *exo*-7-ethyl-5-methyl-6,8-dioxabicyclo [3.2.1] octane
endo-brevicomin: *endo*-7-ethyl-5-methyl-6,8-dioxabicyclo [3.2.1] octane
frontalin: 1,5-dimethyl-6,8-dioxabicyclo [3.2.1] octane
grandisal: 1-methyl-2[1-methylethenyl]cyclobutaneacetaldehyde
grandisol: [1*R*-*cis*]-1-methyl-2-[1-methylethenyl]cyclobutaneethanol
ipsdienol: 2-methyl-6-methylene-2, 7-octadien-4-ol
ipsenol: 2-methyl-6-methylene-7-octen-4-ol
lineatin: 3,3,7-trimethyl-2,9-dioxatricyclo [3.3.1.0^{4,7}]nonane
3,2-MCH: 3-methylcyclohex-2-en-1-one
3,3-MCH: 3-methylcyclohex-3-en-1-one
1,2-MCH-ol: 1-methyl-2-cyclohexen-1-ol
3-methyl-3-buten-1-ol
methylbutanone: 3-hydroxy-3-methylbutan-2-one
methylheptanol: 4-methylheptan-3-ol
methylheptenone: 6-methylhept-5-en-2-one
α-multistriatin: 2,4-dimethyl-5-ethyl-6,8-dioxabicyclo [3.2.1] octane
myrcenol: 2-methyl-6-methylene-2,7-octadien-3-ol
myrtenol: 4,6,6-trimethylbicyclo [3.1.1] hept-3-en-10-ol
trans-pentenol: *trans*-pent-3-en-2-ol
retusol: *S*-(+)-6-methylhept-5-en-2-ol
seudenol (3,2-MCH-ol): 3-methyl-2-cyclohexen-1-ol
sulcatol: (±) 6-methylhept-5-en-2-ol
cis-verbenol: *cis*-4,6,6-trimethylbicyclo [3.1.1] hept-3-en-2-ol
trans-verbenol: *trans*-4,6,6-trimethylbicyclo [3.1.1] hept-3-en-2-ol
verbenene: 6,6-dimethyl-4-methylenebicyclohept-2-ene
verbenone: 4,6,6-trimethylbicyclo [3.1.1] hept-3-en-2-one

(Continued)

Table 1. Semiochemicals isolated and/or identified in Canadian bark and timber beetles: their source and biological activity. *(Continued)*

Family, species, and common name (if any)	Semiochemical[a]	Source and biological activity
		et al. 1985). *Exo*-brevicomin inhibits response by white pine populations (Pitman et al. 1978). *Endo*-brevicomin inhibits response (Rudinsky et al. 1974a). Mated males in lodgepole pine populations produce frontalin which acts as multifunctional pheromone (Ryker and Libbey 1982; Borden et al. 1987a). Verbenone isolated and identified in hindguts of females (Pitman et al. 1969), but is produced primarily by microorganisms (Hunt and Borden 1989), acts as antiaggregation pheromone, (-) enantiomer active (Ryker and Yandell 1983; Borden et al. 1987a). *S*-(+)-ipsdienol detected in abdomens of males exposed to myrcene vapors (Hunt et al. 1986). (±) - and *S*-(+)-ipsdienol inhibit response in field (Hunt and Borden 1988). Myrcene and terpinoline better kairomones than α-pinene in ponderosa or lodgepole pine populations (Billings et al. 1976; Conn et al. 1983; Borden et al. 1983a).
Dendroctonus pseudotsugae Douglas-fir beetle	frontalin *trans*-verbenol seudenol 1,2-MCH-ol verbenone *trans*-pentenol 3,2-MCH 3,3-MCH methylheptenone α-pinene camphene ethanol	Frontalin (Pitman and Vité 1970), *trans*-verbenol (Rudinsky et al. 1972a), seudenol (Vité et al. 1972), verbenol (Rudinsky et al. 1974b), *trans*-pentenol (Ryker et al. 1979) and 1,2-MCH-ol (Libbey et al. 1983) isolated and/or identified from female hindguts or volatiles; 3,3-MCH (Libbey et al. 1976) and methylheptenone (Ryker et al. 1979) from male volatiles; 3,2-MCH from both sexes (Kinzer et al. 1971; Pitman and Vité 1975). Frontalin, seudenol, 1,2-MCH-ol, *trans*-pentenol, α-pinene, camphene, and ethanol all components of attractive mixtures (Pitman and Vité 1970; Furniss and Schmitz 1971; Pitman et al. 1975; Ryker et al. 1979; Libbey et al. 1983). *Trans*-verbenol synergistic with frontalin (Rudinsky et al. 1972a), but role unclear in other tests (Rudinsky et al. 1972b; Furniss et al. 1972). Verbenone (Rudinsky et al. 1974b) and 3,2-MCH (Rudinsky et al. 1972b; Rudinsky 1973; Rudinsky and Ryker 1980) attractive with other components at low concentrations, inhibitory at high concentrations, 3,3-MCH (Rudinsky and Ryker 1979) and methylheptenone (Ryker et al. 1979) inhibit response.
Dendroctonus rufipennis Spruce beetle	frontalin seudenol 1,2-MCH-ol verbenene α-pinene	Frontalin (Pitman and Vité 1970) and seudenol (Vité et al. 1972a) detected in female hindguts. Verbenene detected in volatiles of boring females (Gries et al. 1991). Frontalin induces attack of baited trees (Dyer 1973, 1975), attack enhanced by 1,2-MCH-ol (Wieser et al. 1991). Seudenol with α-pinene more attractive than frontalin in traps (Furniss et al. 1976; Dyer and Lawko 1978). Verbenene in traps attractive alone and in combination with frontalin and α-pinene (Gries et al. 1991).
Dryocoetes confusus Western balsam bark beetle	*exo*-brevicomin *endo*-brevicomin myrtenol	(+)-*exo*-, (+)-*endo*-Brevicomin and myrtenol detected in abdominal extracts of males (Schurig et al. 1983; Borden et al. 1987b). *exo*-Brevicomin acts as aggregation pheromone in traps or as a tree bait (Borden et al. 1987b), while *endo*-brevicomin inhibits response; natural enantiomers of both most active (A.J. Stock, A. Camacho, and J.H. Borden, unpubl. data). Myrtenol attractive in combination with *exo*-brevicomin (Borden et al. 1987b), but activity not upheld (A.J. Stock and J.H. Borden, unpubl. data).
Gnathotrichus retusus	retusol α-pinene ethanol	Retusol [S-(+)-sulcatol] isolated and identified from volatiles of boring males, attractive alone in laboratory and field, antipode inhibitory (Borden et al. 1980a). Synergized in the field by ethanol and α-pinene (Borden et al. 1980c, 1982).

(Continued)

Table 1. Semiochemicals isolated and/or identified in Canadian bark and timber beetles: their source and biological activity. *(Concluded)*

Family, species, and common name (if any)	Semiochemical[a]	Source and biological activity
Gnathotrichus sulcatus	sulcatol α-pinene ethanol	A 65:35 mixture of *S*-(+)- and R-(-)-sulcatol isolated and identified from male boring dust, hindguts, and volatiles (Byrne et al. 1974). Both enantiomers synergistic and essential for response (Borden et al. 1976). Ethanol and α-pinene synergize response in the field (Borden et al. 1980c, 1982).
Ips latidens	ipsenol *cis*-verbenol β-phellandrene α-pinene 3-carene terpinoline myrcene	Ipsenol found in male frass; (+)-, (-)-, and (±)-ipsenol attractive in field traps (Miller et al. 1991). *cis*-Verbenol synergistic with ipsenol in California (Wood et al. 1967), but inhibitory in British Columbia (Miller et al. 1991). β-phellandrene attractive alone. Attraction to ipsenol in British Columbia increased by addition of β-phellandrene, decreased by a mixture of four other monoterpenes (Miller and Borden 1990a).
Ips pini Pine engraver	ipsdienol myrcenol β-phellandrene	Ipsdienol isolated and identified from boring male hindguts, abdomens, or volatiles; populations from California, Idaho, and southeastern British Columbia produce primarily *R*-(-)-ipsdienol (Plummer et al. 1976; Birch et al. 1980; Miller et al. 1989); New York and southwestern British Columbia populations a mixture of enantiomers (Lanier et al. 1980; Miller et al. 1989). *R*-(-)-ipsdienol attractive to California beetles, *S*-(+)-ipsdienol inhibitory (Birch et al. 1980). British Columbia and New York populations respond to a mixture of both enantiomers, except for preference for (-)-ipsdienol in southeastern B.C. (Lanier et al. 1980; D.R. Miller and J.H. Borden, unpubl. data). *E*-myrcenol identified in the volatiles of boring males (Gries et al. 1988), acts as multifunctional pheromone (Miller et al. 1990). β-phellandrene attractive in combination with ipsdienol; (Miller and Borden 1990b).
Polygraphus rufipennis Foureyed spruce bark beetle	3-methyl-3-buten-1-ol	Detected in abdominal extracts and volatiles of boring males. Attractive in field traps and as tree baits (Bowers et al. 1991).
Scolytus multistriatus Smaller European elm bark beetle	methylheptanol α-multistriatin δ-multistriatin α-cubebene	Isolated and identified from frass and volatiles from female-infested elm logs (Pearce et al. 1975; Peacock et al. 1975). Absolute configurations determined from synthetic compounds as 3*S*, 4*S*-(-)methylheptanol and 1*S*,2*R*,4*S*,5*R*-(-)-α-multistriatin (Pearce et al. 1976; Mori 1976, 1977; Elliott et al. 1979), α-cubebene present as (-) enantiomer (Pearce et al. 1975). Ternary mixture necessary for optimal activity (Pearce et al. 1975; Lanier et al. 1977). In Germany (Gerken et al. 1978), but not England (Blight et al. 1980), δ-multistriatin is the active isomer.
Trypodendron lineatum Striped ambrosia beetle	lineatin α-pinene ethanol	Lineatin isolated and identified from female frass (MacConnell et al. 1977). Absolute configuration of active enantiomer (Borden et al. 1980b) determined from synthetic enantiomers 1*R*,4*S*,5*R*,7*R*-(+)-lineatin (Slessor et al. 1980). Highly attractive to both sexes in field (Borden et al. 1979; Vité and Bakke 1979; Klimetzek et al. 1980). Ethanol and α-pinene synergistic with lineatin for European populations (Vité and Bakke, 1979; Klimetzek et al. 1980), but α-pinene not required for maximal response in England (Borden et al. 1982), and inactive in North America (Salom and McLean 1988). Ethanol inhibitory at high concentrations in North America (Borden et al. 1982); either inert (Borden et al. 1982; Salom and McLean 1988) or synergistic (Shore and McLean 1983) at low concentrations.

using traps with dark, vertical silhouettes). Two operational tactics arising from this strategy are: (1) to mass-trap populations of beetles so that few are left to attack hosts, or (2) to induce attack on trap trees or logs before killing the beetles (e.g. with pesticides or by logging and milling, Borden 1989).

The second, major strategy is to exploit the beetles' use of antiaggregation pheromones in resource partitioning and termination of attack (Fig 1). Two practical tactics arising from this strategy are: (1) to preclude a buildup of beetle populations by broadcast applications of antiaggregation pheromones to prevent beetles from attacking a vulnerable resource (e.g. windthrown trees), or (2) to cause beetles to disperse from an already infested area, preserving the remaining trees intact.

Because other means of detecting outbreaks or infestations are well developed, semiochemicals are only rarely used operationally to survey for bark and timber beetles in Canada. However, much developmental research has been done using other tactics (Table 2), and semiochemicals are now used operationally against several species, particularly ambrosia beetles and the mountain pine beetle. The case histories of the two research and development programs leading to semiochemical-based management of these beetles exemplify the advent of a new era in pest management.

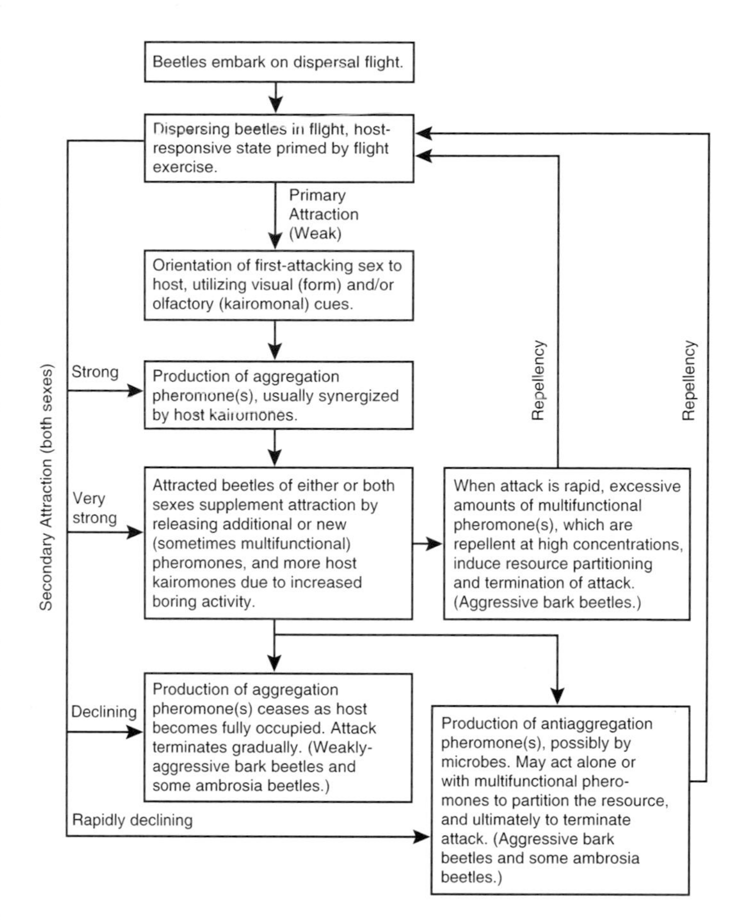

Figure 1. Flow diagram depicting generalized, semiochemical-mediated host selection by bark and timber beetles. Central column describes primary attraction of first attacking beetles, as well as pheromone production and cessation mechanisms. Left-hand loops depict response of dispersing beetles to attractive semiochemicals, contributing to the mass attack of new hosts. Right-hand loops depict role of antiaggregation pheromones in returning host-seeking beetles to the dispersal phase.

Ambrosia Beetles

The Problem

Ambrosia beetles, *Trypodendron lineatum*, *Gnathotrichus sulcatus*, and *G. retusus*, infest the sapwood of felled timber and freshly milled, green lumber. The "pin holes" or tunnels stained dark by the beetles' ambrosia fungus cause degrade of lumber and plywood and the rejection of lumber for export, and result in unacceptable resorting and remanufacturing costs. The estimated impact of $63 million per year in 1985 on the British Columbia coast (McLean 1985) can be increased to an estimated $200 million per year in 1991.

In the years following World War II, conventional chemical insecticides were sprayed from the ground and the air onto decked logs, booms, and unseasoned lumber to control ambrosia beetles (Lejeune and Richmond 1975; Borden and McLean 1981). DDT was used first, but gave way to lindane, which in turn gave way to methyl trithion. Finally, faced with increasing resistance to the use of chemical pesticides by labor, as well as mounting environmental concern, all use of such insecticides was stopped.

The ambrosia beetle problem is particularly acute in dryland sorting areas (dryland sorts), "field warehouses" where logs trucked from the logging site are sorted by species and grade and held for transport to the mill (Borden and McLean 1981). In early spring, *T. lineatum*, which have overwintered in the surrounding forest, invade the sorts and mass attack valuable logs in which they are intent on raising a brood. Then *Gnathotrichus* spp. emerge from stored logs, debris, or newly arrived logs and also attack fresh logs. The risk of attack by the univoltine *T. lineatum* subsides somewhat in early summer, but in August, the main, summer generation flight and attack of *G. sulcatus* begins, extending the ambrosia beetle risk period to 8 months, from March through October (Borden and McLean 1981). In some sawmills, *G. sulcatus* is a severe threat to green lumber, which it attacks freely and in which it can successfully reproduce, constituting a significant threat to lumber-importing countries (McLean and Borden 1975).

Development of Semiochemicals for Pest Management

Pioneering research at Oregon State University and at the Pacific Forestry Centre, Forestry Canada, Victoria, British Columbia, disclosed that mass attack by *T. lineatum* was mediated by a female-produced pheromone (Rudinsky and Daterman 1964a,b; Chapman 1966). This pheromone enables the beetles in nature to locate and to exploit fully the scarce natural resource of windthrown or broken trees (Atkins 1966). Similarly, mass attack by *Gnathotrichus* spp. was later shown to be mediated by male-produced aggregation pheromones (Borden and Stokkink 1973; Borden and McLean 1979).

Table 2. Summary of the status of semiochemicals of bark and timber beetles as pest management tools in Canada.

Species, common name (if any), and pest status	Semiochemical-based pest management tactics according to developmental status		
	Operational use	Operational readiness	Underdevelopment
Dendroctonus brevicomis Western pine beetle Restricted to mature ponderosa pine in British Columbia. A major mortality factor in U.S. Could assume greater importance as greenhouse effect develops, but at present is a minor pest.		Mass-trapping techniques developed in California can cause significant reductions in beetle numbers and infestation rate (Bedard and Wood 1974; DeMars et al. 1980). Could be adapted readily for use in Canada.	Tree baiting in combination with a bark-applied insecticide can induce very high mortality of attracted beetles (Smith 1986). Could be developed for application against small infestations in Canada. Tree baiting alone could be modified to contain and concentrate infestations before logging. Although antiaggregants are known (Table 1), more operational research is required before they could be recommended for use.
Dendroctonus ponderosae Mountain pine beetle Major mortality factor in mature pines in western Canada. Killed over 200 million lodgepole pines in British Columbia from1972 to 1986 that were capable of generating over $20 billion in economic activity.	Attractive semiochemicals are produced commercially and are in operational use as tree baits to contain and concentrate infestations before logging, to mop up residual populations after sanitation-salvage logging, and with MSMA to control small, spot infestations (Borden 1990; Borden and Lindgren 1989). Minor use in survey traps to delineate safe period for hauling infested logs (Stock 1984).		Verbenone can reduce further attacks in heavily infested forests and inhibit them in high hazard stands (Amman et al. 1989; Lindgren et al. 1989a). Underdevelopment in British Columbia as operational pest management tool.
Dendroctonus pseudotsugae Douglas-fir beetle Moderately severe mortality factor in mature Douglas-fir in the interior of British Columbia.		Antiaggregation pheromones operationally tested and ready for use as broadcast granular insecticide in U.S. (Furniss et al. 1981, 1982). Could be adapted for use in British Columbia if registration as pesticide granted. Tree baiting tested to contain and concentrate infestations before logging in U.S. (Ringold et al. 1975) and now used operationally. Could be readily adapted for use in Canada.	Mass trapping in small infestations shows promise (B.S. Lindgren, Phero Tech Inc., Vancouver, B.C., pers. comm.), but problems with spillover attacks on standing trees must be overcome.
Dendroctonus rufipennis Spruce beetle Serious mortality factor of mature spruce in British Columbia and Alberta, and threat to spruces defoliated but not killed by spruce budworm in eastern Canada.	Baiting of trees with frontalin and α-pinene following Dyer's (1973, 1975) methodology is now used operationally in some areas to contain and concentrate infestations before logging.		While MCH will inhibit attraction to hosts (Rudinsky et al. 1974c; Lindgren et al. 1989b) and shows promise for operational use, more work is needed to achieve success similar to that with the Douglas-fir beetle. The discovery of verbenene and 1,2-MCH-ol as aggregation pheromones (Table 1) has stimulated renewed operational research on tree baiting and mass trapping.

(Continued)

Table 2. Summary of the status of semiochemicals of bark and timber beetles as pest management tools in Canada. *(Continued)*

Species, common name (if any), and pest status	Semiochemical-based pest management tactics according to developmental status		
	Operational use	Operational readiness	Underdevelopment
Dryocoetes confusus Western balsam bark beetle The most severe mortality agent of subalpine fir in western Canada.		With known semiochemicals, *exo*-brevicomin can be used as a tree bait to induce attack before logging, and *endo*-brevicomin can be deployed at 10-m centers to reduce the attack rate in stands to be conserved (Stock 1991).	Isolation of an unknown host attractant is under way; this material could be used to improve the efficacy of tree baits (A. Camacho and J.H. Borden, unpubl. data).
Gnathotrichus retusus The least important of the three major ambrosia beetle pests on the British Columbia coast. More prevalent in the interior and U.S. in warm dry regions.	Retusol effective in attracting beetles to traps (Borden et al. 1980c, 1981). Used with host volatiles as a bait in multiple-funnel traps to survey for and mass-trap *G. retusus* (B.S. Lindgren, Phero Tech Inc., Vancouver, B.C., pers. comm.).		
Gnathotrichus sulcatus A major pest of felled timber in the woods and in timber processing areas. Can also attack and degrade unseasoned lumber. A major threat to be introduced in infested logs or lumber into timber-importing countries.	Semiochemical-based, mass-trapping programs developed in sawmills (McLean and Borden 1977, 1979) and dryland sorting areas (Lindgren and Borden 1983) have become the basis of a major semiochemical-based management program for ambrosia beetles offered by Phero Tech Inc., Vancouver, British Columbia.		
Ips pini Pine engraver An occasional killer of sappling to mature trees under stressed conditions. Becoming more important with advent of intensive silviculture (e.g. through buildup of populations in thinnings representing a threat to residual, high-value, leave trees).		Ipsdienol-baited traps could be used to survey for occurrence, if justified by demand.	Interest in mass trapping high in U.S., hence developmental work there. Efficacy of mass trapping could be increased by addition of host kairomones (Table 1). Research under way on use of antiaggregants to inhibit infestation in thinned stands (Borden et al. 1991).
Polygraphus rufipennis Foureyed spruce bark beetle An aggressive, secondary beetle that kills black spruce defoliated by spruce budworm.			Mass trapping being investigated as means for reducing populations in budworm-defoliated stands, allowing trees to recover (W.W. Bowers and J.H. Borden, unpubl. data).
Scolytus multistriatus Smaller European elm bark beetle An introduced vector of Dutch elm disease. Of major importance in eastern Canada, but occurs in the West. Displaced by native elm bark beetle in colder regions.	Semiochemicals used in traps in the Prairie provinces to survey for expansion of range and establishment of this European insect (Ellis 1985).	Operational programs using semiochemical-baited trap trees in which beetles are killed with an arsenical herbicide and/or an insecticide (O'Callaghan et al. 1980; Lanier and Jones 1985) are effective in the U.S. (Lanier and Jones 1984) and could be adapted for use in Canada if desired.	

(Continued)

Table 2. Summary of the status of semiochemicals of bark and timber beetles as pest management tools in Canada. *(Concluded)*

Species, common name (if any), and pest status	Semiochemical-based pest management tactics according to developmental status		
	Operational use	Operational readiness	Underdevelopment
Trypodendron lineatum Striped ambrosia beetle The most important ambrosia beetle pest in Canada as well as elsewhere in the temperate world. Responsible for majority of est. $200 million losses per year on British Columbia coast.	Semiochemical-based, mass-trapping programs developed in dryland sorting areas (Lindgren and Borden 1983) and around booming grounds (Shore 1982) have become the basis of Phero Tech's large semiochemical-based management program in British Columbia.		A search for an illusive antiaggregation pheromone continues (Borden 1988). It could be used to repel beetles from vulnerable logs. Other repellents, such as turpentine (Nijholt 1973) and pine oil, (Nijholt 1980) operationally ineffective.

Intensive research on the isolation and identification of the unknown pheromones soon began. The research culminated (Table 1) in the identification, synthesis, and field testing of lineatin, a single aggregation pheromone for *T. lineatum* (MacConnell et al. 1977; Borden et al. 1979), a 65:35 blend of *S*-(+)- and *R*-(-)-sulcatol for *G. sulcatus* (Byrne et al. 1974), and *S*-(+)-sulcatol (retusol) for *G. retusus* (Borden et al. 1980a). Because *T. lineatum* produced and responded only to the (+) enantiomer of lineatin, while the (-) enantiomer was inert (Borden et al. 1980b), the racemate, a 50:50 mixture of both enantiomers can be used operationally. *Gnathotrichus sulcatus* required both enantiomers of sulcatol, which act synergistically (Borden et al. 1976), again allowing the racemate to be used operationally. However, *G. retusus* uses *S*-(+)-sulcatol and is repelled by the antipode (Borden et al. 1980a, 1981), ensuring species specificity, but demanding that chirally pure *S*-(+)-sulcatol (retusol) be used in pest management programs.

In addition to the aggregation pheromones, ethanol, produced by anaerobic metabolism in cut logs (Graham 1968), was shown to act as a primary, kairomonal attractant (Fig. 1) for ambrosia beetles (Moeck 1970, 1971; Cade et al. 1970). It acts synergistically with sucatol (Borden et al. 1980c) and lineatin[2] to create an enhanced attraction. A second host compound, α-pinene, weakly enhanced the activity of pheromones for *Gnathotrichus* spp. (Borden et al. 1980c), but its role in interacting with lineatin is unclear.

In operational programs, lineatin is released from Hercon® Luretape dispensers (Herculite Products Inc., New York) at a rate of 0.4 mg per day. Sulcatol is released from bubble cap dispensers (Phero Tech Inc., Vancouver, British Columbia) at 10 mg per day, and ethanol from polyvinylchloride tubes at 60 mg per day.

Operational Research

The first operational research was directed against *G. sulcatus* in sawmills. A courageous Ph.D. student, J.A. McLean, undertook the pioneering challenge of mass trapping *G. sulcatus* on sulcatol-baited, wire-mesh, sticky traps in the immense lumber storage yards of a sawmill on Vancouver Island. He worked on the assumption that the total *G. sulcatus* population in the sawmill was represented by those that were caught in traps, or on semiochemical-baited trap-piles of suitably aged lumber, plus those that attacked other lumber piles in the mill site. A weekly survey for new attacks on piles of green lumber was made, and the efficiency of the mass-trapping program was calculated using the following formula:

$$\%\ \text{Suppression} = \frac{\text{No. beetles caught on traps and trap piles} \times 100}{\text{No. beetles caught on traps and trap piles} + \text{Est. no. beetles attacking lumber in the mill site}}$$

Using traps deployed throughout the mill site and applying this formula, McLean and Borden (1977, 1979) achieved an overall efficiency of 65.1% for the 1976 season. In midsummer, when the population was low, the efficiency approached 100%. However, the suppression decreased during the peak flight periods in the spring and late summer, as beetles apparently bypassed the traps and set up their own attraction centers on freshly attacked lumber. This problem can be eliminated partly by integrating mass trapping with inventory control, in which infested logs are neither brought to nor stored at the mill, and in which rough-cut, green lumber is either processed rapidly into finished, kiln-dried lumber, or is rapidly removed from the mill site. A second Ph.D. student, T.L. Shore (1982), later surveyed the entire ambrosia beetle population in the sawmill and an adjacent booming ground and expanded the mass-trapping program to include *T. lineatum* and *G. retusus*.

A third Ph.D. student, B.S. Lindgren, extended the mass-trapping research into dryland sorts. The primary target was *T. lineatum*, with the trapping of *G. sulcatus* as a secondary goal. In one experiment, the size of the overwintering population of *T. lineatum* on the forest floor was estimated in a 60-m-wide band covering 10 ha around a dryland sort on Vancouver Island. Low and high estimates were based on extending the area in which minimal densities of beetles occurred around the base of trees either 0.45 or 0.90 m from the

[2]Borden et al. (1982) found ethanol to be an effective synergist in Europe, but did not verify this activity in North America. Salom and McLean (1988) found no synergistic activity, but recent studies by Phero Tech Inc., Vancouver, British Columbia, have confirmed the activity in North America.

root collar. The experimental strategy of the program was to intercept beetles in the spring as they flew into the sort by capturing them in semiochemical-baited traps placed around the sort perimeter. These traps were supplemented by semiochemical-baited bundles of trap-logs also placed around the sort margin. These were reduced to pulp chips following attack by ambrosia beetles. Taking into account the high and low estimates of overwintering populations, the 2.26 million *T. lineatum* captured in 1981 represented from 44% to 77% of the target population (Lindgren and Borden 1983).

Operational Programs

Mass-trapping programs require a trap that is efficient in capturing the target species, is easy to use, and is durable. In British Columbia, the multiple-funnel trap (Lindgren 1983) rapidly became the preferred device. It consists of a series of vertically aligned black, plastic funnels above a collecting receptacle (Fig. 2). Baits are suspended inside the open column in the middle of the trap. These traps provide a prominent vertical silhouette to which the beetles orient. They eliminate offensive sticky material used in some earlier traps, are collapsible for transport, and can be hung from ropes, wires, or from metal bars driven into the ground. Thus, the trap can move with the wind and return to its original vertical position when the wind subsides. The trap remains effective under dusty conditions, although leaves and spider webs occasionally reduce the trapping power. Finally, the traps capture beetles in great numbers.

Following completion of McLean and Borden's (1979) operational research in the sawmill, the mass-trapping program was taken over for 7 years by MacMillan Bloedel Ltd. until the mill was closed. The year after Lindgren and Borden's (1983) dryland sort research project, the trapping program was taken over by Canadian Pacific Forest Products Ltd., which had the foresight to employ a pest management forester. In 1983, the second year of the company's operational program, the pest management forester reported that virtually no beetles broke through the trap barrier to attack logs in the sort. The success of the program was aided by rigorous inventory control. The program is still run today by the company, and it has been expanded to include three dryland sorts, two booming grounds, and three sawmills with adjacent booming grounds. Both programs are rare examples of continuing semiochemical-based pest management run by enlightened forest products companies.

Most of the semiochemical-based management of ambrosia beetles is now done by Phero Tech Inc., Vancouver, British Columbia, which runs the largest, annual, commercial program ever developed for the mass trapping of any insect. By 1989, many timber processing areas in British Columbia, as well as in Washington, Oregon, and Alaska, were using Phero Tech's trapping technology. Fifty areas were under pest management contract to the company, and 11 additional areas were managed independently, with baits and traps purchased from Phero Tech.

A fundamental precept of Phero Tech's program that ensures success is the company's insistence that a new customer purchase the consulting services of the company's professionals along with the purchase of traps and baits. This policy circumvents failures borne of a lack of comprehension by nonspecialists on the ambrosia beetle problem and the techniques and dedication required to solve it, and avoids the creation of an unjustifiably bad image of semiochemical-based management.

Figure 2. Multiple-funnel trap in use for mass trapping ambrosia beetles in a timber processing area. (R.G. Long, Simon Fraser University)

Phero Tech's approach integrates host-habitat management (i.e. cultural control through manipulation of log inventories in the woods, as well as in processing areas, with suppression of beetle populations by using semiochemicals to intercept and kill host-seeking beetles). Fundamental to such an integrated pest management program is the hypothesis that the size of ambrosia beetle populations is determined primarily by the amount of host-habitat available (Borden 1988). When the population of beetles is reduced by removal of host material, mass trapping has a much greater chance of success than if it were used alone.

In the first year of a typical management program, a timber-processing area is surrounded by survey traps. Thereafter, mass trapping supplemented by trap logs is conducted with emphasis on the "hot" areas of greatest beetle activity. Typically a timber processing area with a 1 000-m perimeter will have about 100 to 120 traps for *T. lineatum* plus 10 to 15 bundles of trap logs. In early spring, there would also be 15 to 18 traps for *G. sulcatus*, and an equivalent number for *G. retusus* if it is a problem. From midsummer until the *G. sulcatus* flight is finished in late October, a 50:50 ratio of *G. sulcatus* to *T. lineatum* traps is maintained.

Every 5 weeks the captured beetles are collected and the traps serviced. The numbers of beetles are rapidly determined by volumetric measure. These counts, as well as a written assessment of the problem and recommendations for integrated management, are provided to the customer within 5 to 10 working days of the collection date.

In evaluating semiochemical-based management, confidence is placed largely on the original research that demonstrated efficacy of the strategy. The number of beetles captured can be related to the number of logs that could be potentially attacked, but this assessment does not incorporate the benefits achieved through host-habitat management. Therefore, much evaluation is visual; instead of the pre-management observations of white sawdust dripping from attacked logs, logs in managed areas remain largely unattacked. Sawmill divisions no longer complain of huge ambrosia beetle problems passed on from the woods, the dryland sort, or the booming ground. When doubting customers have cancelled the pest management program, there have been striking examples of renewed impact of the beetles. Ultimately, it should be possible to construct a realistic simulation model of ambrosia beetle infestation dynamics and impact. The efficacy of an integrated pest management program that removes beetles and protects the product from attack could then be evaluated in terms of its benefit to the final product.

The Mountain Pine Beetle

The Problem

The mountain pine beetle, *Dendroctonus ponderosae*, is a devastating forest pest in western Canada. Between 1972 and 1985, it killed approximately 195.7 million pines in British Columbia, representing an estimated $14.4 to 19.6 billion (Canadian) in potential economic impact (Borden 1990). Even after salvage logging, the real losses were conservatively estimated to range from $4.1 to 5.4 billion.

Silvicultural methods, such as sanitation-salvage logging or falling and burning of infested trees (McMullen et al. 1986), are very effective in reducing populations of the beetle, if they are applied in a thorough and timely fashion. However, these methods are often inadequate for several reasons. Many infestations cannot be harvested before the beetles fly in midsummer; harvesting operations often fail to encompass entire infestations or overlook asymptomatic trees; dispersing beetles routinely establish new infestations that are undetectable until the infested trees turn red the following summer; and numerous trees in environmentally sensitive or inaccessible areas may succumb to beetle attacks while land managers helplessly watch the plunder. While infestations raged in British Columbia in the early 1970s, it became clear that these limitations on conventional silvicultural control methods had emphasized the need for new pest management tools.

Development of Semiochemicals for Pest Management

In large part, semiochemicals have become or are being developed as these new tools. However, "Pondelure," a mixture of the host tree monoterpene α-pinene and its oxidation product, the female-produced aggregation pheromone *trans*-verbenol (Table 1), failed to attract mountain pine beetles in lodgepole pine forests in British Columbia (Conn et al. 1983), despite its attractiveness in ponderosa and white pine forests in the United States (Pitman and Vité 1969; Pitman 1971). In British Columbia and in the United States Pacific northwest, myrcene replaced α-pinene as the most effective synergist with *trans*-verbenol for the attraction of males (Billings et al. 1976; Conn et al. 1983; Borden et al. 1983a). Additional research showed that *exo*-brevicomin and another male-produced compound, frontalin, act as multifunctional pheromones, enhancing attraction at low concentrations (particularly for females) and serving as antiaggregation pheromones at high concentrations (Borden et al. 1987a). Finally, verbenone that is produced by microorganisms associated with the beetles (Hunt and Borden 1989) also serves as an antiaggregation pheromone (Ryker and Yandell 1983; Borden et al. 1987a).

The attractive blend of myrcene, *trans*-verbenol, and *exo*-brevicomin, called "Mountain Do," has proven to be a most effective pest management tool either as tree baits or as trap lures. Effective release rates for each component are as follows: myrcene, 20 mg per 24 h; *trans*-verbenol, 1 mg per day; and *exo*-brevicomin, 0.2 mg per day in tree baits and 0.02 mg per day in trap lures.

In considering semiochemicals to manipulate mountain pine beetle populations, examination was made of points of vulnerability in the complex of semiochemical-mediated events involved in host selection and mass attack by the beetle (Borden 1989). Specifically, the blend of myrcene, *trans*-verbenol, and *exo*-brevicomin was successfully exploited. In addition, the practical use of verbenone to induce beetles to leave an infested zone or to exclude beetles from it is under exploration.

A critical step in the development of Mountain Do was the design of a cheap, durable, "idiot proof," bait receptacle for affixing to trees. The device that was eventually adopted and used operationally was a waterproof cardboard envelope (Fig. 3) that already contained the bait dispensers (two closed 1.9-mL polyethylene tubes for myrcene; a 1.9-mL polyethylene tube for *trans*-verbenol that required opening; and a 1.0-mm inside diameter capillary tube that needed to be broken for *exo*-brevicomin). The envelope was stapled to a tree and the flap folded into a slot on the outside to protect against sun and rain. When affixed to the north side of a tree, this device had an effective life of more than 100 days. In 1989, the cardboard envelope was replaced by Phero Tech's new "baggie" receptacle (Fig. 4) containing *trans*-verbenol in a bubble cap release device, *exo*-brevicomin in a piece of Hercon® Lure tape (Herculite Products Inc., New York), and myrcene in a polyethylene tube. This receptacle is easier to use than the cardboard envelope and is well camouflaged when affixed to a tree.

Exploitation of the Aggregation Response

Survey—Surveys for the location of most *D. ponderosae* infestations are conventionally done from midsummer to early fall by aerial reconnaissance for red trees killed by beetles the previous year. Ground probes (British Columbia Ministry of Forests and Lands 1987) are then conducted to locate freshly attacked green trees that can be subjected to one or more treatments, possibly involving semiochemicals.

Figure 3. Waterproof cardboard envelope containing attractive semiochemicals and affixed to lodgepole pine tree. Pitch tube on the tree is evidence of attack by the mountain pine beetle induced by the semiochemical bait. (R.G. Long, Simon Fraser University)

Figure 4. "Baggie" receptacle containing attractive semiochemicals being stapled to lodgepole pine tree. (R.G. Long, Simon Fraser University)

However, semiochemical-baited, multiple-funnel traps have been used to monitor for the occurrence of peak flights of the beetle (Stock 1984). This knowledge can then be used to impose "no-haul" restrictions on loggers, so that infested logs are not transported when the beetles are emerging and flying, and new infestations are prevented at sites where trucks stop or break down, or around mill sites.

Treatment of small infestations—For small, localized infestations, Mountain Do has been integrated with chemical pesticides when it is too dangerous or too difficult to use fire (i.e. to dispose of infested trees by felling and burning) (McMullen et al. 1986). Two procedures have been adopted (Borden 1989b; Borden and Lindgren 1989).

The first is called **cut, spray, bait, and treat**. In late spring to midsummer, before the beetles emerge and disperse, infested trees are felled and sprayed with a water-based formulation of carbaryl to kill emergent beetles (Fuchs and Borden 1985). Trees surrounding the felled trees are then baited with Mountain Do at the rate of one baited tree per two felled and sprayed infested trees. Any residual beetles not killed by the insecticide are induced to attack the baited trees and those surrounding them. Following attack, the mass-attacked trees are ax-frilled at the base and the arsenical herbicide monosodium methane arsenate is squirted into the ax-frill. The herbicide is translocated up the tree and kills the new brood larvae (Maclauchlan et al. 1988).

A modification of this procedure, **treat, bait, and treat**, has been extensively adopted by the Merritt Forest District in south-central British Columbia (T.E. Lacey, Merritt Forest District, British Columbia Forest Service, Merritt, B.C., pers. comm.). Newly infested trees are located shortly after attack and treated with monosodium methane arsenate. The next spring, a few trees surrounding the treated trees are baited with Mountain Do. Any residual beetles that attack these baited trees are then killed by a second, post-attack, herbicide treatment that prevents further reproduction.

Post-logging mop-up—A second approach to exploiting the aggregation response to semiochemicals is to use them in post-logging mop-up (Borden et al. 1983b). In this case, a sanitation-salvage cut is done in a heavily infested stand to extract as many living beetles as possible in infested trees. Depending on the intensity of the infestation, this may be either a clearcut or selection cut in which only the infested trees are removed. However, as in some types of cancer surgery, it is very difficult to detect and remove all of the cancer (i.e. the infested trees). A chemotherapeutic mop-up is thus done for 1 to 3 years by baiting selected, large-diameter trees within and around a selection cut and around the periphery of a clearcut. Residual beetles are attracted to these trees and those around them. The induced infestations can be easily located, and the beetles therein killed by monosodium methane arsenate or by felling and burning, or the infested trees can be excised and removed from the forest by small-scale logging.

Investigation of this tactic occurred in an infestation in Manning Park, British Columbia (Borden et al. 1983b). By

1982, this infestation had reached a magnitude of between 5 000 to 6 000 green-attacked trees, many large enough to generate 5 000 to 10 000 brood beetles each. In an intensive survey, all infested trees that could be found were marked for extraction. A selection cut was carried out, taking only the marked trees. Their stumps were peeled to kill all possible beetles. Fifty trees well dispersed within the logged area were then semiochemical-baited. The attack on these trees clearly showed that the logging operation had not removed all of the beetles from the park. Forty-eight of the 50 baited trees were attacked, as were 80 trees within 12 m of them, representing 128 attacked trees on an area of 2.2 ha. Only 86 attacked trees were found on the remaining 63 ha, demonstrating the power of the baits to concentrate the attack in a small area. All of the attacked trees were cut and burned.

The next year 10 trees were baited and 29 were attacked (Wood 1983). Again, these were cut and burned. The process was repeated in 1984, with three trees baited, attacked, and cut and burned (Wood 1984). No infestation recurred. While immigrating beetles could create a new infestation, the old one was successfully mopped up. An integrated pest management approach of intensive survey, radical surgery, chemotherapy, re-survey, minor surgery, and repeated re-survey and treatment had solved the problem. The use of semiochemicals in post-logging mop-up is now practised on a moderate scale throughout British Columbia.

Containment and concentration—Under the pressure of mountain pine beetle outbreaks, one of the most frustrating situations encountered by forest pest management staff in the past was the necessity to neglect certain infestations for extended periods of time while attention was focused on others. Often sanitation-salvage cuts could not be done before beetle flight, infestations were not heavy enough to justify immediate logging, or it was too late in the spring to fell and burn because of the danger of fire. Thus, the beetles in these neglected infestations were allowed to emerge and to disperse and attack new trees, often far from their original infestation. New outbreaks flared up continuously, while others expanded. If, however, there were a mechanism whereby infestations could be contained, and possibly concentrated, allowing them to intensify without expanding, then their eventual excision by clearcutting would be much more effective and limited in size. Moreover, the chances of new infestation centers flaring up would be greatly reduced.

A cost-effective procedure was developed (Borden et al. 1983c), in which trees at 50-m centers in infested stands (4 trees per ha) are baited on the north side with Mountain Do, with care taken to bait living pine trees of the largest possible girth. These baited trees serve as olfactory beacons for emergent beetles, which cannot fly out of the stand without passing within 25 m of a baited tree. When infestations are large, the beetles that move from tree to tree within the stand are of little consequence. In this case, containment of dispersing beetles heading out of the infestation can be achieved by baiting the perimeter in two lines, 50 m apart, with the baited trees in the two lines offset by 25 m.

Containment and concentration were shown by three studies comparing green:red ratios (i.e. the ratios of newly attacked green trees to red trees attacked the previous year) in baited versus control stands. In all three instances, there were significantly higher green:red ratios in baited stands than in unbaited control stands (Fig. 5).

The tactic was further validated by Gray and Borden (1989) who compared green:red ratios in a central baited zone A compared to those in concentric 50- and 100-m-wide zones (B and C) surrounding the central zone. In three experimental stands in the Nelson Forest Region of British Columbia, zone A was baited, whereas in two control stands, it was left unbaited. There was a migration of beetles outward from zone A to zone B in the unbaited stands, while in the baited ones, the attack was contained and concentration was primarily within the baited zones A. The attractive range of a baited tree appeared to extend up to 75 m. When the green:red ratios were weighted by attack density and diameter of freshly

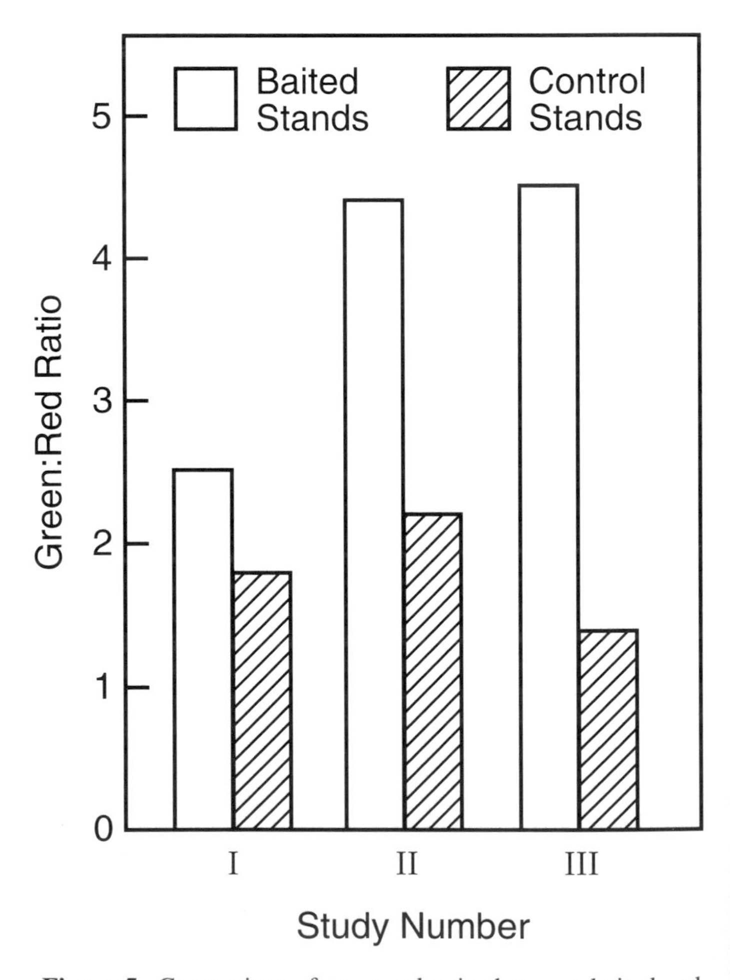

Figure 5. Comparison of green:red ratios between baited and unbaited control stands in three studies evaluating the use of semiochemicals to contain and concentrate infestations of the mountain pine beetle. Study I (Borden et al. 1983c): combined results from five grid-baited and five control stands. Study II (Heath 1986): combined results from five grid-baited and three perimeter-baited stands, each with a paired control stand. Study III (Gray and Borden 1989): combined results from three grid-baited stands compared with two control stands. In each case, chi-square tests ($P<0.01$) indicate that there is a significant difference in the ratios between baited and control stands.

attacked (green) trees to more closely reflect actual beetle populations, the containment and concentration effect was even more pronounced.

Since its operational introduction in 1984 (Borden and Lacey 1985), the containment and concentration tactic has been widely accepted in western Canada. It has made it possible to avoid a chaotic state of beetle chasing by allowing a regulated harvest of an artificially intensified infestation to be conducted after 1 or 2 years of baiting in a relatively small area. If the size of the beetle population has been overestimated, and the induced attack is light, the attacked trees can be easily found and the beetles therein killed by felling and burning or by monosodium methane arsenate treatment. From 1985 to 1989, approximately 12 500 ha per year were baited in British Columbia to contain and concentrate *D. ponderosae* infestations.

Exploitation of the Antiaggregation Response

Two large-scale trials conducted in 1987, one in the United States (Amman et al. 1989) and the other in Canada (Lindgren et al. 1989a), confirmed that verbenone may be promising in management of mountain pine beetle populations.

Two hypotheses were tested in British Columbia (Lindgren et al. 1989a). The first was that verbenone can be used to induce dispersal of beetles from an existing infested area, and the second was that it can be used to exclude beetles from an area, thus preventing new infestations. When verbenone release devices were applied at 10-m centers to infested 4-ha blocks, the mean number of newly attacked trees in treated blocks was 62.1% lower than in the control treatment. The mean number of trees with mass attacks, defined as more than 31.3 attacks/m^2 at 1.3 m above the ground, was 75.2% lower in the treated than in the control blocks. In another experiment verbenone was applied at 10-m centers to 1-ha blocks, which had been baited with aggregation pheromones to determine whether verbenone could counteract the concentration effect of the baits. The mean number of infested trees was 58.5% lower in the verbenone-treated blocks than in the control blocks, and the mean number of mass-attacked trees was reduced by 74.3%.

The use of verbenone in one area combined with aggregation pheromones in another area may ultimately be the most operationally feasible tactic. The potential for moving the locus of an infestation may be enhanced by the simultaneous use of verbenone to push the beetles out of one stand, and aggregation pheromones to draw them into another stand. This tactic has generated considerable interest for use in infested campgrounds, riparian areas, deer and elk winter range, and protection forests on steep slopes where harvesting is not a desirable option.

Conclusion

The successful, operational exploitation of the aggregation response of mountain pine beetles by use of tree baits in conjunction with harvesting and use of systemic pesticides, and the promising results with antiaggregation pheromones, demonstrate clearly that semiochemicals have the potential of becoming routine tools in forest management wherever bark beetles are of concern. However, it should be stressed that semiochemicals may not be effective at the zenith of an outbreak and will probably rarely be useful if used alone. By using several tactics simultaneously, including prevention of population buildup by careful silvicultural planning, maintenance of thrifty forests, and sound harvesting practices, bark beetle populations may be kept at levels where semiochemicals can be useful as supplementary tools, not only to provide information on population trends, but also as primary agents for dealing efficiently with problem situations when they do occur.

The Future

As long as timber is harvested and high-quality wood is coveted, ambrosia beetles will continue to be pests and semiochemicals will have a place in their management. It is less clear if semiochemicals will ever have a practical role in the management of other timber pests, such as those in the families Cerambycidae and Buprestidae, because little is known about their chemical ecology.

There are some who expect bark beetles to become a minor problem as mature and overmature forests give way to young, thrifty forests under intensive silviculture. This expectation is unlikely to materialize for two reasons. Firstly, many forests will never be intensively managed. Secondly, if one looks south to California and the southern United States, or across the Atlantic Ocean to Europe, he or she will observe bark beetles readily infesting trees several generations after the virgin forest was harvested. One might expect the same trend to occur in Canada.

Moreover, more and more stress factors such as acid rain, increased ultraviolet radiation, and higher temperatures caused by the greenhouse effect will test the stability of our forests in the future. The stressed trees will be vulnerable to many pests, including bark beetles, and semiochemicals will certainly have a place in their management.

Continued research and development on the use of semiochemicals to manage major bark and timber beetle pests, such as the mountain pine beetle, the spruce beetle, and the western balsam bark beetle, thus should continue to provide worthwhile dividends. Finally, because of an increased vulnerability of our forests and a corresponding increase in their value, even species now considered to be minor pests, or even nonpests, may assume prominence. Therefore, pioneering research on the semiochemicals of bark and timber beetles should never be abandoned.

References

Amman, G.D.; Thier, R.W.; McGregor, M.D.; Schmitz, R.F. 1989. Efficacy of verbenone in reducing lodgepole pine infestations by mountain pine beetles in Idaho. Can. J. For. Res. 19: 60-64.

Anderson, R.F. 1948. Host selection by the pine engraver. J. Econ. Entomol. 41: 596-602.

Atkins, M.D. 1966. Behavioural variation among scolytids in relation to their habitat. Can. Entomol. 98: 285-288.

Bedard, W.D.; Tilden, P.E.; Wood, D.L.; Lindahl, K.Q., Jr.; Rauch, P.A. 1980a. Effects of verbenone and *trans*-verbenol on the response of *Dendroctonus brevicomis* to natural and synthetic attractant in the field. J. Chem. Ecol. 6: 997-1013.

Bedard, W.D.; Tilden, P.E.; Wood, D.L.; Lindahl, K.Q., Jr.; Silverstein, R.M.; Rodin, J.O. 1980b. Field responses of the western pine beetle and one of its predators to host- and beetle-produced compounds. J. Chem. Ecol. 6:625-641.

Bedard, W.D.; Tilden, P.E.; Wood, D.L.; Silverstein, R.M.; Brownlee, R.G.; Rodin, J.O. 1969. Western pine beetle: field response to its sex pheromone and a synergistic host terpene, myrcene. Science 164: 1284-1285.

Bedard, W.D.; Wood, D.L. 1974. Programs utilizing pheromones in survey or control. Bark beetles — the western pine beetle. Pages 441-449 *in* M.C. Birch (Ed.), Pheromones. North-Holland Pub. Co., Amsterdam.

Billings, R.F.; Gara, R.I.; Hrutfiord, B.F. 1976. Influence of ponderosa pine resin volatiles on the response of *Dendroctonus ponderosae* to synthetic *trans*-verbenol. Environ. Entomol. 5: 171-179.

Birch, M.C.; Light, D.M.; Wood, D.L.; Browne, L.E.; Silverstein, R.M.; Bergot, B.J.; Ohloff, G.; West, J.R.; Young, J.C. 1980. Pheromonal attraction and allomonal interruption of *Ips pini* in California by the two enantiomers of ipsdienol. J. Chem. Ecol. 6: 703-717.

Blight, M.M.; Ottridge, A.P.; Wadhams, L.J.; Wenham, M.J.; King, C.J. 1980. Response of a European population of *Scolytus multistriatus* to the enantiomers of α-multistriatin. Naturwissenschaften. 67: 517-518.

Booth, D.C.; Phillips, T.W.; Claesson, A.; Silverstein, R.M.; Lanier, G.N.; West, J.R. 1983. Aggregation pheromone components of two species of *Pissodes* weevils (Coleoptera: Curculionidae): isolation, identification and field activity. J. Chem. Ecol. 9: 1-2.

Borden, J.H. 1988. The striped ambrosia beetle. Pages 579-596 *in* A.A. Berryman (Ed.), Dynamics of Forest Insect Populations. Patterns, Causes, Implications. Plenum Press, New York.

Borden, J.H. 1989. Semiochemicals and bark beetle populations: exploitation of natural phenomena by pest management strategists. Holarctic Ecology 12: 501-510.

Borden, J.H. 1990. Use of semiochemicals to manage coniferous tree pests in western Canada. Pages 281-315 *in* R. Ridgway, R.M. Silverstein, and M. Inscoe (Eds.), Behavior-modifying chemicals for insect management. Marcel Dekker, New York.

Borden, J.H.; Chong, L.J.; Fuchs, M.C. 1983b. Application of semiochemicals in post-logging manipulation of the mountain pine beetle, *Dendroctonus ponderosae* (Coleoptera: Scolytidae). J. Econ. Entomol. 76: 1428-1432.

Borden, J.H.; Chong, L.; McLean, J.A.; Slessor, K.N.; Mori, K. 1976. *Gnathotrichus sulcatus*: synergistic response to enantiomers of the aggregation pheromone sulcatol. Science 192: 894-896.

Borden, J.H.; Chong, L.J., Pratt, K.E.G.; Gray, D.R. 1983c. The application of behaviour-modifying chemicals to contain infestations of the mountain pine beetle, *Dendroctonus ponderosae*. For. Chron. 59: 235-239.

Borden, J.H.; Chong, L.; Slessor, K.N.; Oehlschlager, A.C.; Pierce, H.D., Jr.; Lindgren, B.S. 1981. Allelochemic activity of aggregation pheromones between three sympatric species of ambrosia beetles (Coleoptera: Scolytidae). Can. Entomol. 113: 557-563.

Borden, J.H.; Conn, J.E.; Friskie, L.M.; Scott, B.E.; Chong, L.J.; Pierce, H.D., Jr.; Oehlschlager, A.C. 1983a. Semiochemicals for the mountain pine beetle, *Dendroctonus ponderosae* (Coleoptera: Scolytidae), in British Columbia: baited tree studies. Can. J. For. Res. 13: 325-333.

Borden, J.H.; Delvin, D.R.; Miller, D.R. 1991. Synomones of two sympatric species deter attack by the pine engraver, *Ips pini* (Say) (Coleoptera: Scolytidae). Can. J. For. Res. (In press)

Borden, J.H.; Handley, J.R.; Johnston, B.D.; MacConnell, J.G.; Silverstein, R.M.; Slessor, K.N.; Swigar, A.A.; Wong, D.T.W. 1979. Synthesis and field testing of 4,6,6-lineatin, the aggregation pheromone of *Trypodendron lineatum* (Coleoptera: Scolytidae). J. Chem. Ecol. 5: 681-689.

Borden, J.H.; Handley, J.R.; McLean, J.A.; Silverstein, R.M.; Chong, L.; Slessor, K.N.; Johnston, B.D.; Schuler, H.R. 1980a. Enantiomer-based specificity in pheromone communication by two sympatric *Gnathotrichus* species (Coleoptera: Scolytidae). J. Chem. Ecol. 6: 445-456.

Borden, J.H.; Hunt, D.W.A.; Miller, D.R.; Slessor, K.N. 1986. Orientation in forest Coleoptera: an uncertain outcome of responses by individual beetles to variable stimuli. Pages 97-109 *in* T.L. Payne, M.C. Birch, and C.E.J. Kennedy (Eds.), Mechanisms in Insect Olfaction. Clarendon Press, Oxford.

Borden, J.H.; King, C.J.; Lindgren, B.S.; Chong, L.; Gray, D.R.; Oehlschlager, A.C.; Slessor, K.N.; Pierce, H.D., Jr. 1982. Variations in response of *Trypodendron lineatum* from two continents to semiochemicals and trap form. Environ. Entomol. 11: 403-408.

Borden, J.H.; Lacey, T.E. 1985. Semiochemical-based manipulation of the mountain pine beetle, *Dendroctonus ponderosae* Hopkins: a component of lodgepole pine silviculture in the Merritt Timber Supply Area of British Columbia. Z. Angew. Entomol. 99: 139-145.

Borden, J.H.; Lindgren, B.S. 1989. The role of semiochemicals in IPM of the mountain pine beetle. Pages 247-255 *in* T.L. Payne and H. Saarenmaa (Eds.), Integrated control of Scolytid Bark Beetles. Virginia Polytechnic Institute and State University, Blacksburg, Va.

Borden, J.H.; Lindgren, B.S.; Chong, L. 1980c. Ethanol and α-pinene as synergists for the aggregation pheromones of two *Gnathotrichus* species. Can. J. For. Res. 10: 290-292.

Borden, J.H.; McLean, J.A. 1979. Secondary attraction in *Gnathotrichus retusus* and cross attraction of *G. sulcatus* (Coleoptera: Scolytidae). J. Chem. Ecol. 5: 79-88.

Borden, J.H.; McLean, J.A. 1981. Pheromone-based suppression of ambrosia beetles in industrial timber processing areas. Pages 133-154 *in* E.R. Mitchell (Ed.), Management of Insect Pests with Semiochemicals. Plenum Press, New York.

Borden, J.H.; Oehlschalger, A.C.; Slessor, K.N.; Chong, L.; Pierce, H.D., Jr. 1980b. Field tests of isomers of lineatin, the aggregation pheromone of *Trypodendron lineatum* (Coleoptera: Scolytidae). Can. Entomol. 112: 107-109.

Borden, J.H.; Pierce, A.M.; Pierce, H.D., Jr.; Chong, L.J.; Stock, A.J.; Oehlschlager, A.C. 1987b. Semiochemicals produced by western balsam bark beetle, *Dryocoetes confusus* Swaine (Coleoptera: Scolytidae). J. Chem. Ecol. 13: 823-836.

Borden, J.H.; Ryker, L.C.; Chong, L.J.; Pierce, H.D. Jr.; Johnston, B.D.; Oehlschlager, A.C. 1987a. Response of the mountain pine beetle, *Dendroctonus ponderosae* Hopkins (Coleoptera: Scolytidae), to five semiochemicals in British Columbia lodgepole pine forests. Can. J. For. Res. 17: 118-128.

Borden, J.H.; Stokkink, E. 1973. Laboratory investigation of secondary attraction in *Gnathotrichus sulcatus* (Coleoptera: Scolytidae). Can. J. Zool. 51: 469-473.

Bowers, W.W.; Gries, G.; Borden, J.H.; Pierce, H.D. Jr. 1991. 3-Methyl-3-buten-1-ol: an aggregation pheromone of the four-eyed spruce bark beetle, *Polygraphus rufipennis* (Kirby) (Coleoptera: Scolytidae). J. Chem. Ecol. 17: 1989-2002.

Brand, J.M; Young, J.C.; Silverstein, R.M. 1979. Insect pheromones: a critical review of recent advances in their chemistry, biology and application. Fortschritte Chem. Org. Naturst. 37: 1-190.

British Columbia Ministry of Forests and Lands. 1987. Pest Management. Protection Manual, Vol. II. Victoria, British Columbia.

Byers, J.A. 1982. Male-specific conversion of the host plant compound, myrcene, to the pheromone, (+)-ipsdienol, in the bark beetle, *Dendroctonus brevicomis*. J. Chem. Ecol. 8: 363-371.

Byers, J.A.; Wood, D.L.; Craig, J.; Hendry, L.B. 1984. Attractive and inhibitory pheromones produced in the bark beetle, *Dendroctonus brevicomis*, during host colonization: regulation of inter- and intraspecific competition. J. Chem. Ecol. 10: 861-877.

Byrne, K.J.; Swigar, A.A.; Silverstein, R.M.; Borden, J.H.; Stokkink, E. 1974. Sulcatol: population aggregation pheromone in the scolytid beetle, *Gnathotrichus sulcatus*. J. Insect Physiol. 20: 1895-1900.

Cade, S.C.; Hrutfiord, B.F.; Gara, R.I. 1970. Identification of a primary attractant for *Gnathotrichus sulcatus* isolated from western hemlock logs. J. Econ. Entomol. 63: 1014-1015.

Chapman, J.A. 1966. The effect of attack by the ambrosia beetle *Trypodendron lineatum* (Olivier) on log attractiveness. Can. Entomol. 98: 50-59.

Conn, J.E.; Borden, J.H.; Scott, B.E.; Friskie, L.M.; Pierce, H.D. Jr.; Oehlschlager, A.C. 1983. Semiochemicals for the mountain pine beetle, *Dendroctonus ponderosae* (Coleoptera: Scolytidae) in British Columbia: field trapping studies. Can. J. For. Res. 13: 320-324.

De Mars, C.J.; Slaughter, G.W.; Bedard, W.D.; Norick, N.X.; Roettgering, B. 1980. Estimating western pine beetle-caused tree mortality for evaluating an attractive pheromone treatment. J. Chem. Ecol. 6: 853-866.

Dyer, E.D.A. 1973. Spruce beetle aggregated by the synthetic pheromone frontalin. Can. J. For. Res. 3: 486-494.

Dyer, E.D.A. 1975. Frontalin attractant in stands infested by the spruce beetle, *Dendroctonus rufipennis* (Coleoptera: Scolytidae). Can. Entomol. 107: 979-988.

Dyer, E.D.A.; Lawko, C.M. 1978. Effect of seudenol on spruce beetle and Douglas-fir beetle aggregation. Environ. Canada, Ottawa. Bi-Mon. Res. Notes 34: 30-32.

Elliott, W.J.; Hromnak, G.; Fried, J.; Lanier, G.N. 1979. Synthesis of multistriatin enantiomers and their action on *Scolytus multistriatus* (Coleoptera: Scolytidae). J. Chem. Ecol. 5: 279-287.

Ellis, R.A. 1985. Annual report on elm bark beetle surveillance and control in Winnipeg—1985. Insect. Cont. Br., Parks Rec. Dept., City of Winnipeg. Unpubl. report.

Francke, W.; Heeman, V.; Gerken, B.; Renwick, J.A.A.; Vité, J.P. 1977. 2-Ethyl-1,6-dioxaspiro[4.4]nonane, principal aggregation pheromone of *Pityogenes chalcographus* (L.). Naturwissenschaften. 64: 590-591.

Fuchs, M.G.; Borden, J.H. 1985. Pre-emergence insecticide applications for control of the mountain pine beetle. *Dendroctonus ponderosae* (Coleoptera: Scolytidae). J. Entomol. Soc. British Columbia. 82: 25-28.

Furniss, M.M.; Baker, B.H.; Hostetler, B.B. 1976. Aggregation of spruce beetles (Coleoptera) to seudenol and repression of attraction by methylcyclohexenone in Alaska. Can. Entomol. 108: 1297-1302.

Furniss, M.M.; Clausen, R.W.; Markin, G.P.; McGregor, M.D.; Livingston, R.L. 1981. Effectiveness of Douglas-fir beetle antiaggregative pheromone applied by helicopter. USDA For. Serv. Washington, D.C. Gen. Tech. Rep. INT-101.

Furniss, M.M.; Kline, L.N.; Schmitz, R.F.; Rudinsky, J.A. 1972. Tests of three pheromones to induce or disrupt aggregation of Douglas-fir beetles (Coleoptera: Scolytidae) on live trees. Ann. Entomol. Soc. Amer. 65: 1227-1232.

Furniss, M.M.; Markin, G.P.; Hager, Y.J. 1982. Aerial application of Douglas-fir beetle antiaggregative pheromone: equipment and evaluation. USDA For. Serv. Washington D.C. Gen. Tech. Rep. INT-137.

Furniss, M.M.; Schmitz, R.F. 1971. Comparative attraction of Douglas-fir beetles to frontalin and tree volatiles. USDA For. Serv. Washington, D.C. Res. Pap. INT-96.

Gerken, B.; Grune, S.; Vité, J.P.; Mori, K. 1978. Response of European populations of *Scolytus multistriatus* to isomers of multistriatin. Naturwissenschaften. 65: 110-111.

Graham, K. 1968. Anaerobic induction of primary chemical attractancy for ambrosia beetles. Can. J. Zool. 46: 905-908.

Gray, D.R.; Borden, J.H. 1989. Containment and concentration of mountain pine beetle (Coleoptera: Scolytidae) infestations with semiochemicals: validation by sampling of baited and surrounding zones. J. Econ. Entomol. 82: 1399-1405.

Gries, G.; Borden, J.H.; Gries, R.; Lafontaine, J.P.; Dixon, E.A.; Weiser, H.; Whitehead, A.T. 1991. 4-Methyl 6, 6-dimethyl-bicyclo [3.1.1] hept-2-ene (verbenene): new aggregation pheromone of the scolytid beetle, *Dendroctonus rufipennis*. Naturwissenschaften. (In press)

Gries, G.; Pierce, H.D., Jr.; Lindgren, B.S.; Borden, J.H. 1988. New techniques for capturing and analyzing semiochemicals for scolytid beetles (Coleoptera: Scolytidae). J. Econ. Entomol. 81: 1715-1720.

Heath, D. 1986. Assessment of operational pheromone-based containment programs for mountain pine beetle control in the Cariboo Region. British Columbia For. Serv. Victoria Int. Rep. PM-C-1.

Hunt, D.W.A.; Borden, J.H. 1988. Response of mountain pine beetle, *Dendroctonus ponderosae* Hopkins, and pine engraver, *Ips pini* (Say), to ipsdienol in southwestern British Columbia. J. Chem. Ecol. 14: 277-293.

Hunt, D.W.A.; Borden, J.H. 1990. Conversion of verbenols to verbenone by yeasts isolated from *Dendroctonus ponderosae* (Coleoptera: Scolytidae). J. Chem. Ecol. 16: 1385-1397.

Hunt, D.W.A.; Borden, J.H.; Pierce, H.D.; Slessor, K.N.; King, G.G.S.; Czyzewska, E. 1986. Sex-specific production of ipsdienol and myrcenol by *Dendroctonus ponderosae* (Coleoptera: Scolytidae) exposed to myrcene vapors. J. Chem. Ecol. 12: 1579-1586.

Kinzer, G.W.; Fentiman, A.F.; Foltz, R.L.; Rudinsky, J.A. 1971. Bark beetle attractants: 3-methyl-2-cyclohexen-1-one isolated from *Dendroctonus pseudotsugae*. J. Econ. Entomol. 64: 970-971.

Kinzer, G.W.; Fentiman, A.F., Jr.; Page, T.F., Jr.; Foltz, R.L.; Vité, J.P.; Pitman, G.B. 1969. Bark bettle attractants: indentification, synthesis and field bioassay of a new compound isolated from *Dendroctonus*. Nature 221: 477-478.

Klimetzek, von D.; Vité, J.P.; Mori, K. 1980. Zur Wirkung und Formulierung des Populationslockstoffes des Nutzholzborkenkafers *Trypodendron (=Xyloterus) lineatum*. Z. Angew. Entomol. 89: 57-63.

Lanier, G.N.; Claesson, A.; Stewart, T.; Piston, J.J.; Silverstein, R.M. 1980. *Ips pini*: the basis for interpopulational differences in pheromone biology. J. Chem. Ecol. 6: 677-687.

Lanier, G.N.; Gore, W.E.; Pearce, G.T.; Peacock, J.W.; Silverstein, R.M. 1977. Response of the European elm bark beetle, *Scolytus multistriatus* (Coleoptera: Scolytidae), to isomers and components of its pheromone. J. Chem. Ecol. 3: 1-8.

Lanier, G.N.; Jones, A.H. 1984. Integrated management of Dutch elm disease in Washington D.C. 1984. State Univ. N.Y., Coll. Environ. Sci. For., Unpubl. rep.

Lanier, G.N.; Jones, A.H. 1985. Trap trees for elm bark beetles. Augmentation with pheromone baits and chlorpyrifos. J. Chem. Ecol. 11: 11-20.

Lejeune, R.R.; Richmond, H.A. 1975. Striped ambrosia beetle. Pages 246-249 *in* M.L. Prebble (Ed.), Aerial Control of Forest Insects in Canada. Can. Dept. Environ., Ottawa, Ont.

Libbey, L.M.; Morgan, M.E.; Putman, T.B.; Rudinsky, J.A. 1976. Isomer of antiaggregative pheromone identified from male Douglas-fir beetle: 3-methylcyclohex-3-en-1-one. J. Insect Physiol. 22: 871-873.

Libbey, L.M.; Oehlschlager, A.C.; Ryker, L.C. 1983. 1-Methylcyclohex-2-en-1-ol as an aggregation pheromone of *Dendroctonus pseudotsugae*. J. Chem. Ecol. 12: 1533-1541.

Libbey, L.M.; Ryker, L.C.; Yandell, K.L. 1985. Laboratory and field studies of volatiles released by *Dendroctonus ponderosae* Hopkins (Coleoptera: Scolytidae). Z. Angew. Entomol. 100: 381-392.

Lindgren, B.S. 1983. A multiple funnel trap for scolytid beetles (Coleoptera). Can. Entomol. 115: 299-302.

Lindgren, B.S.; Borden, J.H. 1983. Survey and mass trapping of ambrosia beetles (Coleoptera: Scolytidae) in timber processing areas on Vancouver Island. Can. J. For. Res. 13: 481-493.

Lindgren, B.S.; Borden, J.H.; Cushon, G.H.; Chong, L.J.; Higgins, C.J. 1989a. Reduction of mountain pine beetle (Coleoptera: Scolytidae) attacks by verbenone in lodgepole pine stands in British Columbia. Can. J. For. Res. 19: 65-68.

Lindgren, B.S.; McGregor, M.D.; Oakes, R.D.; Meyer, H.E. 1989b. Suppression of spruce beetle attacks by MCH released from bubble caps. Western J. Appl. For. 4: 49-52.

MacConnell, J.G.; Borden, J.H.; Silverstein, R.M.; Stokkink, E. 1977. Isolation and tentative identification of lineatin, a pheromone from the frass of *Trypodendron lineatum* (Coleoptera: Scolytidae). J. Chem. Ecol. 3: 549-561.

Maclauchlan, L.E.; Borden, J.H.; D'Auria, J.M.; Wheeler, L.A. 1988. Distribution of arsenic in MSMA-treated lodgepole pines infested by the mountain pine beetle, *Dendroctonus ponderosae* (Coleoptera: Scolytidae), and its relationship to beetle mortality. J. Econ. Entomol. 81: 274-280.

McLean, J.A. 1985. Ambrosia beetles: a multimillion dollar degrade problem of sawlogs in coastal British Columbia. For. Chron. 61: 295-298.

McLean, J.A.; Borden, J.H. 1975. *Gnathotrichus sulcatus* attack and breeding in freshly sawn lumber. J. Econ. Entomol. 68: 605-606.

McLean, J.A.; Borden, J.H. 1977. Suppression of *Gnathotrichus sulcatus* with sulcatol-baited traps in a commercial sawmill and notes on the occurrence of *G. retusus* and *Trypodendron lineatum*. Can. J. For. Res. 7: 348-356.

McLean, J.A.; Borden, J.H. 1979. An operational pheromone-based suppression program for an ambrosia beetle, *Gnathotrichus sulcatus*, in a commercial sawmill. J. Econ. Entomol. 72: 165-172.

McMullen, L.H.; Safranyik, L.; Linton, D.A. 1986. Suppression of mountain pine beetle infestations in lodgepole pine forests. Can. For. Serv., Pac. For. Cen., Victoria, Inf. Rep. BC-X-276.

Miller, D.R.; Borden, J.H. 1990a. The use of monoterpenes as kairomones by *Ips latidens* (Le Conte) (Coleoptera: Scolytidae). Can. Entomol. 122: 301-307.

Miller, D.R.; Borden, J.H. 1990b. β-Phellandrene: kairomone for pine engraver, *Ips pini* (Say) (Coleoptera: Scolytidae). J. Chem. Ecol. 16: 2519-2531.

Miller, D.R.; Borden, J.H.; King, G.G.S.; Slessor, K.N. 1991. Ipsenol: an aggregation pheremone for *Ips latidens* (LeConte) (Coleoptera: Scolytidae). J. Chem. Ecol. 17: 1517-1527.

Miller, D.R.; Borden, J.H.; Slessor, K.N. 1989. Inter- and intrapopulation variation of the pheromone, ipsdienol produced by male pine engravers, *Ips pini* (Say) (Coleoptera: Scolytidae). J. Chem. Ecol. 15: 233-247.

Miller, D.R.; Gries, G.; Borden, J.H. 1990. *E*-myrcenol: a multifunctional pheromone for the pine engraver, *Ips pini* (Say) (Coleoptera: Scolytidae). Can. Entomol. 122: 401-406.

Moeck, H.A. 1970. Ethanol as the primary attractant for the ambrosia beetle *Trypodendron lineatum* (Coleoptera: Scolytidae). Can. Entomol. 102: 985-995.

Moeck, H.A. 1971. Field test of ethanol as a scolytid attractant. Environ. Can., Ottawa, Ont. Bi-Mon. Res. Notes 27(2): 11-12.

Mori, K. 1976. Synthesis of (1*S*: 2*R*: 4*S*: 5*R*)-(-)-α-multistriatin, the pheromone in the smaller European elm bark beetle, *Scolytus multistriatus*. Tetrahedron 32: 1979-1981.

Mori, K. 1977. Absolute configuration of (-)-4-methylheptan-3-ol, a pheromone of the smaller European elm bark beetle, as determined by the synthesis of its (3*R*,4*R*)-(+)- and (3*S*,4*R*)-(+)-isomers. Tetrahedron 33: 289-294.

Nijholt, W.W. 1973. Ambrosia beetle attacks delayed by turpentine oil. Can. Dep. Fish. and For., Ottawa, Ont. Bi-Mon. Res. Notes 29(6): 36.

Nijholt, W.W. 1980. Pine oil and oleic acid delay and reduce attacks on logs by ambrosia beetles (Coleoptera: Scolytidae). Can Entomol. 112: 199-204.

Nordlund, D.A. 1981. Semiochemicals: a review of the terminology. Pages 13-28 *in* D.A. Nordlund, R.L. Jones, and W.J. Lewis (Eds.), Semiochemicals. Their Role in Pest Control. Wiley, New York.

O'Callaghan, D.P.; Gallagher, E.M.; Lanier, G.N. 1980. Field evaluation of pheromone-baited trap trees to control elm bark beetles, vectors of Dutch elm disease. Environ. Entomol. 9: 181-185.

Paine, T.D.; Hanlon, C.C. 1991. Response of *Dendroctonus brevicomis* and *Ips paraconfusus* (Coleoptera: Scolytidae) to combinations of synthetic pheromone attractants and inhibitors verbenone and ipsdienol. J. Chem. Ecol. 17: 2163-2176.

Peacock, J.W.; Cuthbert, R.A.; Gore, W.E.; Lanier, G.N.; Pearce, G.T.; Silverstein, R.M. 1975. Collection on Porapak Q of the aggregation pheromone of *Scolytus multistriatus* (Coleoptera: Scolytidae). J. Chem. Ecol. 1: 149-160.

Pearce, G.T.; Gore, W.E.; Silverstein, R.M. 1976. Synthesis and absolute configuration of multistriatin. J. Org. Chem. 41: 2797-2803.

Pearce, G.T.; Gore, W.E.; Silverstein, R.M.; Peacock, J.W.; Cuthbert, R.A.; Lanier, G.N.; Simeone, J.B. 1975. Chemical attractants for the smaller European elm bark beetle, *Scolytus multistriatus* (Coleoptera: Scolytidae). J. Chem. Ecol. 1: 115-124.

Phillips, T.W.; Lanier, G.N. 1986. Interspecific activity of semiochemicals among sibling species of *Pissodes* (Coleoptera: Scolytidae). J. Chem. Ecol. 12: 1587-1601.

Pitman, G.B. 1971. *Trans*-verbenol and *alpha*-pinene: their utility in manipulation of the mountain pine beetle. J. Econ. Entomol. 64: 426-430.

Pitman, G.B.; Hedden, R.L.; Gara, R.I. 1975. Synergistic effects of ethyl alcohol on the aggregation of *Dendroctonus pseudotsugae* (Coleoptera: Scolytidae) in response to pheromones. Z. Angew. Entomol. 78: 203-208.

Pitman, G.B.; Stock, M.W.; McKnight, R.C. 1978. Pheromone application in mountain pine beetle/lodgepole pine management: theory and practice. Pages 165-173 *in* A.A. Berryman, G.D. Amman, and R.W. Stark (Eds.), Theory and practice of mountain pine beetle management in lodgepole pine forests. Univ. Idaho, Moscow, and USDA For. Serv., Washington, D.C.

Pitman, G.B.; Vité, J.P. 1969. Aggregation behavior of *Dendroctonus ponderosae* (Coleoptera: Scolytidae) in response to chemical messengers. Can. Entomol. 101: 143-149.

Pitman, G.B.; Vité, J.P. 1970. Field response of *Dendroctonus pseudotsugae* (Coleoptera: Scolytidae) to synthetic frontalin. Ann. Entomol. Soc. Amer. 63: 661-664.

Pitman, G.B.; Vité, J.P. 1975. Biosynthesis of methylcyclohexenone by male Douglas-fir beetle. Environ. Entomol. 3: 886-887.

Pitman, G.B.; Vité, J.P.; Kinzer, G.W.; Fentiman, A.F., Jr. 1968. Bark beetle attractants: *trans*-verbenol isolated from *Dendroctonus*. Nature 218: 168-169.

Pitman, G.B.; Vité, J.P.; Kinzer, G.W.; Fentiman, A.F., Jr. 1969. Specificity of population-aggregating pheromones in *Dendroctonus*. J. Insect Physiol. 15: 363-366.

Plummer, E.L.; Stewart, T.E.; Byrne, K.; Pearce, G.T.; Silverstein, R.M. 1976. Determination of the enantiomeric composition of several insect pheromone alcohols. J. Chem. Ecol. 2: 307-331.

Renwick, J.A.A. 1967. Identification of two oxygenated terpenes from the bark beetles *Dendroctonus frontalis* and *Dendroctonus brevicomis*. Contrib. Boyce Thompson Inst. 23: 355-360.

Renwick, J.A.A; Vité, J.P. 1970. Systems of chemical communication in *Dendroctonus*. Contrib. Boyce Thompson Inst. 24: 283-292.

Ringold, G.B.; Gravelle, P.J.; Miller, D.; Furniss, M.M; McGregor, M.D. 1975. Characteristics of Douglas-fir beetle infestation in northern Idaho resulting from treatment with Douglure. USDA For. Serv. Washington, D.C. Res. Note INT-189.

Rudinsky, J.A. 1973. Multiple functions of the Douglas-fir beetle pheromone, 3-methyl-2-cyclohexene-1-one. Environ. Entomol. 2: 579-585.

Rudinsky, J.A.; Daterman, G.E. 1964a. Field studies on flight patterns and olfactory responses of ambrosia beetles in Douglas-fir forests of western Oregon. Can. Entomol. 96: 1339-1352.

Rudinsky, J.A.; Daterman, G.E. 1964b. Response of the ambrosia beetle *Trypodendron lineatum* (Olivier) to a female-produced pheromone. Z. Angew. Entomol. 54: 300-303.

Rudinsky, J.A.; Furniss, M.M; Kline, L.N.; Schmitz, R.F. 1972b. Attraction and repression of *Dendroctonus pseudotsugae* (Coleoptera: Scolytidae) by three synthetic pheromones in traps in Oregon and Idaho. Can. Entomol. 104: 815-822.

Rudinsky, J.A.; Kinzer, G.W.; Fentiman, A.F., Jr.; Foltz, R.L. 1972a. *Trans*-verbenol isolated from Douglas-fir beetle: laboratory and field bioassay in Oregon. Environ. Entomol. 1: 485-488.

Rudinsky, J.A.; Morgan, M.E.; Libbey, L.M.; Putnam, T.B. 1974a. Antiaggregative-rivalry pheromone of the mountain pine beetle, and a new arrestant of the southern pine beetle. Environ. Entomol. 3: 90-98.

Rudinsky, J.A.; Morgan, M.E.; Libbey, L.M.; Putnam, T.B. 1974b. Additional components of the Douglas-fir beetle (Col., Scolytidae) aggregative pheromone and their possible utility in pest control. Z. Angew. Entomol. 76: 65-77.

Rudinsky, J.A.; Ryker, L.C. 1979. Field bioassay of male Douglas-fir beetle compound 3-methylcyclohex-3-en-1-one. Experientia 35: 1302.

Rudinsky, J.A.; Ryker, L.C. 1980. Multifunctionality of Douglas-fir beetle pheromone 3,2-MCH confirmed with solvent dibutyl phthalate. J. Chem. Ecol. 6: 193-201.

Rudinsky, J.A.; Sartwell, C.; Graves, T.M.; Morgan, M.E. 1974c. Granular formulation of methylcyclohexenone: an antiaggregative pheromone of the Douglas-fir and spruce bark beetles. Z. Angew. Entomol. 75: 254-263.

Ryker, L.C.; Libbey, L.M.; 1982. Frontalin in the male mountain pine beetle. J. Chem. Ecol. 8: 1399-1409.

Ryker, L.C.; Libbey, L.M.; Rudinsky, J.A. 1979. Comparison of volatile compounds and stridulation emitted by the Douglas-fir beetle from Idaho and western Oregon populations. Environ. Entomol. 8: 789-798.

Ryker, L.C.; Rudinsky, J.A. 1982. Field bioassay of *exo*- and *endo*-brevicomin with *Dendroctonus ponderosae* in lodgepole pine. J. Chem. Ecol. 8: 701-707.

Ryker, L.C.; Yandell, K.L. 1983. Effect of verbenone on aggregation of *Dendroctonus ponderosae* Hopkins (Coleoptera: Scolytidae) to synthetic attractant. Z. Angew. Entomol. 96: 452-459.

Salom, S.M.; McLean, J.A. 1988. Semiochemicals for capturing the ambrosia beetle, *Trypodendron lineatum*, in multiple funnel traps in British Columbia. J. Entomol. Soc. British Columbia. 85: 34-39.

Schurig, V.; Weber, R.; Nicholson, G.I.; Oehlschlager, A.C.; Pierce, H., Jr.; Pierce, A.M.; Borden, J.H.; Ryker, L.C. 1983. Enantiomer composition of natural *exo*- and *endo*-brevicomin by complexation gas chromatography/selected ion mass spectrometry. Naturwissenschaften. 70: 92-93.

Shore, T.L. 1982. A pheromone-mediated mass-trapping program for three species of ambrosia beetle in a commercial sawmill. Ph.D. Thesis. University of British Columbia, Vancouver, British Columbia.

Shore, T.L.; McLean, J.A. 1983. A further evaluation of the interaction between the pheromone and two host kairomones of the ambrosia beetles, *Trypodendron lineatum* (Olivier) and *Gnathotrichus sulcatus* (LeConte). Can. Entomol. 115: 1-5.

Silverstein, R.M. 1970. Methodolgy for isolation and identification of insect pheromones—examples from Coleoptera. Page 285-299 *in* D.L. Wood, R.M. Silverstein and M. Nakajima (Eds.), Control of Insect Behavior by Natural Products. Academic Press, New York.

Silverstein, R.M.; Brownlee, R.G.; Bellas, T.E.; Wood, D.L.; Browne, L.E. 1968. Brevicomin: principal sex attractant in the frass of the female western pine beetle. Science 159: 889-890.

Silverstein, R.M.; Rodin, J.O.; Wood, D.L. 1966. Sex attractants in frass produced by male *Ips confusus* in ponderosa pine. Science 154: 509-510.

Silverstein, R.M.; Rodin, J.O.; Wood, D.L. 1967. Methodology for isolation and identification of insect pheromones with reference to studies on California five-spined ips. J. Econ. Entomol. 60: 944-949.

Slessor, K.N.; King, G.G.S; Miller, D.R.; Winston, M.L.; Cutforth, T.L. 1985. Determination of chirality of alcohol or latent alcohol semiochemicals in individual insects. J. Chem. Ecol. 11: 1659-1667.

Slessor, K.N.; Oehlschlager, A.C.; Johnston, B.D.; Pierce, H.D., Jr.; Grewal, S.K.; Wickremesinghe, L.K.G. 1980. Lineatin: regioselective synthesis and resolution leading to the chiral pheromone of *Trypodendron lineatum*. J. Org. Chem. 45: 2290-2297.

Smith, R.H. 1986. Trapping western pine beetles with baited toxic trees. USDA For. Serv. Washington, D.C. Res. Note PSW-382.

Stewart, T.E.; Plummer, E.L.; McCandless, L.L.; West, J.R.; Silverstein, R.M. 1977. Determination of enantiomer composition of several bicyclic ketal insect pheromone components. J. Chem. Ecol. 3: 27-43.

Stock, A.J. 1984. Use of pheromone baited Lindgren funnel traps for monitoring mountain pine beetle flights. British Columbia For. Serv. Victoria Int. Rep. PM-PR-2.

Stock, A.J. 1991. The western balsam bark beetle, *Dryocoetes confusus* Swaine: impact and semiochemical-based management. Ph.D. Thesis. Simon Fraser University, Burnaby, British Columbia.

Vité, J.P.; Bakke, A. 1979. Synergism between chemical and physical stimuli in host colonization by an ambrosia beetle. Naturwissenschaften 66: 528-529.

Vité, J.P.; Pitman, G.B. 1969. Aggregation behavior of *Dendroctonus brevicomis* in response to synthetic pheromones. J. Insect Physiol. 15: 1617-1622.

Vité, J.P.; Pitman, G.B.; Fentiman, A.F., Jr.; Kinzer, G.W. 1972. 3-Methyl-2-cyclohexen-1-ol isolated from *Dendroctonus*. Naturwissenschaften 59: 469.

Wieser, H.; Dixon, E.A.; Cerezke, H.F.; Mackenzie, A.A. 1991. Pheromone formulation for attracting spruce beetles. U.S. Patent No. 4,994,268.

Wood, D.L. 1972. Selection and colonization of ponderosa pine by bark beetles. Pages 101-117 *in* H.F. van Emden (Ed.), Insect/Plant Relationships. Blackwell Scientific, London.

Wood, D.L.; Browne, L.E.; Ewing, B.; Lindahl, K.; Bedard, W.D.; Tilden, P.E.; Mori, K.; Pitman, G.B.; Hughes, P.R. 1976. Western pine beetle: specificity among enantiomers of male and female components of an attractant pheromone. Science 192: 896-898.

Wood, D.L.; Browne, L.E.; Silverstein, R.M.; Rodin, J.O. 1966. Sex pheromones of bark beetles—I. Mass production, bio-assay, source, and isolation of the sex pheromone of *Ips confusus* (LeC.). J. Insect Physiol. 12: 523-536.

Wood, D.L.; Stark, R.W.; Silverstein, R.M.; Rodin, J.O. 1967. Unique synergistic effects produced by the principal sex attractant compounds of *Ips confusus* (LeConte) (Coleoptera: Scolytidae). Nature 215: 206.

Wood, R.O. 1983. Mountain pine beetle in Manning Park. A post-control survey. Can. For. Serv., Pac. For. Res. Cen., Victoria, B.C. Pest. Rep., September 1983.

Wood, R.O. 1984. Mountain pine beetle in Manning Park. A post-control survey. Can. For. Serv., Pac. For. Res. Cen., Victoria, B.C. Pest Rep., September 1984.

Chapter 46

Pesticide Formulations

ALAM SUNDARAM

Introduction

Effective protection of forests against defoliating insects requires that pesticides be formulated in an appropriate medium to provide good physical stability and facilitate easy mixing, pumping, and storing capabilities; to cause efficient spray atomization and droplet impaction on targets; to have optimum field stability during the critical period of pest control; and, finally, to rapidly degrade in the environment to innocuous products.

Manufacturers closely guard the details of their formulation techniques to protect their products against the competition from rival companies (Matthews 1979). Methods of formulating a pesticide must be based not only on its effectiveness but also on its biodegradability and overall effect on the environment. In recent years, the awareness of environmental contamination by xenobiotic compounds and economic considerations have encouraged intensive research with specific objectives to reduce dosage rates and, at the same time, to increase effectiveness, and to minimize undue losses of pesticides that may produce adverse side effects on the environment.

Formulation Research Requirements

Formulation technology is a complex science. The behavior of formulation components and their effects on biological matrices and on the environment need to be understood so that pesticides can be delivered to target organisms with maximum efficiency and with minimum impact on the environment. Formulation techniques involve different approaches for agricultural and forestry uses because of the differences in application technologies, particularly in the use of large volume rates in agriculture as opposed to low volume rates in forestry, and types of terrain encountered. In forestry, because of the small droplet sizes that are needed, the physicochemical properties of formulations must be optimized so that the maximum amount of the spray cloud released above the tree canopy can reach the intended target.

With chemical insecticides, low application rates and fine atomization are used; thus low volatility formulations are required to prevent loss of spray droplets into the off-target environment (Sundaram 1987a; Sundaram and Retnakaran 1987; Maas 1971; Sundaram et al. 1987a). With *Bacillus thuringiensis* (Bt) and other biological control agents, the mode of entry is via oral ingestion so concentrated foliar deposits are needed to provide lethal doses with minimum foliage feeding. Spray droplets of some high potency aqueous formulations of Bt do not spread on targets, often resulting in poor adhesion and low deposits (Sundaram 1989a). Suitable adjuvants must be added, therefore, to ensure good adhesion of spray droplets onto target foliage, and also to ensure protection of deposits from ultraviolet light and rain (van Frankenhuyzen and Nystrom 1989).

When a pesticide manufacturer introduces a new formulation to the market for forestry use, it is either (i) a basic formulation which is a concentrated product that can be readily mixed with a carrier liquid such as water or a diluent oil, or (ii) a tank-mix formulation that requires no further dilution by the user. It is necessary to investigate the behavior of both types of product, and to achieve this, physicochemical properties and behavior are often compared with those of a standard (i.e. a product currently in extensive use).

Basic Formulations

Research requirements with the basic formulations involve: (i) storage stability of the pesticide material; (ii) rate of sedimentation and/or crystallization of the pesticide and inerts; (iii) resuspensibility of sediments and redissolution of the pesticide on agitation after storage over the winter; (iv) variations in the mass of pesticide and mass of inerts ratio for different lot numbers; (v) investigations to determine the suitability of a range of carrier liquids; (vi) investigation of the physicochemical properties of a variety of diluent oils and other adjuvants.

Because few formulations are developed by the pesticide industry specifically for forestry use, data on all of the above have to be generated elsewhere so that formulation properties and behavior can be optimized for forestry use.

Tank-Mix Formulations

Research requirements with the tank-mix formulations involve: (i) physical and chemical stability of the active material and inerts (phase separation and degradation) at a wide range of temperatures; (ii) a comparative evaluation of physicochemical properties between the new tank-mix formulation and the currently used one(s); (iii) spray generation, drop size spectra, and deposition patterns on targets, using different types of atomizers; (iv) a comparative evaluation of droplet retention and adhesion on targets.

Manufacturing companies, in general, are unable to provide data on tank-mix formulations because of a lack of knowledge of dosage rates, and the type and amount of diluents to be used for preparing tank-mix formulations for forestry use. Therefore, the properties and behavior of the tank-mix formulations are usually investigated by outside agencies.

To investigate the physicochemical properties and behavior of both basic and tank-mix formulations under forestry use patterns, reliable, accurate, economical, and simple methods are needed. Therefore, methodology development is one of the major roles of formulation research projects.

Types of Basic Formulations

Liquid Formulation Concentrates

Only a few pesticides dissolve readily in water or in acceptable solvents and can be applied as solutions. Consequently, the basic formulations of pesticides are prepared either as emulsifiable concentrates, normal or invert emulsions, oil soluble concentrates, flowable formulations, or microencapsulated formulations. These are usually diluted at the field-mixing site with water or with nonaqueous carrier diluents to form the tank-mix formulations.

Emulsifiable concentrates—An important component in these formulations is the emulsifier, a surface active agent, which is partly hydrophilic and partly lipophilic. A pesticide solution in an organic solvent (such as petroleum distillates) cannot be mixed with water because the two liquids would form separate layers. The addition of an emulsifier produces emulsion particles (i.e. those that are composed of a mixture of the active ingredient, solvent, and emulsifier), usually less than 10 µm. These particles are capable of forming a homogeneous and stable dispersion in water, resulting in a "normal" oil-in-water emulsion. The concentration of the active ingredient in most emulsifiable concentrates is usually 25 to 50% w/v (concentration is expressed as percent weight over volume). Some pesticides, such as carbaryl, cannot be formulated economically as an emulsifiable concentrate because the solvents in which the active ingredient is soluble are too expensive for field use (Matthews 1979). Also, solvents with high solubilizing power are often environmentally unacceptable.

The stability of an emulsion is improved by a mixture of surfactants as the anionics increase in solubility at higher temperatures, whereas the reverse is true of nonionic surfactants (Van Valkenburg 1973). Becher (1973) lists several emulsifiers, together with a numerical value for the hydrophile–lipophile balance. An unstable emulsion "breaks" if it separates and forms a "cream" on the surface, or the globules coalesce to form a separate layer. Creaming is due to differences in specific gravity between the two phases, and also to the presence of surfactants that are incompatible with the organic solvent used.

Agitation of the spray mix normally prevents creaming. Breaking of an emulsion after the spray droplets reach the target is due partly to evaporation of the water present, and leaves the pesticide in a film that may readily penetrate the surface of the target. The stability of the emulsions is affected by the hardness and pH of water used for preparing the tank mix, and also by conditions under which the emulsifiable concentrate is stored. Extreme temperatures can adversely affect an emulsion formulation.

The choice of a solvent may be restricted by its flash point so as to reduce possible risks of fire during transportation and use, especially during aerial application. Emulsifiable concentrates pre-mixed with a small quantity of water to form a mayonnaise-type formulation have been used, but these are likely to deteriorate during storage.

A typical example of an emulsifiable concentrate is Ambush® 500 EC, a formulation of a synthetic pyrethroid, permethrin, although this chemical is now registered only for small-scale uses in forestry involving ground application. Another emulsion formulation used in forestry is fenitrothion, an organophosphorus insecticide. Only one emulsifiable concentrate is commercially available at present (Product Label, Sumithion® 50 EC Emulsifiable Concentrate Insecticide, BASF Canada, 345 Carling Drive, Toronto, Ontario), which is prepared by the spray operators at the mixing site. The active liquid material is readily mixed with defined quantities of a solvent (Dowanol® DPM) and a surfactant (Atlox® 3409F or Triton® X-114) at the mixing site and added to water to form an emulsion. This tank-mix formulation is still being used in eastern Canada for the control of spruce budworm, *Choristoneura fumiferana*.

Invert emulsions—This is a change from an oil-in-water emulsion to water-in-oil. An "invert" emulsion is usually very viscous and is sometimes produced by mixing the two phases at the nozzle tip (Muller and Stovell 1966). Invert emulsions have been used principally for the application of herbicides. Careful formulation is essential to produce large droplets and reduce drift (Ford and Furmidge 1966). Aircraft equipment has to be specially designed for invert emulsions. For example, a hydraulically operated positive displacement pump is needed instead of the more common centrifugal pump so that the emission rate can be controlled (Pearson and Masheder 1969). The use of invert emulsions has not been widely accepted because of the need for specialized equipment (Matthews 1979).

Oil soluble concentrates—These formulation concentrates usually contain high amounts of solvents and/or surfactants to help keep the active material in solution when tank-mixed with a diluent oil. An example of this type of formulation is Matacil® 180D, which contains a carbamate insecticide, aminocarb. This formulation is diluted at the mixing site with an insecticide diluent marketed by Shell as ID585. This tank mix is very easy to use because no particulate matter is present to cause sedimentation or crystallization problems. In spite of these advantages, Matacil® 180D is not widely used because nonyl phenol, the surfactant present in the concentrate, could possibly cause aquatic toxicity.

Flowable formulations—A flowable formulation is a water- or oil-based dispersion (the latter is sometimes called an oil suspensible concentrate) of the active ingredient. Several flowable formulations have recently been developed and commercialized. In most cases, the active ingredients are solids, which are usually ground in the carrier liquid phase during preparation of these formulations (one example is the formulation concentrate, Matacil® 180 flowable, of milled aminocarb particles in a heavy paraffinic oil). However, phase separation (i.e. physical instability) could be a problem; and caking was encountered in some cases. Various kinds of thickening agents were added to overcome these problems (Tsuji and Fuyama 1982).

Matacil® 180 flowable was used for several years during the early 1980s to control the spruce budworm. It was tank-mixed as a flowable emulsion by adding the surfactant, Atlox® 3409F (or Triton® X-114) and water. Alternatively, it was also simply diluted with ID585 to provide a nonaqueous tank mix. However, limited interest has been shown in Matacil® 180 flowable and it is not widely used at present.

Emulsion-type aqueous flowable concentrates use liquid active ingredients (e.g. Sumithion® Flowable, containing fenitrothion) (Tsuji and Fuyama 1982). These concentrates are prepared by adding liquid pesticides to water in the presence of efficient dispersing agents, such as water-soluble polymers like polyvinyl alcohol and gum arabic. These dispersions can be stabilized by adding thickening agents like xanthan gums, sodium carboxymethyl cellulose, and polyacrylic polymers. Tank mixing is usually done simply by diluting the concentrate with water.

Originally, Sumithion® Flowable was found unsuitable for forestry use because of the beading problems of the spray droplets. These beads could possibly be picked up by nontarget invertebrates and cause toxicity. A new formulation, which did not show the beading problem, was introduced and field-tested, and it provided enhanced efficacy compared to the conventional fenitrothion emulsion. At present, Sumithion® Flowable is available for use in Canada for the control of spruce budworm.

Microencapsulated pesticides—Microcapsules about 10 µm in diameter are made by a coacervation (phase separation) process which contains a pesticide or a pesticide solution within a hardened polymer wall to produce capsules down to 3 µm in diameter (Marrs and Middleton 1973; Anon. 1974). A reduced amount of pesticide may be adequate for insect control as release of the toxicant can be controlled to some extent by the thickness and structure of the capsule wall. Nonpersistent chemicals can be applied more effectively under field conditions by using special polymeric materials to filter the effect of ultraviolet light. Specificity can be increased, especially if a suitable attractant is used along with the pesticide (Markin et al. 1972). Retention of microcapsules applied as a suspension in water with stickers is claimed to be excellent even after rain (Phillips and Gillham 1973). The slow release characteristics of the active ingredient from microcapsules are particularly useful for the application of chemicals that affect the behavior of insects, such as pheromones (Campion 1976).

Microencapsulated formulations of conventional chemical pesticides have not been developed for forestry use because of suspected toxicity to nontarget organisms. Nevertheless, this type of formulation technology has great potential for enhancing the pesticidal activity of biorational and biological control agents.

Recently, interest has been increasing in controlled release formulations for agricultural uses, either as microencapsulated products, coated granules, or polymer matrix systems. The benefits achieved are extended longevity of biological effectiveness, reduced phytotoxicity, decreased pesticide levels in the environment, and reduced mammalian toxicity (Seaman 1990). However, the use of such formulations in forestry appears to be limited only to microbial pesticides.

Dry Formulation Concentrates

Dry formulation concentrates have the advantage of longer shelf life for the active ingredient during storage. Because of this, these formulations are usually much less expensive than liquid formulations. Dry formulations can be divided into three main categories.

Wettable powders—These formulation concentrates consist of finely divided pesticide particles together with inert powder materials that enable the pesticide to be mixed with water or other diluents to form a homogeneous suspension. Wettable powders frequently contain 25% to 50% of the active material, together with naturally occurring clays and synthetic silica powders, and surface active agents. The amount of silica is usually low (2% to 5%) because this material is very abrasive. Inert fillers are often used to obtain the required pesticide concentration.

Wettable powders have a high proportion of particles less than 5 µm in diameter, and all of the particles are less than about 40 µm. These products should flow easily to facilitate measuring into the mixing container and, because they are made up of extremely small particles, care must be taken to avoid off-target drift problems. The powder should disperse and wet easily when mixed with diluent carriers, and not form lumps. To ensure good mixing, most powder should be premixed with about 5% of the final volume of water and creamed to a thin paste. When added to the remaining carrier diluent, the pre-mix should disperse easily with stirring and remain suspended for a reasonable time. The surface active agents should prevent the particles from aggregating and settling out in the application tank. The rate of sedimentation in the tank is directly proportional to the size and density of the particles (Matthews 1979). Good suspensibility is important because spray equipment, such as the knapsack sprayers, does not have agitators.

Wettable powders should retain their fluidity, dispersibility, and suspensibility characteristics even after prolonged storage. Containers should be designed so that even if the material is stored in stacks, the particles should not be affected by pressure and excessive heat, which may cause agglomeration (Lindner 1973; Polon 1973). Wettable powders are not usually compatible with other types of formulations, because mixing them with an emulsion frequently causes flocculation or sedimentation owing to interference with the surface active agents in the emulsion.

Carbaryl, azinphos-methyl, and diflubenzuron are typical examples of forestry pesticides formulated as wettable powders. At present, the diflubenzuron wettable powder (trade name, Dimilin® WP-25), containing 25% active material is being studied in experimental spray trials, either as a suspension in water or in low-volatility oils, to investigate its use to control several forest insects.

Wettable powders are generally viewed less favorably by regulatory authorities and forest managers than flowable formulations, because powders are dusty and less safe during mixing and handling. Also, they are bulky and difficult to measure (Seaman 1990).

Dust—Dust is a general term applied to fine dry particles ranging from 10 to 30 µm in diameter (Matthews 1979). A dust formulation concentrate (e.g. carbaryl dust) usually contains about 20% active material, but normally these form the basis for further dilution to 0.5 to 10% by other formulators. Less concentrated dusts have also been commercially available, but these were applied undiluted to crops. A dust formulation concentrate is prepared by impregnating highly sorptive particles with a solution of a pesticide. The concentrate is then mixed with dry diluent fillers, such as attapulgite, kaolinite, silica, talc, etc., to the strength required in the field (Polon 1973). The size of the pesticide-carriers and diluent fillers should not be smaller than 10 µm, otherwise the particles would be highly abrasive to the insect cuticle. Mortalities of up to 85% have been reported when dusts of 1.0 to 10 µm in diameter were used with no insecticides (Hunt 1947). In fact, the possibility of using small dust particles without insecticides has been explored for the control of cockroaches, dry-wood termites, and other household pests (Ebeling 1971).

The use of dusts has several advantages. They can be used readily in areas where water is scarce. Small particles that reach the target adhere better to surfaces than large particles, for example, a force of 500 kg is needed to shake off a 10-µm particle, in contrast to 4 kg for a 100-µm particle (Matthews 1979). Fine particles are generally more toxic than coarse ones. However, there are some disadvantages that override the advantages. Small particles create off-target drift problems, including the human inhalation hazard. Because of this, dusts should not be applied from aircraft. Dusts are useful only for specialized treatments, such as treatment of seeds before sowing, and for protection of stored grain from decay.

Dust formulations, similar to the wettable powders, are not favored at present in forestry, partly because of the potential for human exposure during handling and for off-target drift problems during delivery. However, they may have some potential for use in certain cases, especially in forest nurseries and plantations.

Granules—To overcome off-target drift and concern about the human inhalation hazard, large discrete dry particles (granules) could be used, especially for the most potentially hazardous pesticides. To prepare granules, a carrier filler is impregnated with the pesticides in the same way dusts are prepared. Like dusts, the concentration of active ingredient is usually about 15% to 30%. Because of the large size of the filler particles, the rate of release of pesticides is usually slow, but the period of effectiveness is often longer than that obtained with dusts or liquid formulations. In addition, granules can be coated with various materials, such as starch, polymers, etc., to control the rate of release of the toxicant.

Granules have several advantages. Chemicals with short-term longevity can be protected against loss into the environment by the granular technique. The placement of granules can be very precise and, therefore, low quantities of pesticides can be used. Spot treatment to individual plants is also possible and no on-site mixing is required. In spite of these advantages, there has been a slow acceptance of this method for pesticide application. Forest managers generally prefer broadcast treatment because of the necessity to treat large areas within a short period of time, but the present techniques of broadcast release of granules need improvement to maximize their effectiveness.

Granules of carbofuran and other systemic insecticides may have some potential use in forestry because they can be applied under the soil near the root system for uptake by the tree. Nevertheless, such uses are not employed at present because of the potential exposure to nontarget organisms.

Recently, improvements have been made in granular formulations so that they can be readily dispersed in water (called water-dispersible-granules). These formulations are now receiving increasing attention because they are made to be nondusty, less bulky, and easier to measure than conventional granules (Seaman 1990).

Ultra-Low-Volume Formulations

When pesticides are applied at low volume rates, ranging from 15 to 75 L/ha (Maas 1971), the droplets are sufficiently large (from 200 to 500 µm in diameter) to ensure minimum loss into the off-target environment. However, efficacy of the spray is reduced by the delivery of such large droplets because plant coverage is less (i.e. droplets per unit area). Also, many large droplets bounce off the foliage or shatter because of their high-impact velocity (Johnstone 1973). Large droplets also fall through open areas of a tree canopy onto the forest floor and increase soil residues (Sundaram and Sundaram 1987b).

Ultra-low-volume application techniques using volume rates less than 5 L/ha (Matthews 1979) and involving the release of very small droplets (those having diameters smaller than 100 µm) may increase spray effectiveness because of efficient foliar impaction with minimum bounce-off (Maas 1971). However, off-target losses of the toxicant present in the small droplets is also relatively higher because the emitted drop size range of a spray cloud having a volume median diameter of 150 µm means 25% to 30% of the drops are less than 50 µm in diameter (N.B. Akesson, pers. comm.). If the carrier liquid is highly evaporative, even greater losses of the toxicant will occur, because the droplets will end up at distances several kilometres downwind from the target region (Yates and Akesson 1973). Droplets less than 50 µm must not evaporate markedly in-flight and decrease further in size. Therefore, volatility of the tank- mix formulations should be maintained as low as possible. Moreover, efficient atomization of the formulation must be maintained to ensure that excessively large droplets will not be produced during spray release. Adjustment and fine-tuning of the physicochemical properties of ultra-low-volume formulations are necessary for all of these objectives to be realized.

Ultra-low-volume spray techniques involving concentrated tank-mix formulations became more popular with the advent of synthetic pesticides following World War II. However, ultra-low-volume techniques were adopted much earlier, because aircraft chemical foggers were used to release extremely fine droplets for mosquito and fly control even during the 1940s (Deonier et al. 1949; Brown and Watson 1953; Hocking et al. 1953). In the earliest experiments for locust control in East Africa, concentrated solutions of dieldrin were sprayed in diesel fuel (Sayer 1959). To control the desert locust, vast areas had to be treated rapidly, and with the immense logistic problems involved, any reduction in volume that could be achieved without loss of effect was much exploited. Because of the great mobility of the adult locusts, it was not necessary to have a very dense deposit coverage of the vegetation. This explains why the ultra-low-volume spraying was a great improvement even in a very elementary form.

The development of the technique for food crop and forestry spraying started later. There were isolated trials in the early 1960s in Russia and Germany in which strong solutions of DDT in oil were used for the control of Colorado potato beetles, *Leptinotarsa decemlineata*, on potatoes (Angermann 1961; Bondine 1963). Further development of this technique was encouraged by the good results obtained by American Cyanamid in the application of malathion (Messenger 1963, 1964, 1965), for control of rangeland grasshoppers (Skoog et al. 1965), and for control of the boll weevil *Anthonomus grandis grandis* (Maas 1971).

From 1965 to 1968, a series of aerial spray trials, using fenitrothion and several types of atomizers, was carried out in Canada and the ultra-low-volume method was found to be highly effective for the control of forest defoliators (Randall 1971). Since then, the technique has been examined in detail, and is well accepted in forestry aerial applications for the successful control of several defoliators including the spruce budworm, *Choristoneura fumiferana*; white pine weevil, *Pissodes strobi*; spruce bud moth, *Zeiraphera canadensis*; and hemlock looper, *Lambdina fiscellaria* (Morris et al. 1974; DeBoo 1980; Nigam 1971; DeBoo and Campbell 1974; Cadogan et al. 1984; Cadogan 1986; Randall et al. 1977, 1981; Retnakaran et al. 1988; West et al. 1989; Helson et al. 1989).

Some liquid insecticides such as malathion can be applied without any formulation, although such high concentration is seldom justified. Because the number of such uses is very restricted, suitable solvents are always needed. Therefore, at present, almost all conventional forestry chemical insecticides, and Bt, are sprayed using the ultra-low-volume technique. Volume rates from 1.0 to 2.0 L/ha are being successfully used.

In addition, the use of a hygroscopic solvent, Dowanol® TPM, was recently tested with fenitrothion using volume rates less than 1.0 L/ha (Raske et al. 1989) for the control of the hemlock looper. The results were encouraging and warranted additional field trials using fenitrothion/Dowanol® formulations at rates below 1.0 L/ha. The presence of Dowanol® TPM may have contributed not only to reducing droplet evaporation in-flight but may also have increased the droplet sizes while falling due to absorption of moisture from the ambient air. In a previous laboratory study, a thin film of fenitrothion/ Dowanol® mixture was demonstrated to absorb moisture from the ambient air at high humidity (more than 70%) and at low temperatures (less than 10°C) (Sundaram 1987a).

Influence of Formulation on Pesticide Performance

The types of inert ingredients affect physicochemical characteristics, thus influencing field performance of pesticides. The following provides a description of how optimum formulation techniques can enhance pesticide effectiveness and reduce the adverse side effects that are sometimes encountered.

Storage, Mixing and Pumping, and Flowing in the Application Systems

Viscosity plays an important role in storage stability of tank-mix formulations containing particulate materials, such as wettable powders, milled pesticide particles, Bt, etc. If viscosity is below an optimum value, phase separation will occur, resulting in agglomeration of solid particles, caking, and settling (Sundaram and Sundaram 1989). If viscosity is too high, mixing and pumping may be difficult (if viscosity does not decrease with shear rate experienced during pumping), especially at the cold temperatures encountered in the spring months in Canada at the time of spruce budworm spray operations. Consequently, high-powered pumps may be required. High-viscosity formulations also generally pose problems in controlling flow rates through aircraft pipes and nozzle systems.

The viscosity of a liquid determines its flow behavior in the pumps and application equipment. Therefore, it is important to understand flow behavior of liquids under varied shearing forces that are likely to be encountered during field use. Based on their viscosity values, pesticide formulations (both basic and tank-mix types) can be divided into two major categories, Newtonian and non-Newtonian. The latter can be further divided into five groups as described below.

Newtonian Formulations

The viscosity of a Newtonian formulation is almost constant at a given temperature and is considered to be independent of the shear rate of the force applied. Typical examples are water, organic solvents, mineral oils, solutions of inorganic and organic substances, and many of the dilute formulations of chemical insecticides. Newtonian liquids can be divided into two categories, those of low viscosities (i.e. less than 30 mPa·s) and those of higher viscosities. The low-viscosity liquids are relatively easy to handle, mix, and pump even at cold temperatures, because their viscosities will increase only slightly at low temperatures. High-viscosity Newtonian liquids can pose problems in handling at cold temperatures because their viscosities tend to increase markedly as the temperature decreases. These liquids may often be difficult to mix and pump even at normal temperatures, because their viscosities do not decrease with the increasing shear rates encountered during mixing and pumping.

Non-Newtonian Formulations

Pseudoplastic formulations—To this group of formulations belong suspensions, dispersions, and those containing certain polymers and clays. These formulations are characterized by their high storage viscosity, which helps keep particulate materials in uniform suspension. However, with increasing shear rates the viscosity decreases very rapidly, resulting in a thin formulation at high shear rates. These formulations are easy to handle, mix, and pump because of rapid lowering of viscosity as the shear stress is applied. Typical examples of this group are aqueous Bt formulations, Sumithion® Flowable, thick emulsions, and invert emulsions.

Dilatant formulations—This flow behavior is rare, but its effect on highly concentrated suspension of Bt and on certain chemical insecticide formulations containing silicone additives can be grave. The viscosity of dilatant formulations increases with increasing shear rates. In other words, increasing the rotational speed of a nozzle means a multiplied resistance by the formulation, a phenomenon that can lead to breakage of application equipment. By discovering the extent of dilatant behavior ahead of time under laboratory conditions, such catastrophes can be avoided in the field.

Plastic formulations—Belonging to this group are formulations that are sold as pastes or suspensions with high filler content. Although such thick pastes are not used at present in forestry, knowledge of their flow behavior is important so that any future problems can be solved. Also, certain surfactant adjuvants used in tank-mix formulations do behave similarly and are likely to pose problems during mixing and pumping. Plastic formulations have an initial high viscosity called the "yield point," which may not be overcome easily by the application of shearing force, and these liquids may not start to flow until the yield point has been surpassed. After the yield point, however, the viscosity decreases in almost all cases with increasing shear rates.

Thixotropic formulations—Viscosities of thixotropic formulations decrease with the duration of the applied shear and with increasing shear rates. However, these liquids will regenerate toward their original viscosities once the shearing force is removed. Typical examples of this type are flowable formulations of Bt and those containing bentonite clays or some polymers. These formulations are easy to handle, and because of reversal to their original viscosities, they have an added advantage of providing a reduced proportion of fine droplets in the spray cloud, a phenomenon that is helpful in reducing off-target drift problems in pesticide applications.

Rheopectic formulations—This behavior is rare, and is observed only with some highly filled aqueous suspensions containing polymeric additives. The viscosity of a rheopectic formulation decreases with increasing shear rates but increases with the duration of the applied shear. This property is usually reversible once the shearing force is removed. Being the opposite of thixotropy, it is sometimes described as "anti-thixotropy." Some Bt formulations could behave in this manner and contribute to flow problems in the application equipment. Therefore, viscosity–shear rate relationships of Bt formulations should be investigated in the laboratory before field application trials are undertaken.

The degree to which the active material is soluble in the chosen solvent also plays an important role in phase separation, agglomeration, and settling characteristics of particulate materials–the higher the solubility the less the tendency for settling of the pesticide particles. However, extremely high solubility should be avoided, otherwise the inert ingredients can undergo phase separation.

Viscosities, flow behavior, and other relevant physical properties of some basic formulations of pesticides and inert ingredients registered and/or used in forestry spraying are provided in Tables 1 and 2 at a range of temperatures.

Liquid Atomization and Droplet Formation

Formulation ingredients and physical properties influence spray atomization and droplet formation (Page 1961; Colthurst et al. 1966; Argauer et al. 1968; Butler et al. 1969; Richardson 1974; Yates et al. 1976, 1983, 1985a, 1985b; Akesson and Yates 1984; Akesson and Gibbs 1990; Bode and Zain 1987; Bouse et al. 1988; Hall 1989; Haq et al. 1983; Sundaram 1987b). The influence is the greatest when ultra-low-volume rates are used. When small droplets are released by aircraft, droplet volatility in-flight must be minimized so that off-target loss will be insignificant. Therefore, the physicochemical properties of a tank-mix formulation should be

Table 1. Solubility/miscibility in water and other relevant properties of some basic formulations of pesticides and inert ingredients that are registered and/or used at present in forestry.

Name	Source	Product type	Pour point/ freezing point (°C)	Flash point (°C)	Solubility/ miscibility in water
Fenitrothion (technical)	Sumitomo Chemical (Osaka, Japan)	Pesticide material	Below 0	–	Insoluble
Sumithion® 20% Flowable	Sumitomo Chemical (Osaka, Japan)	Basic formulation	Slightly below 0	–	Miscible
Matacil® 180D	Chemagro Ltd. (Mississauga, Ont.)	"	Below 0	80	Insoluble
Matacil® 180F	Chemagro Ltd. (Mississauga, Ont.)	"	Below 0	93	Insoluble
Dipel® 132 (aqueous)	Abbott Laboratories (Chicago, Ill.)	"	Slightly below 0	–	Miscible
Dipel® 176 (oil-based)	Abbott Laboratories (Chicago, Ill.)	"	Below 0	98	Insoluble
Thuricide® 48LV (aqueous)	Zoecon Corporation (Palo Alto, Calif.)	"	Slightly below 0	–	Miscible
Triton® X-114	Rohm and Haas (Westhill, Ont.)	Surfactant	Pour point −9	>150	Soluble
Triton® X-100	Rohm and Haas (Westhill, Ont.)	"	Pour point 7	>150	Soluble
Atlox® 3409F	Atkemix, Inc. (Burlington, Ont.)	"	Pour point 4	12.2	Soluble
Cyclosol® 63	Shell (Toronto, Ont.)	Solvent	Pour point −39	52	Insoluble
ID585	Shell (Toronto, Ont.)	Diluent oil	Pour point −42	59	Insoluble
Sunspray® 6N	Sun Oil Co. (Philadelphia, Pa.)	Diluent oil	Pour point −15	355	Insoluble
Dowanol® TPM	Dow Chemical (Sarnia, Ont.)	Solvent	Pour point −78	110	Soluble

optimized to accommodate this requirement. To achieve this, the nature of the various atomization processes must be considered together with how a change in the fundamental liquid properties will affect the size of the droplets formed. For Newtonian formulations, surface tension and viscosity are known to influence atomization (Ford and Furmidge 1967; Fraser 1958; Kaupke and Yates 1966; Sundaram and Leung 1984; Sundaram and Retnakaran 1987; Sundaram et al. 1987b; Sundaram 1989b). In contrast, non-Newtonian pseudoplastic formulations are characterized by a viscosity that decreases logarithmically with increasing shear, and the role of viscosity on atomization becomes more complex. Thickeners, which increase the viscosity of tank-mix formulations, have been used to minimize off-target loss of pesticide sprays (Akesson and Yates 1964; Suggitt and Winter 1964). Invert emulsions with high viscosities have also been used to alter droplet sizes during atomization (Courshee 1961). However, the type of nozzle used was also found to influence the atomization processes of the non-Newtonian formulations. There is a need for research to determine the most suitable nozzle types and optimum shear rates, so that the desired droplet spectrum can be obtained.

Viscosity also has an indirect effect on off-target drift and on the droplet size obtained at the target site (Sundaram 1986a) because it affects the rate of evaporation that occurs while the droplet is falling. Several adjuvants such as water-soluble polymers have been introduced in recent years to increase the viscosity of tank-mix formulations to reduce the off-target loss of spray droplets.

Surface tension is a measure of the force required to form a new surface. Liquids with high surface tensions will resist the formation of small droplets during atomization (Yates and Akesson 1973). Therefore, the surface tension of tank-mix formulations should be reduced to ensure the formation of desirable droplet size ranges during atomization.

Studies on the role of physical properties of tank mixes on drop formation have indicated that spray atomization is a highly complex process in which a myriad of factors, including nozzle type, application parameters, and volume rates, interact simultaneously (Sundaram et al. 1987a, 1987c; Picot et al. 1989). Nevertheless, extremely small quantities of certain viscoelastic polymers have been shown to markedly alter droplet size spectra in spite of the small increase in viscosities of the tank-mix formulations (Sparks et al. 1988; Akesson et al. 1989; Akesson and Gibbs 1990). Further research is required to understand the role of formulation ingredients on liquid atomization and droplet size formation.

Table 2. Viscosities of some basic formulations of pesticides and other inert ingredients that are registered and/or used at present in forestry, at temperatures encountered during field use.

Name	Viscosities[a] (mPa·s) at temperatures of				Viscosity rating	Flow behavior
	5°C	10°C	15°C	20°C		
Fenitrothion (technical)	126.0	82.5	53.4	40.0	Medium	Nearly Newtonian
Sumithion® 20% Flowable	570.0	533.3	419.0	256.2	High	Pseudoplastic
Matacil® 180D	5025	3280	1810	750	High	Nearly Newtonian
Matacil® 180F	157.0	111.0	80.0	62.0	Medium	Slightly dilatant
Dipel® 132 (aqueous)	92.5	80.1	70.0	60.3	Medium	Slightly pseudoplastic
Dipel® 178 (oil-based)	65.3	56.5	48.0	43.0	Medium	Newtonian
Thuricide® 48LV (aqueous)	48.5	40.5	31.0	22.4	Medium	Slightly pseudoplastic
Triton® X-114	1470	974	600	380	High	Slightly dilatant
Triton® X-100	Paste	Paste	6880	1010	Very high	Slightly dilatant
Atlox® 3409F	Paste	5660	443	217	High	Slightly dilatant
Cyclosol® 63	1.62	1.47	1.33	1.23	Very low	Newtonian
ID585	2.39	2.12	1.89	1.78	Very low	Newtonian
Dowanol® TPM	20.2	16.3	13.1	10.8	Low	Newtonian
Sunspray® 6N	63.0	43.8	32.5	25.7	Medium	Nearly Newtonian

[a] Viscosities were measured at a shear rate of 695 reciprocal seconds.

Table 3. Application details, droplet spectra, percent recovery of applied material, and physical properties of aminocarb and mexacarbate tank mixes, following treatment in New Brunswick.

Parameters	Aminocarb tank mixes		Mexacarbate tank mixes	
	Aqueous	Oil-based	Aqueous	Oil-based
Date of appln.	June 12/1981	June 18/1981	June 20/1984	June 17/1984
Time of appln.	2100	2023	2112	2105
Dosage rate (g/ha)	70	70	70	70
Volume rate (L/ha)	1.5	1.5	1.5	1.5
Atomizer used	AU3000	AU3000	AU3000	AU3000
Temperature (°C)	10	22.3	15	19.7
Relative humidity (%)	96	58	78	47
Wind speed (km/h)	None	1.5	2.0	8.0
Maximum diameter (μm)	85	105	81	187
Minimum diameter (μm)	16	4	10	12
NMD (μm)[a]	35	55	21	60
VMD (μm)[a]	41	68	37	106
Droplets/cm^2	13	14	5.3	17
Percent recovery	3.3	14.7	2.4	14.7
Viscosity (mPa.s)	4.24	25.0	2.73	5.80
Surface tension (mN/m)	29.2	31.0	28.2	31.0
Non-vol. components (%)	38.0	100.0	14.5	34.0

[a] NMD = number median diameter; VMD = volume median diameter.

Droplet Transport and Deposition

The rate of fall of droplets and the efficiency of inertial impaction are both functions of droplet size. If the droplets are too small (i.e. less than 20 μm in diameter), they do not impact efficiently on targets (Mason 1971). Consequently, the properties of a tank-mix formulation are only relevant in so far as they affect atomization, volatility, and behavior relating to initial deposition. The latter has been examined in some detail by Ford and Furmidge (1966). Because spray formulations intended for ultra-low-volume aerial application are expected to have low volatilities (Maas 1971), the main interest has centered on the use of additives to render the carrier or diluent less volatile (Matthews and Johnstone 1968; Sundaram and Retnakaran 1987; Sundaram et al. 1987a). Much of the work carried out from 1960 to 1970 involved invert emulsions (Murrow and Phillips 1962; Colthurst et al. 1966). Other additives that were used to reduce droplet evaporation in-flight included hygroscopic materials such as polyethylene glycols and molasses (Tunstall 1968). In recent studies, the use of low volatility oils as carriers for Bt has been shown to provide increased deposits on balsam fir foliage during hemlock looper control operations in eastern Canada (West et al. 1989).

Density of a formulation has some effect on the terminal velocity of a droplet as it is related to the gravitational force acting on the droplet. However, pesticide formulations used in agricultural and forestry sprays normally have a low range of density values, ranging from 0.8 g/cm^3 for oil-based formulations to 1.2 g/cm^3 for some technical materials and aqueous tank mixes, while the bulk of the applications are still carried out with aqueous formulations with a density value of 1.0 g/cm^3 (Yates and Akesson 1973). Therefore, the contribution of density to droplet transport and deposition is minimal.

Laboratory and field studies carried out to assess the interrelationships between physical properties of tank mixes and droplet spectra on artificial samplers indicated that at equal volume rates, pseudoplastic and thixotropic formulations of Bt provided droplet spectra different from those of the chemical insecticides tested (Sundaram et al. 1987d, 1988). This was because pseudoplastic formulations of chemical insecticides contained polymers, whereas those of Bt did not. Some field data on droplet spectra and deposits of aqueous and oil-based tank mixes are given in Tables 3 and 4.

Droplet Retention and Spreading on Target Surfaces

The biological activity of pesticidal sprays on foliage is a function of many factors including those relating to the physicochemical interactions of pesticide formulations and plant material. One important aspect of this interaction, which largely determines the initial character and subsequent availability of the deposit, is the behavior of liquid droplets as they impact and adhere to the foliar surface. Not all the droplets that reach the forest tree canopy will impinge on foliage (Davies 1966; Himel and Moore 1969; Bache 1975; Cramer and Boyle 1976; Uk 1977; Barry et al. 1977; Barry 1984; Sundaram et al. 1986), since droplet impingement depends on several factors including the microclimatic conditions in the

Table 4. Application details, droplet spectra, percent recovery of applied material, and physical properties of Sumithion® Flowable, Dimilin® ODC 45, Futura® XLV, and Dipel® 132 formulations following treatment in Newfoundland.

Parameters	Sumithion® Flowable (aqueous)	Dimilin®[a] ODC 45 (oil-based)	Futura®[b] XLV (aqueous)	Dipel® 132 (oil-based)
Date of appln.	July 12/1985	June 25/1988	July 3/1985	June 26/1987
Time of appln.	2105	0645	0715	0540
Dosage rate (per ha)	210 g	70 g	30 BIU	30 BIU
Volume rate (L/ha)	1.5	2.5	2.1	2.36
Atomizer used	AU5000	AU4000	AU5000	AU4000
Temperature (°C)	14.0	10.0	14.5	8.0
Relative humidity (%)	88	85	74	91
Wind speed (km/h)	5.0	4.0	8.3	4.5
Maximum diameter (µm)	215	140	89	165
Minimum diameter (µm)	9	10	15	5
NMD (µm)[c]	88	40	54	51
VMD (µm)[c]	133	85	65	86
Droplets/cm^2	7.6	67	4.1	45
Percent recovery	18.0	41.4	11	37.9
Viscosity (mPa.s)	41.2	38.1	43.0	85.0
Surface tension (mN/m)	46.2	34.1	43.6	34.3
Non-vol. components (%)	20.8	100.0	18.3	100.0

[a] Dimilin ODC is a diflubenzuron formulation.
[b] Futura® XLV is a Bt formulation.
[c] NMD = number median diameter; VMD = volume median diameter.

canopy, canopy density, and size and shape of the plant material. Furthermore, not all droplets that impinge on plant surfaces will be retained there, because retention and adhesion processes depend on several physical phenomena including those specific to the spray liquid and the composition and texture of the foliar surfaces (Furmidge 1959; Corn 1966; Johnstone 1973; Uk 1977). The influence of impact velocity on drop spreading, reflection (or bouncing off), and shatter was considered, but droplets smaller than 1 000 µm in diameter are usually reflected with minimum shatter (Johnstone 1973). Droplets larger than 250 µm can be reflected from a waxy leaf surface if they are made up of a high surface tension liquid (Becher and Becher 1969).

Factors contributing to adhesion of small droplets (40 to 60 µm in diameter) to target surfaces are largely the London–Van der Waals attractive forces (Fowkes 1965; Corn 1966), which primarily depend on the physicochemical nature of the droplet and target surface. The role of electrostatic force was explored because spray electrification invariably occurs whenever certain polar liquids are atomized into very fine droplets. However, the magnitude of the electrostatic forces is about one-thousandth of that of the London–Van der Waals forces and, therefore, the former can be ignored for all practical purposes (Kunkel 1950). The microscopic texture of the target surface plays a significant role, and forces of adhesion can easily be increased by a factor of ten for the same drop sizes just by altering the surface texture (Kordecki and Orr 1960). Roughness, hairiness, solubilizing power of the waxy cuticle for ingredients in the droplet, relative humidity of the ambient air, and hydrophilic–hydrophobic characteristics of the leaf surface all contribute to droplet adhesion (Furmidge 1959).

Laboratory findings (Sundaram 1989a) on the role of physical properties of some oil-based and aqueous tank mixes of Bt on droplet spectra and deposit recovery on artificial samplers, and droplet spreading and retention on conifer foliage are listed in Table 5.

Deposit Availability and Mode of Entry into Insects

The intended mode of action determines the form in which the toxicant should be applied and deposited on the target. If the toxicant acts through direct topical contact, then it will almost certainly be more effective in the form of oily droplets, which readily penetrate the insect cuticle (Johnstone 1971; Sundaram 1988). If the toxicant is an oily liquid it may be applied undiluted, provided it can be finely atomized. On the other hand, if the toxicant acts indirectly, as for instance by pick-up via crawling contact of insecticidal residues from the surface of vegetation, the requirements may be more complicated. For many insects, pick-up occurs most readily from solid residues in finely divided form (e.g. wettable powder). Surface deposits of this type are most easily produced by an aqueous suspension in which the particle sizes of the pesticide

Table 5. Application details, droplet spectra, droplets/fir needle, recovery of applied spray, and physical properties of Bt formulations following atomization under laboratory conditions.

Parameters	Dipel® 132 (aqueous)	Dipel® 132 (oil-based)	Dipel® 176 (aqueous)	Dipel® 176 (oil-based)
Date of appln.	June 12/1988	July 13/1988	June 17/1988	June 5/1988
Time of appln.	1000	1100	1130	0930
Dosage rate (per ha)	30 BIU	30 BIU	30 BIU	30 BIU
Volume rate (L/ha)	2.36	2.36	1.78	1.78
Atomizer used	Spinning disk	Spinning disk	Spinning disk	Spinning disk
Temperature (°C)	22	22	22	22
Wind speed (km/h)	Still air	Still air	Still air	Still air
Relative humidity (%)	65	65	65	65
Maximum diameter (μm)	160	125	170	125
Minimum diameter (μm)	20	7	30	7
NMD (μm)[a]	63	58	80	58
VMD (μm)[a]	80	81	102	84
Droplets/cm^2	60	110	36	70
Percent recovery	63	100	91	100
Droplets/fir needle	4.8	12.8	4.0	11.6
Viscosity (mPa·s)	58.3	38.1	84.2	43.2
Surface tension (mN/m)	38.9	31.7	43.5	32.8
Non-vol. components (%)	55.2	100.0	58.4	100.0

[a]NMD = number median diameter; VMD = volume median diameter.

residue can be controlled during the manufacture of the basic formulation. Maximum mortality of first-instar larvae was observed for certain insects caused solely by pick-up of the fine particles of a wettable powder (Johnstone 1971).

With liquid insecticides such as fenitrothion, surface deposits, relative to residues penetrated into inner layer of the foliar cuticle, were similar for oil-based and emulsion tank mixes (Sundaram 1986b), and surface deposits could only be increased by adding certain water-soluble polymers to the formulation (Sundaram and Sundaram 1987a).

When certain oil-based, ultra-low-volume formulations were used, phytotoxicity was a potential problem (Coutts and Parish 1967). Universal adoption of ultra-low-volume spraying of oily liquids is therefore not recommended (Joyce 1968, 1969). Every carrier liquid should be evaluated, in laboratory studies, for its phytotoxicity using small droplets, before it can be considered for ultra-low-volume treatments.

Droplets of some highly concentrated aqueous formulations (e.g. the high potency Bt aqueous formulations) tend to evaporate in-flight, forming nearly dried particles. These droplets do not spread on target foliage, resulting in poor adhesion (Sundaram et al. 1987e; Sundaram 1989a). To overcome this problem, polymeric stickers are being recommended (Matthews 1979) to bind the residual deposits to the target surface. However, bioavailability should not be impaired because of the use of such stickers. It is relatively easy to render a deposit persistent, but of no avail if this effectively prevents pick-up by a moving insect. Such stickers may be advantageous if the deposit is to be orally consumed along with the foliage, and subsequently act through the insect's stomach, although bioavailability can still be reduced if the active ingredient is entrapped in the polymeric structure of such stickers.

Often hygroscopic adjuvants such as polypropylene glycols, glycerol, sugar, or molasses have been added to the tank-mix formulations to reduce in-flight evaporation and drying of droplets, so that adhesion to target foliage can be improved.

Persistence of Pesticides in Target/Nontarget Matrices

The presence of oil carriers in tank-mix aminocarb formulations contributes to increased longevity on target foliage when compared to that of an aqueous formulation containing surfactants (Sundaram and Szeto 1984). Although this behavior could be considered beneficial for enhancing efficacy, nonvolatile oil carriers have also increased both the initial deposits and persistence of residues in soil (Sundaram et al. 1985; Sundaram and Sundaram 1987b). Thus, the possible adverse environmental effects of certain carrier oils should not be overlooked. In contrast, nonvolatile oil carriers have contributed to low short-term persistence in stream waters, compared to that of an aqueous formulation containing surfactants (Sundaram and Szeto 1985). The data thus indicate that carrier oils and surfactant adjuvants can be chosen to influence the longevity of a pesticide in target and nontarget matrices.

Choice of Formulation

Formulations are traditionally selected on the basis of convenience in handling, low mammalian toxicity, ready availability to the user, as well as the price. In general, the cheapest in terms of the active ingredient are wettable powder formulations, and those with the highest amount of active ingredient per unit weight of formulation. When assessing the costs, the whole application technique needs to be considered because the use of a particular formulation may affect the labor costs, the equipment, and spraying time.

In future spray applications, consideration will have to be given to formulations that have proven increased effectiveness with minimum environmental impact. For example, when stomach poisons are applied, surface deposits are effective against leaf-chewing insects, but less so against borers, which often do not ingest their first few bites of plant tissue. The effectiveness of a stomach poison can be improved by the addition of a feeding stimulant, such as molasses, to the spray mix. Formulations that tend to leave powdery residues on the leaf surface are more readily available to larvae crawling on sprayed leaves than an emulsifiable formulation, perhaps because the emulsifiable concentrate is more readily absorbed into the leaf.

In conclusion, formulation technology entails both an art and a science. The physicochemical properties of a tank-mix formulation affect droplet size formation during spray release and the behavior of droplets from the time of release until their dissipation in the environment, including those processes that occur at the droplet/target interface. Therefore, the physical, chemical, biological, and toxicological properties of spray formulations have to be optimized for every pest control situation to achieve the pre-determined objective of increasing the efficacy and reducing the environmental contamination by xenobiotic compounds.

References

Akesson, N.B.; Bayer, D.E.; Yates, W.E. 1989. Application effects of vegetable oil additives and carriers on agricultural sprays. Pages 121-137 *in* P.N.P. Chow, C.A. Grant, A.M. Hinshalwood and E. Simundsson (Eds.), CRC Press, Inc., Boca Raton, Fla.

Akesson, N.B.; Gibbs, R.E. 1990. Pesticide drop size as a function of spray atomizers and liquid formulations. Pages 170-183 *in* L.E. Bode, J.L. Hazen and D.G. Chasin (Eds.), Pesticide Formulations and Application Systems: 10th Vol., ASTM STP 1078, American Soc. for Test. and Mater., Philadelphia, Pa.

Akesson, N.B.; Yates, W.E. 1964. Problems relating to the application of agricultural chemicals and resulting drift residues. Ann. Rev. Entomol. 9: 285-318.

Akesson, N.B.; Yates, W.E. 1984. Physical parameters affecting spray application. Pages 95-115 *in* W.Y. Garner and J. Harvey, Jr. (Eds.), Chemical and Biological Controls in Forestry. Am. Chem. Soc. Symp. Ser. No. 238, Washington, D.C.

Angermann, R. 1961. Aerial control of the Colorado beetle by ultra-low-volume spraying. Agric. Av. 3: 49-54.

Anon. 1974. Microencapsulated pesticide reaches market. Chem. and Eng. News, 52: 15-16.

Argauer, R.J.; Mason, H.C.; Corley, C.; Higgins, A.H.; Sauls, J.N.; Liljedahl, L.A. 1968. Drift of water-diluted and undiluted formulations of malathion and azinphosmethyl applied by airplane. J. Econ. Entomol. 61: 1015-1020.

Bache, D.H. 1975. Transport of aerial spray, III. Influence of microclimate on crop spraying. Agric. Meteorol. 15: 379-383.

Barry, J.W. 1984. Deposition of chemical and biological agents in conifers. Pages 117-137 *in* W.Y. Garner and J. Harvey, Jr. (Eds.), Chemical and Biological Controls in Forestry. Am. Chem. Soc. Publ., Washington, D.C.

Barry, J.W.; Ciesla, W.M.; Tysowsky, M. Jr.; Ekblad, R.B. 1977. Impaction of insecticide on western spruce budworm larvae and Douglas-fir needles. J. Econ. Entomol. 70: 387-388.

Becher, P. 1973. The emulsifier. Pages 65-92 *in* W. Van Valkenburg (Ed.), Pesticide Formulations. Marcel Dekker, Inc., New York.

Becher, P.; Becher, D. 1969. The effect of hydrophilelipophile balance on contact angle of solutions of nonionic surface-active agents: relation to adjuvant effects. Pages 15-23 *in* R.F. Gould (Ed.), Pesticidal Formulations Research. Physical and Colloidal Chemical Aspects, Am. Chem. Soc. Publ., Washington, D.C.

Bode, L.E.; Zain, S.B., Md. 1987. Spray deposits from low volume applications using oil and water carriers. Pages 93-103 *in* G.B. Beestman and D.I.B. Vander Hooven (Eds.), Pesticide Formulations and Application Systems, 7th vol., ASTM STP 968, Am. Soc. for Test. and Mater., Philadelphia, Pa.

Bondine, M. 1963. Experiments in using aircraft for controlling the Colorado beetle. Agric. Av. 5: 9-11.

Bouse, L.F.; Carleton, J.B.; Janks, P.C. 1988. Effects of water soluble polymers on spray droplet sizes. Trans. ASAE 31: 1633-1641, 1648.

Brown, A.W.A.; Watson, D.L. 1953. Studies on fine spray and aerosol machines for control of adult mosquitoes. Mosquito News 13: 81-95.

Butler, B.J.; Akesson, N.B.; Yates, W.E. 1969. Use of spray adjuvants to reduce drift. Trans. ASAE 12: 182-186.

Cadogan, B.L. 1986. Relative field efficacies of Sumithion® 20% flowable and Sumithion® technical formulations against spruce budworm, *Choristoneura fumiferana* (Lepidoptera: Tortricidae). Can. Entomol. 118: 1143-1149.

Cadogan, B.L.; Zylstra, B.F.; de Groot, P.; Nystrom, C. 1984. The Efficacy of Aerially Applied Matacil to Control Spruce Budworm *Choristoneura fumiferana* (Clem.) in Bathurst, New Brunswick, Environ. Can., Can. For. Serv., For. Pest Manage. Inst. Inf. Rep. FPM-X-64, 33 p.

Campion, D.G. 1976. Sex pheromones for the control of Lepidopterous pests using microencapsulation and dispenser techniques. Pestic. Sci. 7: 636-641.

Colthurst, J.P.; Ford, R.E.; Furmidge, C.G.L.; Pearson, A.J.A. 1966. Water-in-oil emulsions and the control of spray drift. Monogr. Soc. Chem. Ind. 21: 47-60.

Corn, M. 1966. Adhesion of particles. Pages 359-392 *in* C.N. Davies (Ed.), Aerosol Science. Academic Press, London, England.

Courshee, R.J. 1961. Avoidance of drift from airborne sprayers. Outlook on Agric. 111: 76-80.

Coutts, H.H.; Parish, R.H. 1967. The selection of a solvent for use with low volume spraying of cotton plants. Agric. Aviat. 9: 125-126.

Cramer, H.E.; Boyle, D.G. 1976. The micrometeorology and physics of spray particle behavior. Pages 27-39 *in* Pesticide Spray Application, Behavior and Assessment. Workshop Proc., Pac. Southwest For. and Range Exp. Stn., USDA For. Serv. Rep. PSW-15.

Davies, C.N. 1966. Deposition from moving aerosols. Pages 393-446 *in* C.N. Davies (Ed.), Aerosol Science, Academic Press, London, England.

DeBoo, R.F. 1980. Experimental Aerial Applications of Permethrin for Control of *Choristoneura fumiferana* in Quebec, 1976-1977. Environ. Can., Can. For. Serv., For. Pest Manage. Inst. Inf. Rep. FPM-X-41, 24 p.

DeBoo, R.F.; Campbell, L.M. 1974. Plantation Research: XI. Experimental Aerial Applications of Methoxychlor and Carbaryl for Control of White Pine Weevil (*Pissodes strobi*) in Ontario, 1974. Environ. Can., Can. For. Serv., Chem. Cont. Res. Inst. Inf. Rep. CC-X-87, 30 p.

Deonier, C.C.; Sullivan, W.N.; Nottingham, E.; Lindquist A.W. 1949. Control of *Anopheles* mosquitoes with coarse and fine DDT sprays applied by airplane. J. Econ. Entomol. 42: 447-450.

Ebeling, W. 1971. Sorptive dusts for pest control. Ann. Rev. Entomol. 16: 123-158.

Ford, R.E.; Furmidge, C.G.L. 1966. The viscosity and size of invert emulsion droplets used in pesticidal sprays. Proc. 3rd Int. Agric. Aviat. Congr., pp. 172-179.

Ford, R.E.; Furmidge, C.G.L. 1967. Physico-chemical studies on agricultural sprays. VIII. Viscosity and spray drop size of water-in-oil emulsions. J. Sci. Food Agric. 18: 419-428.

Fowkes, F.M. 1965. Attractive forces at interfaces. Pages 1-12 in Sydney Ross (chairman), Chemistry and Physics of Interfaces. Am. Chem. Soc. Publ., Washington, D.C.

Fraser, R.P. 1958. The fluid kinetics of application of pesticide chemicals. Adv. Pest Cont. Res. 2: 1-106.

Furmidge, C.G.L. 1959. Physico-chemical studies on agricultural sprays. General principles of incorporating surface active agents as spray supplements. J. Sci. Food Agric. 10: 267-273.

Hall, F.R. 1989. Effect of formulation, droplet size and spatial distribution on dose transfer of pesticides. Pages 145-154 in D.A. Hovde and G.B. Beestman (Eds.), Pesticide Formulations and Application Systems, 8th vol., ASTM STP 968, Am. Soc. for Test. and Mater., Philadelphia, Pa.

Haq, K.; Akesson, N.B.; Yates, W.E. 1983. Analysis of droplet spectra and spray recovery as a function of atomizer type and fluid physical properties. Pages 67-82 *in* T.M. Kaneko and N.B. Akesson (Eds.), Pesticide Formulations and Application Systems, 3rd vol., ASTM STP 828, Am. Soc. for Test. and Mater., Philadelphia, Pa.

Helson, B.V.; de Groot, P.; Turgeon, J.J. 1989. Toxicity of insecticides to first-instar larvae of the spruce budmoth, *Zeiraphera canadensis* Mut. and Free. (Lepidoptera: Tortricidae): laboratory and field studies. Can. Entomol. 121: 81-91.

Himel, C.M.; Moore, A.D. 1969. Spray droplet size in the control of spruce budworm, boll weevil, bollworm, and cabbage looper. J. Econ. Entomol. 62: 916-918.

Hocking, K.S.; Parr, H.C.M.; Yeo, D.; Robins, P.A. 1953. An experimental attempt to produce a fly-free corridor through a belt of tsetse-infested woodland. Bull. Entomol. Res. 44: 601-609.

Hunt, C.R. 1947. Toxicity of insecticide dust diluents and carriers to larvae of the Mexican bean beetle. J. Econ. Entomol. 40: 215-219.

Johnstone, D.R. 1971. Formulations and atomization. Proc. 4th Int. Agric. Aviat. Congr. Kingston (Ont.) Canada, Aug. 25-29 1969, Int. Agric. Aviat. Publ., pp. 225-233.

Johnstone, D.R. 1973. Spreading and retention of agricultural sprays on foliage. Pages 343-386 *in* W. Van Valkenburg (Ed.), Pesticide Formulations. Marcel Dekker, Inc., New York.

Joyce, R.J.V. 1968. Possible developments in the use of aircraft and associated equipment. Chem. Ind. 117-120.

Joyce, R.J.V. 1969. Recent developments in ultra-low-volume application. J.R. Aeronaut. Soc. 73: 80-83.

Kaupke, C.R.; Yates, W.E. 1966. Physical properties and drift characteristics of viscosity-modified agricultural sprays. Trans. ASAE 9: 797-799, 802.

Kordecki, M.C.; Orr, C. 1960. Adhesion of solid particles to solid surfaces. Arch. Environ. Health 1: 1-9.

Kunkel, W.B. 1950. Static electrification of dust particles on dispersion into a spray cloud. J. Appl. Phys. 21: 820-832.

Lindner, P. 1973. Agricultural formulations with liquid fertilizers. Pages 113-141 *in* W. Van Valkenburg (Ed.), Pesticide Formulations. Marcel Dekker, Inc., New York.

Maas, W. 1971. ULV Application and Formulation Techniques. Philips Duphar, B.V., Crop Protection Division, Amsterdam, the Netherlands, 164 p.

Markin, G.P.; Henderson J.A.; Collins, H.L. 1972. Aerial application of microencapsulated insecticide. Agric. Aviat. 14: 70-75.

Marrs, G.J.; Middleton, M.R. 1973. The formulation of pesticide for convenience and safety. Outlook on Agric. 7: 231-235.

Mason, B.J. 1971. The Physics of Clouds, 2nd ed. Clarendon Press, Oxford, England.

Matthews, G.A. 1979. Pesticide Application Methods. Longman, New York, 336 p.

Matthews, G.A.M.; Johnstone, D.R. 1968. Aircraft and tractor spray deposits on irrigated cotton. Cotton Growing Rev. 45: 207-218.

Messenger, K. 1963. Low volume aerial spraying will be boon to applicators. Agric. Chem. 18: 63-64, 66.

Messenger, K. 1964. Low volume aerial spraying. Agric. Chem. 19: 61-62, 64.

Messenger, K. 1965. Aircraft armament for insect control. Agric. Chem. 20: 47-52.

Morris, O.N.; Armstrong, J.A.; Howse, G.M.; Cunningham, J.C. 1974. A 2-year study of virus-chemical insecticide combination in the integrated control of the spruce budworm, *Choristoneura fumiferana* (Tortricidae: Lepidoptera). Can. Entomol. 106: 813-824.

Muller, J.R.; Stovell, F.R. 1966. Development of the biflon system. Proc. 3rd Int. Agric. Aviat. Congr., pp. 55-59.

Murrow, J.; Phillips, P.J. 1962. Stall bifluid spray system. Proc. 2nd Agric. Aviat. Congr. (Grignon 1962), pp. 305-314.

Nigam, P.C. 1971. Insecticide evaluation for aerial application against forest insect pests. Pages 131-143 *in* Proceedings of the Seventh Northeast Aerial Applicators Conference, Ithaca, N.Y.

Page, G.E. 1961. Influence of nozzle size and type, hardness of water, and wetting agents on spray drift. Weeds 9: 258-263.

Pearson, A.J.A.; Masheder, S. 1969. Bi-flon. Agric. Aviat. 11: 126-129.

Phillips, F.T.; Gillham, E.M. 1973. Persistence to rainwashing of DDT wettable powders. Pestic. Sci. 4: 51-57.

Picot, J.J.C.; van Vliet, M.W.; Payne, N.J. 1989. Droplet size characteristics for insecticide and herbicide spray atomizers. Can. J. Chem. Eng. 67: 752-761.

Polon, J.A. 1973. Formulation of pesticidal dusts, wettable powders and granules. Pages 143-234 *in* W. Van Valkenburg (Ed.), Pesticide Formulations. Marcel Dekker, Inc., New York.

Randall, A.P. 1971. Some aspects of aerial spray experimentation for forest insect control. Proc. 4th Int. Agric. Aviat. Congr. Kingston (Ont.) Canada, Aug. 25-29, 1969, Int. Agric. Aviat. Publ. pp. 308-315.

Randall, A.P.; McFarlane, J.W.; Zylstra, B.; Desaulniers, R. 1981. Assessment of early multiple applications of pesticides on high populations of spruce budworm larvae, *Choristoneura fumiferana* (Clem.), Quebec, 1977. Environ. Can., Can. For. Serv., For. Pest Manage. Inst. Inf. Rep. FPM-X-42, 78 p.

Randall, A.P.; Zylstra, B.F.; McFarlane, J.W. 1977. The effect of fenitrothion and aminocarb on second instar spruce budworm *Choristoneura fumiferana* (Clem.) in Quebec. Environ. Can., Can. For. Serv., For. Pest Manage. Inst. Inf. Rep. FPM-X-5, 24 p.

Raske, A.G.; Sundaram, K.M.S.; Sundaram, A.; West, R.J. 1989. Fenitrothion deposits on simulated and live fir foliage following aerial spraying of two formulations. Pages 233-253 in J.L. Hazen and D.A. Hovde (Eds.), Pesticide Formulations and Application Systems, International Aspects, 9th vol., ASTM STP 1036, Am. Soc. for Test. and Mater., Philadelphia, Pa.

Retnakaran, A.; Raske, A.G.; West, R.J.; Lim, K.P.; Sundaram, A. 1988. Evaluation of diflubenzuron as a control agent for hemlock looper (Lepidoptera: Geometridae). J. Econ. Entomol. 81: 1698-1705.

Richardson, R.G. 1974. Control of spray drift with thickening agents. J. Agric. Eng. Res. 19: 227-231.

Sayer, H.J. 1959. An ultra-low-volume spraying technique for the control of the desert locust *Schistocerca gregaria* (Forsh). Bull. Entomol. Res. 50: 371-386.

Seaman, D. 1990. Trends in the formulation of pesticides—an overview. Pestic. Sci. 29: 437-449.

Skoog, F.E.; Cowan, F.T.; Messenger, K. 1965. Ultra-low-volume aerial spraying of dieldrin and malathion for rangeland grasshopper control. J. Econ. Entomol. 58: 559-565.

Sparks, B.D.; Sundaram, A.; Kotlyar, L.; Leung, J.W.; Curry, R.D. 1988. Physicochemical properties, atomization and deposition patterns of some Newtonian spray mixtures of glyphosate containing two spray modifier adjuvants. J. Environ. Sci. Health B23: 235-266.

Suggitt, J.W.; Winter, J.E.F. 1964. Minimum drift viscous herbicide sprays for helicopter application to woody growth. Ontario Hydro Res. Q. (2nd quarter): 20-36.

Sundaram, A. 1986a. Understanding volatilities of forestry spray mixtures from their viscosities and viscosity-temperature relationships. Pages 37-55 *in* L.D. Spicer and T.M. Kaneco (Eds.), Pesticide Formulations and Application Systems: 5th vol., ASTM STP 915, Am. Soc. for Test. and Mater., Philadelphia, Pa.

Sundaram, A. 1987a. Influence of temperature on physical properties of non-aqueous pesticide formulations and spray diluents: relevance to u.l.v. applications. Pestic. Sci. 20: 105-118.

Sundaram, A. 1987b. Optimizing physical properties of diluent oils for maximum target deposition of ultra-low-volume spray droplets under laboratory conditions. Pages 101-122 *in* D.I.B. Vander Hooven and L.D. Spicer (Eds.), Pesticide Formulations and Application Systems: 6th vol., ASTM STP 943, Am. Soc. for Test. and Mater., Philadelphia, Pa.

Sundaram, A. 1989a. Drop size spectra, spreading, and adhesion and physical properties of eight *Bacillus thuringiensis* formulations following spray application under laboratory conditions. Pages 129-141 *in* J.L. Hazen and D.A. Hovde (Eds.), Pesticide Formulations and Application Systems: International Aspects, 9th vol., ASTM STP 1036, Am. Soc. for Test. and Mater., Philadelphia, Pa.

Sundaram, A. 1989b. Influence of adjuvants on spray atomization, droplet size spectra, and deposits of four fenitrothion formulations. Pages 75-82 *in* P.N.P. Chow, C.A. Grant, A.M. Hinshalwood, and E. Simundsson (Eds.), Adjuvants and Agrochemicals, vol. II, CRC Press, Inc., Boca Raton, Fla.

Sundaram, K.M.S. 1986b. A comparative evaluation of dislodgeable and penetrated residues, and persistence characteristics of aminocarb and fenitrothion, following application of several formulations onto conifer trees. J. Environ. Sci. Health B21: 539-560.

Sundaram, K.M.S. 1988. Cuticular penetration and *in vivo* metabolism of fenitrothion in spruce budworm. J. Environ. Sci. Health B23: 643-659.

Sundaram, A.; van Frankenhuyzen, K.; Meating, J.H.; Howse, G.M. 1987e. Droplet size spectrum and deposit pattern of Thuricide® 48LV sprayed aerially over a spruce plantation in Ontario. Pages 359-363 *in* G.W. Green (Ed.), Proceedings of the Symposium on the Aerial Application of Pesticides in Forestry, Natl. Res. Coun. Can., Ottawa, Ont., AFA-TN-18, NRC No. 29197.

Sundaram, A.; Kotlyar, L.; Sparks, B.D. 1987d. Influence of surfactants and polymeric adjuvants on physicochemical properties, droplet size spectra and deposition of fenitrothion and *Bacillus thuringiensis* formulations under laboratory conditions. J. Environ. Sci. Health B22: 691-720.

Sundaram, A.; Leung, J.W. 1984. Physical properties, droplet spectra and deposits of oils used in pesticide sprays. J. Environ. Sci. Health B19: 793-805.

Sundaram, A.; Leung, J.W.; Curry, R.D. 1987b. Influence of adjuvants on physicochemical properties, droplet size spectra and deposit patterns: relevance in pesticide applications. J. Environ. Sci. Health B22: 319-346.

Sundaram, A.; Retnakaran, A. 1987. Influence of formulation properties on droplet size spectra and ground deposits of aerially-applied pesticides. Pestic. Sci. 20: 241-257.

Sundaram, A.; Retnakaran, A.; Raske, A.G.; West, R.J. 1987c. Effect of application rates on droplet size spectra and deposit characteristics of Dimilin® spray mixtures in an aerial spray trial. Pages 104-115 *in* G.B. Beestman and D.I.B. Vander Hooven (Eds.), Pesticide Formulations and Application Systems: 7th vol., ASTM STP 968, Am. Soc. for Test. and Mater., Philadelphia, Pa.

Sundaram, A.; Sundaram, K.M.S. 1989. Influence of spray diluents on physicochemical properties, and physical and chemical compatibility of eight oil-based tank mixes of aminocarb. J. Environ. Sci. Health B24: 1-22.

Sundaram, K.M.S.; Sundaram, A. 1987a. Influence of formulation on spray deposit patterns, dislodgeable and penetrated residues, and persistence characteristics of fenitrothion in conifer needles. Pestic. Sci. 18: 259-271.

Sundaram, K.M.S.; Sundaram, A. 1987b. Role of formulation ingredients and physical properties on droplet size spectra, deposition, and persistence of aerially sprayed aminocarb and mexacarbate in forest litter and soil samples. Pages 139-151 in G.B. Beestman and D.I.B. Vander Hooven (Eds.), Pesticide Formulations and Application Systems: 7th vol., ASTM STP 968, Am. Soc. for Test. and Mater., Philadelphia, Pa.

Sundaram, A.; Sundaram, K.M.S.; Cadogan, B.L. 1985. Influence of formulation properties on droplet spectra and soil residues of aminocarb aerial sprays in conifer forests. J. Environ. Sci. Health B20: 167-186.

Sundaram, K.M.S.; Sundaram, A.; Nott, R. 1986. Mexacarbate deposits on simulated and live fir needles during an aerial spray trial. Trans. ASAE 29: 382-388, 392.

Sundaram, A.; Sundaram, K.M.S.; Leung, J.W.; Holmes, S.B.; Cadogan, B.L. 1987a. Influence of physical properties on droplet size spectra and deposit patterns of two mexacarbate formulations, following spray application under laboratory and field conditions. Pestic. Sci. 20: 179-191.

Sundaram, K.M.S.; Szeto, S. 1984. Persistence of aminocarb in balsam fir foliage, forest litter, and soil after aerial application of three formulations. J. Agric. Food Chem. 32: 1138-1141.

Sundaram, K.M.S.; Szeto, S. 1985. Distribution and persistence of aminocarb in stream water, sediment and fish after application of three Matacil formulations. J. Environ. Sci. Health B20: 187-200.

Sundaram, A.; West, R.J.; Raske, A.G.; Retnakaran, A. 1988. Spray atomization and deposition patterns of one Newtonian and two pseudoplastic formulations after aerial application over a mature conifer forest in Newfoundland. J. Environ. Sci. Health B23: 85-99.

Tsuji, K.; Fuyama, H. 1982. Development of emulsion-type flowable formulation. Vth Intl. Congr. on Pestic. Chem. (IUPAC), VIIId-3, Kyoto, Japan.

Tunstall, J.P. 1968. Preliminary trials of the addition of molasses to insecticides. Cotton Growing Rev. 45: 198-206.

Uk, S. 1977. Tracing of insecticide spray droplets by size on natural surfaces. The state of the art and its value. Pestic. Sci. 8: 501-509.

van Frankenhuyzen, K.; Nystrom, C. 1989. Residual toxicity of a high-potency formulation of *Bacillus thuringiensis* to spruce budworm (Lepidoptera: Tortricidae). J. Econ. Entomol. 82: 868-872.

Van Valkenburg, W. 1973. The stability of emulsions. Pages 93-112 *in* W. Van Valkenburg (Ed.), Pesticide Formulations. Marcel Dekker, Inc., New York.

West, R.J.; Raske, A.G.; Sundaram, A. 1989. Efficacy of oil-based formulations of *Bacillus thuringiensis* Berliner var. *kurstaki* against the hemlock looper, *Lambdina fiscellaria fiscellaria* (Guen.) (Lepidoptera: Geometridae). Can. Entomol. 121: 55-63.

Yates, W.E.; Akesson, N.B. 1973. Reducing pesticide chemical drift. Pages 275-341 *in* W. Van Valkenburg (Ed.), Pesticide Formulations, Marcel Dekker, Inc., New York.

Yates, W.E.; Akesson, N.B.; Bayer, D. 1976. Effects of spray adjuvants on drift hazards. Trans ASAE 19: 41-46.

Yates, W.E.; Cowden, R.E.; Akesson, N.B. 1983. Nozzle orientation, air speed and spray formulation effects on drop size spectrums. Trans. ASAE 26: 1638-1643.

Yates, W.E.; Cowden, R.E.; Akesson, N.B. 1985a. Drop size spectra from nozzles in high-speed airstreams. Trans. ASAE 28: 405-410.

Yates, W.E.; Cowden, R.E.; Akesson, N.B. 1985b. Effects of Nalco-Trol on atomization. USDA For. Serv., For. Pest Manage. Rep., FPM 85-2, Davis, Calif., 37 p.

Chapter 47

Spray Dispersal, Deposition, and Assessment

N. J. Payne

Introduction

The efficacy, environmental impact, and even the cost of an insecticide spray application are all affected by the initial active ingredient distribution after spraying, which is the result of the three physical processes of atomization, dispersal, and deposition. In general, the maximum amount of active ingredient should be deposited on-target during insecticide applications to give maximum efficacy, minimum off-target drift and environmental impact, with the lowest waste of active ingredient and application costs. In this chapter, the behavior of small drops, with diameters less than 100 µm, is reviewed because this size range is widely used in ultra-low-volume insecticide applications for Canadian forestry.

Spray Dispersal

Spray cloud dispersal is governed by initial drop momentum and aerodynamic, gravitational, and electrostatic forces. The small drop sizes usually employed in aerial forestry insecticide applications quickly lose their initial momentum due to aerodynamic drag, as evidenced by their small stop distances. For example, if projected horizontally at 10 m/s in still air, drops with diameters of 10, 50, and 100 µm travel a horizontal distance of 0.2, 4, and 1 cm (Chamberlain 1975). These stop distances indicate that regardless of their initial velocity, small drops quickly reach a velocity similar to the surrounding air.

The effect of gravity on the dispersal of small drops is also limited. For example, drops with diameters of 10, 50, and 100 µm have terminal velocities of 0.3, 7, and 25 cm/s, respectively (Chamberlain 1975), whereas the standard deviation of vertical wind speed in the planetary boundary layer caused by turbulence is typically 20 to 40 cm/s (Pasquill 1974). Thus, the vertical motion of a small drop can be greatly augmented, and for drop diameters less than 50 µm may be dominated by the effect of turbulence. In most applications electrostatic forces on spray drops are small compared to aerodynamic forces and do not significantly affect dispersal. The dispersal of a spray cloud comprising drops with diameters less than 100 µm is therefore largely determined by the motion of the air layer in which it is dispersed, which in turn is governed by meteorological conditions, and for aerial applications or those using an air-blast sprayer, the air motion caused by the aircraft vortices or sprayer.

Atmospheric Influence

Atmospheric dispersal of a spray cloud comprising small drops is described in terms of the processes of advection and atmospheric diffusion. Advection is the downwind transport of the cloud and occurs at a rate proportional to the local average wind speed. Atmospheric diffusion is the three-dimensional spreading of the cloud about its center of mass caused by turbulence (i.e. the cloud spreads vertically and horizontally acrosswind and alongwind according to the strength of turbulence in these directions). Thus, advection results in a horizontal movement of the cloud, whereas atmospheric diffusion provides both horizontal and vertical movement and therefore acts to bring down the cloud from release height to the forest canopy.

Pesticide spray dispersal occurs in the 500 to 1 000-m-deep planetary boundary layer, the portion of the atmosphere immediately overlaying the earth's surface, in which airflow is affected by friction effects at the surface. Air motion in the boundary layer is primarily driven by the geostrophic wind, which is the result of horizontal pressure gradients in the atmosphere, and the Coriolis force arising from the earth's rotation (Neiburger et al. 1973). In addition to those caused by geostrophic winds, air currents in the layer result from differential heating and cooling of air by the earth's surface (Atkinson 1981). For example, in hilly areas upslope or anabatic winds are observed during the day, while downslope or katabatic winds occur at night, due to the radiative heating and cooling of the hillside, respectively. Sea and lake breezes also arise from differential heating over land and water surfaces.

The vertical profile of average horizontal wind speed in the boundary layer varies according to the geostrophic wind and the diurnal variations in vertical air temperature profile (Yasuda 1988; Bergen 1971). These are illustrated in Figures 1 and 2, which show average wind speed at different heights at various times of day. Another essential feature of winds in the boundary layer is the variation of wind direction with height (Neiburger et al. 1973). The average wind direction at various heights shows a systematic shift (Fig. 1), typically 30° to 50° between the top and bottom of the layer. This wind shift depends on the strength of vertical turbulence, with stronger insolation or geostrophic wind speeds causing a reduction in wind shift. Smith et al. (1972) have measured wind shift through a Scots pine canopy (height 16 m). Their results (Fig. 3) cover a range of wind speeds and turbulence levels and show an average shift of 27° between the top and bottom of the canopy. Thus, the rate and direction of advection of a spray cloud by winds in the boundary layer are widely variable.

Turbulence in the boundary layer is caused by the stirring of air by obstacles at the earth's surface, and by the buoyancy forces arising from air temperature gradients. The resulting air motions are known as forced and free convection, respectively (Thom 1975). In the absence of buoyancy forces, when frictional forces are the only source of turbulence in the boundary layer, the atmosphere is neutrally stable. Under these conditions the vertical air temperature gradient is due to air expansion only and is equal to the adiabatic lapse rate (−0.01°C/m). The horizontal wind speed u (z) at height z (> d) is given by:

$$u(z) = \frac{u_*}{k} \log_e((z-d)/z_o) \quad \mathbf{(1)}$$

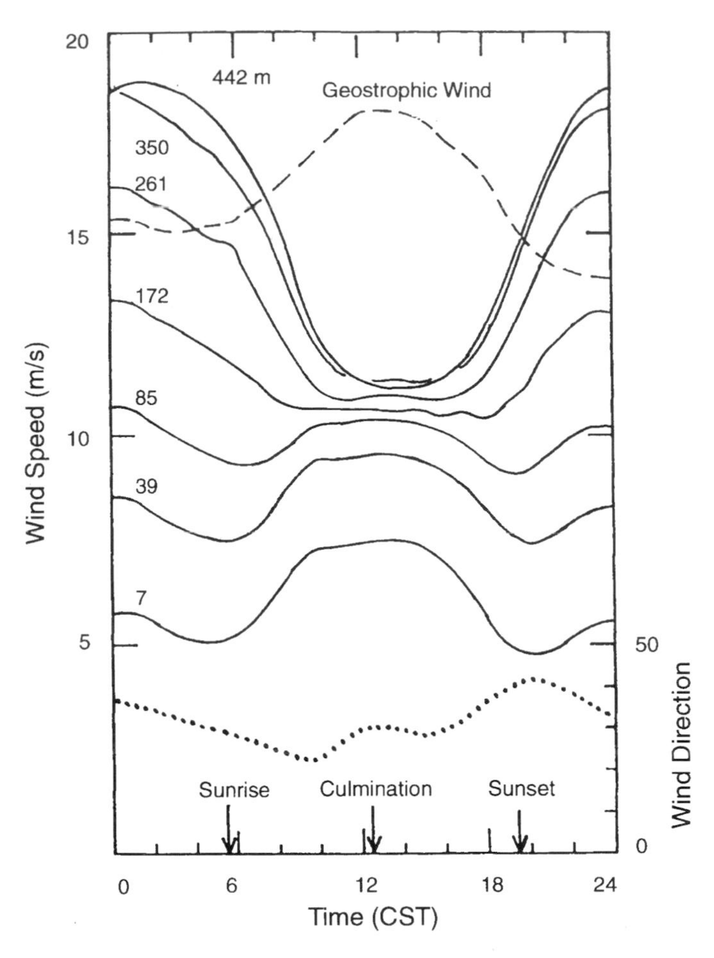

Figure 1. Average wind speed at various heights in the boundary layer at various times of day (Yasuda 1988). Dotted line denotes angle between geostrophic and surface wind.

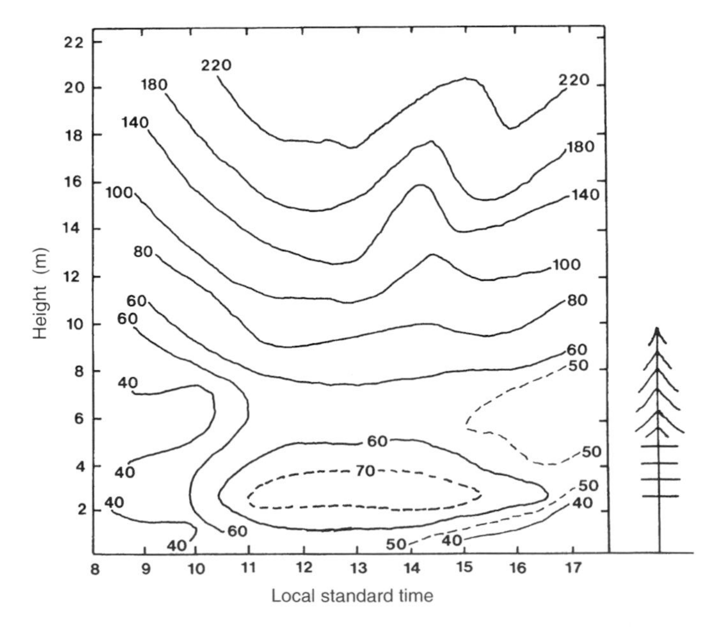

Figure 2. Average wind speed contours (cm/s) in and above a pine canopy at various times of day (Bergen 1971).

where u_* is the friction velocity, k is von Karman's constant, d is displacement height, and z_0 is roughness length (Thom 1975). Friction velocity u_* is proportional to the geostrophic wind speed, and d and z_0 are proportional to the size of roughness elements on the earth's surface. Buoyancy forces arise when the vertical air temperature gradient is different from the adiabatic lapse rate, and enhance or dampen turbulence according to whether the boundary layer is unstable or stable. The air temperature decrease with height in an unstable boundary layer is greater than the lapse rate whereas in a stable boundary layer, the air temperature decrease is less than the lapse rate, or more usually, increases with height. Lapse or inversion conditions, respectively, are said to exist depending on whether the air temperature decreases more or less quickly with height than the adiabatic lapse rate. The wind speed profile in an unstable or stable boundary layer differs from that under near neutrally stable conditions, with higher average wind speeds under unstable conditions and lower wind speeds in stable conditions. Unstable conditions are found during daylight hours, when significant insolation occurs, whereas stable conditions are found at night and around dawn and dusk. Neutral stability occurs in overcast conditions, or with high wind speeds, and for short periods near dawn and dusk.

The Richardson number (Ri) is used to quantify the effect of vertical air temperature gradient on turbulence (Thom 1975) and is calculated from the ratio of kinetic energy supplied or removed by buoyancy forces ($k.e._b$) to that arising from the vertical gradient of horizontal wind speed ($k.e._i$), as a parcel of air changes height.

$$\begin{aligned} Ri &= -(k.e._b)/(k.e._i) \\ &= 2g(T_2-T_1)(z_2-z_1)/(T_2+T_1)(u_2-u_1)^2 \end{aligned} \quad \mathbf{(2)}$$

where g is gravitational acceleration, T_1, T_2 and u_1, u_2 are air temperatures and average horizontal wind speeds at heights z_1 and z_2, respectively. Ri is positive in a stable air layer, and at values above 0.2 turbulence is effectively damped out, whereas in an unstable layer Ri is negative, and for values < -1 convection caused by heating is the dominant process. In near-neutral conditions Ri is close to zero ($-0.01 < Ri < 0.01$). Another measure of air layer stability is the Stability Ratio (SR) (Coutts and Yates 1968), which is given by

$$SR = 10(T_2-T_1)/u^2 \quad \mathbf{(3)}$$

where u is the average horizontal wind speed in the air layer, expressed in metres per second.

The rate of atmospheric diffusion of airborne material is proportional to the strength of turbulence, therefore diffusion rates are greatest in an unstable air layer and lowest in stable conditions. Eddy diffusivity (K) is used to quantify the effectiveness of atmospheric diffusion (Pasquill 1974). Figure 4 shows vertical eddy diffusivities measured at different heights above ground at various times of day (Yasuda 1988). Rates of atmospheric diffusion of a spray cloud are widely variable, according to K values; however, diurnal trends are present. During evening hours, when insolation is slight, the ground cools by radiation, thereby cooling the air above. Under these conditions a stable air layer may develop immediately above the ground, in which vertical atmospheric diffusion is slow or

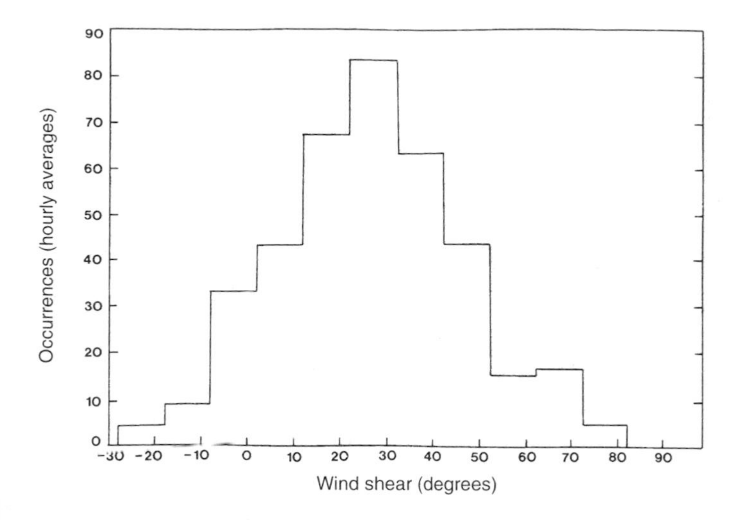

Figure 3. Frequency distribution of wind shift through a pine canopy (Smith et al. 1972).

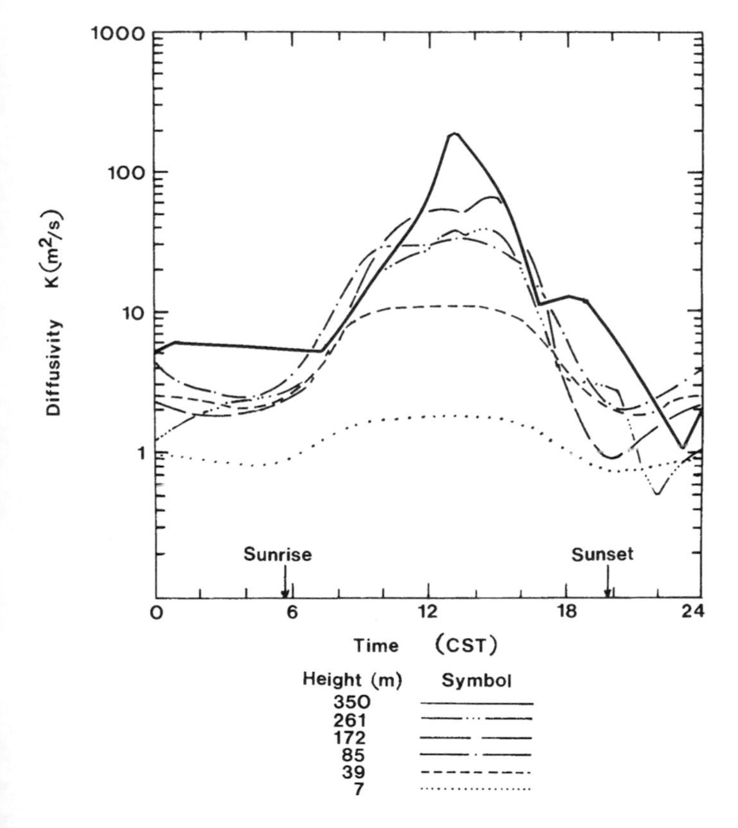

Figure 4. Eddy diffusivity at various heights in the boundary layer at various times of day (Yasuda 1988).

nonexistent, whereas the air above is still unstable or near neutrally stable, thereby supporting quicker diffusion. When a spray cloud comprising small drops is released above a ground or radiation inversion, its downward movement into the stable air layer is slow, and if there is a wind above the inversion, the cloud may be transported for a considerable distance before reaching the canopy, possibly missing the target area altogether. By dawn the inversion layer may have grown to a depth of several hundred metres. After sunrise an unstable air layer begins to develop immediately above the ground, due to radiative ground heating. This is called a mixing layer, and the remainder of the inversion above it is then known as a capping inversion because it limits the upward diffusion of material. As insolation continues, the depth of the mixing layer increases to several hundred metres. A spray cloud released into this mixing layer is dispersed quickly throughout its depth, but upward diffusion from the layer is slow. The effect of vertical air temperature profile on spray cloud dispersal is shown graphically in Figure 5 (Quantick 1985).

Aircraft Wake Influence

An aircraft wake modifies the local state of the atmospheric boundary layer and affects initial spray cloud dispersal. In essence, an aircraft wake at a distance of several wingspans behind the aircraft consists of two downward moving counter-rotating vortices trailing from the wing tips of a fixed-wing aircraft (Fig. 6), or in the case of a rotary wing aircraft, from the rotor downwash. The vortices possess a downward momentum as a consequence of the lift provided to the aircraft (Wickens 1980). The momentum stream-tube theory considers aircraft lift to be provided by the downward momentum imparted to a cylindrical air mass trailing behind the aircraft. Applying Newton's second law, that is, force equals rate of momentum change, gives the following relation:

$$L = \rho VWS \quad \textbf{(4)}$$

where L is the lift force on an aircraft, V is aircraft velocity, W is the downward velocity of the stream-tube or wake velocity, S is the cross-sectional area of the momentum stream-tube, and ρ is air density. For a fixed-wing aircraft the stream-tube is assumed to have a circular cross section with diameter equal to the aircraft wingspan, and the lift force during horizontal flight is equal to the aircraft weight. This gives the following expression for wake velocity (W):

$$W = \frac{4gM}{\pi \rho V b^2} \quad \textbf{(5)}$$

where M is the mass of the aircraft, b is the wingspan, and g is gravitational acceleration. Thus, wake velocity increases with aircraft weight, but decreases with increases in flying speed and wing span. A heavy slow-flying aircraft will therefore give a relatively high vortex descent rate. Wake velocities for aircraft used in pesticide application are typically 0.5 to 1 m/s. For example, a Grumman Avenger traveling at 76 m/s has a wake velocity of 0.6 m/s in a neutrally stable boundary layer. Buoyancy forces also affect vortex motion and in a stable boundary layer the air temperature gradient retards and may arrest the downward vortex motion. In contrast, an unstable boundary layer accelerates the downward vortex motion, and the increased turbulence level found there accelerates vortex decay and shortens vortex lifetimes compared to those in a stable layer.

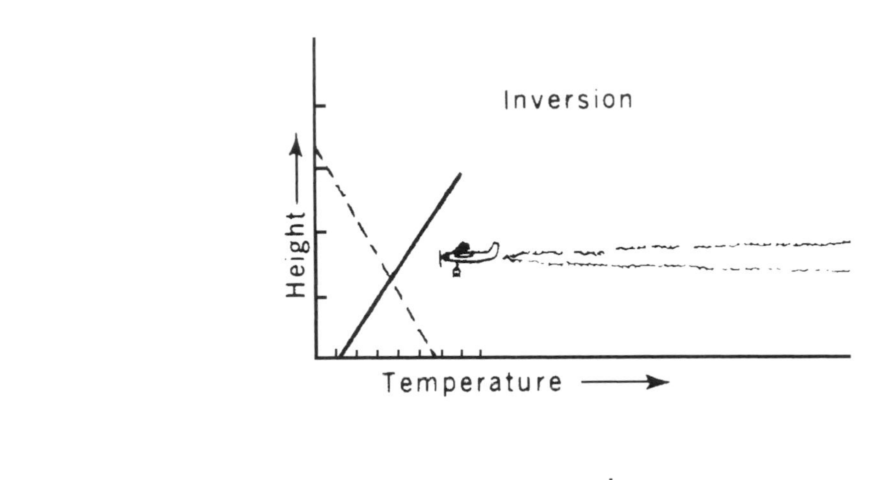

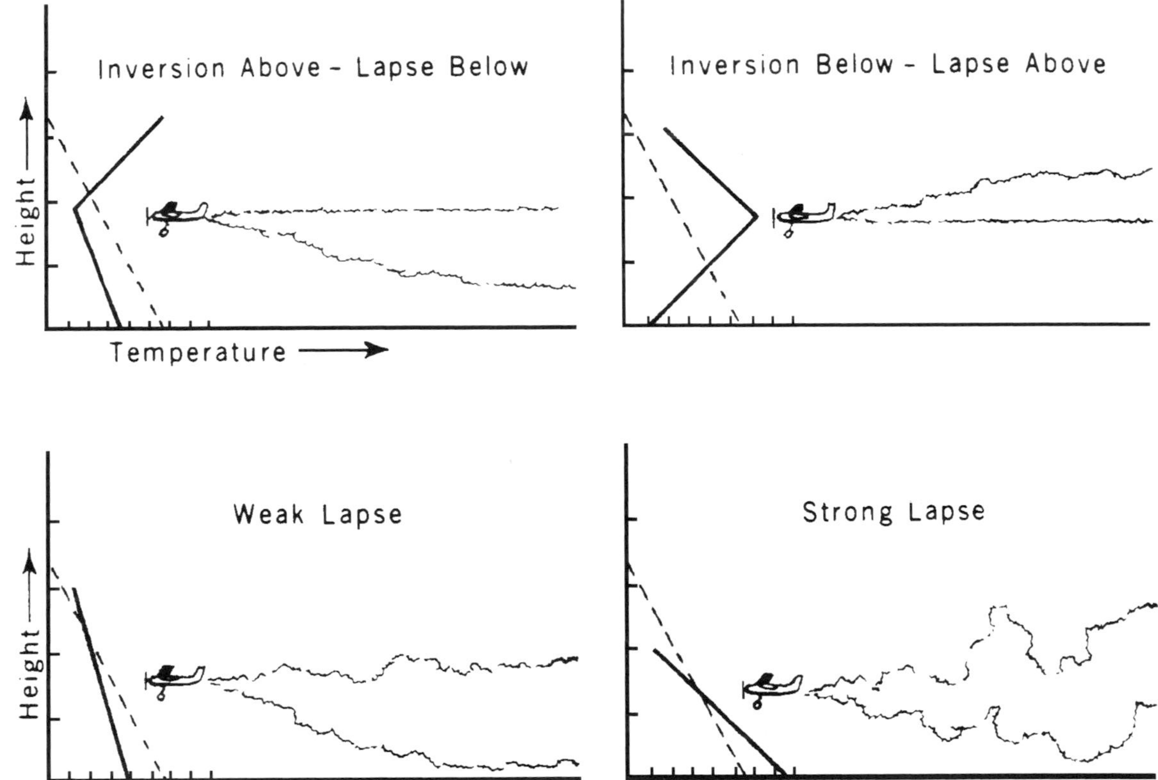

Figure 5. Spray cloud dispersal in boundary layers with various vertical air temperature profiles (Quantick 1985). – – – – Dry adiabatic lapse rate; ——— environmental lapse rate.

Soon after release a substantial portion of a spray cloud comprising small drops becomes entrained in these aircraft vortices, and because of their sedimentation velocity, a large proportion will remain there for an appreciable time, typically 10 to 20 seconds, until the vortices decay sufficiently to release them (Tennankore et al. 1980). The motion of aircraft vortices thus helps to bring drops down from release height to the forest canopy, initially at a rate considerably greater than the drop sedimentation velocity, although the importance of the vortex effect in cloud dispersal is variable and depends on the spray release height, vortex strength, and the vertical air temperature profile in the boundary layer. Unfortunately, the downward vortex motion is retarded when turbulent diffusion is weak, as in a stable boundary layer, and this is when an additional transport mechanism would be most useful to augment atmospheric diffusion. However, the vortex effect during cloud dispersal can be maximized by use of a heavy aircraft with high wind loading, and low release height. A photo-

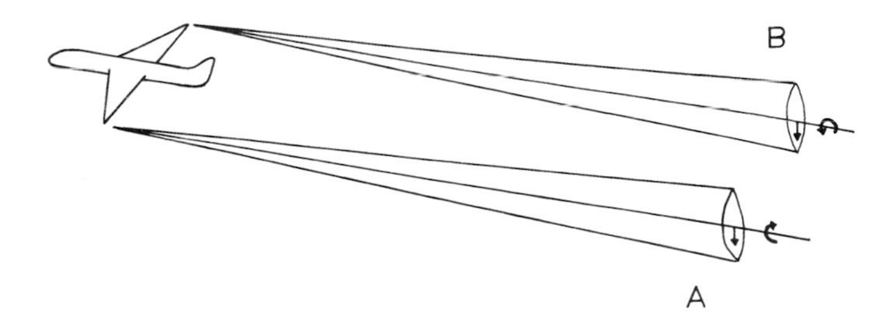

Figure 6. Counter-rotating aircraft vortices (A & B), with similar downward momentum.

graphic record of vortex behavior visualized with smoke has been made by Pelletier (1985), and Drummond (1988) has reviewed aircraft vortex effects on drop motion.

Influence of Electrostatic Force

The process of liquid atomization usually imparts a small electrical charge to the drops (Green and Lane 1964) which can be enhanced by equipment such as the Electrodyne® sprayer (Coffee 1980) that deliberately increases the charge to influence drop dispersal. In most applications the spray cloud generated has nil net charge, however, because electric charges are distributed to provide negatively and positively charged drops, with a mean square charge that increases with liquid polarity. A cloud carrying a net charge, similar in polarity to the atomizer, results from an atomizer held at an electrostatic potential different from that of the earth.

When a drop carrying charge q is placed in an electric field of strength E, a force (F_e) is exerted on the drop where:

$$F_e = Eq. \qquad (6)$$

Electric fields exist around any charged object, and the field strength (E) at a distance r from a point charge Q is given by:

$$E = Q/4\pi \in \in_o r^2 \qquad (7)$$

where $\in$ and $\in_o$ are the relative and absolute permittivities. In air the value of $\in$ is close to unity, and $\in_o$ is 8.8 10^{-2} F/m (Gibson 1969).

A charged drop in a spray cloud dispersing in the atmosphere is subject to electrostatic forces of various origins. If the spray cloud has a net charge, an equal and opposite charge is induced at the earth's surface, causing a downward attractive force. Individual drops are also repelled by the remainder of the similarly charged cloud, causing cloud expansion. In addition, the earth's electric field, typically 120 V/m at the surface (Mason 1971), induces an upward or downward motion of negatively or positively charged drops, respectively. The electrostatic forces on a spray cloud that has not been deliberately charged are typically small compared to aerodynamic and gravitational forces; however, a deliberately charged cloud may be used to affect dispersal (Inculet et al. 1983).

Drop Evaporation

Spray drops evaporate during dispersal, causing a reduction in drop size. When the initial drop size spectrum is small (diameter less than 100 µm), the reduction in drop size by evaporation during flight may significantly reduce the drop impaction efficiency. The evaporation rate depends on drop diameter, because this determines the area of the liquid–air interface. The molecular weights of the spray components also affect evaporation rate, with volatility increasing as molecular weight is reduced. Meteorological conditions also affect evaporation rate; for nonaqueous tank mixes the evaporation rate depends on the air temperature, with higher temperatures causing faster evaporation (Green and Lane 1964), whereas for aqueous tank mixes the evaporation rate depends on both the air temperature and the relative humidity, with increased relative humidity slowing the evaporation rate (Rogers 1979).

The effect of evaporation on the diameter of a falling water drop may be calculated from the following equations (Rogers 1979):

$$D(t) = (D_o^2 - 8ct)^{0.5} \qquad (8)$$

$$c = (1-S)/(F_k+F_D) \qquad (9)$$

where D(t) is drop diameter at time t, D_o is the initial drop diameter, S is the relative humidity, and F_k and F_D are constants dependent on ambient air temperature and pressure. The value of (F_k+F_D) is 8.3 10^7 s/m at 20°C and 101.5 kPa. Figure 7 shows the lifetime of a water drop as a function of diameter under various conditions, calculated from equations 8 and 9. For example, with an air temperature of 20°C, and relative humidity of 60%, a drop 100 µm in diameter takes 25 seconds to evaporate, compared with 6 seconds for a 50-µm drop and 0.5 seconds for a 15-µm drop.

With knowledge of the initial drop size spectrum, these results may be used to estimate the drop size spectrum of an aqueous tank mix at the target. Similarly, if drop lifetimes for a nonaqueous tank mix diluent are known, the drop size spectrum at the target may also be estimated. Otherwise, experimental measurements such as those made by Picot et al. (1981) are required. A comparison of the volatilities of various aqueous and nonaqueous forestry insecticide tank mixes has been made by Sundaram and co-workers (1985, 1986).

The flight time of an aerially applied spray drop is typically 1 to 5 minutes. This appears to exclude the use of small aqueous drops for insecticide applications due to evaporation, and yet they have been widely used. If these drop sizes are to be useful for pesticide applications, aqueous tank mixes must include a low volatility liquid that will not evaporate during the flight of the drop, and in effect puts a lower limit on the drop size. For example, by including only 12.5% by volume of low volatility liquid in a tank mix, the lower drop size limit is 50% of the initial size. Tank mix adjuvants are available for this, for example, the emulsifiable oil Ulvapron (Wodageneh and Matthews 1981). In some cases, a low volatility active ingredient may provide a lower limit to drop size; for example, fenitrothion acts in this way (Green and Lane 1964, Sumitomo Chemical Co. 1980). Drop evaporation depends on the conditions at and around the drop–air interface, and in some tank mixes, for example, those containing the bacterial insecticide Bt, evaporation may be slowed by the formation of a low volatility layer at the drop surface that impedes the evaporation of more volatile spray components.

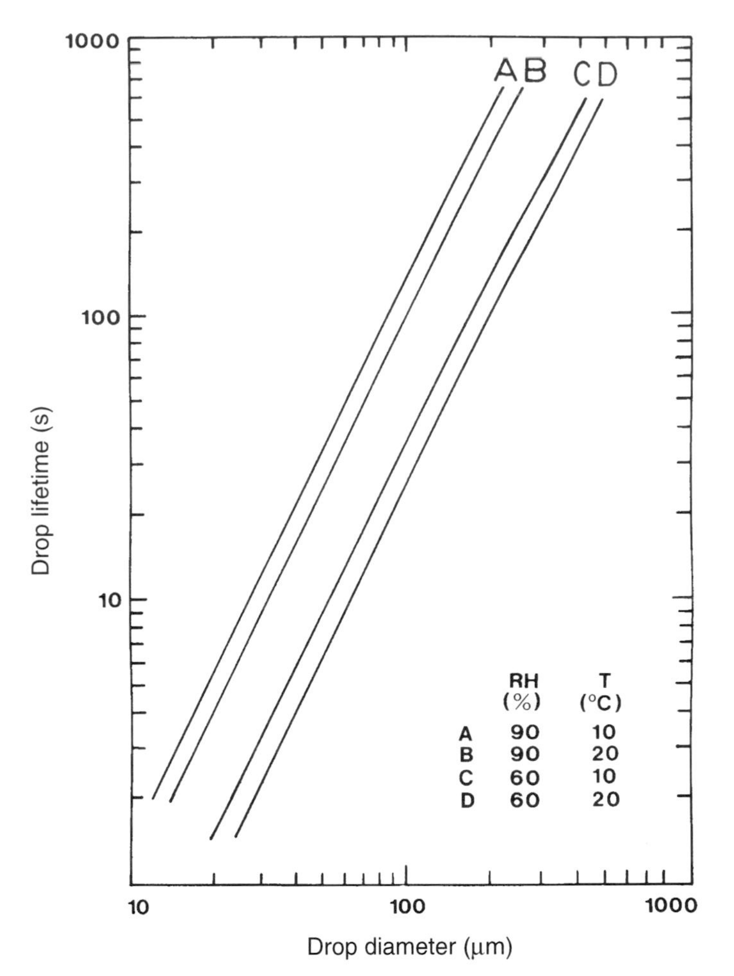

Figure 7. Water drop lifetimes at various relative humidities and air temperatures.

Spray Deposition

Spray cloud deposition occurs by drop sedimentation, inertial impaction, and electrostatic attraction. Drop deposition by sedimentation is the gravitational settling of a drop onto an obstacle. A drop in free fall reaches its terminal velocity when the downward gravitational force is balanced by the upward forces of aerodynamic drag and buoyancy. For insecticide drops in air the buoyancy force is small (compared to aerodynamic drag). The aerodynamic drag on a drop is the sum of inertial and viscous forces, and the nondimensional Reynolds number (Re) indicates the relative magnitude of these forces (inertial/viscous) and the dominant force causing drag. For a drop falling in air Re is defined by:

$$\mathrm{Re} = \frac{uD}{v} \qquad \textbf{(10)}$$

where u is the speed of the drop, D is drop diameter, and v is the kinematic viscosity of air (Chamberlain 1975). For small drops moving at slow speeds the viscous force is larger than the inertial force, i.e. Re << 1, and the vertical component of aerodynamic drag is the viscous force (F_D), given by Stokes law:

$$F_D = 3 \pi \mu D u \qquad \textbf{(11)}$$

where μ is the air viscosity. Equating the gravitational and viscous drag force gives an expression for terminal velocity (V_s), valid for drop diameters up to 60 μm:

$$V_s = \frac{D^2 g \rho}{18 \mu} \qquad \textbf{(12)}$$

and ρ is drop density. Thus, for Re << 1 drop terminal velocity is proportional to the square of its diameter. Above this limit inertial forces may not be ignored and a correction is needed. Figure 8 shows terminal velocity for various drop sizes, taking into account inertial forces for D > 60 μm.

Inertial impaction occurs when an obstacle interrupts the airflow in which drops are dispersed. The airflow deviates around the obstacles, whereas the drop changes direction more gradually due to its greater inertia and may be deposited. Drop impaction efficiency is defined as the deposited portion (percentage) of drops in the airflow directly upward of the obstacle (i.e. deposited portion of drops that would pass through the volume occupied by the obstacle, were it removed). Drop impaction efficiency increases with wind speed and drop size but decreases for increasing obstacle size. Drop impaction efficiency also depends on the pattern of flow around the obstacle. The Reynolds number for the flow pattern,

$$\mathrm{Re} = \frac{UL}{v} \qquad \textbf{(13)}$$

where U is wind speed over obstacle, and L is the width of the obstacle, determines the characteristics of the flow around the obstacle. In the case of potential flow (Re >> 1), the fluid is in effect inviscid, and inertial forces predominate, whereas for viscous flow (Re << 1) inertial forces are small compared to viscous forces. The potential flow regime is most often found for airflow around natural obstacles.

Inertial impaction efficiency in potential flow has been calculated or measured by several investigators, for cylinders, spheres, ribbons, and disks. Figure 9 shows theoretical and experimental impaction efficiencies (E) on a cylinder, plotted against impaction parameter (P) which is calculated by dividing the stop distance of the drop by the obstacle width (May and Clifford 1967). The nondimensional Stokes number, also used to predict particle impaction efficiency, is twice the impaction parameter. Stop distance is the horizontal distance traveled by a drop projected horizontally in still air. For Re < 1 the stop distance may be calculated by applying Stokes law to the horizontal motion of a drop projected in still air to give:

$$s = u D^2 \rho / 18 \mu \qquad \textbf{(14)}$$

where s is the stop distance, u is the velocity of projection, and ρ is the drop density. Thus, the stop distance of a drop in air increases with the velocity of projection, drop diameter, and drop density. Figure 10 shows stop distances of unit density particles in air at various velocities of projection.

For large values of impaction parameter (i.e. when the stop distance is much greater than the obstacle width), drops are deposited with high efficiency, whereas for small values of the impaction parameter when stop distances are much smaller than the obstacle width, drops are deposited with low or zero efficiency. These calculated and experimental results are for laminar flow, but may be used to estimate deposition in turbulent flow if the eddy size is greater than the obstacle width, because under these conditions the flow around the obstacle is locally quasilaminar. In the atmosphere the lower limit of eddy size is about 1 mm, and the predominant eddy sizes are typically greater than 10 cm (Pasquill 1974). However, turbulent airflow sometimes causes drop deposition on the downwind side of an obstacle, a feature not predicted by laminar flow experiments or calculations.

These results have been applied in estimating the impaction efficiency on balsam fir needles of various drop sizes, at various wind speeds (Table 1). These estimates are for an isolated needle; in reality, a needle close to a twig may capture drops somewhat less efficiently because the curvature of the airflow is reduced by the presence of the twig. In other words, the pattern of airflow over the whole twig affects the impaction efficiency on needles attached to it. For example, at speeds less than 1.5 to 2.5 m/s, depending on orientation, airflow over a spruce twig has been found to penetrate between needles, allowing them to act approximately as individual collectors (Grant 1985). With wind speeds above this range, airflow between individual needles is reduced, and more air is deflected around the twig thereby reducing impaction efficiency on individual needles. It is noteworthy that drops impinging on foliage are not necessarily deposited, due to their kinetic energy. For efficient deposition, the evaporated insecticide drop must be sufficiently sticky to prevent it bouncing off foliage after impingement.

Both sedimentation and inertial impaction contribute to canopy deposition of small drops, and their relative importance depends on the drop size spectrum of the cloud and the prevailing meteorological conditions, in particular the average wind speed profile in the canopy. Wind speed increases with height above ground in both coniferous and deciduous canopies (Geiger 1961; Oliver 1971), resulting in higher impaction efficiency near the canopy top. The wind speed profile inside a forest canopy also shows diurnal trends, with highest wind speeds occurring near mid-day when insolation is strongest (Fig. 2). Thus, during the periods normally chosen for insecticide spray applications, wind speeds in the canopy are relatively light, and impaction efficiencies are relatively low.

Table 1. Drop impaction efficiencies on balsam fir needle.

Drop diameter (µm)	Impaction efficiency (%) Wind speed (m/s) 0.25	0.5	1
10	0	0	0
30	5	40	50
50	45	60	75

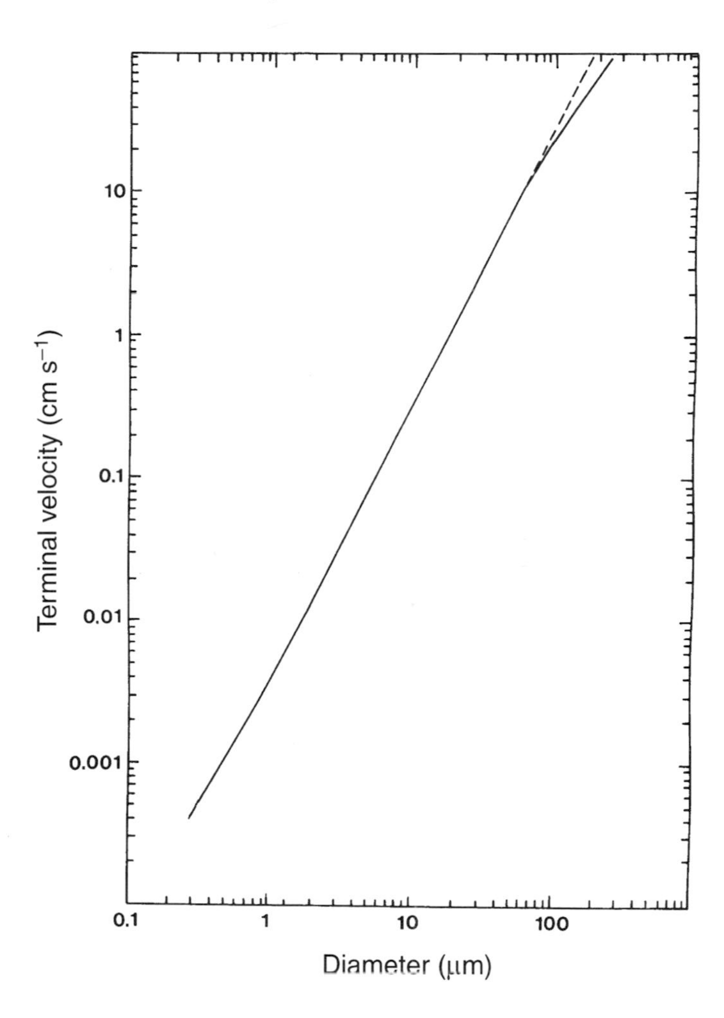

Figure 8. Terminal velocity of water drops of various diameters (Chamberlain 1975).

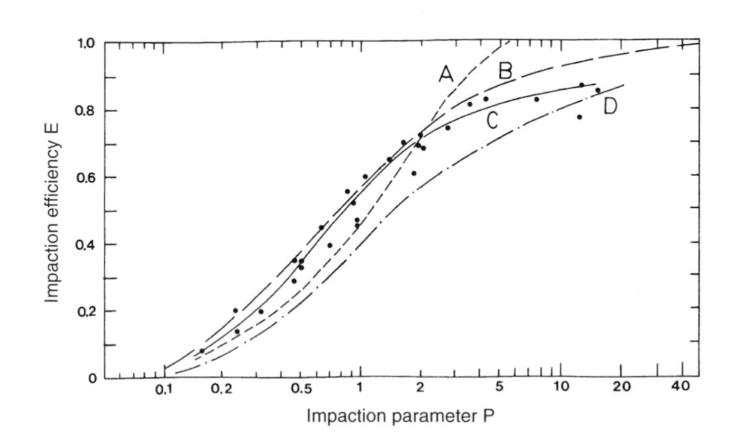

Figure 9. Drop impaction efficiency (E) on cylinder versus impaction parameter (P); curves A, C (fitted to points), and D are from experimental measurements; B is from theoretical calculations (May and Clifford 1967).

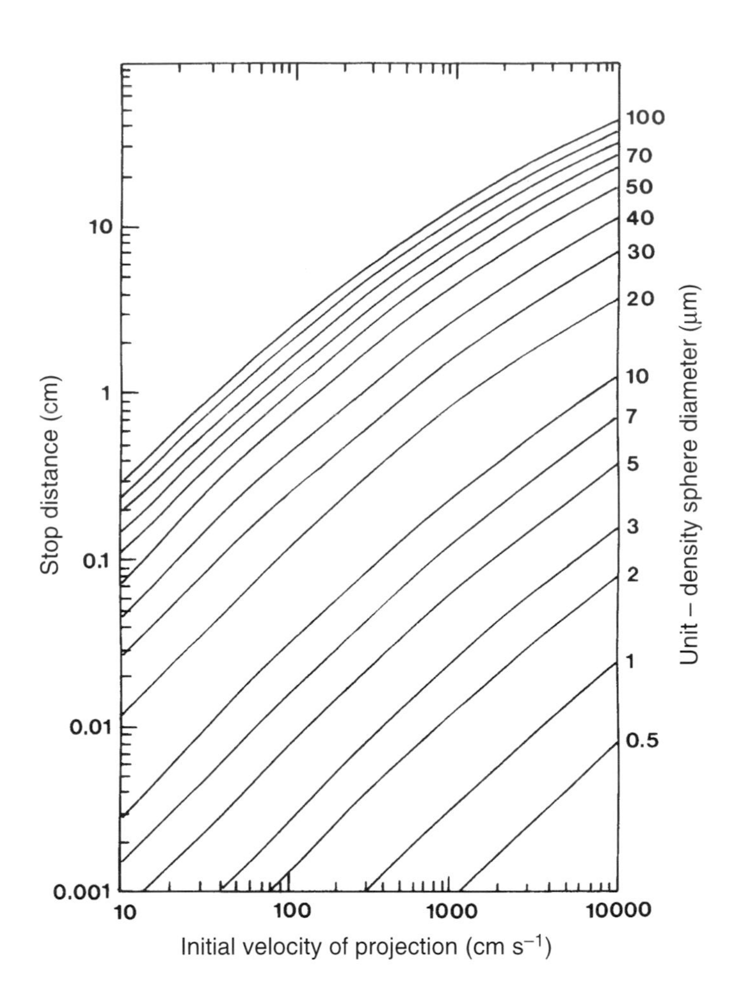

Figure 10. Stop distance of unit density particles in air at various velocities of projection (Chamberlain 1975).

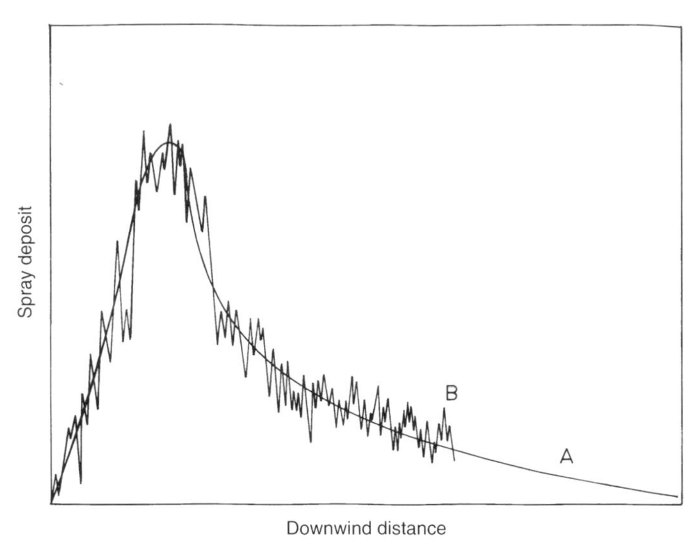

Figure 11. Spray deposit at various downwind distances from a single crosswind swath, averaged (A) and from an individual swath (B).

Drop deposition is also affected by electrostatic forces. A drop approaching an earthed obstacle in the airflow will be attracted to it by induced charges, or for a charged obstacle either attract or repelled, according to the sign of the electrostatic charge. The importance of the electrostatic force in drop deposition depends on its magnitude, as compared to the aerodynamic and gravitational forces on a drop. From equation 4 it is apparent that the electrostatic force increases as a charged drop approaches an obstacle. Thus, the electrostatic force can be negligible during dispersal, but still become important during deposition.

Spray Deposit Pattern

The average spray deposit at various downwind distances from a single aerially applied crosswind swath has a characteristic form (Fig. 11A) that has been measured in many trials (Crabbe and McCooeye 1985; Cadogan et al. 1986; Uk 1987). The general shape describes both foliar and ground deposits, although actual deposit differs according to substrate. The source strength, that is, drops or mass of active ingredient released per unit length of swath, is directly proportional to the average deposit at a given downwind distance from the release line. The peak deposit and its downwind distance from a single crosswind swath depend on cloud release height and vertical intensity of turbulence (standard deviation of vertical wind speed/average horizontal wind speed). Peak deposit falls with increasing release height, and the downwind distance of the peak increases, whereas increasing turbulence causes both peak height and downwind distance to decrease. Wind speed alone is not an influential variable in determining the pattern of deposit from aerial applications using drops with diameters less than 100 µm, but acts through the intensity of turbulence. In aerial applications with small drops the aircraft vortices give an effective release height lower than the aircraft flying height, because the cloud is carried downward in the wake before being released as the wake decays (Picot et al. 1986).

Figure 11B shows typical deposit from a single swath. Spray deposit from a single swath is more variable than the characteristic form (Fig. 11A, which represents an average from multiple swaths overlaid along a single track. This variability is due to the stochastic nature of the principal dispersal mechanism, atmospheric diffusion, and is an inherent deposit characteristic. In operational sprays swaths are often applied at orientations other than crosswind. Bache (1984) has analyzed the effect of orientation on deposit, and except for alongwind swaths, the characteristic deposit pattern is similar regardless of orientation, showing a peak and extended deposit tail, although peak height and downwind distance does change with orientation.

The spray deposit resulting from a multiple swath application is the sum of deposit from individual swaths (Bache 1984). When using sprays comprising drops with diameters of less than 100 µm, the deposit tail extends well beyond the swath region and may contribute significantly to spray deposits in neighboring downwind swaths. Pesticide use-strategies employing small drops are sometimes described as incremental spraying because spray deposit at any position is the sum of contributions from several swaths. In New Brunswick, for example, the typical swath width used for applying the insecticide fenitrothion with Grumman Avenger (TBM) aircraft is 135 m, and with this use-strategy, significant spray deposits have been observed up to 400 m downwind of the spray line (van Vliet 1987).

Several investigators have measured the pattern of spray deposit inside a coniferous canopy resulting from aerial applications made with small drops (Armstrong and Yule 1978; Joyce et al. 1981; Sundaram et al. 1989). The results of these investigations have been reviewed by Payne (1988), who noted several common trends found in the deposit measurements. First, foliar spray deposit was found to increase with height above the ground, for spray applications made in meteorological conditions covering a range of wind speeds and atmospheric stabilities. In addition, spray deposits at canopy top were considerably higher when applications were made in relatively high winds (Joyce et al. 1981). Second, a radial variation was observed with deposit increasing toward the branch tip. Deposit increases were also observed from the downwind to the upwind side of the tree, and from the underside to the top of the twig. In addition to these average trends, foliar deposit variability was large in all these investigations. These vertical and radial foliar deposit distributions are well matched to the distribution of spruce budworm larvae in the canopy during spring and early summer (Prebble 1975), making small drops useful for applying insecticide to control them.

Off-target deposits are concomitant with the use of small drops, due to their low terminal velocities. The proportion of active ingredient deposited outside the treatment area is dependent on many factors, including spray application, meteorological and plant canopy variables, the effect of which has been discussed by Payne et al. (1988). Over recent years experimental trials have been carried out in New Brunswick to quantify the effect of meteorological conditions on spray dispersal during spruce budworm control operations. The measurements made (Crabbe et al. 1984; Crabbe and McCooeye 1985; Elias 1988) show that drift (i.e. the airborne fraction of the spray cloud leaving the treatment area) increases with increasing boundary layer stability (i.e. with decreasing turbulence). This result is due to the important role of atmospheric diffusion in transporting drops down to the canopy. For many years only stable conditions, found in early morning and late evening, were thought suitable for spruce budworm control operations; however, these recent results show that neutrally stable or unstable conditions are preferable for reducing drift. This is an important conclusion because it widens the time window suitable for ultralow-volume insecticide applications. In addition, airborne drop concentrations in and immediately above the canopy were reduced with increased wind speed. This finding also has significant ramifications for pesticide regulation because it shows that bystander exposure is not necessarily increased by spraying in higher wind speeds.

A useful way to ameliorate the effects of off-target deposits from pesticide applications is to use a buffer zone. This is a minimum distance of approach between a treatment area and sensitive area, designed to allow sufficient spray cloud dilution to ensure that off-target deposits are reduced to a tolerable level. The problem of setting scientifically defensible buffer zone widths is complex, because the required width depends on pesticide toxicity, organism sensitivity, the insecticide use-strategy employed, and the prevailing meteorological conditions during spraying.

At present there is limited consensus from province to province on the width of buffer zone needed to provide protection to sensitive areas (Eastern Spruce Budworm Council 1987), and this is due partly to the lack of a scientific method for choosing a buffer width. A useful development in relation to this problem was the publication of a scientific method for estimating buffer widths (Payne et al. 1988). This technique identified the sensitive area to be protected, and using toxicology measurements for the sensitive species, estimated a buffer zone width from mathematical models of spray cloud dispersal based on a worst case scenario. Field measurements of spray cloud dispersal were used as a basis for the model. This technique was developed to set buffer widths around applications of the insecticide permethrin, but it is generally applicable, requiring only the provision of appropriate toxological and spray cloud dispersal measurements.

Spray Dispersal Models

Spray cloud dispersal may be predicted by mathematical modeling, and various theoretical approaches have been developed for this purpose. The simplest of these is the ballistic model that is used to calculate the downwind distance traveled before a drop released in a crosswind is deposited; it is based on the equation,

$$x = u\ H\ /\ V_s \tag{15}$$

where x is the distance traversed by the drop before deposition, H is the release height, u is the average horizontal wind speed in the air layer through which the drop falls, and V_s is the drop terminal velocity (Pasquill 1974). This model applies to large drops, whose terminal velocity is greater than the standard deviation of vertical wind speed in the atmospheric boundary layer. Thus, for drops with diameters less than 50 µm, the distance traveled by the drop is not well predicted by equation 15.

Mathematical models applicable to small drop dispersal have also been developed. The concept of atmospheric diffusion has been applied to provide the gradient-transfer model, in which the rate of transport of airborne material down a concentration gradient is proportional to the magnitude of the gradient, that is,

$$F_x = -K_x \frac{\sigma C}{\sigma x} \tag{16}$$

where F_x is the drop flux parallel to the x axis, K_x is eddy diffusivity parallel to the x axis, and C is the airborne drop concentration. A differential equation taking into account drop advection, diffusion, sedimentation, and deposition is formulated using this concept, and solved analytically or numerically to calculate airborne drop concentrations and other statistics of interest (Ermak 1977).

Another small drop dispersal model uses the statistical theory of dispersion, that relates the spread of drops along any axis to the rms turbulent velocity fluctuations in that direction, giving:

$$\overline{X^2} = 2\ \overline{u'^2} \int_0^T \int_0^t R\ (s)\ ds\ dt \qquad \textbf{(17)}$$

where X is the displacement of a drop due to eddy velocity u′ after time T, and R(s) is the Lagrangian correlation coefficient between the drop velocity induced by the eddy velocity at time t and (t+s) (Pasquill 1974). This theory is applied in a numerical model to predict spray cloud dispersal by calculating the position of many individual drops at successive instants of time, also taking into account drop advection, sedimentation, and deposition.

Reid and Crabbe (1980) compared predictions from two mathematical models of spray cloud dispersal, based on the gradient-transfer and statistical theories of turbulent dispersal, and Picot et al. (1986) compared predictions from a model based on the statistical theory, with experimental measurements of canopy deposit from a small drop forestry insecticide application. The gradient-transfer theory is also the basis for the "tilted plume" model used to calculate spray cloud dispersal in the USDA—Forest Service model (Dumbauld and Bjorklund 1977). Mickle (1988) also compared experimental measurements with predictions from both models.

Spray Assessment

Spray deposits from forestry insecticide applications are assessed for two purposes, to measure on- or off-target deposit. The amount of active ingredient deposited on-target (i.e. in the pest habitat) is often required for correlation with efficacy. Operationally, measurements of spray deposit may be required to demonstrate that a particular insecticide application gave the required deposit for adequate insect control. In research, measurements of drops per unit area may be needed to compare with larval population reduction measurements, to estimate the active ingredient deposit required for adequate pest control (Cadogan et al. 1984). Spray assessments are also used to measure the amount of active ingredient transported off-target, for example, that drifted beyond the swath region (Crabbe et al. 1984) or that deposited on water bodies (Payne et al. 1986), to estimate the potential for environmental impact.

Assessment Techniques

Spray deposit assessments may be made using a variety of techniques, depending on the information required and the resources available. Deposits are usually quantified by chromatography, fluorometry, colorimetry, or by visual inspection, although other techniques such as atomic absorption spectroscopy and neutron activation analysis may also be used (EPPO 1982). In Canadian forestry a bioassay has been developed to quantify Bt deposits from measurements on larval frass drop (Wiesner et al. 1988), and immunoassay techniques to quantify forestry pesticides are also under consideration (K. Sundaram, pers. comm.).

Gas and liquid chromatography are the two chromatographic techniques commonly used to quantify pesticide residues. In gas chromatography, the sample is vaporized and injected into an inert gas stream that carries it through a column containing a solid material. A detector downstream of the column is used to indicate the amount of the pesticide of interest in the gas stream at various times after injection. Various types of detectors may be used; two often employed in quantifying forestry insecticides make use of electrical current measurements across a cavity in which the samples are ionized with beta radiation or in a flame (McNair and Bonelli 1969). The system is calibrated by injecting a known quantity of the pesticide of interest. Gas chromatography is a relatively sensitive technique for quantifying pesticides down to 0.01 to 1.0 ppm. Armstrong and Yule (1978) measured foliar spray deposit on white spruce using gas chromatography to quantify a tracer chemical added to the tank mix, and Sundaram et al. (1986) also used this technique to quantify mexacarbate deposits on balsam fir needles. High pressure liquid chromatography uses a liquid stream to carry the residue sample through the column to the detector. The types of detector usually employed in this method rely on ultraviolet or visible light absorbtion, fluorescence or refractive index measurements, and these techniques typically provide a detection limit of 0.1 to 10 ppm.

Fluorometry involves the measurement of light emitted by an irradiated sample containing a fluorescent tracer. Ultraviolet or visible light is used for irradiation. Deposit fluorescence may be measured on the collection surface (Furness and Newton 1988), or the spray deposit is washed off to quantify the total amount of tracer in the sample (Sharp 1974). This latter technique provides detection down to 0.1 to 10 ppm.

Colorimetry requires a dyed spray. Rhodamine B and Erio Acid Red dyes are often used with aqueous formulations, while Automate Red B or Uvitex OB (fluorescent) are suitable for dyeing oil-based formulations. Deposits are washed from collection surfaces, and light transmission through the washings is measured at a particular wave length to quantify the dye concentration in the sample. The spectrophotometer is calibrated by testing a solution of known concentration. This technique provides detection down to 10 to 100 ppm. The techniques of chromatography, fluorometry, and colorimetry are used to quantify the total amount of active ingredient in a sample, rather than the drop sizes contributing to the deposit.

Visual inspection of spray deposits is a simple and widely used technique for quantifying pesticide spray deposits. Deposits are collected on natural or artificial collectors and visualized with dye or fluorescent particles. Deposits may be manually or semiautomatically counted and sized using a microscope and graticule or image analysis system (e.g. Artek 810, Artek Systems Corporation, Farmingdale, N.Y.). An automatic system has also been developed that is capable of quantifying spray deposits that can be recorded on film (Slack 1972). From measurements of stain density and diameter the drop density and an estimate of the drop size distribution can be obtained, with knowledge of the appropriate stain factor. A stain or spread factor is the ratio of the diameter of the drop and the stain it produces on the collector. Stain factors for drops with diameters less than 100 μm depend on the drop size, its composition, and the composition of the collector and are usually assumed to be independent of the kinetic energy of the drop on landing. Ideally, stain factors should be measured with these conditions as close as possible to those found during the field trials. For example, to provide realistic drop composition after flight requires a reduction in the volatile fraction compared to the original tank mix.

Deposit Collectors

Various collector types are used to sample spray deposits from insecticide applications. If measurements of small drop deposit on natural substrates are required, but artificial collectors are to be used, careful consideration must be given to their design and location. To ensure a representative spray deposit, the artificial collector should be similar in shape, size, and location to the natural substrate of interest. This is due to the importance of inertial impaction in depositing small drops, which depends on local wind speed, airflow pattern, and obstacle size. Artificial collectors should ideally be easy and inexpensive to manufacture, and made of material that doesn't react chemically with any of the tank mix ingredients.

When spraying to control spruce budworm larvae, the target comprises the conifer needles near branch tips, and the most reliable way to measure on-target deposit is therefore to collect foliage samples and quantify the spray deposit found there. But for practical reasons an artificial collector may be desirable, for example, to reduce sample clean-up required for pesticide residue measurements by gas chromatography. Kristmanson et al. (1988) fabricated an aluminum frond similar to coniferous foliage for use in measuring foliar spray deposit, to which were added Kromekote® card strips for drop size spectrum measurements (van Vliet 1987). During spraying the frond is suspended in the forest canopy close to a tree, and deposits are later quantified by gas chromatography or visual inspection. Measured spray deposits from this collector have been compared with those on balsam fir, black spruce, and jack pine to provide a correlation that may be used to calculate actual foliar deposit. In Quebec both natural and artificial conifer foliage have been used to assess operational spray deposits (Lambert 1988). Another example of an artificial collector chosen to imitate a natural surface is that used by Payne et al. (1986). To measure offtarget deposit expected on a water body, a plastic sheet was pegged over an area of ground cleared of vegetation, and a glass plate was placed on the sheet to collect drops from a permethrin spray. This collector is aerodynamically similar to a water surface, and spray deposition is therefore similar to that on water.

The highly calendered and filled Kromekote® card has long been used for collecting and displaying drop deposits for visual assessment (Haliburton et al. 1975). Convenient devices for holding and transporting Kromekote® cards during and after forestry insecticide applications have been developed, including a ground sampling unit (Randall 1980) and a foliar sampling apparatus (Cadogan and Zylstra 1984). It is necessary to consider the effect of collector size when using Kromekote® card collectors to estimate foliar deposit, because of its effect on impaction efficiency. Card strips about 1 cm wide provide convenient handling and an obstacle to airflow more similar to a fir or spruce twig than the 10 × 10 cm card often used and give more representative measurements of foliar spray deposits.

Airborne Spray Assessment

Measurements of airborne insecticide may also be required, for example, to quantify the amount of active ingredient in a drift cloud. A simple to use, inexpensive, and reliable device for collecting airborne pesticide sprays is the Rotorod® (Fig. 12, Ted Brown Associates, Los Altos Hills, Calif.). This instrument uses a small electric motor to translate a narrow sampling surface at high speed (10 m/s), thereby enabling drops with diameters exceeding 10 μm to be efficiently collected from the volume of air swept (Edmonds 1972). Experiments using this sampler include those by Crabbe and McCooeye (1985) and Payne et al. (1987). Total active ingredient collected may be quantified, or the airborne drop size spectrum can be assessed by placing strips of Kromekote® cards on the collection surfaces (van Vliet 1987).

Another technique for airborne drop measurement is the aspirated sampler, in which drop-laden air is collected via an orifice, filtered to remove drops, and the spray deposit then quantified. For accurate sampling, a sharp-edged orifice must be operated isokinetically and isoaxially with the surrounding airflow (Fuchs 1975). Isokinetic sampling implies that the speed of the air entering the orifice should be the same as the surrounding air; this is required to maintain ambient drop concentration as the air sample enters the orifice. When air samples are taken by anisokinetic sampling, the airflow streamlines near the orifice are deviated by its presence, because the flow of air through the orifice is either more or less than would occur if the orifice were not present. Inertia then causes the drop trajectories to deviate from the flow streamlines resulting in atypical drop concentrations in the air sample. The requirement for isoaxial sampling implies that the axis of the sampling probe be parallel to the flow streamlines. The reason for this is that the curvature in the streamlines caused by anisoaxial sampling will, as described previously, result in atypical drop concentrations in the air sample. The requirement for a sharp-edged orifice is again to ensure that the flow streamlines are not deviated on entering the sampler, as they would be with a blunt-edged orifice. Crabbe and McCooeye (1985) and Riley et al. (1989) have used aspirated samplers to quantify airborne insecticide downwind of forestry applications.

The airborne drop size spectrum can also be measured with a cascade impactor. This device separates the spray cloud into different aerodynamic size-classes as it passes through progressively narrower orifices and is deposited on impaction surfaces placed close to these orifices (Cohen 1986; Fig. 13). At each successive stage the impaction surface, which is usually placed at right angles to the local flow, is placed closer to the

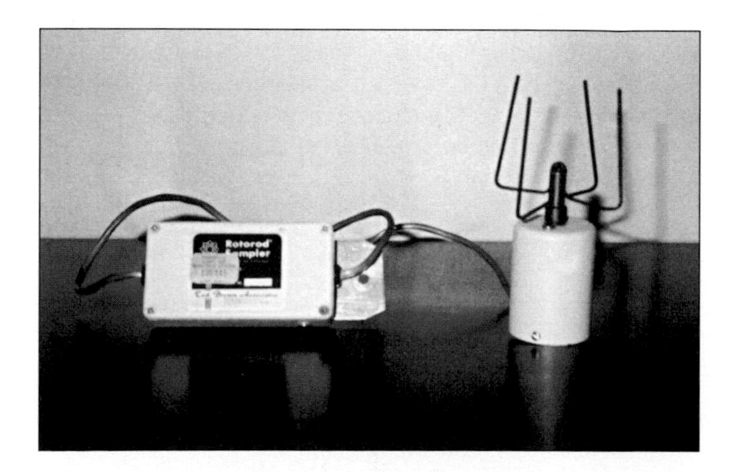

Figure 12. Rotorod® sampler for collecting airborne drops.

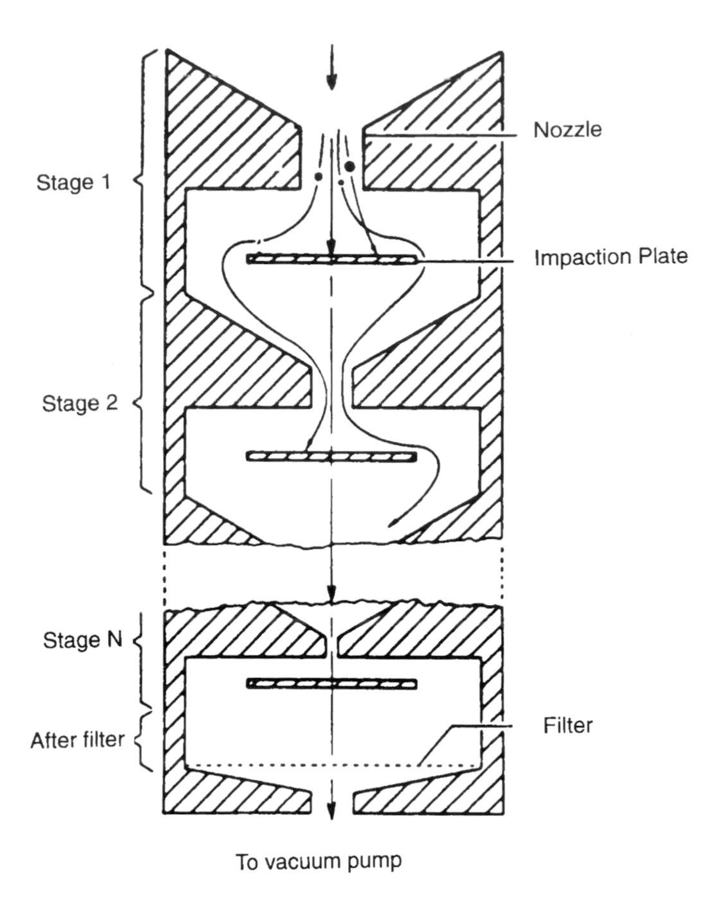

Figure 13. Cross section through a cascade impactor, showing trajectories for three particle sizes.

orifice. This causes progressively greater curvature in the flow streamlines at each stage, causing a different portion of the drop size spectrum to be deposited. The overall result is that the first impactor encountered by the airflow catches the largest airborne drops, and the drop sizes caught on subsequent impactors are progressively reduced, providing sampling of the whole drop size spectrum. A variety of impactors is available, covering different drop size ranges. The upper diameter limit for measurements with this instrument is typically 30 to 50 µm, while the lower limit is about 0.05 to 0.5 µm. Crabbe et al. (1980) made use of this measurement technique in quantifying the drop size spectra in the drifted portion of the spray cloud from forestry insecticide applications.

References

Armstrong, J.A.; Yule, W.N. 1978. The distribution of aerially-applied spray deposits in spruce trees. Can. Entomol. 110: 1259-1267.

Atkinson, B.W. 1981. Meso-scale atmospheric circulations, Academic Press, London, England. 495 p.

Bache, D.H. 1984. A practical scheme for estimating swath width using ULV sprays. Crop Protection 3: 451-468.

Bergen, J.D. 1971. Vertical profiles of wind speed in a pine stand. For. Sci. 17: 314-321.

Cadogan, B.L.; Zylstra, B.F. 1984. An apparatus and technique for sampling aerially applied sprays in conifers. Pestic. Sci. 15: 417-423.

Cadogan, B.L.; Zylstra, B.F.; de Groot, P.; Nystrom, C. 1984. The efficacy of aerially applied Matacil to control spruce budworm *Choristoneura fumiferana* (Clem) in Bathurst, New Brunswick. Can. For. Serv., For. Pest Manage. Inst., Sault Ste. Marie, Ont. Inf. Rep. FPM-X-64. 33 p.

Cadogan, B.L.; Zylstra, B.F.; Nystrom, C.; Pollock, L.B.; Ebling, P.M. 1986. Spray deposits and drop size spectra from a high wing monoplane fitted with rotary atomizers. Trans. ASAE 29: 402-406.

Chamberlain, A.C. 1975. The movement of particles in plant communities. Pages 155-203 *in* J.L. Monteith (Ed.), Vegetation and the Atmosphere, volume 1. Academic Press, London, England.

Coffee, R. 1980. Electrodynamic spraying. Pages 95-107 *in* J.O. Walker (Ed.), BCPC Application Symp. "Spraying Systems for the 1980's." BCPC, Croydon, England.

Cohen, B.S. 1986. Introduction: the first 40 years. Pages 1-21 *in* J.P. Lodge and T.L. Chan (Eds.), Cascade impactor-sampling and data analysis. Am. Ind. Hyg. Assoc., Akron, Ohio.

Coutts, H.H.; Yates, W.E. 1968. Analysis of spray droplet distribution from agricultural aircraft. Trans. ASAE 11: 25-27.

Crabbe, R.S.; Elias, L.; Krzymien, M.; Davie, S. 1980. New Brunswick Forestry Spray Operations: field study of the effect of atmospheric stability on long range pesticide drift. Rep. No. LTR-UA-52. NAE, Natl. Res. Counc., Ottawa, Ont.

Crabbe, R.S.; McCooeye, M. 1985. Effect of atmospheric stability and wind speed on wind drift in aerial spray trials, neutral to unstable conditions. Rep. No. LTR-UA-82. NAE, Natl. Res. Counc., Ottawa, Ont.

Crabbe, R.S.; McCooeye, M.; Elias, L. 1984. Effect of atmospheric stability on wind drift in aerial forest sprays. Neutral to stable conditions. Rep. No. LTR-UA-73. NAE, Natl. Res. Counc., Ottawa, Ont.

Drummond, A.M. 1988. Aircraft vortex effects on droplet motion and vortex stability. Pages 51-68 *in* G.W. Green (Ed.), Proc. Symp. Aerial Application of Pesticides in Forestry AFA-TN-18. Natl. Res. Counc., Ottawa, Ont.

Dumbauld, R.K.; Bjorklund, J.R. 1977. Deposition profile calculations for the State of Maine 1977 spray program. Rep. No. TR-77-310-01. H.E. Cramer Company, Salt Lake City, Utah.

Eastern Spruce Budworm Council. 1987. Buffer zones: their application to forest insect control operations. Proc. Buffer Zone Workshop, Quebec City, April 1986. Can. For. Serv., For. Pest Manage. Inst., Sault Ste. Marie, Ont.

Edmonds, R.L. 1972. Collection efficiency of Rotorod samplers for sampling fungus spores in the atmosphere. Plant. Dis. Rep.56: 704-708.

Elias, L. 1988. Specification of most acceptable spray weather. Rep. No. LTR-UA-97. NAE, Natl. Res. Counc., Ottawa, Ont.

EPPO. 1982. Guidelines for reduced volume pesticide application. 2nd ed. European and Mediterranean Plant Protection Organization, Paris. 48 p.

Ermak, D.L. 1977. An analytical model for air pollutant transport and deposition from a point source. Atmos. Environ. 11: 231-237.

Fuchs, N.A. 1975. Sampling of aerosols. Atmos. Environ. 9: 697-707.

Furness, G.O.; Newton, M.R. 1988. A leaf scanning technique using a fluorescence spectrophotometer for the measurement of spray deposits. Pestic. Sci. 24: 123-137.

Geiger, R. 1961. The climate near the ground. Harvard University Press, Cambridge, Mass. 611 p.

Gibson, W.M. 1969. Basic electricity. Penguin Books, Harmondsworth, Eng. 192 p.

Grant, R.H. 1985. The influence of the physical attributes of a spruce shoot on momentum transfer. Agric. and For. Meterol. 36: 7-18.

Green, H.L.; Lane, W.R. 1964. Particulate clouds, dusts, smokes, and mists. 2nd ed. E. & F.N. Spon, London, England. 471 p.

Haliburton, W.; Hopewell, W.W.; Yule, W.N. 1975. Deposit assessment, chemical insecticides. Pages 59-67 *in* M.L. Prebble (Ed.), Aerial Control of Forest Insects in Canada. Dept. Environ., Ottawa, Ont.

Inculet, I.I.; Hodgson, K.J.; Millward, J.G. 1983. Cross current dual airfoil electrostatic spray nozzle. Pages 1050-1054 *in* Proc. IEEE-IAS Ann. Meeting, Mexico City.

Joyce, R.J.; Schaefer, G.W.; Allsopp, K. 1981. Distribution of spray and assessment of larval mortality at Annabaglish. Pages 15-46 *in* A.V. Holden and D.B. Bevan (Eds.), Aerial Application of Insecticide against Pine Beauty Moth. For. Comm., U.K.

Kristmanson, D.D.; Picot, J.J.C.; van Vliet, S.; Henderson, G.W. 1988. Measuring foliar deposits from aerial spraying of insecticides. Pages 269-273 *in* G.W. Green (Ed.), Proc. Symp. Aerial Application of Pesticides in Forestry AFA-TN-18. Natl. Res. Counc., Ottawa.

Lambert, M. 1988. Quantification of spray deposit in experimental and operational aerial spraying operations. Pages 125-130 *in* G.W. Green (Ed.), Proc. Symp. Aerial Application of Pesticides in Forestry AFA-TN-18. Natl. Res. Counc., Ottawa, Ont.

Mason, B.J. 1971. The physics of clouds. 2nd ed. Clarendon Press, Oxford, Eng. 671 p.

May, K.R.; Clifford, R. 1967. The impaction of aerosol particles on cylinders, spheres, ribbons and discs. Ann. Occup. Hyg. 10: 83-95.

McNair, H.M.; Bonelli, E.J. 1969. Basic gas chromatography. 5th ed. Varian Aerograph, Walnut Creek, Calif. 306 p.

Mickle, R.E. 1988. A review of models for ULV spraying scenarios. Pages 179-188 *in* G.W. Green (Ed.), Proc. Symp. Aerial Application of Pesticides in Forestry AFA-TN-18. Natl. Res. Counc., Ottawa, Ont.

Neiburger, M.; Edinger, J.G.; Bonner W.D. 1973. Understanding our atmospheric environment. W.H. Freeman and Co., San Francisco. 293 p.

Oliver, H.R. 1971. Wind profiles in and above a forest canopy. Quart. J.R. Met. Soc. 97: 548-553.

Pasquill, F. 1974. Atmospheric diffusion: the dispersion of windborne material from industrial and other sources. 2nd ed. Ellis Horwood Ltd., Chichester, England. 429 p.

Payne, N.J. 1988. Canopy penetration and deposition of small droplets. Pages 95-101 *in* G.W. Green (Ed.), Proc. Symp. Aerial Application of Pesticides in Forestry AFA-TN-18. Natl. Res. Counc., Ottawa, Ont.

Payne, N.; Feng, J.; Reynolds, P. 1987. Off-target deposit measurements and buffer zones required around water for various aerial applications of glyphosate. Can. For. Serv., For. Pest Manage. Inst., Sault Ste. Marie, Ont. Inf. Rep. FPM-X-80, 23 p.

Payne, N.; Helson, B.; Sundaram, K.; Fleming, R. 1988. Estimating buffer zones for pesticide applications. Pestic. Sci. 24: 147-161.

Payne, N.; Helson, B.; Sundaram, K.; Kingsbury, P.; Fleming, R.; de Groot, P. 1986. Estimating the buffer required around water during permethrin applications. Can. For. Serv., For. Pest Manage. Inst., Sault Ste. Marie, Ont. Inf. Rep. FPM-X-70, 26 p.

Pelletier, M. 1985. Étude du vortex d'un Piper Pawnee. Québec Ministère de l'Énergie et des Ressources, Quebec City. 40 p.

Picot, J.J.C.; Chitrangad, B.; Henderson, G. 1981. Evaporation rate correlation for atomized droplets. Trans. ASAE 24: 552-554.

Picot, J.J.C.; Kristmanson, D.D.; Basak-Brown, N. 1986. Canopy deposit and off-target drift in forestry aerial spraying: the effects of operational parameters. Trans. ASAE 29: 90-96.

Prebble, M.L. 1975. Spruce budworm: introduction. Pages 77-84 *in* M.L. Prebble (Ed.), Aerial Control of Forest Insects in Canada. Dept. Environ., Ottawa, Ont.

Quantick, H.R. 1985. Aviation in crop protection, pollution and insect control. Collins, London, England. 428 p.

Randall, A.P. 1980. A simple device for collecting aerial spray deposits from calibration trials and spray operations. Can. For. Serv., Ottawa, Ont. Bi-mon. Res. Notes 36: 23.

Reid, J.D.; Crabbe, R.S. 1980. Two models of long-range drift of forest pesticide aerial spray. Atmos. Environ. 14: 1017-1025.

Riley, C.W.; Wiesner, C.J.; Ecobichan, D.J. 1989. Measurement of aminocarb in long-distance drift following aerial application to forests. Bull. Environ. Contam. Toxicol. 42: 37-44.

Rogers, R.R. 1979. A short course in cloud physics. 2nd ed. Pergamon Press, Oxford, England. 235 p.

Sharp, R.B. 1974. Spray deposit measurement by fluorescence. Pestic. Sci. 5: 197-209.

Slack, W.E. 1972. The NAE flying spot scanner/analyzer. DME/NAE Q. Bull. No. 1972(3). Natl. Res. Counc., Ottawa, Ont.

Smith, F.B.; Carson, D.J.; Oliver, H.R. 1972. Mean wind-direction shear through a forest canopy. Boundary-Layer Meteorol. 3: 178-190.

Sumitomo Chemical Company Ltd. 1980, Sumithion Technical Manual. 25 p.

Sundaram, A. 1985. A gravimetric method for determining the relative volatilities of non-aqueous pesticide formulations and spray diluents. Pestic. Sci. 16: 397-403.

Sundaram, A.; Leung, J.W. 1986. A simple method to determine relative volatilities of aqueous formulations of pesticides. J. Environ. Sci. Health B21: 165-190.

Sundaram, K.M.; Millikin, R.; Sundaram, A. 1989. Assessment of canopy and ground deposits of fenitrothion following aerial and ground application in a northern Ontario forest. Pestic. Sci. 25: 59-69.

Sundaram, K.M.; Sundaram, A.; Nott, R. 1986. Mexacarbate deposits on simulated and live fir needles during an aerial spray trial. Trans. ASAE 29: 382-388 and 392.

Tennankore, K.; Picot, J.J.C.; Chitrangad, B.; Kristmanson, D.D. 1980. Aircraft vortex studies in forest aerial spraying. Trans. ASAE 23: 1076-1079 and 1083.

Thom, A.S. 1975. Momentum, mass and heat exchange. Pages 57-109 *in* J.L. Monteith (Ed.), Vegetation and the atmosphere. Vol. 1. Academic Press, London, England.

Uk, S. 1987. Distribution patterns of aerially applied ULV sprays by aircraft over and within the cotton canopy in the Sudan Gezira. Crop Protection 6: 43-48.

van Vliet, S.J. 1987. Aerial spraying of forests: measurements of ground and canopy deposits. B.Sc. Thesis. Univ. New Brunswick, Fredericton, N.B.

Wickens, R.H. 1980. A stream tube concept for lift: with reference to the maximum size and configuration of aerial spray emissions. Can. Aeronaut. Space J. 26: 134-143.

Wiesner, C.J.; Kettela, E.G.; Thomas, A.W.; Riley, C.M. 1988. Field validation of a foliar deposit bioassay for *Bacillus thuringienis*. Pages 329-331 *in* G.W. Green (Ed.), Proc. Symp. Aerial Application of Pesticides in Forestry AFA-TN-18. Natl. Res. Counc., Ottawa, Ont.

Wodageneh, A.; Matthews, G.A. 1981. The addition of oil to pesticide sprays—effect on droplet size. Tropical Pest Manage. 27: 121-124.

Yasuda, N. 1988. Turbulent diffusivity and diurnal variations in the atmospheric boundary layer. Boundary-Layer Meteorol. 43: 209-221.

Chapter 48

Design of Experimental Efficacy Trials in Canadian Forests

B.L. Cadogan and P. de Groot

Introduction

Insecticides were aerially applied to sectors of Canada's forests for the first time in 1927 (Randall 1975). Since then, efforts in eastern Canada have centered on control of forest insect defoliators primarily by developing new efficacious insecticides and by improving application techniques. Field testing is an integral part of the development and registration of insecticides and field trials ultimately confirm or deny the adequacy of the insecticide or control strategy. Ideally, two stages of field trials are required to evaluate efficacy: (1) field experiments, where the effectiveness of insecticides are compared and/or various combinations of dosage (concentration and application rate), timing, and frequency of application are tested, and (2) the pilot test, where the "best" treatments are applied in an operational manner to confirm their performance and assess the extent that operational factors might affect their efficacies. Unfortunately, the latter step has often been ignored or changed by compromises to make a field experiment quasi-operational with inadequate design and implementation. Nevertheless, the principal objective of an insecticide field trial is to test the effectiveness of potentially useful products and strategies against pest insect populations. For a forestry efficacy field trial to achieve its objectives it must be planned, conducted, and evaluated scientifically so that the results will be conclusive and acceptable for a variety of locations and conditions.

In addition to planning and design that are statistically sound, a field trial requires a thorough knowledge of the insect pest; the forest as a dynamic system; the pesticide and its application; and the required logistics in terms of personnel, supplies, and financial costs.

The Insect Pest

It is important for the personnel involved in the spray to have an exact knowledge of the biology, behavior, and epidemiology of the pest species. This knowledge is instrumental in determining when the pest is most susceptible to sprays and what population levels are optimum for testing.

The Forest

Broadly defined there are many classes of forests, but within each class there are two types: a natural forest with tremendous inter- and intra-species heterogeneity, and a managed forest (i.e. one that was nurtured silviculturally where there is more homogeneity in tree age, height, crown structure, and spacing, and significantly less species diversity than in a natural forest). An appreciation of the dynamics of these forests is mandatory because the forest's topography, age, vigor, species composition, and so forth significantly influence the experimental design of the trial.

The Pesticide and Its Application

The efficacy of a pesticide is dependent not only on its inherent properties (e.g. toxicity, sprayability, persistence), but also on how efficiently it is applied. Thus, a thorough knowledge of the pesticide's strengths and weaknesses and detailed attention to its application in the planning and implementation stages are obligatory in a field trial.

Required Logistics

An aerial field trial is an extensive undertaking that requires considerable human, material, and financial resources. They must be carefully planned and used judiciously. It is important therefore to acknowledge at every stage, that the logistical support (e.g. aircraft acquisition, pesticide analyses, insect sampling) will determine whether a trial can be conducted, to what extent, and how well it is done.

Planning

The Experimental Design

The experimental design is the blueprint of the experimental field trial, dictating the terms of how the trial will be conducted. Yet a design must not be regarded merely as a theoretical concept because it affects and is affected by the practical aspects of the trial. A good design confers the integrity that is necessary for a valid operation.

Preparing an experimental design for a field efficacy trial in forests presents special difficulties: forests are vastly different from agricultural crops; what is practically feasible takes precedence over what might be recommended as ideal or desirable; and because of the many difficulties and hazards in treating forests, especially if aircraft are used, safety influences or limits all experimental plans in forests much more than in agricultural settings. Therefore, modern experimental designs, largely developed (by R. A. Fisher and his school) for agricultural field crops, cannot be applied directly to forest situations. In Fisher's concept of the randomized block design (Fisher 1960), treatments are allocated independently, and at random, to plots within replicated blocks (Fig. 1), where each block is more or less a homogeneous entity. This situation rarely, if ever, exists in forests, and together with other practical constraints, usually makes a "true" randomized block design infeasible. Unbiased replication of the treatment units can be accomplished in other ways (e.g. by random allocation to a predetermined number of plots in a generally infested area, without reference to blocks) (Cadogan 1986). A modification of this completely randomized arrangement, which usually is advantageous, is to stratify the plots (i.e. to apply a series of treatments in areas differing in host type, insect population levels, etc.) so that these areas or blocks are not true replicates, but differ in a known way. This provides a selective spectrum of conditions for the trial, and the effect of the different conditions can be accounted for and assessed in the statistical analyses of the data. Latin square, split plot, or other conventional designs are generally impractical or inappropriate for aerial spraying experiments.

Furthermore, during aerial applications there is the potential that aerosols and vapor will probably drift farther than they would during ground applications. Large buffer zones could be required to eliminate interplot contamination. Consequently, inordinately large areas would be required to accommodate a conventional experimental design. However, there are forest-related situations (e.g. nurseries and seed orchards) where ground equipment is used and conventional agricultural designs can be utilized.

The number of replicates that can be incorporated into a valid efficacy trial is determined not only by the statistical obligation to control the size of the experimental error, but also by practical considerations. We have found that in general three replications per treatment will suffice. It is only in exceptional cases that a larger number is possible, and the number of replications n for a given significance level t can be estimated approximately using the equation $n = 2s^2t^2$, where s is the standard deviation.

There may be situations (e.g. using a single treatment), where a randomized block design might be feasible in Canadian forests; however, it is evident that experimental designs for aerial field trials in forests will be dictated by the realities of the forest setting and not vice versa. Nevertheless, it is necessary to always acknowledge that an exact field trial must include untreated check blocks, test compound or treatment blocks and, if possible, standard or so-called positive check blocks as components. The untreated check is used to document natural uninfluenced changes in the pest population over the course of the trial; and the standard treatment is used to compare the test treatments with a known treatment.

Plot Size and Shape

In most trials in forests, the size and shape of the treatment units are dependent on what the forest resources allow, and what type of spray application is to be employed. Regardless of how they are to be replicated (Figs. 1 and 2), the plots should conform to a shape and size that can accommodate the spray and sampling methodologies.

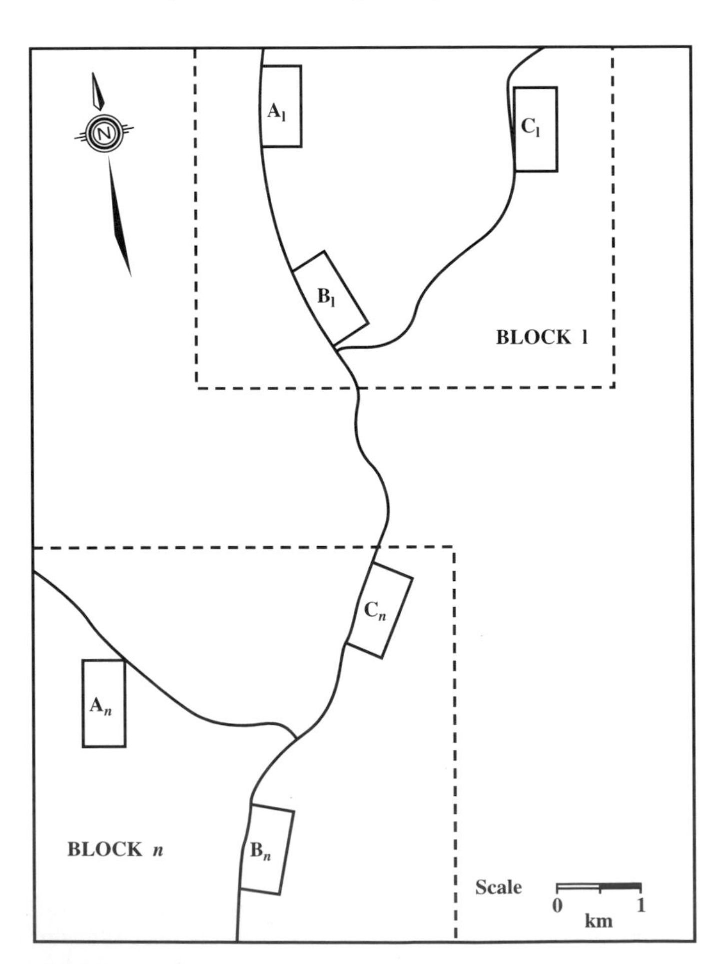

Figure 1. An ideal concept of a randomized block design with three treatments, A, B, and C, where n is the number of replicates. This design requires large areas of forests.

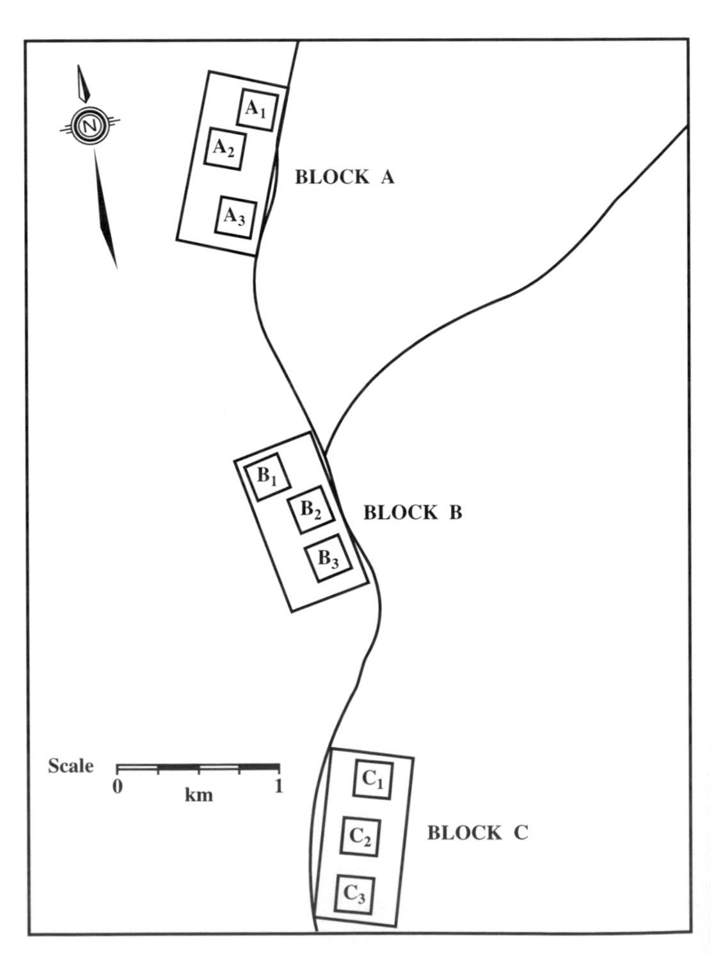

Figure 2. A compromise design that requires relatively small areas. A, B, and C are treatment areas, each with n=3 randomized plots.

In our research we use a rule of thumb which dictates that the minimum plot size is the smallest area that can be sprayed and sampled efficiently and the maximum size is the largest one that can be sampled adequately (i.e. without introducing unacceptable levels of experimental error). Consequently, in our aerial trials, plots range from 20 to 100 ha (1.0 × 0.2 km to 2.0 × 0.5 km). In ground trials, the size and shape of experimental areas is largely determined by the physical nature of the areas and to what extent the spray equipment can enter and maneuver within them. In general most ground plots range from 0.01 to 0.50 ha.

Whenever possible, individual spray units should be rectangular, as this shape best facilitates efficient spraying. This does not imply that other shapes are invalid. If a small, fixed-wing aircraft is to be used to apply the spray, we would suggest that the minimum length of the spray rectangle be 1.0 km, as this distance allows enough time (22.5 seconds) with the aircraft flying at approximately 160 km/hr for the atomizers to stabilize and spray optimally.

Implementation

Execution of an aerial field trial encompasses three general areas of work: selecting and preparing the experimental areas; applying the pesticide; and evaluating the spray and biological (insect and host tree) components of the trial.

Implementation is labor intensive and requires trained, highly conscientious personnel who have a rigorously keen eye for detail and accuracy, for it is the implementation that will ultimately decide if the results from the trial are credible or not.

Plot Selection

Plots must be located and prepared, and generally this is initiated with a preconceived ideal design. However, circumstances might force compromises that result in a practical design, provided the compromises do not affect the integrity of the field trial. The research location of an efficacy trial, besides being constrained by biotic and abiotic requirements, must also satisfy certain guidelines that are prescribed by pesticide regulatory bodies. The location should, among other things, be isolated from permanent human habitation; avoid proximity to productive water bodies and other environmentally sensitive areas; be within reasonable ferrying distance from an airport if fixed-wing aircraft are to be used; and be within reasonable travel time from living quarters of the field personnel. Potentially suitable areas in Canada are best located with the assistance of Forestry Canada's Forest Insect and Disease Survey and provincial forestry personnel. The former provides information on pest status throughout Canada, the latter expertise on regional and district forest resources and facilities. The suitability of these proposed areas is confirmed by aerial and ground reconnaissances, forest resource inventory cruisings, and pest population assessments.

Plot Preparation

Preparation of the plots entails simple surveying of the dimensions agreed upon and defining the boundaries (Cadogan 1987) so that the pesticide can be targeted precisely to the area. Easy access into and throughout a plot should be made possible by removing undergrowth to make networks of discrete pathways. Sample trees should be chosen randomly from trees with similar characteristics (height, DBH, canopy form, etc.). Areas around these sample trees should be prepared to facilitate sampling (Maksymiuk 1963; Cadogan 1987). These procedures will vary according to the objectives and experimental design of the trial, as well as with such factors as forest type, pest species, available resources.

Applying the Pesticide

Spraying is the most influential practical aspect of the aerial or ground efficacy trial. Good application is necessary if the material is to achieve its optimum efficacy, and this requires careful attention to a number of parameters. First, the type and size of the aircraft or ground sprayer should be capable of functioning efficiently in the chosen situation; load capacity, spray requirements, mixing and loading facilities, and ferrying distances must be considered. Second, the atomization equipment, comprising atomizers, pumps, etc., should be calibrated to produce the desired spray spectrum at the precise flow rates. These rates are calculated using the ground sprayer or aircraft speed, its designated swath width, and the volume of the pesticide to be applied per unit area. Ideally, the calibration and spray characteristics should be done using the test pesticide, but if this is not practical, an acceptable alternative formulation that mimics the pesticide must be used. In any situation, the application equipment must be calibrated prior to the actual application.

There are two rules regarding the mixing and loading of pesticides that should be observed. First, knowledge of the pesticides is a prerequisite, as this assists in avoiding problems relating to compatibility or cross-contamination of materials. Second, all personnel should be aware of any special safety regulations or other requirements relating to the pesticide and these should be clearly followed.

Spraying should be conducted under circumstances that allow sprays to be timed and applied optimally. This demands that the application should be made when the target insect is at a developmental stage when it is most vulnerable toxicologically and most accessible ecologically. Sprays should also be applied under weather conditions that maximize spray deposition on the target insect or host plant. For a given aircraft type, there is no agreement among researchers as to which weather conditions create the best environment for efficient droplet deposit. It is generally agreed, however, that some turbulent energy, either from wind or from aircraft vortices, is required to distribute droplets efficiently. Our experience with light aircraft has shown that most sprays are deposited equally well in both stable and turbulent (unstable) weather conditions. Moderate windspeeds (3 to 12 km/hr), however, are an advantage in displacing incremental swaths and thus improving spray coverage.

Evaluation

The evaluation of a field trial is a complex multifaceted task, the aim of which is to collect data that allow the effectiveness of a product or technique to be assessed precisely and dependably (Unterstenhofer 1976). Reliable evaluation requires that appropriate criteria are used for assessments and that these criteria are used optimally; the sampling procedure is biologically and mensurationally sound (i.e. there is a minimum of measurement error); the sample size is adequate to provide estimates of satisfactory precision; and the method of analysis used is appropriate for the data obtained. The choice of "correct criteria" and using an adequate number of samples should be decided by the experimenter as part of the overall experimental design. However, it is generally accepted that too few samples can lead to erroneous conclusions (Payandeh and Beilhartz 1978) and that too many samples can be costly and logistically unmanageable, thus introducing unacceptable measurement errors.

In both aerial and ground field trials, the criteria used to evaluate efficacy in forests vary according to the circumstances surrounding the trial. Generally, three components—the spray, the insect population, and the yield of, or damage to, the host plant—are assessed.

Evaluating the Spray

The product can be effective only if it reaches the target. Therefore, the spray should be monitored and assessed to determine how much of the spray reached the targets, or how much spray the targets were exposed to and how effective it was.

Ideally, spray evaluations should be made at or in close proximity to the target, but this is not always practical in forest settings. In most cases, spray assessments can be made using either artificial substrates [Kromekote® cards and glass slides (Randall 1980; Cadogan and Zylstra 1984), Mylar film (Akesson and Cowden 1978), Millipore filters (Morris 1972), foliage mimics (Sundaram et al. 1986; Lambert 1987), etc.] placed wherever representative samples of the spray deposit can be made. Or they can be made using natural foliage (conifer needles, deciduous leaves, stems, etc.) (Uk 1977; Neisess 1978).

Spray effectiveness is dependent on spray coverage; therefore, it is necessary to determine the number of drops per unit area or the mass deposited as milligrams of pesticide per unit gram, volume, or area of artificial medium or foliage. Drop size is implicit in coverage because the number of drops that can be produced from a given volume is a function of the diameter of the drops (e.g. 1 000 drops each with a diameter of 10 µm can be produced from one 100-µm drop) (Joyce 1969); therefore, drop size distributions of the spray should be assessed.

Because they do not adequately simulate the intended spray target, artificial surfaces should not be expected to attract and receive drops the same as they would be collected on the target. Nevertheless, such substrates offer distinct practical advantages without compromising the status of the assessment. Although foliage could be considered the ideal medium on which spray deposit assessments should be made, it is not the most practical. Certain pesticides tend to be quickly incorporated into foliage, which makes all assessments very difficult. The cuticular waxes, stomata, and trichomes on leaves severely distort surface deposits, thus making the accurate sizing of droplets almost impossible.

Evaluating Pest Responses

Insect population suppression is probably the most widely used criterion employed to assess the efficacy of an insecticidal treatment. The two standard measurements of this criterion are the percent population reduction or percent insect mortality, and the surviving insect population density. The latter is of prime importance because it indicates the population's potential to resurge to intolerable levels. Unfortunately, there is no general consensus as to what levels of survivorship constitute effective control of a particular pest.

A wide variety of sampling methods, dictated by the trial's objectives, the biology and behavior of the pest, and the particular population parameters being estimated, have been used in field efficacy trials. Nevertheless, nearly all use the number of living or dead insects to calculate population changes. Because the data obtained are the basis for analysis and evaluation, the number of samples and samplings, their timing and the time it takes to complete a sampling must be carefully chosen if the resulting data are to be accurate and reliable. In agricultural trials both treated and control plots can be sampled on the same day. The nature of forest trials does not always permit this; therefore, sample evaluations should be carried over the shortest practical period because with some pest species, population density can change substantially within a short time.

Evaluating Host Plant Responses

Damage by the causal pest to the host plant and crop yield are the two criteria most often used to evaluate plant responses in a field trial. However, in most cases only plant damage is appropriate, primarily because yield could be affected by other factors (e.g. fertilizer, rainfall) not associated with the pesticide treatment.

Damage in most field trials is categorized either as damage inflicted (e.g. percent defoliation) or damage prevented (e.g. percent foliage saved); expressed numerically as percentages or scale values (i.e. a range of ratings), or both. Percentages are preferred because they are more objective than ratings, which present the risk of subjective bias.

For the evaluations to be credible, the sampling method must be appropriate and care should be taken to ascertain that the damage was indeed caused by the pest. Once the evaluations have been made with the right criteria, the data must be analyzed to answer the question of whether and to what extent the treatment was effective.

Data Analysis

Statistical analyses are necessary whenever the results are expressed as numerical values regardless of whether they involved counts, estimates, or scale ratings. When statistical analysis is used, the conditions specific to the experiment should be acknowledged. In this context it is very helpful, in the planning stages of the experiment, to consult with a biometrician (who is familiar with forest conditions) about methods of analysis that are appropriate and reliable.

Various methods are used to compute pesticide efficacy (i.e. that percentage of pest population reduction or reduction in crop damage that can be attributed to the treatment in comparison with the untreated check or a standard). The formula most commonly used with experimental treatments for pest population reduction is that proposed by Abbot (1925), but his procedure was intended for laboratory use where variables are easily controlled and both treated and untreated organisms are evaluated simultaneously. The formula can be modified for field data (Sun and Shepard 1947), however, if certain conditions are met. In most aerial forest trials all plots cannot be sampled on the same day. If this reality is ignored, as many forest researchers appear to do, using Abbot's formula without special provisions (Cadogan 1987) might be inappropriate because some insect populations change substantially over a short period.

Conclusions

The forests of Canada range from the mountains and foothills of the western Pacific rim to the Atlantic coast, encompassing extremely diverse climates, topography, and socioeconomic conditions. These conditions require that ground or aerial field trials should be flexible and adaptable.

The concept of standardizing field methods (Walton and Lewis 1982) is optimistically ideal, and although some criteria and methodology can be used uniformly in certain situations (e.g. the 45-cm branch tip as the sample unit to assess larval spruce budworm, *Choristoneura fumiferana*, populations), rigid standardization is impractical and undesirable. Standardization not only overlooks the inherent difficulties imposed by the diversity of Canada's forest conditions, it tends to suppress those innovative approaches that are the foundation of scientific progress.

References

Abbott, W.S. 1925. A method of computing the effectiveness of an insecticide. J. Econ. Entomol. 18: 265-267.

Akesson, N.; Cowden, R. 1978. Metallic salts as tracers for spray applications. Pages 107-112 *in* Barry, J.W., R.B. Ekblad, G.P. Markin and G.C. Trostle (compilers), Methods for sampling and assessing deposits of insecticidal sprays released over forests. USDA For. Serv., Washington, D.C. Tech. Bull. 1696. 161 p.

Cadogan, B.L. 1986. Relative field efficacies of Sumithion flowable 20% flowable and Sumithion technical formulations against spruce budworm *Choristoneura fumiferana* (Lepidoptera: Tortricidae). Can. Entomol. 118: 1143-1149.

Cadogan, B.L. 1987. Experimental aerial application of Matacil flowable insecticide to control spruce budworm (Lepidoptera: Tortricidae). Crop Protection 6(2): 130-135.

Cadogan, B.L.; Zylstra, B.F. 1984. An apparatus and technique for sampling aerially applied sprays in conifers. Pestic. Sci. 15: 417-423.

Fisher, R.A. 1960. The design of experiments. Oliver and Boyd. Edinburgh. 248 p.

Joyce, R.J.V. 1969. Operational frontiers. Pages 66-72 *in* Proc. 4th Int. Agric. Aviation Congr., Kingston. (1969). 622 p.

Lambert, M.C. 1987. Quantification of spray deposit in experimental and operational spraying operations. Pages 125-130 *in* G.W. Green (Ed.), Proc. Symp. on the aerial application of pesticides in Forestry. NRCC AFA-TN-18. Ottawa, 387 p.

Maksymiuk, B. 1963. Screening effect of the nearest tree on aerial spray deposits recovered at ground level. J. For. 61: 143-144.

Morris, O.N. 1972. Deposit assessment, biological insecticides. Pages 68-71 *in* M.L. Prebble (Ed.), Aerial control of forest insects in Canada. Dep. Environ. Ottawa, Ont. 330 p.

Neisess, J. 1978. Assessment of microbial spray deposits. Pages 99-100 *in* J.W. Barry, R.B. Ekblad, G.P. Markin, and G.C. Trostle (compilers), Methods for sampling and assessing deposits of insecticidal sprays released over forests. USDA For. Serv. Washington, D.C. Tech. Bull. 1596. 161 p.

Payandeh, B.; Beilhartz, D.W. 1978. Sample size estimation made easy. Environ. Canada, Can. For. Serv., Great Lakes Forestry Centre, Sault Ste. Marie, Inf. Rep. 0-X-275. 19 p.

Randall, A.P. 1975. Application technology. Pages 35-55 *in* M.L. Prebble (Ed.), Aerial control of forest insects in Canada. Dep. Environ. Ottawa, Ont. 330 p.

Randall, A.P. 1980. A simple device for collecting aerial spray deposits from calibration trials and spray operations. Can. For. Serv., Ottawa, Ont. Bi-month Res. Notes 36(5): 23.

Sun, Y.; Shephard, H.H. 1947. Methods of calculating and correcting the mortality of insects. J. Econ. Entomol. 40: 710-715.

Sundaram, K.M.S.; Sundaram, A.; Nott, R. 1986. Mexacarbate deposits on simulated and live fir needles during an aerial spray trial. Trans. ASAE 29: 382-388, 391.

Uk, S. 1977. Tracing insecticide spray droplets by sizes on natural surfaces. The state-of-the-art and its value. Pestic. Sci. 8: 501-509.

Unterstenhofer, G. 1976. The basic principles of crop protection field trials. Pflanzenschutz Nachrichten 29 (1976, 2): 83-180.

Walton, G.S.; Lewis, F.B. 1982. Spruce budworm core *B.t.* test—1980. Combined summary, CANUSA-USDA-FS Research Paper NE—506, Broomall, Pa. 9 p.

Chapter 49

Sampling Forestry Materials for Chemical Accountability Studies

K.M.S. SUNDARAM

Introduction

Chemicals have been used aerially to control forest insect pests since 1927 (Shea and Nigam 1984). The distribution, persistence, mobility, and fate of the chemicals in various components of Canadian forests at on-target and off-target sites have been studied extensively and documented to meet the registration requirements of new pesticides according to the Pest Control Products Act; to understand the environmental and associated ecological hazards; to improve the efficiency and effectiveness of pesticide use; and to address properly the provincial and public concerns about the use of these forest management tools (Sundaram et al. 1980; Sundaram and Szeto 1984; Sundaram and Nott 1985; Sundaram and Szeto 1987; Sundaram, A. et al. 1987).

The sampling protocol (methods, techniques, and procedures) used in collecting different substrates for pesticide residue analysis are well-established in agriculture, food, environment, and other settings (Lykken et al. 1957; Lykken 1963; ASTM 1973; Anon. 1974; Smith and James 1981; Kratochvil et al. 1984; Park and Pohland 1989; Kratochvil and Peak 1989). However, such sampling processes for the monitoring and analysis of pesticide residues in various forestry substrates are neither well-developed nor properly documented. Most of the sampling techniques used in forestry to date have been derived and modified from those used in agriculture. The myriad of problems encountered in the sampling of forestry substrates from spray application sites, including complex terrain and topography, vast and often inaccessible spray areas, uneven canopy cover, heterogeneity of matrices, and capricious deposition of pesticides over forests, have presented severe difficulties. These problems have necessitated the development of sample collection and handling methods that are uniquely suitable for forestry situations.

The reliability of any research measurement depends on sample quality and the sampling scheme used to obtain representative samples. Sampling methods in pesticide residue analysis are critical to the accuracy and validity of the data generated, especially when the matrix sampled is heterogeneous and contains only trace levels of the chemical. Uncertainties in the sampling process (e.g. site selection and preparation, sampling techniques used, frequency of sampling, size and number of samples), sample handling (compositing, packing, labeling, storing to minimize sample deterioration and transporting to a residue analysis laboratory) and pre-analytical treatment (e.g. homogenization, subsampling of composite samples, solvent extraction, concentration, removal of interferences), will ultimately affect the results. Poor techniques will lead to erroneous results and eventual invalid conclusions. According to Kratochvil et al. (1984), the following factors contribute to the overall uncertainty in analysis: sampling; subsampling and preparation; and analytical protocol and data evaluation. Youden (1967) and Horwitz (1979) considered sampling to be a major source of error in any analytical process: no amount of analytical expertise will result in reliable residue data if this major source of error is not eliminated or at least reduced to a minimum.

In forestry situations, whenever possible, random sampling techniques should be used to reduce sampling errors and remove potential bias in the sampling method. Increasing the size and number of samples and compositing or pooling samples will minimize variations within the matrix in order to obtain more representative sample lots for analysis. Similarly, errors due to sample preparation are decreased by further compositing and homogenizing the field samples to form a homogeneous laboratory sample and using aliquots (test samples) of it in the final residue analysis.

An overview of the sampling protocols used in forestry situations is presented in this chapter, followed by some specific procedures and guidelines used for the collection and processing of various indicator matrices from terrestrial and aquatic components of a forest environment. In addition, the pre-analytical treatments required to process samples for residue analysis are included. Such rigorous and standardized sampling procedures not only serve to generate reliable and meaningful residue data but are also useful in modeling pesticide movement in a forest ecosystem and in evaluating its fate and possible hazard.

The sampling protocols reported here have been developed primarily for insecticide applications; however, these may be applied to other types of pesticides with necessary modifications.

General Considerations

Kratochvil and Peak (1989) defined sampling as the collection from a defined population of a portion that is representative of the population as a whole with reference to the substance of interest. Therefore, the usefulness of any sampling procedure hinges upon the approach taken in obtaining a representative sample of the population under study as a whole. A good sampling program requires personnel who are adequately trained in sample collection and handling procedures and who follow a systematic written protocol that is statistically sound and devoid of any undue bias (Anon. 1983). The following points should be considered as necessary prerequisites to develop meaningful sampling protocols for pesticide analysis in forestry situations:

1. clearly stated objectives of the study, including the type of information required;
2. the experimental design and availability of trained personnel;
3. the type and physical state of the population to be sampled and its location, especially accessibility to spray site;

4. the nature of the substance to be measured, the methodology available for it, and the desired degree of precision and accuracy of the measurements;
5. time and cost constraints involved (both are directly related to accuracy; and
6. the size and number of samples (related to no. 1) and frequency of sampling required to meet the accuracy level.

Emphasis on the statistical considerations necessary for proper sampling (e.g. minimum increment, size, interval, number, replicates), theories covering the sampling process, analytical procedures, and computational details required for data evaluation from the standpoint of quality assurance are not given in this book. Since statistical and experimental designs are specific to the objectives and goals of each study, every effort should be made to obtain sufficient statistical input in the experimental design from practising statisticians and/or from the literature (Youden 1951; Sampford; 1962; Cochran 1963; Snedecor and Cochran 1967; Kratochvil and Taylor 1981; Anon. 1983). A concise, informative account on sampling for chemical analysis, including various theories, is given by Kratochvil et al. (1984) and Kratochvil and Peak (1989). The mathematics of planning, sample design, and environmental sampling are comprehensively discussed by Keith (1988). Also, a recent overview on sample planning, sampling definitions and the principles of sampling is provided by Garfield (1989).

Sampling Plans

The substrates of interest, which serve as indicators for pesticide exposure, are sampled for residue analysis after the aerial application of pesticides. The substrates are air, foliage, soil, litter, water, sediment, and aquatic organisms (animal and plant). Usually, an average residual profile is sufficient for monitoring pesticide residues in forestry situations, and this profile can be provided by a random sampling process accompanied by compositing. The distribution, persistence, dissipation, and fate of the released chemical in a forest environment are assessed from the average pesticide concentrations and their metabolites found in different types of matrices at zero time and at different post-spray intervals.

To ensure that the sample is taken in as random a fashion as possible, a plot of approximately 5 ha, which is fully exposed to the spray cloud, is chosen, preferably from the center of the spray block. The size and location of the plot depends somewhat on the type of substrate (i.e. aquatic or terrestrial) to be sampled. The plot is cleared of unwanted vegetation, ground cover, and so forth, and divided into segments. Samples of each population are collected randomly from these segments so that every unit of the entire plot has an equal chance of being incorporated into the selected sample. The individual samples or increments of each population are processed (if size is a problem), then pooled according to type to form three or four well-integrated composite samples. Although analysis of individual increments provides information on between-sample variability, compositing is done for considerations of time, and human and monetary resources.

The composite samples of each population are packed in suitable containers, labeled, stored at –20°C to prevent sample deterioration, and transported to the residue laboratory for analysis. In the laboratory, all samples are logged and assigned laboratory numbers with field record work sheets. After removing any foreign objects from the composites, laboratory samples of uniform composition are prepared by grinding, mixing, sieving, filtering, and so forth, according to each substrate type. Aliquots of test samples are taken from this homogeneous laboratory sample for residue analysis. The remainder are stored at –20°C for future use.

Pre-spray and control samples are collected and shipped separately to examine possible interferences due to coextractives and to assess recovery levels of spiked analyte. Laboratory blanks are analyzed with each batch of samples to verify that glassware, instruments, and reagents are contaminant-free.

On-site meteorological conditions (temperature, relative humidity, wind speed and direction, sunlight or cloud cover), aircraft parameters (e.g. aircraft type, speed, nozzle type, spray height, swath width, emission and volume rates, dosage) and formulation properties (e.g. tank mix composition, types of ingredients, viscosity, surface tension, volatility) considerably influence deposition patterns of chemicals at the canopy and on forest floor levels (Matthews 1979). Documented information on these are collected to interpret the deposition data correctly.

Sampling and Processing Terrestrial Components

Forest Soils

The forest floor is the major receptor of pesticides during aerial applications; however, these deposits are often not distributed uniformly, making representative sampling difficult. The major problem areas concerning soil residues are persistence, vertical leaching leading to groundwater contamination, and run-off with accompanying residue movement.

The method of sampling soil and the type of equipment used for soil sampling vary according to the objectives of the study. For persistence and vertical mobility studies, three soil plots (ca. 15 m^2) are randomly chosen in the open areas (ca. 5 ha) at the center of a spray block. The plots are prepared by removing the overlying litter, moss, and other organic detritus and exposing the soil layer to the spray cloud. All small objects such as fallen branches, twigs, roots, and stones, are removed (to eliminate possible bias in analysis) and the soil surface is leveled. The surrounding overstory and understory vegetation are cleared to prevent spray interception and to enhance uniform pesticide deposition on the soil surface. The soil plots are marked by corner posts surrounded by a strand of rope containing colored ribbons for easy identification.

To study persistence, samples are taken randomly, at pre-spray and at various intervals of time post-spray (starting from zero time, hourly, daily, and gradually decreasing the sample frequency), from the top 1 to 3 cm (depending on bulk density) of the soil layer as 2.5-cm diameter cores (20 cores per sample per plot) using a tube sampler. The 20 cores from each plot are pooled to form three composite samples, wrapped in aluminum foil, labeled (e.g. date, time, pesticide and dosage, site number, location, sample name, core size, number of cores), sealed in plastic bags with masking tape, stored in Styrofoam coolers equipped with "freezer packs," and transported to the field laboratory, where the samples are stored in a freezer at −20°C (to minimize chemical and microbial degradation) until analyzed.

For vertical mobility studies, a split-tube auger (9.2 cm in diameter, 50 cm long, with markings to indicate depth), consisting of two halves locked together with screw bolts, is used to obtain soil cores. The front half of the auger, which can be opened up lengthwise, is attached at the top by hinges to a closed solid metal base. The open end of the auger is driven into the soil to a depth of 20 or 30 cm by a sledge hammer. The auger is drawn out gently by twisting and pulling. It is then laid flat on a plastic sheet, disassembled, and the soil core resting on the basal half is sliced with a knife into 5-cm-long segments, yielding four or six segments per core. Three soil cores are taken randomly from each plot and the three soil segments from the same depth are pooled to form one composite sample. Each composited 5-cm segment from three separate soil plots (four or six segments per plot) is processed as described above.

To avoid sample contamination, samplers and other tools are thoroughly cleaned after sampling each plot by scrubbing with a clean stiff brush and water, rinsed with acetone or methanol, air-dried and stored in plastic bags. Similarly, vinyl gloves used to handle soils are discarded or washed with methanol and air dried between samples.

If the pesticide is not persistent and residue degradation or soil adsorption leading to bound residues are possibilities, the pesticide should be monitored by fortifying pre-spray soil samples and treating them with the same schedule (freezing, shipping, and storage) as the actual field samples. It is preferable to analyze soil samples as soon as possible rather than storing them for a prolonged period.

Processing of soil samples before analysis is done by first thawing each composite soil. Stones, roots, organic detritus, and so forth, are then removed and the sample is mixed well with a metal spatula. Each soil is blended in a Hobart food chopper for about 8 minutes and passed through a 10-mesh sieve (2-mm openings). Aliquots of test portions from this laboratory sample are used for moisture determination (AOAC 1955); particle size composition; and residue analysis by solvent extraction (serial or sequential), filtration, liquid–liquid partition, cleanup, concentration, eventual quantification and data evaluation.

Forest Litter

Litter samples are collected either for monitoring studies or for evaluating persistence and fate of pesticides. The sampling design for monitoring is similar to the one described for soil. About 20 cores (diameter 5 cm, depth 4 cm), free from fallen branches, twigs, roots, stones, and so forth, are taken randomly in the central area (ca. 5 ha) of the spray block either by using a tube sampler marked at 4 cm or by driving a metal frame (5 cm × 4 cm × 4 cm depth) into the ground and removing the contents with a trowel. The litter cores are then pooled to form three composite samples. A comparison of the two devices (tube sampler versus rectangular metal frame—a device developed at the Forest Pest Management Institute of Forestry Canada)—indicated that both are reliable and produce consistent results, but use of the tube sampler increases the time required for sample collection especially when the litter is moist.

For long-term persistence and fate studies, three fully exposed litter plots (ca. 20 m^2) are established in the central area (5 ha) as described for soil. All small objects such as fallen branches, twigs, small stones, roots and other debris are cleared from the sampling areas to minimize bias in the final analysis. Litter is sampled with a metal frame collector (15.5 × 15.5 × 4.0 cm) at the same frequency as soil at a 1 cm or 4 cm depth (with a surface area of 240 cm^2) depending on the composition, compaction (porosity/density), and depth of the litter. About three to five cores per plot are taken randomly and composited to form a single sample per plot. The tube sampler may also be used for litter sampling but use of the metal frame reduces the time required for sample collection.

Precautions should be taken, as described earlier, to prevent sample contamination. Procedures for packing, handling, storage, transportation to the residue laboratory, and pre-analytical sample preparation are as described for soil.

Foliage

The procedures used to collect conifer (fir, *Abies* sp.; spruce, *Picea* sp.; and pine, *Pinus* sp.) and deciduous (oak, *Quercus* sp.; maple, *Acer* sp.; and birch, *Betula* sp.) foliage samples at desired pre- and post-spray time intervals to study persistence and fate, or at a specific time to monitor residues, are similar. The general sampling plan to study persistence and fate is to select about 12 dominant trees (coniferous and/or deciduous, ca. 15 m high and ca. 15 cm DBH) with fully developed crowns, ample growing space and exposure to sunlight in the central area (covering ca. 5 ha) of the spray block, before spray application, and mark them with colored ribbons for ready identification. Ground vegetation and trees neighboring each sample tree are removed, up to a radius of 5 m, to enhance their exposure to the spray cloud. At each sampling, one 20- to 30-cm-long branch tip is taken at midcrown level from each quadrant of the tree using a pole pruner. If vertical distribution of the chemical in the tree crown is one of the objectives of the study, foliage samples can be similarly collected from top and bottom crown positions as well. The 16 branch tips corresponding to four nearby trees are pooled after removing the shoots (to minimize dilution effects due to growth and associated error in subsequent residue analysis) and three composite samples from the 12 sample trees are obtained. The samples are kept in plastic bags and promptly stored in Styrofoam coolers kept at 0°C and transported to the field laboratory. In the field laboratory, mature foliage or needles are carefully clipped from the branch tips of each composite sample, mixed thoroughly and stored at −20°C in sealed, labeled plastic bags until they are analyzed.

For pesticide monitoring, a similar sampling procedure is used, except that the site preparation and clearing of surrounding understory and overstory vegetation are not required. The sampling trees selected should be in the open and fully exposed to sunlight. If distribution variability rather than average concentration of the chemical is the primary objective in monitoring, then the foliage samples from each tree could be analyzed separately instead of compositing.

Storing foliage in plastic bags may not be ideal due to possible adsorption of surface deposits onto the walls of the bag. Periodic analysis of the bags after proper solvent rinsing and concentration indicated that the error contributed by adsorption is negligible, provided that the analyte was lipophilic. Ideal storage containers are Teflon bags and bottles, but they are expensive.

Sample processing and laboratory subsampling for analysis are done as described for soil except that the Hobart-blended foliage does not require sifting through a 2.0-mm mesh sieve. Also, adequate care, as described earlier, should be taken to avoid sample contamination.

Air

In forestry spraying, airborne pesticides usually exist in aerosol (particulate) and vapor phases. Extraction devices fitted with glass or Teflon fiber filters can be used to sample airborne pesticides. A number of air sampling devices have been reported in the literature (Sherma 1979; Mallet and Cassista 1984). In forestry situations, portable gas-washing bottles (250 ml) (impingers) fitted with fritted glass inlets containing dimethyl formamide, ethylene glycol, or toluene (120 ml) as the trapping medium are practical for sampling total airborne pesticides (Sundaram 1984). Of the three trapping media used, toluene was found to be acceptable for both vapor and aerosol phases, because of its lower boiling point (110.6°C compared with 153°C for dimethyl formamide and 197°C for ethylene glycol) and ready flash-evaporation without much degradative loss of pesticides. Because of the volatility of toluene, a second absorption bottle was used in series to minimize the re-entry of the chemical to the gas stream, and the flow rate of air was also adjusted to a low level of 0.5 L/min. Usually, the gas-washing bottles are covered with aluminum foil to prevent possible degradation of pesticides by sunlight.

The usual air sampling procedure requires 10 random sampling sites (more sites are advisable, if resources permit, due to high variability observed in relative amounts) in a 5-ha plot in the center of the spray block. At each sampling site, air is sucked by battery-operated pumps for an hour at a constant speed (0.5 L/min) into glass impingers containing toluene as the trapping medium. The flow rate of each air sampler is carefully calibrated before use to determine exactly the volume of air sampled during the 60-minute period. The sampler and the collecting reagent are pre-tested in the laboratory, using the intended chemical in vapor phase, to determine sampling efficiency and retention at the planned flow rate to be used in the field. The toluene trapping media in the impingers and the washings of sampling train, glass wool, and so forth, are transferred quantitatively and pooled to get five composite samples, which are stored in clean screw-top (Teflon or foil-lined) amber-colored glass containers, and transported to the field laboratory at 0°C and refrigerated immediately.

In the residue laboratory, the composite samples of toluene are passed through a column of Na_2SO_4, flash-evaporated gently at low pressure to a small volume, cleaned by column adsorption chromatography (if necessary), eluted, and concentrated before analysis.

Spray Deposition

Various types of collectors are used to evaluate spray deposits (Randall 1980). Collectors made of aluminum (10 × 10 cm), Mylar sheets (10 × 10 cm) and glass plates (10 × 10 cm or two 5.0 × 7.5 cm plates fastened together by Tuck Tape and mounted on folding aluminum sheets) are popular. Petri dishes (Pyrex glass, 10 cm in diameter) containing distilled water (to simulate water surface) or binder-free glass fiber filters (Gelman A/E-GFF No. 61633, 10.2 cm diameter or equivalent with high water absorbency), snugly fitted to the petri dish bottom (to simulate the forest floor) to absorb large droplets (encountered while applying high volume rates) and to prevent loss through splashing and run-off due to droplet coalescence, are also used. The use of metal rods, aluminum coils and leaves to simulate natural conifer and deciduous foliage have been used to assess deposits as reported in previous studies (Sundaram, K.M.S. et al. 1986b, 1987, 1989; Sundaram 1987c; Sundaram and Sundaram 1988; Raske et al. 1989). A number of deposit collectors are placed randomly around the central area (ca. 5 ha) of the spray block at a height of 15 cm above the ground level one-half hour before spray application, and removed one hour after treatment. More details on the placement and collection of deposit collectors are given elsewhere (Sundaram et al. 1985a,b). It usually takes about 1 hour for the spray cloud to settle. A longer waiting period should be avoided to reduce the risk of loss of the chemical through volatilization. Care should be taken to ensure that the ground vegetation and/or over-hanging foliage does not obscure the surface of the sampling units during spray application.

Glass plates, Mylar sheets, aluminum plates, and metal rods are repeatedly rinsed at the site itself with suitable solvents. Like samples are pooled to form five or six composite samples (the residues deposited on individual collectors will be too low to quantify accurately) to provide distribution variability and average deposition within the plot. They are stored in sealed (Teflon or foil-lined), labeled, amber-colored bottles at 0°C and transported to the field laboratory and kept frozen until analysis. All artificial foliage samplers and glass-fiber filters (including solvent rinses of the petri dishes) are individually transferred to Teflon bottles (1-L capacity) and repeatedly extracted with suitable solvents. Like samples are pooled to form composite samples and each is handled as described above. Glassware used for sampling, sample processing, and storage should be monitored to determine if any adsorption of pesticides (such as synthetic pyrethroids) has occurred.

In the laboratory, the solvent rinses are filtered through anhydrous Na_2SO_4 and the volume is adjusted to about 2 ml by flash-evaporation before cleanup and analysis.

Sampling and Processing Aquatic Components

Water

Stream and pond waters from forest areas are sampled for monitoring and persistence studies. An essential consideration in a forestry situation is the selection of suitable water-sampling stations that are easily accessible. Sampling method (grab or continuous), frequency of sampling, and number of samples required to minimize variability in population (e.g. pesticide concentration varies with amount of input, runoff, dilution effect, degradation, adsorption) arising from fluctuations (e.g. stream discharge and water velocity, pond and stream depth, width, turbidity, temperature, alkalinity, rainfall) during the sampling period should be determined by the objectives of the study. Generally, the use of plastic (adsorption of pyrethroids and possible breakage in shipment and during freezing) bottles is discouraged over more expensive but durable Teflon bottles. Prolonged storage of water, even at −20°C, is not recommended because of hydrolysis and gradual adsorption of pesticides onto suspended matter.

To collect water samples from a shallow stream, select three to four sampling sites (increasing distances downstream if mobility is one of the objectives) where the water flow is uniform and has good inherent mixing. Each site is identified by colored ribbons attached to trees on the stream bank. At each site, four equidistant sampling stations (two near the bank; one on each side and two in midstream) are established. From each station, water is sampled (grab sample) by dipping a clean wide-mouthed, open, 1-L Teflon bottle just about 1 cm beneath the water surface and allowing the water to flow in until about 40% of the bottle is filled. Care should be taken to avoid skimming of the surface or stirring up the bottom sediment while sampling. Two of the four samples collected at each site are pooled (i.e. one sample near the bank and the other one collected in the middle) to form a composite sample. Similarly, the other two samples are pooled to provide two composite samples per site. The bottle containing each composite sample is closed tightly with a Teflon screw cap, labeled, and stored at 0°C in a cooler. The addition of about 50 ml of pesticide-grade hexane or dichloromethane to each water sample is recommended to stabilize the chemical if prolonged storage is a possibility.

Water samples from ponds in the spray area are collected from six randomly selected sites covering the entire pond area (periphery and center). The samples (one near the bank and the other collected in center) are pooled to form three composite subsamples per pond and brought to the field laboratory at 0°C. All samples (stream and pond) are pH adjusted (pH < 7), if required, with H_2SO_4 (e.g. carbamates), and then kept frozen at -20°C until analysis. The grab sampling technique is also used to collect discrete water samples at different depths of stream and pond water columns, to study the vertical distribution of pesticides, by slowly lowering tightly closed bottles and opening them at the appropriate depth.

Surface water samples are collected by gently touching a thin glass plate (10 × 10 cm) to the surface (pond or stream) layer with the aid of tongs. The plate is held on the surface for only a few seconds and then the adsorbed deposits are immediately rinsed into an amber-colored bottle using 2 × 25 ml of ethyl acetate (or any other suitable solvent). Recent studies have shown that the use of Gelman A/E glass fiber filters (No. 61638) cut into 10-cm squares serve the purpose equally well.

Where practicable, water samples are solvent-extracted, dried (Na_2SO_4), concentrated, and then stored at −20°C. This procedure will considerably minimize the hydrolytic loss of a chemical during storage. The Sep-Pak cartridges (Wolkoff and Creed 1981) and ion-exchange resins (XAD-2) (Mallet et al. 1978) for collection and concentration of low-level pesticide constituents have been tried under field conditions, but the conventional solvent-extraction techniques are equally suitable and give reliable results for compound-specific sampling of pesticides in natural waters.

In the laboratory, the pH of the composited water sample is adjusted to 7.0 by adding drops of $Na_2CO_{3(aq)}$ mixed with 100 ml of 20% $NaCl_{(aq)}$. Aliquots of the sample are partitioned serially or sequentially with the appropriate organic solvent (e.g. CH_2Cl_2, $C_6H_5CH_3$, C_6H1_4, $CHCl_3$, CCl_4). The pooled organic phase is dried through a column of anhydrous Na_2SO_4 and flash-evaporated to a small volume. The crude extract is cleaned by column adsorption chromatography, eluted, concentrated, and finally analyzed.

Sediment

Accurate sampling of sediment from a forest stream bed is complicated because the concentration of pesticide varies with particle size composition (the smaller the particle size, the higher the adsorption) and organic matter (the higher the organic matter, the higher the uptake) (Eidt et al. 1984). Because of the dynamics of a stream environment, the particle size composition varies from location to location; consequently, the concentration of pesticide also varies, and collection of a representative sediment sample from a streambed is often difficult. A variety of core and dredge samplers suitable for fine, compacted sediments have been reported (Ford et al. 1975; Kratochvil et al. 1984). Scooping sediments using collection bottles, as described below, was found to be adequate, especially when the samples encountered are coarse, noncohesive, and shallow.

Usually the sediment samples from streams and ponds are collected from the same vicinity and with the same sampling schedule as the water samples. Because of the expected variability in residue concentration, a larger number of individual samples (six to nine scoops of sediment each weighing about 100 g, drained wet weight), are taken using clean, wide-mouthed amber-colored glass bottles fitted with Teflon-lined screw-caps. At each sampling station, the bottle is gently lowered to the bottom, the lid is unscrewed, and one single scoop of sediment of about 1 cm thickness is taken by moving the bottle around. The bottle is tightly sealed, brought to the surface, and decanted gently to remove all the inflowed water. All six to nine sediment scoops collected from the six stations are pooled after removing all the debris to form three composite samples. The samples are labeled, sealed, and brought to the field laboratory in coolers at 0°C and kept frozen until analysis.

In the laboratory, each composite sediment sample is allowed to thaw and organic debris, stones, roots, twigs, and so forth are removed. The sample is then filtered under suction to remove excess water and mixed thoroughly using a metal spatula. Aliquots of laboratory samples are taken for extraction and analysis using a procedure identical to that for soil samples.

Fish

A problem usually encountered when collecting fish samples for residue analysis is the selection of a standard indicator species. No one species can represent all fishes and the species vary according to environment. Accumulation of pesticides by fish depends upon species, water concentration, and food type and supply. Therefore, collection of representative pre- and post-treatment fish samples requires careful planning.

Brook trout, *Salvelinus fontinalis*; Atlantic salmon, *Salmo salar*; rainbow trout, *Salmo gairdneri*; fathead minnow, *Pimephales promelas*; and creek chub, *Semotilus atromaculatus* are some of the predominant fish species in the streams of eastern Canada. Three to five uniform-sized fish (mean weight about 10 g; mean length about 8 cm) (large or old fish are not representative samples for monitoring purposes), contributing to a composite sample per sampling station, are collected in the same area as the water and sediment samples either by hook and line or by electroshocking. Each composite sample is wrapped in aluminum foil, packed in a labeled polyethylene bag, and chilled immediately before transportation to the field laboratory for storage at –20°C until analysis. Fish samples are collected before the pesticide application to provide an adequate background basis against which exposed samples can be compared.

During analysis, the whole fish in a composite sample is chopped into small pieces using a sharp knife and mixed thoroughly. Aliquots of laboratory samples are homogenized with anhydrous Na_2SO_4 and acetonitrile or any other suitable organic solvent (polar or nonpolar depending on the chemical of interest), sequentially in a Polytron or Sorvall type homogenizer. The extract is filtered and processed as described for sediment.

Aquatic Plants and Insects

The sampling of plant and insect species is, to a large extent, governed by their availability in sufficient amounts for successful analysis. Some common aquatic plants usually available in eastern Canada and sampled for residue analysis (Sundaram et al. 1986a; Sundaram and Nott 1986; Sundaram 1987a,b) are algae, *Draparnaldia* sp.; moss, *Sphagnum* sp. and *Fontinalis* sp.; manna grass, *Glyceria borealis*; watercress, *Nasturtium officinale*; liverwort, *Jungermannia* sp. and *Riccia* sp.; cattails, *Typha latifolia*; and buttercup, *Ranunculus aquatilis*. They are collected one day before spraying and at increasing intervals after spraying for persistence and fate studies. If the goal is to estimate the levels of pesticide pollution, a single post-spray sampling is sufficient. Each sample, according to its type, is collected (usually around the same vicinity where water was sampled), either by scooping or uprooting bunches of plants per sample to form three composite samples. The adsorbed water (submerged and floating species) is squeezed out, roots and dead tissues are chopped and removed, and each sample is wrapped separately in aluminum foil. Samples are packed separately in labeled polyethylene bags, stored in coolers at 0°C in transit, and brought to the field laboratory immediately for storage at –20°C.

Sufficient masses of plankton are collected by using a Schindler-Patalas plankton trap (Schindler 1969) by lowering it repeatedly below the water surface and straining the volume of water (12 L) each time through a net and removing the plankton that had deposited in it. Most aquatic invertebrates and bottom fauna are collected respectively by sweep netting and dip nets (Sundaram et al. 1991). The samples are separated according to species and stored in glass vials at -20°C until analysis.

Before blending, the excess water is removed from the thawed samples by pressing them in folds of water-absorbent paper. The samples are extracted and cleaned as described in the earlier sections.

Conclusions

Pesticides are intentionally released by aircraft into the forest environment to protect forests from insects, diseases, competing vegetation, and so forth. Only careful and restrictive usage will prevent environmental and ecological damages from pesticides. Rigid scientific protocols must therefore be used to monitor these pest control chemicals. In addition, their mobility, persistence, fate, and impact in different components should be studied. These are essential pursuits to alleviate some of the public's concerns about the use of pest control chemicals (Dunster 1987).

During an average aerial application, the amount of chemical that reaches different forestry components is often in the order of parts per million (micrograms per gram) to parts per billion (nanograms per gram) levels with very uneven distribution. At present, sophisticated and precise analytical techniques are available to detect and quantify even lower levels of various pesticide residues present in the forestry environment, and the expertise of practising analytical chemists has grown considerably. Commensurate with these advances, it is critical that every effort be made to eliminate experimental errors during pre-analytical treatments, that is, during sample procurement, processing, storage and transportation to the residue laboratory. All of these operations must be taken seriously and conducted on a scientific and statistical basis by any prospective sample collector who wants to complete sampling with confidence. If these details are overlooked or given scant attention, the residue data generated will be either unreliable or worthless; this inevitably leads to misinterpretation and inaccurate conclusions that create either false security or undue concern and alarm. In addition to crude techniques of sampling, compromising the quality of work for expedience is another major error that should be avoided at all costs if our aims are to maintain environmental quality and ecological safety.

References

Anon. 1974. Guidelines on sampling and statistical methodologies for ambient pesticide monitoring. U.S. Federal Working Group on Pest Management, Washington, D.C. 57 p.

Anon. 1983. Principles of environmental analysis – ACS Committee on Environmental Improvement. Anal. Chem. 55: 2210-2218.

AOAC. 1955. Official methods of analysis. 8th ed. Assoc. Official Anal. Chem., Washington, D.C. 1008 p.

ASTM. 1973. Sampling, standards and homogeneity. ASTM Special Publ. 540. Am. Soc. Testing Materials, Philadelphia, Pa.

Cochran, W.G. 1963. Sampling techniques. 2nd. ed. Wiley and Sons, Inc. New York.

Dunster, J.A. 1987. Chemicals in Canadian forestry: the controversy continues. Ambio 16(2-3): 142-148.

Eidt, D.C.; Sosiak, A.J.; Mallet, V.N. 1984. Partitioning and short-term persistence of fenitrothion in New Brunswick (Canada) headwater streams. Arch. Environ. Contam. Toxicol. 13: 43-52.

Ford, J.H.; McDaniel, C.A.; White, F.C.; Vest, R.E.; Roberts, R.E. 1975. Sampling and analysis of pesticides in the environment. J. Chromatogr. Sci. 13: 291-295.

Garfield, F.M. 1989. Sampling in the analytical scheme. J. Assoc. Off. Anal. Chem. 72(3): 405-411.

Horwitz, W. 1979. The inevitability of variability in pesticide residue analysis. Part 3. Pages 649-655 *in* H. Geissbühler (Ed.), Advances in Pesticide Science. Pergamon Press, New York.

Keith, L.H. 1988. Principles of environmental sampling. Am. Chem. Soc., Washington, D.C. 480 p.

Kratochvil, B.; Peak, J. 1989. Sampling techniques for pesticide analysis. Vol. XVII. Pages 1-33 *in* J. Sherma (Ed.), Analytical Methods for Pesticides and Plant Growth Regulators. Academic Press Inc., New York.

Kratochvil, B.; Taylor, J.K. 1981. Sampling for chemical analysis. Anal. Chem. 53: 924A-938A.

Kratochvil, B.; Wallace, D.; Taylor, J.K. 1984. Sampling for chemical analysis. Anal. Chem. 56: 113R-129R.

Lykken, L. 1963. Important considerations in collecting and preparing crop samples for residue analysis. Residue Rev. 3: 19-34.Lykken, L.; Mitchell, L.E.; Woogerd, S.M. 1957. Sampling crops for residue analysis. J. Agric. Food Chem. 5: 501-505.

Mallet, V.N.; Brun, G.L.; Macdonald, R.N.; Berkane, K. 1978. A comparative study of the use of XAD-2 resin and the conventional serial solvent extraction procedure for the analysis of fenitrothion and some derivatives in water-preservation techniques. J. Chromatogr. 160: 81-88.

Mallet, V.N.; Cassista, A. 1984. Fenitrothion residue survey in relation to the 1981 spruce budworm spray program in New Brunswick, Canada. Bull. Environ. Contam. Toxicol. 32: 65-74.

Matthews, G.A. 1979. Pesticide Application Methods. Longman, London, England, 336 p.

Park, D.L.; Pohland, A.E. 1989. Sampling and sample preparation for detection and quantification of natural toxicants in food and feed. J. Assoc. Off. Anal. Chem. 72(3): 399-404.

Randall, A.P. 1980. A simple device for collecting aerial spray deposits from calibration trials and spray operations. Environ. Can., Can. For. Serv., Ottawa, Ont. Bi-mon. Res. Notes, 36(5): 23.

Raske, A.G.; Sundaram, K.M.S.; Sundaram, A.; West, R.J. 1989. Fenitrothion deposits on simulated and live fir foliage following aerial spraying of two formulations. Pages 233-253 *in* J.L. Hazen and D.A. Houde (Eds.), Pesticide Formulations and Application Systems, Vol. 9, ASTM STP 1036, Am. Soc. Testing Materials, Philadelphia, Pa.

Sampford, M.R. 1962. An introduction to sampling theory. Oliver and Boyd, Edinburgh.

Schindler, D.W. 1969. Two useful devices for vertical plankton and water sampling. J. Fish. Res. Board. Can. 26: 1948-1955.

Shea, P.J.; Nigam, P.C. 1984. Chemical control. Pages 115-132 *in* D.M. Schmitt, D.G. Grimble and J.L. Searcey (tech. coord.), Managing the Spruce Budworm in Eastern North America. USDA For. Serv., Coop. State Res. Serv., Washington, D.C. Agric. Handbook No. 620.

Sherma, J. 1979. Manual of analytical quality control for pesticides and related compounds. U.S. Environ. Protect. Agency, Office Res. Dev., Health Effects Res. Lab., Research Triangle Park, N.C. No. EPA-600/1-79-008, 402 p.

Smith, R.; James. G.V. 1981. The Sampling of Bulk Materials. Royal Soc. Chem., London, England. 191 p.

Snedecor, G.W.; Cochran, W.G. 1967. Statistical methods. 6th ed. Iowa State University Press, Ames, Iowa.

Sundaram, A.; Sundaram, K.M.S.; Cadogan, B.L. 1985a. Influence of formulation properties on droplet spectra and soil residues of aminocarb aerial sprays in conifer forests. J. Environ. Sci. Health B20(2): 167-186.

Sundaram, A.; Sundaram, K.M.S.; Cadogan, B.L.; Nott, R.; Leung, J.W. 1985b. An evaluation of physical properties, droplet spectra, ground deposits and soil residues of aerially applied aminocarb and fenitrothion emulsions in conifer forests in New Brunswick. J. Environ. Sci. Health B20(6): 665-688.

Sundaram, A.; Sundaram, K.M.S.; Leung, J.W.; Holmes, S.B.; Cadogan, B.L. 1987. Influence of physical properties on droplet size spectra and deposit patterns of two mexacarbate formulations, following spray application under laboratory and field conditions. Pestic. Sci. 20: 179-191.

Sundaram, K.M.S. 1984. Residue levels of fenitrothion and aminocarb in the air samples collected from experimental spray blocks in New Brunswick in 1982. J. Environ. Sci. Health B19(4 and 5): 409-426.

Sundaram, K.M.S. 1987a. Persistence characteristics of operationally sprayed fenitrothion in nearby unsprayed areas of a conifer forest ecosystem in New Brunswick. J. Environ. Sci. Health B22(4): 413-438.

Sundaram, K.M.S. 1987b. Distribution, dissipation and persistence of aminocarb in aquatic components of the forest environment after aerial application of two Matacil 180F formulations. Can. For. Serv., For. Pest Manage. Inst., Sault Ste. Marie. Ont. Inf. Rep. FPM-X-78, 19 p.

Sundaram, K.M.S. 1987c. Fenitrothion deposits on simulated and live surfaces during an aerial spray trial in Northern Ontario. Pages 293-297 in Proc. Symp. Aerial Appln. Pestic. For., Ottawa, Ont.

Sundaram, K.M.S.; de Groot, P.; Sundaram, A. 1987. Permethrin deposits and airborne concentrations downwind from a single swath application using a back pack mist blower. J. Environ. Sci. Health B22(2): 171-193.

Sundaram, K.M.S.; Holmes, S.B.; Kreutzweiser, D.P.; Sundaram, A.; Kingsbury, P.D. 1991. Environmental persistence and impact of diflubenzuron in a forest aquatic environment following aerial application. Arch. Environ. Contam. Toxicol. 20: 313-324.

Sundaram, K.M.S.; Millikin, R.; Sundaram, A. 1989. assessment of canopy and ground deposits of fenitrothion following aerial and ground application in a Northern Ontario forest. Pestic. Sci. 25: 59-69.

Sundaram, K.M.S.; Nott, R. 1985. Mexacarbate residues in selected components of a conifer forest following aerial applications of oil-based and aqueous spray formulations. J. Environ. Sci. Health B20(4): 425-444.

Sundaram, K.M.S.; Nott, R. 1986. Persistence of fenitrothion residues in a conifer forest environment. Can. For. Serv., For. Pest Manage. Inst., Sault Ste. Marie, Ont. Inf. Rep. FPM-X-75, 14 p.

Sundaram, K.M.S.; Nott, R.; Holmes, S.B.; Boyonoski, N. 1986a. The distribution and fate of mexacarbate in a forest aquatic ecosystem. Can. For. Serv., For. Pest Manage. Inst., Sault Ste. Marie, Ont. Inf. Rep. FPM-X-73, 23 p.

Sundaram, K.M.S.; Sundaram, A. 1988. Deposition of aerially sprayed mexacarbate on balsam fir canopy and the forest floor: relevance to ULV applications. J. Environ. Sci. Health B23(5): 453-473.

Sundaram, K.M.S.; Sundaram, A.; Nott, R. 1986b. Mexacarbate deposits on simulated and live fir needles during an aerial spray trial. Trans. ASAE. 29(2): 382-388, 392.

Sundaram, K.M.S.; Szeto, S.Y. 1984. Persistence of aminocarb in balsam fir foliage, forest litter, and soil after aerial application of three formulations. J. Agric. Food Chem. 32: 1138-1141.

Sundaram, K.M.S.; Szeto, S.Y. 1987. Distribution and persistence of carbaryl in some terrestrial and aquatic components of a forest experiment. J. Environ. Sci. Health B22(5): 579-599.

Sundaram, K.M.S.; Szeto, S.Y.; Hindle, R.; MacTavish, D. 1980. Residues of nonylphenol in spruce foliage, forest soil, stream water and sediment after its aerial application. J. Environ. Sci. Health B15(40): 403-419.

Wolkoff, A.W.; Creed, C. 1981. Use of Sep-Pak C-18 cartridges for the collection and concentration of environmental samples. J. Liq. Chromatogr. 4: 1459-1472.

Youden, W.J. 1951. Statistical methods for chemists. John Wiley and Sons, Inc., London, England. 126 p.

Youden, W.J. 1967. Role of statistics in regulatory work. J. Assoc. Off. Anal. Chem. 50: 1007-1013.

Chapter 50

Review of the Role of Drop Size Effects on Spray Efficacy

CHARLES J. WIESNER

Introduction

The aerial application of insecticides to control defoliating insects involves a complex, interdependent chain of events or processes as illustrated in Table 1. To optimize the efficiency and effectiveness (efficacy) of the aerial application, while minimizing unwanted or unnecessary impact on both the on-target and the off-target environment, each of these processes must be understood in a detailed mechanistic fashion. The ultimate objective of aerial spray efficacy research must be to develop an accurate, predictive model incorporating all the physical, chemical, and biological factors that usually interact and influence the final result. Such a model will permit the manager or the regulator to predict the spray deposit, the drift, and the biological efficacy of any selected application scenario simply by inputting the appropriate conditions into the computerized model. Several such models exist now (FSCBG, AGDISP, PKBW), although they deal only with the physical aspects of deposit and/or drift; chemical and biological processes have not yet been included. These models have been validated using data from actual spray trials and the predictions have been found to agree fairly well with field measurements.

Although much remains to be done in refining our understanding of the spray efficacy process and the models thereof, the progress in recent years has been remarkable. In 1977 at the Symposium on Long Distance Drift of Forest Insecticides held in Fredericton, New Brunswick, it was generally agreed that drift model predictions available then had a tenfold margin of error (i.e. 1 000%); by contrast, present models are able to predict deposit within 20–50% of actual measured values. This paper discusses the role of droplet size on the transport of the spray cloud from the aircraft to the forest canopy and the deposition of individual droplets on the target foliage.

Table 1. Processes and major influences determining the efficacy of aerial spray operations.

Process	Major influences
A. Droplet spectrum generation	Formulation Atomizer design Aircraft
B. Spray cloud descent and canopy penetration	Aircraft wake Meteorology Terrain Stand characteristics
C. Droplet deposition	Droplet spectrum Meteorology Target characteristics
D. Insect/deposit interaction	Larval development Larval behavior Shoot phenology Larval toxicology Chemical transport (capillary, systemic, vapor)

The first step in the spray process is the breakup of the liquid insecticide formulation (tank mix) into a cloud of droplets. This breakup or atomization is achieved by pumping the tank mix through either spray nozzles or rotary atomizers mounted on a boom attached to the aircraft wing. The nature and potential transport characteristics of the spray cloud are defined by the sizes of the droplets comprising it. The distribution of size and number of droplets within a given size range is combined in the so-called droplet spectrum, a representation that essentially characterizes the physical properties of the spray cloud.

The size of a droplet has a profound influence on several factors, for example, the motion of the droplet in response to air currents. "Small" droplets tend to follow air currents whereas "large" ones tend to respond to gravity rather than air movement and, consequently, fall more vertically. These characteristics reflect the momentum of the droplet and influence both droplet trajectories between aircraft and forest canopy and efficiency of depositing or landing on target foliage.

The droplet spectrum is probably the single most important parameter in the entire spray process. At the same time, the droplet spectrum is, in principle, the most easily manipulated parameter; by changing atomizer type, orientation relative to the airstream and throughput, the range of emitted droplets can be adjusted to coarser or finer sizes. However, in practice, achieving selective production of only the most effective or optimum droplet sizes is not yet feasible with currently available commercial atomizers. Much remains to be done to improve existing atomizer designs and to develop entirely new and more efficient concepts.

The choice of the optimum droplet spectrum for forestry applications has been the subject of long-standing controversy fueled by too much misguided intuition and too little science. However, over the last two decades, evidence has gradually been accumulating to suggest that very small droplets are, in fact, responsible for the bulk of target insect mortality, and that the spray material in the form of large drops is essentially wasted.

During the so-called Dunphy experiments which were carried out by the New Brunswick Spray Efficacy Research Group (NBSERG) from 1981 to 1986, one of the major objectives was determining the optimum drop size for conifer application.

Droplet Spectrum Characterization

The droplet spectrum essentially characterizes the spray cloud by providing a measure of the distribution of small, medium, and large drops comprising the cloud. Generally, a spray cloud contains many more small drops than large ones irrespective of the atomizer used.

The diameter of the drops becomes particularly important when considering the amount of volume of spray material contained within a given size range. This is due to the mathematical relationship between diameter and volume:

$$\text{Volume of drop} = \frac{\pi}{6}\,(\text{diameter})^3$$

The third power relationship between the two means, for example, that doubling the diameter of a drop increases its volume eightfold, and a tenfold increase in diameter results in a thousandfold increase in volume. Consequently, although a spray cloud may consist of many small and very few large drops, the bulk of the active ingredient can be contained in the few large drops.

This simple fact has far-reaching consequences on insecticide efficacy because the prerequisite to killing an insect must be contact or encounter between any individual insect and one or more spray droplets. The probability of an insect encountering a droplet increases in direct proportion to the number of droplets deposited in the immediate vicinity or "microhabitat" of the insect. Studies aimed at determining the optimum droplet size have integrated three considerations: (1) What size most effectively and uniformly reaches and deposits on the insect microhabitat (i.e. foliage)? (2) What size maximizes the probability of encounter? (3) What size is sufficient to produce a toxic effect?

Historical Perspective

The classical view of forestry spraying has been strongly influenced by agricultural crop protection practices together with heavy doses of intuitive thinking. Unfortunately, intuition has proved to be a poor predictor of complex physical phenomena.

Agriculture has traditionally associated good coverage of the crop with very high application rates or volumes of pesticide. Spraying to run-off (i.e. to the point where excess spray liquid actually drips from plant foliage) has been a common practice in both field and orchard crops. High volumes are inevitably associated with large drops which, in turn, were considered beneficial in reducing the potential for off-target drift.

The volumes commonly used in agriculture (20 to 50+L/ha) were entirely impractical for forestry use due to the large acreages and long ferry distances from airstrip to spray block. From necessity, early spray programs against the spruce budworm reduced application volumes to about 5 L/ha. Shortly after the inception of the New Brunswick budworm operation in 1952, studies carried out by the Defense Research Board and the Forest Insect Investigations Unit indicated that spruce budworm mortality correlated with unit area coverage (drops per square centimetre) rather than drop size or volume of spray material deposited. These results provided the first indication that small droplets might have some advantage over traditional high-volume techniques. Over the next decade, Canadian spray tactics gradually evolved, leading to progressively lower spray volumes and the use of crosswind, wide lane applications to provide uniform coverage by successive incremental deposition. In spite of the successful use of this system over millions of acres of Canadian forest, the tactics were far from universally accepted in other jurisdictions and, even in Canada, the detailed mechanisms that made the system work were not well understood.

Field Observations

Most studies of spray deposition have used artificial collection surfaces to sample deposits, primarily because of their convenience. However, it has long been recognized that natural plant surfaces have very different physical and aerodynamic properties and, hence, very different droplet collection efficiencies (Baker et al. 1978; Hadaway and Barlow 1965; Uk 1977).

One of the earliest reports addressing the size of spray droplets affecting spruce budworm larvae and their microhabitat was published in 1967 by Himel and Moore. The authors examined larvae (Himel and Moore 1967; Himel 1969) and foliage following aerial applications of Zectran® against an infestation of western spruce budworm, *Choristoneura occidentalis*, on Douglas-fir in Montana. Spray droplets were made visible using fluorescent particles suspended in the tank mix. Although the emitted sprays had maximum drop sizes of 350 µm and volume median diameters in the order of 150 µm, 99% of all spray droplets observed on larvae were smaller than 70 µm. Larger drops were found primarily on peripheral foliage and on the ground. The authors concluded that only small drops effectively penetrate the forest canopy and impinge within the larval microhabitat. Furthermore, since the large drops cannot, due to their scarcity, contribute significantly to larval mortality, they must be regarded as a waste and an environmental burden. Clearly, since the vast bulk of the spray (and of most conventional sprays) was in the form of large drops, it appeared that 95% to 99% of the applied spray was ineffective and that efficacy was achieved only through the 1% to 5% of the spray which was in the form of small droplets.

In 1975, Barry and Ekblad (1978) conducted extensive field trials with both fixed- wing aircraft and helicopter using the synthetic insecticides carbaryl and trichlorphon as well as *Bacillus thuringiensis* (Bt). Spray droplets were made visible by addition of dyes to the tank mixes. In spite of the broad range of application parameters, the results were amazingly consistent (see Table 2).

Again, it was clear that only small droplets were found on foliage in sufficient numbers to lead to significant larval contact and mortality. Furthermore, the uniformity of these results illustrates convincingly that spray deposition is a physical phenomenon mediated by drop size, target morphology, and air movement, and that an in-flight droplet will behave in response to these influences, not to its chemical composition. The evaporation rate of the droplet is the only factor significantly influenced by the chemical composition of the tank mix. This fact is still widely misunderstood.

Table 2. Foliar deposition of small spray drops.[a]

Insecticide	Emitted vmd[b] (µm)	Volume applied (L/ha)	Foliar deposit (drops/needle)	Proportion of drops on foliage < 81 µm diam.
Bt	350	18.7	2.4	95.4%
Bt	350	18.7	1.8	97.4%
Carbaryl	270	9.3	2.3	92.2%
Trichlorphon	280	4.7	2.3	92.0%

[a]From Barry and Ekblad (1978).
[b]Volume median diameter.

Aerial spray trials conducted in Scotland in 1978 against the pine beauty moth, *Panolis flammea*, a major pest of Scots and lodgepole pine, examined the differences between ultra-low-volume and low-volume application of fenitrothion (Joyce and Spillman 1979). Both treatments received 300 g/ha of active ingredient but the former was sprayed at 1.0 L/ha while 20 L/ha were applied to the latter. Analysis of foliage for fenitrothion showed that the ultra-low-volume treatment resulted in nearly 75% more foliar deposit than did the low-volume treatment. On the other hand, ground deposit or contamination was eight times higher in the low-volume than in the ultra-low-volume plot.

This mounting volume of evidence in support of the role of small drops in forest spray efficacy has obviously raised numerous questions. For example, how does a cloud composed of small drops find its way from aircraft height down to the target? Does the increased drift potential of small drops outweigh their efficacy benefits?

Preliminary studies at the Cranfield Institute of Technology and the CIBA-GIEGY Agriculture Aviation Research Unit in England indicated that air turbulence in the form of eddy-type circulation above the canopy, together with aircraft vortex or wake, may provide the vertical transport vectors necessary for cloud descent. Under all but the most calm conditions, gravity and droplet sedimentation appeared to have little influence on cloud transport.

In 1979, studies carried out in New Brunswick by the National Research Council (Crabbe et al. 1980) examined the effect of meteorological conditions on long-range drift. Crabbe and his co-workers were able to demonstrate significant reductions in drift at 7.5 and 24 km downwind under more turbulent conditions than under stable atmospheric conditions, thereby corroborating the turbulent transport hypothesis.

Clearly, the stage was set for detailed and intensively monitored experiments to define more quantitatively the mechanisms of small droplet transport from aircraft to forest canopy and of impaction in the target microhabitat.

Dunphy Field Trials

In 1981, Forest Protection Ltd. undertook a long-term research program to develop a rigorous definition of the processes involved in aerial spray drift and deposit. Over the next 6 years, the company contracted with several members of the New Brunswick Spray Efficacy Research Group to carry out collaborative spray trials operating out of Dunphy airstrip in central New Brunswick. The collaborators, who included R.S. Crabbe (National Research Council), R.B.B. Dickison and J.J.C. Picot (University of New Brunswick), R.E. Mickle (Atmospheric Environment Service), and C.J. Wiesner (Research and Productivity Council), developed an experimental design based on an intensive sampling and monitoring effort. The experimental objective was to measure downwind foliar deposit and drift under a range of well-characterized meteorological conditions.

The general layout of the Dunphy experiments consisted of a single swath crosswind application with various sampling and measuring devices deployed from the swath line out to 3 600 m downwind. A Grumman Avenger TBM flying approximately 20 m above the forest canopy applied dyed spray with an emitted volume median diameter close to 100 µm. The spray mix, which simulated the operational aqueous fenitrothion emulsion using noninsecticidal chemicals, was applied at a rate equivalent of 1.5 L/ha and had a residual volume of 20% after evaporation of the water. Consequently, the evaporated volume median diameter was approximately 58 µm.

Meteorological measurements were made by cooperators from the National Research Council, Atmospheric Environment Service, and the University of New Brunswick, using an instrumented 46-m mast situated 400 m downwind of the spray line as well as balloon-borne minisondes and tethersondes. Spray drift samplers were deployed at 400, 1 200, and 3 600 m while vertical profiles of spray cloud flux were provided to a height of 200 m by lightweight dosimeters (aspirating samplers) suspended from tethered balloons. This sampling system was designed to provide data necessary to enable calculation of the total mass of the drifting spray cloud past the three downwind distances (Crabbe et al. 1983).

Spray deposits were measured on conifer foliage, on foliage simulators, and on the ground using microscopic examination to determine droplet frequency and size as well as by gas chromatography to determine total mass deposited. Airborne drift was measured by means of three different sampling devices: (a) aspirating dosimeters to determine total mass flux past the sampling point; (b) rotorods also for mass flux measurement; (c) cascade impactors for determination of droplet size distribution.

By comparing the droplet spectra and populations found on the ground, the foliage and in the air under various meteorological conditions against the spectrum emitted by the spray aircraft, and by determining the collection efficiencies of various target elements, Picot et al. (1986) were able to define the optimum droplet spectrum for foliar deposition. The optimum was found to lie between 15 to 55 µm (Fig.1). This range of sizes affords the best mix of foliar impaction efficiency and number frequency and avoids the waste of active ingredient inherent in large drops.

The results of the Dunphy trials are detailed in accompanying chapters, but in summary they have provided highly sophisticated information on the relationships between spray drift and the most significant operational parameters. The complex interaction of aircraft height, aircraft vortex, meteorological conditions, and spray spectrum is now understood in

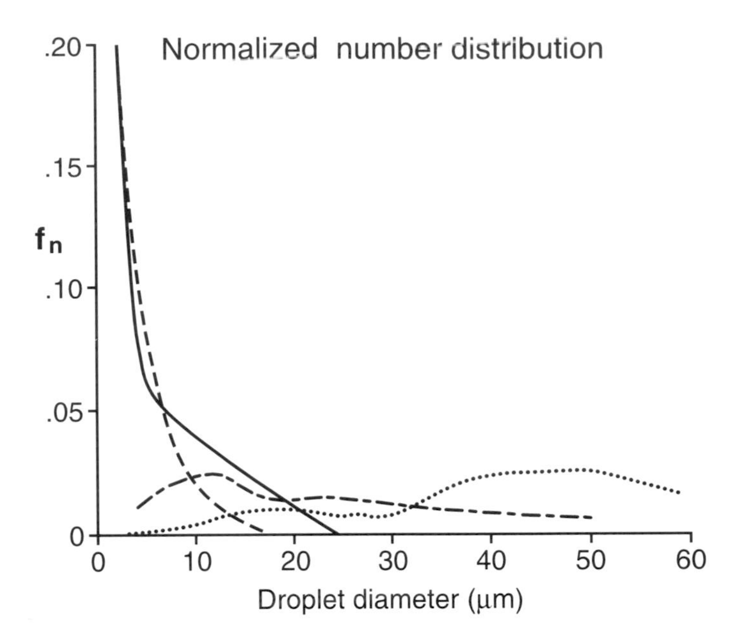

Figure 1. Number distribution of drifting and deposited droplet sizes in Dunphy trials, where: f_n equals number fraction; ——— is emitted evaporated spectrum; ---- is drifting at 400 m; —-— is foliar deposit at 100 m; and ..... is ground deposit at 100 m. (J.J.C. Picot and D.D. Kristmanson, pers. comm.)

sufficient detail to form the basis of the evolution of new operational hardware and tactics.

Although progress in the field has been rapid over recent years and the extensive research is finally coming to fruition via gradual transfer of refined application technologies to operational practices, we still have a long way to go. For example, a detailed understanding of aircraft vortex behavior still eludes us, although the importance of this phenomenon to spray delivery is now accepted. Other parameters such as topography and forest stand structure remain to be addressed in detail and much needs to be done to elucidate the mechanisms of dose transfer from foliar deposit sites to the target pest insects.

References

Baker, B.L.; Thompson J.F., Jr.; Threadgill, E.D. 1978. Spray deposition on a three-dimensional object. Trans. ASAE. 21(5): 806-812.

Barry, J.W.; Ekblad, R.B. 1978. Deposition of insecticide drops on coniferous foliage. Trans. ASAE. 21: 438-441.

Crabbe, R.S.; Elias, L.; Davie, S.J. 1983. Field study of effect of atmospheric stability on target deposition and effective swath widths for aerial forest sprays in New Brunswick — Part II. Rep. No. LTR-UA-65. National Aeronautical Establishment, Natl. Res. Counc. Can., Ottawa, Ont.

Crabbe, R.S.; Elias, L.; Krzymien, M.; Davie, S. 1980. New Brunswick Forestry Spray Operations: field study of the effect of atmospheric stability on long-range pesticide drift. Rep. No. LTR-UA-52. National Aeronautical Establishment, Natl. Res. Counc. of Can., Ottawa, Ont.

Hadaway, A.B.; Barlow, F. 1965. Studies on deposition of oil drops. Ann. Appl. Biol. 55: 267-274.

Himel, C.M. 1969. The optimum size for insecticide spray droplets. J. Econ. Entomol. 62: 919-925.

Himel, C.M.; Moore, A.D. 1967. Spruce budworm mortality as a function of aerial spray droplet size. Science 156: 1250-1251.

Joyce, R.J.V.; Spillman, J.J. 1979. Pages 13-24 *in* A.V. Holdan; D. Bevan (Eds.), Control of pine beauty moth by fenitrothion in Scotland, 1978. For. Comm., Edinburgh.

Picot, J.J.C.; Kristmanson, D.D.; Basak-Brown, N. 1986. Canopy deposit and off-target drift in forestry aerial spraying. The effects of operational parameters. Trans. ASAE. 29: 90-96.

Uk, S. 1977. Tracing insecticide spray droplets by sizes on natural surfaces. The state of the art and its value, 1977. Pestic. Sci. 8: 501-509.

Chapter 51

Effect of Atmospheric Stability on Wind Drift of Spray Droplets from Aerial Forest Pesticide Applications

R.S. CRABBE AND M. MCCOOEYE

Introduction

Large-scale protection of Canadian forests from insect damage is accomplished by ultra-low-volume aerial insecticide applications. The limited spray volume emitted (about 0.2 kg of active ingredient in 2 L/ha of water or light oil) is finely atomized to extend protection to the very large number of foliar elements requiring it. However, because of the low sedimentary deposition of the spray droplets (maximum evaporated diameter approximately 150 µm with sedimentation velocity = 0.5 m/s) and the proclivity of pilots to maintain a safe flying height of several tens of metres above the trees, the resultant off-target wind drift can be significant unless atmospheric turbulence is present to bring the emission into the canopy. Generally, provincial regulations governing ultra-low-volume forest spray programs in Canada have been satisfied by applicators restricting their aerial insecticide applications to the thermally stable (nonturbulent) daytime atmospheric conditions. But at the same time, in many jurisdictions, the spray program has been accompanied by complaints from the public of excessive exposure of nontarget receptors to droplet wind drift.

In 1979, at the request of the New Brunswick forestry spray agency, Forest Protection Limited, the Unsteady Aerodynamics Laboratory of the National Research Council of Canada, henceforth referred to as the Laboratory, in cooperation with the Department of Chemical Engineering of the University of New Brunswick, conducted a controlled field experiment in northern New Brunswick on long-range windborne droplet drift from a Grumman TBM spray plane (a converted World War II Grumman Avenger Torpedo Bomber). The intent of that study was to measure the far-field atmospheric spray droplet concentrations at tree height to enable toxicologists to assess the hazards, if any, to the environment and the public at large from long-range drift from aerial forest spraying. Twenty-four nozzles (Teejet® Number 11010) were fitted to atomize a stimulant of the fenitrothion-in-oil formulation comprising 5% tris-2-ethyl-hexyl phosphate (TEHP) in number 585 fuel oil. When atomized, about 95% of the oil would evaporate (Picot et al. 1981) so that the nonvolatile fraction still in the atmosphere, TEHP and oil residue, could be measured.

That investigation (Crabbe et al. 1980a) revealed the presence of fine spray droplets to a distance of 45 km from the flight line, although only in parts-per-trillion concentrations. A not unexpected discovery was that turbulent atmospheric conditions gave an order of magnitude lower atmospheric droplet concentrations at tree height than did stable conditions. This result led to subsequent field studies to determine whether ultra-low-volume aerial spray applications in neutral to unstable (i.e. turbulent) atmospheric conditions would lead to higher target deposit and lower off-target drift than in stable conditions. This chapter presents results of controlled field experiments in the New Brunswick and Northern Ontario forests, hereafter referred to as the Dunphy and Kapuskasing studies, that demonstrate the beneficial effect of aerial spraying in neutral to unstable surface-layer flows to reduce wind drift and enhance target deposition.

Chemical Sampling and Analysis

A major aspect of these investigations was the development in the Laboratory of chemical samplers with high collection efficiency for windborne spray droplets when deployed in a turbulent atmosphere. In 1979, the Laboratory used a chemical sampler, or dosimeter, comprising a battery-driven pump that draws air through a glass tube at 1 to 2 L/min. This glass tube, or adsorber, contained a 1-cm length of fine nickel mesh coated with a chromatographic liquid phase. Because the Laboratory's sampling stations were more than 7 km from the flight line, where only the small droplets of the emission would still be in the atmosphere, stabilizing devices were not considered for the sampler to align the intake tube into the local wind direction. Indeed, analysis of plate deposits in a commercial cascade impactor at 22 km downwind revealed few droplets larger than 20 µm in diameter in the atmosphere so that the loss of sampling efficiency through misalignment of the tube was probably only nominal, particularly in view of the relatively low wind speeds at tree height.

These samplers were also deployed from a tethered balloon in 1979 but here misalignment of the intake did lead to a significant loss in sampling precision owing to the increase in wind speed with height. Nevertheless, the estimated fraction of nonvolatiles still airborne at 22 km downwind in moderately turbulent conditions, approximately 6%, is close to the prediction of Crabbe and Reid (1980).

In the Dunphy and Kapuskasing studies, the Laboratory undertook to demonstrate that the drift cloud mass decreased with increasing thermal instability. It therefore partly redressed the above sampler deficiency by attaching a monovane tail in an attempt to guide the sampler into the horizontal wind direction when deployed from a tethered blimp. However, early field results from the Dunphy study still revealed lack of sampling efficiency so that the Laboratory continued to improve its dosimeter by streamlining the pump housing, replacing the monovane tail with a bivane tail, and gimbal-mounting and balancing the unit to stabilize its motion about both the pitch and yaw axes.

Rotorods manufactured by Ted Browne & Assoc. of California were also modified for use in a vertical array from a mast in the Dunphy study and from a tethered blimp in the

Kapuskasing study. The Rotorod is acknowledged to be the industry standard for measuring droplet concentration (May et al. 1976), but it does not trap vapor whereas the dosimeter can trap both droplets and vapor. Bystander inhalation exposure to pesticide vapor from volatization of foliar deposits is reported in Crabbe et al. (1980b) to be comparable to that from the droplet drift cloud.

Photographs of the Laboratory's dosimeter and modified Rotorod sampler are shown in Figure 1a and b. Figure 1c shows the comparison of the dosimeter's performance to that of the Rotorod in a coarse aerosol stream in a wind tunnel. The quantity compared in that test was the time-integrated concentration, or dosage, in the aerosol plume.

The Laboratory also designed a prototype cascade impactor for size determination of droplets up to 40 µm in diameter (about 75 µm in diameter before evaporation of volatile components in most aerial insecticide sprays), which separated the size distribution of the drifting droplets into six stages for use in correcting the dosimeter dosages to isokinetic (true) values. The aerodynamic analysis and performance verification of the impactor is described in Crabbe (1984), and it is shown in Figure 2.

Paralleling the development of the droplet samplers in the Laboratory was the development of foliage simulators at the University of New Brunswick (Kristmanson and Picot 1983). The New Brunswick Research and Productivity Council deployed these in the Dunphy study to measure the deposition of spray droplets (e.g. Riley 1985). Obviously, the sum of individual measurements of drift cloud mass and foliar deposition must equal the spray emission. To the extent it does is a measure of the precision of the field data.

For the Kapuskasing field study, deposit measurements were made using a passive droplet sampler (Fig. 3) designed in the Laboratory for use in the forest canopy. This sampler is a cylindrical mesh cage supported by stainless steel end caps with hooks to support the sampler at top and bottom. The cage material is 13 mesh brass with a wire diameter of 0.025 cm. The sampler is 12.8 cm long and has a diameter of 2.3 cm. The cages were deployed at 0.75 m and 1.5 m above ground level. For comparison purposes, some broad leaf foliage simulators designed at the Forest Pest Management Institute of Forestry Canada were deployed at selected deposit measuring stations. Ground deposit was measured using hinged stainless steel plates measuring 7.6 cm × 15.2 cm.

a

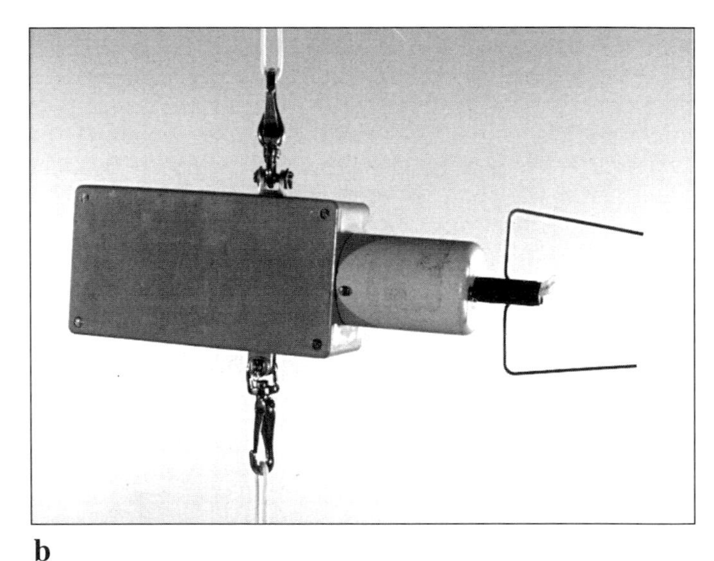

b

c

Figure 1. (a) NRC dosimeter; (b) NRC Rotorod package; (c) comparison of Rotorod to dosimeter performance.

Figure 2. NRC cascade impactor.

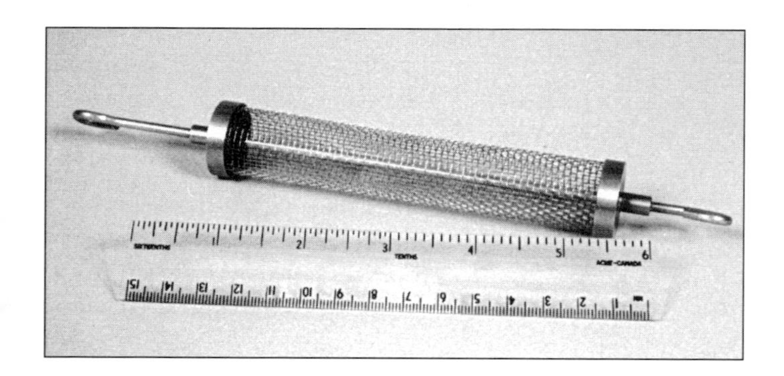

Figure 3. NRC passive droplet sampler.

In all field trials in New Brunswick and Northern Ontario, the spray formulation contained a chemical tracer. Since two aircraft were flown in the Dunphy investigation, two tracers, TEHP and tri-phenyl-phosphate (TPP), were used. In the Kapuskasing investigation only one tracer, TEHP, was required.

Both TEHP and TPP were determined in a single gas-chromatographic analysis for each sample. Analyses were performed on a Varian model 1400 gas chromatograph modified to accept adsorber tubes and provide means of thermal desorption of vapors. The gas chromatograph was equipped with a Varian thermionic flame ionization detector. Details of the procedure for analysis of adsorber tubes are given in Krzymien (1980). A similar protocol developed for the analysis of Rotorod deposits is described in Crabbe and McCooeye (1985). The cascade impactor plates, deposit cages, foliage simulator "leaves," and ground plates were all washed with reagent grade acetone. The washings were adjusted to a constant volume and analyzed by gas chromatography as described in the above report.

The Field Studies

Dunphy

From 1981 through 1983, exploratory field studies of drift and deposit to 1.2 km from the TBM flight line proceeded through the aegis of the New Brunswick Spray Efficacy Research Group. This organization was created to coordinate the activities of all the organizations participating in the research program. A flat forest site at 46.58°N, 65.83°W in east-central New Brunswick was selected for the experiment and the sampling grid at this location (Fig. 4) was identified and progressively refined during this period. Five swaths were overlaid on a (crosswind) flight line in 30 minutes in each spray trial in years 1984 and 1986 of the Dunphy study to obtain ensemble-averaged data (Crabbe and McCooeye, 1985, 1988).

During the field trials, the University of New Brunswick Faculty of Forestry provided 24-hour weather forecasts as well as short-term forecasts of upper and surface wind directions with the aid of pre-spray minisonde ascents at the Upper Blackville, New Brunswick, aerodrome (minisonde—a free-rising balloon instrument package for measuring and transmitting atmospheric state variables such as temperature to a ground station). The airport is also known locally as the Dunphy airstrip.

Dunphy forest site—The forest at the Dunphy site is essentially a mixture of 11 to 15-m-high jack pine and spruce trees on flat terrain. From forest boundary layer measurements described below, values of zero-plane displacement, d, of 11 m and aerodynamic roughness length, z_0, of 1.6 m were determined where d is the vertical displacement of the surface winds by the forest and z_0 is a measure of the aerodynamic roughness of the tree tops (z_0 is about 1 mm for a desert). These parameters appear as $\ln((z-d)/z_0)$ in the expression for the lower portion of the wind speed profile used to compute drift cloud mass, z being the height above ground level.

Picot et al. (1987) give more detailed properties of the vegetation at this site.

Forest boundary layer measurements—To profile the atmospheric surface layer, the Laboratory instrumented a 48-m mast in 1985 and a 63-m mast in 1986 (Fig. 5), each at 400 m from the flight line, with up to six levels of propeller bivanes (trade name: MRI; manufactured by Belfort Inst. Co., Baltimore, Md.) which the Laboratory fitted with microbead thermistors. The performance of the MRI propeller bivane in atmospheric turbulence is documented in Raman and Brown (1976).

The mast instrumentation was used not only to measure wind speed for use in computing drift cloud mass, but also temperature and turbulent flux of heat and momentum for determination of atmospheric thermal stability. To further aid in the assessment of the state of thermal stability, the temperature difference over tower height was measured with an MRI matched thermistor pair to within 0.02°C. Thirty-minute averages were obtained using an on-site microcomputer equipped with analog-to-digital conversion (in 1984, the tower data were logged with an FM recorder as a multiplex time series for later analysis with a minicomputer).

Relative humidity was measured with a capacitance probe (trade name: Humicap, available from Vaisala Inc., Woburn, Ma.) at 25 m above ground level in 1986, approximately midway between the aircraft and the canopy, and at 10 m above ground level in 1984 using a wet and dry bulb psychrometer. The psychrometer data were not measured at actual spray time and so represent approximate values prevailing during the spray trials in 1984. Relative humidity affects wind drift of

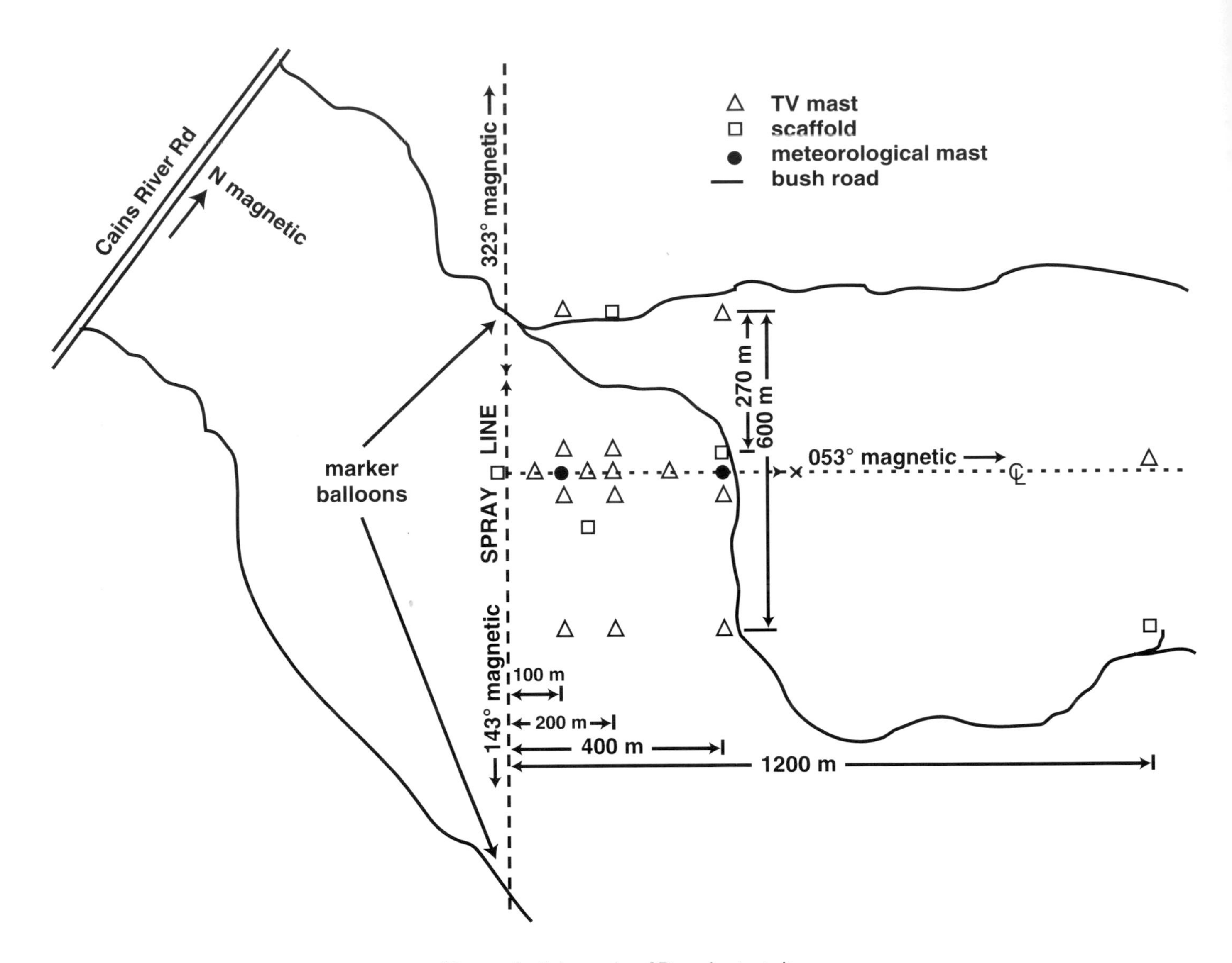

Figure 4. Schematic of Dunphy test site.

Figure 5. Dunphy meteorological mast.

aqueous formulations by influencing the rate of evaporation of the water component and hence the terminal velocity of the droplet.

The extension of the wind profile to 250 m above ground level was obtained from minisonde ascents (Steeves 1986) and tethersonde profiles (R. Mickle, pers. comm.) (tethersonde—an instrument package on a tethered blimp for measuring and transmitting atmospheric flow and state variables such as wind speed and temperature to a ground station). The vector mean wind profile to drift cloud top is required for computing cloud mass.

Spray protocol—In each spray trial, Forest Protection Limited flew a Grumman TBM fitted with 24 Teejet® 11010 atomizers in 1984 and 1986 as well as a Cessna-188 fitted with four Micronair® AU4000 atomizers in 1986. Only one flight line was identified in the forest, but within certain constraints of homogeneity, this was oriented approximately orthogonal to the prevailing wind direction. The average flying heights of the Cessna-188 and TBM were determined photographically by the University of New Brunswick Department of Chemical Engineering to be 42 m and 38 m above ground level, respectively, in 1986 and 49 m above ground level for the TBM in 1984. The Laboratory measured the mass of nonvolatiles in the drift cloud at 400 and 1 200 m from the flight line in 1984 and 400, 600, and 900 m in 1986 (400 m is close to the swath interval used operationally by a team of three TBMs), and the Research and Productivity Council measured deposition.

Spray formulation and application strength—An emulsion of light oil (dibutyl phthalate in place of fenitrothion) and other components in water was used as a spray simulant. The compositions of the Dunphy aerial spray formulations are given in Table 1. The sum of the nonvolatiles (dibutyl phthalate, tracer, Atlox surfactant, Dowanol solvent, and dye) comprised 17% of the spray volume.

The TBM spray simulant in Table 1 was shown by Tomney et al. (1978) to have essentially the same density, viscosity, and interfacial surface tension as the aqueous fenitrothion formulation, thereby ensuring little difference in the droplet size spectra. However, the higher fraction of Atlox in the Cessna-188 spray simulant shown in Table 1 would be expected to lower both the surface tension and droplet sizes relative to the fenitrothion formulation. Dibutyl phthalate has about the same vapor pressure as fenitrothion, approximately 10^4 mm Hg at room temperature, so that further loss of droplet mass following evaporation of the water component is considered negligible in the transport time scale of interest in this study, about 20 minutes.

The line source strength of the TBM spray in grams per metre of flight line was computed from knowledge of the TBM calibrated emission rate (24 U.S. gal/min), air speed (150 kn) and mean wind speed measured at aircraft height on the meteorological mast. In the Cessna-188, the pilot radioed instantaneous readings of the total Micronair® volume flow rate displayed by the calibrated flowmeter installed in the cockpit. The Laboratory recorded and later averaged these to 7.6 U.S. gal/min for computation of the Cessna-188 line source strength at its 95-kn air speed.

Experimental results—Field data on drift cloud mass from the Dunphy, New Brunswick, site encompassed a wide range of wind directions and atmospheric conditions. A complete description of the experimental details in the Dunphy experiment as well as the method of computing drift cloud mass from vertical profiles of droplet dosage and vector wind appears in Crabbe and McCooeye (1988). Selected data on drift cloud mass from years 1984 and 1986 of the Dunphy study (Crabbe and McCooeye 1985, 1988) together with foliar deposit data (van Vliet 1985, 1987) appear in Tables 2 and 4, respectively, and Tables 3 and 5 contain the corresponding 30-minute tower profiles of the atmospheric surface layer flow.

In Tables 3 and 5, L is the Monin-Obukhov length, a measure of atmospheric thermal stability; large values (more than 500 m) of either sign denote near-neutral conditions. Increasing departures from neutral (adiabatic) conditions are identified by decreasing absolute values of L; negative values denote unstable, more turbulent, conditions while positive values denote stable, less turbulent, conditions than in neutral flow. An alternative measurement of thermal stability is the gradient Richardson number, R_i. Neutral conditions are defined by $R_i = 0$, unstable conditions by $R_i<0$, and stable conditions by $R_i>0$.

Table 1. Dunphy spray formulations.

TBM spray formulation		Cessna-188 spray formulation	
Component	% Volume	Component	% Volume
dibutyl phthalate	9.3	dibutyl phthalate	2.3
TEHP	2.6	TPP	2.3
Atlox	2.6	Dowanol	2.5
Dowanol	2.6	Atlox	9.2
water with 0.75% mass of rhodamine B dye	82.9	Erioglaucine A100 dye	0.5
		water	83.2

In Table 5, pwv is the difference in water vapor pressure between the droplet surface and the surrounding air that drives droplet evaporation.

Discussion of 1984 drift data—An inspection of 1984 spray trials 2, 3, and 5 in Table 2 shows essentially the same wind speed at aircraft height but systematically decreasing drift past 400 m downwind with decreasing stability (increasing turbulence) as measured by the turbulent deposition velocity (see below). The spray budget (sum of drift past 400 m and deposit to 400 m) shows closure to within about 25%.

The drift past 1 200 m is also tabulated but because of experimental errors described in Crabbe and McCooeye (1988), the results shown tend to underestimate the cloud mass.

The quantity v_d in Tables 3 and 5 is the deposition velocity of fine spray droplets as measured in terms of atmospheric turbulence. Field measurements described in Wesley et al. (1977) and numerical simulation (Crabbe and Reid 1980) indicate that this is 70% of the turbulent transfer velocity of horizontal momentum, u_*^2/U where u_* is the friction velocity defined as the square root of the turbulent wind stress on the trees divided by the fluid density and U is the mean wind speed. The data in these tables indicate the higher the level of turbulence the larger v_d will be and the less the wind drift will be.

Discussion of 1986 drift data—The Dunphy study was resumed in 1986 with improved instrumentation to cover some gaps in year 1984.

Referring to spray trials 3 and 4 in Table 4 (having nearly the same spray release height and wind speed), the Cessna-188 emission leaves 86% of the nonvolatiles in the atmosphere at 400 m downwind in moderately stable flow (spray trial 3) and 82% in near-neutral flow; the equivalent value from the TBM emission in stable flow is lower, only 56%.

The drift past 600 m in trials 3 and 4 displays more pronounced effects of stability with 66% airborne from the Cessna-188 vs 86% at 400 m in the stable case, but only 56% vs 82% in the near-neutral case. Equivalent values for the TBM are again lower, 52% vs 56% at 400 m downwind in the stable case and only 31% in the near-neutral case (spray trial 4).

The influence of relative humidity on drift appears from a comparison of Cessna-188 trials 6 and 10 (again having approximately the same release height and wind speed) where the slightly stable trial (10) leaves 54% in the atmosphere at 900 m downwind compared to only 30% in the moderately stable trial (6) because the higher relative humidity in trial 6, 83% vs 43%, reduced evaporative loss of droplet mass and terminal velocity. Higher relative humidity also lowers the TBM drift past 400 m downwind as may be seen in a comparison of TBM spray trials 3 and 6 where the less stable of these two (3) left 56% of the nonvolatiles in the atmosphere compared to only 50% in the moderately stable trial (6), because of the lower relative humidity, 36% vs 83%.

The influence of wind speed on drift in turbulent conditions may be seen by comparing the drift past 600 m downwind between sprays 2 and 4. Averaged over both aircraft emissions, 43% of the nonvolatiles drift past 600 m when the wind speed at aircraft height is 3.5 m/s (trial 4) compared to only 54% when the wind speed is 7.1 m/s. Some of this difference can be attributed to the difference in aircraft height.

Table 2. Drift and deposit from TBM Teejet® emission in Dunphy, 1984.

Trial	Date Time ADT	Stability	Aircraft ht (m)	Wind[a] °mag./m·s	Deposit to 400 m (% nonvolatiles)	Drift past 400 m	Drift past 1 200 m	Sunrise or sunset ADT
1	12/6, 1952	near neutral	48	255/5.6			23	2116
2	13/6, 0843	slightly stable	51	240/3.6	25	56	19	0529
3	17/6, 0657	slightly unstable	52	250/3.3	24	48	15	0528
4	17/6, 2123	slightly stable	49	225/6.4	32		27	2119
5	21/6, 1659	moderately unstable	44	269/3.5	44	33	6	2120

[a]Tabulated wind speed and direction are at aircraft height. The direction should be compared to the upwind perpendicular to the flight line, 233° magnetic.

Table 3. Tower profiles from Dunphy, 1984.

Item[a]	Trial 1	Trial 2	Trial 3	Trial 4	Trial 5
Mean wind speed (m/s)					
U(18)	3.2		1.5	3.0	2.2
U(24)	5.6	1.7	1.9	3.6	2.6
U(48)		3.5	3.3	6.4	3.5
Friction velocity (m/s)					
$u_*(18)$			0.41	0.75	0.68
$u_*(24)$	0.74	0.36	0.30	0.68	0.50
$u_*(48)$			0.35	0.68	0.50
Root mean square vertical turbulent velocity (m/s)					
$\sigma_w(18)$			0.41	0.80	0.58
$\sigma_w(24)$	0.86	0.47	0.34	0.76	0.54
$\sigma_w(48)$			0.38	0.81	0.58
Deposition velocity (m/s)					
$v_d(18)$	0.11	0.04	0.08	0.13	0.15
Temperature (°C)					
T(25)	25	17	11	18	21
Relative humidity (%)					
rh(10)	35	75	50		40
Monin-Obukhov length (m)					
L	900	200	–340	410	–94

[a]Numbers in parentheses are heights in metres above ground level.

Nonetheless, as argued in Crabbe and McCooeye (1985), progressive increases in wind speed in turbulent flow raise drift by increasingly smaller amounts, essentially because the mean (negative) slope of ultra-low-volume droplet trajectories approaches a limited value given by the quantity v_d/U (averaged over, approximately, the depth of the layer between aircraft and tree height). In turbulent flow, this ratio is largely independent of wind speed.

A further beneficial effect of turbulence (i.e. of neutral to unstable flow) is that it transfers horizontal fluid momentum, as evidenced by U(h), the wind speed at tree height, into the canopy flow, thereby raising the foliar collection efficiency for ultra-low-volume droplets. As a result, atmospheric droplet dosages in the subcanopy flow decrease with increasing wind speed (Fig. 6) and the ratio of ground to foliar deposit also decreases (van Vliet 1987).

Overall, the TBM spray emission leaves less nonvolatile material in the atmosphere than the Cessna-188 emission. This difference in drift between aircraft is ascribed to larger droplets from the Teejet® 11010 atomizers on the TBM than from the

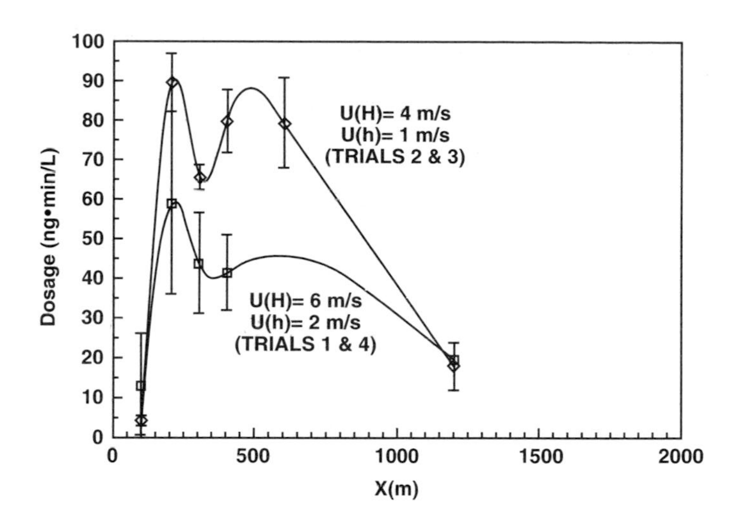

Figure 6. Dosage of nonvolatiles at 5 m above ground level from a single TBM spraying five swaths on a line (Dunphy 1984).

Table 4a. Drift/deposit from TBM Teejet® emission in Dunphy, 1986.

Trial	Date Time ADT	Stability	Aircraft ht (m)	Wind[a] °mag./m·s	Drift/deposit (% nonvolatiles) 400 m	600 m	900 m	Sunrise or sunset ADT
2	07/7, 1034	near neutral	45	229/6.6		58/18		0536
3	08/7, 2055	moderately stable	40	291/3.5	56/44	52/47		2120
4	10/7, 0625	near neutral	36	268/3.5		31/37		0539
6	17/7, 0644	moderately stable	36	265/4.2	50/26		35/36	0548

[a] Tabulated wind speed and direction are at aircraft height. The direction should be compared to the upwind perpendicular to the flight line, 255° magnetic.

Table 4b. Drift/deposit from Cessna-188 Micronair® emission in Dunphy, 1986.

Trial	Date Time ADT	Stability	Aircraft ht (m)	Wind[a] °mag./m·s	Drift/deposit (% nonvolatiles) 400 m	600 m	900 m	Sunrise or sunset ADT
2	07/7, 1034	near neutral	63	229/7.6		50/09		0536
3	08/7, 2055	moderately stable	35	291/3.5	86/25	66/30		2120
4	10/7, 0625	near neutral	37	268/3.5	82/23	56/25		0539
6	17/7, 0644	moderately stable	35	265/4.1	80/27		30/39	0548
10	22/7, 2032	slightly stable	42	232/4.4	88/08		54/20	2112

[a] Tabulated wind speed and direction are at aircraft height. The direction should be compared to the upwind perpendicular to the flight line, 255° magnetic.

Table 5. Tower profiles from Dunphy, 1986.

Item[a]	Trial									
	1	2	3	4	5	6	7	8	9	10
Mean wind speed (m/s)										
U(9)										0.21
U(18)	2.1	3.7	1.5	1.9	1.3	1.4	2.6	0.5	1.8	1.6
U(25)	2.9	4.6	2.2	2.8	2.4	3.0	3.5	1.1	2.8	2.8
U(38)	3.9	6.0	3.2	3.6	3.2	4.4	4.3	1.8	3.2	3.9
U(63)	5.3	7.6	4.9	4.9	4.3	6.6	5.1	4.0	3.7	5.8
Friction velocity (m/s)										
$u_*(9)$										0
$u_*(18)$	0.48	0.87	0.21	0.42	0.25	0.23	0.67	0.11	0.53	0.21
$u_*(25)$	0.44	0.81	0.21	0.43			0.67	0.06	0.48	0.31
$u_*(38)$	0.46	0.80	0.23	0.48			0.73			0.26
$u_*(63)$	0.47	0.71	0.24	0.40	0.26	0.17	0.78	0.05	0.49	0.18
Root mean square vertical turbulent velocity (m/s)										
$\sigma_w(9)$										0.07
$\sigma_w(18)$	0.57	0.96	0.24	0.44	0.28	0.27	0.74	0.07	0.58	0.28
$\sigma_w(25)$	0.57	1.04	0.27	0.55			0.86	0.08	0.72	0.37
$\sigma_w(38)$	0.60	1.02	0.26	0.57			0.93			0.25
$\sigma_w(63)$	0.60	0.96	0.28	0.54	0.33	0.23	1.02	0.07	0.90	0.25
Deposition velocity (m/s)										
$v_d(18)$	0.08	0.14	0.02	0.06	0.03	0.03	0.12	0.02	0.11	0.02
Temperature (°C)										
T(25)	19.3	15.2	24.7	9.3	13.4	13.3	28.3	9.6	24.4	21.6
Relative humidity (%)										
rh(25)	45	78	36	71	92	83	30	90	31	43
Water vapor deficit (mb)										
$P_{wv}(25)$	4.5	1.5	6.2	1.7	0.6	1.1	7.7	0.6	6.9	5.0
Gradient Richardson number										
$R_i(30)$	0.032	−0.013	0.087	0.029	0.032	0.077	−0.051	0.245	−0.356	0.076
Monin-Obukhov length (m)										
L	491	−833	79	540	500	53	−175	39	−42	165

[a] Numbers in parentheses are heights in metres above ground level.

Micronairs on the Cessna-188. Emission size spectra measurements (Picot et al. 1987) show a larger fraction of the Teejet® emission above 100 µm in diameter than of the Micronair® emission. Also, as discussed earlier, the higher fraction of the surfactant, Atlox, in the Cessna-188 emission would lower the droplet sizes relative to the TBM emission.

Further support for this contention of smaller droplets in the Cessna-188 emission arose from an "H-rod" dosage from the Cessna-188 spray exceeding the "U-rod" dosage by 29% in the one case where they were compared, trial 2; the equivalent value from the TBM spray was only 6% (i.e. within experimental error). As explained in Crabbe and McCooeye (1988), the "H" rod collects small droplets more efficiently than the "U" rod.

It is also likely that the higher vortex wake descent speed of the TBM, 0.8 m/s vs 0.5 m/s, contributed to lower drift from the TBM emission.

Averaged over both aircraft, budget closure in these cases selected from the 1986 Dunphy study ranged from 99% at 400 m to 80% at 600 m to 71% at 900 m downwind. The degradation in accuracy with downwind distance resulted partly from sparser foliar sampling with increasing distance and from inaccuracies in the blimp dosimeter readings, particularly at the 900-m station.

Kapuskasing

The Laboratory undertook the 2-year Northern Ontario field study to extend the Dunphy data into the extremes of stable and unstable meteorological conditions and to test its new portable Rotorod packages as a replacement for the dosimeters in measuring the vertical droplet dosage profile in the drift cloud. This site had been used by the Forest Pest Management Institute of Forestry Canada in a field program to establish buffer zones for aerial herbicide releases (e.g. Payne et al. 1986).

Forest site and spray protocol—Vegetation at the Kapuskasing site in Northern Ontario (49.30°N, 83.29°W) is a mixture of 2- to 6-m-high alders and some higher conifers on flat terrain. Tower measurements revealed site-averaged, zero-plane

Figure 7. Kapuskasing test site.

displacement, and roughness lengths of 4 m and 1 m, respectively. A photograph of the site appears in Figure 7.

The higher alders and conifers for most of the roughly 2 km × 2 km site predominated the height and density of the vegetation covering most of the near field grid, which averaged only 1.6 m at 12 300 stems per hectare (N. Payne, pers. comm.). However, some of this vegetation was denuded during the herbicide buffer zone study. Thus, when measuring foliar deposition in 1988 with the simulators described earlier under Chemical Sampling and Analysis, the data were weighted according to whether or not the surrounding vegetation had leaves.

In the first year of the Kapuskasing study, two orthogonal spray lines were identified to accommodate wind directions from the southwest and northwest. A Cessna-188 aircraft fitted with four Micronair® AU5000 atomizers was flown at 22 m above ground level to apply four to six overlaid swaths of the Dunphy TBM formulation at 5-minute intervals. (Repeated applications tend to remove random effects of aircraft motion and atmospheric turbulence by the process of ensemble averaging.) In all cases, the line flown minimized the difference between the wind direction and the flight-line perpendicular. Two atomizer settings were used, 4 535 rpm and 14.5 L/min to generate spray droplets with a volume median diameter about 200 μm and 6 425 rpm and 4.6 L/min to generate droplets with a volume median diameter about 120 μm (50% of the spray volume resides in droplets of diameter greater than the volume median diameter and 50% resides in droplets of diameter less than the volume median diameter). Drift cloud mass was measured at 400 m and 1 200 m downwind.

In the second year of the Kapuskasing study, a third flight line was identified for westerly flow and the measurements were extended to include ground and foliar deposition to 400 m and cloud mass at 400 m and 2 200 m from each flight line. The spray aircraft was fitted with four Micronair® AU4000 atomizers operating at 9 500 rpm and 5 L/min to generate ultra-low-volume aerial sprays, representative of those recommended by the New Brunswick Spray Efficacy Research Group to optimize forest insecticide spray efficacy (i.e. an evaporated emission with a major proportion of its mass from 15 to 55 μm in diameter). Up to six overlaid swaths were applied in each 30-minute trial by the Cessna-188 flying at a targeted height of 24 m above ground level (approximately 20 m above vegetation) to better approximate operational altitudes in forest insecticide applications than was the case in 1987.

The line source strength of the Cessna-188 spray in 1987 and the Micronair® speed were determined from the pilot's observations of flow rate and rotary speed displayed in the cockpit. In 1988, the frequency of the rotating flowmeter turbine was logged, as well as the pressure of a sensitive barometer installed in the cockpit and the outputs of the Micronair® rotary speed magnetic pick-ups. When suitably software-processed, these records yielded volume flow rate (summed over the four Micronairs), actual aircraft height above ground level, and Micronair® revolutions per minute (R. Mickle, pers. comm.).

Drift and deposit samplers—Up to seven levels of the Rotorod packages, described earlier under Chemical Sampling and Analysis, were deployed to heights ranging from 200 m above ground level at 400 m from the flight line to 400 m above ground level at 2 200 m from the flight line. The vertical Rotorod arrays comprised fewer samplers than the equivalent dosimeter arrays in the Dunphy study because the Rotorod package is about 50% heavier than the dosimeter (1 kg vs, 0.7 kg). Fifteen deposit stations were serviced to a distance of 400 m from each flight line using the instrumentation described earlier.

Meteorological measurements—Meteorological data from years 1987 and 1988 of the Kapuskasing study comprised tower profiles of atmospheric surface layer turbulence, vertical heat flux, and vector mean wind speeds using Dunphy instrumentation, as well as tethersonde profiles of temperature, humidity, and vector winds to 300 m above ground level in 1987 and 400 m above ground level in 1988 (R. Mickle pers. comm.). These profiles were extended to 4 km above ground level with an airsonde in 1988. In 1987, a 20-m instrumented meteorological mast was raised on site; a 92-m microwave tower 35 km north of the site was also instrumented in 1987 for temperature profiling (feasible because of large-scale horizontal homogeneity) and in 1988 the Laboratory raised a 26 m instrumented mast on site.

Experimental results—A complete description of the experimental details in the Kapuskasing experiment appears in Crabbe and McCooeye (1989). Selected data on drift cloud mass from year 1987 and year 1988 of the study appear in Tables 6 and 8, respectively, and Tables 7 and 9 contain the corresponding 30-minute tower profiles of the atmospheric surface layer flow.

Discussion of experimental results from Kapuskasing 1987—Referring to Tables 6 and 7, spray trials 1 and 2 featured large contrasts in wind speed and thermal (static) stability with near-neutral conditions and moderate wind speed flow in trial 1 and light wind speed with almost free thermal convection in trial 2. The former corresponds to the most frequently studied atmospheric surface layer flow, and the drift value at 400 m downwind, 43%, may establish the reference for droplet drift from low-volume aerial emissions over moderately rough vegetation for these conditions, although the large difference between the wind direction, 170° magnetic, and the upwind perpendicular to the spray line, 217° magnetic, will have influenced this result somewhat.

The free convection conditions in spray trail 2 raised the cloud top to 700 m above ground level at 1 200 m downwind, but in contrast to widely held perceptions of the hazards of aerial spraying in "thermal" conditions, the mass of nonvolatiles was low, at 19% of the emission. The explanation lies partly in the definition of near-free convection conditions, large turbulent transport of heat from the ground with very light winds, and partly in the effect of large turbulent motions in near-free convection. Light winds lower off-target wind-borne droplet transport and the value of v_d at 3 cm/s was among the higher values measured during the study.

In trials 1 and 2, the application rates were higher and droplet sizes larger than in the remaining trials. The drift figures for these trials would therefore increase slightly for the ultra-low-volume applications used in trial 3 and beyond.

The most elucidating comparison involves spray trials 3 to 5 which featured conditions varying from an unstable mixing layer capped by inversion at 80 m above ground level in trial 3 to an (uncapped) moderately stable layer in trial 4. In the latter, the Cessna-188 emission left 60% of nonvolatiles in the atmosphere at 400 m downwind and 20% at 1 200 m,

Table 6. Drift from Cessna-188 Micronair® emission in Kapuskasing, 1987.

Trial	Date Time EDT	Stability	Aircraft ht (m)[a]	Wind[b] °mag./m·s	Drift (% non-volatiles) past 400 m	1 200 m	Sunrise or sunset EDT	Inversion base ht (m)
1	13/8, 1920	near neutral	22	170/3.5	43		2055	0
2	14/8, 1047	very unstable	22	296/0.9		19	0622	700
3	19/8, 0736	moderately unstable	22	276/4.1	18	10	0631	80
4	19/8, 1957	moderately stable	22	318/2.0	60	20	2043	0
5	20/8, 0708	slightly unstable	22	224/4.0	56	21	0632	18

[a] Actual heights were not measured but were observed to be close to the targeted value of 22 m above ground level.
[b] Tabulated wind speed and direction are at aircraft height.

Table 7. Tower profiles from Kapuskasing, 1987.

Item[a]	Trial 1	2	3	4	5
Mean wind speed (m/s)					
U(9)	2.4	0.7	2.9	1.3	2.6
U(20)	3.4	0.9	4.1	2.0	3.9
Friction velocity (m/s)					
$u_*(9)$	0.30	0.16	0.29	0.00	0.29
$u_*(20)$	0.29	0.24	0.37	0.10	0.23
Root mean square vertical turbulent velocity (m/s)					
$\sigma_w(9)$	0.34	0.26	0.38	0.05	0.34
$\sigma_w(20)$	0.31	0.35	0.47	0.05	0.32
Deposition velocity (m/s)					
$v_d(9)$	0.03	0.03	0.02	0.00	0.02
Temperature (°C)					
T(20)	20	24	10	17	8
Relative humidity (%)					
rh(20)	90	70	100	66	100
Monin-Obukhov length (m)					
L	9999	−2	−52	46	−123
Inversion base height (m)	0	700	80	0	18

[a] Numbers in parentheses are heights in metres above ground level.

whereas the former, which featured higher wind speeds, left only 18% and 10% respectively, because of the higher downward turbulent transfer rate (v_d = 2 cm/s vs 0 cm/s) and confined vertical mixing.

The corresponding values in the slightly unstable trial 5 are higher at 56% and 21% respectively, than those of the moderately unstable spray trial 3 because the Cessna-188 injected the spray material into a stable layer capping the shallow 18-m-thick unstable surface layer. The aircraft vortex wake, which contains the emission, does not readily sink into a stable layer, resulting in relatively large values of drift. Nevertheless, the drift is still less than in the surface-based inversion of trail 4 in spite of higher wind speed.

Vertical profiles of longitudinal droplet transport normalized by line source strength are plotted in Figure 8 for spray trials 3, 4, and 5 of year 1987 of the Kapuskasing study. The area enclosed by curves and the height axis yield the drift values reported in Table 6 for these spray trials. Plotted in this manner, the data clearly reveal the beneficial effect of thermal instability in reducing the fraction that drifts beyond the target boundaries.

Discussion of experimental results from Kapuskasing 1988— The 1988 Kapuskasing field experiment extended the 1987 results into a wider range of atmospheric thermal stability in sufficient detail to elucidate further the role of the turbulent mixing layer depth or inversion base height in influencing drift. An inversion base acts as a cap on vertical droplet dispersion and thereby reduces drift by maintaining higher droplet concentrations near the canopy and higher deposition than in "unlimited" vertical mixing. Referring to Table 8, the drift cloud mass at 400 m downwind in the unstable spray trials varied from 56% of the nonvolatiles when the inversion base was 750 m above ground level (trial 10) to 44% when it was 410 m above ground level (trial 3), and to 38% when the inversion base was only 155 m above ground level (trial 7). Further reduction in mixing layer depth to 80 m lowered the drift cloud mass at 400 m downwind to only 18% (trial 3 in year 1987). The exception to this trend is trial 11 where the aircraft height was 3 to 4 m higher than in the other unstable trials. However, the drift past 2 200 m downwind in this trial, as in the other unstable trials, was less than 30% of nonvolatiles.

The highest off-target drift of nonvolatiles past 400 m downwind, 77%, occurred in the slightly stable spray trial 4 due to the moderately high wind speed, 5.1 m/s, but deceased to only 27% at 2 200 m because of the relatively large value of downward turbulent transfer, v_d = 6 cm/s (see Table 9). The highest drift of nonvolatiles past 2 200 m downwind, 50%,

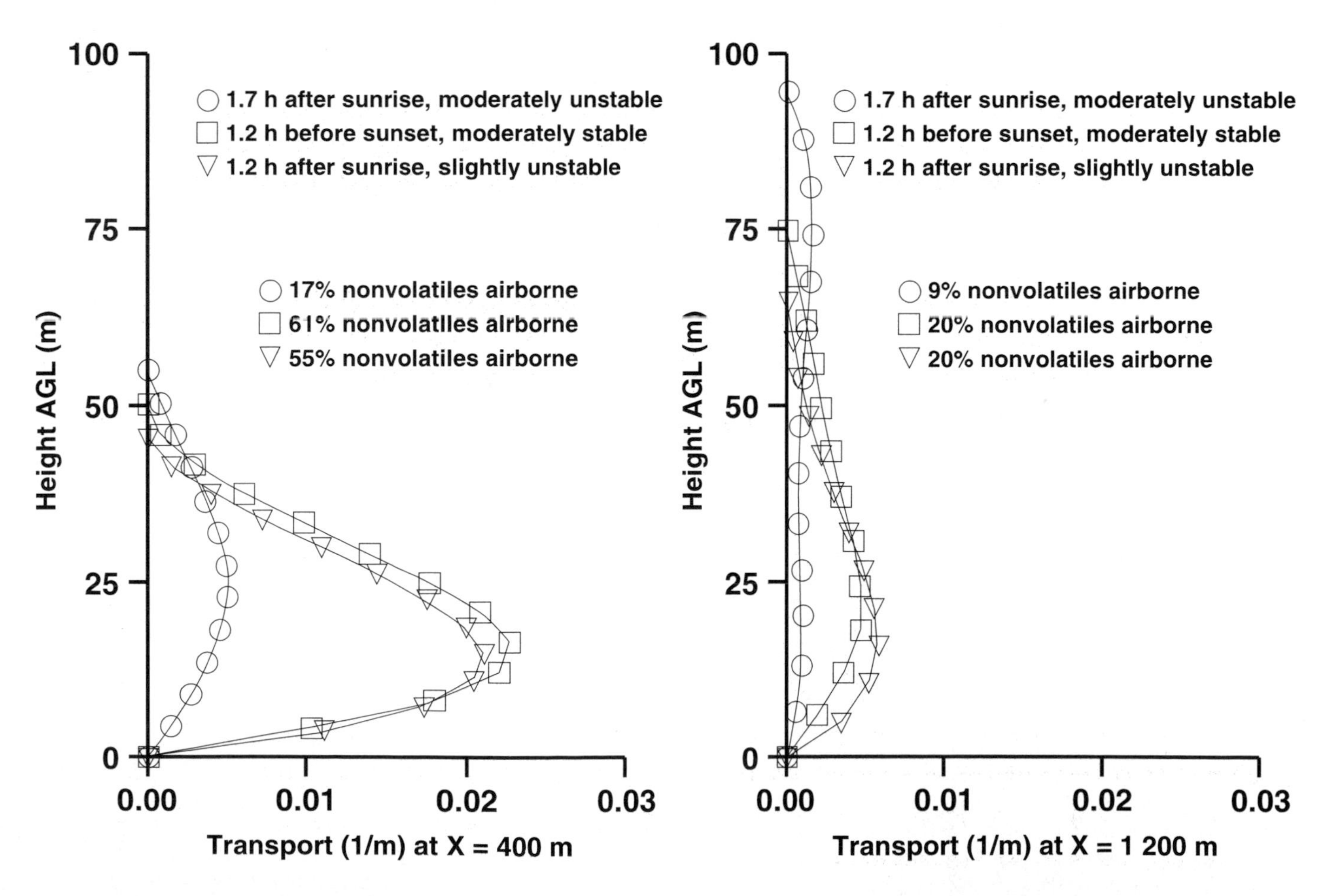

Figure 8. Fractional wind drive per metre of height from Micronair® AU5000 atomizers on a Cessna-188.

Table 8. Drift from Cessna-188 Micronair® emission in Kapuskasing, 1988.

Trial	Date Time EDT	Stability	Aircraft ht (m)	Wind[a] °mag./m·s	Drift (% non-volatiles) past 400 m	Drift (% non-volatiles) past 1 200 m	Sunrise or sunset EDT	Inversion base ht (m)
2	02/7, 0700	moderately stable	28	228/3.0	71	50	0530	0
3	03/7, 1038	moderately unstable	25	285/3.3	44	24	0531	410
4	03/7, 2020	slightly stable	27	255/5.1	77	27	2143	0
7	09/7, 0920	slightly unstable	26	292/2.9	38	29	0537	155
8	09/7, 1837	moderately stable	24	311/1.9	44	31	2139	0
10	14/7, 1642	moderately unstable	26	290/4.2	56		2135	750
11	17/7, 0907	slightly unstable	29	248/3.8	50	27	0546	315

[a] Tabulated wind speed and direction are at aircraft height.

Table 9. Tower profiles from Kapuskasing, 1988.

Item[a]	Trial 1	2	3	4	5	6	7	8	9	10	11
Mean wind speed (m/s)											
U(10)	2.1	1.5	2.4	2.9	1.3	2.2	2.1	1.2	2.8	3.1	2.4
U(18)	2.4	2.0	2.9	4.2	2.4	3.2	2.6	1.6	3.6	3.9	3.2
U(26)	2.7	2.8	3.3	5.0	3.1	4.4	2.9	2.0	4.2	4.2	3.7
Friction velocity (m/s)											
$u_*(10)$	0.40	0.13	0.43	0.50	0.10	0.31	0.40	0.12	0.42	0.52	0.43
$u_*(18)$	0.42	0.12	0.41	0.52	0.00	0.37	0.44	0.12	0.53	0.48	0.43
$u_*(26)$	0.39	0.18	0.48	0.59	0.00	0.34	0.46	0.16	0.54	0.51	0.45
Root mean square vertical turbulent velocity (m/s)											
$\sigma_w(10)$	0.56	0.17	0.52	0.55	0.10	0.41	0.43	0.17	0.55	0.61	0.52
$\sigma_w(18)$	0.64	0.17	0.54	0.59	0.10	0.48	0.48	0.18	0.61	0.67	0.53
$\sigma_w(26)$	0.71	0.19	0.60	0.62	0.10	0.43	0.53	0.20	0.63	0.73	0.52
Deposition velocity (m/s)											
$v_d(10)$	0.05	0.01	0.05	0.06	0.00	0.03	0.05	0.01	0.04	0.06	0.05
Temperature (°C)											
T(10)	18	11	24	26	16	29	22	28	19	15	17
Relative humidity (%)											
rh(10)	40	64	44	35	61	51	53	37	52	57	84
Monin-Obukhov length (m)											
L	≈−30	24	−64	255	0	159	−118	62	2666	−65	−131
Inversion base height (m)	100	0	410	0	0	0	155	0	9999	750	315

[a] Numbers in parentheses are heights in metres above ground level.

occurred in the moderately stable lighter wind speeds of trial 2 due to the buoyant retardation of the wake vortex, which contains most of the emission, and the relatively low value of v_d = 1 cm/s. However, very light winds in moderately stable flow do lower drift as in trial 8 where 44% drifted past 400 m downwind and 31% past 2 200 m downwind in a wind speed of only 1.9 m/s. Nevertheless, these values also exceed those in the unstable trials with similar aircraft flying heights, provided vertical mixing is limited.

General Discussion

Not every spray trial in the field studies returned useful data. Drift data from trials 1 and 5 of year 1986 of the Dunphy study and trial 9 of year 1988 of the Kapuskasing study were discarded because of rain contamination. Drift data from Dunphy trials 8 and 9 in 1986 and Kapuskasing trials 1 and 6 in 1988 are omitted from discussion because of the largely unknown effect of the highly oblique winds in these trials.

Table 10. Deposit from Cessna-188 Micronair® emission in Kapuskasing, 1988.

	Trial							
	2	3	4	6	7	8	10	11
Deposit to 400 m (% nonvolatiles)	13	4	5	11	6	15	5	4

Results from Kapuskasing trial 5 in 1988 where blimp sampler spacing was too coarse to define the shape of the relatively compact vertical dosage profiles in the very stable conditions of this trial were also discarded. Drift cloud data from trial 7 of year 1986 of the Dunphy study are also not reported because the vertical dosimeter arrays did not encompass the entire depth of the drift cloud in the unstable conditions of this trial.

The measured values of deposit to 400 m from the flight line in year 1988 of the Kapuskasing study (Table 10) tend to be too low inasmuch as they do not generally effect budget closure. More deposit stations would have helped, but the nature of the vegetation, 1- to 2-m largely leafless alders and willows, raised difficulties in relating the cage and leaf sampler deposits to actual values because a significant fraction of the vegetation comprised stems rather than leaves. The deposit measurements were preliminary and were undertaken to assess the performance of a new sampler design.

This widespread absence of foliage surface area is expected to have further influenced the mass balance in that it would not only increase ground deposit at the expense of foliar deposit, but also increase the subcanopy wind speeds over the very low values typically observed in denser vegetation. Thus, it is unlikely that the horizontal deposit plates in sparse grass 20 cm or so above ground level would have efficiently collected the small droplets comprising these ultra-low-volume emissions. Support for this hypothesis arises in budget closure values in the stable trials, where the subcanopy wind speeds would be low due to the absence of significant momentum transfer, exceeding those in the unstable trials. Average values are 75% in the stable flows but only 52% in the unstable flows.

Summary and Conclusions

The results of the controlled field studies in New Brunswick and Northern Ontario forests described in this chapter, particularly from the Kapuskasing study where improved droplet samplers were used, clearly demonstrate that neutral to unstable (i.e. turbulent) conditions can reduce off-target drift relative to stable conditions. This conclusion would be particularly relevant to the aerial application of ultra-low-volume formulations comprising (evaporated) droplets 15 to 55 μm in diameter which the New Brunswick Spray Efficacy Research Group advocates for forest protection. However, it also applies to any pesticide spray application where droplets smaller than about 100 μm in diameter are present.

Generally, neutral to unstable conditions occur between about 2 hours after sunrise and about 2 hours before sunset. However, since the inversion base has a characteristic rate of rise following sunrise (approximately 50 m/h), forest pesticide applicators, to reduce off-target drift, should fly their spray lines in the morning turbulent mixing layer beneath a low, well-defined capping inversion. The height of the inversion base can be determined from a vertical temperature profile.

References

Crabbe, R.S. 1984. Aerodynamic analysis of a prototype cascade impactor. Natl. Res. Counc., Ottawa, Ont. NAE LTR-UA-74.

Crabbe, R.S.; Elias, L.; Krzymien, K.; Davie, S. 1980a. New Brunswick forestry spray operations: field study of the effect of atmospheric stability on long-range pesticide drift. Natl. Res. Counc., Ottawa, Ont. NAE LTR-UA-52.

Crabbe, R.; Krzymien, M.; Elias, L.; Davie, S. 1980b. New Brunswick forestry spray operations: measurement of atmospheric fenitrothion concentrations near the spray area. Natl. Res. Counc., Ottawa, Ont. NAE LTR-UA-56.

Crabbe, R.S.; McCooeye, M. 1985. Effect of atmospheric stability and windspeed on wind drift in aerial forest spray trials, neutral to unstable conditions. Natl. Res. Counc., Ottawa, Ont. NAE LTR-UA-82.

Crabbe, R.S.; McCooeye, M. 1988. The Dunphy field study relating atmospheric stability to wind drift from aerial forest spray operations. Natl. Res. Counc., Ottawa, Ont. NAE LTR-UA-98.

Crabbe, R.S.; McCooeye, M. 1989. The Kapuskasing field study relating atmospheric stability to wind drift from aerial forest spray operations. Natl. Res. Counc., Ottawa, Ont. NAE LTR-UA-99.

Crabbe, R.S.; Reid, J.D. 1980. Results from two models for the long-distance drift and dry deposition of an insecticide cloud onto a forest. Second Joint Conference on Applications of Air Pollution Meteorology. March 24-27, 1980, New Orleans, La.

Kristmanson, D.D.; Picot, J.J.C. 1983. Dunphy drift trials-1982. Measurement of airborne drops and Kromekote card deposits at ground level; foliage simulator tests. Univ. of N.B., Dept. of Chem. Eng. Rep., Fredericton, N.B.

Krzymien, M. 1980. Determination of fenitrothion andtris (2-ethyl3 hexyl) phosphate vapor pressures. Natl. Res. Counc., Ottawa, Ont. NAE LTR-UA-19.

May, K.R.; Pomeroy, N.P.; Hibbs, S. 1976. Sampling techniques for large windborne particles. J. Aerosol. Sci. 7: 53-62.

Payne, N.; Helson, B.; Sundaram, K.; Kingsbury, P.; Fleming, L.; de Groot, P. 1986. Estimating the buffer required around water during permethrin applications. For. Can., For. Pest Manage. Inst., Sault Ste. Marie, Ont. Inf. Rep. FPM-X-70.

Picot, J.C.; Chitrangad, G.; Henderson, G. 1981. Evaporation rate correlation for atomized droplets. Transaction of the ASAE, 24, No. 3, pp 552-554.

Picot, J.J.C.; Wallace, D.J.; Kristmanson, D.D. 1987. The PKBW model for prediction of aerial spraying deposition and drift. ACAFA Symp. on the aerial application of pesticides in forestry, Ottawa, Canada, Oct. 1987.

Raman, S.; Brown, R.M. 1976. A comparison of turbulence measurements made by a hot-film probe, a bivane, and a directional vane in the atmospheric surface layer. J. Appl. Meteorol 15: 138-144.

Riley, C.M. 1985. Deposition of spray droplets within the forest canopy, Dunphy-1984. Research and productivity council report number C/84/006, Fredericton, N.B.

Steeves, B.G. 1986. Meteorological analysis pertaining to the FPL/NRC experimental spray trials of 1983, 1984 and 1986. Univ. of N.B., Faculty of Forestry Rep., Fredericton, N.B.

Tomney, T.D.; Smedley, J.B.; Kristmanson, D.D.; Picot, J.J.C. 1978. The characteristics of drop size distributions generated by aerial spray atomizers. Univ. of N.B., Dept. of Chem. Eng. Rep., Fredericton, N.B.

van Vliet, M.W. 1985. Deposition of aerial insecticide sprays. Univ. of N.B., Dept. of Chem. Eng., B.Sc. thesis, Fredericton, N.B.

van Vliet, S. 1987. Summary of results for Dunphy 1986 spray trials. Univ. of N.B., Dept. of Chem. Eng. Rep., Fredericton, N.B.

Wesley, M.L.; Hicks, B.B.; Dannevik, W.K.; Fusella, S.; Husar, R.B. 1977. An eddy correlation method of particulate deposition from the atmosphere. Atmos. Environ. 11.

Chapter 52

Spray Atomizer Droplet Characterization

J.J.C. PICOT, D.D. KRISTMANSON, AND M.W. VAN VLIET

Introduction

The control of forest pests is usually done by atomizing a pesticide (as a liquid or slurry) and introducing the droplets into the air over the forest canopy. If the objective is to apply the pesticide to the needles of conifers, the important mechanism is the aerodynamic capture of droplets by the needles as the droplets are carried past in the airstream (assuming that the droplets are less than 100 μm in diameter and have very low settling velocities). Wallace (1988), in his description of a simulation model, shows how to calculate the probability of deposition of droplets on foliage using information on weather parameters and the aerodynamic collection efficiency of foliage shoots on coniferous species, as well as data on the droplet size emission characteristics of the atomizers used. Kristmanson et al. (1987) describe field trial results from a simulated spruce budworm, *Choristoneura fumiferana*, spray which show that droplet collection efficiency drops off sharply for droplets smaller than about 20 μm. They note also that droplets larger than about 50 μm are not present in sufficient numbers to provide effective coverage in a conventional application rate of 1.4 L/ha of emulsion containing 13.8% active ingredient by volume (210 g/ha of active ingredient), with boom and nozzle spray gear. Picot et al. (1985) have claimed that the optimal droplet size range arriving at the needles for spruce budworm control is from 15 to 55 μm. For a different pest and target, the governing mechanisms for optimal droplet size are the same (including gravitational settling), but the optimal range may well be different, according to the specifics of the problem. The most efficient atomizer will be that which generates all of its droplets in the desired size range. Smaller droplets produce undesirable drift off-target, and larger droplets are wasteful of pesticide. This paper reports on methods used to characterize atomizers and gives measured data on several commercially available atomizers used in aerial spraying.[1]

Measurement Techniques

Because of the extreme difficulty of modeling atomizer performance, tests of atomizer droplet size emission characteristics are carried out at full scale. Actual atomizers are placed in a wind tunnel and tested with the wind tunnel air speed equal to the aircraft operating speed. This assumes that atomizers on aircraft are mounted in an undisturbed flow location in the aircraft slipstream, which is not always the case. In this work, droplets were measured in flight using two single-particle laser droplet spectrometers constructed by Particle Measuring Systems, Boulder, Colorado. One of these is a light scattering probe (the FSSP-100 probe) to measure droplets in the range (nominal) 0.5 to 45.5 μm, and the other is a light imaging probe (the OAP-260X) for droplets in the range 7.1 to 625 μm. Figure 1 shows details of the wind tunnel used, while Figures 2 and 3 give details of the probe systems. Some modifications of the laser probe tips were required and these are shown in Figure 4. Additional information on wind tunnel and probe systems is given below.

Wind Tunnel

The test section is 1.02 m in diameter and 3.05 m long. The overall length of the tunnel is 7.73 m. Tunnel air speeds up to 75 m/s can be generated by the 153-kW (205-hp) diesel engine which drives a 1.37-m diameter axial flow fan. This

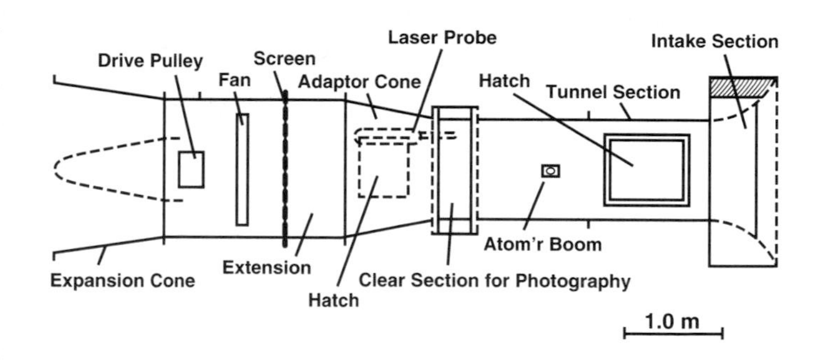

Figure 1. Wind tunnel.

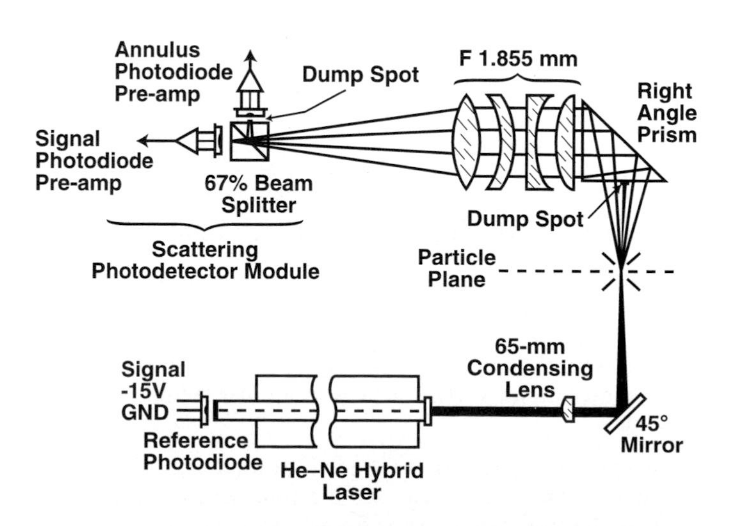

Figure 2. FSSP-100 probe optics.

[1] A correlation for volume median diameter and spectrum span is given by the authors for a range of atomizers and operating conditions in Can. J. Chem. Eng., 1989, 67: 752–761, Droplet size characteristics for insecticide and herbicide spray atomizers, by J.J.C. Picot, M.W. van Vliet, and N.J. Payne.

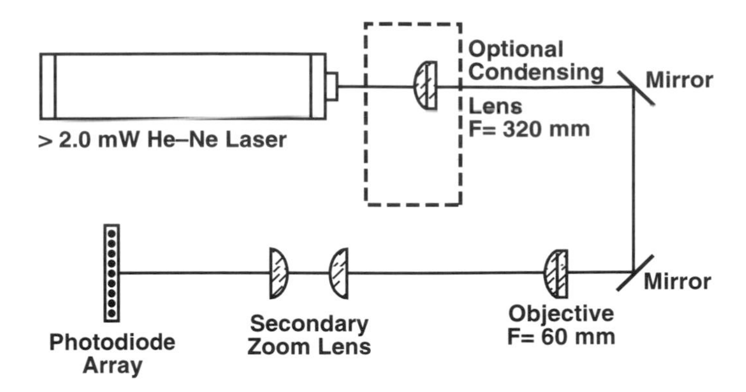

Figure 3. OAP-260X probe optics.

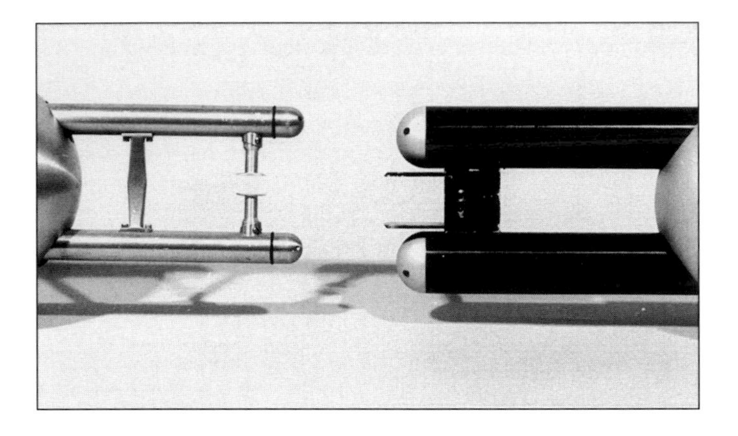

Figure 4. Laser probe tips, OAP-260X (*left*) and FSSP-100 (*right*).

whole apparatus is mounted on a trailer. Actual pesticide formulations may be tested by moving the trailer to a remote location.

The atomizer being characterized is mounted in the test section 1 m upstream of the sensing probe. The atomizer is supported by an airfoil-shaped boom which can be moved laterally across the diameter of the tunnel. Metered fluid is supplied via a conduit inside the boom. The sensing probes can be traversed vertically at a controlled rate by a screw drive assembly coupled to a 560-W (0.75-hp) electric motor.

Forward Scattering Spectrometer Probe

Figure 2 shows the optics of the Forward Scattering Spectrometer Probe, FSSP-100. Particles passing through the beam within the sample area are measured by an electronic comparison of the relative quantity of forward scattered light to predetermined voltage levels in a pulse height analyzer module.

Probe tip modifications—Figure 4 shows the probe tip modification. The cylindrical sample tube was replaced by a rectangular tube where two sides are extended sufficiently far ahead of the optical turrets to prevent droplets rebounding off the curved surface into the probe sample area. This modification was required because the original sample tube created some masking of larger droplets, preventing them from arriving at the sample area. Downstream turbulence created by rotating atomizers was sufficient to create droplet velocity vectors that were not parallel to the wind tunnel axis. For the original sample tube, the maximum permissible off-axis angle for large droplets was 18°. This was increased to 28° on the sides adjacent to the turrets, and to much greater values on the other two sides. Special inserts were constructed for calibration with glass beads, permitting aspiration of the beads past the sample area.

Instrument calibration—In this work, the light scattering probe is used as the standard for droplet flux (droplets per unit time and droplet volume per unit time for each size-class). This requires extensive calibration of the instrument and continuous monitoring of the calibration. The instrument provides for a printed output of droplets counted in 15 size-classes covering the range from 0.5 to 45.5 μm (nominal) with the internal software programed for the refractive index of water. Picot et al. (1985) give the corrections needed to convert the output sizes for liquids of different refractive index. Cerni (1983) and Dye and Baumgardner (1984) have studied the measurement accuracy of the instrument and have both indicated a size reduction effect at particle transit velocities greater than 50 m/s. This has not been observed by the authors of this paper. However, an elaborate calibration procedure was found necessary to calibrate probe sample cross section in output size ranges, as explained below. The sample cross section is that portion of the laser beam in which particles in transit will be "seen" and processed as acceptable particles by the probe.

A Berglund-Liu (TSI, St. Paul, Minnesota) monodisperse droplet generator shown in Figure 5 was mounted on the horizontal traversing boom in the wind tunnel approximately 25 cm upstream of the FSSP-100 probe. This atomizer was initially set to generate peak numbers of droplets approximately 30 μm (using a nonvolatile, filtered oil). With the wind tunnel air speed at 25 m/s, a droplet spectrum was measured using a tight (1.0-cm intervals) sampling grid to encompass all emitted droplets. Fluid flow was measured with an electronic gravimetric balance. The procedure was repeated with the atomizer set to generate peak numbers of droplets less than 10 μm. In addition, two calibrated glass bead samples were measured with the probe. A calibration curve of the form shown in equation 1 was postulated for instrument calibration based on instrument outputs adjusted for refractive index:

$$D_{ci} = a + b \bullet D_m + c \bullet D_m^2 \quad (1)$$

where

D_{ci} = corrected particle size
a, b, c = calibration constants
D_m = instrument output particle size corrected for refractive index only

The data on fluid flow, instrument number frequency output, sample cross section, and bead calibration are all related by equation 1 and the following two equations:

$$V^j = \pi \cdot \sum_i n_i^j \cdot (D_{ci}^j)^3 \cdot L \cdot W/(t^j \cdot Sa \cdot 6) \quad (2)$$

where
V^j = measured volumetric flow rate to the mondisperse atomizer, for test j
n_i^j = droplet count per second in channel i, for test j
D_{ci}^j = instrument output average size for channel i, for test j, corrected for refractive index and instrument calibration
t^j = sample period for test j
Sa = probe sample area

$$\text{and } D_{bc}^j = a + b \cdot D_{bm}^j + c \cdot (D_{bm}^j)^2 \quad (3)$$

where
D_{bc}^j = microscope calibrated bead size for sample j (j = 1 or 2), based on maximum frequency size or $D_{v,0.5}$ of the sample (volume median diameter)
D_{bm}^j = instrument output bead size for sample j, corrected for refractive index only, corresponding to D_{bc}^j

Equations 1, 2, and the two equations 3, one for each calibration bead sample, provide a solution for *Sa*, *a*, *b*, and *c*, which are the sample cross section and the size calibration equation for the output ranges. Adequate accuracy requires that the two Berglund-Liu atomizer tests be repeated three times for the small and large droplet emissions, respectively. (Note that the Berglund-Liu atomizer in these tests should be operated in an untuned mode to generate a broad spectrum, but droplets must all be within the probe measurement range.)

Results from these calibration tests show that the probe can seem to be adequately calibrated from glass bead sample measurements alone, but still be sufficiently out of optical alignment to cause a serious error in sample cross section. Therefore, calibration as described here is required after each realignment of the optical components. Typical values of *Sa*, *a*, *b*, and *c* are:

Sa = 0.282 mm^2 (manufacturer's spec'n is 0.211 mm^2)
a = 0.376 μm
b = 0.4226
c = 160.6×10^{-4} μm^{-1}

The FSSP-100 probe contains a circuit that rejects particles that do not transit the central portion of the beam. This is done by comparing each particle transit time with the sample average particle transit time. Particles with times shorter than the average are electronically rejected. The assumption is that the particles are all moving at the same velocity, and short transit times arise from transit paths near the edges of the circular cross section laser beam, causing undersizing. The reject ratio for the instrument is nominally 0.50. Tests show that this value is achieved within 2% when 3 000 or more particles are processed for a single sample. Monitoring of this circuitry is necessary to ensure that it is not malfunctioning.

The probe is equipped with a clock that measures the time taken up in processing particle transit pulses. This time is greater than the actual particle transit time through the laser beam. Accordingly, since new particle transits are ignored during this time, it is necessary to account for this because resulting

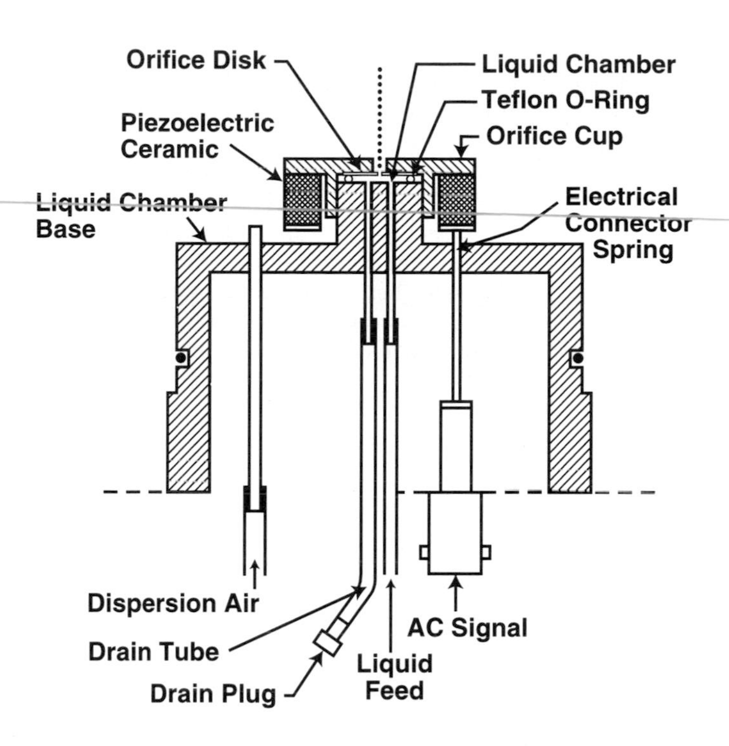

Figure 5. Berglund-Liu monodisperse droplet atomizer.

counts will be based on the actual sample period minus probe dead time. The following formula is recommended:

$$\frac{n_{ci}}{n_{ri}} = \frac{1}{1 - A \cdot K} \quad (4)$$

where
n_{ri} = droplet count for channel i, instrument output
n_{ci} = droplet count as above corrected for probe dead time
A = activity in percent, an instrument output value
K = constant, 0.007 for the probe in this work

An electronic method was developed to verify the above equation and the factor *K*. This is described in detail by van Vliet and Picot (1989) and results show that equation 4 gives values within ± 3.5% of the correct value for indicated activities up to 91%. The manufacturer gives a simple method to check this circuit.

Due to the small sample cross section of this instrument, the probability of occupancy of this cross section by more than one droplet is of the order of 10 parts per million or less for any of the atomizer tests mentioned herein.

Light Imaging Probe

Probe tip modifications—The tip modifications for the OAP-260X probe are shown in Figure 4. Two aluminum turrets were positioned over each laser beam aperture. These were terminated with flat, oval disks defining the sample area. A 4-mm diameter central hole concentric with the laser beam was provided. These openings were connected to the "purge air" openings at the apertures to provide a constant air flow from inside to outside, preventing particles from getting trapped into eddies formed at the opening on the disk surface. The spacing between disks is 11 mm. The resulting sample areas are shown in Tablc 1.

This modification required no electronic adjustment. Data processing software was modified to account for the change in sample areas. This probe tip modification prevented droplet recirculation in the vicinity of the probe laser beam apertures, a problem with the factory version of the probe. Beam sampling length was reduced from the original 61 mm to decrease sample area droplet frequency which was originally much greater than the probe's processing capacity for the large rotating atomizers studied. The optical configuration for this probe is shown on Figure 3.

Instrument calibration—Hovenac (1986) and Hovenac et al. (1985) have studied the sizing accuracy of the OAP-260X probe. The former mentions spectrum broadening as a problem when sizing glass bead samples. Probe response broadening is also mentioned in the latter where calibrations were done using deposited circular spots on a rotating glass disk. Several graphs depicting probe outputs versus microscope sizing data for glass beads are shown in Figure 6. Sample bead velocities were approximately 20 m/s for the smallest particle size studied; a very noticeable spectrum broadening by the probe is evident. This instrument functions by counting the number of photodiodes (in a photodiode strip with 64 elements) that are covered by a particle shadow and multiplying by the width per element to get particle size. Edge effects are eliminated by rejecting events where either or both end photodiodes are covered. The spectrum broadening is an instrument artifact that is also evident when uniform size droplets are tested, as from the Berglund-Liu instrument. Its treatment is discussed below.

Tests to map out the sample area using a monodisperse atomizer effectively corroborate Table 1. As mentioned by Hovenac et al. (1985), probe size output depends on axial location of the particle in the laser beam field. However, in practice, a random distribution of particle transit positions occurs, and this matches the calibration situation expressed in Figure 6.

Experiments with a monodisperse atomizer show that the upper limit of response for the instrument is approximately

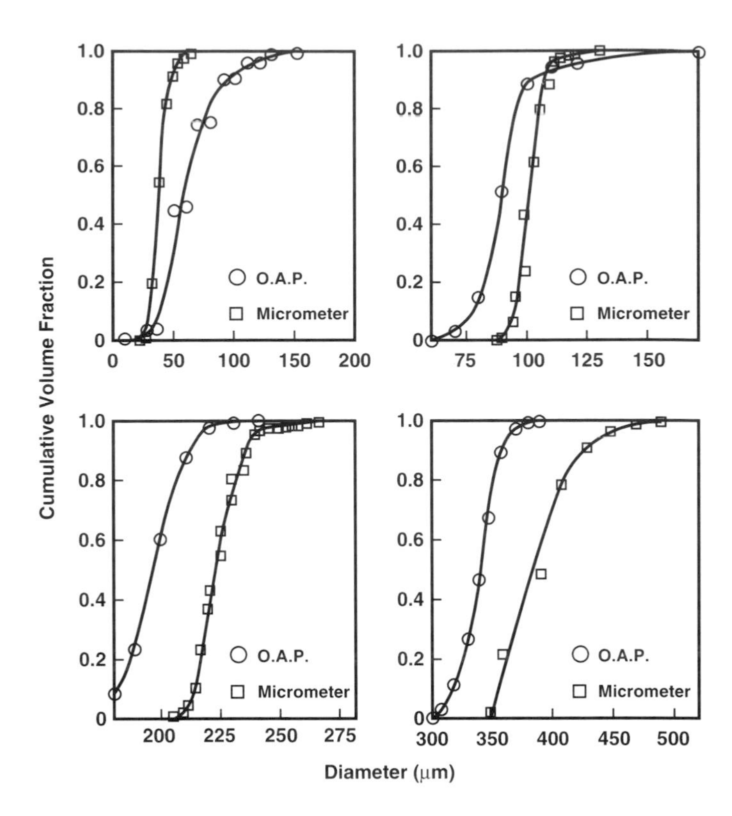

Figure 6. Calibration curves for the OAP-260X probe using microscope-measured glass bead samples.

35 000 particles per second when particle speed is about 10 m/s. This illustrates the dead time problem. The probe requires time to process the electronic data. This then creates a probability of temporal coincidence in tests with atomizers at flow rates of 2 L/min and higher. The probe ignores an unknown fraction of the droplet transit events. Spectrum shape is not distorted except in the case of continuous traversing through a structured plume. Spatial coincidence is supposedly eliminated by a rejection circuit that detects the simultaneous presence of more than one particle shadow on the photodiode strip.

Table 1. Sample areas for a modified OAP-260X probe.

Channel (i)	Sample area (mm^2)
1	0.059
2	0.456
3	1.354
4	2.571
5	3.868
6	5.39
7	6.00
8	8.25
9 to 62	8.10-[(i-9)•(0.15)]

Method for Spectrum Characterization of Atomizers

The FSSP-100 instrument is used as the basis for droplet counts per micrometres of size per unit sample area per unit time. Traverses through the complete plume are done, and probe calibration and functions are checked before and after each sequence of measurements. Secondary flow patterns caused by inlet instabilities create reasonably good mixing in the wind tunnel when continuous vertical traverse times are of the order of 10 seconds. This ensures that the probe activity represents an average activity for the overall traverse. When an inlet flow straightener is used, the plume remains inhomogeneous as it passes by the probe due to the absence of the secondary flows. This forces vertical traverses to be conducted in a point-wise fashion; for a continuous traverse, the activity output would be incorrect because it would represent time occupied in processing pulses in the central region of the plume divided by total sampling time. In these tests, continuous traverses were done at 5-cm horizontal spacing, and point-wise traverses at 5-cm horizontal and vertical spacing.

The OAP-260X output in droplets per micrometre per unit sample area per unit time is used only at sizes greater than the upper limit size of the FSSP-100 probe. An appropriate OAP-260X channel is fitted to the top end of the FSSP distribution by interpolation. The sizes and counts are then reduced to a volume flow rate as follows:

$$\text{Tot. Vol. Flux} = \sum \frac{n_{ci} \bullet L \bullet W}{Sa \bullet t} \bullet \frac{\pi}{6} \bullet D_{ci}^{3} \qquad \textbf{(5)}$$

where
L = distance of vertical spacing of probe traverse
W = distance of horizontal spacing of atomizer traverse
Since the OAP-260X suffers from unknown dead time, the total flow rate given by the FSSP-100 (equation 5) is subtracted from the flow rate supplied to the atomizer being tested. This is then compared with the volume given by the OAP-260X counts above the FSSP-100 upper size limit, using equation 5. This comparison produces a ratio by which the reported volumetric flux of the OAP-260X must be multiplied to produce the correct atomizer total flux. This same ratio is used to correct the individual OAP-260X channel counts. This assumes that OAP-260X errors do not distort spectrum shape. This assumption is valid except in the case of a structured plume when using a continuous traverse, where the probe can move from a region of large numbers of small droplets to one of small numbers of large droplets in the same sample period. A larger fraction of the smaller droplets would be ignored, producing spectrum distortion.

A droplet spectrum prepared as mentioned above is shown in Figure 7a. The mismatch in the number distribution at the interface between the two instruments is evident. This is due to the spectrum broadening effect of the OAP-260X mentioned earlier. The cumulative volume fraction curve for the same test is shown in Figure 7b. In this case, the mismatch creates a minimal problem.

Results of Atomizer Characterization Tests

The measurements described previously produce droplet spectra at generation since negligible evaporation occurs between atomizer and probe. However, a sufficient downstream distance is maintained between the two to ensure that droplet breakup is complete. In addition, because of the extreme dilution of liquid droplets by the moving airstream (concentration of droplets in air is of the order of 10 ppm), a negligible degree of droplet recombination or agglomeration exists. The properties and compositions of the test formulations used are summarized in Table 2. Results are taken from papers by van Vliet and Picot (1987) and van Vliet et al. (1987). The formulations used (except for Dipel® 132 which is a commercial Bt formulation from Abbott Labs, Figure 8) are simulants for a range of chemical insecticides with the same physical properties.

Complete droplet number frequency distribution spectra and cumulative volume fraction for a Teejet® 800050 nozzle, the Micron® X-1 rotating atomizer, and the Micronair® AU4000 atomizer are shown in Figures 9, 10, and 11, respectively. (A photograph of these and other atomizers is shown in Figure 12.) These plots all show increasing droplet frequency with decreasing droplet size, a characteristic of almost all of the atomizers tested by the authors. Some discontinuity is shown where the two sets of probe data meet, but overall accuracy is adequate for purposes of evaluating the atomizers for use in applying pesticides to coniferous foliage. The criterion for selection herein is a maximum volume fraction in the size range 15 to 55 μm at generation. Table 3 gives comparative results of atomizer performance based on this criterion for the commercial atomizers shown in Figure 12. The Micronair® AU4000 rotating at 14 000 rpm is seen to be the best performer

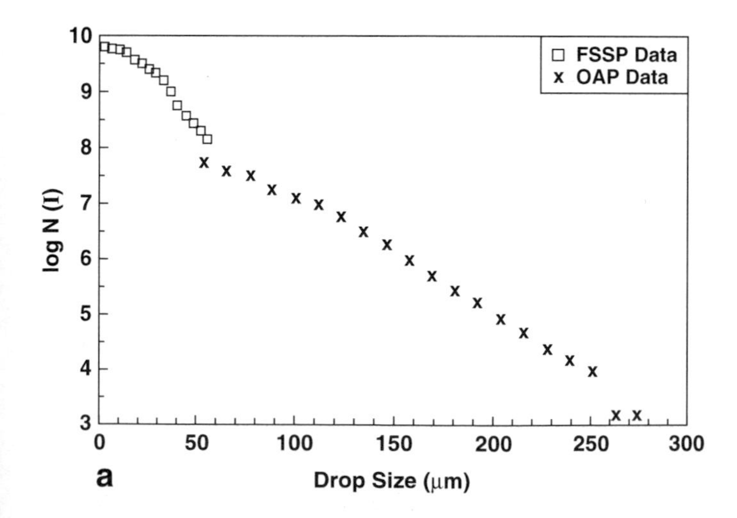

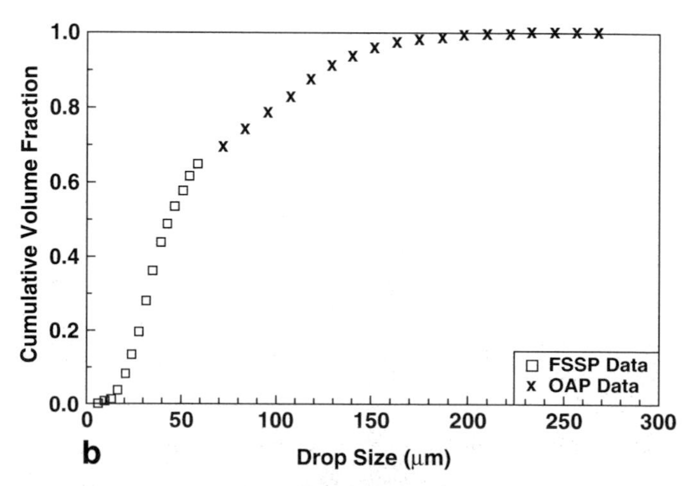

Figure 7. Number distribution curve (a) and cumulative volume fraction (b) for a Micronair® AU4000 atomizer at 9 000 rpm (150 Hz) at an air speed of 49 m/s for fluid no. 6 at 1.83 L/min. N(I) is drops/(μm•min) for size range I, for the total spray plume.

Table 2. Atomizer test fluids.

Fluid number	Composition (% volume)	Surf. tension (mN/m)	Viscosity (mN•s/m^2)	T[a] (°C)
1	91% Water 9% 2-Propanol	48	4	12.5
2	De-Icing Fluid	45.0	35.9	–8.5
3	80% Water 20% 2-Propanol + 10% NaCl (wt)	38.0	5.1	–5
4	50% 2-Propanol 50% De-Icing Fluid	30.4	16.7	–3
5	60 wt% Sucrose 40 wt% Water	76.4	260	–5
6	68 wt% Sucrose 32 wt% Water	74.1	270	22
7	80% Water 20% 2-Propanol	39.8	2.3	18
8	Water	72.8	1	20
9	100% Dipel® 132 (Abbott Labs)	29.1	Fig. 8	23

[a] Properties measured at test temperature.

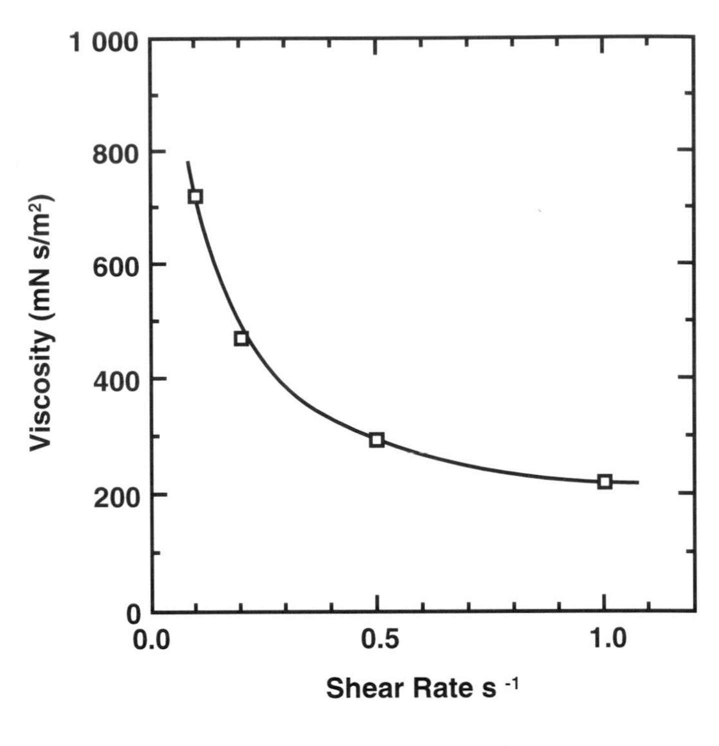

Figure 8. Viscosity versus shear rate, Dipel® 132 (fluid no. 9).

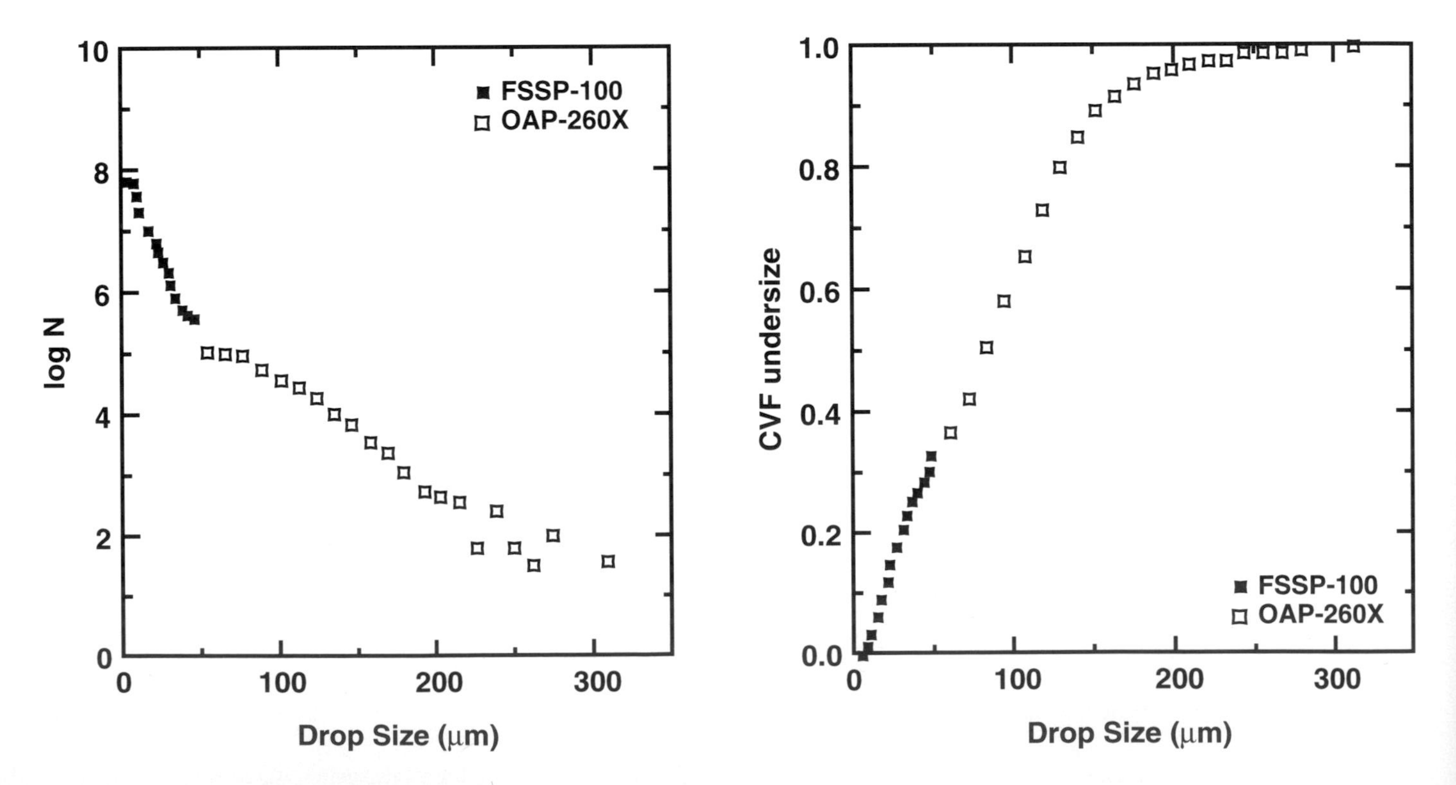

Figure 9. Droplet number distribution and cumulative volume fraction versus drop size, Teejet® 800050 (test no. 3, fluid no. 1). N is drops/(μm•min) for total plume.

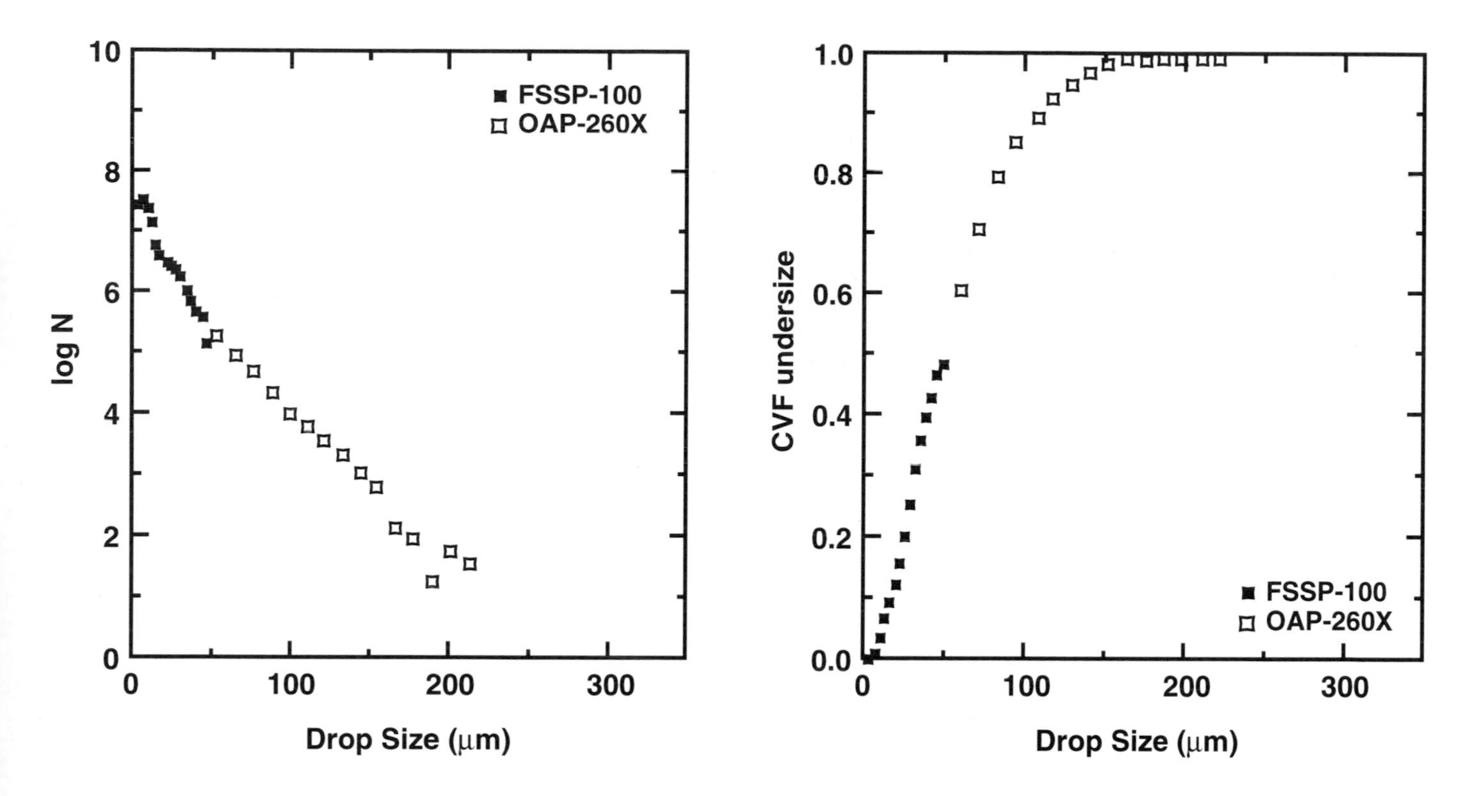

Figure 10. Droplet number distribution and cumulative volume fraction, Micron® X-1 atomizer (test no. 4, fluid no. 1). N is drops/(μm•min), total plume.

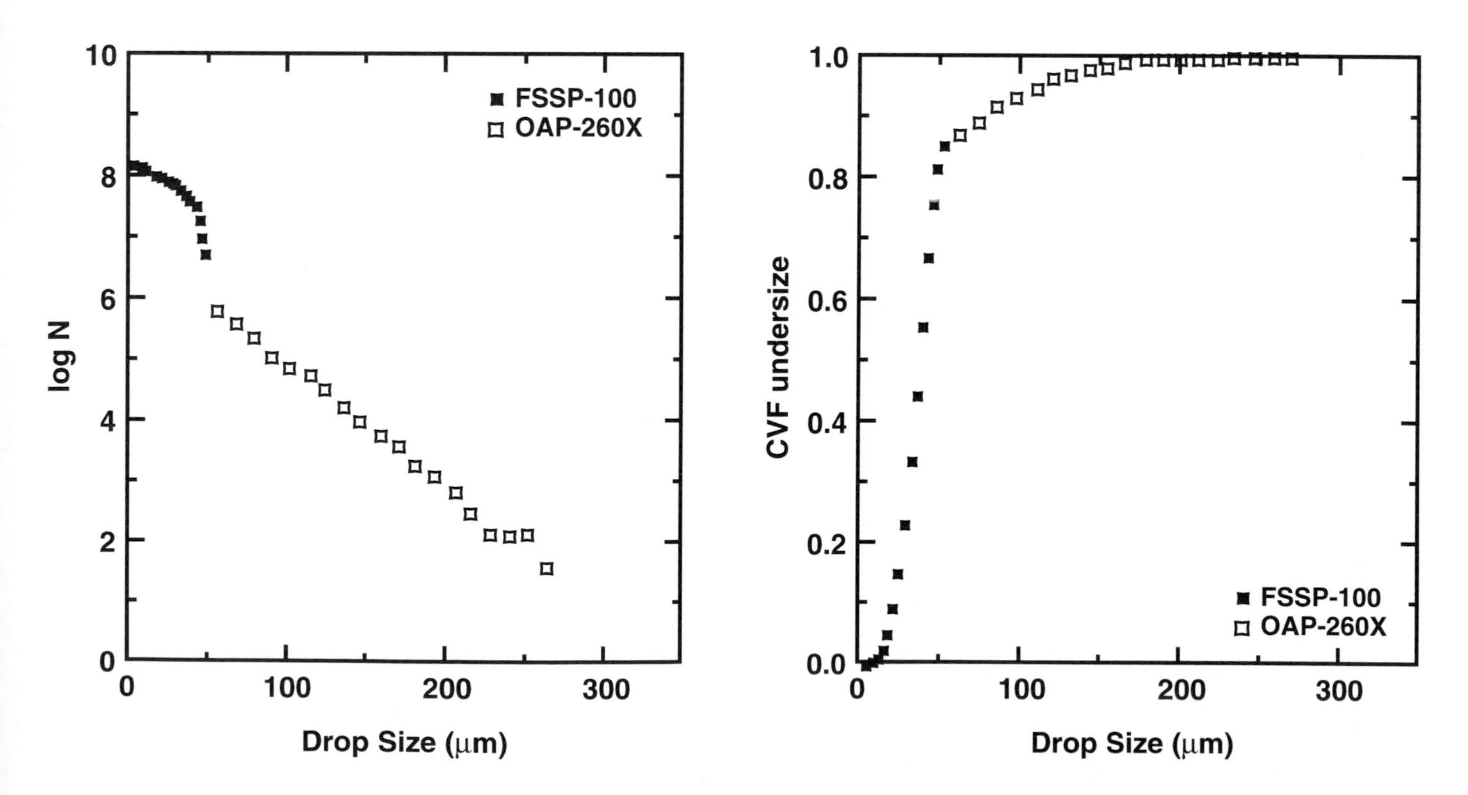

Figure 11. Droplet number distribution and cumulative volume fraction, Micronair® AU4000 atomizer (test no. 7, fluid no. 1). N is drops/(μm•min), total plume.

Figure 12. Atomizers (*left to right*): Teejet®, Micron® X-1, Micron® X-1VP, Micronair® AU4000, and Beecomist® 360.

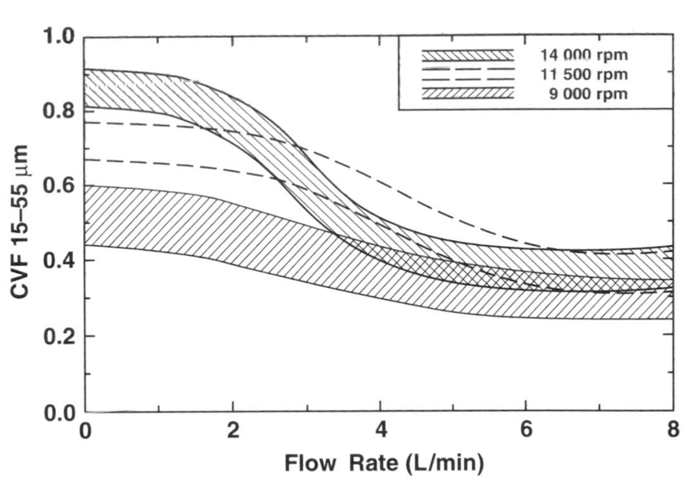

Figure 13. Performance plot for Micronair® AU4000. Volume fraction from 15 to 55 µm (CVF_{15-55}) plotted against fluid flow rate for three operating speeds, 233 Hz, 192 Hz, and 150 Hz. Fluids no. 3 to no. 8 used to evaluate performance bands.

Table 3. Comparative atomizer performance.

Atomizer	Fluid number	Flow (ml/s)	Air velocity (m/s)	$D^a_{v,0.1}$ (µm)	$D^a_{v,0.5}$ (µm)	$D^a_{v,0.9}$ (µm)	CVF^b (µm)
1-Teejet® 11010 906-18°	1	63.5	64	41	124	222	.167
2-Teejet® 8002 90°	1	13.2	63	37	113	180	.19
3-Teejet® 800050 90°	1	3.2	62	20	85	150	.28
4-Micron® X-1 378 Hz	1	1.5	45	20	55	110	.46
5-Micron® X-1VP 378 Hz	1	1.5	45	18	54	113	.44
6-Beecomist®360 40 µm SS Sleeve 233 Hz	2	12.7	42	24	47	123	.54
7-Micronair® AU4000 20 Mesh 233 Hz	3	30.7	49	23	37	78	.83
8-AS no. 7	9	31.5	49	24	42	108	.73

[a] $D_{v,0.1}$: 10% of the spray volume is in the diameters smaller than this value.
[b] CVF is the spray cumulative volume fraction between the droplet sizes indicated.

of the devices tested. This atomizer was extensively studied in the van Vliet and Picot (1987) work, and a summary of operating performance is given in Figure 13. There, the 15 to 55 µm volume fraction is plotted against fluid flow rate supplied to the atomizer, with rotational speed as a parameter, for fluids ranging in viscosity from 1 to 270 mN•s/m^2 and in surface tension from 30.4 to 76.4 mN/m at a wind tunnel air velocity of 49.2 m/s. The versatility of this unit is shown by the narrow performance bands associated with atomization of fluids with a wide range of viscosity and surface tension. Atomization performance according to the stated criterion above deteriorates sharply at flow rates greater than 2 L/min per unit. Recent tests have shown that performance also deteriorates at speeds above 14 000 rpm, perhaps due to bulging of the wire mesh basket causing fluid to accumulate in the central portion of the rotating wire mesh. In spring 1989, the manufacturers supplied the authors with a stamped metal sleeve basket that seemed to eliminate this problem. Although the manufacturer does not recommend operating speeds greater than 12 000 rpm, no difficulty has been noted in operation, provided the propeller blades are not clamped too tightly (this can cause blade root cracking and subsequent failure at speed).

Effect of Operating Variables on Atomizer Performance

No general quantitative correlation exists for the prediction of the performance of spray atomizers. Manufacturer's data cannot generally be relied upon for accuracy, and experiments such as those reported herein must therefore be done to evaluate performance. General qualitative trends for dependency on operating variables have been observed. The most important factor, for the single-fluid atomizers described here, is the relative velocity between the emitted fluid and the airstream. Increased speed means smaller droplets: this means aircraft speed for Teejet®-type atomizers and rotational speed for rotary atomizers. For Teejet® atomizers, Tomney's (1979) results indicate a decreasing dependency on increasing air speed above 70 m/s. For rotary atomizers, increased rotational speed also causes decrease in droplet size, except for the limitations due to distortion of the wire mesh (AU4000 type) by centrifugal force, mentioned previously.

For jet atomizers, fluid surface tension is of secondary importance in determining droplet size. For rotary atomizers, fluid flow rate is the secondary factor, and fluid properties of third order importance. When the Beecomist® 360 (with the 40-µm stainless steel porous sleeve) was fed with a fluid having a viscosity of 7.0 centipoise and 26 dyne/cm, Pang (1985) reported that 75% of the volume was from 15 to 55 µm, for the same flow rate as for test 6, Table 3. Viscosity is very important in this case, apparently due to its effect on flow through the small pores in the porous mesh. Pang (1985) also reports an increase in performance when 5 ppm of Polyox, a drag-reducing additive, was added to the above solution.

Conclusions

Along with Matsumoto et al. (1985), the authors conclude that detailed study of each individual atomizer type is necessary for an adequate knowledge of performance characteristics. Measurements of the type described herein are required. As mentioned earlier, this work has concentrated on atomizers for deposition of pesticide on coniferous foliage. The choice of atomizers for deposition of droplets on other targets requires an understanding of the capture mechanism and a knowledge of the optimal size for capture.

Acknowledgments

The authors are grateful for the support of the Natural Sciences and Engineering Research Council; of Forest Protection Ltd., Fredericton, New Brunswick; and of Forestry Canada. The excellent work by the technicians of the University of New Brunswick Chemical Engineering Department in constructing the wind tunnel and other ancillary devices is also hereby acknowledged.

References

Cerni, T.A. 1983. Determination of size and concentration of cloud drops with an FSSP. J. Clim. Appl. Met. 22: 1346-1355.

Dye, J.E.; Baumgardner, D. 1984. Evaluation of the forward scattering probe, part I: electronic and optical studies. J. Atmos. Ocean Tech. 1: 329-344.

Hovenac, E.A. 1986. Calibration of droplet sizing and liquid water content instruments: survey and analysis. NASA Contract Rep. 175099, Lewis Research Center, Cleveland, Ohio.

Hovenac, E.A.; Hirleman, E.D.; Ide, R.F. 1985. Calibration and sample volume characterization of PMS optical spray probes. Int. Conf. Liq. Atom. and Spray Systems, Vol. 2, July 1985. Inst. of Energy, London, England.

Kristmanson, D.D.; Picot, J.J.C.; van Vliet, S.; Henderson, G.W. 1987. Measuring foliar deposits from aerial spraying of insecticides. Proc. Symp. Aerial Appl. of Pest. in For., Ottawa, Oct. 20-22, 1987. Natl. Res. Counc., Ottawa, Ont.

Matsumoto, S.; Belcher, D.W.; Crosby, E.J. 1985. Rotary atomizers, performance understanding and prediction. Int. Conf. Liq. Atom. and Spray Systems, Vol. 1, July 1985. Inst. of Energy, London, England.

Pang, M.K.H. 1985. M.Sc. thesis, Univ. of N.B., Dept. of Chem. Eng., Fredericton, N.B.

Picot, J.J.C.; Bontemps, X.; Kristmanson, D.D. 1985. Measuring spray atomizer droplet spectrum down to 0.5 µm size. Trans. A.S.A.E. 28: 1367-1370.

Picot, J.J.C.; van Vliet, M.W.; Payne, N.J. 1989. Droplet size characteristics for insecticide and herbicide spray atomizers. Can. J. Chem. Eng. 67: 752-761.

Tomney, T.D. 1979. Effect of polymer additives on jet breakup. M.Eng. thesis, Univ. of N.B., Dept. of Chem. Eng., Fredericton, N.B.

van Vliet, M.W.; Henderson, G.W.; Picot, J.J.C. 1987. Droplet spectra of atomizers used in pest control. Proc. Symp. Aerial Appl. of Pest. in For., Ottawa, Oct. 20-22, 1987. Natl. Res. Counc., Ottawa, Ont.

van Vliet, M.W.; Picot, J.J.C. 1987. Calibrating light imaging and light scattering probes for atomizer characterization. 8th ASTM Symp. on Pesticide Formulations and Application Systems, Bal Harbour, Fla. 1987.

van Vliet, M.W.; Picot, J.J.C. 1989. Drop spectrum characterization for the Micronair AU4000 aerial spray atomizer. Atom. Spray Tech. 3: 123-134.

Wallace, D.J. 1988. Aircraft vortex effects in forestry aerial spraying. Ph.D. thesis, Univ. of N.B., Dept. of Chem. Eng., Fredericton, N.B.

Chapter 53

Principles of Atomization and Atomizer Selection

N.J. PAYNE

Introduction

Most insecticides presently used in Canadian forestry are aerially applied in sprays. Atomization is the process by which a spray is formed from a tank mix, and it results in a spray cloud comprising drops of various sizes. The drop size spectrum employed in an insecticide application affects the fate of the spray cloud and the resulting efficacy and environmental impact. The choice of drop size spectrum, and the atomizer and settings to provide it, is consequently an important one.

Spray Atomization

Physical Process

A requirement of any atomizer is that it results in forces being applied to the liquid that break it up into small particles. This may be achieved in many ways, but a universal requirement in the atomization process is the input of energy to break the cohesive intermolecular forces present in the liquid. Surface tension is a manifestation of these intermolecular forces, which resist the creation of new liquid surface area. The energy E required to create the new surface area in forming drops with diameter D from a liquid volume V is given by:

$$E = 6\ \sigma V/D \qquad (1)$$

where σ is the liquid surface tension. Thus, the finer the spray the greater the new surface area and the energy required to create it. The required energy may come from aerodynamic, inertial, or electrostatic forces which usually act in combination to provide atomization. Aerodynamic forces occur when the speed or direction of the liquid stream differ from the airstream into which it is released. The relative motion of the airstream over the liquid causes the force and energy input. Inertial forces result from the kinetic energy of the liquid stream itself and are used when the atomizer forms a liquid stream whose momentum assists its breakup, as in a flat-fan hydraulic nozzle. Electrostatic forces arise from the mutual repulsion of similarly charged portions of the liquid stream. The intermolecular or surface tension charged portions of the liquid stream. The intermolecular or surface tension force opposes these disintegrating forces during liquid atomization.

The function of an atomizer is to form the liquid stream into sheets, ligaments, or drops, which are then subjected to the forces that complete atomization. The intermediate liquid form is an important step in atomization because it affects the final drop size spectrum generated. In addition, an atomizer usually accelerates the liquid, thereby increasing the aerodynamic and inertial forces acting on the liquid stream. Liquid sheets are formed by design in various types of hydraulic nozzle, including the flat-fan and hollow cone nozzles. Several modes of liquid sheet disintegration are found. First, the sheet may perforate; these perforations grow in extent and number leaving a network of liquid ligaments (Fraser et al. 1962; Fig. 1). The perforations are caused by the action of turbulence in the liquid stream, or by buffeting from the airstream into which it is released, and grow by surface tension. Second, wave propagation on the sheet may cause it to separate into ligaments (Fraser et al. 1962; Fig. 2). These waves may be created by hydrodynamic or aerodynamic disturbances in the liquid flow and grow in amplitude until the sheet fragments. Third, thick sheet disintegration occurs at relatively high liquid velocities (100 m/s) when liquid is removed directly from wave crests by interaction with the surrounding air (Mayer 1961). Finally, rim disintegration occurs at the edge of the liquid sheet (Fig. 3), again forming ligaments (Clark and Dombrowski 1972). Rim disintegration also results from wave propagation; however, the drop sizes produced are generally larger than those produced by sheet disintegration.

Liquid ligaments are formed directly in hydraulic jet nozzles and rotary atomizers (e.g. a spinning disk or cup), as

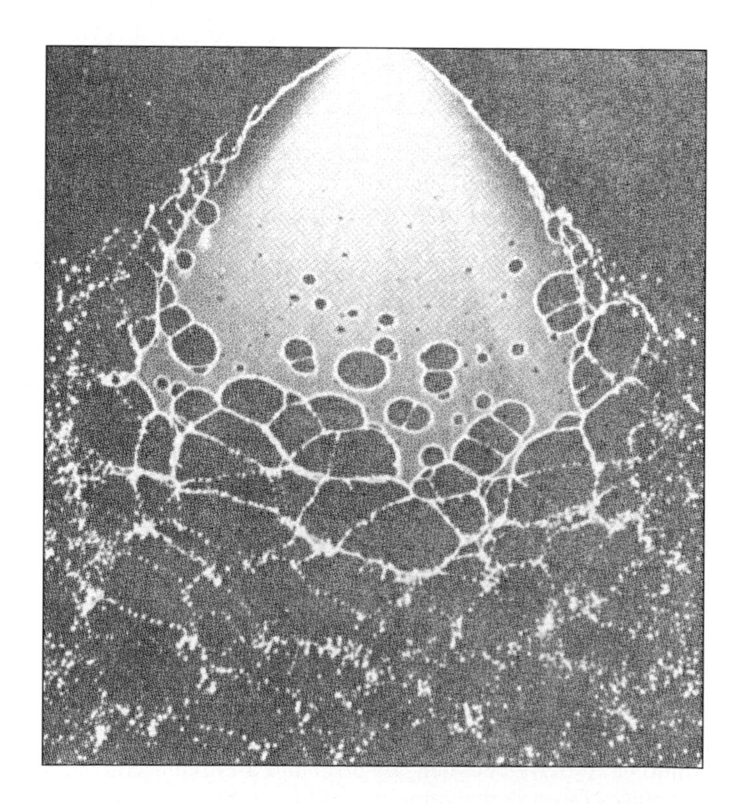

Figure 1. Liquid sheet disintegration by perforation (Matthews 1979).

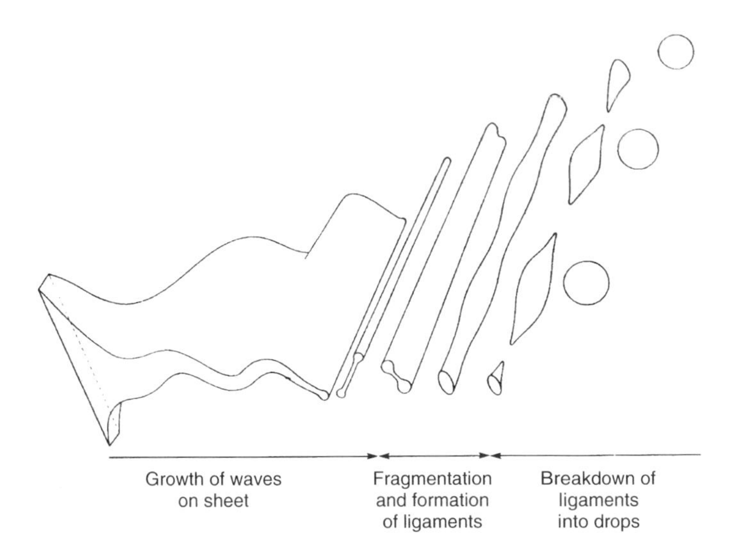

Figure 2. Liquid sheet disintegration by wave propagation (Dombrowski and Johns 1963).

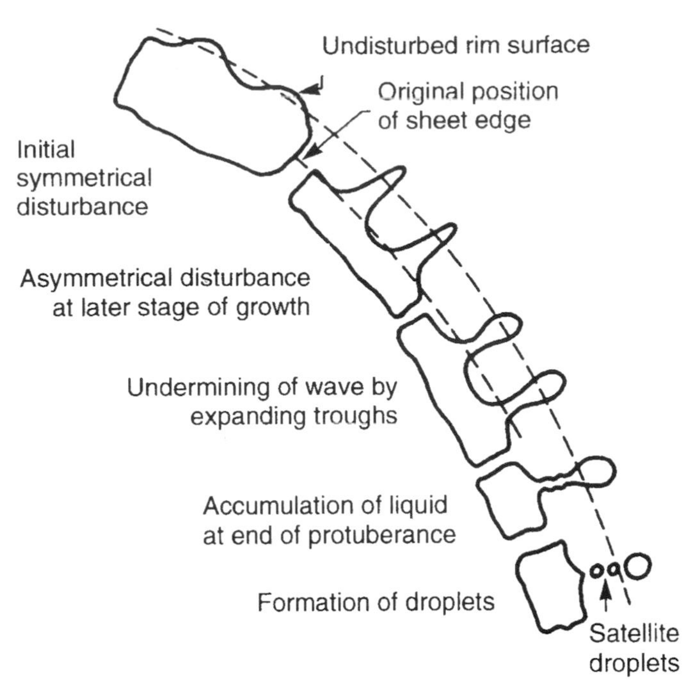

Figure 3. Rim disintegration on a liquid sheet (Clark and Dombrowski 1972).

well as from sheet breakup. Ligament disintegration also occurs in various modes. Waves on the ligament are caused by hydrodynamic or aerodynamic disturbances and these perturbations grow and cause fragmentation. Axisymmetric disturbances are known as Rayleigh instability (Clift et al. 1978; Fig. 4a). At certain wavelengths this type of disturbance can lead to the formation of a narrow drop size spectrum with a main drop diameter about twice that of the ligament, together with satellite drops. Another mode of breakup results from the propagation of sinuous waves (Clift et al. 1978; Fig. 4b).

Drop disintegration is caused by aerodynamic and electrostatic forces. A drop disintegrates when the surface tension force is insufficient to maintain its integrity. Aerodynamic forces arise from the relative velocity between the air and drop, and the diameter limit above which drops are fragmented by these forces may be calculated by comparing the magnitudes of surface tension and aerodynamic forces, using the Weber number criterion (Hinze 1953) given by:

$$D_m = k\ \sigma/v_m^2 \qquad (2)$$

where D_m is the diameter limit, v_m is the relative velocity between the drop and air, σ is the surface tension, and k is a constant. Thus D_m decreases with surface tension. The diameter limit for water drops has been measured by Lane (1951), who found:

$$D_m = 0.612/v_m^2 \qquad (3)$$

This implies a diameter limit of about 320 and 1 200 µm, respectively, at relative velocities of 45 and 22 m/s (162 and 79 km/h). Two modes of drop disintegration have been observed (Lane 1951). Bag breakup (Fig. 5) occurs where the relative velocity between the drop and airstream gradually increases through the velocity limit (v_m). Stripping breakup occurs when the relative velocity between the drop and air is abruptly changed, causing liquid to be stripped from the drop surface. Drops also disintegrate when the repulsive force resulting from electrical charge on the drop exceeds the surface tension force holding the drop together (Rayleigh 1982). The Rayleigh charge limit Q_m for a drop of diameter D is given by:

$$Q_m = (2\pi\sigma D^3)^{1/2} \qquad (4)$$

where σ is the liquid surface tension. The charge limit reduces with surface tension and drop diameter; thus, as a drop evaporates it may become unstable because of the charge already present. After a drop disintegrates due to electrostatic force, the fragments are stable because the charge is distributed over a larger surface area.

Atomizer Types

Many different types of atomizer are available for use in insecticide applications, providing drops with diameters from 20 to 2 000 µm depending on the equipment and settings chosen. Atomizers may be classified according to the principal mode of atomization, including hydraulic, rotary, airblast, electrostatic, vibrating, and thermal fogging.

The hydraulic or pressure nozzle is the simplest type of atomizer and is therefore inexpensive to manufacture and maintain. In its basic form the hydraulic nozzle comprises an orifice or outlet that forms the liquid flow into a thin stream moving at a relatively high speed. For example, the liquid speed relative to an 8006 flat-fan nozzle, with a flow rate of 2.3 L/min, is 15 m/s (54 km/h). The liquid pressure provides the kinetic energy of the liquid stream that is used in the atomization process. When employed for ground-based applications the kinetic energy of the liquid is the principal source of atomization energy, whereas for aerial applications the kinetic energy of the air moving past the aircraft also contributes to atomization.

Hydraulic nozzles are classified according to the shape of the liquid stream emitted. For example, a plain jet nozzle produces a cylindrical jet from a circular orifice, whereas the elliptical orifice in a flat-fan nozzle gives a flat liquid sheet, and the hollow and full cone nozzles impart a tangential motion to the liquid stream to produce a cone-shaped stream, also using a circular orifice. An example of hydraulic nozzle operation is given in Figure 6, which shows a cross section through a Raindrop® hydraulic nozzle, designed to minimize the proportion of small drops generated (Delevan Catalog). In this atomizer, liquid is admitted to the primary swirl chamber via slots, causing tangential acceleration of the liquid stream, which after passing through the primary orifice enters a second whirl chamber where air becomes mixed with the liquid before it passes through another orifice to form a hollow cone spray, with drops moving at several metres per second. hydraulic nozzles can be used to generate a wide range of drop sizes. A plain jet, such as a Microfoil® nozzle can provide a very coarse spray (Picot et al. 1989), whereas a flat-fan nozzle used on an aircraft provides a relatively fine spray (Yates et al. 1984a). In general, the average drop size produced by a hydraulic nozzle increases with the size of orifice used.

Airblast or pneumatic atomizers, including two-fluid nozzles, use a high-speed airstream to cause atomization. Two-fluid nozzles are classed as internal or external mixers according to whether the gas and liquid streams are mixed inside or outside the nozzle body. The liquid stream mixed with the airstream is broken up principally by aerodynamic forces, and the atomization energy is provided by the kinetic energy of the air. A knapsack mistblower is an example of an airblast atomizer. A small internal combustion engine is used to provide an airstream with a velocity of typically 60 to 80 m/s, into which the tank mix is injected from a simple orifice, or for finer atomization from an air-driven spinning disk. The airstream providing atomization is sometimes used to carry the spray to the target, as in a mistblower. A hydraulic nozzle used on an aircraft may effectively operate as an airblast atomizer, depending on the air speed and orientation. Figure 7 shows a cross section of an internally mixing two-fluid nozzle in which air is introduced tangentially into the nozzle chamber and perpendicular to the liquid flow, to create an air–liquid mixture that is further atomized by the deflector assembly as it leaves the nozzle. Airblast or pneumatic atomizers generate a relatively fine drop size spectrum, and the average drop size decreases with increasing air speed.

Rotary atomizers dispense liquid over a rotating part that divides the flow into thin streams, which are thrown off by centrifugal force and atomized by inertial and aerodynamic forces. The relative importance of the aerodynamic forces increase when this type of atomizer is used on an aircraft. The rotating part can take several forms, including a porous cylinder, a bristled cylinder, and serrated-edged disk, and may be driven electrically, hydraulically, or aerially. For example, the wind-driven Micronair® atomizers (Fig. 8) employ a porous cylinder, and the electrically driven Micron® atomizers (Fig. 9) use a serrated-edged disk. In the Micronair® atomizer, tank mix is admitted close to the longitudinal axis of the rotating cylinder, and initial breakup is effected by the diffuser tube. Under centrifugal force the liquid is flung out to a gauze cylinder, which further atomizes the liquid before the airblast past the atomizer completes the process. The gauze speed is typically 60 m/s. By contrast, in the Micron® atomizer, tank mix is fed to the center of the spinning disk and spreads to the angled portion where it is channeled by fine grooves to the teeth at the disk periphery and thrown off as ligaments, which break up under the influence of inertial and aerodynamic forces. The peripheral speed

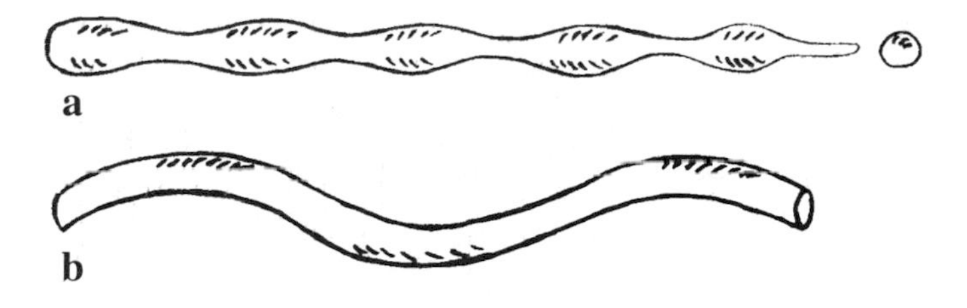

Figure 4. Waves on liquid ligaments: (a) axisymmetric, (b) sinuous.

Figure 5. Bag breakup of drop (Lane 1951).

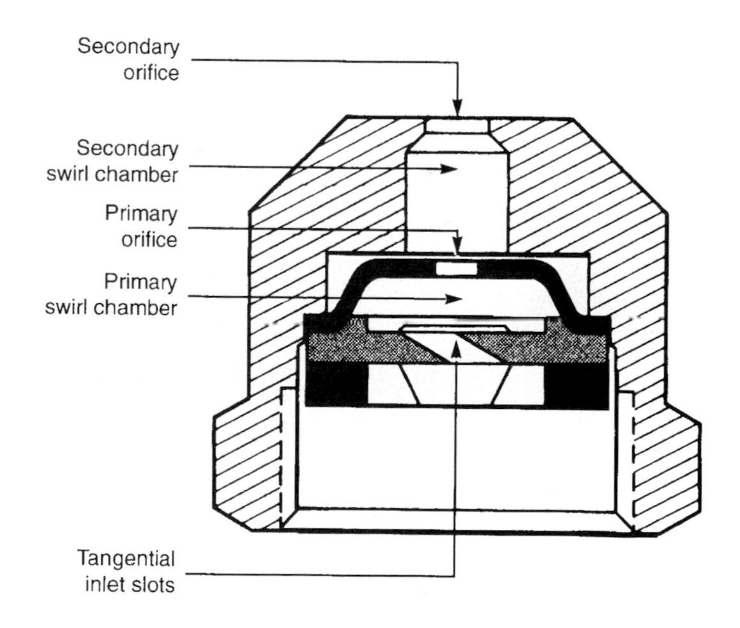

Figure 6. Cross section of Raindrop® hydraulic nozzle (Delevan Catalog).

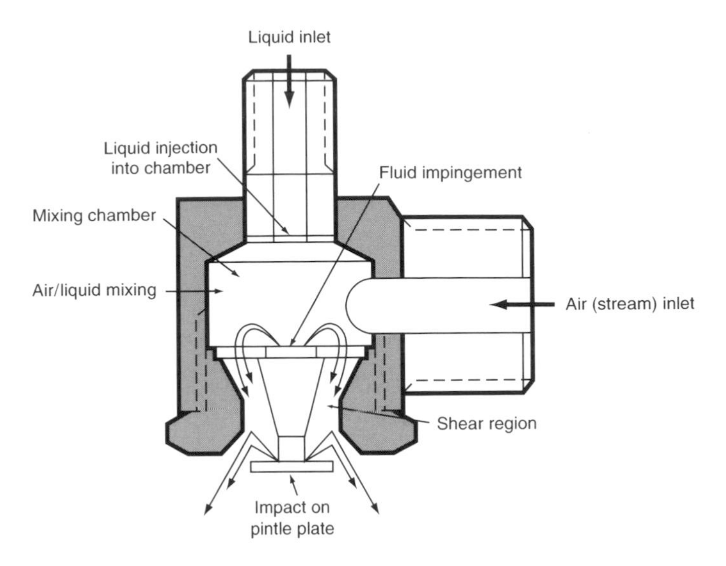

Figure 7. Cross section of two-fluid nozzle (Delevan Catalog).

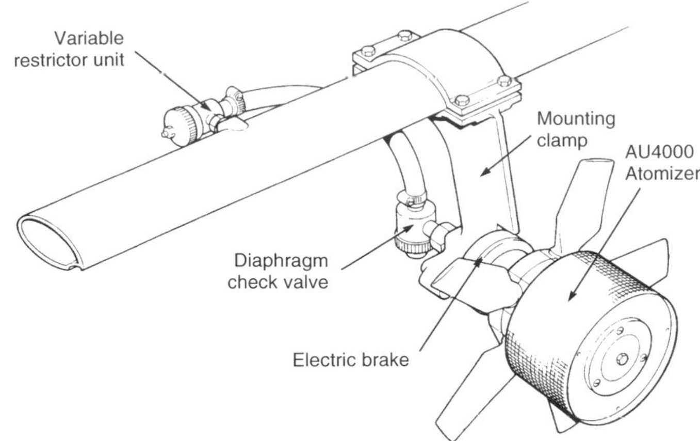

Figure 8. Typical installation of Micronair® AU4000 atomizer (Micronair handbook).

a

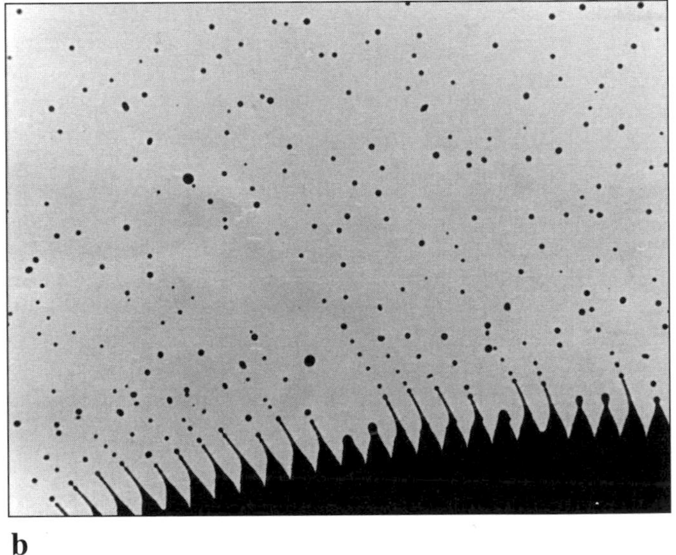
b

Figure 9. (a) Serrated-edged disk from Micron® Ulva and (b) drop formation from ligaments at rim.

is typically 5 m/s, and the liquid flow rate is controlled to provide ligamented output for optimum performance. The rotary atomizer is used to generate medium and fine drop size spectra (e.g. Yates et al. 1984a; van Vliet and Picot 1987; Picot et al. 1989), and the average drop size produced by this atomizer type decreases with increasing atomizer rotation rate.

Electrostatic sprayers are divided into two categories, those that use electrostatic forces as the primary means of liquid atomization, and those that use nonelectrostatic forces for liquid atomization but electrostatically charge drops in the process. In both cases, the atomizers generate an electrostatically charged cloud to enhance drop deposition. The Electrodyne® atomizer (Fig. 10) uses a gravity feed to provide a thin liquid stream through a nozzle maintained at high voltage. Electrodynamically amplified waves on the liquid surface cause atomization, with charged drops emitted from the wave crests. This produces a relatively narrow drop size spectrum (VMD/NMD less than 1.2). The average drop size decreases with increasing nozzle voltage, and for optimum atomizer performance the liquid resistivity must be within a certain range (Coffee 1980, 1981). An example of an electrostatic atomizer using nonelectrostatic forces for atomization is that developed by Law (1978). This atomizer employs airblast or pneumatic atomization and charges drops by induction. Arnold and Pye (1980) have also developed an electrostatic atomizer using nonelectrostatic forces, in which liquid is atomized using a spinning disk and charged by induction. Matthews (1989) has reviewed the electrostatic spraying of pesticides.

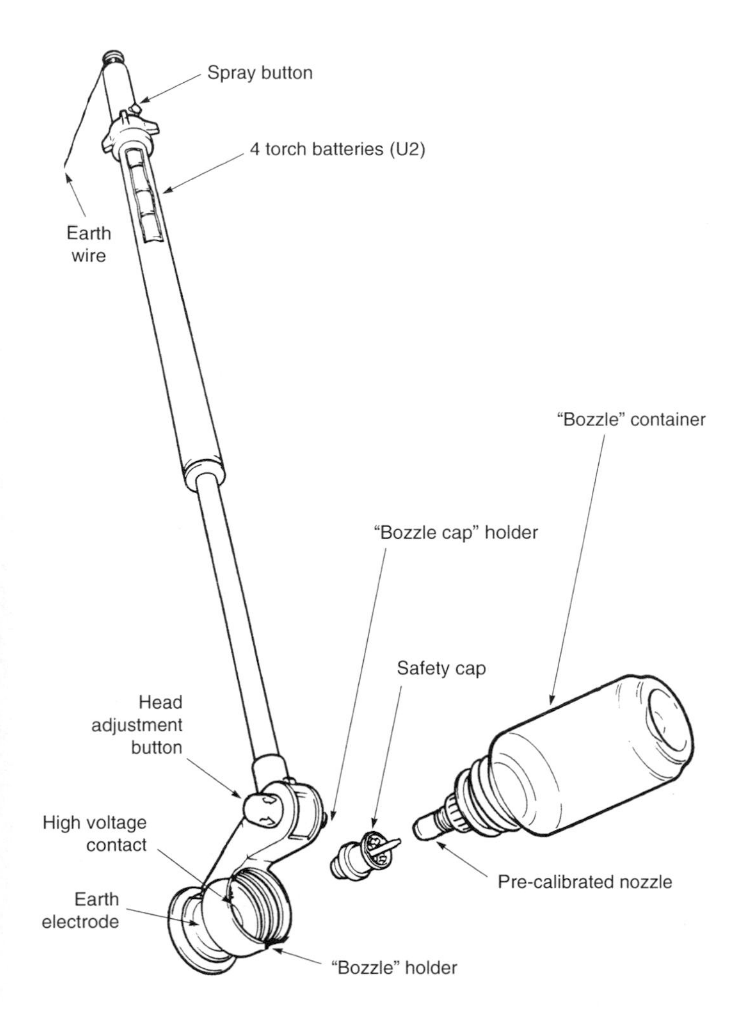

Figure 10. Electrodyne® sprayer (ICI handbook).

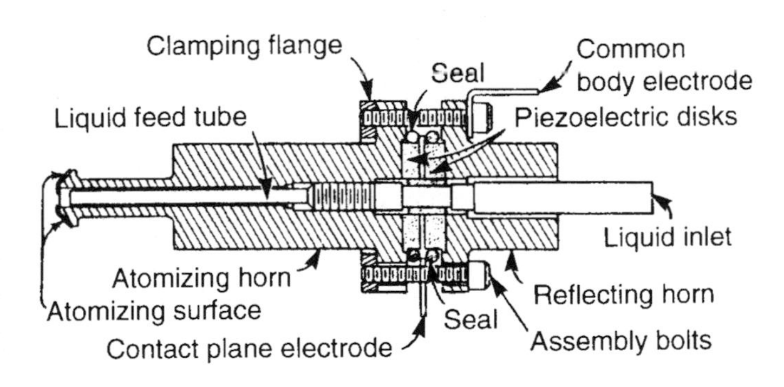

Figure 11. SONO-TEK ultrasonic nozzle (Sonotek Corp.).

Vibrating atomizers include acoustic, ultrasonic, and vibrating types, all of which use vibrations to cause liquid atomization. An example of an ultrasonic atomizer is the SONO-TEK nozzle (Sonotek Corp., Poughkeepsie, N.Y.; Fig. 11) in which liquid is dispersed in a thin film over a rapidly vibrating surface. The vibrations (40 to 1 000 kHz) cause capillary waves to form at the liquid surface, the wavelength of which is inversely proportional to the vibration frequency. When the amplitude of the vibrations is sufficient, drops are emitted from the wave crests. Drop size is inversely proportional to vibration frequency, and this type of atomizer produces relatively fine sprays, with drop sizes from 20 to 50 µm in diameter. This atomizer type is effectively uncloggable, making it very suitable for use with pesticide formulations with a high solid content; however, the maximum attainable flow rate with this nozzle is relatively low, thereby limiting its usefulness for aerial insecticide applications. Another example of a vibrating atomizer is the vibrating orifice generator (Berglund and Liu 1973), in which liquid is atomized by forcing it through a small hole in a diaphragm vibrating at a high frequency. This atomizer is limited to low flow rates and is therefore not suitable for large-scale insecticide applications.

Thermal fogging devices are also used for insecticide dispersal. A fog refers to an aerosol cloud largely comprised of drops from 1 to 10 µm in diameter. In this type of atomizer, the pesticide is mixed with a suitable oil and the mixture is vaporized by injecting it into a hot airstream with a temperature of 200–300°C. A fog is formed by condensation when the vapor-laden airstream mixes with the atmosphere (Matthews 1979). The high temperature of the airstream into which the tank mix is injected may result in the deactivation of some insecticides (e.g. Bt), thereby limiting the usefulness of this atomization process for forestry insecticide application.

Formulation Effects

The physical characteristics of a tank mix, in particular the surface tension, viscosity, and density, all effect the drop size spectrum generated by an atomizer. Surface tension is caused by forces of attraction between liquid molecules, which balance out for any molecule in the body of the liquid, but for molecules at the liquid surface result in a net inward force, causing the liquid to behave as though it were covered by a thin elastic membrane. The surface tension of a liquid is defined as the force exerted on a cut of unit length in its surface (Massey 1983). For pesticide tank mixes, surface tension typically ranges from 20 to 70 mN/m and decreases with increasing temperature. Sundaram (1987) has reported surface tension measurements for a variety of nonaqueous insecticide tank mixes and ingredients. Because surface tension opposes the formation of new surface area, the average drop size generated by an atomizer increases with the surface tension of the tank mix.

Viscosity is a measure of the resistance offered by a liquid to a shearing force (i.e. one that causes liquid molecules to move over one another) and is proportional to the internal liquid "friction." If the viscosity is independent of the rate of shear, the liquid (e.g. water) is Newtonian, whereas liquids whose viscosity is shear rate-dependent are described as non-Newtonian. Dilatant tank mixes show an increase in viscosity with increasing shear rate, where as the viscosity of pseudoplastic mixes decreases with increasing shear rate. The shear rate is proportional to the velocity difference between liquid parcels at adjacent locations in the flow (Massey 1983). Viscosity opposes liquid deformation and therefore the average drop size generated by the atomizer increases with the viscosity of the tank mix. The values of viscosity of pesticide tank mixes are typically 1 to 20 mPa·s, although component liquids may have greater viscosities. Sundaram (1987) has reported measured viscosities of various nonaqueous tank mixes and ingredients.

The mass per unit volume of liquid is defined as its density. For pesticide tank mixes and their ingredients the value of this parameter usually lies between 800 and 1 200 km/m^3. This property has a minor effect on atomization, with the average drop size generated by an atomizer increasing with tank mix density. The effect of these liquid properties on the average drop size from various different atomizers has been summarized by Lapple et al. (1967).

Atomizer Characterization

To select the most suitable atomizer and settings for an insecticide application, it is necessary to know the drop sizes generated and their frequency, information that has to be obtained by experimental measurement. To obtain realistic measurements of drop size spectra for aircraft-mounted atomizers, it is necessary to use an airstream velocity similar to that near the atomizer, which results in rapid dispersal of the cloud as the airstream carries away the drops. If the drop size spectrum at release is to be accurately measured, the sampling protocol used must recover all the drop sizes in their original proportion. Remote-sensing devices are used to obtain state-of-the-art drop size spectra measurements for aircraft-mounted atomizers. These measurements are made close to the atomizer, but far enough away for atomization to be complete. Investigations using laser spectrometers or photographic techniques to measure drop size spectra about 1 m downstream of the atomizer placed in a wind tunnel to provide realistic air speeds have been made by Picot et al. (1989, 1990) and by Yates et al. (1985). Earlier techniques for measuring drop size spectra from pesticide atomizers involved sampling the cloud with multiple collectors several metres from the atomizer and quantifying stain diameters (Coutts and Yates 1968).

Figure 12 provides an example of typical drop size spectrum measurements, including a volume and number distribution of the type used in atomizer selection. These results are plotted as ogives or cumulative frequency distributions on log-probability scales, enabling the volume or number proportion of drops in a chosen size range to be easily estimated.

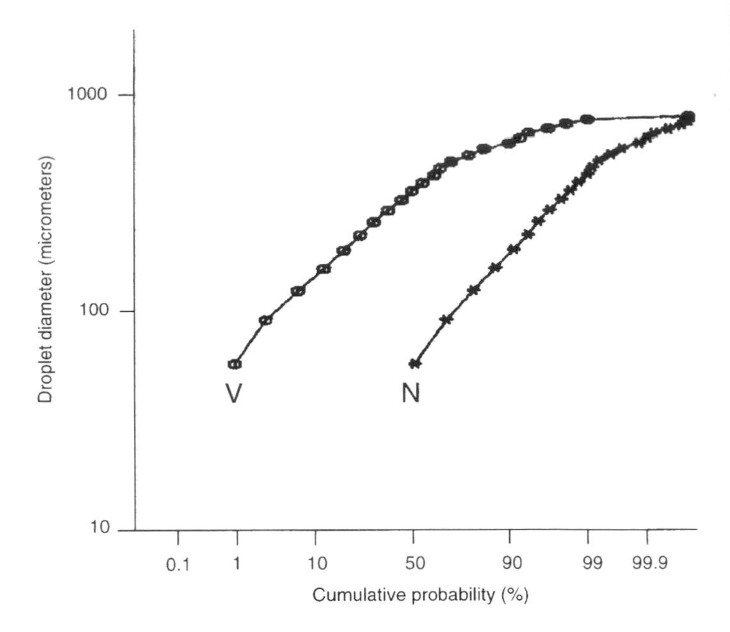

Figure 12. Ogives describing volume and number distributions from hydraulic nozzle (D8-jet, Yates et al. 1984a).

Drop size measurements are usually made in a series of ten or more size-classes; however, the measured spectrum may be typified by the volume and number median diameters (VMD and NMD), that is, the diameters above (or below) which 50% of the spray volume and drop numbers are found. The length, surface, and volume mean diameters are also used to describe drop size spectra (Hinds 1982). The volume mean or volume average diameter (VAD) is the diameter of the drop having the average volume and is used to calculate the number of drops generated per second by the dispersal system. A measure of the drop size range is the difference between diameters defining the 10th and 90th percentile in the volume spectrum. This difference, divided by the VMD, is the relative span of the distribution. The ration VMD/NMD also describes spectrum width. The various mathematical forms and statistics used to describe drop size distributions have been discussed by Johnstone (1978) and Kristmanson (1988).

Various atomizers suitable for aerial insecticide applications in forestry have been characterized with laser spectrometers by Picot and co-workers (Picot et al. 1985, 1989, 1990; van Vliet and Picot 1987) and Yates and his co-workers (Yates et al. 1984a–c, 1985). Many types of hydraulic and rotary atomizers have been characterized, and in the latter category Micronair® atomizers, Micronair (Aerial) Ltd., Bembridge Fort, Sandown, Isle of Wight, England, are probably the most widely tested. Because of their effect on the atomization process, the physical properties of the tank mix to be sprayed are also a consideration in atomizer characterization. This is particularly relevant for Bt tank mixes, which are suspensions having a viscosity that varies with shear rate. Yates and Cowden (1986a–c) have characterized various rotary atomizers using Bt formulations for use in forestry.

Atomizer, Setting, and Volume Rate Selection

To select an atomizer and setting suitable for a particular insecticide application, the required drop size range must be know. Screening may then be carried out to select the equipment, flow rate, etc., that provide the greatest proportion of drop sizes in that range. The drop size spectrum of the cloud has a significant effect on the fate of the applied insecticide. For example, a coarse spectrum results in high ground deposits, thereby wasting insecticide (Sundaram et al. 1987; Sundaram and Retnakaran 1987), whereas a very fine spectrum causes a large proportion of the cloud to drift out of the treatment area, reducing on-target deposit and again wasting insecticide. These facts lead naturally to the concept of an optimum drop size range, which results in a high proportion of the active ingredient being deposited on-target. The drop size spectrum found in the habitat of the target life-stage must be quantified and compared to the spectrum at release to discover which sizes are deposited most frequently and are therefore best for delivering active ingredient to that target. Knowledge of the optimum drop size spectrum is the key to designing an efficient insecticide use-strategy that maximizes efficacy and minimizes environmental impact and application costs.

This concept has been applied in New Brunswick for spruce budworm, *Choristoneura fumiferana*, control operations, for which drops with diameters from 15 to 55 μm are considered optimum (Picot et al. 1985). The spruce budworm larval habitat during insecticide applications comprises the conifer needles near branch tips. Drops with diameters less than 15 μm are not intercepted in sufficient numbers by the needles due to the low inertial impaction efficiency of the drops, while those with diameters greater than 55 μm are too frequently deposited beneath the canopy. In addition, large drops account for a significant proportion of the applied insecticide, while contributing only slightly to coverage (drops per square centimetre). Although this drop size range is suitable for spruce budworm larviciding, it will also be useful for other pests that have a similar habitat (i.e. conifer foliage near the branch tips), because it maximizes the proportion of active ingredient reaching this target. The optimum drop size range for controlling a specific pest depends on the habitat of the target life-stage; thus, a ground-dwelling target insect would require the use of large drops to maximize the proportion of active ingredient deposited there.

Drop coverage is another aspect of insecticide use-strategy design. To ensure that an application is efficacious, the active ingredient distribution in the pest habitat must be such that there is a high probability that the majority of the pest population will receive a lethal dose. This is achieved by ensuring adequate coverage of the target (i.e. sufficient drops per unit area) with adequate active ingredient content. For example, controlling fourth-instar *C. fumiferana* larvae with Bt requires a lethal dose per 10 mm^2 of needle surface (Fast 1981). However, this foliar coverage requirement has a concomitant requirement for a minimum number of drops to be applied per hectare, and with the drop size spectrum determines the required volume application rate for the insecticide application. The required volume application rate is calculated from the chosen drop size spectrum and drop coverage as follows:

$$V = v\ I\ J\ 10^8 \qquad (5)$$

where V is the volume application rate (litres per hectare), v is release volume of the volume average drop size in litres, I is the leaf area index, and J is the coverage requirement (drops per square centimetre), adjusted to account for drops lost by drift and ground deposit (Joyce et al. 1981). For example, if 25% of the drops in the spray cloud are lost, the adjusted coverage requirement must exceed the actual coverage by 33%. This calculation requires an estimate of the leaf area index of the canopy to be treated (i.e. the area of foliage per unit of ground area).

The drop size spectrum chosen for an insecticide application affects the application cost through its effect on the volume application rate. As the diameter of the drop of average volume is reduced, so the required volume application rate is decreased, while maintaining the number of drops generated. The latter increases the area able to be sprayed per sortie, which reduces the ferrying time and consequently the application cost. There has been a tendency to decrease the volume application rates used in aerial applications of forestry insecticides to take advantage of this opportunity. Volume application rates in Canadian forestry were typically about 1.5 L/ha in the early 1980s. In 1988, operational applications were carried out in New Brunswick using a total volume of 0.5 L/ha. Because of the potential cost benefit, efforts to reduce the proportion of diluent used in the tank mix are likely to continue. However, drop impaction efficiency is reduced by decreasing drop size and this puts a lower limit on the useful drop size range and the volume application rate.

Drop evaporation is a consideration when screening atomizers using drop size measurements made soon after spray generation, because of the difference it causes between the drop size spectrum of the spray cloud on reaching the target and that at generation. If the drop size spectrum required at the target is known, then the spectrum at release can be estimated from the droplet flight time, tank-mix volatility, and air temperature (see Chapter 47). For water-based sprays, the relative humidity must be considered. Picot et al. (1981) described a technique for measuring the tankmix evaporation rate and used it in estimating the evaporation of a multicomponent drop, and Sundaram (1985, 1986) has compared the volatilities of various aqueous and nonaqueous insecticide tank mixes. When dealing with drops having diameters less than 100 μm, the drop size spectrum at the target may sometimes be obtained from the low-volatility fraction of the tank mix. For example, water drops with diameters of 60 μm or less have a lifetime of less than 1 minute (Hinds 1982), shorter than their flight time in a typical forestry insecticide application. Thus, for a water-based tank mix and drop sizes less than 60 μm, the drop diameter at the target can be estimated by assuming that all the water, and any other component with similar or greater volatility, evaporates during flight.

The atomizer and settings selected for an application must be capable of providing not only the required drop size spectrum, but also the drop output. The number of drops per second N required from the spray aircraft may be calculated as follows:

$$N = u\ w\ J \qquad (6)$$

where u is the air speed, w is the swath width, and J is the adjusted coverage requirement. From the number of atomizers mounted on the aircraft, the drops per second required from each individual atomizer may then be calculated. The number of drops generated per second (n) by an atomizer may be calculated from equation 7:

$$n = Q/v \qquad (7)$$

where Q is the flow rate through the atomizer in litres per second and v is the release volume of the VAD drop in litres. Aircraft speed and swath width can be adjusted to compensate for limited atomizer output, although with some atomizers (e.g. Micronairs), the effect of air speed changes on the drop size spectrum must also be considered.

The comparison of atomizer performance given in Table 1 was prepared by Picot (pers. comm.) for use in selected atomizers

Table 1. Atomizer performance comparison.

Atomizer	Flow rate (mL/s)	Air speed (m/s)	Volume fraction (15–55 μm)
Flat-fan nozzles			
Teejet 11010	64	64	0.16
Teejet 8002	13	63	0.19
Teejet 800050	3.1	62	0.29
Rotary atomizers			
Beecomist® 360 (40-μm pore sleeve)	20	40	0.97
Micron X-1	1.5	45	0.50
Micronair® AU3000	190	47	0.35
Micronair® AU4000	33	49	0.79
Micronair® AU4000	120	45	0.40
Unimizer No. 10	16	45	0.18

and settings that can be used to generate drop size spectra closest to the optimum, while providing an adequate flow rate for spruce budworm larviciding in New Brunswick. The tank mix used in these tests was a 9% (v/v) solution of isopropyl alcohol in water, except for the Micronair® AU4000 tests, which employed a 50% solution, and the Beecomist® test for which water was used. The rotary atomizers were all set for maximum rotation rates, and the liquid flow from the hydraulic nozzles was oriented at 90° to the crosswind, to obtain fine drop size spectra. On the basis of spectrum alone, the Beecomist® atomizer appears preferable, having 95% of the spray volume within the required size range. However, this atomizer has a relatively low maximum flow rate and comprises a porous polyethylene sleeve that is easily clogged, and these factors together with the power requirement precluded its use. This exemplifies the point that the dispersal system selected must be capable of providing the required number of drops per unit length of spray line to satisfy the coverage requirement.

References

Arnold, A.J.; Pye, B.J. 1980. Spray application with charged rotary atomizers. Pages 109-117 *in* J.O. Walker (Ed.), Proceedings of British Crop Protection Council Symposium Spraying Systems for the 1980s. Monogr. No. 24, BCPC, Croydon, England.

Berglund, R.N.; Liu, B.Y. 1973. Generation of monodisperse aerosol standards. Environ. Sci. Technol. 7: 147-153.

Clark, C.J.; Dombrowski, N. 1972. On the formation of drops from the rims of fan spray sheets. Aerosol Sci. 3: 173-183.

Clift, R.; Grace, J.R.; Weber, M.E. 1978. Bubbles, drops and particles. Academic Press, New York. 380 p.

Coffee, R.A. 1980. Electrodynamic spraying. Pages 95-107 *in* J.O. Walker (Ed.), Proceedings of British Crop Protection Council Symposium Spraying Systems for the 1980s. Monogr. No. 24, BCPC, Croydon, England.

Coffee, R.A. 1981. Electrodynamic crop spraying. Outlook on Agriculture 10: 350-356.

Coutts, H.H.; Yates, W.E. 1968. Analysis of spray droplet distributions from agricultural aircraft. Trans. ASAE 1: 25-27.

Dombrowski, N.; Johns, W.R. 1963. The aerodynamic instability and disintegration of viscous liquid sheets. Chem. Eng. Sci. 18: 203214.

Fast, P.G. 1981. Measurement of needle area consumed by larvae of *Choristoneura fumiferana*. Can. For. Serv., Ottawa. Ont. Res. Notes 1: 28-29.

Fraser, R.P.; Eisenklam, P.; Dombrowski, N.; Hasson, D. 1962. Drop formation from rapidly moving liquid sheets. A. I. Ch. E. J. 8: 672-680.

Hinds, W.C. 1982. Aerosol technology: properties, behavior, and measurement or airborne particles. John Wiley and Sons, New York. 424 p.

Hinze, J.O. 1953. Fundamentals of the hydrodynamic mechanism of splitting in dispersion processes. A. I. Ch. E. J. 1: 289-295.

Johnstone, D.R. 1978. Statistical description of spray drop size for controlled drop application. Pages 35-42 *in* Proceedings of British Crop Protection Council Symposium Controlled Drop Application. Monogr. No. 22, BCPC, Croydon, England.

Joyce, R.J.; Schaefer, G.W.; Allsopp, K. 1981. Distribution of spray and assessment of larval mortality at Annabalish. Pages 1546 *in* A. Holden and D. Bevan (Eds.), Aerial application of insecticide against pine beauty moth. For. Comm., U.K.

Kristmanson, D.D. 1988. Principles of spray droplet formation: the role of emission systems. Pages 15-20 *in* G.W. Green (Ed.), Proc. NRC-ACAFA Symp. Aerial Application of Pesticides in Forestry AFATN-18. Natl. Res. Counc., Ottawa, Ont.

Lane, W.R. 1951. Shatter of drops in streams of air. J. Ind. and Eng. Chem. 43: 1312-1317.

Lapple, C.E.; Henry, J.P.; Blake, D.E. 1967. Rep. No. AD 821-314, Stanford Res. Inst., CA.

Law, S.E. 1978. Embedded-electrode electrostatic-induction spraycharging nozzle: theoretical and engineering design. Trans. ASAE 21: 1096-1104.

Massey, B.S. 1983. Mechanics of fluids. 5th ed. Van Nostrand Reinhold, Wokingham, England, 625 p.

Matthews, G.A. 1979. Pesticide application methods. Longman, London, 334 p.

Matthews, G.A. 1989. Electrostatic spraying of pesticides: a review. Crop Protection 8: 3-15.

Mayer, E. 1961. Theory of liquid atomization in high velocity gas streams. ARS July 31: 1783-1785.

Picot, J.J.; Bontemps, X.; Kristmanson, D.D. 1985. Measuring spray atomizer droplet spectrum down to 0.5 μm size. Trans. ASAE 18: 1367-1370.

Picot, J.J.; Chitrangad, B.; Henderson, G. 1981. Evaporation rate correlation for atomized droplets. Trans. ASAE 24: 552-554.

Picot, J.J.; van Vliet, M.W.; Payne, N.J. 1989. Dropsize spectra from insecticide and herbicide atomizers. Can. J. Chem. Eng. 67: 752-761.

Picot, J.J.C.; van Vliet, M.W.; Payne, N.J.; Kristmanson, D.D. 1990. Characterization of Aerial Spray Nozzels with Laser LightScattering and Imaging Probes and Flash Photography. Pages 142150 *in* E.D. Hirleman, W.D. Bachalo, and P.G. Felton (Eds.), American Soc. for Testing and Materials, Philadelphia.

Rayleigh, Lord. 1982. On the equilibrium of liquid conducting masses charged with electricity. Phil. Mag. 14: 184-186.

Sundaram, A. 1985. A gravimetric method for determining the relative volatilities of non-aqueous pesticide formulations and spray diluents. Pestic. Sci. 16: 397-403.

Sundaram, A. 1986. A simple method to determine relative volatilities of aqueous formulations of pesticides. J. Environ. Sci. Health. B, 21: 165-190.

Sundaram, A. 1987. Influence of temperature on the physical properties of non-aqueous pesticide formulations and spray diluents: relevance to ULV applications. Pestic. Sci. 20: 105118.

Sundaram, A.; Retnakaran, A. 1987. Influence of formulation properties on droplet size spectra and ground deposits of aerially-applied pesticides. Pestic. Sci. 20: 241-257.

Sundaram, A.; Sundaram, K.M.; Leung, J.W.; Holmes, S.B.; Cadogan, B.L. 1987. Influence of physical properties on droplet size spectra and deposit patterns of two mexacarbate formulations, following spray application under laboratory and field conditions. Pestic. Sci. 20: 179-191.

van Vliet, M.W.; Picot, J.J. 1987. Drop spectrum characterization for the Micronair AU4000 aerial spray atomizer. Atomization and Spray Technol. 3: 123-134.

Yates, W.E.; Akesson, N.B.; Cowden, R.E. 1984a. Measurement of dropsize frequency from nozzles used for aerial applications of insecticides in forests. USDA, For. Pest Manage., Washington, D.C. Rep. No. 8434 2804.

Yates, W.E.; Akesson, N.B.; Cowden, R.E. 1984b. Dropsize spectra for applications of Thuricide with Micronair and Unimizer atomizers. USDA, For. Pest Manage., Washington, D.C. Rep. No. 842.

Yates, W.E.; Akesson, N.B.; Cowden, R.E. 1984c. Dropsize spectra for application of Thuricide 32LV with a Beecomist atomizer. USDA, For. Pest Manage., Washington, D.C. Rep. No. 84-3.

Yates, W.E.; Cowden, R.E. 1986a. Dropsize spectra of rotary atomizers with *Bacillus thuringiensis* tank mixes. USDA, For. Pest Manage., Washington, D.C. Rep. No. 86-1.

Yates, W.E.; Cowden, R.E. 1986b. Dropsize spectra of Dipel 8L and Thuricide 48LV atomized with a Micronair. USDA, For. Pest Manage., Washington, D.C. Rep. No. 86-2.

Yates, W.E.; Cowden, R.E. 1986c. Dropsize spectra for Beecomist, Micronair and 8006 flat fan nozzle with Dipel 8L and Thuricide 48LV. USDA, For. Pest Manage., Washington, D.C. Rep. No. 86-4.

Yates, W.E.; Cowden, R.E.; Akesson, N.B 1985. Dropsize spectra from nozzles in high-speed airstreams. Trans. ASAE 28: 405-410 and 414.

Chapter 54

Equipment for Ground Application of Insecticides

PETER DE GROOT

Introduction

Various types of ground-based equipment for the application of insecticides are important tools in the protection of Canada's forests. For small areas, ground spray equipment may be better than aerial equipment for efficient delivery of the insecticide to the target. In some circumstances, ground-based spray equipment is the only equipment that can be used because aerial applications do not give the desired targeting and coverage, are too expensive, or may not be politically or legally permitted. Choosing correct and properly calibrated ground application equipment is essential to ensure accurate targeting of an insecticide at the correct application rate.

Several types of ground spray equipment are used in insect pest management programs in Canada. The following sections briefly discuss the primary considerations in choosing this equipment and describe their principal components, the types of equipment available, as well as their basic operating principles, where they are used in forestry, and the advantages and disadvantages regarding safety and use pattern.

Choosing Ground Application Equipment

Choosing the correct spray application equipment is essential to the effective, safe, and economical use of insecticides. Once a piece of equipment is chosen, it must be properly calibrated because this is just as essential as choosing the right equipment. Techniques for the calibration are usually provided in the instruction manual and are not discussed further here.

There are five questions that should be asked before deciding which type and model of ground spray equipment to use: (1) What are the types, sizes, and conditions of the forest stands to be protected? (2) What are the habits of the target insects? (3) Will the equipment be compatible with the insecticide formulations and their physical and biological requirements for biorational insecticides? (4) What are the local meteorological conditions that will influence the deposit, coverage, and loss of the insecticide, and hence affect the performance of the equipment? (5) Will the equipment be cost effective, provide maximum efficiency, and apply insecticides in an environmentally acceptable manner? Choosing a piece of application equipment can be a complex decision and is seldom dictated by a single factor. Because forest pest management is often faced with limited budgets and labor, and a variety of pests and forest conditions, no single piece of equipment will do all tasks equally well and a compromise is often necessary.

Plantations are better suited for ground spray equipment than natural stands because they generally contain only one species, the trees are more evenly spaced and they have a narrow range of tree heights. Furthermore, greater expenditure on equipment is warranted in plantations because of their relatively higher value. For some types of equipment, tree height, crown size and shape, density of foliage, and spacing may impose limitations. The size of the stand is a major factor; in particular, the area the equipment can cover per unit time and the time available to treat the area are important.

Topography, ground cover, and soil conditions affect the rate of operation of the equipment as does the type of mover (i.e. wheels, tracks, or skids). In recently cutover stands, the amount and location of logging debris can pose a problem. The environmental sensitivity of the stand (or the surrounding area) and the meterological conditions may dictate the choice of equipment. For example, airblast sprayers may produce an unacceptable amount of insecticide drift in parklands or in areas adjacent to water.

Effective control of an insect pest requires an adequate knowledge of its biology and habits. After the most effective and economical control measure, location, timing, and formulation have been determined, then the type of ground application equipment can be chosen. For example, a contact insecticide for open-feeding sawflies can be easily applied with most types of ground spray equipment, but an insect feeding on seeds within a cone may only be controlled with equipment that can inject or implant a systemic insecticide. The equipment must be able to deliver the insecticide within the "spray window" or the time in which the spray must be applied to be effective. The equipment must also be capable of delivering the formulation of the insecticide chosen, and in the case of biological insecticides such as viruses, it must not affect their viability.

There are many factors to consider when evaluating the equipment including ability to change spray direction to account for tree or target differences; flexibility of changing flow rate, spray pattern, droplet sizes, and the distance the insecticide is to be carried; rate of work; simplicity of operation; and operating and maintenance costs. Design features of the equipment such as ease of loading, cleaning, and storage; accessible controls; number and placement of filters; weight; and operator comfort should be considered. The equipment should be safe, durable, and easy to maintain. Dealer service and availability of parts should also be considered before purchase.

Components of Ground Spray Equipment

Three basic components of ground spray equipment are a tank to hold the insecticide, a system to transfer the insecticide from the tank to the spray nozzle(s), and a nozzle system to emit and disperse the insecticide.

There are a variety of shapes and sizes, construction materials, and accessories for spray tanks. The most suitable tank will depend on the needs of the spray program. Some desirable features of tanks are corrosion resistance, an opening large enough for easy filling and cleaning, a tight-fitting cap to prevent insecticide leaks, a liquid level indicator, a filter at the opening to remove debris, an agitation system, a shape and drainage system to allow easy clean-up, and a construction material that is easy to repair.

The insecticide transfer system (tank to nozzle) may be a simple gravity-fed tube (e.g a spinning disk ultra-low-volume sprayer) or it may consist of a series of hoses, valves, filters, and pumps. The pump is the most important apparatus and is subject to the most problems. There are a variety of pumps (gear, roller, piston, centrifugal, and diaphragm) with a wide range of specifications relating to resistance to corrosion, wear characteristics, operating speed, output capacity, and pressure. The choice of pump depends on the volume and pressure required and, to some extent, on the formulation of the insecticide. Most spray equipment sold is matched with the correct pump design and capacity, but the abrasiveness of the insecticide must be considered before using the equipment. For example, roller pumps are subject to rapid wear when using wettable powders, so centrifugal or diaphragm pumps would be better choices. The pump must be capable of producing the correct operating pressure for the insecticide formulation used. For the large hydraulic or airblast sprayers, other essential spray transfer equipment includes control valves, agitators, suction lines, return lines, a pressure regulator and relief lines, a readily visible pressure gauge, and line strainers.

The basic functions of a nozzle system are control of the flow rate, production of droplets of a desired size, and dispersal of droplets in a specific pattern. Correct selection, use, and maintenance of nozzles are critical to the proper and safe application of insecticides. There are many different types of nozzles which may be categorized by the type of energy source: hydraulic, gaseous, centrifugal, kinetic, and thermal. Hydraulic nozzles commonly used in forestry are of the flat fan, hollow cone, or solid type. Airblast sprayers and mist blowers use gaseous energy or twin-fluid nozzles, while spinning disk ultra-low-volume sprayers use centrifugal energy to produce droplets. Further information on nozzles can be found in the preceding chapter.

Types of Ground Spray Equipment Used in Forestry in Canada

Backpack Sprayers

Backpack sprayers are light, portable, operator-carried machines designed to apply small quantities of concentrated or diluted spray. They are mainly used for spot treatments of individual trees and for broadcast treatments in small stands. They are useful where accurate targeting is essential.

The sprayer consists of an insecticide tank, sometimes pressurized by air. The air pressure, generated by a diaphragm or piston pump, forces the insecticide through a hose to a trigger. The spray is directed through a wand or lance by squeezing the trigger. The wand extends the reach of the applicator and can be fitted with a variety of nozzle types to create different droplet sizes and spray patterns. There are basically two types of sprayers available: pressurized containers that are carried by hand or with a shoulder strap and knapsack sprayers that are carried on the back (Fig. 1). The knapsack sprayers have a frame that provides comfort for the operator.

Backpack sprayers offer the advantage of precision targeting of an insecticide, particularly where there are environmental concerns. Their initial low purchase price, low operating costs, light weight, simple operation, and easy cleaning, maintenance, repair, and storage make them attractive for small-scale operations.

One of the disadvantages of backpack sprayers is their uneven application rate. The operator controls the rate of forward movement and tank pressure; therefore, accurate application rates are a problem, especially over rough terrain. Rough and slippery terrain also increases operator fatigue and may increase the risk of injury or insecticide exposure should a fall occur. Problems with clogged nozzles and leaky pistons are not uncommon with some sprayers. With some models, occasional leakage at the filler-cap or from around the pump seal can expose the neck and back of the operator to the insecticide. On other models, the hose is too short for certain applications. Some of the sprayers also lack agitation systems, thus wettable powders are difficult to keep properly mixed. Continuous operation of the hand lever to maintain an adequate spray pressure may also become difficult and uncomfortable after extended use. The risk of exposure to the insecticide is a concern when using backpack sprayers because the operator tends to walk forward into the sprayed areas.

Since backpack sprayers have small tank capacities, they require frequent refilling and may be uneconomical for use in large areas or where high volumes are applied to individual trees. Trees up to 3 to 4 m can be treated efficiently with most backpack sprayers, but motorized backpack mistblowers are preferred for tall trees or large areas that need to be treated quickly.

Motorized backpack mistblowers—Backpack mistblowers use forced air to produce and move spray droplets. They are typically used for the control of insects in young and small plantations (usually less than 10 ha). Individual trees 8 m high and plantations with trees 3 to 4 m high can be treated with most mistblowers. Many small-scale research trials use mistblowers to evaluate the efficacy of different products and dosages or the timing of insecticide applications. Backpack mistblowers are the most frequently used type of ground spray application equipment in Canada's forests.

A typical mistblower consists of an air-cooled two-stroke engine mounted on a knapsack frame, with a pressurized tank, hose, and nozzle (Fig. 2). The engine drives a fan to produce an airstream. The insecticide, on injection into the airstream, is sheared into a fine spray or mist. The velocity of the airstream is controlled by the fan speed, which in turn is controlled by the engine speed using a throttle: a valve on the nozzle regulates the flow rate of the chemical to the nozzle. Droplet size can be regulated by changing the flow rate, air velocity, or nozzle. Most backpack mistblowers use a twin-fluid nozzle and some models can be fitted with a centrifugal (spinning disk) nozzle. The nozzle is mounted on a flexible hose which, together with the airstream, helps direct the insecticide mist to the target area.

Mistblowers are simple to operate, maintain, and calibrate. They are more expensive to purchase and operate than backpack hydraulic sprayers. They are versatile machines that can be used to treat individual trees or stands quickly and efficiently, and many models have a variety of accessories to diversify their use.

The small-sized droplets produced by the mistblowers have both positive and negative attributes. Small droplets are usually preferable for optimizing coverage and insecticidal efficacy, but under the influence of the airstream they are subject to increased evaporation and drift compared to larger droplets. Drift may be used to advantage under the right conditions, particularly where incremental spraying by overlapping swaths is practised. Under improper meteorological conditions, however, drift may increase environmental contamination. For some viscous pesticide formulations, it may be difficult to get a uniform spray deposit.

The engine noise and vibration produced by the machine can be uncomfortable. When used for extended periods of time, the heavy mistblower becomes a burden. The protective clothing for a mistblower operator is often uncomfortable in hot weather because the operator carries and walks with the heavy mistblower. As with backpack hydraulic sprayers, the small tank capacity (usually 10 to 15 L) limits their use to treating small areas (usually less than 10 ha). The size of area where mistblowers become inefficient in terms of cost and efficacy will depend on the size and spacing of the trees, how much of the tree is being treated, the volume of spray being applied, and the work rate of the operator. In large forest stands or for large trees, where there is good access to the trees and fairly level topography, airblast or hydraulic sprayers may have cost advantages over backpack mistblowers.

Airblast sprayers—An airblast sprayer uses the energy of an airstream to transport and distribute the spray droplets (Fig. 3). Nozzles dispense the spray into a high velocity air current generated by a fan. The fans are powered by an engine or they can be driven by the power take-off from a tractor. Airblast sprayers may be mounted on a trailer or directly on a tractor.

These sprayers are large, expensive, and require good access and maneuverability within the stands to be treated, and thus their use in forestry is limited to high-value plantations and seed orchards.

It is important to match the sprayer with the size of trees to be sprayed. The operation of the sprayer must be adjusted to the volume of spray required, which depends on the size of the tree, foliage density, and the coverage required for the insect pest. Properly used, airblast sprayers provide good penetration and coverage of the tree with the spray. These sprayers produce a high number of small droplets that are prone to drift; therefore, they should not be used in environmentally sensitive areas.

Powered Hydraulic Sprayers

Powered hydraulic sprayers are designed to deliver large volumes of dilute spray under high or low pressure. There are two configurations used in forestry: those equipped with a boom, and those equipped with hand-held, trigger-type spray guns. Boom sprayers (also referred to as row, crop sprayers) are often used in nurseries, whereas gun-type sprayers are used to treat trees in shelterbelts, seed orchards, and roadside plantations (Fig. 4), or to treat individual high-value trees. Hydraulic sprayers can either be trailer- or tractor-mounted or mounted on a truck (Fig. 5).

In hydraulic sprayers, the liquid is put under pressure by a pump, then transferred to the nozzles which break up the stream of liquid into droplets. The size of the droplets depends on the pressure applied to the liquid and the type of nozzle selected. These sprays are often applied to the point of run-off to ensure thorough coverage and are excellent for penetrating dense foliage. The trend in forestry has been away from hydraulic gun-type sprayers to airblast sprayers, where conditions permit, because airblast sprayers give just as good coverage of trees, with a greatly reduced liquid volume and in less time.

Ultra-Low-Volume Sprayers

There are several types of ultra-low-volume (ULV) sprayers. The type commonly used in forestry is the battery-powered, hand-held, spinning-disk type applicator (Fig. 6). In Canada, their use is usually confined to herbicide programs, but they are suited for certain insecticide applications in a nursery or for treating young individual trees or small plantations (usually less than 5 ha). They are also useful for experimental applications of insecticides.

In a typical ULV sprayer, the insecticidal liquid is fed by gravity from a small bottle onto a spinning-disk atomizer mounted on a hollow tube. The hollow tube serves as the handle and as the container for batteries that power the spinning disk. An air bleed to the bottle provides a small constant pressure head to the liquid. The flow rate from the bottle is controlled by interchangeable restrictors. Atomization takes place at the teeth of the disk as it rotates, and droplet size can be adjusted by varying the rotational speed of the disk. Wind is used as an agent for dispersal of the insecticide. The distance spray droplets are carried also depends on the droplet size, and the release height above the trees.

One of the advantages of this type of sprayer is that it produces very uniform small drops with little or no pesticide carrier. The ULV sprayer is very light, easy to maintain, and relatively easy to use once the operator has learned how to use the wind effectively for dispersal and deposition of the insecticide. It may lack the durability needed for forestry applications, but extensive testing to assess durability has not been done under Canadian conditions. Battery consumption may be a problem, particularly if rechargeable batteries are not used. Because highly concentrated insecticides are used, there is potentially an increased hazard to the operator. The availability of suitable insecticide formulations for ULV application may also be a problem for some insecticides.

Tree Implants and Tree Injectors

The implantation or injection of systemic insecticides is practised on high-value, isolated trees such as "plus trees," "seed trees," or those in seed production areas. These techniques are suited for trees in rough terrain or where access is remote or limited during parts of the year, and for areas

Figures 1–7

Figure 1. Knapsack sprayer. **Figure 2.** Backpack mistblower. **Figure 3.** Airblast sprayer. **Figure 4.** Hydraulic sprayer with trigger spray gun spraying roadside trees. **Figure 5.** Truck-mounted hydraulic sprayer. **Figure 6.** Spinning-disk ultra-low-volume sprayer. **Figure 7.** Tree injectors for systemic insecticides.

5

6

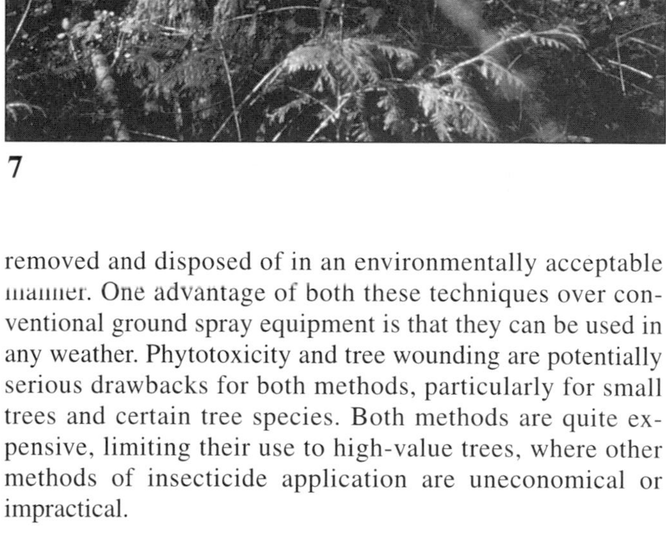

7

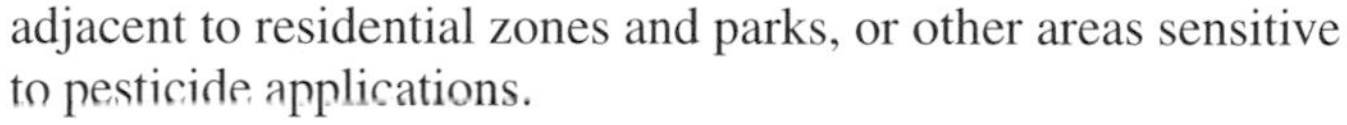

adjacent to residential zones and parks, or other areas sensitive to pesticide applications.

The implantation method involves the use of plastic perforated shells that encase a gelatin capsule containing the insecticide. The cartridges are placed in evenly spaced holes drilled into the trunk of the tree, usually 15 to 45 cm above the ground, and preferably in the root-flare area. The number of cartridges per tree is based on the trunk circumference. The tree's sap dissolves the gelatin capsule and distributes the insecticide.

Tree injectors (Fig. 7) have a plastic reservoir that holds the insecticide and a tube that feeds the insecticide into a hole drilled into the trunk of the tree. The number of units per tree is also based on the tree circumference. Tree injection usually results in faster translocation of the insecticide to the desired site (e.g. foliage at midcrown or cones and seeds) than implantation.

Both implantation and injection are closed insecticide application systems that offer greater protection for the user against insecticide spills and contamination, and minimize risk to nontarget organisms. Nevertheless, pesticide safety equipment is essential because leaks can expose the applicator to the concentrated insecticide, and injectors must be removed and disposed of in an environmentally acceptable manner. One advantage of both these techniques over conventional ground spray equipment is that they can be used in any weather. Phytotoxicity and tree wounding are potentially serious drawbacks for both methods, particularly for small trees and certain tree species. Both methods are quite expensive, limiting their use to high-value trees, where other methods of insecticide application are uneconomical or impractical.

Bibliography

Fisher, H.H.; Deutsch, A.E. 1985. Lever operated knapsack sprayers. International Plant Protection Centre, Oregon State University, Corvallis, Ore. 32 p.

Henigman, J.F.; Beardsley, J.D. (Eds.) 1985. Forest Pesticide Handbook of British Columbia. 2nd ed. Vol. 4. Council of Forest Industries and British Columbia Ministry of Forests, Victoria, B.C.

Matthews, G.A. 1979. Pesticide application methods. Longman Group Ltd. London and New York. 334 p.

Chapter 55

Aerial Control Equipment

B.L. CADOGAN

Introduction

Modern forestry practices, like modern farming methods, require the use of pesticides. Aircraft have always been one of the vehicles used to apply these pesticides, especially if large tracts are to be treated.

In the early days of aerial pesticide application, the aircraft used were converted armed forces, training, sport, and even commercial airplanes with makeshift tanks and booms fitted to conventional airframes. Today, most aerial applications are done with specialized agricultural airplanes that are designed for aerial spraying so that pesticides can be applied with precision.

Equipment for aerially dispersing pesticides was adapted from ground spray equipment and was initially designed for applying dusts, from which practice the term "crop-dusters" probably arose. However, as ground and aerial operations shifted predominantly to liquid pesticides, equipment was modified accordingly (Randall 1975). Even today, most equipment for applying pesticides aerially is generally modified ground equipment, primarily because this is where regular advances in new and innovative spray equipment are being made.

In 1989, there are still only two classes of pesticide spray systems (namely booms and nozzles, and rotary atomizers) that are used to apply pesticides aerially to Canadian forests. These two systems have been in use for more than 50 years (Randall 1975) and their principles have been thoroughly discussed (Akesson and Yates 1979; Matthews 1979). Although no new system of aerial spray equipment has emerged over the last five decades, both of the currently used systems have undergone major improvements over that time. The use of both systems has changed significantly, however, in the last 20 years. Today, most insecticides are applied to forests with rotary atomizers, in contrast to the 1960s and 1970s, when boom and nozzle applications were predominant. This development was precipitated by the trend to apply insecticides in volumes of less than 2.0 L/ha and to accommodate small spray drops less than 100 µm in diameter that are believed to be more effective against forest defoliators than large drops 200 µm or larger in diameter.

Rotary Atomizers

The principles of rotary atomization have been discussed elsewhere in this publication and by Bals (1969, 1977). Although the principles might have remained unchanged over the last 15 years, there have been significant technical and performance improvements in modern rotary atomizers.

Two brand name rotary atomizers are currently being used to apply pesticides regularly to Canadian forests and an additional two have the potential for regular use.

Micronair Atomizers

The Micronair rotary atomizer is the most widely used rotary atomizer in Canada. Almost 100% of the treatments to Ontario Crown land are applied with aircraft equipped with these atomizers. In other provinces, large aircraft equipped with booms and nozzles are sometimes used to spray insecticides, but when light aircraft are required, Micronairs are used almost exclusively. The atomizer consists of a cylindrical gauze cage that rotates around a fixed spindle attached to a mounting bracket on the wing of the aircraft.

Changing the pitch of the atomizer blades varies the rotational speed of the atomizer cage which in turn varies the size of the drops produced. The pesticide enters through the hollow spindle and is deflected evenly over the rotating cage, where atomization is completed when the liquid is thrown clear of the cage (Micronair [Aerial] Ltd. 1986).

There are 11 Micronair models that can be used for aerial spraying (Micronair [Aerial] Ltd. 1986), but only four are currently being used operationally in Canada. One of the four models is hydraulically driven. The other are wind-driven directly from the airflow (i.e. aircraft slipstream) through the atomizer blades. Changes in the speed of the aircraft result in variable power being supplied to the atomizers and consequent fluctuations in the rotational speed of the units. This inevitably leads to the production of a wider droplet size spectrum. On the other hand, hydraulically driven atomizers maintain constant rotational speeds that are independent of an aircraft's air speed, but the power to drive these units at the rotational speeds necessary for efficient atomization may require major modifications to the average spray aircraft.

The big advantage of Micronairs over booms and nozzles is their ability to atomize low- or ultra-low-volume spray emissions efficiently into very small and relatively uniform droplets 20 to 100 µm in diameter. Production of small drops is necessary if effective spray coverage of the prescribed target area is to be achieved. At present, Micronairs appear to be used almost exclusively on light aircraft such as Cessna Ag Trucks, Ayers Thrush, etc., which suggests, albeit incorrectly, that rotary atomizers in general are not well suited to withstand the high stresses created by the speeds of large aircraft (Douglas DC-6, etc.). In the past, early models of wind-driven rotary atomizers did suffer premature structural failure when fitted to large aircraft because the atomizers were often not optimally located on the aircraft's wings; consequently, the powerful slipstreams caused the atomizer cages to rotate unrestrictively at speeds higher than they were designed for. Nevertheless, rotary atomizers were used successfully on large aircraft as early as 1975 (Table 1). During the 1980s, the performance of Micronairs on medium and large aircraft was investigated with the emphasis on determining the number of atomizers per aircraft, the location on the wing, the atomizer blades, and the angle settings that would minimize stress on the atomizer and

Table 1. Field calibrations in Canada with Beecomist rotary atomizers between 1975 and 1988.

Location	Year	Aircraft	Equipment configuration	Air speed (km/h)	Study objective
Manitoba	1975	DC-6	16 Beecomists	352	Product efficacy
Quebec	1978	DC-6	16 Beecomists	352	Products & flow rate
Ottawa	1978	Hughes 500D	4 Beecomists	96, 64, 32	Air speed on deposit
New Brunswick	1985	Bell 206	Beecomists	120, 40	Air speed on deposit
Ontario	1988	Bell 212	Beecomists		Product efficacy

be optimum for spraying with these aircraft (Québec Énergie et des Ressources, unpublished data). Consequently, in 1985, Micronairs were successfully fitted to an Avenger TBM aircraft (Fig. 1) (no undue atomizer fatigue induced by stress was evident) and acceptable atomization was achieved. The four Micronair models that are most commonly used for aerial applications of pesticides in Canada are discussed individually.

AU3000—The compact Micronair® AU3000 was the first Micronair model to be used operationally on aircraft in Canada (Randall 1975). This model is wind-driven and is still in use on some aircraft, which is testimony to its mechanical performance and overall reliability. The AU3000 was introduced to more effectively apply spray volumes that range from ultra-low to high.

Fitting the AU3000 to an aircraft requires the installation of special support booms (Figs. 2, 3). This requirement is considered a major constraint to use of the AU3000 because the procedure is costly and requires the services of a skilled technician and regulatory certification.

AU4000—The AU4000 (Fig. 4) is a streamlined updated version of the AU3000 with a cage (gauze) that is reduced to half of its original length. Unlike the AU3000, it does not require special supports for fitting, but can be attached to either standard or streamlined spray booms. This is a distinct advantage because it allows for easy interchange of the atomizer with conventional booms and nozzles. This model is also capable of rotating significantly faster than any other Micronair atomizer. With an aircraft flying at 160 km/h, a dry (i.e. without a dampening liquid flow) wind-driven AU4000 can rotate at approximately 12 000 rpm, and it is well-suited for producing small droplets less than 60 μm), which are necessary in the application of ultra-low spray volumes. One AU4000 model can be driven hydraulically, which allows constant rotational speeds that are independent of the aircraft's speed. This allows the production of relatively uniform spray spectra. However, the hydraulic model requires considerably more power than is normally available in spray planes to drive it.

AU5000—The AU5000 mini-atomizer (Figs. 5, 6) is a small lightweight (1.8-kg) direct-drive unit that was developed before to the AU4000 to meet the needs of operators who required the advantages of a Micronair system without the constraint of the special support brackets that are required for the AU3000 (Micronair [Aerial] Ltd. 1982). The AU5000 is a versatile unit that can be fitted to several different spray aircraft and is comparatively simple to attach to most standard spray booms without the need for structural modifications to the aircraft. This adaptability allows an aircraft to be prepared quickly to use either Micronairs or booms and nozzles. The mini-atomizer will handle "flow rates" up to 22 L/min, thus accommodating a wide range of application volumes without significantly losing its atomization efficiency. In general, these units do not rotate as fast as the larger models; consequently, they produce droplet spectra that might include a significant complement of drops in the 250-μm diameter range. Trials have been conducted since 1983, primarily in Quebec, to investigate the operation of the AU5000, and how to optimize its performance. During these trials, the positioning of the atomizers on the aircraft, the length and pitch of the atomizer blades, the physicochemical characteristics of the formulation, and how flow rates influence atomizer performance (Table 1) were studied. Consequently, the mini-atomizers are now used widely in control operations.

AU7000—The AU7000 atomizers are small units that are presently used predominantly on ground equipment but show some potential for use on aircraft. These units can be powered either hydraulically or by direct drive with the former allowing the atomizers to rotate at constant yet acceptably high speeds. This facilitates the fitting of these atomizers to slow-moving aircraft without seriously compromising their ability to atomize efficiently. The direct-drive model can be attached to spray aircraft that fly at approximately 100 km/h; however, these direct-driven units do not appear to have any advantages over the AU5000 model for aerial spraying.

The Beecomist Rotary Atomizer

The Beecomist® Rotary atomizer gained prominence during the 1960s, when there was a proliferation of ultra-low-volume spray equipment (Randall 1975). The Beecomist spray head uses sleeves driven at high speeds by either an electric or a hydraulic motor. The electric motor weighs 3.6 kg and operates on either a 12- or 24-V DC power supply. The model draws 24 A (amps) at 12 V or 14 A at 24 V when atomizing a highly viscous material. These units, therefore, require a powerful electrical source to operate efficiently.

The hydraulic model weighs 1.54 kg and thus is considerably lighter than the electric model, but it requires significant hydraulic power (13 L/min of hydraulic fluid at 5860.5 kPa) to achieve the high rotational speeds (10 000 to 15 000 rpm) that are the greatest advantage of these units.

Figure 1. Micronair® AU4000 rotary atomizers fitted to a Grumman Avenger (TBM). (H.J. Irving)

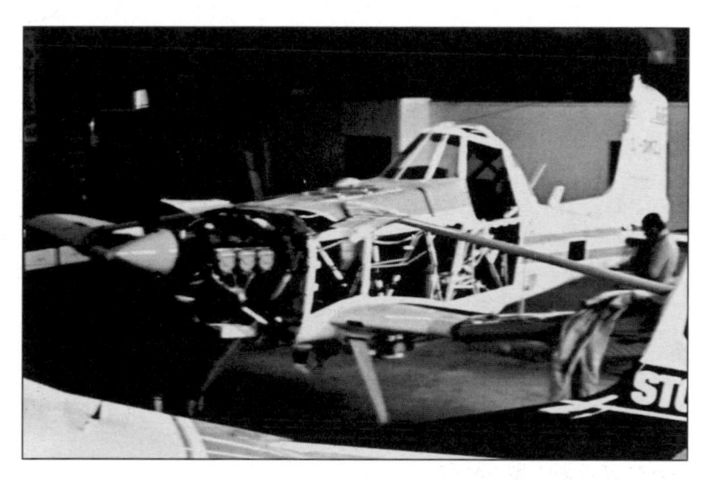

Figure 2. Installing support booms to a Cessna-188 Ag Truck to accommodate Micronair® AU3000 rotary atomizers.

Figure 3. Micronair® AU3000 rotary atomizers installed on support booms in a Cessna-188 Ag Truck.

Figure 4. Micronair® AU4000 rotary atomizer.

Figure 5. Micronair® AU5000 atomizers fitted to a Piper Pawnee. (M. Pelletier)

Figure 6. Micronair® AU5000 atomizers fitted to an Ayers Thrush.

The power sources of both models enable them to maintain uniform rotational velocities independent of aircraft speed and both are highly capable of producing uniform droplets. Because atomizer rotational speeds are dampened by the volumetric flow rate and the viscosity of the spray-mix, Beecomists tend to compromise the ability to use high flow rates to achieve high rotational speeds and are, therefore, most efficient using ultra-low-volume sprays. Consequently, Beecomists are suited to aircraft that can supply high power sources and are recommended for slow aircraft that are unable to generate high atomizer speeds and small drops when using direct-drive atomizers.

In 1975, Beecomists were used successfully on large (DC-6) aircraft to spray mosquitoes (Table 1). Since then, the atomizers have been used experimentally on both rotary wing and fixed-wing aircraft (Figs. 7, 8, 9), primarily to investigate deposit patterns, product efficacy, and the influence of flow rates and air speed on atomizer output (Table 2). They have also been used occasionally in operations against spruce budworm, *Choristoneura fumiferana*, and spruce bud moth, *Zeiraphera canadensis*. Although Beecomists can atomize ultra-low spray volumes most efficiently, these units are not used widely in operational pest control in forestry, probably because of their significant power requirements.

The Unimizer

The Unimizer no. 10 rotary atomizer is a lightweight, direct-drive unit that is considered an alternative to the Micronair® AU5000 mini-atomizer. However, this unit has not been used experimentally or operationally in Canadian forests and no performance data are available. The Unimizer mounts simply on any aircraft with a trailing edge boom, but it does not rotate at speeds that are required for efficient ultra-low-volume spraying. This is perhaps the main reason why it has not been considered as a serious competitor for other rotary atomizers in Canada.

Airbi Rotary Atomizer

The Airbi is a French-made rotary atomizer that has been introduced recently, intended primarily for light, fixed-wing aircraft and helicopters. The Airbi is electrically propelled by a 12- or 24-V electromotor. It differs from the conventional rotary atomizer in that it has a magnet and a watertight casing equipped with a joint that operates by centrifugation and an overpressure by air inside the head. The rotary head is perforated with holes to prevent clogging and is equipped with conventional blades or with 10-mm long bristles.

The unit is lightweight (1.17 kg) with peak consumption of 15 A at 12 V or 10 A at 24 V, and is reportedly capable of

Table 2. Field calibrations with Micronair® AU4000 and AU5000 rotary atomizers between 1983 and 1988.

Location	Year	Aircraft	Equipment	Air speed (km/h)	Study objective
Ontario	1983[a]	Cessna-188 Ag Truck	6-AU5000	160	Flow rates, blade pitch
Quebec	1983[b]	Piper Pawnee	6-AU5000	160	Products, diluents, and flow rates
Quebec	1984[c]	DC-4	Booms & nozzles	–	Products, atomization, aircraft type
		Turbo Thrush	6-AU5000	210	
		Piper Pawnee	6-AU5000	160	
Quebec	1985[b]	Piper Pawnee	6-AU5000	160	Atomizer position and aircraft vortices
Quebec	1986[d]	Turbo Thrush	6-AU5000	209	Atomizer position, blade length, blade pitch, and flow rates
		Bull Thrush		209	
		Thrush Commander		177	
		Ag Cat		177	
		Piper Pawnee		161	
Quebec	1986[b]	Piper Pawnee	6-AU5000	160	Atomizer position
Quebec	1986[b]	Piper Pawnee	6-AU5000	160	Atomizer position and blade pitch
Quebec	1987[e]	Piper Pawnee	8-AU5000	160	Products and flow rates
N.B.	1988[f]	Avenger TBM	AU4000		Spray distribution

[a] Calibrations by Forest Pest Management Institute.
[b] Calibrations by Émile Aubin, Québec Énergie et des Ressources.
[c] Calibrations by Émile Aubin and M. Barras, Québec Énergie et des Ressources.
[d] Calibrations by M.-C. Lambert, Québec Énergie et des Ressources.
[e] Calibrations by M. Pelletier, Québec Énergie et des Ressources.
[f] Calibrations by Forest Protection Limited.

Figure 7. Beecomists fitted to a DC-6 wing. (M. Pelletier)

Figure 8. Beecomists fitted to a DC-6. (M. Pelletier)

Figure 9. Beecomists fitted to a Bell 207 helicopter. (R. Johnston)

rotating at approximately 2 000 to 12 000 rpm. At present, no reports are available on field performances of this new atomizer.

Booms and Nozzles

Booms and nozzles have long been the mainstay of agricultural and forestry pesticide applications. Even with the increased use of rotary atomizers for applying insecticides to forests, a variety of booms and nozzles are still widely used, primarily for aerial spraying of herbicides. Although considerable improvements have been made to the design, manufacture, and performance of booms and nozzles over the past 10 years, their use to control forest insects has been waning since 1979. It is only occasionally, and even then, primarily on large fixed-wing aircraft and helicopters, that booms and nozzles are used for insecticide applications in Canada.

Booms that used to be made of heavy metal pipes have been replaced largely by booms made of lightweight alloys, hardened plastics, and polyvinyl chloride. Many of the new booms are streamlined (Fig. 10), which makes them more aerodynamic than the standard booms.

Over the past 20 years, a wide variety of nozzles has been available for use. Nozzles are now made from almost 50 different materials (Spraying Systems Co.), many of them synthetic. These advancements have greatly enhanced the performance of nozzles.

The Thru Valve Boom

Perhaps the one novel development in boom and nozzle technology since 1975 that is relevant to the aerial application of pesticides is the Thru Valve Boom. It is streamlined with multiple nozzles and valves faired into the boom (Fig. 10) and is mounted to create a minimum of air turbulence in the area of the nozzle orifice. With an aircraft traveling at high speeds (160 to 248 km/h), the boom is reportedly the first spray device that allows liquid to be discharged in laminar flow that forms sprays that are free of aerosols. Although it was intended primarily for use on helicopters applying high-volume herbicides, the system can also be fitted to small aircraft (Figs. 11, 12). The nozzle design and configurations show that the system can be adapted to apply low or ultra-low spray volumes and is capable of producing droplet spectra that are similar to those produced by the Micronair® AU5000s and therefore applicable to insecticide treatments.

Acccessory Equipment

The atomizers have always been the main focus of a spray dispersal system; however, the accessories that complement them (i.e. pumps, regulatory valves, etc.) are vitally important. Sprays can be applied efficiently only if the entire dispersal system is operating optimally as a unit.

Randall (1975) described the basic components of a spray dispersal system. Whereas the generic composition of a spray system has not changed significantly since then, the components have been improved dramatically. Significant advances have been made, particularly to the regulatory

Figure 10. A Thru Valve Boom showing multiple nozzles.

Figure 11. A Thru Valve Boom fitted to an Ayers Turbo Thrush.

Figure 12. A Thru Valve Boom fitted to a Cessna-188 Ag Truck.

components of dispersal systems. These improvements have resulted in improved accuracy in the aerial application of pesticides.

Pumps and Regulators

The liquid pump has long been considered the heart of a spray machine (Akesson and Yates 1979). In boom and hydraulic pressure nozzle systems, where the pump supplies the pressure to atomize the liquid into spray drops, and rotary atomization, where the function of the pump is to force the spray mixture to the atomizers, pump improvements have led to more efficient application of pesticides.

Positive displacement pumps are now equipped with flow sensors that monitor pump performance and output. These provide a uniform flow to the atomizers. To further ensure precise flow rates, most rotary atomizers are equipped with a variable restrictor unit that regulates the flow through each atomizer.

At present, most hydraulic packages for rotary atomizers largely consist of an independent hydraulic motor that drives the spray pump and include a hydraulic filter and hoses with "quick-disconnects" for easy removal. Electrically driven spray pumps are used primarily on helicopters and are designed to improve flow rates.

Chemical Injection Spraying

Some of the most radical advances in accessories have been in the area of electronics. One of these is a chemical injection spraying system. It differs from conventional spray dispersal in one fundamental way: instead of mixing the pesticide with the diluent in the sprayer or mixing tank, a precise amount of the pesticide from one tank is metered into a line carrying the diluent from another tank to the atomizers. The output of the metering pump is monitored and controlled electronically and an in-line mixer blends the diluent and pesticide before the mixture is atomized.

This system eliminates surplus spray-mix and manual mixing and substantially reduces cleaning of mixing equipment. However, because current trends emphasize spraying undiluted insecticides, these benefits might not be readily available for forest applicators.

Other Electronic Accessories

Digital flow meters display information in the cockpit on flow rates in litres per minute or litres per hectare and provide read-outs of the volume of spray remaining in the spray tank.

Application computers and electronic Sprayer Monitors keep track of the spraying systems, and provide read-outs to the pilot on aspects such as pass time, volume pumped per pass, total volume pumped from the tank, area covered, minutes of actual spraying time, L/ha, etc. Micronair Ltd. has developed a digital instrument that measures the rotational speed of Micronair rotary atomizers, enabling the operator to set them to produce desirable droplet patterns.

Data loggers have recently been incorporated into these systems so that permanent hard copies of the data are available. This development is very important when assessing the performance of a spray dispersal system.

Summary and Prognosis

Although no novel atomization processes have evolved to replace or even supplement rotary atomizers and hydraulic boom and nozzle systems, the atomization of aerially applied liquid pesticides has improved significantly since 1975. Rotary atomizers are now better designed to withstand stresses than they were 15 years ago and can rotate at higher speeds than earlier models. Because drop size is determined by the rotational speed of the atomizers, modern rotary units are capable of producing high numbers of droplets from substantially reduced application volumes. New synthetic materials and other lightweight alloys of high tensile strength have largely replaced the steel in the older and heavier conventional booms. Additionally, many of these new booms are more aerodynamically streamlined to reduce drag. Nozzles are also better designed, specifically for efficient dispersal of liquids. These improvements have resulted from a universal requirement to apply pesticides more efficiently.

It is not clear whether demand will continue to dictate the path that future developments in pesticide dispersal equipment will take or if development will limit how efficiently pesticides can indeed be applied. The present trend in insecticide applications to forests is ultra-low-volume spray emissions which, by their nature, implies the production of very small drops and very precise application techniques if the target area is to be covered effectively. This suggests that in the future, research on rotary atomizers will be directed toward achieving much higher speeds than the present atomizers are capable of.

However, it appears that the public's perceptions of pesticides will influence whatever development there will be; their concerns for clean pristine forest environments will probably dictate not only what pesticides are used, but also how they are applied.

Regardless, it is envisaged that future spray equipment will be largely electronically controlled. This implies precise functioning and thus suggests very efficient insecticide applications to control forest pests.

References

Akesson, N.B.; Yates, W.E. 1979. Pesticide application equipment and techniques. FAO Services Bulletin 38. 257 p.

Bals, E.J. 1969. Design of rotary atomizers. Pages 156-165 *in* Proc. 4th Internat. Agric. Aviation Congr. Kingston, Ont.

Bals, E.J. 1977. A new rotary disc nozzle. Pages 523-525 *in* Proc. 1977 British Crop Protection Conference (Pests and Diseases).

Matthews, G.A. 1979. Pesticide application methods. Longman, N.Y. 334 p.

Micronair (Aerial) Ltd. 1982. Micronair AU 5000 mini-atomizer handbook. Isle of Wight, England. 34 p.

Micronair (Aerial) Ltd. 1986. Micronair AU atomizer handbook. Isle of Wight, England. 58 p.

Randall, A.P. 1975. Application technology. Pages 34-55 *in* M.L. Prebble (Ed.), Aerial control of forest insects in Canada. Dept. of Environ. Ottawa, Ont. 330 p.

Spraying Systems Co. 1987. Industrial spray nozzles and accessories. Catalog 25. Ill. 64 p.

Part III
Application of Technology and Equipment

Introduction

Environmental Concerns Related to Forest Insect Control Programs 1973–1988

PETER D. KINGSBURY

The past decade and a half of forest insect control operations has been a period of consistent and increasing public concern over the health and environmental effects of pesticides. Environmental concerns first popularized by Rachel Carson's landmark book *Silent Spring* (Carson 1962) have expanded into widespread public anxiety over pesticide effects on the environment, bolstered by media attention of issues such as dioxin contamination of defoliants used in Vietnam. These concerns have created political responses that have had tremendous impacts on forest insect control programs in Canada. At the same time, a great deal of scientific effort has been applied to determine what actual impacts forest spraying has had on the environment, including nontarget organisms. The articles that follow this introduction will expand on the nature and findings of these studies.

The Nature of Environmental Impact Studies

Environmental impact studies relevant to the effects of forest insect control operations have been carried out by a wide range of federal, provincial, academic, and private agencies for a variety of purposes (Table 1). Although these can generally be differentiated into the categories of product development, monitoring, or research studies, these differences are not always clear-cut and there is considerable overlap between these three activities. Monitoring activities have often focused on new use patterns of insecticides in evaluating the effects of previously unstudied timing, dosage, formulation, application system, or sequence of applications. Product development studies often provide considerable new information on insecticide ecology because they tend to involve more in-depth studies under more controlled conditions than monitoring studies. All kinds of studies may contribute to the development and testing of appropriate impact study methodologies. Each type of study has built-in advantages and limitations, and the best assessment of environmental effects comes from drawing on a variety of studies. Just as laboratory and field studies provide different balances between controlled experimental and real world conditions, so various types of studies cover a range of situations that complement our understanding of insecticide effects.

Impact studies have focused on a broad range of topics ranging from effects on microorganisms to impacts on vertebrate communities. Residue monitoring has ranged from verification of active ingredient concentrations in tank mixes, to residue levels within a wide range of substrates within spray blocks, to residues found long distances offtarget in air, water, and shellfish. Terrestrial invertebrate studies have focused their attention on invertebrate-dependent ecological processes, pollination and parasitism in particular. Aquatic studies have concentrated on food chains in forest streams, although some work has also been done in bogs, ponds, and lakes. Terrestrial vertebrate studies have focused primarily on the effects of insecticides on forest songbird populations.

Evolution of Federal and Provincial Regulatory and Monitoring Agencies

From 1973 to 1988, there was a gradual shift in the regulation and monitoring of forest insect control programs from federal to provincial agencies. The information on environmental effects of forest insect control operations summarized in Prebble (1975) was almost exclusively generated by the Canadian Forestry Service, Canadian Wildlife Service, and Department of Fisheries — all federal agencies. Although forest insect control programs were planned and conducted by provincial agencies, they played little role in evaluation of side effects. As provincial legislation and regulations serving the purposes of the Pest Control Products Act have come into being during the 1970s and 1980s, the authority to approve spray programs and the mandate to evaluate their effects have been taken over by provincial agencies. This process has evolved at a different pace in different jurisdictions and is continuing to evolve in some provinces. Details of this transition for several eastern Canada provinces can be found in Kingsbury and Trial (1987). Current provincial regulations range from review of spray programs by a Pesticides Advisory Board that makes recommendations to the provincial environment minister (e.g. Newfoundland, New Brunswick), to formal environmental assessment processes involving submission of an impact study and public hearings (Quebec).

As each province has taken over the regulation of their spray program, there has tended to be an attempt to generate a local spray monitoring initiative within that province's research communities. In many instances this has been accomplished by conditions of the licensing process which make the spray proponent provide funding for monitoring studies (e.g. Newfoundland). In other cases, the spray agency itself has developed internal expertise and monitoring capability (e.g. Quebec). Private consultants have done the majority of monitoring work in some provinces (e.g. Ontario), while in others (e.g. New Brunswick), monitoring has involved an extensive group of federal, provincial, university, and private consultant groups. The provinces of New Brunswick, Newfoundland, and Quebec have each had ongoing activities in the area of environmental monitoring of forest insect control operations for over a decade (Table 2). In general, the amount of field monitoring has declined substantially since the early 1980s as the scale and diversity of forest spray programs have declined.

Insecticide research activities have tended to remain the domain of federal agencies, with occasional contributions

Table 1. Various types of environmental impact and health studies carried out on forest insect control.

Type of study	Purpose	Comments
Pre-registration or Product Development Studies	- Generate information to be used in the federal registration process to evaluate the potential hazard of allowing the insecticide being studied to be used for forest insect control	- Generally carried out by pesticide developers (industry or Forest Pest Management Institute) - Includes acute and chronic toxicology studies on a range of organisms as well as special studies such as carcinogenicity and mutagenicity evaluations
Monitoring Studies	- Evaluate the effects of a given spray program (or portion of it) in a given year	- Often required as a condition of licensing by the provincial regulatory body - May identify new areas of concern not previously identified
Exposure (residue) monitoring	- Evaluate the magnitude and extent to which organisms are exposed to insecticides in various substrates (air, water, foliage, soil, food) during and after spray programs	- May be independent of or combined with effects (impact) monitoring - Must be combined with simultaneous or previous study on the biological consequences of the exposure levels found to determine their significance
Effects (impact) monitoring	- Evaluate the biological response of organisms or communities exposed to insecticides	- Interpretation of results may be limited by inadequate knowledge of the influences of other factors such as natural stresses (e.g. weather) on the phenomenon studied
Research Studies	- Evaluate the nature of insecticide organism–ecosystem interactions	- May direct monitoring studies into the most appropriate areas of concern - Also follow up on unforeseen or unusual results from monitoring studies
Methodology research	- Develops methods for evaluating exposure levels and organism response to insecticides	- Degree to which methodologies are developed limits monitoring that can be carried out
Impact research	- Determine the nature and ramifications of cause–effect relationships between insecticides and organisms or ecosystems	- May define pathways of effect applicable to a variety of insecticides and determine significance of process disruption or dysfunction

from university researchers who have been federally or provincially funded. The Forest Pest Management Institute and Maritimes laboratories of Forestry Canada, the Fredericton office of the Canadian Wildlife Service, and the Atlantic regional office of the Environmental Protection Service have been the most active research groups in the field over the past decade. Several organizations and meetings have served as a focus for reporting and discussing environmental effects of forest insect control programs (Table 3).

Insecticide Risk Evaluation

The interactions between an insecticide and the forest environment are complex and dynamic. Toxicity of a given insecticide varies widely among the enormous diversity of organisms that may be exposed to it. Toxicity also varies widely depending on the routes of exposure (e.g. oral, dermal, inhalation) by which organisms contact the toxicant. Numerous ecological factors such as life history, trophic relationships, and behavior influence the inherent risk an organism may be at, and these in turn are influenced by environmental factors such as weather. Exposure of the insecticide to an organism is itself influenced by a host of physical, meteorological, topographic, and biological factors. Although we can simplify the interaction to a relationship such as:

$$\text{Environmental response} = f\left[\left(\begin{matrix}\text{Toxicological and}\\ \text{ecological risk}\end{matrix}\right)\left(\begin{matrix}\text{Environmental}\\ \text{exposure}\end{matrix}\right)\right],$$

we can never hope to fully quantify or understand it completely.

To compensate for the inadequacies in our ability to predict the precise interactions between an insecticide and the

Table 2. Provincial monitoring groups that have produced extensive literature on monitoring of forest insect control operations.

Province	Group and responsible agency	Period active	Comments
New Brunswick	EMOFICO (**E**nvironmental **M**onitoring of **F**orest **I**nsect **C**ontrol **O**perations) Committee	1976–present	- Evolved from PERG (**P**esticide **E**cology **R**esearch **G**roup), an informal organization of researchers
	Environment Canada University of New Brunswick New Brunswick Department of the Environment	1976–78 1978–80 1980–present	- Have issued annual reports since 1976 summarizing extensive biological and residue studies
Newfoundland	Environmental Assessment Division Department of Environment Government of Newfoundland and Labrador	1977–present	- Have issued annual reports summarizing residue and biological monitoring carried out on spruce budworm (1977 to 1985 except for 1981) and hemlock looper (1985 to 1987) spraying
Quebec	Service des Études environnementales Ministère de l'Énergie et des Ressources Gouvernement du Québec	1979–present	- Carried on monitoring program initially conducted under the Sous-comité de l'Environnement, comité de Surveillance Écologique des Pulvérisations Aériennes - Have issued numerous reports principally dealing with deposits and residues of chemical and biological insecticides but including some biological studies

Table 3. Organizations and meetings that have regularly dealt with environmental effects of forest insect control programs from 1973 to 1988.

Organization or meeting	Period active	Comments
Annual Forest Pest Control Forum	1973–present	- Held annually under the aegis of Forestry Canada to provide a forum for the review and discussion of forest pest control operations in Canada and related research
Eastern Spruce Budworm Council	1977–present	- A committee of forest resource department deputy ministers from eastern North America - Have standing health and environmental committees that report biannually on current issues in these areas
Annual Eastern Spruce Budworm Research Work Conference	1977–present	- An annual meeting of researchers and forest managers concerned with the spruce budworm and its control
National Research Council of Canada Associate Committee on Scientific Criteria for Environmental Quality	1973–1985	- Established expert panels to evaluate and produce documents on environmental risks associated with the use of several insecticides including fenitrothion, aminocarb, and Bt
Canada–United States Spruce Budworm Program (CANUSA)	1977–1983	- An international program of research into spruce budworm control and management of susceptible forests - The program included an environmental impact working group that dealt with issues in this field - Funded several annotated bibliography and methodology reports in the field of environmental effects

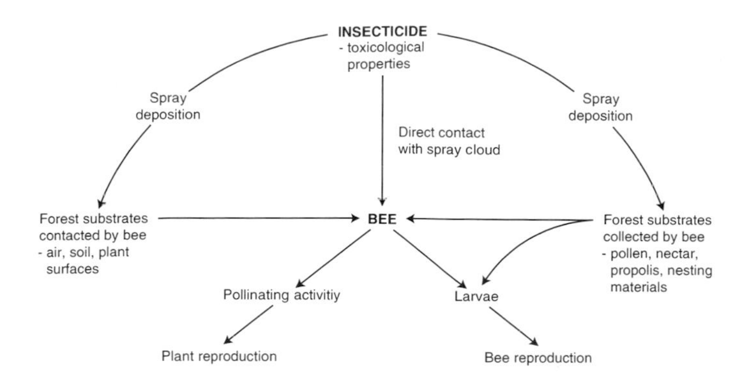

Figure 1. Important factors contributing to the ecotoxicity of forestry insecticides to bees.

environment, risk evaluation processes have focused on selected indicator species or processes that can be evaluated under conditions felt to represent worst case situations to indicate the maximum level of effect likely to occur in the real world. This approach is taken with assessments of risks to both health and the environment. The potential effects are conceptualized, as is done for insecticide–bee interactions in Figure 1, to guide toxicological and ecological studies required to clarify the levels at which significant effects will be expressed. The probability of direct toxic effects occurring can then be predicted by comparing levels of the insecticide-inducing effects in toxicological studies with measurements of insecticide found in the appropriate substrates after spraying.

Where direct effects are predicted, ecological studies help predict the possible secondary effects that might result. For instance, studies showing the dependence of a forest plant on a given group of pollinating insects for successful seed set suggest the potential impact of reductions in populations of that pollinator on reproduction of that plant. Although evidence for predicting these types of effects can be accumulated, it is necessary to evaluate the extent to which they actually occur under operational use patterns because of our inability to understand all of the factors that might enhance or mitigate the predicted effects. Sublethal effects on organisms not detected under bioassay conditions may produce unpredicted effects in the real world. Conversely, compensatory mechanisms in ecological systems may reduce or eliminate effects that are predicted. Monitoring programs guided by research studies continue to be important to our understanding of the real impacts of insecticide spray programs.

References

Carson, R. 1962. Silent Spring. Houghton Mifflon, Boston. 368 p.

Kingsbury, P.D.; Trial, J.G. 1987. Buffer zones: their application to forest insect control operations. Proceedings of the Buffer Zone Workshop sponsored by the Eastern Spruce Budworm Council's Environmental Committee. Quebec, Que., 16-17 April 1986. Forest Pest Management Institute, Sault Ste. Marie, Ont.

Prebble, M.L. (Ed.). 1975. Aerial control of forest insects in Canada. Dep. Environ., Ottawa, Ont. 330 p.

Chapter 56

Fate of Insecticides in the Forest Environment

K.M.S. SUNDARAM

Introduction

An array of insecticides has been used by perceptive Canadian forest managers since 1927 to control epidemic populations of insects in the orders Lepidoptera and Coleoptera (Shea and Nigam 1984). Because of their efficacy, economy, and availability, the use of insecticides has become very widespread in recent decades and has enhanced the output of forest lands by protecting the forest resource from some of the most destructive organisms affecting it. Before 1967, DDT [1,1,1-trichloro-2,2-bis-(*p*-chlorophenyl) ethane] was used extensively (Fettes and Buckner 1976). Due to environmental concerns about this chemical (Bitman et al. 1970; Kenaga 1972), it was phased out and replaced by a variety of short-lived, readily degradable but potent organophosphorus and carbamate insecticides. Table 1 lists some of the registered or experimental insecticides used at present in Canadian forestry spray programs. Insecticides registered for minor use, such as forest nursery, seed orchards, ornamental and shade trees, and woodlots, are not included in the list.

Insecticides, by definition, are toxic to living organisms. Broadcast application and subsequent distribution of insecticides in the environment could have adverse effects on the other nontarget organisms including humans. Realization of such adverse effects depends on the inherent toxicity of the compound and the potential for nontarget organisms to be exposed. Considerable public and regulatory concerns have therefore been expressed about the chemical control practices used in forestry and controversies have arisen within the last decade (Dunster 1987). The Forest Pest Management Institute of Forestry Canada initiated research on the distribution, persistence, movement, and fate of insecticide residues in different compartments of the forest environment to gain some insight into their potential hazards or damages and to possibly alleviate some of the concerns. The task is extremely complex, especially if one must systematically follow each insecticide and its additives released into the forest environment, the metabolites formed and their ultimate fate and consequences, in the four environmental compartments of the forest, namely atmosphere (air), hydrosphere (water), lithosphere (soil), and biosphere (biota). Data gaps exist concerning the environmental fate of most of the forestry insecticides listed in Table 1. The presentation of a holistic picture is hampered by lack of information on the persistence and degradative pathway of each chemical for even a single environmental compartment under actual conditions of use. Bearing this in mind, the objectives of this chapter will be twofold: (1) to present an overview on the general principles governing the fate of chemicals in the forest environment and (2) to compile and discuss briefly the fate and persistence of some of the specific insecticides used in forestry spray programs in Canada, during the post-DDT era, either on an operational or on an experimental basis.

Environmental Fate and Persistence of Insecticides in the Forest Environment: An Overview

Aerial broadcast of insecticides is the most prevalent application method in forestry. Ground application methods are used if the area sprayed is small and thorough target coverage is essential. In all forestry sprays, the target is the tree crown. The rate of application or dosage varies with the insecticide and in most cases it is less than 1 kg/ha. (All dosages refer to the amount of active ingredient applied, unless otherwise stated.) The dosages for aminocarb and fenitrothion, the two principal insecticides used in recent years, are, respectively, 70 and 280 g/ha per application.

Once the insecticide is released over the forest canopy, it is distributed among the four compartments (air, water, soil, and biota) in a way that depends on the mode of application, nature and thickness of the forest canopy, and prevailing meteorological conditions (NRCC 1975; Armstrong 1977). Insecticide distribution within the forest is quite varied (Sundaram 1989b) and as little as 20% or less of the applied dosage reaches the forest floor (Sundaram 1990a). The rest is distributed unevenly among the other three compartments (Norris 1971). Figure 1 illustrates the pathways by which an aerially released insecticide is distributed among the forestry compartments and the possible transformation processes it could undergo. The residues in the environmental compartments are subjected to physical, chemical, and biological transformations (Lichtenstein 1972a; Glotfelty 1985).

The movement of the chemical among the four compartments and its persistence and fate in each are influenced by (1) the intrinsic properties of the chemical such as its structure, polarity, reactivity (photolysis, hydrolysis, etc.), vapor pressure (volatilization), water solubility (soil mobility and adsorption), hydrophobicity (K_{OW}, partitioning), and bioactivity (permeability, bioconcentration, etc.); and (2) the extrinsic factors such as dosage, formulation, equipment, application method, topography of spray block, nature and thickness of canopy, plant surface, soil type and composition, and weather (wind, temperature, relative humidity, rain, sunlight, etc.) (Crosby 1973; Glotfelty 1985; NRCC 1986; Seiber 1987). The key physical properties of an insecticide, their interrelationships and how they affect movement of the chemical between compartments (phase transfers) are shown in Figure 2 (Seiber 1985). These physical properties are becoming the basis for making predictions about insecticide persistence in aquatic and, to a lesser extent, terrestrial forest ecosystems. The various phase transfer processes that operate among the four forestry compartments to facilitate the movement of insecticides are given in Figure 3. The magnitude of each process depends on the intrinsic properties of the chemical and the matrix under consideration in each compartment, and various extrinsic factors acting on them.

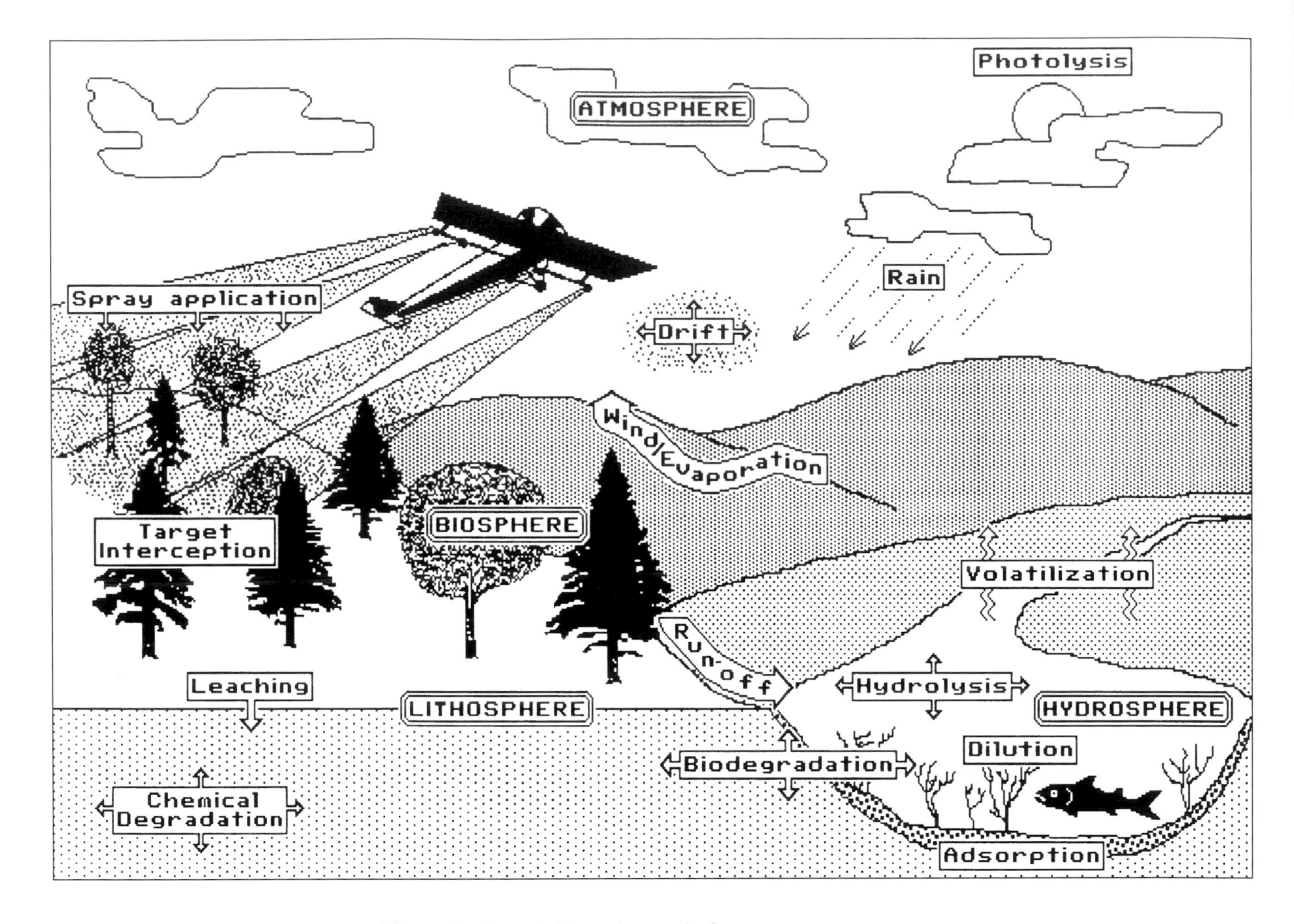

Figure 1. Insecticide pathways in forestry compartments.

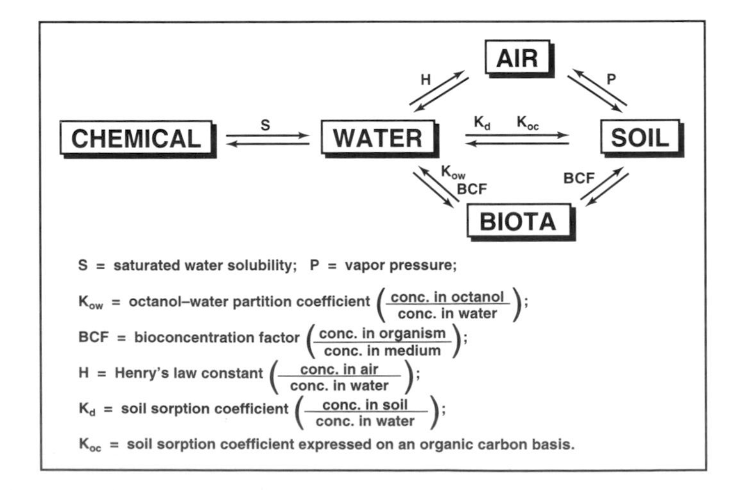

Figure 2. Role of physical properties in the movement of insecticides between forestry compartments (Seiber 1985).

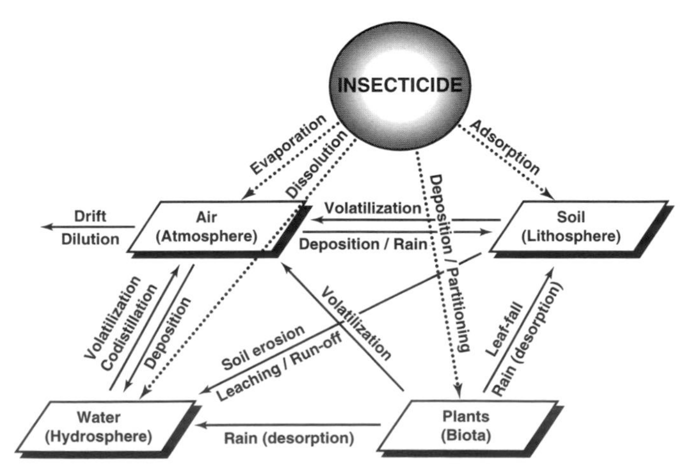

Figure 3. Various transfer processes operating on the movement of insecticides in the four forestry compartments.

Table 1. Registered and experimental insecticides used in Canadian forest insect control programs.

Common name	Trade name	Chemical name	Structural formula	Important physical properties
Acephate	Orthene	O,S-dimethyl acetyl-phosphoramidothioate	$(H_3CS)(H_3CO)P(=O)-N(H)COH_3$	Colorless solid, m.p. 82–89°C, v.p. 0.267×10^{-3} mPa at 24°C, solubility in water 790 g/L at 20°C
Aminocarb	Matacil	4-dimethylamino-*m*-tolyl N-methylcarbamate	$(H_3C)_2N-C_6H_3(CH_3)-O-C(=O)-N(H)CH_3$	Colorless to beige crystals, m.p. 93–94°C, v.p. *ca* 665 mPa at 25°C solubility in water 2 g/L at 25°C[a]
Carbaryl	Sevin	1-napthyl N-methylcarbamate	$C_{10}H_7-O-C(=O)-N(H)CH_3$	White crystals, m.p. 142°C, v.p. *ca* 665 mPa at 26°C, solubility in water 120 mg/L at 30°C
Chlorpyrifos -methyl	Reldan	O,O-dimethyl O-3,5,6 -trichloro-2-pyridyl phosphorothioate	$Cl_3C_5HN-O-P(=S)(OCH_3)_2$	Colorless crystals with odor of S, m.p. 45.5–46.5°C, v.p. 5.6 mPa at 25°C, solubility in water 4 mg/L at 24°C
Diflubenzuron	Dimilin	1-(4-chlorophenyl)-3-(2,6-difluorobenzoyl) urea	$F_2C_6H_3-C(=O)-N(H)-C(=O)-N(H)-C_6H_4-Cl$	Colorless crystals, m.p. 230–232°C, v.p. < 1.3×10^{-2} mPa at 50°C, solubility in water 0.1 mg/L at 20°C

(Continued)

[a]NRCC (1982) value is high. K. Sundaram's value is about 80 mg/L at 25°C.

Table 1. Registered and experimental insecticides used in Canadian forest insect control programs. *(Continued)*

Common name	Trade name	Chemical name	Structural formula	Important physical properties
Fenitrothion	Accothion, Folithion, Sumithion	O,O-dimethyl O-4-nitro-*m*-tolyl phosphorothioate		Yellowish brown liquid, b.p. 140–145°C at 13.3 × 10^{-3} mPa, v.p. 18 mPa at 20°C, solubility in water 14 mg/L at 30°C
Methoxychlor	Marlate	1,1,1-trichloro-2,2-bis (4-methoxyphenyl) ethane		Colorless crystals, m.p. 89°C, v.p. very low, solubility in water 0.1 g/L at 25°C
Mexacarbate	Zectran	4-dimethylamino-3,5-xylyl N-methylcarbamate		White solid, m.p. 85°C, v.p. < 0.013 mPa at 139°C, solubility in water 0.1 mg/L at 25°C
Permethrin	Ambush, Pounce	3-phenoxybenzyl (1*RS*,3*RS*;1*RS*,3*SR*)-3-(2,2-dichlorovinyl)-2,2-dimethylcyclo-propanecarboxylate		*Pure form*—colorless crystals, m.p. 34–35°C, v.p. 261 mPa at 30°C, solubility in water 0.2 mg/L at 30°C *Technical*—yellowish brown liquid, b.p. *ca* 200°C at 1000 mPa Pure permethrin is *ca* 40% *cis* and 60% *trans* isomers; *cis* is more toxic and stable compared to *trans* m.p. of pure *cis*-form 63–65°C, m.p. of pure *trans*-form 44–47°C

(Continued)

Table 1. Registered and experimental insecticides used in Canadian forest insect control programs. *(Concluded)*

Common name	Trade name	Chemical name	Structural formula	Important physical properties
Phosphamidon	Dimecron	2-chloro-2-diethyl-carbamoyl-1-methylvinyl dimethyl phosphate	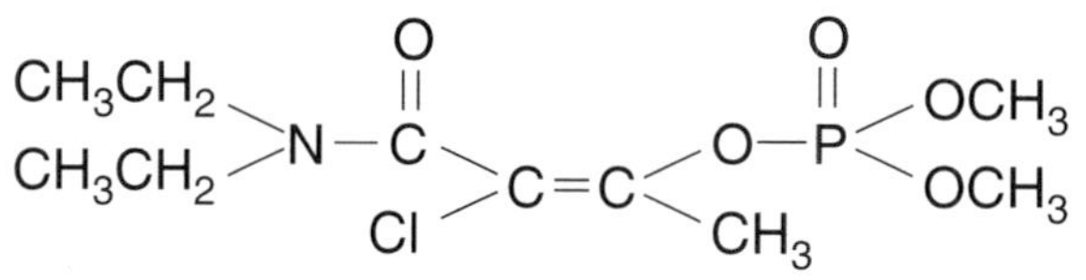 *cis* or E or α-isomer *ca* 30% (less toxic; fast dissipation) 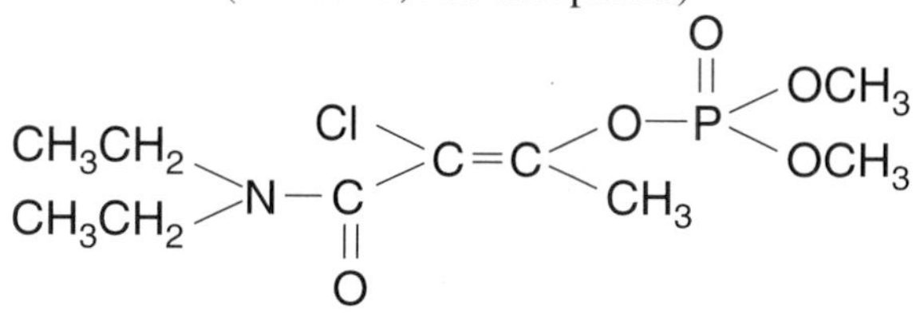 *trans* or Z or β-isomer *ca* 70% (more toxic; slow dissipation)	*cis* and *trans* mixture is yellowish liquid, b.p. 94°C at 5330 mPa, v.p. 3.3 mPa at 20°C, completely miscible in water
Tetrachlor-vinphos	Gardona	2-chloro-1-(2,4,5-tri-chlorophenyl) vinyl dimethyl phosphate	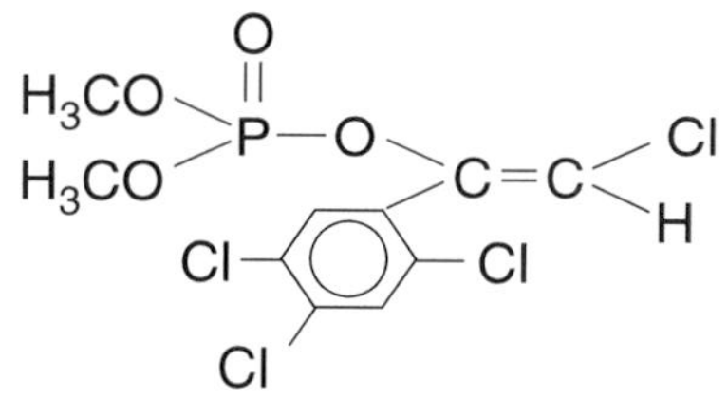 (*trans* or Z-isomer)	Colorless crystals, m.p. 95–97°C, v.p. 5.6×10^{-3} mPa at 20°C, solubility in water 11 mg/L at 20°C
Trichlorfon	Dylox	dimethyl 2,2,2-tri-chloro-1-hydroxy-ethylphosphonate	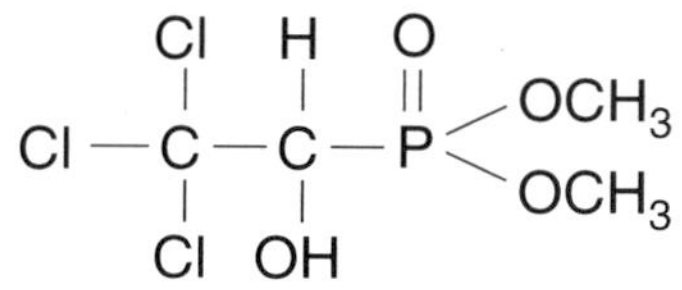	Colorless crystals, m.p. 83–84°C, v.p. 1.0 mPa at 20°C, solubility in water 154 g/L at 25°C

Within each forestry compartment, various physical, chemical, and metabolic factors, summarized in Table 2, will affect the insecticide and lead to its inactivation either by altering its structure to form innocuous compounds or depleting its concentration or both. Most of the factors acting on the chemical are interrelated and therefore a combined action will be exerted on the chemical to cause its eventual breakdown. Considering the low rates of application (70–280 g/ha), the residue levels of sprayed insecticides reaching different forestry compartments are very low—often below submicrogram quantities. At these prescribed dosages and assuming that 20% reaches the forest floor, the amount of sprayed insecticide found would be 0.14 to 0.56 $\mu g/cm^2$. In most cases, the residues are lost rapidly and often exponentially, obeying the decay equation $Y = A + Be^{-Ct}$ (where Y = residues at time, t; A = residues at $t = \infty$; B = residues dissipated; and C = rate constant of dissipation) (Sundaram 1986a,b). An inverse relationship is usually found between the rate of loss of an insecticide and its initial deposit in the matrix (Sundaram 1990b). Most of the currently available data show that the chemicals now used in Canadian forestry spray programs seldom cause definite adverse effects on the quality of the forest environment, because of low residue levels, poor mobility among the components, short persistence, and poor bioaccumulation potential.

Fate in Air

All currently used forestry insecticides (Table 1) have very low vapor pressures and are therefore nonvolatile and not readily transported to the gaseous phase from the deposits on litter, soil, plant, and water surfaces through volatilization. Consequently, vapor phase transportation and movement to other areas, with accompanying photodegradation, is not a major pathway of dissipation, except for the airborne residues present during and immediately following spray application. Although the magnitudes of some of the processes are not well understood, studies have shown (Sundaram 1984a) that offsite drift, dilution, and precipitation reduced airborne residue levels over forest canopies to very low levels and completely removed them from the air above the application site within a few hours after application. The potential for offsite drift depends largely on formulation, mode of application, and climatic and environmental factors (van Valkenburg 1973; Wheatley 1973). In general, the initial fate of the insecticide in the air is primarily related to release height, wind speed, atomizer type, droplet-size spectrum, etc., which control the drift and deposition of the spray cloud, and other factors such as photolysis and dilution influence the residual concentration in the atmosphere.

Fate in Soil

Forestry insecticides (Table 1) are polar and nonionic. Except for acephate, phosphamidon, and trichlorfon, they have very low solubility in water. During an aerial spray, only minute amounts of insecticide reach each square centimetre of the forest floor, depending on the nature and thickness of the forest canopy. Initial ground deposits will be lower under a dense canopy than in thinly forested areas (NRCC 1986). The various factors listed in Table 2 will influence the loss of

Table 2. Factors affecting the fate of insecticides in forestry compartments.

Compartments			
Air	Soil	Biota (plants)	Water
Physical	Physical	Physical	Physical
Drift	Solubilization	Migration	Dilution
Dilution	(leaching)	(translocation, penetration, etc.)	Volatilization/temp.
Precipitation	Adsorption	Solubilization	Codistillation
	Volatilization/temp.	Volatilization/temp.	Adsorption
Chemical	Root uptake	Growth dilution	pH
Photolysis	(translocation)	Wash-off	
	pH	(dew, rain, wind)	Chemical
	Soil texture		Molecular alteration
	Moisture		(as in biota)
		Chemical	Photolysis
	Chemical	Molecular alteration	
	Chemisorption	(reduction, oxidation, hydrolysis,	Metabolic
	Ion-exchange	conjugation, etc.)	Microbes
	Complex formation	Photolysis	
	Photolysis		
	Hydrolysis		
	Oxidation	Metabolic	
	Reduction	Enzymatic action	
		(molecular alteration)	
	Metabolic		
	Microbial action		
	(molecular alteration)		

the chemical from the environment. The extent that each of these factors influence degradation or persistence is dictated by: (1) the various intrinsic properties of the chemical discussed earlier; (2) soil characteristics such as particle size, mineral and organic content, pH, and type and amount of microbial population; and (3) environmental factors such as temperature, moisture content, precipitation, sunlight, air movement, etc. (Edwards 1975). Microbial degradation—not chemical hydrolysis, photolysis, or volatilization—is the major route for dissipation of insecticides from soil. Photodegradation is not significant in the forest soil/litter system because of the lack of intense sunlight and extended exposure (Ghassemi et al. 1981). The mobility of an insecticide in soil is inversely related to its affinity for adsorption (Nicholls 1988). In general, insecticides with low water solubility, such as diflubenzuron, are immobile and are not leached in the soils (Sundaram and Nott 1989). Adsorption is greater in soils containing higher organic content, hydrogen ion concentration, and temperature (Ghassemi et al. 1981). Water soluble insecticides, such as trichlorfon and acephate (Table 1), are also rarely mobile and seldom reach water bodies by overland run-off or leaching due to their sorption and fixation by soil colloids and eventual chemical and microbial degradations (Ghassemi et al. 1981; Boberschmidt et al. 1989). The DT_{50} values (the time required for 50% of the active ingredient to decay) of most insecticides used in forestry range from a few hours to days and none persist for an extended time.

Fate in Foliage

The initial deposit of aerially applied insecticides on a forest canopy depends on its nature and thickness (NRCC 1975; Armstrong 1977). A wide variation has been observed in deposits on foliage: deposits were higher on foliar samples collected from the upwind side of a tree canopy than from the downwind side (Sundaram 1990a). The crown foliage contains higher initial levels of insecticides than the lower foliage (Sundaram 1974b; Sundaram et al. 1986b; Sundaram 1987b; Sundaram and Sundaram 1988). Higher initial levels were also found in young shoots than in older foliage, probably due to its exposure to the spray cloud (Sundaram et al. 1987c; Sundaram and Varty 1989). In general, higher initial levels were also found in deciduous foliage than in conifer needles (Yule and Varty 1975; Sundaram et al. 1989; Sundaram 1991b; refer to “Permethrin” and “Phosphamidon” which follow), showing that foliar morphology and surface characteristics play important roles in droplet reception and retention.

The various factors listed in Table 2 act on the foliar residues degrading them to polar, hydrophilic terminal products. Their influence on degradation is extremely complex and depends on: (1) the different intrinsic properties of the chemical, especially its structural stability, volatilization, and solubility in foliar lipids; (2) extrinsic factors such as tree species (namely, nature and type of foliage), method of application, dosage, formulation, temperature, rain, sunlight, and air movement; and (3) tree factors such as morphology, canopy penetration of spray droplets and its uptake by foliage, growth rate, translocation and storage, and metabolism (type and amount of plant enzymes) (Edwards 1975). Absorption of spray droplets and further retention of residues are influenced largely by the intrinsic properties of the insecticide formulation additives and the nature of the foliage (Sundaram 1986a). The bulk of the terminal products formed by metabolic activity are transitory and innocuous; however, a small fraction of these are conjugated and held in the plant tissues as bound residues (Sundaram et al. 1977).

Usually, the bulk of the insecticide that deposits on the plant surface suffers an initial rapid loss due to various physical factors (Table 2). A smaller fraction of the deposited insecticide, depending on its lipophilicity (indicated by its K_{OW}), is absorbed by waxy surfaces of leaves, especially conifer needles, penetrates and retreats into subcuticular layers, and is retained at submicrogram levels for some time (Sundaram 1987b). Unlike the organochlorines, insecticides currently used in forest protection seldom cause any adverse effects because they do not persist for long periods or in sizable amounts in the biota. The DT_{50} in most cases varies from a few hours to less than a week.

Fate in Water

Released insecticides are deposited on water bodies in the forest environment, which act as reservoirs for the contaminants. In addition, rainfall, soil erosion, and surface run-off also contribute to their presence. Studies conducted to date show that the amounts reaching the water surface are extremely small and in parts per billion (micrograms per litre) level. Water functions both as a reaction medium for the chemical and also as a nucleophilic reactant hydrolyzing and deactivating the material as per the following reaction:

$$R{-}X + H_2O \rightarrow R{-}OH + X^-_{(aq)} + H^+_{(aq)}$$

where R–X is the insecticide in molecular form. Since most of the insecticides have low water solubility and do not exist as true solutions, the insecticide on the surface layer will be lost rapidly by dilution and to a lesser extent by volatilization and codistillation. Little work has been done on quantifying the role of volatilization in the dissipation process of insecticides in aquatic environments of forests. Another contributory factor for the loss could be photolysis, which is influenced largely by the depth of water column, amount of sunlight, and presence of interfering substances. So far little quantitative data are available on the contribution of photodegradation to the dissipation of insecticides from water bodies in forest areas. Recent field studies (Sundaram 1991a) have shown that, in addition to dilution, a significant loss of chemical from segments of the hydrosphere is attributable to adsorption and partition to particulates, sediments, and other aquatic substrates, possibly accompanied by microbial action. Apart from gross predictions, quantitative data on microbial degradation of insecticides and their degradative pathways in the aqueous environment of forests are not yet available. According to Ghassemi et al. (1981), microbial degradation is one of the most prevalent routes. Intrinsic properties of the chemical, additives in the spray-mix, dosage, mode of application, water quality, and hydrodynamics would play a considerable role in the dissipation process, as would pH and temperature of the water. Nearly all insecticides (Table 2) are degraded under alkaline conditions. Despite many years of forestry insecticide spraying, very little uptake or biomagnification and associated deleterious effects in aquatic organisms have been reported.

Environmental Persistence and Fate of Individual Insecticides

Fate and persistence of individual insecticides used in forestry spray programs, especially those conducted by the Forest Pest Management Institute during the post-DDT era are presented briefly in the following pages. As previously mentioned, apart from the persistence of active ingredient, the total fate of any insecticide or even its fate in a single environmental compartment cannot be thoroughly evaluated because of analytical difficulties encountered in the extraction, identification, and quantification of the very low levels of metabolites present in forestry matrices. Therefore, possible degradative pathways of insecticides in forestry matrices will be given in outline only or as gross predictions along with their persistence.

Acephate

Acephate, an organophosphorus insecticide, has been used periodically since 1974 on an experimental basis to control spruce budworm at a dosage of about 500 g/ha.

The documented information that is available on the fate of acephate in air is sparse. Szeto et al. (1978) found 1.16 $\mu g/cm^2$ (open area) and 0.62 $\mu g/cm^2$ (under tree canopy) of acephate in litter following an aerial application at 1 120 g/ha over a conifer forest. Residues dissipated rapidly due to soil microorganisms and persisted only for 10 days (open area) and 30 days (under tree canopy). The DT_{50} of acephate in soil is about 1 to 3 days depending on soil type, moisture content, and initial acephate concentration. The DT_{50} was lower in soils with little organic matter than in clay loams. Methamidophos was the primary metabolite. Little or no acephate decomposition took place in sterile soil. No photodegradation of acephate was observed in soils (Boberschmidt et al. 1989).

Peak concentrations (27 hours post-spray) in the upper, middle, and lower crown needles of Douglas-fir, *Pseudotsuga menziesii*, were, respectively, 9.18, 6.83, and 5.71 µg/g following an aerial spray at 1 120 g/ha. Concentrations declined to 50% in 3 days. No residues were detected after 60 days. Low concentration levels of transient methamidophos were found in both substrates (Szeto et al. 1978). During a simulated aerial application at 280 g/ha, the initial deposit found in white spruce foliage, *Picea glauca*, was 31.45 µg/g and it decreased, primarily through physical processes (Table 2), to 2.12 µg/g within 5 days and disappeared completely after 32 days with a DT_{50} of about 2 days (Sundaram and Hopewell 1976). Only trace levels of methamidophos were found. Following a trunk injection study using C-14 acephate, Sundaram et al. (1977) found rapid apoplastic migration of the chemical to the upper and growing parts of spruce trees. The rapid translocation is due to the chemical's water solubility. Methamidophos was the primary product due to the rupture of N-C bonds in the needle. Some persistent derivatives of phosphoric acid were also found through the rupture of P-N, P-O, and P-S bonds of the molecule by plant enzymes.

Stability of acephate in water is pH dependent. The DT_{50} (21°C) values are, respectively, 65.5, 46.7, and 16.1 days at pH 3, 7, and 9 (Boberschmidt et al. 1989). Acephate appears to be more susceptible to hydrolysis at high pH. The DT_{50} of the chemical decreased considerably in natural water containing sediment due to concerted action of hydrolysis and microbial activity. Because of the low vapor pressure, acephate did not escape from water into the atmosphere and no photodegradation was observed either (Szeto et al. 1979). Acephate is short-lived in stream water after its injection at 1 100 µg/L and dissipated completely within 96 hours primarily due to dilution. No methamidophos was found in water. Although 4- to 8-hour sediment samples contained about 0.13 µg/g, this too disappeared within 96 hours (Geen et al. 1981). Following an aerial spray, stream waters showed a peak concentration (1 hour post-spray) of 140 µg/L, which was reduced to 9 µg/L 2 days later due to chemical breakdown and flushing downstream (Rabeni and Stanley 1979). Pond waters spiked with 100 µg/L showed DT_{50} of 3 to 5 days in water and 1 to 3 days in sediment (Boberschmidt et al. 1989). Neither acephate nor its toxic metabolite methamidophos persisted in the forestry components at the recommended use patterns.

Aminocarb

Aminocarb marketed as Matacil® has been used extensively since 1970 to control budworm at an operational dosage of 2×70 g/ha.

In the vapor phase, aminocarb was found to undergo rapid photodegradation (NRCC 1982); therefore, it is not expected to persist in air. Sundaram and Sundaram (1989) found that the chemical is less volatile than fenitrothion or mexacarbate and attributed the lack of volatility to its low vapor pressure. The peak aminocarb concentrations found in air inside the spray block after application at 70 g/ha were 211 and 1 201 ng/m^3 for oil-based and emulsion formulations, respectively. The residues decreased rapidly with time due to dilution, offsite movement, and photolysis. Small amounts of demethylated products and aminophenols were found in the air (Sundaram 1984a).

Aerial spray trials of aminocarb at a dosage of 70 g/ha were conducted using oil-based and emulsion formulations (Sundaram et al. 1976; Sundaram 1981; Sundaram and Szeto 1984; Sundaram et al. 1985a; Sundaram and Nott 1985a). Generally, the amount of aminocarb reaching the forest floor was extremely low (range <0.01–0.2 µg/g) and was influenced largely by existing meteorological conditions, application parameters, and the type of formulation used. Deposits were usually higher with formulations containing denser, highly viscous, nonvolatile oils than with ones containing either light oils or emulsions. They all dissipated rapidly with approximate DT_{50} ranging from 2 days in litter to 12 hours in soil. In a typical spray operation using three formulations, one containing heavy oil, another with light oil, and the third an emulsion (Sundaram and Szeto 1984), the average peak residue concentrations (1 or 2 hours post-spray) found in litter were, respectively, 0.20, 0.08, and 0.02 µg/g, and these persisted for 12 to 21 days. (All concentrations are expressed on a fresh or wet weight basis, unless otherwise noted.) The corresponding soil concentrations were 0.05, 0.02, and 0.01 µg/g and they persisted for about 5 days. Following a simulated aerial application at 59 g/ha, Sundaram and Hopewell (1977a) found peak concentrations of aminocarb (1 hour post-spray) ranging from 4.08 (under tree cover) to 5.22 µg/g (open area) with corresponding DT_{50} of 3 and 2 days. The residues persisted up to 40 days. Volatilization, photolysis, and microbial action may be the primary causes for aminocarb's dissipation from soil and litter matrices. The difference in persistence observed between litter and soil could be attributed to the higher lipid content in litter than in soil (Armson 1977), formation of cationic species by the chemical due to a lower pH in litter, and ready adsorption to litter matrices minimizing the bioavailability of the chemical.

Foliar painting of ^{14}C-aminocarb on potted spruce trees showed that the insecticide is not systemic (Sundaram and Hopewell 1977b). Different types of aminocarb formulations were sprayed onto single conifer trees as simulated aerial sprays to examine the influence of additives on deposition and subsequent penetration of the active ingredient (Sundaram and Hopewell 1977a: Sundaram and Sundaram 1981; Sundaram 1986a). The 1-hour post-spray peak concentration varied from 10.0 to 29.0 µg/g depending on the type of formulation used. The amount of tissue or penetrated and surface or dislodgeable residues varied according to the solubilizing power and lipophilic nature of the adjuvants. Usually both type of residues increased with formulations containing dense, viscous oils as additives compared to hydrophilic components. The maximum and minimum DT_{50} for penetrated and dislodgeable residues were, respectively, 26.0 and 12.3 days for lipophilic adjuvants and 13.5 and 8.8 days for hydrophilic ones.

Aminocarb residues in conifers have been studied extensively following aerial application at 70 g/ha (Sundaram et al. 1976; Sundaram and Szeto 1984; Sundaram 1985; Sundaram and Nott 1985a). The initial peak concentration in the needles varied widely from 0.79 to 2.76 µg/g according to the formulation sprayed. The dissipation of the chemical was exponential: within 2 days the bulk of the material was lost. Thereafter, the residues persisted at lower levels (<1.0 µg/g) for nearly 30 days and then dissipated to below the detection limit (0.005 µg/g). Volatilization, weathering, photolysis, and to a certain extent plant enzymes would have been responsible for the original rapid decrease (Abdel-Wahab et al. 1966; Abdel-Wahab and Casida 1967). Then, as the residues were gradually absorbed by the lipophilic terpenoids of the foliage, they degraded relatively slowly.

In aquatic model systems containing natural water, Sundaram et al. (1984) found rapid loss of the chemical in water due to hydrolysis and volatilization. The DT_{50} found was about 35 hours but increased with decreasing pH (NRCC 1982), probably due to the formation of a stable protonated cation (Sundaram et al. 1984). In the water/sediment system, the chemical moved readily from the water to sediment due to adsorption by the colloidal materials present in sediment. The adsorbed insecticide was lost primarily by microbial action forming different demethylated aminocarb moieties and phenols, which eventually converted into CO_2, NH_3, and H_2O. Of these, 3-methyl-4-(methylamino)phenyl-N-methylcarbamate (MAA), 3-methyl-4-(amino)phenyl-N-methylcarbamate (MAC), and 4-(dimethylamino)-3-methylphenol (AP) were predominant (Sundaram et al. 1984).

Distribution, persistence, and partition of aminocarb in forestry streams were studied extensively by applying the material to water surfaces (Sundaram et al. 1984; Sundaram and Szeto 1985; Eidt et al. 1988). The initial concentration of the chemical on the surface layer varied enormously, ranging from less than 1 µg/L to more than 5 µg/L, depending on dosage and method of application. Disappearance of the chemical was very rapid primarily due to downstream transport and dilution, movement into other substrates, and to a lesser extent by physicochemical and biological processes. The insecticide loss varied according to the type of formulation used, mode of application, stream discharge, and other site conditions. Stream particulates, sediments, aquatic plants, algae, fish, and aquatic invertebrates accumulated residues to varying extents and showed a wide range of retention times. Sundaram and Szeto (1985) found that levels of accumulation were higher for emulsion formulations than oil solutions. For example, the highest concentrations of aminocarb found in sediment (5 minutes post-spray) for emulsion and oil formulations were, respectively, 20.2 and 7.6 ng/g. Similarly, fish collected 2 hours after exposure to emulsion and oil formulations contained, respectively, 106.0 and 31.6 ng/g of aminocarb. Very small amounts of MAA and AP were found occasionally in some of the stream substrates.

Fate and persistence of aminocarb in aquatic components of a forest environment after aerial application at 70 g/ha were studied on two occasions, Sundaram et al. 1976 and Sundaram 1987a. In Sundaram et al. 1976, the peak concentrations of the chemical in pond and stream waters, sampled after application of an oil formulation, were 2.1 and 2.6 µg/L, respectively. The rate of disappearance was slow and uneven probably due to low pH (5.2–5.4) of the waters wherein the insecticide existed as protonated cation resisting volatilization, hydrolysis, adsorption, and microbial action. Aminocarb is stable under acidic conditions with DT_{50} of 90 days in a pH 5 buffer solution held in the dark at 20°C (NRCC 1982). Foliar washings by rain and run-off waters from land would have contributed further to the residues in natural waters. The DT_{50} ranged from 4.4 to 8.7 days for pond and stream waters, respectively. In Sundaram 1987a, the 1 hour post-spray peak concentrations in stream waters ranged from 3.06 to 22.67 µg/L. The residues diminished logarithmically to below detectable levels (0.1 µg/L) within 24 hours. Sediment samples contained a peak level of 5.8 µg/g, which also dissipated rapidly. Concentrations of aminocarb found in different aquatic plants (approx. 880 ng/g) and moss (210 ng/g) were much higher than in ambient water and gradually dissipated. Trace levels of MAA were occasionally found in some of these samples. The peak residue levels in fish ranged from 6.1 to 13.8 ng/g but they were eliminated within 24 to 72 hours. In addition to MAA and MAC, sporadic presence of the phenol AP was found in some of the samples, indicating that demethylation and possibly enzymatic hydrolysis may have played some role in the dissipation of this chemical in fish (Sundaram and Szeto 1979; Szeto and Holmes 1982).

At the dosage rate used, it is apparent that aminocarb is labile and does not persist in different terrestrial and aquatic components of the forest environment.

Carbaryl

Carbaryl is a broad spectrum carbamate insecticide extensively used (ground and aerial) in the United States and to a more limited extent in Canada at dosage ranges of 0.50 to 1.25 kg/ha to control primarily the spruce budworm, *Choristoneura fumiferana*, and the western spruce budworm *C. occidentalis*.

During a forest spray operation, average carbaryl air residues at 1 hour post-spray were found to be 3.4 $\mu g/m^3$ which decreased rapidly to 0.02 $\mu g/m^3$ within 24 hours (Sundaram, unpubl. data). Data on the decomposition pattern of the insecticide in air are nonexistent. However, it is possible to postulate that the particulate and gaseous forms of carbaryl in the air will be dispersed, diluted, degraded by the sun, and eventually deposited.

In an experimental aerial spray at 280 g/ha, Sundaram and Szeto (1987) found peak residue levels of 1.121 and 0.59 µg/g in forest litter and soil, respectively. The residues dissipated rapidly by chemical and biological means and the DT_{50} value for both matrices was 1.5 days. Lower levels (0.06 µg/g) persisted in the litter than in the soil beyond the 10-day sampling period, probably due to higher acidity of the former causing the basic insecticide (R) to be protonated, physically adsorbed, and stabilized as $RH^+_{(aq)}$ as per the equation:

$$R + H^+_{(aq)} = RH^+_{(aq)}$$

Adsorption and persistence of the insecticide are influenced by soil texture and organic matter content. A detailed account on the persistence and fate of carbaryl in field soils is given by Boberschmidt et al. (1989).

Several studies have addressed the degradation of carbaryl in agricultural and food crops (Boberschmidt et al. 1989). Kuhr and Casida (1967) demonstrated that the chemical is metabolized and hydroxylated into several water-soluble moieties. Sundaram and LeCompte (1975) measured the carbaryl residues in eastern white pine, *Pinus strobus*, shoots following aerial application at 1.12 kg/ha. At 1 hour post-spray the maximum residue level (as sampled) was 19.1 µg/g (dislodgeable 12.9 µg/g, penetrated 6.2 µg/g). Fifteen days later, the residue levels had declined to 4.8 µg/g (dislodgeable 2 µg/g, penetrated 2.8 µg/g) and at 90 days, there was no dislodgeable residue of carbaryl in pine shoots and only a trace level (<1 µg/g) of penetrated residue was found in the plant tissues. The DT_{50} values for the total, dislodgeable, and penetrated residues were 6.5, 5, and 12 days, respectively. Pieper (1979) measured carbaryl residues in forest foliage and found that initial residue levels were 30 and 138 µg/g for aspen and Douglas-fir, respectively, with the corresponding DT_{50} of 8 and 4.5 days. Sixty-three days after application, the residues in the two substrates declined to 0.5 and 3.8 µg/g. In a recent experimental spray application at 280 g/ha, Sundaram and Szeto (1987) found a peak concentration, 1 hour after application, of 4.20 µg/g, which dissipated rapidly and curvilinearly to 0.48 µg/g after 5 days. Residues persisted at low levels (0.09 µg/g) up to 10 days post-spray. The DT_{50} value obtained from the depletion curve was 2.3 days. From the data collected, it can be concluded that carbaryl has a relatively short persistence in a terrestrial environment.

Carbaryl is relatively stable in acidic waters due to the formation of protonated cation (RH^+) but readily hydrolyzed to 1-naphthol under basic conditions. Szeto et al. (1979) found that in pond and creek waters held for 42 and 52 days, respectively, at 9°C, carbaryl degraded to about 80% in the former and to about 60% in the latter, primarily due to microbial activity. During an aerial application of the chemical, the 3-hour post-spray concentration in stream waters ranged from 2 to 160 µg/L (Pieper 1979). Stanley and Trial (1980) found 0.4 to 0.9 µg/L of the chemical in streams and brooks with a DT_{50} ranging from 23 to 28 hours. However, Gibbs et al. (1984) found initially lower levels in water (0.25 µg/L) that persisted for 14 months at 0.06 µg/L. Sediment samples contained 54 µg/g of carbaryl as a peak concentration that persisted as long as 16 months. Recent studies by Sundaram and Szeto (1987) showed that at the dosage level of 0.28 kg/ha, the maximum residue level found in stream water was 314 µg/L and more than 50% of it had dissipated within 1 hour. Low but detectable levels (1 µg/L) of the chemical persisted in water until the end of the 10-day sampling period. Sediment samples contained a maximum level of 0.04 µg/g, which dissipated below the detection limit within 5 hours. Brook trout, *Salvelinus fontinalis*, and slimy sculpins, *Cottus cognatus*, captured in the stream 1 day after the spray, contained on average about 0.04 µg/g of carbaryl and none of it was found in 3-day post-spray samples. Under normal use patterns, carbaryl does not seem to persist in terrestrial and aquatic environments of forests.

Chlorpyrifos-methyl

Chlorpyrifos-methyl is a broad spectrum organophosphorus insecticide, sprayed once only over a conifer forest at 70 g/ha (Szeto and Sundaram 1981) to suppress spruce budworm.

Very little is known about the fate and persistence of the chemical in the air mass over the spray area. Since the air was in a dynamic state, it is likely that the chemical would have dissipated rapidly by drift and dilution.

The maximum residue levels found in litter and soil (as sampled) were, respectively, 0.254 µg/g (1.5 hour post-spray) and 0.253 µg/g (1 hour post-spray). Residues persisted in quantifiable levels (>0.006 µg/g) up to 42 days in both substrates. The DT_{50} in both substrates was about 3 days. After 25 days, trace amounts (<0.006 µg/g) were still found in litter, due to either its lipoid content (Armson 1977), or low microbial activity, or concerted influence of both, but were undetected in soil. The chemical degraded initially to 3,5,6-trichloro-2-pyridinol, the principal metabolite that subsequently degraded to organic chlorine compounds and CO_2.

The residue in balsam fir foliage was highest 1 hour after spraying and rapidly declined to about 30% within 1 day giving a DT_{50} value of about 5 hours. Low levels of residue (0.03 µg/g) persisted for 125 days probably due to the absorption, dissolution, and entrapment of a fraction of the chemical into foliar waxes resisting further weathering and enzymatic activity.

The maximum initial residue levels found in stream water ranged from 5.5 to 221 μg/L and they dissipated very rapidly: more than 90% disappeared within 3 hours primarily due to dilution and probably hydrolysis, and none was detected after 4 days. Sporadic presence of the chemical at a range of 5 to 67 ng/g was found in bottom sediment and persisted for 10 days at 8 ng/g, confirming the movement of chlorpyrifos-methyl from water to sediment (Szeto and Sundaram 1982). The average maximum residue levels found (2 days post-spray) in brook trout and slimy sculpins were 34.4 ng/g and 10.1 ng/g, respectively. However, the bulk of the residues was metabolized and lost within 4 days and only trace levels were found 9 days post-spray. No insecticide was detected after 47 days. Chlorpyrifos-methyl appears to be nonpersistent in the aquatic components of the forest environment.

Diflubenzuron

Diflubenzuron is an insect growth regulator that inhibits chitin synthesis in insects during molting.

Studies on the fate of diflubenzuron in air have not been reported. Schaefer and Dupras (1979) found very slow photodegradation of the chemical. Microcosm studies showed that volatilization of the chemical from treated conifer seedlings and forest soil was low and the chemical diminished with time (Sundaram, unpubl. data), probably due to its low vapor pressure.

Sundaram and Nott (1989) found that mobility of diflubenzuron in forest soil columns was low and did not increase with dosage. No residues were found below the 10-cm level in soil columns or in the leachates, even at 630 g/ha. Mobility is influenced by soil type and additives present in the formulation. In a laboratory soil study using clay loam, Sundaram and Nott (unpubl. data) found a DT_{50} of 39 days, which is longer than the average values of about a week and confirmed the dependence of the DT_{50} on particle size. Furthermore, they found that hydrolysis and microbial action were responsible for the degradation of diflubenzuron to 4-chlorophenylurea, 2,6-difluorobenzoic acid, and CO_2. No free 4-chloroaniline was found in the soils. In a simulated aerial spray at 90 g/ha, Sundaram (1986b) found 1-hour post-spray peak concentration levels in forest soil and litter ranging from 1.87 to 3.20 μg/g and 3.08 to 4.60 μg/g, respectively. The residues dissipated rapidly with DT_{50} ranging from 6.52 to 7.49 days in soil and 7.34 to 8.36 days in litter. In an experimental aerial spray program at 70 g/ha, Sundaram (1991b) found peak residue levels of 1.03 μg/g in soil and 0.46 μg/g in litter, and they disappeared below the detection limit within 10 days after application.

In a simulated aerial spray study, Sundaram (1986b) found peak concentration levels ranging from 20.9 to 36.3 μg/g in spruce needles 1 hour after application. The residues decreased curvilinearly with DT_{50} values ranging from 9.30 to 12.8 days, indicating that the chemical is relatively stable on the needles. Sloughing, volatilization, photolysis, hydrolysis and weathering factors probably played major roles in its dissipation. Sundaram (1991b) studied the persistence and fate of diflubenzuron in eastern white pine, *Pinus strobus*, and sugar maple, *Acer saccharum*, foliage after an aerial application of the chemical as a wettable-powder formulation, Dimilin® WP-25 at 70 g/ha using 10, 5, and 2.5 L/ha over three blocks in a mixed forest. Samples were collected before application and at intervals of time after application. The peak (1-hour post-spray) residue levels varied according to the foliar type and volume rates of application. In addition, stability and persistence of the chemical in the two foliar species were influenced by the volume rates. The maximum 1-hour post-spray concentrations (micrograms per gram) in the foliar samples were:
white pine: 13.07 (10 L/ha), 6.90 (5 L/ha), and 4.67 (2.5 L/ha)
sugar maple: 19.09 (10 L/ha), 4.10 (5 L/ha), and 3.36 (2.5 L/ha)
The chemical was persistent in the foliage of both species, and dissipation was gradual and followed first order kinetics. The chemical persisted for 120 days in foliage sampled from the block that received 10 L/ha compared to a persistence of about 40 days in the other two blocks treated with 5 L/ha and 2.5 L/ha. The DT_{50} (days) values for the pine and maple foliage, depending on the volume rates applied, were:
white pine: 4.6 (10 L/ha), 3.3 (5 L/ha), and 3.9 (2.5 L/ha)
sugar maple: 8.7 (10 L/ha), 3.5 (5 L/ha), and 4.3 (2.5 L/ha)
The present findings on the longevity of diflubenzuron in pine and maple foliage agree with those observed earlier in hardwood and conifer foliage (van den Berg 1986), spruce needles (Sundaram 1986b), and oak foliage (Martinat et al. 1987).

Stability of diflubenzuron in natural water varies with pH, its physicochemical characteristics, and climatic conditions. It degrades rapidly through hydrolysis in neutral or alkaline waters with a DT_{50} of about 7 days (Ivie et al. 1980). Schaefer and Dupras (1979) found photolytic degradation of diflubenzuron in water with a DT_{50} of about 3.4 days. Sundaram (1989a) studied the fate and persistence of the chemical in an aquatic environment after aerial application at 70 g/ha in 10, 5, and 2.5 L/ha over a boreal forest containing ponds and streams. Water, sediment, and aquatic plants were collected from pond and stream environments at intervals of time up to 30 days post-treatment for determining diflubenzuron residues. The peak 1-hour post-spray residues of the chemical in the pond and stream waters for the three application rates were 13.82, 3.25, and 1.59 μg/L, respectively, indicating that initial water concentration was influenced by the volume rate applied. The concentration of diflubenzuron in pond water decreased gradually, probably due to dilution, adsorption, and subsequent microbial degradation, hydrolysis, and photolysis, but after 20 days none was found. However, the rate of loss of the material was rapid in streams due to mixing and downstream mobility and no residues were found after 2 days. The approximate DT_{50} in pond and stream waters were 1.3 and 0.2 days, respectively. The concentration of diflubenzuron in pond sediments increased gradually from 0.13 μg/g at 6 hours post-spray to a maximum level of 0.24 μg/g after 1 day and then decreased to nondetectable levels (<0.05 μg/g) within 5 days. None of the stream sediments contained any diflubenzuron. Apperson et al. (1978) also found no residues in lake sediments following a 5-μg/L application. Aquatic plants in the pond acted as a transient sink for the chemical, building irregularly the concentration of diflubenzuron from traces (0.05–0.10 μg/g) to 0.29 μg/g (6 hours post-spray) and then to a peak level of 0.36 μg/g (1 day post-spray): within 10 days the levels were below the detection limit (Sundaram 1989a). Sporadic presence of the chemical at trace levels were found in aquatic plants from streams up to 5 days after application. Booth and Ferrell (1977) found that diflubenzuron does not accumulate to significant amounts in aquatic plants. Caged brook trout from the pond and streams did not contain measurable levels (>0.05 µg/g) of diflubenzuron (Sundaram 1989a). Diflubenzuron did not exhibit unusually high persistence in aquatic components of the forest environment, and its bioaccumulation potential in forestry matrices appears to be minimum.

Fenitrothion

Fenitrothion has been the choice insecticide for controlling spruce budworm in eastern Canada since 1967. The normal operational dosage is two applications each at 210 g/ha or one at 280 g/ha.

During an aerial spray, part of the material release from the aircraft remains airborne and is distributed within and outside the application site. Sundaram (1984a) found that the air sampled within the spray block 1 hour post-spray contained a peak concentration of 1 997 ng/m^3, which diminished to about 85% within 24 hours, due to dilution and drift. Air samples collected at intervals of time outside the spray blocks contained varying amounts of fenitrothion as particulates and vapor, ranging from 40 to 80 ng/m^3 (Mallet and Volpé 1982; Mallet and Cassista 1984). Yule et al. (1971) reported that total phosphorus content in the air, soon after application, rose to a maximum of 3 $\mu g/m^3$ per day. The airborne fenitrothion is degraded by sunlight (Ohkawa et al. 1974; Addison 1981), adsorbed onto particulates, deposited or diluted by atmospheric transport, and eventually degraded to innocuous products.

Degradation patterns of fenitrothion in sandy and clay loam forest soils and leaf litter collected from Northern Ontario forests were studied in laboratory microcosms (Sundaram 1990b). The chemical was lost rapidly through microbial action with DT_{50} of 6.34 days in sandy loam, 7.98 days in clay loam, and 16.42 days in leaf litter. Major degradation products were 4-nitrocresol and CO_2 with occasional presence of aminofenitrothion, as observed earlier by others (Takimoto et al. 1976; Miyamoto 1977; Spillner et al. 1979a,b). Soil properties and incubation conditions influenced the DT_{50} values and the types and amounts of degradation products formed. The degradative pathway was hydrolysis of the P-O-Ar bond, ring cleavage, oxidation, and conjugation of fragments to soil and litter matrices. Yule and Duffy (1972) found that aerial application of fenitrothion at 280 g/ha gave only 0.04 µg/g as the peak soil residue concentration, which dissipated to trace levels (<0.01 µg/g) within 64 days. Sundaram (1974b) and Symons (1977) found a similar pattern of loss; the insecticide content in soils ranged from trace levels to 0.1 µg/g, lingered for a while around 0.02 µg/g (probably due to reversible adsorption to soil mineral and organic matter), and disappeared within 45 days due to physical (volatilization), chemical (photolysis), and biological means. The initial deposit was too small to determine any DT_{50} and possible degradative pathways. In a later in-depth study, Sundaram and Nott (1984) found that at 210 g/ha dosage, the maximum range of residue levels reaching the forest soil and litter layers were only 0.13 to 0.48 µg/g and 0.13 to 0.29 µg/g, respectively, and the corresponding DT_{50} value for both substrates was about 2.5 days. Although the forest floor acts as a reservoir for sprayed chemicals, accumulation of fenitrothion and its persistence, after consecutive years of application, have seldom been observed in forest soil and litter (Yule and Duffy 1972; Sundaram 1974c). The exception is a single observation of very low level buildup of about 0.06 µg/g in litter and 0.03 µg/g in soil after five yearly applications of the chemical to a plantation forest (Sundaram 1984b). This is probably due to the transfer of fenitrothion-contaminated leaves to the soil stratum during rain and autumn leaf-fall.

Conifer seedlings grown in nutrient solutions fortified with fenitrothion gradually accumulated the chemical in needles, stem, and roots. It was metabolized eventually by hydrolysis and hydroxylation rather than by oxidative desulfurization (Sundaram and Prasad 1975). However, when fenitrothion was applied to conifer needles, the bulk of the material was lost within 2 days and, apart from cuticular penetration, little translocation to other parts had occurred (Sundaram et al. 1975). Using simulated aerial application of different fenitrothion formulations at 340 g/ha, Sundaram and Sundaram (1982) and Sundaram (1986a) observed initial rapid loss of the chemical from conifer needles, decreasing gradually afterwards to less than 2% of the initial deposit beyond 150 days. The DT_{50} values for the dislodgeable (surface) and penetrated (tissue) residues were influenced by formulation additives and were in the range of 13.2 to 15·1 days for dislodgeable and 16.0 to 23.5 days for penetrated residues. The reason for the higher DT_{50} of penetrated residues is that fenitrothion is lipophilic (K_{OW} = 1 862) and is absorbed into the cuticular waxes of the subsurface layer, delaying bio- and photodegradations, and hydrolysis. The dislodgeable residues, however, are still susceptible to loss by volatilization and degradation, resulting in a faster rate of disappearance than those that penetrated into the subsurface regions of the foliage. In a similar study at a lower dosage (270 g/ha), Sundaram and Sundaram (1987) found, under laboratory conditions, a distinct difference in the rate of loss of fenitrothion in the initial and later stages of decay obeying two types of exponential decay equations, and the overall DT_{50} ranged from 9.3 to 12.4 days.

Under operational conditions (280 g/ha) for spruce budworm control, Yule and Duffy (1972) found peak fenitrothion residues in conifer needles ranging from 2.25 to 2.50 µg/g, which decreased by about 50% within 4 days and 70 to 85% within 2 weeks. About 10% of the initial deposit persisted for most of the year. In an experimental spray over a plantation forest, Sundaram (1974b) found that maximum initial residues in conifers ranged from 3.47 to 5.45 µg/g with an approximate DT_{50} of 2 days. Nearly 95% of the chemical intercepted by the needles was lost within 10 days, but detectable amounts persisted up to 155 days. Oxon was found, infrequently, near trace levels in some samples. The mechanisms of dissipation are primarily volatilization, photolysis, and weathering. In a later field trial at 210 g/ha, Sundaram and Nott (1984) found that the highest post-spray concentrations ranged from 0.66 to 0.82 µg/g for the oil-based and 1.72 to 1.82 µg/g for aqueous formulations with the corresponding average DT_{50} of 2.9 and 2.6 days. The hardwood species within mixed forests, such as maple, collected three to four times higher deposits on their foliage than conifers during operational sprays; however, the residues decreased rapidly (Yule 1974; Yule and Varty 1975; LaPierre 1982;

Sundaram 1984b; LaPierre 1985). At the current use pattern fenitrothion is lost rapidly from various forestry compartments and does not persist for prolonged periods. However, there is considerable evidence in the literature to show that conifer foliage acts as a microsink for the chemical (Yule 1974; Sundaram 1974c; Sundaram and Sundaram 1982; Ayer et al. 1984; Sundaram 1984b; Eidt and Pearce 1986; Sundaram 1987c; Ernst et al. 1989). Fenitrothion is lipophilic; therefore, a small fraction (<10%) of the material intercepted by conifer needles is absorbed, transported and stored as intact molecules in cuticular waxes of the foliage, resisting subsequent volatilization, weathering action, photo- and biodegradations.

The fate of fenitrothion in aquatic systems has attracted considerable attention. Numerous laboratory studies, which are summarized in the NRCC (1975) review, show that hydrolysis of fenitrothion is largely influenced by the pH and temperature of the medium. The chemical is relatively stable in acidic waters but hydrolyzes rapidly with an increase in alkalinity, forming predominantly 3-methyl-4-nitrophenol (cresol). The DT_{50} at pH 5.4 and 8.2 were 60 and 22 hours, respectively (Sundaram 1973a). Photolysis plays a major role in the removal of fenitrothion from standing waters (Lockhart et al. 1973) forming oxidative products involving the P=S and *m*-CH_3 groups and by the hydrolysis of P-O-Ar bond (Miyamoto 1977). The fate of fenitrothion in natural waters in model ecosystems was studied with and without sediment. The DT_{50} was approximately 22 hours in the aerated system and nearly twice that in the nonaerated sample. In the water–sediment model, fenitrothion moved from water to sediment and the adsorbed residues in sediment were degraded by the microbes, forming different amounts of the cresol, aminofenitrothion, and demethylated products of the parent material (Sundaram et al. 1984). The DT_{50} of the chemical in pond and lake waters varied from 18 minutes to 2 hours. The loss was attributed to volatilization, photolysis, hydrolysis, and adsorption. Cresol and aminofenitrothion were found in small quantities (Metcalfe et al. 1980; Maguire and Hale 1980; Malis and Muir 1984). Aerial application of the chemical at 280 g/ha gave 9 µg/L to 26 µg/L as peak concentrations in pond waters and the corresponding DT_{50} varied from a few hours to 2 days (Sundaram 1974b). In stream waters the concentration and DT_{50} ranges were rather low, 0.25 to 6.7 µg/L and 1.5 to 5 hours, respectively. Moody et al. (1978) found a peak level of 1.32 µg/L in creek water and within 10 hours the concentration decreased to 0.57 µg/L. Ground applications usually produced higher initial concentrations (4.1–22 µg/L) in stream water than aerial spray, and the rate of dissipation was slow (Sundaram et al. 1983). Disappearance of fenitrothion from streams was rapid due to downstream transport and dilution and adsorption to stream substrates such as stream particulates, moss, aquatic plants, stream insects, fish, and bottom sediment, and the adsorbed material persisted for only a short while (Sundaram 1974b; Eidt and Sundaram 1975; Morrison and Wells 1981; Sundaram et al. 1984). Studies by Moody et al. (1978) and Eidt et al. (1984) agree with these observations. The data accumulated so far under laboratory microcosms and in experimental or operation spray programs show that fenitrothion, under the current use patterns, does not persist in natural waters and accumulate in aquatic flora and fauna for a long time even after consecutive treatment of the chemical for a 5-year period (Sundaram 1987c).

Methoxychlor

Methoxychlor was used to control the white pine weevil, *Pissodes strobi*, the native elm bark beetle, *Hylurgopinus rufipes*, and the smaller European elm bark beetle, *Scolytus multistriatus*, during the 1970s by aerial and ground applications as aqueous emulsions or oil-based formulations. Unlike DDT, the chemical has little tendency to be stored in the body fat or accumulated in the food chain; however, the high dosage (>2 kg/ha) required was neither economical nor viewed favorably by the public. Consequently, the product has seldom been used since the mid-1970s.

The initial residues in pine shoots and bark and the corresponding DT_{50} values (Table 3) varied according to the dosage applied, formulation type, and mode of application. Usually ground application and emulsion formulations gave higher initial deposits than aerial spray and oil formulations. No detailed metabolic studies have been done on the dissipation mechanisms and fate of the chemical in pine shoots and elm bark. The regular presence of methoxychlor ethene and sporadic formation of phenolic moieties (Sundaram 1975c, 1976) indicate that in addition to various physical processes such as weathering by rain and wind, abrasion, and hydrolysis, photo- and enzymatic degradations would have played some role (Kapoor et al. 1970).

Mexacarbate

Mexacarbate is a carbamate insecticide that is very effective against budworm larvae. In an aerial application at 70 g/ha, Sundaram et al. (1986a) found a peak concentration range of 110 to 440 ng/m^3 in air inside the spray block. The residue levels fluctuated but decreased with time and after 10 hours post-spray, the amount approached the minimum detection limit of 50 ng/m^3. After 12 hours post-spray, none was detected. Spray chamber studies (Sundaram and Sundaram 1989) showed that mexacarbate is more volatile than aminocarb and fenitrothion, and at 210 g/ha, the peak residues at 0.25 hour post-spray were 21.7 µg/L for mexacarbate, 12.2 µg/L for fenitrothion, and 6.3 µg/L for aminocarb. Studies showed that degradation of the chemical in air is primarily through demethylation of the N-dimethylamino group with little hydrolysis of the carbamate ester bond (Kuhr and Dorough 1976). Dissipation processes, in addition to photolysis (Silk and Unger 1973), included mixing of aerosol with air, atmospheric transportation, and eventual sedimentation.

Using simulated rainfall, Sundaram et al. (1985b) showed that very little mobility and leaching of mexacarbate occurred in soil columns containing sandy and clay loam forest soils. The bound insecticide was degraded rapidly by soil microbes to demethylated carbamates, phenols, and unidentified metabolites. Mexacarbate is rapidly degraded from forest litters in laboratory microcosms. The DT_{50} for aerobic and submerged litters ranged from 7.3 to 8.8 days and 7.7 to 9.3 days, respectively. Litter microbes played a major role in the degradation of the chemical and formation of CO_2. Metabolites detected included demethylated carbamates, xylenols, polar and other unknown products. The principal degradative pathway for the

Table 3. Methoxychlor treatment for the control of the white pine weevil, *Pissodes strobi*, and two species of elm bark beetles, *Hylurgopinus rufipes* and *Scolytus multistriatus*.

Formulation	Dosage	Mode of application	Matrix type	Initial conc. (μg/g) (fresh wt.)	DT_{50} (days)	Reference
Emulsion	2.25 kg/ha	Hydraulic sprayer (truck mounted)	W. pine needles W. pine shoots	520 113	21.5 26.0	Sundaram et al. 1972
Oil (Fuel oil no. 2)	2.53 kg/ha	Aircraft—four Micronair® AU 2000	W. pine leader	12.8	5.0	Sundaram 1973b
Oil (Fuel oil no. 2)	2.80 kg/ha	Aircraft—boom and nozzle	W. pine leader	6.32 4.79[a] 1.53[b]	5.0 1.5 7.5	Sundaram 1975b, 1975c
Emulsion	2.24 kg/ha	Aircraft—four Micronair® AU 3000	W. pine leader	40.8	6	Sundaram 1977
Oil (Arotex)	2.80 kg/ha	Aircraft—four Micronair® Au 3000	W. pine leader	91.5	8	Sundaram 1977
Emulsion 2%	92 L/elm tree	Hydraulic sprayer (truck mounted)	Elm bark	223	28	Sundaram 1975c, 1976
Emulsion 12.5%	9.2 L/elm tree	Mistblower (truck mounted)	Elm bark	272	32	Sundaram 1975c, 1976

[a] Dislodgeable or surface residues.
[b] Penetrated or tissue residues.

chemical was N-demethylation, hydrolysis, ring-cleavage, and conjugation to litter matrix (Sundaram et al. 1987a).

Residues of mexacarbate, formulated as emulsion and oil solutions, were measured in the forest soil and litter after application at the rate of 70 g/ha (Sundaram and Nott 1985b). The peak concentrations for the two formulations ranged from 0.02 to 0.11 μg/g in litter and from 0.01 to 0.06 μg/g in soil. The residue levels in both matrices decreased rapidly to below the detection limit (0.005 μg/g) within 1 day.

The behavior of mexacarbate in conifers was evaluated by the above authors in the same study. The peak 1-hour post-spray concentrations observed in fir needles were 0.51 and 0.19 μg/g for the oil-based and emulsion formulations, respectively. The residue levels decreased rapidly with DT_{50} of approximately 5 hours. No residue levels were found 8 days after application. Pieper and Miskus (1967) also observed similar rapid dissipation of the chemical in Douglas-fir needles. The rapid dissipation of mexacarbate in conifers appears to be due to volatilization, photolysis, wash-off by rain and weathering, and to a lesser extent by biological degradation.

The fate and persistence of aerially applied mexacarbate were studied (Sundaram et al. 1986a) in a forest aquatic environment after spraying by aircraft twice at a dosage of 70 g/ha. The maximum 1-hour post-spray concentrations in stream and pond waters were 0.73 and 18.74 μg/L, respectively. Concentrations fell rapidly to below detection limits (0.01 μg/L) in stream waters within 12 hours and within 3 days in pond water. Different species of aquatic plants in the pond sampled 1 hour after application contained 81 to 720 μg/g of mexacarbate. The residue levels decreased rapidly with DT_{50} values ranging from 2 to 8.5 hours. Aquatic plants collected from the stream did not contain measurable levels of mexacarbate. None of the fish, insects, and other aquatic animals or sediment samples collected from the stream and pond contained any mexacarbate. Various demethylated products and xylenols were found as metabolites in water and aquatic plants. Mexacarbate did not persist in forest aquatic environments because of its rapid loss through various physical, chemical (photolysis), and metabolic processes involving demethylation and hydrolysis (Crosby et al. 1965; Abdel-Wahab and Casida 1967; Meikle 1973). Mexacarbate when sprayed at 70 g/ha per application was lost rapidly from the terrestrial and aquatic components of the forest environment and did not persist in them.

Permethrin

Permethrin is a synthetic pyrethroid insecticide effective against a broad range of forestry pests at a dosage range of 17.5 to 70 g/ha.

Little is known about the fate and persistence of permethrin aerosols over the forest canopy. Ruzo and Casida (1980) found that the chemical is degraded by prolonged exposure to sunlight. Depending on climatic and environmental factors we can expect that permethrin droplets will be diluted, volatilized, drifted offsite (Sundaram et al. 1987b; Payne et al. 1988), degraded to innocuous products, and eventually deposited.

Some information is available on the environmental fate of permethrin relevant to Canadian forest situations (Kingsbury 1983; NRCC 1986). Permethrin residues measured in forest soil and litter after aerial application of the chemical at 17.5 g/ha ranged, respectively, from 0.07 to 0.12 μg/g and 0.157 to 0.750 μg/g and were relatively stable for 2 months (Kingsbury 1983). In contrast, studies (Sundaram, unpubl. data) conducted using 70 g/ha, at a high wind speed with small droplets, gave peak residue levels of 0.01 μg/g in soil and 0.13 μg/g in litter. The residues dissipated to below detection limits within 2 weeks. Degradation of permethrin in soil is due to hydrolysis of the ester bond followed by microbial action, and the rate of loss is influenced by soil type, climate, and microbial population (NRCC 1986; Leahey 1985). Therefore, extrapolation of the results from one study to another is not possible.

Permethrin residues measured following applications of 17.5 g/ha were higher for deciduous foliage with peak values ranging from 0.78 to 1.55 μg/g than for coniferous foliage whose peak value ranged from 0.24 to 0.32 μg/g. Residues gradually declined and approached detection limits (0.01 μg/g) within 2 months in both coniferous and deciduous foliage.

Sundaram (unpubl. data) demonstrated that at a dosage of 70 g/ha, trembling aspen, *Populus tremuloides*, foliage contained 4.542 μg/g of permethrin at 1 hour post-spray; that is, on average nearly 50% more than the needles of jack pine, *Pinus banksiana*, (3.466 μg/g), white pine (3.125 μg/g), and white spruce (2.573 μg/g), probably because of the greater surface-to-volume ratio of the aspen compared to conifers. Also, the morphology of foliage and its surface characteristics play key roles in droplet reception and retention during spray applications. The initial bark residue level was also much higher in aspen (1.646 μg/g) than in the barks of jack pine (0.818 μg/g), white pine (0.240 μg/g), and white spruce (0.409 μg/g). Residues in all samples decreased rapidly at the beginning and then gradually: measurable levels persisted over winter including the fallen aspen foliage collected from the forest floor. The residue levels (micrograms per gram) in samples collected 363 days post-spray were: aspen foliage from forest floor 0.661; aspen bark 0.029; jack pine needle 0.037; jack pine bark 0.048; white pine needle 0.021; white pine bark 0.021; white spruce needle 0.039; white spruce bark 0.017.

The DT_{50} and DT_{90} values (days) for these samples were:

	DT_{50}	DT_{90}
aspen		
foliage	6.95	40.4
bark	5.02	31.7
jack pine		
needles	7.13	25.0
bark	4.23	17.8
white pine		
needles	6.54	21.4
bark	6.93	24.6
white spruce		
needles	5.72	24.7
bark	5.19	22.7

The amount of initial deposits found in conifers was influenced largely by the geometry of tree crown, needle density, and configuration. Furthermore, the collection efficiency of conifer needles and the impaction potential of the droplets in the vicinity of the targets would have played some role. Dissipation patterns and residual levels in needles, apart from the initial permethrin levels found in them, are to some degree influenced by the type and amount of lipoidal content of the matrices.

The degradative pathway of permethrin loss from forestry foliage has not been studied, but it is reasonable to assume from other studies (Ohkawa et al. 1977; Gaughan and Casida 1978) that volitilization, photolysis, weathering, and metabolic breakdown by plant enzymes are responsible for its dissipation. Gaughan and Casida (1978) found that the disappearance of the *cis* isomer in nearly all forestry matrices was less than that of the *trans* isomer. The prolonged persistence of small amounts of the chemical in plant matrices is due to the lipophilicity (NRCC 1986) of permethrin (log K_{OW} = 3.5 – 6.5) leading to its absorption and retention on the waxy layers of foliage and bark.

Peak permethrin residues of 2.5 μg/L and 147.0 μg/L have been found in stream and pond waters, respectively, after application of 17.5 g/ha (Kingsbury and Kreutzweiser 1980). Residues in water fell rapidly to below detectable levels within 6 hours and no DT_{50} values could be established because of a lack of sufficient residue data. Maximum residue levels of 0.04 μg/g in sediment and 0.12 μg/g in fish were found (Kingsbury 1983). Rawn et al. (1982) reported ready absorption and retention of permethrin in a model aquatic ecosystem by aquatic plants, fish, and sediment. The loss of permethrin in another model aquatic ecosystem study was attributed to the walls of the container (Solomon et al. 1985). In a recent stream injection study, Sundaram (1991a) found nonuniform movement of the chemical downstream (12 μg/L at 30 m and 0.1 μg/L at 730 m from treatment site) with rapid dissipation primarily due to dilution and absorption. Aquatic plants, stream detritus, fish, and other aquatic organisms including invertebrate drift acted as a sink by absorbing, partitioning, and retaining the chemical from ambient water (Sundaram 1991a). Eventually the residues in them were lost by various physicochemical and metabolic processes (Table 2).

Phosphamidon

Phosphamidon is an organophosphorus insecticide that was used for spruce budworm control at the end of the DDT era. It was phased out and is withdrawn now because of its avian toxicity.

No information is available on the fate and persistence of the chemical in air following forestry spray application.

After an operational spray application of phosphamidon at 210 g/ha, Varty and Yule (1976) found the highest concentration of 0.311 μg/g (as sampled) in 2-day soil samples. The chemical dissipated rapidly and none was found after 4 days. In an in-depth experimental study using the same dosage, the highest concentrations of the chemical found in litter and soil 1 hour after the spray application were, respectively, 0.42 μg/g (*cis* 0.11, *trans* 0.31) and 0.34 μg/g (*cis* 0.08, *trans* 0.26) (Sundaram, unpubl. data). The residues dissipated rapidly to below the detection level (0.005 μg/g) after 12 days. The DT_{50} values for litter and soil samples were, respectively, 3.53 days (*cis* 2.01, *trans* 4.05) and 2.30 days (*cis* 2.14, *trans* 2.59).

The fate and persistence of phosphamidon in red and white spruce, *Picea rubens* and *P. glauca*, buds and needles, balsam fir, *Abies balsamea*, buds and needles, and red maple, *Acer rubrum*, foliage were studied by Varty and Yule (1976). The peak residue levels in the 6-hour post-spray samples were 3.358, 2.288, 2.375, 5.250, and 9.350 μg/g, respectively. The residues dissipated rapidly by various physical processes (Table 2) to below the detection limit within 8 to 15 days, except for spruce buds and shoots in which the residues lingered up to 64 days after application. Sundaram (1975a) measured the initial deposits of phosphamidon in conifers after aerial application of the chemical as an adulticide at a dosage of 70 g/ha to control spruce budworm moths. The residue levels varied enormously, ranging from 0.16 to 7.60 ng/g (average 1.75) in white spruce and 0.27 to 10.63 ng/g (average 3.04) in balsam fir, probably due to nonuniform distribution of the chemical during application.

In the experimental spray program conducted by the author (refer to soil and litter), the residue levels in 1-year-old needles of balsam fir, white spruce, white pine, red pine, *Pinus resinosa*, and jack pine, and in fresh foliage of white birch, *Betula papyrifera*, and red maple were determined at intervals after application. The 1-hour post-spray peak residue levels and the corresponding DT_{50} values are given in Table 4. The data reaffirmed the earlier findings in Sundaram 1990a and 1991b (also refer to the discussions on permethrin) that morphology of foliage and its surface characteristics played key roles in the droplet reception and retention.

Residues were lost rapidly and exponentially, except for jack pine. Other conifers did not contain any residues beyond 40 days post-spray. Jack pine needles and the foliage from the two deciduous species contained detectable levels beyond 40 days. Various physical factors (Table 2) and a myriad of metabolic processes such as dealkylation, hydrolytic cleavage of the ester bond, and dechlorination would have operated together on the chemical to reduce its concentration (Geissbühler et al. 1971). The less toxic *cis* form is lost more rapidly than the *trans* isomer (Table 4), as was observed previously by Bull et al. (1967) and Westlake et al. (1973). The persistence of a chemical depends on the nature of the matrix and the initial concentration found in it (Lichtenstein 1972b). This is illustrated by the higher DT_{50} values for phosphamidon in birch and maple foliage than in conifers (Table 4). Within the conifers, the initial deposition, retention, and subsequent degradation patterns of the chemical are largely influenced by many interrelated factors such as the geometry and density of tree crown, needle configuration, and the type and amount of cuticular waxes present in the needles. Apparently, needle configuration plays a vital role in the droplet reception either by sedimentation or impaction or by both processes.

Only limited information is available on the degradation of phosphamidon in forest aquatic systems. Hydrolytic cleavage of the bond between the methoxy group and phosphorus is enhanced by increase in pH. After an experimental aerial spray at 280 g/ha over a conifer forest, Sundaram (unpubl. data) found maximum residue levels of 5.33 μg/L (*cis* 1.71 and *trans* 3.62) and 8.79 μg/L (*cis* 2.63 and *trans* 6.16) in stream (pH 6.8) and pond (pH 6.4) waters, respectively. The residues dissipated within 8 and 18 days, respectively, in the stream and pond and the corresponding DT_{50} values were 1.11 days (*cis* 0.76 and *trans* 1.29) and 2.57 days (*cis* 1.84 and *trans* 2.92). The rapid decline was attributed to dilution and transport in the stream water and adsorption followed by chemical and microbial breakdown in pond water.

Tetrachlorvinphos

Tetrachlorvinphos is an organophosphorus insecticide used only once on an experimental basis to control white pine weevil. An aqueous formulation containing wettable powder of the insecticide and adjuvant was aerially applied to a white pine plantation at 1.125 kg/ha. The initial insecticide concentration in pine leaders was 4.20 μg/g (Sundaram 1974a). It decreased rapidly with time following first-order kinetics with DT_{50} of about 5 days. The residue level on the last sampling day (50 days) was 0.17 μg/g. Because of its low persistence, caused by its intrinsic instability, weevil control was found to be marginal and the chemical was withdrawn. Its metabolism and breakdown in the plantation forest ecosystem are therefore unknown.

Trichlorfon

Trichlorfon is an organophosphorus insecticide used in spruce budworm control at uneconomic dosage levels (>1 kg/ha), usually at the late-instar stages when foliar damage has already occurred. For these reasons, the insecticide has fallen into disfavor and very little information is available on its fate in different components of the forest environment.

Table 4. Initial residues and DT_{50} values of phosphamidon isomers in forestry foliage.

	Conc. (μg fresh wt.)			DT_{50} (days)		
Foliar type	*cis*	*trans*	total	*cis*	*trans*	total
Balsam fir	0.830	2.030	2.860	1.74	3.53	2.94
White spruce	0.317	0.839	1.156	2.95	3.30	3.20
White pine	0.364	0.931	1.295	1.58	1.78	1.72
Red pine	0.406	1.102	1.508	2.39	2.40	2.39
Jack pine	0.677	1.706	2.383	1.89	2.45	2.24
White birch	1.446	3.906	5.352	3.39	4.00	3.84
Red maple	1.974	4.706	6.680	3.88	5.63	5.06

Sundaram and Varty (1989) sprayed a boreal forest with trichlorfon at an application rate of 1.14 kg/ha and monitored the residue levels in various forestry components. They found a peak residue level of 2.84 µg/g in forest soil that dissipated, probably due to physicochemical and metabolic factors, to below the detection limit (0.05 µg/g) within 2 weeks. The DT_{50} is about 6 days. Trichlorfon is subjected to leaching and wash out from soil due to its high water solubility. The maximum initial residue levels in conifer needles and buds ranged from 14.32 to 17.26 µg/g and 10.83 to 11.80 µg/g, respectively. They dissipated rapidly in 3 days to 6.08 to 6.48 µg/g in needles and 2.63 to 4.20 µg/g in buds. Detectable levels (0.15 µg/g) were found in conifers until 32 days post-spray. The initial residue in maple foliage was 14.03 µg/g, in 3 days it decreased to 3.57 µg/g, and none was found at 32 days. The authors hypothesized that the rapid dissipation was due to a combination of physical (especially volatilization and weathering), chemical (hydrolysis), and biological (dehydrochlorination to dichlorvos followed by hydrolysis) factors. Pieper and Richmond (1976) reported nearly similar dissipation patterns of the chemical in conifer and deciduous foliage samples. It is not certain whether degradation in conifers involves production of dichlorvos.

The stream water (pH 6.3) showed (Sundaram and Varty 1989) an initial concentration of 95.6 µg/L with DT_{50} less than 1 day and the residue levels reached below the detection limit (0.05 µg/L) within 2 weeks post-treatment. Dilution, transportation, hydrolysis, and adsorption (Table 2) effects would have played major roles in the disappearance of the chemical. Pieper and Richmond (1976) observed a similar pattern of decline and no residues were found in water samples collected 7 days after application. Both studies indicated a rapid dissipation of the chemical in soil, foliage, and water components of the forest environment.

Conclusions

Chemical insecticides have played a vital role in Canadian forestry for many years in the suppression of various insect pests and in enhancing the production of fiber and lumber. A sound knowledge of their fate and persistence in various forestry compartments and their potential impacts on nontarget organisms is essential to assess environmental acceptability, especially when constant assaults are being made against their broadcast application, although most of the assaults are at odds with available facts. Compilation and review of the data on the persistence and fate of insecticides currently used in forest protection have shown that pursuant to the established use patterns, little damage has occurred, contrary to public perception, to the quality of our environment.

Until now, undue emphasis has been placed on efficacy assessment of sprayed chemicals over forest lands and their environmental fate and persistence have been relegated to the sidelines. Current societal demands and public concerns emphasize the necessity to understand the distribution, behavior, persistence, and fate of sprayed materials over forests and to be accountable to maintain the quality of our forest environment. We should act quickly to meet these challenges.

References

Abdel-Wahab, A.J.; Casida, J.E. 1967. Photooxidation of two 4-dimethylaminoaryl methylcarbamate insecticides (Zectran®and Matacil®) on bean foliage and of alkylaminophenyl methyl carbamates on silica gel chromatoplates. J. Agric. Food Chem. 15(3):479-487.

Abdel-Wahab, A.J.; Kuhr, R.J.; Casida, J.E. 1966. Fate of ^{14}C-carbonyl-labeled insecticide chemicals in and on bean plants. J. Agric. Food Chem. 14:290-298.

Addison, J.B. 1981. Vapor phase photochemistry of fenitrothion and aminocarb. Bull. Environ. Contam. Toxicol. 27: 250-255.

Apperson, C.S.; Schaefer, C.H.; Colwell, A.E.; Warner, G.H.; Anderson, N.L.; Dupras, E.F., Jr.; Longanecker, D.R. 1978. Effects of Diflubenzuron on *Chaoborus astictopus* and nontarget organisms and persistence of diflubenzuron in lentic habitats. J. Econ. Entomol. 71(3): 521-527.

Armson, K.A. 1977. Forest soils: properties and processes. Univ. of Toronto Press, Toronto, Ont. 390 p.

Armstrong, J.A. 1977. Relationship between the rates of pesticide application and the quantity deposited on the forest. Pages 183-202 *in* Proc. Symp. on Fenitrothion: the long-term effects of its use in forest ecosystems. Associate Committee on Scientific Criteria for Environmental Quality, Natl. Res. Counc., Ottawa, Ont. NRCC No. 16073.

Ayer, W.C.; Brun, G.L.; Eidt, D.C.; Ernst, W.R.; Mallet, V.N.; Matheson, R.A.; Silk, P.J. 1984. Persistence of aerially applied fenitrothion in water, soil, sediment and balsam fir foliage. Can. For. Serv., Mar. For. Cent. Inf. Rep. M-X-153, 14 p.

Bitman, J.; Cecil, H.C.; Fries, G.F. 1970. DDT-induced inhibition of avian shell gland carbonic anhydrase: a mechanism for thin eggshells. Science 168: 594.

Boberschmidt, L.; Saari, S.; Sassaman, J.; Skinner, L. 1989. Pesticide background statements, Vol. IV, Insecticides. USDA, For. Serv., Washington, D.C. Agric. Handbook No. 685, pp. AC 1-99.

Booth, G.M.; Ferrell, D. 1977. Degradation of Dimilin® by aquatic foodwebs. Pages 221-243 *in* M.A.Q. Khan (Ed.), Pesticides in Aquatic Environments. Plenum Press, New York.

Bull, D.L.; Lindquist, D.A.; Grabbe, R.R. 1967. Comparative fate of the geometric isomers of phosphamidon in plants and animals. J. Econ. Entomol. 60(2): 332-349.

Crosby, D.G. 1973. The fate of pesticides in the environment. Ann. Rev. Plant Physiol. 24: 467-492.

Crosby, D.C.; Leitis, E.; Winterlin, W.L. 1965. Photodecomposition of carbamate insecticides. J. Agric. Food Chem. 15(3): 479-487.

Dunster, J.A. 1987. Chemicals in Canadian forestry: the controversy continues. Ambio 16(2-3): 142-148.

Edwards, C.A. 1975. Factors that affect persistence of pesticides in plants and soils. Pure and Appl. Chem. 42: 39-56.

Eidt, D.C.; Bacon, G.B.; Degraeve, G.M.; Mallet, V.N. 1988. Fate and short-term persistence of the insecticide aminocarb in a New Brunswick (Canada) headwater stream. Arch. Environ. Contam. Toxicol. 17: 817-829.

Eidt, D.C.; Pearce, P.A. 1986. The biological consequences of lingering fenitrothion residues in conifer foliage—a synthesis. For. Chron. (Aug.) pp. 246-249.

Eidt, D.C.; Sosiak, A.J.; Mallet, V.N. 1984. Partitioning and short-term persistence of fenitrothion in New Brunswick (Canada) headwater streams. Arch. Environ. Contam. Toxicol. 13: 43-52.

Eidt, D.C.; Sundaram, K.M.S. 1975. The insecticide fenitrothion in headwater streams from large-scale forest spraying. Can. Entomol. 107: 735-742.

Ernst, W.R.; Pearce, P.A.; Pollock, J.L. 1989. Environmental effects of fenitrothion use in forestry. Environ. Can., Conservation and Protection Br., Dartmouth, N.S. 166 p.

Fettes, J.J.; Buckner, C.H. 1976. Historical sketch of the philosophy of spruce budworm control in Canada. Pages 57-60 *in* W.H. Klein (Ed.), Proc. Symp. Spruce Budworm. USDA, For. Serv., Washington, D.C. Misc. Publ. No. 1327.

Gaughan, L.C.; Casida, J.E. 1978. Degradation of *trans*- and *cis*-permethrin on cotton and bean plants. J. Agric. Food Chem. 26: 525-528.

Geen, G.H.; Hussain, M.A.; Oloffs, P.C.; McKeown, B.A. 1981. Fate and toxicity of acephate (Orthene®) added to a coastal B.C. stream. J. Environ. Sci. Health B16(3): 253-271.

Geissbühler, H.; Voss, G.; Anliker, R. 1971. The metabolism of phosphamidon in plants and animals. Pages 39-60 *in* F.A. Gunther and J.A. Gunther (Eds.), Residue Reviews—Phosphamidon. Springer-Verlag, New York.

Ghassemi, M.; Fargo, L.; Painter, P.; Quinlivan, S.; Scofield, R.; Takata, A. 1981. Environmental fates and impacts of major forest use pesticides. U.S. Environ. Protection Agency, Washington, D.C. Publ. No. PB 83-124552, 436 p.

Gibbs, K.E.; Mingo, T.M.; Courtemauch. 1984. Persistence of carbaryl (Sevin-4-Oil) in woodland ponds and its effects on pond macroinvertebrates following forest spraying. Can. Entomol. 116: 203-213.

Glotfelty, D.E. 1985. Pathways of pesticide dispersion in the environment. Pages 425-435 *in* J.L. Hilton (Ed.), Agricultural chemicals of the future—BARC Symp. No. 8. Rowman and Allanheld, Totowa, N.J.

Ivie, G.W.; Bull, D.L.; Velch, J.A. 1980. Fate of diflubenzuron in water. J. Agric. Food Chem. 28(2): 330-337.

Kapoor, I.P.; Metcalf, R.L.; Nystrom, R.F.; Sangha, G.K. 1970. Comparative metabolism of methoxychlor, methiochlor and DDT in mouse, insects and in a model ecosystem. J. Agric. Food Chem. 18(6): 1145-1152.

Kenaga, E.E. 1972. Guidelines for environmental study of pesticides: Determination of bioconcentration potential. Residue Rev. 44: 73.

Kingsbury, P.D. 1983. Permethrin in New Brunswick salmon nursery streams. Environ. Can., Can. For. Serv., For. Pest Manage. Inst., Sault Ste. Marie, Ont. Inf. Rep. FPM-X-52, 192 p.

Kingsbury, P.D.; Kreutzweiser, D.P. 1980. Environmental impact assessment of a semi-operational permethrin application. Environ. Can., Can. For. Serv., For. Pest Manage. Inst., Sault Ste. Marie, Ont. Inf. Rep. FPM-X-30, 47 p.

Kuhr, R.J.; Casida, J.E. 1967. Persistent glycosides of metabolites of methylcarbamate insecticide chemicals formed by hydroxylation in bean plants. J. Agric. Food Chem. 15: 814-825.

Kuhr, R.J.; Dorough, H.W. 1976. Carbamate insecticides: chemistry, biochemistry and toxicology. CRC Press, Inc., Cleveland, Ohio. 301 p.

LaPierre, L.E. 1982. The persistence of fenitrothion insecticide in red maple (*Acer rubrum* L.) and white birch (*Betula papyfifera* Marsh) deer browse. J. Range Manage. 35(1): 65-67.

LaPierre, L.E. 1985. Persistence of fenitrothion insecticide in poplar (*Populus tremuloides*) and gray birch (*Betula populifolia*). Bull. Environ. Contam. Toxicol. 35: 471-475.

Leahey, J.P. 1985. Metabolism and environmental degradation. Pages 263-342 *in* J.P. Leahey (Ed.), The Pyrethroid Insecticides. Taylor and Francis, London, England.

Lichtenstein, E.P. 1972a. Environmental factors affecting fate of pesticides. Pages 190-205 *in* Conference Proceedings on Degradation of Synthetic Organic Molecules in the Biosphere. Natl. Acad. Sci., Washington, D.C.

Lichtenstein, E.P. 1972b. Persistence and fate of pesticides in soils, water and crops: significance to humans. Pages 1-22 *in* A.S. Tahori (Ed.), Fate of Pesticides in Environment. Gordon and Breach Science Publishers, New York.

Lockhart, W.L.; Metner, D.A.; Grift, N. 1973. Biochemical and residue studies on rainbow trout, *Salmo gairdneri*, following field and laboratory exposures to fenitrothion. Manit. Entomol. 7: 26-36.

Maguire, R.J.; Hale, E.J. 1980. Distribution of fenitrothion sprayed on a pond: kinetics of its distribution and transformation in water and sediment. J. Agric. Food Chem. 28: 372-378.

Malis, G.P.; Muir, D.C.G. 1984. Fate of fenitrothion in shaded and unshaded ponds. Pages 277-295 *in* W.Y. Garner and J. Harvey, Jr. (Eds.), Chemical and biological controls in forestry. Am. Chem. Soc., Symp. Ser. No. 238, Am. Chem. Soc., Washington, D.C.

Mallet, V.N.; Cassista, A. 1984. Fenitrothion residue survey in relation to the 1981 spruce budworm spray program in New Brunswick, Canada. Bull. Environ. Contam. Toxicol. 32: 65-74.

Mallet, V.N.; Volpé, G. 1982. A chemical residue survey in relation to the 1980 spruce budworm spray program in New Brunswick (Canada). J. Environ. Sci. Health B17(6): 715-736.

Martinat, P.J.; Christman, V.; Cooper, R.J.; Dodge, K.M; Whitmore, R.C.; Booth G.M.; Seidel, G.E. 1987. Environmental fate of Dimilin® WP-25 in a central Appalachian forest. J. Agric. Food Chem. 39: 142-149.

Meikle, R.W. 1973. Metabolism of 4-dimethylamino-3,5-xylyl methylcarbamate (mexacarbate, active ingredient of Zectran® insecticide): a unified picture. Bull. Environ. Contam. Toxicol. 10: 29-36.

Metcalfe, C.D.; McLeese, D.W.; Zitko, V. 1980. Rate of volatilization of fenitrothion from fresh water. Chemosphere 9: 151-156.

Miyamoto, J. 1977. Degradation of fenitrothion in terrestrial and aquatic environments including photolytic and microbial reactions. Pages 105-134 *in* J.R. Roberts, R. Greenhalgh and W.K. Marshall (Eds.), The long-term effects of its use in forest ecosystems. Associate Committee on Scientific Criteria for Environmental Quality, Natl. Res. Counc., Ottawa, Ont. NRCC No. 16073.

Moody, R.P.; Greenhalgh, R.; Weinberger, P. 1978. The fate of fenitrothion in an aquatic ecosystem. Bull. Environ. Contam. Toxicol. 19: 8-14.

Morrison, B.R.S.; Wells, D.E. 1981. The fate of fenitrothion in a stream environment and its effect on the fauna following aerial spraying of a Scottish forest. Sci. Total Environ. 19: 233-252.

Nicholls, P.H. 1988. Factors influencing entry of pesticides into soil water. Pestic. Sci. 22: 123-137.

Norris, L.A. 1971. The behavior of chemicals in the forest. Pages 90-106 *in* J. Capizzi and J.M. Witt (Eds.), Pesticides, pest control and safety of forest and range lands. Continuing Education Publications, Corvallis, Ore.

NRCC. 1975. Fenitrothion: the effects of its use on environmental quality and its chemistry. Associate Committee on Scientific Criteria for Environmental Quality, Natl. Res. Counc. Can., Ottawa, Ont. NRCC No. 14104, 162 p.

NRCC. 1982. Aminocarb: the effects of its use on the forest and the human environment. Associate Committee on Scientific Criteria for Environmental Quality, Natl. Res. Counc. Can., Ottawa, Ont. NRCC No. 18979, 253 p.

NRCC. 1986. Pyrethroids: their effects on aquatic and terrestrial ecosystems. Associate Committee on Scientific Criteria for Environmental Quality, Natl. Res. Counc. Can., Ottawa, Ont. NRCC No. 24376, 303 p.

Ohkawa, H.; Kaneko, H.; Miyamoto, J. 1977. Metabolism of permethrin in bean plants. J. Pestic. Sci. 2: 67-76.

Ohkawa, H.; Mikami, N.; Miyamoto, J. 1974. Photodecomposition of Sumithion® [0,0-dimethyl-0-(3-methyl-4-nitrophenyl)-phosphorothioate. Agric. Bio. Chem. 38(11): 2247-2255.

Payne, N.J.; Helson, B.V.; Sundaram, K.M.S.; Fleming, R.A. 1988. Estimating buffer zone widths for pesticide applications. Pestic. Sci. 24: 147-161.

Pieper, G.R. 1979. Residue analysis of carbaryl on forest and in stream water using HPLC. Bull. Environ. Contam. Toxicol. 22: 167-171.

Pieper, G.R.; Miskus, R.P. 1967. Determination of Zectran® residues in aerial forest spraying. J. Agric. Food Chem. 15(5): 915-916.

Pieper, G.R.; Richmond, C.E. 1976. Residues of trichlorfon and lauroye trichlorfon in Douglas-fir, willow, grass, aspen foliage and in creek water after aerial application. Bull. Environ. Contam. Toxicol. 15: 250-256.

Rabeni, C.F.; Stanley, J.G. 1979. Operational spraying of acephate to suppress spruce budworm has minor effects on stream fishes and invertebrates. Bull. Environ. Contam. Toxicol. 23: 327-334.

Rawn, G.P.; Webster, G.R.B.; Muir, D.C.G. 1982. Fate of permethrin in model outdoor ponds. J. Environ. Sci. Health B17: 463-486.

Ruzo, L.O.; Casida, J.E. 1980. Pyrethroid photochemistry: mechanistic aspects in the reaction of the (dihalovinyl) cyclo-propanecarboxylate substituent. J. Chem. Soc. Perkin Trans. 3: 728-732.

Schaefer, C.H.; Dupras, E.F., Jr. 1979. Factors affecting the stability of SIR-8514 (2-chloro-N-[[[4-(trifluoromethoxy) phenyl]amino]carbonyl] benzamide) under laboratory and field conditions. J. Agric. Food Chem. 27(5): 1031-1034.

Seiber, J.N. 1985. General principles governing the fate of chemicals in the environment. Pages 389-402 *in* J.L. Hilton (Ed.), Agricultural Chemicals of the Future—BARC Symp. No. 8. Rowan and Allanheld, Totowa, N.J.

Seiber, J.N. 1987. Principles governing environmental mobility and fate. Pages 88-105 *in* N.N. Ragsdale and R.J. Kuhr (Eds.), Pesticides: Minimizing the Risks. Am. Chem. Soc. Symp. Ser. No. 336. Am. Chem. Soc., Washington, D.C.

Shea, P.J.; Nigam, P.C. 1984. Chemical control. Pages 116-132 *in* D.M. Schmitt, D.G. Grimble and J.L. Searcey (Eds.), Spruce budworm handbook: managing the spruce budworm in eastern North America. USDA, For. Serv., Washington, D.C. Agric. Handbook No. 620.

Silk, P.J.; Unger, I. 1973. The photochemistry of carbamates I: the photodegradation of Zectran®. Int. J. Environ. Anal. Chem. 2: 213-220.

Solomon, K.R.; Yoo, J.Y.; Lean, D.; Kaushik, N.K.; Day, K.E.; Stephenson, G.L. 1985. Dissipation of permethrin in limnocorrals. Can. J. Fish. Aquat. Sci. 42: 70-76.

Spillner, C.J.; Thomas, V.M.; DeBaun, J.R. 1979a. Effect of fenitrothion on microorganisms which degrade leaf-litter and cellulose in forest soils. Bull. Environ. Contam. Toxicol. 23: 601-606.

Spillner, C.J.; DeBaun, J.R.; Menn, J.J. 1979b. Degradation of fenitrothion in forest soil and effects on forest soil microbes. J. Agric. Food Chem. 27(5): 1054-1060.

Stanley, J.G.; Trial, J.G. 1980. Disappearance constants of carbaryl from streams contaminated by forest spraying. Bull. Environ. Contam. Toxicol. 25: 771-776.

Sundaram, A.; Sundaram, K.M.S.; Cadogan, B.L. 1985a. Influence of formulation properties on droplet spectra and soil residues of aminocarb aerial sprays in conifer forests. J. Environ. Sci. Health B20(2): 167-186.

Sundaram, K.M.S. 1973a. Degradation dynamics of fenitrothion in aqueous systems. Environ. Can., Can. For. Serv., Chem. Cont. Res. Inst. Inf. Rep. CC-X-44, 19 p.

Sundaram, K.M.S. 1973b. Persistence studies of insecticides: I. Aerial application of methoxychlor for control of white pine weevil in Ontario, 1973. Environ. Can., Can For. Serv., Chem. Cont. Res. Inst. Inf. Rep. CC-X-57, 34 p.

Sundaram, K.M.S. 1974a. Persistence studies of insecticides: II. Degradation of Gardona® on white pine leaders (*Pinus strobus* L.) after aerial application for control of white pine weevil (*Pissodes strobi* Peck) in Ontario, 1973. Environ. Can., Can. For. Serv., Chem. Cont. Res. Inst. Inf. Rep. CC-X-62, 32 p.

Sundaram, K.M.S. 1974b. Distribution and persistence of fenitrothion residues in foliage, soil and water in Larose Forest. Environ. Can., Can. For. Serv., Chem. Cont. Res. Inst. Inf. Rep. CC-X-64, 43 p.

Sundaram, K.M.S. 1974c. Persistence studies of insecticides: III. Accumulation of fenitrothion and its oxygen analog in foliage, soil and water in Larose Forest. Environ. Can., Can. For. Serv., Chem. Cont. Res. Inst. Inf. Rep. CC-X-65, 21 p.

Sundaram, K.M.S. 1975a. Phosphamidon isomers in coniferous foliage. Environ. Can., Can. For. Serv., Chem. Cont. Res. Inst. Inf. Rep. CC-X-95, 27 p.

Sundaram, K.M.S. 1975b. Persistence studies of insecticides: VI. Degradation of methoxychlor on white pine leaders (*Pinus strobus* L.) after aerial application for control of white pine weevil (*Pissodes strobi* Peck) in Ontario, Spring 1974. Environ. Can., Can. For. Serv., Chem. Cont. Res. Inst. Inf. Rep. CC-X-99, 19 p.

Sundaram, K.M.S. 1975c. Gas chromatographic analysis of methoxychlor formulations and spray mixtures. Environ. Can., Can. For. Serv., Chem. Cont. Res. Inst. Inf. Rep. CC-X-94, 17 p.

Sundaram, K.M.S. 1976. Persistence and fate of methoxychlor used for elm bark beetle control in the urban environment of the national capital area. Environ. Can., Can. For. Serv., Chem. Cont. Res. Inst. Inf. Rep. CC-X-118, 45 p.

Sundaram, K.M.S. 1977. A study on the comparative deposit levels and persistence of two methoxychlor formulations used in white pine weevil control. Environ. Can., Can. For. Serv., Chem. Cont. Res. Inst. Inf. Rep. CC-X-142, 14 p.

Sundaram, K.M.S. 1981. Distribution, persistence and fate of Matacil® formulations in a forest ecosystem. Environ. Can., Can. For. Serv., For. Pest Manage. Inst., Sault Ste. Marie, Ont. File Rep. 20, 32 p.

Sundaram, K.M.S. 1984a. Residue levels of fenitrothion and aminocarb in the air samples collected from experimental spray blocks in New Brunswick in 1982. J. Environ. Sci. Health B19: 409-426.

Sundaram, K.M.S. 1984b. Fenitrothion residues in some forestry samples from a plantation forest following experimental spray application for five consecutive years. J. Environ. Sci. Health B22(4): 409-426.

Sundaram, K.M.S. 1985. Foliar residues of aminocarb at tree canopy and ground levels in an experimental spray operation in conifer forests. Pestic. Sci. 16: 463-471.

Sundaram, K.M.S. 1986a. A comparative evaluation of dislodgeable and penetrated residues, and persistence characteristics of aminocarb and fenitrothion, following application of several formulations onto conifer trees. J. Environ. Sci. Health B21(6): 539-560.

Sundaram, K.M.S. 1986b. Persistence and degradation of diflubenzuron in conifer foliage, forest litter and soil, following simulated aerial application. Can. For. Serv., For. Pest Manage. Inst., Sault Ste. Marie, Ont. Inf. Rep. FPM-X-74, 19 p.

Sundaram, K.M.S. 1987a. Distribution, dissipation and persistence of aminocarb in aquatic components of the forest environment after aerial application of two Matacil® 180F formulations. Can. For. Serv., For. Pest Manage. Inst., Sault Ste. Marie, Ont. Inf. Rep. FPM-X-78, 19 p.

Sundaram, K.M.S. 1987b. Fenitrothion deposits in simulated and live surfaces during an aerial spray trial in northern Ontario. Pages 293-297 *in* G.W. Green (Ed.), Symposium on the aerial application of pesticides in forestry. The Associate Committee on Agriculture and Forestry Aviation, Natl. Res. Counc. Can., Ottawa, Ont. Ref. No AFA-IN-18, NRCC No. 29197.

Sundaram, K.M.S. 1987c. Persistence characteristics of operationally sprayed fenitrothion in nearby unsprayed areas of a conifer forest ecosystem in New Brunswick. J. Environ. Sci. Health B22(4): 413-438.

Sundaram, K.M.S. 1989a. Deposition, persistence and dissipation of diflubenzuron in pond and stream environments following aerial application of Dimilin® WP-25. Can. J. Plant Sci. 69: 259-260.

Sundaram, K.M.S. 1989b. Fenitrothion deposits on different components of a forest ecosystem during an aerial spray trial. Pages 244-253 *in* J.L. Hazen and D.A. Hovde (Eds.), Pesticide formulations and application systems: international aspects, vol. 9, ASTM 1036. Am. Soc. Test. Materials, Philadelphia, Pa.

Sundaram, K.M.S. 1990a. Foliar and ground deposits of fenitrothion following aerial application over a plantation forest. J. Environ. Sci. Health B25 (5): 643-663.

Sundaram, K.M.S. 1990b. Persistence and metabolic fate of fenitrothion in northern Ontario forest soils and leaf litter under laboratory conditions. J. Environ. Sci. Health B25(6): 743-766.

Sundaram, K.M.S. 1991a. Fate and short-term persistence of permethrin insecticide injected in a northern Ontario (Canada) headwater stream. Pestic. Sci. 31: 281-294.

Sundaram, K.M.S. 1991b. Spray deposit patterns and persistence of diflubenzuron in some terrestrial components of a forest ecosystem after application at three volume rates under field and laboratory conditions. Pestic. Sci. 32: 275-293.

Sundaram, K.M.S; Boyonoski, N.; Feng, C. 1987a. Degradation and metabolism of mexacarbate in two types of forest litters under laboratory conditions. J. Environ. Sci. Health B22(1): 29-54.

Sundaram, K.M.S.; de Groot, P.; Sundaram, A. 1987b. Permethrin deposits and airborne concentrations downwind from a single swath application using a back pack mist blower. J. Environ. Sci. Health B22(2): 171-193.

Sundaram, K.M.S.; Feng, C.; Boyonoski N.; Manniste-Squire, V. 1985b. Leaching, degradation and fate of ^{14}C-mexacarbate in columns packed with forest soils. Can. For. Serv., For. Pest Manage. Inst., Sault Ste. Marie, Ont. Inf. Rep. FPM-X-71, 34 p.

Sundaram, K.M.S.; Feng, J.; Nott, R.; Feng, C. 1983. Distribution, dynamics, persistence and fate of aminocarb and fenitrothion formulations containing Triton® X-100 in a New Brunswick forest environment. Environ. Can., Can. For. Serv., For. Pest Manage. Inst., Sault Ste. Marie, Ont. File Rep. 44, 83 p.

Sundaram, K.M.S.; Hopewell, W.W. 1976. Distribution, persistence and translocation of Orthene® in spruce trees after simulated aerial spray application. Environ. Can., Can. For. Serv., Chem. Cont. Res. Inst. Inf. Rep. CC-X-121, 25 p.

Sundaram, K.M.S.; Hopewell, W.W. 1977a. Fate and persistence of aminocarb in conifer foliage and forest soil after simulated aerial spray application. Environ. Can., Can. For. Serv., For. Pest Manage. Inst., Sault Ste. Marie, Ont. Inf. Rep. FPM-X-6, 41 p.

Sundaram, K.M.S.; Hopewell, W.W. 1977b. Penetration, translocation and fate of C-14 aminocarb in spruce trees. Environ. Can., Can. For. Serv., Chem. Cont. Res. Inst. Inf. Rep. CC-X-140, 25 p.

Sundaram, K.M.S.; Hopewell, W.W.; Lafrance, G. 1977. Uptake, translocation and metabolism of C-14 acephate in spruce trees. Environ. Can., Can. For. Serv., Chem. Cont. Res. Inst. Inf. Rep. CC-X-139, 31 p.

Sundaram, K.M.S.; LeCompte, P.E. 1975. Persistence studies of insecticides: V. Degradation of carbaryl on white pine leaders (*Pinus strobus* L.) after aerial application for control of white pine weevil (*Pissodes strobi* Peck) in Ontario, 1974. Environ. Can., Can. For. Serv., Chem. Cont. Res. Inst. Inf. Rep. CC-X-98, 16 p.

Sundaram, K.M.S.; Kingsbury, P.D.; Holmes, S.B. 1984. Fate of chemical insecticides in aquatic environments: Forest spraying in Canada. Pages 253-276 *in* W.Y. Garner and J. Harvey, Jr. (Eds.), Chemical and biological controls in forestry. Am. Chem. Soc., Symp. Ser. No. 238, Am. Chem. Soc., Washington, D.C.

Sundaram, K.M.S.; Millikin, R.; Sundaram, A. 1989. Assessment of canopy and ground deposits of fenitrothion following aerial and ground application in a northern Ontario forest. Pestic. Sci. 25: 59-69.

Sundaram, K.M.S.; Nott, R. 1984. Fenitrothion residues in selected components of a conifer forest following aerial application of tank mixes containing Triton® X-100. Environ. Can., Can. For. Serv., For. Pest Manage. Inst., Sault Ste. Marie, Ont. Inf. Rep. FPM-X-65, 14 p.

Sundaram, K.M.S.; Nott, R. 1985a. Distribution and persistence of aminocarb in terrestrial components of the forest environment after semi-operational application of two mixtures of Matacil® 180F. Can. For. Serv., For. Pest Manage. Inst., Sault Ste. Marie, Ont. Inf. Rep. FPM-X-67, 10 p.

Sundaram, K.M.S.; Nott, R. 1985b. Mexacarbate residues in selected components of a conifer forest following aerial applications of oil-based and aqueous spray formulations. J. Environ. Sci. Health B20(4): 425-444.

Sundaram, K.M.S.; Nott, R. 1989. Mobility of diflubenzuron in two types of forest soils. J. Environ. Sci. Health B24(1): 65-86.

Sundaram, K.M.S.; Nott, R.; Holmes, S.B.; Boyonoski, N. 1986a. The distribution and fate of mexacarbate in a forest aquatic ecosystem. Can. For. Serv., For. Pest Manage. Inst., Sault Ste. Marie, Ont. Inf. Rep. FPM-X-73, 23 p.

Sundaram, K.M.S.; Prasad, R. 1975. Persistence studies of insecticides: IV. Root penetration, translocation and metabolism of C-14 labeled fenitrothion in young spruce trees. Environ. Can., Can. For. Serv., Chem. Cont. Res. Inst. Inf. Rep. CC-X-86, 15 p.

Sundaram, K.M.S.; Raske, A.G.; Retnakaran, A.; Sundaram, A.; West, R.J. 1987c. Effect of formulation properties on ground and foliar deposits of two insecticides in flushed and one-year-old balsam fir needles following aerial application. Pestic. Sci. 21: 105-118.

Sundaram, K.M.S.; Smith, G.G.; O'Brien, W.; Bonnet, D. 1972. A preliminary report on the persistence of methoxychlor for the control of white pine weevil in plantations. Environ. Can., Can. For. Serv., Chem. Cont. Res. Inst. Inf. Rep. CC-X-31, 27 p.

Sundaram, K.M.S.; Sundaram, A. 1981. Effect of additives on the persistence of aminocarb in conifer foliage. Environ. Can., Can. For. Serv., Ottawa, Ont. Res. Notes 1(3): 18-21.

Sundaram, K.M.S.; Sundaram, A. 1982. Influence of formulation on droplet size, deposit concentration and persistence of fenitrothion in conifers following a simulated aerial application. Environ. Can., Can. For. Serv., Ottawa, Ont. Res. Notes 2(1): 2-5.

Sundaram, K.M.S.; Sundaram, A. 1987. Influence of formulation on spray deposit patterns, dislodgeable and penetrated residues, and persistence characteristics of fenitrothion in conifer needles. Pestic. Sci. 18: 259-271.

Sundaram, K.M.S.; Sundaram, A. 1988. Deposition of aerially sprayed mexacarbate on balsam fir canopy and the forest floor: relevance to ULV applications. J. Environ. Sci. Health B23(5): 453-473.

Sundaram, K.M.S.; Sundaram, A. 1989. Relative volatilization of three insecticides from deposits on fir foliage following spray application under laboratory conditions. J. Environ. Sci. Health B24(2): 167-182.

Sundaram, K.M.S.; Sundaram, A.; Nott, R. 1986b. Mexacarbate deposits on simulated and live fir needles during an aerial spray trial. Trans. ASAE 29(2): 382-388 and 392.

Sundaram, K.M.S.; Szeto, S.Y. 1979. A study on the lethal toxicity of aminocarb to freshwater crayfish and its *in vivo* metabolism. J. Environ. Sci. Health B14(6): 589-602.

Sundaram, K.M.S.; Szeto, S.Y. 1984. Persistence of aminocarb in balsam fir foliage, forest litter, and soil after aerial application of three formulations. J. Agric. Food Chem. 32: 1138-1141.

Sundaram, K.M.S.; Szeto, S.Y. 1985. Distribution and persistence of aminocarb in stream water, sediment and fish after application of three Matacil® formulations. J. Environ. Sci. Health B20(2): 187-200.

Sundaram, K.M.S.; Szeto, S.Y. 1987. Distribution and persistence of carbaryl in some terrestrial and aquatic components of a forest environment. J. Environ. Sci. Health B22(5): 579-599.

Sundaram, K.M.S.; Varty, I.W. 1989. Distribution and persistence of trichlorfon in a forest environment. J. Environ. Sci. Health B24: 647-659.

Sundaram, K.M.S.; Volpé, Y.; Smith, G.G.; Duffy, J.R. 1976. A preliminary study on the persistence and distribution of Matacil® in a forest environment. Environ. Can., Can. For. Serv., Chem. Cont. Res. Inst. Inf. Rep. CC-X-116, 44 p.

Sundaram, K.M.S.; Yule, W.N.; Prasad, R. 1975. Studies of foliar penetration, movement and pesistence of C-14 labeled fenitrothion in spruce and fir trees. Environ. Can., Can. For. Serv., Chem. Cont. Res. Inst., Inf. Rep. CC-X-85, 17 p.

Symons, P.E.K. 1977. Dispersal and toxicology of the insecticide fenitrothion: predicting hazards of forest-spraying. Residue Reviews 38: 1-36.

Szeto, S.Y.; Holmes, S.B. 1982. The lethal toxicity of Matacil® 1.8F to rainbow trout and its *in vivo* metabolism. J. Environ. Sci. Health B17(1): 51-61.

Szeto, S.Y.; MacCarthy, H.R.; Oloffs, P.C.; Shepherd, R.F. 1978. Residues in Douglas-fir needles and forest litter following an aerial application of acephate (Orthene). J. Environ. Sci. Health B13(2): 87-103.

Szeto, S.Y.; MacCarthy, H.R.; Oloffs, P.C.; Shepherd, R.F. 1979. The fate of acephate and carbaryl in water. J. Environ. Sci. Health B14(6): 635-654.

Szeto, S.Y.; Sundaram, K.M.S. 1981. Residues of chlorpyrifos-methyl in balsam fir foliage, forest litter, soil, stream water, sediment and fish tissue after double aerial application of Reldan®. J. Environ. Sci. Health B16: 743-766.

Szeto, S.Y.; Sundaram, K.M.S. 1982. Behavior and degradation of chlorpyrifos-methyl in two aquatic models. J. Agric. Food Chem. 30: 1032-1035.

Takimoto, Y.; Hirota, M; Inui, H.; Miyamoto, J. 1976. Decomposition and leaching of radioactive Sumithion® in 4 different soils under laboratory conditions. J. Pestic. Sci. 1: 131-143.

van den Berg, G. 1986. Dissipation of diflubenzuron residues after application of Dimilin® WP-25 in a forestry area in North Carolina (USA) and some ecological effects. Duphar, B.V., Crop Protection Division, The Netherlands, Rep. No. 56637/47/1986, 111 p.

van Valkenburg, W. (Ed.). 1973. Pesticide Formulations. Marcel Dekker Inc., New York. 341 p.

Varty I.W.; Yule, W.N. 1976. The persistence and fate of phosphamidon in a forest environment. Bull. Environ. Contam. Toxicol. 15(3): 257-264.

Westlake, W.E.; Ittig, M.; Ott, D.E.; Gunther, F.A. 1973. Persistence of residues of the insecticide phosphamidon on and in oranges, lemons and grapefruit, and on and in orange leaves and in dried citrus pulp cattle feed. J. Agric. Food Chem. 21(5): 846-850.

Wheatley, G.A. 1973. Pesticides in the atmosphere. Pages 365-408 *in* C.A. Edwards (Ed.), Environmental pollution by pesticides. Plenum Press, New York.

Yule, W.N. 1974. The persistence and fate of fenitrothion insecticide in a forest environment. II. Accumulation of residues in balsam fir foliage. Bull. Environ. Contam. Toxicol. 12(2): 249-252.

Yule, W.N.; Cole, A.F.W.; Hoffman, I. 1971. A survey of atmospheric contamination following forest spraying with fenitrothion. Bull. Environ. Contam. Toxicol. 6: 289-296.

Yule, W.N.; Duffy, J.R. 1972. The persistence and fate of fenitrothion insecticide in a forest environment. Bull. Environ. Contam. Toxicol. 8: 10-17.

Yule, W.N.; Varty, I.W. 1975. The persistence and fate of fenitrothion insecticide in a forest environment. III. Deposit and residue studies with black spruce and red maple. Bull. Environ. Contam. Toxicol. 13(6): 678-680.

Chapter 57

Monitoring Insecticide Residues After Operational Forest Pest Control Programs

PIERRE-M. MAROTTE, LUC MAJOR, AND CHARLES MAISONNEUVE

Introduction

Several types of environmental monitoring programs were implemented throughout Canada from 1973 to 1988 following insecticide spraying operations. The main purpose of the follow-up was to make sure that the insect control operations were environmentally safe. Consequently, there have been a number of studies, some of which have involved the use of living organisms that are particularly susceptible to the effects of the insecticides.

In this period, growing public concern over environmental issues forced agencies involved in pest control operations to initiate follow-up studies to determine the fate of pesticide residues. The development of laboratory techniques that allowed detection of very low levels has enabled researchers to detect changes in insecticide behavior, to determine accumulation of these substances after repeated applications, and to validate some palliative measures designed to protect sensitive areas, and so forth. These findings have also been used within the framework of risk analysis to predict the effects of insecticides on some organisms.

Over the last 15 years, several environments have been examined. We have concentrated, however, on four of these: i.e. water, foliage, soil, and air. In this survey, we shall deal with only three major insecticides: fenitrothion, aminocarb, and *Bacillus thuringiensis* (Bt). These products have been used primarily as a means of controlling the spruce budworm, *Choristoneura fumiferana*.

In this paper, we discuss the main follow-up studies conducted in Canada on the fate of residues in the environment after spraying operations. The most important results are given, by study, in the tables. Maximum values are mentioned more often than mean values, which were only occasionally provided by these investigations. Furthermore, it was necessary to consult technical reports since most of the follow-up studies were not published in scientific journals.

Water

Water samples were taken in most of the environmental follow-up studies of insecticide spraying programs. From a scientific point of view, water is an important substrate because it harbors a multitude of organisms (Figs. 1 and 2).

Since water is directly linked to human needs because of its role in our survival and recreational activities, we have become particularly aware of this environment. As an indication of its importance, protection zones have been established around bodies of water in most provinces (Kingsbury and Trial 1987).

Because of the diversity of habitat studies, a synthesis of the environmental follow-up studies proved to be a highly complex matter. There were references to lakes, pools, ponds, swamps, streams, rivers, estuaries, fish hatcheries, and groundwater tables as well as mention of a wide range of sampling delays. For the sake of simplification, the results were grouped into two types of environment, lotic (flowing water) and lentic (still water) and only certain significant sampling delays were included.

Figure 1. As part of the environmental monitoring program, samples were collected in various environments to assess the residual quantities of insecticides.

Figure 2. Water sampling in a lake after insecticide spraying.

Table 1a. Maximum fenitrothion concentrations in water samples collected within treated areas following spraying operations.

Location	Year	Sampling delay (hr)	Concentration (μg/L)	Reference
		Lotic Environment		
Newfoundland	1977	3-6	49.0	Environ. Monit. Committee 1979b
New Brunswick	1973	1	6.38	Eidt and Sundaram 1975
	1976	24	42.0	Lord et al. 1978
	1979	12	0.64	Montreal Engineering 1980
		48	0.11	
	1980	0-408	20.0	Mallet and Volpé 1982
	1981	24	1.10	Mallet and Cassista 1982
		120	<0.01	
	1982	1	1.84	Sundaram et al. 1982
		24	0.33	
Quebec	1975	2	8.72	Mamarbachi and St-Jean 1976
		24	3.50	
	1978	1	2.36	Mathieu et al. 1979
		96	<0.01	
	1979	1-4	127.0	Morin et al. 1986
	1980	1-4	124.0	
	1981	1-4	14.2	
	1982	1-4	55.1	
	1983	1-4	5.93	Dostie et al. 1985
	1983	1-2	9.08	Major 1983
	1984	5	7.43	Mamarbachi et al. 1985
		168	0.055	
		Lentic Environment		
New Brunswick	1980	?	11.8	Mallet and Volpé 1982
Quebec	1979	1-4	13.9	Morin et al. 1986
	1980	1-4	74.9	
	1981	1-4	583.0	
	1982	1-4	1114.0	
	1983	6	6.68	Dostie et al. 1985
	1984	36	202.0	Mamarbachi et al. 1985
		168	0.033	
		456	<0.010	

Fenitrothion

The mean concentrations detected in the waterways within the areas treated in Quebec between 1979 and 1983 ranged from 4.04 to 20.5 μg/L, a few hours after spraying (Morin et al. 1986; Dostie et al. 1985). These mean values constitute a more accurate reflection of the concentrations generally encountered in a lotic environment than the maximum values of 124 and 127 μg/L reported in one of these studies. These maximum values were based on samples taken at the edge of waterways where the absence of a current provided the conditions typical of a lentic environment. If the latter values are not taken into account, the maximum concentrations reported in the literature ranged from 1.84 to 55.1 μg/L. One day after application, the amount of fenitrothion detected in one sample was still 49 μg/L. Levels decreased rapidly thereafter, with the result that the maximum value after 1 week was 0.055 μg/L (Table 1a).

In the waterways located outside the treated areas, maximum concentrations were detected immediately after spraying. They were lower, however, than levels found in the sprayed sites, since the highest value, recorded 3.5 km from the treatment site, was 11.9 μg/L (Table 1b).

Concentrations were generally higher in a lentic than in a lotic environment. The maximum values of 202 to 1 114 μg/L

Table 1b. Maximum fenitrothion concentrations in water samples collected outside treated areas following spraying operations.

Location	Year	Sampling delay (hr)	Concentration (μg/L)	Distance from treated area (km)	Reference
			Lotic Environment		
New Brunswick	1981	<24	0.91	4.0	Mallet and Cassista 1984
		288	<0.01		
Quebec	1978	1	2.63	1-42	Mathieu et al. 1979
		96	0.019	1-42	
	1981	<24	3.79	1.2	Morin et al. 1983
	1982	1	11.9	0-3.4	Morin et al. 1984
	1984	12	0.137	35.0	Mamarbachi et al. 1985
	1985	24	0.356	21.5-35.0	Dostie and Parent 1986
		168	<0.01	21.5-35.0	
	1986	6	6.03	1.5-3.0	Marotte and Cimon 1987
			Lentic Environment		
Quebec	1981	<24	1.72	2-4.5	Morin et al. 1983
	1982	2	5.92	2-3	Morin et al. 1984
	1983	1-2	0.175	4.6	Dostie et al. 1985
	1986	12	<0.01	8.8	Delisle et al. 1986b

recorded immediately after spraying operations in Quebec were probably based on surface samples and are not considered representative of the body of water in the lakes affected. The mean values of 7.29 to 109.1 μg/L obtained in Quebec from 1979 to 1982 (Morin et al. 1986) are in all likelihood a more accurate reflection of the amount of fenitrothion that may normally be found in a lentic environment. Furthermore, concentrations drop rapidly and maximum values of 0.033 μg/L were recorded after 1 week. Less than 5 km from the areas treated, maximum concentrations were still lower, ranging from 0.175 to 5.92 μg/L in the hours following spraying.

Aminocarb

The mean concentrations encountered in the waterways within the areas treated from 1979 to 1983 ranged from 1.22 to 2.40 μg/L (Quebec), a few hours after spraying (Morin et al 1986; Dostie et al 1985). Maximum concentrations were normally detected within hours of spraying and rose as high as 34.0 μg/L (Newfoundland). They dropped rapidly thereafter, and the maximum values 1 and 5 days later were 0.998 μg/L (Quebec) and 0.1 μg/L (Ontario), respectively (Table 2a).

Values were generally higher in a lentic environment. The mean values recorded in this type of environment within the areas treated in Quebec between 1979 and 1983 ranged from 3.0 to 16.2 μg/L a few hours after the operations (Morin et al. 1986; Dostie et al. 1985). Maximum concentrations shortly after spraying were 7.95 to 331.0 μg/L. In addition, the product persisted longer in a lentic environment and concentrations of up to 9.54 μg/L were recorded after 3 days, with traces still present after 1 month.

Aminocarb was generally present, although in smaller concentrations, in bodies of water and waterways outside the areas treated (Table 2b). The maximum values recorded immediately after spraying ranged from 2.11 to 13.5 μg/L in a lotic environment and from 0.133 to 1.62 μg/L in a lentic environment. After 1 week, they were under 0.01 μg/L in a lotic environment but as high as 0.9 μg/L in a lentic environment.

Bt

Although Bt has only recently been introduced on an operational scale, it has already been the focus of several monitoring programs. One of the first studies (Beak Consultants Ltd. 1980) expressed results in colony-forming units (CFU) per square centimetre, and the others, in CFUs per litre of water, making any comparison difficult.

As with chemical insecticides, the highest concentrations were obtained in the hours following application. Thus, in the waterways within the areas treated in Quebec, maximum and minimum concentrations ranged from 330×10^4 to 860×10^4 CFU/L and from 33×10^4 to 140×10^4 CFU/L respectively (Table 3). Even though values subsequently dropped, the spores seemed to persist since concentrations of up to 4.6×10^4 CFU/L were detected 11 to 50 days after application and were still present after 11 months (0.329×10^4 CFU/L).

Concentrations found in the waterways outside the treated areas were generally lower than within the areas. The maximum levels recorded were 250×10^4 CFU/L, 12 km from an area treated less than 1 day before. The spores persist for some period of time since concentrations of up to 44×10^4 CFU/L were found after 6 days in a sample taken 18 km from the blocks.

The values obtained in a lentic environment were generally higher than those recorded in a lotic environment. Maximum values in lentic waters within the treatment area were $3\,530 \times 10^4$ CFU/L 1 hour after the application. Concentrations subsequently decreased to a maximum of 5.45×10^4 CFU/L after 11 months. Contamination was lower outside these blocks. The maximum value reported (i.e. 10.4×10^4 CFU/L) was obtained in a sample collected less than 5 km from the treated areas. After 11 months, concentrations of up to 0.12×10^4 CFU/L could still be detected at this distance.

Table 2a. Maximum aminocarb concentrations in water samples collected inside treated areas following spraying operations against the spruce budworm, *Choristoneura fumiferana*.

Location	Year	Sampling delay (hr)	Concentration (µg/L)	Reference
		Lotic Environment		
Newfoundland	1978	6	34.0	Environ. Monit. Committee 1979a
New Brunswick	1979	12	1.0	Montreal Engineering 1980
Ontario	1980	1	<0.1	Beak Consultants Ltd. 1980
		24	0.18	
		120	0.1	
Quebec	1978	5	3.34	Mathieu et al. 1979
		48	0.075	
	1979	1-4	8.92	Morin et al. 1986
	1980	1-4	6.95	
	1981	1-4	18.4	
	1981	0.5	0.639	Marotte 1982
		7.5	0.211	
	1982	1	8.63	Morin et al. 1984
	1982	1-4	6.46	Morin et al. 1986
	1983	1-2	9.31	Dostie et al. 1985
	1984	<1	30.5	Mamarbachi et al. 1985
		24	0.998	
		96	0.066	
		Lentic Environment		
Newfoundland	1978	36	81.0	Environ. Monit. Committee 1979a
Quebec	1978	1	12.7	Mathieu et al. 1979
		48	1.99	
	1979	1-4	25.0	Morin et al. 1986
	1980	1-4	7.95	
	1981	1-4	39.1	
	1982	1-4	331.0	
	1983	4	18.6	Dostie et al. 1985
		72	9.54	
	1984	1	35.9	Mamarbachi et al. 1985
		120	1.09	
		720	0.011	

Table 2b. Maximum aminocarb concentrations in water samples collected outside treated areas following spraying operations against the spruce budworm, *Choristoneura fumiferana*.

Location	Year	Sampling delay (hr)	Concentration (μg/L)	Distance from treated area (km)	Reference
			Lotic Environment		
Quebec	1978	5	2.11	1-42	Mathieu et al. 1979
		48	0.071	1-42	
		96	<0.01	1-42	
	1982	1	13.5	1.0-3.4	Morin et al. 1984
	1984	24	0.013	35.0	Mamarbachi et al. 1985
		144	<0.01	7.5 and 35.0	
			Lentic Environment		
Newfoundland	1977	168	0.9	7.2	Environ. Monit. Committee 1979b
Quebec	1979	2	0.133	5.4	Gaboury 1980
	1982	2	1.62	1.8-3.6	Morin et al. 1984

Table 3. Concentrations of the biological insecticide Bt in water samples collected following spraying operations.

Location	Year	Sampling delay (hr)[a]	Concentration (CFU × 10^4/L)[b] Min.	Max.	Distance from treated area (km)	Reference
			Lotic Environment			
Ontario	1980	0.5	360/cm^2	2 100/cm^2		Beak 1980
		72	460/cm^2	718/cm^2		
		32 days	44/cm^2	95/cm^2		
Quebec	1984	0.5	51.5	500.0		Dugal et al. 1985
		24	1.5	110.0		
		11 months	0.0034	0.329		
	1985	0.33	140.0	860.0		Delisle et al. 1986a
		2	33.0	330.0		
		11-50 days	0.022	4.6		
	1985	9-14	0.04	37.0	0.1-5.0	Cimon et al. 1986
		24	0.044	0.63	0.1-5.0	
		11 months	N.D.[c]	0.0035	2.8-4.8	
	1985	<24		250.0	12.0	Dostie et al. 1986a
		24		50.0	18.0	
		144		44.0	18.0	
			Lentic Environment			
Quebec	1984	1	48.5	3 530		Dugal et al. 1985
		24	9.5	1 005		
		1 month	1.67	61.5		
		11 months	0.0135	5.45		
	1985	48-72	7.6	380.0		Dostie et al. 1986b
		11-40 days	0.079	8.2		
		11 months	<0.01	0.37		
	1985	11-40 days	<0.01	0.47	2.7	
	1985	9-14	0.011	10.4	1.2-4.9	Cimon et al. 1986
		24	0.045	4.2	0.8-4.9	
		11 months	0.0069	0.12	8-4.9	
	1986	72-84	N.D.[c]	0.72	8.6	Delisle et al. 1986b

[a] The delay between spraying and sampling is given in hours unless otherwise noted.
[b] The concentration is expressed as the number of colony-forming units (CFU) per litre, unless otherwise noted.
[c] The CFU counts were below detectable levels.

Table 4. Fenitrothion concentrations in coniferous foliage (unless otherwise noted) following spraying operations against the spruce budworm, *Choristoneura fumiferana*.

Location	Year	Sampling delay	Concentrations (μg/g d.w.)[a] Mean	Maximum	Reference
Quebec	1975	A few hours	2.8 (f.w.)		Coulombe et al. 1976
		1-1.5 months	0.23 (f.w.)		
	1979	A few hours	2.58		Gaboury 1980
	1979	A few hours	2.87	7.97	Morin et al. 1986
	1980	A few hours	12.01	111.0	
	1980	1 year		3.53	Morin et al. 1983
	1981	A few hours	6.71	37.1	Morin et al. 1986
	1981	A few hours		5.01	Morin et al. 1984
		4 months		1.53	
		1 year		1.48	
		A few hours		29.0 (decid.)	
New Brunswick	1973	A few hours	6.25		Yule and Varty 1975
		128 days	0.03		
	1981	A few hours		0.53 (f.w.)	Mallet and Cassista 1984
		2 weeks		0.43 (f.w.)	
	1982	8 months		0.41 (f.w.)	Ayer et al. 1984
		20 months		0.26 (f.w.)	
		32 months		0.32 (f.w.)	
	1983	A few hours	0.76		Eidt and Mallet 1986
		1 year	0.63		
		2 years	0.27		

[a]Concentrations are expressed as micrograms per gram of dry foliage (μg/g d.w.) unless otherwise noted; decid. = deciduous, f.w. = fresh weight.

Foliage

Insecticide concentrations in the foliage and their persistence after treatment are useful indicators of environmental contamination following spraying operations. Several animal species may ingest or come in contact with these residues and suffer certain effects that, in turn, may have an impact on the food chain. McNeil and McLeod (1977), for instance, observed acute susceptibility in the Swaine jack pine sawfly, *Neodiprion swainei*, with the result that very low concentrations of fenitrothion on jack pine foliage produced a high mortality rate.

The following studies were conducted mainly in balsam fir stands (Fig. 3), from 1979 to 1983. They were carried out in New Brunswick, Quebec, Ontario, and Newfoundland.

Fenitrothion

Table 4 shows that mean concentrations of fenitrothion in the foliage a few hours after spraying ranged from 0.76 to 12.01 μg/g, with a maximum value of 111.0 μg/g. Values decreased substantially with time, but small amounts still persisted for more than 1 year. In fact, residues of up to 1.48 and 1.69 μg/g were detected 1 year after operations. The study involving deciduous trees included in Table 4 seems to indicate the same type of product behavior in broadleaf trees (i.e. initially high concentrations [29.0 μg/g] dropping significantly in the months that follow).

Aminocarb

In the case of aminocarb, the mean values ranged from 1.01 to 9.82 μg/g a few hours after application and maximum values rose to 37.3 μg/g (Table 5). As with fenitrothion, concentrations decreased dramatically in the months that followed, and very low concentrations were found 1 year after application (maximum: 0.09 μg/g). In deciduous trees, the initial concentration was high (32.2 μg/g), but after 4 months, this value dropped to a maximum of 8.9 μg/g.

Bt

The only available study on Bt residues in the foliage was conducted in Ontario in 1980 (Beak Consultants Ltd. 1980). A few hours after application, the maximum count was 1.56×10^6 CFU/cm^2. One month later, the maximum concentration was only 1 290 CFU/cm^2.

Soil

Forest soil generally consists of an organic horizon on top of mineral soil. The litter and humus layers forming the organic horizons described in this section are primarily made up of conifer needles. They contain most of the microbial population responsible for breaking down the organic material.

Table 5. Aminocarb concentrations in coniferous foliage (unless otherwise noted) following spraying operations against the spruce budworm, *Choristoneura fumiferana*.

Location	Year	Sampling delay	Concentrations (μg/g d.w.)[a] Mean	Maximum	Reference
Quebec	1979	A few hours	1.01	5.5	Morin et al. 1986
	1980	A few hours	9.82	37.3	
	1980	1 year		0.03	Morin et al. 1983
	1981	A few hours	2.21	26.2	Morin et al. 1986
	1981	4 months		0.61	Morin et al. 1984
		1 year		0.09	
		A few hours		32.2 (decid.)	
		4 months		8.9 (decid.)	
	1982	1 3 hr	2.14	20.0	Morin et al. 1986
	1983	1-2 hr	1.92	13.6	Dostie et al. 1985
Newfoundland	1978	A few hours		5.4 (f.w.)	Environ. Monit. Committee 1979a
		22 days		1.3 (f.w.)	
	1978	1 year		0.04	Gov. Nfld.-Labr. 1980
Ontario	1980	A few hours		2.24 (f.w.)[b]	Beak 1980
		60 days		<0.01 (f.w.)[b]	

[a] Concentrations are expressed as micrograms per gram of dry foliage (μg/g d.w.) unless otherwise noted; decid. = deciduous, f.w. = fresh weight.
[b] A mixture of coniferous and deciduous foliage.

Irrespective of whether insecticides have drifted directly to the ground from the air or have been borne by contaminated leaves falling to the ground, they may upset mineral element cycles and affect soil fertility. Furthermore, insecticides may penetrate the mineral soil and contaminate the groundwater tables.

Consequently, it is important to know the level of persistence of the various insecticides and the risks of accumulation in the soil. Several programs monitoring spraying operations included sampling of forest soils (Fig. 4). The samples were usually collected with a corer and generally contained only the organic horizon. In some instances, however, a sample of the mineral soil was also taken.

Insecticide residues in the soil are often expressed in terms of the amount of insecticide per unit of soil weight. Thus, for chemical insecticides, the results are given in micrograms of active ingredient per gram of soil. For biological insecticides, the results of the initial studies, which tended to be based on deposits, were given in colony-forming units per unit of area. Subsequently, determinations were expressed in colony-forming units per unit of soil weight in the interests of achieving a better understanding of the residues present in the environment.

When interpreting the results, it must be remembered that the data were sometimes expressed on a fresh weight basis and at other times, on a dry weight basis, and in some instances no mention was made of this factor. Concentrations may vary considerably, therefore, depending on the amount of water contained in the sample.

Fenitrothion

Only two studies provided maximum fenitrothion values less than 2 weeks after spraying. These studies were conducted in New Brunswick and Quebec and yielded values of 0.19 and 0.093 μg/g, respectively, in the organic portion. There was only one instance of follow-up after 4 months and it revealed a maximum value of 0.066 μg/g for the same horizon. This concentration was higher than the maximum values recorded 1 year after spraying (i.e. 0.018 μg/g) (Table 6).

Analysis of mineral soil revealed either the absence of detectable levels of fenitrothion or that levels were lower than found in the organic horizon. The studies conducted in the two provinces concurred in their finding that the accumulation of fenitrothion in the soil did not seem to be related to the number of treatments (Morin et al. 1983; Ayer et al. 1984).

Aminocarb

Less than 2 weeks after spraying, maximum aminocarb concentrations in the organic horizon ranged from nondetectable levels to 0.93 μg/g (Table 7). Mean concentrations reached 0.34 μg/g. Values generally tended to decrease thereafter with the result that after 1 year, no traces of aminocarb were found in the soil in some studies, whereas in others, very small maximum concentrations (i.e. less than 0.1 μg/g) were detected. In one exceptional case, values rose to 0.631 μg/g. One study revealed a maximum value of 0.24 μg/g in the mineral portion 24 hours after the application. In another

Table 6. Fenitrothion concentrations in the soil of coniferous forests following spraying operations against the spruce budworm, *Choristoneura fumiferana*.

Location	Year	Sampling delay	Concentrations (µg/g d.w.)[a] Minimum	Maximum	Soil horizon	No. seasons treated	Reference
New Brunswick	1981	2 weeks	0.04 (f.w.)	0.19 (f.w.)	organic		Mallet and Cassista 1984
	1983	1 year	<0.005	0.008	mineral	3	Ayer et al. 1984
		2 years	<0.005	0.007	mineral	2	
		3 years	0.005	0.01	mineral	1	
Quebec	1979	<1 day		0.093	organic		Gaboury 1980
	1981	1 year	0.005	0.007	organic	1	Morin et al. 1983
		1 year	0.003	0.014	organic	2	
		1 year	0.003	0.018	organic	3	
		1 year	<0.001	<0.001	mineral	1	
		1 year	<0.001	<0.001	mineral	2	
		1 year	<0.001	<0.001	mineral	3	
	1981	4 months	0.020	0.066	organic		Morin et al. 1984
	1982	1 year	0.015	0.060	organic		

[a]Concentrations are expressed as micrograms per gram dry weight (µg/g d.w.) unless otherwise noted; f.w. = fresh weight.

Table 7. Aminocarb concentrations in the soil of coniferous forests following spraying operations against the spruce budworm, *Choristoneura fumiferana*.

Location	Year	Sampling delay	Maximum concentrations (µg/g)[a]	Soil horizon	No. seasons treated	Reference
Newfoundland	1978	2 days	0.14 (f.w.)	organic		Environ. Monit. Committee 1979a
		14 days	0.01 (f.w.)	organic		
	1979	1 year	0.011	organic		Gov. Nfld.-Labr. 1980
Ontario	1979	2 months	0.01 (f.w.)	organic		MacCallum 1980
	1980	1 day	0.19	organic		Beak 1980
		5 days	0.93	organic		
		1 month	0.47	organic		
		2 months	0.03	organic		
		1 day	0.24	mineral		
		5 days	<0.01	mineral		
Quebec	1979	A few hours	0.05 (d.w.)	organic		Gaboury 1980
	1981	1 year	<0.001 (d.w.)	organic	1	Morin et al. 1983
		1 year	0.631 (d.w.)	organic	3	
		1 year	<0.001 (d.w.)	organic	5	
		1 year	<0.001 (d.w.)	mineral	1	
		1 year	<0.001 (d.w.)	mineral	3	
		1 year	<0.001 (d.w.)	mineral	5	
	1981	4 months	0.131 (d.w.)	organic		Morin et al. 1984
		1 year	0.096 (d.w.)	organic		

[a]Maximum concentrations are expressed as micrograms per gram of soil (µg/g). In some cases it was not specified if this was on a fresh or dry weight basis; f.w. = fresh weight, d.w. = dry weight.

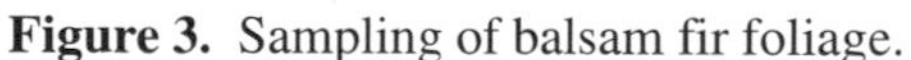

Figure 3. Sampling of balsam fir foliage.

Figure 4. Forest soil sampling using a metal core drill.

study, no insecticide concentrations could be detected in the mineral soil samples after 1 year despite 5 years of treatment.

The studies of Morin et al. (1983) suggest that the product does not tend to accumulate in the soil as a result of the number of treatments.

Bt

Maximum Bt concentrations in the organic horizon less than 2 weeks after applications were 7.42×10^6 CFU/cm^2 when expressed on a unit surface basis and 1.4×10^6 CFU/g when expressed in terms of unit of soil weight. With respect to mean concentrations, available findings mention values of 1.22×10^5 CFU/cm^2 and 2.8×10^5 CFU/g for the same interval and horizon (Table 8).

For the same period, maximum concentrations were lower (i.e. 7.4×10^4 CFU/g) in the underlying mineral soil. In the 3 months following spraying operations, concentrations in the organic layer seemed to remain stable, whereas they increased slightly in the mineral soil. Despite a significant decrease in Bt concentrations in the soil after 1 year, concentrations were still higher than pre-treatment levels. This suggests that spores may accumulate in the soil if spraying operations are repeated over time (Cardinal and Marotte 1987). Even though Cardinal and Marotte (1987) have mentioned that the pesticide can drift over a distance of 31 km, a maximum Bt concentration of only 1.4×10^4 CFU/g was detected 1.5 km outside the treated areas.

Air

Since the publication of the book edited by Prebble (1975), several studies have been carried out to determine the degree of air contamination resulting from aerial spraying of insecticides. Virtually all of these studies have been conducted within the framework of spruce budworm treatment programs in Quebec and New Brunswick. Environmental monitoring operations have been conducted during applications of fenitrothion, aminocarb, and Bt.

Sampling methods for determining insecticide concentrations in the air generally include the use of various types of suction pumps equipped with collectors (Figs. 5 and 6). In the case of chemical insecticides, these collectors may be bubblers (containing dimethylformamide or ethylene glycol) or absorbent filters (Florisil, Amberlite XAD-2, polyurethane foam). For bacteriological insecticides containing Bt, the

Table 8. Biological insecticide (Bt) concentrations in the soil following spraying operations against the spruce budworm, *Choristoneura fumiferana.*

Location	Year	Sampling delay	Concentrations (CFU)[a] Minimum	Mean	Maximum	Soil horizon	Reference
Nova Scotia	1980	3.5 days			7.5×10^3/g	organic	McBride and Coyle 1980
		8 days			19.0×10^3/g	organic	
		18 days			440.0×10^3/g	organic	
		3.5 days			<1000/g	mineral	
		8 days			74×10^3/g	mineral	
		18 days			290×10^3/g	mineral	
Ontario	1980	1 day			$7.42 \times 10^6/cm^2$	organic	Beak 1980
		5 days			$1.08 \times 10^6/cm^2$	organic	
		1 month			$1000/cm^2$	organic	
		1 day			$0/cm^2$	mineral	
		5 days			$0/cm^2$	mineral	
Quebec	1984	1-2 days	$1.06 \times 10^4/cm^2$	$1.22 \times 10^5/cm^2$	$3.80 \times 10^5/cm^2$	organic	Cardinal and Marotte 1987
		2-3 months	$9.54 \times 10^3/cm^2$	$1.07 \times 10^5/cm^2$	$4.87 \times 10^5/cm^2$	organic	
		1 year	$7.60 \times 10^3/cm^2$	$0.87 \times 10^5/cm^2$	$3.64 \times 10^5/cm^2$	organic	
	1985	1-3 days	1.6×10^4/g d.w.	2.8×10^5/g d.w.	1.4×10^6/g d.w.	organic	Cardinal and Marotte 1988
		12-42 days	3.6×10^4/g d.w.	2.3×10^5/g d.w.	1.2×10^6/g d.w.	organic	
		1 year	1.2×10^4/g d.w.	1.4×10^5/g d.w.	5.5×10^5/g d.w.	organic	
		1 month	$<0.1 \times 10^4$/g d.w.	1.3×10^3/g d.w.	6.5×10^3/g d.w.	mineral	
		1 year	$<0.1 \times 10^4$/g d.w.	4.3×10^3/g d.w.	2.3×10^4/g d.w.	mineral	

[a]Concentrations were expressed as the number of colony-forming units (CFU) per gram (g) of soil, per square centimetre (cm^2) of soil surface, or per gram of dry soil (g d.w.).

Table 9. Fenitrothion concentrations in the air following spraying operations against the spruce budworm, *Choristoneura fumiferana.*

Location	Year	Sampling delay	Concentrations (µg/m³) Mean	Maximum	Distance from treated area (km)	Reference
Quebec	1982	0-30 min	23.3	58.7	within area	St-Louis et al. 1983
	1983	0-120 min	9	26.0	within area	Major et al. 1984a
	1984	0-120 min		12.1	within area	Major et al. 1985a
	1985	~16 hr		1.23	within area	Dugal and Major 1986
		~2.5 days		0.35	within area	
New Brunswick	1978	0-120 min	5.8		within area	Wood 1980
		36 hr	0.04		within area	
	1980			12.0[a]	within area	Mallet and Volpé 1982
	1982	0-60 min		2.0	within area	Sundaram 1984
Quebec	1982	0-30 min	5.26	13.5	2	St-Louis et al. 1983
	1984	0-120 min		0.6	3	Major et al. 1985a
	1984			<0.01	>8[b]	Major et al. 1984b
	1985			0.007	>25[b]	Major et al. 1986b
	1986			0.044	>8[b]	Dugal and Major 1987
New Brunswick	1980			0.05	3-7[b]	Mallet and Cassista 1984
	1981			0.08	variable	Wood 1981
	1981			441	1	Wood 1982

[a]The residue consisted of aminofenitrothion, a metabolite of fenitrothion.
[b]These figures represent the shortest distance between the sampling station and the closest treated area. Part of each concentration, however, may be the result of treatments carried out much farther away.

Table 10. Aminocarb concentrations in the air following spraying operations against the spruce budworm, *Choristoneura fumiferana*.

Location	Year	Sampling delay (min)	Maximum concentrations (µg/m³)	Distance from treated area (km)	Reference
Quebec	1982	0-30	15.3	within area	St-Louis et al. 1983
	1983	0-30	24.6	within area	Major et al. 1984a
	1984	0-120	18	within area	Major et al. 1985a
New Brunswick	1982	180	1.20	within area	Sundaram 1984
Quebec	1982	0-30	3.92	2[a]	St-Louis et al. 1983
	1983	30-60	1.94	2	Major et al. 1984a
	1984		<0.01	>8	Major et al. 1984b
	1984	0-120	0.47	3	Major et al. 1985a
Ontario	1980		0.14	outside area	Carrow et al. 1980

[a]This figure represents the shortest distance between the sampling station and the closest treated area. Part of the concentration, however, may be the result of treatments carried out much farther away.

Table 11. Biological insecticide (Bt) concentrations in the air following spraying operations against the spruce budworm, *Choristoneura fumiferana*.

Location	Year	Sampling delay	Concentrations (CFU/m³)[a]		Distance from treated area (km)	Reference
			Minimum	Maximum		
Quebec	1984	2 months		presence	within area	Bastille et al. 1985
	1984	26 days		presence	nearby	Bastille et al. 1984
	1984			94.0	>6[b]	Major et al. 1985b
	1984		84% < 0.8	938	1-10[b]	Major et al. 1986a
	1984			6.7	2-13[b]	Dugal and Major 1987
	1984		95% < 1.0	59.5	>5[b]	Dugal 1988
	1985	<1 day		>100	within area	Bastille et al. 1985
	1985	0-120 min	38	2 316	within area	Dugal et al. 1986

[a]Concentrations are expressed as the number of colony-forming units per cubic metre of air (CFU/m³).
[b]These figures represent the shortest distance between the sampling station and the closest treated area. Part of each concentration, however, may be the result of treatments carried out much farther away.

collectors used are Millipore® filter membranes (pore size: 0.45 µm) or nutrient agars. During aerial applications, the insecticide is sprayed in the form of a cloud made up of fine droplets. Ideally, the aerosol should be sampled according to isokinetic laws, especially near the spray area where an abundance of the largest droplets is found. According to this principle, the rate at which the air is drawn into the sampling tube must be equal to the rate at which the cloud is displaced (wind velocity) with the tube facing the wind. When these conditions are met, results are representative of the concentrations in the air. As it is extremely difficult to carry out this type of sampling procedure during follow-up work, virtually all of the studies described here were conducted with permanently oriented fixed-flow-volume equipment. Even so, they reliably indicate the degree of contamination of the air following spraying operations.

Fenitrothion and Aminocarb

Tables 9 and 10 provide a synopsis of chemical insecticide monitoring studies. Examination of these tables shows that within the areas treated, maximum concentrations shortly after application (less than 120 minutes) ranged from 2 to 59 µg/m³ air in the case of fenitrothion and from 15 to 24.6 g/m³ in

Figure 5. Apparatus used to assess air contamination in sectors treated for spruce budworm.

Figure 6. Pump used to measure the concentration of insecticide in the air outside the treated sectors.

the case of aminocarb. It is interesting that Mallet and Volpé (1982) found 12 µg/m^3 aminofenitrothion (a fenitrothion metabolite) within the treated area.

Inside the treated areas, most of the product was detected within the first half hour. Thereafter, concentrations dwindled rapidly (Wood 1982; Major et al. 1984a; Dugal et al. 1986).

Outside these areas, the concentrations detected were much lower but highly variable. However, Wood (1982) recorded very high maximum values of 441 µg/m^3 fenitrothion 1 km from the treated area.

Bt

There were far fewer studies on contamination of the air by biological insecticides (Table 11). Within the treated areas, Dugal et al. (1986) obtained maximum Bt counts of 2 316 CFU/m^3 within the first 2 hours of application, and Bastille et al. (1985) confirmed the presence of Bt spores in the air 2 months after the end of the spraying program in Quebec. Outside the treated areas concentrations were generally very low, as with the chemical insecticides. In studies designed to determine the degree of contamination of the air in municipalities close to the treated areas in Quebec, under 1 CFU/m^3 air was reported in 63, 87, 91, and 95 percent of the results obtained in 1984, 1985, 1986, and 1987, respectively (Dugal 1988). Major et al. (1986a), however, found 938 CFU/m^3 air in a sample that was collected during one of these studies. These same authors also found traces of Bt more than 50 km from the treated areas.

Discussion

Regardless of the type of insecticide used, concentrations in the air vary greatly after aerial spraying operations. This variability is observed not only between treated plots but also within the same plots (Dugal and Major 1986). Significant correlations, moreover, have been recorded between concentrations in the air and deposits on the ground in the same area (Major et al. 1984a, 1985a; Dugal and Major 1986). This variability may be caused by a number of factors. The most commonly mentioned include temperature, moisture content, wind, local topography, forest cover, location of the station, type of plane used, altitude of flight, area treated, spraying equipment, formulation of the product, and time of treatment. Sundaram et al. (1983) suggested that water-based formulations yielded higher concentrations in the air than oil-based formulations. The same authors also noted greater concentrations with evening applications. They attributed this phenomenon to an increase in volatilization with higher temperatures. Dugal and Major (1987) found that the use of single-engine planes instead of four-engine ones seemed to promote lower Bt concentrations in the air in municipalities close to the treated sites.

Studies also showed that storing and handling insecticides at operational bases caused only very slight contamination of the air in nearby inhabited areas (Major et al. 1984b, 1986b).

References

Ayer, W.C.; Brun, G.L.; Eidt, D.C.; Ernst, W.; Mallet, V.N.; Matheson, R.A.; Silk, P.J. 1984. Persistence of aerially applied fenitrothion in water, soil, sediment and balsam fir foliage. Can. For. Serv., Marit. For. Res. Cent., Fredericton, N.B. Inf. Rep. M-X-153. 14 p.

Bastille, A.; Boily, Y.; Boudreau, R.; Bouliane, G.; Nadeau, A.; Patry, L. 1984. Programme de surveillance médico-environmentale des pulvérisations aériennes d'insecticides biologiques *Bacillus thuringiensis* var. *kurstaki* contre la tordeuse des bourgeons de l'épinette. D.S.C. Rivière-du-Loup. Rapport synthèse. 68 p.

Bastille, A.; Laferrière, M.; Leclerc, J.C.; Nadeau, A. 1985. Programme de surveillance médico-environmentale des pulvérisations aériennes d'insecticides biologiques *Bacillus thuringiensis* var. *kurstaki* contre la tordeuse des bourgeons de l'épinette: Territoire du département de santé communautaire de Rivière-du-Loup, Québec. D.S.C. Rivière-du-Loup. Rapport 1985. 66 p.

Beak Consultants Ltd. 1980. Environmental monitoring of the Ontario Ministry of Natural Resources Spruce Budworm Spraying Program, Kirkland Lake, 1980. Ont. Minist. Nat. Resour. Cochrane, Ont. 107 p.

Cardinal, P.; Marotte, P.-M. 1987. Persistance des spores de *Bacillus thuringiensis* dans le sol forestier suite aux pulvérisations aériennes contre la tordeuse des bourgeons de l'épinette au Québec en 1984. Gouvernement du Québec, ministère de l'Énergie et des Ressources, Direction de la conservation, Service des études environnementales. 40 p.

Cardinal, P.; Marotte, P.-M. 1988. Persistance du *Bacillus thuringiensis* dans le sol forestier au Québec. Gouvernement du Québec, ministère de l'Énergie et des Ressources, Direction de la conservation, Service des études environnementales. 63 p.

Carrow, J.R.; Nicholson, S.A.; Howse, G.M. 1980. Spruce budworm spraying in Ontario, 1980. Ont. Minist. Nat. Resour., Forest Res. Branch, Pest Control Rep. No. 10. 32 p.

Cimon, A.; Marotte, P.-M.; Dostie, R. 1986. Surveillance des prises d'eau potable situées à proximité des aires traitées au *Bacillus thuringiensis* contre la tordeuse des bourgeons de l'épinette en 1985. Gouvernement du Québec, ministère de l'Énergie et des Ressources, Service des études environnementales. 17 p.

Coulombe, L.; Noel, S.; Mamarbachi, G.; St. Jean, R. 1976. Lutte contre la tordeuse des bourgeons de l'épinette. 1. Distribution et persistance des résidus de fénitrothion dans le feuillage - arrosage été 1975. Gouvernement du Québec, Service de protection de l'environnement. 29 p.

Delisle, S.; Dostie, R.; Marotte, P.-M. 1986a. Comportement du *Bacillus thuringiensis* en milieu lotique à l'intérieur des aires traitées. Gouvernement du Québec, ministère de l'Énergie et des Ressources, Service des études environnementales. 16 p.

Delisle, S.; Dostie, R.; Parent, G. 1986b. Surveillance des prises d'eau potable réalisée dans le cadre du suivi environnemental des pulvérisations aériennes d'insecticides contre la tordeuse des bourgeons de l'épinette au Québec en 1986. Gouvernement du Québec, ministère de l'Énergie et des Ressources, Direction de la conservation, Service des études environnementales. 20 p.

Dostie, R.; Marotte, P.-M.; Parent, G. 1985. Résidus de fénitrothion et d'aminocarbe dans l'eau et le feuillage. Gouvernement du Québec, Direction de la conservation, Service des études environnementales. 36 p.

Dostie, R.; Parent, G. 1986. Comportement du fénitrothion (Sumithion[MD]) en milieu lotique à l'extérieur des aires traitées. Gouvernement du Québec, ministère de l'Énergie et des Ressources, Service des études environnementales. 14 p.

Dostie, R.; Parent, G.; Delisle, S.; Marotte, P.-M. 1986a. Comportement du *Bacillus thuringiensis* en milieu lotique à l'extérieur des aires traitées. Gouvernement du Québec, ministère de l'Énergie et des Ressources, Service des études environnementales. 43 p.

Dostie, R.; Parent, G.; Marotte, P.-M.; Delisle, S. 1986b. Concentration de *Bacillus thuringiensis* dans l'eau des lacs situés à l'intérieur et à l'extérieur des aires traitées contre la tordeuse des bourgeons de l'épinette au Québec en 1985. Gouvernement du Québec, ministère de l'Énergie et des Ressources, Direction de la conservation, Service des études environnementales. 21 p.

Dugal, J. 1988. Concentration d'insecticides biologiques dans l'air de certaines municipalités à la suite des pulvérisations aériennes contre la tordeuse des bourgeons de l'épinette en 1987. Gouvernement du Québec, ministère de l'Énergie et des Ressources, Direction de la conservation, Service des études environnementales. 22 p.

Dugal, J.; Major, L. 1986. Concentrations résiduelles de fénitrothion dans l'air à l'intérieur des secteurs traitées contre la tordeuse des bourgeons de l'épinette en 1985. Gouvernement du Québec, ministère de l'Énergie et des Ressources, Direction de la conservation, Service des études environnementales. 27 p.

Dugal, J.; Major, L. 1987. Résidus d'insecticides dans l'air de certaines municipalités suite aux pulvérisations aériennes contre la tordeuse des bourgeons de l'épinette en 1986. Gouvernement du Québec, ministère de l'Énergie et des Ressources, Direction de la conservation, Service des études environnementales. 28 p.

Dugal, J.; Major, L.; Cardinal, P.; Delisle, S. 1985. Étude de la persistance des insecticides biologiques dans les eaux naturelles en milieu forestier. Gouvernement du Québec, ministère de l'Énergie et des Ressources, Service d'entomologie et pathologie. 29 p.

Dugal, J.; Major, L.; Rousseau, G. 1986. Concentrations d'insecticides biologiques dans l'air à l'intérieur des aires traitées. Gouvernement du Québec, ministère de l'Énergie et des Ressources, Service des études environnementales. 19 p.

Eidt, D.C.; Mallet, V.N. 1986. Accumulation and persistence of fenitrothion in needles of balsam fir and possible effects on abundance of *Neodiprion abietis* (Harris) (Hymenoptera: Diprionidae). Can. Entomol. 118: 69-77.

Eidt, D.C.; Sundaram, K.M.S. 1975. The insecticide fenitrothion in headwaters streams from large-scale forest spraying. Can. Entomol. 107: 735-742.

Environmental Monitoring Committee. 1979a. 1978 Environmental monitoring of the spruce budworm spraying program in Newfoundland. Gov. Newfoundland & Labrador, Dep. Consumer Affairs Environ. Inf. Rep. RA-79-2. 29 p.

Environmental Monitoring Committee. 1979b. 1977 Environmental monitoring of the spruce budworm spraying program in Newfoundland. Gov. Newfoundland & Labrador, Dep. Consumer Affairs Environ. Inf. Rep. RA-79-1. 21 p.

Gaboury, G. 1980. Surveillance environnementale des pulvérisations aériennes 1979 contre la tordeuse des bourgeons de l'épinette au Québec. Gouvernement du Québec, Rapport préliminaire, ministère de l'Énergie et des Ressources, Direction de la conservation, Service d'entomologie et de pathologie. 13 p.

Government of Newfoundland and Labrador. 1980. 1979 Environmental monitoring of the spruce budworm spraying program in Newfoundland. Dep. Consumer Affairs Environ. Inf. Rep. RA-80-1. 18 p.

Kingsbury, P.D.; Trial, J.G. 1987. Buffer zones: their application to forest insect control operations. Proc. Buffer Zone Workshop sponsored by the Eastern Spruce Budworm Council's Environmental Committee, Quebec City, 16-17 April 1986. 124 p.

Lord, D.A.; Matheson, R.A.F.; Stuart, L.; Swiss, J.J.; Wells, P.G.; 1978. Environmental monitoring of the 1976 spruce budworm spraying program in New Brunswick. Environ. Protection Serv., Halifax, Atlantic Region. EPS-5-AR-78-3. 161 p.

MacCallum, M.E. 1980. The short-term environmental impact of the spruce budworm spray program in Ontario 1979. Ont. Minist. Nat. Resour., For. Res. Branch, Pest Control Sect. 58 p.

Major, L. 1983. Surveillance des pulvérisations aériennes d'insecticides contre le Diprion de Swaine au Québec en 1983: résidus de fénitrothion dans l'eau et les bleuets; évaluation du dépôt. Gouvernement du Québec, ministère de l'Énergie et des Ressources, Direction de la conservation, Service d'entomologie et pathologie. 24 p.

Major, L.; Mamarbachi, G.; Rousseau, G.; St-Louis, N.; Adley, R. 1984a. Surveillance des pulvérisations aériennes d'insecticide contre la tordeuse des bourgeons de l'épinette au Québec en 1983: échantillonnage de l'air. Gouvernement du Québec, ministère de l'Énergie et des Ressources et Eco-Recherches Inc. 41 p.

Major, L.; Rousseau, G.; Mamarbachi, G. 1984b. Résidus d'insecticides chimiques dans l'air de certaines agglomérations urbaines de la région 01. Gouvernement du Québec, ministère de l'Énergie et des Ressources, Service d'entomologie et pathologie. 14 p.

Major, L.; Rousseau, G.; Mamarbachi, G. 1985a. Concentrations d'insecticides chimiques dans l'air à l'intérieur et à l'extérieur (3 km) des aires traitées. Gouvernement du Québec, ministère de l'Énergie et des Ressources, Service d'entomologie et pathologie. 41 p.

Major, L.; Rousseau, G.; Lamontagne, B. 1985b. Concentrations d'insecticides biologiques dans l'air de deux agglomérations urbaines de la région 01. Gouvernement du Québec, ministère de l'Énergie et des Ressources, Service d'entomologie et pathologie. 23 p.

Major, L.; Rousseau, G.; Cardinal, P. 1986a. Concentrations d'insecticides biologiques dans l'air de sept municipalité situées au voisinage d'aires traitées. Gouvernement du Québec, ministère de l'Énergie et des Ressources, Direction de la conservation, Service des études environnementales. 32 p.

Major, L.; Rousseau, G.; Mamarbachi, G. 1986b. Résidus de fénitrothion dans l'air de trois municipalités de la région Bas St-Laurent-Gaspésie. Gouvernement du Québec, ministère de l'Énergie et des Ressources, Direction de la conservation, Service des études environnementales. 16 p.

Mallet, V.N.; Cassista, A. 1982. A chemical residue survey in relation to the 1982 Spruce Budworm Spray Program in New Brunswick (Canada). *In* Environmental Monitoring of Forest Insect Control Operations - Interim Report 1982. 39 p.

Mallet, V.N.; Cassista, A. 1984. Fenitrothion residue survey in relation to the 1981 Spruce Budworm Spray Program in New Brunswick, Canada. Bull. Environ. Contam. Toxicol 32: 65-74.

Mallet, V.N.; Volpé, G. 1982. A chemical residue survey in relation to the 1980 Spruce Budworm Spray Program in New Brunswick, Canada. J. Environ. Sci. Health B17: 715-736.

Mamarbachi, G.; Dostie, R.; Marotte, P.-M. 1985. Persistence des résidus d'aminocarbe et de fénitrothion dans les eaux naturelles en milieu forestier. Gouvernement du Québec, ministère de l'Énergie et des Ressources, Service d'entomolgie et pathologie. 19 p.

Mamarbachi, G.; St-Jean, R. 1976. Lutte contre la tordeuse des bourgeons de l'épinette. II. Résidus de fénitrothion, diméthoate, phosphamidon et trichlorphon dans l'eau de certaines ruisseaux du Québec et de fénitrothion dans les aiguilles du Sapin baumier—Arrosage été 1975. Gouvernement du Québec, Service de protection de l'environnment, Recherche et planification, Laboratoire des pesticides, 15 p.

Marotte, P.-M. 1982. Evaluation des impacts sur les invertébrés aquatiques d'eau courante. Gouvernement du Québec, ministère de l'Énergie et des Ressources, Service d'entomologie et de pathologie. 29 p.

Marotte, P.-M.; Cimon, A. 1987. Environmental monitoring and quality control of aerial spraying of insecticides against the spruce budworm in Quebec in 1986. Pages 44-54 *in* Environmental Monitoring of Forest Insect Control Operations (EMOFICO) - 1986 Report.

Mathieu, P.; Larochelle, P.; Dostie, R. 1979. Comportement des insecticides dans les milieux aquatiques lors des pulvérisations aériennes contre la tordeuse des bourgeons de l'épinette, région de St-Pascal-de-Kamouraska—Printemps 1978. Gouvernement du Québec, ministère des Richesses

Naturelles, Direction domaine hydr., Groupe conseil en écologie. 57 p.

McBride, R.P.; Coyle, A.M. 1980. The persistence of aerially applied *Bacillus thuringiensis kurstaki* in Nova Scotia forest soils. Nova Scotia Dep. Lands For. and R.W. Welsford Research Group Ltd. 25 p.

McNeil, J.N.; McLeod, J.M. 1977. Apparent impact of fenitrothion on the Swaine jack pine sawfly, *Neodiprion swainei* Midd. Pages 203-215 *in* CNRC 1977a.

Montreal Engineering Company Ltd. 1980. Effects on fish of the 1979 New Brunswick spruce budworm control program — aerial spraying. Montreal Engineering Company Ltd., Fredericton., N.B.

Morin, R.; Gaboury, G.; Dostie, R. 1984. Résidus de fénitrothion et d'aminocarbc dans l'environnement. Gouvernement du Québec, ministère de l'Énergie et des Ressources, Service d'entomologie et de pathologie. 56 p.

Morin, R.; Gaboury, G.; Mamarbachi, G. 1986. Fenitrothion and aminocarb residues in water and balsam fir foliage following spruce budworm spraying programs in Quebec, 1979 to 1982. Bull. Environ. Contam. Toxicol. p. 662-628.

Morin, R.; Gaboury, G.; Mamarbachi, G.; Dostie, R. 1983. Surveillance des pulvérisations aériennes d'insecticides contre la tordeuse des bourgeons de l'épinette au Québec en 1981: Résidus de fénitrothion et d'aminocarbe dans l'environnement. Gouvernement du Québec, ministère de l'Énergie et des Ressources, Direction de la conservation. 60 p.

Prebble, M.L. (Ed.). 1975. Aerial control of Forest Insects in Canada. Environ. Can., Ottawa, Ont.

St-Louis, M.; Fung, I-Fu.; Gaboury, G.; Rousseau, G.; Mamarbachi, G. 1983. Surveillance des pulvérisations aériennes d'insecticides contre la tordeuse des bourgeons de l'épinette au Québec en 1982; Echantillonnage de l'air. Gouvernement du Québec, Eco-Recherches Inc., ministère de l'Énergie et des Ressources et ministère de l'Environnement. 40 p.

Sundaram, K.M.S. 1984. Residuc levels of fenitrothion and aminocarb in the air samples collected from experimental spray blocks in New Brunswick in 1982. J. Environ. Sci. Health B19 (4-5): 409-426.

Sundaram, K.M.S.; Feng, J.; Nott, R.; Feng. C. 1982. Chemical accountability of forest pest control products. Can. For. Serv., For. Pest Manage. Inst., Sault Ste. Marie, Ont. Study Ref. No. FP-16. 14 p.

Sundaram, K.M.S.; Feng, J.; Nott, R.; Feng. C. 1983. Distribution, dynamics, persistence and fate of aminocarb and fenitrothion containing Triton X-100 in a New Brunswick forest environment. Can. For. Serv., For. Pest. Manage. Inst., Sault-Ste-Marie, Ont., File Rep. 44, 88 p.

Wood G.W. 1980. Fenitrothion movement in air. Page 68 *in* Environmental Surveillance in New Brunswick 1978-1979. Effects of Spray Operations for Forest Protection. EMOFICO, Fredericton, N.B. 68 p.

Wood, G.W. 1981. Monitoring of insecticide drift from forest spray operations and its effect on low bush blueberry in New Brunswick in 1980. Agric. Can., Frederic., N.B.

Wood, G.W. 1982. Duration of Fenitrothion spray cloud. *In* D.R. Townsend, (Ed.). Environmental Surveillance of Spray Operations for Forest Protection in New Brunswick, 1980-1981. Washburn and Gillis Associates Ltd., Fredericton, N.B.

Yule, W.N.; Varty, I.W. 1975. The persistence and fate of fenitrothion insecticide in a forest environment. III. Deposit and residue studies with black spruce and red maple. Bull. Environ. Contam. Toxicol. 13 (6): 678-680.

Chapter 58

Impacts of Forest Aerial Spray Programs on Nontarget Terrestrial Invertebrates

PETER D. KINGSBURY AND KEVIN N. BARBER

Introduction

Invertebrates comprise a diverse and ubiquitous component of virtually all terrestrial ecosystems. In terms of biomass, numbers of individuals, and species, the arthropods comprise the most dominant group of terrestrial invertebrates. The earthworms represent a significant group of invertebrates that has received attention in agricultural systems. However, they are not widely distributed in the post-glacially forested areas of Canada, and many are presumed to be introduced European species. The reader is referred to the reviews of Edwards and Lofty (1977), Reynolds (1977), and Stenersen (1984) for more information. Arthropods, on the other hand, are virtually synonymous with "terrestrial invertebrates" when discussing impacts of pesticides in forestry and further discussion will be principally restricted to this group.

The terrestrial arthropods are tremendously diverse and still, to a significant extent, replete with undescribed species (Danks 1979). Although this situation presents the opportunity to study a wide range of ecological relationships, it also hinders the formulation of broad generalizations regarding the potential for pesticide-induced impacts based upon the study of a few populations. Lehmkuhl et al. (1984) and Rosenberg et al. (1986) describe the suitability of insect populations in environmental impact assessment studies (examples of applications primarily in aquatic systems) and emphasize the necessity to focus these studies at the species level—an approach with increasingly more potential in terrestrial systems with the accumulation of identification skills, sampling techniques, and ecological information.

In the context of environmental impact investigations in forestry, the terrestrial arthropods occupy an unenviable position. Virtually all pestiferous animals of forests that are targeted by chemical pesticides are arthropods, chiefly insects. Therefore, of all the terrestrial forest biota, the nontarget arthropods are considered the most vulnerable to impacts at the individual level caused by wide-spectrum synthetic insecticides. At the same time, these animals are generally the most adept at recovering from such disturbances, at least at the population level and beyond. Thus, the situation arises that short-term effects are observed to be both prevalent and sometimes extensive, but the ecological significance of these disturbances, interpreted or speculated upon over the long-term, is generally considered to be slight.

Nontarget insecticidal impacts can be reduced by developing insecticides with a greater selectivity of action (e.g. Bt, *Bacillus thuringiensis*). Alternatively, no-spray, or buffer, zones can be observed to avoid highly sensitive areas occupied by vulnerable animals. This last approach is often the exception since the ubiquity of arthropods places them throughout the milieu of the forest environment, sometimes even within (internal parasitoids) the targeted pest, and aesthetic/ecological value judgments are not as well developed for arthropods as for vertebrates. The commercial blueberry fields of New Brunswick are a classic example, where buffers have been established and/or a narrower spectrum insecticide (trichlorfon) applied to forested perimeters to reduce impacts on pollinating insects.

Role of Arthropods in the Terrestrial Forest Ecosystem

Terrestrial arthropods exhibit a wide range of habit, habitat, and phenology. Predators and parasitoids actively search or lie in wait for prey in the soil, litter, rock crevices, bark, foliage (Fig. 1), carrion, and air. The many species of prey may feed externally or internally on vascular plants, fungi, lichen, carrion, or detritus. All of these, particularly larval insects, may in turn be utilized as food by other animals, including vertebrates, often comprising a substantial portion of their diet. Ectoparasites of invertebrates and vertebrates are abundant and widespread and may serve as vectors of disease.

Where insects are concerned, those that undergo a complete metamorphosis may display widely divergent habits as larvae, compared with adults. Caterpillars and sawflies generally feed externally upon vascular plants, while the adults feed upon nectar. The adults of parasitoid wasps, flies, and beetles are generally winged and feed mainly on nectar and/or pollen. However, their larvae are not free-living but feed on or within host individuals, sometimes within the confines of a burrow. Some winged adult insects encountered in terrestrial ecosystems have an aquatic larval stage (e.g. dragonflies, biting flies such as black flies and mosquitoes). Some beetles, however, have similar feeding habits as larvae and adults (e.g. predaceous ground beetles, herbivorous or coprophagous scarab beetles, aphidophagous ladybird beetles) except for the qualification that adult insects do not grow, but feed principally to mature eggs and to supply energy for maintenance and activity.

Besides their important role in cycling energy among animals, arthropods assist in the fragmentation of forest floor litter and facilitate the microbial humification and mineralization of organic material in addition to transforming some of this energy into animal tissue. Particularly abundant groups of litter- and soil-inhabiting arthropods are the springtails and mites.

Pollination has become the leading process-approach to the study of insecticidal impacts on terrestrial invertebrates. Such concerns began when it was determined that low crop yields in commercial blueberry fields in New Brunswick were attributable to reductions in populations of pollinators (Kevan 1975). Since pollination is a mutualistic relationship between flowering plants and pollinating animals (often insects), whenever one partner in the relationship is detrimentally affected, the other is placed at risk. Thus, there exists the hypothetical potential for a spiral decline of both, unlike other symbiotic relationships such as predation. More importantly, bees, especially social bees, are animals that have a relatively lower intrinsic rate of increase and provide care and food for their larvae. The dependency of the developing generation upon provisions supplied by a relatively long-lived parent or by siblings serves to limit the capacity of social bees to recover from insecticidal impacts, notwithstanding differentials in their physiological detoxification systems. Similar concerns can be raised regarding carnivorous Hymenoptera with some degree of social (provisioning) development (e.g. sphecoid and vespoid wasps).

Reductions in rates of pollen transport, thus affecting pollination, could result in subsequent reductions in plant recruitment and availability of forage for pollinators. This is in addition to the immediate reductions in fruit and seed available as food for other animals such as birds and small mammals. To date, no investigation of the significance of insecticide-induced reductions in fruit- or seed-production in the forest has been undertaken. Most of these concerns of long-term ecological disruption are based on extrapolation and speculation from sound, but short-term, impact research and is discussed more fully in the following chapter of this book.

Methods of Investigation

Field assessment of insecticidal impact has been approached in various ways, and Southwood (1978) provides detailed, technical discussion of applicable ecological techniques. The many options can be grouped as measures of density or biomass in individual populations or communities, the level or rate of some process or activity, or the response of sentinel animals exposed to aerial sprays. Responses can also be measured in the short term/acute or in the longer term. Generally, a monitoring program should employ a comparison of sprayed and unsprayed areas of comparable habitat over a period of time including the collection of pre-spray and post-spray data derived from replicated collections. This design provides an evaluation of the contribution that natural oscillations in population parameters (e.g. density and activity) make to the overall fluctuations observed over time, whether or not the actual causes are weather, habitat assemblages, or recruitment to subsequent developmental stages. The remaining difference is then attributed to the insecticidal treatment.

The real-life situation is often quite different from the ideal because of constraints of limited resources and availability and suitability of control sites. All too often, studies are open to criticisms concerning uncontrolled or unmeasured variables that may influence or even obscure treatment effects. Most arthropods are classical poikilotherms (bumblebees are notable exceptions) and temperature, wind, and rain can exert significant influence on their activity in terms of time (diurnal and seasonal) and space (cover and aspect). Longer term studies seldom include pre-spray data because they are usually implemented with operational spray programs and not with more highly controlled field experiments. Sometimes, this can be overcome by increasing the number of blocks studied. The development of a sufficiently focused study to provide meaningful information, unobscured by excessive variability introduced by low numbers in samples or extensive pooling of samples, is a goal that will continue to be pursued.

Measures of Density or Biomass

Immediate effects can be seen in terms of the "knockdown" of arthropods (Varty 1982; references in Table 1). This is achieved by placing buckets, trays (Fig. 2), or cloths of known area under individual trees, and sometimes in a variety of cover types. When compared over time, including pre- and post-spray periods, and against a control site, effects of the spray are often reflected in abrupt increases in the natural fall of living, dead, and dying insects or "insect rain." The relative proportions of the various types of arthropods collected can provide a cursory indication of those groups immediately killed or suffering a delayed kill over the first few days only. The length of time that a spray event continues to have an effect can be roughly estimated, but the long-term consequences cannot be determined. Primarily a coarse indexing technique, the data derived with this method have been rarely analyzed statistically.

An extension of this technique involves a post-spray, ground-applied insecticidal drench or wash (e.g. permethrin) of the trees above the drop trays in an attempt to dislodge the remaining or surviving portion of the external arboreal fauna (Varty 1975, 1980; Kreutzweiser 1983). This permits estimation of the total community and further estimates of proportional impacts can be made of individual populations or ecological groupings. Some bias attributable to the criteria used to position the collection devices may still remain, however, because deflection by wind and amount of foliage in the column above each device affects catch size.

The use of caged insects facilitates estimation of the direct influence of the spray on individuals (Kingsbury et al. 1985; unpubl. data, Forest Pest Management Institute, Forestry Canada). After exposure to the spray, the insects are maintained in a controlled environment for a period of time while mortality is recorded and compared with similar data gathered for individuals caged on control sites. Problems with this technique include potentially differential penetration of droplet size through the restraining-cage material, and a confounding of routes of exposure since these animals are held in the same contaminated cages after the spray event. However, this approach does provide preliminary estimates of the risk that various kinds of arthropods face. An alternative method would use tethered or relatively sedentary animals, obviating the need for a potential spray interception surface surrounding them, but as yet this has not been attempted. These data are best used when compared with companion studies of laboratory toxicology and field residues. A more refined extension of caged sentinels is the use of artificial colonies of honey bees and bumble bees which will be discussed below.

To obtain an insight into the disruption caused by insecticidal stress at the population level and over longer periods of time, a more extensive sampling program must be implemented. This involves tailoring a sampling design to best fit the population or community studied. This represents the generalized approach to field studies of populations whereas the previous techniques have been developed primarily for study of insecticidal impacts.

What is often involved is a destructively sampled unit of habitat such as a length of tree branch (Millikin 1987; Woodworth-Lynas et al. 1987; Buckner et al. 1974; Anon. 1980, 1987), a whole plant (Anon. 1980, 1987), an area of forest floor litter or a core of litter/soil (Environmental Monitoring Committee 1979b; Shepherd 1980). The samples are then processed in such a way that the invertebrates are separated from the extraneous material or matrix. This is variously accomplished with sifting, dislodgement, differential flotation, or by forcing the living individuals to move from a heat stress (Berlese–Tullgren funnel) or by taking advantage of a taxic response to light or gravity. Density estimates can then be derived on the basis of habitat units of linear measure, area, volume, or weight.

A less refined sampling procedure that sometimes provides similar information is a standardized, direct collection of arthropods in the field. Two very commonly used tools are the use of sweep nets and beat-sheets. Sweep nets are used to sample arthropods on low vegetation and the numbers of arthropods can be estimated for a given number of swings or sweeps of the net or length of transect (Wan 1974; Environmental Monitoring Committee 1979a; Bendell et al. 1986; Shepherd 1980; Buckner et al. 1974). However, the method is prone to individual collector bias and is not always replicable or comparable from study to study, particularly in heterogeneous habitats. Just the same, it can provide information readily and cheaply and could be improved if used as a defaunating tool for a standard area of habitat. Beat-sheets are similar to drop cloths except a stick is used to dislodge living animals, usually from woody plants (Environmental Monitoring Committee 1979a,b; Varty 1978; Shepherd 1980). Two disadvantages of using beat-sheets are the difficulty in determining how much of the habitat is being sampled and the differential escape behavior of the arthropods present. This latter approach is best left as a qualitative inventory tool for such groups as arboreal beetles, spiders, and aphids.

Measures of Activity or Process Rates

Estimates of levels of activity of litter- and ground-dwelling arthropods (e.g. ground beetles, spiders, centipedes, millipedes) have been made using fine-sand transects to provide counts of arthropod tracks per unit time and length of transect (Bracher and Bider 1982). More frequently, pitfall traps are used (Carter and Brown 1973; Kingsbury and Kreutzweiser 1980; Kreutzweiser 1982; Shepherd 1980; Buckner et al. 1974). Most often they are destructive, incorporating a killing/preserving agent, but can be operated as simple dry canisters to provide live traps for arthropods such as ground beetles if emptied frequently to minimize damage. Mark–recapture techniques can be applied in conjunction with this type of livetrapping or other census techniques to estimate local populations. Pitfall trapping provides perhaps the best and most readily implemented way to study ground- and litter-dwelling macro-arthropods. Smaller arthropods such as springtails and mites are likely to be studied better with habitat collections obtained with soil/litter cores followed by Berlese–Tullgren funnel extraction of living invertebrates (Carter and Brown 1973; Environmental Monitoring Committee 1979b; Shepherd 1980).

Malaise traps (Environmental Monitoring Committee 1979b), sticky traps, pan traps, and light traps (Wan 1974; Shepherd 1980) are used primarily to intercept flying insects. They may be operated conventionally, or a bait (e.g. food, pheromone) may be incorporated to increase the efficiency or selectivity of the trap for specific types of arthropods. These trapping techniques have been used sparingly or not at all in insecticide impact studies, mostly because they are not always amenable to quantification and reproducibility, with the possible exception of sticky traps.

Process-oriented monitoring is another means of assessing impacts of pesticides. Good examples are such processes as pollination, litter decomposition, parasitism, and to a lesser extent, predation. They provide a means of monitoring a process with a standardized technique that circumvents the necessity to track specific populations or communities. Companion assessments can be made of particular components of the community involved to provide an opportunity to correlate measurements and support cause–effect hypotheses.

Litter decomposition rates have been estimated by weight loss from known amounts of litter or foliage held in mesh bags (Eidt and Weaver 1985, 1986). This technique cannot discriminate among the relative contributions that various biotic components make to the total process. Thus, any recorded impacts cannot be attributed to arthropod or microbial disturbance, for example, without the accompanying sampling of these communities. Secondary investigations must pursue the identity of the impacted components if this is warranted.

Parasitism rates can be estimated by sampling the host and either rearing or dissecting out the internal parasitoids (e.g. Kettela and Varty 1972; Gesner and Varty 1973; Otvos and Raske 1980). The parasites can then be identified and additional information gained regarding the relative contribution each makes to the parasite load that the host carries in sprayed and unsprayed areas. Virtually all of this type of monitoring has concentrated on parasites of the targeted pest because this is of immediate concern, more is known of the parasitic complex attacking the host, and sampling procedures have already been developed.

Predation rates are more difficult to compartmentalize since the opportunity to encounter the predator with a sampling method is more fleeting and predators are usually much less selective of their prey. Rate of removal of prey from a given habitat would be a generalized approach to consider, but as in litter decomposition, impacts cannot be compartmentalized without accompanying or subsequent study of the populations/communities involved (but see Sunderland 1988). This is particularly important when there are vertebrates utilizing the same prey as arthropods. To date, only suggestions of impacts on predation by arthropods have been made as a projection from directly measured reductions in populations of known predators (Varty 1975, 1977, 1978, 1980; Varty and Carter 1974; Carter and Brown 1973).

Table 1. Summary of invertebrate knockdown studies conducted in forest spray blocks in eastern Canada during June and early July, 1972–1986.

Material tested	Application rate	Spray formulation	Tree species	Area of drop tray (m^2)	Mean drop tray catch per day: Pre-spray	Days post-spray[a]: +0	+1	+2	Family groups[b] in descending order of abundance	Reference
Fenitrothion	2 × 140 g/ha	Aerotex/Atlox/H_2O	fir/spruce	0.37	2.9	7.1,15.9	8.3,5.3	3.0,1.1	Lep., Hym., Hem.	Miller et al. 1973[c]
	2 × 140 g/ha	Aerotex/Atlox/H_2O	open	0.37	0.3	6.8,19.3	16.2,6.6	3.3,1.2	Hym., Lep., Hem.	Miller et al. 1973[c]
	210 g/ha	Cyclosol/Triton X-100/H_2O	fir	0.13	2.1	15.7	17.9	9.1*	Lep. larvae	McLeod 1982
	280 g/ha	Dowanal/Atlox/H_2O	fir	0.25	1.8	17.5	6.6	5.2	Dip., Hom., Hym.	Millikin 1987; unpubl. data
	280 g/ha	Dowanal/Atlox/H_2O	birch	0.25	2.7	42.4	7.0	4.7	Hom., Dip., Hym.	Millikin 1987; unpubl. data
Aminocarb	2 × 70 g/ha	Matacil 1.8D/ID585	spruce	0.13	3.0	14.0,4.4	9.4,4.0	1.8,3.4	Dip.	Kingsbury et al. 1981a
	2 × 70 g/ha	Matacil 1.8D/ID585	willow	0.13	3.1	6.3,3.3	0.7,0.7	0.7,–	Dip., Hym.	Kingsbury et al. 1981a
	2 × 70 g/ha	Matacil 1.8D or 180 F in ID585, Sunspray 6N oil, or Atlox/H_2O	fir	0.12	4.2	15.6,17.2	14.0,12.1	9.4,8.1	Dip., Lep. larvae	Millikin 1981
	70 g/ha	Matacil 1.80F/Triton X-100/H_2O	fir	0.13	1.6	8.7	9.3	10.4*	Lep. larvae	McLeod 1982
	2 × 70 g/ha	Matacil 1.8D/ID585	fir/alder/open	1.00	3.8	31.5.,11.6	24.1,10.8	NC,5.3	Dip., Hym., Hom.	Woodworth-Lynas et al. 1987
	175 g/ha	Matacil 1.4 OSC/ID585	fir	0.13	1.6	28.5	7.0	2.6	Dip., Aran., Coll.	Kingsbury et al. 1981a
	175 g/ha	Matacil 1.4 OSC/ID585	alder	0.13	1.4	15.6	16.0	6.0	Dip., Hom., Hym.	Holmes & Kingsbury 1980
Permethrin	2 × 17.5 g/ha	ID585	spruce	0.05	1.4	5.8,1.8	8.4,0.8	3.4,–	Lep. larvae	Kingsbury & McLeod 1979
	2 × 17.5 g/ha	ID585	willow	0.05	1.8	21.1,0.0	3.6,2.4	2.2,2.6	Hom., Col., Lep.	Kingsbury & McLeod 1979
	2 × 17.5 g/ha	ID585	nannyberry/ choke cherry	0.05	0.9	2.5,0.7	2.3,1.8	1.3,1.4	Hym., Dip., Lep., Hom.	Kingsbury & McLeod 1979
	2 × 17.5 g/ha	ID585	open	0.05	0.5	0.8,1.2	0.4,1.4	–,0.4	Dip.	Kingsbury & McLeod 1979
	17.5 g/ha	ID585	coniferous	0.12	0.6	6.3	2.0	1.3	Dip.	Kingsbury & Kreutzweiser 1980
	17.5 g/ha	ID585	deciduous	0.12	1.3	6.0	0.8	2.0	Dip.	Kingsbury & Kreutzweiser 1980
	2 × 17.5 g/ha	ID585	fir/pin cherry	0.12	2.0	4.0,6.0	2.0,6.5	2.7,2.1	Dip.	Kreutzweiser 1982

(Continued)

Table 1. Summary of invertebrate knockdown studies conducted in forest spray blocks in eastern Canada during June and early July, 1972–1986. *(Concluded)*

Material tested	Application rate	Spray formulation	Tree species	Area of drop tray (m^2)	Mean drop tray catch per day: Pre-spray	Days post-spray[a]: +0	+1	+2	Family groups[b] in descending order of abundance	Reference
Permethrin	2 × 17.5 g/ha	ID585	fir	0.12	1.9	13.1,5.5	5.3,3.4	2.0,1.5	Dip.	Kreutzweiser 1982
	2 × 17.5 g/ha	ID585	fir	0.10	3.2	6.4,8.2	4.2,5.4	3.2,3.1*	Dip., Lep. larvae	Kreutzweiser 1983
	2 × 17.5 g/ha	ID585	pin cherry	0.10	2.4	3.1,6.2	3.5,NC	4.4,2.9	Dip., Hym., Lep. larvae	Kreutzweiser 1983
	2 × 17.5 g/ha	ID585	fir	1.00	2.6[d]	10.5,25.5	35.0,31.0	15.5,25.0	Dip., Lep. larvae, Hom.	Kreutzweiser 1983
Azamethiphos	2 × 70 g/ha	Cellosolve	fir/spruce	0.13	1.5	24.9,10.1	7.0,9.1	0.9,7.6*	Dip., Hom.	Kingsbury et al. 1980
	2 × 70 g/ha	Cellosolve	alder/hazel/fir	0.13	2.0	6.0,4.8	10.4,3.1	2.4,*10.9	Dip., Arach., Hem., Coll.	Kingsbury et al. 1980
Carbaryl	2 × 280 g/ha	SEVIN-2-OIL in ID585	fir	0.13	0.4	0.8,2.4	3.7,1.4	1.1,0.8	Dip., Col., Lep. larvae	Holmes et al. 1981
	2 × 280 g/ha	SEVIN-2-OIL in ID585	alder	0.13	3.3	12.0,10.2	13.8,6.6	7.4,*15.0	Dip., Col.	Holmes et al. 1981
Chlorpyrifos-methyl	2 × 70 g/ha	ID585	fir	0.13	0.8	10.4,1.9	1.4,1.0	0.3,0.8	Dip.	Holmes & Millikin 1981
	2 × 70 g/ha	ID585	alder	0.13	2.2	6.4,4.6	1.6,2.2	3.4,1.4	Dip.	Holmes & Millikin 1981
	725 or 825 g/ha	undiluted RELDAN	alder	0.13	2.1	25.3	5.9	5.6*	Dip., Hem., Coll., Col.	Kingsbury & Holmes 1980
Nonylphenol	470 mL/ha	1:3 with ID585	fir	0.13	0.8	1.2	1.7	0.8	–	Kingsbury et al. 1981a
	470 mL/ha	1:3 with ID585	willow	0.13	0.6	1.9	2.9	1.0	Dip., Aran.	Kingsbury et al. 1981a
	467 mL/ha	1:2 with ID585	spruce/alder/open	0.13	2.2	6.2	3.0	1.6	Hym., Dip., Col.	Holmes & Kingsbury 1980
Atlox	2 × 40 mL/ha	3% in H_2O	fir	0.12	3.2	14.1,12.1	7.3,3.7	4.9,–	Dip.	Millikin 1981

[a] Double numbers are post-spray catches following each of double spray application 3 or more days apart; dashes indicate that catches have returned to pre-spray levels; NC indicates collections not made; and asterisks indicate that catches did not return to pre-spray levels until 4 days after spraying.

[b] Family groups are as follows: Arach. — Archnida; Aran. — Araneida; Col. — Coleoptera (Staphylinidae sometimes not reported, as they were apparently attracted to preservative in knockdown buckets); Coll. — Collembola; Dip. — Diptera; Hem. — Hemiptera; Hom. — Homoptera; Hym. — Hymenoptera (mostly parasites of SBW); and Lep. — Lepidoptera (usually spruce budworm, SBW, on fir or spruce).

[c] Adulticide trials; all others larvicide trials.

[d] Average post-spray (6 days) on control site.

Response of Sentinel Animals

Monitoring of pollination impacts employs an array of different techniques and is discussed more fully in the next chapter. Nondestructive sampling usually involves the measurement of the number of bees visiting a certain area (Fig. 3) or transect of bloom over a period of time (Larson et al. 1987; Fudge, Lane and Associates Ltd. 1987). In addition to caged pollinators, artificial colonies of bumblebees and honey bees (Fig. 4) have been used to estimate effects of insecticidal stress. The artificial colony approach is very similar to the population techniques already discussed but provides certain advantages. Sentinel colonies are amenable to control of placement, replication, age, and physiological condition. They provide avenues to monitor effects on behavior, activity level, production of brood and honey, and accumulations of insecticidal residues in wax and honey. Honey bee hives can be fitted with an electronic activity counter, pollen trap, and dead bee trap to measure the activity at the entrance, the rate of pollen collection (by weight), and mortality occurring in the hive (see references in Table 4).

Measurements of the indirect effects of insecticides on pollination can be made by focusing on plants that require insect visitation to achieve normal rates of fertilization (Bouchard 1981; SOMER 1985). This can be achieved by sampling flowers that have opened during or shortly after insecticidal treatment. If a significant reduction in fruit-set or seed-set is attributable to the treatment, a survey of the relative activity of the visiting insects over time could provide an indication as to which pollinator group has been affected. This may require additional background information regarding the relative pollinating efficiency of each taxon since it is not always the most abundant visitors that are the most effective pollinators (Kevan and Baker 1983; Barber 1988).

As more attention is drawn to the importance of invertebrate communities, and as more of their ecological relationships become better understood, there will be a progressive refinement of associated impact studies. Perpetual problems in the past have been associated with the vast array of organisms encountered in generalized sampling designs and how to deal with these data. These studies serve the purpose of demonstrating abrupt, acute impacts but provide little means of drawing unequivocal conclusions, especially regarding long-term implications, as pointed out by Varty (1975). By concentrating on key ecological processes and component species, hymenopterous pollinators in particular, there is the potential for the development of a truly rigorous and informative study of insecticidal impacts on arthropods, on par with those conducted with vertebrate populations.

Ecological Effects on Nontarget Arthropods

Insecticide-Induced Knockdown

A steady natural "rain" of living, moribund, and dead immature and adult arthropods falls from forest canopies over the growing season. Varty (1982) reports the aggregate tallies of this natural fallout to be quite large by mid-June, equal to about one-quarter of the total population present on the trees at this time. Applications of broad spectrum chemical insecticides can induce large short-term increases, referred to as knockdown, in this natural fallout, which can total many thousand individuals of various pest and nonpest species per hectare sprayed (Kettela and Varty 1972).

A large number of insecticide formulations and spray ingredients have been studied for their ability to induce invertebrate knockdown from a variety of forest canopies (Table 1). Results show that spray-induced knockdown is very common and can even be induced by "inert" formulation components such as the emulsifier Atlox® 3409. Knockdown may involve a combination of toxic action of the spray and physical interaction with spray components leading to dislodgement from the canopy. Evidence for the latter can be seen in short-term increases of more than one hundredfold in adult mayfly and midge catches in drift nets set in a stream during aerial application of Bt formulations (Buckner et al. 1974).

In almost all the studies reviewed (Table 1), knockdown consisted primarily of adult flies (Diptera) or the target pest species (Lepidoptera larvae or adults). Adult bees and wasps (Hymenoptera) were frequently major components of knockdown, especially from sites with flowering trees and shrubs. Bugs (Homoptera and Hemiptera), beetles (Coleoptera), springtails (Collembola), and spiders (Araneida) were also important in some knockdown collections. There are few indications of different materials selectively affecting different groups of organisms, not because the materials are not selective but because the monitoring method is not sufficiently rigorous. The variety of organisms collected seems to be influenced more by the canopy-type they were collected from, the presence or absence of significant pest populations, and the magnitude of knockdown. In general, heavier knockdowns also encompassed a wider variety of organisms.

Differences in the extent to which different insecticides induced knockdown were modest, or confounded by variability between different studies with the same compound. Fenitrothion generally induced knockdown, peaking at less than 10 times pre-spray levels (less than 4 times at 140 g/ha) and lasting for 2 or 3 days. The most noticeable exception was up to 60-fold increases in knockdown in forest clearings with no overhead canopy after spruce budworm adulticide applications of fenitrothion in mid-July, while knockdown under white birch peaked at about 15 times pre-spray levels. Knockdown effects of aminocarb applications were similar, but usually produced increases of less than 6 times pre-spray levels. A seasonal maximum (175 g/ha) application of aminocarb caused 10- to 18-fold increases in knockdown, but even at this high rate, effects were only noticeable for 3 days post-spray. Knockdown induced by permethrin and chlorpyrifos-methyl sprays, although similar in initial magnitude to that from other insecticides, tended to be of shorter duration with little effect beyond the second day after spraying. In contrast, limited studies on azamethiphos and carbaryl indicated relatively prolonged periods of elevated knockdown levels after applications of these compounds. The insect growth regulator BAY SIR 8514 (Millikin and Mortensen 1980) and various formulations of Bt (Morris and Hildebrand 1974) did not appear to elevate drop-tray catches during the sampling period. These compounds would be expected to have a delayed knockdown effect because they must be

Figures 1–4
Figure 1. Crab spider feeding on a hover fly visiting flowers of American mountain-ash. **Figure 2.** Drop tray (0.25 m^2) erected to collect falling "insect rain" within an experimentally treated forest. **Figure 3.** Forestry technician observing visitation rates of bees to wildflowers within a fixed area and time period. **Figure 4.** Honey bee hives being tended before insecticidal treatment of forested area.

ingested to be effective and are selective for susceptible life stages or species of organisms.

There is little suggestion from knockdown studies encompassing double applications that the initial spray depleted canopy invertebrate communities for extended periods. Knockdown after second applications was often as great as or greater than after first sprays. However, virtually all of these studies do not provide sufficient discrimination of individual species populations. Thus, the relative contributions of pesticide tolerance, and recolonization and/or recruitment (of the same or related species, including different life stages) cannot be assessed.

Effects on Arthropod Communities

Most impact studies evaluating the effects of forest spraying on terrestrial invertebrate communities have focused on arthropods resident on various forest plants, with a smaller number of studies looking at soil- and/or litter-inhabiting or aerial insect communities. Most studies have looked for short-term effects within the season of application, but some longer-term studies have been carried out in areas treated with fenitrothion and aminocarb.

Fenitrothion and aminocarb applications sometimes depress arthropod communities immediately after treatment, with effects appearing greater and more consistently in studies carried out in fenitrothion blocks (Table 2). From the reports summarized in Table 2, it appears fenitrothion primarily affects populations of Diptera and Homoptera. The effect on spiders in Wan's (1974) study represents a smaller increase in sweep-net collections on the sprayed sites as compared with the untreated sites. Aminocarb impacts seem to involve primarily spiders, various Hymenoptera, and Lepidoptera. Caution should be applied, however, to such generalizations, as they are highly subject to the particular habitat, fauna, and conditions of the studies available. Other chemical insecticides have been less well studied, but appear to impact to some extent on arboreal and ground-dwelling arthropod communities. Applications of Bt, on the other hand, have little effect on nontarget arthropods except for limited evidence for impact on nontarget lepidopteran larvae (Table 2).

The most comprehensive studies of long-term effects of forest sprays on terrestrial arthropods have been carried out in study plots in New Brunswick with up to 9 consecutive years of fenitrothion treatment in initial studies (Varty 1977), followed by additional evaluations in subsequent years (Varty 1980, 1982). Results indicate that the arthropod community on balsam fir is not perceptibly destabilized by perennial or intermittent fenitrothion applications. Most predatory, phytophagous, and fungivorous arthropods show patterns of population cycling related to prey abundance or habitat change. Thinning of populations by insecticide treatments does not appear to be profoundly influential on natural population regulatory mechanisms. A fir-dwelling ladybird beetle, *Mulsantina hudsonica*, has become scarcer during the period of fenitrothion usage (Varty 1982), but natural fluctuations may be responsible. There is no evidence to suggest biocontrol mechanisms have been disrupted or that minor-pest eruptions have resulted from fenitrothion impacts on predatory and parasitic arthropods.

Spider densities studied in aminocarb-treated, fenitrothion-treated, and untreated plots over a 4-year period showed similar annual patterns of increase or decrease (Varty 1982). Both density and diversity (number of species) of spiders and their fluctuations were similar in the aminocarb and untreated plots over the 4 years. Spider densities were consistently higher and species diversity consistently lower in the fenitrothion plot. Densities of predaceous insects in the same plots also showed parallel trends over the 4 years, apparently in response to pulses in aphid populations (Varty 1982). Mean densities of insect predators were consistently lower in the fenitrothion plot, intermediate in the aminocarb plot, and highest in the untreated plot, leaving open the hypothesis that the insecticide regimes depressed populations of predaceous insects.

The arthropod fauna of 12 forest plant species in treated as compared with untreated areas was evaluated 1 and 2 years after aminocarb sprays in Newfoundland (Anon. 1980, 1987). There were few indications of long-term effects of aminocarb on the total arthropod fauna or specific taxonomic groups. Groups for which densities at sprayed sites were considered significantly lower than control sites on more than one plant species were limited to three of the 30 indicator-groups evaluated, namely Araneida, Collembola, and Hymenoptera: Formicidae (Anon. 1987).

Varty and Carter (1974) compared the arthropod communities of the forest floor sampled with pitfall traps on two plots, one with 4 consecutive years of fenitrothion treatment preceded by a long history of DDT sprays, the other with a single, current year of fenitrothion treatment. For the most part, the faunas of the two plots were quite similar. Overall abundance and diversity of the fauna was lower on the plot with the longer history of insecticide treatment, primarily due to fewer species of spiders and substantially lower numbers of harvestmen (phalangids) on that plot. When this same plot was resampled in 1977 after a further 5 years of fenitrothion treatment, the numbers of carabid beetles and spiders captured over a 10-week period were one-third and one-half the numbers sampled in 1972 (Varty 1978). Total numbers of organisms trapped from this plot with 9 consecutive years of fenitrothion treatment were 50% of those from a plot that had never been treated. A plot treated with aminocarb in the year of sampling and sporadically with fenitrothion in earlier years yielded 81% of the numbers trapped on the untreated plot (Varty 1978). Most striking in these data sets was the low number of springtails (Collembola) in the long-term fenitrothion plot, but millipedes, spiders, and beetles all showed suggestions of some population-depressing influences related to long histories of insecticidal spraying.

In Newfoundland, total soil arthropods (primarily springtails and mites) were 40 to 50% lower in sites the year after aminocarb treatment than in untreated sites (Anon. 1980). Caution must be taken in drawing conclusions from these data, however, in light of a lack of indications of effects on soil arthropod populations in one of these same aminocarb treatment blocks in the year of treatment (Environmental Monitoring Committee 1979b). Further, results from 2 years post-spray indicated that although densities of springtails and mites were still significantly lower on two treated mature forest sites, springtails were more plentiful on treated regeneration sites compared with similar control sites (Anon. 1987).

Eidt and Weaver (1985, 1986) examined effects of fenitrothion and aminocarb on activity of soil arthropods by studying decomposition rates of conifer foliage in litter bags within small hand-sprayer treated plots. No effects on decomposition rates were found over a year-long period in plots treated with up to double the conventional single emission rate (i.e. 140 g/ha aminocarb; 420 g/ha fenitrothion) for double aerial applications or 10 times the double rate (i.e. 1 400 and 4 200 gm/ha) of either insecticide (Eidt and Weaver 1986).

One long-term effect of fenitrothion for which there is considerable evidence is the ability of persistent low-level fenitrothion residues in conifer foliage to cause mortality of insects feeding on previously sprayed foliage. The Swaine jack pine sawfly, *Neodiprion swainei*, has been shown to suffer mortality in the larval stage and sublethal effects such as smaller cocoons and reduced fecundity when reared on foliage from areas sprayed with fenitrothion (McNeil and McLeod 1977; McNeil et al. 1979). The near disappearance of the balsam fir sawfly, *Neodiprion abietis* complex, from New Brunswick over the period of fenitrothion spraying may reflect the sensitivity of this insect to fenitrothion residues in balsam fir foliage (Eidt and Mallet 1986). The concerns about fenitrothion residues in conifer foliage are reviewed by Eidt and Pearce (1986).

Effects on Parasitism

Varying amounts of information are available on the effects of insecticide applications on the level of parasitism in spruce budworm, *Choristoneura fumiferana*, populations surviving treatments. Parasitism levels could be affected through higher mortality of parasitized versus unparasitized hosts or through direct mortality of adult parasites present at the time of spray application.

Early studies in New Brunswick suggest that there are large differences in the magnitude of kills of adult parasites associated with spray applications at different stages of the target insect's development (Table 3). Far more adult parasites, particularly ichneumonoid wasps, are killed by insecticides applied to kill budworm moths (adulticides) than larviciding treatments applied earlier in the year. Hymenopterous parasites (particularly two parasites of spruce budworm, *Apanteles fumiferanae* and *Glypta fumiferanae*) suffered contact toxicity more frequently while foraging for sugar foods among ground cover than while searching trees for host larvae, as judged from their distribution, after adulticide sprays, on drop cloths set out under fir and spruce and in open areas (Gesner and Varty 1973).

Larvicide sprays seem to have little influence on parasites of small budworm larvae, such as *Apanteles* and *Glypta*, which are concealed within the host's body at the time of spray. Spray-related reductions in these populations would be expected only if parasitized larvae are more sensitive to the sprays than are unparasitized larvae. Comparisons of the percentages of parasitized budworm from fenitrothion- and aminocarb-treated plots and untreated plots in the year of treatment have consistently shown similar or higher levels in the sprayed areas (EMOFICO 1976; Varty 1978; Sarrazin 1978; Magasi et al. 1980; Otvos and Raske 1980). Varty (1977) presented data on annual percent parasitism by *Apanteles* and *Glypta* in a number of New Brunswick plots with spray histories ranging from never treated to eight annual treatments. All the parasitism values fall within the norms of abundance on epidemic hosts in untreated stands, and there is little evidence of an insecticide impact on the parasitism process.

There are more suggestions of impact on the parasites of large budworm larvae and pupae (e.g. ichneumonoid wasps such as *Meteorus, Ephialtes* (= *Apec[h]this*), *Itoplectis* and *Phaeogenes* and tachinid flies such as *Actia*, *Eumea* and *Winthemia*). These parasites are sometimes present as adults at spray time, and consequently can be directly exposed to treatments. Kettela and Varty (1972) reported less than 1% parasitism by these groups in plots sprayed with fenitrothion compared with levels around 10% in untreated epidemic budworm populations. Other suggestions of lower incidences of pupal parasitism in sprayed blocks as compared with untreated blocks are reported in Varty (1978) and Sarrazin (1978), the latter referring specifically to *Phaeogenes*. There is little evidence from studies on parasitism of spruce budworm eggs by *Trichogramma minutum* that insecticide treatments have adverse effects on egg parasitism (Kettela and Varty 1972; Varty 1978; Magasi et al. 1980).

As previously illustrated (Table 3), adulticide treatments can kill large numbers of parasites of larval budworm, especially *Apanteles* and *Glypta*, present as adults at the time of spraying. Three successive applications of 70 g phosphamidon/ha at 2 to 3 day intervals in New Brunswick in 1977 were estimated to kill 80% to 90% of these budworm parasites, for the most part before they could reproduce (Varty 1978). Despite this high rate of kill, parasitism of overwintering larvae from the treated blocks was 22%, only slightly lower than levels of 26% and 30% in two adjacent untreated areas. In earlier studies in New Brunswick, rates of parasitism were only 2.6% to 5.3% and 6.2% to 10.3% in larvae from fenitrothion and phosphamidon blocks respectively, the year after budworm adulticide treatments caused heavy mortality of adult parasites (Gesner and Varty 1973). This suggests considerable but not total depression of the parasitism process when compared with the normal levels of about 20% reported for unsprayed plots.

Conversely, percent parasitism of Swaine jack pine sawfly increased after feeding on conifer foliage containing aged residues of fenitrothion. This was explained as resulting from increased mortality of the host while parasites were unaffected and capable of attacking a higher proportion of this smaller host population, apparently as the result of an increased parasite to host ratio alone (McLeod 1975). However, independent observations indicated a tendency for the sawfly larvae to disperse when feeding on residue-bearing foliage in contrast to the gregarious feeding of colonies on untreated foliage (McNeil and Mcleod 1977; McNeil et al. 1979). This may represent the more important mechanism for increased parasitism if a mutual benefit (parasite deterrence) to the sawflies is derived from gregarious feeding. Studies with Bt, alone or in combination with other chemicals, have revealed no observable effects on parasites or impacts on parasitism in treated plots (Buckner et al. 1974; Morris 1977; Otvos and Raske 1980).

Table 2. Summary of studies on short-term effects of insecticide spray on arthropod communities.

Insecticide	Dosage	Community studies	Sampling methods	Sampling intensity	Results of treatment on populations	Reference
Fenitrothion	140 g/ha	Arthropods on grass and brush	Sweep nets	Moderate for short period	Diptera and Arachnida declined in treated areas	Wan 1974
	140 g/ha	Night-flying insects	Light traps	Light for short period	No detectable effects	Wan 1974
	280 g/ha	Arthropods on fir and birch	Branch samples	Intensive for 3 weeks	About 39% reduction on fir and 13% on birch	Millikin 1987
	140 + 210 g/ha	Predaceous arthropods in soil and litter	Pitfall traps, Tullgren funnels	Prolonged and intensive	Harvestmen and spiders declined in year of treatment	Carter and Brown 1973
	2 × 210 g/ha	Spiders on fir	Insecticide wash	Not known	No evident effects	Varty 1980
	2 × 210 g/ha	Arthropods on shrubs and herbs	Sweep nets	Intensive for 2 weeks	Caterpillars, beetles, wasps, and flies declined due to treatments	Bendell et al. 1986
Fenitrothion + Aminocarb	2 × 210 + 88 g/h	Arthropods on trees and shrubs	Beating, sweep nets	Moderate for 2 months	Adults primarily affected but no prolonged effects	Environmental Monitoring Committee (EMC) 1979a
Aminocarb	70 g/ha	Predaceous arthropods on fir	Beating	Not known	Spiders and chalcids affected more than beetles and ichneumonoids	Varty 1978
	70-87 g/ha	Spiders on fir	Insecticide wash	Not known	Marked reductions in number of spiders but no lasting effects	Varty 1980
	2 × 70 g/ha	Flying and foliage insects	Sweep nets, Malaise traps	Intensive for several weeks	Adult Lepidoptera only group demonstrated to be affected	EMC 1979b
	2 × 70 g/ha	Soil arthropods	Berlese funnels	Moderate	No detectable effects	EMC 1979b
	2 × 70 g/ha	Arthropods on shrubs and herbs	Sweep nets	Intensive for 2 weeks	Beetle, spider, and ant catches declined in treated areas	Bendell et al. 1986
	2 × 70 g/L	Arthropods on forest trees, shrubs	Branch samples	Moderate for 2 months	Possible reductions of flies and spiders	Woodworth-Lynas et al. 1987
	3 × 70 and 2 × 88 g/ha	Arthropods on trees and shrubs	Beating, sweep nets	Light for 2 months	Small initial population reduction, but no lasting effects	EMC 1979a
	175 g/ha	Arthropods activity on ground surface	Sand transect	Intensive over 2-yr period	Spiders + harvestmen and Lepidoptera less active for 10 days after spraying	Bracher and Bider 1981
Permethrin	17.5 g/ha	Ground-dwelling arthropods	Pitfall traps	Moderate for several weeks	Moderate short-term effects	Kingsbury and Kreutzweiser 1980
	2 × 17.5 g/ha	Ground-dwelling arthropods	Pitfall traps	Moderate for 2–3 weeks	No detectable effects	Kreutzweiser 1983

(Continued)

Table 2. Summary of studies on short-term effects of insecticide spray on arthropod communities. *(Concluded)*

Insecticide	Dosage	Community studies	Sampling methods	Sampling intensity	Results of treatment on populations	Reference
Permethrin	17.5 and 2×17.5 g/ha	Arthropods on fir	Insecticide wash	Light, 5 days after spraying	Budworm larvae and beetles reduced in treated areas	Kreutzweiser 1983
Acephate	1120 g/ha	Various arthropod habitats	Sweep nets, beatings, pitfalls, light traps	Intensive for short period	May have reduced number of parasitic wasps, ants, and spiders	Shepherd 1980
Diflubenzuron	280 g/ha	Various arthropod habitats	Sweep nets, beating, pitfalls, light traps, Berlese funnels	Intensive for several weeks	Possible reduction in parasite wasps, spiders, ants, and some soil organisms	Shepherd 1980
Bt	10 BIU/ha	Various arthropod habitats	Sweep nets, pitfalls, branch samples	Intensive for several weeks	No detectable effects	Buckner et al. 1974
	20 BIU/ha	Arthropods on 12 plant species	Branch or whole plant samples	Intensive for 3 months	No detectable effects	Anon. 1980
	30 BIU/ha	Arthropods on 12 plant species	Branch or whole plant samples	Intensive for 3 months	No detectable effects	Anon. 1987
	20-30 BIU/ha	Arthropods on shrubs and herbs	Sweep nets	Intensive for 2 weeks	Lepidopterous larval populations reduced	Bendell et al. 1986

Table 3. Estimates from drop-cloth catches of parasites killed per hectare by various spruce budworm control treatments (adapted from Kettela and Varty 1972 and Gesner and Varty 1973).

	Budworm life stage at time of treatment			
	Third-instar larvae	Fifth-instar larvae	Adults	Adults
Date treated	5 June	16 June	11–14 July	12–16 July
Insecticide treatment	210 g/ha fenitrothion	210 g/ha fenitrothion	2 × 140 g/ha fenitrothion	2 × 140 g/ha phosphamidon
Kill/ha of				
Apanteles	–	–	239 500	300 300
Glypta	–	–	11 400	30 100
Other ichneumonoids	–	–	93 700	92 600
Total ichneumonoids	12 400[a]	7 400[a]	344 600	423 000
Tachinids	500	1 000	1 200	3 800

[a] Not broken down into lower taxa.

Effects on Honey Bees

Studies on the effects of forest sprays on honey bee colonies were initiated in 1971 by the Chemical Control Research Institute with the cooperation of Agriculture Canada. Early studies demonstrated mortality of adult bees after spraying with fenitrothion and aminocarb, while Bt applications were found to have no effect on honey bees (Buckner et al. 1974, 1975a). Throughout the mid- and late-1970s, staff at the Institute continued to evaluate the effects of forest sprays on honey bees in small and large experimental spray plots in Ontario as well as in large operational spray blocks in Quebec and New Brunswick. The majority of these studies evaluated the effects of broad spectrum chemical insecticides (Table 4). Results show that, in general, applications of the organophosphate insecticides, fenitrothion and acephate, cause more extensive and prolonged mortality among honey bee colonies than do applications of the carbamates aminocarb and mexacarbate. Synthetic pyrethroids such as permethrin, although highly toxic to bees, have repellent properties which reduce their actual effects in the field (NRCC 1981).

Weather and foraging activity at the time of application can greatly influence the extent to which honey bees are directly affected by spraying applications. Late afternoon and evening sprays generally have little impact on honey bees until the next day. For most insecticides, this significantly lowers the overall impact from the spray, but fenitrothion seems to be capable of causing substantial mortality even when honey bees first encounter residues the day after application. When the spray application occurs mid-day under conditions where bees are actively foraging (e.g. Table 4, aminocarb, 52 g/ha, Buckner et al. 1975a) initial impact can be considerably greater than generally expected. However, most forest spraying occurs early in the morning or in the evening during periods when honey bees tend to be relatively inactive, thereby reducing exposure to the spray cloud. Early season applications (April–May) tend to have less impact than later applications, reflecting generally lower foraging activity and exposure to insecticide residues. The residual effects of fenitrothion on honey bees appear to be considerably greater than for other insecticides, with mortality commonly still occurring 2 to 4 days after spray application. In one instance, fenitrothion was still causing mortality 6 days after an evening application. Aminocarb and the other chemical insecticides tested did not cause mortality beyond the second day post-spray.

In many instances where considerable mortality occurred, pollen collection was depressed for 2 to 3 days after treatment (Table 4). Activity measurements at the hive entrance indicate that affected colonies were characterized by considerable activity removing dead bees from the colony, sometimes accompanied by observations of apparent disorientation of foragers entering and leaving the hive. Continued monitoring of colony strength and honey production in hives exposed to sprays in May and/or June did not produce evidence of lasting effects. In all instances treated hives were similar to untreated control hives by season's end. Shea (1978) reported complete destruction of honey bee colonies due to experimental forest sprays in Oregon with acephate (1.12 and 2.24 kg/ha) and carbaryl (2.24 kg/ha). Extensive damage to a commercial apiary of 44 hives was documented near a 140 g/ha phosphamidon spray block in Quebec in 1975 (Buckner et al. 1975c). Examination of the hives 2 weeks after spraying revealed large numbers of dead bees, few nurse bees, little pollen, nectar, eggs, or brood and abnormal behavior of surviving bees. Eight samples of dead bees contained average phosphamidon residues of 1.14 ppm.

Honey bee colonies set out in forest plots treated with the insect growth regulators diflubenzuron (350 g/ha) and BAY SIR 8514 (280 g/ha) did not suffer abnormal mortality of adults or development of brood (Buckner et al. 1975d; Mìllikin and Mortensen 1980). Applications of the nuclear polyhedrosis viruses of spruce budworm and redheaded jack pine sawfly, *Neodiprion rugifrons*, were also found to have

Table 4. Summary of honey bees studied carried out during May and June in forest spray operations in eastern Canada, 1972–1979.

Insecticide tested	Application rate	Average mortality per colony per day[a]					Comments	Reference
		0	+1	+2	+3	+4		
Fenitrothion	140 g/ha	-	-	-	-	-	No effects observed	Buckner and Sarrazin 1975
	140 g/ha	-	50	11	-	-	Modest depression of pollen collection for several days	Buckner and Sarrazin 1975
	2 × 140 g/ha	43	28	17	-	-	Pollen collection ceased for 3 days after second spray. Experiment disrupted by bear predation	Unpublished data from CCRI's[b] Environmental Impact Project files
		184	65	68	48	12		
	2 × 210 g/ha	-	26	32			Pollen collection ceased for 3 days after second spray. Honey production normal at end of summer	Buckner and McLeod 1977
		-	122	118	118	93*		
	280 g/ha	14	16	-	-	-	Little evidence of effects	Unpublished data, CCRI
	280 g/ha	225	119	46	32	14	Pollen collection depressed for 2 days after spray	Unpublished data, CCRI
	280 g/ha	-	284	56	20	56	Cool rainy weather depressed pollen collection by treatment and control hives several days post-spray	Buckner et al. 1975a
	2 × 280 g/ha	94	-	-	-	-		Unpublished data, CCRI
		159	21	14	-	-		
Aminocarb	52 g/ha	505	46	-	-	-	Bees actively foraging when spray applied. Pollen collection depressed for 2 days. Honey production normal at end of summer	Buckner et al. 1975b
	70 g/ha	13	-	-	-	-	No foraging activity in AM when block partly treated, little foraging when spraying completed in PM	Buckner et al. 1975b
	2 × 70 g/ha	-	85	14	-	-	Pollen collection reduced for 2 days despite good foraging weather after first spray. Wet cold weather after second spray	McLeod 1978; unpublished data CCRI
		-	-	-	-	-		
	2 × 70 g/ha	-	12	-	-	-	No effects on pollen collection, capped brood production or activity	Kingsbury et al. 1981a
		-	-	-	-	-		
Mexacarbate	52 g/ha	-	16	-	-	-		Buckner and Sarrazin 1975
Acephate	560 g/ha	242	87	20	-	-	Pollen collection sharply reduced for 3 days. Honey production normal at end of summer	Buckner and McLeod 1975
Permethrin	2 × 17.5 g/ha	11	-	-	-	-	Bees not foraging until at least 2 hr after both applications	Kingsbury and McLeod 1979
		11	-	-	-	-		

[a] Only given for period over which mortality exceeded that documented for control hives; dash indicates normal level of mortality on that day; asterisk indicates that mortality did not drop to normal levels until 6 days after spraying.
[b] Chemical Control Research Institute.

no effect on honey bee colonies (Buckner et al. 1975e; Kingsbury et al. 1978). Treatments with Bt also had no detectable effects on honey bees (Buckner et al. 1974).

Effects on Wild Pollinators and Pollination of Forest Plants

One of the major areas of research and monitoring studies over the past 15 years concerning forest spraying impacts of terrestrial invertebrates has been effects on wild pollinators and pollination of forest plants. The initial impetus for this research came out of concerns that forest sprays were severely reducing native wild bee populations in commercial blueberry-growing areas of New Brunswick, leading to severe reductions in fruit-set and crops available for harvest (Kevan 1975). The research conducted in New Brunswick to determine the nature and extent of effects of fenitrothion and aminocarb spraying on wild bees and fruit-set of forest plants is described in some detail in the next chapter. In this chapter, studies carried out in other provinces and with other insecticides will be summarized.

The relative toxicities of six chemical insecticides used in forestry to four bee species has recently been evaluated in cooperative studies between the Forest Pest Management Institute's Environmental Impact and Insecticide Toxicology projects. Using a microdoser, acetone-based solutions of the insecticides were topically applied to the thorax of bees anaesthetized with CO_2. Results (Table 5) indicate a broad range of toxicities with different insecticides with the synthetic pyrethroid, permethrin, the most toxic compound to all species. Fenitrothion, aminocarb, and mexacarbate are all fairly similar in toxicity, ranging from 2 to 10 times less toxic than permethrin to various bee species. Carbaryl is considerably less toxic to bees by topical application, and trichlorfon an order of magnitude less toxic than carbaryl. Comparison of the relative sensitivities of different bee species to each insecticide (Table 6) indicates that workers of the honey bee, *Apis mellifera*, are consistently the most susceptible. A native solitary bee, *Andrena erythronii*, is only slightly (1 to 4 times) less susceptible than the honey bee, the domesticated Eurasian alfalfa leafcutting bee, *Megachile rotundata*, generally 3 to 10 times less sensitive, and workers of the native bumblebee, *Bombus terricola*, are generally 10 to 150 times less susceptible. The notable exception to this pattern is the narrow (1 to 3 times) range of relative susceptibility of the four bee species to fenitrothion. Of particular significance is the greater toxicity of fenitrothion (in relation to aminocarb) to the bumblebee (Table 5) compared with the slightly lower relative toxicity of fenitrothion to the honey bee and andrenid bee.

Bouchard (1981) conducted limited caged bumblebee studies in Quebec spray blocks treated with 210 g/ha fenitrothion and 53 g/ha aminocarb. There was complete mortality within 24 hours of treatment among bumblebees exposed to the fenitrothion spray in cages with contaminated foliage. Over half the bumblebees from the aminocarb block survived past 24 hours in the presence of contaminated foliage; however, these bees had been collected 2 hours after the spray. Mortality estimates of bumblebees set out in cages under 280 g/ha carbaryl sprays in New Brunswick were only around 10%, similar to or lower than among control bees over 7 to 10 days after spraying (Holmes et al. 1981). In 1985, a 280 g/ha fenitrothion application was made to an experimental block in the Icewater Creek research area near Sault Ste. Marie and cages of honey bees, leafcutting bees, and bumblebees were examined for spray-induced mortality (Kingsbury et al. 1985; unpubl. data, Forest Pest Management Institute, Forestry Canada). Twenty-four-hour mortalities of *Apis mellifera*, *Megachile rotundata*, and *Bombus terricola*, corrected for control mortality, were 96%, 98%, and 77% respectively, while overall mortality in the same period of 65 bumblebees representing eight species was 93%. Wild pollinators were still active the afternoon of treatment and were found to have picked up fenitrothion residue levels in the range of 0.4 to 0.5 ng/bumblebee and 0.05 ng/solitary bee in the course of their foraging activities. These compare with residues of 0.4 to 1.6 ng/bumblebee and 0.02 to 0.2 ng/honey bee and leafcutting bee found in caged samples collected one hour after spray application.

Bouchard (1981) reported no apparent effects on bumblebee density but reduced diversity of bumblebee species after 210 g/ha fenitrothion and 52 g/ha aminocarb sprays in

Table 5. Relative toxicity of six insecticides to each of four species of bees.

Insecticide	Relative toxicity[a]			
	Apis mellifera	*Andrena erythronii*	*Megachile rotundata*	*Bombus terricola*
Permethrin	1.00	1.00	1.00	1.00
Mexacarbate	0.38	0.21	0.26	0.30
Aminocarb	0.22	0.16	0.27	0.10
Fenitrothion	0.15	0.10	0.48	0.34
Carbaryl	0.07	0.02	0.03	<0.01
Trichlorfon	<0.01	<0.01	<0.01	<<0.01

[a] Based on parallel line probit analysis of 48-hour mortality data (in units of μgAI/g body weight) relative to permethrin (comparisons within columns only).

Table 6. Relative susceptibility of four species of bees to each of six insecticides.

Insecticide	Relative susceptibility[a]			
	Apis mellifera	*Andrena erythronii*	*Megachile rotundata*	*Bombus terricola*
Permethrin	1.00	1.00	0.28	0.10
Mexacarbate	1.00	0.53	0.18	0.08
Aminocarb	1.00	0.71	0.29	0.05
Fenitrothion	1.00	0.67	0.91	0.31
Carbaryl	1.00	0.26	0.11	0.01
Trichlorfon	1.00	0.45	0.10	0.01

[a] Based on parallel line probit analysis of 48-hour mortality data (in units of μgAI/g body weight) relative to *Apis* (comparisons within rows only).

Quebec. He reported that later emerging long-tongued species appeared to profit from and compensate for impacts on earlier emerging short-tongued species. McLeod and Kingsbury (1982) also could not find any indications of reductions of bumblebee numbers after 52 g/ha aminocarb flowable sprays in Quebec. Knockdown of pollinators (mostly flies but including a few bumblebees) was reported from small plots of forest wildflowers in Ontario hand-sprayed with 350 g/ha applications of Matacil® 1.8D and Matacil® 1.8F, two commercial formulations of aminocarb (Kingsbury et al. 1981b). Effects on wild pollinator activity in the plots were slight, and primarily limited to reductions in dipteran visitation to flowers the day of treatment. Numbers of pollinators standardized for the quantity of bloom available, pollinator visitation patterns to bloom, and seasonal changes in pollinator populations were similar in blocks treated twice with 70 g/ha aminocarb in Newfoundland in 1985 and in untreated blocks, indicating no effect of the treatments on insect pollinators (Larson et al. 1987).

Attempts to monitor the impacts of 210 g/ha fenitrothion applications to control hemlock looper, *Lambdina fiscellaria*, in Newfoundland have been limited by low and highly variable pollinator abundance at the time of spray application, but have produced little evidence of spray impact (Strong 1987; Fudge Lane and Associates Ltd. 1987). Fruit-set studies with raspberries *Rubus* sp. (Strong 1987) and buttercups, *Ranunculus* spp. (Fudge Lane and Associates Ltd. 1987) were limited by methodology and spray-timing problems, but did not indicate any spray related impacts.

Studies in New Brunswick (Thaler and Plowright 1980; Thomson et al. 1985) have shown that fenitrothion, and to a lesser extent aminocarb, can reduce wild pollinator activity enough to cause reduced fruit-set in forest plants dependent upon insect visitors to transfer pollen between plants. Seed-set in *Clintonia borealis*, a species sensitive to fenitrothion spraying, did not appear to be affected in a 400-ha forest block treated with two 280 g/ha applications of carbaryl (Holmes et al. 1981). Studies in Quebec spray blocks treated with fenitrothion and aminocarb report fruit-set reductions in *Clintonia borealis*, *Cornus canadensis*, *Maianthemum canadense*, and *Aralia nudicaulis*, and reduced seed production in *Clintonia borealis* (SOMER 1985). Recent studies at the Forest Pest Management Institute have evaluated a number of other forest wildflowers for their usefulness as indicators of spray impacts on pollinators, with *Polygonatum pubescens*, emerging as a useful species (Barber and Kingsbury 1987; Barber 1988).

References

Anon. 1980. 1979 environmental monitoring of the spruce budworm spray program in Newfoundland. Dep. Consumer Affairs Environ., Res. and Assess. Br., St. John's, Nfld. Inf. Rep. RA-80-1. 18 p.

Anon. 1987. Environmental monitoring of the 1980 spruce budworm spraying program in Newfoundland. Dept. Environ., Environ. Assess. Div., St. John's, Nfld. Inf. Rep. PM-1. 21 p.

Barber, K.N. 1988. Development of a monitoring protocol for using the fecundity of *Polygonatum pubescens* as an indicator of impacts of forest sprays on bumblebees. Can For. Serv., For. Pest Manage. Inst., Sault Ste. Marie, Ont. File Rep. No. 92. 59 p.

Barber, K.N.; Kingsbury, P.D. 1987. Developing monitoring techniques to measure insecticidal impacts on solitary bees and the fecundity of selected forest wildflowers. Can. For. Serv., For. Pest Manage. Inst., Sault Ste. Marie, Ont. File Report No. 82. 50 p.

Bendell, J.F.; Naylor, B.J.; Szuba, K.J.; Innes, D.G.L.; James, R.D.; Smith, B.A. 1986. Jack pine plantations, outbreaks of jack pine budworm, birds, and mammals, and impacts of *Bacillus thuringiensis* (*Bt.*), Matacil and fenitrothion. Pages 56-63 *in* Jack Pine Budworm Information Exchange. Dep. Nat. Resour., Winnipeg, Man. 95 p.

Bouchard, Y. 1981. Impact des pulverisations aériennes de fénitrothion et d'aminocarb sur les populations de bourdons au Québec en 1991. Département de biologie, Université Laval. 42 p.

Bracher, G.A.; Bider, J.R. 1982. Changes in terrestrial animal activity of a forest community after an application of aminocarb (Matacil®). Can. J. Zool. 60: 1981-1997.

Buckner, C.H.; Gochnauer, T.A.; McLeod, B.B. 1975a. The impact of aerial spraying of insecticides on bees. Pages 276-279 *in* M.L. Prebble (Ed.), Aerial Control of Forest Insects in Canada. Dep. Environ., Ottawa, Ont. 330 p.

Buckner, C.H.; Kingsbury, P.D.; McLeod, B.B.; Mortensen, K.L.M.; Ray, D.G.H. 1974. Section F: Impact of aerial treatment on non-target organisms, Algonquin Park, Ontario and Spruce Woods, Manitoba. 72 p. *In* Evaluation of commercial preparations of *Bacillus thuringiensis* with and without chitinase against spruce budworm. Can. For. Serv., Chem. Control Res. Inst., Ottawa, Ont. Inf. Rep. CC-X-59.

Buckner, C.H.; McLeod, B.B. 1975. Impact of aerial applications of Orthene® upon non-target organisms. Can. For. Serv., Chem. Control Res. Inst., Ottawa, Ont. Inf. Rep. CC-X-104. 48 p.

Buckner, C.H.; McLeod, B.B. 1977. Ecological impact studies of experimental and operational spruce budworm (*Choristoneura fumiferana* Clemens) control programs on selected non-target organisms in Quebec, 1976. Can. For. Serv., Chem. Control Res. Inst., Ottawa, Ont. Inf. Rep. CC-X-137. 81 p.

Buckner, C.H.; McLeod, B.B.; Kingsbury, P.D. 1975b. Studies of the impact of the carbamate insecticide Matacil® on components of forest ecosystems. Can. For. Serv., Chem. Control Res. Inst., Ottawa, Ont. Inf. Rep. CC-X-91. 59 p.

Buckner, C.H.; McLeod, B.B.; Kingsbury, P.D. 1975c. Accident investigation activities of the ecological impact team in forest areas treated with insecticide in 1975. Can For. Serv., Chem. Control Res. Inst., Ottawa, Ont. Inf. Rep. CC-X-103. 22 p.

Buckner, C.H.; McLeod, B.B.; Kingsbury, P.D. 1975d. The effect of an experimental application of Dimilin® upon selected forest fauna. Can. For. Serv., Chem. Control Res. Inst., Ottawa, Ont. Inf. Rep. CC-X-97. 26 p.

Buckner, C.H.; McLeod, B.B.; Kingsbury, P.D. 1975e. The effect of an experimental application of nuclear polyhedrosis virus upon selected forest fauna. Can. For. Serv., Chem. Control Res. Inst., Ottawa, Ont. Inf. Rep. CC-X-101. 25 p.

Buckner, C.H.; Sarrazin, R. 1975. Studies of the environmental impact of the 1974 spruce budworm control operation in Quebec. Can. For. Serv., Chem. Control Res. Inst., Ottawa, Ont. Inf. Rep. CC-X-93. 103 p.

Carter, N.E.; Brown, N.R. 1973. Seasonal abundance of certain soil arthropods in a fenitrothion-treated red spruce stand. Can. Entomol. 105: 1065-1073.

Danks, H.V. (Ed.). 1979. Canada and its insect fauna. Mem. Entomol. Soc. Can. 108. 573 p.

Edwards, C.A.; Lofty, J.R. 1977. The biology of earthworms. 2nd ed. Chapman and Hall, London. 333 p.

Eidt, D.C.; Mallet, V.N. 1986. Accumulation and persistence of fenitrothion in needles of balsam fir and possible effects on abundance of *Neodiprion abietis* (Harris) (Hymenoptera: Diprionidae). Can. Entomol. 118: 69-77.

Eidt, D.C.; Pearce, P.A. 1986. The biological consequences of lingering fenitrothion residues in conifer foliage — a synthesis. For. Chron. Aug. 1986: 246-249.

Eidt, D.C.; Weaver, C.A.A. 1985. Effects of fenitrothion and adjuvants on the decomposition rate of white spruce foliage in litter bags on the forest floor. Can. J. For. Res. 15: 174-176.

Eidt, D.C.; Weaver, C.A.A. 1986. Effects of aminocarb and fenitrothion formulations on decomposition rate of balsam fir foliage in litter bags on the forest floor. Can. J. For. Res. 16: 145-146.

EMOFICO. 1976. Environmental effects of the spruce budworm spray program in New Brunswick - 1976. Can For. Serv., Marit. For. Res. Cent., Fredericton, N.B. Inf. Rep. M-X-67. 21 p.

Environmental Monitoring Committee. 1979a. 1977 environmental monitoring of the spruce budworm spray program in Newfoundland. Dep. Consumer Affairs Environ., Res. and Assess. Br., St. John's, Nfld. Inf. Rep. RA-79-1. 21 p. + app.

Environmental Monitoring Committee. 1979b. 1978 environmental monitoring of the spruce budworm spray program in Newfoundland. Dep. Consumer Affairs Environ., Res. and Assess. Br., St. John's, Nfld. Inf. Rep. RA-79-2. 29 p. + app.

Fudge, Lane and Associates Ltd. 1987. A study to monitor the effects of aerially applied fenitrothion on the activity of Newfoundland pollinating insects. Dep. Environ., Environ. Assess. Div., St. John's, Nfld. 11 p. + figs. and tables.

Gesner, G.N.; Varty, I.W. 1973. Side effects of spray program: mortality of arthropod predators and parasites. Pages 28-35 *in* C.A. Miller, J.F. Stewart, D.E. Elgee, D.D. Shaw, M.G. Morgan, E.G. Kettela, D.O.Greenbank, G.N. Gesner, and I.W. Varty (Eds.), Aerial spraying against spruce budworm adults in New Brunswick. A compendium of reports on the 1972 test program. Can. For. Serv., Marit. For. Res. Cent., Fredericton, N.B. Inf. Rep. M-X-38. 35 p.

Holmes, S.B.; Kingsbury, P.D. 1980. The environmental impact of nonyl phenol and the Matacil® formulation. Part 1: Aquatic ecosystems. Can. For. Serv., For. Pest Manage. Inst., Sault Ste. Marie, Ont. Report FPM-X-35. 52 p. + app.

Holmes, S.B.; Millikin, R.L. 1981. A preliminary report on the effects of a split application of Reldan® on aquatic and terrestrial ecosystems. Can. For. Serv., For. Pest Manage. Inst., Sault Ste. Marie, Ont. File Rep. No. 13. 34 p. + app.

Holmes, S.B.; Millikin, R.L.; Kingsbury, P.D. 1981. Environmental effects of a split application of Sevin®-2-oil. Can. For. Serv., For. Pest Manage. Inst., Sault Ste. Marie, Ont. Rep. FPM-X-46. 58 p. + app.

Kettela, E.G.; Varty, I.W. 1972. Assessment of the 1971 spruce budworm aerial spraying program in New Brunswick and forecast for 1972. Can. For. Serv., Marit. For. Res. Cent., Fredericton, N.B. Inf. Rep. M-X-29. 30 p.

Kevan, P.G. 1975. Forest application of the insecticide fenitrothion and its effect on wild bee pollinators (Hymenoptera: Apoidea) of lowbush blueberries (*Vaccinium* spp.) in southern New Brunswick, Canada. Biol. Conserv. 7: 301-309.

Kevan, P.G.; Baker, H.G. 1983. Insects as flower visitors and pollinators. Ann. Rev. Entomol. 28: 407-453.

Kingsbury, P.D.; Holmes, S.B. 1980. A preliminary report on the impact on stream fauna of high dosage applications of Reldan®. Can. For. Serv., For. Pest Manage. Inst., Sault Ste. Marie, Ont. File Rep. No. 2. 27 p. + app.

Kingsbury, P.D.; Holmes, S.B.; Millikin, R.L. 1980. Environmental effects of a double application of azamethiphos on selected terrestrial and aquatic organisms. Can. For. Serv., For. Pest Manage. Inst., Sault Ste. Marie, Ont. Rep. FPM-X-33. 32 p. + app.

Kingsbury, P.D.; Kreutzweiser, D.P. 1980. Environmental impact assessment of a semi-operational permethrin application. Can. For. Serv., For. Pest Manage. Inst., Sault Ste. Marie, Ont. Rep. FPM-X-30. 47 p. + app.

Kingsbury, P.D.; McLeod, B.B. 1979. Terrestrial impact studies in forest ecosystems treated with double applications of permethrin. Can. For. Serv., For. Pest Manage. Inst., Sault Ste. Marie, Ont. Report FPM-X-28. 54 p. + app.

Kingsbury, P.D.; McLeod, B.B.; Millikin, R.L. 1981a. The environmental impact of nonyl phenol and the Matacil® formulation. Part 2: Terrestrial ecosystems. Can. For. Serv., For. Pest Manage. Inst., Sault Ste. Marie, Ont., Rep. FPM-X-36. 47 p. + app.

Kingsbury, P.; McLeod, B.; Mortensen, K. 1978. Impact of applications of the nuclear polyhedrosis virus of the red-headed pine sawfly, *Neodiprion lecontei* (Fitch), on non-target organisms in 1977. Can. For. Serv., For. Pest Manage. Inst., Sault Ste. Marie, Ont. Rep. FPM-X-11. 27 p.

Kingsbury, P.D.; McLeod, B.B.; Mortensen, K.L. 1981b. Preliminary report on studies of the impact of Matacil® formulations on honeybees and wild pollinators. Can. For.

Serv., For. Pest Manage. Inst., Sault Ste. Marie, Ont. File Rep. No. 18. 38 p.

Kingsbury, P.D.; Millikin, R.L.; Sundaram, K.M.S. 1985. Preliminary report on the 1985 experimental fenitrothion spray in the Icewater Creek research area. Can. For. Serv., For. Pest Manage. Inst., Sault Ste. Marie, Ont. File Rep. No. 69. 6 p.

Kreutzweiser, D.P. 1982. The effects of permethrin on the invertebrate fauna of a Quebec forest. Can. For. Serv., For. Pest Manage. Inst., Sault Ste. Marie, Ont. Rep. FPM-X-50. 44 p. + app.

Kreutzweiser, D.P. 1983. Terrestrial invertebrate knockdown studies. Pages 96-102 *in* P.D. Kingsbury (Ed.), Permethrin in New Brunswick salmon nursery streams. Can. For. Serv., For. Pest Manage. Inst., Sault Ste. Marie, Ont. Rep. FPM-X-52. 192 p.

Larson, D.J.; Constantine, J.; McCarthy, M. 1987. Impact of 1985 forest insecticide spraying on Newfoundland pollinating insects. Dep. Environ., Environ. Assess. Div., St. John's, Nfld. 68 p.

Lhemkuhl, D.M.; Danks, H.V.; Behan-Pelletier, V.M.; Larson, D.J.; Rosenberg, D.M.; Smith, I.M. 1984. Recommendations for the appraisal of environmental disturbance: some general guidelines, and the value and feasibility of insect studies. Suppl. Bull. Entomol. Soc. Can. 16(3): 8 p.

Magasi, L.P.; Titus, F.A.; Kettela, E.G. 1980. Parasitism and biological control. Pages 46-47 *in* EMOFICO 1980, Environmental surveillance in New Brunswick, 1978-79. Effects of spray operations for forest protection against spruce budworm. Dep. For. Resour., Univ. New Brunswick, Fredericton. 76 p.

McLeod, B.B. 1978. Effects of aminocarb on honeybees. Page 11 *in* EMOFICO 1978, 1977 environmental surveillance of insecticide spray operations in New Brunswick's budworm-infested forests. Can. For. Serv., Marit. For. Res. Cent., Fredericton, N.B. Inf. Rep. M-X-87. 24 p.

McLeod, B.B. 1982. Terrestrial impact studies on experimental aqueous formulations of spruce budworm, *Choristoneura fumiferana* Clem., control agents containing Triton® X-100, New Brunswick 1982. Can. For. Serv., For. Pest Manage. Inst., Sault Ste. Marie, Ont. File Rep. No. 37. 21 p.

McLeod, B.B.; Kingsbury, P.D. 1982. Terrestrial impact studies of Matacil® Flowable formulated in vegetable oil, Quebec 1982. Can. For. Serv., For. Pest Manage. Inst., Sault Ste. Marie, Ont. File Rep. No. 41. 29 p.

McLeod, J.M. 1975. Possible residual effect of fenitrothion on Swaine jack pine sawfly following aerial applications against the spruce budworm in Quebec. Ann. Entomol. Soc. Que. 20: 82-85.

McNeil, J.N.; Greenhalgh, R.; McLeod, J.M. 1979. Persistence and accumulation of fenitrothion residues in jack pine foliage and their effect on the Swaine jack pine sawfly, *Neodiprion swainei* (Hymenoptera: Diprionidae). Environ. Entomol. 8: 752-755.

McNeil, J.N.; McLeod, J.M. 1977. Apparent impact of fenitrothion on the Swaine jack pine sawfly, *Neodiprion swainei*, Midd. Pages 203-215 *in* R.J. Roberts, R. Greenhalgh and W. Marshall (Eds.), Fenitrothion: long-term effects of its use in forest ecosystems. Associate Committee on Scientific Criteria for Environmental Quality, Natl. Res. Counc. Canada, Ottawa, Ont. NRCC No. 16073. 628 p.

Miller, C.A.; Stewart, J.F.; Elgee, D.E.; Shaw, D.D.; Morgan, M.G.; Kettela, E.G.; Greenbank, D.O.; Gesner, G.N.; Varty, I.W. 1973. Aerial spraying against spruce budworm adults in New Brunswick: a compendium of reports on the 1972 test program. Can. For. Serv., Marit. For. Res. Cent., Fredericton, N.B. Inf. Rep. M-X-38. 35 p.

Millikin, R.L. 1981. Preliminary report on the impact of Matacil® formulations on terrestrial invertebrates, Bathurst, New Brunswick, 1981. Can. For. Serv., For. Pest Manage. Inst., Sault Ste. Marie, Ont. File Rep. No. 15. 20 p.

Millikin, R.L. 1987. Sublethal effects of fenitrothion on forest passerines. M.Sc. Thesis. Univ. British Columbia. Vancouver, B.C. 126 p.

Millikin, R.L.; Mortensen, K.L. 1980. A preliminary report on the environmental impact of an experimental application of the insect growth regulator Bay Sir 8514, Wawa, Ontario. 1979. Can. For. Serv., For. Pest Manage. Inst., Sault Ste. Marie, Ont. File Rep. No. 3 28 p. + app.

Morris, O.N. 1977. Long term study of the effectiveness of aerial application of *Bacillus thuringiensis* – acephate combinations against the spruce budworm, *Choristoneura fumiferana* (Lepidoptera: Tortricidae). Can. Entomol. 109: 1239-1248.

Morris, O.N.; Hildebrand, M.J. 1974. Section E: Assessment of effectiveness of aerial application, Algonquin Park, Ontario. 53 p. *In* Evaluation of commercial preparations of *Bacillus thuringiensis* with and without chitinase against spruce budworm. Can. For. Serv., Chem. Control Res. Inst., Ottawa, Ont. Inf. Rep. CC-X-59.

NRCC. 1981. Pesticide–pollinator interactions. Associate Committee on Scientific Criteria for Environmental Quality, Natl. Res. Counc. Canada, Ottawa, Ont. NRCC No. 18471. 190 p.

Otvos, I.S.; Raske, A.G. 1980. The effects of fenitrothion, Matacil and *Bacillus thuringiensis* plus Orthene on larval parasites of the spruce budworm, *Choristoneura fumiferana* (Lepidoptera: Tortricidae). Can. For. Serv., Nfld. For. Res. Cent., St. John's, Nfld. Inf. Rep. N-X-184. 24 p.

Reynolds, J.W. 1977. The earthworms (Lumbricidae and Sparganophilidae) of Ontario. Life Sci. Misc. Publ., R. Ont. Mus., Toronto, Ont. 141 p.

Rosenberg, D.M.; Danks, H.V.; Lhemkuhl, D.M. 1986. Importance of insects in environmental impact assessment. Environ. Manage. 10(6): 773-783.

Sarrazin, R. 1978. Effects environnementaux des pulvérisations aériennes contre la tordeuse des bourgeons de l'épinette au Québec - 1978. Rapport préliminaire du Sous-comité de l'Environnement du Comité de Surveillance Écologique des Pulvérisations Aériennes. Rapport de Recherche Faunique No. 36. Ministère du Tourisme, de la Chasse et de la Pêche, Québec, P.Q.

Shea, P.J. 1978. Environmental safety. Pages 122-129 *in* M.H. Brookes, R.W. Stark and R.W. Campbell (Eds.), The Douglas-fir tussock moth: a synthesis. U.S. Dep. Agric., Washington, D.C. Tech. Bull. 1585.

Shepherd, R.F. (Ed.). 1980. Operational field trails against the Douglas-fir tussock moth with chemical and biological insecticides. Can. For. Serv., Pac. For. Res. Cent., Victoria, B.C. Rep. BC-X-201, 19 p.

SOMER. 1985. Effets des insecticides fénitrothion et aminocarbe sur la fructification de quatre espèces végétales en milieu forestier. Gouv. du Québec, Min. Énergie et Ressources, Dir. de la Conservation. 208 p.

Southwood, T.R.E. 1978. Ecological methods with particular reference to the study of insect populations. 2nd ed. Chapman and Hall, London, Eng. 524 p.

Stenersen, J. 1984. Detoxification of xenobiotics by earthworms. Comp. Biochem. Physiol. 78C: 249-252.

Strong, K.W. 1987. A study to monitor the impact of the 1985 hemlock looper insecticide spray program on pollinators and berry production. Dep. Environ., Environ. Assess. Div., St. John's, Nfld. 33 p.

Sunderland, K.D. 1988. Quantitative methods for detecting invertebrate predation occurring in the field. Ann. Appl. Biol. 112: 201-224.

Thaler, G.R.; Plowright, R.C. 1980. The effect of aerial insecticide spraying for spruce budworm control on the fecundity of entomophilous plants in New Brunswick, Canada. Can. J. Bot. 58: 2022-2027.

Thomson, J.D.; Plowright, R.C.; Thaler, G.R. 1985. Matacil insecticide spraying, pollinator mortality, and plant fecundity in New Brunswick forests. Can. J. Bot. 63: 2056-2061.

Varty, I.W. 1975. Side effects of pest control projects on terrestrial arthropods other than the target species. Pages 266-275 *in* M.L. Prebble (Ed.), Aerial control of forest insects in Canada. Dep. Environ., Ottawa, Ont. 330 p.

Varty, I.W. 1977. Long-term effects of fenitrothion spray programs on non-target terrestrial arthropods. Pages 343-375 *in* R.J. Roberts, R. Greenhalgh and W. Marshall (Eds.), Fenitrothion: long-term effects of its use in forest ecosystems. Associate Committee on Scientific Criteria for Environmental Quality, Natl. Res. Counc. Canada, Ottawa, Ont. NRCC No. 16073. 628 p.

Varty, I.W. 1978. Hazards to non-target insects. Pages 14-15 *in* EMOFICO 1978. 1977 environmental surveillance of insecticide spray operations in New Brunswick's budworm-infested forests. Can. For. Serv., Marit. For. Res. Cent., Fredericton, N.B. Inf. Rep. M-X-87. 24 p.

Varty, I.W. 1980. Effects on predaceous arthropods in trees. Pages 44-46 *in* EMOFICO 1980, Environmental surveillance in New Brunswick, 1978-79. Effects of spray operations for forest protection against spruce budworm. Dep. For. Resour., Univ. New Brunswick, Fredericton. 76 p.

Varty, I.W. 1982. Environmental impact of aerial spraying on predaceous arthropods in fir stands. Pages 70-72 *in* EMOFICO 1982. Environmental surveillance of spray operations for forest protection in New Brunswick, 1980-81. Environ. New Brunswick, Fredericton. 130 p.

Varty, I.W.; Carter, N.E. 1974. Inventory of litter arthropods and airborne insects in fir-spruce stands treated with insecticides. Can. For. Serv., Marit. For. Res. Cent., Fredericton, N.B. Inf. Rep. M-X-48. 32 p.

Wan, M.T.K. 1974. Effect on non-target arthropods. Pages 40-45 *in* J.R. Carrow (Ed.), Aerial spraying operations against black-headed budworm on Vancouver Island – 1973. Can. For. Serv., Pac. For. Res. Cent., Victoria, B.C. Inf. Rep. BC-X-101. 56 p. + app.

Woodworth-Lynas, C.; Bennett, G.; Knoechel, R. 1987. Environmental monitoring of the 1982 spruce budworm spraying program in Newfoundland. Dep. Environ., Environ. Assess. Div., St. John's, Nfld. 139 p.

Chapter 59

Impact of Pesticides on Forest Pollination

P.G. Kevan and R.C. Plowright

Introduction

Although pollination can be defined simply as the transfer of pollen from the microsporophylls (e.g. anthers) of one plant to the megasporophylls (e.g. pistil) of the same or another plant (Figs. 1 and 2), the total process, together with the events that precede it and those that follow, is deceivingly complex. Not only that, pollination is a keystone process in all biomes of the world and inextricably links, both directly and indirectly, the vast majority of the floral and faunal biomass. Thus, the vital importance of pollination should not be overlooked or denigrated, as it has so often been, in considerations of the impact of environmental perturbations like the application of pesticides. In fact, the value of insect pollination of crops in Canada has been estimated to be more than $1.2 billion (Winston and Scott 1984) and nearly $20 billion in the United States (Levin 1983).

In this chapter, we intend to explore the ramifications of pesticidal impact on pollination and related processes. Unfortunately, information on the pollination requirements of much of the Canadian flora is seriously lacking, and the importance of anthophily (flower visiting) in the lives of both pestiferous and beneficial insects is hardly recognized, and at best, only in passing. Understanding the role of pollination in the rete of ecological communities, including agricultural and especially natural ones, is in its infancy (Kevan and Baker 1983, 1984).

From the foregoing, one might reasonably ask, Is there enough information to discuss the impact of pesticides on forest pollination? Answering that question requires delving into a diverse scientific literature and relating it to what we know about pollination and community ecology in general. In Canada, a prime example of the effects of pesticides on pollination at community levels, including forests, is the use of fenitrothion and other insecticides to attempt control of the spruce budworm, *Choristoneura fumiferana*. This chapter relies heavily on the lessons of that example and uses the results of many diverse studies as they apply. Before venturing into the effects of pesticides on pollination, it is important to itemize the components of the system in general from both botanical and zoological viewpoints. Figure 3a shows the variety of rewards and attractants that flowers offer their visitors, including animal pollinators. On the botanical phase can be seen the reproductive requirements of the plants. Insect pollination is generally accepted as basic to the flowering plants (Angiospermae), and self-pollination, asexual processes, and wind pollination as derived characters. In the Gymnospermae (conifers in Canada), wind pollination is the general rule. Pollination does not really apply to the cryptogamic flora, although an analogous process mediated by arthropods may have been important to very early tracheophytes and predisposed evolving higher plants and insects to mutualistic interrelationships (Kevan and Baker 1983; Crepet 1983).

The Canadian flora is young by any standards, having originated since the retreat of the Wisconsin ice sheets. However, it should not be assumed, therefore, that only generalized, opportunistic, and resilient biotic interactions apply. The symbiotic relationships of organisms in Canadian environments likely did not evolve *in situ* (as is certainly implied by the paucity of endemism in Canada [Danks 1979]), and many symbionts in Canada likely have come together in combinations somewhat different from those in which the individual partners occurred before their post-glacial migrations. Nevertheless, pollination systems exemplify a wide range of associations, from highly evolved and precise one-on-one relationships

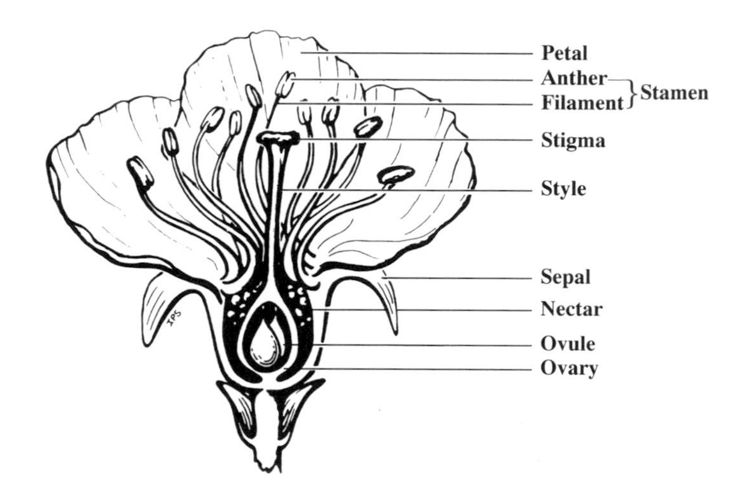

Figure 1. Parts of a flower. Nectar is secreted in droplets from nectaries; pollen is shed as microscopic grains from the anthers.

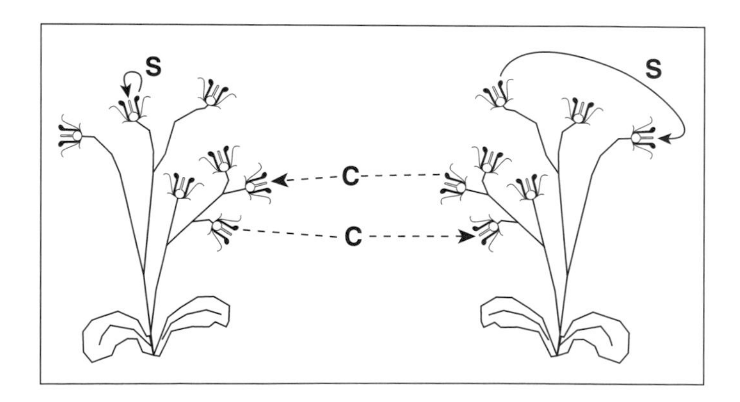

Figure 2. Cross- and self-pollination. Note that self-pollination is very frequent but will not result in fertilization unless the plant is a self-fertile species or variety.

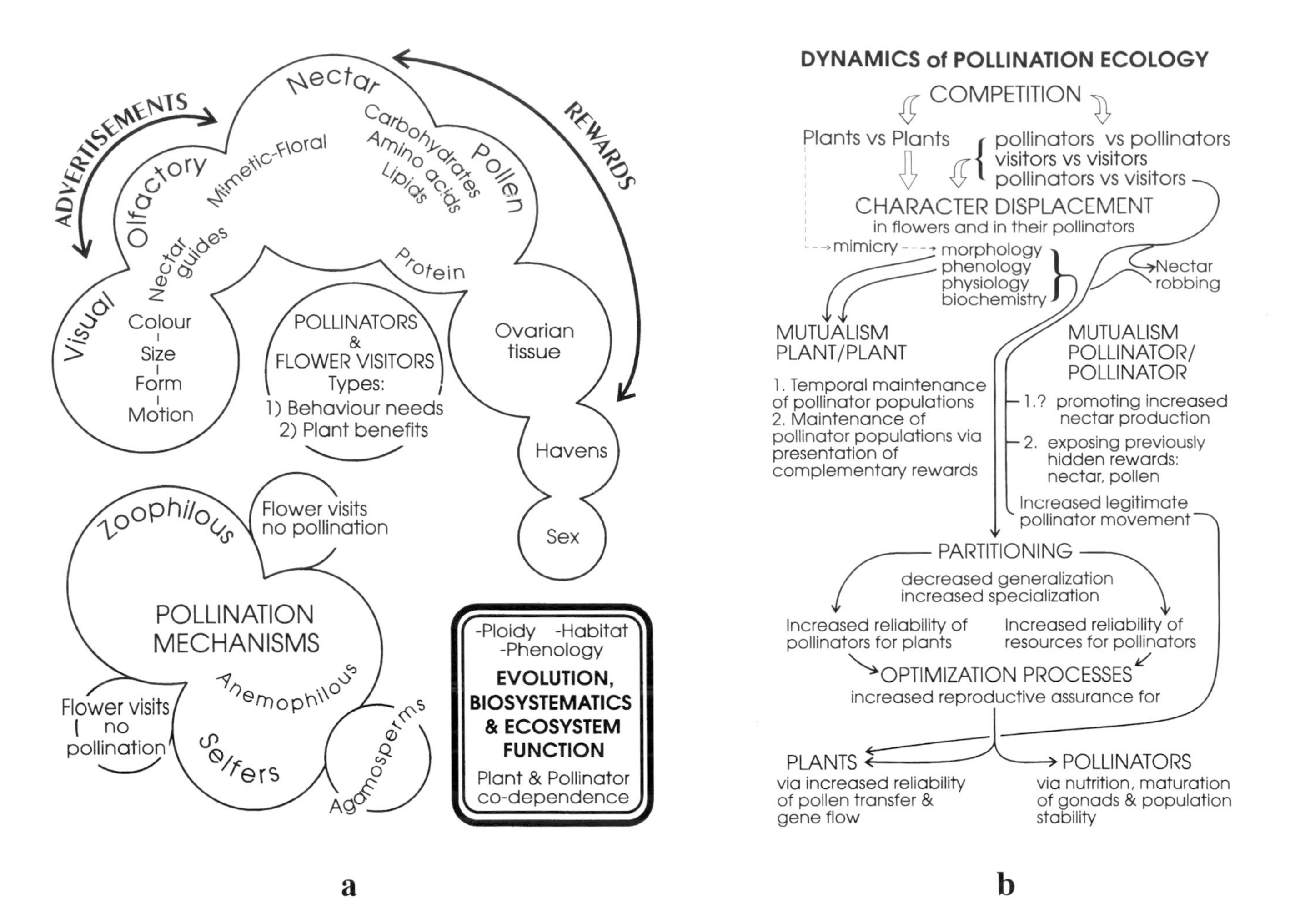

Figure 3. Components of pollination. (a) The dimensions of pollination. This entomocentric view shows the relationships of floral attractants and rewards on pollinators and insect visitors and the roles of the latter on plant reproduction. The processes involved lead to understandings of evolution, ecology, and biosystematics. (b) The dynamics of pollination starts with aspects of competition in pollination systems and follows their consequences through mainstream processes of character displacement, partitioning, and optimization, all of which heighten the mutualism basic to evolution in pollination systems. Side issues of mutualisms between plants and between anthophiles may heighten the effectiveness of the basic mutualism in assuring the reproductive success of both plants and pollinators. (Adapted with permission from Kevan and Baker 1984)

(monolectic pollinators and monophilous plants) to others that involve polylectic pollinators that visit the flowers of many species of plant and polyphilous plants that use a wide variety of pollinators. Figure 3b shows the way in which the community dynamics function.

In short, pollination must be regarded as a keystone process in ecology generally, one that is vulnerable to environmental perturbations, especially as caused by pesticides. The disruption of pollination can be expected to have far-reaching consequences of direct concern to human welfare, and the situation in Canadian forests is no different (Kevan 1991) (Fig. 4).

Components of Pollination and Pesticides

Figure 5 lists about 19 components of pollination that can be influenced by toxic chemicals. The diagram attempts to place the components in a logical sequence, starting with the pollinators and the plant, and proceeding through the process of fertilization in plants and eventuating seed-set as that may influence human affairs and the economy of nature. The process is cyclic in that the demise of pollination reduces the reproductive output

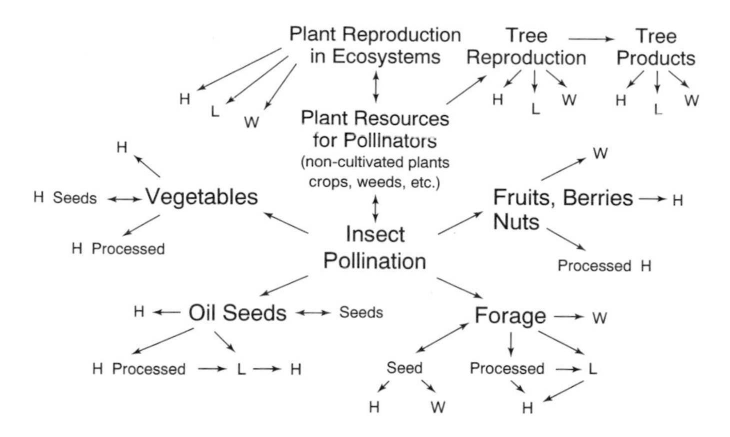

Figure 4. Insect pollination and the consumer (human, H; livestock, L; and wildlife, W).

of plants, which in turn eventually adversely affects the reliability of pollination. Few of the components given in Figure 5 have been investigated with respect to Canadian forests; nevertheless, we draw attention to the general body of relevant literature.

Toxicology and Pollinators

General Considerations

Toxicological information on pesticides is most complete for the target organism and in reference to human health. Nontarget organisms are generally ignored and the greatest body of knowledge available is that for honey bees, *Apis mellifera*. The greatest danger of pesticide poisonings to nontarget organisms exists in agricultural settings. There, the value of honey bees for pollination is in debate, but it far exceeds the value of hive products (Robinson et al. 1989; Southwick and Southwick 1989). Several reference books on pesticides include data on toxicity to honey bees. There are compendia of existing information for honey bees and some other pollinators (e.g. *Megachile centruncularis* and a few others) (National Research Council Canada 1981; Johansen and Mayer 1990). The motivation for obtaining that sort of information has been the protection of crop pollinators, and ultimately the crop itself. However, within the natural environment, and in natural pollination systems, there is little information.

Despite the state of general ignorance, some generalizations about the effects of pesticides on pollinators can be made. The data on the toxicity of pesticides to honey bees can be used as general guidelines for other bees. The formulations

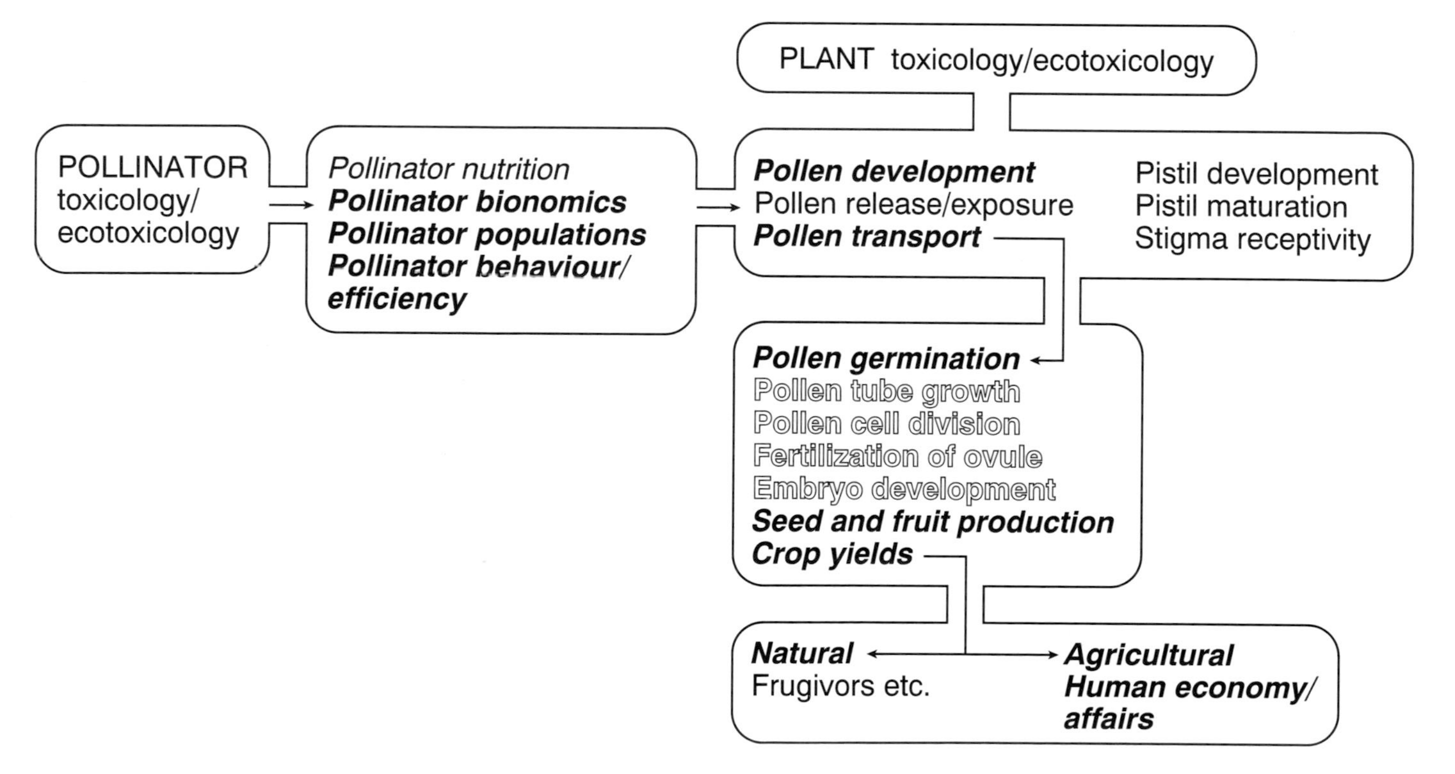

Figure 5. Components of pollination directly affected by insecticides used in forestry in Canada (bold italic), indirectly affected (italic), and presumably affected (outline) together with processes known to be directly affected (arrow.) (Adapted with permission from Kevan and Plowright [1989], p. 26, fig. 2.5)

can be ranked in terms of their relative hazard. Granular formulations are the safest, and fine sprays are safer than coarse ones, which are safer than dusts. Emulsifiable and water soluble concentrates quickly become safe for bees after drying. Adjuvants such as solvents or oily substances tend to make sprays safer for bees, and stickers provide additional protection. The latter is true even when microencapsulated formulations, which are extremely hazardous to bees that collect the microcapsules with pollen, are used. Insecticides are the most toxic, followed by miticides, herbicides and fungicides. All these generalizations can be applied to other nontarget insects, provided that the manner in which the organisms would come into contact with the pesticide is similar to that for honey bees.

The principle by which pesticides are developed and used is that of selective toxicity, and so it is difficult to generalize about the classes of insecticides. Only the biological insecticides (Bt toxin, viruses) are known to not affect honey bees and probably constitute little or no threat to other Hymenoptera. However, Bt is known to affect a wide variety of Lepidoptera, some of which are important pollinators, even in Canadian forests and on Canadian trees (e.g. Kentucky coffee tree is nocturnally pollinated by Noctuidae) (Kevan and Ambrose, unpubl. data).

Among the bees, a generalization that is widely cited is that pesticides are a greater hazard to smaller bees than to larger ones. This principle must be applied with great care, however, because many exceptions are known (Torchio 1973).

Table 1 lists pesticides that have actual or potential use in Canadian forestry and indicates their relative toxicity to honey bees. Also included in the list is the "hazard ratio," which Felton et al. (1986) explain is a useful guideline for assessing the likely hazard a pesticide may present to honey bees. The ratio is calculated as follows:

$$H = a / LD_{50}$$

where H is the hazard ratio, a is the application rate in grams active ingredient per hectare, and LD_{50} is the LD_{50} in micrograms active ingredient per bee for oral and/or contact toxicity. If the value for H is equal to or exceeds 2 500, the compound should be regarded as dangerous to honey bees. If the value for H is less than 50, the compound can be considered nondangerous. If the value for H falls between 50 and 2 500, cage trials, field trials, or both are required before conclusions can be made. After trials, even if the conclusion is that the compound is nondangerous, investigators must be aware of special problems arising, including those of sublethal effects ("Sublethal Effects" section follows).

Table 1. Pesticides used in Canadian forestry with their relative toxicity[a] (H = high; M = medium; L = low; N = relatively nontoxic), slope probit, and hazard ratio to honey bees for application rates noted in Nigam (1975) with notes on other pollinators.

						Bee species[b]			
Pesticides	Class	Relative toxicity	LD_{50}	Slope probit	Hazard ratio	*Meg. cen.*	*Nom. mel.*	*Bombus* spp.	Other
Aminocarb	Carb.	H	1.12	3.61	31-1000	H[c]		M-L[d]	
Carbaryl	Carb.	H	1.54	3.04	364-910	H	H	H	
Fenitrothion	OP	H	0.176	5.75	204-8363	H[c]	H	H[d]	H[d]
DDT[e]	CHC	M	6.29	4.74	45-905	H	M	M-L	
Dimethoate	OP	H	0.191	5.84	1173-5870	H	H	H	
Mexacarbate	Carb.	H	0.302	4.87	59-1485				
Phosphamidon	OP	H	1.45	12.74	25-1546	H	H		
Trichlorfon	OP	L	>11	-	14-76	M-L[f]	M-L	M-L	
Bt	Bact.	N							
2-4-D[g]	Herb.	N							
Glyphosate	Herb.	N							

[a] Source: National Research Council Canada (1981), unless otherwise noted.
[b] *Megachile centruncularis* (*Meg. cen.*); *Nomia melanderi* (*Nom. mel.*); H means that the material is hazardous to pollinators at any time; M-L means that the material should be applied only while pollinators are not foraging.
[c] Thorpe (1980) measured the LD_{50} of aminocarb and fenitrothion on males and females of *Megachile centruncularis* at 0.23 and 0.45 (µg/g aminocarb for m and f, respectively) and 0.17 and 0.27 (µg/g fenitrothion for m and f, respectively.
[d] Details in Kevan and Plowright (1989).
[e] The ages of honey bees had been shown to influence the toxicity of DDT (Ladas 1972).
[f] But has sublethal effects (Torchio 1983).
[g] The nutritional states and ages of honey bees have been shown to influence the toxicity of 2,4-D (Wahl 1976).

Pollinator Poisonings in Forestry

In the context of Canadian forests, most of the available information comes from New Brunswick, where the application of pesticides to control spruce budworm has been made annually since 1952 (except for 1959). The potential for impact on forest pollinators was not recognized during the 1950s and 1960s, but it is highly likely that some pollinators were adversely affected throughout those decades. In about 1970, some growers of lowbush blueberries in New Brunswick claimed that the change in the pesticide from DDT (only moderately toxic to honey bees, Table 1) to fenitrothion (highly toxic, Table 1) had caused losses of native pollinating bees and had reduced fruit-set and crops. The ensuing legal action contributed to a greater sensitivity and awareness about the likelihood of wide-ranging effects of pesticides in the pollination of natural and seminatural communities (Kevan and Collins 1974; Kevan 1975a), an idea that when first proposed was denigrated.

Now, by far the greatest amount of information is available for fenitrothion. That information has been recently reviewed by Kevan and Plowright (1989). Even for fenitrothion and honey bees, LD_{50} values are quite variable, depending on the investigator, the mode of application to the bee, and probably other factors. Nevertheless, fenitrothion is highly toxic to all the bees on which it has been tested (*Apis mellifera*, *Megachile centruncularis*, *Nomia melanderi*, *Andrena erythronii*, *Bombus terricola*). Thus, it is not surprising that populations of bees in areas sprayed with fenitrothion have been depressed. Butterflies of the genus *Erynnis* are also killed by field applications of fenitrothion (K. Barber, pers. comm.). Matacil®, also used in Canadian forests, is not as toxic to honey bees as fenitrothion. In field tests in the forests of New Brunswick, Plowright and Rodd (1980) showed that it did not appear to affect caged bumblebees, but that it may have had an indirect effect in killing solitary bees caged together in groups.

Residue analyses of dead bees generally agree with the pesticide treatment to which they were exposed, either by experimental intent or by accidental circumstances.

Cage Tests in Forestry

To offset the problem that counts of dead bees at a hive do not accurately estimate the mortality of foraging individuals that die while still foraging, and because that technique cannot be used for wild pollinators, Plowright and co-workers used "exposure cages" for assessing pollinator mortality in the field. Their method consisted of exposing captive insects in screen cages in areas sprayed with fenitrothion at 0.206 kg/ha or with aminocarb (Matacil®) at 0.07 kg/ha, while setting out other cages with control individuals in nearby unsprayed areas. The effects of fenitrothion on all the bees and wasps and other insect taxa that were studied were similar. Mortality was high (up to 100% in cages located in the open and directly under the flight path of the spray planes). For insects that survived, it seemed that the cages were partially protected by forest canopy or were some distance (at least 100 m) from the closest spray pass (Plowright 1977; Plowright et al. 1978a; Plowright and Rodd 1980). Under the protection of a dense canopy, the mortality of bumblebees was reduced to 47% (Plowright et al. 1978a). Similar results were obtained by Helson et al. (cited in Kevan and Plowright 1989) in a more recent series of similar, but not identical, experiments with exposure cages. Mortality of caged bees was more than 84%, and mostly total. Matacil® did not appear to affect caged bumblebees, but there was significantly greater mortality in solitary bees exposed to Matacil® than in the controls. However, Plowright and Rodd (1980) cast doubt on the direct effect of the insecticide, suggesting instead that agonistic behavior among the solitary bees caged together resulted in their deaths.

Sublethal Effects

Sublethal doses of some pesticides are known to have adverse effects on arthropod behavior because they interfere with neural and muscular coordination and learning. These sorts of effects have been demonstrated for honey bees treated with parathion. Schricker (1969) first described the effect of this organophosphorous insecticide as impairing the abilities of honey bees to communicate distance by their dance language. Stephen and Schricker (1970) and Schricker (1974b) found that directionality was also upset. Later, Schricker noted adverse effects of parathion on chronological memory (Schricker 1972) and sense of time (Schricker 1974a) in honey bees. Synthetic pyrethroids upset conditioned responses in honey bee learning (Taylor et al. 1987). The effect of a few other insecticides on the behavior of honey bees has also received attention. Cox and Wilson (1984, 1987) discussed the generally erratic behavior of honey bees affected by permethrin. Taylor et al. (1987) noted that the behavioral conditioning of honey bees is impaired by sublethal doses of synthetic pyrethroid insecticides.

Other sublethal effects on the life cycles of bees have been documented. Smirle et al. (1984) showed that malathion and diazinon at sublethal doses reduced the longevity of honey bee workers. Torchio (1983) discovered that Dylox®, used at field recommended doses on alfalfa, did not affect the survival of alfalfa leafcutting bees, but did reduce significantly their cell-building activity. Dylox® is considered to be relatively safe for use around bees (Table 1) and is used in forestry in buffer zones in the vicinity of blooming crops, notably blueberries (Trial 1987).

The sublethal effects on pollinators of pesticides commonly used in forestry have not been documented. Plowright and Rodd (1980) hint at the possible heightened agonistic behavior of solitary bees caged together and exposed to Matacil®. Certainly, the effects of Dylox® on leafcutting bees (Torchio 1983) and the lack of data on fenitrothion, Matacil®, etc., point to a serious gap in information.

Pesticides and Pollinator Populations

Population Reductions

Some of the first strong evidence for the demise of pollinator populations caused by applications of insecticides to forests came from blueberry fields. Kevan (1974) and Kevan and LaBerge (1979) showed strong and significant correlations between the proximity of fields of native blueberry to blocks of the forest treated with fenitrothion, and the populations and diversity of blueberry pollinators, almost all of which were native, wild bees. However, some other early studies reported either negative or equivocal findings on the reductions of pollinator populations. For example, Varty and Carter (1974) concluded on the basis of a program of insect trapping in the forest that the abundance and diversity of wild bees have not been measurably reduced by the perennial use of fenitrothion. Their study was based on a comparison of two sites, one sprayed with fenitrothion for 4 years from 1969 to 1972, the other sprayed in the year of the investigation, 1972. Because both sites were treated in the year of the investigation, it is not surprising that the results were inconclusive. Moreover, the study relied on counts of insects collected at window-traps, which are not effective for collecting bees and give variable results for other insects. In another Canadian forestry project, Buckner (1975) indicated 50% reductions in populations of bumblebees in one study but not in another.

Nevertheless, the great preponderance of documentation on the effects of fenitrothion on populations of pollinators indicates that drastic reduction is the general rule in all habitats studied (Kevan and Plowright 1989). Much of the work of relevance to Canadian forests concerns native bee pollinators of lowbush blueberries in New Brunswick. The blueberries are wild plants encouraged to grow on otherwise abandoned farms surrounded by forest and the bees are mostly native Andrenidae, Halictidae, and *Bombus* spp. which can also be found in the adjoining forest. Kevan (1975b) and Kevan and LaBerge (1979) noted that after 1971, populations of blueberry pollinators were significantly lower in areas when and where fenitrothion was used than in unsprayed areas. Wood (1977) also documented similar findings. In one field where the activity of bees could be measured before and after an application of fenitrothion, he reported a greater than 90% reduction in both solitary bee and bumblebee activity.

Several studies have examined changes in the abundance of pollinators in other parts of the forest ecosystem in New Brunswick. Several authors (Plowright et al. 1977, 1978a,b; Plowright and Rodd 1980; Plowright 1977, 1980; Plowright and Thaler 1979; Plowright and Pendrel 1978; and Thomson et al. 1985) have concluded that populations of bumblebees and other pollinators have been reduced in forests treated with fenitrothion. Moreover, the demographic details revealed in those studies tend unequivocally to link the observed reductions in populations to applications of the insecticide. The late-emerging species of bumblebees (which are still in hibernation when fenitrothion is being applied) generally do not show reductions in their populations. In fact, by late summer, they tend to be more common in the treated forest than in the untreated forest. That may be the outcome of "ecological release" because the late-emerging species benefit from the relative lack of competition for floral resources from earlier-emerging species (Plowright and Rodd 1980).

Aminocarb (Matacil®) did not appear to affect populations of either early-emerging or late-emerging bumblebees foraging at flowering goldenrod in New Brunswick (Plowright and Rodd 1980).

In the western United States, Robinson and Johansen (1978) reported reduced numbers of foraging wild bees in plots treated with Sevin® and Orthene® for control of Douglas-fir tussock moth, *Orgyia pseudotsugata*. Diflubenzuron did not have that effect. Those results agree with the relative toxicities of the pesticides to honey bees. The use of carbaryl, which is highly toxic to honey bees, and trichlorfon, which is only moderately toxic to honey bees, against western spruce budworm, *Choristoneura occidentalis*, in Montana reduced the populations of some of the small, solitary bees (USDA 1978). Miliczky and Osgood (1979) reported more than 50% mortality of native bees in some areas of Maine sprayed with carbaryl against spruce budworm.

The magnitude of the reductions of populations of pollinators is difficult to assess and depends partly on the frame of reference being used. Table 2 presents some of the findings from various studies that are relevant to pollinators of the forests.

The difficulties in assessing the overall impact of such reductions can be illustrated by examples. Plowright (unpubl. data) noted that on the day after the application of fenitrothion to a block of forest in New Brunswick, no queen bumblebees were to be found. This indicates that the population that had been active at the time of the spray was almost entirely poisoned. Near the edges of sprayed blocks, and even in the middle of the blocks several days after the treatment, at least a few live and apparently healthy queen bees were usually seen. Furthermore, populations of bumblebees within sprayed blocks gradually recovered throughout the summer so that by the time goldenrod bloomed in early August, populations of bumblebees had returned to at least 20% of their normal levels for that area (Plowright et al. 1978a). The recovery can be explained by immigration of bees early in the season and their relatively high success in rapid expansion of their colonies in the comparative absence of competition.

Other complications in interpreting the effects of insecticides on pollination may result from the uneven application of insecticide. Protective habitats beneath dense canopy and other naturally protective niches all contribute to confusing spatial and temporal complexity.

Population Recoveries

The amount of time required for populations of pollinators to recover is not really known in either the short term (that is, within the same year) or in the long term.

In the short term, Plowright et al. (1977, 1978a) found that pollinating bees were nearly completely absent for at least 10 days from the central areas of spray blocks treated with fenitrothion. They also reported that bees that were still in hibernation at the time of the spray applications were apparently unaffected. These bees then became abundant, so masking the immediate effects of the pesticide on the general population of all pollinating bees. This effect is also reported by Bouchard (1981) working with bumblebees in fenitrothion- and aminocarb-treated forest plots in Quebec.

The long-term effects of the reduction of pollinator populations in the forest are not known. Kevan (1977) and Kevan and LaBerge (1979) made preliminary estimates of the time taken for populations of blueberry pollinators to recover as being from as little as 3 to 4 years to more than 10 years, depending on chronicity of spray applications and the severity of the poisoning. Kevan and LaBerge (1979) documented

Table 2. Reductions in pollinator populations attributed to insecticide usage in forestry.

Pesticide	Insect affected	Effect[a]	Region	Reference
Fenitrothion	Wild pollinators of blueberries	-ve	N.B.	Kevan 1975
		-ve		Kevan and LaBerge 1979
		-ve		Wood 1974, 1977
		-ve		Thorpe 1980
		-ve		New Brunswick Supreme Court 1976
	Bumblebees	-ve	N.B.	Plowright et al. 1978
				Plowright 1977, 1980
	Solitary bees	-ve	N.B.	Plowright and Rodd 1980
		-ve	N.B.	Thorpe 1976
	Bumblebee diversity	Δ	Que.	Bouchard 1981
Aminocarb	Bumblebee diversity	Δ	Que.	Bouchard 1981
	Solitary bees	-ve	N.B.	Plowright and Rodd 1980
		-ve	N.B.	Thorpe 1976
	Pollinators incl. Diptera	-ve	N.B.	Thomson et al. 1985
	Bumblebees	o	N.B.	Plowright 1980
				Plowright and Pendrel 1978
				Plowright and Rodd 1980
Carbaryl	Wild bees	o	Montana	USDA 1978
		-ve	Maine	Miliczky and Osgood 1979
Carbaryl + acephate	Wild bees	-ve	U.S. Pacific N.W.[b]	Robinson and Johansen 1978
Diflubenzuron	Wild bees	o	U.S. Pacific N.W.	Robinson and Johansen 1978

[a] Negative, -ve; changed, Δ; no effect recorded, o.
[b] United States Pacific Northwest.

5 years of continued recovery on previously chronically affected blueberry fields, and W. Bridges (pers. comm.) indicated that he felt that complete recovery of all but one of those fields had taken place by 1985, 12 years after the discontinuation of the use of fenitrothion near the blueberry fields. Wood (1979) concluded from data gathered on a field that was sprayed only once that full recovery took place in 1 to 3 years, but in 1982a wrote about full recovery "after 3 years." Both Wood and Kevan arrived at much the same conclusion for the same field after independent assessments, despite Wood's (1982a) implication of a difference of opinion.

The matter of recovery of pollinator populations is not simple. First, those insects that are vagile (e.g. bumblebees) are likely to show more rapid recolonization of sprayed forest than those that form nesting aggregations, often mediated by the presence of pheromones of founding individuals, in the same place year after year (e.g. many species of solitary bees). Also, in some groups of pollinating insects, ecological replacement by later emerging species provides a mechanism for partial and rapid recovery of populations, albeit with differences in the fauna (Plowright and Rodd 1980). Finally, given that the full recovery of populations is likely an asymptotic process, the concepts of "acceptable" and "effective" levels of recovery need definition.

Pesticides and Plant Reproduction

Pesticide Toxicity to Pollen Sporogenesis

There appears to be little information on the effects of pesticides on sporogenesis of pollen. Our examination of the literature produced data indicating that all pesticides tested caused increasing aberrations in the development of pollen mother cells with increasing dose. The insecticides examined were aldicarb, BHC, dichlorvos, dimethoate, dipterex, dursban, fenitrothion, monocrotophos, nuvacron, and phosphamidon. The fungicides benomyl and maneb have also been tested. Curiously, we found no literature on the effects of herbicides.

Pesticides and Pollen Viability

Pollen has been used for toxicological bioassay. From our review of about 30 papers published between 1940 and 1987, it is clear that most pesticides and adjuvants reduce the viability and germinability of pollen grains; cause aberrations in their growth, meioses, and mitoses; reduce their rates of growth; and even cause their death after germination. Fungicides have received the greatest amount of attention. However, only a few compounds have been tested on a few plants,

Table 3. Effects of insecticides applied in forests on the fruit-set of entomophilous plants.[a]

Plant species	Insecticide			Location	Reference
	Fen.[b]	Amino.[b]	Sevin		
Vaccinium angustifolium and *V. myritilloides*	-ve			N.B.	Kevan 1977 Wood 1977 Kevan 1975 Thorpe 1980 Kevan and LaBerge 1979
Rhodora canadensis	-ve	-ve	x	N.B.	Thorpe 1980
Kalmia angustifolia	-ve	o	x	N.B.	Thorpe 1980
	-ve	o	x	N.B.	Thaler and Plowright 1980
Maianthemum canadensis	-ve	o	x	Que.	SOMER 1985
	x	-ve	x	N.B.	Thomson et al. 1985
	-ve	x	x	N.B.	Plowright and Thaler 1979 Thaler and Plowright 1980
Cornus canadensis	-ve	o	x	Que.	SOMER 1985
	x	o	x	N.B.	Thomson et al. 1985
	-ve	x	x	N.B.	Plowright and Thaler 1979 Thaler and Plowright 1980
C. stolonifera	-ve	x	x	N.B.	Plowright and Thaler 1979 Thaler and Plowright 1980
	x	(-ve)	x	N.B.	Thomson et al. 1985
C. alternifolia	-ve	x	x	N.B.	Plowright and Thaler 1979 Thaler and Plowright 1980
Aralia nudicaulis	-ve	-ve	x	Que.	SOMER 1985
	-ve	x	x	N.B.	Plowright and Thaler 1979 Thaler and Plowright 1980
Clintonia borealis	o	o	x	Que.	SOMER 1985
	-ve	o	x	N.B.	Thaler and Plowright 1980
Trifolium pratense	-ve	x	x	N.B.	Hartling 1977 in Plowright et al. 1978 Thaler and Plowright 1980
Viburnum cassinoides	x	x	-ve	Maine	Miliczky and Osgood 1979
V. trilobum	-ve	x	x	N.B.	Plowright and Thaler 1979 Thaler and Plowright 1980
Cypripedium acaule	-ve	x	x	N.B.	Plowright et al. 1980
Prunus pensylvanicus	o	o	o	Ont.	Buckner 1975
Ranunculus	o	x	x	Nfld.	Fudge et al. 1986-87

[a] Significant negative effect, -ve; negative effect, but insignificant, (-ve); no significant effect, o; not tested in this study, x.
[b] Fenitrothion; aminocarb.

and the scant literature available makes it difficult to arrive at any general conclusions.

We know of only seven studies in which pesticide applications have been related to pollination success as measured by seed-set. Again, most of the compounds studied were fungicides and irrelevant to forestry. We found only two studies relevant to forest trees. The first is on the pollen of Douglas-fir and white spruce treated with dimethoate, oxdemeton-methyl, potassium coconate, and ferbam. All of those pesticides, except ferbam, which sometimes enhanced pollen germination for Douglas-fir, depressed pollen germinability (Sutherland et al. 1984). The second, by DeBarr and Matthews (1985) working with loblolly pine in seed orchards in the southern United States, indicated no adverse effects of guthion, pydrin, or Ambush® on pollen germinability and eventual seed-set.

From the available information, apparently the application of pesticides generally has an adverse effect on pollen, but the effect is not usually manifest through reduced fecundity. In some instances (e.g. *Vaccinium* spp.), floral form is important. In blueberries, the corolla protects the sporophylls, but in cranberries, whose pollen can be affected by pesticides, it does not. Also, field trails to examine the effect of pesticides on fecundity are difficult to design. Pollen may be released over several days and only the pollen coming directly in contact with a pesticide at the time of its application would be expected to be affected. Thus, the observed lack of or minimal effects on seed-set may well reflect fertilization by pollen that was naturally unavailable for pesticide treatment. Given the strong evidence that most types of pollen so far tested, including those of trees important to Canadian forests, are adversely affected by pesticides, the warning of DeBarr and Matthews (1985) that "it still appears prudent to avoid all unnecessary pesticide use during this sensitive [flowering] period" should be heeded.

Pollinator Populations and Reproduction of Forest Plants

The suspicion that forest pesticides were having a detrimental effect on the reproduction success of forest plants was first broached in about 1970 regarding blueberries and fenitrothion in New Brunswick. The arguments associated with that idea were published by Kevan and Collins (1974) and Kevan (1975a,b). The first empirical investigation of the question (Buckner 1975 in National Research Council Canada 1975) reported negative results, but the conclusion is invalid and no data are provided. Further, the choice of plant, pin cherry, presents problems as it often blooms before the insecticide, fenitrothion, is sprayed; also, it is not known to be a obligate entomophile and it certainly does not depend on bees. Similarly, the choice of buttercup (*Ranunculus* sp.) in Newfoundland by Fudge et al. (1986–87) was inappropriate as many common species are agamospermous. The conclusions are similar in all studies in which the activities of insects, particularly bees, are not at least partially required for pollination. The reproductive output of a plant can be depressed by reduction in populations of pollinators only if the following criteria are met: (1) the pollinators of the plant must be vulnerable to the pesticide in question; (2) the blooming period of the plant must coincide with the period in which the reduction in the pollinator force is most intense; and (3) the breeding system of the plant must be such that insect visitation is required for fruit production. Table 3 lists plants that fit those criteria and have been investigated regarding the effects of insecticides on their fruit production. All have shown depressed reproductive output when pollinator populations have also been shown to have been reduced.

Investigating the long-term effects of reduced seed-set on plant populations is not easy. Most plants of the Canadian forest, even the herbaceous ones, are long-lived perennials. Further, many are at least partially self-fertile (Barrett and Helenurm 1987; Helenurm and Barrett 1987; Kevan et al. 1992). Even so, one might expect to see reduced recruitment into the next generation of plants or reduced dispersal into newly opened habitats. However, for most of the plant species in question, seedling establishment is generally rare.

Demonstrating the effect of pesticide-reduced plant fecundity on populations of forest species would be difficult, even with a small army of trained botanists. The effects could be seen to take place only over a long period. Short-term changes would be too subtle to measure even in a monolectic species of plant for which the pollinator had been utterly extirpated. It is noteworthy that the decline in populations of certain rare plants in Europe has been related to the absence of their pollinators, and the lack of fruiting in some tropical trees is thought to be the consequence of two few pollinators brought about by habitat destruction (Kevan et al. 1990). In Carolinian Canada, some rare trees (*Magnolia acuminata*, *Gymnocladus dioicus*, *Asimina triloba*, *Castanea dentata*) and shrubs (*Rosa setigera*, *Smilax rotundifolia*) do not fruit where distances to pollen-donor plants are too great (Ambrose and Kevan 1990). Thus, because the effects of pesticides in reducing pollinator populations and in reducing plants' fruit- and seed-set have not been documented in the long term in Canadian forests, and the long-term effects on plant populations would be almost impossible to demonstrate, it would be prudent to recognize the short-term evidence to support the scenario presented. Eventually, the effects of the continued removal of pollinators from the ecosystem would become evident.

Plant Fecundity and the Forest Ecosystem

Kevan and Collins (1974) suggested the following chain of events could result from the detrimental effects of insecticides on pollinators. Assuming a reduction in seed and fruit production by plants for which pollinators were severely reduced, one might expect a response from animals that normally depend on those seeds and fruit for food (Fig. 4). The exact nature of such a response is a speculative matter because of the enormous research effort required to document cause and effect. In New Brunswick, Kevan and Collins (1974) suggested that frugivorous birds turned their attentions to commercial blueberry fields to substitute for the lack of fruit in the forest. Certainly the blueberry fields growers suffered unprecedented losses to birds during the period when fenitrothion was used around blueberry fields. Further, it was suggested by Laird (in Kevan and Collins 1974) that fenitrothion reduced populations of bloodsucking insects that transmit avian blood diseases. That resulted in greater reproductive success for the birds, so exacerbating the problem. In the northern forests, boreal or mixed woodland, berries are an important component of the diets of black bears, other wildlife, and a wide variety of birds. Small crops of berries in the natural habitat of these animals probably affect their reproductive success, migration, overwintering capabilities, and propensity to become nuisances.

Economic and Legal Issues

The economic impact of pesticides on pollinators has been assessed mostly on the basis of honey bees and agriculture. In the context of forestry, the issues surrounding crop losses to blueberry farmers in New Brunswick are germane. The blueberries are a native forest plant and the pollinators mostly native forest insects. The blueberry fields are regenerating forest on abandoned farm lands, or cleared forest, held at an exaggerated and early successional stage. While fenitrothion was in use in the vicinity of blueberry fields in New Brunswick, and for some time after, estimated losses were

665 tonnes per year (Kevan and Plowright 1989). Since the recovery of pollinator populations, blueberry crops appear to have returned to near those expected by comparison with crops in Nova Scotia.

The events in New Brunswick led to lengthy litigation, resulting in a landmark in Canadian environmental law (New Brunswick Supreme Court 1976) by which the company applying fenitrothion to the forest was required to pay a settlement to blueberry farmers for some of their crop losses. In general, there are laws to protect beekeepers from losses caused by pesticide application (Crane and Walker 1983), which coupled with the decision in New Brunswick protect pollinators of crops.

Conservation

Although most of Canada's forests are well separated from agricultural areas, the encouragement of agroforestry, woodlot management, and sustainable agriculture are bringing forestry pests into close proximity with cultivation and fields. Particularly important in this light is the gypsy moth, *Lymantria dispar*. The impact that gypsy moth control programs are likely to have on pollinators has not been assessed. However, if a general policy of encouraging the use of biological pesticides such as Bt in its various strains and formulations is followed, the populations of important hymenopterous and dipterous pollinators probably will not be affected. Lepidopterous pollinators may be adversely affected, but the consequences of that are unknown because so little is known about moth pollination in Canadian settings. The Kentucky coffee tree, *Gymnocladus dioicus*, an endangered plant of the Carolinian forest remnants of Ontario, is pollinated by night-flying moths (Ambrose and Kevan 1990). Flowers of many species of plants are visited by moths that may be just as important to pollination as their diurnal visitors. This has been demonstrated for milkweeds, *Asclepias* spp., in southern Ontario (Punchihewa 1985). The role of moths outside pollination ranges from a few being destructive agricultural and forestry pests, to their general importance as food for wild birds and bats.

The economic impact of disrupting natural ecosystems is difficult to assess. The effects are often subtle, evident only over the long term, and usually unforeseen or dismissed by planners. The lessons of bioaccumulation of pesticides in the food chain, acid rain, tropical deforestation, erosion of the ozone layer, and global warming are now being heeded. All are major environmental disruptions that started rather inconspicuously. The demise of pollinator populations, even of honey bees by exotic diseases, has been shown to have had far-reaching implications in agriculture and nature in particular circumstances (Kevan et al. 1990). We contend that continued neglect of the keystone ecological interplay of animals and plants in pollination will affect the diversity and abundance of life in Canadian forests and, indeed, on earth.

References

Ambrose, J.D.; Kevan, P.G. 1990. Reproductive biology of rare Carolinian plants with regard to conservation management. Pages 57-63 *in* G.M. Allen, P.F.J. Eagles, and S.D. Price (Eds.), Conserving Carolinian Canada. University of Waterloo Press, Waterloo, Ont.

Barrett, S.C.H.; Helenrum, K. 1987. The reproductive biology of boreal forest herbs. I. Breeding systems and pollination. Can. J. Bot. 65: 2036-2046.

Bouchard, Y. 1981. Impacte des pulvérisations aériennes de fénitrothion et d'aminocarbe sur les populations de bourdons au Québec. Unpubl. Rep., Dépt. Biologie, Univ. Laval, Québec. 42 p.

Buckner, V.H. 1975. Supporting documents supplied to the expert panel on fenitrothion. Natl. Res. Counc. Can. Publ. No. NRCC 14104, Ottawa, Ont.

Cox, R.R.; Wilson, W.T. 1984. Effects of permethrin on the behaviour of individually tagged bees, *Apis mellifera* Hymenoptera: Apidae. Environ. Entomol. 13: 375-378.

Cox, R.R.; Wilson, W.T. 1987. The behaviour of insecticide-exposed honey bees. Am. Bee J. 127: 118-119.

Crane, E.; Walker, P. 1983. The impact of pest management: bees and pollination. Intl. Bee Res. Assoc. and Tropical Dev. and Res. Inst., London, England. ix + 129 p., + 73 p., + 9 p.

Crepet, W.I. 1983. The role of insect pollination in the evolution of the Angiosperms. Pages 31-48 *in* L. Real (Ed.), Pollination Biology. Academic Press, Orlando, Fla.

Danks, H.V. (Ed.). 1979. Canada and its insect fauna. Mem. Entomol. Soc. Can., No. 108. 573 p.

Debarr, G.L.; Matthews, F.R. 1985. Insecticide applications during pollination period do not adversely affect seed-yields of loblolly pine, *Pinus taeda*. South. J. Appl. For. 9: 240-243.

Felton, J.C.; Oomen P.A.; Stevenson, J.H. 1986. Toxicity and hazard of pesticides to honey bees: harmonization of test methods. Bee World 67: 114-124.

Fudge, S.M.; Pratt, J.; Anstey, V.; Meron, S.; Hicks, B.; Gilkinson, K. 1986-87. A study to monitor the effects of aerially applied fenitrothion on the activity of Newfoundland pollinating insects. Prepared for Nfld. and Lab. Dept. of For. Res. and Land, Corner Brook, Nfld. Fudge, Lane and Associates Limited, St. John's, Newfoundland. 11 p. + tables.

Helenurm, K.; Barrett, S.C.H. 1987. The reproductive biology of boreal forest herbs. II. Phenology of flowering and fruiting. Can. J. Bot. 65: 2047-2056.

Johansen, C.A.; Mayer, D.F. 1990. Pollinator protection: a bee and pesticide handbook. Wicwas Press, Cheshire, Conn.

Kevan, P.G. 1974. Effects of the 1973 forest spray program on the wild been pollinators of lowbush blueberries in southern New Brunswick. (Preliminary report). Unpubl. Rep. to Bridges Bros., Ltd., Commercial Blueberry Farms, St. Stephen, N.B. 22 p.

Kevan, P.G. 1975a. Pollination and environmental conservation. Environ. Conserv. 2: 293-298.

Kevan, P.G. 1975b. Forest application of the insecticide fenitrothion and its effect on the wild bee pollinators (Hymenoptera: Apoidea) of lowbush blueberries (Vaccinium spp.) in southern New Brunswick, Canada. Biol. Conserv. 7: 301-309.

Kevan, P.G. 1977. Blueberry crops in Nova Scotia and New Brunswick. Pesticides and crop reduction. Can. J. Agric. Econ. 25: 61-64.

Kevan, P.G. 1991. Pollination: Keystone process in sustainable global productivity. Acta Hortic. 288: 103-110.

Kevan, P.G.; Baker, H.G. 1983. Insects as flower visitors and pollinators. Ann. Rev. Entomol. 28: 407-453.

Kevan, P.G.; Baker, H.G. 1984. Insects on flowers: pollination and floral visitations. Pages 607-631 *in* C.B. Huffaker and R.C. Rabb (Eds.), Ecological Entomology. J. Wiley, New York.

Kevan, P.G.; Clark, E.A.; Thomas V.G. 1990. Insect pollinators and sustainable agriculture. Am. J. Alternative Agric. 5: 13-22.

Kevan, P.G.; Collins, M. 1974. Bees, blueberries, birds, and budworms. Osprey (Nfld. Natural Hist. Soc. Newsletter) 5: 54-72.

Kevan, P.G.; LaBerge, W.E. 1979. Demise and recovery of native pollinator populations through pesticide use and some economic implications. Proc. of the 4th Intl. Symp. on Pollination, Maryland Agric. Exp. Stn., Spec. Misc. Publ. 1: 489-508.

Kevan, P.G.; Plowright, R.C. 1989. Fenitrothion and insect pollinators. Pages 13-42 *in* W.R. Ernst, P.A. Pearce and T.C. Pollock (Eds.), Environmental effects of fenitrothion use in forestry. Environ. Can., Dartmouth, N.S.

Kevan, P.G.; Usui, M.; Tikhuenev, E.A. 1992. Insects and plants in the pollination ecology of the boreal zone. Proc. Entomol. Soc. Ont. Vol. 122. (In press)

Ladas, A. 1972. Der Einfluss Verschiedener Konstitutions-und Umweltfaktoven auf die Anfalligkeit der Honigbiene (*Apis Mellifera* L.). Gegenüber zwei Insektiziden Pflanzenschutzmitteln. Apidologie 3: 55-78.

Levin, M.D. 1983. Value of bee pollination to U.S. agriculture. Bull. Entomol. Soc. Am. 29: 50-51.

Miliczky, E.R.; Osgood, E.A. 1979. The effects of spraying with Sevin®-4-oil on insect pollinators and pollination in a spruce-fir forest. University of Maine, Orono. Life Sciences Agric. Exp. Stn. Tech. Bull. No. 90. 21 p.

National Research Council Canada. 1975. Fenitrothion: the effects of its use on environmental quality and its chemistry. Natl. Res. Counc. Can. Publ. No. NRCC 14104, Ottawa, Ont. 162 p.

National Research Council Canada. 1981. Pesticide-pollinator interactions. Natl. Res. Counc. Can. Publ. No. NRCC 18471, Ottawa, Ont. 190 p.

New Brunswick Supreme Court. 1976. Between: Bridges Brothers Ltd., Plantiff, and Forest Protection Ltd., Defendant. Queen's Bench Division, Fredericton, N.B.

Nigam, P.C. 1975. Chemical insecticides. Pages 8-24 *in* M.L. Prebble (Ed.), Aerial Control of Forest Insects in Canada. Environ. Can., Ottawa, Ont.

Plowright, R.C. 1977. The effect of fenitrothion on forest pollinators in New Brunswick. Pages 335-341 *in* Proceedings of a symposium on fenitrothion. Natl. Res. Counc. Can. Publ. No. NRCC 16073, Ottawa, Ont.

Plowright, R.C. 1980. Annual and seasonal trends in bumblebee abundance and flowering plant reproduction, associated with aerial spray operations. Pages 37-40 *in* I.W. Varty (Ed.), Environmental Surveillance in New Brunswick, 1978-79. Effects of Spray Operations for Forest Protection against Spruce Budworm. Committee for Environmental Monitoring of Forest Insect Control Operations (EMOFICO). Univ. of New Brunswick.

Plowright, R.C.; Pendrel, B.A. 1978. A comparison of the effects of aminocarb and fenitrothion on forest pollinators in New Brunswick. *In* V.N. Mallet (Ed.), Proceedings of the symposium on aminocarb: effects of its use on environmental quality. Univ. of Moncton, N.B. 145 p.

Plowright, R.C.; Pendrel B.A.; McLaren, I.A. 1978a. The impact of aerial fenitrothion spraying upon the population biology of bumble bees (Bombus Latr.: Hym.) in southern New Brunswick. Can. Entomol. 110: 1145-1156.

Plowright, R.C.; Pendrel B.A.; Thaler, G.R. 1978b. The impact of spruce budworm control operations on forest pollination in New Brunswick during 1977. Report to the Maritimes For. Res. Cent., Fredericton, N.B. (Contact No. 07SU3KL003-7-0010 Dept. Supply and Services, Gov't. of Can.). 82 p.

Plowright, R.C.; Pielou, D.P.; McLaren I.A.; Pendrel, B.A. 1977. The impact of fenitrothion spraying upon the population biology of bumble bees (Bombus Latr.: Hym.) in southwest New Brunswick. Report to Dept. Fisheries and the Environment, Gov't. of Can. 57 p.

Plowright, R.C.; Rodd, F.H. 1980. The effect of spruce budworm control operations on Hymenoptera in New Brunswick. Can. Entomol. 112: 259-270.

Plowright, R.C.; Thaler, G.R. 1979. The effects of biocides on forest pollination in New Brunswick. Proceedings of the 4th International Symposium on Pollination, Maryland Agric. Exp. Stn. Spec. Misc. Publ. 1: 483-487.

Plowright, R.C.; Thomson, J.D.; Thaler, G.R. 1980. Pollen removal in Cypripedium acaule (Orchidaceae) in relation to aerial fenitrothion spraying in New Brunswick. Can. Entomol. 112: 765-769.

Punchihewa, R.W.K. 1984. Anthophilous insects and the pollination ecology of *Asclepias Syriaca* L. and *Asclepias incarnata* L. in southern Ontario. Unpubl. M.Sc. dissertation, Univ. of Guelph, Guelph, Ont. 126 p.

Robinson, W.S.; Johansen C.A. 1978. Effects of control chemicals for Douglas-fir tussock moth, *Orgyia pseudotsugata* (McDunnough) on forest pollination (Lepidoptera: Lymantriidae). Melanderia 30: 9-57.

Robinson, W.S.; Nowogrodzki R.; Morse R.A. 1989. Value of honey bees as pollinators of U.S. crops. Am. Bee J. 129: 411-423; 477-487.

Schricker, B. 1969. The effect of sublethal doses of parathion on the indication of distance in the honey bee. XXII Intl. Beekeeping Congr., Munich. p. 568. Apimondia Publishing House, Bucharest, Rumania.

Schricker, B. 1972. The effect of sublethal doses of parathion E-605 on the chronological memory of the honey bee. Insect. Soc. 19: 411.

Schricker, B. 1974a. Der Einfluss subletalen Dosen von Parathion (E 605) auf das Zeitgeächtnis der Honigbiene. [The effect of sublethal doses of parathion (E 605) on the time sense in the honey bee.] Apidologie 5: 385-398.

Schricker, B. 1974b. Der Einfluss subletalen Dosen von Parathion (E 605) auf die Entfernungsweising bei der Honigbiene. [The effect of sublethal doses of parathion on the indication of distance in honey bees.] Apidologie 5: 149-175.

Smirle, M.J.; Winston M.L.; Woodward K. 1984. Development of a sensitive bioassay for evaluating sublethal pesticide effects on the honey bee (Hymenoptera: Apidae). J. Econ. Entomol. 77: 63-67.

SOMER. 1985. Effets des insecticides fénitrothion et aminocarbe sur la fructification de quatre espèces végétales en milieu forestier. Final Report to Gouvernement du Québec, ministère de l'Énergie et des Ressources. SOMER, Société Multidisciplinaire d'Études et de Recherches de Montréal Inc., Montréal. xiv + 208 p.

Southwick, L. Jr.; Southwick E.E. 1989. A comment on "Value of honey bees as pollinators of U.S. crops." Am. Bee J. 129: 805-807.

Stephen, W.P.; Schricker B. 1970. The effect of sublethal doses of parathion. II. Site of parathion activity, and signal integration. J. Apic. Res. 9: 155-164.

Sutherland, J.R.; Woods T.A.D.; Miller G.E. 1984. Effect of selected insecticides and fungicides on germination of Douglas-fir and white spruce pollen. Tree Planters' Notes 34(1): 22-24.

Taylor, K.S.; Waller G.D.; Crowder L.A. 1987. Impairment of a classical conditioned response of the honey bee *Apis mellifera* L. by sublethal doses of synthetic pyrethroid insecticides. Apidologie 18: 243-252.

Thaler, G.R.; Plowright R.C. 1980. The effect of aerial insecticide spraying for spruce budworm control on the fecundity of entomophilous plants in New Brunswick. Can. J. Bot. 50: 2022-2027.

Thomson, J.D.; Plowright R.C.; Thaler G.R. 1985. Matacil insecticide spraying, pollinator mortality, and plant fecundity in New Brunswick forests. Can. J. Bot. 63: 2056-2061.

Thorpe, E.J. 1980. Some aspects of the effect of forest spraying on solitary bee (Hymenoptera: *Apidae*) populations in New Brunswick. Unpubl. M.Sc. dissertation, Univ. of New Brunswick, Fredericton, N.B. 95 p.

Torchio, P.F. 1973. Relative toxicity of insecticides to the honey bee, alkali bee, and alfalfa leafcutting bee. Kans. Entomol. Soc. 46: 446-453.

Torchio, P.F. 1983. The effects of field applications of naled and trichlorfon on the alfalfa leafcutting bee, *Megachile rotundata* (Fabricius). J. Kansas Entomol. Soc. 51: 62-68.

Trial, J.G. (Ed.) 1987. Buffer zones: their application to forest insect control operations. Proc. of the Buffer Zone Workshop, Quebec City, April 16-17, 1986. Ministry of Supply and Services, Ottawa, Ont. Catalogue No. FO18-7/1987E. 124 p.

U.S. Dept. Agric. 1978. Western spruce budworm, a pilot project with carbaryl and trichlorfon, 1975. USDA For. Serv. Rep. No. 78-5. Missoula, Mont. (Prepared by D. Schmidt et al., Olson-Elliott & Assoc.)

Varty, I.W.; Carter N.E. 1974. Inventory of litter arthropods and airborne insects in fir-spruce stands treated with insecticides. Can. For. Serv., Maritimes For. Res. Cent., Inf. Rep. M-X-48, 32 p.

Wahl, O. 1976. Zur frage nichtbienengefährlicher Pflanzenschutzmittel. Zeitschrift für angewandte Entomologie 82: 82-86.

Winston, M.L.; Scott C.D. 1984. The value of bee pollination to Canadian apiculture. Can. Beekeeping 11; 134.

Wood, G.W. 1974. Information supplied to the expert panel on fenitrothion. Natl. Res. Counc. Can. Publ. No. NRCC 141104, Ottawa, Ont.

Wood, G.W. 1977. The effects of the fenitrothion spray program, on bees and pollination. Pages 321-333 *in* Proceedings of a Symposium of Fenitrothion. Natl. Res. Counc. Can. Publ. No. NRCC 16073, Ottawa, Ont.

Wood, G.W. 1979. Recuperation of native bee populations in blueberry fields exposed to drift of fenitrothion from forest spray operations in New Brunswick. J. Econ. Entomol. 72: 36-39.

Wood, G.W. 1982a. A review of forest spraying and its effect on pollinators. Pages 66-68 *in* D.R. Townsend (Ed.), Environmental surveillance of spray operations for forest protection in New Brunswick, 1980-81. Committee for Environmental Monitoring of Forest Insect Control Operations (EMOFICO). Environment, N.B.

Chapter 60

Impacts of Forest Aerial Spray Programs on Aquatic Ecosystems

D.C. Eidt, W.R. Ernst, and S.B. Holmes

Introduction

Aerially broadcast insecticides may fall directly on water surfaces, may be carried in from terrestrial deposits, or may settle on water after drifting in air. Waters both inside and outside sprayed areas may be contaminated. Thus, there may be a level of risk to each of the organisms that live in the pesticide-contaminated waters.

The risk to any organism is a function of the amount of insecticide to which it is exposed, the duration of the exposure, the organism's tolerance, and its ability to detect and avoid the contaminant. Each of these factors is very variable, but easier to measure and assess for aquatic organisms, which are immersed in a contaminated matrix, than for terrestrial organisms such as mammals and birds, which usually are not.

It is commonly perceived that in flowing water, the hazard increases as pesticide-laden waters converge and combine their loads. The opposite is true because the farther surface water travels, the greater its load is diluted because of clean water entering from tributaries and groundwater. Further losses in concentration occur because the toxicant volatilizes, may be absorbed or adsorbed, and decomposes through both physical and biological means.

The amount of aerially applied insecticide impinging on water is inversely related to the type and amount of screening vegetative cover, which is related to the width of the watercourse or waterbody. The vertical distribution of insecticide in water may also vary. The concentration in well-mixed water is a function of the water depth (Eidt et al. 1984) and in running water, mixing is rapid whether the formulation is oil- or water-based. In standing water, concentration differences with depth last much longer (Maguire and Hale 1980). High surface concentration could be very hazardous to vertically migrating animals such as plankton and surface-breathing water bugs and beetles.

The amount of insecticide in water can be measured directly by chemical analysis or bioassay. It changes rapidly and predictably because of chemical and biological degradation, volatilization, dilution, dispersion, and in streams, flushing. Thus, exposure of aquatic organisms is to rapidly diminishing concentrations, not static concentrations as are used in laboratory toxicity tests. The matter is further complicated because many insecticides partition to suspended particles (fenitrothion) or are themselves particulate (Bt) or suspensions ("flowable" formulations in which particles persist until dissolved). The result is that filter-feeding organisms such as black fly larvae, net-spinning caddisfly larvae, and clams may encounter more risk than do grazers and predators.

Sorption of toxicants to both living and dead substrates has the potential to be hazardous to organisms that feed on them or live in or on these substrates. Fenitrothion is known to be sorbed by organic sediments (Zitko and Cunningham 1974; Eidt et al. 1984) and water plants (Weinberger et al. 1982).

Aquatic organisms may encounter insecticides in several ways. The most obvious is through the body wall from the water. This is usual with higher plants, algae, and microorganisms, but also occurs with animals, particularly through thin places such as blood or tracheal gills. Animals may also take in pesticides by ingesting contaminated microorganic, plant, animal, or detrital food. Living aquatic organisms are said to bioaccumulate pesticides when they take any from their environment into their tissues. If they concentrate the pesticide above the concentration in the water, they bioconcentrate it. Biomagnification occurs when predators accumulate greater concentrations than those of the prey they feed upon.

There are many references to studies on the effects of forest insecticides on aquatic fauna. There are more than 400 references on fenitrothion alone (Fairchild et al. 1989). These references describe three types of studies, and their interpretation requires close attention to the method of investigation and its meaning. They are laboratory toxicity studies, field monitoring, and field experimentation.

Laboratory studies are used to determine toxicities of insecticides in terms of controlled concentrations for, usually, 48 or 96 hours. Actually, organisms in nature are exposed to rapidly changing concentrations, and laboratory toxicities are only rough guides to what might be expected in the field. Peak concentrations observed during field monitoring are rarely the peaks that actually occurred, but decay and dissipation is exponential. Use of microcosms in the laboratory is an improvement because it simulates field conditions, but microcosms invariably lack many of the components that buffer field situations.

Field monitoring describes the technique of observing what happens in water bodies within experimental or operational spray areas. Many of the studies based on this technique are seriously flawed by inadequate controls, inadequate water analysis, inadequate sampling techniques for biota, and by problems of sample timing. Aquatic insect emergence peaks in late June, and it is essential that benthos depletion by emergence be distinguished from that due to the insecticide.

Field experiments are attempts to overcome all the problems of laboratory and monitoring studies, but of course they do not. Perennial problems are lack of sufficient samples, replication, and adequate temporal and spatial controls, which are problems difficult to overcome in aquatic work, particularly in streams. The advantage is that the experimenter has control of dosage and timing.

From the large body of data available from each kind of approach, we understand the hazards and risks associated with fenitrothion, Bt, aminocarb, and permethrin very well. For others, such as acephate, mexacarbate, and diflubenzuron, we know much less because the database is much smaller, and viruses are so specific that the hazard to aquatic organisms is generally discounted.

The insecticides of concern since 1973 are Bt, fenitrothion, aminocarb, and diflubenzuron which are currently registered for use against one insect defoliator or another. Trichlorfon was used aerially in forestry before 1977, but it is no longer used and the database on environmental impacts is now out of date. Permethrin is registered for ground applications only against certain pests such as the black army cutworm, *Actebia fennica*. Mexacarbate has been used experimentally for aerial application against forest defoliators, but its future is uncertain.

Fenitrothion

Fenitrothion is a broad-spectrum insecticide that can affect some aquatic organisms when used in operational forest sprays. In spite of the use of buffer zones, it is not possible to prevent some getting into natural water bodies. Maximum recorded concentrations of fenitrothion in streams after operational spraying range from approximately 1.4 μg/mL to 127 μg/mL (Mallet and Cassista 1984; Morin et al. 1986). Estimated half-lives of fenitrothion are 13 hours in streams and 18 hours in lakes (Fairchild et al. 1989). Contributing to a rapid decline in concentration are hydrolysis, volatilization, biological degradation, and sorption to suspended particles, organic sediments, and some aquatic plants (Eidt et al. 1984); in streams, dilution and flushing are also important.

Some organisms absorb and accumulate fenitrothion above ambient concentrations, but the residues usually persist only for short times (Lakshminarayana and Bourque 1980; Lockhart et al. 1973). Acidic bog ponds are unusual because fenitrothion residues in water, although undetectable within several days after spraying, returned to measurable concentrations the following spring (Fairchild 1990). This suggests that fenitrothion sequestered in some component of the system, and that the fenitrothion was somehow subsequently released.

If fenitrothion concentration in water becomes high enough it can have dramatic effect on drift of aquatic arthropods (Eidt 1975). Increased drift, especially of particularly sensitive arthropods, is a common consequence of fenitrothion spraying (Montreal Engineering Co. 1976; Rabeni and Gibbs 1976; Riche and Coady 1969). Results of some studies suggested to their authors that benthic biomass depletion occurred (Dimond and Malcolm 1971; Buckner et al. 1976a; Coady 1978), as indeed must occur when dead benthic insects are found in spray-induced drift. Other monitoring studies in Canadian streams showed that the effect of increased drift, even if composed largely of dead insects, had no substantial impact on the benthic invertebrate community (Eidt 1975, 1977; Montreal Engineering Co. 1976). In Japan, under a spray regime at seven times the rate permitted in Canada, macrobenthos was severely depleted, but most recovered quickly, probably through recolonization from upstream (Hatakeyama et al. 1990). Density of stoneflies was still low after 11 months. No one has been able to detect changes in community structure or the elimination of rare or sensitive species, either because they haven't occurred or because the techniques have not been sensitive enough.

Using the results of monitoring studies, the proportion of habitat affected in sprayed areas can be estimated, but the estimates would not necessarily reflect what is happening under current use patterns. Current effects should be less than those summarized and judged to be environmentally tolerable by Symons (1977a,b), because of lower application rates, improved formulations, and modified application strategies, particularly the observance of buffer zones. This does not mean there will be no effects on aquatic organisms because aquatic habitats have many and various characteristics that make them particularly sensitive. Isolated instances of kill of aquatic invertebrates undoubtedly could still be detected, but these probably would be fewer. However, there is good evidence that invertebrates in some types of standing water habitats, which have been little studied, are sensitive to both past and present spray practices. For example, Fairchild et al. (1987) noted risk to specialized mosquito and midge inhabitants of pitcher plants. Fairchild and Eidt (1992) discovered that nutrient flow patterns in acidic bog ponds could be temporarily interrupted. This was demonstrated by eliminating some functional groups of organisms with fenitrothion treatment at concentrations that according to Ernst et al. (1991) and Sundaram (1991) occur from direct operational sprays.

Direct effects on fish under current operational conditions have not been detected, although fenitrothion concentrations ranging from 0.13 to 13.7 μg/g wet weight (Hatfield and Riche 1970; Lockhart et al. 1973) have been found in their tissues. Sensitivity decreases with development stage from embryo or fingerling to adult (Klavercamp et al. 1977; Takimoto et al. 1984b). Acetylcholinesterase inhibition, the primary mechanism of fenitrothion toxicity, has been measured in field-collected fish, but no clear relationship has been established with physiological impairment (Witty 1985). Many species of fish engorge themselves on the increased number of arthropods that drift after sprays (Ernst and Julien 1984; Kingsbury 1978), but this apparently does not affect the fish. Fish regurgitated contaminated food that was force fed in the laboratory (Scherer 1975; Symons 1973) and have been observed to avoid fenitrothion-contaminated water in laboratory choice tests (Scherer 1975).

All attempts to identify effects on fish populations in the field have been handicapped by the inability to control fish immigration and emigration. Notwithstanding this handicap, and lack of statistical significance, MacDonald and Penney (1969) and Montreal Engineering Co. (1978) recorded reductions in fish biomass within operational spray blocks. In a controlled stream experiment, Symons and Harding (1974) injected fenitrothion for 24 hours at 10 to 100 times peaks measured after forest spraying, which killed most aquatic insects, and estimated a real reduction in fish biomass of 45 ± 17%. It is generally accepted that reduction in fish growth due to depletion of food supply may occur after operational sprays, but it has not been measured because it is small, and it is deemed negligible.

Hiraoka and Okuda (1984) noted vertebral centrum deformities were three to six times as numerous in fish embryos from a stream draining an operational block sprayed at 1 800 g/ha than in those from untreated streams. The maximum single dose rate permitted in Canada is 280 g/ha. Other morphological effects, or physiological effects other than acetylcholinesterase inhibition, have not been noted in the field.

Several effects on fish have been demonstrated in the laboratory using sustained concentrations well above those encountered in the field. Some are biological changes indicative of increased anaerobic metabolism and hyperglycemia (Koundinya and Ramamurthi 1979a); reduction of cholinesterase activity (Klavercamp et al. 1977); inhibition of oxygen consumption (Murty et al. 1983; Koundinya and Ramamurthi 1979b); reduced protein-bound iodine (Saxena and Mani 1985); cellular damage to liver, kidney, and intestine (Mandal and Kulshrestha 1980); teratogenic effects (Hiraoka and Okuda 1984; Kamaldeep and Toor 1977; Takimoto et al. 1984a); reduced hatch (Miyashita 1984); and inhibition of gonadal development and fecundity of viviparous fish (Saxena et al. 1986).

Except for the field studies of Fairchild and his associates in New Brunswick bogs (Fairchild 1990; Fairchild and Eidt 1992; Fairchild et al. 1987), there has been little research on or monitoring of aquatic impacts of fenitrothion since 1986 in Canada. This is because fenitrothion has been used for forest spraying only in New Brunswick and Newfoundland since then. Streams and lakes in areas sprayed most often (year-to-year frequency) have fauna apparently unaltered in species composition or populations. One remaining matter of concern, however, is the impact on the fauna of small woodland pools with limited forest cover and flow through, and the subsequent impact on organisms at higher trophic levels of temporarily reduced arthropod populations in ponds. The latter question has been posed by wildlife biologists concerned about declines in duck brood success, but in spite of circumstantial evidence the answer has proved elusive.

Bt

Bt is classed as a biological insecticide, and the type usually used in forestry is highly specific to Lepidoptera (Anon. 1987b). Because of this specificity, Bt is considered environmentally safer than the broad-spectrum chemical insecticides. An experimental application of Bt at 9.9 BIU/ha (billion international units per hectare) in Ontario had no adverse effect on stream insects (Buckner et al. 1974; Kingsbury 1975). Applications at 7.13 BIU/ha in Quebec (Kingsbury and Sarrazin 1975) and 10 and 20 BIU/ha in Nova Scotia also had no detectible effects on stream benthic invertebrates (MacDonald 1980). Eidt (1985) estimated the risk to aquatic insects from Bt spraying based on the results of laboratory tests using representative aquatic insects and a knowledge of expected exposures in water after forest spraying. He concluded that there is no hazard to stream insects of Bt aerial sprays at rates adequate for spruce budworm control, but indicated a possibility of effects in ponds on mosquito larvae and insect larvae with similar feeding strategies and high gut pH. Bt as used in forestry is generally regarded as not hazardous to any nontarget aquatic organisms, although the toxicity to aquatic Lepidoptera, of which there are several species of aquatic Pyralidae in Canada, has not been investigated.

Aminocarb

Aminocarb is the insecticide of choice for spruce budworm control when effectiveness is weighed against impacts on aquatic organisms. At the operational rate of application it has been economically competitive, more reliable, and causes no known impacts. The matter may be academic, because aminocarb, although registered for use, is no longer available for reasons regarding the economy of production.

The field and laboratory monitoring and toxicology studies published up to and including most of 1981 were reviewed in considerable detail by a panel convened by the National Research Council of Canada (NRCC 1982). It was observed that peak concentrations of aminocarb rarely exceeded 10 µg/L, which was well below concentrations known to cause biological effects in static tests in the laboratory. The principal concern was with nonylphenol, which was shown to be as toxic as aminocarb and constituted a greater part of the formulation than did aminocarb. McLeese et al. (1980a,b, 1981) clearly demonstrated that the formulation using nonylphenol was more toxic to fish than was aminocarb alone. Eidt and Kingsbury (1981) concluded that this did not pose any significant hazard to the integrity of aquatic environments under the then current forestry use patterns, but it was subsequently accepted that the margin of safety was too narrow.

When the NRCC report was published, a "flowable" formulation that did not contain nonylphenol was being introduced and subsequently completely replaced formulations containing nonylphenol. This overcame any concerns about direct hazard to fish from operational sprays. Since that time, a few studies have clarified questions about the fate and field toxicity of aminocarb in the flowable formulation. The only derivative commonly found in aqueous matrices, albeit erratically and in small concentrations, has been aminocarb phenol. In a clear, cold, headwater trout stream, aminocarb concentrations in suspended particles, sediments, periphyton, higher plants, insects, and brook trout were not higher than ambient concentrations, and they dissipated quickly (Eidt et al. 1988). Nonetheless, some water plants, notably hornwort, *Ceratophyllum demersum*, have a high absorptive capacity for aminocarb (Weinberger and Greenhalgh 1985). Eidt and Weaver (1984) confirmed this laboratory discovery in the field but failed to find evidence that the distribution of hornwort had been influenced by the distribution of forest spraying with aminocarb. More perplexing was finding aminocarb residue in hornwort the year after and downstream of a spray 20 to 40 km distant.

In a single purpose field experiment, using the flowable formulation, Eidt and Weaver (1983) determined a threshold of 30 µg/L for induction of drift of invertebrates in a cool headwater stream. This confirmed a similar threshold determined for the nonylphenol formulation by Ahern (1980) and Ahern and Leclerc (1981).

The greatest concern expressed by the NRCC (1982) panel was that high localized concentrations of aminocarb might occur in unprotected standing water; the panel recommended that such occurrences and their biological impacts be investigated. There has been only one field experiment in a pond using the flowable formulation. Fairchild and Eidt (unpubl. data) treated a warm shallow bog pond of pH 4 by spraying it from the ground with a mist blower. The concentrations in the water were 25 and 7 µg/L after two consecutive applications, greater than three times that expected from an operational spray at 70 g/ha, assuming the unlikelihood that deposition was equal to emission. Only midges of the families

Ceratopogonidae and Cecidomyidae were affected and recovery was complete in less than 6 weeks as determined by population density. A greater impact should be expected in more alkaline water because Sanders et al. (1983) found increasing aminocarb toxicity to two species of freshwater fish with increasing temperature and pH.

Permethrin

From 1976 to 1981, the synthetic pyrethroid, permethrin, was subjected to extensive environmental impact and efficacy testing as a potential new spruce budworm, *Choristoneura fumiferana*, larvicide. The results of field studies on stream fish and invertebrates are available in detail in reports issued by the Forest Pest Management Institute of Forestry Canada and are summarized in Kingsbury and Kreutzweiser (1987) and Kreutzweiser and Kingsbury (1987). These studies showed that at aerial application rates between 8.8 and 70 g/ha, permethrin always resulted in major increases in invertebrate drift in cold-water forest streams in the sprayed areas and usually caused measurable depletions of benthos. Increases in drift were observed as far as 7.6 km downstream from treated areas and depletions of benthos as far as 4.2 km downstream. Ephemeroptera was generally affected most. Recovery, in terms of total benthos density, required from 1 to 18 months, with the rate of recovery inversely related to the initial exposure (i.e. application rate). Effects on species composition and diversity persisted longer. Direct effects on fish food organisms resulted in altered diets for brook trout, *Salvelinus fontinalis*, and Atlantic salmon, *Salmo salar*, for several months to a year. Indirect effects on fish populations included temporary reduction in growth rate and emigration out of treated areas, presumably in response to increased competition for decreased food availability. None of the applications studied caused observed mortality among caged or free fish.

Substantial effects on fish and fish food organisms have also been documented in lakes sprayed with permethrin (Kingsbury 1976). At an aerially emitted rate of 35 g/ha, zooplankton (Cladocera and Copepoda) and aquatic insect (Chironomidae and Chaoboridae) populations were reduced, leading to changes in diets of white sucker, *Catostomus commersoni*, and brown bullhead, *Ictalurus nebulosus*. Minnows (Cyprinidae), pumpkinseeds, Lepomis gibbosus, and brown bullheads held in cages near the surface of the lake suffered almost complete mortality although little or no mortality occurred among uncaged fish or fish held in cages at depths of 2 to 3 m. Effects were much more severe in a lake sprayed at 140 g/ha, including large reductions in zooplankton and aquatic benthos populations, and some mortality of native fish. Burrowing mayfly nymphs (Ephemeroptera: Ephemeridae), dragonfly nymphs (Odonata), chironomid larvae, caddisfly larvae (Trichoptera), and amphipods (Amphipoda) were most affected and post-treatment changes in white sucker and smallmouth bass, *Micropterus dolomieui*, diets were dramatic. Effects lasted up to several months.

Kingsbury and Kreutzweiser (1979) documented permethrin impacts in forest ponds. They observed large numbers of dead water boatmen (Hemiptera: Corixidae), predacious diving beetles (Coleoptera: Dytiscidae), and whirligig beetles (Coleoptera: Gyrinidae) and population reductions of caddisfly larvae, mosquito larvae (Diptera: Culicidae), and chironomid larvae in ponds treated twice at 17.5 g/ha. Giant water bugs, *Belostoma* sp., and dragonfly nymphs were affected to a lesser extent. There was no pesticide-related mortality among fish (bullheads, *Ictalurus* spp., Cyprinidae, and mudminnows, Umbridae) or green frog tadpoles, *Rana clamitans*, held in cages in the treated ponds.

Because of its severe effects on stream invertebrate populations and resulting effects on fish production, permethrin is not an acceptable treatment for large-scale forestry use. Aerial application should be restricted to situations where aquatic systems can be effectively buffered. Permethrin has been studied intensively as a larvicide against spruce bud moth, *Zeiraphera canadensis* and *Zeiraphera* spp., in white spruce plantations in New Brunswick and Quebec. To this end, considerable work has been directed toward establishing standards for buffer zones between spray areas and surface waters (Payne et al. 1986).

Acephate

Acephate has been used occasionally in Canada in forest pest control operations against spruce budworm and Douglas-fir tussock moth, *Orgyia pseudotsugata*. Acephate applied at 1 121 g/ha in British Columbia had no obvious effect on pond invertebrates, and a decline in numbers of Amphipoda sp. was attributed to factors other than the effects of the insecticide (Buckner et al. 1976a). However, the techniques used were probably not sensitive enough to detect anything other than gross effects (Anon. 1980). Moderate increases in the numbers of drifting aquatic invertebrates (about twice prespray numbers) occurred in Ontario streams in forests sprayed with acephate at 560 g/ha (MacCallum 1980). Chironomidae appeared to be the group most affected. Although it was concluded that benthic invertebrate populations were unaffected by treatment, sample sizes were probably too small to adequately assess effects on this environmental component. Buckner and McLeod (1975) observed no mortality of wood frogs, *Rana sylvatica*, held in cages in a pond treated experimentally with acephate at 560 g/ha. Free-living wood frogs, spring peepers, *Hyla crucifer*, and blue-spotted salamanders, *Ambystoma laterale*, were similarly unaffected. Lyons et al. (1976) modeled exposure and susceptibility of green frog tadpoles to acephate and concluded that there is little potential for direct toxic effects on tadpole populations when acephate is applied at recommended dosage rates.

Mexacarbate

Mexacarbate is a carbamate insecticide with toxicity and field effects on aquatic animals similar to those of aminocarb. Kingsbury and Sarrazin (1975) found that two aerial applications at 52 g/ha had no apparent effect on benthic invertebrate populations in a stream based on Surber samples. After two aerial applications of mexacarbate at 70 g/ha each, Holmes (1986) described only a slight effect on aquatic fauna. Black fly (Simuliidae) and midge (Chironomidae) larvae drifted in a treated stream after the second application, but there was no detectable effect on benthos populations measured by Surber sampling. *Callibaetis* larvae (Ephemeroptera: Baetidae), *Trichocorixa* (Corixidae), and Oligochaeta declined in numbers in a treated pond after the second application, but it is doubtful whether the reductions were related to the spray. Brain cholinesterase activities of wild brook trout and Atlantic salmon, caged in the stream, were not inhibited. Mortality rates of caged green frog tadpoles did not differ between treated and control ponds. The candidacy of mexacarbate as a spruce budworm insecticide was withdrawn by the manufacturer for economic reasons in 1987, before it was registered for use.

Diflubenzuron

Diflubenzuron acts by interfering with the synthesis of chitin, an important component of insect cuticle (Anon. 1987a) and thus has little potential for harmful effects on vertebrates, which do not synthesize chitin. Aquatic arthropods, which synthesize chitin, are susceptible.

After an early experimental aerial application at 350 g/ha, populations of amphipods and Coleoptera larvae were reduced in a small stream, but not significantly (Buckner et al. 1975a). The observation lacked a suitable control for comparison, and samples were not collected later than 3 days after spray, which allowed little time for effects. In a pond treated with 140 g/ha, Kingsbury (pers. comm.) found that amphipods, damselfly nymphs (Odonata: Zygoptera), midge larvae, and caddisfly larvae (Trichoptera) declined in numbers. He collected samples as late as 23 days after the treatment, but the reductions could not be positively attributed to the insecticide application because of differences in fauna between the treated and control ponds. Direct application to two large beaver ponds resulted in dramatic declines in zooplankton populations (Kingsbury et al. 1987). Cladocera had virtually disappeared from both ponds by 5 days after spray and did not reappear until 2 to 3 months later. Copepod populations (particularly nauplii) were also affected but had begun to recover about 3 weeks after treatment. Water boatman nymphs (Hemiptera: Corixidae) were affected but benthic invertebrates were not noticeably affected by the treatment. There was some indication of an increased mortality rate among caged amphipods in the treated pond. In operational pest control programs, where an attempt is made to avoid introducing spray products into water bodies, the impact would be less.

Viruses

Nuclear polyhedrosis viruses are naturally occurring pathogens of many insect species, a few of which have been commercially developed and are registered for forest management use in Canada (Anon. 1987c). Because these viruses are extremely host-specific they present little hazard to the environment (Anon. 1987c). Nuclear polyhedrosis viruses of spruce budworm and redheaded pine sawfly, *Neodiprion lecontei*, have been tested in Ontario for effects on nontarget fauna (Buckner et al. 1975b; Kingsbury et al. 1978). Stream invertebrate populations in these tests were unaffected.

Other Insecticides

Other forest insecticides that have had only limited experimental or operational use but do not have a current Canadian registration, and that have been tested in Canada for impacts on aquatic organisms follow:

Insecticide	Applications × Rate (g/ha)	Reference
azamethiphos	2 × 70	Kingsbury et al. 1980
carbaryl	2 × 280	Holmes et al. 1981
chlorpyrifos-methyl	1 × 140	Kingsbury 1979
	1 × 725	Kingsbury and Holmes 1980
	1 × 858	Kingsbury and Holmes 1980
	2 × 70	Holmes and Millikin 1981
dimethoate	1 × 140	Kingsbury and McLeod 1979
phosphadmidon	2 × 70	Kingsbury 1975
	1 × 174	Eidt 1976
	3 × 70	Buckner et al. 1976b
	2 × 210	Buckner et al. 1976b
	2 × 140	Buckner and McLeod 1977
	1 × 280	Buckner and McLeod 1977
	2 × 70	McLeod 1977

References

Ahern, A. 1980. Mesures biologiques de l'impact du Matacil 1.4 OSC en eaux courantes. Rep. Bio-Conseil Inc. Québec, Que. 151 p.

Ahern, A.; Leclerc. 1981. Étude comparative des effets du Matacil et de fénitrothion sur les populations benthiques en eaux courantes. Rep. Bio-Conseil Inc., Québec, Que. 181 p.

Anon. 1980. Operational field trails against the Douglas-fir tussock moth with chemical and biological insecticides. Can. For. Serv., Pac. For. Res. Cent., Inf. Rep. BC-X-201, Victoria, B.C. 19 p.

Anon. 1987a. Diflubenzuron—technical reference. Canadian Pulp and Paper Association, Montreal, Que. 3 p.

Anon. 1987b. *Bacillus thuringiensis*—technical reference. Canadian Pulp and Paper Association, Montreal, Que. 4 p.

Anon 1987c. Nuclear polyhedrosis virus—technical reference. Canadian Pulp and Paper Association, Montreal, Que. 4 p.

Buckner, C.H.; Kingsbury, P.D.; McLeod, B.B.; Mortensen, K.L.; Ray, D.G. H. 1974. Impact of aerial treatment on non-target organisms, Algonquin Park, Ontario and Spruce Woods, Manitoba. *In* Evaluation of Commercial Preparations of *Bacillus thuringiensis* With and Without Chitinase Against Spruce Budworm. Can. For. Serv., Chem. Cont. Res. Inst., Rep. CC-X-59, Ottawa, Ont. 26 p.

Buckner, C.H.; McLeod, B.B. 1975. Impact of aerial applications of Orthene® upon non-target organisms. Can. For. Serv., Chem. Cont. Res. Inst., Rep. CC-X-104, Ottawa, Ont. 48 p.

Buckner, C.H.; McLeod, B.B. 1977. Ecological impact studies of experimental and operational spruce budworm (*Choristoneura fumiferana* Clemens) control programs on selected non-target organisms in Quebec, 1976. Can. For. Serv., Chem. Cont. Res. Inst., Rep. CC-X-137, Ottawa, Ont. 81 p.

Buckner, C.H.; McLeod, B.B.; Kingsbury, P.D. 1975a. The effect of an experimental application of Dimilin® upon selected forest fauna. Can. For. Serv., Chem. Cont. Res. Inst., Rep. CC-X-97, Ottawa, Ont. 26 p.

Buckner, C.H.; McLeod, B.B.; Kingsbury, P.D. 1975b. The effect of an experimental application of nuclear polyhedrosis virus upon selected forest fauna. Can. For. Serv., Chem. Cont. Res. Inst., Rep. CC-X-101, Ottawa, Ont. 25 p.

Buckner, C.H.; McLeod, B.B.; Lidstone, R.G. 1976a. Effects of experimental applications of Orthene® on forest songbirds and aquatic fauna in New Brunswick and British Columbia. Can. For. Serv., Chem. Cont. Res. Inst., File Rep. 67, Ottawa, Ont. 28 p.

Buckner, C.H.; McLeod, B.B.; Lidstone, R.G. 1976b. Environmental impact studies of spruce budworm (*Choristoneura fumiferana* Clemens) control programs in New Brunswick in 1976. Can. For. Serv., Chem. Cont. Res. Inst., Rep. CC-X-135, Ottawa, Ont. 141 p.

Coady, L.W. 1978. Immediate chemical and toxicological influences of aerially applied fenitrothion and aminocarb in selected Newfoundland streams. Environ. Can. Surv. Rep. EPS-5-AR-78-1. 83 p.

Dimond, J.B.; Malcolm, S.A., 1971. Accothion and aquatic insects: monitoring of stream populations. Pages 60-68 *in* Environmental Studies of Accothion for Spruce Budworm Control. A cooperative study by the State of Maine, U.S. Department of Agriculture and U.S. Department of Interior.

Eidt, D.C. 1975. The effect of fenitrothion from large-scale forest spraying on benthos in New Brunswick headwaters streams. Can. Entomol. 107: 743-760.

Eidt, D.C. 1976. Effects of fenitrothion and phosphamidon on stream benthos—1975. Can. For. Serv., Marit. For. Res. Cent., Inf. Rep. M-X-65, Fredericton, N.B. 17 p.

Eidt, D.C. 1977. Effects of fenitrothion on benthos in the Nashwaak project study streams in 1976. Can. For. Serv., Marit. For. Res. Cent., Inf. Rep. M-X-70, Fredericton, N.B. 22 p.

Eidt, D.C. 1985. Toxicity of *Bacillus thuringiensis* var. *kurstaki* to aquatic insects. Can. Entomol. 117: 829-837.

Eidt, D.C.; Bacon, G.B.; DeGraeve, G.M.; Mallet, V.N. 1988. Fate and persistence of the insecticide aminocarb in New Brunswick (Canada) headwater stream. Arch. Environ. Contam. Toxicol. 17: 817-829.

Eidt, D.C.; Kingsbury, P.D. 1981. Toxicity to fish and fish-food organisms. Pages 148-152 *in* J. Hudak and A.G. Raske (Eds.), Review of the Spruce Budworm Outbreak in Newfoundland—Its Control and Management Implications. Nfld. For. Res. Cent., Inf. Rep. N-X-205.

Eidt, D.C.; Sosiak, A.J.; Mallet, V.N. 1984. Partitioning and short-term persistence of fenitrothion in New Brunswick (Canada) headwater streams. Arch. Environ. Contam. Toxicol. 13: 43-52.

Eidt, D.C.; Weaver, C.A.A. 1983. Threshold concentration of aminocarb that causes drift of stream insects. Can. Entomol. 115: 715-716.

Eidt, D.C.; Weaver, C.A.A. 1984. The fenitrothion and aminocarb content and the distribution of the aquatic plant *Ceratophyllum demersum* relative to forest spraying in New Brunswick and Nova Scotia. Naturaliste can. 111: 235-239.

Ernst, W.R.; Julien, G.R. 1984. Effects on brook trout (*Salvelinus fontinalis*) feeding behaviour and brain cholinesterase activity of experimental fenitrothion and aminocarb formulations containing Triton® X-100 and Cyclosol 63. Environ. Can. Surv. Rep. EPS-5-AR-84-9. 28 p.

Ernst, W.R.; Julien G.; Hennigar, P. 1991. Contamination of ponds by fenitrothion during forest spraying. Bull. Environ. Contam. Toxicol. 46:815-821.

Fairchild, W.L. 1989. Perturbation of the aquatic community of acid bog ponds by the insecticide fenitrothion. Ph.D. thesis, University of New Brunswick, Fredericton, N.B.

Fairchild, W.L.; Eidt, D.C. 1992. Perturbation of the aquatic invertebrate community of acidic bog ponds by the insecticide fenitrothion. Arch. Environ. Contam. Toxicol. (In press)

Fairchild, W.L.; Eidt, D.C.; Weaver, C.A.A. 1987. Effects of fenitrothion insecticide on inhabitants of leaves of the pitcher plant, *Sarracenia purpurea* L. Can. Entomol. 119: 647-642.

Fairchild, W.L.; Ernst, W.R.; Mallet, V.N. 1989. Fenitrothion effects on aquatic systems. Pages 113-166 *in* Environmental Effects of Fenitrothion in Forestry. Environ. Can., Conserv. and Pro., Atlantic Region.

Hatakeyama, S.; Shiraishi, H.; Kobayashi, N. 1990. Effects of spraying of insecticides on nontarget macrobenthos in a mountain stream. Ecotoxicol. Environ. Safety 19: 254-270.

Hatfield, C.T.; Riche, L.G. 1970. Effects of aerial Sumithion spraying on juvenile Atlantic salmon (*Salmo salar* L.) and brook trout (*Salvelinus fontinalis* Mitchill) in Newfoundland. Bull. Environ. Contam. Toxicol. 5: 440-442.

Hiraoka, Y.; Okuda, H. 1984. A tentative assessment of water pollution by the medaka, *Oryzias latipes*, egg stationing method aerial application of fenitrothion emulsion. Environ. Res. 34: 262-267.

Holmes, S.B. 1986. A preliminary report on the impact of an experimental aerial application of Zectran® (mexacarbate) insecticide on aquatic ecosystems in New Brunswick, Canada. Can. For. Serv., For. Pest Manage. Inst., Sault Ste. Marie, Ont. File Rep. No. 70, 13 pp + figure and tables.

Holmes, S.B.; Millikin, R.L. 1981. A preliminary report on the effects of a split application of Reldan on aquatic and terrestrial ecosystems. Can. For. Serv., For. Pest Manage. Inst., Sault Ste. Marie, Ont. File Rep. No. 13. 88 p.

Holmes, S.B.; Millikin, R.L.; Kingsbury, P.D. 1981. Environmental effects of a split application of Sevin-2-Oil. Can. For. Serv., For. Pest Manage. Inst., Sault Ste. Marie, Ont. Rep. FPM-X-46. 58 p. + appendices.

Kamaldeep, K.; Toor, H.S. 1977. Toxicity of pesticides to embryonic stages of *Cyprinus carpio communis* Linn. Ind. J. Exp. Biol. 15: 193.

Kingsbury, P.D. 1975. Monitoring aquatic insect populations in forest streams exposed to chemical and biological insecticide applications. Proc. Entomol. Soc. Ont. 106: 19-24.

Kingsbury, P.D. 1976. Studies of the impact of aerial applications of the synthetic pyrethroid NRDC-143 on aquatic ecosystems. Can. For. Serv., For. Pest Manage. Inst., Ottawa, Ont. Rep. CC-X-127. 111 p.

Kingsbury, P.D. 1978. A study of the distribution, persistence and biological effects of fenitrothion applied to a small lake in an oil formulation. Can. For. Serv., For. Pest Manage. Inst., Sault Ste. Marie, Ont. Rep. FPM-X-113. 85 p.

Kingsbury, P.D. 1979. Preliminary field assessment of the effects of chlorpyrifos methyl on stream fauna. A report on a preliminary field study carried out by FPMI's Environmental Impact Section at Ste-Anne-des Monts, Quebec in 1977. Can. For. Serv., For. Pest Man. Inst., Sault Ste. Marie, Ont. Unpubl. report 9 p.

Kingsbury, P.D.; Holmes, S.B. 1980. A preliminary report on the impact on stream fauna of high dosage applications of Reldan. Can. For. Serv., For. Pest Man. Inst., Sault Ste. Marie, Ont. File Rep. No. 2, 27 p. + appendix.

Kingsbury, P.D.; Holmes, S.B.; Millikin, R.L. 1980. Environmental effects of a double application of azamethiphos on selected terrestrial and aquatic organisms. Can. For. Serv., For. Pest Manage. Inst., Sault Ste. Marie, Ont. Rep. FPM-X-33. 32 p. + appendices.

Kingsbury, P.D. Kreutzweiser, D.P. 1979. Impact of double applications of permethrin on forest streams and ponds. Can. For. Serv., For. Pest Manage. Inst., Sault Ste. Marie, Ont. Rep. FPM-X-27. 42 p. + appendices.

Kingsbury, P.D.; Kreutzweiser, D.P. 1987. Permethrin treatments in Canadian forests. Part 1: impact on stream fish. Pestic. Sci. 19: 35-48.

Kingsbury, P.D.; McLeod, B.B. 1979. Ecological monitoring of the sequential application of insecticides in forest protection programs. A summary of studies carried out by FPMI's Environmental Impact Section. Can. For. Serv., For. Pest Manage. Inst., Sault Ste. Marie, Ont. Unpubl. report 13 p.

Kingsbury, P.D.; McLeod, B.B.; Mortensen, K. 1978. Impact of applications of the nuclear polyhedrosis virus of the redheaded pine sawfly, *Neodiprion lecontei* (Fitch), on nontarget organisms in 1977. Can. For. Serv., For. Pest Manage. Inst., Sault Ste. Marie, Ont. Rep. FPM-X-11. 27 p.

Kingsbury, P.D.; Sarrazin, R. 1975. Aquatic fauna. Pages 80-100 *in* C.H. Buckner and R. Sarrazin (Eds.), Studies of the Environmental Impact of the 1974 Spruce Budworm Control Operation in Quebec. Can. For. Serv., Chem. Cont. Res. Inst., Ottawa, Ont. Rep. CC-X-93.

Kingsbury, P.D.; Sundaram, K.M.S.; Holmes, S.B.; Nott, R.; Kreutzweiser 1987. Aquatic fate and impact studies with Dimilin®. Can. For. Serv., For. Pest Manage. Inst., Sault Ste. Marie, Ont. File Rep. No. 78. 45 p.

Klaverkamp, J.F.; Duangsawasdi, M.; Macdonald, W.A.; Majewski, H.S. 1977. An evaluation of fenitrothion toxicity in 4 life stages of rainbow trout, *Salmo gairdneri*. Pages 231-240 *in* F.L. Mayer and J.L. Kaelink (Eds.), Aquatic Toxicology and Hazard Evaluation. ASTM Spec. Tech. Publ. No. 634, Philadelphia, Pa.

Koundinya, P.R.; Ramamurthi, R. 1979a. Tissue respiration in *Tilapia mossambica* exposed to lethal (LC50) concentration of Sumithion and Sevin. Ind. J. Environ. Health. 20: 426-428.

Koundinya, P.R.; Ramamurthi, R. 1979b. Effect of organophosphate pesticide Sumithion (Fenitrothion) on some aspects of carbohydrate metabolism in a freshwater fish, *Sarotherodon* (Tilapia) *mossambicus* (Peters). Experientia (Basel). 35: 1632-1633.

Kreutzweiser, D.P.; Kingsbury, P.D. 1987. Permethrin treatments in Canadian forests. Part 2: impact on stream invertebrates. Pestic. Sci. 19: 49-60.

Lakshminarayana, J.S.S.; Bourque, H. 1980. Absorption of fenitrothion by plankton and benthic algae. Bull. Environ. Contam. Toxicol. 24: 389-396.

Lockhart, W.L.; Metner, D.A.; Grift, N. 1973. Biochemical and residue studies on rainbow trout, *Salmo gairdneri*, following field and laboratory exposures to fenitrothion. Manit. Entomol. 7: 26-36.

Lyons, D.B.; Buckner, C.H.; McLeod, B.B.; Sundaram, K.M.S. 1976. The effects of fenitrothion, Matacil® and Orthene® on frog larvae. Can. For. Serv., Chem. Cont. Res. Inst., Rep. CC-X-129, Ottawa, Ont. 86 p.

MacCallum, M.E. 1980. The short-term environmental impact of the spruce budworm spray program in Ontario—1979. Report prepared under contract to the Pest Cont. Sec., For. Res. Group, Ont. Ministry of Nat. Res. 59 p. + appendices.

MacDonald, P. 1980. Effect of *Bacillus thuringiensis* spraying on aquatic fauna. *In* T.D. Smith, C.A. Miller, K.R. Rozee, A.S. Menon, D. MacKay, P. MacDonald, P. Germain, D.G. Embree, R.M. Bulmer, and R. Wheeler (Eds.), An Experiment with Thuricide 16B for the Reduction of Population Densities of Spruce Budworm (*Choristoneura fumiferana* (Clemens, 1865)) (Lepidoptera: Tortricidae), Nova Scotia, 1979. Rep. to the Minister of Lands and Forests, N.S.

MacDonald, J.R.; Penney, G.H. 1969. Preliminary report on the effects of the 1969 New Brunswick forest spraying on juvenile salmon and their food organisms. Dep. Fish. For. Can., Resource Dev. Branch. 17 p.

Maguire, R.J.; Hale, E.J. 1980. Distribution of fenitrothion sprayed on a pond: kinetics of its distribution and transformation in water and sediment. J. Agric. Food Chem. 28: 372-378.

Mallet, V.N.; Cassista, A. 1984. Fenitrothion residue survey in relation to the 1981 spruce budworm spray program in New Brunswick, Canada. Bull. Environ. Contam. Toxicol. 32: 65-74.

Mandal, P.K.; Kulshrestha, A.K. 1980. Histo-pathological changes induced by the sublethal Sumithion in *Clarias batrachus*. Indian J. Exp. Biol. 18: 547-552.

McLeese, D.W.; Zitko, V.; Metcalfe, C.D.; Sergeant, D.B. 1980a. Lethality of aminocarb and the components of the aminocarb formulation to juvenile Atlantic salmon, marine invertebrates and freshwater clams. Chemosphere 9: 79-82.

McLeese, D.W.; Sergeant, D.B.; Metcalfe, C.D.; Zitko, V.; Burridge, L.E. 1980b. Uptake and excretion of aminocarb, nonyl phenol, and pesticide diluent oil 585 by mussels (*Mytilus edulis*). Bull. Environ. Contam. Toxicol. 24: 575-581.

McLeese, D.W.; Zitko, V.; Sergeant, D.B.; Burridge, L.; Metcalfe, C.D. 1981. Lethality and accumulation of alkylphenols in aquatic fauna. Chemosphere 10: 723-730.

McLeod, B.B. 1977. Environmental monitoring of experimental and operational forest spray programmes in New Brunswick in 1977. A preliminary report to New Brunswick Department of Natural Resources and Committee for Environmental Monitoring of Forest Insect Control Operations (EMOFICO). Can. For. Serv., For. Pest Manage. Inst., Sault Ste. Marie, Ont. Unpubl. report 6 p.

Miyashita, M. 1984. Effects of a short period exposure of fenitrothion on the reproduction of pregnant females of the guppy, *Poecilia reticulata*. Jpn. J. of Hyg. 39: 647-650.

Montreal Engineering Company, Limited. 1976. Assessment of effects of selected insecticide application on drift of stream macroinvertebrates. Report for New Brunswick Department of Natural Resources, Montreal Engineering Company, Limited, Fredericton, N.B. 18 p.

Montreal Engineering Company, Limited. 1978. Assessment of the effects of the 1977 New Brunswick spruce budworm control program on fish food organisms and fish growth. Report for New Brunswick Department of Natural Resources, Montreal Engineering Company, Limited, Fredericton, N.B. 40 p.

Morin, R.; Gaboury, G.; Mamarbachi, G. 1986. Fenitrothion and aminocarb residues in water and balsam fir foliage following spruce budworm spraying programs in Quebec, 1979 to 1982. Bull. Environ. Contam. Toxicol. 36: 622-628.

Murty, A.S.; Rajabhushanam, B.R.; Ramani, A.V.; Christopher, K. 1983. Toxicity of fenitrothion to the fish *Mystus cavasius* and *Lambeo rahita*. Environ. Pollut. Ser. A. Ecol. Biol. 30: 225-232.

NRCC. 1982. Aminocarb: the effects of its use on the forest and the human environment. Associate Committee on Scientific Criteria for Environmental Quality, National Research Council of Canada. NRCC No. 16740. 246 p.

Payne, N.; Helson, B.; Sundaram, K.; Kingsbury, P; Fleming, R.; de Groot, P. 1986. Estimating the buffer required around water during permethrin applications. Can. For. Serv., For. Pest Manage. Inst., Sault Ste. Marie, Ont. Rep. FPM-X-70. 26 p.

Rabeni, C; Gibbs, K.E. 1976. Effects of Dylox, Matacil and Sumithion on non-target stream insects in Maine, 1975. Pages 64-74 *in* U.S. Forest Service, 1975 Cooperative Pilot Control Project of Dylox, Matacil, and Sumithion for spruce budworm control in Maine, U.S. Forest Service, Upper Darby, Pa.

Riche, L.G.; Coady, L. 1969. Effects of aerial spraying with Sumithion on salmon brooks in western and central Newfoundland—1969, with special reference to aquatic insects. Can. Dep. Fish. For., Res. Dev. Branch. Prog. Rep. 61: 43 p.

Sanders, H.O.; Findley, M.T.; Hunn, J.B. 1983. Acute toxicity of 6 forest insecticides to 3 aquatic invertebrates and 4 fishes. U.S. Fish. Wildl. Serv. Tech. Pap. No. 110, 5 p.

Saxena, P.K.; Mani, K. 1985. Protein-bound iodine levels in the blood plasma of freshwater teleost, *Channa punctatus* (Bl.) exposed to subtoxic pesticide concentrations. Toxicol. Lett. 24(1): 33-36.

Saxena, P.K.; Mani, K.; Kondal, J.K. 1986. Effect on safe application rate concentrations of some biocides on the freshwater murrel, *Channa punctatus*, a biochemical study. Ecotoxicol. Environ. Safety 12: 1-14.

Scherer, E. 1975. Avoidance of fenitrothion by goldfish, *Carassius auratus*. Bull. Environ. Contam. Toxicol. 13: 492-496.

Sundaram, K.M.S. 1991. Foliar and ground deposits of fenitrothion following aerial application over a plantation forest. J. Environ. Sci. Health, Part B. 25:643-663.

Symons, P.E.K. 1973. Behavior of young Atlantic salmon (*Salmo salar*) exposed to or force-fed fenitrothion, and organophosphate insecticide. J. Fish. Res. Board Can. 30: 651-655.

Symons, P.E.K. 1977a. Assessing and predicting effects of the fenitrothion spray program on aquatic fauna. Natl. Res. Counc. Can., Assoc. Comm. Sci. Criteria Environ. Qual., Publ. 16073: 391-413.

Symons, P.E.K. 1977b. Dispersal and toxicology of the insecticide fenitrothion; predicting hazards of forest spraying. Residue Rev. 38: 1-36.

Symons, P.E.K.; Harding, G.D. 1974. Biomass changes of stream fishes after forest spraying with the insecticide fenitrothion. Fish. Res. Board Can., Tech. Rep. 432: 47 p.

Takimoto, Y.; Hugino, S.; Yamanda, H.; Miyamoto, J. 1984a. The acute toxicity of fenitrothion to killifish (*Oryzias latipes*) at twelve different stages of its life history. J. Pestic. Sci. 9: 463-470.

Takimoto, Y.; Ohshima, M.; Yamada, M.; Miyamoto, J. 1984b. Fate of fenitrothion in several developmental stages of the killifish (*Oryzias latipes*). Arch. Environ. Contam. Toxicol. 13: 579-587.

Weinberger, P.; Greenhalgh, R. 1985. The sorptive capacity of an aquatic macrophyte for the pesticide aminocarb. J. Environ. Sci. Health. B20: 263-273.

Weinberger, P.; Greehalgh, R.; Moody, R.P.; Boulton, B. 1982. Fate of fenitrothion in aquatic microcosms and the role of aquatic plants. Environ. Sci. Technol. 16: 470-473.

Witty, M.J. 1985. Cholinesterase activities in brook trout: a bioassay study of fenitrothion containing Triton X-114 on brook trout. Unpublished, Sault College of Applied Arts and Technology. 24 p.

Zitko, V.; Cunningham, T.D. 1974. Fenitrothion, derivatives and isomers: hydrolysis, absorption and biodegration. Fish. Mar. Serv. Tech. Rep. 458.

Chapter 61

Impacts of Forest Aerial Spray Programs on Terrestrial Vertebrates

PETER D. KINGSBURY AND RHONDA L. MILLIKIN

Introduction

Forests are inhabited by a wide variety of amphibians, reptiles, birds, and large and small mammal species. Of all these animals, forest songbirds and small mammals have been considered at greatest risk from the effects of insecticides sprayed on the forest and are the subjects of most impact studies.

Birds are a critical component of the forest ecosystem. By responding to changes in food availability and forest structure, they can signal the state of the ecosystem. Furthermore, they have an important influence on insect populations at an endemic level and may suppress or delay insect population increases to epidemic levels (Otvos 1979; Takekawa et al. 1982). This predation effect of birds has been verified through exclusion experiments (e.g. Dowden et al. 1953; Holmes et al. 1979; Torgersen and Campbell 1982; Kelly and Régnière 1985). They respond through changes in their density (numerical response: MacArthur 1958; Gage and Miller 1978) and by increasing the number of insects taken per bird (functional response: Holling 1959; Zach and Falls 1975). The demand for insects is greatest during the late larval and pupal periods, a critical time for the survival of spruce budworm, *Choristoneura fumiferana* (Crawford et al. 1983).

Predation by small animals, on the other hand, has been reported to have little influence on spruce budworm population dynamics (Morris et al. 1958). The predation potential of certain species, particularly the red squirrel, *Tamiasciurus hudsonicus* (Jennings and Crawford 1985), suggests further research is merited.

Both exposure (vulnerability) and response (susceptibility) are important factors contributing to the toxicity of pesticides. These factors can be separated into biological factors (habitat preference, timing of migration, feeding strategy of the species, nesting activities, territorial habits, singing habits, metabolism of species, body size) and physicochemical factors (toxicity of the chemical, application method, timing and number of applications, pesticide formulation, droplet size, weather). It is generally accepted that the most vulnerable and most susceptible avian species are the smaller, more active, insectivorous birds that feed and nest in the upper half of the canopy (Germain 1980a; Peakall and Bart 1983; Busby et al. 1987a). For most of these birds, applications occur when insects are essential to their maintenance and reproduction. Small mammal exposure to insecticides is limited by their largely nocturnal habits and the protection of underground burrows (Buckner et al. 1977a).

Exposure may be direct (inhalation, dermal, ingestion via food or preening) or indirect (food shortage, residues in food, habitat changes). For an insecticide emitted at 70 g/ha, Peakall and Bart (1983) calculated the potential exposure routes to be 33 μg pesticide per 10-g bird orally, through contaminated food; about 3 μg dermally, by adsorption and preening; and extremely small amounts through inhalation. Most passerines do not hatch their young until after the application and, therefore, a direct effect on fledglings would be negligible. It is more likely that lethal or sublethal effects on the adults would lead to reduced parental care.

Duration of exposure depends on persistence of the pesticide in the environment and on the metabolic capabilities of the exposed animals. Secondary poisoning through the consumption of dead or dying insects is possible (DeWeese et al. 1979) and may present a major route of exposure for nestlings (EMOFICO 1982).

Methods of Investigation

Birds

Current census methodology originates from Kendeigh's (1947) application of the territorial mapping technique (Williams 1936). The singing activity of territorial males is monitored before and after treatment on both treated and untreated control plots. A relative decrease in singing activity on the treated plots is assumed to indicate a detrimental impact. Early studies using this method were designed to detect immediate lethal effects. These have since been replaced (in the late 1970s and the 1980s) with programs using a broader range of methods more sensitive to sublethal effects.

Canadian research has taken a variety of directions subsequent to Prebble's (1975) compilation. Some censuses conducted in the mid-1970s concentrated on singing activity without territory mapping (Pearce et al. 1976; Pearce et al. 1979). Other researchers used mist-netting to indicate reproductive success (Ouellet 1981a; Bennet et al. 1987).

Other studies have continued monitoring singing activity using the territory mapping technique. Censuses are conducted, however, in combination with mist-netting for both behavioral observations of marked individuals and an indirect measure of reproductive success (Millikin 1982). Recent emphasis has been on long-term effects as well as immediate behavioral responses to a depletion of food (Millikin 1987). Laboratory toxicology studies have been initiated using zebra finches for immediate and short-term effects on behavior and reproductive success (Holmes 1988).

Since the late 1970s, researchers of the Canadian Wildlife Service at Fredericton, New Brunswick, have taken an indicator species approach, concentrating on the white-throated sparrow, its nesting success and nestling growth (EMOFICO 1982, 1984). Additionally, they use the cholinesterase assay, developed by Hill and Fleming (1982), as a monitoring tool to measure exposure of selected species of songbirds to forestry insecticide sprays. In their interpretation of changes in cholinesterase activity levels after treatment, they have adopted the standards set by Ludke et al. (1975), that a 20% cholinesterase depression indicates exposure, and by Zinkl et al. (1979), that a 50% cholinesterase depression signifies a life-threatening situation.

Researchers outside of Canada have concentrated mainly on cholinesterase depression and studies on feeding behavior (e.g. Grue et al. 1982; Graber and Graber 1983).

There are a variety of techniques available, each having specific benefits and limitations (Table 1). The optimum monitoring program combines some of these methods, the selection depending on the spray program and resource logistics in question. A measure of bird abundance should be combined, however, with some index of reproductive success.

Small Mammals

Impact studies on small mammal populations have used three main techniques (Table 2). Bracher et al. (1986) compared kill-trapping to the sand-transect technique, and found the sand transect to be more reliable because a greater number of species could be analyzed with less disturbance to the small mammal community. However, the sand-transect method could not provide data on breeding status or age structure.

Comments and Recommendations on Census Methodology for Songbirds

Singing activity is generally measured over an area ranging from 4 to 10 ha, the configuration of which varies from a plot to a strip transect (one line with stations flagged along it), and, at the other extreme, a single "spot" census monitored for a set amount of time (usually between 0.8 and 5 km). The progression from plot to transect and spot has allowed more area to be sampled with less time and personnel, but this has been accomplished at the expense of precision and certainty about individual territories mapped and, consequently, the number of birds and their behavior (Germain 1980a; Millikin 1988). With smaller plots a larger proportion of the territories will lie partly outside the plot.

With line transects, the census area varies with both species and hearing abilities. There is a tendency to record a greater proportion of singing birds when fewer birds are present, thereby masking any decline. Much research has been conducted on optimum hearing distance (usually between 0.8 and 5 km) so that the exact census area can be calculated for a measure of relative abundance (Ralph and Scott 1981). A short-term impact is measured as a change from before treatment. Therefore, a reliable index of the population is all that is necessary. Long-term studies, on the other hand, require as accurate a measure of absolute densities as possible.

Lehoux et al. (1982) compared various methods for their ability to detect an impact and found banding efforts and searches for sick or dead birds to be more sensitive than territory mapping or line transects. They favored mapping over line transects because of its increased accuracy. Due to the problem of variable spray deposit, Peakall and Bart (1983) recommend the transect method to cover more spray swaths. Territory mapping along a transect can approximate the accuracy of the plot method (Millikin 1988), but only the plot method provides a close observation of individual birds. With more detailed observations of individual birds and their movement, results are more comparable between observers and between years. Therefore, long-term studies should use the plot method. Plots should be 10 to 15 ha (an equal number of treatment and control, each pair censused by one observer), and only 10 to 15 selected species censused (Germain 1977). Nonetheless, it is unclear what change in singing should be expected following spray-induced nest failure. If the nest fails, the male may renest or be replaced by another male. In both cases, singing becomes more emphatic. Furthermore, censusing is restricted to effects on males. On the other hand, mist-netting provides data on juveniles and females in addition to males and can be used throughout the season. Together, these two methods provide information on both survival and reproduction. Netting should be restricted to species potentially exposed to the treatment.

Behavioral observations require considerable expertise but can assess short-term impacts such as (1) response to changes in food availability, and (2) movements related to insecticide deposit (exposure). For behavioral changes to be detected, observations must be taken within 5 days after treatment; samples designed to measure the effect of a decreased food source or continued exposure should be extended to at least 5 days after spraying. The area must be large enough to minimize movement of birds across plot boundaries. This approach is not restricted to cholinesterase-inhibiting compounds.

In light of the destructive nature of cholinesterase sampling, and the difficulties in interpreting the data, this method should not be recommended. Rather, it should be used only as a tool in situations when it is necessary to verify exposure.

Intense studies of sublethal effects on birds should be accompanied by extensive studies on regional bird populations. If individuals are replaced locally by the immigration of birds from surrounding habitats, population effects may only be visible on a larger scale.

For a review of census methodology related to pesticide impacts, see Bart and Hunter 1978, Germain 1980a, Grue et al. 1982, Lehoux et al. 1982, Peakall and Bart 1983, Peakall 1985, and Mineau and Peakall 1986. For a general discussion of census methodology, see Svennson 1970, Emlen 1971, and Ralph and Scott 1981.

Effect on Songbirds

Phosphamidon

Phosphamidon (Dimecron®) was brought into operational use in spruce budworm control programs in the early 1960s as a far less impactive material on fish than DDT (Kingsbury 1975). Studies in the mid- to late-1960s suggested that 140 g/ha was the threshold application rate for phosphamidon at which symptoms of bird poisoning could be expected. Dosages marginally tolerated by birds did not provide acceptable foliage protection. Consequently, phosphamidon was replaced by fenitrothion in the early 1970s (Pearce 1975). In the mid-1970s, however, large increases in the scale of spruce budworm control programs coupled with limitations in the quantities of fenitrothion available led to renewed use of phosphamidon in larviciding programs. In addition, phosphamidon had been found to be a better spruce budworm

Table 1. Census methods for monitoring the impact of forest insecticide application on songbirds.

Method	Unit of measure	Uses	Limitations	Factors influencing the method	Example references
(1) territory mapping	no. breeding pairs	- change in activity (usually 3 days before to 5 days after treatment; treated vs control) by species or foraging guild - no. territories occupied - fate of individual territories after treatment	- large sample variability due to weather and migration which may mask insecticide-induced effects - inability to distinguish between emigration, mortality, and behavioral changes that influence conspicuousness - immigration of birds from untreated areas - seasonal changes in singing activity - only applicable when birds are defending territories - small sample size	weather, observer, time of day, season, habitat, interpretation errors, bird density, bird species, environmental noise	Germain and Morin 1979; Kingsbury and McLeod 1981
(2) line transect	no. individuals (their position is not recorded nor is there any attempt to follow them over the study)	- relative counts only - requires less effort than territory mapping	- same as above - less sensitive to movement of individuals, therefore less precise - masking of distant individuals by other close and more conspicuous individuals, so that a greater proportion of singing males are recorded when fewer birds are present, thereby masking any decline	- same as above	Lehoux et al. 1982; Pearce et al. 1979
(3) mist-netting	- population structure (sex ratio and no. young per breeding pair) - population stability (return of birds the year following treatment)	- indirect measure of reproductive success - includes effects on females and young as well as males - marking of individuals (can detect movement in and out of treatment area) - long-term population trends	- inadequate sampling of crown-inhabiting species - limited data for effort expended	weather, time of day, season, habitat, net avoidance	Millikin 1987; Ouellet 1981a; Bennet et al. 1987
(4) brain cholinesterase analysis	cholinesterase activity before and after treatment; treated vs control	- index of exposure to organophosphates and carbamates - physiological response	- singing birds are collected, possibly biasing the sample toward healthier individuals - sampled individuals may have immigrated into the treated area after treatment - method is destructive - biological significance of different levels of cholinesterase depression is not well known	age, sex, species, sampling technique, postmortem conditions (field and laboratory), sample size	Busby et al. 1987b
(5) nest observations	- no. successful nests - weight and growth of fledglings	reproductive success	- ability to locate enough nests for statistical analysis, and to locate enough nests of crown-inhabiting species - effect of observer	ability of observer	Peters 1979; EMOFICO 1980
(6) behavioral observations	amount of time spent in various behaviors and locations, before and after treatment, treated vs control	- indices of sublethal effects (activity and movement) - behavioral changes of marked individuals; symptoms of pesticide poisoning and indirect effects on foraging behavior - movement of birds within and across spray boundaries	- time consuming (little data per effort) - sample size - immigration of invertebrates from outside treated area - effects manifested over short time period	weather, observer, habitat, season, time of day	Millikin 1987
(7) search for dead and sick birds	no. per unit effort, treated vs control	- evidence of toxic effect (mortality or symptoms of poisoning) - can be used at any time of season (early or late applications)	- little return for effort - not sensitive to subtle impacts	habitat, observer, weather, scavengers	Lehoux et al. 1982
(8) sand transect	activity based on animal tracks	- change in behavior	- highly labor intensive and costly - method monitors ground-inhabiting species less exposed to aerial sprays - for songbirds, tracks can be identified only to family	observer, habitat, weather	Bracher and Bider 1982

Table 2. Census methods for monitoring the impact of forest insecticide application on small mammals.

Method	Unit of measure	Uses	Limitations	Factors influencing the method	Example references
(1) capture/ recapture (live-trapping)	no. individuals	- marking of individuals - estimate of population before and after treatment compared to control - mortality and movement out of sprayed area	- low population numbers before treatment	- immigration, weather, location of trap sites	Bracher et al. 1986
(2) kill-trapping	no. embryos and placental scars capture rate	- reproductive success (mean litter size and trends in litter size) on treatment vs control - carcasses for residue analysis	- sample size; species must be combined which may mask any impact	- greater problem of immigration due to removal of individuals and repopulation - weather, location of traps, size and type of trap and bait	Kingsbury et al. 1979
(3) sand transect	amount of activity at discrete instances over a period of time	- resampling of individuals (i.e. no disturbance to the population) - data on a wider range of species - predator–prey interactions	- severely affected by rain - expensive and labor intensive - no specimens for residue analysis or information on breeding status and age structure of population	- immigration, weather, location of transect with respect to habitat (i.e. increased species diversity with increased habitat diversity), sampling time period	Bracher et al. 1986

adulticide than fenitrothion (Miller et al. 1973). Larviciding application rates were generally 140 to 175 g/ha, often in sequential applications with other insecticides. Phosphamidon was used fairly extensively as an adulticide, generally in multiple applications of 70 g/ha.

From 1974 to 1976, phosphamidon larviciding treatments of 140 to 280 g/ha were monitored by territory mapping (Buckner and Ray 1975; McLeod 1976a; Buckner and McLeod 1977a; Buckner et al. 1977a) and singing male counts on line transects (Pearce et al. 1976, 1979). Results of these studies, partly summarized by Peakall and Bart (1983), provided conclusive evidence that phosphamidon applications of 140 g/ha or greater consistently affected forest songbird populations, particularly canopy, midcrown, or wide-ranging species such as kinglets and several warblers. In a few instances, some of the most sensitive species, the ruby-crowned kinglet, *Regulus calendula*, in particular, virtually disappeared after spraying.

Although considerable evidence of songbird mortality was gathered, it was impossible to determine the extent of the overall impact with the monitoring techniques used. Extensive bird kills in 1974 and 1975 caused by unseasonably cold weather during migration (Pearce 1974; Buckner et al. 1975a) made it difficult to measure the extent of phosphamidon-related mortality. Even where evidence of mortality existed, effects were masked by rapid replacement of affected individuals. An experimental application of 1 200 g/ha phosphamidon (designed to study songbird impact methodology) caused a significant lowering of singing activity and mortality among at least 15 species representing possibly 25% of the birds in the treated area, but effects could not be measured beyond 3 days (Lehoux et al. 1982).

Adulticide applications of 70 g/ha in New Brunswick were monitored in 1974 by singing male counts (Pearce 1974), and from 1974 to 1977 by mapping techniques (McLeod and Millikin 1982). Little impact on songbirds is evident in any of the data collected, even though the latter studies were carried out in blocks receiving double, triple, or quintuple applications of phosphamidon within 4 to 8 day periods. Some small population reductions were occasionally observed among kinglet and warbler populations (McLeod and Millikin 1982). The lack of observed effects may reflect (1) limited impact of phosphamidon at this application rate; (2) less sensitivity among bird populations at this stage of their breeding cycle (most breeding activity is complete by early to mid-July when adulticide treatments occur); or (3) inability of the census techniques to detect changes in bird populations when breeding territories have naturally broken down.

The use of phosphamidon for spruce budworm control ceased entirely after 1977 because of the documented effects on songbirds at larviciding rates. The lack of significant markets for this product led the manufacturer to choose not to respond to requests for replacement data for suspect information in the original registration package; phosphamidon was subsequently deregistered and withdrawn from use in Canada in the early 1980s.

Fenitrothion

Fenitrothion (Folithion®, Novathion®, Sumithion®) has been the most extensively used forestry insecticide since the late 1960s. Early studies suggested that operational spruce budworm larviciding with 140 or 210 g/ha presented no significant hazard to eastern forest bird populations. Nonetheless, occasional reports of avian casualties in spray areas suggested a rather narrow margin of safety (Pearce 1975).

From 1973 to 1982, many census studies were carried out to monitor songbird populations in areas treated with fenitrothion. Treatments were generally single or double applications between 140 and 280 g/ha but included sequential applications with other chemical insecticides. Reports are available detailing monitoring studies in New Brunswick (Buckner and Ray 1975; Buckner et al. 1977a; Germain 1977; Germain and Tingley 1980; McLeod 1976a, 1982; Pearce 1974; Pearce et al. 1976, 1979), Quebec (Buckner and McLeod 1977a; Buckner and Sarrazin 1975; Kingsbury and McLeod 1980, 1981; Ouellet 1981b; Sarrazin 1977), British Columbia (Carrow 1974), Newfoundland (Buckner and McLeod 1977b), and Manitoba (Buckner et al. 1973). In the majority of these reports, no dead or sick birds were found or observed during censusing. Peakall and Bart (1983), in their review of these studies and earlier studies carried out at higher application rates, conclude that although some species show decreases after 280 g/ha applications, many canopy feeding species did not show significant decreases in areas treated with up to 560 g/ha. Kinglets and some warblers show the greatest responses to 210 and 280 g/ha treatments, but in most instances the evidence for impact at these applications is not statistically valid (e.g. Germain 1977; Peakall and Bart 1983). Lehoux et al. (1982) could not find significant differences in changes on a 1 400 g/ha fenitrothion-treated plot and untreated control plot in either the numbers of individual birds counted by transect census or the numbers of breeding territories identified by the mapping method.

In addition to short-term impact studies, census studies have been used to draw conclusions about long-term changes in songbird numbers in areas repeatedly treated with fenitrothion. Germain and Morin (1979) found comparable numbers of birds in plots sprayed with fenitrothion and/or aminocarb annually for 3 years and in unsprayed plots. Pearce et al. (1979) reviewed census data for 10 forest songbird species in New Brunswick from 1967 to 1976 when fenitrothion was sprayed extensively. They concluded that population trends closely parallel trends of the Maritime provinces as a whole and suggested that any short-term impacts of fenitrothion have been compensated for by the natural resilience of songbird populations.

By the late 1970s, it was apparent that conventional census studies were not able to consistently detect effects of operational fenitrothion applications on populations of forest songbirds. There was evidence of occasional avian mortality, but both the extent and significance of this mortality could not be determined with the methods used. Consequently, experiments with brain cholinesterase determination as a method for evaluating fenitrothion's impact on songbirds were begun. Findlay et al. (1974) had previously used this method with caged Japanese quail during spruce budworm spraying in Manitoba and had shown relatively little effect (13% inhibition of brain cholinesterase activity).

Field studies of cholinesterase levels in forest passerines shot after fenitrothion applications in New Brunswick and Newfoundland have shown a broad range of responses that do not always correlate well with the application rates studied (Table 3). Brain cholinesterase studies therefore appear to be largely descriptive of the individual spray event monitored,

Table 3. Forest songbird brain cholinesterase responses following forest spraying with fenitrothion.

Application rate	Province and year	Application equipment[a]	No. species studied	Percent of birds with ChE[b] depression of: 20%	50%	Reference
Untreated control	NB 84	–	7	11	0	Busby et al. 1987a
	Nfld 86	–	6	9	0	Fudge, Lane & Assoc. 1986
140 g/ha	NB 87	RA	4	70	18	Busby et al., unpubl. data
210 g/ha	NB 78	B+N	1	58	0	Busby et al. 1983b
	NB 78	B+N	1	50	7	"
	NB 79	B+N	1	83	6	"
	Nfld 85	B+N	1	29	6	Strong and Wells 1987
	Nfld 85	B+N	1	37	0	"
	NB 86	RA	4	16	1	Busby et al. 1987b
	Nfld 86	B+N	3	76	55	Fudge, Lane & Assoc. 1987
	Nfld 86	B+N	3	73	19	"
	NB 87	RA	4	55	17	Busby et al., unpubl. data
280 g/ha	NB 80	B+N	4	30	5	Busby et al. 1981
	NB 80	RA	4	16	5	"
	Nfld 85	B+N	1	40	20	Strong and Wells 1987
350 g/ha	Nfld 86	B+N	2	71	43	Fudge, Lane & Assoc. 1986
420 g/ha	NB 79	B+N	1	95	25	Busby et al. 1983b

[a] Boom and nozzle; rotary atomizer (Micronair®).
[b] Cholinesterase.

and caution should be taken before making broad predictions to other fenitrothion sprays. Many factors such as weather conditions, topography, height of spray release, nature of the forest cover, activity of the songbird population at the time of and following spraying, and aspects (e.g. time, location) of the specimen collecting procedure may influence the results obtained from cholinesterase studies in the field as much, or more than, the application rate. For this reason, laboratory studies have been initiated with zebra finches, *Peophila guttata*, a small cage bird, to better define under controlled conditions the causes and consequences of fenitrothion-induced cholinesterase inhibition in small songbirds.

Researchers in New Brunswick have carried out several field studies looking at effects of aerial sprays and/or oral dosing on behavior, growth, and survival of white-throated sparrows, *Zonotrichia albicollis*. Results are summarized in reports of the Committee for Environmental Monitoring of Forest Insect Control Operations (EMOFICO 1980, 1982, 1984). Normal operational sprays of 210 g/ha did not affect the singing frequency of males or clutch size, hatching success, or fledging rates in nests studied, but a 10% reduction in nestling growth was reported. Markedly different results were reported after a contrived over-spray of 420 g/ha followed by a 210 g/ha application several days later. This spray regime resulted in the disappearance of 26% of the parent birds, reduced parental care, and territorial defence behaviors, and severely affected reproductive success. Oral dosing studies with nestlings show that younger birds are more sensitive than older ones, and gradual exposure is less harmful than ingestion of a single higher dose. Nestlings lost weight the day after dosing, but survivors then showed weight gains greater than control nestlings, although weight at fledging remained lower. Oral dosing of adult males, at a level inducing 40% brain cholinesterase depression, resulted in reduced feeding of nestlings and a 3% to 5% reduction in nestling growth, despite increased feeding activity by the undosed females.

Effects of a very high dosage (1 400 g/ha) experimental fenitrothion application on forest songbirds were evaluated by several techniques evaluating responses of known individuals (Lehoux et al. 1982). Nest observations revealed partial mortality of nestlings in one of eight nests observed over the spray period. Twenty-six percent of 38 marked birds resident to the spray plot were not observed after spraying. When birds that disappeared long enough before spraying to doubt their presence at the time of treatment were excluded from the analysis, only 13% of the marked birds could not be accounted for. Netting studies revealed no differences in the population structure, capture rate, or appearance of fledglings, between the treated and an untreated control plot. All these results indicate a surprisingly modest level of impact from this extremely high dosage of fenitrothion (five times the registered maximum rate).

A recent study (Millikin 1987) combining census, mist-netting, and time-budget analysis of marked individuals found little evidence of impact following 280 g/ha fenitrothion applications. Behavioral changes were related to measurements of food availability and insecticide distribution within the habitat. No significant changes were found in the time allotted to social, maintenance, or feeding behaviors in the three indicator species (chestnut-sided warbler, *Dendroica pensylvanica*; magnolia warbler; *D. magnolia*; white-throated sparrow, *Zonotrichia albicollis*) studied, although there was some indication the latter two species moved away from portions of trees that received the highest foliar deposits. Both depression of food availability, and songbird behavioral responses to spraying were short-lived.

Aminocarb

The period of greatest use of aminocarb (Matacil®) in forestry was in the late 1970s and early 1980s, when large areas were treated with single or double applications between 52 and 85 g/ha as a spruce budworm larvicide. During this period, aminocarb was applied as an oil-soluble concentrate in a commercial formulation (Matacil® 1.8D) containing the solvent nonylphenol. These spray regimes were extensively monitored in Quebec (Buckner et al. 1975b; Sarrazin 1977, 1978; Kingsbury and McLeod 1979a, 1980), New Brunswick (Pearce et al. 1976, 1979; Germain and Morin 1979; Germain and Tingley 1980; McLeod and Millikin 1982), Newfoundland (Buckner and McLeod 1977b; Environmental Monitoring Committee 1979), and Ontario (MacCallum 1980; Kingsbury et al. 1981). Summaries and reviews of these and other similar studies are contained in Kingsbury et al. 1981, NRCC 1982, and Peakall and Bart 1983. In those studies reviewed by the National Research Council of Canada (1982), no dead birds were found and only one exhibiting apparent symptoms of acute poisoning was sighted. The latter was an immature purple finch, *Carpodacus purpureus*, from an experimental spruce budworm adulticide block found just after the second 70 g/ha application in a 2-day period (McLeod and Millikin 1982). Only relatively minor impacts on the activity of forest songbirds are reported in monitoring reports, and these are found only when aminocarb was applied twice or in regimes involving prior application of fenitrothion or phosphamidon (NRCC 1982). Peakall and Bart (1983) conclude that the critical dosage of aminocarb required to cause detectable effects on songbirds is greater than conventional forestry application rates.

Higher than normal aminocarb application rates have been evaluated only occasionally for their impact on songbirds. An inadvertent double treatment of a portion of a spray block in Quebec with two 52 g/ha applications within hours of each other did not appear to damage avian populations (Buckner and McLeod 1977a). Kingsbury et al. (1981) studied the effects of seasonal maximum (175 g/ha) applications of aminocarb and the equivalent amount of the solvent nonylphenol applied in a season. Songbirds censused by territorial mapping showed no adverse response to either treatment, and mist-netting studies in the 175 g/ha block also failed to find evidence of impact on the resident songbird population (Ouellet 1981a).

In the early 1980s a flowable formulation (Matacil® 180 Flowable) of aminocarb was developed that did not contain nonylphenol. This product has been evaluated for its impact on songbirds when applied in single or double applications of 52 or 70 g/ha in both oil- and water-based spray mixes (Millikin 1982; Major and Ouellet 1982; McLeod 1982; McLeod and Kingsbury 1982). These studies found little evidence for effects on songbirds from either census or netting results. A single sick Tennessee warbler, *Vermivora peregrina*, was observed in a portion of a spray block that had received heavy deposits during an evening spray, possibly followed by further deposits the next morning when the rest of the block was treated (McLeod and Kingsbury 1982).

Forestry aminocarb applications appear to have a rather limited effect on songbird brain cholinesterase activity. Less than 10% of the songbirds collected from experimental or semi-operational aminocarb flowable treatment plots in New Brunswick exhibit cholinesterase depression greater than 20%, and no birds had 50% depression (Busby et al. 1982, 1983a). Mist-netting studies in aminocarb-treated and untreated control areas in Newfoundland in 1983 indicated that the production and/or dispersal of immature birds was not affected by two applications of 70 g/ha (Bennett et al. 1987).

Other Chemical Control Agents

During the 1970s, a large number of neurotoxins, growth regulators, and mating disruption chemicals were field tested for their efficacy against forest defoliators. Potential effects on songbirds were evaluated using census techniques during many of these field efficacy trials. Although most of these agents were only briefly or never used in operational programs, some are still registered and/or potential candidates for future use.

Acephate—Acephate (Orthene®) has been evaluated for its impact on birds over a wide range of application rates. Reduced warbler activity was apparent after each of two applications of 56 g/ha acephate combined with 20 BIU/ha (billion international units per hectare) Bt in Newfoundland in 1977, but overall impact was minimal (Buckner and McLeod 1977b). Two applications of 280 g/ha in New Brunswick in 1976 had no effect on bird populations (Buckner et al. 1977a). Applications of 560 and 1 400 g/ha in Ontario and New Brunswick have been accompanied by increases (MacCallum 1980) and decreases (Buckner and McLeod 1975a; Buckner et al. 1977a) in bird activity, but these are either statistically nonsignificant or occurred before spraying (Peakall and Bart 1983). An application of 1 129 g/ha in British Columbia in 1976 was accompanied by a 1-day reduction in bird activity, but breeding territories remained occupied and no effect was evident among blackbird nestlings monitored over the spray (McLeod 1976b).

Carbamates—The carbamate insecticides carbaryl (Sevin®) and mexacarbate (Zectran®) have both been used fairly extensively for forest insect control in the United States, but only sparingly in Canada. Single applications of 670 and 1 121 g/ha carbaryl and double applications of 280 and 560 g/ha carbaryl did not harm songbird populations when applied as spruce budworm larvicides in Manitoba and New Brunswick (Buckner et al. 1973, 1977b; Holmes et al. 1981). An application of 52 g/ha mexacarbate to a spray block in Quebec previously treated twice with fenitrothion did not appear to damage resident songbird populations (Buckner and Sarrazin 1975). Recent field studies evaluating the effects of 70 and 140 g/ha applications of mexacarbate on songbird brain cholinesterase found limited effect, with similar proportions (11%) of the birds sampled from the 70 g/ha and control areas exhibiting cholinesterase activity more than 20% lower than the mean for unexposed birds (Busby et al. 1987a). Nineteen percent of the birds from the 140 g/ha block showed this level of cholinesterase activity.

Organophosphates—The organophosphate insecticides azamethiphos (Alfacron®) and chlorpyrifos-methyl (Reldan®) were evaluated for their potential for spruce budworm control in the late 1970s but never registered for forestry use. Double applications of 70 g/ha azamethiphos had little effect on forest songbirds, but territorial abandonment by Traill's flycatchers, *Empidonax traillii*, possibly related to treatment, occurred after the first treatment (Kingsbury et al. 1980). Double applications of 70 g/ha chlorpyrifos-methyl had little effect on songbirds (Holmes and Millikin 1981). Another insecticide evaluated for forestry use in the late 1970s which has been registered for ground sprays and applied aerially under experimental permits is the synthetic pyrethroid permethrin (Ambush®). Double applications of 70 g/ha permethrin did not affect breeding songbirds inhabiting three different forest types (Kingsbury and McLeod 1979b).

Synthetic hormones—Several chemical control agents other than neurotoxins have been evaluated for their effects on songbirds by census techniques during experimental field trials. Applications of 70 and 210 g/ha of juvenile hormone analog (Altosid®) for hemlock looper, *Lambdina fiscellaria*, control had no adverse effect on songbirds (Buchner et al. 1975c). Forest songbirds were not affected by applications of the insect growth regulators diflubenzuron (Dimilin®) at 280 or 350 g/ha (McLeod 1976b; Buckner et al. 1975c) or BAY SIR 8514 at 280 g/ha (Millikin and Mortensen 1980). The extended application of spruce budworm pheromone encapsulated in Conrel fibers and emitted at an active ingredient rate of 13 mg/ha over a 7-day period did not cause any measurable disturbance to the normal activity patterns of resident avifauna (Kingsbury and McLeod 1979c).

Biological Control Agents

Both bacterial and viral insect pathogens have been assessed for their effects on forest songbirds when aerially applied for spruce budworm control. Commercial Bt preparations (Thuricide® 16B, Dipel® WP) applied at approximately 10 BIU/ha in Ontario and Manitoba did not cause statistically significant reductions in bird populations (Buckner et al. 1973, 1974). Germain (1980b) monitored 10 and 20 BIU/ha applications of Bt in Nova Scotia in 1979 by two variants of the territorial mapping method and could not demonstrate any effects of the spraying on songbirds. Applications of up to 50g/ha of the poxvirus of spruce budworm 25 × 1010 PIB/ha (polyhedral inclusion bodies per hectare) of the nuclear polyhedrosis virus of spruce budworm and 55 × 108 PIB/ha of the nuclear polyhedrosis virus of redheaded pine sawfly, *Neodiprion lecontei*, have all been found not to have any effects on forest avifauna (Buckner and Cunningham 1972; Buckner et al. 1975e; Kingsbury et al. 1978).

Effects on Other Terrestrial Vertebrates

Amphibians

Relatively few studies have been carried out on the effects of forest sprays in Canada on larval or adult amphibians. Although direct mortality of both life stages was found in early studies of DDT, less impact was observed with lower dosages, and early studies with organophosphates and carbamates indicated they had little effect on amphibians (Pearce and Price 1975). Static bioassays with larvae of the green frog, *Rana clamitans*, gave 24-hour LC_{50} values of 10, 247, and 6 433 mg/L for fenitrothion, aminocarb, and acephate, respectively (Lyons et al. 1976). These values are all substantially higher than toxicity values for these insecticides to fish and aquatic invertebrates (Eidt et al. 1989), indicating that at least for these materials, larval amphibians tend to be less sensitive than other aquatic organisms. Field studies with caged tadpoles have failed to indicate spray-induced mortality in forest ponds within areas treated with 70 g/ha aminocarb (Buckner et al. 1975b), 560 g/ha acephate (Buckner and McLeod 1975a), or 70 g/ha mexacarbate (Holmes 1986).

Aminocarb applied at 760 g/ha did not affect caged frogs or toads, and no effects on populations were found by adult census or pitfall trapping studies (Buckner et al. 1975b). Sand-transect studies in an area treated with 175 g/ha aminocarb did not show any changes in frog activity after treatment, but did indicate a reduction in toad activity presumed to be related to reduced prey availability (Bracher and Bider 1982).

Small Mammals

Small mammals are used extensively in the toxicological evaluations undertaken before the registration and use of insecticides. Because of their use as experimental animals in these studies, there is considerable information available on their responses to forest insecticides in the laboratory (e.g. NRCC 1975, 1976, 1982). In contrast, small mammal populations are difficult to study in the field, and the data obtained may be difficult to interpret (Buckner and McLeod 1975b).

Most of the monitoring of small mammal populations in insecticide-sprayed forests during the 1970s has compared the populations, age-class structure, and evidence of breeding success of small mammals trapped in treated and untreated control sites some period after spraying. In many instances, low catches have limited the success of these studies in indicating the presence or absence of insecticide effects. Examination of several studies where reasonable sample sizes were obtained shows that in general no effects were apparent from trapping conducted in insecticide-treated forests. Examples are studies carried out in spray operations in Newfoundland (Environmental Monitoring Committee 1979), New Brunswick (Buckner et al. 1975b), Quebec (Buckner and Sarrazin 1975), Ontario (Buckner and Cunningham 1972; Buckner et al. 1974; Kingsbury and McLeod 1979b), Manitoba (Buckner et al. 1973), and British Columbia (Carrow 1974; Shepherd 1980) involving applications of fenitrothion, aminocarb, acephate, carbaryl, mexacarbate, permethrin, diflubenzuron, spruce budworm poxvirus, and Bt. One exception to the general lack of evidence of impact is the report of low population density and limited evidence of breeding success compared to untreated or phosphamidon-treated plots from a fenitrothion (2 × 210 g/ha) block in New Brunswick (Buckner et al. 1977a).

Spray chamber studies with fenitrothion indicated thresholds for lethal effects equivalent to deposits of 560 g/ha for the masked shrew, *Sorex cinereus*; 840 g/ha for the red-backed vole, *Clethrionomys gapperi*; and 1 120 g/ha for the white-footed mouse, *Peromyscus maniculatus* (Buckner et al. 1977b). Breeding success was suppressed for a short period in both voles and mice that survived treatments of 280 g/ha or greater. The same study found that shrews and voles consuming 700 mg/kg of fenitrothion or greater in contaminated mealworm larvae failed to survive. Adult animals rejected fenitrothion-contaminated food material, as did juveniles with previous exposure experience. Tschaplinski and Buckner (1978) found the 96-hour LD_{50} of fenitrothion intraperitoneally administered to red-backed voles was 1 330 mg/kg. Fenitrothion was rapidly metabolized and excreted, with symptoms of intoxications and mortality limited to within 48 hours of dose application. There is little indication from small mammal studies on plots treated annually with 280 g/ha for up to 5 years that fenitrothion depressed small mammal populations (Buckner et al. 1977b).

A major study using both sand-transect and kill-trapping techniques was carried out in Quebec in 1978 and 1979 to evaluate the effects of 175 g/ha aminocarb applied in a single treatment on small mammals. Both methods indicated the aminocarb treatment had little effect on small mammal populations (Bracher et al. 1986). The activity of nearly all species or groups of mammals increased or did not significantly change over the immediate or medium term post-spray period (Bracher and Bider 1982). Ermine, *Mustela* spp., activity increased greatly on the spray plot between the 5th and 8th day post-spray, possibly due to increased avian activity on the ground in response to reduced food availability in the forest canopy. Increased ermine activity is believed to have suppressed shrew and vole activity. Trapping results indicated normal breeding activity had continued among small mammal populations during the treatment period (Kingsbury et al. 1981).

References

Bart, J.; Hunter, L. 1978. Ecological impacts of forest insecticides: an annotated bibliography. New York Cooperative Wildlife Research Unit. Ithaca, NY. 128 p.

Bennet, G.; Woodworth-Lynas, C.; Knoechel, R. 1987. Environmental monitoring of the 1983 spruce budworm spraying program in Newfoundland. Newfoundland and Labrador Department of Environment. Environmental Assessment Division. 112 p.

Bracher, G.A.; Bider, J.R. 1982. Changes in terrestrial animal activity of a forest community after an application of aminocarb (Matacil). Can. J. Zool. 60: 1981-1997.

Bracher, G.A.; Kingsbury, P.D.; Bider, J.R. 1986. A comparison of two techniques for assessing the impact of pesticides on small mammals. Can. Field-Nat. 100 (1): 52-57.

Buckner, C.H.; Cunningham, J.C., 1972. The effect of the poxvirus of the spruce budworm, *Choristoneura fumiferana*

(Lepidoptera: Tortricidae), on mammals and birds. Can. Entomol. 104: 1333-1342.

Buckner, C.H.; Kingsbury, P.D.; McLeod, B.B.; Mortensen, K.L.M.; Ray, D.G.H. February 1974. Section F: Impact of aerial treatment on non-target organisms. Algonquin Park, Ontario and Sprucewoods, Manitoba. 72 p. In Evaluation of commercial preparations of *Bacillus thuringiensis* with and without chitinase against spruce budworm. Chem. Cont. Res. Inst. Inf. Rep. CC-X-59.

Buckner, C.H.; McLeod, B.B. 1975a. Impact of aerial applications of Orthene upon non-target organisms. Can. For. Serv., Chem. Cont. Res. Inst., Ottawa, Ont. Inf. Rep., Ottawa, Ont. CC-X-104. 38 p.

Buckner, C.H.; McLeod, B.B. 1975b. The impact of insecticides on small forest mammals. Pages 314-318 *in* M.L. Prebble (Ed.), Aerial Control of Forest Insects in Canada. Dep. Environ. Ottawa, Ont.

Buckner, C.H.; McLeod, B.B. 1977a. Ecological impact studies of experimental and operational spruce budworm control programs on selected non-target organisms in Quebec in 1976. Can. For. Serv., Chem. Cont. Res. Inst., Inf. Rep. Ottawa, Ont. CC-X-137. 81 p.

Buckner, C.H.; McLeod, B.B. 1977b. Impact of experimental spruce budworm suppression trials upon forest dwelling birds in Newfoundland in 1977. Can. For. Serv., Forest Pest Management Institute, Sault Ste. Marie, Ont. Report FPM-X-9. 80 p.

Buckner, C.H.; McLeod, B.B.; Kingsbury, P.D. 1975a. Accident investigation activities of the ecological impact team in forest areas treated with insecticide in 1975. Can. For. Serv., Chem. Cont. Res. Inst., Ottawa, Ont. Inf. Rep. CC-X-103. 21 p.

Buckner, C.H.; McLeod, B.B.; Kingsbury, P.D. 1975b. Studies of the impact of the carbamate insecticide, Matacil, on components of forest ecosystems. Can. For. Serv., Chem. Cont. Res. Inst., Ottawa, Ont. Inf. Rep. CC-X-91. 61 p.

Buckner, C.H.; McLeod, B.B.; Kingsbury, P.D. 1975c. Insecticide impact and residue studies on Anticosti Island, Quebec, in 1973. Can. For. Serv., Chem. Cont. Res. Inst., Ottawa, Ont. Inf. Rep. CC-X-102. 38 p.

Buckner, C.H.; McLeod, B.B.; Kingsbury, P.D. 1975d. The effects of an experimetnal application of Dimilin upon selected forest fauna. Can. For. Serv., Chem. Cont. Res. Inst., Ottawa, Ont. Inf. Rep. CC-X-97. 26 p.

Buckner, C.H.; McLeod, B.B.; Kingsbury, P.D. 1975e. The effect of an experimental application of nuclear polyhedrosis virus upon selected forest fauna. Can. For. Serv., Chem. Cont. Res. Inst., Ottawa, Ont. Inf. Rep. CC-X-101. 25 p.

Buckner, C.H.; McLeod, B.B.; Lidstone, R.G. 1977a. Environmental impact of spruce budworm control programs in New Brunswick in 1976. Can. For. Serv., Chem. Cont. Res. Inst., Ottawa, Ont. Inf. Rep. CC-X-135. 141 p.

Buckner, C.H.; Ray, D.G.H. 1975. The effects of aerial applications of the larvicides phosphamidon and fenitrothion on populations of songbirds and small mammals in New Brunswick in 1974. Can. For. Serv., Chem. Cont. Res. Inst., Ottawa, Ont. Inf. Rep. Ottawa, Ont. Unpublished report. 11 p.

Buckner, C.H.; Ray, D.G.H.; McLeod, B.B. 1973. The effects of pesticides on small forest vertebrates of the Spruce Woods Provincial Forest, Manitoba. Man. Entomol. 7: 37-45.

Buckner, C.H.; Sarrazin, R. 1975. Studies of the environmental impact of the 1974 spruce budworm control operation in Quebec. Can. For. Serv., Chem. Cont. Res. Inst., Ottawa, Ont. Inf. Rep. CC-X-93. 106 p.

Buckner, C.H.; Sarrazin, R.; McLeod, B.B. 1977b. The effects of the fenitrothion spray program on small mammals. Pages 377-390 *in* J.R. Roverts, R. Greenhalgh, and W.K. Marshall (Eds.), Proceedings of a symposium on fenitrothion: the long-term effects of its use in forest ecosystems. Associate Committee on Scientific Criteria for Environmental Quality. Nat. Res. Counc. Can., Ottawa, Ont. NRCC No. 16073. 628 p.

Busby, D.G.; Holmes, S.B.; Pearce, P.A.; Fleming, R.A. 1987a. The effect of aerial application of Zectran on brain cholinesterase activity in forest songbirds. Arch. Environ. Contam. Toxicol. 16: 623-629.

Busby, D.G.; Pearce, P.A.; Garrity, N.R. 1981. Brain cholinesterase response in songbirds exposed to experimental fenitrothion spraying in New Brunswick, Canada. Bull. Environ. Contam. Toxicol. 26: 401-406.

Busby, D.G.; Pearce, P.A.; Garrity, N.R. 1982. Brain cholinesterase inhibition in forest passerines exposed to experimental aminocarb spraying. Bull. Environ. Contam. Toxicol. 28: 225-229.

Busby, D.G.; Pearce, P.A.; Garrity, N.R.; 1983a. Brain ChE response in forest songbirds exposed to aerial spraying of aminocarb and possible influence of application methodology and insecticide formulation. Bull. Environ. Contam. Toxicol. 31: 125-131.

Busby, D.G.; Pearce, P.A.; Garrity, N.R.; Reynolds, L.M. 1983b. Effect of an organophosphorus insecticide on brain cholinesterase activity in white-throated sparrows exposed to aerial forest spraying. J. Appl. Ecol. 20: 255-263.

Busby, D.G.; Pearce, P.A.; Garrity, N.R. 1987b. Effect of Ultra ULV fenitrothion spraying on brain cholinesterase activity in forest songbirds. Bull. Environ. Contam. Toxicol. 39: 304-311.

Carrow, J.R. (Ed.). 1974. Aerial spraying operations against blackheaded budworm on Vancouver Island, 1973. Can. For. Serv., Pac. For. Res. Cent., Victoria, B.C. Inf. Rep. BC-X-101. 56 p.

Crawford, H.S.; Titterington, R.W.; Jennings, D.T. 1983. Bird predation and spruce budworm populations. J. For. 81(7): 433-435.

DeWeese, L.R.; Henny, C.J.; Floyd, R.L.; Bobal, K.A.; Schultz, A.W. 1979. Response of breeding birds to aerial sprays of trichlorfon (Dylox) and carbaryl (Sevin-4-Oil) in Montana forests. U.S. Fish Wildl. Serv., Spec. Sci. Rep. Wildl. 224, 29 p.

Dowden, P.B.; Jaynes, H.A.; Carolin, V.M. 1953. The role of birds in a spruce budworm outbreak in Maine. J. Econ. Entomol. 46: 307-312.

Eidt, D.C.; Hollebone, J.E.; Lockhart, W.L.; Kingsbury, P.D.; Gadsby, M.C.; Ernst, W.R. 1989. Pesticides in forestry and agriculture: effects on aquatic habitats. Pages 245-284 *in* J.A. Nriagu (Ed.), Aquatic toxicology and water quality management.

Emlen, J.T. 1971. Population densities of birds derived from transect counts. Auk 38: 323.

EMOFICO 1980. Environmental surveillance in New Brunswick, 1978-1979—effects of spray operations for forest protection against spruce budworm. Department of Forest Resources, University of New Brunswick, Fredericton, N.B. 76 p.

EMOFICO 1982. Environmental surveillance of spray operations for forest protection in New Brunswick, 1980-81. Environment New Brunswick, Fredericton, N.B. 130 p.

EMOFICO 1984. Environmental monitoring of forest insect control operations, 1983. Environment New Brunswick, Fredericton, N.B. 40 p.

Environmental Monitoring Committee. 1979. 1978 environmental monitoring of the spruce budworm spray program in Newfoundland. Department of Consumer Affairs and Environment, Research and Assessment Branch, Inf. Rep. RA-79-2. St. John's, Nfld. 29 p.

Findlay, G.M.; Howe, G.J.; Lockhart, W.L. 1974. Impact of fenitrothion upon Japanese quail (*Coturnix coturnix japonica*) in a forest ecosystem. Man. Entomol. 8: 10-15.

Fudge, Lane and Associates Ltd. 1987. A study to monitor the impacts of the 1986 hemlock looper insecticide spray program on songbirds. Newfoundland and Labrador Dept. of Environment. Environmental Assessment Division. 51 p.

Gage, S.H.; Miller, C.A. 1978. A long-term bird census in spruce budworm-prone balsam fir habitats in northeastern New Brunswick. Can. For. Serv., Mar. For. Res. Cent., Fredericton, N.B., Inf. Rep. M-X-84, 6 p.

Germain, P. 1977. Songbird census in fenitrothion regime and 1977 forest spray program in New Brunswick. Final report to Forest Protection Ltd. and New Brunswick Dept. of Natural Resources. University of Moncton, Moncton, N.B. 30 p. + appendices.

Germain, P. 1980a. Monitoring songbirds in the context of forest insect pest control. A review of the methodology with special reference to the use of "census" techniques. Final Report, DSS contract no. 0SS79-00169, 46 p.

Germain, P. 1980b. Impact study of B.t. spraying of forest songbirds, Nova Scotia, 1979. A final report to Nova Scotia Dep. of Lands and Forests. Avifauna Ltd. Moncton, N.B. 20 p.

Germain, P.; Morin, G. 1979. Forest songbird population 1978 monitoring program in relation to aerial spray of insecticides against spruce budworm. A report to Forest Protection Ltd. and N.B. Dep. Nat. Resour. 31 p. + appendices.

Germain, P.; Tingley, S. 1980. Population responses of songbirds to 1979 forest spray operations in New Brunswick. A report to Forest Protection Ltd. 32 p. + appendices.

Graber, J.W.; Graber, R.R. 1983. Feeding rates of warblers in spring. Condor 85(2): 139-150.

Grue, C.E.; Powell, G.V.N.; McChesney, M.J. 1982. Care of nestlings by wild female starlings exposed to an organophosphate pesticide. J. Appl. Ecol. 19: 327-335.

Hill, E.F.; Fleming, W.J. 1982. Anticholinesterase poisoning of birds: field monitoring and diagnosis of acute poisoning. Environ. Toxicol. Chem. 1: 27-38.

Holling, C. 1959. Some characteristics of simple types of predation and parasitism. Can. Entomol. 91: 385.

Holmes, S.B. 1986. A preliminary report on the impact of an experimental aerial application of ZECTRAN (mexacarbate) insecticide on aquatic ecosystems in New Brunswick, Canada. Can. For. Serv., Forest Pest Management Institute, Sault Ste. Marie, Ont. File Rep. No. 70. 35 p.

Holmes, S.B. 1988. Behavioural and reproductive correlates of reduced levels of cholinesterase activity in zebra finches following acute exposures to fenitrothion. Queen's University M.Sc. thesis (Dep. of Biology).

Holmes, S.B.; Millikin, R.L. 1981. A preliminary report on the effects of a split application of Reldan on aquatic and terrestrial ecosystem. Can. For. Serv., Forest Pest Management Institute, Sault Ste. Marie, Ont. File Rep. No. 13. 88 p.

Holmes, S.B.; Millikin, R.L.; Kingsbury, P.D. 1981. Environmental effects of a split application of SEVIN-2-OIL. Can. For. Serv., Forest Pest Management Institute, Sault Ste. Marie, Ont., Rep. FMP-X-46. 112 p.

Holmes, R.T.; Schultz, J.C.; Nothnagle, P. 1979. Bird predation on forest insects: an exclosure experiment. Science 206: 462-463.

Jennings, D.T.; Crawford, H.S. 1985. Predators of the spruce budworm. Spruce budworms handbook. USDA, Washington, D.C. Handbook No. 644. 77 p.

Kelly, B.; Régnière, J. 1985. Predation on pupae of the spruce budworm (Lepidoptera: Tortricidae) on the forest floor. Can. Entomol. 117: 33-38.

Kendeigh, S.C. 1947. Bird population studies in the coniferous forest biome during a spruce budworm outbreak. Ont. Dep. Lands and Forests Biol. Bull. 1: 100 p.

Kingsbury, P.D. 1975. Effects of aerial forest spraying on aquatic fauna. Pages 280-292 *in* M.L. Prebble (Ed.), Aerial control of forest insects in Canada. Dep. Environ. Ottawa, Ont.

Kingsbury, P.D.; Holmes, S.B.; Millikin, R.L. 1980. Environmental effects of a double application of azamethiphos on selected terrestrial and aquatic organisms. Can. For. Serv., Forest Pest Management Institute, Sault Ste. Marie, Ont. Rep. FPM-X-33. 75 p.

Kingsbury, P.D.; McLeod, B.B. 1979a. The impact of the 1978 spruce budworm, *Choristoneura fumiferana* Clemens, larvicide timing trials in Quebec upon forest avifauna. Can. For. Serv., Chem. Cont. Res. Inst., Sault Ste. Marie, Ont. Rep. FPM-X-29. 107 p.

Kingsbury, P.D.; McLeod, B.B. 1979b. Terrestrial impact of studies in forest ecosystems treated with double applications of Permethrin. Can. For. Serv., Forest Pest Management Institute, Sault Ste. Marie, Ont. Rep. FPM-X-28. 65 p.

Kingsbury, P.D.; McLeod, B.B. 1979c. The impact of an experimental application of spruce budworm pheromone upon forest birds. Can. For. Serv., Forest Pest Management Institute, Sault Ste. Marie, Ont. Unpublished report. 19 p.

Kingsbury, P.D.; McLeod, B.B. 1980. The impact of spruce budworm, *Choristoneura fumiferana* Clemens, control operations involving sequential insecticide applications upon forest avifauna in the lower St. Lawrence region of Quebec. Can. For. Serv., Forest Pest Management Institute, Sault Ste. Marie, Ont. Rep. FPM-X-34. 64 p.

Kingsbury, P.D.; McLeod, B.B. 1981. Fenitrothion and forest avifauna—studies on the effects of high dosage applications. Can. For. Serv., Forest Pest Management Institute, Sault Ste. Marie, Ont. Rep. FPM-X-43. 54 p.

Kingsbury, P.D.; McLeod, B.B.; Millikin, R.L. 1979. A seasonal maximum aerial application of Matacil to a forest ecosystem and its impact upon small mammals, birds, and insects. Can. For. Serv., Forest Pest Management Institute, Sault Ste. Marie, Ont. Preliminary Rep. 12 p.

Kingsbury, P.D.; McLeod, B.B.; Millikin, R.L. 1981. The environmental impact of nonyl phenol and the MATACIL formulation. Part 2: Terrestrial ecosystems. Can. For. Serv., Forest Pest Management Institute, Sault Ste. Marie, Ont. Rep. FPM-X-36. 88 p.

Kingsbury, P.D.; McLeod, B.B.; Mortensen, K.L.M. 1978. Impact of applications of the nuclear polyhedrosis virus of the red-headed pine sawfly, *Neodiprion lecontei* (Fitch), on non-target organisms in 1977. Can. For. Serv., Forest Pest Management Institute, Sault Ste. Marie, Ont. Rep. FPM-X-11. 27 p.

Lehoux, D.; Ouellet, R.; Laporte, P. 1982. A comparative study of various methods for monitoring forest bird populations following experimental insecticide sprays. Canadian Wildlife Service and Québec Ministère du Loisir, de la Chasse et de la Pêche. 77 p. + appendices. (Translated from the French original)

Ludke, J.L.; Hill, E.F.; Dieter, M.P. 1975. Cholinesterase (ChE) response and related mortality among birds fed ChE inhibitors. Arch. Environ. Contam. Toxicol. 3: 1-21.

Lyons, D.B.; Buckner, C.H.; McLeod, B.B.; Sundaram, K.M.S. 1976. The effects of fenitrothion, MATACIL and Orthene on frog larvae. Can. For. Serv., Chem. Cont. Res. Inst., Ottawa, Ont. Inf. Rep. CC-X-129. 86 p.

MacArthur, R.H. 1958. Population ecology of some warblers of northern coniferous forests. Ecology 39: 487-494.

MacCallum, M.E. 1980. The short-term environmental impact of the spruce budworm spray program in Ontario, 1979. Pest Control Section, Forest Resources Branch, Ontario Ministry of Natural Resources. 58 p. + appendices.

Major, L; Ouellet, R. 1982. Évaluation d'impact sur les populations d'oiseaux forestiers lors de pulvérsations expérimentales d'une fin suspension d'aminocarbe (Matacil 180F). Gouvernement du Québec, Ministère de l'Énergie et des Ressources, Service d'entomologie et de pathologie, Division environnement et sécurité, 22 p.

McLeod, B.B. 1976a. Environmental monitoring on New Brunswick forests, 1975-76. Sault Ste. Marie, Ont. No. 49. 54 p.

McLeod, B.B. 1976b. Monitoring avian populations during experimental insect control operations in British Columbia in 1976. Can. For. Serv., Chem. Cont. Res. Inst., Ottawa, Ont. File Rep. No. 60. 18 p.

McLeod, B.B. 1982. Terrestrial impact studies on experimental aqueous formulations of spruce budworm control agents containing TRITON X-100, New Brunswick 1982. Can. For. Serv., Forest Pest Management Institute, Sault Ste. Marie, Ont. File Rep. No. 37. 21 p.

McLeod, B.B.; Kingsbury, P.D. 1982. Terrestrial impact studies of MATACIL flowable formulated in vegetable oil, Quebec 1982. Can. For. Serv., Forest Pest Management Institute, Sault Ste. Marie, Ont. File Rep. No. 41. 29 p.

McLeod, B.B.; Millikin, R.L. 1982. Environmental impact assessment of experimental spruce budworm adulticide trials: effects on forest avifauna. Can. For. Serv., Forest Pest Management Institute, Sault Ste. Marie, Ont. Rep. FPM-X-54. 109 p.

Miller, C.A.; Stewart, J.E.; Elgee, D.E.; Shaw, D.D.; Morgan, M.G. 1973. Aerial spraying against spruce budworm adults in New Brunswick. A compendium of reports on the 1972 test program. Can. For. Serv., Marit. For. Res. Cent., Fredericton, N.B. Inf. Rep. M-X-38. 35 p.

Millikin, R.L. 1982. Songbird studies in New Brunswick forests treated with semi-operational applications of MATACIL Flowable formulations in 1982. Can. For. Serv., Forest Pest Management Institute, Sault Ste. Marie, Ont. File Rep. No. 39. 71 p.

Millikin, R.L. 1987. Sublethal effects of fenitrothion on forest passerines. M.Sc. thesis. University of British Columbia. Vancouver, B.C. 98 p.

Millikin, R.L. 1988. A comparison of spot, transect and plot methods for measuring the impact of forest pest control strategies on forest songbirds. Can. For. Serv., Forest Pest Management Institute, Sault Ste. Marie, Ont. Rep. FPM-X-83. 23 p. + appendices.

Millikin, R.L.; Mortensen, K.L. 1980. Preliminary report on the environmental impact of an experimental application of the insect growth regulator Bay Sir 8514, Wawa, Ontario. 1979. Can. For. Serv., Forest Pest Management Institute, Sault Ste. Marie, Ont. File Rep. No. 3. 39 p.

Mineau, P., Peakall, D.B. 1986. An evaluation of avian impact assessment techniques following broad-scale forest insecticide sprays. Environ. Toxicol. Chem. 6: 781-791.

Morris, R.F.; Cheire, W.F.; Miller, C.A.; Mott, D.G. 1958. The numerical response of avian and mammalian predators during a gradation of the spruce budworm. Ecology 49(4): 487-494.

NRCC. 1975. Fenitrothion: the effects of its use on environmental quality and its chemistry. Associate Committee on

Scientific Criteria for Environmental Quality, Natl. Res. Counc. Can., Ottawa, Ont. NRCC No. 14104. 162 p.

NRCC. 1976. *Bacillus thuringiensis*: its effects on environmental quality. Associate Committee on Scientific Criteria for Environmental Quality, Natl. Res. Counc. Can., Ottawa, Ont. NRCC No. 15385. 134 p.

NRCC. 1982. Aminocarb: the effects of its use on the forest and the human environment. Associate Committee on Scientific Criteria for Environmental Quality, Natl. Res. Counc. Can., Ottawa, Ont. NRCC No. 18979. 253 p.

Otvos, I.S. 1979. The effects of insectivorous bird activities in forest ecosystems: an evaluation. Pages 341-374 *in* J.G. Dickson, R.N. Connor, R.R. Fleet, J.C. Kroll and J.A. Jackson (Eds.), The role of insectivorous birds in forest ecosystems. Academic Press.

Oucllet, R. 1981a. Étude d'impact sur les populations aviennes lors d'un arrosage experimental d'aminocarb (Saint-Donat-de-Montcalm), 1979. Québec Ministère du Loisir, de la Chasse et de la Pêche. Direction de la recherche faunique RRF63. 18 p.

Ouellet, R., 1981b. Études des populations d'oiseaux forestiers lors de deux arrosages au fenitrothion (1.4 L/ha) dans la forêt domainale des Appalaches. Québec Ministère du Loisir, de la Chasse et de la Pêche. Direction de la recherche faunique. Unpublished report. 28 p.

Peakall, D.B. 1985. Behavioral responses of birds to pesticides and other contaminants. Residue Reviews 96: 46-77.

Peakall, D.B.; Bart, J.R. 1983. Impact of aerial applications of insecticides on forest birds. CRC Critical Reviews in Environmental Control. 13(2): 117-165.

Pearce, P.A. 1974. Bird responses to forest sprays in New Brunswick, 1974. Unpublished paper presented at Canadian Forest Pest Control Forum. 11 p. (Annotated in Bart and Hunter 1978)

Pearce, P.A. 1975. Effects on birds. Pages 306-313 *in* M.L. Prebble (Ed.), Aerial control of forest insects in Canada. Dep. Environ. Ottawa, Ont.

Pearce, P.A.; Peakall, D.B.; Erskine, A.J. 1976. Impact on forest birds of the 1975 spruce budworm spray operation in New Brunswick. Canadian Wildlife Service Progress Notes No. 62. 7 p.

Pearce, P.A.; Peakall, D.B.; Erskine, A.J. 1979. Impact on forest birds of the 1976 spruce budworm spray operation in New Brunswick. Canadian Wildlife Service Progress Notes No. 97. 15 p.

Pearce, P.A.; Price, I.M. 1975. Effects on amphibians. Pages 301-305 *in* M.L. Prebble (Ed.), Aerial control of forest insects in Canada. Dep. Environ. Ottawa, Ont.

Peters, D.C. 1979. Growth rates of white-throated sparrow chicks on a fenitrothion sprayed and unsprayed plot in northeastern New Brunswick, Ms report, Faculty of Forestry, University of New Brunswick, 22 p.

Prebble, M.L. (Ed.). 1975. Aerial control of forest insects in Canada. Dep. Environ., Ottawa, Ont.

Ralph, C.J.; Scott, J.M. (Eds.). 1981. Estimating numbers of terrestrial birds. Studies in Avian Biology No. 6, Cooper Ornithological Society. 630 p.

Sarrazin, R. 1977. Effets environnementaux des pulvérisations aériennes contre la tordeuse des bourgeons de l'épinette au Québec—1977. Rapport préliminaire du Sous-Comité de l'Environnement du Comité de Surveillance Ecologique des Pulvérisations Aériennes. Rapport de Recherche Faunique No. 15. Ministère du Tourisme, de la Chasse et de la Pêche. Québec, Qué.

Sarrazin, R. 1978. Effets environnementaux des pulvérisations aériennes contre la tordeuse des bourgeons de l'épinette au Québec—1978. Rapport préliminaire du Sous-Comité de l'Environnement du Comité de Surveillance Ecologique des Pulvérisations Aériennes. Rapport de Recherche Faunique No. 36. Ministère du Tourisme, de la Chasse et de la Pêche. Québec, Qué.

Shepherd, R.F. (Ed.). 1980. Operational field trials against the douglas-fir tussock moth with chemical and biological insecticides. Can. For. Serv., Pac. For. Res. Cent., Victoria, B.C. Rep. BC-X-201. 19 p.

Strong, K.W.; Wells, J. 1987. A study to monitor the impact of the 1985 hemlock looper insecticide spray program on songbird populations. Newfoundland and Labrador Dep. of Environ. Environmental Assessment Division. 26 p.

Svennson, S. 1970. Bird census work and environmental monitoring. Ecol. Res. Comm. Bull. 9: 92 p.

Takekawa, J.Y.; Garton, E.O.; Langelier, L.A. 1982. Biological control of forest insect outbreaks: the use of avian predators. Transactions of the North American Wildlife and Natural Resources Conference 47: 393-409.

Torgerson, T.R.; Campbell, R.W. 1982. Some effects of avian predators on the western spruce budworm in north central Washington. Environ. Entomol. 11(2): 429-431.

Tschaplinski, P.J.; Buckner, C.H. 1978. Toxicity and metabolism of fenitrothion in red-backed voles, *Clethrionomys gapperi*. Can. For. Serv., Forest Pest Management Institute, Sault Ste. Marie, Ont. Rep. FPM-X-17. 215 p.

Williams, A.B. 1936. The composition and dynamics of a beech-maple climax community. Ecol. Monogr. 6: 317-408.

Zach, R.; Falls, J. 1975. Response of the ovenbird (Aves: Parulidae) to an outbreak of the spruce budworm. Can. J. Zool. 53: 1669-1672.

Zinkl, J.G.; Henry, C.J.; Shea, P.J. 1979. Brain cholinesterase activities of passerine birds in forests sprayed with cholinesterase-inhibiting insecticides. Pages 356-365 *in* Animals as monitors of environmental pollutants. National Academy of Sciences. Washington, D.C.

Part IV
Operational Control Programs

Chapter 62

Insect Control in Newfoundland, 1973–1989

H. CRUMMEY

Introduction

The forest resource of Newfoundland is composed predominantly of softwoods, with balsam fir and black spruce being the major species used by the forest industry. Because white spruce and larch seldom occur as large homogeneous tracts of timber, they are only used incidentally. The remaining white pines and the few scattered native red pine stands are more of ecological than economic importance. Of the hardwood species, only white and yellow birch and trembling aspen occur in volumes of any consequence. Nevertheless, these species do not at present support any significant industrial use, although there is an increasing demand for fuel wood.

Besides its relatively undiversified species complex, another feature worth noting about the forest resource is its age-structure. Newfoundland's forests have a significant proportion of overmature timber.

Understanding the species and age-structure of the forest provides a good basis for appreciating the magnitude of pest problems that have occurred.

When the forest industry began in the early 1900s in Newfoundland, accessibility was largely limited and the resource was virtually untapped, leading to an apparent surplus situation that lasted into the early 1970s. Consequently, pest outbreaks and volume losses in earlier years were largely regarded as natural phenomena that could be tolerated.

Major Forest Insect Species

Two native insects, the spruce budworm, *Choristoneura fumiferana*, and the hemlock looper, *Lambdina fiscellaria*, have periodically become primary pests of the forest.

Spruce Budworm

Before 1972 outbreaks of the spruce budworm, *C. fumiferana*, in Newfoundland were generally of short duration and collapsed naturally without causing significant tree mortality (Otvos and Moody 1978). The outbreak that began in 1972 and declined in the late 1970s was unprecedented in duration, intensity, and devastation (Hudak and Raske 1981); scattered pockets of infestation still remained in 1989. Host trees in Newfoundland include balsam fir, white and black spruce, and occasionally larch. Although mature and overmature stands of balsam fir suffer the most mortality, stands of regeneration and immature trees may be killed during an epidemic. Nonfatal attack often produces top-kill and reduced volume increment in attacked trees. Balsam fir suffers higher mortality than black spruce, although the latter species may exhibit reduced cone production and also be weakened to the point of being more susceptible and vulnerable to attack by secondary insects and diseases such as the four-eyed spruce bark beetle and armillaria root rot.

Hemlock Looper

Outbreaks of the hemlock looper, *L. fiscellaria*, in Newfoundland have been recognized since 1912, but losses were not recorded until 1947 (Carroll 1956; Otvos et al. 1971; Otvos et al. 1979). During two outbreaks that occurred between 1944 and 1963 it was estimated that 2.651 million m^3 of wood (all species) were killed. In the next outbreak, between 1966 and 1972, another 12.021 million m^3 of merchantable wood (all species) were estimated to have been killed with less than 20% being salvaged (Otvos et al. 1979). This outbreak, however, resulted in the first large-scale aerial chemical spray programs being conducted in Newfoundland in 1968 and 1969. It was estimated that these protection programs saved about 24.1 million m^3 of merchantable forest (Otvos et al. 1979).

Evidence suggests that before 1970 hemlock looper outbreaks of varying intensity have been cyclical, lasting for 5 to 7 years with 3 to 4 years in between. Outbreaks tend to develop in about 2 years, especially during warm, dry periods. High population levels seldom persist for more than 2 years in any particular stand, and declines are associated with cool, wet weather, starvation, and disease caused by native fungi (*Entomophthora* spp.). Nonetheless, major damage to mature and overmature stands can occur in a short period of time. Looper larvae are very wasteful feeders in that they seldom consume all attacked needles. In sufficient numbers, the loopers can cause severe defoliation throughout the crown of the tree leading to tree mortality in 1 or 2 years (Otvos et al. 1971; Otvos et al. 1979).

An outbreak of hemlock looper was expected to occur in the mid-1970s. However, the spruce budworm epidemic, which began in 1972 and continued throughout the decade, probably depleted so much of the food supply that looper numbers were unable to reach their potential during that period (Otvos et al. 1979). At the end of the spruce budworm outbreak, an epidemic of hemlock looper was again expected. Given the extremely weakened conditioned of much of the island's timber resource, such an uncontrolled looper outbreak would have brought disaster to the forest industry of the province. In fact, during 1984, the hemlock looper population did increase dramatically (Clark and Carew 1985), and by 1985 a major outbreak of hemlock looper covered much of central and eastern Newfoundland. By 1989 the outbreak had progressed over most of the island but due to major control programs and natural factors, collapsed, except for pockets of infestation on the Northern and Avalon peninsulas.

Control Policy

The need to protect the forest resource against insects became obvious in the late 1960s during an outbreak of the hemlock looper (Otvos et al. 1971). Because of the potential impact of this pest on the resource, the provincial government (henceforth referred to as the government) decided to carry out the first large-scale forest protection program in 1968 and 1969, using aerially applied chemical insecticides. In the early 1970s, an outbreak of spruce budworm started in Newfoundland (Otvos and Moody 1978). Initially, no control action was contemplated because prior outbreaks had lasted only a few years (Otvos and Moody 1978) and because control programs in other jurisdictions at the time were quite controversial and not well understood. In 1974, the government and the forest industry felt that some measure of action should be undertaken to protect the forest from this potential threat (Carter 1979). However, spraying was considered the last choice.

As a result of the outbreak, a 5-year Federal/Provincial Forest Subsidiary Agreement, in place at the time, was amended to allow for salvaging of spruce budworm-affected timber to minimize expected losses. The significance of this effort was minimal compared with the magnitude of the problem (Anon. 1980). By 1977, 90% of the productive forest area of the island was expected to be infested, but a large-scale spray program was not initiated because government policy toward forest spraying was being shaped. The government decided instead to carry out an experimental program to gather additional information under Newfoundland conditions, including determining the efficacy of various insecticides and monitoring environmental side-effects, and to gain experience in conducting large-scale operations (Carter 1977). Based on the success of this program, the first large-scale budworm spray program was carried out in 1978 (Carter 1979). But in 1979 and 1980, because of local public concern and opposition to the 1978 operation, only experimental and/or semi-operational programs using Bt (*Bacillus thuringiensis*) were permitted (Carter 1979; Carter 1980); in 1980 the government appointed a Royal Commission on Forest Protection. Part of its mandate was to review all applicable current information and, in particular, to recommend options to the government on forest protection against insect pests. The royal commission confirmed the magnitude of the insect problem and recommended that the government adopt a long-term policy on protection, particularly as it related to investment in expensive silvicultural practice to renew the forest resource (Poole 1981). This recommendation, along with many others, was adopted by the government and has provided the basis for forest spraying policy within the province. Control programs since 1980 have become an integral part of forest management. Particular emphasis has been placed on protecting silviculture areas. The current position of the Newfoundland Department of Forestry and Agriculture is that the forest resource will be protected against insect pests, using federally registered pesticides, subject to annual environmental assessment and review processes within the province, as deemed necessary. The government will decide on the nature and extent of a program based on all available information and recommendations.

Without such a policy on protection, the royal commission recommended that silvicultural prescriptions not be undertaken.

Insect Control Operations

Before 1977, no expertise existed within the provincial government to carry out all the necessary aspects of forest insect control operations. In 1977, the Forest Protection Division of the Newfoundland Department of Forestry and Agriculture assumed responsibility for any control programs conducted against insect and disease pests, and to date have planned and supervised major insect control programs.

The insect population forecast is generated by Forestry Canada, based on cooperative fall surveys with provincial forestry departments, and this determines the need for a control program and provides the outline to initiate proposed treatment areas.

Since 1977, the Newfoundland Department of Forestry and Agriculture has carried out all other aspects of the aerial programs (apart from the actual application of the insecticide and aircraft maintenance), including the transportation, handling, mixing, loading, and decontamination of insecticides and related products at spray bases, as well as the loading of aircraft. The department also oversees the actual spraying by the contractor to ensure that the proper areas are treated under the appropriate weather conditions. From early in the season, it observes insect and host-tree shoot development and larval numbers to determine the ideal application date(s) and priorities of areas. Monitoring of insecticide efficacy continues throughout the spray program, and the final assessment is made after insect feeding has ended. All necessary ground, communication, and sampling equipment is supplied and owned by the department.

Staff for spray programs include entomologists and technicians of the Forest Protection Division, as well as sufficient temporary personnel to carry out the programs efficiently and effectively. This usually involves about 30 people, but can reach 60 or more depending on the size of the program. In the past, the department's Regional Services have also provided personnel, as required, to assist in various capacities. As procedures are complex, workers who are familiar with the operation are desirable. Mistakes can be serious and costly in both time and money.

Operational Tactics

There have been few changes in operational tactics over the past 12 years, other than using available equipment and technology as it develops and complying with current regulatory guidelines. The most significant changes in operations have been the utilization of improvements in insecticide application technology and the better formulations and knowledge of how Bt products can be used to maximize efficacy.

Spray aircraft—In the past, spraying has been done with various types of aircraft, both single- and/or multi-engine (Tables 1 and 2). Generally, three or four single-engine aircraft form a "team," spraying together in formation. For practical and safety reasons, fewer aircraft would be employed to

Table 1. Summary of various spray regimes used for spruce budworm, *Choristoneura fumiferana*, control and foliage protection in Newfoundland, 1977–1989.

Year	Spray regime[a]	Area (ha) treated by aircraft type		
		Single-engine[b]	Multi-engine[c]	Total
1977[d]	1. 2 appl. 210 g Sum/1.46 L/ha+ 1 appl. 88 g Mat/1.46 L/ha	–	16 180	16 180
	2. 3 appl. 88 g Mat/1.46 L/ha	–	16 180	16 180
	3. 2 appl. 210 g Sum/1.46 L/ha	–	16 180	16 180
	4. 2 appl. 88 g Mat/1.46 L/ha	9 600	–	9 600
	5. 3 appl. 70 g Mat/1.46 L/ha	–	16 180	16 180
	6. 1 appl. 20 BIU[e] Th16B + 56 g Ort/5.84 L/ha and 1 appl. 20 BIU Th16B/5.84 L/ha	–	2 590	2 590
	Annual Total	**9 600**	**67 310**	**76 910**
1978	1. 2 appl. 70 g Mat/1.46 L/ha	9 929	340 003	349 932
	2. 1 appl. 70 g Mat/1.46 L/ha	102	26 570	26 672
	Annual Total	**10 031**	**366 573**	**376 604**
1979[d]	1. 1 appl. 20 BIU Th16B/5.84 L/ha	5 650	–	5 650
	2. 1 appl. 20 BIU Th24B/5.84 L/ha	140	–	140
	3. 1 appl. 20 BIU N45/5.84 L/ha	80	–	80
	Annual Total	**5 870**	**–**	**5 870**
1980[d]	1. 1 appl. 20 BIU Th16B/5.84 L/ha	7 345	–	7 345
	2. 1 appl. 20 BIU Th24B/4.67 L/ha	400	–	400
	3. 1 appl. 20 BIU Th32B/2.34 L/ha	400	–	400
	4. 1 appl. 20 BIU N45/4.67 L/ha	400	–	400
	5. 1 appl. 20 BIU N3/4.67 L/ha	1 820	–	1 820
	6. 1 appl. 20 BIU D88/4.67 L/ha	600	–	600
	7. 2 appl. 20 BIU Th16B/5.84 L/ha	96	–	96
	8. 2 appl. 20 BIU D88/4.67 L/ha	700	–	700
	Annual Total	**11 761**	**–**	**11 761**
1981	1. 1 appl. 70 g Mat/1.46 L/ha	–	17 990	17 990
	2. 1 appl. 86 g Mat/1.46 L/ha	3 200	70 478	73 678
	3. 2 appl. 70 g Mat/1.46 L/ha	2 180	51 943	54 123
	4. 2 appl. 86 g Mat/1.46 L/ha	–	13 208	13 208
	5. 1 appl. 70 g Mat/1.46 L/ha + 86 g Mat/1.46 L/ha	–	79 064	79 064
	6. 2 appl. 20 BIU Th16B/7 L/ha	720	–	720
	7. 1 appl. 20 BIU D88/7 L/ha	1 200	–	1 200
	Annual Total	**7 300**	**232 683**	**239 983**
1982	1. 1 appl. 70 g Mat/1.46 L/ha	1 293	–	1 293
	2. 2 appl. 70 g Mat/1.46 L/ha	41 816	–	41 816
	3. 1 appl. 20 BIU Th16B/7 L/ha	4 725	–	4 725
	Annual Total	**47 834**	**–**	**47 834**

[a] Dipel (Bt)(D); Matacil (aminocarb)(Mat); Matacil Flowable (aminocarb)(MatF); Novabac (Bt)(N); Orthene(acephate)(Ort); Sumithion (fenitrothion)(Sum); and Thuricide (Bt)(Th).
[b] Cessna AgTruck (1979); Cessna AgWagon (1977, 1978); Grumman AgCat (1980, 1981, 1982, 1983, 1984, 1985); Piper Pawnee (1980).
[c] Douglas DC-6B (1977, 1978, 1983, 1985); Douglas DC-4 (1986); Super Constellation (1981).
[d] Designated years were experimental or semi-operational.
[e] Billion international units (BIU).

(Continued)

Table 1. Summary of various spray regimes used for spruce budworm, *Choristoneura fumiferana*, control and foliage protection in Newfoundland, 1977–1989. *(Concluded)*

Year	Spray regime[a]	Area (ha) treated by aircraft type		
		Single-engine[b]	Multi-engine[c]	Total
1983	1. 1 appl. 70 g Mat/1.46 L/ha	–	622	622
	2. 2 appl. 70 g Mat/1.46 L/ha	15 749	50 183	65 932
	3. 2 appl. 70 g MatF/1.46 L/ha	2 591	3 615	6 206
	Annual Total	**18 340**	**54 420**	**72 760**
1984	1. 1 appl. 70 g Mat/1.46 L/ha	1 410	–	1 410
	2. 2 appl. 70 g Mat/1.46 L/ha	21 816	–	21 816
	3. 1 appl. 20 BIU Th16B/7 L/ha	720	–	720
	4. 1 appl. 20 BIU D132/1.57 L/ha	1 680	–	1 680
	5. 2 appl. 20 BIU D132/1.57 L/ha	710	–	710
	Annual Total	**26 336**	**–**	**26 336**
1985	1. 1 appl. 20 BIU D132/1.57 L/ha	3 450	–	3 450
	2. 2 appl. 70 g Mat/1.46 L/ha	2 945	–	2 945
	Annual Total	**6 395**	**–**	**6 395**
1986–1989	No programs			
	Total area treated	**143 467**	**720 986**	**864 453**

[a] Dipel (Bt)(D); Matacil (aminocarb)(Mat); Matacil Flowable (aminocarb)(MatF); Novabac (Bt)(N); Orthene(acephate)(Ort); Sumithion (fenitrothion)(Sum); and Thuricide (Bt)(Th).
[b] Cessna AgTruck (1979); Cessna AgWagon (1977, 1978); Grumman AgCat (1980, 1981, 1982, 1983, 1984, 1985); Piper Pawnee (1980).
[c] Douglas DC-6B (1977, 1978, 1983, 1985); Douglas DC-4 (1986); Super Constellation (1981).
[d] Designated years were experimental or semi-operational.
[e] Billion international units (BIU).

Table 2. Summary of various spray regimes used for hemlock looper, *Lambdina fiscellaria*, control and foliage protection in Newfoundland, 1985–1989.

Year	Spray regime[a]	Area (ha) treated by aircraft type			
		Single-engine[b]	Multi-engine[c]	Combination[d]	Total
1985	1. 1 appl. 210 g Fen/1.5 L/ha	4 864	10 382	3 964	19 210
	2. 1 appl. 280 g Fen/1.5 L/ha	2 920	36 894	–	39 814
	3. 2 appl. 210 g Fen/1.5 L/ha	12 889	33 232	17 583	63 704
	4. 1 appl. 30 BIU[e] D132/2.36 L/ha	2 365	–	–	2 365
	Annual Total	**23 038**	**80 508**	**21 547**	**125 093**
1986	1. 1 appl. 210 g Fen/1.5 L/ha	64	–	1 120	1 184
	2. 2 appl. 210 g Fen/1.5 L/ha	9 818	38 655	29 371	77 844
	3. 1 appl. 30 BIU D132/2.36 L/ha	5 420	–	–	5 420
	Annual Total	**15 302**	**38 655**	**30 491**	**84 448**
1987	1. 1 appl. 210 g Fen/1.5 L/ha	6 226	7 216	1 108	14 550
	2. 2 appl. 210 g Fen/1.5 L/ha	81 392	54 759	13 711	149 862
	3. 1 appl. 30 BIU D132/2.36 L/ha	3 308	–	–	3 308
	4. 2 appl. 30 BIU D132/2.36 L/ha	875	–	–	875
	Annual Total	**91 801**	**61 975**	**14 819**	**168 595**

(Continued)

Table 2. Summary of various spray regimes used for hemlock looper, *Lambdina fiscellaria*, control and foliage protection in Newfoundland, 1985–1989. *(Concluded)*

Year	Spray regime[a]	Area (ha) treated by aircraft type			
		Single-engine[b]	Multi-engine[c]	Combination[d]	Total
1988	1. 2 appl. 210 g Fen/1.5 L/ha	45 138	–	–	45 138
	2. 1 appl. 30 BIU D176/1.8 L/ha	6 964	–	–	6 964
	3. 2 appl. 30 BIU D176/1.8 L/ha	1 111	–	–	1 111
	4. 1 appl. 30 BIU D176/1.8 L/ha + 1 appl. 30 BIU D132/2.36 L/ha	1 042	–	–	1 042
	5. 1 appl. 30 BIU D176/1.8 L/ha + 1 appl. 30 BIU F-XLV/2.0 L/ha	261	–	–	261
	6. 1 appl. 30 BIU D176/1.8 L/ha + 1 appl. 210 g Fen/1.5 L/ha	9 557	–	–	9 557
	7. 1 appl. 30 BIU F-XLV/2.0 L/ha	3 492	–	–	3 492
	8. 1 appl. 30 BIU F-XLV/2.0 L/ha + 1 appl. 30 BIU D176/1.8 L/ha	1 361	–	–	1 361
	Annual Total	**68 926**	**–**	**–**	**68 926**
1989	1. 1 appl. 30 BIU F-XLV/2.0 L/ha	1 624	–	–	1 624
	2. 2 appl. 30 BIU F-XLV/2.0 L/ha	3 738	–	–	3 738
	Annual Total	**5 362**	**–**	**–**	**5 362**
	Total area treated	**204 429**	**181 138**	**66 857**	**452 424**

[a] Futura (Bt)(F); fenitrothion (either Folithion or Sumithion)(Fen); and Dipel (Bt)(D).
[b] Ayres Bull Thrush (1987, 1988, 1989); Grumman AgCat (1985, 1986, 1987, 1988).
[c] Douglas DC-6B (1985, 1987); Douglas DC-4 (1986).
[d] Large aircraft and small aircraft combined to spray this block area.
[e] Billion international units (BIU).

treat blocks of smaller size. The government has contracted out this aspect of the program to companies that have the necessary application equipment and experienced aircraft personnel. Usually proposals for aircraft are requested, received and evaluated, and a decision made on the numbers and kinds of aircrafts that could do the job. Therefore, over the years, an assortment of aircraft have been used, depending upon the size of the program, extent of treatment areas, distance from operating bases, and past performance.

In the last 4 years, larger single-engine aircraft (i.e. Ayers Bull Thrush) have been desirable because of their faster ferrying and spray speeds and larger hopper capacity, thus maximizing the productivity of each spray mission.

Application equipment—Up until 1984, boom and nozzle equipment was employed for both multi- and single-engine spray aircraft. As a result of research and experience in other jurisdictions, rotary atomizers were first used in 1984 to apply Bt for spruce budworm control. Since that time, only rotary atomizers have been accepted for both chemical and Bt insecticide application using single-engine aircraft. Initially, Micronair® AU5000 units were employed but this has changed, and preference is now given to Micronair® AU4000 units. Multi-engine spray hardware has not changed over the years.

Guidance—Guidance of multi-engine aircraft has been accomplished using on-board navigation systems (Litton Inertial Navigation System). This has been a standard for multi-engine aircraft for many years and is effective when used by experienced pilots. With single-engine aircraft, navigation has been provided by departmental personnel in helicopters that lead spray aircraft along predetermined flight lines marked on 1:50 000 scale topography maps or 1:12 500 scale color photographs, depending on the size of the applicable treatment area(s). In addition, an aerial supervisor in a fixed-wing aircraft checks the accuracy of the navigation and performance of the spray aircraft and initiates corrective action, as necessary. This applies to each multi-engine aircraft and to each team of single-engine aircraft. The supervisor also assesses weather parameters before and during spray missions. The supervisory aircraft operates well above the spray formation. The navigator and supervisor configuration is employed for each team of single-engine aircraft and, although somewhat expensive, has proven to be effective.

Mixing–loading systems—Spray bases have been equipped with appropriate-size steel tanks for chemical insecticides and, recently, polyethylene tanks for Bt preparations, with the associated plumbing to facilitate the mixing and agitation of insecticide and loading of aircraft. These set-ups are designed to ensure environmental safety by using approved containment dyking and currently acceptable safety and emergency equipment and materials.

Financial—For spray programs, the financial considerations (based on actual costs) from 1977 to 1989, have ranged from as little as $270 000 to as high as $3 million, depending on the size of the program. An average figure would be from $1 million to $2 million per year for a modest program. Control projects have generally been conducted under a prearranged financial agreement with two paper companies and the provincial government. In 1978 and 1979, the ratio was 50:50. From 1980 to the present, the government has funded one-third of the cost of the operational control programs and the two paper companies the remaining two-thirds.

Safety precautions during spray operations—To minimize the risk of exposure of people to insecticide spray, there is generally a 1.6-km (multi-engine aircraft) or a 0.8-km (single-engine aircraft) buffer zone left around known places of permanent human habitation and, where possible, around areas such as cabin development and camp and day-use areas in parks. Likewise, a 1.6-km buffer zone is left around identifiable intakes to known community water supplies; however, it may be feasible to decrease buffers in particular cases, especially with Bt. These are dealt with in consultation with the provincial Department of Environment and Lands on an individual basis as and when identified. If, during the course of a spray mission, unauthorized personnel are detected in or near a treatment area, the aerial supervisor will instruct the spray aircraft pilot(s) to provide extra buffers or to terminate the mission, as applicable in the circumstance.

Before spraying, hospitals and regional public health officials in the vicinity are notified concerning which products are to be used, general areas of spray blocks, timing of spray season, and so forth. This allows them to prepare should an incident occur that would require medical assistance.

Safety precautions for project personnel—Each mixer/loader is required to wear a hooded rubber suit, rubber gloves, rubber boots, goggles, and a face mask during the mixing of the insecticide formulation, the filling of loading and holding tanks and aircraft, and the decontamination of insecticide drums. Pilots and navigators/supervisors are not permitted to be involved in the handling of insecticides.

All mixer/loader personnel are instructed not to eat, smoke, or chew while handling insecticides or rub their eyes or mouths, to wash their hands with soap and water before eating, smoking, or using the toilet, and to shower or bathe completely after each day's operation. Gloves must be washed in neutralizing solution before removal. Contaminated clothing must not be reused until it has been laundered.

If contact with an insecticide occurs, the person is required to immediately remove the contaminated clothing and to wash thoroughly. Should symptoms of illness occur during or shortly after handling of any insecticide or mix, a doctor (who has received prior notification of the program, its insecticides, exposure symptoms, and antidotes) is contacted immediately to arrange the necessary medical attention. Symptoms of poisoning are reviewed and this information is given to those supervising and handling the insecticides and mix. Hospital and emergency telephone numbers are also posted in a conspicuous place to be used in the event of an accident.

All stipulations in the licence issued by the provincial Department of Environment and Lands concerning environmental safety are followed. These include the reporting of any incidents, such as spills, to the appropriate authorities. In connection with this, the Department of Forestry and Agriculture has a contingency plan that is annually updated and submitted for approval before receiving an operator's licence. The plan outlines procedures for spill reporting, emergency first aid for exposure, insecticide spill only, aircraft crash in the bush, aircraft accident on or near an airport, jettisoned aircraft load, drum decontamination and disposal, and other general regulations and instructions as necessary.

As part of the program, the public is informed and kept aware of details of the operation and the progress. A phone-in information line is set up and the general public can call collect to find out where and approximately when spraying will be carried out and the ongoing status of all blocks. Similarly, since 1977 telefacsimile messages are sent twice a day to the news media with information indicating what blocks are ready to be treated and the status of blocks that have been treated since the last update.

Regional offices of the Department of Forestry and Agriculture and the Department of Environment and Lands are provided with maps showing spray blocks. These maps are available for viewing by the general public during regular office hours. Also, forestry unit offices of the department are made aware of spray blocks in their area and are provided with applicable detailed maps to show the public.

Specific environmental concerns—The government also made the commitment that as part of the control programs and in cooperation with the Department of Environment and Lands, environmental monitoring would be carried out as required. In 1985, studies on the impact of fenitrothion on birds and bird behavior, on pollinators and pollination, and on deposits/residues were done. Additional studies on songbird cholinesterase depressions as a result of applications of fenitrothion for hemlock looper control were conducted in 1986, 1987, and 1988.

Operational Programs

Major aerial control efforts were mounted against the spruce budworm up to 1985 (the last program required against this pest) and against the hemlock looper from 1985 to 1989. As silvicultural efforts continue to increase, the need to protect these investments against losses to insects and disease becomes more apparent. A future wood supply for the industry is dependent on a vigorous, healthy, growing stock, which can reach rotation age free from insect and disease infestation.

The purpose of the insect control programs is to reduce populations of these insects during the larval feeding stages to minimize feeding pressure on infested trees and thereby (a) prevent additional tree mortality; (b) promote growth in younger stands to aid in offsetting projected future wood deficits; (c) protect seed supplies for natural regeneration and potential production areas; (d) protect existing and/or proposed silvicultural treatment areas; and (e) help to suppress population levels in areas of declining infestation as well as in potential areas of population buildup.

Selected areas for treatment include those forecast to receive moderate to severe defoliation as well as silvicultural areas (both existing and proposed) where the forecast is for light defoliation. The latter category is of major importance since the government and the paper companies have made, or will make, a substantial investment in forest management, and this investment in the future wood supply for the island must be protected.

Spruce budworm—Control programs against the spruce budworm developed into an integral part of forest protection and management (Table 1). In conjunction with these programs, from 1977 to 1985, the provincial Department of Environment and Lands monitored the spray programs and studied the impact of chemical insecticides, particularly Matacil®, on the environment.

In 1977, the department conducted an experimental spray program to test the efficacy of various insecticides, monitor environmental side-effects, and obtain operational experience in the conduct of such programs. The area treated was approximately 77 000 ha. The next year, an operational program was successfully conducted using the chemical insecticide aminocarb (Matacil®) over 376 000 ha. In 1979 and 1980, spraying was limited to small experimental and/or semi-operational programs (approximately 6 000 and 12 000 ha respectively) using the biological insecticide Bt. In 1980, the government appointed a Royal Commission on Forest Protection and Management to look at all aspects of control options. During the following years, up to and including 1985 (the last year when a spruce budworm program was required), control programs were conducted using chemical and biological insecticides.

Hemlock looper—Based on the experience with the spruce budworm epidemic in the 1970s, and the government's commitment to protect the resource as a result of adopting the recommendations of the royal commission, immediate action was taken to control the sudden, but not unexpected, outbreak of the hemlock looper. This decision was even more critical in view of the great loss of timber from the budworm outbreak and the projected wood supply deficit. An uncontrolled looper outbreak would have meant disaster to the resource. Therefore, in 1985, the government approved a large-scale aerial protection program against the looper using the only registered product available, the chemical insecticide, fenitrothion. Approval for this was based on the 1980 royal commission report, which examined the product for spruce budworm control and was satisfied that it was acceptable for use under existing regulations, stipulations, and ongoing procedures associated with aerial control programs. The Department of Forestry and Agriculture also encouraged the manufacturers of the biological insecticide Bt to obtain a temporary registration for their product for use on areas where chemical insecticide was not preferred.

Aerial control programs were successfully carried out from 1985 to 1989 (Table 2). Also, Forestry Canada, in cooperation with the department and the forest industry, carried out experimental trials on various chemical and biological insecticides to assess their efficacy against the hemlock looper. The objectives were to improve control measures and techniques and to have alternatives to the one registered product.

These experimental programs were continued from 1985 to 1988, with the result that Bt application parameters have been better defined and one Bt product was registered for use against the looper. Experimental programs are an integral part of operational programs and essential to better manage pest problems in an effective and efficient manner.

References

Anon. 1980. Brief to the Newfoundland and Labrador Royal Commission on forest protection and management. Gov. Nfld. and Lab., Dep. For. Resour. Lands, St. John's, Nfld. 230 p.

Carroll, W.J. 1956. History of the hemlock looper, *Lambdina fiscellaria fiscellaria* (Guen.), (Lepidoptera: Geometridae) in Newfoundland, and notes on its biology. Can. Entomol. 88(10): 587-599.

Carter, N.E. 1977. The 1977 experimental spruce budworm spray program in Newfoundland. Dep. For. Agric., For. Prot. Div., St. John's, Nfld. Inf. Rep. FP-1, 32 p.

Carter, N.E. 1979. The 1978 aerial forest protection program against spruce budworm in Newfoundland. Dep. For. Agric., For. Prot. Div., St. John's, Nfld. Inf. Rep. FP-2, 19 p.

Carter, N.E. 1979. The 1979 Bt spray trails in Newfoundland. Dep. For. Resour. Lands, For. Prot. Div., St. John's, Nfld. Inf. Rep. FP-3, 22 p.

Carter, N.E. 1980. The 1980 Bt spray program in Newfoundland. Dep. For. Resour. Lands, For. Prot. Div., St. John's, Nfld. Inf. Rep. FP-4, 37 p.

Clark, L.J.; Carew G.C. 1985. Forest insect and disease conditions in Newfoundland and Labrador in 1984. Environ. Can., Can. For. Serv., Nfld. For. Cent., St. John's, Nfld. Inf. Rep. N-X-229, 31 p.

Hudak, J.; Raske, A.G. (Eds.). 1981. Review of the spruce budworm outbreak in Newfoundland—its control and forest management implications. Environ. Can., Can. For. Serv., Nfld. For. Cent., St. John's, Nfld. Inf. Rep. N-X-205, 280 p.

Otvos, I.S.; Clark R.C.; Clarke L.J. 1971. The hemlock looper in Newfoundland; the outbreak from 1966 to 1971 and aerial spraying in 1968 and 1969. Environ. Can., Can. For. Serv., Nfld. For. Cent., St. John's, Nfld. Inf. Rep. N-X-68, 62 p.

Otvos, I.S.; Clark L.J.; Durling D.S. 1979. A history of recorded hemlock looper outbreaks in Newfoundland. Environ. Can., Can. For. Serv., Nfld. For. Cent., St. John's, Nfld. Inf. Rep. N-X-179, 46 p.

Otvos, I.S.; Moody B.H. 1978. The spruce budworm in Newfoundland: history, status and control. Environ. Can., Can. For. Serv., Nfld. For. Cent., St. John's, Nfld. Inf. Rep. N-X-150, 76 p.

Poole, C.F. (Chairman). 1981. Report of the Royal Commission on Forest Protection and Management. Part I. Gov. Nfld. and Lab. St. John's, Nfld. 114 p.

Chapter 63

Insect Control in Nova Scotia, 1979–1987

T.D. Smith

Introduction

The forests of Nova Scotia are both susceptible and vulnerable to the spruce budworm, *Choristoneura fumiferana* (Blais 1983). Usually, expanses of tree mortality are limited to the North and South Mountains, northern mainland Nova Scotia, and Cape Breton Island (Fig. 1). Visible defoliation for the recent epidemic was first noted in 1969 on the mainland (Forbes et al. 1975) (Table 1). At times the spruce budworm acts in concert with the hemlock looper, *Lambdina fiscellaria*, and the eastern blackheaded budworm, *Acleris variana* (Forbes et al. 1961).

The latest spruce budworm infestation persisted from 1974 to 1981 on Cape Breton and from 1969 to 1987 on the northern mainland. On Cape Breton Island, the population declined with the exhaustion of the food supply and on the mainland from a combination of environmental factors, including a high rate of larval infection (66–73%) by the gut parasite, *Nosema fumiferanae*. The larvae contained debilitating numbers of microsporidia, 18 to 31 ($\times 10^5$) per milligram of tissue, which severely weakened the surviving adults (Smith and Georgeson 1987).

The impact of the spruce budworm on the timber resources of Cape Breton can be inferred by comparing the pre-infestation (1970) inventory of 40.5 million m^3 of wood to the post-infestation (1984–85) inventory of 14.3 million m^3 of wood. An extensive wood salvage program was undertaken, however, and the overall loss of timber (18.0 million m^3) was less than indicated by the inventory figures above. Also, these inventory figures do not contain estimates of losses in volume for the Cape Breton Highlands National Park. Approximately 3.3 million m^3 were lost on the mainland. The combined losses are equivalent to a 10-year wood supply to the forest industry of Nova Scotia.

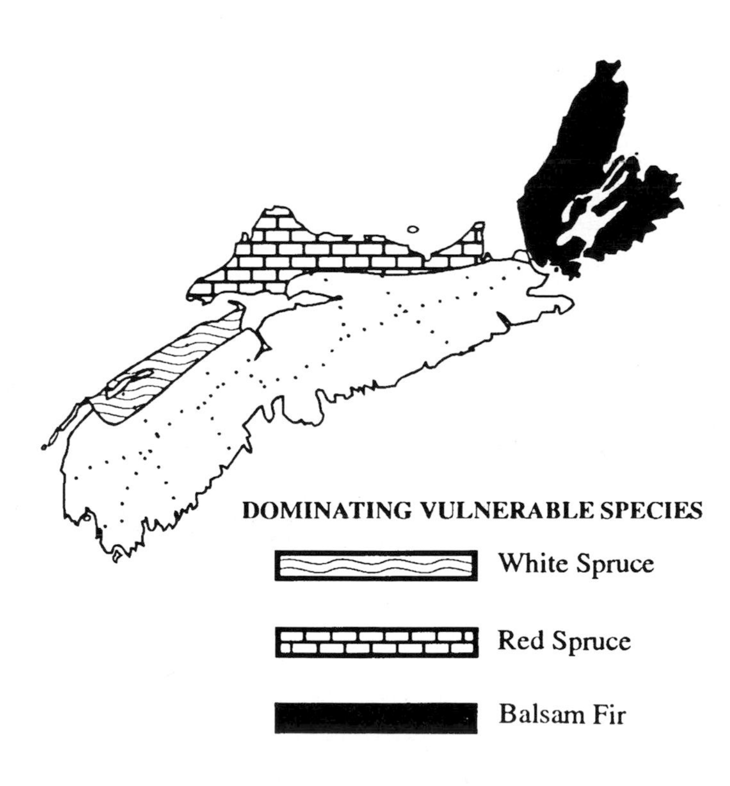

Figure 1. Spruce–fir forests vulnerable to spruce budworm (*Choristoneura fumiferana*).

Factors Affecting Control Operations

During the period under consideration, several factors influenced the amount and types of insecticides used to control the spruce budworm in Nova Scotia: changes in legislation, aerial foliage protection program, blueberries in forested areas, land ownership, insecticide formulation, and buffer zones in designated watersheds.

Table 1. Gross area of defoliation (×10 000 ha) by the spruce budworm, *Choristoneura fumiferana*, in Nova Scotia, 1968 to 1987.

Year	Low	Moderate	Severe	Total
1968	0.0	0.0	0.0	0.0
1969	0.0	0.5	0.0	0.5
1970	0.2	3.1	0.3	3.6
1971	4.6	2.1	16.1	22.8
1972	1.6	2.5	11.5	15.6
1973	0.0	3.6	5.3	8.9
1974	3.0	15.4	3.5	21.9
1975	6.5	12.2	20.4	39.1
1976	2.0	5.8	1.0	8.8
1977	11.5	4.6	8.1	24.2
1978	1.4	3.2	7.6	12.2
1979	5.6	3.8	10.3	19.7
1980	10.8	15.4	13.2	39.4
1981	10.1	8.5	11.8	30.4
1982	3.7	12.5	5.0	21.2
1983	6.3	5.2	24.2	35.7
1984	2.6	3.3	2.6	8.5
1985	2.6	3.2	28.7	34.5
1986	14.3	28.3	0.6	43.2
1987	0.0	0.0	0.0	0.0

Figure 2. Results of foliage protection efforts in 1979 and 1980 on a balsam fir tree that supported high to moderate populations of spruce budworm larvae from 1977 to 1980.

Changes in Legislation

The spruce budworm outbreak on Cape Breton Island and the proposal to initiate an Aerial Foliage Protection Program sparked considerable public controversy (Brett-Crowther 1981). The devastation of the timberlands on Cape Breton Island and the persistence of the infestation on the mainland, were among the reasons for a review of the status of Nova Scotia forestry. A Royal Commission on Forestry was formed in 1982 and submitted its report in 1984 (Connor et al. 1984). It recommended a comprehensive program for the restoration, conservation, and improvement of Nova Scotia forests, and protection of the forest was identified as a key component of forest management. In early 1986, the government released a comprehensive forest policy that addressed the question of protection. Two new pieces of legislation were subsequently developed, the Forests Act and the Forest Enhancement Act. These acts came into effect on May 26, 1986, and enable the Crown to use the most appropriate methods for the protection of forests from insects and pathogens.

Aerial Foliage Protection Program

In 1978, the department initiated an investigation into the use of the microbial agent *Bacillus thuringiensis* (Bt) against the spruce budworm. In the spring of 1979, 5 553 ha were sprayed with Bt and acceptable foliage protection was achieved (Smith et al. 1980) (Fig. 2). The Aerial Foliage Protection Program therefore became operational in 1980. From 1979 to 1981 the principal host species selected for protection was balsam fir on the Cape Breton Highlands. Emphasis then

Table 2. Land ownership of areas treated against the spruce budworm, *Choristoneura fumiferana*, under the Aerial Foliage Protection Programs in Nova Scotia, 1979 to 1987.

	Ownership area (ha/%)						No. small owners applied	Program area (ha/%)
	Crown lands			Private lands				
Year	Prov.	Federal	Total	Large	Small	Total		
1979	236.0	0.0	236.0	5 537.0	0.0	5 537.0	0	5 537.0
	4.1	0.0	4.1	95.9	0.0	95.9		100.0
1980	2 827.9	0.0	2 827.9	22 842.1	0.0	22 842.1	0	25 670.0
	11.0	0.0	11.0	89.0	0.0	89.0		100.0
1981	2 321.5	0.0	2 321.5	28 874.5	0.0	28 874.5	0	31 196.0
	7.4	0.0	7.4	92.6	0.0	92.6		100.0
1982	3 415.3	0.0	3 415.3	15 103.2	634.7	15 737.9	1	19 153.2
	17.8	0.0	17.8	78.9	3.3	82.2		100.0
1983	9 239.2	0.0	9 239.2	10 852.6	634.7	11 487.3	1	20 726.5
	44.6	0.0	44.6	52.4	3.1	55.4		100.0
1984	6 389.0	0.0	6 389.0	2 368.0	11 780.0	14 148.0	625	20 537.0
	31.1	0.0	31.1	11.5	57.4	68.9		100.0
1985	8 300.7	0.0	8 300.7	14 245.8	27 174.1	41 419.9	800	49 720.6
	16.7	0.0	16.7	28.7	54.7	83.3		100.0
1986	6 603.2	282.6	6 885.8	5 410.6	43 858.4	49 269.0	2200	56 154.8
	11.8	0.5	12.3	9.6	78.1	87.7		100.0
1987	4 109.2	90.0	4 199.2	1 115.5	25 765.6	26 881.1	3000	31 080.3
	13.2	0.3	13.5	3.6	82.9	86.5		100.0

shifted to protecting high-value red spruce stands on both Crown and private land on the mainland following the collapse of the infestation on the Cape Breton Highlands (Table 2).

Blueberries in Forested Areas

Blueberries are the leading export fruit crop of Nova Scotia. The total area under production is 10 500 ha, involving about 1 000 growers. Blueberry fields are distributed rather uniformly through the spruce budworm-infested area of the northern mainland (Fig. 1). The presence of blueberry fields effectively curtails the use of chemical pesticides currently registered for aerial application for foliage protection against spruce budworm on the mainland. Thus, on the mainland, Bt is the pesticide of choice for use against lepidopteran forest defoliators.

Land Ownership

In Nova Scotia, the landowner must request foliage protection for spruce budworm-infested trees on private land. The applicant may submit requests to protect any size parcel of land, but if the area is to be protected, it must either be greater than 5 ha or be merged with adjoining areas to form a treatment area that is equal to or greater than 5 ha. In 1982, only one small landowner applied for inclusion in the Aerial Foliage Protection Program. However, by 1987, over 3 000 small farm woodlot owners had submitted applications (Table 2). The review of the applications and the collation of areas is coordinated by the department. Areas for which the owner has applied for foliage protection and sensitive no-spray areas are noted on a 1:10 000 scale orthortopographic map. Proposed treatment areas are designed on the basis of requested foliage protection areas and timber hazard from the spruce budworm. Each landowner whose land is to be sprayed is contacted to review the status of the application. The applications are valid for the duration of the infestation.

Insecticide Formulation

Two changes in the formulations of Bt had a marked influence on spray operations. The introduction of the high potency Bt formulations prompted use of Micronair® AU5000 mini-atomizers in 1984. The use of oil formulations decreased evaporation rates and allowed a widening of the relative humidity limits for effective spraying in 1986 (Table 3).

Buffer Zones

There have been significant changes in the buffer zone restrictions with the introduction of Bt as the insecticide used in spruce budworm control. The most dramatic was the elimination of the spray prohibition within designated watersheds. Before 1983, spraying could not occur within 1.5 km of the boundary of a designated watershed. In 1984, this was changed to a buffer zone of 0.5 km on each side of a stream, upstream from the extraction point for a town, for a designated distance plus 60 m. The designated upstream distance is determined from thc rate of stream discharge at the point of extraction. In the 1987 spray program, 2 877 ha were treated within the Tatamagouche watershed.

Treatment areas not involving watersheds are designed with the boundary as a flight line. There is a 60-m buffer zone, a treatment area, and an adjoining area. There is no special buffer zone with the use of Bt near blueberry fields in Nova Scotia. Provincial parks are closed 12 hours before spraying and opened 24 hours after spraying. Portable water sources are covered with plastic bags and picnic tables are turned over.

Aerial Control Operations

Large- and small-scale aerial control operations are best treated separately, because the procedures differ in several aspects. All these control treatments were operational, and no records were kept on spray effectiveness.

Large-Scale Operations

These operations were under the control of a ground crew at a central airport and used a team of spray and guidance planes.

The ground crew is the operational nucleus of the spray program. At the main airport the ground crew is made up of the program director, one airport manager, one assistant airport manager, one air traffic controller, one information officer, and three mixers and loaders. While spraying is ongoing, the airport manager is responsible for running the operation. The ground crew maintains contact with field and air crews via aircraft and departmental radio systems.

The mix crew was trained in the use and care of pumps and meters. Clothing for the mix crew and ground crew included a wide-brimmed hard hat, goggles, rubber gloves,

Table 3. Increase in area treated due to the introduction of oil formulations of Bt which allowed lowering the relative humidity (RH) levels for effective spraying from 65% to 35% in 1986.

Base of operations	Area below 65% RH		Area above 65% RH		Total area	
	ha	%	ha	%	ha	%
Parrsboro	10 060	17.9	30 380	54.1	40 440	72.0
Trenton	3 316	5.9	12 398	22.1	15 715	28.0
Total	13 377	23.8	42 778	76.2	56 155	100.0

Figure 3. A team of Grumman AgCats spraying Bt on mature red spruce and Norway spruce seedlings planted in a recent strip-cut at Moose River, Cumberland County, Nova Scotia, 1981.

Table 4. Types of spray aircraft used on the Aerial Foliage Protection Programs in Nova Scotia, 1979 to 1987.

Year	Spray aircraft Type	Number	Spray system[a]
1979	Stearman	1	B & N
	Cessna AgTruck	1	B & N
	Grumman AgCat Model A	4	B & N
1980	Grumman AgCat Model A	12	B & N
1981	Grumman AgCat Model A	12	B & N
1982	Grumman AgCat Model A	9	B & N
1983	Hughes 500 helicopter	1	B & N
	Grumman AgCat Model A	9	B & N
1984	Hughes 500 helicopter	1	B & M
	Grumman AgCat Model A	6	B & M
1985	Bell 206 B helicopter	1	B & M
	Cessna Pawnee	1	B & M
	Grumman AgCat (Models A & B)	9	B & M
	Thrush Commander	5	B & M
1986	Cessna Pawnee	3	B & M
	Grumman AgCat (Models A & B)	6	B & M
	Ayers Thrush	3	B & M
1987	Grumman AgCat	7	B & M

[a] Boom and nozzle (B&N); Boom and Micronair® AU5000 mini-atomizer (B&M).

cotton overalls, and steel-toed boots (Weekman 1978; Singer 1980). Wash basins with water, soap, and industrial cleansers were provided as per Nova Scotia Department of Health norms. If emergency treatment were required, the nearest hospital was 2 km from the airport.

The principal spray aircraft used against the spruce budworm in large-scale operations was the Grumman AgCat (Table 4). This aircraft usually flies 30 m above the tree canopy at ground speeds of 150 to 160 km per hour.

The spray aircraft usually fly in teams of three (Fig. 3) and are accompanied by two guidance aircraft, usually Cessna 172. The first guidance aircraft, flying 100 m to 200 m above the canopy and about 300 m in front, guides the spray aircraft along the correct azimuth. The second, flying 500 m to 700 m above the spray team, supervises the spraying. The areas sprayed with Bt are shown in Table 5.

Small-Scale Operations

Three different types of aircraft were used to treat small areas of 5–50 ha: (1) a spray helicopter accompanied by a guidance helicopter; (2) three Cessna Pawnee accompanied by two guidance aircraft; and (3) one Ayers Thrush accompanied by a guidance aircraft (Table 6). The best combination for treating these small areas was a single spray aircraft such as a Grumman AgCat or an Ayers Thrush, accompanied by a guidance helicopter, and is the system currently used. The

Table 5. Area treated with different formulations of Bt in Nova Scotia from 1979 to 1987.

Year	Formulation	Area (ha)	Application Dose (BIU/ha)[a]	Application Rate (L/ha)
1979	Thuricide 16B	5 773.0	10-20	4.7-9.4
1980	Thuricide 16B	25 670.0	20	9.4
1981	Thuricide 16B	26 626.0	20	7.1
	Dipel 88	4 570.0	20	5.9
1982	Thuricide 16B	1 455.6	20	7.1
	Dipel 88	536.3	20	5.9
	Thuricide 32LV	17 161.3	20	5.9
1983	Dipel 88	994.1	20	5.9
	Thuricide 32LV	3 981.7	20	5.9
	Novabac 3	15 650.9	20	4.7
	Futura	109.8	20	2.5
1984	Dipel 176	171.3	20	1.1
	Dipel 132	176.0	30	2.4
	Dipel 132	20 189.7	20	1.6
1985	Dipel 132	8 570.7	20	1.6
	Dipel 132	41 149.9	30	2.4
1986	Dipel 132	56 154.8	30	2.4
1987	Dipel 132	31 080.3	30	2.4

[a] Billion international units per hectare.

Table 6. Comparison of mean area flown by three sets of aircraft used to treat small areas (5–50 ha) in Nova Scotia, 1985 to 1987.

Spray aircraft	No.	Mean area (ha)	Total area (ha)	Guidance no.	Aircraft type	Year
Bell 206 B[a]	1	2 955.8	2 955.8	1	Hughes 500[a]	1985
Cessna Pawnee	3	1 824.3	5 473.0	2	Cessna 172	1986
Ayers Thrush	1	5 717.1	5 717.1	1	Hughes 500	1986
Grumman AgCat	1	4 552.4	4 552.4	1	Bell 206/ Hughes 500	1987

[a]Helicopters.

spray aircraft operated from an airport whereas the helicopter was serviced by a mobile crew that stayed near the proposed treatment areas. The helicopter hovered 100 m to 150 m above the first flight line of the treatment area and the spray aircraft passed underneath. Spraying was supervised from the guidance helicopter. If more than one pass was required, the process was repeated on the second spray line.

In addition to the small-scale spraying operations to control the spruce budworm, a 136-ha stand of red spruce was sprayed in 1986 with a single application of Bt at the rate of 30 BIU/ha (billion international units per hectare). This stand had high and equal populations of spruce budworm and hemlock looper. The spray was applied late for spruce budworm and early for hemlock looper. The looper populations in the treated area were lower in 1987 than in 1986. Looper larvae were noted feeding in adjacent blueberry fields.

References

Blais, J.R. 1983. Trends in the frequency, extent, and severity of spruce budworm outbreaks in eastern Canada. Can. J. For. Res. 13(4): 539-547.

Brett-Crowther, M.R. 1981. The spruce budworm controversy in Nova Scotia. Science and Public Policy—February 1981. 55-76.

Connor, J.; MacKinnon, G.A.; Matheson, D.L. 1984. Forestry: report of the Nova Scotia Royal Commission on Forestry. N.S. Gov't. Misc. Publ. xx + 113 p.

Forbes, R.S.; Underwood, G.R.; Cumming, F.G.; Eidt, D.C. 1961. Maritime Provinces. Pages 17-30 *in* B.M. McGuan, Annual report of the Forest Insect and Disease Survey. 1960. Dep. For., For. Ent. and Path. Br. Ottawa, Ont. 121 p.

Forbes, R.S.; Underwood, G.R.; Van Sickle, G.A. 1970. Maritimes Region. Pages 20-36 *in* R.M. Prentice and A.G. Davidson (Eds.), Annual report of the forest Insect and Disease Survey. 1969. Dep. Fish. and Oceans, Can. For. Serv., Ottawa, Ont. 125 p.

Forbes, R.S.; Underwood, G.R.; and Van Sickle, G.A. 1975. Maritimes Region. Pages 12-36 *in* A.C. Molnar and A.G. Davidson (Eds.), Annual report of the Forest Insect and Disease Survey. 1974. Dep. Environ., Can. For. Serv., Ottawa, Ont. 109 p.

Singer, J. 1980. Pesticide safety guidelines for personal protection. USDA, For. Ser., FIDM-MAG, Davis California. Misc. Publ. Contract No. 53-9158-9-6257, 45 p.

Smith, T.D.; Georgeson, E. 1987. Aerial foliage protection program, Nova Scotia, 1987. Ent. Serv. Lands and Forests, Truro, N.S. Misc. Publ., 34 p.

Smith, T.D.; Miller, C.A.; Rozee, K.R.; Menon, A.S.; Mackay, D.; MacDonald, P.; Germain, P.; Embree, D.G.; Bulmer, R.M.; Wheeler, R. 1980. An experiment with Thuricide 16B for the reduction population densities of spruce budworm (*Choristoneura fumiferana* (Clemens 1985)) (Lepidoptera: Tortricidae), Nova Scotia, Canada. Misc. Publ., 120 p.

Weekman, G.T. (Ed.) 1978. Apply pesticides correctly. USDA and USEPA. U.S. Gov't. Printing Office., Washington, D.C. 0-275-518. 40 p.

Chapter 64

Insect Control in New Brunswick, 1974–1989

E.G. KETTELA

Introduction

The spruce budworm, *Choristoneura fumiferana*, remained the most serious threat to forest management in eastern North America during the period covered by this review. In 1975, approximately 54 026 million ha were seriously defoliated in Canada from Ontario east to Newfoundland (Kettela 1981, 1983a). The ever-increasing economic importance of the fir/spruce resource and the serious threat that the spruce budworm outbreak posed throughout the affected regions were responsible for the creation of several organizations to cope with the problem. The Eastern Spruce Budworm Council was an international organization with the primary function of lobbying for improved usable pest management options. It initiated a host of task force studies of which the Baskerville Task Force for the Evaluation of Budworm Control Alternatives (Baskerville 1976) was the most important to New Brunswick. The Canada–United States (CANUSA) Spruce Budworms Program was ratified in 1977 and facilitated the coordination of research and the development of environmentally acceptable strategies for managing spruce budworm outbreaks and the forests susceptible to their attacks. The New Brunswick Spray Efficacy Research Group was established in 1980 and is still functioning. Its goal is to improve spray efficacy through multidisciplinary research and development.

Finally, a key development in 1976 was the formation of the Environmental Monitoring of Forest Insect Control Operations (EMOFICO), a group dedicated to review the environmental aspects of forest spray operations and identify areas of environmental concern.

Factors Influencing Operational Spray Programs

Resurgent infestations of the spruce budworm, and their associated control programs, coincided with a new and vigorous concern from the general public. Special interest groups began to question all aspects of forest protection and forest practices generally, and in the mid-1970s expressed grave concerns for human health and threats to the environment in the sprayed areas. This concern led to several court cases, such as Bridges Brothers versus Forest Protection Limited and Abram Friesen et al. versus Forest Protection Limited. Information was also laid before the courts by the Concerned Parents Incorporated (1977) alleging that spraying was the plausible cause of Reye's syndrome. Policy on spray matters was revolutionized by court decisions, which greatly influenced politicians and forest managers alike. These decisions have been responsible for changes in use patterns of these formulations of insecticides, even though some of the newer insecticides were apparently less effective than the ones they have replaced. It was recognized that the goals of protection programs were part of an intensive forest management strategy. This meant that infestations would in some circumstances be allowed to kill forest stands and that wood supply analysis drove the forest management/budworm management system.

Starting in the early 1970s, buffer zones (where any spraying was excluded) became a feature of spray programs as politicians, by edict, set down arbitrary non-spray areas. To minimize the adverse effects of chemical sprays, two types of buffer zones or setbacks evolved. Areas near human habitation were left unsprayed, to reduce contamination of the air and water, and this often meant that the forests, almost all in private woodlots, were sacrificed to the budworm and salvage operations. In addition, however, complete avoidance of spray effects in shut-off areas was not always achieved, either due to spray drift or to error in shut-off. This point is illustrated by an incident that occurred in 1981 in an area that was being treated with fenitrothion by a DC-6 aircraft. Two areas in the middle of a spray block were omitted from treatment because people and trucks were seen along the same sections of road during both spray applications. Fortunately, 120 sample trees were located in this area, and inspection of the spray team reports showed which areas had been omitted from direct spray. The results show that although there were some reductions in spruce budworm survival and in defoliation on trees in the shut-off zones, both were considerably less than in the properly treated areas (Table 1). The second type of buffer zone employed an alternative, often less effective, insecticide. For example, the practice of using phosphamidon along streams in DDT spray blocks was developed in 1963 and used until 1969 when fenitrothion completely replaced DDT. Initial dosage trials in 1963 showed that the most effective dosage of phosphamidon for foliage protection was very toxic to birds. Consequently, the dosage rate for phosphamidon was reduced

Table 1. Results of spray efficacy in a DC-6 spray block in sections treated with fenitrothion and shut-off zones in that block, 1981.

Treatment	Percent reduction	
	Budworm survival	Defoliation
Treated sections		
A	85	45
B	80	70
C	85	55
Shut-off zones		
D	42	15
E	35	10

to lessen the impact on birds, but it also reduced the effectiveness of the insecticide in foliage protection. Phosphamidon, used on large blocks and in buffers along streams, was less effective than DDT (Table 2).

The search for an alternative insecticide for use near blueberry fields and adjusted use patterns for fenitrothion began in 1971 as a result of litigation involving blueberry growers who claimed that fenitrothion affected pollinators and the blueberry crop. Because of injunctions and policy decisions, spraying fenitrothion within 3.2 km of blueberry fields was curtailed and a temperature restriction was placed on fenitrothion use that necessitated early spring spraying, before pollinator emergence (Table 3). The 3.2-km buffer encompassed about 100 000 ha of forest and Dylox® was tried operationally to provide protection of trees within this area. The protection provided by Dylox® was lower than that provided by fenitrothion (Table 3). One reason for the difference is that Dylox® is ineffective against early instar, bud-mining larvae and must be applied precisely against late-instar larvae on expanded foliage when the chance of spray delay is critical. The temperature injunction dictated the application of fenitrothion very early in the spray season (second-instar spray timing), resulting in spray efficacy that was lower than desired. These spray practices were pursued for several years because some protection was considered better than none. Formulations of Bt were sometimes used near potable water supplies, or to minimize potential damage to fish, because the danger of contaminating the water supplies was considered to be minimal (Table 4). For example, Bt was used as a buffer along a stream that forms a potable water supply approximately 16 km downstream, and also along a major salmon stream. The objective of these sprayed buffers was to reduce the fenitrothion contamination of the streams. Along the salmon stream, Bt provided adequate protection. However, the trees along the water supply suffered more defoliation than trees sprayed with fenitrothion. In this instance there was a lower level of protection because the Bt was sprayed after most of the defoliation had occurred. Therefore, this usage of Bt probably still has promise, particularly in situations where trees along water courses need protection and chemical contamination is unacceptable.

Adequate protection was a particular problem with farm woodlots that were often small and irregularly shaped. Formerly, these woodlots were in the original setback zone decreed by the provincial Cabinet in 1977, but they were targeted for protection in 1980, even though this posed severe logistical

Table 2. Comparison of efficacy of DDT and phosphamidon used at operational dosages in New Brunswick, 1968.

Treatment	Percent reduction	
	Budworm survival	Defoliation
DDT	85	70
Phosphamidon	65	30

Table 3. Comparison of the results of early and optimal timing of fenitrothion spray treatments and fenitrothion and Dylox® spray treatments against the spruce budworm, *Choristoneura fumiferana*, in New Brunswick.

Treatment	Percent reduction	
	Budworm survival	Defoliation
1972		
Fenitrothion regular timing L3–L4[a]	82	50
Fenitrothion early L2–L3	60	34
1974		
Dylox® L5–L6	60	12
Fenitrothion L3–L4	85	65

[a] The designations L2, L3, L4, L5, L6 indicate larval instar at time of treatments.

Table 4. Spray efficacy (percent reduction in survival and defoliation) for two spray scenarios in which Bt was used along streams in New Brunswick to reduce fenitrothion contamination.

Treatment	Percent reduction	
	Budworm survival	Defoliation
	Potable water supply stream	
1981		
Bt	80	10
Fenitrothion	85	65
	Salmon-producing stream	
1984		
Bt	90	70
Fenitrothion	95	70

Table 5. Results of spraying in terms of percent reduction in survival and defoliation in irregular and large rectangular-shaped treatment areas in New Brunswick, 1970 to 1974.

	Percent reduction in irregular-shaped spray blocks		Percent reduction in large rectangular spray blocks	
	Survival	Defoliation	Survival	Defoliation
	TBM spray aircraft			
1970	45	25	65	40
1971	20	10	80	65
1972	35	0	75	60
	Small agricultural aircraft			
1974	30	5	81	45

problems for the spray company and created difficulty in obtaining effective control. Similarly, between 1970 and 1974, data collected from irregularly shaped spray blocks and large contiguous blocks show that the percent reduction in survival and in defoliation was much less in the irregularly shaped spray blocks than in the large rectangular spray blocks, irrespective of the type of aircraft used (Table 5).

Surveys and Assessment

Forestry Canada was responsible for surveys and assessment of spruce budworm infestations and for evaluating spray programs from 1952 to 1982. The resources for these activities came from Forestry Canada and Forest Protection Ltd. The intensity of the surveys was increased in 1975 to meet the need for more information on the timing of sprays, for the intermediate assessment of results to eliminate unnecessary spraying, and to provide estimates of predicted and actual defoliation. A full-time staff of five technicians and one professional was employed by Forest Protection Ltd., under the direction of Forestry Canada, to provide this information.

Costs to the company ranged from about $100,000 in 1975 to about $300,000 in 1977 and reflected the need for more information to assist decision making. In 1983, Forest Protection Ltd. assumed responsibility for all aspects of operational surveys and spray assessments under the guidance of Forestry Canada. These duties were transferred to the Forest Management Branch, New Brunswick Department of Natural Resources, in the latter half of 1983–84.

The development and planning of realistic protection programs largely depend on information provided by key surveys to forecast infestations and to predict and assess the level of damage in each area. Information from these surveys is integrated and a map depicting hazard (risk) is prepared. Although the procedures used in preparing this map have been modified over the years to reflect changing needs and forest conditions, the procedure is essentially the same as that devised by Webb et al. (1956). An example of some of the criteria that must be met to derive a spray program is shown in Table 6.

Effective spray operations are contingent on the spray being applied at the correct time to achieve maximum effectiveness. A key service provided to the spray operators is information on the phenological development of the host trees and of the pest insect (Dorais and Kettela 1982). The phenological map devised by Webb (1958) is still used as a general guide in planning spray operations and in locating reference points for development. Auger's classification (Dorais and Kettela 1982) of shoot development indices and larval development is used to aid in the timing of spray operations in New Brunswick, except when court injunctions have dictated when spray operations may take place. Normal procedures for applying chemical insecticides to balsam fir called for spraying when the buds started to show green and all the larvae had moved out of the old needles to the buds, usually during the peak of the third instar. If Bt was being used, spraying normally began at the start of shoot flush and when the larvae were mostly in the fourth instar.

Key surveys have been egg-mass and overwintering larval surveys to assess spruce budworm abundance and aerial sketch map surveys to assess the severity and extent of defoliation. Counts of spruce budworm egg masses and overwintering larvae have proven to be reliable predictors of spruce budworm abundance and defoliation for the coming year. Before 1985, the overwintering larval surveys were used to provide additional information on infestations by cross-checking with egg-mass surveys to evaluate the infestation status of selected forest stands. The egg-mass survey technique developed by Webb et al. (1956) also included estimates of defoliation, tree condition, and past damage for each plot, and thus enabled hazard ratings for specific sites to be prepared. Although aerial defoliation surveys have several limitations (Kettela 1982; Dorais and Kettela 1982; Miller and Kettela 1975), they nevertheless provide key information on extent and severity of damage. The introduction of Loran C navigation equipment has improved these surveys by eliminating many of the navigational errors that resulted from dead reckoning navigation.

Before 1983, assessments of the effectiveness of the spray program were based on evaluation of defoliation on sample trees in treated and untreated areas and on estimates of population densities (spruce budworms per 45-cm branch tip) before treatment and at 25% adult emergence. Calculation of spray efficacy, percent reduction in defoliation, and percent reduction in survival is based on graphs showing the relation between population density and defoliation and between survival and defoliation. This method of calculating efficacy was devised by MacDonald (1963) and was the standard measure of efficacy until 1983.

Table 6. Criteria used to determine larval spray areas in 1978.[a]

A. 1st draft

1. The hazard map as compiled by Forestry Canada was used as a base.
2. Low hazard areas were eliminated.
3. Nonsusceptible areas were eliminated.
4. Municipal water supplies were eliminated.
5. Areas within 1 mile of known year-round habitation were eliminated.
6. Areas within 2 miles of blueberry fields were eliminated.
7. All federal lands were eliminated.
8. Remaining areas less than 12 000 acres were eliminated.

B. 2nd draft reflected the New Brunswick Department of Natural Resources wish that the proposed area be reduced to conform with monies available.

Therefore:

a) Dylox® was removed from consideration and a 4-mile buffer was established around all blueberry fields,
b) further revision of the nonsusceptible types through aerial observation was made,
c) all small isolated blocks were deleted, and
d) enough moderate hazard areas were deleted to reduce the plan to an affordable size.

[a] From Forest Protection Ltd. year-end report, 1978.

The transfer of survey responsibility from one agency to another in the early 1980s has resulted in several changes in the procedures used. Overwintering larval surveys (developed by Forestry Canada) are now used as the primary forecasting tool and a change in the procedure for calculating hazard has been introduced. In addition to plotting points and contouring areas of similar hazard, the province also derives hazard by simply overlaying forecast maps and defoliation maps. Hazard areas are considered to be those areas where overlap occurs (Carter and Lavigne 1986, 1987). Foliage retention has been used as the baseline estimate of spray efficacy since 1983, and efficacy is expressed as the percentage of trees with foliage retention above the minimum set limit (Table 7). The minimum acceptable foliage retention level is 60% for balsam fir and 50% for spruce. Graphical information showing defoliation as a function of population density provides continuity with earlier methods and usually demonstrated whether there is a difference between treated and untreated areas.

Although methods for assessing spray deposit have existed since the early 1950s, routine spray deposit assessment of New Brunswick aerial control operations was not normally conducted. In the late 1980s, however, techniques were developed for assessing foliar deposits (Lambert 1987; Kettela, unpubl. data), and these showed that the apparent failure of a spray operation on a block of land can be due to poor spray deposition caused by physical factors at the time of spraying.

Table 7. Synopsis of spray results for balsam fir in New Brunswick, 1974 to 1988.

Year	Percent reduction[a]		Percent trees meeting foliage retention objective	
	Survival	Defoliation	Mean	Range
1974	70	25	–	–
1975	75	20 (0–50)	–	–
1976	85–99	40–95	–	–
1977	20–95	0–78	–	–
1978	30–95	40–85	–	–
1979	10–80	20–85	–	–
1980	10–99	0–86	–	–
1981	85–99	40–90	–	–
1982	43–99	10–99	–	–
1983	–	–	70	
1984	–	–	85	
1985	–	–	85	
1986	–	–	80	69–94
1987	–	–		27–99 (fenitrothion) 73–100 (Bt)
1988	–	–		37–88 (fenitrothion) 18–37 (Bt)

[a] These expressions of efficacy were not used after 1982.

Aircraft and Insecticides

Aircraft

The workhorse of the spray fleets deployed in New Brunswick was the Grumman Avenger TBM aircraft. In 1975, Forest Protection Ltd. purchased 22 TBMs to assure that New Brunswick would have an affordable spray fleet that would meet acceptable levels of performance. In addition to TBMs, DC-6 aircraft with boom and nozzle spray systems and inertial navigation guidance systems were used. Small agricultural type spray aircraft used were M-18s, Grumman AgCats, Thrush Commanders, Cessna-188s, Piper Pawnees, and Bell 206 helicopters. Guidance for this variety of aircraft was provided by navigator aircraft (usually a Cessna-172) using the system developed by Forest Protection Ltd. (Dixon and Irving 1985). Standard spray equipment on the TBMs was boom and nozzle while rotary atomizers were used on the smaller aircraft during the 1980s.

Insecticides

From 1974 to 1989, the primary insecticide used was fenitrothion and the secondary was aminocarb (Matacil®). Generally there were two applications, in any one season, usually at 210 g/ha for fenitrothion and 70 g/ha for Matacil®. From 1974 to 1977, trichlorfon (Dylox®) was used on a limited scale to protect the forest close to blueberry fields, trichlorfon being

Table 8. Percentage of different insecticides used in spray operations in New Brunswick by Forest Protection Ltd., 1978 to 1989.

Year	Percent of area treated with each insecticide		
	Aminocarb	Fenitrothion	Bt
1978	48	52	0
1979	97	3	<1
1980	15	84	1
1981	0[a]	100	0[b]
1982	3	96	1
1983	7	90	3
1984	50	45	5
1985	65	24	11
1986	39	39	22
1987	3	80	17
1988	0	53	47
1989	0	82	18

[a] No oil-based sprays permitted, aminocarb formulation diluted with 585 insecticide diluent.

[b] No Bt used because pilot trials in woodlots in 1980 yielded poor to mediocre results.

Table 9. Summary of Bt usage and effectiveness in New Brunswick, 1979 to 1989.[a]

Year	Hectares treated	Comments
1979	1 500	Pilot project to develop Bt for woodlot protection in setback zone. Results mediocre.
1980	10 500	Woodlot protection program in setback zone. Results poor to mediocre.
1981	0	
1982	4 000	Demonstration trials with a variety of Bt formulations. Results acceptable.
1983	10 300	Use of Novabac-3 in woodlot protection program. Results mostly acceptable.
1984	37 300	Operational use of Thuricide® 48LV in woodlot protection. Results acceptable. Budworm populations tend to be lower in woodlots than industrial forest. Some industrial forest treated as well.
1985	81 000	Operational use of Dipel® 132 & Thuricide® 48LV in woodlot protection. Results acceptable. Some industrial forest treated as well.
1986	111 500	Operational use of Dipel® 132 in a woodlot protection program. Results acceptable. Populations of budworm lower in these areas before treatment. Small fixed-wing and helicopter aircraft used.
1987	81 700	Dipel® and Futura® FC used in woodlot program. 6 100 ha treated with reduced dose, enhanced atomization in test program. All Bt effective for foliage protection. Results acceptable.
1988	210 000	Dipel® 132 & Futura® XLV used in industrial program. Applied with TBMs and DC-6. Results poor to mediocre due to cold wet weather.
1989	105 000	Futura® XLV and Dipel® 176 used in industrial program. Results acceptable, as effective as fenitrothion for foliage protection. Single and double applications. Most applied by enhanced atomization.

[a] Source is a variety of Forest Protection Ltd. and Department of Natural Resources reports.

deemed to be relatively nontoxic to pollinating bees. Phosphamidon (Dimecron®) was used in spray operations in 1974–75, usually at a dosage of 140 g/ha per application.

The standard application rate for formulated insecticides was 1.46 L/ha. For fenitrothion, both emulsion and oil formulation were used. Generally, TBM aircraft were used to apply the emulsions and DC-6 aircraft to apply the oil-based sprays, but the TBMs also applied oil sprays. Phosphamidon and trichlorfon were diluted with water, but aminocarb could only be applied as an oil-based spray. Aminocarb was not used operationally in 1981 and 1982 but it was used from 1983 to 1987 after the development of a water-based formulation (Matacil® 180F). However, after 1987 Matacil® 180F was not made available due to low sales potential. The percentages of aminocarb, fenitrothion, and Bt used from 1978 to 1989 are shown in Table 8.

A major change in forest protection operations has been the increased use of Bt. Increasing technological development in the 1980s resulted in some radical changes in potency and formulation technology, all in response to the needs of forest managers and the resulting increased research on Bt use (Wiesner et al. 1984). Political decrees in Ontario, Quebec, and Nova Scotia led to a Bt-only use policy for all budworm control programs. However, its use in New Brunswick (Table 9) has been promoted but not totally accepted, and fenitrothion is still the principal chemical insecticide used (Table 8). By 1987, New Brunswick was the only province using chemical insecticides as part of its forest management program.

Spray Efficacy and Attainment of Objectives

Operational Spraying

The 1989 spruce budworm infestation in New Brunswick, which now appears to be in a receding phase, peaked in 1974–75 (Table 10). The initial 5 822-million-ha infestation for 1975 resulted in moderate to severe defoliation of 3.567 million ha in spite of a massive spray program. The shock to the forest was so great, and the threat of extensive forest loss so overwhelming, that the decision was made to change the very conservative minimum dose/minimum cost approach of protection to a more aggressive one. For the first time since 1957, enough resources were found to tackle the infestation on an unprecedented scale. The results were spectacular, aided by suitable spray weather throughout the operation (Table 10). The trend of the previous 3 years was reversed and only 4% of the treated area had moderate to severe defoliation in 1976, compared with 55% in 1975 and 75% in 1974.

Woodlots were included in spray operations from 1980 to 1987, but not in 1988 or 1989 because of the low spruce budworm populations in these areas (Table 11). The poor results obtained with Bt in the pilot projects led to the decision to use only chemicals in the woodlot protection program in 1981. The restrictions regarding spraying within the 1.6-km buffer zone allowed the use of small aircraft to within 300 m of habitations. The woodlot program from 1980 to 1985 called for treatment of irregular-shaped blocks of forest. From 1986, the woodlot blocks tended to be larger and rectangular, facilitating

Table 10. Summary of infestations, areas treated, and defoliation caused by the spruce budworm, *Choristoneura fumiferana*, in New Brunswick, 1974 to 1989.

Year	Forecast of moderate to severe infestation for (ha × 1000)	Hectares treated (× 1000)	Percent of infestation treated	Moderate–severe defoliation (ha × 1000)	Percent of treated area moderate–severe defoliation
1974	5125	1601	31	3403	75
1975	5822	2781	47	3567	55
1976	5494	3933	67	943	4
1977	3690	1705	46	492	4
1978	4722	1574	30	669	5
1979	4674	1601	34	1337	14
1980	4700	1625	35	673	6
1981	2600	1900	73	1221	14
1982	4000	1693	42	1202	16
1983	5300	1741	33	2028	20
1984	4100	1245	31	730	–
1985	3570	725	20	1070	19
1986	3150	545	17	927	19
1987	1706	583	34	430	16
1988	1505	547	36	500	39[a]
1989	1643	611	37	396	21

[a] Average percent of treated area with moderate to severe defoliation. In this year, the forest in the Bt operational area was 60% defoliated while that in the fenitrothion area was only 19% defoliated.

more accurate spray treatment. From 1987, the ultra-ultra-low-volume spray usage pattern developed in New Brunswick was used for the chemical treatment of all small blocks, or any area treated with slow flying aircraft equipped with Micronair® atomizers.

Variability of Results

The extensive size of spray programs and high proportion of infested area treated have demonstrated the will of the province to cope with the spruce budworm problem. Except for 1981 and 1976 and perhaps 1975 and 1977, the tendency has been to spray conservatively (Table 10). The concept that the proportion of the treated area defoliated provides a reasonable measure of overall efficacy seems to have some merit but shows that results of conservative spraying have been variable. Even in years with good overall foliage protection results varied, but clearly the levels of foliage protection in 1974, 1975, and 1988 were less than those of the other years. In 1974 to 1975, low doses of insecticide (a cost-cutting measure) were used against a vigorous population of insects, and in 1988 the overall lower level of efficacy was due to the poor results with Bt which was used on 47% of the spray program.

Spray efficacy for fenitrothion and aminocarb tend to be similar, while results with phosphamidon and trichlorfon were generally lower (Kettela, unpubl. data). Results from 1983 to 1989 are presented in a series of reports prepared by Hartling (1984) and Carter and Lavigne (1985 to 1989), which support the results shown in Table 10 and show variability in spray efficacy. The apparent overall variability of efficacy is often the result of some areas being treated only once. Steel (unpubl. data) (under contract to Forest Protection Ltd.) examined efficacy of single applications and concluded such treatments produced more highly variable results than double applications, either because intermediate surveys often showed falsely optimistic results or because spruce budworm development was too far advanced.

In the mid-1970s, there was considerable debate over the merits of oil-based versus water-based sprays. Ultimately the debate was settled with the analysis of 5 years (1975 to 1979) of data of oil- and water-based sprays applied from TBM and DC-6 aircraft, which showed no significant difference between either type of formulation (Kettela 1979).

Formulations of Bt were initially used sparingly from 1980 to 1984. The cautious approach reflected the dismal re-

Table 11. Summary of woodlot protection program in New Brunswick, 1979 to 1989.[a]

Year	Hectares treated	Comments
1980	10 500	Trial program with Bt. Targeted spraying.[b] Odd-shaped blocks. Results poor to mediocre.
1981	30 000	Only fenitrothion used. Odd-shaped blocks. Inclusion at owners request. Results—generally good foliage protection.
1982	45 000	Mostly chemical except for 4 000 treated with Bt. Odd-shaped and odd-sized blocks. Results variable, fair to good foliage protection.
1983	82 000	Fenitrothion on 71 700 and Bt on 10 300 ha. Odd-shaped blocks. Results—foliage protection variable. Decision of areas to be treated madc by Dep. of Natural Resources. Exclusion at owners request.
1984	92 078	Bt and chemicals. Results—acceptable foliage retention.
1985	105 357	Bt and chemicals. Results—protection met Dep. of Natural Resources objectives.
1986	48 562	Bt and chemicals. Results—protection met Dep. of Natural Resources objectives of 60% foliage retention on fir and 50% on spruce. Spray blocks rectangularized and made generally larger. Setback zone for Bt reduced to 155 m from habitations.
1987	50 000	Chemicals and Bt. Some chemicals applied with ultra-ultra-low-volume technology. Results met Dep. of Natural Resources objectives.
1988	6 000	Treatment of seed production areas only. Infestation had receded to beyond 1.6 km setback for the most part.
1989	0	No woodlot program as such due to the recession of the infestation.

[a] All treatments were applied with small agricultural-type aircraft.
[b] The term targeted spraying refers to the practice of spraying only selected portions of a stand.

sults attained in the pilot project in 1980 (Table 11). However, from 1984 to 1989, considerable advances have been made in the potency and formulations of Bt, and in the greater understanding of the dose–response relationships (Wiesner et al. 1984, 1985). As a result, Bt usage has increased through to 1988 when it was used in 47% of the industrial spray program; unfortunately, the results were poor and reflected the impact of weather factors on its performance. The poor result in 1988 led to decreased usage of Bt in 1989. However, in 1989, Bt provided a level of protection similar to fenitrothion, a result that may have been due to the combination of two applications to much of the area and to the use of enhanced atomization, which has been shown to provide improved results with one-half the dose.

Conclusion

The responsibility for the success or failure of spruce budworm spraying ultimately rests with individuals or groups not directly involved with spray operations. Regulatory bodies, pressure groups, and politicians have all had major impacts on the conduct of forest protection practices. Concern for the environment has been a priority during spruce budworm spray operations in the province of New Brunswick, and buffer zones around sensitive areas have been used since the mid-1950s to mitigate the environmental impacts of all chemical and biological insecticides used against the spruce budworm. In many instances, this buffering has led to a reduced ability to provide adequate protection to the forest resource. Nevertheless, forest protection practices in New Brunswick have provided significant protection to a substantial proportion of the merchantable wood in the province. However, progress in improving the safety and effectiveness of control procedures has been slow, and it is only because of the systematic exploration of systems that affect spraying that incremental improvements have been made. Recent advances have been limited to the more effective usage of Bt formulations and to the development of an elite wood supply analysis system (Table 12). How to intervene early before outbreaks develop is the challenge of the future.

Table 12. Chronology of key events affecting spray operations against the spruce budworm, *Choristoneura fumiferana*, in New Brunswick, 1974 to 1989.[a]

Year	Events
1974	• First spraying in northwestern New Brunswick (N.B.) since 1958 reflecting the northerly spread of the infestation. Protection to the forest appears minimal.
1975	• Modification of spray planning as dictated by a court injunction. • Setbacks from blueberry fields and alternative insecticides for spraying near these fields (Dylox®). • Worldwide shortage of insecticides. • Forest Protection Ltd. (FPL) acquired TBM fleet (consisting of Grumman Avenger carrier-borne aircraft used in the Second World War) to ensure affordability and quality of spray aircraft. • Defoliation reached maximum level. • Start of coordinated funding by Forest Protection Ltd. for monitoring and research projects. Funding of research and development work with University of New Brunswick (UNB) Dep. of Chemical Engineering. • FPL and Forestry Canada sponsor field trials with two Bt products—positive results.
1976	• Last use of Dimecron® (phosphamidon) in operational spraying. • Formation of EMOFICO (Environmental Monitoring of Forest Insect Control Operations), with Dr. I.W. Varty as Chairman. • Children's Hospital in Halifax suggests link between Reye's syndrome and N.B. Forest Spraying. • Schneider Panel Report on Reye's syndrome. • FPL carries out largest spray program ever in N.B. Foliage protection is excellent. • Law suit—Abram Friesen et al. vs FPL. • Report of Task Force for Evaluation of Budworm Control Alternatives—Dr. G. L. Baskerville, Chairman. FPL Board of Directors unanimously agrees to early intervention philosophy but opts for conservative approach to protection in 1977. • Province elected majority of Board of Directors FPL, including five cabinet ministers.
1977	• CANUSA (Canada—USA Spruce Budworms Program) introduced. • Last year for operational use of Dylox®. • Mile setback from habitations decreed. Thus, protection removed from about 40% of province and virtually all small private woodlots. • Research on the effects of organophosphate insecticides on cholinesterase depression and nerve conduction velocity. A cooperative project between the federal Dep. of Health and Welfare, the N.B. Dep. of Health, the Bio-Engineering Institute of UNB, and FPL. • Litigation—Stanley Lewis et al. vs The Town of St. Stephen and FPL. • Concerned Parents Inc. lays 35 charges against FPL.
1978	• As a condition of the 1978 spray permit, FPL sponsors a Task Force on Long Distance Drift, with I.W. Varty as Chairman. Included were scientists from UNB, Dep. of Chemical Engineering, the N.B. Research and Productivity Council, the federal Dep. of Health and Welfare, and a Steering Committee, consisting of members from federal and provincial regulatory agencies. This eventually led to the Dunphy drift trials and the formation of the N.B. Spray Efficacy Research Group.
1979	• No operational use of fenitrothion permitted while health aspects were being resolved. However, the Canadian Wildlife Service requested a double application on 20 200 ha to assess impact of fenitrothion on birds. Matacil® used for operational treatments. • The N.B. Dep. of Environment took over administration of the N.B. Pesticides Control Act from the Dep. of Agriculture and Rural Development. • Bt operational pilot project in woodlots successfully conducted. • FPL supports environmental monitoring of spraying, including work on pollinators, songbirds, secondary insects, and red squirrels, and finances EMOFICO. • Four films produced by Kingsley Brown for FPL.
1980	• The N.B. Dep. of Environment took over EMOFICO and environment monitoring. • Use of Matacil® (aminocarb) restricted to inventory carry-over from previous year, while it was reevaluated by the federal and provincial Departments of Health. Concerns about Matacil® nonylphenol 585 oil formulation raised by the Dep. of Fisheries and Oceans. Manufacturers of Matacil® develop new formulation. • Fenitrothion used operationally. • Small woodlot spraying pilot project conducted over 10 000 ha in setback zone with Bt.
1981	• Use of fenitrothion permitted in the 1-mile setback zone, but only with small aircraft. • Fenitrothion used operationally.

(Continued)

Table 12. Chronology of key events affecting spray operations against the spruce budworm, *Choristoneura fumiferana*, in New Brunswick, 1974 to 1989.[a] *(Concluded)*

Year	Events
1982	• The Spitzer Task Force on Reye's syndrome reports in April and the use of diluents from fossil fuels was prohibited (i.e. no oil formulations.) • Large-scale field testing of new emulsifiers for fenitrothion application. • Search for alternative formulations of fenitrothion. • Matacil® 180F successfully tested.
1983	• Bt used operationally in small woodlots. • Matacil® and fenitrothion used in operational spraying. • FPL directed to take over surveys and assessment from Forestry Canada. • Special task forces on cancer and birth defects. • Trials with Bt demonstrate that deposit on foliage and efficacy are directly linked.
1984	• Original formulations of fenitrothion approved for continued use. • New formulations of Bt tested successfully. • Spray trials with fenitrothion and Matacil® demonstrate efficacy of ultra-ultra-low-volume (UULV) reduced diluent sprays. This leads to registration of this spray pattern for small plane spraying. • The N.B. Dep. of Natural Resources takes over surveys and assessment from FPL.
1985	• Oil formulations of Bt approved for use. • Large-scale testing of reduced diluent spraying carried out. • Further testing to determine toxic pathways of Bt.
1986	• Budworm infestation appears to be receding. • Increased use of Bt. • Operational measurements of spray deposits show the high variability of spray deposition in treatment blocks. • Reduced diluent demonstration trials with fenitrothion are successful, leading to registration of this use pattern for small aircraft.
1987	• Research and development results begin to be transferred to operational use. TBM aircraft equipped with Micronair atomizers provide enhanced deposition with reduced dosage and volume. • Reduced dosage-enhanced atomization trials with Bt are very successful. • High potency Bt products provide excellent results. • Matacil®, the environmentally preferred insecticide, not available for purchase—remaining inventory used.
1988	• Enhanced atomization use of Bt tested further; results are very positive. • Province calls for substantially increased use of Bt; 47% of plan treated with Bt. Results are poor to mediocre. This causes retrenchment on use of Bt.
1989	• Reduced use of Bt (18% of operation). • FPL funds $300 000 study on Bt optimum use. • Results with Bt in 1989 same as with fenitrothion. • Optimum use trials show enhanced efficacy with double applications and enhanced atomization. This has influenced plans for 1990.

[a] Information provided by Mr. H.J. Irving, Managing Director and President of Forest Protection Ltd., Fredericton, New Brunswick.

Bibliography

Baskerville, G.L. 1976. Report of the task force for evaluation of budworm control alternatives. New Brunswick Dep. Natural Resources, Fredericton, N.B. 210 p.

Carrow, J.R. (Ed.). 1983. B.T. and the spruce budworm—1983 proceeding of a seminar held in Fredericton, New Brunswick, on September 8, 1983. New Brunswick Dep. of Natural Resources.

Carter, N.E.; Lavigne D.R. 1985. Protection spraying against spruce budworm in New Brunswick, 1984. Year-end report. Timber Management Branch, New Brunswick Dep. Natural Resources.

Carter, N.E.; Lavigne D.R. 1986. Protection spraying against the spruce budworm in New Brunswick, 1985. Year-end report. Timber Management Branch, New Brunswick Dep. Forests, Mines and Energy.

Carter, N.E.; Lavigne D.R. 1987. Protection spraying against spruce budworm in New Brunswick, 1986. Year-end report. Timber Management Branch, New Brunswick Dep. Natural Resources and Energy.

Carter, N.E.; Lavigne D.R. 1988. Protection spraying against spruce budworm in New Brunswick, 1986. Year-end report. Timber Management Branch, New Brunswick Dep. Natural Resources and Energy.

Carter, N.E.; Lavigne D.R. 1989. Protection spraying against spruce budworm in New Brunswick, 1986. Year-end report. Timber Management Branch, New Brunswick Dep. Natural Resources and Energy.

Carter, N.E.; Lavigne D.R. 1990. Protection spraying against spruce budworm in New Brunswick, 1986. Year-end report. Timber Management Branch, New Brunswick Dep. Natural Resources and Energy.

Dixon, C; Irving, H.J. 1985. Aerial navigation of spruce budworm spraying in New Brunswick. Pages 406-408 *in* C.J. Sanders; R.W. Stark; E.J. Mullins; J. Murphy (Eds.), Recent Advances in Spruce Budworms Research. Proceedings of the Spruce Budworms Research Symposium, Bangor, Me., 16-20 September, 1984.

Dorais, L; Kettela, E.G. 1982. A review of entomological survey and assessment techniques used in regional spruce budworm, *Choristoneura fumiferana* (Clem.) surveys and in assessment of operational spray programs. A report of the Committee for Standardization of Survey and Assessment Techniques. Eastern Spruce Budworm Council. Ministère de l'Énergie et des Ressources, Québec, Que. 43 pp.

Ernst, W.R.; Pearce, P.A.; Pollock, J.L. (Eds.). 1989. Environmental effects of fenitrothion use in forestry: impacts on insect pollinators, songbirds and aquatic organisms. Conservation and Protection, Environment Canada, Atlantic Region.

Greenbank, D.O.; Schaefer, G.W.; Rainey, R.C., 1980. Spruce budworm (Lepidoptera: Tortricidae) moth flight and dispersal: new understanding from canopy observations, radar, and aircraft. Memoirs of the Entomological Society of Canada—No. 110, 1980.

Hartling, L.K. 1984. Spruce budworm protection program in New Brunswick, 1983. Forest Management Branch, New Brunswick Dep. Natural Resources, N.B.

Irving, H.J. 1983. Operational use of Bt in New Brunswick: Part 1. *In* J.R. Carrow (Ed.), Costs, operational and environmental constraints in Bt and the spruce budworm—1983, Forest Protection Limited.

Irving, H.J. 1987. Buffer zones—an operator's perspective in buffer zones: their application to forest insect control operations. Canadian Forestry Service (Proceedings of the buffer zone workshop sponsored by the Eastern Spruce Budworm Council's Environmental Committee, Québec, Que., 16-17 April 1986).

Kettela, E.G. 1979. Summary statement on the efficacy of spruce budworm defoliation abatement spray operations in New Brunswick in 1979 and the status of infestations in the Maritimes, 1979-1980. Can. For. Serv., Mar. For. Res. Cent. (a report prepared for the Seventh Forest Pest Control Forum, Ottawa, Ont. November 27-28, 1979).

Kettela, E.G. 1980. A synopsis of results of budworm spraying in New Brunswick in 1980 and a forecast of infestations and hazard for 1981. Can. For. Serv., Mar. For. Res. Cent., Fredericton, N.B. Technical Note No. 18.

Kettela, E.G. 1981. Overview of forest protection operations and spruce budworm surveys in New Brunswick, 1981. Can. For. Serv., Mar. For. Res. Cent., Fredericton, N.B. Technical Note No. 44.

Kettela, E.G. 1981. Status of spruce budworm infestations in eastern North America. 62nd Annual Meeting Woodlands Section Canadian Pulp and Paper Association, 1981.

Kettela, E.G. 1982. Results of aerial surveys for current spruce budworm defoliation in New Brunswick and a review of the methodology. Can. For. Serv., Mar. For. Res. Cent., Fredericton, N.B. Technical Note No. 66.

Kettela, E.G. 1983a. Status of spruce budworm infestations in New Brunswick in 1983 and a forecast of conditions for 1984. Can. For. Serv., Mar. For. Res. Cent., Fredericton, N.B. Technical Note No. 100.

Kettela, E.G. 1983b. Operational use of Bt in New Brunswick: Part 2. *In* J.R. Carrow (Ed.), Efficacy in Bt and the spruce budworm—1983. Mar. For. Res. Cent., Fredericton, N.B.

Kettela, E.G. 1987. The impact of forest spray buffer zones on the quality of forest protection in buffer zones: their application to forest insect control operations. Canadian Forestry Service. (Proceedings of the buffer zone workshop sponsored by the Eastern Spruce Budworm Council's Environmental Committee, Québec, Que. 16-17 April 1986.)

Kettela, E.G.; Easton, R.W.; Craig, M.B.; Van Raalte, G.D. 1977. Results of spray operations against spruce budworm in New Brunswick 1977 and a forecast of conditions in the Maritimes for 1978. Can. For. Serv., Mar. For. Res. Cent., Fredericton, N.B. Information Rep. M-X-81.

Kettela, E.G.; Steel, V. 1984. Results of the 1980 woodlot protection project with *Bacillus thuringiensis kurstaki* in New Brunswick. Can. For. Serv., Mar. For. Res. Cent., Fredericton, N.B. Information Rep. M-X-150.

Knox, S. 1981. Aerial evaluation of damage from spruce budworm attack in New Brunswick, New Methodology—1980. Forest Protection Ltd., Fredericton, N.B. Technical Rep. 80/T-3.

MacDonald, D.R. 1963. Studies of aerial spraying against the spruce budworm in New Brunswick: XVIII operational summaries and assessments of immediate results—1960, 1961, 1962. Parts 1 and 2). Forest Entomology and Pathology Laboratory Fredericton, N.B. Information Rep. 95 pp.

MacFarlane, M.D. 1987. The impact of forest spray buffer zones on the quality of forest protection in buffer zones: their application to forest insect control operations. Canadian Forestry Service. (Proceedings of the buffer zone workshop sponsored by the Eastern Spruce Budworm Council's Environmental Committee, Québec Que., 16-17 April 1986.)

Magasi, L.P. 1977. Forest pest conditions in the Maritimes in 1977 with an outlook for 1978. Can. For. Serv., Mar. For. Res. Cent., Fredericton, N.B. Information Rep. M-X-82.

Magasi, L.P. 1982. Forest pest conditions in the Maritimes in 1981. Can. For. Serv., Mar. For. Res. Cent., Fredericton, N.B. Information Rep. M-X-135.

Magasi, L.P. 1983. Forest pest conditions in the Maritimes in 1982. Can. For. Serv., Mar. For. Res. Cent., Fredericton, N.B. Inf. Rep. M-X-141.

Magasi, L.P. 1984. Forest pest conditions in the Maritimes in 1983. Can. For. Serv., Mar. For. Res. Cent., Fredericton, N.B. Information Rep. M-X-149.

Magasi, L.P. 1985. Forest pest conditions in the Maritimes in 1984. Can. For. Serv., Mar. For. Res. Cent., Fredericton, N.B. Inf. Rep. M-X-154.

Magasi, L.P. 1986. Forest pest conditions in the Maritimes in 1985. Can. For. Serv., Mar. For. Res. Cent., Fredericton, N.B. Inf. Rep. M-X-159.

Magasi, L.P. 1987. Forest pest conditions in the Maritimes in 1986. Can. For. Serv., Mar. For. Res. Cent., Fredericton, N.B. Inf. Rep. M-X-161.

Magasi, L.P. 1988. Forest pest conditions in the Maritimes in 1987. Can. For. Serv., Mar. For. Res. Cent., Fredericton, N.B. Inf. Rep. M-X-166.

Magasi, L.P. 1989. Forest pest conditions in the Maritimes in 1988. Can. For. Serv., Mar. For. Res. Cent., Fredericton, N.B. Inf. Rep. M-X-174.

Miller, C.A.; Greenbank, D.O.; Thomas, A.W.; Kettela, E.G.; Volney, W.J.A. 1977. Spruce budworm adult spray tests, 1976. Can. For. Serv., Mar. For. Res. Cent., Fredericton, N.B. Inf. Rep. M-X-75.

Miller, C.A.; Kettela, E.G. 1975. Aerial control operations against the spruce budworm in New Brunswick, 1952-73. Pages 94-112 *in* M.L. Prebble (Ed.), Aerial control of forest insects in Canada. Eviron. Can., Ottawa, Ont.

Miller, C.A.; Steward, J.F.; Elgee, D.E.; Shaw, D.D.; Morgan, M.G.; Kettela, E.G.; Greenbank, G.O.; Gesner, G.N.; Varty, I.W. 1973. Aerial spraying against spruce budworm adults in New Brunswick—a compendium of reports on the 1972 test program. Can. For. Serv., Mar. For. Res. Cent., Fredericton, N.B. Inf. Rep. M-X-38.

Miller, C.A.; Varty, I.W.; Thomas, D.O.; Greenbank, D.O.; Kettela, E.G. 1980. Aerial spraying of spruce budworm moths, New Brunswick, 1972-1977. Can. For. Serv., Mar. For. Res. Cent., Fredericton, N.B. Inf. Rep. M-X-110.

Prebble, M.L. (Ed.). 1975. Aerial control of forest insects in Canada. Environ. Can., Ottawa, Ont.

Thomas, A.W., Miller, C.A.; Greenbank, D.O. 1979. Spruce budworm adult spray tests, New Brunswick, 1977. Can. For. Serv., Mar. For. Res. Cent., Fredericton, N.B. Inf. Rep. M-X-99.

Varty, I.W. 1978. Environmental surveillance of insecticide spray operations in New Brunswick's budworm-infested forests, 1977. Can. For. Serv., Mar. For. Res. Cent., Fredericton, N.B. Inf. Rep. M-X-87.

Webb, F.E. 1958. Biological assessment of aerial forest spraying against the spruce budworm in New Brunswick. II. A review of the period 1952–1956. Proc. Tenth International Congress of Entomology (1956). 4: 303-316.

Webb, F.E., Cameron, D.G.; MacDonald, D.R. 1956. Studies of aerial spraying against the spruce budworm in New Brunswick. V. Techniques for large-scale egg and defoliation ground surveys 1953-55. Interim Rep. 1955-8, Forest Biology Laboratory, Fredericton, N.B.

Wiesner C.J.; Kettela, E.G.; Fast, P.G. 1984. *Bacillus thuringiensis* foliar deposition and efficacy against the eastern spruce budworm. RPC. Rep. No. C/83/221.

Wiesner C.J.; Kettela, E.G.; Fast, P.G. 1985. Assessment of the influence of concentration and foliar deposition on the efficacy of *Bacillus thuringiensis*. 1984 field trial. New Brunswick Spray Efficacy Research Group. Rep. No: C/85/047.

Chapter 65

Insect Control in Quebec, 1974–1987

LOUIS DORAIS, MICHEL AUGER, MICHEL PELLETIER,
MICHEL CHABOT, CLÉMENT BORDELEAU, AND JEAN CABANA

Introduction

At the beginning of the 1970s, Quebec suffered a serious infestation of the spruce budworm, *Choristoneura fumiferana*. In 1973, the Quebec Ministry of Forests, at that time called the Ministry of Lands and Forests, launched a large-scale aerial spraying program to minimize the expected damage. A brief historical survey of the infestation and an explanation of the treatments conducted are presented in this chapter.

Areas Infested

The spruce budworm, *Choristoneura fumiferana*, is considered to be the most serious defoliator of softwood forest species in northeastern North America. It mainly attacks balsam fir, *Abies balsamea*, and white spruce, *Picea glauca*.

The insect causes damage during its larval period by feeding on the annual shoots of fir and spruce. A severe attack repeated over several years can cause a tree to die. Mortality in fir and white spruce was estimated, respectively, at 65% and 20% of the stems during the last infestation in western Quebec (McLintock 1955).

The spruce budworm is native to North America. The current infestation is the tenth of its kind reported in Quebec since the beginning of the 18th century, and the third since the beginning of the present century (Blais 1964). The budworm returns cyclically about every 30 years and moves across the province from west to east in greater or lesser numbers (Brown 1970). The mechanisms determining its appearance or disappearance are still unknown, and forest planners are still ill-prepared to fight the insect because they are unable to predict when infestations will begin or how long they will last.

The upsurge in budworm populations during the current infestation was first observed in 1966 in spruce plantations located near Grand-Mère in the center of Quebec. An infestation covering 1 000 ha was reported in 1967 at Low in western Quebec, and over the following years, it spread rapidly from west to east. The infested areas totaled more than 2.4 million ha by the end of 1970 (Fig. 1).

In eastern Quebec, the budworm first appeared at two different areas: Témiscouata in 1968 and the southern part of the Gaspé Peninsula in 1971 (Baie des Chaleurs). By the end of 1975, the infestation covered about 35 million ha (Table 1). By that time, the entire balsam fir zone in Quebec was affected. Starting in 1976, the infestation began to recede in western Quebec, then 2 years later in the east (1978).

The infestation reappeared in central and eastern Quebec at the beginning of the 1980s, after a remission of 3 years, following the same pattern observed throughout North America during the previous infestation (Fig. 2). This upsurge continued until 1983, the year that marked the second peak in the infestation, then it gradually declined until 1988.

The infestation began to recede in 1986 and was confined to the Gaspé region. Mortality assessments were conducted periodically to identify sectors that should be withdrawn from spraying programs and reserved for salvage. There was no significant extension of the high-mortality sectors after 1986, and the numerous salvage programs and the evolution of the residual stands rendered it problematical to continue province-wide stem mortality assessment, so it was abandoned.

The impact of the budworm infestation that raged from 1974 to 1987 is difficult to assess, but the following facts are clear:

- 13 million ha suffered various degrees of mortality;
- mortality rates, applied to forest volumes, indicated by 1985 that total losses of 235 million m^3 could be expected;
- large-scale salvage programs have been set up and insofar as the market allows, it should be possible to absorb these losses;
- the mortality rate was low (less than 25%) in many stands, and in these cases it is especially difficult to draw conclusions. The denser stands were not degraded by such a low mortality rate; in fact, it produced a thinning effect that was

Table 1. Areas (ha) infested by the spruce budworm, *Choristoneura fumiferana*, in Quebec from 1967 to 1987.

Year	Annual defoliation (light–moderate–severe)	Mortality	Total
1967	1 044	-	1 044
1968	151 401	-	151 401
1969	803 504	-	803 504
1970	2 422 170	-	2 422 170
1971	5 276 329	-	5 276 329
1972	10 301 480	79 650	10 381 130
1973	10 069 711	248 285	10 317 996
1974	29 597 282	1 145 565	30 742 847
1975	32 300 722	3 020 984	35 321 706
1976	29 335 216	3 814 319	33 149 535
1977	27 116 467	5 672 386	32 788 853
1978	16 257 796	6 296 423	22 554 219
1979	5 311 110	8 290 141	13 601 251
1980	6 222 282	9 254 931	15 477 213
1981	6 935 278	10 164 410	17 099 688
1982	9 848 506	11 190 109	Noncomplementary
1983	13 211 876	11 964 783	areas (recovery)
1984	11 043 238	12 633 425	"
1985	9 259 816	12 863 014	"
1986	2 831 947	not available	"
1987	1 041 654	-	"

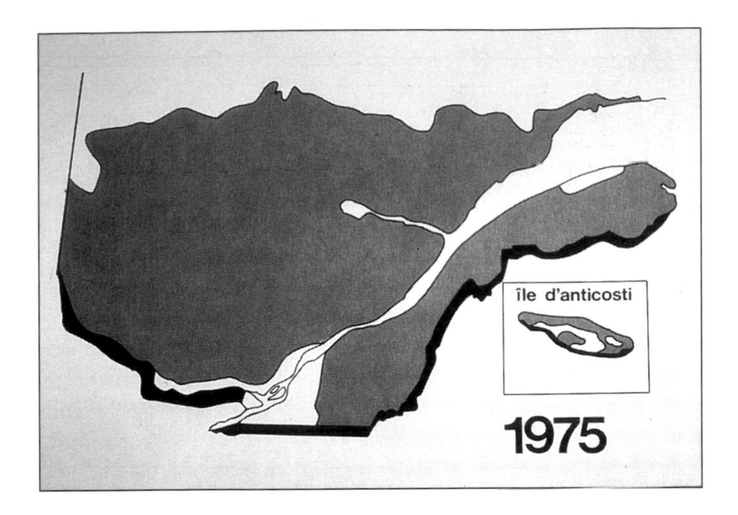

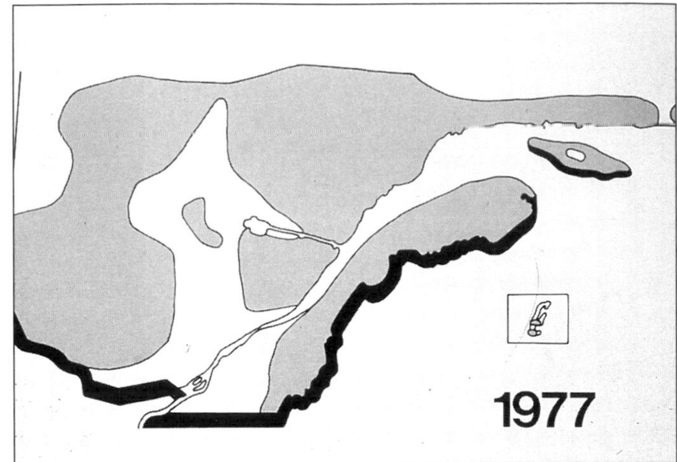

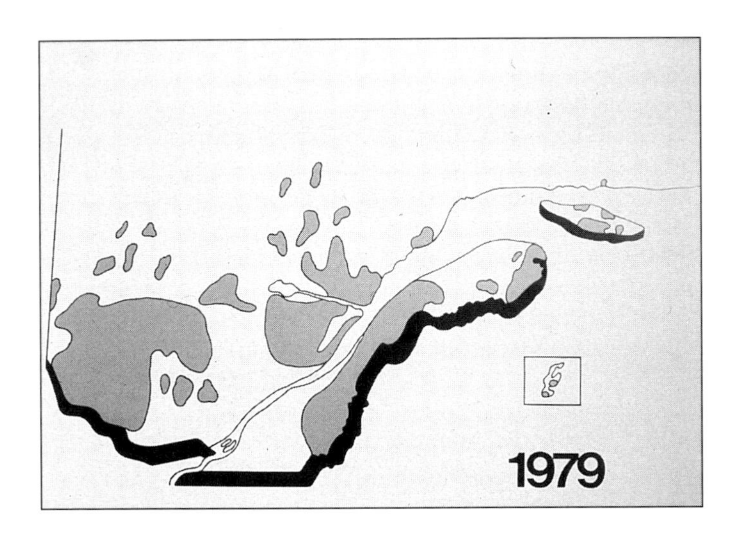

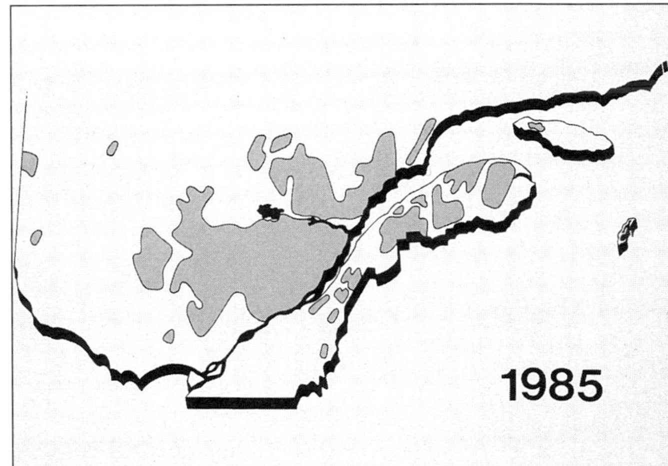

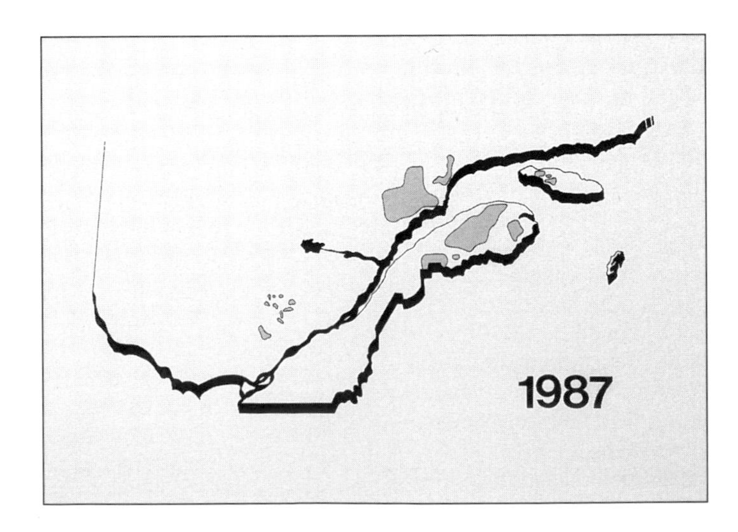

Figure 1. Distribution of the spruce budworm, *Choristoneura fumiferana*, infestation in Quebec from 1975 to 1987.

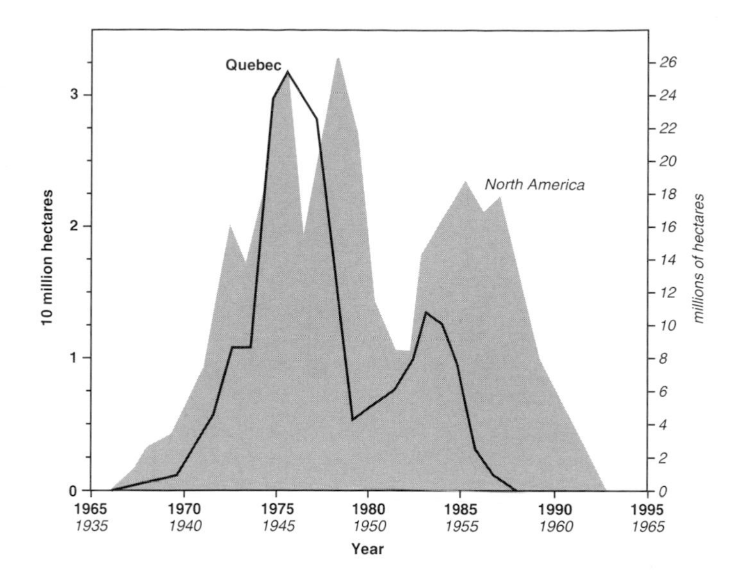

Figure 2. Behavior of the spruce budworm, *Choristoneura fumiferana*, infestation from 1935 to 1963 in North America and from 1967 to 1987 in Quebec.

beneficial to the less vulnerable residual stems. However, the same low mortality in thinner stands tended to reduce the merchantable volume of timber below the threshold of profitability (50 m^3/ha).

Knowledge of Spraying Operations

Our knowledge of how to combat harmful forest insects on a large scale was rather limited in 1973. Extensive operations against the spruce budworm were conducted from 1954 to 1962 in the Gaspé region with DDT sprayed from single-engine planes (Stearman-type). These operations were organized by the Quebec Forest Industries Association and carried out by Forest Protection Limited of New Brunswick.

A single application at a rate of 0.5 pounds of DDT diluted in 0.5 gallons per acre and directed against the fifth-larval instar resulted in larva mortality of about 99%. The treatment reduced the population to such an extent that the sectors sprayed did not require a second treatment the following year. The treatment was usually performed after the third year of a severe attack and was intended to prevent tree mortality. However, when DDT was banned in 1965, a substitute had to be found and a new approach adopted to combat the insect.

The Laurentian Forest Research Centre and the Chemical Control Research Institute of the Canadian Forestry Service, along with the Faculty of Forestry and Surveys of Laval University, acted as technical advisors to the Ministry of Lands and Forests planning these initial programs.

1974–1976

From 1974 to 1976, the infestation reported in western Quebec and the two smaller pockets of infestation detected in the east (Témiscouata and Baie des Chaleurs) gradually crept closer together as the budworm spread throughout Quebec. By the end of 1975, the insect was present in about 35 million ha.

The Operations

The Ministry of Lands and Forests launched a 10-year aerial spraying program to combat the spruce budworm. Mixing and loading centers were set up at existing airports, and landing strips were extended to facilitate spraying in selected territories (Fig. 3). From 2.5 to 3.6 million ha were treated annually from 1974 to 1976, yet these areas never amounted to more than one-quarter of the total area infested (Tables 1 and 2).

Various types of four-engine and twin-engine aircraft (Super Constellation L-1049, Constellation L-749, DC-6, DC-4, DC-3, and CL-215) were used in the operations (Fig. 4) to cover such a vast territory during the brief period suitable for treatment. The number of spray aircraft varied from year to year, depending on the characteristics of each aircraft (Table 3) and the size of the area to be sprayed. On average, a four-engine airplane could treat 250 000 ha/season (double application).

Because spray aircraft operated at a higher altitude (100 m), oil was used instead of water as the insecticide spray medium to reduce evaporation. The higher air speed (300 km/h) also tended to break up the mixture into fine droplets and produced more even coverage by the solution allowing a reduction in the quantity of mixture to be sprayed per hectare. Because of the higher spraying altitude and finer misting, the spray cloud was also widened. The quantity of solution found at ground level resulted from the three to four successive swaths by the aircraft at intervals of 1 000 m.

After various tests, a self-contained inertial navigation system (Litton LTN-51) was adopted. Twelve units were purchased and installed in the spray aircraft under an annual contract. Spraying was conducted with booms and fan nozzles.

Precise standards for spraying were developed in 1974. Treatments had to be conducted with winds at less than 15 km/h and at an altitude of 50 to 200 m depending on wind speed. Each of the different spray regimes used was verified at ground level using test papers (Kromekote®) The drops produced had a diameter of approximately 70 µm, and 10 drops per square centimetre was considered an adequate deposition.

Figure 3. Insecticide mixing and loading plan implemented at the Bonaventure operation base.

Table 2. Treatments conducted in Quebec from 1974 to 1976 against the spruce budworm, *Choristoneura fumiferana*.

Year	Area (ha)	Insecticides and doses	Volume (L/ha)	Synchronization	Airplane (no.)	Cost ($/ha)
1974	2 567 859	Fenitrothion 140-210 g AI/ha[a] Aminocarb 52 g AI/ha Mexacarbate 52 g AI/ha	0.9 to 1.12	50% L2 + L3[d]	(3) L-1049 (1) L-749 (5) DC-6 (2) CL-215	2.87
	(including 3 237)	Bt[b] 20 BIU/ha[c]	4.6	L3		
1975	2 927 518	Fenitrothion 140 g AI/ha Aminocarb 52 g AI/ha Mexacarbate 52 g AI/ha	0.9 to 1.12	50% L2 + 25% L4	(3) L-1049 (2) L-749 (5) DC-6 (2) DC-4	4.03
	(including 95 932)	Bt 20 BIU/ha	2.3 to 4.6	L3		
1976	3 655 841	Fenitrothion 140 g AI/ha Aminocarb 52 g AI/ha Dimethoate 140 g AI/ha	0.9	10% L2 + 25% L4 1% L2 + 10% L2 + 25% L4 10% L2 + 25% L4 + L5	(3) L-749 (7) DC-6	3.18

[a]Grams of active ingredient per hectare.
[b]*Bacillus thuringiensis*.
[c]Billion international units per hectare.
[d]Larval instar (L2–L5).

Table 3. Characteristics of spray aircraft used operationally in Quebec from 1974 to 1987 to combat the spruce budworm, *Choristoneura fumiferana*.

Type		Useful load (L)	Speed (km/h)	Width of spray (m)	Radius of operational action (km)
Super Constellation L-1049		16 600	336	915	100
Constellation L-749	(CH)[a]	13 600	320	915	100
	(Bt)[b]	12 100	320	305	100
DC-6	(CH)	13 600	368	915	100
	(Bt)	12 100	368	305	100
DC-4	(CH)	11 000	280	915	100
	(Bt)	8 500	280	305	100
CL-215		5 200	240	305	100
Thrush	(CH)	1 500	232	122	40
	(Bt)	1 500	232	122	40
AgCat	(CH)	750	176	122	20
	(Bt)	750	176	122	20
Pawnee	(Bt)	450	176	122	20

[a]Chemical preparations.
[b]Biological preparations.

Figure 4. Aerial insecticide spraying using a four-engine airplane.

Products Used

Three chemical insecticides were used in most spray programs on an operational basis from 1974 to 1976: one was an organophosphate, fenitrothion (Sumithion® and Folithion®), and the other two were carbamates, aminocarb (Matacil®) and mexacarbate (Zectran®). Doses and volumes varied from year to year (Table 2) but the standard treatment consisted of two applications of one of these products. Where insect populations were especially heavy (more than 25 larvae per branch), three applications were sometimes made.

The insecticide was sprayed at a volume of 0.9 to 1.12 L/ha. Because only a slight difference in annual foliage protection was detected, the total volume sprayed was decreased to 0.9 L/ha, so that a larger surface could be sprayed and treatment costs reduced accordingly.

Synchronization of Treatments

To reduce the insect populations before they enter the buds, the first application was as the larvae were emerging (L2—larval instars are sometimes referred to as L2, L3, etc.), and the second was after the buds had opened (L4). This approach was designed to take the best advantage of a migration period for second-instar larvae (beginning of the fourth instar) when the budworm stops feeding on old needles and moves into the bud. By synchronizing the applications in this way, a part of the population was destroyed before it could attack the annual foliage. If three applications were planned, the first two were usually applied while the larvae were emerging and the final one while the buds were opening (Table 2).

Environmental Monitoring

Environmental monitoring of aerial spraying was also begun in 1974 and was conducted mainly by the Chemical Control Research Institute and the Ministry of Recreation, Hunting and Fishing. Surveys were made of aquatic insects, birds, and small mammals before and after treatments for each spray regime to detect any impact on nontargeted organisms.

Spraying Strategy

At the end of 1974, the Ministry of Lands and Forests, in cooperation with Laval University, developed a technique that used the balsam fir's capacity to resist repeated insect attacks to measure the risk of fir mortality in forests attacked by the budworm (Hardy and Dorais 1976). This technique provided a basis for a general policy to combat the budworm with chemical insecticides and guided entomologists in selecting areas for treatment. Essentially, the approach consisted of starting treatments after 1 year of severe attack (moderate risk) and continuing treatments for the duration of the infestation, so long as the forests had not yet reached a level of extreme risk. Stands exhibiting such a level of risk were then withdrawn from the spraying programs because the trees were considered to be too heavily damaged to benefit from treatment.

Costs of Treatments

The cost of treatments (airplanes, products, and related assessments) amounted to nearly $3/ha. However, beginning in 1975, the cost rose to more than $4/ha due to the large number of treatments applied on a semi-operational basis using a biological insecticide (Table 2). Two-thirds of these costs were paid by the Ministry of Lands and Forests and the rest by the forest industry.

Effectiveness of Treatments

The treatments performed from 1974 to 1976 were particularly difficult due to the intensity (population of more than 25 larvae per branch) and the extent of the infestation. Protection of annual foliage was judged adequate (defoliation varying from light to moderate) on 78, 64, and 68%, respectively, of the territory treated during these 3 years (Table 4).

This period also led to an adjustment in insecticide spraying volumes, synchronization of treatments, and a treatment strategy based on new products.

1977–1980

After the decline in populations reported in western Quebec in 1976, programs to combat the spruce budworm were concentrated in the Lower St. Lawrence–Gaspé region from 1977 to 1980. This region was selected because of the importance of its forest industry and because timber shortages were expected in the area unless steps were taken to protect supplies.

The Operations

Aerial spraying operations were less extensive, varying from 1.4 to 0.2 million hectares, but were relatively similar to those performed during the previous period. They were carried out mostly by four-engine planes (L-749, DC-4, and DC-6) working out of existing airports. Litton LTN-51

Table 4. Annual effectiveness of treatments conducted in Quebec from 1974 to 1976 against the spruce budworm, *Choristoneura fumiferana*.

Year	Populations before treatment (larvae/branch)	Larva mortality (%)	Annual defoliation (%)	Proportion of the area adequately protected (%)
1974				
Treated areas	25.7	78	59	78
Control areas	22.4	69	68	-
1975				
Treated areas	33.8	87	73	64
Control areas	37.4	86	87	-
1976				
Treated areas	37.3	89	44	68
Control areas	35.3	85	75	-

systems were still used for navigating the aircraft, but starting in 1978, each spray aircraft was accompanied by a surveillance plane. The planes continued to use a system of booms and fan nozzles. Due to the rugged terrain in the Gaspé region, the spraying volume was increased from 0.9 L/ha to 1.4 L/ha (Table 5).

Products Used

Fenitrothion and aminocarb continued to be the main chemical products used. They were delivered in bulk or in barrels, as needed, and mixed with oil in a closed circuit (Fig. 5) at one location (Rivière-du-Loup) before being transported to the spraying base in their final form. To ensure an even mixture, the preparations were agitated a second time before being loaded onto the planes. All the products used during this period kept very well from one season to the next and posed no residue problem. The oil used to rinse the pumps, meters, and tubes was reused in preparations the following year.

Table 5. Treatments conducted in Quebec from 1977 to 1980 against the spruce budworm, *Choristoneura fumiferana*.

Year	Area (ha)	Insecticides and doses	Volume (L/ha)	Synchronization	Airplane type (no.)	Cost ($/ha)
1977	1 395 666	Phosphamidon 140 g AI/ha[a] Fenitrothion 210 to 280 g AI/ha Aminocarb 70 g AI/ha	0.9	20% L2 + 50% L2 + 25% L4[c]	DC-6 (7)	4.71
1978	1 255 329	Fenitrothion 210 g AI/ha Aminocarb 52 to 70 g AI/ha	1.12 to 1.4	L4 50% L2 + L4 L4 and L5 50% L2 + L4 + L5	DC-6 (2) DC-3 (3) L-749 (3)	4.88
	(including 7 326)	Bt 20 to 40 BIU/ha[b]	4.6 to 9.36	L4	DC-6 (1) AgCat (3)	17.59
1979	582 965	Fenitrothion 210 g AI/ha Aminocarb 52 g AI/ha	1.12	L4 L4 + L5 L3, L4 + L5	L-749 (2)	5.96
	(including 17 030)	Bt 20 to 30 BIU/ha	4.6 to 8	L4	L-749 (1)	19.54
1980	188 511	Fenitrothion 210 g AI/ha Aminocarb 52 g AI/ha	1.4	L4 + L5	L-749 (2) AgCat (2)	15.07
	(including 22 180)	Bt 30 BIU/ha	7.0	L4		37.80

[a] Grams of active ingredient per hectare.
[b] Billion international units per hectare.
[c] Larval instar (L2–L5).

Figure 5. Closed-circuit siphoning of insecticide barrels.

A biological insecticide, Bt, began to be used in 1978 in certain parts of the operational program. The product formed a suspension that had to be diluted with water before application and agitated to prevent sedimentation until it was loaded on board the planes. The spraying volumes were much higher (4 to 8 L/ha), and the product had to be protected from freezing during the winter.

Synchronization of Treatments

In 1977, the infestation reached its maximum intensity in eastern Quebec. The insect populations encountered there were much heavier, and treatment could not be delayed. Two to three applications were systematically applied depending on population levels as projected from insect egg surveys and tree mortality risk. Early applications (20% L2 and 50% L2) proved effective in reducing larva populations before they entered the buds. Although insufficient, these applications were indispensable in providing a certain degree of protection for annual foliage against excessive populations (Table 5).

When the populations declined to less critical levels in 1978, standard treatments were resumed, with one or two applications after the buds opened.

Eastern Spruce Budworm Council

Some Canadian provinces and northeastern states joined forces in 1977 to combat the spruce budworm problem and created the Eastern Spruce Budworm Council. The council provided a forum where forest managers could discuss both the technical and administrative aspects of their operations. Ontario, Quebec, New Brunswick, Newfoundland, Nova Scotia, and the state of Maine were active members. Forestry Canada and its American counterpart (USDA-FS) were also represented. One of the council's main accomplishments was to standardize, from province to province, the entomological techniques used during operations (Dorais and Kettela 1982).

Use of Bt

Beginning in 1978, larger and larger areas were being treated with Bt. This product, which had been under development since 1970, was by then better known and more effective, provided that its use was subjected to certain precise restrictions (Dorais et al. 1980):

a) Population level and synchronization of treatments—Because Bt is an ingestion insecticide, it has to be applied after the bud opens to allow contamination of annual foliage. In a heavy infestation, a large proportion of the annual foliage is already destroyed even before Bt can be applied. For this reason, Bt is reserved for sectors where the expected population levels do not exceed 25 larvae per branch.
b) Insecticide deposition and meteorological conditions associated with spraying—Deposition of the biological insecticide is assessed on nutritive agar in Petri dishes by counting the number of drops forming colonies. From 20 to 25 colonies per cm^2 are required for adequate protection. Deposition is conditioned by meteorological conditions at the time of spraying, the ideal conditions being a relative humidity of more the 50%, winds less than 12 km/h, and the absence of thermal turbulence.
c) Dose and volume—The volume of the mixture sprayed varies with the type of aircraft used. A single-engine plane flying at 30 m above the trees can spray the product at a rate of 4.7 L/ha, whereas a four-engine plane flying at 100 m above the trees can spray 7 L/ha. The dose sprayed varies from 20 BIU/ha for single-engine planes to 30 BIU/ha for four-engine planes, depending on the quantity of preparation used.
d) Spraying system and droplet spectrum—The droplet spectrum is still assessed using Kromekote® test paper. The ideal droplet varies from 85 to 150 μm in diameter. The spraying system consequently has to be adjusted to the aircraft's speed. Single-engine planes are equipped with 40 fan nozzles, while four-engine planes have 168 open nozzles.

Effectiveness of Treatments

The operations conducted from 1977 to 1980 made it possible to protect annually more than 90% of the territory treated (annual defoliation varying from light to moderate) (Table 6).

Cost of Treatments

The cost of chemical treatments varied from \$5 to \$6/ha during this period. This increase was considered normal and was attributed to oil price increases. The cost of biological treatments, by contrast, fluctuated between \$18 and \$20/ha, three times more expensive than chemical treatments (Table 5). For operational and financial reasons, Bt spraying was therefore limited to sectors that were especially environmentally sensitive.

To summarize, this was a period in which operation and assessment techniques were refined and adjusted to declining populations. Four-engine aircraft and chemical insecticides were used much more rigorously, and a much safer biological insecticide was introduced.

Table 6. Annual effectiveness of treatments conducted in Quebec from 1977 to 1980 against the spruce budworm, *Choristoneura fumiferana*.

Year	Populations before treatment (larvae/branch)	Larva mortality (%)	Annual defoliation (%)	Proportion of the area adequately protected (%)
1977				
Treated areas	30.7	93	52	89
Control areas	29.1	81	78	-
1978 LSL[a]				
Treated areas	15.9	94	33	
Control areas	15.2	70	67	
1978 GASP[b]				98
Treated areas	5.2	85	26	
Control areas	5.0	62	43	
1979				
Treated areas	13.3	89	45	96
Control areas	14.9	69	76	-
1980				
Treated areas	10.4	80	32	99
Control areas	8.9	45	57	-

[a]Lower St. Lawrence Region.
[b]Gaspé region.

1981–1983

The spruce budworm infestation, which had been declining since 1976, resumed in eastern Quebec in 1980. The spraying program was continued in the Lower St. Lawrence–Gaspé region to maintain a viable forest and was extended to sectors of the North Shore, the Saguenay, and the Laurentian Reserve to slow forest deterioration and allow salvage operations in stands already damaged (mortality rate: 25 to 50%).

The Operations

The areas treated from 1981 to 1983 vary from 0.7 to 1.3 million ha (Table 7). The majority of the treatments continued to be applied by four-engine planes, while single-engine planes (AgCat, Pawnee) were introduced to treat smaller areas with specialized forest purposes (seed orchards, parks, etc.). Spraying techniques (booms and nozzles) and navigation (Litton LTN-51) remained the same, except for single-engine planes, which were guided by a spotter plane.

In addition, after the Environment Quality Act and related regulations were passed at the beginning of the 1980s, every agency planning to conduct aerial pesticide spraying on more than 600 ha was required to submit an impact study to the Quebec Ministry of the Environment and to defend it in public hearings. The first such study was filed in 1982. Without actually changing the nature of the operations, this new process for conducting spraying programs gave rise to an

Table 7. Treatments conducted in Quebec from 1981 to 1983 against the spruce budworm, *Choristoneura fumiferana*.

Year	Area (ha)	Insecticides and doses	Volume (L/ha)	Synchronization	Airplane type (no.)	Cost ($/ha)
1981	705 164	Fenitrothion 210 g AI/ha[a] Aminocarb 52 g AI/ha	1.4	L4 + L5[c]	L-749 (2) DC-4 (6) AgCat (2)	9.43
	(including 15 001)	Bt 20 to 30 BIU/ha[b]	7.0	L4		31.13
1982	1 298 495	Fenitrothion 210 g AI/ha	1.4	L4 + L5	L-749 (2)	8.15
		Aminocarb 2 × 52 g AI/ha 1 × 87 g AI/ha	2.34	L4	DC-4 (6) AgCat (3)	
	(including 31 877)	Bt 20 to 40 BIU/ha	3.5 to 8.8	L4	Pawnee (3)	20.52
1983	1 253 605	Fenitrothion 210 g AI/ha	1.4	50% L2 + L4	L-749 (2) DC-6 (3)	11.28
		Aminocarb 2 × 52 g AI/ha 1 × 87 g AI/ha	2.34	L4 + L5 L4	DC-4 (3)	
	(including 45 627)	Bt 20 to 30 BIU/ha	2.5 to 4.6	L4	AgCat (3)	27.15

[a]Grams of active ingredient per hectare.
[b]Billion international units per hectare.
[c]Larval instar.

Figure 6. Balsam fir bud flare classes used in synchronizing treatments.

annual public information campaign; a follow-up program for workers assigned to mixing and spraying the products; preparation of an emergency plan; a rigorous stock management program, including residue follow-up; and a product quality control program (researching the presence of pathogenic contaminants and testing for insecticidal potential).

Products Used

Chemical insecticides continued to be the most widely used products during this period, while the biological insecticide Bt was sprayed by four-engine planes on ever larger areas (15 000 to 45 000 ha). At the same time, the great quantities of spray formulations required to obtain the desired results prompted manufacturers to develop ever more concentrated products to reduce logistic constraints and treatment costs.

Synchronization of Treatments

Treatments conducted while the larvae were emerging (50% L2) were applied only in critical cases where populations were excessive and the trees were heavily damaged (Table 7). The use of Bt, which acts by ingestion, had to be synchronized with bud development, which had to be sufficiently flared to receive the insecticide (Dorais and Kettela 1982) (Fig. 6).

Effectiveness of Treatments

During this period, spraying was recommended as a means of prolonging the salvage period, but it is difficult to quantify the actual effect of such treatments. Many stands did reach the point of no-return. It was projected that by the end of this period, the budworm infestation would destroy about 235 million m^3. Integrating heavily damaged sectors into the program had an impact on the annual effectiveness of sprayings: 89, 57, and 78%, respectively, of the territory treated received adequate protection during these 3 years (Table 8).

Cost of Treatments

The cost of chemical treatments during this period was $10/ha and the cost of biological treatments was between $20 and $30/ha (Table 7).

1984–1987

A remission in the infestation in eastern Quebec and the development of more concentrated Bt solutions that could be used undiluted allowed spraying programs to be gradually converted to exclusive use of biological insecticides.

Under the Environment Quality Act a second impact study was filed in 1984 by the Ministry of Energy and Resources for the period 1985 to 1989. This study proposed that standard chemical insecticides gradually be replaced by the biological insecticide Bt. This decision was made at a time when the infestation had been receding since 1983 and when forest managers were learning to use Bt more effectively. It had a decisive impact on the development of Bt, prompting several producers to pursue research to improve their product or develop new ones. In addition, Ontario, New Brunswick, Newfoundland, and Maine followed Quebec's lead in calling for extended use of Bt in their respective programs.

The Operations

About 700 000 ha were treated in 1984 and 1985. The infestation declined in 1986, and the areas treated were decreased to between 50 000 ha and 200 000 ha (Table 9). As

Table 8. Annual effectiveness of treatments conducted in Quebec from 1981 to 1983 against the spruce budworm, *Choristoneura fumiferana*.

Year	Populations before treatment (larvae/branch)	Larva mortality (%)	Annual defoliation (%)			Proportion of the area adequately protected (%)
			LSL[a]	GASP[a]	NS[a]	
1981						
Treated areas	24.6	92	28	78	65	89
Control areas	21.8	73	52	90	90	-
1982						
Four-engine planes						
Treated areas	25.2	88	57			
Control areas	21.1	61	94			
Single-engine planes						57
Treated areas	24.3	95	35			-
Control areas	22.2	66	78			
Bt						
Treated areas	16.5	82	54			
Control areas	13.0	51	86			
1983						
Treated areas	15.6	84	38			78
Control areas	15.5	59	71			-

[a] Lower St. Lawrence Region; Gaspé region; North Shore.

the treatment areas were reduced, they were divided into smaller and smaller blocks, and single-engine planes (Thrush, AgCat) had to be used for greater efficiency.

The spraying technique was adapted to the new, more concentated Bt preparations, and the spraying volumes were reduced to 2.37 L/ha and no longer required dilution immediately before the operation began. Under these conditions, a four-engine aircraft could treat from 20 000 to 65 000 ha/year. A single-engine aircraft, equipped with Micronair® AU-5000 atomizers to ensure finer misting of spray products, could treat from 5 000 to 10 000 ha/year.

Insecticide deposition was now assessed directly on fir needles or on paper foliage simulators at the time of calibration (Fig. 7). From 0.7 to 1 drop per needle, with each drop measuring between 30 and 50 μm in diameter, deposition was sufficient to ensure adequate protection of annual foliage (Lambert 1987). To guarantee the necessary minimum deposition for adequate protection, the spraying dose per hectare was increased from 20 to 30 BIU/ha.

The equipment used in airports had to be meticulously cleaned after each operation because major problems with corrosion and scale deposits had been discovered on aluminum parts in the meters, pumps, and planes.

The intensive quality control that was exercised starting in 1985 (Fig. 8) revealed that Bt preparations also contain microcontaminants of species other than *Bacillus thuringiensis*. In 1987, certain species such as enterococci were detected in Dipel® 132 in such massive numbers that concerns were voiced for human health (Cabana 1989). In response to this problem, the Pesticides Directorate of Agriculture Canada issued temporary standards or regulations to limit and even ban the presence of certain microcontaminants.

In addition, insecticides were tested before take-off to detect any unacceptable variations in insecticidal potential in certain products. This problem was remedied by mixing low-potential and high-potential lots (Cabana 1986). These discoveries highlighted the need for quality control of preparations supplied by producers before spraying begins.

Table 9. Treatments conducted in Quebec from 1984 to 1987 against the spruce budworm, *Choristoneura fumiferana*.

Year	Area (ha)	Insecticides and doses	Volume (L/ha)	Synchronization	Airplane type (no.)	Cost ($/ha)
1984	708 482	Fenitrothion 210 g AI/ha[a] Aminocarb 52 g AI/ha	1.4	50% L2 + L4[c] L4 + L5	L-749 (2) DC-6 (3) DC-4 (5) AgCat (2)	19.24
	(including 323 948)	Bt Dipel 132 Thuricide 32LV-48LV Futura FC 20 to 30 BIU/ha[b]	2.34 to 4.68	L4		27.70
1985	667 418	Fenitrothion 210 g AI/ha	1.4	L4 + L5	L-749 (1) DC-6 (3)	22.74
	(including 482 963)	Bt Thuricide 48LV Novabac 3 Futura FC 20 to 30 BIU/ha	2.37 to 3.55	L4	DC-4 (5) Thrush (7) Pawnee (3)	32.27
1986	51 155	Fenitrothion 210 g AI/ha	1.4	L4 + L5	Thrush (3) AgCat (1)	32.25
	(including 18 117)	Bt Thuricide 48LV 30 BIU/ha	2.37	L4		
1987	197 992	Bt Dipel 132 30 BIU/ha	2.37	L4	DC-4 (3) Thrush (5)	26.80
	(including 927)	Dipel 176 30 BIU/ha	1.77	L4		

[a] Grams of active ingredient per hectare.
[b] Billion international units per hectare.
[c] Larval instar (L2–L5).

Figure 7. Foliage simulator used to evaluate insecticide deposition for calibrating spraying airplanes.

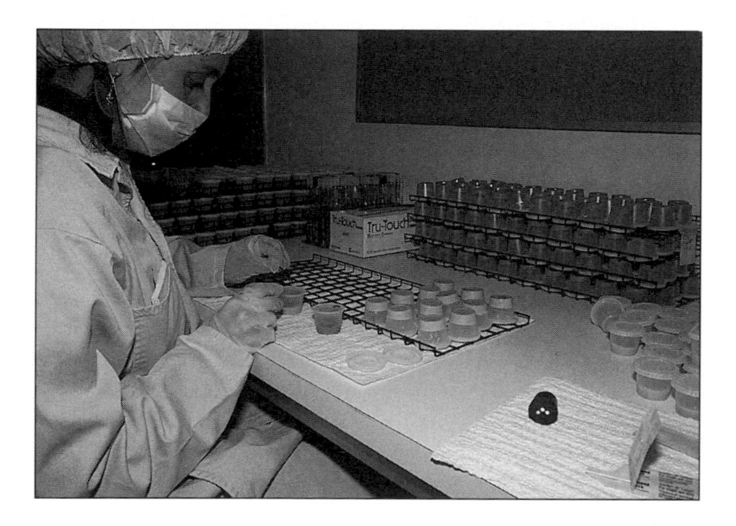

Figure 8. Product quality control.

Treatment areas were also selected more carefully, according to the following forestry, entomological, and operational criteria: balsam fir and white spruce should make up 50% of stand volume; stands should be age 20 or older; the maximum gradient should be no greater than 40%; only the areas where moderate to severe defoliation is predicted (based on surveys of larvae in hibernation) should be considered; no cutting is planned for the next 5 years; the tree mortality rate should be less than 50%; the maximum area should be 250 ha forming a single block. Using these criteria, it was possible to identify territories suitable for spraying (target territories) even before they became infested and to better coordinate spraying and salvage operations. This procedure also made it possible to avoid unnecessary treatment of stands that were inaccessible, or were not targeted for timber harvesting, and that might otherwise act as reservoirs for insects for reinfesting protected territories.

Vulnerability Grid

The damage caused by the budworm was studied to characterize the mortality reported in certain sectors according to stand type (species, age, density), quality of the station (slope, deposition, drainage), and persistence of infestation. A stand vulnerability grid was developed based on the trees' capacity to resist insect attacks. Used as a planning tool by forest managers, this grid made it possible to orient cutting operations toward the most vulnerable stand concentrations, thereby reducing the treatment areas in the short term and the impact of future infestations in the long term (Gagnon and Chabot 1988).

Spraying came to represent a new management tool that contributes to bringing the forest back to normal in the medium and long term. Combined with salvage operations, spraying makes it possible to prolong the life of a stand and to achieve a better distribution of different age-classes over time and space.

Products Used

Standard chemical products continued to be used until 1986. Beginning in 1985, four different Bt preparations were introduced, three of them water-based (Thuricide,® Novabac® 3, and Futura® FC). The concentration of these products varied between 8 and 12 BIU/L, which entailed a variation from 1.77 to 4.68 L/ha in the quantity of mixture that had to be sprayed to ensure a dose of 30 BIU/ha (Table 9).

Effectiveness and Cost of Treatments

The new Bt preparations were used without jeopardizing the health of the forests, which were maintained in good condition following sustained treatments in the Gaspé region. It is estimated that from 1984 to 1987, more than 90% of the areas treated each year enjoyed adequate protection (Table 10).

Table 10. Annual effectiveness of treatments conducted in Quebec from 1984 to 1987 against the spruce budworm, *Choristoneura fumiferana*.

Year	Populations before treatment (larvae/branch)	Larva mortality (%)	Annual defoliation (%)	Proportion of the area adequately protected (%)
1984				
Treated areas (CH)[a]	10.7	81	28	96
Treated areas (Bt)[b]	9.0	77	26	-
Control areas	7.1	45	32	-
1985				
Treated areas	8.5	84	22	94
Control areas	7.1	70	30	-
1986				
Treated areas	14.0	89	32	95
Control areas	14.3	80	50	-
1987				
Treated areas	13.7	93	36	92
Control areas	8.6	80	55	-

[a] Chemical preparations.
[b] Biological preparations.

The cost of treatments was approximately $20/ha for a double application of chemical insecticide and $30/ha for a single application of Bt. Treatments were judged to be effective if infestation levels did not exceed 25 larvae per branch. Since 1987 (Forest Act), the government has assumed 50% and industry 50% of the spraying costs in public forests. In small private forests, the government assumes the full cost.

Conclusion

The spruce budworm infestation that ravaged Quebec's forests from 1974 to 1987 required spraying 23.1 million ha at a total cost of $137 million. One positive outcome of this fight against the budworm was the development of aerial spraying as a forest management tool that is acceptable both environmentally and socially. In the short term, aerial spraying will help to maintain a viable forest by reducing the annual defoliation caused by spruce budworm (80% of treated areas were adequately protected 10 years out of 14). In the long term, it will contribute to normalizing the forest by helping to distribute the different age-classes of trees more evenly over time and space.

Judging from past experience, we can already assume that the spruce budworm will one day return to infest Quebec's forests and that the products, doses, and volumes now in use will be outdated by then. However, the approach that has been developed can help forest managers in their struggle against the insect.

Bibliography

Blais, J.R. 1964. History of spruce budworm outbreaks in the past two centuries in southeastern Quebec and northern Maine. Department of Forestry, Canadian Forest Entomology and Pathology Branch, Ottawa. Bi-monthly Program Rep. 20 (5) 1-2.

Brown, C.E. 1970. A cartographic representation of spruce budworm (*Choristoneura fumiferana*) infestation in eastern Canada, 1909-1966. Can. For. Serv., Ottawa, Ont. Rep. No. 1263.

Cabana, Jean. 1989. Contrôle de la qualité effectué en 1987 sur les préparations à base *Bacillus thuringienses* utilisées lors des pulvérisations aériennes contre la tordeuse des bourgeons de l'épinette. Québec, ministère de l'Énergie et des Ressources. 23 p.

Cabana, Jean. 1986. Contrôle de la qualité effectué, en 1986, sur les préparations à base de *Bacillus thuringiensis* utilisées lors des pulvérisations aériennes contre la tordeuse des bourgeons de l'épinette. Québec, ministrère de l'Énergie et des Ressources. 23 p.

Dorais, L.; Kettela, E.; 1982. A review of entomological survey and assessment techniques used in regional spruce budworm, *Choristoneura fumiferana*, surveys and in the assessment of operational spray programs—Report of the Committee for Standardization of Survey and Assessment Techniques. Eastern Spruce Budworm Council, Fredericton, N.B.

Dorais, L.; Pelletier, M.; Smirnoff, W.; 1980. Aerial spraying applications of *Bacillus thuringiensis* conducted on Quebec forest (Canada) from 1971 to 1979 against the spruce budworm, *Choristoneura fumiferana* (Lepidoptera Tortricidae). VIth International Agricultural Aviation Congress, Turin, Italy.

Gagnon, R.; Chabot, M.; 1988. Système d'évaluation de la vulnérabilité des peuplements à la tordeuse des bourgeons de l'épinette : ses fondements, son implantation et son utilisation en aménagement forestier. L'Aubelle, No. 67, pp. 7-14.

Hardy, Y-J.; Dorais, L-G.; 1976. Cartographie du risque de retrouver de la mortalité dans les forêts de sapin baumier attaquées par la tordeuse des bourgeons de l'épinette. Can. J. For. Res., 6: 262-267.

Lambert, M.C. 1987. Proceedings of the Symposium on Aerial Application of Pestcides in Forestry, Natl. Res. Counc. Can., Ottawa, Ont. No. 29197.

McLintock, F.F. 1955. How damage to balsam fir develops after a spruce budworm epidemic. USDA For. Serv., Northeast For. Exp. Sta., Upper Darby, Pa. Res. Pap. 75.

Chapter 66

Insect Control in Ontario, 1974–1987

G.M. Howse, J.H. Meating, and J.J. Churcher

Introduction

This review describes operational control programs that were carried out against six insects including spruce budworm, *Choristoneura fumiferana*, 1974 to 1987; gypsy moth, *Lymantria dispar*, 1982 and 1985 to 1987; jack pine budworm, *Choristoneura pinus*, 1985 to 1987; forest tent caterpillar, *Malacosoma disstria*, 1975, 1977, and 1978; oak leafshredder, *Croesia semipurpurana*, 1977 to 1980 and 1983; and Saratoga spittlebug, *Aphrophora saratogensis*, 1976. In addition, the effects of spray treatments aimed at spruce budworm were also assessed for linden looper, *Erannis tiliaria*, in 1975 and spruce coneworm, *Dioryctria reniculelloides*, in 1978 to 1980 when high populations of these insects occurred in situations that were being treated for spruce budworm.

In Ontario, protection of Crown forests that make up 84% of the total productive forest land base of the province is the responsibility of the Ontario Ministry of Natural Resources (henceforth referred to as the Ministry). However, since 1967, Forestry Canada, Ontario Region, Forest Insect and Disease Survey Unit (henceforth referred to as the Unit) has provided advice and assistance to the Ministry in the conduct of its operational control programs against destructive forest insect pests. Forestry Canada assists the Ministry in planning control programs but final decisions about what is done are the Ministry's. Forestry Canada provides the biological information required for planning, timing, and evaluating the effectiveness of these operations. In addition, surveys for insect population densities, defoliation, damage, tree condition, and forecasts based on egg-mass and L_2 population samples that are necessary for planning future protection action are conducted by the Unit. The Ministry is responsible for the logistics of the spraying operations including insecticides, aircraft, and landing strips and provides assistance to the Unit in the form of aircraft time for surveys and funding for casual help for foliage mills, field crews, and vehicles.

Starting in 1971, the Unit became involved in planning and assessing cooperative experimental trials involving biological and chemical insecticides at the request of the Insect Pathology Research Institute, the Chemical Control Research Institute, and the Ministry. These research trials, which included Bt, virus, chemical insecticides, insect growth regulators, pyrethroids, and parasites, have continued to the present time (1981–1988), largely in collaboration with the Forest Pest Management Institute of Forestry Canada, the successor to the two previously mentioned institutes, the Ministry, and in the case of aerial releases of *Trichogramma*, the universities of Guelph and Toronto.

Since 1980, the planning and execution of insect control operations using aerial application of insecticides have been conducted in accordance with a policy, procedure, and a manual prepared by the Ministry. The procedure requires the formation of regional planning committees with representation from the Unit, and public notification of proposed insect control projects, and it provides opportunity for public input during the planning process through open houses or other procedures. Options to spraying such as accelerated or reallocated harvests, salvage cuts, prescribed burns, and no action (i.e. allowing the infestation to run its course) must be considered.

The Ministry has used only Bt in its aerial spray programs since 1985. Although this was a decision by the cabinet of the Ontario government in the spring of 1985, it should be realized that the policy of the Ministry is "where alternatives to chemical insecticides are commercially available, reasonably cost effective and approved federally and provincially for use, the Ministry will use such alternatives in preference to chemical insecticides." This policy was introduced in 1980 and the Ministry's use of chemical insecticides has declined substantially since then.

Spruce Budworm

Introduction

The earliest attempts to control spruce budworm in Ontario involved experimental aerial dusting of calcium arsenate near Westree, Gogama District, in 1928 and 1929.

Aerial spraying against the spruce budworm in Canada began in 1944 and continued in 1945 when DDT in oil solution was applied to experimental plots in Algonquin Provincial Park, Ontario. In 1945, the Ministry sprayed two blocks of budworm-infested forest southwest of Lake Nipigon with DDT. One block about 16 000 ha was located at Little Sturge Lake, Nipigon District; the second block about 9 830 ha was located at Circle Lake, Thunder Bay District. In addition, a strip-spraying operation totaling 320 ha was also carried out in 1945 in the vicinity of Sioux Lookout and Hudson in the Sioux Lookout District. This was an attempt to contain a 7 700-ha infestation by spraying a 120-metre strip of forest around the periphery of the infestation mapped in 1944.

In 1946, additional spraying totaling about 11 650 ha with DDT was conducted by the Ministry in the vicinity of Eaglehead Lake about 80 km north of the city of Thunder Bay, Thunder Bay District. The spraying in 1946 involved a routine spray area (about 9 600 ha), several test plots, and several wildlife assessment plots.

Results of spraying on the spruce budworm in 1944 to 1946 were studied by members of the Forest Insect Investigation Unit, Canada Department of Agriculture, Sault Ste. Marie and Ottawa, which was one of the predecessor agencies that now make up Forestry Canada, Ontario Region.

Following the spraying at Eaglehead Lake in 1946, no control action was taken during the next 20 years despite extensive, damaging outbreaks that occurred throughout northwestern Ontario. The outbreaks collapsed in 1963 and populations remained extremely low until 1967 when a small, severe infestation was found near Burchell Lake, 100 km west of Thunder Bay. The Ministry decided to attempt elimination of the infestation center or suppress it sufficiently to prevent spread to the surrounding area. Subsequent control actions from 1968 to 1973 are described by Howse and Sippell in Prebble (1975).

Throughout this chapter, geographic locations of spraying operations within Ontario are Ministry administrative regions and districts. Ontario is subdivided into 8 such regions and each region is further divided into 5 to 8 districts for a total of 47 districts. The regions and districts are depicted in Figure 1 of Chapter 4.

Northwestern Ontario—In northwestern Ontario, spruce budworm infestations were relatively minor from 1967 to 1976 due to aerial control operations that were designed to eliminate infestations or to suppress populations sufficiently to prevent buildup and spread into adjacent areas. These comprehensive control operations were accompanied by intensive surveys designed to detect and delineate areas of infestation as early as possible.

In 1974, two blocks totaling 9 955 ha were sprayed in Quetico Provincial Park in the Atikokan District, North Central Region. Details on dates, dosages, formulation, type of aircraft, and spray systems in this and subsequent operations in northwestern Ontario are provided in Table 1. The purpose of the operation was the same as the past 3 years (1971–1973) (i.e. to prevent the spread of budworm into susceptible forests to the east of Quetico). Zectran® was applied at a rate of 84 g in 1.4 L/ha. The spray mixture consisted of 1 part Zectran® to 2 parts Arotex.

On aerial defoliation surveys, pupal counts, and egg-mass counts, the results appeared to be good. Several small pockets of defoliation totaling about 280 ha were mapped in one spray block. No defoliation was evident in the other spray block. Pupal counts and egg-mass counts from the sprayed areas showed that budworm populations were reduced by more than 95%.

In 1975, 8 094 ha consisting of two blocks, 7 689 ha and 405 ha, were sprayed with a single application of Dylox® used in undiluted form at an application rate of 840 g in 1.68 L/ha near Bennet Lake, Fort Frances District, Northwestern Region. This was the only significant infestation known to exist in northwestern Ontario, and it appeared to be quite discrete and well-defined. The spray aircraft were delayed in getting to Bennet Lake, consequently sprays were applied to fifth- and sixth-instar larvae.

In early July, aerial observers mapped 10 900 ha of defoliation which included virtually all of the treated area. Larval mortality due to the Dylox® treatment was calculated to be at least 60% and post-spray pupal populations on balsam fir were reduced to slightly less than 1.0 per 46-cm branch tip. Egg-mass counts for 1975 increased throughout the area and other infestations were found in several locations in Atikokan and Thunder Bay districts.

In 1976, 35 270 ha were sprayed in the vicinity of Bennet Lake, Fort Frances District, Northwestern Region. Matacil® mixed with fuel oil was applied to 17 824 ha at a rate of 105 g in 0.94 L/ha. Fenitrothion in Arotex was sprayed on the remaining 17 446 ha at a rate of 280 g in 0.94 L/ha. Much of this area, 23 452 ha, received a second application of fenitrothion at the same rate as the first application. A block 6 475 ha centering on an infestation at Fluker Lake on the eastern edge of the Atikokan District was not sprayed owing to advanced

Table 1. Summary of spray operations against the spruce budworm, *Choristoneura fumiferana*, in northwestern Ontario, 1974–1976.

Year and date	Location	Area sprayed (ha)	No. applic.	Insecticide and applic. rate/ha	Carrier	Corrected population reduction (balsam fir) (%)	Aircraft, spray system, and contractor
1974	**Quetico Prov. Park**						
June 13–July 2	Poohbah Lake	8 741	1	Zectran 84 g/1.4 L	Arotex	95[a]	Stearman (1) Micronair Au 3000
	Basswood Lake	1 214	1	Zectran 84 g/1.4 L	Arotex	95	General Airspray
1975 June 12–17	Bennet Lake	7 689	1	Dylox 840 g/1.75 L	none-undiluted	60–70 (approx.)	AgCat (1) Micronair
	Manion Lake Road	405	1	Dylox 840 g/1.75 L	none-undiluted	60–70 (approx.)	General Airspray
1976	Bennet Lake	1 409	1	Fenit.[b] 280 g/0.94 L	Arotex	n.a.[c]	
June 6–20		16 037	2	Fenit. 280 g/0.94 L	Arotex	62	Stearman (2) AgCat (2)
		10 409	1	Matacil 105 g/0.94 L	fuel oil	80	Micronair
		7 415	2	Matacil 105 g/0.94 L plus fenit. 280 g/0.94 L	fuel oil Arotex	93	General Airspray

[a] Population reduction in sprayed areas, uncorrected for natural mortality.

[b] Fenitrothion.

[c] Data not available.

development of the spruce budworm. Overall, spraying conditions were not favorable as windy weather prevailed.

Results varied according to the treatment, but overall larval mortality averaged nearly 80%. The highest larval reduction of 93% was recorded for the area treated with Matacil® and fenitrothion. The single application of Matacil® resulted in 80% mortality whereas two applications of fenitrothion caused only 62% mortality. In early July, about 27 520 ha of defoliation consisting of numerous, small pockets were located within an area of 182 000 ha between Bennet Lake and Rainy Lake, Fort Frances District. Some of these pockets of defoliation were within the area sprayed in 1976, but most were outside. Numerous pockets of defoliation totaling 30 530 ha were also mapped within an area of 202 000 ha between Kawnipi Lake in the Atikokan District, and Lower Shebandowan Lake and Aldina Township in Thunder Bay District, North Central Region. In addition, 3 640 ha of defoliation were mapped along the Pigeon River south of Thunder Bay, and several other small pockets of defoliation totaling a few hundred hectares recurred at Bayley Bay on the Ontario–Minnesota border.

In the fall of 1976, it was determined that attempted suppression of spruce budworm outbreaks would require a control operation of at least 400 000 ha in 1977. An operation of this size was rejected on environmental and economic grounds and spraying against the spruce budworm in this part of the province was abandoned until 1985, when spraying to keep trees alive in high-value stands was resumed.

Reflecting on the events of the 10-year period from 1967 to 1977, the question of what was achieved deserves discussion. One conclusion that can be drawn from these events is that an early detection, delineation, and suppression approach may be successful in stopping or slowing outbreaks. This may provide the rationale for a strategy to manage budworm in the future, but other factors such as migrant moths from Minnesota or Manitoba must also be considered. Certainly the actions of the late 1960s and early 1970s bought 10 years, or so, of time that enabled the forest industry and the province to get better prepared for the inevitable outbreak and its consequences by building roads and accessing remote areas in order to harvest or salvage susceptible stands depending on the eventual course of events.

Southern Ontario, 1974 to 1976, 1980 to 1981—Following several years of moderate to severe defoliation, a total of 726 ha in Algonquin Provincial Park were aerially sprayed in 1974 to protect foliage on host trees in high-value areas along Highway 60. Many major camping or recreational areas such as those located at Mew Lake, Lake of Two Rivers, Killarney Lodge, Pog Lake, Whitefish Lake, and Opeongo Lake were sprayed. Materials used were Zectran® FS 15 mixed with Arotex applied at a rate of 84 g in 1.4 L/ha and Bt (Thuricide® 16B) at a rate of 10 BIU (billion international units) in 4.7 L/ha. A variety of treatments involving aerial applications of Zectran® and Bt and ground application of Bt were used. Details are summarized in Table 2.

The basic approach was to spray one application of Zectran® at 84 g/ha. Some areas, such as campgrounds that had been heavily damaged previously and where a high level of protection was currently required received a second application of Zectran,® and in addition, Bt was applied to accessible trees from a mistblower, diluted 7:1 with water. The 40-ha aerial application of Bt was an experimental trial located at the south end of Opeongo Lake where the Harkness Fisheries Laboratory and a campground were located. Foliage protection obtained in any treatment involving Zectran® was good or excellent and the level of protection is enhanced when followed by Bt from a mistblower. Differences in results obtained from aerial, aerial and mistblower, and mistblower applications only on the two host species indicate that the better results on white spruce are probably due to improved coverage by the mistblower. The aerial application of Bt provided moderate foliage protection although the data do not show a population reduction on white spruce. The mistblower application of Bt would probably have provided more effective foliage protection, if applied a few days earlier.

In 1975, 862 ha were sprayed in Algonquin Provincial Park, again to protect foliage on host trees in high-value areas along Highway 60. Most of the major camping or recreational areas were sprayed. In addition, the provincial tree nursery at Midhurst was sprayed with fenitrothion from a mistblower. In Algonquin Provincial Park, fenitrothion (Sumithion® 10E, Niagara Chemicals) was applied from an aircraft and Bt (Thuricide® 16B) was sprayed from the ground using a mistblower. Treatment details are summarized in Table 2. The basic approach this year was to spray one application at 140 g/ha. Some areas, such as campgrounds, where a high level of protection was required, received a second application of fenitrothion at 140 g/ha and a mistblower application of Bt. Some areas received mistblower applications of Bt only, diluted 7:1 with water.

All treatments provided satisfactory results although a single application of fenitrothion was the poorest of the four. Excellent results were achieved with mistblower applications of fenitrothion at Midhurst.

In 1976, 777 ha in Algonquin Provincial Park were sprayed with Bt from an aircraft, and another 40 ha were treated with Bt from a mistblower to protect foliage on host trees in high-value areas, namely organized campgrounds, and scenic or hiking areas along Highway 60. The provincial tree nursery at Midhurst was sprayed with fenitrothion from a mistblower. Bt was applied at a rate of 10 BIU in 4.7 L/ha. Pre-spray larval counts were, overall, lower than expected. However, it was decided to proceed with the operation because there were high counts in some locations. Results were satisfactory, with substantial larval reduction and foliage protection recorded for most of the sample locations. Spruce budworm populations continued to decline and there was no further spraying in Algonquin Provincial Park.

In 1980, an 850-ha portion of a deer yard in Spence Township, Parry Sound District, was aerially sprayed. This was the first time that a wildlife management area had been protected against spruce budworm damage. The area was treated with two Bt formulations: 567 ha were treated with Dipel® 88 under experimental permit and 283 ha were sprayed with Thuricide® 16B. Both formulations were mixed with water and applied at a rate of 20 BIU in 9.4 L/ha. Moderate levels of larval mortality ranging from 34% to 72% resulted from the treatments. Foliage protection was fair or poor. Foliage protection was slightly better in the Dipel®-treated areas;

however, pre-spray budworm densities were twice as high in the Thuricide®-treated area.

In 1981, the Spence Township deer yard was sprayed again, with 620 ha treated with two different Bt formulations. One was Dipel® 88 applied at 24 BIU in 7.0 L/ha, the other was Thuricide 32® BX at 20 BIU per 4.7 L/ha applied under experimental permit. Results of these treatments were generally very good (Table 2). Spruce budworm populations collapsed in southern Ontario and control has not been necessary since 1981.

Northern and Northeastern Ontario, 1974 to 1986— Spraying for spruce budworm in northeastern Ontario started in 1970 and until 1973 was aimed exclusively at protecting fir/spruce stands in provincial parks in Chapleau and Wawa districts. A total of 16 277 ha or approximately 4 000 ha a year was sprayed with Zectran® or fenitrothion.

In 1974, the policy of park protection was continued with the spraying of 8 863 ha in two provincial parks; 8 175 ha were treated in Lake Superior Provincial Park, Wawa District, and 688 ha in the Shoals Provincial Park, Chapleau District. All applications consisted of Zectran® at 84 g in 1.4 L/ha, but the results were inconsistent. Spraying along Highway 17 and Mijin Lake road in Lake Superior Provincial Park gave excellent protection, but a sprayed corridor along the Sand River in Lake Superior Provincial Park was almost completely defoliated. Similarly, results from Shoals Provincial Park were poor, with defoliation reaching 75% to 80% in several

Table 2. Summary of spray operations against the spruce budworm, *Choristoneura fumiferana*, in southern Ontario, 1974–1976, 1980–1981.

Year and date	Location	Area sprayed (ha)	No. applic.	Insecticide and applic. rate/ha	Carrier	Corrected population reduction (%)		Reduction in current defoliation[a] (%)		Aircraft, spray system, and contractor
						wS[b]	bF[c]	wS	bF	
1974	Algonquin Prov.	546	1	Zectran 84 g/1.4 L	Arotex	36	92	57	71	Stearman (1)
aerial June 6–8	Park									Micronair
ground June 5–10		8	1	Zectran 84 g/1.4 L	Arotex	83	99	82	87	General Airspray
			1	+ Bt mistblower	water					
		81	2	Zectran 84 g/1.4 L	Arotex	n.a.[d]	n.a.	n.a.	n.a.	John Bean Roto-
		51	2	Zectran 84 g/1.4 L	Arotex	79	100	80	88	Mistblower
										OMNR[e]
			1	Bt mistblower						
		40 (exper.)[f]	2	Bt 10 BIU[g]/4.7 L	water	0	64	52	56	
		26	1	Bt mistblower	water	86	78	39	55	
1975	Algonquin Prov.	259	1	Fenit.[h] 140 g/1.75 L	water	64	78	65	46	AgCat (1)
May 28–30	Park									Micronair
		563	2	Fenit. 140 g/1.75 L	water	49	90	74	73	General Airspray
		40	2	Fenit. 140 g/1.75 L	water	66	89	52	79	
			1	Bt mistblower	water					
		20	2	Bt mistblower	water	78	87	83	73	
1976	Algonquin Prov.	777	1	Bt 10 BIU/4.7 L	water	21	78	20	28	Piper Pawnee
aerial June 8–9	Park									Micronair
ground June 5–9			1	Bt 10 BIU/4.7 L	water	52	83	56	61	Central AgAir
		40	1	+ Bt mistblower	water					
			2	Bt mistblower	water	56	60	31	76	
1980	Spence Twp	567 (exper.)	1	Dipel 88 20 BIU/9.4 L	water	34	56	17	32	AgCat (1), b & n[i]
May 27–28	Parry Sound Dist.	283	1	Thur.[j] 16B 20 BIU/9.4 L	water	72	68	0	33	Pawnee (1), b & n
										Can Ag Spray
1981	Spence Twp	296	1	Dipel 88, 24 BIU/7.0 L	water	81	100	76	94	
June 2	Parry Sound Dist.	324	1	Thur. 32, BX,	water	30	63	14	90	
		(exper.)		20 BIU/4.7 L						

[a] Difference in percentage defoliation between treated and untreated plots.
[b] White spruce.
[c] Balsam fir.
[d] Data not available.
[e] Ontario Ministry of Natural Resources.
[f] Experimental spray.
[g] Billion international units.
[h] Fenitrothion.
[i] Boom and nozzle.
[j] Thuricide 16B.

locations throughout the sprayed area. The focus on provincial parks in this part of Ontario continued in 1975 when a total area of 4 504 ha constituting one park in Wawa District, four parks in Chapleau District, and a white spruce plantation near Searchmont, Sault Ste. Marie District, were sprayed. In addition, provincial nurseries at Chapleau and Swastika were sprayed by aircraft and with ground equipment. Fenitrothion was applied to 2 359 ha in single applications at a rate of 280 g in 1.75 L/ha or in two applications, the first at 280 g in 1.75 L/ha, the second at 175 g in 1.75 L/ha. Dylox® was applied to 2 145 ha at either single or double applications, each at a rate of 840 g in 1.75 L/ha. Results were quite variable. A single application of Dylox® caused heavy larval mortality but foliage protection was poor. A double application of Dylox® along the Sand River afforded excellent foliage protection. Fenitrothion was responsible for high larval mortality and good foliage protection in Lake Superior Provincial Park, but results with fenitrothion in parks in Chapleau and the plantation near Searchmount were poor.

Spraying in 1976 broadened to include provincial nurseries at Chapleau and Swastika, a spruce seed production area in Triquet Township, Chapleau District, as well as four provincial parks in Chapleau District and one park, Lake Superior, in Wawa District. A total of 4 158 ha was treated. Matacil® was applied to 4 058 ha at a rate of 105 g in 0.94 L/ha in single or double applications. One hundred hectares were treated with a single application of fenitrothion at 280 g in 0.94 L/ha. Another 732 ha were treated with Orthene® in single or double applications of 280 or 560 g/ha under an experimental permit. Operations were extended beyond optimum limits by poor weather. Results were variable but generally poor.

A program similar to that in 1976 was conducted in 1977. Areas sprayed included four provincial parks, a nursery, and two seed production areas in Chapleau District; a nursery and a seed production area in Kirkland Lake District; and various locations in Lake Superior Provincial Park, Wawa District. In all, 3 855 ha were aerially sprayed with Matacil®. Operational trials totaling 405 ha comparing various dosages and application rates of Orthene® were carried out as well as operational protection spraying of Matacil® in Lake Superior Provincial Park. Matacil® mixed with fuel oil was applied to 3 855 ha in single (664 ha) or double (3 191 ha) applications of 70 g in 1.46 L/ha. In three of the provincial parks in Chapleau District, Matacil® was sprayed throughout the parks except for a buffer strip of approximately 100 m along lake edges and around campgrounds. Bt (Thuricide® 16B) was applied from a mistblower along shorelines and campgrounds. About 77 ha were treated in this manner. As in 1976, the operations were prolonged beyond optimum limits by poor spraying weather. The results for double applications of Matacil® in the Chapleau District were inconsistent, ranging from very poor to good. Mistblower applications of Bt produced generally good results. In Lake Superior Provincial Park, two applications of Matacil® generally provided good results.

In 1978, the operational aerial spray program was reduced to only 200 ha with a seed production area and tree nursery in Kirkland Lake District and a white spruce plantation in Hearst District being protected. Matacil® and Orthene® were the materials used. Operational trials with various Bt formulations were conducted in Lake Superior Provincial Park. A total of 350 ha was aerially sprayed at various dosages. Another 100 ha, including campgrounds in four provincial parks and a tree nursery in Chapleau District, and plantations in the Bonner Tree Improvement Centre, Kapuskasing District, were sprayed from the ground with Bt (Dipel® WP) using a hydraulic sprayer. A white spruce seed production area in Chapleau District was also ground sprayed with Orthene® 85 using the hydraulic unit. Results with aerially applied Matacil® and Orthene® in Kirkland Lake District were good in the seed production area and excellent in the nursery. Ground spraying with Bt produced generally fair to good results in Chapleau but poor foliage protection at the Bonner Centre in Kapuskasing District. Orthene® applied hydraulically produced fair results in the Chapleau white spruce seed production area.

The amount of aerial spraying increased considerably in 1979 to a total of 20 248 ha. The majority of the increase involved protection spraying of high-value commercial forest to minimize damage until harvested in Lamplugh and Elliott townships, Kirkland Lake District, and Clavet Township, Geraldton District. A total of 15 263 ha was treated with Matacil®, 1 936 ha with Orthene® (including 809 ha that received a second application of Novabac®), 2 651 ha with Bt (Thuricide® 16B), and 398 ha with Novabac®. In addition, other high-value stands (tree nurseries, seed production areas, or plantations) were sprayed in the Kirkland Lake, Chapleau, Gogama, Cochrane, Kapuskasing, and Hearst districts. Campgrounds in two provincial parks totaling 54 ha in Chapleau District were treated with Bt (Dipel® WP) using a mistblower. Results with Matacil® and Orthene® were quite variable but would be considered fair on an overall basis. Novabac® results were generally poor, whereas Thuricide® produced good or excellent results. Matacil® and Orthene® were much more effective protecting black spruce than white spruce or balsam fir.

In 1980, 10 068 ha were aerially sprayed to reduce budworm damage. Of this total, about 1 300 ha were high-value forests in six districts in the Northern Region such as seed production areas (16), one nursery (Chapleau), and spruce plantations. The intent was to improve application and deposit in these locations (small blocks), compare insecticides with systemic action, Orthene® and Cygon®, and compare three application regimes. All spraying was done with a Hughes 300 helicopter. Poor weather in June and other technical problems made it impossible to complete all the proposed applications. Orthene® provided good to excellent foliage protection; Cygon® results were very poor. In Kirkland Lake District (Elliott and Lamplugh townships), 8 742 ha of commercial forest, primarily balsam fir, were sprayed. About 4 684 ha were treated with Matacil® and 4 058 ha with Bt (Novabac®). Good to excellent foliage protection was achieved with the various Bt applications as well as Bt plus Matacil® and with double applications of Matacil®.

From 1981 to 1984, 20 266 ha in northeastern Ontario were aerially sprayed to protect balsam fir, white spruce, and black spruce stands from the spruce budworm. The largest operation was in 1981 when 9 613 ha were treated. In the following 3 years, aerial spraying programs were considerably smaller, averaging about 3 500 ha per year. Despite the small total area treated in these programs, they were important years in the development of control materials, especially the biological insecticide Bt. Aerial spraying operations conducted during this period were generally targeted to relatively small discrete blocks of infested forest, thus presenting an ideal opportunity to carry out several experimental trials annually. Also, until 1984, it was possible to use both chemical and biological insecticides to determine the relative merits of each. After 1984, Bt was used exclusively, and chemical insecticides were eliminated from aerial spraying programs.

In 1981 and 1982, aerial spraying operations were conducted in several locations in Gogama, Kapuskasing, Hearst, and Temagami districts to protect female flowers in black spruce and white spruce seed production areas from budworm feeding, a practice that started in 1976 in this part of Ontario. There was a severe shortage of white and black spruce seed in Ontario at that time and, during an outbreak, the spruce budworm could have a severe impact on the developing flowers and cones. Therefore, in 1981, an early or pre-budworm emergence spray was planned using chemical insecticides. Originally, Orthene® was to be applied at the first sign of flowering on spruce followed by an application of either permethrin or Matacil® at the peak of spruce budworm emergence. However, permits for the use of Orthene® and permethrin were not issued, so Matacil® was used alone. Technical problems delayed the first application of Matacil® until after budworm emergence, but a very light flower crop in 1981 prevented assessment of the efficacy of this program. In 1982, the first application was scheduled for the first sign of spruce budworm emergence, but again technical problems resulted in serious delays. Two seed production areas in Temagami District, Friday Lake and Matabitchuan, were treated with a double application of Orthene,® while two others in the Hearst District, Arnott and Hanlan, were treated with an application of Orthene® followed by an application of Matacil® 5 to 7 days later. Despite the abundant flower crop that was observed in 1982, the effectiveness of flower protection provided by these treatments was not determined because of the lateness of the operation.

In 1981, three high-value areas totaling 108 ha in the Northern Region were treated with a nuclear polyhedrosis virus provided by Dr. J.C. Cunningham of the Forest Pest Management Institute. The virus was applied at the peak of the fifth instar and, despite encouraging levels of larval mortality, was not effective in limiting defoliation. A complicating factor found in the Reeves Township seed production area, Chapleau District, was the presence of a high population of spruce coneworm larvae. A follow-up of these treatments in 1982 showed that there was apparently very little carryover of the virus, although low budworm populations in two of the three plots made assessment very difficult.

In 1982, wildlife habitat, specifically a moose yard in Chelsea Township, Hearst District, was aerially treated in an attempt to preserve balsam fir foliage while studies of effects of experimental harvesting techniques on moose behavior could be assessed. A single application of Matacil® 1.8D (90 g in 4.7 L/ha) was applied to 180 ha. In 1983 and 1984, the moose yard was treated with Matacil® 1.8F (90 g in 3.0 L/ha) and in 1985 a single application of Thuricide® 48LV (20 BIU in 1.6 L/ha) was used. Foliage protection from these treatments was generally not very satisfactory because the area was usually one of the last to be treated each year and often had already suffered significant levels of defoliation.

In 1981, a protection spraying program was begun in Nagagamisis Provincial Park, Hearst District. Each year, until the budworm infestation subsided in 1986, about 600 ha of campgrounds, shoreline, and road corridors were treated with a variety of Bt formulations including Thuricide® 16B (20 BIU in 7.2 L/ha) in 1981, Dipel® 88 and Thuricide® 32B (20 BIU in 5.9 L/ha) in 1982, Dipel® 88 (20 BIU in 5.9 L/ha) in 1983 and 1984, and Thuricide® 48LV (20 BIU in 1.6 L/ha) in 1985 and 1986. Another park in Hearst District, Fushimi Provincial Park, was also treated from 1982 to 1986. Similar Bt formulations were used except in 1983 when Novabac® 3 was applied at a rate of 20 BIU in 5.9 L/ha. Overall, these programs appeared to be very effective in reducing budworm populations and protecting foliage.

From 1981 to 1985, several white spruce plantations in Rogers Township, Hearst District, were treated with a wide variety of operational and experimental Bt formulations as well as chemical insecticides. The plantations were 10 to 20 years old and contained a substantial balsam fir component. In 1981, three materials—Matacil® (86 g in 3.0 L/ha), Dipel® 88 (20 BIU in 5.9 L/ha), and nuclear polyhedrosis virus (741×10^9 PIB [polyhedral inclusion bodies] in 9.4 L/ha)—were used to treat nearly 1 600 ha. Only 136 ha were treated in 1982 with Matacil® 1.8D (90 g in 3.0 L/ha). The area treated with Matacil® in 1981 and 1982 was used in a study to assess the impact of continued spruce budworm defoliation on growth in white spruce plantations. Chemical insecticides were used on this plantation to hopefully minimize budworm impact and thus provide a "control" for the study. Nearly 500 ha were sprayed in 1983. Dipel® 88, 6L and 8L were applied at 30 BIU/ha each at two rates—neat and in a 50:50 dilution with water. Results, in terms of foliage protection, were variable but generally poor. Pre-spray larval populations were high in most plots, ranging from 24 to 56 per 45-cm branch tips. The plantation was treated with Matacil® 1.8F (90 g in 3.0 L/ha). In 1984, Matacil® 1.8F (90 g in 3.0 L/ha) and Sevin®-4-Oil (0.85 kg in 2.3 L/ha) were used very successfully in the Rogers Township plantations. Another experimental program was also carried out using five Dipel® formulations (Dipel® 6L, 8L, 132, 6158, and 6167) and Thuricide® SAN-415. Despite delays and heavy rainfall following several treatments, results of the 1984 experimental trials were very encouraging with significant foliage protection on balsam fir and white spruce in most plots. The Rogers Township plantations were treated for the final time in 1985 with Thuricide® 48LV (20 BIU in 1.6 L/ha). No experimental spraying was carried out in Hearst District that year.

An important component of spruce budworm control programs from 1981 to 1985 was the protection of commercial forests that were scheduled for harvest within 5 years of the initiation of spraying. Each forest could be treated a maximum of 3 years within that period. In 1986, the criteria were expanded to include areas scheduled for harvest within 10 years, with a maximum of 6 years of treatment.

From 1981 to 1984, the total area of commercial forests treated was relatively small, totaling less than 12 000 ha. In 1981, 5263 ha in Elliott Township, Kirkland Lake District, were treated with Matacil® (86 g in 3.0 L/ha), Dipel 88® (20 BIU in 5.1 L/ha), and Thuricide® 16B (16 BIU in 6.2 L/ha). From 1982 to 1984, the only commercial stands treated were in Hearst District where an average of 2 100 ha were sprayed annually. The materials used in these programs included Dipel® 88, Dipel® 132, and Novabac® 3. All were applied at a rate of 20 BIU in 5.9 L/ha.

In 1985, 5 013 ha were treated in Hearst District with a single application of Thuricide® 48LV at the rate of 20 BIU in 1.6 L/ha. Both commercial (2 874 ha) and high-value (2 139 ha) forests including plantations, provincial parks, and wildlife habitat were sprayed in mid-June. The spraying was a week to 10 days late in terms of insect development which undoubtedly reduced the effectiveness of the treatments.

In 1986, the provincial spruce budworm spray program totaled 150 633 ha. Of this, 51 084 ha were located in northeastern Ontario in Wawa District, Northeastern Region, and Hearst District, Northern Region. Treatment was a single application of Thuricide® 48LV at the rate of 20 BIU per 1.6 L/ha. Results in Hearst District were good although prespray populations were not as high as expected and high levels of natural mortality occurred during the larval and pupal stages in much of this area. In fact, a decline of more than 2.5 million ha of infestation occurred in 1986 in Hearst and Wawa districts. Protection was not necessary in 1987.

A summary of area treated and insecticides used each year in northeastern Ontario is presented in Table 3.

North Central and Northwestern Ontario, 1985 to 1987— The spruce budworm infestation in the northwestern part of the province followed a much different course than in northeastern or southern Ontario. Extensive aerial spraying operations conducted in northwestern Ontario from 1968 to 1976 appeared to have been effective in delaying the rapid buildup and expansion of the outbreak that had occurred farther east. After 1976, the northwestern outbreak followed a similar course to that in the northeast, expanding each year until more than 7 million ha were affected by 1985.

In the fall and winter 1983 and 1984, a control program to protect commercial stands that were considered critical to local mill supplies in the Thunder Bay District was proposed. Of particular concern were 6 000 to 7 000 ha in the Burchell Lake area that were part of the Canadian Pacific Forest Products limits. However, this project was canceled in the spring of 1984 due to political concern about the public's perception of such operations.

In the fall of 1984, a communications plan designed to inform and educate the public about forest pest situations, options, and environmental impacts and to give the public an opportunity to comment was developed. Open houses were held during the winter in districts proposing an aerial spraying program to provide the public with an opportunity to view and comment on proposals that included reallocation of harvest, accelerated harvest, salvage, clearcutting, aerial spraying, or doing nothing.

In 1985, 24 357 ha in Thunder Bay and Fort Frances districts were aerially sprayed to protect foliage from the spruce budworm. Nearly 90% of the spraying occurred on commercial stands using Bt (Dipel® 132) applied at rates of 20 or 30 BIU/ha. No chemical insecticides were used in 1985 or subsequent years.

Table 3. Summary of spray operations against the spruce budworm, *Choristoneura fumiferana*, in northeastern Ontario, 1974–1986.

Year	Insecticide	Area treated (ha)	
1974	Zectran	8 863	
1975	Fenitrothion	2 359	
	Dylox	2 145	
		4 504	
1976	Matacil	4 058	
	Fenitrothion	100	
	Orthene (exper.)[a]	732	
		4 890	
1977	Matacil	3 855	
	Orthene (exper.)	405	
		4 260	
1978	Matacil	160	(81 ha initially treated with Matacil received a 2nd application of Orthene)
	Orthene	40	
	Novabac-3 (exper.)	77	
	Dipel WP (exper.)	88	
	Thuricide 16B (exper.)	185	
		550	
1979	Matacil	15 263	
	Orthene	1 936	(809 ha initially treated with Orthene received a 2nd application of Novabac)
	Thuricide 16B	2 651	
	Novabac 32B	398	
		20 248	
1980	Matacil	4 684	(247 ha initially treated with Matacil received a 2nd application of Bt)
	Orthene	740	
	Cygon	182	
	Novabac	4 058	(182 ha initially treated with Novabac received a 2nd application of Matacil)
	Permethrin (exper.)	279	
	Matacil (exper.)	125	
		10 068	
1981	Matacil	3 225	
	Dipel 88	2 807	
	Thuricide 16B	3 473	
	Virus (exper.)	108	
		9 613	
1982	Dipel 88	2 744	
	Matacil	316	
	Novabac	247	
	Thuricide 32B	77	
	Orthene	45	(23 ha initially treated with Orthene received a 2nd application of Matacil)
	Thuricide 48B (exper.)	25	
		3 454	
1983	Dipel 88	1 513	
	Novabac	1 250	
	Matacil	399	
	Dipel 88 (exper.)	70	
	Dipel 8L (exper.)	110	
	Dipel 6L (exper.)	100	
	Bactospeine (exper.)	60	
		3 502	
1984	Dipel 88	2 324	
	Matacil	400	
	Dipel 132	364	
	Sevin-4-Oil	200	
	Bt (six materials [exper.])	409	
		3 697	
1985	Thuricide 48LV	5 013	
1986	Thuricide 48LV	51 084	

[a] Experimental.

In the fall of 1985, district and regional working committees met to begin planning for 1986. Previously, these committees had been primarily made up of representatives from provincial and federal governments. However, by 1985, industry was taking a much greater interest in the budworm problem and the spraying operations in Ontario. The various forest products companies were invited to participate on the district and regional working committees and two representatives from the Ontario Forest Industries Association were included on the provincial Budworm '86 Committee. As a result of industries' interest and participation in the planning process, about 227 200 ha overall were proposed for treatment in 1986.

Budworm larval surveys were conducted during the winter to assess populations in several areas proposed for treatment in 1986. Based on the results of this survey, many blocks were deleted and in May and June 1986, about 150 633 ha were aerially treated in the four northern regions with 99 549 ha of the total in northwestern Ontario. Most of the area treated in 1986 was commercial forest. Thuricide® 48LV was the only material used in the program and was applied undiluted at either 20 or 30 BIU/ha. Results were quite variable but rapid larval development and technical problems that led to delays in the programs may have accounted for some of the poorer results.

Budworm defoliation and egg-mass information collected in 1986 indicated that populations were declining in the area east of Lake Nipigon and in parts of Thunder Bay District. This information enabled the district and regional committees to eliminate many of the areas treated in 1986 and it was proposed that about 93 000 ha of commercial and high-value forest be sprayed in 1987. As in 1985 and 1986, most of the stands in the proposal were commercial timber.

During the winter and spring 1987, spruce budworm larval surveys were conducted to further assess populations in some of the areas included in the 1987 spray proposal. The results of these surveys led to the cancellation of several blocks in Geraldton and Terrace Bay districts. In May and June, OMNR treated 76 689 ha in six districts in the North Central and Northwestern regions with Dipel® 132, applied undiluted at either 20 or 30 BIU/ha. Results were generally very good with defoliation rates in sprayed plots averaging 50% less than in unsprayed plots.

As the spruce budworm outbreak developed in northwestern Ontario, aerial spraying operations were begun to protect high-value stands in the North Central Region. In 1985, about 22 white spruce plantations and seed production areas in Thunder Bay District were treated with a single application of Dipel® 132 or 176 at either 20 or 30 BIUs per hectare. The following year, this effort was essentially repeated using Thuricide® 48LV at 20 BIU in 1.6 L/ha. Low spruce budworm populations and low defoliation levels were observed in both treated and untreated check plots in 1986. In 1987, the number of plantations included in the control program was greatly reduced because of the low larval populations encountered the previous year and low egg-mass counts. Those that were treated received a single application of Dipel®132 at 20 or 30 BIU/ha.

In 1986 and 1987, protection spraying was extended to other provincial parks affected by the budworm in the North Central and Northwestern regions. Sibley Provincial Park, Thunder Bay District, was treated with Dipel® 132 in 1985 and 1987 and Thuricide® 48LV in 1986. All three treatments were applied at a rate of 20 BIU in 1.6 L/ha. In 1986, Ojibway Provincial Park, Sioux Lookout District, and Sandbar Lake Provincial Park, Ignace District, were aerially sprayed with Thuricide® 48LV (30 BIU in 2.4 L/ha). The following year, 1987, Sandbar Lake Provincial Park was treated with a single application of the experimental formulation Dipel® 176 at a rate of 30 BIU in 1.6 L/ha. Lake Nipigon Provincial Park, Nipigon District, and Ojibway Provincial Park received single applications of Dipel® 132 (20 BIU in 1.6 L/ha) in 1987. All of these treatments successfully protected foliage.

A summary of area treated and insecticides used each year is presented for northwestern Ontario in Table 4. A summary of spruce budworm spray operations for Ontario from 1974 to 1987 is presented in Table 5.

During the last decade or so in Ontario, aerial spraying programs conducted against the spruce budworm have undergone some significant changes. There have been technological changes leading to improved spray delivery systems and navigational aids. Experimental programs, particularly involving Bt formulations, have led to the development of new high-potency/low-volume materials that have greatly improved the efficacy and cost effectiveness of these products. This last development has taken on even greater significance in light of the recent practice of excluding chemical insecticides from provincial programs. At one time, chemical insecticides were an integral part of pest control operations in the province. However, since 1980 their use has declined to the point where they are no longer considered to be an operational alternative. Another important development has been the growing involvement of the forest industry in all phases of pest control in the province. Individual companies and, indeed, the industry as a whole have become much more active in the development of control policy and programs. The large aerial spraying operations carried out over commercial forest stands during the last 3 years (1985, 1986, 1987) are undoubtedly a result of this new effort by the industry.

Table 4. Summary of spray operations against the spruce budworm, *Choristoneura fumiferana*, in north central and northwestern Ontario, 1985–1987.

Year	Insecticide	Area treated (ha)
1985	Dipel 132	24 266
	Dipel 176 (experimental)	91
		24 357
1986	Thuricide 48LV	99 549
1987	Dipel 132	76 689
	Dipel 176 (experimental)	130
		76 819

Table 5. Summary of spruce budworm spray operations[a] against the spruce budworm, *Choristoneura fumiferana*, in Ontario, 1974–1987.

Year	Area treated (ha)				Insecticides
	Northwestern	Northeastern	Southern	Total	
1974	9 955	8 863	686	19 504	Zectran
1975	8 094	4 504	862	13 460	Dylox, fenitrothion
1976	35 270	4 158	777	40 205	Fenitrothion, Matacil, Thuricide 16B
1977		3 855		3 885	Matacil
1978		200		200	Matacil, Orthene
1979		20 248		20 248	Matacil, Orthene, Thuricide 16B, Novabac 32B
1980		9 664	283	9 947	Matacil, Orthene, Cygon, Thur. 16B, Novabac
1981		9 505	296	9 801	Matacil, Dipel 88, Thuricide 16B
1982		3 429		3 429	Dipel 88, Matacil, Novabac, Thuricide 16B, Orthene
1983		3 162		3 162	Dipel 88, Novabac, Matacil
1984		3 288		3 288	Dipel 88, Dipel 132, Matacil, Sevin-4-Oil
1985	24 266	5 013		29 279	Dipel 132, Thuricide 48LV
1986	99 549	51 084		150 633	Thuricide 48LV
1987	76 689			76 689	Dipel 132

[a] Does not include experimental sprays.

Gypsy Moth, 1982–1987

During the 15 years from 1973 to 1987, the gypsy moth, *Lymantria dispar*, situation in Ontario has undergone some dramatic changes. First discovered in Ontario in 1969 on Wolfe Island, near the city of Kingston, gypsy moth populations remained relatively low for more than a decade, based on annual surveys conducted to monitor the spread of this introduced forest pest in southeastern Ontario. Starting in 1977, an annual pheromone trapping survey was initiated in Northern Ontario. Population suppression and/or containment operations, involving ground and aerial applications of Sevin®-4-Oil and Dimilin® were conducted by Agriculture Canada from 1970 to 1981. A total of about 20 500 ha were treated during this period.

Despite the control attempts, gypsy moth populations continued to slowly build and spread in eastern Ontario during the 1970s. Then in 1981, aerial surveys detected approximately 1 450 ha of defoliation associated with the gypsy moth (Table 6) in about 23 pockets throughout the Eastern Region. Of most concern to local provincial authorities were four pockets of moderate to severe defoliation totaling about 1 000 ha in the Kaladar area of Tweed District. This discovery in the Kaladar area started a new chapter in the gypsy moth story in Ontario.

In 1981, several recreational, commercial, and environmental values, such as tourism, cottaging, hunting, and timber, that could be severely affected by a gypsy moth outbreak were identified. Options were discussed and preparations made to inform the public about the situation. Intensive egg-mass surveys were conducted to delineate the four pockets of infestation and assess population densities. In the spring of 1982, three municipalities in the Kaladar area requested a spray program to control gypsy moth on private land. Shortly thereafter, such a program was initiated using the chemical insecticide Sevin®-4-Oil along with small-scale trials involving the experimental biological insecticides Gypchek (a nuclear polyhedrosis virus) and Bt.

Initial plans called for the treatment of about 2 000 ha in the Kaladar area that would include the four pockets of gypsy moth defoliation as well as buffer zones around each. Several weeks before the anticipated spray date, public controversy erupted over the proposed use of Sevin®-4-Oil and the program was reduced in size to 416 ha. The area treated with Sevin®-4-Oil was limited to a 90-ha block of Crown land. Bt (Dipel® 88) was used on about 263 ha and Gypchek was applied to 63 ha (Table 7). All treatments appeared to be effective in reducing gypsy moth populations and limiting defoliation.

Table 6. Gypsy moth, *Lymantria dispar*, infestation and aerial spraying programs in Ontario, 1981–1987.

Year	Gross area of moderate to severe defoliation (ha)	Area treated operationally (ha)
1981	1 450	0
1982	4 800	416
1983	40 954	0
1984	80 624	0
1985	246 342	170
1986	167 776	103 094
1987	12 678	40 249

In 1982, aerial surveys showed that the area of moderate to severe defoliation had expanded, affecting 4 800 ha in the Kaladar area, and it was generally felt that future efforts to control the gypsy moth would be extremely difficult. Large operational aerial spraying programs to control the pest were considered impractical in view of the large proportion of private land in the region and the diversity of opinions concerning the use of pesticides. The identification of high-value areas such as parks, plantations, and wildlife habitat that would warrant protection in the event they were affected by the gypsy moth was therefore begun.

No operational or experimental aerial spraying occurred in 1983 or 1984. Surveys during these 2 years showed substantial increases in the size of the infestation from 40 954 ha in 1983 to 80 624 ha in 1984.

In 1985, a small protection program was carried out in three provincial parks: Frontenac, Sharbot Lake, and Silver Lake. A total of 170 ha were aerially treated with Dipel® 88 at 40 BIU in 12.0 L/ha. Because of an unusually prolonged egg-hatch period, the efficacy of these treatments was somewhat reduced, especially at Sharbot Lake Provincial Park. Second applications were made to Sharbot Lake and Frontenac parks.

A total of 246 342 ha of forest was infested by the gypsy moth in the Eastern Region in 1985 and it was now affecting more urban and residential areas. Gypsy moth began to receive considerable attention from the media and the province was receiving requests to take appropriate action from large numbers of municipal, county, and regional governments as well as individual citizens. A large-scale egg-mass survey was conducted in the fall. The results of this survey and the aerial survey conducted in July were the basis for the largest aerial spraying program ever proposed in Canada to protect forests from the gypsy moth, and in 1986, about 103 094 ha in eastern Ontario were aerially sprayed with Bt.

In 1986, more than 50% (57 417 ha) of the total gypsy moth program was on private land. This part of the program

Table 7. Summary of aerial spraying programs against the gypsy moth, *Lymantria dispar*, in Ontario, 1982–1987.

Year	Treatment	Area (ha)	Dosage	Type of spraying[a]	No. applications	Aircraft[b]	Equipment[c]
1982	Sevin-4-Oil	90	1.1 kg/2.4 L•ha^{-1}	O	1	RW	
	Dipel 88[d]	119	20 BIU[e]/5.9 L•ha^{-1}	E	2	FW (Stearman)	
	Dipel 88[d]	144	30 BIU/8.8 L•ha^{-1}	E	2	FW (Stearman)	
	Gypchek (NPV)[f]	36	2.5×10^{11} PIB[g]/18.8 L•ha^{-1}	E	2	FW (Stearman)	
	Gypchek (NPV)[f]	27	2.5×10^{11} PIB/18.8 L•ha^{-1}	E	2	FW (Stearman)	
1983	No spray						
1984	No spray						
1985	Dipel 88[d]	170	40 BIU/12 L•ha^{-1}	O	2	RW	B & N
1986	Dipel 132[d]	103 094[h]	30 BIU/6.0 L•ha^{-1}	O	2-3	RW/FW	M
	SAN-415[d]		30 BIU/6.0 L•ha^{-1}	E	2		M
	NPV (wild)[i]	87	2.7×10^{11} PIB/9.4 L•ha^{-1}	E	2	RW	M
	NPV (wild)[i]	10	2.2×10^{12} PIB/9.4 L•ha^{-1}		2	RW	M
1987	Dipel 132[d]	40 249[j]	30 BIU[k]	O	2-3	RW/FW	M
	Thuricide 48LV[d]		30 BIU/6 L•ha^{-1}	O	2-3	RW/FW	M
	Futura XLV[d]		30 BIU/6 L•ha^{-1}	O	2-3	RW/FW	M
	Dipel 176[d]	82	50 BIU/3.0 L•ha^{-1}	E	1	RW	M
	Dipel 176[d]	146	30 BIU/4.4 L•ha^{-1}	E	2	RW	M

[a] Operational (O) or experimental (E).
[b] Rotary wing (RW) or fixed wing (FW).
[c] Boom and nozzle (B & N) or Micronair (M).
[d] Formulations of Bt.
[e] Billion international units.
[f] A commercial preparation of nuclear polyhedrosis virus.
[g] Polyhedrosis inclusion bodies.
[h] This includes 45 677 ha of Crown land and 57 417 ha of private land.
[i] A wild strain of virus isolated by the Forest Pest Management Institute.
[j] This includes 26 310 ha of Crown land and 13 939 ha of private land.
[k] Volumes varied from 2.1 L to 6.0 L per ha.

was carried out with the cooperation of nine counties in the affected area and was based on a cost-sharing formula with the individual landowners. All areas, Crown and private, received two applications of Dipel® 132 at a rate of 30 BIU in 6.0 L/ha. Several outlying "hot spots" involving about 10 000 ha received a third application in an attempt to further suppress gypsy moth populations and thus reduce the chances for further expansion or spread. A variety of fixed-wing aircraft (Dromadier M-18's, Turbo Thrushes, Piston Thrushes, AgCats, AgTrucks, Piper Pawnees, and Piper Braves) and rotary-wing aircraft (Bell 47's, Bell 212's, and Hughes 500's) were used. A small experimental program was conducted involving double applications of the nuclear polyhedrosis virus, produced by Dr. J.C. Cunningham of the Forest Pest Management Institute, and the Bt formulation SAN-415, produced by Sandoz Agro Canada Inc.

The results of the 1986 operational and experimental programs were difficult to assess, based on egg-mass reduction and foliage protection, because of a somewhat unexpected natural decline in gypsy moth populations in many areas. Overall, the program appeared to be effective in reducing defoliation rates.

Aerial surveys conducted in July 1986 revealed that for the first time in 5 years, the gypsy moth infestation had decreased in size. Overall, about 167 776 ha of hardwood forest were defoliated, a reduction of 32% from 1985. Results of egg-mass surveys indicated that a further reduction would likely occur in 1987, especially in the older parts of the infestation. However, several areas of high egg-mass densities still remained and several new areas were detected.

Based on results of the aerial and egg-mass surveys, a plan to treat approximately 68 000 ha of Crown and private land with Bt in 1987 was formulated. However, a reduction in the number of private landowner requests for protection and the exclusion, for operational reasons, of many small private land blocks reduced the actual size of the 1987 program to 40 249 ha. Crown land spraying accounted for 26 310 ha and the private land component was 13 939 ha. Three Bt formulations (Dipel® 132, Thuricide 48LV and Futura XLV) were used in the program and, as in 1986, a variety of fixed-wing and rotary-wing aircraft were employed. All blocks received two treatments, approximately 5 to 10 days apart, and several blocks were treated a third time. Each application was at a rate of 30 BIU/ha and both Thuricide® 48LV and Futura® XLV were applied at a volume of 6.0 L/ha. Volumes of the Dipel® 132 treatments varied from 2.1 to 6.0 L/ha because of problems encountered during mixing.

Two experimental trials were conducted as part of the 1987 operation. A single treatment of Dipel® 176 at a rate of 50 BIU in 3.0 L/ha was applied to an 82-ha block in Carleton Place District and 146 ha in Pembroke District received a double application of Dipel® 176 at a rate of 30 BIU in 4.4 L/ha. A Bell 206 helicopter was used in both cases.

On May 14, the same day that the 1987 gypsy moth spray operation began, the entire Ontario Bt spray program was put on hold until reports of contaminants in the Bt formulations could be investigated. The program did not resume until a week later when it was reported that the contaminants did not pose a human health hazard. This delay, at a critical period in insect and host development, undoubtedly influenced the effectiveness of the program. However, as in 1986, gypsy moth populations declined throughout much of the infestation, and it appears that a combination of natural factors such as weather, disease, and parasitism were the main causal agents. Defoliation rates were, on average, only slightly higher in unsprayed check plots than in spray blocks. Egg-mass reductions were, on the other hand, approximately 28% greater in spray blocks, indicating some additional reductions in gypsy moth populations as a result of spraying.

The gypsy moth infestation in eastern Ontario continued to decline in 1987. Only 12 678 ha of moderate to severe defoliation were mapped, a reduction of 92%. Despite this good news, the gypsy moth continued to expand its range in southern Ontario. New pockets of defoliation appeared around the fringe of the old infestation in the east and, for the first time, a pocket of moderate to severe defoliation, totaling 115 ha, was mapped in Simcoe District, Southwestern Region. Larvae were found in parks in about 10 districts in southern Ontario and moths were trapped throughout this area and on Manitoulin Island and along the North Channel of Lake Huron in Northern Ontario. The final distribution of gypsy moth in Ontario is impossible to predict but it will take a coordinated effort from each of the agencies involved to manage this pest in Ontario.

Jack Pine Budworm, 1985–1987

The jack pine budworm, *Choristoneura pinus*, is considered to be the most serious insect pest of jack pine in Ontario. Earlier outbreaks of this pest occurred in northwestern Ontario, but during the last outbreak, 1967 to 1972, high populations occurred in central and northeastern Ontario. Tree mortality up to 30% to 40% of the jack pine component in a stand may occur following 2 to 3 years of severe defoliation. Top killing is a common result also, thus half or more of the stand may be damaged. Outbreaks of this insect appear suddenly, last for 2 or 3 years in any given location, and then collapse abruptly. Spraying operations totaling about 5 600 ha were conducted in 1968, 1969, and 1972 to protect high-value areas such as provincial parks, plantations, and stands of high quality timber. Fenitrothion was the insecticide used in 1968 and 1969 and Zectran® in 1972.

The current outbreak of this insect in Ontario started in 1982. By 1984, infestations had expanded to more than 1 million ha throughout the northern regions as well as the Algonquin Region. Large infestations totaling more than 600 000 ha were present in northeastern Ontario, an unusual situation in that there was no known historical occurrence of infestations, particularly of this magnitude, in this part of the province. Because of the increasing importance of jack pine, several forest industry companies in Northern Ontario expressed great concern about the impact this outbreak would have and strongly advocated that a large-scale protection spraying program be initiated.

During the winter, protection plans involving the use of biological insecticides and other nonspray options such as reallocation of harvest, accelerated harvest, and salvage operations were developed and a temporary registration for the operational use of Bt in 1985 was granted.

In late June and early July 1985, about 220 000 ha of jack pine forest were aerially treated in Chapleau, Gogama, Kirkland Lake, Blind River, Espanola, and Sudbury districts. Single applications of Dipel® 132 or Thuricide® 48LV were made with both fixed-wing and rotary-wing aircraft equipped with Micronair atomizers (Table 8). All treatments were at 20 BIU/ha. A small experimental program involving six different Bt treatments was conducted in Gogama District (Table 8).

Overall, the 1985 operational spraying program was effective in reducing defoliation. In most cases, defoliation levels were 40% to 50% lower in treated plots than in untreated check plots and the two Bt formulations appeared comparable. Because of delays in carrying out the experimental program, results were not as good as they might have been with an earlier application.

Aerial surveys conducted in July 1985 showed that the jack pine budworm infestations had again increased substantially. Overall, about 3.66 million ha of moderate to severe defoliation were mapped, an increase of 2.51 million ha from 1984. These increases occurred in each of the three geographical regions previously affected.

Egg-mass surveys indicated that in northeastern Ontario populations would decline considerably and that a population collapse was possible over extensive areas in 1986. In northwestern Ontario, despite an overall decline in egg-mass densities, moderate to severe defoliation was expected to persist within areas defoliated in 1985 in Red Lake, Fort Frances, Kenora, Dryden, Ignace, and Atikokan districts and could, in fact, expand in 1986.

Table 8. Summary of spraying operations against the jack pine budworm, *Choristoneura pinus*, in Ontario, 1985–1987.

Year	Treatment[a]	Area treated (ha)
1985	**Operational**	
	Dipel 132, 20 BIU/1.6 L•ha^{-1}	140 000
	Thuricide 48LV, 20 BIU/1.6 L•ha^{-1}	80 000
		220 000
	Experimental	
	Dipel 132, 30 BIU/2.4 L•ha^{-1}	140
	Dipel 176, 30 BIU/1.8 L•ha^{-1}	140
	Dipel 176, 30 BIU/3.6 L•ha^{-1}	140
	Thuricide 48LV, 10 BIU/1.6 L•ha^{-1}	140
	Thuricide 64LV, 20 BIU/1.2 L•ha^{-1}	140
	Thuricide 64LV, 30 BIU/1.8 L•ha^{-1}	140
		840
1986	**Operational**	
	Dipel 132, 20 BIU/1.6 L•ha^{-1}	n.a.[b]
	Thuricide 48LV, 20 BIU/1.6 L•ha^{-1}	n.a.
		488 032
	Experimental	
	Foray 48B, 20 BIU/1.6 L•ha^{-1}	n.a.
	Dipel ABG-61, 20 BIU/1.6 L•ha^{-1}	n.a.
1987	**Operational**	
	Dipel 132, 20 BIU/1.6 L•ha^{-1}	105 463

[a] Various formulations of Bt were used, at rates of 20 or 30 billion international units per hectare.

[b] Areas treated were not available.

Since it was feared that any additional defoliation in many stands would result in unacceptable levels of damage, about 240 135 ha in northeastern Ontario were aerially sprayed between June 9 and June 25, 1986, with Dipel® 132 or Thuricide® 48LV at a rate of 20 BIU in 1.6 L/ha using both fixed-wing and rotary-wing aircraft equipped with Micronair atomizers. Two experimental treatments involving Foray 48B (20 BIU in 1.6 L/ha) and Dipel® ABG-61 (20 BIU in 1.6 L/ha) were carried out in Sudbury District. Generally, low pre-spray larval populations were encountered throughout the region and so defoliation levels were light in both spray plots and checks.

The situation in northwestern Ontario in 1985 was very similar to that experienced in northeastern Ontario the previous year, and in June 1986 about 247 897 ha were aerially sprayed in six districts—Red Lake, Kenora, Dryden, Sioux Lookout, Fort Frances, and Atikokan. As in the northeast, Dipel® 132 and Thuricide® 48LV were the operational materials used (Table 8). Overall, results of the 1986 operations in northwestern Ontario were good in terms of foliage protection.

Efficacy data from the 1985 jack pine budworm aerial spraying led to the full registration of two Bt formulations, Dipel® 132 and Thuricide® 48LV, for application rates of 30 BIU/ha.

The 1986 aerial surveys showed that the infestation in the Northern and Northeastern regions had essentially collapsed with only 135 196 ha of defoliation scattered over six districts. In the northwest, the infestation was concentrated in the Northwestern Region where 1.55 million ha were affected, a slight increase from 1985. An additional 31 391 ha of defoliation were mapped in the North Central Region, down significantly from the 285 406 ha observed in 1985.

Egg-mass surveys indicated that defoliation was not likely to occur in the northeast in 1987 and in the northwest the outbreak was expected to decline significantly or even collapse over a widespread area. This placed the Northwestern Region in the same position faced by the northeast in 1985. Additional overwintering larval surveys were conducted to provide information on populations and parasitism levels in several proposed spray blocks. This information, along with egg-mass and defoliation records, was used to identify the commercial and high-value stands most likely to be affected by the jack pine budworm in 1987. As a result, 105 463 ha were aerially sprayed in four districts—Red Lake, Kenora, Dryden, and Fort Frances—in the Northwestern Region in 1987. The entire area was treated with a single application of Dipel® 132 at a rate of 20 BIU in 1.6 L/ha using fixed-wing aircraft. No experimental trials were conducted against the jack pine budworm in 1987. On a plot-to-plot basis, spray efficacy was typically quite variable. However, the overall results of the 1987 program indicate substantial foliage protection at all population densities.

Based on results of the 1987 egg-mass survey and historical records of earlier jack pine budworm outbreaks, the current outbreak was expected to decline in Ontario in 1988. Populations in the Northern, Northeastern, and Algonquin regions had returned to endemic levels. In the Northwestern

Region, small scattered pockets of light to moderate defoliation were expected in some areas, but large extensive areas of moderate to severe defoliation were not expected to occur.

Forest Tent Caterpillar, 1975–1978

In response to an outbreak of forest tent caterpillar, *Malacosoma disstria*, that occurred in Ontario from 1973 to 1979, the Ministry conducted a total of 2 727 ha of operational and 405 ha of experimental aerial spraying in 1975, 1977, and 1978 (Table 9).

In general, the purpose of the operational spraying was to reduce larval populations in provincial parks to a level that would prove not to be a nuisance to park users and, secondly, to minimize foliage loss. In addition to the operational spraying each year, various field trials involving aerial applications of chemical and biological insecticides were conducted. The materials tested were registered insecticides used against forest insects such as spruce budworm but were not specifically registered for use against forest tent caterpillar. As of 1973, commercial preparations of Bt had not been tested in aerial applications against the forest tent caterpillar in Canada and it was recommended that "such tests should be carried out to determine whether this biological insecticide can be used to protect foliage during periods of infestation" (Prebble 1975).

In 1975, the Ministry aerially sprayed about 506 ha in Grundy Lake Provincial Park, Parry Sound District (Table 10), with two applications of fenitrothion at a rate of 210 g in 1.75 L/ha for each application. In addition to the main operation at Grundy Lake, four field trials involving aerial applications of fenitrothion and Thuricide® 16B were carried out in heavily infested aspen stands at the Key River along Highway 69 about 3 km south of Grundy Lake (Table 10). All sprays were applied May 22 and 23 with an AgCat equipped with Micronair atomizers. The Unit assisted with the timing and evaluation of results. Although assessments of the spraying were not conducted in the park, the field trials showed that fenitrothion 420 g/ha (two applications of 210 g in 1.75 L/ha each) was most effective in reducing larval populations, followed in descending order by Bt at 20 BIU/ha (two applications of 10 BIU in 4.7 L/ha each), a single 210 g in 1.75 L/ha application of fenitrothion, and lastly, a single application of Bt at 10 BIU in 4.7 L/ha. Larval mortalities due to each treatment were 99.7, 92, 86, and 71% respectively (Table 10). The field trials were carried out too late to achieve any significant degree of foliage protection, since most of the larvae had reached third or fourth instar at the time of treatment.

In 1976, the Ministry planned to spray a total of 648 ha with Bt in three heavily infested provincial parks. The proposed operations were 486 ha in Grundy Lake Provincial Park, Parry Sound District, and 162 ha in Arrowhead and Mikisew provincial parks, Bracebridge District. Warm weather in late April followed by cool, wet conditions resulted in poor hatch of egg bands combined with low survival of larvae from those eggs that did hatch. Significant larval populations could not be found by Unit personnel within the proposed spray boundaries of the three parks; consequently, the Ministry acting on a recommendation from Forestry Canada canceled the operations.

Table 9. Summary of area treated (ha) by operational and experimental aerial spraying against forest tent caterpillar, *Malacosoma disstria*, in Ontario 1975–1978.

	Area treated (ha)			
	Operational		Experimental	
	Bt	Chemical	Bt	Chemical
1975		506	81	81
1976		no spraying		
1977	1 788			162
1978	433 (OMNR)[a]		81	
	253 (private)			

[a] Ontario Ministry of Natural Resources.

In May 1977, the Ministry sprayed a total of 1 788 ha in six provincial parks in Parry Sound and Bracebridge districts and a research area in Huronia District with Thuricide® 16B. In addition, aerial trials totaling 162 ha involving Dylox®, Orthene®, and Dimilin® were carried out near the Key River in Parry Sound District. Two Cessna AgTrucks equipped with Micronair atomizers were used for both the field trials and the operational spraying. Details on the dates, materials, dosages, formulations, type of aircraft, spray systems, and results are provided in Table 10. Larval development at the time of treatment was primarily second instar, and trembling aspen leaves, which had just flushed from buds, were approximately 2 to 3 cm in diameter. The parks were sprayed with two applications of Bt at a dosage of 10 BIU in 4.7 L/ha each. There was a 2- to 3-day interval between applications on the same area.

The double applications of Bt in the parks provided results that were somewhat variable but, overall, reasonably effective. Larval populations were lowered to non-nuisance levels and good foliage protection was generally achieved. The contrasts evident between treated and untreated stands in several locations were very striking. Results of the trials at Key River showed that Dimilin® at 70 g/ha and Orthene® at 560 g/ha were the most effective treatments, with high larval mortality and excellent foliage protection. However, there was a strong probability that the Dimilin® plot was either sprayed directly or contaminated with drift from the Orthene® 560 g/ha treatment. The two plots were about 183 m from each other. Orthene® at 280 g/ha provided excellent results and Dylox® at 280 g/ha provided acceptable results, particularly since most of the defoliation was caused by high populations of large aspen tortrix, *Choristoneura conflictana*, in this plot. This species is a relatively late developing insect and probably not particularly susceptible to sprays timed for forest tent caterpillar.

In 1978, the Ministry sprayed a total of 433 ha which included three provincial parks, one each in Parry Sound, Bracebridge, and Huronia districts, a research area in Huronia District, the Frost Centre in Minden District, and a county forest in Owen Sound District (Table 10). Thuricide® 16B was applied at 10, 12, and 15 BIU/ha. A group of private landowners, advised by the Ministry, Owen Sound District, sprayed 253 ha with Dipel® WP at 6 BIU/ha. Two trials totaling 81 ha

Table 10. Summary of spray operations against the forest tent caterpillar, *Malacosoma disstria*, in Ontario, 1975–1978.

Year and date	Location	Area sprayed (ha)	No. applic.	Insecticide and application rate/ha	Carrier	Corrected population reduction (%) tA[a]	Corrected population reduction (%) sM[b]	Reduction in current defoliation[c] (%) tA	Reduction in current defoliation[c] (%) sM	Aircraft, spray system, and contractor
1975	Grundy Lake P.P.[d]	506	2	Fenit.[e] 210 g/1.75 L	Water					AgCat (1)
May 22–23	Key River	40.5 (exper.)[f]	2	Fenit. 210 g/1.75 L	Water	99.7				Micronair atomizers
	Key River	40.5 (exper.)	1	Fenit. 210 g/1.75 L	Water	86				General Airspray
	Key River	40.5 (exper.)	2	Thur.[g] 16B, 10 BIU[h]/4.7 L	Water	92				
	Key River	40.5 (exper.)	1	Thur. 16B, 10 BIU/4.7 L	Water	71				
1977	Grundy Lake P.P.	314	2	Thur. 16B, 10 BIU/4.7 L	Water	69		85		Cessna AgTrucks (2)
May 14–18	Killbear P.P.	587	2	Thur. 16B, 10 BIU/4.7 L	Water					Micronair atomizers
	Oastler P.P.	21	2	Thur. 16B, 10 BIU/4.7 L	Water					Modern Air Spray
May 16–18	Mikisew P.P.	69	2	Thur. 16B, 10 BIU/4.7 L	Water					
	Arrowhead P.P.	513	2	Thur. 16B, 10 BIU/4.7 L	Water	79	69	95	85	
	BRMC[i]	264	2	Thur. 16B, 10 BIU/4.7 L	Water	30	38	53	86	
	Research Plot	20	1	Thur. 16B, 10 BIU/4.7 L	Water					
May 12–13	Key River	40.5 (exper.)	1	Dimilin[j], 70 g/4.7 L	Water	93		94		
	Key River	40.5 (exper.)	1	Orthene, 560 g/4.7 L	Water	92		90		
	Key River	40.5 (exper.)	1	Orthene, 280 g/4.7 L	Water	70		94		
	Key River	40.5 (exper.)	1	Dylox, 280 g/4.7 L	Water	48		68		
1978										
May 19 and 23	Grundy Lake P.P.	131	2	Thur.16B, 7.5 BIU/4.7 L	Water	99				Cessna AgTruck (1)
May 24 and 29	Mikisew P.P.	71	2	Thur.16B, 7.5 BIU/4.7 L	Water					Stearman (1)
May 25 and 30	Six Mile Lake P.P.	61	2	Thur.16B, 7.5 BIU/4.7 L	Water	79				Micronair atomizers
n.a.	Research plot	20	1	Thur.16B, 10 BIU/4.7 L	Water					Zimmer Air Services
May 25 and 30	Frost Centre	69	2	Thur.16B, 7.5 BIU/4.7 L	Water					
May 29	Country Forest	81	1	Thur.16B, 12 BIU/4.7 L	Water		100			
May 19	Key River	40.5 (exper.)	1	Thur.16B, 7.5 BIU/4.7 L	Water	83				
May 19 and 24	Key River	40.5 (exper.)	2	Thur.16B, 7.5 BIU/4.7 L	Water	89				
Private Spraying										
May 26–29	Owen Sound	253	1	Dipel WP, 6 BIU/18.7 L	Water	98				Piper Pawnee (1) Micronair atomizers Crop Protection Services

[a] Trembling aspen.
[b] Sugar maple.
[c] Difference in percentage defoliation between treated and untreated plots.
[d] Provincial Park.
[e] Fenitrothion.
[f] Experimental spraying.
[g] Thuricide® 16B.
[h] Billion international units.
[i] Bracebridge Resource Management Centre.
[j] Plot may have been sprayed directly or contaminated with drift from Orthene® 560 g/ha treatment.

comparing 7.5 BIU/ha and 15 BIU/ha of Thuricide® 16B were conducted by the Unit at Key River. All spraying was done between May 19 and 30. Results were inconclusive because forest tent caterpillar late larval and pupal populations were very low due to parasitism and disease in treated and check areas.

Oak Leafshredder, 1977–1980 and 1983

Between 1973 and 1983, nearly 3 600 ha of oak forest in southern Ontario, specifically Huronia and Maple districts, were aerially treated with chemical insecticides to reduce populations of the oak leafshredder (Table 11). This insect has been a persistent pest in red oak stands in Ontario in recent years, particularly in the late 1970s and early 1980s when higher than normal populations prevailed. The spraying operations described in the following section were conducted by the Ministry. Larval and host development (spray timing), treatment effectiveness, and special surveys to obtain the biological information required to determine areas requiring protection were carried out by the Unit with assistance from the Ministry.

Severe defoliation of oak stands for 3 consecutive years in parts of Dufferin County, Huronia District, prompted the Ministry to carry out spray trials in 1973. About 526 ha were sprayed with Sevin®-4-Oil. These 1973 trials were described by Sippell in Prebble (1975).

In 1977, 612 ha of oak–maple forest at Awenda Provincial Park, Huronia District, were sprayed to prevent as much defoliation as possible. The main operation consisted of 486 ha treated with Sevin®-4-Oil at a dosage and rate of 1 120 g in 9.4 L/ha. Operational field trials using Orthene® at 840 g in 9.4 L/ha or 1 120 g in 9.4 L/ha totaling 126 ha were also conducted. All spraying was done between May 19 and May 24. Details on the dates, materials, dosages formulations, type of aircraft, spray systems, and results are provided in Table 12. All treatments proved to be effective in reducing larval populations. Sevin®-4-Oil achieved satisfactory results, although defoliation was slightly higher in some of the Sevin®-treated blocks compared to early Orthene® treatments. Much of the Sevin®-treated area was sprayed 2 or 3 days past the date for optimum foliage protection. Two Orthene® treatments carried out early, May 20 or 21, when larvae were in second instar, provided excellent foliage protection. The late Orthene® treatment, May 24, applied when larvae were in third instar, provided much less foliage protection.

In 1978, a total of 758 ha of oak–maple forest in Huronia District was sprayed to minimize defoliation by the oak leafshredder. Spraying took place in three locations: Midhurst, Wildman, and Dufferin (Table 12). Twenty-eight ha were treated at Midhurst, a Ministry nursery, 195 ha were sprayed at Wildman Tract, part of the Simcoe County Forest, and 535 ha in Dufferin County Forest. The county forests were under Ministry management. Based on the results of the 1977 spraying at Awenda Provincial Park, Orthene® 85 mixed with water was used as the operational insecticide in 1978 applied at a dosage of 840 g in 9.4 L/ha by fixed-wing aircraft with a boom and nozzle delivery system. Several trials involving reduced dosage and application rate and different delivery systems were incorporated into the 1978 operational program. Thus, the various trials carried out were helicopter equipped with Beecomists, Orthene® 840 g in 4.7 L/ha, and fixed wing with Micronair atomizers, Orthene® 840 g and 560 g in 4.7 L/ha. Sprays were applied May 25 to 28 with the larvae primarily in second instar. All treatments were highly effective in reducing larval populations and protecting foliage (Table 12).

In 1979, the Ministry aerially sprayed a total of 303 ha in Huronia District which basically involved areas treated in 1977 or 1978 at Awenda, Wildman, and Dufferin. Based on the results of trials conducted in 1977 and 1978, Orthene® 85 SP mixed with water was used as the operational insecticide in 1979 applied at a dosage and rate of 840 g in 9.4 L/ha by an AgCat equipped with boom and nozzles. Larval development at time of spraying, which was May 22 to 24, was second and third instars. Results for 1979 (Table 12), particularly Awenda and Wildman with 94% to 95% larval mortality and good foliage protection, were equivalent to previous years.

In 1980, 767 ha of oak–maple forest in Huronia District were sprayed to minimize defoliation. Another 81 ha consisting of equal parts of private and Crown land were sprayed at Uxbridge, Maple District. Orthene® 85 SP mixed with water was applied at 840 g in 9.4 L/ha by AgCat or Piper Pawnee aircraft equipped with boom and nozzles. Spraying was conducted May 19 to 23 when the insects were in second and third instar. All areas received good foliage protection except Uxbridge where larval mortality and foliage protection were considerably less than the other three areas that were sprayed. Reasons for the poorer results at Uxbridge are not clear except that the spraying in Maple District was done by a different contractor and aircraft than the operation in Huronia District.

On May 24 and 25, 1983, the Ministry aerially treated 579 ha of oak–maple forest in Dufferin County Forest and adjacent areas in Simcoe County Forest, Huronia District. Two Piper Pawnees equipped with Micronair atomizers were used to apply single applications of both chemical and biological insecticides. Orthene® 97 SP was applied at two dosages, 840 g and 560 g in 9.4 L/ha to approximately 195 ha. The lower dosage of

Table 11. Summary of area (ha) aerially sprayed for the oak leafshredder, *Croesia semipurpurana*, in Ontario, 1973, 1977–1980, and 1983.

	Area treated	
	Operational	Experimental
1973[a]	526	
1977	486	126
1978	758	
1979	303	
1980	848	
1983	515	64[b]

[a] 1973 operation described by Sippell in Prebble (1975).
[b] Chemical insecticide was used for all spraying except for the 64 ha of experimental spraying in 1983 when Bt was tested.

Table 12. Summary of spray operations against the oak leafshredder, *Croesia semipurpurana*, in Ontario, 1977–80 and 1983.

Year and date	Location	Area sprayed (ha)	No. applic.	Insecticide and application rate/ha	Carrier	Corrected population reduction (%)	Reduction in current defoliation (%)[a]	Aircraft, spray system, and contractor
1977								
May 21–23	Awenda P.P.[b]	486	1	Sevin-4-Oil, 1120 g/9.4 L	fuel oil	80	60	Piper Pawnee (1)
May 21 (early)	Awenda P.P.	49 (exper.)[c]	1	Orthene, 840 g/9.4 L	water	78	90	b & n[d]
May 24 (late)	Awenda P.P.	16 (exper.)	1	Orthene, 840 g/9.4 L	water	87	31	Crop Protection
May 20 (early)	Awenda P.P.	61 (exper.)	1	Orthene, 1120 g/9.4 L	water	84	89	Services
1978								
May 25	Midhurst	28	1	Orthene, 840 g/4.7 L	water	90	70	Hughes 500 C Beecomists (4) Viking Helicopters Ltd.
May 26	Wildman	81	1	Orthene, 560 g/4.7 L	water	71	89	Stearman, Micronair atomizers
May 26	Wildman	114	1	Orthene, 840 g/4.7 L	water	80	86	AgTruck, Micronair atomizers
May 26–28	Dufferin	535	1	Orthene, 840 g/9.4 L	water	95	61	Stearman-AgTruck, b & n Zimmer Air Service
1979								
May 22	Awenda P.P.	81	1	Orthene, 840 g/9.4 L	water	94	85	AgCat, b & n
May 23	Wildman	121	1	Orthene, 840 g/9.4 L	water	95	82	General Airspray
May 24	Dufferin	101	1	Orthene, 840 g/9.4 L	water	55	74	
1980	**Huronia District**							
	Awenda P.P.	688	1	Orthene, 840 g/9.4 L	water	83	79	AgCat, b & n
May 20–23	Hendrie Forest	51	1	Orthene, 840 g/9.4 L	water	100	79	General Airspray
	Midhurst	28	1	Orthene, 840 g/9.4 L	water	98	70	
	Maple District							
May 19	Uxbridge	81	1	Orthene, 840 g/9.4 L	water	25	36	Pawnee, b & n Sandham Air Service
1983								
May 24	Dufferin and Simcoe	320	1	Sevin-4-Oil,1120g/9.4 L	n.a.[e]	99	65	Piper Pawnees (2)
May 25	County Forests,	132	1	Orthene, 840 g/9.4 L	water	98	57	Micronair atomizers
May 24	Huronia District	63	1	Orthene, 560 g/9.4 L	water	97	47	
May 24		64 (exper.)	1	Dipel 88, 40 BIU[f]/4.7 L	water	49	32	

[a] Difference in percentage defoliation between treated and untreated plots.
[b] Provincial Park.
[c] Experimental spraying.
[d] Boom and nozzles.
[e] Data not available, probably fuel oil.
[f] Billion international units.

Orthene® was applied in response to a request from the Canadian Wildlife Service. Sevin®-4-Oil was applied to 320 ha at 1 120 g in 9.4 L/ha. Dipel® was sprayed on 64 ha at 40 BIU in 4.7 L/ha and it was the first time that Bt had been used against the oak leafshredder in Ontario. Insect development at time of spraying was first and second instar. Spraying began the evening of May 24 and was completed the following morning. Just as spraying was completed, a light rain began to fall and continued for the rest of the day. As well, temperatures decreased considerably and it remained cool for 3 days following the spray. Sevin® and Orthene® (at both dosages) were very effective in reducing larval numbers and limiting defoliation. The Dipel® treatment produced rather modest results but it is likely that the cool, wet weather following the spray reduced the effectiveness of this treatment.

Saratoga Spittlebug, 1976

In 1975, severe damage in the form of branch and top mortality caused by Saratoga spittlebug, *Aphrophora saratogensis*, was discovered in parts of 13 red pine plantations near Pembroke, Ontario. Special surveys in the fall of 1975 to determine the number of feeding punctures and eggs indicated that populations would be high in 1976. The Ministry decided to spray six of these plantations, totaling 143 ha, to try to prevent further damage.

Malathion 50% E.C. mixed with water was applied at 560 g in 4.7 L/ha using a Piper Super Cub from Sandham Air Service, equipped with boom and nozzles. Nymphal development and adult emergence was followed closely in the field by a Unit technician. Adult emergence appeared to be nearly complete by mid-July and sprays were applied the evening of July 22.

Population counts were made before and after spraying in each of the six plantations slated for treatment and in four privately owned plantations selected for check plots. Trees were swept using a heavy duty insect net. A sweep consisted of a quick swing of the net upward along a tree's foliage from the lowest branch to as high as the individual could reach or to the top of the tree. Two sweeps were made per tree, each on opposite sides. Adults captured in the net were counted and released. Pre-spray surveys were made July 13 to 15; post-spray surveys were made July 26 and 27. Average tree height in each plantation ranged from 1 to 4 m.

Results are presented in Table 13. The difference in the population levels between the treated plantations and the check plots after spraying is very pronounced. The insecticide and method of application would seem to be an extremely effective method of controlling this pest. Surveys carried out during the fall of 1976 showed a large decrease in feeding damage by this insect in treated plantations (Table 14).

Linden Looper, 1975

The linden looper, *Erannis tiliaria*, is widely distributed in Canada and the United States and feeds on a wide variety of deciduous tree species. The most recent outbreak of this insect in Ontario occurred in 1988 when nearly 9 000 ha of defoliated white birch were mapped in the Terrace Bay District, North Central Region. Previous to this, an outbreak occurred in Ontario from 1974 to 1977, peaking in 1975 when 200 000 ha of white birch and yellow birch were severely defoliated in Lake Superior Provincial Park, Wawa District.

Table 13. Summary of pre- and post-spray counts and percent change of Saratoga spittlebug, *Aphrophora saratogensis*, population in red pine plantations sprayed evening of July 22, 1976, with Malathion at 560 g in 4.7 L/ha in Pembroke District.

	Pre-spray (July 13-15)			Post-spray (July 26-27)			
	No. trees swept	Total no. insects trapped	Average insects per tree	No. trees swept	Total no. insects trapped	Average insects per tree	% change
Sprayed plantations	825	544	0.6594	1 800	3	0.0017	−99.7
Unsprayed plantations	635	166	0.2614	1 273	266	0.2090	−20.0

Table 14. Summary of Saratoga spittlebug, *A. saratogensis*, feeding scars on samples of 1974 and 1975 growth of red pine taken from 10 plantations (6 sprayed, 4 not treated) in the Pembroke District.

		Average no. feeding scars per 10 cm of twig		Average no. eggs per sample[a]	
	No. twigs examined	1975	1976	1975	1976
Sprayed plantations	80	25.4	1.7	17.1	0.6
Unsprayed plantations	50	26.3	7.4	12.0	6.1

[a] Average number of eggs based on examination of one bud cluster from each of four trees.

Table 15. Population reduction and foliage protection due to application of Dylox®, fenitrothion, and Bt against third- and fourth-instar larvae of linden looper, *Erranis tiliaria*, on white birch in Lake Superior Provincial Park in 1975.

Treatment	Pre-spray larvae/61-cm branch tip	Surviving larvae/61-cm branch tip	Days after spray	% population reduction due to spray	% 1975 defoliation
Fenitrothion, 280 g/1.75 L·ha^{-1}, June 4–10					
Rabbit Blanket	50.4	2.9	9	64	36
Check	42.1	6.7		–	29
Mijin Road	15.2	.15	12	97	23
Check	19.6	6.55		–	22
Dylox, 840 g/1.75 L·ha^{-1}, June 10					
Dylox	36.5	.2	11	98	17
Check	31.7	7.6		–	24
Bt materials					
16B — 19BIU/5.8 L·ha^{-1}, June 4	15.0	.2	16	96	12
Check	19.6	6.55		–	22
24B — 24 BIU/5.3 L·ha^{-1}, June 7	25.1	.08	12	99	18
32B — 21 BIU/3.1 L·ha^{-1}, June 8	25.1	0	11	100	20
Check	31.7	7.6		–	24

Table 16. Population reduction of third- and fourth-instar spruce coneworm, *Dioryctria reniculelloides*, larvae on white spruce due to various treatments in Chapleau and Kirkland Lake districts, 1978.

Treatment	Pre-spray larvae per 46-cm branch tip	Surviving pupae per 46-cm branch tip	% population reduction due to treatment
Aerial application—Kirkland Lake District			
Matacil, 2 applications, 87.5 + 87.5 g in 4.67 L/ha			
Burt Twp Seed Production Area	14.5	4.79	38
Check	16.2	8.06	
Matacil, 1st application, 87.5 g in 4.67 L/ha			
Orthene, 2nd application, 0.56 kg in 4.67 L/ha			
Swastika Nursery	5.0	1.24	78
Check	5.7	6.50	
Hydraulic application—Chapleau District			
Orthene, 0.45 kg/378.5 L			
Reeves Twp Seed Production Area	23.2	3.64	55
Check	23.2	8.10	
Dipel, 0.45 kg/378.5 L			
Wakami Lake	9.4	4.7	0
Check	9.5	4.0	
5 Mile Lake	7.7	2.5	55
Check	7.4	5.4	
Shoals	5.3	1.1	82
Check	5.7	6.5	
Missinaibi Lake	4.3	3.4	31
Check	5.7	6.5	

Table 17. Population reduction of spruce coneworm, *D. reniculelloides*, larvae on white spruce due to various treatments in six districts in Northern Ontario, 1979.

Treatment	Pre-spray larvae per 46-cm branch tip	Surviving pupae per 46-cm branch tip	% population reduction due to treatment
Orthene, 560 g in 4.7 L/ha	9.06	1.84	62
Check	4.87	2.58	
Orthene, 280 g in 1.5 L/ha–Novabac 32B, 20 BIU/4.7 L•ha^{-1}	11.24	.76	85
Check	11.50	5.08	
Matacil, 63 g in 1 L/ha	8.54	1.90	0
Check	8.94	1.60	
Matacil, 87.5 g in 1.5 L/ha	9.12	2.45	41
Check	8.15	3.70	
Thuricide 16B, 10 BIU in 4.7 L/ha	3.80	.20	96
Check	4.00	5.93	
Thuricide 16B, 10 BIU in 4.7 L/ha, 2 applications	7.15	1.74	63
Check	4.60	3.03	
Novabac 32B, 20 BIU in 4.7 L/ha, 2 applications	11.64	3.35	56
Check	6.62	4.32	

Operational spraying totaling about 3 300 ha was conducted for spruce budworm in Lake Superior Provincial Park in 1975. Insecticides used were fenitrothion and Dylox®. Also, three Bt products, Thuricide® 16B, 24B, and 32B, were tested against the spruce budworm in Lake Superior Provincial Park. In the course of budworm larval development activities and pre-spray sampling, it was realized that heavy infestations of linden looper were present on white birch throughout those stands selected for operational and experimental aerial spraying. White birch occurs commonly in association with balsam fir and white spruce in this part of the province; hence, it was decided to assess the effects of the treatments on larval populations of linden looper.

The results of the various treatments in terms of population reduction and foliage protection are presented in Table 15. Densities ranged from 15 to 50 larvae per 61-cm branch tip before spraying. All treatments were responsible for high larval mortality except for fenitrothion at Rabbit Blanket Lake Campground. Some foliage protection occurred in four of the six sites evaluated but severe defoliation was unlikely to have occurred because many loopers became diseased by a nuclear polyhedrosis virus and an entomophthora fungus. Afterward, it was concluded that an epizootic of polyhedrosis virus caused the collapse of the Lake Superior Provincial Park infestation. Finally, the timing of the sprays that were aimed at spruce budworm may not have been suitable for linden looper.

Spruce Coneworm, 1978–1980

The spruce coneworm, *Dioryctria reniculelloides*, is an insect associated with spruce budworm and is reported to be occasionally important as a cone-feeder and defoliator on white spruce.

Populations of spruce coneworm increased in many locations in northeastern Ontario in 1976 and 1977, and in some instances the numbers of coneworms alone were capable of causing significant damage. For example, in 1978 in a white spruce seed production area in Reeves Township, Chapleau District, pre-spray numbers of coneworm averaged 23.2 per 46-cm branch tip or about 25% of the total larval population (budworm and coneworm) feeding on white spruce in this location. An opportunity existed to assess the effects of various treatments planned for spruce budworm control on spruce coneworm populations. Consequently, in 1978, 1979, and 1980, when white spruce sample branches were examined for budworm larvae or pupae, counts were also made of the numbers of coneworm present.

In 1978, several treatments in Chapleau and Kirkland Lake districts including aerial applications of Matacil® and Matacil® + Orthene® and ground applications of Orthene® and Dipel® were assessed for effectiveness against spruce coneworm (Table 16). Average density of pre-spray larvae in treated areas was 9.9 per 46-cm branch tip. All treatments had

Table 18. Population reduction of spruce coneworm, *D. reniculelloides*, larvae on white spruce due to various aerial treatments in three districts in Northern Ontario in 1980.

Treatment	Pre-spray larvae per 46-cm branch tip	Surviving pupae per 46-cm branch tip	% population reduction due to treatment
Kapuskasing District			
Cygon, 560 g in 18.8 L/ha	1.33	.16	15
Check	1.26	.20	
Hearst & Kirkland Lake District			
Orthene, 560 g or 1.12 kg in 9.4 L/ha	7.74	.33	66
Check	9.83	1.25	
Kirkland Lake District			
Novabac, 25 BIU in 4.7 L/ha	1.60	.42	76
Check	1.03	1.15	
Matacil, 86 g in 4.7 L/ha	1.80	.65	29
Check	2.18	1.12	
Permethrin, 17.5 g in 1.4 L/ha	3.60	.48	60
Check	3.33	1.10	
Novabac, 25 BIU in 4.7 L/ha + Matacil, 86 g in 1.4 L/ha	1.80	1.10	45
Check	1.03	1.15	

some effect, although hydraulically applied Dipel® produced extremely variable results ranging from 0% to 82% larval mortality.

Again in 1979, significant coneworm populations were present with the average number of pre-spray larvae in spray areas being 8.6 per 46-cm branch tip. In 1979, much of the budworm spraying was carried out in seed production areas, so for the second consecutive year the various treatments to control spruce budworm were also assessed for their effects on spruce coneworm populations. Treatments, pre-spray larval densities, surviving densities, and percent mortality due to treatment for 18 locations in six districts are presented in Table 17. Seven different treatments (aerial applications) were involved. Based on the 1979 results, Orthene® and Bt were more effective than Matacil®.

In 1980, six treatments in three districts in the Northern Region were evaluated for effectiveness against spruce coneworm on white spruce. Two treatments, Cygon® and Orthene®, were applied by helicopter to seed production areas. The other four treatments were applied to commercial forests by fixed-wing spray aircraft. Average pre-spray density of spruce coneworm in treated areas that were evaluated was 3.0 larvae per 46-cm branch tip. Treatments, pre-spray numbers of coneworm, surviving numbers of coneworm, and percent population reduction due to treatment are presented in Table 18. Cygon® and Matacil® were the least effective treatments. Novabac®, Orthene®, and permethrin were the most effective, and a double application of Novabac® and Matacil® was only moderately effective. The number of coneworm larvae sprayed was very low in several treatments, which could bring into question the validity of the assessment.

Nevertheless, 3 years of data indicate what works and what doesn't. Thuricide® and Orthene® provided consistently good results whereas Novabac® and Matacil® gave variable amounts of control. With only 1 year of test data and low populations, it can only be suggested that permethrin may be a promising material against coneworm, but the effectiveness of Cygon® cannot be adequately evaluated from the data. The situation with Matacil® is interesting because the dosages used are consistently effective against spruce budworm but much less effective against spruce coneworm. There is some indication that Bt and permethrin are more effective against coneworm than budworm, and Orthene® is equally effective against the two insects.

Bibliography

Howse, G.M.; Carrow, J.R. 1977a. Forest tent caterpillar aerial spraying operations, Ontario, 1977. Report to the 5th Annual Forest Pest Control Forum, Dec. 6-7, 1977. Ottawa, Ont. 8 p.

Howse, G.M.; Carrow, J.R. 1977b. Oak leafshredder aerial spraying operations, Ontario, 1977. Report to the 5th Annual Forest Pest Control Forum, Dec. 6-7, 1977, Ottawa, Ont. 6 p.

Howse, G.M.; Carrow, J.R. 1978a. Forest tent caterpillar aerial spraying operations, Ontario, 1978. Report to the 6th

Annual Forest Pest Control Forum, Nov. 28-29, 1978. Ottawa, Ont. 6 p.

Howse, G.M.; Carrow, J.R. 1978b. Oak leafshredder aerial spraying operations, Ontario, 1978. Report to the 6th Annual Forest Pest Control Forum, Nov. 28-29, 1978, Ottawa, Ont. 11 p.

Howse, G.M.; Carrow, J.R. 1979. Oak leafshredder aerial spraying operations, Ontario, 1979. Report to the 7th Annual Forest Pest Control Forum, Nov. 27-28, 1979, Ottawa, Ont. 15 p.

Howse, G.M.; Carrow, J.R. 1980. Oak leafshredder aerial spraying operations, Ontario, 1980. Report to the 8th Annual Forest Pest Control Forum, Nov. 18-20, 1980, Ottawa, Ont. 7 p.

Howse, G.M.; Harnden, A.A. 1978. The 1977 spruce budworm situation in Ontario. Can. For. Serv., Sault Ste. Marie, Ont. Inf. Rep. 0-X-280. 72 p., illus.

Howse, G.M.; Harnden, A.A. 1979. The 1978 spruce budworm situation in Ontario. Can. For. Serv., Sault Ste. Marie, Ont. Inf. Rep. 0-X-300. 72 p., illus.

Howse, G.M.; Harnden, A.A.; Carrow, J.R. 1980. The 1979 spruce budworm situation in Ontario. Can. For. Serv., Sault Ste. Marie, Ont. Inf. Rep. 0-X-310. 89 p., illus.

Howse, G.M.; Harnden, A.A.; Sippell, W.L. 1975. The 1974 spruce budworm situation in Ontario. Can. For. Serv., Sault Ste. Marie, Ont. Inf. Rep. 0-X-228. 49 p., illus.

Howse, G.M.; Harnden, A.A.; Sippell, W.L. 1976. The 1975 spruce budworm situation in Ontario. Can. For. Serv., Sault Ste. Marie, Ont. Inf. Rep. 0-X-250. 73 p., illus.

Howse, G.M.; Harnden, A.A.; Sippell, W.L. 1977. The 1976 spruce budworm situation in Ontario. Can. For. Serv., Sault Ste. Marie, Ont. Inf. Rep. 0-X-260. 82 p., illus.

Howse, G.M.; Harnden, A.A.; Meating, J.H.; Carrow, J.R. 1981. The 1980 spruce budworm situation in Ontario. Can. For. Serv., Sault Ste. Marie, Ont. Inf. Rep. 0-X-336. 93 p., illus.

Howse, G.M.; Sippell, W.L. 1976. Forest tent caterpillar and Saratoga spittlebug aerial spraying operations, Ontario 1976. Report to the 4th Annual Forest Pest Control Forum, Nov. 23-24, 1976, Ottawa, Ont. 3 p.

Meating, J.H.; Lawrence, H.D.; Howse, G.M.; Carrow, J.R. 1982. The 1981 spruce budworm situation in Ontario. Can. For. Serv., Sault Ste. Marie, Ont. Inf. Rep. 0-X-343. 92 p., illus.

Meating, J.H.; Lawrence, H.D.; Howse, G.M.; Carrow, J.R. 1983a. The 1982 spruce budworm situation in Ontario. Can. For. Serv., Sault Ste. Marie, Ont. Inf. Rep. 0-X-349. 75 p., illus.

Meating, J.H.; Lawrence, H.D.; Cunningham, J.C.; Howse, G.M. 1983b. The 1982 gypsy moth situation in Ontario: general surveys, spray trials and forecasts for 1983. Can. For. Serv., Sault Ste. Marie, Ont. Inf. Rep. 0-X-352. 14 p., illus.

Meating, J.H.; Howse, G.M.; McGauley, B.H. 1983c. Spruce budworm in Ontario, 1983. Can. For. Serv., Sault Ste. Marie, Ont. Report to the 11th Annual Forest Pest Control Forum, Nov. 15-17, 1983, Ottawa, Ont. 20 p., illus.

Meating, J.H.; Howse, G.M.; McGauley, B.H. 1983d. Oak leafshredder aerial spraying operations in Ontario, 1983. Report to the 11th Annual Forest Pest Control Forum, Nov. 15-17, 1983, Ottawa, Ont. 7 p.

Meating, J.H.; Howse, G.M.; McGauley, B.H. 1984. Spruce budworm in Ontario, 1984. Can. For. Serv., Sault Ste. Marie, Ont. Report to the 12th Annual Forest Pest Control Forum, Nov. 27-29, 1984, Ottawa, Ont. 19 p.

Meating, J.H.; Howse, G.M.; McGauley, B.H. 1985a. Spruce budworm in Ontario, 1985. Can. For. Serv., Sault Ste. Marie, Ont. Report to the 13th Annual Forest Pest Control Forum, Nov. 19-21, 1985, Ottawa, Ont. 16 p.

Meating, J.H.; Howse, G.M.; McGauley, B.H. 1985b. Jack pine budworm in Ontario, 1985. Can. For. Serv., Sault Ste. Marie, Ont. Report to the 13th Annual Forest Pest Control Forum, Nov. 19-21, 1985, Ottawa, Ont. 12 p., illus.

Meating, J.H.; Howse, G.M.; McGauley, B.H. 1985c. Gypsy moth in Ontario, 1985. Can. For. Serv., Sault Ste. Marie, Ont. Report to the 13th Annual Forest Pest Control Forum, Nov. 19-21, 1985, Ottawa, Ont. 8 p., illus.

Meating, J.H.; Howse, G.M.; McGauley, B.H. 1986a. Spruce budworm in Ontario, 1986. Can. For. Serv., Sault Ste. Marie, Ont. Report to the 14th Annual Forest Pest Control Forum, Nov. 18-20, 1986, Ottawa, Ont. 15 p.

Meating, J.H.; Howse, G.M.; McGauley, B.H. 1986b. Jack pine budworm in Ontario, 1986. Can. For. Serv., Sault Ste. Marie, Ont. Report to the 14th Annual Forest Pest Control Forum, Nov. 18-20, 1986, Ottawa, Ont. 17 p., illus.

Meating, J.H.; Howse, G.M.; McGauley, B.H. 1986c. Gypsy moth in Ontario, 1986. Can. For. Serv., Sault Ste. Marie, Ont. Report to the 14th Annual Forest Pest Control Forum, Nov. 18-20, 1986, Ottawa, Ont. 15 p., illus.

Meating, J.H.; Howse, G.M.; Lessard, R.A. 1987a. Spruce budworm in Ontario, 1987. Can. For. Serv., Sault Ste. Marie, Ont. Report to the 15th Annual Forest Pest Control Forum, Nov. 17-19, 1987, Ottawa, Ont. 14p.

Meating, J.H.; Howse, G.M.; Lessard, R.A. 1987b. Jack pine budworm situation in Ontario, 1987. Can. For. Serv., Sault Ste. Marie, Ont. Report to the 15th Annual Forest Pest Control Forum, Nov. 17-19, 1987, Ottawa, Ont. 13 p., illus.

Meating, J.H.; Howse, G.M.; Lessard, R.A. 1987c. Gypsy moth in Ontario, 1987. Can. For. Serv., Sault Ste. Marie, Ont. Report to the 15th Annual Forest Pest Control Forum, Nov. 17-19, 1987, Ontario, Ont. 18 p., illus.

Prebble, M.L. (Ed.). 1975. Aerial control of forest insects in Canada. Dep. Environ., Ottawa, Ont. 330 p.

Sippell, W.L. 1975. Oak leafshredder. *In* M. L. Prebble (Ed.), Aerial control of forest insects in Canada. Dep. Environ., Ottawa, Ont.

Weir, H.J.; Jansons, V. 1977. Forest Insect and Disease Surveys in the Algonquin Region of Ontario, 1976. Can. For. Serv., Sault Ste. Marie, Ont. 20 p.

Chapter 67

Insect Control in Manitoba, 1973–1989

K. Knowles and A.R. Westwood

Introduction

Aerial insecticide applications in Manitoba have been implemented for forest pest control for a number of insect species over the last 20 years. These programs were first initiated because of public pressure concerning pests such as the forest tent caterpillar or because of efforts by government agencies to reduce impacts on high-use recreational areas in the province. In more recent years, control programs have also been implemented against jack pine and spruce budworms, which have threatened extensive timber volumes in many areas of the province.

The Forestry Branch of the Manitoba Department of Natural Resources has been working to develop forest pest impact data that can be used to guide pest management programs. Many of the early application projects were reactive, being implemented only after significant foliage loss, top-kill, or outright tree mortality had occurred. By further refining its survey skills and assessing the potential volume loss and the economic viability of taking action, the Forestry Branch intends to develop "thresholds" above which various levels of forest pest control action can be implemented before the damage occurs, or at least during the early stages of infestation. Much of this work is being done in conjunction with Forestry Canada and in cooperation with other provincial governments.

In Manitoba, infestations of jack pine budworm, forest tent caterpillar, and spruce budworm have warranted aerial insecticide applications in recent years. The applications can be split into two periods. Those of the 1970s were generally characterized by the use of small single-engine aircraft, boom and nozzle equipment, conventional application volumes, and a fairly unsophisticated means of controlling aircraft flight paths. Application technology implemented for operations during the late 1980s included larger single-engine aircraft (e.g. Dromader M18), Micronair® rotary atomizers, ultra-low volume applications, and the use of pointer aircraft for delineating flight paths. During the 1970s, the coordination of spray programs was carried out by several agencies, including the Manitoba Parks Branch, the Manitoba Forestry Branch, Forestry Canada, and the Manitoba Department of Agriculture. In 1980, the Forest Protection Section was established within the Forestry Branch to deal with forest pest management. Forest Protection has coordinated forest insect control programs since its inception.

Figure 1 shows Manitoba's wooded areas and the specific locations referred to in the text.

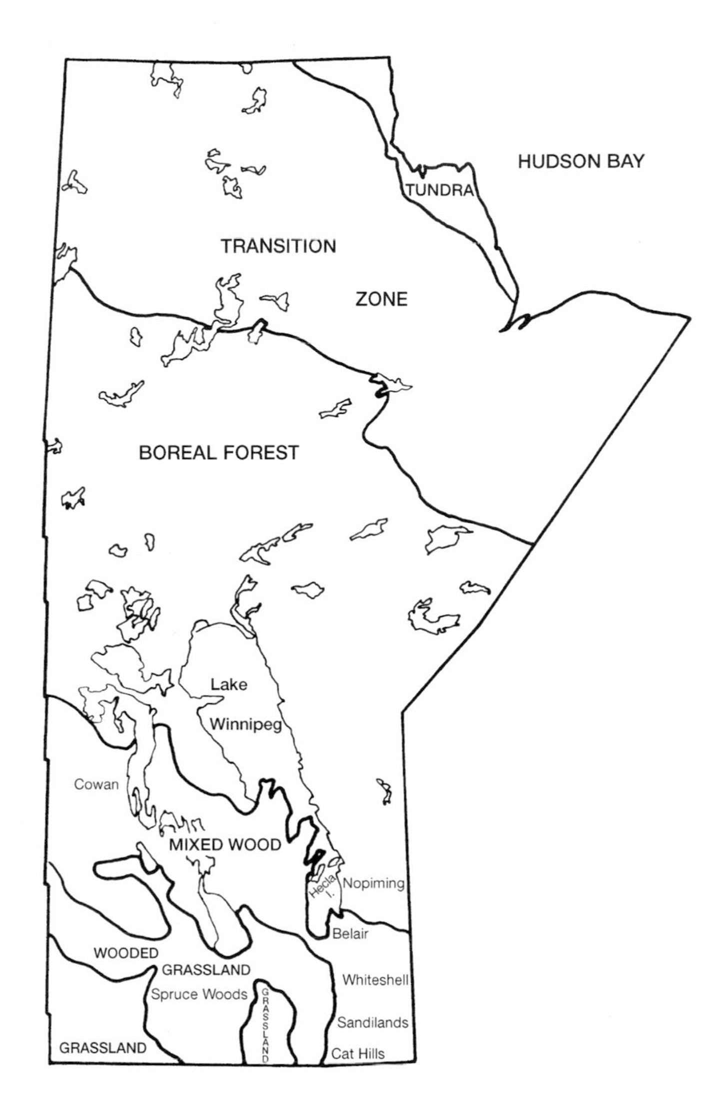

Figure 1. Forest areas of Manitoba, including provincial forest.

Jack Pine Budworm

Jack pine budworm, *Choristoneura pinus*, outbreaks occurred in southern Manitoba from 1973 to 1979. Infested areas included Spruce Woods, Belair, Whiteshell, and Sandilands provincial forests. With the exception of the Spruce Woods outbreak, which persisted for 7 years, the duration of these outbreaks was generally 2 years. In Spruce Woods, control measures were carried out in 6 of the 7 years that the outbreak persisted. Fenitrothion was applied as an operational treatment from 1974 to 1977 inclusive (Table 1). The insecticide was mixed with Atlox® 3409 emulsifier, Texaco Aerotex® 3470 oil (in the ratio 77.5:11.6:10.9), and water. Target E and Atplus 526 were also added to the 1976 and 1977 applications, at the rates of 140 ml/9.35 L and

Table 1. Aerial applications of insecticides to control the jack pine budworm, *Choristoneura pinus*, in Manitoba forests, 1974–1986.

Year	Location	Insecticide	Hectares	Dosage (AI/ha)[a]	Spray vol. (L/ha)
1974	Spruce Woods Provincial Forest	fenitrothion	65	280 ml	9.35
1975	Spruce Woods Provincial Forest	fenitrothion	520	280 ml	9.35
1976	Spruce Woods Provincial Forest	fenitrothion	520	280 ml	9.35
1977	Spruce Woods Provincial Forest	fenitrothion	1 400	280 ml	9.35
1977	Belair Provincial Forest	fenitrothion	720	280 ml	9.35
1979	Spruce Woods Provincial Forest	Bt	25	10 BIU[b]	4.70
1986	Cat Hills Provincial Forest	Bt	409	20 BIU	1.57
	Sandilands Provincial Forest	Bt	19 473	20 BIU	1.57
	Belair Provincial Forest	Bt	3 822	20 BIU	1.57
	Kississing Lake area	Bt	541	20 BIU	1.57
1986	Cowan area	Bt	2 286	20 BIU	3.40

[a] Active ingredient per hectare.
[b] Billion international units.

178 ml/389 L of solution respectively. Cost of the application increased from $4.08/ha in 1974 to $5.25/ha in 1977. Pre- and post-spray larval counts were not done in 1974, 1975, or 1976. In 1977, 68% larval reduction was achieved. Novabac®3, a Bt (*Bacillus thuringiensis*) formulation, mixed with Chevron Sticker in the ratio 1:1 600 and Erio Acid Red at the rate of 0.25 g/L, was used in a trial spray block in 1979 at the cost of $5.20/ha for the application. The application rate was 10 billion international units (BIU)/ha. Despite the low dosage, 83% larval reduction was achieved. However, foliage protection was poor as defoliation was similar in both the treated and untreated areas (77% and 75% respectively). Control was carried out in Belair Provincial Forest in 1977 following severe defoliation in 1976, using the same formulation as the 1977 Spruce Woods operation. The Belair population had collapsed by 1978. No insecticide applications were implemented in either the Sandilands or Whiteshell areas. All jack pine budworm treatments in Belair and Spruce Woods during this outbreak period were applied by Cessna AgWagon aircraft equipped with booms and nozzles. Helium-filled meterological balloons controlled by fishing reels were used to provide aircraft guidance. Timing of all applications coincided with peak fourth-instar larval development.

Jack pine budworm outbreaks were absent from the province in 1980 and 1981. In 1982 severe defoliation was detected north of Grand Rapids. This infestation spread slightly to the northwest in 1983. Considerable expansion occurred in 1984 to the northwest, northeast, and southward. New infestations also appeared east of Lake Winnipeg. The infestation peaked in 1985, covering much of the range of jack pine throughout the province. The following year, a spray program was implemented in high-priority jack pine forests in four separate regions of the province. In total, 26 531 ha were treated (Table 1). Bt was applied undiluted to all spray blocks, at a total cost of $20.63/ha. Two formulations of Bt were used, Thuricide® 48 LV and Thuricide® 32F. Application was carried out by a team of three Dromader M18 aircraft equipped with Micronair® AU5000 rotary atomizers; a Cessna-172 was used as a pointer aircraft. Timing of application coincided with peak third- to fourth-instar larval development. Larval reduction due to treatment was variable, ranging from a low of 31% to a high of 86%.

During the 1986 spray program the original spray blocks were based on an egg-mass survey conducted in August and September of 1985. This resulted in a planned spray program of just under 34 000 ha. Pre-spray larval counts done just prior to opening each spray block indicated a significant reduction in population in some areas. Based on the pre-spray survey, the actual area sprayed was reduced by approximately 20% to the 26 531 ha indicated.

Although the precise relationship between male flower production and jack pine budworm larval survival is not completely understood, the Forestry Branch is attempting to incorporate this variable into the jack pine budworm survey program.

Forest Tent Caterpillar

Localized defoliation of trembling aspen by the forest tent caterpillar, *Malacosoma disstria*, became noticeable in southern Manitoba in 1971. The outbreak that followed persisted until 1978; at its peak, in 1977, the majority of the aspen parkland of southern Manitoba was severely defoliated. Following a Manitoba Clean Environment Commission Hearing held in April 1975, permission was granted for the aerial control of forest tent caterpillar (Table 2). Foliage protection programs were carried out by the Manitoba Department of Agriculture and the Manitoba Parks Branch. Treated areas included farm woodlots, provincial park campgrounds and picnic areas, as well as cottage subdivisions. The farm woodlot

Table 2. Aerial applications of the insecticide Malathion to control the forest tent caterpillar, *Malacosoma disstria*, in Manitoba forests, 1974–1978.

Year	Location	Hectares	Dosage (AI/ha)[a]	Spray vol. (L/ha)
1974	Manipogo Provincial Park	50	350 ml	9.35
1975	Whiteshell area	240	350 ml	9.35
	Grand Beach Provincial Park	300	350 ml	9.35
	Interlake parks	300	350 ml	9.35
	Western parks	375	350 ml	9.35
	Farm woodlots	22 700	350 ml	9.35
1976	Whiteshell area	835	350 ml	9.35
	Grand Beach Provincial Park	300	350 ml	9.35
	Interlake parks	300	350 ml	9.35
	Western parks	440	350 ml	9.35
	Farm woodlots	29 100	350 ml	9.35
1977	Whiteshell area	950	350 ml	9.35
	Grand Beach area	400	350 ml	9.35
	Interlake parks	620	350 ml	9.35
	Western parks	460	350 ml	9.35
	Birds Hill Provincial Park	780	350 ml	9.35
	Farm woodlots	50 000	350 ml	9.35
	Cottage subdivisions	1 800	350 ml	9.35
1978	Whiteshell Provincial Park	400	350 ml	9.35
	Birds Hill Provincial Park	145	350 ml	9.35
	Western parks	265	350 ml	9.35
	Eastern parks	130	350 ml	9.35
	Farm woodlots	14 800	350 ml	9.35

[a] Active ingredient per hectare.

program was coordinated by the Department of Agriculture in conjunction with participating municipalities. Private woodlot owners funded the program on a per hectare basis. Control in campgrounds and picnic areas was carried out by the Parks Branch. The cottage subdivision spray program was coordinated by the Parks Branch and funded by the Whiteshell Cottage Owner's Association. In the initial year of control, 1974, one campground of 50 ha was treated. In 1977, at the peak of infestation, 55 000 ha were aerially sprayed. By 1978, the outbreak had declined substantially, and only 15 740 ha were treated. All applications were with malathion (50 EC) applied with Cessna AgWagon, Cessna AgTruck, and Piper Pawnee aircraft equipped with conventional boom and nozzle systems. Aircraft guidance, where deemed necessary, was achieved through the use of helium-filled meterological balloons. No systematic method of assessing larval reduction or foliage protection was implemented during any of the tent caterpillar spray programs. General visual observation, however, indicated that good control was achieved when application was made during the appropriate stage in larval development (peak third instar) and under favorable weather conditions.

Spruce Budworm

Spruce budworm, *Choristoneura fumiferana*, infestations in Manitoba have persisted considerably longer than those of the jack pine budworm. An infestation in the Spruce Woods Provincial Forest, which was treated three times in the early 1970s, lasted for 13 years, from 1967 to 1979. The Spruce Woods infestation is unique, in that balsam fir, the spruce budworm's primary host, is absent from this area. White spruce is the only host species in Spruce Woods. The current infestation in eastern Manitoba, which has been treated on six different occasions, commenced in the mid-1970s. Despite the longevity of these outbreaks, the spruce budworm is only considered a serious forest pest on a very localized basis due to the lack of abundance of the host species. The white spruce and balsam fir components of Manitoba's commercial forests are 7% and 1% by volume respectively.

In 1973 (Table 3) over 3 000 ha were operationally treated in the Spruce Woods with fenitrothion in solution with an emulsifier, Aerotex oil and water. Two Cessna AgWagons equipped with Micronair® AU2000 and 3000 rotary atomizers

Table 3. Aerial applications of insecticides to control the spruce budworm, *Choristoneura fumiferana*, in Manitoba forests, 1973–1989.

Year	Location	Insecticide	Hectares	Dosage (AI/ha)[a]	Spray vol. (L/ha)
1973	Spruce Woods Prov. Park and Forest	fenitrothion	3 238	280 ml	9.35
	Spruce Woods Prov. Park and Forest	carbaryl	50	1.123 L	2.93
	Spruce Woods Prov. Park and Forest	Bt	80	10 BIU[b]	37.4, 18.7
1974	Spruce Woods Prov. Park and Forest	fenitrothion	800	280 ml	9.35
1975	Farm woodlot next to Spruce Woods Prov. Park	fenitrothion	20	280 ml	9.35
1980	Whiteshell Prov. Park and Hecla Island Golf Course[c]	fenitrothion	77	–	–
1981	Whiteshell Prov. Park campgrounds	Bt	37	–	–
1981	Hecla Island Prov. Park	Bt	365	20 BIU	5.6
1986	Hecla Island Prov. Park	Bt	423	20 BIU	1.57
1986	Bird Lake area	Bt	300	20 BIU	1.57
1987	Whiteshell Prov. Park	Bt	536	30 BIU	3.0
1988	Whiteshell and Nopiming Prov. Park	Bt	1 182	30 BIU	2.4
1989	Whiteshell and Nopiming Prov. Park	Bt	1 003	30 BIU	1.77
1989	Abitibi-Price Forest Management License	Bt	3 981	30 BIU	1.77

[a] Active ingredient per hectare.
[b] Billion international units.
[c] Ground sprayed to run off in a 0.25% solution or suspension in water.

and a Piper Pawnee with boom and nozzle were used. Generally, greater success was achieved with the boom and nozzle equipment at this application volume. Larval reduction due to treatment with fenitrothion ranged from 28% to 86% (Hildahl and DeBoo 1973). Carbaryl was also applied with a Piper Pawnee aircraft fitted with booms and nozzles. Larval reduction due to carbaryl was 86% (Hildahl and DeBoo 1973). Bt was applied to 80 ha at a rate of 10 BIU/ha with a Cessna AgWagon aircraft fitted with Micronair® AU2000 rotary atomizers. Applications contained equal parts molasses and water at volumes of 18.7 and 37.4 L/ha. Larval reductions with Bt were 43% and 61% for the light and heavy applications respectively (Morris et al. 1975). In 1974, several blocks in the Spruce Woods were sprayed with fenitrothion by a Cessna AgWagon aircraft fitted with booms and nozzles. A farm woodlot adjacent to Spruce Woods Park was treated with fenitrothion in 1975 as part of a stream deposition study carried out by Dr. L. Lockhart of the Freshwater Institute, Environment Canada.

In 1980, the Hecla Island golf course and two campgrounds in Whiteshell Provincial Park were ground sprayed with fenitrothion. High pressure 200-gallon tankers were used. Larval reductions in the three areas were 61%, 78%, and 87%. The two Whiteshell campgrounds were ground sprayed again in 1981 with Bt, with a resulting larval reduction of 70%. The high-use recreational area at Hecla Island was aerially sprayed with Bt (Dipel®:water at the proportion 40:60) in 1981. The treated area included a golf course, campground, picnic area, cottage subdivision, and a resort complex. A Cessna AgWagon aircraft equipped with Micronair® AU2000 rotary atomizers was used for the application. Larval reduction due to treatment was 68%.

No control projects for spruce budworm were implemented from 1982 to 1985 inclusive. In 1986, the recreational area at Hecla Island and an area at Bird Lake (campground and cottage subdivision) in Nopiming Provincial Park were sprayed with Bt (Thuricide® 48 LV). A Dromader M18 aircraft equipped with Micronair® AU5000 rotary atomizers was used for the application and a Cessna-172 as a pointer aircraft. Larval reduction due to treatment was 20% at Hecla Island and 36% at Bird Lake.

Two locations in Whiteshell Provincial Park were aerially sprayed for spruce budworm in 1987. At Falcon Lake a 165-ha spray block consisting of a golf course and the townsite was treated. In the Dorothy Lake area 371 ha of campgrounds and cottage subdivisions were treated. Thuricide® 48 LV was applied undiluted at a rate of 3.0 L/ha; the increased volume was to compensate for loss of potency during 1 year of storage. A Cessna AgTruck aircraft fitted with Micronair® AU5000 rotary atomizers and a Cessna-172 pointer aircraft carried out the project. The larval development index was 3.5 at the time of application and shoot development was greater than Auger's class IV on both balsam fir and white spruce. Temperatures at the time of application were 12° to 14°C; relative humidity was 60% and wind speed was calm to 5 km. Despite what appeared to be ideal conditions, in terms of insect and host development and weather, no larval reduction was achieved. Severe defoliation occurred in both treated and untreated blocks.

In 1988, high-use recreation areas were treated in both Whiteshell and Nopiming provincial parks. A total of 1 182 ha were sprayed with Bt (Dipel® 132) undiluted at a strength of 30 BIU/ha. The application rate was 2.4 L/ha. A Thrush Commander (single-engine spray aircraft) equipped with six Micronair® AU5000 rotary atomizers was used for the application and a Cessna-172 was again used to guide the spray aircraft. The operation was hampered by above normal temperatures causing very rapid larval development, strong winds, fog, and thunder showers. Larval reduction due to treatment ranged from 47% to 70%. Defoliation in the treated areas ranged from 25% to 63%, compared with 50% to 86% in untreated areas. The cost of the application was $29.96 per ha.

The spruce budworm infestation in eastern Manitoba increased in 1989. Approximately 58 000 ha of white spruce and balsam fir suffered moderate to severe defoliation within the Abitibi-Price Forest Management License, and in Nopiming and Whiteshell provincial parks. An aerial spray program was implemented over an area of 4 984 ha, of which 1 003 ha was high-use recreational forest in Whiteshell and Nopiming parks and 3 981 ha was commercial forest within the Abitibi-Price Forest Management License.

Bt (Dipel® 8L) was applied undiluted at 30 BIU/ha at a rate of 1.77 L/ha in a single application. Two Thrush Commanders equipped with six Micronair® rotary atomizers were employed for the application. Swath width was 70 m and a Cessna-172 was employed for navigation. Application costs were $21.22/ha.

Larval reduction due to insecticide averaged 27%. Defoliation was 28% in treated blocks and 50% in untreated areas. Unseasonably cool weather causing a decrease in feeding activity is believed responsible for the less than desirable results.

Conclusion

From 1973 to 1989, Manitoba underwent a significant change in the approach to forest pest control in general and in the aerial application of pesticides in particular. Both chemical and biological pesticides have become integral tools of the forest manager as a result of the increased investment in forest management. Forest pest management operations are now implemented as required to suppress pest populations below economic thresholds when conditions warrant. The Forestry Branch works toward all aspects of integrated pest management taking into account biological and economic "values at risk." This philosophy extends to cooperative research efforts with both Forestry Canada and governments of other provinces to better predict population dynamics, forest pest impact, and the value to be gained by taking an appropriate control action.

Acknowledgments

A number of individuals have been involved in these insect control projects. Although not all can be named here, we recognize and appreciate their efforts. A special acknowledgment must be given to the following individuals: J. Bissinger (deceased), Provincial Parks Forester; Y. Beaubien, Forestry Branch; A.E. Campbell (retired), Canadian Forestry Service; V. Hildahl (retired), Canadian Forestry Service, Forestry Branch; L. Matwee, Forestry Branch; J. McCullough, Manitoba Department of Agriculture; M. Slivitzky, Forestry Branch; W. Louttit, Forestry Branch; W. Keller, formerly with Provincial Parks; and G. Munro, now Ontario Ministry of Natural Resources.

References

Forest Protection, Manitoba Natural Resources Report 87-2. Jack Pine Budworm Control Program 1986.

Forest Protection, Manitoba Natural Resources Report 1-88. Spruce Budworm Report 1988 and Predictions for 1989.

Forest Protection, Manitoba Natural Resources Report 2-90. Spruce Budworm Report 1989 and Predictions for 1990.

Hildahl, V.; DeBoo, R.F. 1973. Aerial applications of chemical insecticides against the spruce budworm in Manitoba, 1973. Man. Entomol. 7: 6-14.

Morris, O.N.; Angus, T.A.; Smirnoff, W.A. 1975. Field trials of *Bacillus thuringiensis* against the spruce budworm, 1960-1973. Pages 129-131 *in* M.L. Prebble (Ed.), Aerial control of forest insects in Canada. Dep. Environ., Ottawa, Ont.

Chapter 68

Insect Control in Alberta and Saskatchewan, 1979–1989

MURRAY L. ANDERSON, HIDEJI ONO, AND MADAN PANDILA

Introduction

The most conspicuous forest pests in Alberta and Saskatchewan have been the defoliators of trembling aspen. The forest tent caterpillar, *Malacosoma disstria*, frequently reaches outbreak proportions over large areas of both provinces, but no major control operations have been undertaken. Land owners have organized local spray programs to protect the aspen foliage around their homes, but no estimates of the areas involved are available.

Both provinces have been fortunate in having relatively low populations of forest insect pests that could detrimentally affect our coniferous forests. Although overall population levels have been low, some insect pest outbreaks have occurred in several areas. The jack pine budworm, *Choristoneura pinus*, has caused extensive damage in Saskatchewan. Salvage cutting was undertaken after the outbreak collapsed, but this could not be considered a control measure. The mountain pine beetle, *Dendroctonus ponderosae*, and the spruce budworm, *Choristoneura fumiferana*, however, have necessitated some control action. Both Alberta and Saskatchewan have used sanitation and salvage harvesting where possible as a primary tool to control these outbreaks and use the affected timber.

Mountain Pine Beetle

Although a short-lived outbreak of the mountain pine beetle occurred in the Banff area of Alberta in the early 1940s, it was generally considered that the climate was too severe for the insect to become a major pest east of the divide. A major outbreak in southwestern Alberta and in the Cypress Hills region of southeastern Alberta and southwestern Saskatchewan in the late 1970s to early 1980s proved that this assumption was false.

Table 1. Trees treated in a control program against the mountain pine beetle, *Dendroctonus ponderosae*, in Alberta, 1979–1986.

Beetle development year August–July	No. trees treated	
	Lodgepole pine	Limber pine
1979–1980	15 589	–
1980–1981	24 702	–
1981–1982	7 222	–
1982–1983	7 990	–
1983–1984	8 792	23 250
1984–1985	5 030	13 704
1985–1986	282	1 034

Alberta

The mountain pine beetle was detected in southwestern Alberta in 1977 when nine infested spots were identified in the province. Their origin was suspected to have been Glacier National Park in northern Montana and the Flathead River Valley of southeastern British Columbia. The outbreak spread quickly into Waterton National Park in southwestern Alberta, and subsequently to the Castle River area south of Crowsnest Pass. The exceptionally mild winters in 1978 and 1979 allowed the population to expand very rapidly. By the early 1980s, more than 40 townships in Alberta's Castle River Valley had become infested. In addition, spot infestations were identified in the Porcupine Hills and in the Cypress Hills of Alberta and Saskatchewan.

Lodgepole pine was the major commercial species affected. Infested stands south of Highway 3 in the Crowsnest Pass area were salvage logged since most of this timber had already become heavily infested. All identified beetle-infested sites north of Highway 3 were sanitation logged to retard and control the outbreak (Table 1). Limber pine was also found to have become infested with mountain pine beetle. Affected trees from this species and any debris from the sanitation logging operation were burned to kill the beetle. This process of sanitation logging and burning of the remaining debris was also used to control infestations in the Cypress Hills.

The infestation reached an endemic level by 1986 after which no further major control efforts were undertaken. Sex pheromone traps continue to be deployed to monitor beetle populations.

Saskatchewan

The mountain pine beetle was found in the Cypress Hills region in the late 1970s and early 1980s. It had probably been transported from the Waterton National Park area.

The primary control efforts consisted of the use of pheromone baits, burning of affected trees, and salvage logging (Table 2). The pheromone baits were effective at controlling

Table 2. Trees treated in control program against the mountain pine beetle, *D. ponderosae*, in Saskatchewan, 1980–1986.

Year	No. trees	No. pheromone baits
1980	128	–
1981	767	–
1982	2 222	–
1983	905	335
1984	84	1 000
1985	3	800
1986	0	500

beetle movement during its flight period. By 1987, the population was at its lowest point in 9 years. The long-term plan is to break up the even-age pine forest into a variety of age-classes to reduce the risk of mountain beetle outbreaks in the future. Monitoring of this area will continue.

Spruce Budworm

The spruce budworm is found wherever stands of its host trees occur, but it has been of major concern primarily in the northern and western parts of Alberta and in northern Saskatchewan, as this is where the largest stands of commercial coniferous timber are located. Even in these locations, there have been no extensive spraying operations on the scale of those conducted in eastern Canada.

Alberta

Endemic populations of spruce budworm have been present in Alberta, mostly along river valleys. Periodically populations reach epidemic levels where some control action has been needed. Control efforts by farmers, home owners, towns, cities including Edmonton, and provincial park agencies have been undertaken to maintain trees around homes and parks. Estimates of the total areas treated are not available.

For minor outbreaks, timber salvage and accelerated timber harvesting have been used. In one instance, an epidemic population of spruce budworm was controlled by weather factors where severe, freezing temperatures followed an extended period of warm conditions, which had prompted the budworm larvae to emerge from the protective bud caps.

In 1989, Dipel® 176 (*Bacillus thuringiensis*) at a rate of 1.8 L/ha (undiluted, ultra-low volume) was aerially sprayed over 1 000 ha of forest in the Eaglesham area, northeast of Grande Prairie, by a Grumman AgCat aircraft equipped with Micronair® Au3000 spray nozzles. This site was not conducive to harvesting or salvage logging. Monitoring of budworm levels on Eaglesham and other sites such as Chinchaga River, House River, and Hawk Hills where budworm levels are increasing will continue.

Table 3. Summary of spruce budworm, *Choristoneura fumiferana*, defoliation and timber harvesting for control in Saskatchewan, 1982–1988.

	Area defoliated (ha)			Area harvested (ha)
Location	1982	1984	1988	1984–1988
Tall Pines	2 000	4 800	16 600	5 150
Red Earth	–	7 900	15 000	4 520
Torch River	–	1 000	–	550
Total	2 000	13 700	31 600	10 220

Saskatchewan

Outbreaks of spruce budworm have been recorded in several areas of Saskatchewan. Although the effect in each site has been significant, the total area defoliated has been relatively minor. Progressive timber salvage harvesting of heavily infested areas apparently slowed down the spread of the spruce budworm infestation in Saskatchewan (Table 3).

Chapter 69

Insect Control in British Columbia, 1974–1988

R.F. DeBoo and S.P. Taylor

Introduction

In the previous compendium on aerial applications of insecticides for forest pest control in Canada (Prebble 1975), British Columbia species targeted and time frames for intermittent sprays were: western blackheaded budworm, *Acleris gloverana* (1956–1973); western hemlock looper, *Lambdina fiscellaria lugubrosa* (1929–1964); western false hemlock looper, *Nepytia freemani* (1948–1973); phantom hemlock looper, *Nepytia phantasmaria* (1956–1957); green-striped forest looper, *Melanolophia imitata* (1959–1964); saddleback looper, *Ectropis crepuscularia* (1960–1961); pine butterfly, *Neophasia menapia* (1961); Douglas-fir tussock moth, *Orgyia pseudotsugata* (1962); hemlock needleminer, *Epinotia tsugana* (1965); cone and seed insects (1971); white pine weevil, *Pissodes strobi* (1971); and striped ambrosia beetle, *Trypodendron lineatum* (1958–1970).

Most of these operations were conducted with the deep involvement and interest of Forestry Canada research staff. Aircraft used included a specially fitted Boeing Flying Boat, a Canso Model A, the versatile Stearman, small Bell and Hiller helicopters, Grumman Avengers, and conventional agricultural spray aircraft. Materials used included calcium arsenate, DDT, BHC, phosphamidon, fenitrothion, and commercial formulations of Bt (*Bacillus thuringiensis*). These materials were usually applied because of concern for defoliation, growth reduction, and mortality of trees in commercial forests; degradation of stored logs; and loss of amenity values in parks. A variety of organizations collaborated with Forestry Canada staff to plan, implement, and analyze the projects. These included the British Columbia Forest Service, the Council of Forest Industries of British Columbia, and both federal and provincial parks services. Cost-sharing arrangements and assignment of staff specialists for key roles during applications were variable over the years, but usually involved Forestry Canada research staff for field leadership, with funding for materials, supplies, and support staff from the collaborating organizations.

In retrospect, none of these aerial spray programs could be considered truly operational because the mandate to treat was confounded by the perceived urgent nature of the operation, limited prior experience, and the lack of policy and operational resources. As a result, the plan usually involved a temporary pooling of available human resources and funding, a dominant research core, and poor follow-up after treatment.

Few projects continued for more than one or two seasons. The operations against western blackheaded budworm, for example, included research trials near Port McNeill on Vancouver Island in 1956 and a DDT spray project over about 63 000 ha in 1957. Both DDT and Bt applications were made in 1960 on the Queen Charlotte Islands. Another project, this time using fenitrothion, was conducted over about 12 000 ha, again on northern Vancouver Island, in 1973. Thus, over the period 1956 to 1973, three insecticides were evaluated on three separate occasions, based on the perceptions and needs of researchers and forest managers (Richmond 1986). In fact, these applications and all others reported for British Columbia in Prebble (1975) really constituted pilot studies or operational trials and not spray operations per se.

Not much has changed in British Columbia since 1973. Of the 30 projects listed (Tables 1–3), most were research studies or operational trials by design (e.g. Canadian Forestry Service 1980). Only the cooperative sprays against the Douglas-fir tussock moth in the Kamloops Region during 1983 and the gypsy moth eradication program (1984–1988) might be considered intentionally operational. That is, these projects were structured with clear objectives, defined roles and responsibilities for collaborators, supporting legislation (provincial Forest Act, federal Plant Quarantine Act) and policies (British Columbia Ministry of Forests, 1984, 1988), established procedures, and, most importantly, with an operational agency (British Columbia Forest Service, Agriculture Canada—Plant Protection) solidly in the lead role.

This chapter will summarize these few operational experiences, present the current rationale for dealing with defoliators (which is based largely on experiences gained from the two recent western spruce budworm outbreaks), and outline the decision-making process likely to be followed during future operations with aircraft in British Columbia. Reference should be made to other chapters of this book for details of particular British Columbia research studies and operational trials.

Douglas-fir Tussock Moth

British Columbia Forest Service policy recognizes and encourages the development and implementation of pest management agreements with other government agencies, industrial concerns, and private individuals. This policy stems from Sections 125 to 127 of the British Columbia Forest Act, which require pest control actions when deemed appropriate.

In 1980 to 1984 the Douglas-fir tussock moth, *Orgyia pseudotsugata* (Fig. 1), a cyclical pest, attacked dry-belt Douglas-fir forests in south-central British Columbia. During this outbreak, the British Columbia Forest Service implemented a program that accelerated the development of the preferred treatment, a potent nuclear polyhedrosis virus, and provided relief to woodlot owners, ranchers, municipalities, and individual homeowners (Fig. 2). The primary objective for these spray trials was to develop the ability to mitigate damage due to this native defoliator (Fig. 2) before the next outbreak, which is likely to occur in the early 1990s. Included in the plan was an appraisal of wood volume and property value losses caused by this pest (Ross and Taylor 1983; Taylor 1986). The project (Fig. 3) combined three major components to achieve

Figure 1. Mature larva of the Douglas-fir tussock moth, *Orgyia pseudotsugata*.

Figure 2. Rural residence near Kamloops, British Columbia, surrounded by forest killed by the Douglas-fir tussock moth, *O. pseudotsugata*. In addition to wood loss, property value was estimated to have depreciated 10%.

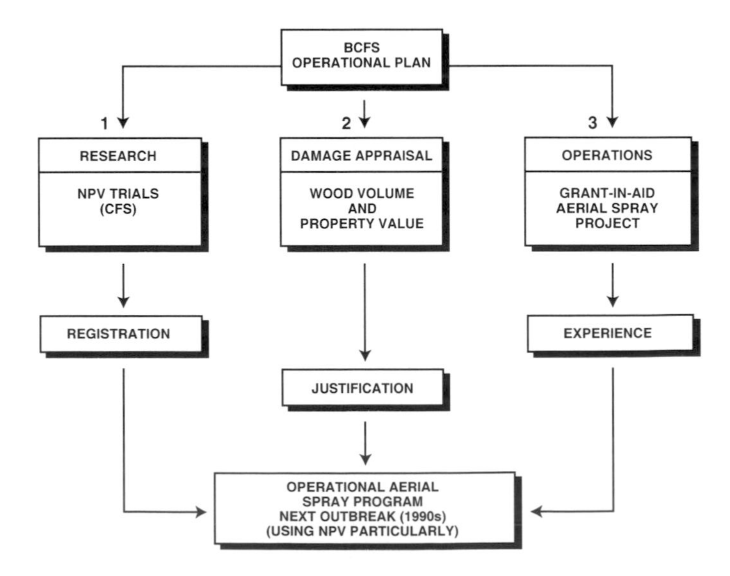

Figure 3. Objectives of the 1981–1983 Douglas-fir tussock moth, *O. pseudotsugata*, program: full operational capability for the next outbreak.

one purpose: An operational capability to aerially apply a registered viral insecticide with an array of conventional chemical spray alternatives in reserve.

The first component was achieved with the temporary registration of Douglas-fir tussock moth nuclear polyhedrosis virus in 1983. The outstanding results of the trials (Shepherd et al. 1984) also enabled the British Columbia Forest Service to negotiate the purchase of a significant supply of the virus from the United States Forest Service for application during the next outbreak.

The work involved Forestry Canada evaluation of several dosages of a virus spray formulation of interest in both Canada and the United States. Resources for the work included funding and support staff from the Protection Branch and the Kamloops Region, British Columbia Forest Service, respectively.

The second aspect of this plan concerned the development of a standard method to assess impacts of Douglas-fir tussock moth defoliation in order to analyze the aerial option for spray treatment. The estimates (Taylor 1986), based on timber volume losses of 308 100 m^3 and a property value reduction factor of 10% (Ross and Taylor 1983), indicated a combined loss of

$3.1 million due to this short-lived outbreak. A more detailed analytical method (M. Stemeroff, pers. comm.) is now available to permit a more precise determination of the need for and net economic benefits of aerial treatments.

Thirdly, at the instruction of the Minister, an operational project was conducted in 1983 to assist individuals and municipalities in protecting woodlots and trees of amenity value. The project included British Columbia Forest Service staff from Kamloops, regional-district staff of the Ministry of Municipal Affairs, and a local aerial applicator. The procedure included on-site inspection, treatment recommendation and project coordination from the British Columbia Forest Service, project approval and administration by the local regional-district office, and delivery of spray treatment by the experienced aerial contractor.

Altogether, some 99 different applications of the insecticide acephate (at the active ingredient rate of 1.1 kg/ha applied by AgTruck) were made to woodlots, townsites, and rural developments. Some results were poor, most likely due to lateness of application, the volatility of the aqueous mixture of acephate spray, and the dry Kamloops climate. Few of the treatments achieved the same superior level of protection (only 8% defoliation) as the trial application of carbaryl (Sevin®-4-oil at the active ingredient rate of 1.1 kg/ha by Cessna AgTruck) on Neskainlith Indian land near Chase in 1982 (Chorney 1982). By comparison, defoliation ranged from 40% to 100% in the acephate-treated stands (Tables 1 and 2).

The project did, however, satisfy the needs of most individuals requesting treatment while enabling them to recover 50% of their expenses through a cost-sharing agreement. Forest Service staff at the Kamloops regional office, as well as at several district offices, served the public well in their role as technical and extension personnel. This comprehensive liaison function in pest management was the first experience for the Forest Service in this capacity during an aerial spray operation. It involved public meetings, on-site inspections, and communication with regional-district personnel on

Table 1. Aerial applications of insecticides to control the Douglas-fir tussock moth, *Orgyia pseudotsugata*, in British Columbia forests, 1975–1988.[a]

Year	Insecticide	Hectares treated	Dosage(s) (AI/ha)[b]	Spray volume (L/ha)	Remarks
1975	Bt	15 260	16.8 and 25.2 BIU[c]	3.5–9.3	Four Micronair® AU3000 atomizers were used in some areas
	NPV[d]	1 210	24.7×10^{10} PIB[e]	9.3, 18.7	
	diflubenzuron	300	0.28 kg	9.3	
	acephate	180	1.12 kg	3.8	
1976	acephate	8 490	0.58–1.12 kg	1.9, 3.8	
	Bt	810	17.0 BIU	4.7	
	diflubenzuron	100	0.035–0.141 kg	3.8	
1981	pheromone[f]	50	0.008, 0.025 kg		
	NPV	20	22.0×10^{10} PIB	11.3	Boom mounted on a Bell 206B helicopter
1982	carbaryl	320	1.1 kg	9.5	Operational application under direction of BCFS
	NPV	40	1.6–25.0 ($\times 10^{10}$) PIB	9.4	
	acephate	20	1.1 kg	9.0	Private operational application in Nelson Forest Region
1983	acephate	1 070	1.1 kg	9.0	BCFS/private operational application

[a] Unless otherwise noted, the outbreaks were in the Kamloops Forest Region, the applications were experimental, the sprays were applied with conventional booms fitted with flat-fan jet nozzles (Teejet nos. 8005-8015) mounted on Cessna AgTruck or AgWagon aircraft, and Forestry Canada was the lead agency.

[b] Active ingredient per hectare.

[c] Billion international units.

[d] Nuclear polyhedrosis virus.

[e] Polyhedral inclusion bodies.

[f] Dry applications of hollow Conrel fibers filled with 2-6 heneicosen-11-one hexane, dispensed under the direction of the United States Forest Service and Albany International Company, using specialized equipment designed with a rotating cone, similar to a fertilizer spreader for dry applications, and mounted on a Bell 206B helicopter.

the status of infestation, recommendation for treatment, and approvals for funding. Detailed technical and professional support included ground support assistance for the operator during the applications, monitoring of Douglas-fir tussock moth populations, timing the treatments, and follow-up inspections and assessments.

All aspects of the three-pronged objective were successfully met, and much valuable experience was gained:

- The virus was registered because of Forestry Canada and United States Forest Service research data; as a result, the British Columbia Forest Service acquired a supply, which has been placed in storage for application early during the next outbreak as a fully registered aerial treatment.
- The British Columbia Forest Service's damage appraisal, including losses in timber and property values, suggested that a realistic method for decision-making is feasible. Decisions on treating an area will be made after the review and analysis of the environmental risks and of the economic benefits of treatment using either nuclear polyhedrosis virus or a conventional synthetic organic insecticide.
- The strictly operational component of the plan satisfied public concerns during the final year of the outbreak phase. Although optimal protection was not achieved with all applications, extremely valuable experience and information was obtained by the British Columbia Forest Service from its leadership role in this component of the program.

Gypsy Moth

The first Canadian collection of gypsy moth, *Lymantria dispar*, eggs was made by a provincial inspector in Vancouver from imported European nursery stock in 1911 (Schmidt 1986). Since then, other collections and infestations were noted at various locations in Canada. It was not until 1978, however, that the first plans were made for controlling a new gypsy moth threat in British Columbia (Powell 1986). This landmark experience, often referred to as the "Kitsilano Battle," included a major confrontation with environmental activist groups and irate citizens (Arrand 1986; Czerwinski and Isman 1986; Cram 1989), and a severely modified ground control operation using carbaryl and insecticidal soap in 1979.

The first aerial applications of an insecticide for eradication of gypsy moth occurred at Fort Langley in 1984 following detection of eggs and male moths the previous two summers (Table 2). A single helicopter application was made in conjunction with three ground spray treatments. All of the 1984 sprays were aqueous mixtures of Bt. Since then, additional aerial treatments of Bt have been made at Courtenay (1985), Chilliwack (1985–1987), Colwood (1988), and Kelowna (1988). The 1978 to 1987 treatments have been deemed successful, as no male moths have been trapped in subsequent years (Sheffield 1988).

The gypsy moth is considered a major threat to the British Columbia forest products industry, particularly since the majority of wood products are delivered out of province. As such, the primary concern has been the establishment of gypsy moth populations, concurrent with the strict enactment of domestic quarantine regulations, and the threat of embargoes on shipments of these wood products by importing countries. Defoliation of shade trees, ornamentals, and forest stands is of

Figure 4. Aerial application of Bt for eradication of gypsy moth, *Lymantria dispar*.

secondary importance at this time. Thus, British Columbia has had to develop a unique approach to dealing with the gypsy moth threat which involves interagency collaboration, pooling of resources to optimize the efficacy of Bt treatments (Fig. 4), and efficient annual surveys. The primary program objective is to prevent establishment, by means of early control intervention, to ensure continuation of export shipments.

This approach (DeBoo 1986), involves federal legislation (Plant Quarantine Act) and a lead role for Agriculture Canada as long as eradication is considered feasible. The work is organized and coordinated largely through the Gypsy Moth Committee of the British Columbia Plant Protection Advisory Council. The 1988 plan included intensive surveys and applications of up to three ground and three aerial Bt sprays. It had direct involvement for personnel in Forestry Canada, the British Columbia Forest Service, the Crop Protection Branch of the British Columbia Ministry of Agriculture and Fisheries, aerial and ground applicators, the Canadian Armed Forces, several municipal governments, as well as Research and Plant Protection units of Agriculture Canada.

The British Columbia experience from 1978 to 1988 can be summarized as follows:

- As a result of the gypsy moth threat, a strong interagency working relationship has evolved in British Columbia at the operational level. Planning and coordination of field operations occurs largely through the British Columbia Plant Protection Advisory Council.
- As a result of the Kitsilano Battle and the lessons learned in public relations, Agriculture Canada and the cooperating members of the Plant Protection Advisory Council have evolved a very successful public relations program to inform residents, elected officials, and special-interest groups of all proposed Bt treatments. To date, this activity, which is time consuming and expensive, has ensured public support and trust and has proven effective in dealing with a plant pest emergency in a difficult setting.
- As a result of the multiple-treatment and interagency approach, all evidence to date has suggested that gypsy moth infestations may be eliminated if areas of attack are small and isolated. The strategy includes both aerial and ground applications of Bt (and synthetic organic materials such as carbaryl and acephate, if feasible) in conjunction with intensive sanitation practices and pheromone trapping of

Table 2. Aerial applications of insecticides to control the gypsy moth, and other defoliators in British Columbia forests, 1975–1988.[a]

Year	Forest region	Insecticide (AI/ha)[b]	Hectares treated	Dosage(s) (AI/ha)[b]	Spray volume (L/ha)	Remarks
			Gypsy moth, *Lymantria dispar*			
1984	Vancouver	Bt	10	30 BIU[c]	29.0	
1985	Vancouver	Bt	160	30 BIU	9.6	
1986	Vancouver	Bt	25	30 BIU	9.6	
1987	Vancouver	Bt	25	30 BIU	5.9	
1988	Vancouver and Kamloops	Bt	150	30 BIU	2.4	
			Black army cutworm, *Actebia fennica*			
1975	Nelson	trichlorfon	100			Boom mounted on Bell 47G3B1 helicopter
			Western false hemlock looper, *Nepytia freemani*			
1975	Kamloops	acephate	60			Boom mounted on Cessna AgTruck or AgWagon aircraft
			Spruce cone maggot, *Strobilomyia neanthracina*			
1979	Prince George	dimethoate	10			Boom mounted on Bell 206B helicopter
			Spruce budworm, *Choristoneura fumiferana*			
1988	Prince George	Bt	150			Piper Brave PA-36 aircraft fitted with Micronair® atomizers

[a] Insecticide applications to control the gypsy moth were operational, under the direction of Agriculture Canada, and used Bell 206B helicopters fitted with four Micronair® AU4000 atomizers. The remaining applications were experimental, and the booms were fitted with flat-fan jet nozzles (Teejet nos. 8005–8015).
[b] Active ingredient per hectare.
[c] Billion international units.

male moths. At the time of this writing, British Columbia remains in the introductory phase of gypsy moth establishment (DeBoo 1986), largely because of the assertiveness of Agriculture Canada's staff in this province, the strong support offered via the Plant Protection Advisory Council, the surprisingly excellent results of all spray treatments and, undoubtedly, a large dose of good luck during the past decade.

Western Spruce Budworm

British Columbia, like most other Canadian provinces, has experienced extensive and damaging outbreaks of budworms (*Choristoneura* and *Acleris* spp.). The most noteworthy, mainly because of its distribution in populated and accessible areas of south-central British Columbia, has been the western spruce budworm, *Choristoneura occidentalis* (Fig. 5). Two outbreaks have occurred in the recent past. Each, according to survey records, surpassed previous levels.

The outbreak in the 1970s occurred primarily in the Vancouver and Kamloops forest regions with most of the infestation located in stands along the Fraser River and its tributaries from Hope north to Pemberton and Cache Creek. The infestation grew in extent and severity of attack from 1970 until it reached nearly 200 000 ha of moderate to severe defoliation in 1976. This serious threat to stand growth and survival was of concern to the British Columbia Forest Service and the forest industry in the Fraser Canyon area between Hope and Boston Bar.

Accordingly, a plan was developed to treat about 100 000 ha of the most seriously affected stands in 1977. The plan included the application of carbaryl and acephate insecticides via helicopters. However, early in the planning stages, local residents voiced their concern and opposition to the aerial spray program directly to the provincial government. Organized groups from Vancouver, including Greenpeace and the Canadian Scientific Pollution and Environmental Control Society, soon became involved. As a result of this vocal and active opposition, Forests Minister Tom Waterland convened a special conference of experts, proponents, and complainants about 1 month before the beginning of the spray period. This hearing (British Columbia

Figure 5. Severe defoliation of young (understory) Douglas-fir by western spruce budworm, *Choristoneura occidentalis*, larvae. This defoliation results from invasion from adjacent mature and semimature trees intensively managed under an uneven-aged stand management regime known as "Faller's Selection."

Ministry of Forests 1977) resulted in a modified plan to treat about 50 000 ha, which was submitted for the consideration and approval of the British Columbia Cabinet. The Cabinet turned down the plan and the spray program was cancelled. That year, 1977, budworm populations collapsed over most of the infested area and remained low to nil for the balance of the decade.

This unique experience, captured in part by Doliner (1984), was a turning point in Canadian forest pest management. For the first time, a provincial government systematically reviewed and considered detailed social and environmental questions, in addition to technical, biological, and economic factors concerning a widespread outbreak of a damaging insect pest on forests. The Cabinet decided that the anticipated benefits to be derived from the aerial spray treatments were surpassed by questions pertaining to toxicological risks to humans and to other elements of the forest environment. Similar decisions have been rendered by other provincial governments on aerial spray proposals since 1977.

The western spruce budworm has again returned to British Columbia in numbers and at locations not previously reported. By 1987, the budworm had infested about 800 000 ha of Douglas-fir stands, mostly in the dry-belt east and northeast of the Fraser Canyon forests. In this case, many young trees in selectively logged stands have been severely damaged or killed (Fig. 4). Thus, managed stands, and the millions of dollars invested for improvements in these easily accessible harvesting and recreational areas, are at a risk. This situation is unlike that of the 1970s because the Fraser Canyon forest that was under attack was mostly mature and even-aged.

The lessons learned during the 1970s coupled with the emergence of Bt as the preferred treatment have changed the basic parameters for dealing with defoliating insects in British Columbia. As a result, Forest Service staff responsible for pest management have adopted a series of required and sequential steps to determine the best treatment option. This exercise in decision making, which can be applied to all defoliators and certain other pests, includes:

- Delineation of areas of primary concern following detailed aerial and ground surveys.

Table 3. Aerial applications of insecticides to control the western spruce budworm, *Choristoneura occidentalis*, in British Columbia forests, 1976–1988.[a]

Year	Insecticide	Hectares treated	Dosage(s) (AI/ha)[b]	Spray volume (L/ha)
1976	NPV[c]	20	25.0×10^{10} PIB[d]	9.4
1978	Bt	160	20.0 BIU[e]	9.4
	NPV	60	75.0×10^{10} PIB	9.4
1982	NPV	180	5.4×10^{11} PIB	9.4
	Granulosis virus	180	1.7×10^{14} CAP[f]	9.4
1986	Bt	200	30 BIU	5.9
1987	Bt	940	30 BIU	2.8, 3.1
1988	Bt	1 550[g]	30 BIU	2.03

[a] All but one of the outbreaks were in the Kamloops Forest Region, and the insecticide applications were experimental. The 1976 to 1982 trials were directed by Forestry Canada and were applied with conventional booms fitted with flat-fan jet nozzles (Teejet nos. 8005–8015) mounted on Cessna AgTruck or AgWagon aircraft. The 1986 to 1988 trials were directed by the British Columbia Forest Service and were applied with four Micronair® AU4000 atomizers mounted on a Cessna AgTruck aircraft.

[b] Active ingredient per hectare.

[c] Nuclear polyhedrosis virus.

[d] Polyhedral inclusion bodies.

[e] Billion international units.

[f] Capsules.

[g] Some infestations were in the Nelson Forest Region.

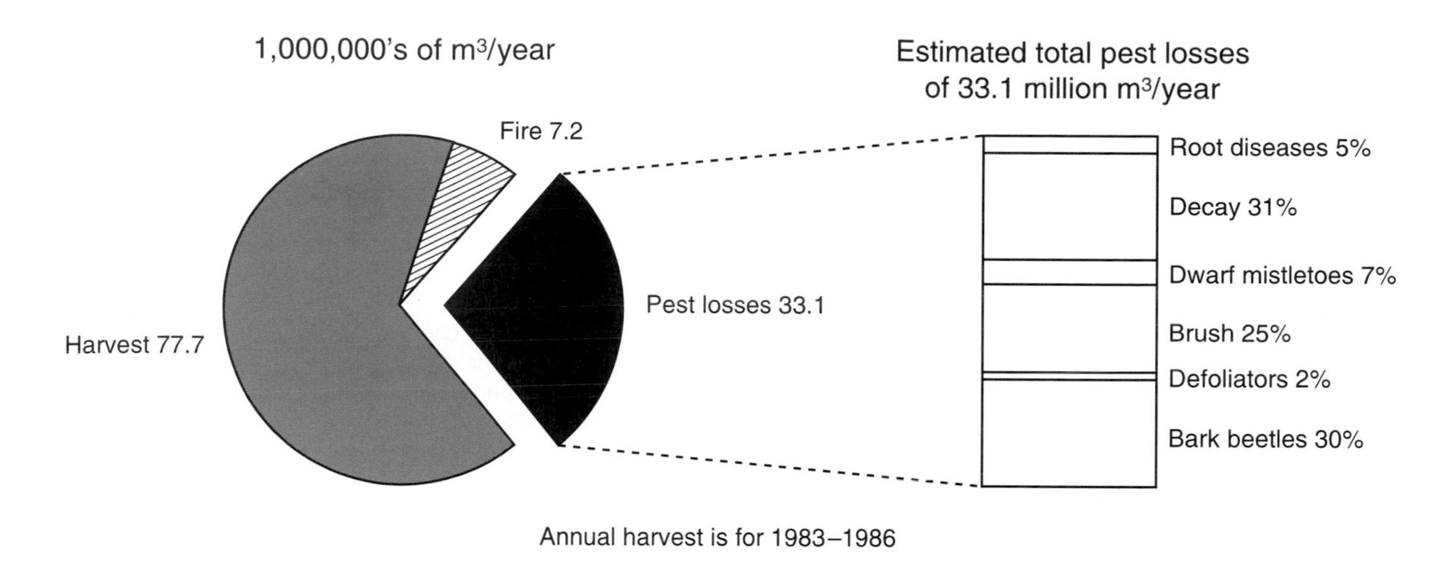

Figure 6. Estimates of gross average annual wood volume depletion in British Columbia, 1983–1987, due to harvesting, fire, and pests. (Canadian Forestry Service and British Columbia Forest Service)

- A damage appraisal procedure, preferably based on current data from permanent sample plots located in infested areas.
- A decision-making profile requiring detailed consideration of treatment options (e.g. aerial spray, priority harvesting and silvicultural prescriptions change). This profile would involve evaluation and comparison of costs, risks (human health, environmental), difficulty, efficacy, reoccurrence probability (thus, need for retreatment), stand characteristics, concerns of residents and special interest groups, as well as the consequences of not treating.

Without completion of all of these tasks, a convincing argument to fund and conduct an aerial spray program in British Columbia will not be possible. At present, Pest Management staff are actively collecting information and reviewing options. In the meantime, operational trials have been conducted since 1986 (Table 3) to evaluate Bt as a potential treatment to afford foliage protection to study plots and a provincial park, and to gain operational experience in planning and implementing a small-scale aerial spray program. To date, results have been variable using single applications at 30 BIU (billion international units)/ha, perhaps due to inexperience, terrain, climate, and the inherent and basically unknown limitations of this insecticide.

Future Programs in British Columbia

Forest insect defoliators will continue to cause losses in valuable growing stock and consternation for responsible forest managers. This concern will grow with the advances in and expectations for forest management in British Columbia. Aerial treatment, with materials like Bt and nuclear polyhedrosis virus, will most likely continue to be a dominant control option. However, resource managers will also be considering harvesting and silvicultural treatments (British Columbia Ministry of Forests 1987), especially for native Douglas-fir defoliators such as tussock moth and budworms. Silvicultural solutions will include the production of mixed forests, including larger quantities of nonsusceptible species such as pine.

Finally, decisions to treat pest problems must be related to available resources and management priorities, as well as the latest available technology. Current estimates for British Columbia suggest that impacts due to insect defoliators are less than for brush problems, several diseases, or bark beetles (Fig. 6). As a result, protection programs for defoliators may be limited in scope because of the many tasks that are required to undertake a broad operational front within a forest management unit. In British Columbia, the primary consideration is always prevention of pest problems through multidisciplinary activities wherever possible. Pest suppression projects will occur following detailed reviews and systematic evaluations of treatment options. With the development of improved formulations of Bt and registration of narrow-spectrum materials such as nuclear polyhedrosis virus, aerial application of an insecticide will remain an important and viable option in British Columbia.

References

Arrand, J. 1986. PPAC and gypsy moth sub-committee. Pages 57-58 *in* Understanding the gypsy moth threat. Proc. Inf. Symp. 5 Nov. 1985, Vancouver. B.C. Plant Prot. Advisory Counc., Agric. Can., New Westminster, B.C.

British Columbia Ministry of Forests. 1977. Hansard report on conference on Fraser Canyon budworm spraying programme. Queen's Printer, Victoria, B.C. 410 p.

British Columbia Ministry of Forests. 1984. Ministry policy. Pest management—administration. Ministry Policy Manual III - PRO-014. 11 p.

British Columbia Ministry of Forests and Lands. 1987. Drybelt fir timber management: a review for practitioners. Proc. Symp. 6-7 Oct., 1987, Williams Lake. North. Silvic. Comm., South. Internal Silvic. Comm., B.C. Min. For. Lands. 106 p.

British Columbia Ministry of Forests and Lands. 1988. Ministry Policy. Forest and range pest management—pesticide regulation and use. Ministry Policy Manual III - PRO-015, 13 p. (revised).

Canadian Forestry Service. 1980. Operational field trials against the Douglas fir tussock moth with chemical and biological insecticides. Can. For. Serv., Pac. For. Cent. Inf. Rep. BC-X-201, 19 p.

Chorney, F.J. 1982. Evaluation of an aerial application of Sevin-4-oil against the Douglas-fir tussock moth, 1982. B.C. For. Serv., Victoria, B.C. Internal Rep. PM-K-3, 9 p.

Cram, W.A. 1989. Gaining supporting for British Columbia's gypsy moth wars 1978-1988. A case study in public relations. B.C. Min. For. Pest Manage. Rep. No. 12, 35 p.

Czerwinski, C.; Isman, M.B., 1986. Urban pest management: decision-making and social conflict in the control of gypsy moth in west-coast cities. Bull. Entomol. Soc. Am. 32(1): 36-41.

DeBoo, R.F. 1986. Developing a strategic plan. Pages 59-66 *in* Understanding the gypsy moth threat. Proc. Inf. Symp. 5 Nov. 1985, Vancouver, B.C. Plant Prot. Advisory Counc., Agric. Can., New Westminster, B.C.

Doliner, L.H. 1984. The political-technical interface in pest management: a historical analysis of the proposed 1977 Fraser Canyon budworm spray program. Simon Fraser Univ. Pest Manage. Pap. No. 27, 219 p.

Powell, G. 1986. The B.C. situation. 1. British Columbia gypsy moth update. Pages 45-46 *in* Understanding the gypsy moth threat. Proc. Inf. Symp. 5 Nov. 1985, Vancouver. B.C. Plant Protect. Advisory Counc., Agric. Can., New Westminster, B.C.

Prebble, M.L. (Ed.). 1975. Aerial control of forest insects in Canada. Dep. Environ., Ottawa, Ont. Cat. No. Fo23/19/1975, 330 p.

Richmond, H.A. 1986. Forest entomology: from pack horse to helicopter. B.C. Minist. For. Lands Pest Manage., Victoria, B.C. Rep. No. 8, 44 p.

Ross, D.W.; Taylor, S.P. 1983. Effects of the current (1980-1984) Douglas-fir tussock moth outbreaks on forest resources and other values. B.C. For. Serv., Victoria, B.C. Internal Rep. PM-PB-9, 45 p.

Schmidt, A.C. 1986. The gypsy moth and Agriculture Canada past, present and future. Pages 47-52 *in* Understanding the gypsy moth threat. Proc. Inf. Symp. 5 Nov. 1985, Vancouver. B.C. Plant Protect. Advisory Counc., Agric. Can., New Westminster, B.C.

Sheffield, S. 1988. B.C. gypsy moth control programs, 1978-1987. Unpublished summary report, Agric. Can., New Westminster, B.C., 4 p.

Shepherd, R.F.; Otvos, I.S.; Chorney, R.J.; Cunningham, J.C. 1984. Pest management of Douglas-fir tussock moth (Lepidoptera: Lymantriidae). Prevention of an outbreak through early treatment with a nuclear polyhedrosis virus by ground and aerial applications. Can. Entomol. 116: 1533-1542.

Taylor, S. 1986. A revised estimate of the impacts of the 1980-84 Douglas-fir tussock moth outbreak. B.C. For. Serv., Victoria, B.C. Internal Rep. PM-DB-23, 9 p.

Part V

Future Management of Forest Pests

Chapter 70

Strategies for the Future

J.R. Carrow

Introduction

Forest management in Canada, including forest pest management, is taking place in an environment of social and economic change. (This manuscript was prepared in 1990.) Around the world, restructuring of social, political, and economic systems have become realities, rather than intellectual musings. We live in a time of globalization of trade, investment, and technology. The popularization of the concept of sustainable development has brought worldwide attention to the prospect of a doubling of the world's human population in the next 50 years, with resulting exhaustion of our natural resources, and continuing degradation of the environment. Although these processes appear distant to many Canadians, they will all affect, directly and indirectly, our country and the way in which we practise forest management in the future.

On a national scale, future management of forest pests will be influenced by several factors. First and foremost, a continuing supply of wood for industrial use, both in the short term and indefinitely into the future, must be assured. As more of the productive forest land base is allocated to non-timber uses, the land base available for production forests will continue to decrease. Public interest in the health of the forest, as well as demand for increased use of the forest environment, will result in more public involvement, which in turn will affect forest resource policies, and more regulatory controls in the interests of environmental protection. Traditional technologies for forest pest control, for example, chemical pesticides, will disappear as older control products are systematically reevaluated by regulatory agencies. As well, the interest of pesticide manufacturers in developing and registering new control technologies for forestry use is highly uncertain. The evolution and use of new pest control technologies will also be greatly influenced by the attitudes of Canadians, the majority of whom disapprove of the use of chemicals in the forest environment. Complicating the future even further is the uncertainty created by climate change, and the effects this will have on forest pests and their associated damage (Borden 1988).

Sustainable Development

The report of the World Commission on Environment and Development, "Our Common Future" (Brundtland Commission 1987), ranks as one of the seminal books of this century. Several of the arguments presented in the report bear repeating, because they will define the social, political, and economic climate within which forest pest management will be practised in the future. The principle of sustainable development is commonly expressed as the ability of humanity " to ensure that it meets the needs of the present without compromising the ability of future generations to meet their own needs." It implies limits that are imposed by the present state of technology and the ability of the biosphere to absorb the effects of human activities. The report criticizes the historic institutional separation of environmental protection and economic development, and points to the need for greater integration of environmental accountability and economic development in government agencies and the private sector.

In a recent report, the Science Council of Canada (1988) focused on the role of science and technology in achieving sustainable development in Canada. The Council pointed out that "the countries that are quick to develop clean technologies to meet their own needs are also the ones whose industries are likely to grow." The Council made specific recommendations for:

1. an environmental assessment capability in all corporations and government agencies whose activities impinge on the environment;
2. the Department of Industry, Science and Technology to provide leadership in promoting technologies that are less environmentally damaging; and
3. the establishment of a national database on the quality of ecosystems.

Public Participation

The past decade has seen a steady growth in public involvement in all aspects of forest management, particularly on Crown land. This is not surprising as 92% of the forest land base is owned by the people of Canada; after decades of apathy, the owners are finally taking an interest in the condition of their property. The Forestry Canada (1989) national public opinion poll revealed a high level of concern about the state of our forests. Among other things, the majority of Canadians:

- view our forests as a "national treasure to be held in trust" for future generations;
- consider that potential environmental impact is more important than economic value and job creation in making land use decisions;
- believe that it is more important to preserve special forest areas than to ensure jobs; and
- disapprove of the use of chemicals in the forest environment and favor biological alternatives.

Many of these attitudes translate into reality when citizens, or advocacy groups, successfully challenge government-approved forestry operations, such as harvesting of old-growth forests and aerial spraying of pesticides. The current environmental assessment of timber management on Crown lands in Ontario represents the most comprehensive public examination of forestry in Canadian history. When complete, the

Ontario Environmental Assessment Board is expected to table a set of "Terms and Conditions" that will regulate how future forestry operations are planned and carried out in Ontario. At this point, it seems clear that the process of timber management will become more restrictive, more transparent to the public, and more accountable. Relative to forest pest management, one of the parties to the environmental assessment hearings, Forests for Tomorrow (1990), a coalition of environmental groups, submitted some interesting draft terms and conditions for pesticide use and forest protection. Among other things, the coalition proposes that:

- the Ontario Ministry of Natural Resources (henceforth referred to as the Ministry) adopt an integrated pest management policy, with the declared objective of reducing pesticide use; and
- where feasible, nonchemical alternatives shall be favored; where alternatives are not available, pesticide use shall continue only while a search for alternatives is initiated, and nonchemical alternative research needs are identified and met.

More specifically, Forests for Tomorrow proposes that "no chemical insecticide use shall be permitted for the purposes of timber management within the area of the undertaking," and that "no aerial spraying of herbicides shall be permitted for the purposes of timber management within the area of the undertaking."

Recently, the issue of unresolved native land claims has surfaced as a major factor in forest management planning and operations in many regions. In the Ontario environmental assessment hearings, native peoples have also proposed terms and conditions that could affect forest pest management (Nishnawbe-Aski Nation 1990). For aerial spraying projects, they propose the use of buffer zones around potable water supplies; fish spawning areas; areas used for fishing; homes, cottages, logging camps, trappers cabins; berry-picking areas; endangered or rare species habitats; and trap lines. Nishnawbe-Aski Nation also proposes the following specific buffer zones:

a) human habitation
   1000 m for herbicides,
   1500 m for aminocarb, fenitrothion, and carbaryl,
   300 m for *Bacillus thuringiensis* (Bt);
b) water supplies—double the width of the zones for human habitation; and
c) sensitive areas
   240 m for herbicides,
   300 m for aminocarb, fenitrothion, and carbaryl.

If the Ministry plans to use aminocarb, carbaryl, or fenitrothion, Nishnawbe-Aski Nation proposes that the Ministry be required to submit an environmental assessment of the proposed spray program, as well as to monitor the effects of operational spraying on bees, raptors, furbearers, frogs, snakes, and aquatic invertebrates in ponds and streams. Although these conditions may not be accepted by the Board, they do reflect a growing tendency to severely restrict how pesticides are used in the forest environment.

The depth of public interest in pesticides is clearly evident in the current Pesticides Registration Review, initiated by the Minister of Agriculture in early 1989. The Review Team consists of representatives from every stakeholder group with an interest in pesticides: environmentalists, regulatory agencies, agriculture, forestry, labor, chemical industry, consumers, and public health.

The negotiations of the Review Team, as well as the hundreds of briefs and submissions from the public, clarify that there is widespread apprehension among Canadians about the use of pesticides, and a general and strong desire to reduce our use of conventional chemical pesticides. At the same time, producer groups are increasingly concerned about loss of pest control technology, brought about by suspension of registrations, and very slow development and registration of new technology.

Productive Forest Land Base

What society expects from the forest environment has changed dramatically in the past decade. We are placing increasing demands on the forest for both industrial and nonindustrial uses. In addition to the traditional uses such as parks, recreation areas, and wildlife management units, there is a growing demand to protect old-growth forests, habitat for endangered species (e.g. northern spotted owl), and sites with heritage and spiritual value.

Decisions reached in recent cases involving the South Moresby Islands, the Carmanah Valley, and Temagami have resulted in significant withdrawals from the industrial forest land base. Globally, the Brundtland Commission (1987) called for a protected-area network representing 12% of the land and water base. In Canada, the World Wildlife Fund is actively promoting completion of our national network by the year 2000; at present, only 3.4% is protected from industrial activity (World Wildlife Fund 1990), so the pressure to withdraw forest land from industrial use will continue. As an example, protected areas now represent 5.24% of the provincial land base in British Columbia; the Valhalla Society (1988) has identified several additional areas that would increase the protected area land base to just over 13%. In addition to the more contentious withdrawals, the Crown forest land base is steadily shrinking due to a variety of demands for other uses including agriculture, commercial use, transportation and utilities, parks and recreation, settlement. Some examples of total withdrawal from the forest land base from 1980 to 1990 are Newfoundland, 9 214 ha; Manitoba, 411 621 ha; Alberta, 103 400 ha; and British Columbia, 105,731 ha (1983–1989).

This trend will continue, and future industrial wood requirements will have to come from a shrinking land base. This in turn will put increased demands on the productive forest land base, and management will become more intense. Historically, Canada has had the luxury of being able to accept much of the growth loss and mortality associated with forest insect attack, largely because the fiber losses were surplus to our industrial needs. However, because of steadily increasing provincial timber production targets, and a diminishing land base, these losses will become less acceptable.

Timber Supply Planning

Protection of the Natural Forest

Historically, the majority of forest insect control programs in Canada have been carried out to protect commercially valuable species that made up part of the short- and long-term timber supply. This forest consisted primarily of older age-class stands that have developed naturally, with little or no silvicultural activity, and industry derived virtually all of its wood supply from these stands. Because the new forests, both natural regeneration and plantations, will not be commercially operable for several decades yet, it will be essential to continue protection of these older stands, particularly those that are included in long-term management plans, and 5-year operating plans. In 1980, the Canadian Council of Resource and Environment Ministers (henceforth referred to as the Council) endorsed a national wood production target of 210 million m^3 by the year 2000—a 35% increase in 20 years. Some provinces have adopted provincial timber production targets. For example, Ontario established a target of 25.8 million m^3 by the year 2020. The balance of extensive and intensive management required to meet these targets in the distant future is under debate (Benson 1988; Sedjo 1990). Regardless, protection of the wood supply from unacceptable losses due to insects (and other agents) for the next few decades will become increasingly essential, largely due to the escalating demand for wood and a shrinking land base to provide that wood. Failure to adequately protect these areas will result in localized shortages in wood supply (Baskerville 1983). Thus, for the near future, the challenge will be to continue protecting older stands to keep them alive for eventual harvest. The objective of such programs will be to protect trees and prevent mortality, rather than preserve growth.

Protection of the New Forests

In the past, most of the forests that received protection tended to be homogeneous in species and age-class, and remote from other values such as habitation. However, the new forests are developing largely because of harvesting activities and fires, and the pattern that is being created presents a different and more difficult challenge; they will take the form of a mosaic of smaller stands, with a range of age-classes, and they will be more intermixed with values such as habitation, water supplies, and aquatic habitat, which need to be protected from pesticides.

A review of provincial harvesting and renewal data (Table 1) gives an appreciation of the structure and extent of the new forests in each province.

Considering the average cost of preparing and planting 1 ha of harvested or burned forest land (approx. $800), as well as the cost of tending operations, Tables 1 and 2 illustrate the enormous investment in Canada's new forests. Between 1977 and 1988, annual expenditures on silviculture increased from $100 million to $700 million, and for protection, annual expenditures increased from $100 million to $400 million (Smyth et al. 1989).

This investment must be protected from pests and fire, both to ensure a predictable wood supply and to justify the continuation of major silvicultural programs, such as those carried out under the Forest Resource Development Agreements. Because of a combination of factors including decadence, fire and pest losses, harvesting, and withdrawals, the industry will be able to rely on the existing mature forest for its wood supply for only another 20 to 40 years, depending on the region. After that, industry will have to shift its harvesting to the new forest that is currently being grown, the majority of which will have been silviculturally treated. It is critically important that this new forest be commercially operable as forecast in timber supply models, if we are to avoid a serious shortfall in supply. Silvicultural activities are expected to continue and in many cases expand (Table 2).

Because of a combination of more effective fire suppression and the imposition of limits on harvest block size (Table 1), the new forest is becoming more diverse and fragmented. This structure will pose some new technical challenges, but it will also present some new opportunities for insect pest management.

Harvest block design—The size and shape of new plantations and natural stands will be determined by harvesting activities, as influenced by provincial guidelines on harvest block design. Several provinces have established upper limits on the size of clearcuts (Table 1), with the smallest limit at 32 ha; however, in many provinces there is some flexibility in the application of block size guidelines, which allows for larger blocks under certain circumstances. Thus, size alone is unlikely to constrain future insect control operations, most of which involve aerial application of insecticides. It will require, however, that the technology shift to smaller fixed-wing aircraft and helicopters. Because the new forest is more fragmented, there will be a requirement for more targeted treatments, that is, application of pest control agents will be restricted to the specific areas that need treatment. Again this will require smaller aircraft or effective and practical ground equipment. One factor that will complicate aerial application is the trend to use landscape design in laying out harvest blocks. This will produce many harvest blocks, and new stands, of random size and irregular shapes, often using physiographic contours as boundaries. Some may argue that harvest blocks should be designed to meet the needs of aerial spraying, but considering current public opinion about harvesting and the use of pesticides, this position would be politically untenable. Rather, aerial spraying technology will have to adapt to this new operational setting.

On the other hand, the well-defined boundaries associated with harvest blocks and young stands should make guidance of spray aircraft easier than the older extensive areas of natural forest that are currently the target of most aerial spraying programs.

Public concern about aerial spraying in the forest environment has persisted for about 15 years, and this concern is unlikely to dissipate. Some years ago, this concern began to deter the development and registration of new active ingredients for forestry use in Canada. The problem became sufficiently serious that the Council established a task force on forestry pesticides, which was directed to identify and

Table 1. Harvesting and renewal activities (hectares) in Canada, 1980–1989.[a]

	Maximum harvest block	Area harvested	Area planted[b]	Area tended[c]
Newfoundland	n/a	176 137	22 887 (0)[d]	41 015
Prince Edward Island	n/a	23 358	5 963 (0)	4 470
Nova Scotia	50	336 800	68 300 (0)	123 200
New Brunswick	125	362 130	119 448 (30)	62 423
Quebec	250	2 360 535	315 000 (2)	152 290
Ontario	n/a	2 001 646	454 655 (-)	542 344
Manitoba	100	105 483	35 185 (0)	16 449
Saskatchewan	40-120	211 695	52 344 (20)	7 779
Alberta	32-60	324 000	160 816 (0)	39 381
British Columbia	n/a	1 797 862	962 528 (2.1)	368 736
Total		**7 699 646**	**2 225 044**	**1 360 467**

[a] All data provided by provincial forestry agencies.
[b] Planting is only one of several artificial and natural techniques used to regenerate harvested and burned areas over time.
[c] Includes spacing, thinning, cleaning, fertilizing.
[d] Percentage of planting done with improved nursery stock in parentheses.

Table 2. Projected harvesting and renewal activities (hectares) in Canada, 1990–1995.[a]

	Harvesting	Planting[b]	Tending[c]
Newfoundland	97 500	30 000 (0)[d]	80 000
Prince Edward Island	15 000	4 500 (49-65)	3 250
Nova Scotia	180 000	54 000 (3-5)	82 500
New Brunswick	250 000	46 900 (80)	38 125
Quebec	1 410 000	110 000 (15)	429 800
Ontario	1 277 878	425 000 (65-75)	491 000
Manitoba	71 500	37 000 (3)	14 000
Saskatchewan	n/a	30 000 (25-30)	10 000
Alberta	400 000	228 000 (5)	42 500
British Columbia[e]	1 185 000	458 216	533 297

[a] All data provided by provincial forestry agencies.
[b] Planting is only one of several artificial and natural techniques used to regenerate harvested and burned areas over time.
[c] Includes spacing, thinning, cleaning, fertilizing.
[d] Percentage of planting done with improved stock in parentheses.
[e] Includes only Ministry of Forests programs.

overcome the obstacles to development and registration of several pesticides. During that review, the task force found obstacles for forestry use among all the players in the process: the pesticide manufacturers, the federal regulatory agencies, the provincial regulatory agencies, and in some cases, the provincial forestry departments (Carrow 1985b). The task force concluded that the problem was the process itself that acted as a major disincentive to the development of new control products. More recently, the federal regulatory agencies and the pesticide industry have been reluctant to register forestry pesticides for aerial applications, even though they have approved ground application. Unless the current regulatory climate changes, the forestry sector will be forced to rely more on ground application, particularly in regenerated areas. Thus, in designing harvest blocks, which determine the size and shape of stands that will comprise the new forest, we should consider that many future pest control operations against insects, competing vegetation, and diseases may have to be carried out from the ground, rather than the air.

Given the widespread use of buffer zones around features such as habitation, water sources, aquatic habitat, ecological reserves, and agricultural operations to protect them from the adverse effects of aerial pesticide spraying, it seems unwise to plan any silvicultural activities within these buffer zones. Certainly, protection of stands within these zones will be very costly, and effectiveness could be limited.

Tree species—The tendency to establish single-species plantations within harvested or burned areas can set the stage for enormous losses. Good examples of this are the white spruce plantations established by J.D. Irving Ltd. in New Brunswick. In the late 1970s, about 6000 ha of white spruce plantation was severely attacked by the spruce bud moth, *Zeiraphera canadensis*. There was no control agent known or registered for this insect, and as a result it caused widespread damage and loss of commercial value in these stands, which were being grown for sawlog production. Because the plantations were one species only, white spruce, their commercial value was largely lost. Had the plantations contained two or more species, the commercial loss would have been less, and possibly the bud moth would not have had such a serious impact on the white spruce component (Carrow 1985a). For operational and logistic reasons, there is a tendency to plant, or regenerate, one species on a block. On sites where other species are present through natural regeneration, the result will be a multi-species stand, which should be able to better tolerate pest attack without significant commercial losses. But in cases where natural regeneration is limited, it is generally imprudent to establish extensive single-species stands.

High-value stands—In addition to new forests that are intended to provide a commercial wood supply, there is an increasing number of "high-value" stands that are being grown or managed for special purposes. These include parks, recreation areas, wildlife habitat, ecological reserves, Christmas tree farms, biomass plantations, seed orchards, seed production areas, and urban forests. All are special challenges for forest pest managers.

They often represent very high value, and therefore the damage threshold is low. Because they are often composed of noncommercial species, they attract insect pests for which there are no available controls. Furthermore, their location close to human habitation, or other values such as water supplies, makes it difficult to use conventional insecticides, and indeed may preclude the use of aircraft. Yet the need for control is often critical, for example, seed and cone insects, and forest tent caterpillar, *Malacosoma disstria*, in parks. In these situations, there is potential for the development and use of "alternative" technologies including parasites, predators, pheromones, and nematodes, and indeed, in many cases, the sometimes elusive concept of integrated pest management should become a reality.

Genetically improved stands—Comparing Tables 1 and 2 shows a clear increase in the use of genetically improved stock in provincial planting programs. Use of this stock will increase the value of these plantations; indeed, because this stock has been selected primarily on the basis of improved growth and form, these plantations will be particularly important in meeting the wood requirements of the industry in the mid-21st century. However, this stock has not generally been selected for insect resistance, and so insect pests will pose a major threat to these plantations.

Economic threshold—Recommendation no. 9 of the National Forest Sector Strategy (Canadian Council Forest Ministers 1987) states that the forestry sector should "ensure that all pest management operations are ecologically and economically justified." Historically, the forestry sector has not addressed the question of economic justification with much rigor, largely because pest control operations were generally carried out during epidemic phases of insect pests such as spruce budworm, *Choristoneura fumiferana*, hemlock looper, *Lambdina fiscellaria*, and the Douglas-fir tussock moth, *Orgyia pseudotsugata*. In those situations, the consequences of not protecting infested forests were usually evident—the trees died. Because the value of the stands greatly exceeded the cost of the control program, there was little need to systematically carry out an economic justification.

Nevertheless, the Strategy points to the need to economically justify control programs on a project basis. If the area proposed for protection falls outside provincial long-term management (or timber supply) planning, then it may be difficult to justify a major control program. However, if the area is included within long-term planning, then the cost of finding an alternative wood supply and re-directing operations, as an option to protection, should be considered. Insect control in the new, younger age-class forests presents a different situation. In some cases, a control program is carried out to prevent mortality, for example, hemlock looper, but in other cases, the objective is to preserve annual growth, or volume increment. As we move towards more intense management of the new forest, insect control to preserve annual growth will become more common. Indeed, as we exhaust the supply of old-growth natural forests, it will become more important to ensure that our new forests are developing as predicted in provincial timber supply projections. Although the principle of preserving annual volume growth in younger developing forests is sound, we have not yet addressed the question of how much of that annual growth must be preserved to satisfy the supply models. Until we do, it will be difficult to economically justify insect control for many of the programs whose objective is to preserve annual growth. Of course, this deficiency is not restricted to young commercial forests; a similar effort is required regarding the economic justification for insect control programs in many other forests, for example, recreation areas and parks.

Looking to the Future

As previously mentioned, the National Forest Sector Strategy (Canadian Council Forest Ministers 1987) contains one recommendation that deals specifically with forest pest management: recommendation no. 9. Its wording reflects the judgment and perspective of all the major stakeholders in Canada's forests, and because of this, it will likely influence the direction of pest management operations and research in the coming decade. The recommendation states that all elements of the forest sector should "recognize that pesticides are

among the legitimate means for effective forest management in specific areas, that their use continue to be regulated, and:
- ensure that all pest management operations are ecologically and economically justified;
- encourage development and use of effective alternative methods of pest control, including integrated pest management;
- accelerate research into the environmental effects of pesticides; and
- ensure that the process for registration of pesticides for forest use is not cost-prohibitive and is open to public scrutiny."

Three years have passed since the Strategy was endorsed, and a survey of provincial forestry agencies shows that, not surprisingly, response to recommendation no. 9 has been variable. All provinces accept the need for pesticide use in some situations, but their stance on chemical pesticides varies. New Brunswick and Newfoundland endorse their use, and others, for example, Alberta and Nova Scotia, maintain that only Bt may be used in the natural forest environment. Regarding ecological and economic justification, British Columbia appears to be the only province to have undertaken a systematic assessment of the economic impact of various forestry pests, and the economic value offered by pesticides, compared with other practical techniques. Other provinces still approach economic and ecological justification on a provincial scale, and in more theoretical terms. Most of the provinces have developed a strong interest in "alternatives," and are funding major research projects on control agents such as insect growth regulators, pheromones, nematodes, fungi, and improved spray technology. There is broad support for and interest in integrated pest management, and clearly, many provinces are using techniques such as adjusting harvest schedules, delaying planting on harvested sites, pheromone monitoring, and trap logging to minimize damage by various pests.

Government Policy

The policies adopted by provincial forestry agencies, or by provincial governments, regarding forestry use of pesticides are important for two reasons: they determine what technology is available to the forest manager for operational use; and they indirectly affect pesticide research and development activities carried out by pesticide manufacturers and research agencies. Although provincial governments generally acknowledge the need to use pesticides for protection of forests from insect outbreaks, the availability of pesticide technology to forest managers varies widely. Spruce budworm control programs, which have been the largest and most frequent forest insect control programs in Canadian history, have relied historically on a combination of microbial (Bt) and chemical insecticides.

However, recently, provinces have shifted steadily away from chemicals and substituted Bt. Although formal policy statements on the use of chemicals versus Bt are difficult to find, many provinces have adopted a "no-chemical" position for their major operations. In recent years, Nova Scotia, Quebec, Ontario, Manitoba, and British Columbia have used only microbial insecticides in their major insect control programs. In some provinces, this position has been developed by senior managers within provincial forestry departments; in others, it has been a ministerial decision. Elsewhere, the sensitivity of insecticide use is illustrated in some provinces (Newfoundland, Nova Scotia, New Brunswick), where the provincial Cabinet, or Executive Council, makes the final decision about what area will be treated and what control products will be used. Thus, the issue of chemical use in the forest environment is politically sensitive, and because of trends in public opinion, the pressure to further restrict chemical use and to find control "alternatives" will continue to mount. This should lead to more emphasis on research and development on alternative methods of control, and it will likely result in a requirement for more rigorous justification (both economic and ecological) of insect control programs, especially those using conventional chemical insecticides.

Future Needs in Forest Pest Management

Our future needs in forest pest management will be determined by four factors:

a) the need to publicly justify insect control programs that involve the introduction of control agents or products into the forest environment;

b) the need to preserve the integrity of existing inventories of old age-class stands for 20-40 years until the new forest becomes mature enough to harvest;

c) the need to preserve growth and prevent mortality in younger stands that make up the new forest, as well as special forest areas with unique value;

d) the need to develop control technology for new pests that arise because of the changed structure of the new forest, as well as climate change.

Public justification—The National Forest Sector Strategy (Canadian Council Forest Ministers 1987) recommends that all pest management operations be ecologically and economically justified. To satisfy this recommendation, control projects should incorporate three attributes:

a) they should have a clearly defined purpose;

b) where possible, they should include a "performance standard" to be achieved; and

c) there should be clear accountability.

In Canada, forest insect control programs are generally undertaken for one of three purposes: outbreak control, outbreak containment, or tree protection (Carrow et al. 1979; Ontario Ministry of Natural Resources 1985). In outbreak control, the purpose is to eliminate an insect infestation, or suppress it to endemic levels, while it is still relatively small and localized. Outbreak containment is the prevention of expansion of an established outbreak into valuable areas presently free of the outbreak. Where an outbreak has become widespread, because it was neither controlled nor contained, the purpose of a control program is usually protection of trees, or some portion of trees, for example, foliage. Thus, to credibly justify a control program, a purpose must be defined.

In most contemporary forestry operations, performance standards are used to assess the degree of operational success; this applies to nursery stock production, planting, regeneration (stocking, free-to-grow), and harvesting. Yet performance standards are uncommon in forest insect control programs. A review of provincial reports from the Annual Forest Pest Control Forums during the 1980s shows that results are usually characterized as "satisfactory," or "variable," or "unsatisfactory." If we are to achieve public credibility, and justify large-scale programs, we must define and adopt numerical standards to be achieved, and report on the degree of success in achieving those standards. For example, New Brunswick currently uses a foliage protection standard of 60% on balsam fir and 50% on spruce in its annual spruce budworm spray program. (Carter 1988). Quebec uses similar standards for "foliage protection" (Dorais 1985). Given the lack of knowledge about what level of protection or control is required to achieve the objective, many pest managers are reluctant to choose a standard. Nevertheless, the absence of a standard for measuring performance leaves the manager in a very vulnerable position, and it is better to adopt an arbitrary standard than to have no standard at all. Research is clearly needed to establish realistic standards for various pest control operations.

Mature forests—In most insect control programs in mature forests, the purpose is to simply keep the trees alive, not to preserve growth. Our experience has developed largely on the basis of protection spraying programs against the spruce budworm, and in provinces where protection has been sustained over time, the health of the forest has been preserved. However, with growing public concern about large-scale programs, the forestry sector should be exploring other tactics to keep the trees alive. The first, and perhaps most difficult question, concerns the level of foliage protection that is required to keep mature trees alive when under attack by a defoliator. Through experience, we know that 50%–60% protection of spruce and balsam fir is effective against spruce budworm, but perhaps a lower level of protection would also be effective. Could microbials be combined with chemical insecticides to achieve a much higher level of control and therefore reduce the need for annual treatment?

Spray scheduling also needs careful examination. Protection spraying of mature forests is done to maintain the trees until harvest, but it is not known whether annual spraying is necessary to keep trees alive. Perhaps the same objective could be attained with less frequent spraying. Again, because of costs and a growing need to justify spray programs, protection spraying should also be linked directly to harvesting plans; stands whose harvest date is farthest into the future warrant protection before stands that are scheduled for harvest in the near future, for example, within 5 years.

In the next decade, provincial pest managers anticipate that most of the existing major pests will persist as problems. The list includes spruce budworm, Choristoneura fumiferana; western spruce budworm, *C. occidentalis*; jack pine budworm, *C. pinus*; gypsy moth, *Lymantria dispar*; forest tent caterpillar, *Malacosoma disstria*; hemlock looper, *Lambdina fiscellaria*; balsam woolly adelgid, *Adelges piceae*; eastern blackheaded budworm, *Acleris variana*; oak leafshredder, *Croesia semipurpurana*; mountain pine beetle, *Dendroctonus ponderosae*; spruce beetle, *D. rufipennis*; Douglas-fir beetle, *D. pseudotsugae*; and Douglas-fir tussock moth, *Orgyia pseudotsugata*. However, several additional pests are expected to emerge as well, for example, larch sawfly, *Pristiphora erichsonii*; large aspen tortrix, *Choristoneura conflictana*; *Ips* spp.; and western balsam bark beetle, *Dryocoetes confusus*. Of the foregoing list, there are no registered controls for eight (Canadian Pulp and Paper Association 1987).

The new forests—In our new production forests, as well as our high-value forests, there is a growing need to protect annual growth, seed and cone crops, aesthetics, and form. These pose a difficult challenge, both in developing effective technology and in defining realistic performance standards. Indeed, for some, it may be impossible to define a credible standard, for example, aesthetic value. As these new forests are managed, new pests commonly appear, either because they move into an area, or because their damage assumes unacceptable proportions. These stands will be susceptible to many of the common pest species such as the spruce budworm, tussock moths, jack pine budworm, sawflies, and gypsy moth, but they will also be severely attacked by a complex of regeneration pests, such as the spruce bud moth, *Zeiraphera canadensis*; Zimmerman pine moth, *Dioryctria zimmermani*; white pine weevil, *Pissodes strobi*; sawflies; black army cutworm, *Actebia fennica*; and root weevils. It is noteworthy that for several of these new pests, there are no registered control products at present (Canadian Pulp and Paper Association 1987).

Other categories of forest will bring with them different insect pests, which heretofore have been generally unimportant economically. Seed and cone pests, such as coneworms, midges, and maggots, have assumed increasing importance as the forestry sector relies more heavily on production of genetically improved seed in seed orchards. The damage caused by these pests, and the general lack of effective control technology, represent major obstacles to the widespread operational use of improved nursery stock. Again, effective, registered control products are generally lacking (Canadian Pulp and Paper Association 1987).

As mentioned earlier, the prospect of climate change further complicates forest pest management. Borden (1988) has raised a series of questions that reveal the types of changes that may occur because of temperature change. For example, it is not known: (a) whether insects with temperature-induced diapause will become year-round pests; (b) whether insects that tolerate high temperatures will displace those with lower tolerance; (c) whether specific host plants will be more or less susceptible to insects at higher temperatures and carbon dioxide levels; (d) whether temperature change will affect the virulence of microbial pathogens; (e) the manner in which temperature change will affect beneficial insects. Although these questions cannot be answered with any certainty, they do raise several factors that have the potential to affect insect–host interactions, the severity of damage to important species, and the effectiveness of and need for insect control programs.

Pest control technology—As we enter the 1990s, Canada's forest managers face a daunting challenge regarding forest pests. The need for protection against pest damage is growing annually, for both the aging natural forest and the young forest. Yet the technology available to control forest insects is extremely limited. In Canada, there are 27 active ingredients (19 insecticides) registered for forest and woodlands use against forest pests (insects, diseases, vegetation) (Canadian Pulp and Paper Association 1987, 1990), whereas in the United States, there are more than 130 active ingredients registered for forestry. Even with this limited technology, provincial governments have imposed further restrictions, in some cases prohibiting the use of chemical insecticides. This type of provincial action severely affects the research and development required for the registration of new control agents (chemical, biochemical, and biological), and the commercial availability of registered products. In recent years, the carbamate insecticide aminocarb (Matacil®) has been withdrawn from production due to lack of provincial demand, even though the product was the most cost-effective and environmentally benign chemical insecticide available for spruce budworm control. As well, in 1988, May and Baker suspended development and registration of another carbamate insecticide, Zectran®, due to the influence of environmental pressure groups and the projected limited provincial demand. Equally disturbing is the recent regulatory opposition to the registration of new insecticides, for example, permethrin (and herbicides) for aerial application.

Canadian pesticide sales represent about 3% of world sales, and the Canadian forestry sector uses less than 5% of the pesticides sold in Canada, that is, less than 0.15% of world use. With such a limited market, along with a strict, uncertain regulatory environment, and unpredictable provincial policies regarding forestry pesticides, it is no wonder that the pesticide industry has little interest in developing new pest control technology for Canada's forestry sector. Nevertheless, the sector has been able to provide adequate protection against major forest pests, but there is an increasing number of pests for which no control technology is available. In 1987, Forestry Canada listed 15 major forest insect pests in Ontario (annual report, Forest Insect and Disease Survey 1987); there are insecticides registered for 13. However, the ministerial ban on the use of chemical insecticides in Ontario precludes the use of most registered products, so that control agents (Bt and sawfly virus) are available for only 3 of the 15 major pests.

Buffer zones have become an integral part of planning and carrying out forest pest control programs. All provinces involved in aerial application of pesticides use buffer zones to protect designated areas or features from unintentional pesticide drift. The design of buffer zones varies widely across Canada, and not surprisingly, the provinces with the longest experience with large-scale programs have developed the most sophisticated criteria. For example, buffer zones in New Brunswick vary with the size of aircraft, the pesticide used, and the area to be protected (Table 3).

In comparison, for chemical insecticides, Ontario uses buffer zones of 120 m to protect lakes and rivers, and 240 m to protect habitation and sensitive areas, such as fish habitat and endangered species habitats. No buffer zones are required for Bt.

Buffer zones, while necessary to protect other features, can affect a substantial land base, and as mentioned earlier, managers should be conscious of this in planning forest renewal operations. For example, the 3200-m buffer used to protect municipal water supplies in New Brunswick precludes the aerial application of insecticide on more than 1000 ha.

Table 3. Buffer zones (metres) for aerial application of insecticides in New Brunswick.

Sensitive area	Chemical (water-based)		Bacterial (Bt)	
	LSP[a]	SSP[b]	LSP	SSP
Habitation	1600	300	155	155
Lakes (40+ ha)	400			
Designated rivers	400			
Open water bodies		65		0
Ecological reserves	400	65	400	65
Blueberry fields	3200	3200	3200	0
Municipal water (point of extraction)	3200	3200	3200	3200

[a]Large spray plane (TBM or larger).
[b]Small spray plane (agricultural-type spray plane).

Despite the discouraging scenario for pest control technology throughout the past decade, there are signs that the regulatory system and the pesticide industry are responding to the needs of the sector. A comprehensive review of the federal pesticide registration system has been completed and recommendations for a new system have been proposed (Pesticide Registration Review 1990). The system is designed to incorporate six principles, all of which have been identified as important by the forestry sector. Concerning future control technology, two of these principles are:

- to provide increased access to pest management strategies that may reduce risk of harm to health, safety, and the environment; and
- to provide support for the development of policies that assist economic viability/competitiveness of farming/forestry/fisheries.

How will this be achieved? The system provides for forestry representation on the Canadian Pest Management Advisory Council, which will, among other things, monitor the performance of the new system to ensure that it is operating efficiently and fairly. The system also provides for several classes of registration that will be important to forestry: User-requested Minor Use Registration; User-requested Minor Use Label Expansion; Critical Need; and Provisional Registration. Because of small market demand, these options should prove useful in registering new technology, including the many alternative technologies that are currently being actively pursued by the sector.

Clearly, however, risk will play an important role in the future registration process, and indeed there is provision to deny registration to a new product without any assessment of value if the risk is considered unacceptable. Forestry scientists must acquaint themselves with this new regulatory environment, before undertaking long-term research on a technology that has little chance of progressing through the registration system. This judgment should be based on the likelihood of public acceptance of the technology. Clearly, because more than 70% of Canadians disapprove of the use of chemicals in the forest environment (Forestry Canada 1989), this should send a strong signal to scientists working on new control technology. But most Canadians cannot distinguish between conventional chemicals (carbamates, organophosphates, pyrethroids) and the newer classes of chemicals, and generally, they are unaware of semiochemicals and insect growth regulators. Hopefully, this technology will be accepted more readily when it has been explained to the public. Looking to the new "pest management strategies that may reduce the risk to health, safety, and the environment," and considering such factors as public acceptability, likelihood of registration, operational feasibility, and commercial interest, the technologies that appear most promising are microbials; semiochemicals; insect biocontrol agents (parasites, predators); insect growth regulators; and short-residual chemicals, for example, pyrethroids. Given the widespread uncertainty and public apprehension about the introduction of genetically altered organisms into the environment, the prospect of registering this technology for use in the forest environment seems particularly dim. Canadians have grown to regard their forests as one of the last natural terrestrial ecosystems left, and they are unlikely to approve the release of an "unnatural" organism into that environment.

It is encouraging to observe the recent interest in some of these new technologies by the pesticide industry. Throughout the 1980s, Bt became the insecticide of choice in many provinces, and commercial interest increased dramatically: there are now 16 registered products, distributed by six companies, a factor that has helped to close the price gap between Bt and chemical insecticides and to improve product performance. Commercial interest in insect growth regulators and pheromones, though limited, has persisted and hopefully will increase. Perhaps the most notable recent development has been the entry of some of the large pesticide manufacturers into the insect biocontrol agent field, that is, the large-scale production and mass release of parasites and predators. This is encouraging because it indicates that the industry is able and willing to adapt to new nontraditional technologies that are not part of their history.

Nevertheless, Canada's forestry sector remains a very small user of pesticides, and that is unlikely to change very much. Limited demand will continue to be a major obstacle to the development of new insect control technology, and it may be that the sector will have to develop cooperative mechanisms for the research, development, and registration of new technology. This would require a provincial–federal–forest industry shared-funding agreement, designed to bring new products forward quickly, and if necessary, produce them for operational use.

Summary

As Canada approaches the 21st century, forest pest managers face the most difficult technical, social, and political challenge in Canada's history. The rapidly diminishing inventory of "old-growth" forest, the requirement to grow the new forest to meet future wood demands, and the shrinking productive forest land base are all combining to make pest-caused losses less tolerable. The need to protect the old forest from major insect pests such as spruce budworm, hemlock looper, mountain pine beetle, and Douglas-fir tussock moth will continue for several decades, until that forest is exhausted through harvesting and natural decline. In addition, there will be the steadily growing need to protect the new forest from a variety of new pests that cause growth loss, mortality and deformation, and the secretive pests of high-value areas, such as seed orchards. This will translate into more complex pest management programs, probably covering more area.

Three major factors, lack of control technology, a restrictive regulatory environment, and public involvement in forest management, will make forest pest management in Canada very difficult. Although there is general enthusiasm among resource managers for the use of alternatives to conventional chemical pesticides, the harsh reality is that very few government-approved alternatives are commercially available now. Of all the insecticides registered for forestry in Canada, there are only three "alternatives": Bt, the redheaded pine sawfly virus, and the Douglas-fir tussock moth virus. In total, only 10 insects can be controlled with these agents. Integrated pest management is a reality with several important insect pests, including the Douglas-fir tussock moth and several western bark beetles. A combination of pheromones, salvage harvesting, and targeted application of pesticides (chemical or microbial) provides an effective management system. But for the major defoliators, the manager must still rely heavily on pesticide technology to adequately protect the resource. Apart from adjusted harvest schedules, few alternatives are available for operational use.

In those provinces where pesticide applicator training is not in place, it will likely become mandatory for the forestry sector in the near future. This will apply to both ground and aerial application. And it would not be surprising to see a provincial requirement for a permit for every forestry pesticide application, particularly on Crown land.

Just as the public commonly has an opportunity for review of and comment on forest management plans, this will increasingly become a feature of planning forest pest management projects. Pest managers will have to be prepared to adjust plans to accommodate the legitimate concerns of local residents and to protect local values from pesticide contamination.

However challenging the future may seem, it will be a stimulating environment for pest managers. Shifting human values, relative to the management of our natural environment, have overtaken science and technology. Forest managers will require an unprecedented level of technical skill and resourcefulness; they will also have to be more responsive and adaptive than ever before. The forest pest manager will have to constantly remember that his/her primary task is not to use pesticides, but rather to use the full range of technology and skills needed to protect Canada's most valuable natural resource from unacceptable damage due to pests.

References

Baskerville, G.L. 1983. Good forest management: a commitment to action. N.B. Dep. Nat. Resour., Fredericton, N.B.

Benson C. 1990. The potential for integrated resource management with intensive or extensive forest management: Reconciling vision with reality - the extensive management argument. For. Chron. 66(5):457-460.

Borden J.H. 1988. The acid test. Bull. Entomol. Soc. Can. 20(3):14-16.

Brundtland Commission. 1987. Our common future. World Commission on Environment and Development.

Canadian Council Forest Ministers. 1987. A national forest sector strategy for Canada. Can. Coun. Forest Ministers, Ottawa.

Canadian Pulp and Paper Association. 1987. Insecticides registered for forest and woodlands management. Can. Pulp and Paper Assoc. Montreal. Tech. Ref.

Canadian Pulp and Paper Association. 1990. Herbicides registered for forest and woodlands management. Can. Pulp and Paper Assoc. Montreal. Tech. Ref.

Carrow J.R. 1985a. Spruce budmoth - a case history: opportunities. For. Chron. 61(3):247-251.

Carrow J.R. 1985b. Improving the process. Proc. Public Affairs & Forest Management Pesticides in Forestry Seminar, Toronto.

Carrow J.R.; Flowers, J.F.; Howse, G.M. 1979. A rationale for spruce budworm spraying in Ontario: present and future. Ont. Min. Nat. Resour. Pest Cont. Sect. Rep. 6.

Carter N.E. 1988. Protection spraying against the spruce budworm in New Brunswick 1988. Dep. Nat. Resour. and Energy, Fredericton, N.B.

Dorais L. 1985. Effectiveness of aerial spraying programs against the spruce budworm. Eastern Spruce Budworm Research Work Conference. Fredericton, N.B.

Forest Insect and Disease Survey. 1987. Annual report Forest Insect and Disease Survey, Ontario Region. Forestry Canada, Sault Ste. Marie.

Forestry Canada. 1989. 1989 National survey of Canadian public opinion on forestry issues. Forestry Canada, Ottawa.

Forests For Tomorrow. 1990. Draft terms and conditions submitted by Forests For Tomorrow for the Class Environmental Assessment of Timber Management in Ontario, 28 November 1990. Toronto.

Nishnawbe-Aski Nation. 1990. Draft terms and conditions submitted by the Nishnawbe-Aski Nation and Windigo Tribal Council for the Class Environmental Assessment of Timber Management in Ontario, 28 November 1990. Toronto.

Ontario Ministry of Natural Resources. 1985. Aerial application of insecticides for forest management in Ontario. Ont. Min. Nat. Resour. Policy and Procedure FR 04 10 10. 1 November 1985. Toronto.

Pesticide Registration Review. 1990. Recommendations for a revised federal pest management regulatory system. Pesticide Registration Review, December 1990, Ottawa.

Science Council Canada. 1988. Environmental peacekeepers: science, technology and sustainable development in Canada. Sci. Counc. of Canada, Ottawa.

Sedjo R. 1990. Comments on the "The potential for integrated resource management with intensive or extensive forest management: Reconciling vision with reality." For. Chron. 66(5):461-463.

Smyth, J.H.; Campbell, K.L.; Lapointe, G.; Martel, J.P. 1989. Forest management expenditures in Canada, 1983-1985. Can. Pulp and Paper Assoc. - Forestry Canada Joint Rep. 14, Ottawa, Ont.

Valhalla Society. 1988. B.C.'s endangered wilderness: a comprehensive proposal for protection. Valhalla Society, New Denver, B.C.

World Wildlife Fund. 1990. Endangered spaces progress report. World Wildlife Fund, Toronto.

Directory of Authors

A Natural Resources Canada, Canadian Forest Service affiliation is indicated by one of the following numbers referring to a Canadian Forest Service research establishment:

(1) Natural Resources Canada
Canadian Forest Service
Newfoundland and Labrador Region
St. John's, Newfoundland
A1C 5X8

(2) Natural Resources Canada
Canadian Forest Service
Maritimes Region
Fredericton, New Brunswick
E3B 5P7

(3) Natural Resources Canada
Canadian Forest Service
Quebec Region
Sainte-Foy, Quebec
G1V 4C7

(4) Natural Resources Canada
Canadian Forest Service
Petawawa National Forestry Institute
Chalk River, Ontario
K0J 1J0

(5) Natural Resources Canada
Canadian Forest Service
Ontario Region
Sault Ste. Marie, Ontario
P6A 5M7

(6) Natural Resources Canada
Canadian Forest Service
Forest Pest Management Institute
Sault Ste. Marie, Ontario
P6A 5M7

(7) Natural Resources Canada
Canadian Forest Service
Northwest Region
Edmonton, Alberta
T6H 3S5

(8) Natural Resources Canada
Canadian Forest Service
Pacific and Yukon Region
Victoria, British Columbia
V8Z 1M5

(9) Natural Resources Canada
Canadian Forest Service
Headquarters
Ottawa, Ontario
K1A 1G5

Anderson, M.L.
Alberta Forestry, Lands and Wildlife
Alberta Forest Service
Edmonton, Alberta T5K 2M4

Auger, M.
Quebec Ministry of Forests
Forest Insect and Disease Protection Service
Quebec City, Quebec G1N 2C9

Barber, K. **(6)**

Bordeleau, C.
Quebec Ministry of Forests
Forest Insect and Disease Protection Service
Quebec City, Quebec G1N 2C9

Borden, J.H.
Simon Fraser University
Department of Biological Sciences
Burnaby, British Columbia V5A 1S6

Bowers, W.W. **(1)**

Butterworth, E.W.
Research and Productivity Council
Fredericton, New Brunswick E3B 6C2

Cabana, J.
Quebec Ministry of Forests
Forest Insect and Disease Protection Service
Quebec City, Quebec G1N 2C9

Cadogan, B.L. **(6)**

Carrow, J.R.
University of Toronto
Faculty of Forestry
Toronto, Ontario M5S 3B3

Cerezke, H.F. **(7)**

Chabot, M.
Quebec Ministry of Forests
Forest Insect and Disease Protection Service
Quebec City, Quebec G1N 2C9

Churcher, J.J.
Ontario Ministry of Natural Resources
Forest Health and Protection Section
Sault Ste. Marie, Ontario

Crabbe, R.S.
National Research Council Canada
Ottawa, Ontario K1A 0R6

Crummey, H.
Newfoundland Department of Forestry and Agriculture
Corner Brook, Newfoundland A2H 6J8

Cunningham, J.C. **(6)**

DeBoo, R.F.
British Columbia Ministry of Forests and Lands
British Columbia Forest Service
Victoria, British Columbia V8W 3E7

de Groot, P. **(6)**

Dorais, L.
Quebec Ministry of Forests
Forest Insect and Disease Protection Service
Charlesbourg, Quebec G1H 6R1

Drouin, J.A. **(7)**

Eidt, D.C. **(2)**

Embree, D.G. **(2)**

Ernst, W.R.
Environment Canada
Conservation and Protection Service
Dartmouth, Nova Scotia

Finney-Crawley, J.R.
Memorial University of Newfoundland
St. John's, Newfoundland A1C 5S7

Grant, G.G. **(6)**

Gray, T.G. **(8)**

Harris, J.W.E. **(8)**

Helson, B.V. **(6)**

Holmes, S.B. **(6)**

Howse, G.M. **(5)**

Hudak, J. **(1)**

Hulme, M.A. **(8)**

Humble, L.M. **(8)**

Jobin, L. **(3)**

Kaupp, W.J. **(6)**

Kettela, E.G. **(2)**

Kevan, P.G.
University of Guelph
Ontario Agricultural College
Guelph, Ontario N1G 2W1

Kingsbury, P.D. **(6)**

Knowles, K.
Manitoba Department of Natural Resources
Forestry Branch
Winnipeg, Manitoba R3N 1Z4

Kristmanson, D.D.
University of New Brunswick
Department of Chemical Engineering
Fredericton, New Brunswick E3B 5A3

Lachance, D. **(3)**

Lyons, D.B. **(5)**

Lysyk, T.J. **(6)**

Magasi, L.P. **(2)**

Maisonneuve, C.
Department of Energy and Resources
Environmental Studies Service
Charlesbourg, Quebec G1H 6R1

Major, L.
Department of Energy and Resources
Environmental Studies Service
Charlesbourg, Quebec G1H 6R1

Marotte, P.-M.
Quebec Ministry of Forests
Environmental Evaluation Branch
Charlesbourg, Quebec G1H 6R1

McCooeye, M.
National Research Council Canada
Ottawa, Ontario K1A 0R6

McLean, J.A.
University of British Columbia
Vancouver, British Columbia V6T 1Z4

Meating, J.H. **(5)**

Miller, G.E. **(8)**

Millikin, R.L. **(6)**

Moeck, H.A. **(8)**

Nealis, V.G. **(5)**

Nigam, P.C. **(2)**

Ono, H.
Alberta Forests, Lands and Wildlife
Alberta Forest Service
Edmonton, Alberta P5G 0S8

Otvos, I.S. **(8)**

Pandila, M.
Saskatchewan Parks and Renewable Resources
Prince Albert, Saskatchewan

Payne, N.J. **(6)**

Pelletier, M.
Quebec Ministry of Forests
Forest Insect and Disease Protection Service
Quebec City, Quebec G1N 2C9

Pendrel, B.A. **(2)**

Perry, D.F. **(3)**

Picot, J.J.C.
University of New Brunswick
Department of Chemical Engineering
Fredericton, New Brunswick E3B 5A3

Plowright, R.C.
University of Toronto
Department of Zoology
Toronto, Ontario M5S 1A1

Raske, A.G. **(1)**

Régnière, J. **(3)**

Rentz, C.L. **(7)**

Retnakaran, A. **(6)**

Safranyik, L. **(8)**

Sanders, C.J. **(5)**

Shepherd, R.F. **(8)**

Shore, T.L. **(8)**

Silk, P.J.
Research and Productivity Council
Fredericton, New Brunswick E3B 6C2

Smith, S.M.
University of Toronto
Faculty of Forestry
Toronto, Ontario M5S 3B3

Smith, T.D.
Nova Scotia Department of Lands and Forests
Truro, Nova Scotia B2N 5B8

Strongman, D.
Saint Mary's University
Biology Department
Halifax, Nova Scotia B3H 3C3

Strunz, G.M. **(2)**

Sundaram, A. **(6)**

Sundaram, K.M.S. **(6)**

Syme, P.D. **(5)**

Taylor, S.P.
British Columbia Ministry of Forests and Lands
British Columbia Forest Service
Victoria, British Columbia V8W 3E7

Turgeon, J.J. **(6)**

Tyrrell, D. **(6)**

van Frankenhuyzen, K. **(6)**

Van Sickle, G.A. **(8)**

van Vliet, M.W.
University of New Brunswick
Department of Chemical Engineering
Fredericton, New Brunswick E3B 5A3

Volney, W.J.A. **(7)**

Wallace, D.R. **(5)**

West, R.J. **(1)**

Westwood, A.R.
Manitoba Department of Natural Resources
Forestry Branch
Winnipeg, Manitoba R3N 1Z4

Wiesner, C.J.
Research and Productivity Council
Fredericton, New Brunswick E3B 6C2

Wilson, G.G. **(6)**

Wong, H.R. **(7)**